MOLECULAR SYSTEMATICS OF PLANTS

DNA SEQUENCING

MOLECULAR SYSTEMATICS
OF PLANTS II
DNA SEQUENCING

EDITED BY
DOUGLAS E. SOLTIS
PAMELA S. SOLTIS
Washington State University, Department of Botany
JEFF J. DOYLE
Cornell University, Department of Botany

KLUWER ACADEMIC PUBLISHERS
Boston / Dordrecht / London

Distributors for North, Central and South America:
Kluwer Academic Publishers
101 Philip Drive
Assinippi Park
Norwell, Massachusetts 02061 USA
Telephone (781) 871-6600
Fax (781) 871-6528
E-Mail <kluwer@wkap.com>

Distributors for all other countries:
Kluwer Academic Publishers Group
Distribution Centre
Post Office Box 322
3300 AH Dordrecht, THE NETHERLANDS
Telephone 31 78 6392 392
Fax 31 78 6546 474
E-Mail <services@wkap.nl>

 Electronic Services <http://www.wkap.nl>

Library of Congress Cataloging-in-Publication Data

Molecular systematics of plants II: DNA sequencing / edited by
    Douglas E. Soltis, Pamela S. Soltis, Jeff J. Doyle.
        p.   cm.
    Includes bibliographical references and index.
    1. Botany -- Classification -- Molecular aspects. 2. Nucleotide
    sequence. I. Soltis, Douglas E. II. Soltis, Pamela S.
    III. Doyle, Jeff J.
    QK95.6.M6525   1998
    572.8'6332-- dc21   97-17374
                                                                CIP

    ISBN 0-412-11121-7 (Casebound edition)
    ISBN 0-412-11131-4 (Paperback edition)

Contents

Preface

In the 6 years that have transpired since the publication of *Molecular Systematics of Plants,* the field of molecular systematics has advanced at an astonishing pace. *Molecular Systematics of Plants* provided an overview of the advances in systematic botany that took place during the first decade of molecular systematics. Hence, from an empirical standpoint, the book dealt primarily with restriction site data, the principal DNA technique used just prior to the publication of *Molecular Systematics of Plants.* Not a single chapter of *Molecular Systematics of Plants* involved comparative DNA sequencing, although one chapter involved RNA sequencing and a second chapter discussed the great potential of DNA sequence data for phylogeny reconstruction. Shortly after the publication of *Molecular Systematics of Plants,* comparative DNA sequencing, facilitated by the amplification of DNA via the polymerase chain reaction (PCR), became the tool of choice for molecular systematics. Comparative DNA sequencing was embraced so widely and so quickly that in just a few years it surpassed restriction site analysis as *the* DNA technique in systematics. As a result, some portions of *Molecular Systematics of Plants* became outdated due to the amount of new empirical data, while a number of important theoretical and analytical issues related to DNA sequence data were simply not addressed there at all.

The primary objectives of *Molecular Systematics of Plants II* are to address those issues and to summarize some of the recent achievements of plant molecular systematics. As was our goal with *Molecular Systematics of Plants,* we hope to illustrate the potential of DNA markers for addressing a wide variety of phylogenetic and evolutionary questions. We attempt to provide guidance in choosing an appropriate technique, as well as an appropriate gene for sequencing, for a given level of systematic inquiry. We endeavor not merely to review available techniques and previous work, but to provide a stimulus for devising future research in this rapidly evolving field. We sincerely hope that the book is deemed valuable by both beginning and experienced researchers.

The first two chapters (D. E. Soltis and P. S. Soltis; Wolfe and Liston) review the diverse array of DNA approaches available and also indicate the types of evolutionary and systematic questions that can be addressed with a given technique. The impetus for these chapters came from the training of our own students. With the development of new techniques and the identification of new genes for sequencing, it became difficult, if not impossible, to direct graduate students to just one or a few sources in the literature for reviews of this material. In the first chapter, D. E. Soltis and P. S. Soltis review the various DNA approaches available for inferring phylogeny; they then concentrate on comparative DNA sequencing and provide an overview of the sequences currently in widespread use, as well as those that appear to have phylogenetic potential. Chapter 2 by Wolfe and Liston takes a similar approach, but concentrates on the contributions of PCR other than for DNA sequencing (e.g., RAPDs, AFLPs). Chapter 2 should be of great interest to both plant systematists and population biologists.

Chapter 3 (Jansen et al.) suggests that restriction site analysis of the chloroplast genome should not be abandoned in favor of comparative sequencing. Their comparisons indicate that cpDNA restriction site data from a reasonable number of enzymes often have less homoplasy and can generate a larger number of phylogenetically informative characters than any single chloroplast gene sequence. Particularly at lower taxonomic levels, their work indicates that restriction site analysis is superior to sequencing multiple genes or noncoding regions of the chloroplast genome.

Chapters 4 and 5 (Doyle and Davis; Lewis) contribute a theoretical perspective, dealing with issues of data analysis (using parsimony and maximum likelihood, respectively). Doyle and Davis also consider the critical issues of character homology and alignment of DNA sequences. Lewis provides a thoughtful, easy-to-follow review of maximum likelihood methods in phylogeny reconstruction.

Three chapters deal with the broad topic of molecular evolution. Chapters 6 and 7 (Olmstead et al.; P. S. Soltis and D. E. Soltis) are complementary in that they deal with molecular evolution and character weighting of DNA sequences representing the chloroplast (*rbcL, ndhF*) and nuclear (18S rDNA) genomes, respectively. These chapters illustrate how knowledge of patterns of gene evolution can be used to augment molecular phylogenetic analyses. After extensive experimentation with weighting strategies, the authors of both chapters independently conclude that standard unweighted parsimony is a good initial approach in phylogeny reconstruction (a conclusion also drawn by Doyle and Davis). In Chapter 8 Nickrent et al. use parasitic plants as a model system for the study of molecular evolution. Because these plants often lack the ability to photosynthesize, they provide important insights into the working and evolution of the plastid genome when functional constraints are relaxed.

Chapter 9 by Sanderson contributes an additional theoretical perspective to the volume. He summarizes the approaches available for estimating rates of evolution and divergence times in molecular phylogenies. Using boxes describing concrete examples, he explains several methods in detail, providing an accessible overview of the topic.

With the widespread use of comparative sequencing and the availability of numerous chloroplast, nuclear, and mitochondrial DNA regions, the issue of phylogenetic congruence between gene trees and data sets is now of practical and theoretical significance. In Chapter 10 Wendel and Doyle provide an overview of the causes of incongruence and also illustrate that discordance can provide novel insights into genome evolution and species history. Complementing this chapter, Johnson and Soltis in Chapter 12 provide a hands-on approach to the topic using multiple DNA data sets for Polemoniaceae and Saxifragaceae. A number of tests for assessing congruence are illustrated in detail (boxes are used to assist those interested in applying these tests to their data), and the results of these congruence tests are discussed in light of the possible causes of incongruence reviewed by Wendel and Doyle.

The years since the publication of *Molecular Systematics of Plants* have seen tremendous advances in our understanding of the molecular basis of morphology, particularly of the angiosperm flower. In Chapter 12 Albert et al. review this information, integrating recent molecular developmental genetic models with data from molecular phylogenies to construct hypotheses for the evolution of petals and petal-like structures in diverse angiosperm taxa. This example provides a glimpse into one direction molecular systematics may take in the future.

The remaining chapters review various applications of DNA markers to a diversity of questions in plant systematics and evolutionary biology. Palmer and Delwiche review the molecular data pertaining to the origin of plastids and reveal that the evolution of plastids and their genomes largely involves lateral transfer, or horizontal evolution. An impressive amount of data indicates that the three well-characterized groups of primary plastids are monophyletic, arising from a single cyanobacterial symbiosis. It is also clear that secondary symbiosis has played a major role in plastid evolution, with many of the major groups of algae having acquired their plastids in this manner.

Baldwin et al. review the important evolutionary insights that have been garnered via the application of DNA approaches to plants endemic to island systems. DNA phylogenetic methods have proved valuable for discerning the ancestry and resolving relationships of oceanic island plants. In addition, phylogenetic inferences of island groups and close relatives have allowed the estimation of the ancestral characteristics of founder species, morphological and ecological change in insular lineages, dispersal to and within archipelagos, and even the timing and rates of diversification.

Arnold and Emms provide a review of the types of systematic and evolutionary questions that can be addressed at the population level. This chapter complements the overview of methods provided by Wolfe and Liston in that it illustrates how various DNA approaches have actually been applied at the population level. Rieseberg provides a thought-provoking review of the application of gene mapping to the study of plant speciation. Most of the initial period of DNA sequencing in plants involved a single gene, *rbcL*. Hence, Chapter 17 (Chase and Albert) is devoted to discussing the development of the large *rbcL* database now available for angiosperms and the issues of data analysis that have arisen as a consequence of this large data set.

Chapters 18 and 19 (Chapman et al.; Wolf et al.) review recent advances in our understanding of phylogeny in groups of green plants that have received less attention than have angiosperms. Chapman et al. provide a valuable overview of the application of DNA sequencing and phylogeny reconstruction to green algae. DNA-based phylogenies have revolutionized our knowledge of algal phylogeny, provided insights into the possible sister group of land plants, and elucidated trends in the evolution of morphological and ultrastructural characters. Wolf et al. provide a timely overview of pteridophytes, the most poorly understood group of

vascular plants from a phylogenetic perspective. Wolf et al. also show the value of combining data sets (both molecular and morphological) in phylogenetic analyses of enigmatic groups, such as pteridophytes. Analyses of combined data sets provide increased resolution, increased internal support, and shortened run times compared to analyses of the individual data sets, a point also stressed by Chase and Albert for angiosperms.

For two multiauthored phylogenetic analyses involving *rbcL* and 18S rDNA sequences, we have followed the suggestion of Mark Chase and used the citation format: Chase, Soltis, Olmstead et al., 1993 and Soltis, Soltis, Nickrent et al., 1997, respectively. The goal of this format is to provide credit to the primary organizers of each paper.

We want to thank some of the many people who have assisted with this book. Our students, Dirk Albach, Linda Cook, Matthew Gitzandanner, Michael Hardig, Jason Koontz, Robert Kuzoff, Mark Mort, Andrea Rabe, Melinda Trask, and Michael Zanis read every chapter of this volume and provided valuable comments. Thanks also to Stephen Downie, Brandon Gaut, Paul Keim, Kevin Nixon, and Elizabeth Kellogg for reviewing portions of the book. The following persons not only contributed chapters, but also reviewed one or more chapters: Mark Chase, Jerrold Davis, Daniel Nickrent, Michael Arnold, Paul Lewis, Richard Olmstead, Loren Rieseberg, Jonathan Wendel, and Paul Wolf. We also thank Lynn Druffel, who spent many hours assisting with the reformatting of chapters. Lastly, we thank the contributors for their cooperation in producing what we hope will be a valuable reference for those interested in plant molecular systematics.

Contributors

Victor A. Albert
The New York Botanical Garden
Bronx, NY 10458-5126

Michael L. Arnold
Department of Genetics
University of Georgia
Athens, GA 30602

Bruce Baldwin
University of California
Department of Integrative Biology
Berkeley, CA 94720

Mark A. Buchheim
Faculty of Biological Science
The University of Tulsa
600 S. College Ave.
Tulsa, OK 74104-3189

Russell L. Chapman
Department of Plant Biology
Louisiana State University
Baton Rouge, LA 70803

Mark W. Chase
Royal Botanical Gardens
Kew, Richmond
Surrey TW93AB, England
United Kingdom

Alison E. Colwell
Department of Botany
University of Washington
Seattle, WA 98195

Daniel J. Crawford
Department of Plant Biology
Ohio State University
1735 Neil Ave.
Columbus, OH 43210-1293

Jerrold I. Davis
L. H. Bailey Hortorium
Cornell University
467 Mann Library
Ithaca, NY 14853

Charles F. Delwiche
University of Maryland
Department of Plant Biology
H. J. Patterson Hall
College Park, MD 20742-5815

Claude W. dePamphilis
Department of Biology
Vanderbilt University
Nashville, TN 37235

Laura DiLaurenzio
The New York Botanical Garden
Bronx, New York 10458-5126

Jeff J. Doyle
L. H. Bailey Hortorium
Cornell University
467 Mann Library
Ithaca, NY 14853

R. Joel Duff
Department of Plant Biology
Southern Illinois University
Carbondale, IL 62901-6509

Simon K. Emms
Department of Genetics
University of Georgia
Athens, GA 30602

Javier Francisco-Ortega
Jardin de Aclimatacion de Orotava
Calle Retama Num. 2, E-38400
Puerto de La Cruz, Tenerife
Canary Islands, Spain

Thomas Friedl
Department of Plant Ecology and Systematics
Universitaet Bayreuth
Bayreuth, Germany

Mats H. G. Gustafsson
The New York Botanical Garden
Bronx, New York 10458-5126

Mitsuyasu Hasebe
National Institute for Basic Biology
38 Nishigounaka
Myo-daiji-cho
Okazaki 444
Japan

Volker A. R. Huss
Universitaet Erlangen
Institut fur Botanik
Staudtstrasse 5
D-91058 Erlangen
Germany

Robert K. Jansen
Department of Botany
University of Texas
Austin, TX 78713

Leigh A. Johnson
Department of Botany
North Carolina State University
Raleigh, NC 27695-7612

Kenneth G. Karol
Laboratory of Molecular Systematics
National Museum of Natural History
Smithsonian Institution
Washington, D.C. 20560

Seung-Chul Kim
Department of Plant Biology
Ohio State University
1735 Neil Avenue
Columbus, OH 43210-1293

Louise A. Lewis
Department of Biology
University of New Mexico
Albuquerque, NM 87131

Paul O. Lewis
Department of Biology
University of New Mexico
Albuquerque, NM 87131

Aaron Liston
Department of Botany and Plant Pathology
Oregon State University
Corvallis, OR 97331-2902

Jim Manhart
Department of Biology
Texas A & M University
College Station, TX 77843-3258

Richard M. McCourt
Botany
Academy of Natural Sciences
1900 Benjamin Franklin Parkway
Philadelphia, PA 19118

David Millie
Southern Regional Research Center
Agricultural Research Service, USDA
New Orleans, LA 70124

Daniel L. Nickrent
Department of Plant Biology
Southern Illinois University
Carbondale, IL 62901

Richard G. Olmstead
Department of Botany
University of Washington
Seattle, WA 98195

Jeanine L. Olsen
Department of Marine Biology
University of Groningen
9750 AA Haren
The Netherlands

Jeffrey D. Palmer
Department of Biology
Indiana University
Bloomington, IN 47405

Kathleen M. Pryer
Department of Botany
The Field Museum
Roosevelt Road at Lake Shore Drive
Chicago, IL 60605-2496

Patrick A. Reeves
Department of Botany
University of Washington
Seattle, WA 98195

Loren H. Rieseberg
Department of Biology
Indiana University
Bloomington, IN 47405

Michael J. Sanderson
Department of Evolution and Ecology
University of California
Davis, CA 95616

Alan R. Smith
University Herbarium
University of California
Berkeley, CA 94720

Douglas E. Soltis
Department of Botany
Washington State University
Pullman, WA 99164-4238

Pamela S. Soltis
Department of Botany
Washington State University
Pullman, WA 99164-4238

Kim E. Steiner
National Botanical Institute
Private Bag X7
Claremont 7735
Republic of South Africa

Tod F. Steussy
Institute for Botany
University of Vienna
Rennweg 14, A-1030
Vienna, Austria

Debra A. Waters
Department of Plant Biology
Louisiana State University
Baton Rouge, LA 70803

James L. Wee
Department of Biological Sciences
Loyola University
New Orleans, LA 70118

Jonathan F. Wendel
Department of Botany
Iowa State University
Ames, IA 50011

Paul G. Wolf
Department of Biology
Utah State University
Logan, UT 84322

Andrea D. Wolfe
Department of Plant Biology
Ohio State University
1735 Neil Ave.
Columbus, OH 43210-1293

Alan C. Yen
Department of Botany
University of Washington
Seattle, WA 98195

Nelson D. Young
Department of Biology
Vanderbilt University
Nashville, TN 37235

1

Choosing an Approach and an Appropriate Gene for Phylogenetic Analysis

Douglas E. Soltis and Pamela S. Soltis

A diverse array of molecular approaches is now available to the plant systematist for use in phylogenetic inference, including restriction site analysis, comparative sequencing, analysis of DNA rearrangements (e.g., inversions) and gene and intron loss, and various PCR-based techniques. Although these various methodologies present systematists with unparalleled opportunities for elucidating relationships and evolutionary processes, the sheer number of molecular approaches available, as well as the number of proven DNA regions for use in comparative sequencing, may seem overwhelming to those new to the field of molecular systematics. This chapter provides a general review of the various molecular techniques currently available to the plant systematist. Our primary goal is to review the types of molecular data sets that can presently be obtained; we discuss the advantages and disadvantages of each and the most appropriate approach for studying a given taxonomic level or type of evolutionary question. Given the current emphasis on DNA sequence data for phylogeny estimation, we then concentrate on the choice of an appropriate gene for comparative sequencing and also briefly discuss the pros and cons of manual versus automated sequencing. A description of selected PCR-mediated techniques for systematic and population-level studies is presented in Chapter 2.

Throughout its short history, plant molecular systematics has relied primarily on the chloroplast genome. This is currently changing as investigators turn to nuclear gene sequences (e.g., ITS, 18S rDNA), often to compare nuclear topologies with existing chloroplast-based topologies. As reviewed in Chapter 10, comparing trees based on nuclear and chloroplast markers can be particularly valuable at lower taxonomic levels, providing a window into evolutionary processes that could not be achieved with either genome alone. Because molecular systematists have relied so heavily on the chloroplast genome, and continue to do so, we first review briefly the general features of this molecule before proceeding with specific molecular approaches.

THE CHLOROPLAST GENOME

The general structural features and advantages, as well as disadvantages, of using the chloroplast genome (cpDNA) have been well reviewed (numerous chapters in P. S. Soltis et al., 1992a; Whitfeld and Bottomley, 1983; Palmer, 1985a, 1985b, 1986, 1987, 1991; Zurawski and Clegg, 1987); we briefly summarize these here.

We thank Jeff Doyle, Dick Olmstead, Paul Wolf, and Mark Chase for helpful comments on the manuscript. This work was supported in part by NSF grant DEB 9307000.

A "typical" land plant chloroplast genome (Fig. 1.1) is a circular molecule characterized by two inverted repeat segments that separate the remainder of the molecule into a large and a small single-copy region, respectively. This general structure is found throughout land plants, with several noteworthy exceptions (conifers [but see Wakasugi et al., 1996], Fabaceae, and others; see below in structural rearrangements in the chloroplast genome). The advantages of the chloroplast genome for phylogeny reconstruction include the fact that the chloroplast genome is small (typically between 120 and 200 kb), making it relatively easy to examine the entire genome via restriction site analysis (see numerous chapters in P. S. Soltis et al., 1992a). Most genes in the chloroplast genome are essentially single-copy (reviewed in Palmer, 1985a, 1985b, 1986). In contrast, most nuclear genes are members of multigene families, which can compromise the phylogenetic utility of these genes. The chloroplast genome is considered to be conservative in its evolution; it is structurally conservative and evolves fairly slowly at the nucleotide sequence level (Palmer, 1985b, 1991; Downie and Palmer, 1992a). Nonetheless, as is evident below, different portions of the chloroplast genome evolve at different rates. As a result, a wide range of possibilities exists for resolving relationships using data from the chloroplast genome, from the level of species and genus to family and even higher levels.

The conservative evolution of the chloroplast genome has also been considered one disadvantage of this molecule for inferring phylogeny, limiting its applicability among closely related species and at the population level. Although it has not been possible to use cpDNA variation at the population level to the degree mtDNA variation has been used in animals (e.g., Avise, 1986, 1994; Mitton, 1994), several studies have successfully used cpDNA variation to examine population-level relationships and evolutionary processes within species (reviewed in D. E. Soltis et al., 1992, 1997a; see also Soltis et al., 1989, 1991; Lavin et al., 1992; Liston et al., 1992; Ferris et al., 1993; Sewell et al., 1996). Several investigations have used cpDNA variation to assess intraspecific phylogeography (sensu Avise et al., 1987), linking patterns of population-level cpDNA variation to past geologic events, most notably glaciation (reviewed in Soltis et al., 1997a).

A second disadvantage of cpDNA for phylogeny estimation at lower taxonomic levels involves the potential occurrence of chloroplast transfer: the movement of a chloroplast genome from one species to another by introgression. This process and the problems it raises in phylogenetic studies have been well reviewed (e.g., Harris and Ingram, 1991; Rieseberg and Soltis, 1991; Rieseberg and Brunsfeld, 1992). Although chloroplast capture, if undetected, will bias estimates of phylogeny, it can, when recognized, be very informative about evolutionary processes (see Chapter 10).

Four main approaches employ the chloroplast genome to infer relationships: (1) restriction site analysis; (2) structural changes in the chloroplast genome, including inversions, large deletions, and the loss of specific introns and genes; (3) comparative DNA sequencing; and (4) PCR-based approaches. The first two of these topics were the primary focus of many of the chapters in P. S. Soltis et al. (1992a); we review these only briefly (see also Chapter 3 and reviews by Palmer et al., 1988a; Clegg, 1993; Doyle, 1993; Olmstead and Palmer, 1994; Sytsma and Hahn,

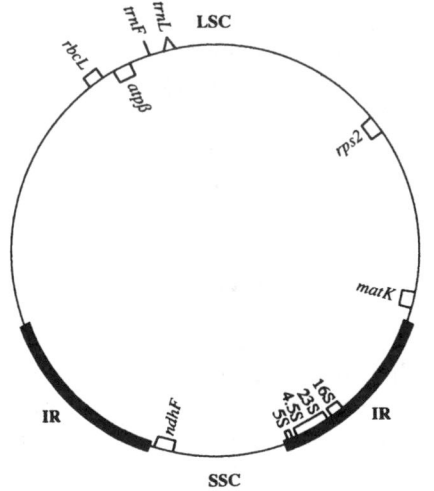

Figure 1.1.   Diagram of chloroplast genome (representative of most land plants) illustrating location of many of the chloroplast regions discussed in text. IR = inverted repeat; SSC = small single-copy region; LSC = large single-copy region.

1994, 1996; and P. S. Soltis and D. E. Soltis, 1995a). PCR-based approaches are described in Chapter 2. In this chapter, we concentrate instead on comparative DNA sequencing.

Restriction Site Analysis

Restriction site analysis of the chloroplast genome was the molecular tool of choice for inferring phylogenetic relationhips for nearly a decade (see P. S. Soltis et al., 1992a), with applications ranging from comparisons of populations (reviewed in D. E. Soltis et al., 1992, 1997a) to phylogenetic analysis of congeneric species and genera. Restriction site analysis of particularly conserved regions of cpDNA (e.g., the inverted repeat) has even been employed at higher taxonomic levels (e.g., Downie and Palmer, 1992b, 1994). It is in the study of congeneric species and related genera, however, that cpDNA restriction site analysis has been most widely employed (for reviews see Palmer et al., 1988a; P. S. Soltis et al., 1992a; Doyle, 1993; P. S. Soltis and D. E. Soltis, 1995a; Sytsma and Hahn, 1994).

In recent years, DNA sequencing has steadily replaced cpDNA restriction site analysis for phylogenetic inference, even at lower taxonomic levels (Olmstead and Palmer, 1994). However, cpDNA restriction site data can provide resolution comparable to, or better than, that achieved with sequences of the internal transcribed spacers of nuclear ribosomal DNA. Thus, particularly for analyses of species or genera, it should not be assumed a priori that sequence data are superior to restriction site data (see Chapter 3). However, sequence data do hold several advantages over restriction sites, including the easy exchange and comparability of data and the rapidity with which data can be garnered. Furthermore, restriction site mapping is a time consuming and laborious process that is not required in comparative DNA sequencing. With the widespread use of automated sequencing, the advantages of DNA sequencing will only be further enhanced. With the exploration of new DNA regions and the discovery of genes or DNA segments that are suitable for inferring phylogeny at the specific and generic levels, the use of restriction site data will continue to de-

crease. For now, however, cpDNA restriction sites continue to be an important source of data, particularly for comparisons of closely related species and for population-level analyses (see Chapter 3).

Structural Rearrangements in the Chloroplast Genome

Structural changes in the chloroplast genome are relatively rare and, when present, have been considered strong evidence of monophyly (reviewed by Downie and Palmer, 1992a). Inversions, large deletions, and the loss of specific genes and introns have provided phylogenetic markers at a variety of taxonomic levels within the land plants. Below we describe the past applications of cpDNA structural rearrangements to the reconstruction of plant phylogeny and discuss levels of homoplasy in structural characters.

Inversions The potential value of cpDNA inversions as phylogenetic markers was first demonstrated by Jansen and Palmer (1987a, 1987b), who reported that a 22-kb inversion in the large single-copy region of the chloroplast genome marked a major divergence in the Asteraceae. The chloroplast genomes of representatives of subtribe Barnadesiinae of tribe Mutisieae have the gene order typical of most land plants, but the remainder of the family, including the rest of Mutisieae, are characterized by a rearranged cpDNA. The phylogenetic inference from the distribution of the inversion is that Barnadesiinae are the sister group to the remainder of the family, a result also supported by phylogenetic analyses of restriction site variation (Jansen and Palmer, 1988) and *rbcL* nucleotide substitutions (Michaels et al., 1993).

Other cpDNA inversions have similarly marked major evolutionary splits in land plants. A 30-kb inversion differentiates the chloroplast genomes of bryophytes from those of most other land plants (Ohyama et al., 1986); the lycopsids have the cpDNA structure of bryophytes and have therefore been regarded as the sister to the remaining tracheophytes (Raubeson and Jansen, 1992a). Structural rearrangements have also identified distinct lineages of ferns (Stein et al.,

1992). A set of three inversions indicates hierarchical relationships among the grasses and their relatives (Doyle et al., 1992b): A 28-kb inversion unites Restionaceae, Ecdeiocoleaceae, and Joinvilleaceae with the Poaceae, a 6-kb inversion identifies Joinvilleaceae as the sister to the grasses, and a third inversion, involving the *trnT* gene, is a synapomorphy for the grasses. Two inversions have been reported in the Fabaceae (Palmer et al., 1987, 1988b): The 78-kb inversion unites 10 genera of subtribe Phaseolinae of tribe Phaseoleae (Bruneau et al., 1990) and defines a group that largely agrees with morphological and anatomical data and with the current taxonomy (Doyle et al., 1992a); the 50-kb inversion provides a synapomorphy for a clade that includes most of Papilionoideae (Doyle et al., 1996a).

Deletions and Loss of Genes and Introns

Deletions of introns, genes, and even entire regions of cpDNA have also been considered as synapomorphies. The most spectacular of the cpDNA deletions are the independent losses of one copy of the inverted repeat (IR) from the chloroplast genomes of six tribes of Fabaceae (Palmer et al., 1987; Lavin et al., 1990), some conifers (Strauss et al., 1988; Raubeson and Jansen, 1992b; *Pinus thunbergii* has an intact, although small, inverted repeat; Wakasugi et al., 1996), *Erodium* and *Sarcocaulon* (Geraniaceae) (Downie and Palmer, 1992a), and *Conopholis* (Orobanchaceae) (Downie and Palmer, 1992a). The loss of this 20–25-kb region and its component rRNA and other genes is tolerated by the plant because one set of functional genes remains in each chloroplast genome. It was initially suggested that one consequence of this major change in the cpDNA structure is a loss of stability of the chloroplast genome: Those cpDNAs that lack the inverted repeat structure are typically highly rearranged (e.g., Palmer et al., 1987; Strauss et al., 1988). However, not all cpDNAs with only one IR region have undergone rearrangement. The rearranged cpDNAs typical of Pinaceae and several legumes may be due to the presence of numerous dispersed repeated sequences that permit recombining and rearrangement, rather than the loss of one copy of the IR (Milligan et al., 1989).

Specific genes have also been occasionally lost from the chloroplast genome, with the sequence and sometimes its expression being transferred to the nuclear genome. For example, the gene *tufA*, which encodes elongation factor Tu (a major factor in protein synthesis), resides in the chloroplast genomes of algae, but occurs in the nuclear genome of all land plants (Baldauf and Palmer, 1990; Delwiche et al., 1995). The chloroplast gene *rpl22* is absent in all legumes surveyed (Downie and Palmer, 1992a; Doyle et al., 1995), and the genes ORF184 and *rps16* have been lost from the chloroplast genomes of most members of subfamily Papilionoideae of the family (Downie and Palmer, 1992a; Doyle et al., 1995). Even introns may be excised, with their loss interpreted as a synapomorphy. For example, the intron of the chloroplast gene *rpl2* has been lost at least six times over the course of angiosperm evolution: It is missing from the cpDNAs of Caryophyllales (Zurawski et al., 1984), Convolvulaceae, *Cuscuta,* Menyanthaceae, some Geraniaceae, Saxifragaceae s.s., and *Drosera* (Downie et al., 1991) and has been lost several times in Fabaceae (Doyle et al., 1995).

Homoplasy in Structural Characters

Although structural rearrangements have typically been regarded as providing strong evidence of monophyly (e.g., Downie and Palmer, 1992a), increasing evidence has revealed that these molecular characters, like restriction sites and nucleotide substitutions, are also prone to homoplasy. The loss of the IR in the legumes and some conifers is obviously convergent, and numerous cases of parallel structural mutations have recently been reported. For example, a 6-kb inversion in the cpDNAs of *Atriplex* and *Chenopodium* (Chenopodiaceae) also characterizes all members of Cactaceae surveyed (Downie and Palmer, 1994); given the relationships of these two families within the Caryophyllales, this inversion has apparently arisen in parallel in Chenopodiaceae and Cactaceae. Parallel inversions have also been detected in *Anemone* and related genera of Ranunculaceae (Hoot and Palmer, 1994), and a parallel deletion of 0.9 kb in the *clpP* gene has occurred in *Lobelia holstii* and *Monopsis lutea* (Lobeliaceae) (Knox et al., 1993).

Both gene and intron losses in the chloroplast genomes of legumes appear homoplasious (Doyle et al., 1995). Losses of both ORF184 and *rps16* have recurred several times even within subfamily Papilionoideae. The *rpl2* intron has been lost at least four times within the Fabaceae, with one of these losses a synapomorphy for the tribe Desmodieae of subfamily Papilionoideae. Therefore, all types of structural mutations, that is, inversions, IR loss, and loss of genes and introns, may be homoplasious, and caution should of course be used when inferring phylogenetic relationships from the distribution of structural rearrangements.

DNA SEQUENCING

In just the past 5 years, the number of genes and DNA regions used for phylogeny estimation has grown rapidly. In the early 1990s, most molecular phylogenetic studies relied on *rbcL* sequences, and to a much smaller degree on 18S ribosomal RNA or DNA (rDNA). In contrast, it is now difficult to review adequately all of the possible candidates for DNA sequencing studies. The field of plant molecular systematics has progressed so rapidly, for example, that several of the genes mentioned only recently (P. S. Soltis and D. E. Soltis, 1995b, and papers in *Annals of the Missouri Botanical Garden* 82:149–321) as new "alternatives" to *rbcL* for comparative sequencing are now widely sequenced (e.g., the rDNA internal transcribed spacer (ITS), *ndhF, matK;* the entire 18S rRNA gene). Undoubtedly additional genes, of which we are not aware at the time of this writing, will be in wide use by the time this volume is published. The number of DNA regions available for comparative sequencing will continue to increase rapidly as researchers seek multiple genes to test existing topologies, to provide additional resolution, and to obtain additional markers for the entire spectrum of taxonomic levels.

In this section we first provide a brief comparison of automated versus manual sequencing. We then present an overview of the many genes and DNA regions presently available for comparative sequencing. We discuss both those genes/DNA regions that are widely used, as well

as those DNA regions that seem to display potential utility. We provide this overview by genome (chloroplast, nuclear, mitochondrial) following relevant introductory comments and references for each of these genomes. For each gene or DNA region, general information is given regarding location, size, appropriate PCR and sequencing primers, and general rate of evolution; also provided is a brief review of exemplar studies.

Readers should keep in mind that rates of evolution for any specific gene or DNA region may vary among, and even within, groups (e.g., Doebley et al., 1990: Bousquet et al., 1992). We stress that for individual genes it is difficult to formulate strict rules regarding rate of evolution and taxonomic level of utility, particularly across the scope of all green plants. The rate of sequence variation for a given gene may vary from group to group, for example, even within a single phylum, such as the angiosperms. A number of factors may potentially contribute to this rate variation, including generation time, extinction, episodic changes of rates of sequence divergence, and lineage-specific rate variation. Furthermore, the relative rate of evolution of two genes may vary greatly among groups. For example, *ndhF* provides approximately three times as many parsimony-informative characters as does *rbcL* in Acanthaceae, Scrophulariaceae, and Asteraceae, but in Solanaceae *ndhF* provides only 1.5 times as many such characters as does *rbcL* (Olmstead and Sweere, 1994; Olmstead and Reeves, 1995; Kim and Jansen, 1995). Similarly, *rbcL* generally evolves about three times faster than 18S rDNA in angiosperms; in Orchidaceae, in contrast, 18S rDNA evolves two times faster than *rbcL* (M. Chase, pers. comm.). Thus, whereas each gene has a "typical" taxonomic range of application, that range of applicability may vary considerably from group to group. Also, the entire scale may slide dramatically in some instances, particularly in rapidly evolving groups or in groups of great antiquity. For example, *rbcL* sequence data are typically used at higher taxonomic levels (family and above) in angiosperms, with some potential to resolve generic-level relationships in some angiosperm families (see below in Chloroplast Sequences-*rbcL*). In ferns, however, *rbcL* sequences have been of great value at the generic and even specific levels (e.g., Wolf, 1995; see Chapter 19). Similarly, in the red algae (Rhodophyta), levels of *rbcL* sequence divergence are typically much higher than those reported for angiosperms at the same taxonomic levels (Freshwater et al., 1994). These caveats should be kept in mind when choosing a gene for phylogeny reconstruction.

Even though we refer to genes as rapidly or slowly evolving, and useful at lower versus higher taxonomic levels, genes are actually a mosaic of individual characters (nucleotides) that may vary considerably in evolutionary rate. Thus, rapidly evolving regions such as the ITS may nonetheless have highly conserved areas that can be aligned over a broad taxonomic scale (Hershkovitz and Zimmer, 1996). That is, all nucleotide positions in a gene sequence are not created equal. Just as plant morphologists or phyleticists using morphological features become well acquainted with individual morphological characters, plant molecular systematists must begin to consider and become more familiar with individual nucleotide positions. Following this approach, for example, a phylogenetic study aimed at elucidating higher-level relationships could use not only conserved genes, but conserved nucleotide positions in genes generally described as rapidly evolving. This topic is discussed in more detail in Chapter 6.

Automated Versus Manual Sequencing

Automated sequencing is becoming more common, and its use will obviously increase in the future, either in individual labs, shared lab facilities, or commercial centers. In many regards, automated sequencing is ideal for the needs of the molecular systematist, for the strength of this approach lies in taking a well-characterized DNA region and sequencing it over and over again. In exploring the phylogenetic potential of poorly known genes or DNA regions, however, manual sequencing may still be preferable in some cases.

One misconception with automated sequencing is that all fluorescence signals (resulting in a chromatogram of nucleotide calls) are foolproof

and can be used immediately in phylogenetic studies. Based on our experience with the automated sequencers of Applied Biosystems Inc., each sequence must be carefully examined and edited base by base, comparing the fluorescence signals with the bases called. An excellent program for the assembly and editing of contigs (e.g., sequence fragments generated by different primers) is Sequencher (Gene Codes Corp., Inc., Ann Arbor, Michigan). This program provides additional control in the final editing of the sequence and gives the user the opportunity to invoke ambiguity codes, to override weak calls, to delete or insert bases in one strand, and to make other changes as necessary; the result is a much more accurate sequence.

The advantages of automated sequencing are several. Because each sample prepared contains all four nucleotides, a gel with 36 lanes will provide 36 primers worth of sequences, whereas with manual sequencing separate sequencing reactions are needed for each nucleotide. Thus, we routinely generate with an automated sequencer, three to four complete 18S rDNA sequences (approximately 1,800 bp in length) or nine complete ITS-1 and ITS-2 sequences (approximately 500 bp in length) per gel, with each base position sequenced at least twice. In addition, the newer automated sequencers have faster run times due to improvements in laser technology; a gel can be completed in four hours, providing 500 or more nucleotides per lane. With manual sequencing, in contrast, it is often necessary to do two separate loadings to obtain that much sequence. Another potential advantage of automated sequencing involves the use of cycle sequencing reactions in the preparation of samples. This approach, which may also be used in manual sequencing, reduces or eliminates secondary structure and therefore facilitates the sequencing of G+C rich areas (see also Soltis and Soltis, 1997; Soltis, Soltis, Nickrent et al., 1997).

One of the biggest time savers with automated sequencers involves the processing of data. Sequences no longer are read by eye from autoradiographs and then entered into a data matrix. With automated sequencing, once a sequence has been edited, it is simply exported to a data file.

In terms of expense, per sample or per sequence, the cost of automated and manual sequencing is comparable, provided the investigator runs the samples personally. Using a commercial facility to obtain sequences greatly increases the price of automated sequencing versus manual sequencing.

Chloroplast Sequences

rbcL   The gene *rbcL* is located in the large single-copy region of the chloroplast genome (Fig. 1.1) and encodes the large subunit of ribulose 1,5-bisphosphate carboxylase/oxygenase (RUBISCO); the small subunit of RUBISCO is encoded by *rbcS,* a nuclear gene. *rbcL* is typically 1,428, 1,431, or 1,434 bp in length; indels (insertions or deletions) are extremely rare.

The enormous phylogenetic utility of *rbcL* sequences has been emphasized and reviewed previously (e.g., Clegg, 1993; Palmer et al., 1988a; Chase, Soltis, Olmstead et al., 1993; see Chapter 17). At the family level and above, *rbcL* has, by far, been the preferred gene for inferring phylogeny. Not only have *rbcL* sequences been widely applied in analyses of extant taxa, but they have also been obtained from fossil leaf samples of seed plants from the Miocene, with some perhaps 17–20 million years old (Golenberg et al., 1990; P. Soltis et al., 1992b; Manen et al., 1995). The gene is among the most frequently sequenced segments of DNA in either plants or animals. *rbcL* sequence data have been so widely used that this single gene is the subject of an entire chapter of this volume (see Chapter 17). For this reason, our coverage of *rbcL* here is brief. Sequence data derived from *rbcL* have been used to address phylogenetic relationships not only in the angiosperms (see Chapter 17 for review), but also in ferns (e.g., Hasebe et al., 1993, 1995; Wolf, 1995; see Chapter 19), bryophytes (B. Mishler, pers. comm.), and various groups of algae (e.g., Freshwater et al., 1994; McCourt et al., 1995, 1996; see Chapter 18 for review of green algae). The application of *rbcL* sequence data spans a very wide taxonomic range. Although perhaps most widely used at the family level, *rbcL* sequences have been employed in very broad phylogenetic analyses involving all major groups of green algae (see Chapter 18) and all major lineages of ferns (Hasebe et al., 1995; see Chapter 19). Perhaps most impressive of these broad analyses is the use of 500 *rbcL* sequences to address phylogenetic relationships primarily within the angiosperms and secondarily among extant seed plants (Chase, Soltis, Olmstead et al., 1993). The broadest taxonomic application of *rbcL* sequence data is that of Manhart (1994), who attempted to elucidate relationships among all lineages of land plants. Although Manhart (1994) stated that this analysis may have overextended the application of *rbcL* sequence data, it is actually unclear as to whether some of the anomalous relationships recovered are indeed due to saturation of *rbcL* sequences or to insufficient taxon density.

As RUBISCO is a critical photosynthetic enzyme, *rbcL* was one of the first plant genes to be sequenced (e.g., Zurawski et al., 1981, 1986). The first papers using *rbcL* sequences to elucidate relationships include Doebley et al. (1990), Soltis et al. (1990), and Les et al. (1991); these studies preceded the use of PCR, relying instead entirely on cloned products for sequencing. With the advent of PCR, the *rbcL* sequencing explosion began, a process facilitated by G. Zurawski (DNAX, Inc.), who made available at no cost a complete set of sequencing primers. These primers were designed based on angiosperm sequences and work throughout the flowering plants, as well as seed plants in general. Given the enormous number of *rbcL* sequences that have accumulated, various aspects of the molecular evolution of *rbcL* have also been studied, including its general rate of evolution and evolution by codon position (e.g., Albert et al., 1994; see also Chapters 6 and 17).

Amplification and sequencing primers for *rbcL* vary among phyla. The following references provide technical information for various groups of algae and land plants: Rhodophyta (Freshwater and Rueness, 1994; Hommersand et al., 1994); Chlorophyta (see Chapter 18); Polypodiophyta (Hasebe et al., 1993, 1995; see Chapter 19); seed plants (Chase, Soltis, Olmstead et al., 1993; Gadek and Quinn, 1993; Brunsfeld et al., 1994). In most green plants, amplification primer Z1, which corresponds to the first 30 bases of *rbcL* of *Zea mays,* is widely used. At the 3′ end of *rbcL,* the amplification primer "3′ rbcL", which is downstream of the *rbcL* terminus, is widely used in angiosperms (Olmstead et al., 1992). In most other groups of plants, a different amplification primer located just inside the 3′ terminus of *rbcL* must be used.

The lower limit of applicability of *rbcL* sequences typically extends to the generic level, but in some groups reaches the specific level. Analyses of *rbcL* sequences have been used to resolve generic relationships within several families of flowering plants (e.g., Morgan et al., 1994; Xiang et al., 1993; Conti et al., 1993; Soltis et al., 1993; Kim and Jansen, 1996). In a few cases, *rbcL* sequence analyses have even clarified relationships among congeneric species: in *Cornus* (Xiang et al., 1993), *Saxifraga* (Soltis et al., 1996), and *Drosera* (Williams et al., 1994). In gymnosperms, examples of the use of *rbcL* at the generic level include Taxodiaceae/Cupressaceae (Brunsfeld et al., 1994; Gadek and Quinn, 1993). In ferns, application of *rbcL* sequences to resolve intergeneric and even interspecific relationships will likely be commonplace. The work of Wolf (1995, unpubl.; Wolf et al., 1994) and Haufler et al. (1996) illustrates the use of *rbcL* sequences at this low level, in Dennstaedtiaceae and Polypodiaceae, respectively (see also Chapter 19). That *rbcL* sequences can be used routinely to resolve relationships among congeneric species of ferns should not be surprising, given that allozyme surveys indicated that on average the genetic differentiation among congeneric fern species is comparable to that among related genera of angiosperms (Soltis and Soltis, 1989). Applica-

tions of *rbcL* at low taxonomic levels is also generally possible in the red algae (Rhodophyta) given that sequence divergence values are much higher than for angiosperms (Freshwater et al., 1994). Applying *rbcL* sequences to lower taxonomic levels may be enhanced somewhat by sequencing past the 3′ end of the gene. This is easily achieved in angiosperms given that the most widely used 3′ amplification primer (3′ *rbcL*) is actually located roughly 80 bp downstream of the terminus of *rbcL*. This region between the terminus of *rbcL* and the 3′ amplification primer has a general rate of base substitution comparable to that of *rbcL* and can provide additional parsimony-informative sites (e.g., Xiang et al., 1993; Morgan et al., 1994; Soltis et al., 1993, 1996).

atpB   *atpB* encodes the β subunit of ATP synthase, an enzyme that couples proton translocation across membranes with the synthesis of ATP (Zurawski et al., 1982; Gatenby et al., 1989; reviewed in Hoot et al., 1995a). The other subunits are encoded in either the chloroplast or nuclear genomes. *atpB* is located in the large single-copy region of the chloroplast genome just downstream from *rbcL* (Fig. 1.1); the two genes are separated by an intergenic spacer that is approximately 900 bp in length (Fig. 1.2) and that has phylogenetic utility itself (see below).

Ritland and Clegg (1987) first noted the phylogenetic potential of sequence data derived from *atpB*. Their analyses suggested that *atpB* has a rate of evolution virtually identical to that of *rbcL*. Additional investigations further confirmed the similar properties of *atpB* and *rbcL*

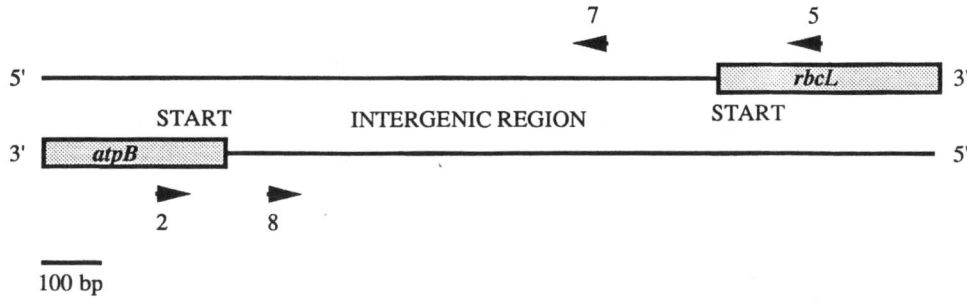

Figure 1.2.   Diagram showing close proximity of *atpB* and *rbcL* genes. Also illustrated is the *atpB–rbcL* intergenic spacer region and the general location of primers that can be used in its amplification and sequencing.

(Hoot et al., 1995a). The length of *atpB* (1,497 bp) is also very similar to that of *rbcL* (typically 1,428, 1,431, or 1,434 bp in angiosperms). As recently reviewed (Hoot et al., 1995a), no indels have been reported for *atpB,* and the gene does not contain introns; thus, *atpB* sequences are easily aligned. The work of Hoot et al. (1995a) indicates that a segment of DNA containing most of *atpB* (1,474 or 1,497 bp) is readily amplified via PCR. Given the conservative nature of *atpB,* the amplification and sequencing primers given in Hoot et al. (1995a) (Fig. 1.3) work broadly across the angiosperms (Savolainen et al., 1996).

The sequencing of *atpB* has greatly increased recently as investigators seek additional base pairs to resolve relationships further and turn to alternative genes to test *rbcL*-based topologies (see Hoot et al., 1995a, 1995b; Savolainen et al., 1996; Wolf, 1996). For example, Savolainen et al. (1996, unpubl.) are sequencing *atpB* across all major lineages of angiosperms for ultimate comparison with the *rbcL* (Chase, Soltis, Olmstead et al., 1993) and 18S rDNA (Soltis, Soltis, Nickrent et al., 1997) trees for angiosperms. The work of Savolainen et al. confirms that *atpB* and *rbcL* have similar rates of evolution in angiosperms and also suggests that *atpB* seems to provide better resolution of relationships. Similarly, *atpB* sequences have great utility in ferns at the same taxonomic levels as *rbcL* (see Chapter 19).

matK Among protein-coding regions in the chloroplast genome, *matK* (formerly ORFK) is one of the most rapidly evolving (Wolfe, 1991). *matK* is located in the large single-copy region of the chloroplast genome (Fig. 1.1); it is approximately 1,550 bp in length and encodes a maturase involved in splicing type II introns from RNA transcripts (Neuhaus and Link, 1987; Wolfe et al., 1992). In all photosynthetic land plants so far examined, *matK* is located within an intron of approximately 2,600 bp positioned between the 5' and 3' exons of the transfer RNA gene for lysine, *trnK* (Fig. 1.4). *matK* is present even in the reduced chloroplast genome of the nonphotosynthetic parasite *Epifagus* (Orobanchaceae), although neither the *trnK* coding regions nor associated intron are present (Wolfe et

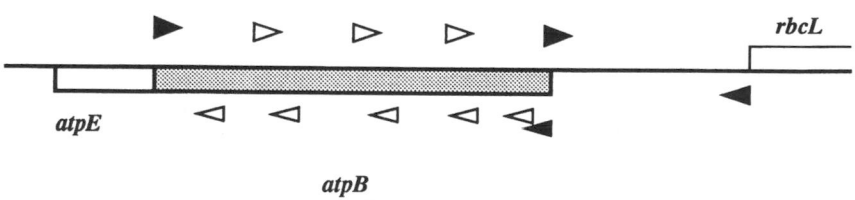

Figure 1.3. General map of *atpB* and location of PCR and sequencing primers. Black arrows indicate the location and direction of the amplification primers. White arrows indicate the location and direction of the internal sequencing primers. Modified from Hoot et al. (1995a).

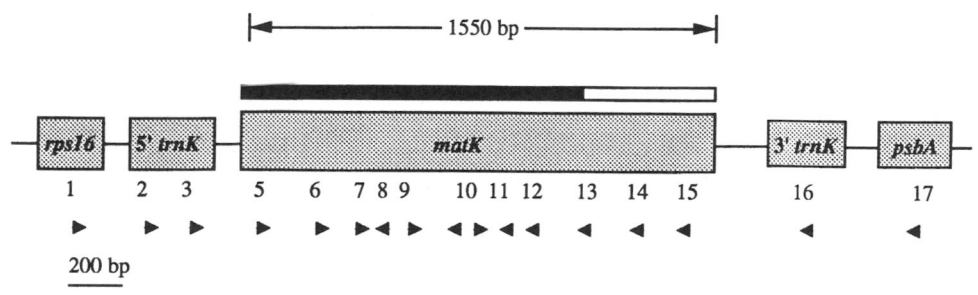

Figure 1.4. Relative location of amplification and sequencing primers used for sequencing *matK* and the *trnK* intron. Black portion of bar above *matK* gene indicates the approximately 1,000-bp region that is typically sequenced. Numbers refer to primers given in Johnson and Soltis (1995). Modified from Johnson and Soltis (1995).

al., 1992). Additional background information is given in Steele and Vilgalys (1994) and Johnson and Soltis (1994, 1995).

The gene *matK*, as well as the noncoding regions that flank it, are easily amplified using the highly conserved flanking coding regions that include the *trnK* exons and the genes *rps16* and *psbA* (Fig. 1.4). Using *matK* sequences from Saxifragaceae, Polemoniaceae, and Brassicaceae (Neuhaus and Link, 1987), Johnson and Soltis (1994, 1995) developed a suite of sequencing primers for the entire gene. Several of these sequencing primers (1412F, 1470R, 1235R) have broad applicability and have been used successfully throughout the dicots. However, because of the relatively rapid rate of evolution of *matK* compared to more conservative genes, such as *rbcL*, it will likely be necessary to design additional sequencing primers for the study group of interest. *matK* is widely used in angiosperms, but it has rarely been used in other plant groups, although P. Gadek and C. Quinn (unpubl.) have successfully employed *matK* in studies of Taxodiaceae/Cupressaceae.

The rate of evolution of *matK* makes this gene appropriate for resolving intergeneric or interspecific relationships in seed plants. Based on data for Saxifragaceae (Johnson and Soltis, 1995), Cornaceae (Xiang et al., 1998), and Taxodiaceae/Cupressaceae (P. Gadek and C. Quinn, unpubl.; Johnson and Soltis, 1995), the average number of nucleotide differences per site in pairwise comparisons for *matK* are 3.2, 2.4, and 3.4 times higher, respectively, than for *rbcL*. Comparisons of several members of Araliales suggest that the average number of nucleotide differences is 1.3 times greater for *matK* than for the chloroplast gene, *ndhF*, although the information content of the two genes may be roughly comparable in Araliales given that *ndhF* is 1.4 times longer than *matK* (Johnson and Soltis, 1995). In Polemoniaceae, the number of nucleotide differences is on average 1.9 times greater for *matK* than ITS. The information content of *matK* is similar to or greater than that of ITS, however, given that *matK* is 3.1 times longer. *matK* sequences may therefore be informative at the generic and even species levels (Johnson and Soltis, 1995; Soltis et al., 1996). Investigations are just beginning to explore the upper limits of the phylogenetic utility of *matK*. Ongoing studies indicate the utility of *matK* sequences for resolving familial relationships within Saxifragales and for identifying the closest relatives of Cornaceae and Apiaceae.

Other aspects of the molecular evolution of *matK* have also been reported. In Polemoniaceae, Saxifragaceae, and Cornaceae, the numbers of transitions and transversions are essentially identical (Johnson and Soltis, 1995; Xiang et al., 1998); in contrast, *rbcL* is biased toward transitions. In all three families, the frequency of mutations in *matK* at the first and second codon positions closely approaches the frequency of mutations in the third codon position. In contrast, in *rbcL* the number of substitutions in the third codon position is much higher than the numbers of substitutions at the first and second positions, as is typical of protein-coding genes with strong functional constraints.

Indels are likely to be present in a *matK* data matrix of any taxonomic breadth (e.g., Polemoniaceae, Saxifragaceae s.s., Apiales, Cornaceae, and Cupressaceae, reviewed in Johnson and Soltis, 1995). However, the number of indels is generally small; a 76-taxon *matK* data set for Apiales contains 12 indels (Plunkett et al., 1997a). All of the indels in the families noted are small, either 3, 6, or 9 bp. Hence, alignment of *matK* sequences is easily accomplished by eye. Although for any DNA region, the homology of indels may be difficult to ascertain, a posteriori mapping of *matK* indels onto trees based on nucleotide substitutions reveals that these indels are often parsimony-informative, providing additional support for monophyletic groups identified by base substitutions (Johnson and Soltis, 1995).

Comparative sequencing of *matK* has great potential for retrieving phylogeny within families and genera of land plants. Most studies have obtained well resolved phylogenies using approximately two-thirds (~ 1,000 bp) of the 1,550-bp *matK* gene; some studies have used considerably less (Steele and Vilgalys, 1994). In Saxifragaceae s.s., *matK* sequences provide a level of resolution comparable to that achieved with cpDNA restriction sites; it has also been possible to use *matK* sequences to discern the maternal parent of allopolyploids in *Saxifraga*

(Johnson and Soltis, 1995). Well-resolved generic and species-level phylogenies have been obtained using *matK* sequences in Saxifragaceae s.s. (Johnson and Soltis, 1995; Soltis et al., 1996), Polemoniaceae (Johnson and Soltis, 1995; Johnson et al., 1996), Apiales (Plunkett et al., 1997a, 1997b), Cornaceae (Xiang et al., 1997), Annonaceae (M. Chase et al., unpubl.), Ericaceae (Kron, 1997; K. Kron et al., unpubl.), and Cupressaceae/Taxodiaceae (P. Gadek and C. Quinn, unpubl.). Many other families are the focus of ongoing studies and further exemplify this applicability within families (e.g., Crassulaceae, Rubiaceae). In several angiosperm families, *rbcL* and *matK* data have been combined, providing enhanced resolution and internal support compared to either gene alone (e.g., Soltis et al., 1996; Xiang et al., 1998; Plunkett et al., 1997b). To date, phylogenetic analyses using the *trnK* region have concentrated on *matK*. However, the *trnK* intron regions flanking *matK* may provide additional useful phylogenetic information. In Saxifragaceae the 5′ flanking region is slightly less variable than *matK,* but the 3′ flanking region is considerably more variable (Johnson and Soltis, 1994, 1995).

ndhF   The gene *ndhF* is located in the small single-copy region of the chloroplast genome close to the junction with the inverted repeat (Figs. 1.1, 1.5). In fact, the termination codon of *ndhF* in tobacco is located only 43 bp from the boundary between the inverted repeat and the small single-copy region. In tobacco, *ndhF* is 2,233 bp in length and hence approximately 1.5 times longer than *rbcL.* The gene encodes a protein that appears to be a subunit of chloroplast NADH dehydrogenase (see Kim and Jansen, 1995; Olmstead and Reeves, 1995).

Comparison of *ndhF* sequences of rice and tobacco suggests that *ndhF* has a nucleotide substitution rate approximately two times higher than that of *rbcL* (reviewed in Olmstead and Reeves, 1995). In studies of intergeneric relationships, *ndhF* provided roughly three times as many characters as *rbcL* in both Acanthaceae (Scotland et al., 1995) and Asteraceae (Kim and Jansen, 1995). In Solanaceae and Scrophulariaceae, however, *ndhF* provided only 60% and 50% more characters, respectively, than *rbcL* (Olmstead and Sweere, 1994; Olmstead and Reeves, 1995).

As previously noted (e.g., Kim and Jansen, 1995; Olmstead and Reeves, 1995), *ndhF* consists of two very different regions. The 5′ region of the gene (1,380 bp) is more similar to *rbcL* in both rate and pattern of nucleotide substitution. In contrast, the 3′ portion of *ndhF* (855 bp) is more A+T rich, has higher levels of nonsynonymous base substitutions, and shows a greater transversion bias at all codon positions (Kim and Jansen, 1995). Kim and Jansen (1995) suggest that the distinct evolutionary patterns in the 5′ and 3′ portions of *ndhF* likely reflect different

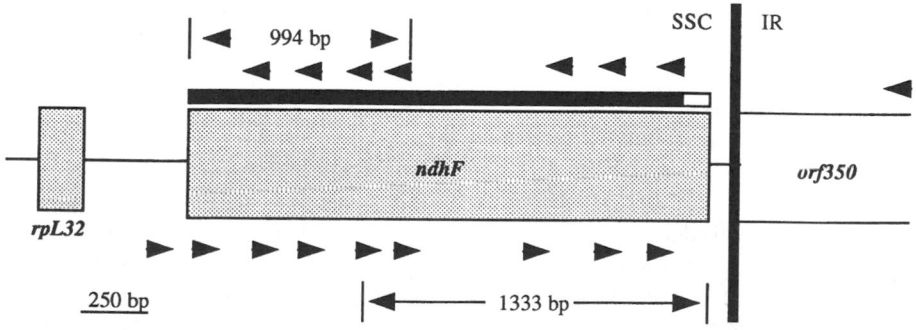

Figure 1.5.   Map of *ndhF* and adjoining regions of cpDNA in tobacco. Boxes indicate reading frames; lines connecting boxes indicate noncoding DNA. Vertical bar at left end of *orf350* indicates junction between inverted repeat (IR) and small single-copy region (SSC). Arrows indicate location of PCR and sequencing primers. Black portion of bar above *ndhF* indicates the region that is typically sequenced—all but approximately 120 bp at the 3′ end, which is not amplified. Overlapping regions marked above and below the map indicate the two fragments in which the gene was amplified. Modified from Olmstead and Reeves (1995).

functional constraints. Kim and Jansen (1995) also note that the presence of these two different patterns of evolution within the same gene may be advantageous for phylogenetic reconstruction. That is, the conserved and more variable portions may be useful for inferring relationships in older and more recently evolved groups, respectively. The more variable domains, for example, have made it possible to infer relationships in recently evolved groups, such as Poaceae subfamily Pooideae (Catalan et al., 1997).

The gene *ndhF* is typically amplified in two overlapping segments. The amplification and sequencing primers described in Olmstead and Sweere (1994) appear to have very broad applicability in the angiosperms. As with other protein-coding genes of the chloroplast genome (*rbcL, atpB, matK*), alignment of *ndhF* sequences is easily accomplished by eye. As was observed for *matK,* indels appear to be a fairly regular feature of *ndhF* sequences. The insertions and deletions detected for Asteridae are mostly confined to a 250-bp region between positions 1,443 and 1,697 of the tobacco gene. This portion of *ndhF* also appears to have an accelerated rate of base substitution (Olmstead and Sweere, 1994). Some indels in *ndhF* (in Solanaceae, Olmstead and Sweere, 1994; and Acanthaceae, Scotland et al., 1995) are phylogenetically informative, whereas others (in Acanthaceae, Scotland et al., 1995; and Poaceae, Clark et al., 1995) are clearly homoplasious. As noted by Olmstead and Reeves (1995), the parallel occurrence of *ndhF* indels is not surprising given their localized distribution. It may be that indels in *ndhF* may be of less phylogenetic utility than in genes such as *matK,* where indels appear to be more evenly distributed across the gene.

The utility of *ndhF* for resolving generic-level relationships has been well demonstrated within a number of angiosperm families, including Acanthaceae (Scotland et al., 1995), Solanaceae (Olmstead and Sweere, 1994), Scrophulariaceae (Olmstead and Reeves, 1995), Asteraceae (Kim and Jansen, 1995), Orchidaceae (Neyland and Urbatsch, 1996), and Poaceae (Clark et al., 1995). At higher levels, *ndhF* sequences have helped to resolve relationships in Asteridae s. 1.

(Kim and Jansen, 1995). For those families for which *rbcL* and *ndhF* sequences are available for a similar suite of taxa, trees based on *ndhF* are more fully resolved than those based on *rbcL.*

16S rDNA   The chloroplast 16S rRNA gene, located within the inverted repeat (Fig. 1.1), is approximately 1,600 bp in length. A complete set of amplification and sequencing primers is provided by Manhart (1995). The gene for the small subunit (SSU) of ribosomal RNA is highly conserved and has been used primarily to reconstruct phylogeny at deep levels. For example, chloroplast 16S rRNA sequences were used along with other SSU sequences to infer the deepest branches of life (e.g., Wolters and Erdmann, 1986; Pace et al., 1986), with chloroplast (as well as mitochondrial) 16S rRNA sequences nested among bacterial sequences. Other investigators have conducted broad phylogenetic analyses of cyanobacteria using l6S rDNA sequences and have included chloroplast sequences (Woese, 1987; Giovannoni et al., 1988; Urbach et al., 1992).

Support for the phylogenetic utility of chloroplast 16S rDNA sequences within plants is still preliminary. Mishler et al. (1992) used 370 bp of both the chloroplast 16S and 23S rRNA genes from 11 bryophytes and concluded that these sequences may be able to resolve the deep branches of land plant phylogeny. Manhart (1995) recently used nearly complete 16S rDNA sequences representing over 20 land plants to assess relationships among the ferns and fern allies and similarly concluded that with the inclusion of more taxa, 16S rDNA sequences may contribute to our understanding of phylogenetic relationships among land plants.

atpB–rbcL Intergenic Region   The noncoding intergenic region between the 3′ ends of *atpB* and *rbcL* is approximately 900 bp in length (Figs. 1.1, 1.2), although the length may vary considerably due to insertions and deletions (Manen et al., 1994a). Nonetheless, some highly conserved areas are present that correspond to the promoter elements of the two genes. The amplification primers typically used for this region are highly conserved given that they rest within the genes *atbB* and *rbcL.* The utility of the *atpB–rbcL* intergenic region has been explored primarily in angiosperms (e.g., Golenberg et al., 1993; Manen et al., 1994b; Natali et al., 1995; Savolainen et al., 1994, 1995a). These studies suggest that this noncoding region may be particularly useful within and between genera. In some instances, species-level relationships were also resolved (Manen et al., 1994b; Natali et al., 1995). Although variation was low, *atpB–rbcL* intergenic sequences revealed the possible maternal parent of *Malus* × *domestica,* as well as the possible maternal parentage of some apple cultivars (Savolainen et al., 1995a).

Indels appear to be frequent in the *atpB–rbcL* intergenic region, and some indels may be phylogenetically informative. Natali et al. (1995) found two large deletions that agreed with base substitutions in defining monophyletic groups in Rubiaceae: A deletion of 204 bp defines subfamily Rubioideae, and a deletion of approximately 50 bp unites members of tribe Rubieae. In contrast, in a study of nine grasses, Golenberg et al. (1993) found that indels in this region apparently recurred at a small number of labile sites and were homoplasious; they cautioned against the use of indels from noncoding regions in general.

There are many different intergenic regions in the chloroplast genome that could potentially be used for phylogenetic inference at lower taxonomic levels. Particularly useful will be those intergenic regions that are flanked by nearby, highly conserved genes. Such regions will be easy to amplify in a wide range of plants and will also provide short segments of DNA that are easily sequenced. It is therefore likely that many additional intergenic regions will be exploited in the next 5 years.

Other Noncoding Chloroplast Sequences Other noncoding sequences of the chloroplast genome also have phylogenetic potential, including the *trnL* (UAA) intron and the intergenic spacer between the *trnL* (UAA) 3′ exon and the *trnF* (GAA) gene (Taberlet et al., 1991; Gielly and Taberlet, 1994, 1996) (Fig. 1.1). Initial comparisons suggest that these regions may evolve at rates similar to that of *rbcL* to as much as three times faster than *rbcL,* depending on the study group. These regions are easily amplified and sequenced (see Taberlet et al., 1991). They are also relatively small, with the *trnL* intron ranging from 350–600 bp and the *trnL–trnF* spacer ranging from roughly 120 to 350 bp in the monocots and dicots initially sampled. Sequences of the *trnL* intron resolved relationships among eight species of *Gentiana* (Gielly and Taberlet, 1994, 1996). Several recent studies have employed both the *trnL* (UAA) intron and the *trnL–trnF* (GAA) intergenic spacer. The level of variation in this *trnL–trnF* data set is comparable to that of *rbcL* for Themidaceae (formerly Brodiaeeae of Alliaceae) (Fay et al., 1996) and Orchidaceae (Whitten et al., 1996). In contrast, the *trnL* regions appear to be evolving about three times faster than *rbcL* in Iridaceae (M. Chase, pers. comm.). In the fern family Ophioglossaceae, the *trnL–trnF* intergenic spacer is about one-third the size of *rbcL,* with sequence divergences for the spacer three to five times higher than sequence divergence for *rbcL* for the same taxa (Hauk et al., 1996). Separate analyses of *rbcL* and *trnL–trnF* spacer sequences produced nearly identical topologies with similar branch lengths.

Goremykin et al. (1996) developed amplification and sequencing primers for a noncoding region in the inverted repeat of the chloroplast genome that spans from the 3′ end of the 23S rRNA gene to the 5′ end of the 5S rRNA gene (Fig. 1.6). This highly conserved region includes

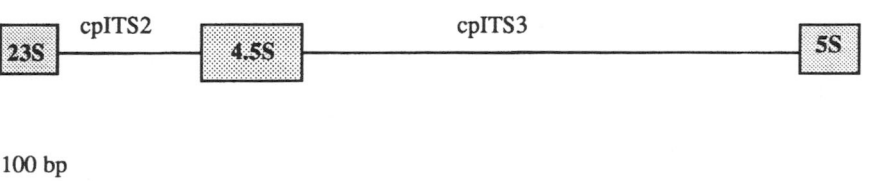

100 bp

Figure 1.6.   Schematic representation of a portion of the inverted repeat in the chloroplast genome showing the two chloroplast intergenic transcribed spacer region (cpITS2 and cpITS3) employed by Goremykin et al. (1996).

the 4.5S rRNA gene and two chloroplast inter-genic transcribed spacer regions (cpITS2 and cpITS3). Goremykin et al. (1996) employed se-quences from these spacer regions (cpITS se-quences) to examine relationships among vas-cular plants. Despite the small size of these sequences (approximately 500 bp), Goremykin et al. concluded that cpITS sequences are useful at higher taxonomic levels; their data suggest, for example, that Gnetales are part of a gymnosperm clade and are not sister to the angiosperms, a finding in agreement with recent analyses of the much longer 18S rDNA sequences (Chaw et al., 1997; P. Soltis et al., unpubl.).

Other Chloroplast Sequences   A number of other chloroplast genes have phylogenetic po-tential. For example, dePamphilis et al. (1997) used sequence data for *rps2,* which is located in the large single-copy region (Fig. 1.1) and en-codes the S2 subunit of the plastic ribosome, to examine the relationship of the angiosperm fam-ilies Orobanchaceae and Schrophulariaceae. Us-ing 613 bp of sequence data, they produced trees suggesting that Orobanchaceae are derived from within Scrophulariaceae. Thus, this gene may have potential for reconstructing phylogeny at the familial and generic levels. More informa-tion on amplification and sequencing primers may be obtained from C. dePamphilis.

Another gene with phylogenetic potential is *rps4,* which encodes the S4 subunit of the plastic ribosome. Nadot et al. (1994, 1995) first demon-strated the phylogenetic utility of this gene in studies of intergeneric relationships in Poaceae and the monocots. The region is short (approxi-mately 600 bp) and appears to exhibit little size variation. For example, a 78-taxon data set in-cluding monocots, dicots, a conifer, cycad, fern, and liverwort was easily aligned by eye (Nadot et al., 1995). The PCR and sequencing primers de-veloped by Nadot et al. (1994) seem to have wide applicability and have been employed success-fully in a diverse array of monocots and dicots (M. Chase, pers. comm.). Based on preliminary data for Iridaceae, Annonaceae, and Orchi-daceae, the level of variation of *rps4* sequences is comparable to, or greater than, that of *rbcL* (M. Chase, pers. comm.). By itself, the *rps4* gene may not be long enough to address many phylo-

genetic questions, but because it is easily ampli-fied and sequenced, it appears to be an efficient means of obtaining an additional cpDNA data set (M. Chase, pers. comm.)

Combining Chloroplast Sequences   Be-cause the chloroplast genome is inherited as a unit and is not subject to recombination, chloro-plast DNA sequences and restriction sites can be readily combined (e.g., Soltis et al., 1993, 1996; Johnson and Soltis, 1995; Olmstead and Sweere, 1994; Olmstead and Reeves, 1995; Plunkett et al., 1997b). Several investigators have, for ex-ample, combined three chloroplast DNA data sets. Olmstead and Sweere (1994) combined cpDNA restriction sites and *rbcL* and *ndhF* se-quences for Solanaceae; Johnson and Soltis (1995) combined cpDNA restriction sites and *rbcL* and *matK* sequences for Saxifragaceae s. s. The advantages of combining chloroplast DNA data sets, as well as data sets in general, include: (1) greatly shortened run times, (2) enhanced resolution, (3) increased internal support for clades, and (4) the presence of uniquely sup-ported clades when compared to the separate data sets (see Olmstead and Sweere, 1994; Soltis et al., 1997b).

Combining data sets will be particularly use-ful in the analysis of very large data sets. The simulation studies of Hillis (1996; using the 232-taxon data set of 18S rDNA sequences for angiosperms of Soltis, Soltis, Nickrent et al., 1997), as well as recent empirical studies (Soltis et al., 1997b) demonstrate that large data sets are tractable in terms of phylogenetic analysis given sufficient data. Soltis et al. (1997b) combined *rbcL* and 18S rDNA sequences for 230 an-giosperms and observed all four of the features noted above. Significantly, parsimony searches of the combined data sets swapped to comple-tion in several days; searches of the individual data sets did not swap to completion in an entire month. In fact, phylogenetic analyses of a com-parably sized data set of only 18S rDNA se-quences did not swap to completion despite over two years of computer time (Soltis, Soltis, Nick-rent et al., 1997). When large data sets of *atpB, rbcL,* and 18S rDNA sequences are combined, run times are further shortened, and resolution and internal support of clades are further enhanced (M. Chase et al., unpubl.; D. Soltis et al., unpubl). As these studies also demonstrate, combining data is not limited to chloroplast genes; nuclear, cpDNA, mtDNA, and morphological, anatomical, and chemical data sets can all be combined, given that the individual data sets are essentially congruent (see Chapters 10 and 11). These and other recent studies in which data sets have been combined set the stage for the next generation of phylogenetic analyses in plants in which multiple data sets representing several genes or DNA regions, as well as other characters, are routinely combined in phylogenetic analyses conducted at all taxonomic levels.

Nuclear Sequences

Despite the large size of the nuclear genome and the large number and diversity of genes that it

includes, most attempts to infer phylogeny with nuclear gene sequences have involved the nuclear ribosomal DNA cistron (rDNA). Several other nuclear genes (e.g., *Pgi, gapA,* and *adh*) appear to have potential as well and are discussed at the end of this section. One obvious by-product of this primary reliance on ribosomal gene regions as a source of nuclear sequence data is a phylogenetic gap in coverage that can be seen clearly in Fig. 1.7. Highly conserved coding regions (18S, 26S rDNA) are useful primarily at the family level and above, whereas rapidly evolving regions such as ITS are often best suited for comparing species and closely related genera. Lacking is a DNA region or regions that will resolve intergeneric relationships across moderate to large families and among closely related families. ITS and 18S–26S rDNA have some applicability at the lower and

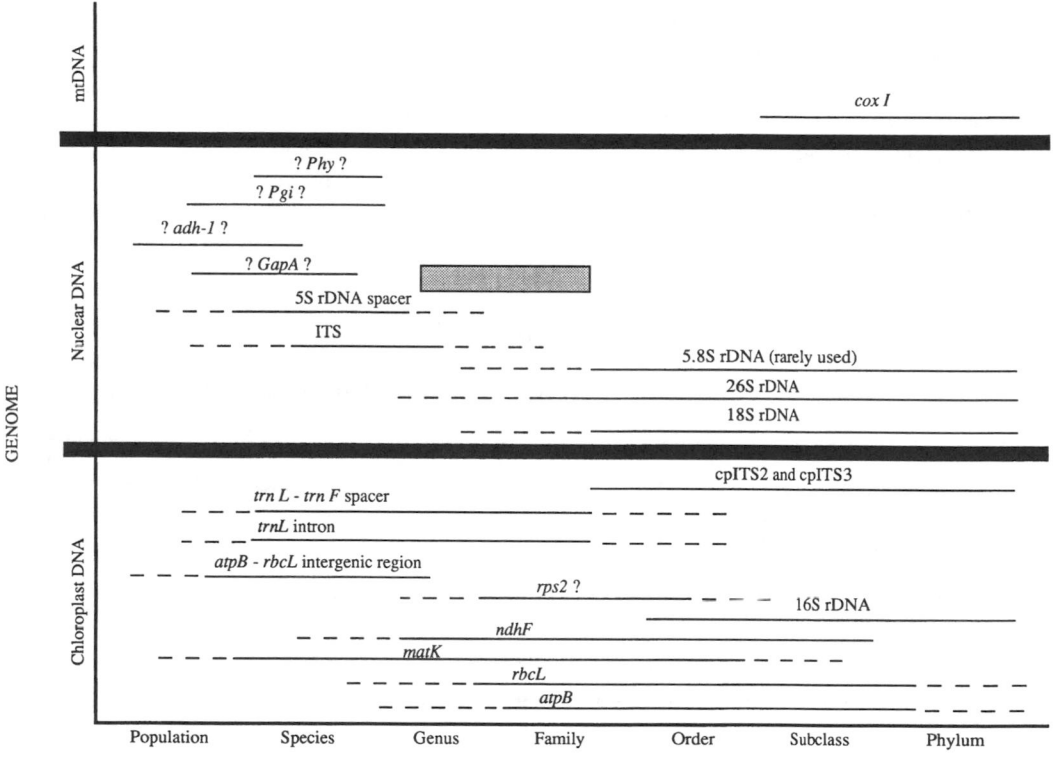

Figure 1.7.   Approximate taxonomic level of utility of various chloroplast, nuclear, and mitochondrial genes and DNA regions used in phylogenetic reconstruction based on angiosperms. The shaded box represents the taxonomic zone that is not presently well covered by nuclear gene sequences. ? refers to genes that have been rarely used; – – – – designates the approximate upper or lower limits of applicability (i.e., the region may work at this level in some groups).

upper end of this range, respectively, but additional nuclear genes are clearly needed to cover this taxonomic level adequately.

The basic structure of rDNA is illustrated in Fig. 1.8, in which a single repeat unit is depicted. Each such repeat is reiterated thousands of times within most plant genomes (Appels and Dvorak, 1982; Appels and Honeycutt, 1986), with rDNA representing as much as 10% of total plant DNA (Hemleben et al., 1988). The number of rDNA repeat units is highly variable (Rodgers and Bendich, 1987; Jorgensen and Cluster, 1988; Appels and Honeycutt, 1986; Riven et al., 1986; Bobola et al., 1992; Govindaraju and Culli, 1992). As reviewed by others (e.g., Hamby and Zimmer, 1992), each repeat consists of a transcribed region that comprises an external transcribed spacer (ETS), followed by the 18S gene, an internal transcribed spacer (ITS-1), the 5.8S gene, a second internal transcribed spacer (ITS-2), and finally the 26S gene. Each such repeat is separated from the next repeat by an intergenic spacer (IGS). Each of these regions is covered in more detail below.

The approximate lengths of the three coding regions are very similar throughout plants: The 18S gene equals 1,800 bp (Nickrent and Soltis, 1995); the 26S gene equals 3,300 bp (Burt et al., 1995); the 5.8S gene equals 160 bp (Takaiwa et al., 1985; Tanaka et al., 1980). In contrast, the length of the IGS varies considerably (from 1 to 8 kb; Jorgensen and Cluster, 1988). This variation in IGS length is the major contributor to the large range of variation in total length of the repeat unit in plants, ranging from approximately 8 to 15 kb (Hemleben et al., 1988). Variation in the length of the ITS-1 and ITS-2 regions is also noteworthy. Whereas the length of these two regions in angiosperms is very similar and falls within a fairly narrow window, considerable variation in the length of ITS-1 has been noted in conifers (see below in ITS section).

The reasons for the emphasis on the rDNA cistron as a source of nuclear sequence data are well reviewed (e.g., Appels and Honeycutt, 1986; Hamby and Zimmer, 1992; Mindell and Honeycutt, 1990; Hillis and Dixon, 1991; Sanderson and Doyle, 1992; Nickrent and Soltis, 1995) and include the tandem repeat structure and extremely high copy number of rDNA (Rodgers and Bendich, 1987), which make it easy to amplify or clone. These genes are also found in all organisms, a feature of particular import to those interested in the deep branches of the history of life or to those who study parasitic plants in which the chloroplast

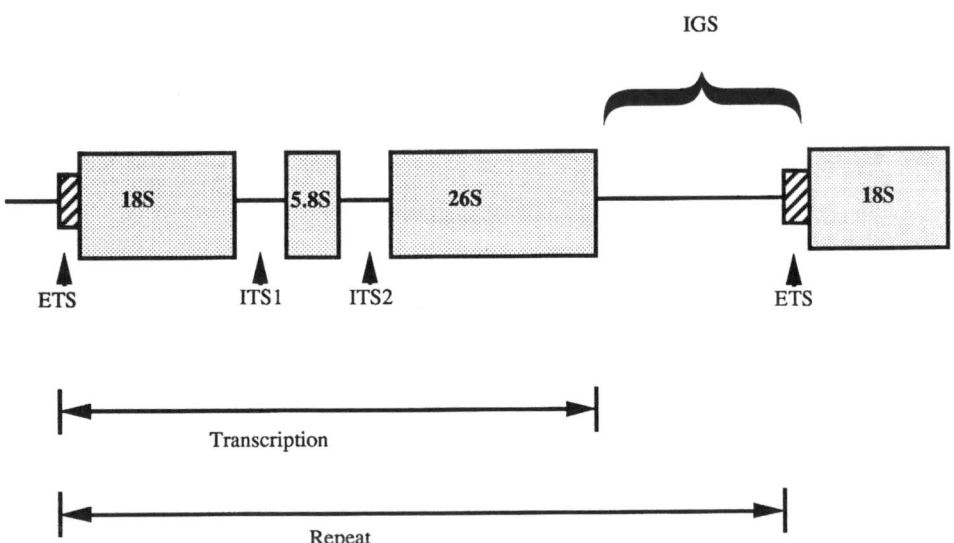

Figure 1.8.   Schematic diagram of rDNA repeat in plants. 18S, 5.8S, and 26S refer to the ribosomal rRNA genes. ITS-1 and ITS-2 are the two internal transcribed spacer regions. IGS is the intergenic spacer; ETS is the external transcribed spacer.

genome has been greatly reduced (see Chapter 8). The rDNA gene family comprises very conserved regions (i.e., the 18S and 26S genes) that can be used to infer phylogeny at higher taxonomic levels, as well as more rapidly evolving segments (i.e., ITS, IGS) that may be useful at the generic, specific, and even (in the case of IGS) the populational levels.

Ribosomal DNA cistrons typically are located in the nucleolar organizing region (NOR) and may be present on several different chromosomes (Thompson and Flavell, 1988). Within a single organism, sequence similarity between individual cistrons is extremely high; this homogeneity of cistrons has been referred to as horizontal evolution (Brown et al., 1972) and coincidental evolution (Hood et al., 1975), but it is typically known as concerted evolution (Arnheim et al., 1980; Zimmer et al., 1980). Possible mechanisms for concerted evolution include unequal crossing over or unequal exchange, gene conversion, transposition, and slippage (e.g., Arnheim et al., 1980; Dover, 1982; Arnheim, 1983). The essentially complete homogenization of ribosomal loci appears to make them advantageous for reconstructing deep phylogenetic events (Sanderson and Doyle, 1992). The structure, function, gene organization, evolution, and phylogenetic utility of ribosomal DNA have been summarized (e.g., Jorgensen and Cluster, 1988; Hillis and Dixon, 1991; Hamby and Zimmer, 1992; Nickrent and Soltis, 1995).

The fact that this gene family undergoes rapid concerted evolution (Arnheim et al., 1980; Zimmer et al., 1980; Arnheim, 1983; Appels and Dvorak, 1982) permits the use of these genes, despite their high copy number, for inferring phylogeny (Sanderson and Doyle, 1992). However, although concerted evolution homogenizes rDNA repeats, recent work not only indicates that the evolution of these regions may be more complex than thought initially, but also suggests caution in phylogeny reconstruction, particularly at lower taxonomic levels. At least some plant genomes exhibit rDNA diversity that includes multiple functional loci, pseudogenes, and recombinants (Buckler et al., 1997; Buckler and Holtsford, 1996a, 1996b). The potential impact of this rDNA diversity on phylogeny reconstruction is particularly great with more rapidly

evolving regions, such as internal transcribed spacer sequences, but likely is of little concern with the more conserved rRNA coding regions (see below).

Nuclear Ribosomal Gene Sequences (18S, 26S, 5.8S, 5S rDNA)   In plants, ribosomes occur in the chloroplasts, mitochondria, and cytoplasm; each is composed of a large and a small subunit containing rRNA and proteins. Thus, there are three potential genomic sources of large and small subunit rDNA sequences in plants; this chapter concentrates on nuclear ribosomal sequence data; the potential of chloroplast rDNA sequence data was noted previously, and the mitochondrial rDNA has not been explored extensively in plants. The nuclear small and large subunit rDNAs are referred to here, respectively, as 18S rDNA and 26S rDNA, although sedimentation coefficients vary. General background for both the 18S and 26S genes are provided; the small 5S and 5.8S genes, which contribute to the large ribosomal subunit, are discussed only briefly.

Prior to 1990, most nuclear ribosomal sequences were RNA sequences, determined using cloned material or via reverse transcriptase with rRNA templates and Sanger dideoxynucleotide reactions (e.g., Lane et al., 1985). Beginning in the early 1990s, the 18S and 26S genes were amplified via PCR and sequenced directly. The main reasons for this shift (given in Nickrent and Soltis, 1995) are: (1) rRNA is difficult to extract, purify, and store, due to its lability; (2) the secondary structure inherent in rRNA causes polymerase stalling during sequencing, which produces what are termed *hard stops* on sequencing gels, and ultimately results in ambiguous sequence (this secondary structure also makes direct sequencing of rDNA difficult, although to a lesser extent; see Soltis, Soltis, Nickrent et al., 1997); (3) only RNA genes can be sequenced from RNA extracts, whereas, using total genomic DNA extracts, any gene from the nucleus, chloroplast, or mitochondrion can, in theory, be sequenced. For example, many of the same DNAs initially used to amplify and sequence *rbcL* are now being used for the amplification and sequencing of *atpB,* 18S rDNA, *ndhF, matK,* and other genes; (4) both strands are available for sequencing DNA, providing the opportunity to double-check each base position, a benefit not possible with RNA. Because of the secondary structure inherent in rRNAs, investigators need to be aware that manual sequencing of rDNA is more difficult than chloroplast genes. However, preparation of sequencing reactions by cycle sequencing followed by automated gel runs seems to alleviate these problems (Soltis, Soltis, Nickrent et al., 1997).

Both small subunit (18S) and large subunit (26S) rRNA and/or rDNA sequences have been used to address a wide variety of higher-level phylogenetic questions. 18S and 26S rRNA or rDNA sequences have been used to examine the deepest branches of life, revealing the domains Eukarya, Bacteria, and Archaea (Wolters and Erdmann, 1986; Olsen, 1987; Woese, 1987; Embley et al., 1994). They have also been employed at high taxonomic levels (family and above) within many groups, including animals (e.g., Sogin et al., 1986; Field et al., 1988; Wada and Satoh, 1994; Pawlowski et al., 1996), protozoa (e.g., Schlegel et al., 1991), fungi (e.g., Förster et al., 1990; Swann and Taylor, 1993; Hinkle et al., 1994; Chapela et al., 1994), and lichens (Gargas et al., 1995). The use of 18S and 26S rRNA and rDNA sequences for phylogenetic reconstruction of plants is reviewed below in 18S rDNA and 26S rDNA (see also Chapters 18 and 19).

Initial applications of rRNA or rDNA sequence data often involved only portions of the 18S and 26S genes. One of the early attempts to use rRNA data to resolve phylogenetic relationships in vascular plants used a total of 1,701 base positions per taxon, 1,097 (of approximately 1,800 bp) from the 18S rRNA and 604 (of approximately 3,300 bp) from the 26S rRNA (Hamby and Zimmer, 1992). Partial 18S sequences were also used in phylogenetic studies of Onagraceae (Bult and Zimmer, 1993) and Poaceae (Hamby and Zimmer, 1988). Mishler et al. (1994) used partial sequences to study relationships among land plants. Early studies of green algae similarly employed partial sequences of both the 18S and 26S genes (e.g., Buchheim et al., 1990; Buchheim and Chapman, 1991, 1992; Chapman and Buchheim, 1991; Kantz et al., 1990). As DNA sequencing has become routine (and also expedited by automated approaches), complete 18S rDNA sequences have become quite easy to obtain, and generating complete, or nearly complete, 26S rDNA sequences is also straightforward.

The importance of integrating a knowledge of secondary structure models to optimize alignment of homologous nucleotide positions for rDNA sequences has been noted by a number of investigators (e.g., Mishler et al., 1988; Hillis and Dixon, 1991; Kranz et al., 1995; Nickrent and Soltis, 1995; Hershkovitz and Zimmer, 1996); compilations of ribosomal RNA structures are available for both SSU (18S) and large subunit (LSU) (26S) rRNAs (Gutell, 1993; Gutell et al., 1993; Neefs et al., 1993). Determining the impact of secondary structure of the 18S rRNA transcript on phylogeny reconstruction may also be important. As reviewed by several investigators (e.g., Mishler et al., 1988; Wheeler and Honeycutt, 1988; Dixon and Hillis, 1993), major questions remain regarding not only the evolution, but also the phylogenetic analysis, of rRNA and rDNA sequences (see Chapter 7). For example, should loop bases (nonpairing bases) and stem bases (pairing bases) both be used in phylogenetic reconstruction, and, if so, should bases from stems and loops be considered equally informative? The prevalence of compensatory changes in stem regions and the potential for nonindependence of characters should also be considered. These topics are covered in more detail in Chapter 7.

18S rDNA   18S rDNA/rRNA sequences have been much more extensively used than have 26S rDNA/rRNA sequences. Although the general taxonomic range of application of the two regions appears to be very similar, the sheer size of the 26S gene (over 3,000 bp) has deterred investigators, particularly with regard to complete sequencing. In contrast, the size of 18S rDNA (SSU; approx. 1,800 bp) has made it much more amenable to PCR amplification and sequencing. The advantage of obtaining complete, rather than partial, sequences of these and other genes is that this approach also provides a database of sequences for the study of molecular evolution. For example, the large data set of complete 18S rDNA sequences obtained by Soltis, Soltis,

Nickrent et al. (1997) afforded the opportunity for the first detailed analyses of molecular evolution of 18S rRNA genes in angiosperms (see Chapter 7).

Only one 18S rDNA sequence type is typically found in an organism, but intraindividual 18S rDNA variation has been detected in some cases. Carranza et al. (1996) found two types of 18S rDNA in the flatworm *Dugesia mediterranea;* although both types appear functional, only one is transcribed. In the North American sturgeon, *Acipenser transmontanus,* four different 18S rDNA sequence types were detected, the presence of which may be due, in part, to polyploidy (J. Krieger et al., unpubl.). However, other related fish also presumed to be of polyploid origin do not exhibit such polymorphism. Regardless, the different 18S sequences of *A. transmontanus* form a clade, and the relationships among sturgeon species remain unchanged in phylogenetic analyses, regardless of the sequence of *A. transmontanus* used. The multiple rDNA types in *A. transmontanus* appear to be paralogues, having diverged since the origin of the species (see discussion of paralogues and orthologues in Chapter 4). That occasional multiple 18S sequence types do not distort phylogenetic relationships is significant in applications of 18S rDNA sequences in plant groups such as ferns and angiosperms, which are noted for polyploidy. In addition, the fact that 18S rDNA topologies are highly congruent with *rbcL*-based topologies at a diverse array of taxonomic levels in ferns (Wolf, 1995; unpubl.) and angiosperms (e.g., Soltis, Soltis, Nickrent et al., 1997; Soltis et al., 1997b; Soltis and Soltis, 1997; Kron, 1996) further suggests that polyploidy has little or no topological impact.

As noted previously, many initial studies involved portions of the 18S region; recent studies have typically used the entire 18S gene (e.g., Wilcox et al., 1992; Lewis et al., 1992; Huss and Sogin, 1990; Henriks et al., 1991; Nickrent and Soltis, 1995; Kranz et al., 1995; Capesius, 1995; Soltis, Soltis, Nickrent et al., 1997). 18S rDNA sequences have been widely employed in algae (e.g., Chapman and Buchheim, 1991; Bakker et al., 1994; Cavalier-Smith et al., 1994; Bhattacharya and Medlin, 1995; Medlin et al., 1996; see Chapter 18), bryophytes (e.g., Mishler et al.,

1994; Hedderson et al., 1996), ferns (Wolf, 1995; Kranz and Huss, 1996; see Chapter 19), gymnosperms (e.g., Chaw et al., 1993, 1995, 1997), and angiosperms (see Zimmer et al., 1989; Hamby and Zimmer 1992; Nickrent and Soltis, 1995; Soltis, Soltis, Nickrent et al., 1997). Several broad analyses of entire 18S rDNA sequences have been conducted that have implications for the origin of land plants, monophyly of gymnosperms, and sister group of the angiosperms (Kranz et al., 1995; Chaw et al., 1997). Because 18S rDNA is so highly conserved, PCR and sequencing primers have broad applicability across all plants. Details regarding PCR and sequencing primers and sequencing strategies are provided in Bult et al. (1992) and Nickrent and Starr (1994).

Despite its potential for reconstructing plant phylogeny, 18S rDNA has, until recently, been underused in angiosperm systematics. Several factors may be responsible, most notably (1) the reliance on *rbcL* sequences for inferring angiosperm phylogeny and (2) misconceptions regarding the evolution of 18S rDNA. The most common criticisms of 18S rDNA as a source of phylogenetic information have been that it is not sufficiently variable for phylogenetic reconstruction within the angiosperms and that it is highly prone to insertion and deletion, making sequence alignment difficult. However, recent studies (e.g., Nickrent and Soltis, 1995; Soltis, Soltis, Nickrent et al., 1997) have demonstrated that 18S rDNA provides a sufficient number of characters for broad-scale phylogenetic reconstruction of the angiosperms. In fact, although 18S rDNA typically evolves at only one-third to one-half the rate of *rbcL* (Nickrent and Soltis, 1995), 18S rDNA sequences have been used to infer relationships within specific angiosperm clades (e.g., Kron, 1996; Soltis and Soltis, 1997; L. Johnson et al., unpubl.; Hoot et al., 1995a, 1995b; Rodman et al., 1998). Within some groups, such as Orchidaceae, 18S rDNA actually evolves faster than *rbcL* (M. Chase, pers. comm.). The relationships recovered in these studies are largely congruent with those inferred from *rbcL* analyses.

Insertion and deletion events in 18S rDNA are largely confined to a few specific regions that correspond to the termini of several helices on the proposed rRNA secondary structure. Because these regions can confound alignment over a broad taxonomic scale, they should not be used when alignment is ambiguous (cf. Soltis, Soltis, Nickrent et al., 1997). The prevalence of insertion and deletion in angiosperm sequences has been overestimated, due in large part to sequencing errors that create "false" indels (reviewed in Soltis, Soltis, Nickrent et al., 1997). In fact, insertions and deletions in highly conserved portions of the 18S gene may be parsimony informative. Alignment of 18S sequences is therefore straightforward and easily accomplished by eye in angiosperms (Soltis, Soltis, Nickrent et al., 1997) and even across most land plants (P. Soltis et al., unpubl.).

Interest in using 18S rDNA sequences will likely continue to increase as researchers seek to test *rbcL*-based phylogenetic hypotheses with nuclear characters. Due, however, to the generally lower information content of 18S rDNA compared to *rbcL*, 18S-based topologies will typically not provide the same degree of resolution possible with *rbcL* (see Hoot et al., 1995b; Nickrent and Soltis, 1995; Soltis and Soltis, 1997; Kron, 1996; Rodman et al., 1998). To obtain nuclear-based trees with similar resolution to that of *rbcL*, it may be necessary to include portions of, if not the entire, 26S rDNA.

26S rDNA   The large size of the 26S rDNA (LSU; over 3,300 bp) has until recently deterred complete sequencing of this region. Typically, sequences derived from several portions of the 26S rRNA gene, rather than the entire gene, have been used to retrieve phylogeny in plants, as well as in other groups of organisms. In several instances, portions of the 26S rRNA gene have been used in concert with portions of the 18S rRNA region (e.g., Kantz et al., 1990, Buchheim and Chapman, 1991; Hamby and Zimmer, 1992, Bult and Zimmer, 1993, and Misher et al., 1994). Significantly, phylogenetic analysis of portions of LSU provided the same general topology as did SSU sequence data for a similar suite of termite and fungal taxa (compare Chapela et al., 1994 and Hinkle et al., 1994). Similarly, initial comparisons in plants suggest that 26S rDNA sequence data reveal the same

relationships revealed by 18S rDNA and *rbcL* sequences (Kuzoff et al., 1998).

The 26S rRNA gene is often noted as a candidate for sequencing as either an alternative or a supplement to 18S rDNA. Assessing the phylogenetic utility and molecular evolution of the entire 26S rDNA in plants has been made difficult, however, by the paucity of sequences. Until recently, only seven complete 26S rDNA sequences (all angiosperms) were available for comparison (Burt et al., 1995). In searching for additional base pairs from the nucleus to elucidate higher-level relationships, and with the increased use of automated sequencing, investigators have recently developed PCR and sequencing primers for the entire 26S rDNA (e.g., Kuzoff et al., 1998; D. Nickrent, unpubl.). Kuzoff et al. (1998) also provide 12 additional complete 26S sequences representing a diversity of angiosperms, plus one species of Gnetales. Their comparisons further confirm the higher-level phylogenetic potential of entire 26S rDNA sequences and demonstrate that entire 26S rDNA sequences evolve 1.6 to 2.2 times faster and provide over three times as many parsimony-informative characters as does 18S rDNA.

Although the phylogenetic potential of portions of 26S rDNA is clear, the patterns of nucleotide change in this gene are more complicated than in 18S rDNA (Bult et al., 1995). The 26S rDNA is a mosaic of highly conserved core regions interspersed with 12 regions of variable size called *expansion segments* (Clark et al., 1984) or *divergent domains* (Hassouna et al., 1984). The number and relative positions of these expansion segments appear to be highly conserved over a wide range of taxa. These expansion segments have elevated rates of sequence variation and are subject to some length variation (Burt et al., 1995). Most of the fluctuations in the length of the expansion segments reflect the gain and loss of short, directly repetitive sequence motifs, apparently as the result of slippage mechanisms (Hancock and Dover, 1988). Tautz et al. (1986, 1988) coined the term "cryptic sequence simplicity" to describe the sequence footprints of the motifs after they are shuffled among themselves by repeated slippage events. Because of their elevated rates of sequence variation, there is debate as to whether these regions are functional (reviewed in Bult et al., 1995). In addition, the possibility of motif shuffling and molecular coevolution among expansion segments may compromise the phylogenetic utility of these regions (Burt et al., 1995). Recent work, however, not only suggests that these factors may not be a concern, but also indicates that expansion segments may have phylogenetic potential (Kuzoff et al., 1998).

5.8S rDNA   The 5.8S rRNA gene is easily amplified and sequenced using primers located in the 18S and 26S rRNA genes (Fig. 1.9). The 5.8S gene has only rarely been used to infer phylogeny both because it is highly conserved (e.g., Hamby and Zimmer, 1992; Troitsky and Bobrova, 1986; Troitsky et al., 1991; Suh et al., 1993; Hershkovitz and Lewis, 1996) and because of its small size (only 164–165 bp; Hershkovitz and Lewis, 1996). An analysis of eight 5.8S sequences representing fungi, green algae, mosses, and angiosperms produced an unresolved trichotomy among land plants, green algae, and fungi, suggesting the region was too conserved and too small to be informative at even deep taxonomic levels (Troitsky and Bobrova, 1986). A larger analysis of thirty-five 5.8S sequences representing land plants, algae, and fungi has shown that 5.8S sequences are readily aligned across land plants, algae, and fungi with few indels present, although a variable-length region (base positions 120–150) is

more readily aligned within than between major taxonomic groups (Hershkovitz and Lewis, 1996). The proportion of potentially informative sites in seed plant sequences (0.14) is very similar to that determined for 18S rDNA sequences (0.18; Nickrent and Soltis, 1995), and phylogenetic analysis of these 5.8S sequences revealed a number of groupings similar to those obtained in comparable analyses with 18S rDNA sequences.

The pattern of sequence divergence of the 5.8S region is in striking contrast to that of the 5S rRNA gene (see below in 5S rRNA Genes and Spacer), another small DNA region. Among green plants, for example, the 5S gene has little or no phylogenetic utility not only because of its small size, but also because of apparent mutational saturation and constraints imposed by rRNA secondary structure (Mishler et al., 1988; Halanych, 1991; Steele et al., 1991—see below in 5S rRNA Genes and Spacer). In contrast, mutational saturation and constraints of secondary structure are not factors when considering 5.8S rDNA comparisons across green plants (Hershkovitz and Lewis, 1996).

ITS   The internal transcribed spacer (ITS) regions (Figs. 1.8, 1.9) of 18S–26S nuclear rDNA have become a major focus of comparative sequencing at the specific and generic levels in angiosperms (see review by Baldwin et al., 1995). Rather than review the various uses of ITS sequencing in plant systematics, we summarize general information for the ITS regions, provide

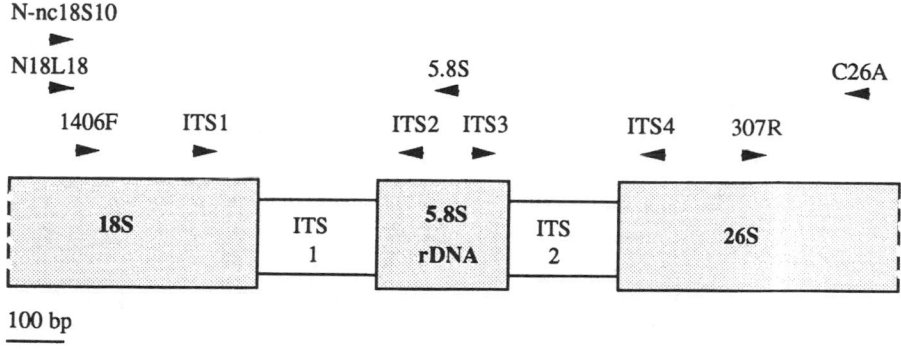

Figure 1.9.   Schematic diagram of the repeat unit of 18S and 26S nuclear ribosomal DNA showing the ITS-1 and ITS-2 regions; the intergenic spacer (IGS) and much of the 26S region are not illustrated. Arrows indicate the relative locations of PCR and sequencing primers available for amplification and sequencing of ITS-1 and ITS-2, and also the 5.8S region (see White et al., 1990; Soltis and Kuzoff, 1995; Wen and Zimmer, 1996; Nickrent et al., 1994 for base composition of primers).

an overview of the structure and utility of these regions in plants other than angiosperms, and review recent work that suggests caution in the use of these sequences for inferring phylogeny.

The internal transcribed spacers (ITS-1 and ITS-2) are part of the nuclear rDNA transcript but are not incorporated into ribosomes. They appear to play a role in the maturation of nuclear rRNAs, bringing the large and small subunits into close proximity within a processing domain (reviewed in Baldwin et al., 1995). This function concomitantly suggests that ITS-1 and ITS-2 are under some evolutionary constraint in structure and sequence, a view that is further supported by recent angiosperm-wide comparisons of ITS-1 and ITS-2 sequences (Hershkovitz and Zimmer, 1996, unpubl.).

ITS sequences are proving valuable for phylogenetic reconstruction in angiosperms (Baldwin et al., 1995), algae (Bakker et al., 1995; Coleman et al., 1994), and ferns (D. B. Stein et al., unpubl.). ITS sequences have also been used to infer phylogeny at lower taxonomic levels in a diverse array of organisms, including fungi (e.g., Vilgalys and Sun, 1994) and insects (e.g., Campbell et al., 1993; Fritz et al., 1994; Schlötterer et al., 1994; Vogler and DeSalle, 1994).

Recent work (Hershkovitz and Zimmer, 1996; Hershkovitz and Lewis, 1996) indicates that ITS-1 and ITS-2 sequences are inherently G+C rich and that portions of these regions are quite conserved among angiosperms. For example, 40% of ITS-2 is conserved across all angiosperms sequenced. Hershkovitz and Zimmer (1996) present an alignment in which as much as 50% of the ITS-2 sequence is alignable above the family level in angiosperms. Thus, the ITS regions not only possess high information content at lower taxonomic levels, but also exhibit conserved sequence patterns and high alignability across angiosperms.

The ITS regions are easily amplified in most angiosperms (but see caveats following), due in part to their small size and the fact that the PCR primers represent highly conserved flanking sequences from the rRNA genes. The PCR and sequencing primers widely used were initially described by White et al. (1990) for fungi and have broad applicability. Additional primers have also shown wide applicability (Soltis and Ku-

zoff, 1995; Nickrent et al., 1994; Wen and Zimmer, 1996).

Several strategies have been employed to amplify and sequence the ITS regions (Baldwin, 1992; Wen and Zimmer, 1996; Soltis and Kuzoff, 1995). Sequencing of the ITS regions can be difficult, however, because the region is G+C rich and prone to secondary structure (Baldwin et al., 1995). The degree of sequencing difficulty encountered can vary greatly from group to group. Adding DMSO to PCR and cycle sequencing reactions is a simple, yet often important, aid (Burt et al., 1992; Winship, 1989). The use of cycle sequencing (e.g., Wen and Zimmer, 1996) also seems to facilitate the sequencing of the ITS, just as was found for 18S rDNA, another region that is difficult to sequence due to both G+C-rich areas and inherent secondary structure (Soltis, Soltis, Nickrent et al., 1997). Other modifications that may facilitate amplification and sequencing are given in Zhang et al. (1992), Bechmann et al. (1990), Kim and Jansen (1994), and Bult et al. (1992).

Perhaps the most extreme example of ITS diversity within individuals is provided by Portulacaceae where individual plants of many species possess numerous nonfunctional or degenerate copies. Determining the actual intact ITS sequences can only be accomplished via laborious cloning and sequencing methods (M. Hershkovitz and E. Zimmer, unpubl.). Although the total length of the ITS regions, plus the intervening 5.8S gene, is fairly short and relatively uniform (600–700 bp) across angiosperms, it is much longer in Gnetales, *Ginkgo,* conifers, and cycads (Liston et al., 1996). For example, representatives of Pinaceae exhibit enormous variation in ITS length (1,550–3,125 bp), with most of the variation occurring in ITS-1 (Bobola et al., 1992; B. LePage, pers. comm., S. Brunsfeld, pers. comm.). As a result, ITS sequences have rarely been used successfully to infer phylogeny in these groups. However, ongoing work (B. LePage, pers. comm.) indicates that a portion of ITS-1 and the entire ITS-2 region can be readily employed in phylogentic reconstruction in conifers. Although ITS has been little studied in ferns, initial studies indicate that ITS regions of ferns may be slightly longer than those of most angiosperms. For example, in tree ferns ITS (including the 5.8S gene) is approximately 800 bp in length (D. Stein et al., unpubl.).

Because amplification primers for ITS are located in highly conserved coding regions of the rRNA genes, it is possible to amplify and sequence inadvertently the ITS of a foreign DNA source rather than of the intended target organism. Hershkovitz and Lewis (1996) demonstrated, for example, that the 5.8S/ITS sequences reported for eight species of the *Mimulus guttatus* complex (Scrophulariaceae) (Ritland and Straus, 1993; Ritland et al., 1993) are likely a chlorophycean algal contaminant. Chlorophyceae are ubiquitous, occurring in air, water, soil, and also environments such as laboratories, growth chambers, and greenhouses. Other possible sources of algal contaminants exist; for example, some tropical and subtropical angiosperms harbor algae in leaf epidermal cells (reviewed in Hershkovitz and Lewis, 1996). A similar example involves the reported ITS sequences of some conifers (e.g., Bobola et al., 1992; Smith and Klein, 1994); these appear to be fungal contaminants (Liston et al., 1996). Fungal contaminant sequences were also amplified and sequenced in work on the fern genus *Botrychium* (W. Hauk, pers. comm.). These examples dramatically illustrate the caution required in interpreting ITS sequences; these contaminant sequences are a direct result of the broad taxonomic applicability of the amplification and sequencing primers for these regions. Particularly when investigating a new group of plants, it is prudent to compare the ITS and 5.8S sequences to those available in GenBank (as recommended by Hershkovitz and Lewis, 1996).

The popularity of ITS sequence analysis relates to the great need for sequence data from the nuclear genome at lower taxonomic levels. The chloroplast genome has been the major focus of molecular phylogenetic analyses over the past 10 years, either through chloroplast DNA restriction site analysis (see chapters in P. Soltis et al., 1992a) or the sequencing of several chloroplast genes or DNA regions. As cpDNA data have rapidly accumulated, the concomitant need for nuclear sequence data has also became clear. Pitfalls and potential problems exist when estimating organismal phylogeny based on a single

organellar genome or a single nuclear gene (e.g., Pamilo and Nei, 1988; Rieseberg and Soltis, 1991; Doyle, 1992). The prevalence of chloroplast capture (Rieseberg and Soltis, 1991; Rieseberg, 1995) may result in incorrect estimates of organismal phylogeny (see Chapter 10). Incongruence between cpDNA-based and ITS-based trees (e.g., Soltis and Kuzoff, 1995) is typically attributed to the cpDNA trees incorrectly reflecting organismal phylogeny because of chloroplast capture. Another potential factor at lower levels is lineage sorting (e.g., Avise, 1986; Pamilo and Nei, 1988), which would also lead to an incorrect hypothesis of relationship. Because lineage sorting could occur for either cpDNA or nuclear markers, the underlying message is that it is imperative to use several markers, preferably representing both the nuclear and chloroplast genomes, in phylogenetic reconstruction. This is particularly critical at taxonomic levels that are most prone to hybridization and lineage sorting. Examination of phylogenetic relationships at lower levels with both cpDNA restriction sites and/or sequences and ITS sequences is not only critical for providing a more accurate picture of relationships, but it also affords the opportunity to garner insights into evolutionary processes (see Chapter 10).

Aspects of ITS evolution must also be considered when using ITS sequences for phylogeny reconstruction. The dynamics of rDNA concerted evolution are not well understood, and simple models of ITS evolution may lead to incorrect phylogenetic estimates. For example, interlocus concerted evolution of rDNA repeats has been reported in allopolyploid species of cotton (Wendel et al., 1995). Phylogenetic analysis of five alloploid (AD genome) cotton species plus their A and D genome progenitors suggests that the rDNA from four polyploid lineages has been homogenized to a D genome repeat type, whereas in one polyploid, concerted evolution fixed an A genome type. Bidirectional interlocus concerted evolution following hybridization or polyploidization could lead to erroneous estimates of organismal relationship in phylogenetic studies employing ITS sequences at lower taxonomic levels. In contrast, interlocus concerted evolution is likely not a problem at higher taxonomic levels when 18S and 26S rDNA sequences are employed because these genes are so conserved that congeners capable of hybridization and forming allopolyploids are likely to be identical or nearly so in sequence. Concerted evolution has not homogenized the ITS regions of all polyploids examined. Both parental ITS sequences are present in polyploid species of *Paeonia* that are approximately one million years of age (Sang et al., 1995); ITS sequences were therefore used to document reticulate evolution in much the same way that allozymes have been used. Similarly, concerted evolution has apparently not occurred in allotetraploid *Tragopogon* species, which are likely no more than 80 years old (J. Koontz et al., unpubl.).

In addition to multiple functional ITS loci, both nonfunctional ITS (what have been termed ITS *pseudogenes,* although ITS is not a gene) and recombinants may also be present in individual plants (Buckler and Holtsford 1996a, 1996b; Buckler et al., 1997; M. Hershkovitz and E. Zimmer, unpubl.; R. Kuzoff et al., unpubl.). Buckler et al. (1997) identified recombinants and putative pseudogenes in the ITS sequences of *Gossypium, Nicotiana, Tripsacum,* Winteraceae, and *Zea.* The putative ITS pseudogenes could be identified by high substitution rates and the presence of many deamination-type substitutions ($C \rightarrow T$ and $G \rightarrow A$). In addition, unlike functional ITS sequences, the pseudogenes had low predicted secondary structure stability. Buckler et al. suggest that this low secondary structure stability may actually enhance the amplification of these pseudogenes with PCR. They suggest that highly denaturing PCR conditions be used (i.e., add Dimethyl Sulfoxide, [DMSO]) to prevent the preferential amplification of pseudogenes. Many investigators have in fact routinely used DMSO in PCR amplification of portions of the ribosomal RNA cistron, following Bult et al. (1992). There is also evidence of ITS pseudogenes in ferns (D. Stein et al., unpubl.). Several studies have documented the presence of recombinant ITS sequences that combine portions of two distinct ITS types (Buckler et al., 1997; R. Kuzoff et al., unpubl.). Although these recombinants may be present in low frequency, the preferential amplification and sequencing of a recombinant (if undetected)

could produce errors in estimates of phylogeny. Thus, concerted evolution has not homogenized rDNA repeats in all taxa.

IGS   The intergenic spacer (IGS) (Figs. 1.8, 1.10) ranges from roughly 1 to 8 kb in plants (Jorgensen and Cluster, 1988); as noted, much of the length variation in the 18S–26S rDNA traces to this region. Much of the middle portion of the IGS consists of a subrepeat structure, with intraspecific variation in IGS length the result of varying numbers of subrepeats (Fig. 1.10). The length of the subrepeats varies, typically ranging from 100–200 bp (e.g., Appels and Dvorak, 1982; Saghai-Maroof et al., 1984; Yakura et al., 1984). Because of this variable subrepeat structure, the IGS varies extensively in length, not only among related species, but also within species and populations, and even within individual plants (reviewed in Rodgers and Bendich, 1987; Schaal and Learn, 1988). Because of its large size and variability, the IGS has been the focus of restriction site or restriction fragment analyses, particularly at lower levels (e.g., among closely related species and at the populational level—reviewed in Schaal and Learn, 1988). However, some plant species are remarkably constant, showing neither length nor restriction site variation in the IGS (e.g., Doyle and Beachy, 1985; Sytsma and Schaal, 1985).

Complete nucleotide sequences of the IGS are known for a diverse array of organisms, including mammals, amphibians, insects, fungi, protozoa (e.g., reviewed in Simeone et al., 1985; Labhart and Reeder, 1987; Umthun et al., 1994; Yang et al., 1994), and several plants (e.g., Lassner et al., 1987; Gruendler et al., 1991; Borisjuk and Hemleben, 1993; Cordesse et al., 1993; King et al., 1993; Polanco and Vega, 1994; Torres-Ruiz and Hemleben, 1994). Despite large variation in primary sequence, these diverse IGSs are architecturally similar and usually contain functionally similar regulatory elements, including the core promoter, upstream control element, transcriptional terminators, and spacer promoters (e.g., Borisjuk and Hemleben, 1993).

In plants, the IGS in *Cucurbita pepo, C. maxima,* and *Cucumis sativus* (Cucurbitaceae) is mainly composed of three different repeated elements interspersed in unique sequences:

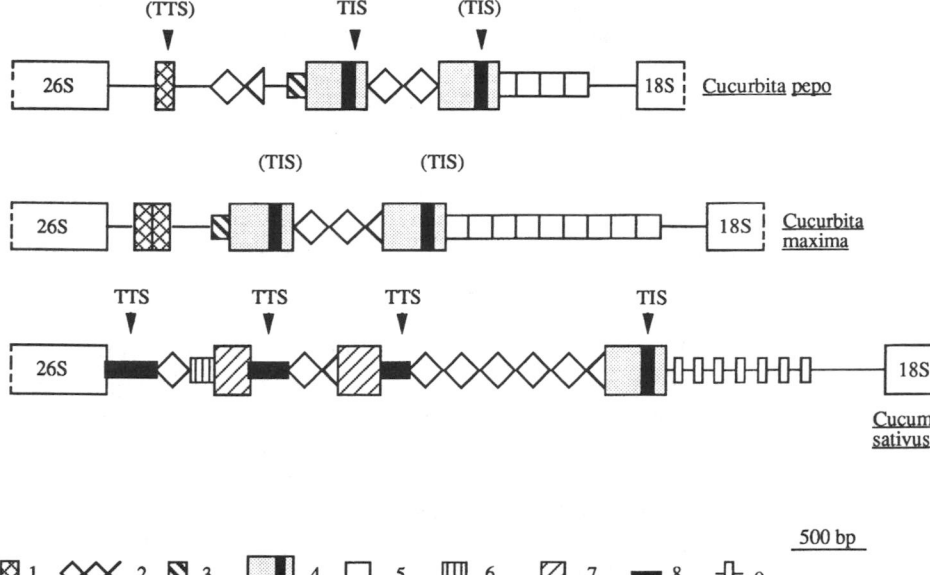

Figure 1.10.   Comparison of the structure of the IGS in three members of Cucurbitaceae (modified from King et al., 1993). TTS = Transcription termination site; TIS = transcription initiation site; ETS = external transcribed spacer. Code for various regions (labeled 1 to 9 below diagram) is as follows: (1) 70-bp element including TTS, (2) GC-rich clusters, (3) 188-bp AT-rich region, (4) 422-bp element including the TIS and an AT-rich adjacent region, (5) 260-bp D repeats, (6) 90-bp elements, (7) duplication of the 3′ end of the 26S rRNA gene, (8) duplication of the 470 bp to the 26S rRNA gene, and (9) 119-bp elements. Differences in the lengths of the two *Cucurbita* IGS regions are mainly due to the different number of 260-bp repeats (see symbol 5) located in the ETS subsequent to the second TIS.

(1) G+C-rich clusters, (2) a 422-bp A+T-rich element that includes the transcription initiation site (TIS) for RNA polymerase I, and (3) 260-bp repeats in the 5′ external transcribed spacer (King et al., 1993) (Fig. 1.10). Comparison of the three IGS sequences revealed 75–100% sequence similarity between the two species of *Cucurbita,* depending on the region compared. In contrast, obvious sequence identity between *Cucumis* and *Cucurbita* species occurs only around the A+T-rich element that includes the TIS.

IGS sequences are composed of both nonrepeated segments as well as tandem arrays of repeated sequences (Fig. 1.10). The evolutionary history of these arrays of repeated sequences may be complex. In rice, for example, successive amplification and deletion events are proposed to explain the distribution of repetitive arrays (Cordesse et al., 1993). This complexity, plus the great length of the IGS, have deterred comparative sequencing for phylogenetic purposes. Nonetheless, comparison of the IGS se-

quences of several cereals shows that some regions are highly conserved and may represent areas of utility for phylogenetic reconstruction (Cordesse et al., 1993).

Important insights into IGS sequence evolution have been obtained recently by D. Hayworth (pers. comm.) for *Arabidopsis.* His work suggests that the entire structure of the IGS is "loose," with even the control elements (TIS, transcription termination site or [TTS], promoter) usually repeated. Hence, without considerable investigation, one cannot assume that a region of IGS chosen for phylogenetic analysis is not repeated. Failure to recognize a repetitive element for what it is may confound assessments of sequence homology. Furthermore, in each new study group, repetitive elements can occur in any region of the IGS. In addition, because there is often intragenomic variation in this region (as reviewed previously), direct sequencing of PCR products yields multiple sequence. Hence, it may be necessary to clone individual copies of the IGS; this would make the use of this region very time consuming and perhaps not cost effective. The most reliable region of IGS to use to avoid these problems appears to be the ETS (Fig. 1.10), which comprises several regions (King et al., 1993; Volkov et al., 1996). One portion contains the putative RNA polymerase I transcription initiation site. A second region consists of a variable number of subrepeats; these repeated elements also appear to have a peculiar base composition, with a very high number of TG and TNG motifs (King et al., 1993). This second portion of the ETS may not have phylogenetic utility. The third portion of the ETS, which is approximately 500 bp in length and immediately adjacent to the 18S rDNA, may hold the greatest phylogenetic potential. However, based on the few sequences available, this region may be no more variable than the ITS regions. It might be a useful supplement to ITS sequences, to obtain additional nuclear characters (D. Hayworth, pers. comm.).

5S rRNA Genes and Spacer In all green plants and animals, the 5S rRNA genes are small (typically 120 bp) and occur in tandem arrays separate from the 18S–5.8S–26S rDNA cistron (Long and Dawid, 1980). The number of such 5S rRNA arrays varies considerably, from typically one or two, to many (reviewed in Schneeberger and Cullis, 1992); the number of copies within an array is also highly variable, ranging from several hundred to several thousand. The sequence of this gene is highly conserved, which initially suggested that it might be used for resolving relationships among the deep branches of the eukaryotes (e.g., Hori and Osawa, 1987; reviewed in Steele et al., 1991). A detailed analysis of the utility of 5S rRNA sequence data for inferring phylogeny in green plants (Steele et al., 1991) revealed, however, that many 5S sites do not vary, and of those that do, many accumulate mutations so rapidly that they are subject to multiple hits. Thus, the small number of phylogenetically informative sites in the 5S rRNA gene, as well as its short length, suggest that estimates of phylogeny based only on analyses of 5S rRNA sequences must be viewed with caution.

Several authors have examined the phylogenetic value of 5S rRNA spacer region sequences (Scoles et al., 1988; Baum and Appels, 1992;

Kellogg and Appels, 1995; Cronn et al., 1996). Recent attempts to use the 5S rRNA spacer region suggest phylogenetic utility within genera. In *Gossypium* there is considerable congruence between gene trees based on 5S intergenic sequences and earlier phylogenetic studes (Cronn et al., 1996). Other recent studies have resolved interspecific relationships within *Cypripedium, Phoenix, Silene,* and *Miscanthus* (M. Chase, pers. comm.). In general, the 5S spacer appears more variable than do ITS sequences. Hence, although short, 5S spacer sequences offer the potential for resolving interspecific and intergeneric relationships for comparison with ITS-based topologies. However, lack of concerted evolution (Kellogg and Appels, 1995; Cronn et al., 1996) among repeat units and arrays makes these spacers difficult to isolate and sequence; cloning of individual repeats is typically required.

Intraspecific and interspecific variation in 5S rRNA gene and spacer sequences appears to be decoupled (Kellogg and Appels, 1995; Cronn et al., 1996). For example, in members of the wheat tribe (Triticeae) nucleotide diversity within a 5S gene array is high; in fact, it is not significantly different from that of the spacers. Rates of concerted evolution are therefore insufficient to homogenize the entire array. Between species, however, there are significantly fewer fixed differences in the gene than would be expected given the high within-species variation. In contrast, however, the amount of variation in the spacer is the same or greater between species as within individuals. Cronn et al. (1996) also showed that duplicated 5S rDNA arrays in allopolyploid *Gossypium* species have retained their identity since polyploid formation, suggesting that interlocus concerted evolution has not played a significant role in these plants.

Other Nuclear Genes Several other nuclear genes appear to have phylogenetic potential but require further investigation. These include the small heat shock genes (Waters, 1995), which, based on a preliminary study in Brassicaceae, may be useful at the generic level. Of the four small heat shock gene families, the gene coding for the chloroplast-localized protein is the most promising.

The phytochrome gene (PHY) family is also a potential source of phylogenetic information. In *Arabidopsis,* five loci are present (PHYA–PHYE). Phytochrome sequence data provide a high degree of resolution within Fabaceae, suggesting that these sequences may be most useful below the family level (Mathews et al., 1995).

The nuclear gene for chloroplastic glyceraldehyde 3-phosphate dehydrogenase (*gapA*) is approximately 1,900 bp in length and is present in a single copy in *Arabidopsis thaliana* (Shih et al., 1991). Adachi et al. (1995) used partial sequences (661 bp) in a phylogenetic analysis of Berberidaceae. The relationships retrieved are comparable to those obtained in a much more extensive phylogenetic analysis of the family based on *rbcL* and ITS sequences (Kim and Jansen, 1996), suggesting that this gene may have potential at the specific and generic levels. Martin et al. (1993) explored the phylogentic utility of glycolytic glyceraldehyde 3-phosphate dehydrogenase (*gapC*) in angiosperms and concluded that the rate of sequence evolution of *gapC* is conservative and comparable to that of *rbcL*. Phylogenetic analyses using full-sized *gapC* sequences (approximately 1,300 bp) for a small number of angiosperms revealed relationships very similar to those obtained with *rbcL* sequences. One potential difficulty with using *gapC* sequences involves the presence of gene duplication; whereas *Arabidopsis* has a single gene for *gapC,* there are three in *Zea* (Martin et al., 1993).

Most higher plants have two alcohol dehydrogenase genes (*adh*) that, based on work with maize (Dennis et al., 1984), are derived from the duplication of a single ancestral gene. Sequences of both *adh1* and *adh2* have been obtained for several higher plants, with many studies focusing on *adh1*. These sequences are approximately 3,100 bp in length and are interrupted by nine introns (Fig. 1.11). Kosuge et al. (1996) amplified *adh* fragments from 17 genera of Ranunculaceae. Using sequence data for exon 3 and a portion of exon 4 (273 bp of exon sequence), they found *adh1* and *adh2* sequences for Ranunculaceae to be easily distinguishable. Concentrating on *adh1* sequences, they estimated phylogenetic relationships among genera that largely correspond to tribes based on cytology and fruit morphology (Kosuge et al., 1996). In a study of *adh1* variation across the geographic range of *Zea mays* and two species of teosinte (*Z. luxurians* and *Z. diploperennis*), Gaut and Clegg (1993) found that the distance between some maize and teosinte alleles is much less than the distance between some maize alleles. This suggests either introgression between maize and teosinte populations, or an ancient polymorphism at *adh1* that predates the separation of these three *Zea* species. R. Small and J. Wendel (pers. comm.) have used PCR primers specific to exon 2 and exon 9 to amplify *adh* genes from *Gossypium;* these products were subsequently cloned and sequenced. Exon sequences appear to have phylogenetic utility at low taxonomic levels. Small and Wendel compared the phylogenetic utility of sequences from various noncoding regions in the chloroplast to *adh* exon sequences in a group of recently diverged species of *Gossypium*. They found that 1.7 kb of *adh* exon sequence provides more phylogenetically informative characters than over 5 kb of noncoding sequence from the chloroplast genome.

Sequences for the nuclear phosphoglucose isomerase genes (*Pgi*) may also have value in phylogeny reconstruction (Gottlieb and Ford, 1996). Paralogous genes (*PgiC1* and *PgiC2*) resulting from a single duplication event encode

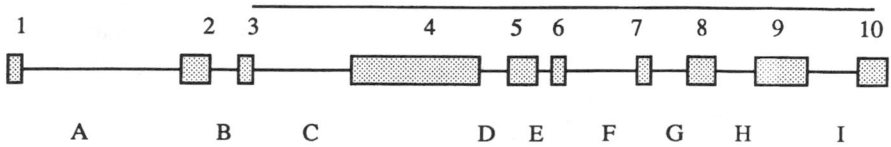

Figure 1.11.   Map of *adh1*. The middle line (with stipled boxes) represents the *adh1* gene. Stipled boxes are exons, numbered 1–10; lines between boxes are intervening sequences, labeled A–I. The line above the gene represents the region cloned and sequenced by Gaut and Clegg (1993). Modified from Gaut and Clegg (1993).

the cytosolic isozymes of PGI in *Clarkia* (Onagraceae). *Clarkia* has been the focus of a series of model studies by Gottlieb and coworkers, including the study of duplicated isozymes for cytosolic PGI (reviewed in Gottlieb and Ford, 1996). All *Clarkia* species examined have both genes, *PgiC1* and *PgiC2;* both genes are expressed in species with two active cytosolic PGI isozymes, but species with a single cytosolic PGI always have an active *PgiC1* and a silenced *PgiC2*. Phylogenetic analyses indicate that two groups of sequences can be recognized, corresponding to *PgiC1* and *PgiC2*. The phylogenetic results for the two genes are in complete accord and show the same sectional relationships. Furthermore, these studies reveal that *PgiC2* has been silenced at least four times independently. *Pgi* sequences may also be of phylogenetic utility in other groups. However, the gene is very large (the sequences examined by Gottlieb and Ford averaged 6,226 bp for *PgiC1,* for example) and structurally complex, consisting of 23 exons and 22 introns.

In most plants, histone H3 loci form a large multigene family; some of these loci possess introns whose sequences may be of phylogenetic utility at the species level (Doyle et al., 1996b). In *Glycine,* phylogenetic analysis of two closely related H3-B loci revealed a complex pattern of evolution; these loci do not have phylogenetic value. In contrast, sequences derived from the single H3-D locus provided gene trees similar to other hypotheses of relationship at the species level in *Glycine* (Doyle et al., 1996b). The PCR primers used were designed based on sequences known from soybean (*G. max*); the applicability of these primers outside of *Glycine* is not known.

The multigene family for glutamine synthetase (GS) in flowering plants includes one member that encodes a chloroplast-expressed isozyme and that represents an ancient divergence in the gene family (Doyle, 1991). The coding region is approximately 1 kb in length and is interrupted by 11 introns in alfalfa (Tischler et al., 1986); intron positions appear to be conserved throughout the gene family. Oligonucleotide primers constructed against conserved regions have been used to amplify one or more introns; sequences of these introns are providing phylogenetic data at the species level in legumes (D. Bailey et al., unpubl.) and *Oxalis* (E. Emshwiller and J. Doyle, unpubl.).

Mason-Gamer and Kellogg (1996) have explored the phylogenetic utility of the starch synthase gene *waxy*. This gene is presumed to be single copy, is approximately 3 kb in length, and contains numerous introns. Mason-Gamer and Kellogg (1996) developed a set of primers that permit amplification and sequencing of a 1.3-kb portion of *waxy* that includes five introns. Based on initial work in Poaceae, the introns may be phylogenetically useful within genera, whereas exon sequences are useful at the family level.

Denton and Hall (1996) have explored the phylogenetic potential of the RNA polymerase II gene (RPB2). RPB2 is found in all eukaryotes and occurs as a single copy in most organisms. A 2.2-kb region of RPB2 containing an intron of 600 bp has been PCR-amplified, cloned, and sequenced in over 25 members of Ericaceae. Analysis of the intron sequences suggests phylogenetic potential at the intrafamilial level (Demon and Hall, 1996).

Several investigators are investigating the phylogenetic utility of MADS-box genes (e.g., D. Bailey and J. Doyle, pers. comm.; B. Baldwin and M. Purugganan, pers. comm.). In both Cruciferae and in the Hawaiian silverswords, these sequences appear to have phylogenetic utility between closely related species. Preliminary investigation of structure and variation of the MADS-box regulatory gene family (see Purugganan et al., 1995) across genetically similar members of the Hawaiian silversword alliance (Madiinae, Asteraceae) has revealed that the organization of these loci is similar to that documented in *Arabidopsis* (Brassicaceae). Moreover, there appears to be sufficient nucleotide sequence and insertion/deletion variation in the relatively conserved domains of these regulatory loci to allow for phylogenetic resolution. Two distinct loci orthologous to the *Arabidopsis* APETALA3 (AP3) gene (see Jack et al., 1992) are present in the tetraploid Hawaiian silverswords and a single locus occurs in the diploid Californian relative, *Madia bolanderi*. Extensive sequence divergence between paralogous copies of AP3 and close similarity of the *M.*

bolanderi sequence to one copy are consistent with either (1) an allopolyploid origin of the silversword alliance or (2) divergent rates of sequence evolution across paralogous gene lineages.

Mitochondrial Genome

Molecular study of the mitochondrial genome involving either restriction site analysis or sequencing has been a major focus of phylogenetic studies in animals (see reviews by Moritz et al., 1987; Harrison, 1989; Mitton, 1994; Avise, 1986, 1994; Avise et al., 1987; chapters in Hillis et al., 1996). In contrast, however, the mitochondrial genome has been little used in studies of plant phylogeny. This lack of emphasis on the plant mitochondrial genome is due to the following (Palmer, 1992): (1) plant mtDNAs are very large and highly variable in size, which contrasts with the small size and stability of animal mtDNAs; (2) many foreign DNA sequences, particularly chloroplast DNA sequences, are present in plant mitochondrial genomes; (3) large duplications (repeats) appear to be frequently created and lost; (4) recombination can occur among repeats with high frequency, creating a very complex genome structure; (5) small (1–11 kb) unstable, extrachromosomal plasmids of unknown origin and function are common in plant mitochondria and may be transmitted differently than the main mitochondrial genome; (6) plant mtDNAs are characterized by many short (50–1,000 bp), dispersed repeats that appear scattered throughout the mitochondrial genome; such repeats are considered rare in other mtDNAs and in cpDNA; (7) plant mtDNAs rearrange very quickly, the result being that even closely related species do not possess the same order of mtDNA genes; (8) the rate of nucleotide substitution is very low compared to the mitochondrial genome of other organisms, as well as compared to the chloroplast and nuclear genomes of plants.

In summary, the genome structure and dynamics of mtDNA in plants are very different from those of other genomes. The rapid changes highlighted above in genome structure, size, and configuration result in a genome that is very difficult to analyze. Palmer, pg. 46 (1992) concludes, for example, that these features "make the molecule essentially worthless for the restriction site-based reconstructions of intrafamilial phylogeny for which cpDNA is so well suited." Compounding this difficulty is the fact that mtDNA is much less abundant than cpDNA in leaf tissue.

The plant mitochondrial genome does have several potential applications, however. First, rearrangements are frequent in plant mtDNA and relatively easy to distinguish by study of restriction patterns (Palmer, 1992). Thus, at lower taxonomic levels, rapid surveys of genome "types" may be valuable (reviewed in Palmer, 1992; Palmer and Herbon, 1988). A second possible application of the mitochondrial genome involves the occasional loss of mitochondrial genes and introns. These could be exploited for systematic purposes in much the same way as in the chloroplast genome (see foregoing; see also Downie et al., 1991; Doyle et al., 1995). At this point, however, gene and intron losses in mtDNA have been little explored as phylogenetic markers. The greatest utility of the plant mitochondrial genome at this point involves comparative gene sequencing. As noted, the rate of nucleotide substitution in plant mtDNA is very low. In fact, it is 3–4 times slower than in cpDNA, roughly 12 times slower than in plant nuclear DNA, and 40–100 times slower than in animal mtDNA (Palmer, 1992). This low rate of base substitution in plant mitochondrial genes makes them potentially valuable in phylogenetic reconstruction at higher taxonomic levels. Several mtDNA genes have recently been analyzed and seem to have potential for inferring phylogeny at higher levels.

coxI and atpA   C. dePamphilis and coworkers (pers. comm.) have developed amplification and sequencing primers and also compiled preliminary data sets for two plant mitochondrial genes, *coxI* and *atpA*. PCR amplification of these genes has been successful with virtually all angiosperms, as well as a wide diversity of gymnosperms. Only smaller fragments can be reliably amplified from the ferns and fern allies; general success with these groups will require further development of PCR primers and/or protocols (L. Bowe and C. dePamphilis, pers.

comm.). The two genes are comparable not only in length (*coxI* is 1,574 to 1,592 bp, *atpA* is 1,521 to 1,536 bp), but also in rate of evolution. Overall sequence divergence (estimated from averages of multiple pairwise comparisons for comparable taxa for *rbcL* and both mitochondrial genes) is 0.43 and 0.45, respectively, that of *rbcL* for *coxI* and *atpA*. For both mitochondrial genes, this slower rate is due to a decreased rate of synonymous substitution relative to *rbcL*.

Phylogenetic analyses of over 50 diverse species of gymnosperms and angiosperms for *coxI* and for a dozen diverse angiosperms for *atpA* indicate that these genes may be applicable over a wide taxonomic range from suprafamilial through ordinal and subclass. However, sequence divergence within families is only a few percent or less, which limits resolution at this level. Strongly conflicting phylogenetic hypotheses have been observed between *rbcL* and *coxI* in some cases (C. dePamphilis, pers. comm.), but these appear to be attributable to specific (and correctable) effects of RNA editing in *coxI*. For the most part, however, results are either congruent with trees drawn from large numbers of *rbcL* sequences, or with slightly less parsimonious trees for one of the genes. In direct comparisons with *rbcL*, homoplasy was considerably lower for *coxI* and *atpA* than with *rbcL*, due to many fewer multiple substitutions in the mitochondrial sequences. This suggests that accurate phylogenetic reconstruction with mitochondrial genes may not require taxon densities as high as needed with more rapidly evolving genes. However, these and all other mitochondrial genes are subject to RNA editing, which may pose special problems for phylogeny estimation unless editing sites are identified and removed prior to phylogenetic analysis (Bowe and dePamphilis, 1996; Malek et al., 1996). Primer sequences and more information are available for *coxI* and *atpA* from C. dePamphilis.

Sequence data derived from several other plant mitochondrial genes, including *coxIII* (see Malek et al., 1996), *atp9, nad2,* and *cob,* may also be applicable to plant phylogenetic research. However, it is difficult to evaluate the potential for most other mitochondrial genes because so few sequences are available at present.

SOURCES OF DNA FOR SYSTEMATIC STUDIES

Simple methods of DNA extraction from fresh plant tissue yield large quantities of high-molecular-weight DNA (e.g., Doyle and Doyle, 1987, 1990), suitable for restriction site analysis, sequencing, and various PCR-mediated techniques. However, fresh tissue is not always available for the taxa of interest. Effective alternatives to fresh material are (1) tissues quick-dried in silica gel (e.g., Chase and Hills, 1991) and (2) herbarium specimens (e.g., Taylor and Swann, 1994; Loockerman and Jansen, 1995; Savolainen et al., 1995b). Improved "microprep" extraction procedures, requiring only milligrams of dried tissue, provide adequate quantities of DNA for PCR amplifications (e.g., Cullings, 1992). Although these approaches are not suitable for large-scale restriction mapping analyses, they do yield sufficient amounts of DNA for comparative sequencing, RAPDs, and RFLP studies of amplified fragments.

The application of these micro-prep techniques to tissue samples removed from herbarium specimens has tremendous potential, providing literally millions of plant samples housed in herbaria worldwide for molecular systematic study. However, as many studies of DNA preservation from both recent and ancient materials have demonstrated, extraction and analysis of DNA from preserved material are hampered by a number of factors, and success rates are neither predictable nor always satisfactory (e.g., Paabo, 1989; Golenberg et al., 1990; Golenberg, 1991; P. S. Soltis et al., 1992b, 1995; P. S. Soltis and D. E. Soltis, 1993; Lindahl, 1993; Cano and Poinar, 1993). Reports of successful DNA extraction and analysis from 100+-year-old herbarium specimens (e.g., Savolainen et al., 1995b) indicate that success is not always correlated with age but may depend more on mode of preservation and drying and on species-specific compounds that affect DNA preservation or inhibit PCR amplification. Potential sources of PCR inhibition in herbarium samples are discussed further by Savolainen et al. (1995b), who also demonstrated that some types of amplification failures can be overcome by simple modifications of existing extraction techniques. Therefore, although the use of herbarium specimens as a source of DNA for systematic studies should perhaps not yet be considered routine (e.g., Savolainen et al., 1995b), we encourage the supervised sampling of herbarium specimens for molecular studies. The potential of herbarium specimens to provide molecular information makes herbaria genetic storehouses and further demonstrates the value of preserved collections for a diversity of scientific disciplines.

VOUCHERS IN MOLECULAR SYSTEMATIC STUDIES

As molecular systematists depend increasingly on botanical gardens, preserved material, and the collections of colleagues as sources of material for analysis, the need for appropriate voucher specimens and information is greater than ever. However, in our zeal to reconstruct plant phylogney or discern the evolutionary dynamics of populations or species, we molecular systematists have been too often irresponsible about documenting the sources of material used in a study or providing specimens to serve as permanent records of the plants included in the study. As noted by Chase, Soltis, Olmstead et al. (1993), misidentified plant material may become apparent because of unexpected placement in phylogenetic trees; however, errors may go undetected, and without appropriate vouchering, the true identity of a misidentified specimen cannot be determined. We echo the call for responsible vouchering made by Goldblatt et al. (1992) and commend journals such as the *Annals of the Missouri Botanical Garden* and *Systematic Botany* for requiring citation of voucher information as a condition of publication. Likewise, all sequences should be accessioned in a major repository.

LITERATURE CITED

Adachi, J., K. Kosuge, T. Denda, and K. Watanabe. 1995. Phylogenetic relationships of the Berberidaceae based on partial sequences of the *gapA* gene. Plant Systematics and Evolution, 9(Suppl.):351–353.

Albert, V. A., A. Backlund, K. Bremer, M. W. Chase, J. R. Manhart, B. D. Mishler, and K. D. Nixon. 1994. Functional constraints and *rbcL* evidence for land plant phylogeny. Annals of the Missouri Botanical Garden **81**:534–567.

Appels, R., and J. Dvorak. 1982. The wheat ribosomal DNA spacer region: its structure and variation in populations among species. Theoretical and Applied Genetics **63**:337–348.

Appels, R., and R. L. Honeycutt. 1986. rDNA: evolution over a billion years. In DNA Systematics, Vol. II, Plants, ed. S. K. Dutta, pp. 81–135. CRC Press, Boca Raton, Florida.

Arnheim, N. 1983. Concerted evolution of multigene families. In Evolution of Genes and Gene Proteins, eds. M. Nei and R. K. Koehn, pp. 38–61. Sinauer Associates, Sunderland, Massachusetts.

Arnheim, N., M. Krystal, R. Schmickel, G. Wilson, O. Ryder, and E. Zimmer. 1980. Molecular evidence for genetic exchanges among ribosomal genes on nonhomologous chromosomes in man and apes. Proceedings of the National Academy of Sciences U.S.A. **77**:7323–7327.

Avise, J. C. 1986. Mitochondrial DNA and the evolutionary genetics of higher animals. Philosophical Transactions of the Royal Society of London B **312**:325–342.

Avise, J. C. 1994. Molecular Markers, Natural History, and Evolution. Chapman & Hall, New York.

Avise, J. C., J. Arnold, R. M. Ball, E. Berminham, T. Lamb, J. E. Neigel, C. A. Reeb, and N. C. Saunders. 1987. Intraspecific phylogeography: the mitochondrial DNA bridge between population genetics and systematics. Annual Review of Ecology and Systematics **18**:489–522.

Bakker, F. T., J. L. Olsen, W. T. Stam, and C. Van Den Hoek. 1994. The *Cladophora* complex (Chlorophyta): new views based on 18S rDNA gene sequences. Molecular Phylogentics and Evolution **3**:365–382.

Bakker, F. T., J. L. Olsen, and W. T. Stam. 1995. Evolution of nuclear rDNA ITS sequences in the *Cladophora albida/sericea* clade (Chlorophyta). Journal of Molecular Evolution **40**:640–651.

Baldauf, S. L., and J. D. Palmer. 1990. Evolutionary transfer of the chloroplast *tuf*A gene to the nucleus. Nature **344**:262–265.

Baldwin, B. G. 1992. Phylogenetic utility of the internal transcribed spacers of nuclear ribosomal DNA in plants: an example from the Compositae. Molecular Phylogenetics and Evolution **1**:3–16.

Baldwin, B. G., M. J. Sanderson, J. M. Porter, M. F. Wojciechowski, C. S. Campbell, and M. J. Donoghue. 1995. The ITS region of nuclear ribosomal DNA: a valuable source of evidence on angiosperm phylogeny. Annals of the Missouri Botanical Garden **82**:247–277.

Baum, B. R., and R. Appels. 1992. Evolutionary change at the *5S Dna* loci of species in the Triticeae. Plant Systematics and Evolution **183**:195–208.

Bechmann, B., W. Luke, and G. Hunsmann. 1990. Improvement of PCR amplified DNA sequencing with the aid of detergents. Nucleic Acids Research **18**:1309.

Bhattacharya, D., and L. Medlin. 1995. The phylogeny of plastids: a review based on comparisons of small-subunit ribosomal RNA coding regions. Journal Phycology **31:**489–498.

Bobola, M. S., D. E. Smith, and A. S. Klein. 1992. Five major nuclear ribosomal repeats represent a large and variable fraction of the genomic DNA of *Picea rubens* and *P. mariana.* Molecular Biology and Evolution **9:**125–137.

Borisjuk, N., and V. Hemleben. 1993. Nucleotide sequence of the potato rDNA intergenic spacer. Plant Molecular Biology **21:**381–384.

Bousquet, J., S. H. Strauss, A. D. Doerksen, and R. A. Price. 1992. Extensive variation in evolutionary rate of *rbcL* gene sequences among seed plants. Proceedings of the National Academy of Sciences U.S.A. **89:**7844–7848.

Bowe, L. M., and C. W. dePamphilis. 1996. Effects of RNA editing and gene processing on phylogenetic reconstruction. Molecular Biology and Evolution **13:**1159–1166.

Brown, D. D., P. C. Wensink, and E. Jordan. 1972. Comparison of the ribosomal DNAs of *Xenopus laevis* and *Xenopus mulleri:* the evolution of tandem genes. Journal of Molecular Biology **63:**57–73.

Bruneau, A., J. J. Doyle, and J. D. Palmer. 1990. A chloroplast DNA structural mutation as a subtribal character in the Phaseoleae (Leguminosae). Systematic Botany **15:**378–386.

Brunsfeld, S. J., P. S. Soltis, D. E. Soltis, P. A. Gadek, C. J. Quinn, D. D. Strenge, and T. A. Ranker. 1994. Phylogenetic relationships among the genera of Taxodiaceae and Cupressaceae: evidence from *rbcL* sequences. Systematic Botany **19:**253–262.

Buchheim, M. A., and R. L. Chapman. 1991. Phylogeny of the colonial green flagellates: a study of 18S and 26S rRNA sequence data. BioSystems **25:**85–100.

Buchheim, M. A., and R. L. Chapman. 1992. Phylogeny of *Carteria* (Chlorophyceae) inferred from molecular and organismal data. Journal of Phycology **28:**362–374.

Buchheim, M. A., M. Turmel, E. A. Zimmer, and R. L. Chapman. 1990. Phylogeny of *Chlamydomonas* (Chlorophyta) based on cladistic analysis of nuclear 18S rRNA sequence data. Journal of Phycology **26:**689–699.

Buckler, E. S., IV and T. P. Holtsford. 1996a. *Zea* ribosomal repeat evolution and substitution patterns. Molecular Biology and Evolution **13:**623–632.

Buckler, E. S., IV and T. P. Holtsford. 1996b. *Zea* systematics: ribosomal ITS evidence. Molecular Biology and Evolution **13:**612–622.

Buckler, E. S., IV, A. Ippolito, and T. P. Holtsford. 1997. Evolution of plant ribosomal DNA: pseudogenes, recombination, and phylogenetic implications. Genetics, **145:**821–832.

Bult, C. J., and E. A. Zimmer. 1993. Nuclear ribosomal RNA sequences for inferring tribal relationships within Onagraceae. Systematic Botany **18:**48–63.

Bult, C. J., M. Kallersjo, and Y. Suh. 1992. Amplification and sequencing of 16/18S rDNA from gel-purified total plant DNA. Plant Molecular Biology Report **10:**273–284.

Bult, C. J., J. A. Sweere, and E. A. Zimmer. 1995. Cryptic sequence simplicity, nucleotide composition bias, and molecular coevolution in the large subunit of ribosomal DNA in plants: implications for phylogenetic analyses. Annals of the Missouri Botanical Garden **82:**235–246.

Campbell, B. C., J. D. Steffen-Campbell, and J. H. Werren. 1993. Phylogeny of *Nasonia* species complex (Hymenoptera: Pteromalidae) inferred from an internal transcribed spacer (ITS2) and 28S rDNA sequences. Insect Molecular Biology **2:**225–237.

Cano, R. J., and H. N. Poinar. 1993. Rapid isolation of DNA from fossil and museum specimes suitable for PCR. Biotechnology **15:**432–435.

Capesius, I. 1995. A molecular phylogeny of bryophytes based on the nuclear encoded 18S rRNA genes. Journal of Plant Physiology **146:**59–63.

Carranza, S., G. Giribet, C. Ribera, J. Baguña and M. Riutort. 1996. Evidence that two types of 18S rDNA coexist in the genome of *Dugesia* (*Schmidtea*) *mediterranea* (Platyhelminthes, Turbellaria, Tricladid). Molecular Biology and Evolution **13:**824–832.

Catalan, M. P., E. A. Kellogg, and R. G. Olmstead. 1997. Phylogeny of Poaceae subfamily Pooideae based on chloroplast *ndhF* gene sequences. Molecular Phylogenetics and Evolution, **8:**150–166.

Cavalier-Smith, T., M. T. E. P. Allsopp, and E. E. Chao. 1994. Chimeric conundra: are nucleomorphs and chromists monophyletic or polyphyletic? Proceedings of the National Academy of Sciences U.S.A. **91:**11368–11372.

Chapela, I. H., S. A. Rehner, T. R. Schultz, and U. G. Mueller. 1994. Evolutionary history of the symbiosis between fungus-growing ants and their fungi. Science **266:**1691–1694.

Chapman, R. L., and M. Buchheim. 1991. Ribosomal RNA gene sequences: analysis and significance in the phylogeny and taxonomy of green algae. Critical Reviews in Plant Sciences **10:**343–368.

Chase, M. W., and H. H. Hills. 1991. Silica gel: an ideal material for field preservation of leaf samples for DNA studies. Taxon **40:**215–220.

Chase, M. W., D. E. Soltis, R. G. Olmstead, D. Morgan, D. H. Les, B. D. Mishler, M. R. Duvall, R. A. Price, H. G. Hills, Y.-L. Qiu, K. A. Kron, J. H. Rettig, E. Conti, J. D. Palmer, J. R. Manhart, K. J. Sytsma, H. J. Michaels, W. J. Kress, K. G. Karol, W. D. Clark, M. Hedrén, B. S. Gaut, R. K. Jansen, K.-J. Kim, C. F. Wimpee, J. F. Smith, G. R. Furnier, S. H. Strauss, Q.-Y. Xiang, G. M. Plunkett, P. S. Soltis, S. M. Swensen, S. E. Williams, P. A. Gadek, C. J. Quinn, L. E. Eguiarte, E. Golenberg, G. H. Learn, Jr., S. W. Graham, S. C. H. Barrett, S. Dayanandan, and V. A. Albert. 1993. Phylogenetics of seed plants: an analysis of nucleotide sequences from the plastid gene *rbcL.* Annals of the Missouri Botanical Garden **80:**528–580.

Chaw, S.-M., H. Long, B.-S. Wang, A. Zharkikh, and W.-H. Li. 1993. The phylogenetic position of Taxaceae based on 18S rRNA sequences. Journal of Molecular Evolution **37:**624–630.

Chaw, S.-M., H. -M. Sung, H. Long, A. Zharkikh, and W.-H. Li. 1995. The phylogenetic positions of the conifer genera *Amentotaxus, Phyllocladus,* and *Nageia* inferred from 18S rRNA sequences. Journal of Molecular Evolution **41**:224–230.

Chaw, S.-M., Sung, H.-M., H. Long, A. Zharkikh, W.-H. Li and T. C. Lau. 1997. Molecular phylogeny of gymnosperms and seed plant evolution: analysis of 18S rRNA sequences. Journal of Molecular Evolution, **14**:56–68.

Clark, G. B., B. W. Tague, V. C. Ware, and S. A. Gerbi. 1984. *Xenopus laevis* 28S ribosomal RNA: a secondary structure and its evolutionary and functional implications. Nucleic Acids Research **12**:6197–6220.

Clark, L. G., W. Zhang, and J. F. Wendel. 1995. A phylogeny of the grass family (Poaceae) based on *ndhF* sequence data. Systematic Botany **20**:436–460.

Clegg, M. T. 1993. Chloroplast gene sequences and the study of plant evolution. Procedings of the National Academy of Sciences U.S.A. **90**:363–367.

Coleman, A. W., A. Suarez, and L. J. Goff. 1994. Molecular delineation of species and syngens in volvocacean green algae (Chlorophyta). Journal of Molecular Evolution **30**:80–90.

Conti, E., A. Fischbach, and K. J. Sytsma. 1993. Tribal relationships in Onagraceae: implications from *rbcL* sequence data. Annals of the Missouri Botanical Garden **80**:672–685.

Cordesse, F., R. Cooke, D. Tremousaygue, F. Grellet, and M. Delseny. 1993. Fine structure and evolution of the rDNA intergenic spacer in rice and other cereals. Journal of Molecular Evolution **36**:369–379.

Cronn, R. C., X. Zhao, A. H. Patterson, and J. F. Wendel. 1996. Polymorphism and concerted evolution in a multigene family: 5S ribosomal DNA in diploid and allopolyploid cottoms. Journal of Molecular Evolution **42**:685–705.

Cullings, K. W. 1992. Design and testing of a plant-specific PCR primer for ecological and evolutionary studies. Molecular Ecology **1**:233–240.

Delwiche, C. F., M. Kuhsel, and J. D. Palmer. 1995. Phylogenetic analysis of *tuf*A sequences indicates a cyanobacterial origin of all plastids. Molecular Phylogenetics Evolution **4**:110–128.

Dennis, E. S., W. L. Gerlach, A. J. Pryor, J. L. Bennetzen, A. Inglis, D. Llewellyn, M. M. Sachs, R. J. Ferl, and W. J. Peacock. 1984. Alcohol dehydrogenase (*adh1*) gene of maize. Nucleic Acids Research **12**:3983–4000.

Denton, A. L., and B. D. Hall. 1996. Evolution of RNA polmerase II intron sequences in *Rhododendron* and the Ericaceae. American Journal of Botany (suppl.) **83**:150.

dePamphilis, C. W., N. D. Young, and A. D. Wolfe. 1997. Evolution of plastic gene *rps*2 in a lineage of hemiparasitic and holoparasitic plants: many losses of photosynthesis and complex patterns of rate variation. Proceedings of the National Academy of Sciences U.S.A. **94**:7367–7372.

Dixon, M. T., and Hillis, D. M. 1993. Ribosomal RNA secondary structure—compensatory mutations and implications for phylogenetic analysis. Molecular Biology and Evolution **10**:256–267.

Doebley, J. F., M. Durbin, E. M. Golenberg, M. T. Clegg, and D. P. Ma. 1990. Evolutionary analysis of the large subunit of carboxylase (*rbcL*) nucleotide sequence among the grasses (Gramineae). Evolution **44**:1097–1108.

Dover, G. A. 1982. Molecular drive: a cohesive mode of species evolution. Nature **299**:111–117.

Downie, S. R., and J. D. Palmer. 1992a. Use of chloroplast DNA rearrangements in reconstructing plant phylogeny. **In** Molecular Systematics of Plants, eds. P. S. Soltis, D. E. Soltis, and J. J. Doyle, pp. 14–35. Chapman & Hall, New York.

Downie, S. R., and J. D. Palmer. 1992b. Restriction site mapping of the chloroplast DNA inverted repeat: a molecular phylogeny of the Asteridae. Annals of the Missouri Botanical Garden **79**:266–283.

Downie, S. R., and J. D. Palmer. 1994. A chloroplast DNA phylogeny of the Caryophyllales based on structural and inverted repeat restriction site variation. Systematic Botany **19**:236–252.

Downie, S. R., R. G. Olmstead, G. Zurawski, D. E. Soltis, P. S. Soltis, J. C. Watson, and J. D. Palmer. 1991. Six independent losses of the chloroplast DNA *rpl*2 intron in dicotyledons: molecular and phylogenetic implications. Evolution **45**:1245–1259.

Doyle, J. J. 1991. Evolution of higher plant glutamine synthetase genes: tissue specificity as a criterion for predicting orthology. Molecular Biology and Evolution **8**:366–377.

Doyle, J. J. 1992. Gene trees and species trees: molecular systematics as one-character taxonomy. Systematic Botany **17**:144–163.

Doyle, J. J. 1993. DNA, phylogeny, and the flowering of plant systematics. BioScience **43**:380–389.

Doyle, J. J., and R. N. Beachy. 1985. Ribosomal gene variation in soybean (*Glycine max*) and its relatives. Theoretical Applied Genetics **70**:369–376.

Doyle, J. J., and J. L. Doyle. 1987. A rapid DNA isolation procedure for small quantities of fresh leaf tissue. Phytochemical Bulletin **19**:11–15.

Doyle, J. J., and J. L. Doyle 1990. Isolation of plant DNA from fresh tissue. Focus **12**:13–15.

Doyle, J. J., M. Lavin, and A. Bruneau. 1992a. Contributions of molecular data to papilionoid legume systematics. **In** Molecular Systematics of Plants, eds. P. S. Soltis, D. E. Soltis, and J. J. Doyle, pp. 223–251. Chapman & Hall, New York.

Doyle, J. J., J. I. Davis, R. J. Soreng, D. Garvin, and M. J. Anderson. 1992b. Chloroplast DNA inversions and the origin of the grass family (Poaceae). Proceedings National Academy of Science U.S.A. **89**:7722–7726.

Doyle, J. J., J. L. Doyle, and J. D. Palmer. 1995. Multiple independent losses of two genes and one intron from legume chloroplast genomes. Systematic Botany **20**:272–294.

Doyle, J. J., J. L. Doyle, J. A. Ballenger, and J. D. Palmer. 1996a. The distribution and phylogenetic significance of a 50-kb chloroplast DNA inversion in the flowering plant family Leguminosae. Molecular Phylogenetics and Evolution **5**:429–438.

Doyle, J. J., V. Kanazin, and R. C. Shoemaker. 1996b. Phylogenetic utility of histone H3 sequences in the perennial relatives of soybean (*Glycine*: Leguminosae). Molecular Phylogenetics and Evolution, **6**:438–447.

Embley, M. T., R. P. Hirt, and D. M. Williams. 1994. Biodiversity at the molecular level: the domains, kingdoms and phyla of life. Philosophical Transactions of the Royal Society of London B **345**:21–31.

Fay, M. F., W. S. Davis, L. Hufford, and M. W. Chase. 1996. A combined cladistic analysis of Themidaceae. American Journal of Botany (suppl.) **83**:155.

Ferris, C., R. P. Oliver, A. J. Davy, and G. M. Hewitt. 1993. Native oak chloroplast reveal an ancient divide across Europe. Molecular Ecology **2**:337–344.

Field, K. G., G. J. Olsen, D. J. Lane, S. J. Giovannoni, M. T. Ghiselin, E. C. Raff, N. R. Pace, and R. A. Raff. 1988. Molecular phylogeny of the animal kingdom. Science **239**:748–753.

Förster, H., M. D. Coffey, H. Elwood, and M. L. Sogin. 1990. Sequence analysis of the small subunit ribosomal RNAs of three zoosporic fungi and implications for fungal evolution. Mycologia **82**:306–312.

Freshwater, D. W., S. Fredericq, B. S. Butler, M. H. Hommersand, and M. W. Chase. 1994. A gene phylogeny of the red algae (Rhodophyta) based on plastic *rbcL*. Proceedings of the National Academy of Sciences U.S.A. **91**:7281–7285.

Freshwater, D. W., and J. Rueness. 1994. Phylogenetic relationships of some European *Gelidium* (Gelidiales, Rhodophyta) species, based on *rbcL* nucleotide sequence analysis. Phycologia **33**:187–194.

Fritz, G. N., J. Conn, A. Cockburn, and J. Seawright. 1994. Sequence analysis of the ribosomal internal transcribed spacer 2 from populations of *Anopheles nuneztovari* (Diptera: Culidae). Molecular Biology and Evolution **11**:406–416.

Gadek, P., and C. J. Quinn. 1993. An analysis of relationships within the Cupressaceae sensu stricto based on *rbcL* sequences. Annals of the Missouri Botanical Garden **80**:581–586.

Gargas, A., P. T. De Priest, M. Grube, and A. Tehler. 1995. Multiple origins of lichen symbioses in fungi suggested by SSU rDNA phylogeny. Science **268**:1492–1495.

Gatenby, A. A., S. J. Rothstein, and M. Nomura. 1989. Translational coupling of the maize chloroplast *atpB* and *atpE* genes. Proceedings of the National Academy of Sciences U.S.A. **86**:4066–4070.

Gaut, B. S., and M. T. Clegg. 1993. Molecular evolution of the *Adh*1 locus in the genus *Zea*. Proceedings of the National Academy of Sciences U.S.A. **90**:5095–5099.

Gielly, L., and P. Taberlet. 1994. The use of chloroplast DNA to resolve plant phylogenies: non-coding versus *rbcL* sequences. Molecular Biology and Evolution **11**:769–777.

Gielly, L., and P. Taberlet. 1996. A phylogeny of the European gentians inferred from chloroplast *trnL* (UAA) intron sequences. Botanical Journal of the Linnean Society. **120**:57–75.

Giovannoni, S. J., S. Turner, G. J. Olsen, S. Barns, D. J. Lane, and N. R. Pace. 1988. Evolutionary relationships among cyanobacteria and green chloroplasts. Journal of Bacteriology **170**:3584–3592.

Goldblatt, P., P. C. Hoch, and L. M. McCook. 1992. Documenting scientific data: the need for voucher specimens. Annals of the Missouri Botanical Garden **79**:969–970.

Golenberg, E. M. 1991. Amplification and analysis of Miocene plant fossil DNA. Philosophical Transactions of the Royal Society of London B **333**:419–427.

Golenberg, E. M., D. E. Giannasi, M. T. Clegg, C. J. Smiley, M. Durbin, D. Henderson and G. Zurawski. 1990. Chloroplast DNA sequence from a Miocene *Magnolia* species. Nature **344**:656–658.

Golenberg, E. M., M. T. Clegg, M. L. Durbin, J. Doebley, and D. P. Ma. 1993. Evolution of a noncoding region of the chloroplast genome. Molecular Phylogenetics and Evolution **2**:52–64.

Goremykin, V., V. Bobrova, J. Pahnke, A. Troitsky, A. Antonov and W. Martin. 1996. Noncoding sequences from the slowly evolving chloroplast inverted repeat in addition to *rbcL* data do not support Gnetalean affinities of angiosperms. Molecular Biology and Evolution **13**:383–396.

Gottlieb, L. D., and V. S. Ford. 1996. Phylogenetic relationships among the sections of *Clarkia* (Onagraceae) inferred from the nucleotide sequences of *PgiC*. Systematic Botany **21**:45–62.

Govindaraju, D. R., and C. A. Culli. 1992. Ribosomal DNA variation among populations of a *Pinus rigida* Mill. (pitch pine) ecosystem: I. Distribution of copy numbers. Heredity **69**:133–140.

Gruendler, P., I. Unfried, K. Pascher, and D. Schweizer. 1991. rDNA intergenic region from *Arabidopsis thaliana*: structural analysis, intraspecific variation and functional implications. Journal of Molecular Biology **221**:1209–1222.

Gutell, R. R. 1993. Collection of small subunit (16S-and 16S-like) ribosomal RNA structures. Nucleic Acids Research **21**:3051–3054.

Gutell, R. R., M. W. Gray, and M. Schnare. 1993. A comilation of large subunit (23S and 23S-like) ribosomal RNA structures. Nucleic Acids Research **21**:3055–3074.

Halanych, K. M. 1991. 5S ribosomal RNA sequences inappropriate for phylogenetic reconstruction. Molecular Biology and Evolution **8**:249–253.

Hamby, K. R., and E. A. Zimmer. 1988. Ribosomal RNA sequences for inferring phylogeny within the grass family (Poaceae). Plant Systematics and Evolution **160**:29–37.

Hamby, K. R., and E. A. Zimmer. 1992. Ribosomal RNA as a phylogenetic tool in plant systematics. In Molecular Systematics of Plants, eds. P. Soltis, D. Soltis and J. Doyle, pp. 50–91. Chapman & Hall, New York.

Hancock, J. M., and G. A. Dover. 1988. Molecular coevolution among cryptically simple expansion segments of eukaryotic 26S/28S rRNAs. Molecular Biology and Evolution **5**:377–391.

Harris, S. A., and R. Ingram. 1991. Chloroplast DNA and biosystematics: the effects of intraspecific diversity and plastid transmission. Taxon **40**:393–412.

Harrison, R. G. 1989. Animal mitochondrial DNA as a genetic marker in population and evolutionary biology. Trends in Ecology and Evolution **4**:6–11.

Hasebe, M., M. Ito, R. Kofuji, K. Ueda, and K. Iwatsuki. 1993. Phylogenetic relationships of ferns deduced from *rbcL* gene sequence. Journal of Molecular Evolution **37**:476–482.

Hasebe, M., P. G. Wolf, K. M. Pryer, K. Ueda, M. Ito, R. Sano, G. J. Gastony, J. Yokoyama, R. J. Manhart, N. Murakami, E. H. Crane, C. H. Haufler and W. D. Hauk. 1995. Fern phylogeny based on *rbcL* nucleotide sequences. American Fern Journal **85**:134–181.

Hassouna, N., B. Michot, and J. Bachellerie. 1984. The complete nucleotide sequence of mouse 28S rRNA gene: implications for the process of size increase of the large subunit rRNA in higher eukaryotes. Nucleic Acids Research **12**:3563–3574.

Haufler, C. H., T. A. Ranker, and L. Jianwei. 1996. Using *rbcL* sequences to test hypotheses of phylogenetic relationship in the fern families Polypodiaceae and Grammitidaceae. American Journal of Botany (suppl) **83**:126.

Hauk, W. D., C. R. Parks, and M. W. Chase. 1996. A comparison between *trnL-F* intergenic spacer and *rbcL* DNA sequence data: an example from Ophioglossaceae. American Journal of Botany (suppl.) **83**:126.

Hedderson, T. A., R. L. Chapman, and W. L. Rootes. 1996. Phylogenetic relationships of bryophytes inferred from nuclear-encoded rRNA gene sequences. Plant Systematics and Evolution **200**:213–224.

Hemleben, V., M. Ganal, J. Gerstner, K. Schiebel, and R. A. Torres. 1988. Organization and length heterogeneity of plant ribosomal RNA genes. **In** Architecture of Eukaryotic Genes, ed. G. Kahl, pp. 371–383. VCH, Weinheim, Germany.

Henriks, L., R. Debaere, Y. Van De Peer, J. Neefs, A. Goris, and R. De Wachter. 1991. The evolutionary position of the rhodophyte *Porphyra umbilicalis* and the basidiomycete *Leucosporidium scottii* among other eukaryotes as deduced from complete sequences of small ribosomal subunit RNA. Journal of Molecular Evolution **32**:167–177.

Hershkovitz, M. A., and L. A. Lewis. 1996. Deep level diagnostic value of the rDNA-ITS region: the case of an algal interloper. Molecular Biology and Evolution **13**:1276–1295.

Hershkovitz, M. A., and E. A. Zimmer. 1996. Conservation patterns in angiosperm rDNA-ITS2 sequences. Nucleic Acids Research **24**:2857–2867.

Hillis, D. M. 1996. Inferring complex phylogenies. Nature **383**:130.

Hillis, D. M., and M. T. Dixon. 1991. Ribosomal DNA: molecular evolution and phylogenetic inference. Quarterly Review Biology **66**:411–453.

Hillis, D. M., C. Moritz, and B. K. Mable. 1996. Molecular Systematics, second edition. Sinauer Associates, Sunderland, Massachusetts.

Hinkle, G., J. K. Wetterer, T. R. Schultz, and M. L. Sogin. 1994. Phylogeny of the attine ant fungi based on analysis of small subunit ribosomal RNA gene sequences. Science **266**:1695–1697.

Hommersand, M. H., S. Fredericq, and D. W. Freshwater. 1994. Observations on the systematics and biogeography of the Gigartinaceae (Gigartinales, Rhodophyta) based on *rbcL* sequence analysis. Botanica Marina **73**:192–203.

Hood, L., J. H. Campbell, and S. C. R. Elgin. 1975. The organization, expression, and evolution of antibody genes and other multigene families. Annual Review of Genetics **9**:305–353.

Hoot, S. B., and J. D. Palmer. 1994. Structural rearrangements, including parallel inversions, within the chloroplast genomes of *Anemone* and related genera. Journal of Molecular Evolution **38**:274–281.

Hoot, S. B., A. Culham, and P. R. Crane. 1995a. The utility of *atpB* gene sequences in resolving phylogenetic relationships: Comparison with *rbcL* and 18S ribosomal DNA sequences in the Lardizabalaceae. Annals of the Missouri Botanical Garden **82**:194–208.

Hoot, S. B., A. Culham, and P. R. Crane. 1995b. Phylogenetic relationships of the Lardizabalaceae and Sargentodoxaceae: chloroplast and nuclear DNA sequence evidence. Plant Systematics and Evolution **9**:195–199.

Hori, H., and S. Osawa. 1987. Origin and evolution of organisms as deduced from SS ribosomal RNA sequences. Molecular Biology and Evolution **4**:445–472.

Huss, V. A. R., and M. L. Sogin. 1990. Phylogenetic position of some *Chlorella* species within the Chlorococcales based upon complete small-subunit ribosomal RNA sequences. Journal of Molecular Evolution **31**:432–442.

Jack, T., L. L. Brockman, and E. M. Meyerowitz. 1992. The homeotic gene APETALA3 of *Arabidopsis thaliana* encodes a MADS Box and is expressed in petals and stamens. Cell **68**:683–697.

Jansen, R. K., and J. D. Palmer. 1987a. Chloroplast DNA from lettuce and *Barnadesia* (Asteraceae): structure, gene localization, and characterization of a large inversion. Current Genetics **11**:553–564.

Jansen, R. K., and J. D. Palmer. 1987b. A chloroplast DNA inversion marks an ancient evolutionary split in the sunflower family (Asteraceae). Proceedings of the National Academy of Sciences U.S.A. **84**:5818–5822.

Jansen, R. K., and J. D. Palmer. 1988. Phylogenetic implications of chloroplast DNA restriction site variation in the Mutisieae (Asteraceae). American Journal of Botany **75**:753–766.

Johnson, L. A., and D. E. Soltis. 1994. *matK* DNA sequences and phylogenetic reconstruction in Saxifragaceae s. s. Systematic Botany **19**:143–156.

Johnson, L. A., and D. E. Soltis. 1995. Phylogenetic inference in Saxifragaceae sensu stricto and *Gilia* (Polemoniaceae) using *matK* sequences. Annals of the Missouri Botanical Garden **82**:149–175.

Johnson, L. A., J. L. Schultz, D. E. Soltis, and P. S. Soltis. 1996. Monophyly and generic relationships of

Polemoniaceae based on *matK* sequences. American Journal of Botany **83**:1207–1224.

Jorgensen, R. A., and P. D. Cluster. 1988. Modes and tempos in the evolution of nuclear ribosomal DNA: new characters for evolutionary studies and new markers for genetic and population studies. Annals of the Missouri Botanical Garden **75**:1238–1247.

Kantz, T. S., E. C. Theriot, E. A. Zimmer, and R. L. Chapman. 1990. The Pleurastrophyceae and Micromonadophyceae: a cladistic analysis of nuclear rRNA sequence data. Journal of Phycology **26**:711–721.

Kellogg, E. A., and R. Appels. 1995. Intraspecific and interspecific variation in 5S RNA genes are decoupled in diploid wheat relatives. Genetics **140**:325–343.

Kim, K.-J., and R. K. Jansen. 1994. Comparisons of phylogenetic hypotheses among different data sets in dwarf dandelions (*Krigia*): additional information from internal transcribed spacer sequences of nuclear ribosomal DNA. Plant Systematics and Evolution **190**:157–185.

Kim, K.-J., and R. K. Jansen. 1995. *ndhF* sequence evolution and the major clades in the sunflower family. Proceedings of the National Academy of Science U.S.A. **92**:10379–10383.

Kim, Y.-D., and R. K. Jansen. 1996. Phylogenetic implications of *rbcL* and ITS sequences variation in the Berberidaceae. Systematic Botany, **21**:381–396.

King, K., R. A. Torres, U. Zentgraf, and V. Hemleben. 1993. Molecular evolution of the intergenic spacer in the nuclear ribosomal RNA genes of Cucurbitaceae. Journal of Molecular Evolution **36**:144–152.

Knox, E. B., S. R. Downie, and J. D. Palmer. 1993. Chloroplast genome rearrangements and the evolution of giant lobelias from herbaceous ancestors. Molecular Biology and Evolution **10**:414–430.

Kosuge, K., K. Swada, T. Denda, J. Adachi, and K. Watanabe. 1997. Phylogenetic relationships of some genera in the Ranunculaceae based on alcohol dehydrogenase genes. Plant Systematics and Evolution, in press.

Kranz, H. D., and V. A. R. Huss. 1996. Molecular evolution of pteridophytes and their relationship to seed plants: evidence from gene sequences. Plant Systematics and Evolution **202**:1–11.

Kranz, H. D., D. Miks, M.-L. Siegler, I. Capesius, C. W. Sensen, and V. A. Huss. 1995. The origin of land plants: phylogenetic relationships between Charophytes, Bryophytes, and vascular plants inferred from complete small subunit ribosomal RNA gene sequences. Journal of Molecular Evolution **41**:74–84.

Kron, K. 1996. Phylogenetic relationships of Empetraceae, Epacridaceae, and Ericaceae: evidence form nuclear ribosomal 18S sequence data. Annals of Botany **77**:293–303.

Kron, K. 1997. Phylogenetic relationships of Rhododendroideae (Ericaceae). Systematic Botany, in press.

Kuzoff, R. K., J. A. Sweere, D. E. Soltis, and E. A. Zimmer. 1997. The phylogenetic potential of entire 26S rDNA sequences in plants. Molecular Biology and Evolution, in press.

Labhart, P., and R. H. Reeder. 1987. DNA sequences for typical ribosomal gene spacers from *Xenopus laevis* and *Xenopus borealis*. Nucleic Acids Research **15**:3623–3624.

Lane, D. J., B. Pace, G. J. Olsen, D. A. Stahl, M. L. Sogin, and N. R. Pace. 1985. Rapid determination of 16S ribosomal RNA sequences for phylogenetic analyses. Proceedings of the National Academy of Sciences U.S.A. **82**:6955–6959.

Lassner, M., O. Anderson, and J. Dvorak. 1987. Hypervariation associated with a 12 nucleotide direct repeat and inferences on intergenomic homogenization of ribosomal RNA gene spacers based on the sequence of a clone from the wheat nor-d3 locus. Genome **29**:770–777.

Lavin, M., J. J. Doyle, and J. D. Palmer. 1990. Systematic and evolutionary significance of the loss of the large chloroplast DNA inverted repeat in the family Leguminosae. Evolution **44**:390–402.

Lavin, M., S. Mathews, and C. Hughes. 1992. Chloroplast DNA variation in Gliricidia *sepium* (Leguminosae): intraspecific phylogeny and tokogeny. American Journal of Botany **78**:1576–1585.

Les, D. H., D. K. Garvin, and C. F. Wimpee. 1991. Molecular evolutionary history of ancient aquatic angiosperms. Proceedings of the National Academy of Science U.S.A. **88**:10119–10123.

Lewis, L. A., L. W. Wilcox, P. A. Fuerst, and G. L. Floyd. 1992. Concordance of molecular and ultrastructural data in the study of zoosporic chlorococcalean green algae. Journal of Phycology **28**:375–380.

Lindahl, T. 1993. Instability and decay of the primary structure of DNA. Nature **362**:709–715.

Liston, A., L. H. Rieseberg, and M. A. Hanson. 1992. Geographic partitioning of chloroplast DNA variation in the genus *Datisca* (Datiscaceae). Plant Systematics and Evolution **181**:121–132.

Liston, A., W. A. Robinson, J. M. Oliphant, and E. R. Alvarez-Buylla. 1996. Length variation in the nuclear ribosomal DNA internal transcribed spacer region of non-flowering seed plants. Systematic Botany **21**:109–120.

Long, E. O., and I. B. Dawid. 1980. Repeated genes in eukaryotes. Annual Review of Biochemistry **49**:727–764.

Loockerman, D. J., and R. K. Jansen. 1995. The use of herbarium material for DNA studies. **In** Sampling the Green World, eds. T. F. Stuessy and S. H. Sohmer, pp. 205–219. Columbia University Press, New York.

Malek, O., K. Lattig, R. Hiesel, A. Brennicke, and V. Knoop. 1996. RNA editing in bryophytes and a molecular phylogeny of land plants. EMBO Journal **15**:1403–1411.

Manen, J.-F., V. Savolainen, and P. Simon. 1994a. The *atpB* and *rbcL* promoters in plastid DNAs of a wide dicot range. Journal of Molecular Evolution **38**:577–582.

Manen, J.-F., A. Natali, and F. Ehrendorfer. 1994b. Phylogeny of Rubiaceae-Rubieae inferred from the sequence of a cpDNA intergene region. Plant Systematics and Evolution **190**:195–211.

Manen, J.-F., V. Savolainen, S. De Marchi, and B. Rion. 1995. Chloroplast DNA sequences from a Miocene diatomite deposit in Ardeche (France) Comptes Rendu Academy of Science Paris **318**:971–975.

Manhart, J. R. 1994. Phylogenetic analysis of green plant *rbcL* sequences. Molecular Phylogeny and Evolution 3:114–127.

Manhart, J. R. 1995. Chloroplast 16S rDNA sequences and phylogenetic relationships of fern allies and ferns. American Fern Journal 85:182–192.

Martin, W., D. Lydiate, H. Brinkmann, G. Forkmann, H. Saedler, and R. Cerff. 1993. Molecular phylogenies in angiosperm evolution. Molecular Biology and Evolution 10:140–162.

Mason-Gamer, R. J., and E. A. Kellogg. 1996. Potential utility of the nuclear gene *waxy* for plant phylogenetic analysis. American Journal of Botany (suppl.) 83:178.

Mathews, S., M. Lavin, and R. A. Sharrock. 1995. Evolution of the phytochrome gene family and its utility for phylogenetic analyses of angiosperms. Annals of the Missouri Botanical Garden 82:296–321.

McCourt, R. M., K. G. Karol, S. Kaplan, and R. W. Hoshaw. 1995. Using *rbcL* sequences to test hypotheses of chloroplast and thallus evolution in conjugating green algae (Zygnematales, Charophyceae). Journal of Phycology 31:989–995.

McCourt, R. M., K. G. Karol, M. Guerlesquin, and M. Feist. 1996. Phylogeny of extant genera in the family Characeae (Charales, Charophyceae) based on *rbcL* sequences and morphology. American Journal of Botany 83:125–131.

Medlin, L. K., WH. C. F. Kooistra, R. Gersonde, and U. Wellbrock. 1996. Evolution of the diatoms (Bacillariophyta). II. Nuclear-encoded small-subunit rRNA sequence comparisons confirm a paraphyletic origin for the centric diatoms. Molecular Biology and Evolution 13:67–75.

Michaels, H. J., K. M. Scott, R. G. Olmstead, T. Szaro, R. K. Jansen, and J. D. Palmer. 1993. Interfamilial relationships of the Asteraceae: insights from *rbcL* sequence variation. Annals of the Missouri Botanical Garden 80:742–765.

Milligan, B.G., J. N. Hampton, and J. D. Palmer. 1989. Dispersed repeats and structural reorganization in subclover chloroplast DNA. Molecular Biology and Evolution 6:355–368.

Mindell, D. P., and R. L. Honeycutt. 1990. Ribosomal RNA in vertebrates: evolution and phylogenetic applications. Annual Review of Ecology and Systematics 21:541–566.

Mishler, B. D., K. Bremer, C. J. Humphries, and S. R. Churchill. 1988. The use of nucleic acid sequence data in phylogenetic reconstruction. Taxon 37:391–395.

Mishler, B. D., P. H. Thrall, J. S. Hopple Jr., E. De Luna, and R. Vilgalys. 1992. A molecular approach to the phylogeny of bryophytes: cladistic analysis of chloroplast-encoded 16S and 23S ribosomal RNA genes. Bryologist 95:172–180.

Mishler, B. D., L. A. Lewis, M. A. Buchheim, K. S. Renzaglia, D. J. Garbary, C. F. Delwiche, F. W. Zechman, T. S. Kantz, and R. L. Chapman. 1994. Phylogenetic relationships of the "green algae" and "bryophytes." Annals of the Missouri Botanical Garden 81:451–483.

Mitton, J. B. 1994. Molecular approaches to population biology. Annual Review of Ecology and Systematics 25:45–69.

Morgan, D. R., D. E. Soltis, and K. R. Robertson. 1994. Genetic variation within and among populations of the narrow endemic, *Delphinium viridescens*. American Journal of Botany 81:890–903.

Moritz, C., T. E. Dowling, and W. M. Brown. 1987. Evolution of animal mitochondrial DNA: relevance for populational biology and systematics. Annual Review of Ecology and Systematics 18:269–292.

Nadot, S., R. Bajon, and B. Lejeune. 1994. The chloroplast gene *rps4* as a tool for the study of Poaceae phylogeny. Plant Systematics and Evolution 191:27–38.

Nadot, S., G. Bittar, L. Carter, R. Lacroix, and B. Lejeune. 1995. Phylogenetic analysis of Monocotyledons based on the chloroplast gene *rps4* using parsimony and a new numerical phenetics method. Molecular Phylogenetics and Evolution 4:257–282.

Natali, A., J.-F. Manen, and F. Ehrendorfer. 1995. Phylogeny of the Rubiaceae-Rubioideae, in particular the tribe Rubieae: evidence from a non-coding chloroplast DNA sequence. Annals of the Missouri Botanical Garden 82:428–439.

Neefs, J.-M., Y. Van de Peer, P. De Rijk, S. Chapelle, and R. De Wachter. 1993. Compilation of small subunit RNA structures. Nucleic Acids Research 21:3025–3049.

Neuhaus, H., and G. Link. 1987. The chloroplast tRNA (UUU) gene from mustard (*Sinapsis alba*) contains a class II intron potentially coding for a maturase-related polypeptide. Current Genetics 11:251–257.

Neyland, R., and L. E. Urbatsch. 1996. Phylogeny of subfamily Epidendroideae (Orchidaceae) inferred from *ndhF* chloroplast gene sequences. American Journal of Botany 83:1195–1206.

Nickrent, D. L., and D. E. Soltis. 1995. A comparison of angiosperm phylogenies from nuclear 18S rDNA and *rbcL* sequences. Annals of the Missouri Botanical Garden 82:208–234.

Nickrent, D. L., and E. M. Starr. 1994. High rates of nucleotide substitution in nuclear small-subunit (18S) rDNA from holoparasitic flowering plants. Journal of Molecular Evolution 39:62–70.

Nickrent, D. L., K. Schuette, and E. Starr. 1994. A molecular phylogeny of *Arceuthobium* (Viscaceae) based on nuclear ribosomal DNA internal transcribed spacer sequences. American Journal of Botany 81:1149–1160.

Ohyama, K., H. Fukuzawa, T. Kohchi, H. Shirai, T. Sano, S. Sano, K. Umesono, Y. Shiki, M. Takeuchi, Z. Chang, S. Aota, H. Inokuchi, and H. Ozeki. 1986. Chloroplast gene organization deduced from complete sequence of liverwort *Marchantia polymorpha* chloroplast DNA. Nature 322:572–574.

Olmstead, R. G., and J. D. Palmer. 1994. Chloroplast DNA systematics: a review of methods and data analysis. American Journal of Botany 81:1205–1224.

Olmstead, R. G., and P. A. Reeves. 1995. Evidence for the polyphyly of the Scrophulariaceae based on chloroplast *rbcL* and *ndhF* sequences. Annals of the Missouri Botanical Garden 82:176–193.

Olmstead, R. G., and J. A. Sweere. 1994. Combining data in phylogenetic systematics: an empirical approach using three molecular data sets in the Solanaceae. Systematic Biology **43**:467–481.

Olmstead, R. G., H. J. Michaels, K. M. Scott, and J. D. Palmer. 1992. Monophyly of the Asteridae and identification of their major lineages inferred from DNA sequences of *rbcL.* Annals of the Missouri Botanical Garden **79**:249–265.

Olsen, G. J. 1987. Earliest phylogenetic branchings: comparing rRNA-based evolutionary trees inferred with various techniques. Cold Spring Harbor Symposium Quantative Biology **52**:825–837.

Paabo, S. 1989. Ancient DNA: extraction, characterization, molecular cloning, and enzymatic amplification. Proceedings of the National Academy of Sciences U.S.A. **86**:1939–1943.

Pace, N. R., G. J. Olsen, and C. R. Woese. 1986. Ribosomal RNA phylogeny and the primary lines of evolutionary descent. Cell **45**:325–326.

Palmer, J. D. 1985a. Evolution of chloroplast and mitochondrial DNA in plants and algae. In Molecular Evolutionary Genetics, ed. R. J. MacIntyre, pp. 131–240. Plenum Press, New York.

Palmer, J. D. 1985b. Comparative organization of chloroplast genomes. Annual Review of Genetics **19**:325–354.

Palmer, J. D. 1986. Isolation and structural analysis of chloroplast DNA. Methods in Enzymology **188**:167–186.

Palmer, J. D. 1987. Chloroplast DNA evolution and biosystematic uses of chloroplast DNA variation. American Naturalist **130**:S6–S29.

Palmer, J. D. 1991. Plastid chromosomes: structure and evolution. In Cell Culture and Somatic Cell Genetics in Plants, Vol. 7, The Molecular Biology of Plastids, eds. L. Bogorad and I. K. Vasil, pp. 5–53. Academic Press, New York.

Palmer, J. D. 1992. Mitochondrial DNA in plant systematics: applications and limitations. In Molecular Systematics of Plants, eds. P. S. Soltis, D. E. Soltis, and J. J. Doyle, pp. 36–49. Chapman & Hall, New York.

Palmer, J. D., and L. A. Herbon. 1988. Plant mitochondrial DNA evolves rapidly in structure, but slowly in sequence. Journal of Molecular Evolution **28**:87–97.

Palmer, J. D., B. Osorio, J. Aldrich, and W. F. Thompson. 1987. Chloroplast DNA evolution among legumes: loss of a large inverted repeat occurred prior to other sequence rearrangements. Current Genetics **11**:275–286.

Palmer, J. D., B. Osorio, and W. F. Thompson. 1988b. Evolutionary significance of inversions in legume chloroplast DNAs. Current Genetics **14**:65–74.

Palmer, J. D., R. K. Jansen, H. J. Michaels, M. W. Chase, and J. R. Manhart. 1988a. Chloroplast DNA variation and plant phylogeny. Annals of the Missouri Botanical Garden **75**:1180–1206.

Pamilo, P., and M. Nei. 1988. Relationships between gene trees and species trees. Molecular Biology and Evolution 5:568–583.

Pawlowski, J., I. Bolivar, J. F. Fahrni, T. Cavalier-Smith, and M. Gouy. 1996. Early origin of foraminifera suggested by SSU rRNA gene sequences. Molecular Biology and Evolution **13**:445–450.

Plunkett, G. M., D. E. Soltis, and P. S. Soltis. 1997a. Evolutionary patterns in Apiaceae: Inferences based on *matK* sequence data. Systematic Botany **21**:477–495.

Plunkett, G.M., D. E. Soltis, and P. S. Soltis. 1997b. Clarification of the relationship between Apiaceae and Araliaceae based on *matK* and *rbcL* sequence data. American Journal of Botany **84**:565–580.

Polanco, C., and M. Perez De La Vega. 1994. The structure of the rDNA intergenic spacer of *Avena sativa* L.: a comparative study. Plant Molecular Biology **25**: 751–756.

Purugganan, M. D., S. D. Rounsley, R. J. Schmidt, and M. F. Yanofsky. 1995. Molecular evolution of flower development: diversification of the plant MADS-box regulatory gene family. Genetics **140**:354–356.

Raubeson, L. A., and R. K. Jansen. 1992a. Chloroplast DNA evidence on the ancient evolutionary split in vascular land plants. Science **255**:1697–1699.

Raubeson, L. A., and R. K. Jansen. 1992b. A rare chloroplast DNA structural mutation is shared by all conifers. Biochemistry and Systematic Ecology **20**:17–24.

Rieseberg, L. H. 1995. The role of hybridization in evolution: old wine in new skins. American Journal of Botany **82**:944–953.

Rieseberg, L. H., and S. J. Brunsfeld. 1992. Molecular evidence and plant introgression. In Molecular Systematics of Plants., eds. P. S. Soltis. D. E. Soltis, and J. J. Doyle, pp. 151–176. Chapman & Hall, New York.

Rieseberg, L. H., and D. E. Soltis. 1991. Phylogenetic consequences of cytoplasmic gene flow in plants. Evolutionary Trends in Plants **5**:65–84.

Ritland, K., and M. T. Clegg. 1987. Evolutionary analysis of plant DNA sequences. American Naturalist **130**: S74–S100.

Ritland, C. E., and N. A. Straus. 1993. High evolutionary divergence of the 5.8S ribosomal DNA in *Mimulus glaucescens* (Scrophulariaceae). Plant Molecular Biology **22**:691–696.

Ritland, C. E., K. Ritland, and N. A. Straus. 1993. Variation in the internal transcribed spacers (ITS1 and ITS2) among eight taxa of the *Mimulus guttatus* species complex. Molecular Biology and Evolution **10**:1273–1278.

Riven, C. J., C. A. Cullis, and V. Walbot. 1986. Evaluating quantitative variation in the genome of *Zea mays.* Genetics **113**:1009–1019.

Rodgers, S. O., and A. J. Bendich. 1987. Ribosomal RNA genes in plants: variability in copy number and in the intergenic spacer. Plant Molecular Biology **9**: 509–520.

Rodman, J. E., P.S. Soltis, D. E. Soltis, and K. J. Sytsma. 1998. Parallel evolution of glucosinolate biosynthesis inferred from congruent nuclear and plastid gene phylogenies. American Journal of Botany, in press.

Saghai-Maroof, M. A., K. M. Soliman, R. A. Jorgensen and R. W. Allard. 1984. Ribosomal DNA spacer-length polymorphism in barley: Mendelian inheritance, chromosomal location, and population dynamics. Proceedings of the National Academy of Sciences U.S.A. **81**:8014–8018.

Sanderson, M. J., and J. J. Doyle. 1992. Reconstruction of organismal gene phylogenies from data on multigene families: concerted evolution, homoplasy, and confidence. Systematic Biology **41**:4–17.

Sang, T., D. J. Crawford, and T. F. Stuessy. 1995. Documentation of reticulate evolution in peonies (*Paeonia*) using internal transcribed spacer sequences of nuclear ribosomal DNA: Implications for biogeography and concerted evolution. Proceedings of the National Academy of Sciences U.S.A. **92**:6813–6817.

Savolainen, V., J.-F. Manen, E. Douzery, and R. Spichiger. 1994. Molecular phylogeny of families related to Celastrales based on *rbcL* 5′ flanking sequences. Molecular Phylogentics and Evolution **3**:27–37.

Savolainen, V., R. Corbaz, C. Moncousin, R. Spichiger, and J.-F. Manen. 1995a. Chloroplast DNA variation and parentage analysis in 55 apples. Theoretical and Applied Genetics **90**:1138–1141.

Savolainen, V., P. Cuenoud, R. Spichiger, M. D. P. Martinez, M. Crevecoeur, and J.-F. Manen. 1995b. The use of herbarium specimens in DNA phylogenetics: evaluation and improvement. Plant Systematics and Evolution **197**:87–98.

Savolainen, V., C. M. Morton, S. B. Hoot, and M. W. Chase. 1996. An examination of phylogenetic patterns of plastid *atpB* gene sequences among eudicots. American Journal of Botany (suppl.) **83**:541.

Schaal, B. A., and G. H. Learn, Jr. 1988. Ribosomal DNA variation within and among plant populations. Annals of the Missouri Botanical Garden **75**:1207–1216.

Schlegel, M., H. J. Elwood, and M. L. Sogin. 1991. Molecular evolution in hypotrichous ciliates: sequence of the small subunit ribosomal RNA genes from *Onychodromus quadricornutus* and *Oxytricha granulifera* (Oxytrichidae, Hypotrichida, Ciliophora). Journal of Molecular Evolution **32**:64–69.

Schlötterer, C., M.-T. Hauser, A. Von Haesleler, and D. Tautz. 1994. Comparative evolutionary analysis of rDNA ITS regions in *Drosophila*. Molecular Biology and Evolution **11**:513–522.

Schneeberger, R. G., and C. A. Cullis, 1992. Intraspecific 5S rRNA gene variation in flax, *Linum usitatissimum* (Linaceae). Plant Systematics and Evolution **183**:265–280.

Scoles, G. J., B. S. Gill, Z.-Y. Xin, B. C. Clarke, C. L. McIntyre, C. Chapman, and R. Appels. 1988. Frequent duplication and deletion events in the 5S RNA genes and the associated spacer regions of the *Triticeae*. Plant Systematics and Evolution **160**:105–122.

Scotland, R. W., J. A. Sweere, P. A. Reeves, and R. G. Olmstead. 1995. Higher-level systematics of Acanthaceae determined by chloroplast DNA sequences. American Journal of Botany **82**:266–275.

Sewell, M. M., C. R. Parks, and M. W. Chase. 1996. Intraspecific chloroplast DNA variation and biogeography of North American *Liriodendron* L. (Magnoliaceae). Evolution **50**:1147–1154.

Shih, N.-C., P. Heinrich, and H. M. Goodman. 1991. Cloning and chromosomal mapping of nuclear genes encoding chloroplast and cytosolic glyceraldehyde-3-phosphate-dehydrogenase from *Arabidopsis thaliana*. Gene **104**:133–138.

Simeone, A., A. La Volpe, and E. Boncinelli. 1985. Nucleotide sequence of a complete ribosomal spacer of *D. melanogaster*. Nucleic Acids Research **13**:1089–1102.

Smith, D. E., and A. S. Klein. 1994. Phylogenetic inferences on the relationship of North American and European *Picea* species based on nuclear ribosomal 18S sequences and the internal transcribed spacer 1 region. Molecular Phylogenetics and Evolution **3**:17–26.

Sogin, M. L., H. J. Elwood, and J. H. Gunderson. 1986. Evolutionary diversity of eukaryotic small-subunit rRNA genes. Proceedings of the National Academy of Sciences U.S.A. **83**:1383–1387.

Soltis, D. E., and R. K. Kuzoff. 1995. Discordance between molecular and chloroplast phylogenies in the *Heuchera* group (Saxifragaceae). Evolution **49**: 727–742.

Soltis, D. E., and P. S. Soltis. 1989. Polyploidy, breeding systems, and genetic differentiation in homosporous pteridophytes. In Isozymes in Plant Biology, eds. D. E. Soltis and P. S. Soltis, pp. 241–258. Dioscorides Press, Portland, Oregon.

Soltis, D. E., and P. S. Soltis. 1997. Phylogenetic relationships among Saxifragaceae sensu lato: a comparison of topologies based on 18S rDNA and *rbcL* sequences. American Journal of Botany **84**:504–522.

Soltis, D. E., P. S. Soltis, T. A. Ranker, and B. D. Ness. 1989. Chloroplast DNA variation in a wild plant, *Tolmiea menziesii*. Genetics **121**:819–826.

Soltis, D. E., P. S. Soltis, M. T. Clegg, and M. Durbin. 1990. *rbcL* sequence divergence and phylogenetic relationships in Saxifragaceae sensu lato. Proceedings of the National Academy of Sciences U.S.A. **87**:4640–4644.

Soltis, D. E., M. Mayer, P. S. Soltis, and M. Edgerton. 1991. Chloroplast-DNA variation in *Tellima grandiflora* (Saxifragaceae). American Journal of Botany **78**: 1379–1390.

Soltis, D. E., P. S. Soltis, and B. Milligan. 1992. Intraspecific chloroplast DNA variation. In Molecular Systematics of Plants, eds. P.S. Soltis, D. E. Soltis, and J. J. Doyle, pp. 177–201. Chapman & Hall, New York.

Soltis, D. E., D. R. Morgan, A. Grable, P. S. Soltis, and R. Kuzoff. 1993. Molecular systematics of Saxifragaceae sensu stricto. American Journal of Botany **80**: 1056–1081.

Soltis, D. E., R. K Kuzoff, E. Conti, R. Gornall, and K. Ferguson. 1996. *matK* and *rbcL* gene sequence data indicate that *Saxifraga* (Saxifragaceae) is polyphyletic. American Journal of Botany **83**:371–382.

Soltis, D. E., M. A. Gitzendanner, D. D. Strenge, and P. S. Soltis. 1997a. Chloroplast DNA intraspecific phylogeography of plants from the Pacific Northwest of North America. Plant Systematics and Evolution **206**:353–373.

Soltis, D. E., C. Hibsch-Jetter, P. S. Soltis, M. W. Chase, and J. S. Farris. 1997b. Molecular phylogenetic relationships among angiosperms: an overview based on rbcL and 18S rDNA sequences. Journal of Plant Research, in press.

Soltis, D. E., P. S. Soltis, D. L. Nickrent, L. A. Johnson, W. J. Hahn, S. B. Hoot, J. A. Sweere, R. K. Kuzoff, K. A. Kron, M. W. Chase, S. M. Swensen, E. A. Zimmer, S.-M. Chaw, L. J. Gillespie, W. J. Kress, and K. J. Sytsma. 1997. Angiosperm phylogeny inferred from 18S ribosomal DNA sequences. Annals of the Missouri Botanical Garden **84:**1–49.

Soltis, P. S., and D. E. Soltis. 1993. Ancient DNA: prospects and limitations. New Zealand Journal of Botany **31:**203–209.

Soltis, P. S., and D. E. Soltis. 1995a. Plant molecular systematics: inferences of phylogeny and evolutionary processes. Evolutionary Biology **28:**139–194.

Soltis, P. S., and D. E. Soltis. 1995b. Alternative genes for phylogentic reconstruction in plants: introduction. Annals of the Missouri Botanical Garden **82:**147–148.

Soltis, P. S., D. E. Soltis, and J. J. Doyle, eds. 1992a. Molecular Systematics of Plants. Chapman & Hall, New York.

Soltis, P. S., D. E. Soltis, and C. J. Smiley. 1992b. An *rbcL* sequence from a Miocene *Taxodium* (Bald Cypress). Proceedings of the National Academy of Sciences U.S.A. **89:**449–451.

Soltis, P. S., D. E. Soltis, S. J. Novak, J. L. Schultz, and R. K. Kuzoff. 1995. Fossil DNA: Its potential for biosystematics. In Experimental and Molecular Approaches to Pant Biosystematics, eds. P. C. Hoch and A. G. Stephenson pp. 1–14. Missouri Botanical Garden, St. Louis.

Steele, K. P., and R. Vilgalys. 1994. Phylogenetic analyses of Polemoniaceae using nucleotide sequences of the plastic gene *matK*. Systematic Botany **19:**126–142.

Steele, K. P., K. E. Holsinger, R. K. Jansen, and D. W. Taylor. 1991. Assessing the reliability of SS rRNA sequence data for phylogenetic analysis. Molecular Biology and Evolution **8:**240–248.

Stein, D. B., D. S. Conant, M. E. Ahearn, E. T. Jordan, S. A.Kirch, M. Hasebe, K. Iwatsuki, M. K. Tan, and J. A. Thomson. 1992. Structural rearrangements of the chloroplast genome provide and important phylogenetic link in ferns. Proceedings of the National Academy of Sciences U.S.A. **89:**1856–1860.

Strauss, S. H., J. D. Palmer, G. T. Howe, and A. H. Doerksen. 1988. Chloroplast genomes of two conifers lack a large inverted repeat and are extensively rearranged. Proceedings of the National Academy of Sciences U.S.A. **85:**3898–3902.

Suh, Y., L. B. Thien, H. E. Reeve, and E. A. Zimmer. 1993. Molecular evolution and phylogenetic implications of internal transcribed spacer sequences of ribosomal DNA in Winteraceae. American Journal of Botany **80:**1042–1055.

Swann, E. C., and J. W. Taylor. 1993. Higher taxa of Basidiomycetes: an 18S rRNA gene perspective. Mycologia **85:**923–936.

Sytsma, K. J., and W. J. Hahn. 1994. Molecular systematics: 1991–1993. Progress in Botany **55:**307–333.

Sytsma, K. J., and W. J. Hahn. 1996. Molecular systematics: 1994–1995. Progress in Botany **58:**470–499.

Sytsma, K. J., and B. A. Schaal. 1985. Phylogenetics of the Lisianthius skinneri (Gentianaceae) species complex in Panama utilizing DNA restriction fragment analysis. Evolution **39:**594–608.

Taberlet, P., L. Gielly, G. Pautou, and J. Bouvet. 1991. Universal primers for amplification of three non-coding regions of chloroplast DNA. Plant Molecular Biology **17:**1105–1109.

Takaiwa, F., K. Ooono, and M. Sugiura. 1985. Nucleotide sequence of the 17S-25S spacer region from rice rDNA. Plant Molecular Biology **4:**355–364.

Tanaka, Y., T. A. Dyer, and G. G. Brownlee. 1980. An improved direct RNA sequence method: its application to *Vicia faba* 5.8S ribosomal RNA. Nucleic Acids Research **8:**1259–1272.

Tautz, D., M. Trick, and G. A. Dover. 1986. Cryptic simplicity in DNA is a major source of genetic variation. Nature **322:**652–656.

Tautz, D., J. M. Hancock, D. A. Webb, and G. A. Dover. 1988. Complete sequences of the rRNA genes of *Drosophila melanogaster.* Molecular Biology and Evolution **5:**366–376.

Taylor, J. W., and E. C. Swann. 1994. DNA from herbarium specimens. In Ancient DNA, eds. B. Hermann and S. Hummel, pp. 167–181. Springer-Verlag, Berlin.

Thompson, W. F., and R. B. Flavell. 1988. DNase I sensitivity of ribosomal RNA genes in chromatin and nucleolar dominance in wheat. Journal of Molecular Biology **204:**535–548.

Tishler, E., S. Dassarma, and H. M. Goodman. 1986. Nucleotide sequence of an alfalfa *Medicago sativa* glutamine synthetase gene. Molecular and General Genetics **203:**221–229.

Torres-Ruiz, R. A., and V. Hemleben. 1994. Pattern and degree of methylation in ribosomal RNA genes of *Cucurbita pepo* L. Plant Molecular Biology **26:**1167–1179.

Troitsky, A. V., and V. K. Bobrova. 1986. 23S rRNA-derived small ribosomal RNAs: their structure and evolution with references to plant phylogeny. In DNA Systematics, Vol. 2: Plants, ed. S. K. Dutta, pp. 137–170. CRC Press, Boca Raton, Florida.

Troitsky, A. V., Y. F. Melekhovets, G. M. Rakhimova, V. K. Bobrova, K. M. Valiejo-Roman, and A. S. Antonov. 1991. Angiosperm origins and early stages of seed plant evolution deduced from rRNA sequences. Journal of Molecular Evolution **32:**253–261.

Umthun, A. R., Z. Hou, Z. A. Sibenaller, W.-L. Shaiu, and D. L. Dobbs. 1994. Identification of DNA-binding proteins that recognize a conserved Type I repeat sequence in the replication origin region of *Tetrahymena* rDNA. Nucleic Acids Research **22:**4432–4440.

Urbach, E., D. L. Robertson, and S. W. Chisholm. 1992. Multiple evolutionary origins of prochlorophytes within the cyanobacterial radiation. Nature **355:** 267–270.

Vilgalys, R., and B. L. Sun. 1994. Ancient and recent patterns of geographic speciation in the oyster mushroom *Pleurotus* revealed by phylogenetic analysis of ribosomal DNA sequences. Proceedings of the National Academy of Sciences U.S.A. **91:**4599–4603.

Vogler, A. P., and R. DeSalle. 1994. Evolution and phylogenetic information content of the ITS-1 region in the tiger beetle Cicindela dorsalis. Molecular Biology and Evolution **11:**393–405.

Volkov, R., S. Kostishin, F. Ehrendorfer, and D. Schweizer. 1996. Molecular organization and evolution of the external transcribed rDNA spacer region in two diploid relatives of *Nicotiana tabacum* (Solanaceae). Plant Systematics and Evolution **20:**117–129.

Wada, H., and N. Satoh. 1994. Phylogenetic relationships among extant classes of echinoderms, as inferred from sequences of 18S rDNA, coincide with relationships deduced from the fossil record. Journal of Molecular Evolution **38:**41–49.

Wakasugi, T., J. Tsudzuki, S. Ito, M. Shibata, and M. Sugiura. 1996. A physical map and clone bank of the black pine (*Pinus thunbergii*) chloroplast genome. Plant Molecular Biology Reporter **12:**227–241.

Waters, E. R. 1995. An evaluation of the usefulness of the small heat shock genes for phylogenetic analysis in plants. Annals of the Missouri Botanical Garden **82:**278–295.

Wen, J. and E. A. Zimmer. 1996. Phylogeny of *Panax* L. (the ginseng genus, Araliaceae): inference from ITS sequences of nuclear ribosomal DNA. Plant Systematics and Evolution **6:**167–177.

Wendel, J. F., A. Schnabel, and T. Seelanan. 1995. Bidirectional interlocus concerted evolution following allopolyploid speciation in cotton (*Gossypium*). Proceedings of the National Academy of Sciences U.S.A. **92:**280–284.

Wheeler, W. C., and R. L. Honeycutt. 1988. Paired sequence difference in ribosomal RNAs: evolutionary and phylogenetic implications. Molecular Biology and Evolution **5:**90–96.

White, T. J., T. Bruns, S. Lee, and J. Taylor. 1990. Amplification and direct sequencing of fungal ribosomal RNA genes for phylogenetics. In PCR Protocols: A Guide to Methods and Applications, eds. M. Innis, D. Gelfand, J. Sninsky, and T. White, pp. 315–322. Academic Press, San Diego.

Whitfeld, P. R., and W. Bottomley. 1983. Organization and structure of chloroplast genes. Annual Review Plant Physiology **34:**279–310.

Whitten, W. M., M. W. Chase and W. L. Stern. 1996. Molecular systematics of *Stanhopeinae* (Orchidaceae). American Journal of Botany (suppl.) **83:**201.

Wilcox, L. W., L. A. Lewis, P. A. Fuerst, and G. L. Floyd. 1992. Assessing the relationships of autosporic and zoosporic chlorococcalean green algae with 18S rDNA sequence data. Journal of Phycology **28:**381–386.

Williams, S. E., V. A. Albert, and M. W. Chase. 1994. Relationships of Droseraceae: a cladisitic analysis of *rbcL* sequence and morphological data. American Journal of Botany **81:**1027–1037.

Winship, P. R. 1989. An improved method for direct sequencing PCR amplified material using dimethyl sulphoxide. Nucleic Acids Research **17:**1266.

Woese, C. R. 1987. Bacterial evolution. Microbiology Review **51:**221–271.

Wolf, P. G. 1995. Phylogenetic analysis of *rbcL* and nuclear ribosomal RNA gene sequences in Dennstaedtiaceae. American Fern Journal **85:**306–327.

Wolf, P. G. 1996. Nucleotide sequences from *atpB* provide useful data for fern systematists. American Journal of Botany (suppl.) **83:**133.

Wolf, P. G., P. S. Soltis, and D. E. Soltis. 1994. Phylogenetic relationships of dennstaedtioid ferns: evidence from *rbcL,* sequences. Molecular Phylogenetics and Evolution **3:**383–392.

Wolfe, K. H. 1991. Protein-coding genes in chloroplast DNA: compilation of nucleotide sequences, data base entries and rates of molecular evolution. In Cell Culture and Somatic Cell Genetics of Plants, Vol. 7BI, ed. K. Vasil, pp. 467–482. Academic Press, San Diego.

Wolfe, K. H., C. W. Morden, and J. D. Palmer. 1992. Function and evolution of a minimal plastid genome from a nonphotosynthetic parasitic plant. Proceedings of the National Academy of Sciences U.S.A. **89:**10648–10652.

Wolters, J., and V. A. Erdmann. 1986. Cladistic analysis of SS rRNA and 16S rRNA secondary and primary structure—the evolution of eukaryotes and their relation to Archaebacteria. Journal of Molecular Evolution **24:**152–166.

Xiang, Q.-Y., D. E. Soltis, D. R. Morgan, and P. S. Soltis. 1993. Phylogenetic relationhips of *Cornus* s. 1. and putative relatives inferred from *rbcL* sequence data. Annals of the Missouri Botanical Garden **80:**723–734.

Xiang, Q.-Y., D. E. Soltis, and P. S. Soltis. Su. 1998. Phylogenetic relationships of Cornaceae and close relatives inferred from *matK* and *rbcL* sequences. American Journal of Botany, in press.

Yakura, K., A. Kato, and S. Tanifuji. 1984. Length heterogeneity of the large spacer of *Vicia faba* is due to the differing number of a 325 bp repetitive sequence element. Molecular and General Genetics **193:**400–405.

Yang, Q., M. G. Zwick, and M. R. Paule. 1994. Sequence organization of the *Acanthamoeba* rRNA intergenic spacer: identification of transcriptional enhancers. Nucleic Acids Research **22:**4798–4805.

Yeh, L.-C., and J. C. Lee. 1991. Higher-order structure of the 5.8S rRNA sequence within the yeast precursor ribosomal RNA synthesized in vitro. Journal of Molecular Biology **217:**649–659.

Zhang, W., C. Reading, and A. B. Deisseroth. 1992. Improved PCR sequencing with formamide. Trends in Genetics **8:**332.

Zimmer, E. A., R. K. Hamby, M. L. Arnold, D. A. LeBlanc, and E. E. Theriot. 1989. Ribosomal RNA phylogenies and flowering plant evolution. In The Hierarchy of Life, eds. B. Fernholm, K. Bremer, and J. Jörnvall, pp. 205–214. Elsevier Science, Amsterdam.

Zimmer, E. A., S. L. Martin, S. M. Beverley, Y. W. Kan, and A. C. Wilson. 1980. Rapid duplication and loss of genes coding for the a chains of hemoglobin. Proceedings of the National Academy of Sciences U.S.A. **77:**2158–2162.

Zurawski, G., and M. Clegg. 1987. Evolution of higher plant chloroplast DNA-coded genes: implications for structure-function and phylogenetic studies. Annual Review of Plant Physiology **38**:391–418.

Zurawski, G., W. Bottomley, and P. R. Whitfeld. 1984. Junctions of the large single copy region of the inverted repeats in *Spinacia oleracea* and *Nicotiana debneyi* chloroplast DNA: sequence of the genes for tRNA[HIS] and the ribosomal proteins S19 and L2. Nucleic Acids Research **12**:6547–6558.

Zurawski, G., B. Perrot, W. Bottomley, and P. R. Whitfeld. 1981. The structure of the gene for the large subunit of ribulose 1,5 bisphosphate carboxylase from spinach chloroplast DNA. Nucleic Acids Research **9**:3251–3270.

Zurawski, G., B. Perrot, W. Bottomley, and P. R. Whitfeld. 1982. Structures of the genes for the B and E subunits of spinach chloroplast ATPase indicate a dicistronic mRNA and an overlapping translation stop/start signal. Proceedings of the National Academy of Sciences U.S.A. **79**:6260–6264.

Zurawski, G., P. R. Whitfeld, and W. Bottomley. 1986. Sequence of the gene for the large subunit of ribulose 1, 5-bisphosphate carboxylase from pea chloroplasts. Nucleic Acids Research **14**:3975.

2

Contributions of PCR-Based Methods to Plant Systematics and Evolutionary Biology

Andrea D. Wolfe and Aaron Liston

In less than a decade, the polymerase chain reaction (PCR) has evolved into the common denominator of molecular biology for nucleic acid research. The most important impact of PCR in plant molecular systematics has been rapid amplification of DNA regions, facilitating nucleotide sequencing (see Chapter 1). However, there are many other techniques aside from DNA sequencing in which PCR is appropriate and extremely useful. In this chapter we review these applications of PCR in plant systematics and highlight areas of research that have been underused in plant molecular systematics laboratories.

THE POLYMERASE CHAIN REACTION

PCR developed from the manipulation of DNA polymerases to study basic processes of cell biology such as DNA replication, repair, and inheritance. The "chain reaction" of the technique refers to an exponential increase in the amount of target DNA obtained through successive cycles of amplification. Although in vitro DNA amplification has been known for nearly three decades (Kleppe et al., 1971), it took the isola-

tion, purification, and marketing of DNA polymerases from thermophilic bacteria (Mullis et al., 1986; Mullis and Faloona, 1987; Saiki et al., 1988) to launch the technique into ubiquity in molecular biology labs today. Before the advent of PCR, molecular biologists relied on cloning DNA into plasmid vectors and bacterial-mediated amplification of the DNA of interest. In fact, the development of these in vivo DNA amplification techniques in the early 1970s (Cohen et al., 1973) virtually terminated research into the development of in vitro methods until Kary Mullis's "California buckeye-scented Eureka! moment" in 1983 (Mullis, 1994).

The basic protocol of PCR is simple (see fig. 1 in both Cha and Thilly, 1995 and Palumbi, 1996): (1) double-stranded DNA is denatured at high temperature to form single strands (templates); (2) short oligonucleotide primers bind at a lower annealing temperature to the single-stranded complementary templates at ends flanking the targeted sequence; (3) the temperature is raised for synthesis, by primer extension, of target sequences; (4) the newly synthesized double-stranded DNA target sequences are denatured at high temperature, and the cycle is repeated. The

We thank K. Bachmann, S. Kephart, and J. Leebens-Mack for comments on an early draft of this manuscript, and those who kindly provided reprints and unpublished data. AL would like to thank Sara Liston and Bill Robinson for assistance with manuscript preparation. This work was supported by NSF DEB-9123080 and new faculty set-up funds from Ohio State University (ADW).

amplification of target DNA can be exponential in that every cycle has the potential to double the amount of target DNA from the previous cycle if there is a sufficient amount of polymerase, primers, and nucleotides in the solution.

Although the basic protocol of PCR is straightforward, each application requires optimization (Innis and Gelfand, 1990). Parameters that can affect the quantity and quality of amplified DNA include the concentration and quality of template DNA, as well as the concentration of primer DNAs, deoxynucleotide triphosphates (dNTPs), and magnesium chloride. The rate of change between steps of the protocol and the annealing temperature are also parameters that can have a large effect on the success of PCR experiments. The materials and methods sections of published papers using PCR (e.g., Baum and Sytsma, 1994; Doyle et al., 1995; Wolfe et al., 1997; see also the references listed in Tables 2.1–2.4) are good starting points for PCR protocols for specific genic regions.

PCR resource material is available through articles (Abrol and Chaudhary, 1993; Nickrent, 1994; Palumbi, 1996), books (Dieffenbach and Dveksler, 1995; Innis et al., 1995), and on the Internet. One Worldwide Web site of general interest is the PCR Jump Station. This Web site contains information on PCR general theory, protocols, software, equipment, applications, references, sequencing, and primers; it offers links to many different pages dealing with molecular biology and genetics.

The heat-stable polymerase originally isolated for PCR came from *Thermus aquaticus*. *Taq* DNA Polymerase is available in native and recombinant forms from most molecular biology supply companies. Other heat-stable polymerases have been isolated from *T. thermophilus, T. flavus, T. brockianus, Thermococcus litoralis, Thermotoga maritima, Pyrococcus furiosus,* and *P. woesei* (Dieffenbach and Dveksler, 1995). *Taq* DNA polymerase does not have a $3' \rightarrow 5'$ proofreading exonuclease capability, but DNA polymerases from *Thermococcus litoralis,* and *Pyrococcus* spp. do. Proofreading exonuclease activity increases the fidelity of PCR. Estimates of error incorporation for *Taq* DNA polymerase range from 7.2×10^{-5} to 2×10^{-4} (reviewed in Cha and Thilly, 1995), but this does not seem to pose a problem for generation of nucleotide sequences because the large supply of homogeneous template sequence at the early stages of PCR amplification minimizes the impact of small errors in DNA amplification.

Long-distance PCR, or long-PCR, is the process of amplifying DNA sequences up to 50 kb (Cheng et al., 1994, Estep, 1995; Foord and Rose, 1995). This technique uses a combination of two thermostable polymerases: a nonproofreading polymerase at high concentration and the secondary proofreading polymerase at a lower concentration. Although applications in plant biology or plant systematics are limited as of yet, the potential for research in plant systematics is clearly great. For example, there are complete chloroplast genome sequences in GenBank for the liverwort *Marchantia* (Ohyama et al., 1986), the gymnosperm *Pinus thunbergii* (Wakasugi et al., 1994), and the angiosperms *Nicotiana tabacum, Oryza sativa, Epifagus virginiana,* and *Zea mays* (Shinozaki et al., 1986; Hiratsuka et al., 1989; Wolfe et al., 1992; Maier et al., 1995). Many genes and genic regions are conserved among these taxa, facilitating the design of universal primers. A 7.5-kb segment of the chloroplast genome has recently been amplified in taxa from Poaceae, Aizoaceae, Boraginaceae, Malvaceae, Solanaceae, and Asteraceae using primers from within the *petA* and *rbcL* genes (V. Symonds and K. Schierenbeck, pers. comm.). If enough universal primers were designed to cover the entire chloroplast genome, it is conceivable that whole genome restriction-site mapping could be done without the traditional blotting and probe hybridization methods (see reviews in P. Soltis et al., 1992a; Olmstead and Palmer, 1994; Soltis et al., 1997). Similarly, DNA sequencing of chloroplast genomes would be possible using long-PCR instead of cloning methods (e.g., Shinozaki et al., 1986).

MAJOR IMPACTS OF PCR ON PLANT SYSTEMATICS

PCR has had the greatest impact in plant systematics in the following areas: nucleotide sequencing (see Chapter 1), analysis of randomly amplified polymorphic DNA (RAPD) markers (see also Chapters 15 and 16), and the detection of structural rearrangements (see Chapter 1). Prior to the introduction of PCR, nucleotide sequencing of plant genes involved cloning the gene of interest and employing overlapping single-stranded DNA sequencing runs (Zurawski et al., 1981, 1984; Coruzzi et al., 1984; Clegg and Durbin, 1990). Early plant systematics papers using PCR amplification to facilitate nucleotide sequencing include Golenberg et al. (1990) and Manhart and Palmer (1990). Since then the number of studies employing PCR for DNA amplification and subsequent sequencing has grown tremendously (see Chapter 1).

PCR has also facilitated the automation of DNA sequencing (Mcbride et al., 1989; Swerdlow and Gesteland, 1990; Tracy and Mulcahy, 1991; Bock and Slightom, 1995). In most automated systems, the gene of interest is amplified by PCR with the inclusion of fluorescent labels specific to each dideoxynucleotide triphosphate (ddNTP) used in the Sanger et al. (1977) sequencing protocol. The labeled fragments are separated on denaturing polyacrylamide gels according to size. The fluorescent labels are detected via a scanning mechanism as the fragment passes through a light beam, and the terminal ddNTP is recorded in a computer file. The amount of nucleotide sequence data currently being generated via automated sequencing is increasing rapidly (see Chapter 1).

One of the more dramatic impacts of PCR on DNA sequencing is the utility of the method in recovering DNA sequences from fossil material or from herbarium specimens. Several studies of DNA recovered from Miocene plant fossil material have been presented in recent years (Golenberg et al., 1990; Golenberg, 1991; P. Soltis et al., 1992b). Compression fossils from the Clarkia, Idaho, site have yielded amplifiable DNA for the chloroplast gene *rbcL* from the extinct *Magnolia latahensis* and a fossil *Taxodium*. Sequences from these two taxa were compared to extant relatives and results of phylogenetic analyses placed the fossil DNA sequences as close relatives to living specimens. However, recent studies have found that most amino acids found in Clarkia fossils have racemized or degraded, suggesting that endogenous nucleic acids are not recoverable (Poinar et al., 1996).

These results underscore the importance of taking steps to ensure that contaminant DNA is not amplified. Standard precautions include the use of UV sterilized reagents and supplies, barrier pipette tips, taxon-specific primers, and negative controls (e.g., all PCR components except target DNA) in all reactions. Even with rigorous controls, the "wrong" DNA can be amplified, for example from endosymbiotic organisms (Liston et al., 1996). Therefore all sequence results should be screened against public DNA sequence databases at an early stage of data collection.

Herbarium specimens are also becoming a resource for plant molecular systematics (see also Chapter 1). DNAs have successfully been extracted and/or amplified via PCR from herbarium specimens since the mid-1980s (e.g., Rogers and Bendich, 1985; Liston, 1992a; Savolainen et al., 1995). Doyle et al. (1995) surveyed 392 genera of legumes for the loss of introns of *rpl2* and *trnI* with approximately half of the taxa represented by DNAs extracted from herbarium specimens. In the Doyle et al. (1995) survey, extracted DNA from herbarium specimens ranged in quality from severely degraded to "normal," i.e., of high molecular weight. Savolainen et al. (1995) found that amplification of DNA from some herbarium samples may be difficult because oxidized material coprecipitates with the DNA, but that addition of certain additives can overcome the inhibiting activities of some herbarium extracts.

RAPD MARKERS

Description of the Method

As a direct consequence of the polymerase chain reaction, DNA analysis has become accessible to a broad range of researchers. PCR can be said to have enabled the "practice of molecular biology without a permit" (Erlich et al., 1991) in reference to the dictum of Chargaff that molecular biology itself was "the practice of biochemistry without a license" (quoted in Erlich et al., 1991). To extend the analogy, the generation of RAPD markers can be considered "the practice of PCR without a clue." This is because the method requires no prior knowledge of DNA sequence to generate genetic markers. Due to the ease of obtaining RAPD data, the method has been employed in hundreds of studies. Unfortunately the method has many limitations, and researchers must be aware of these to avoid being clueless when it comes to the appropriate application and analysis of RAPD markers.

Two related methods of producing genetic markers via the PCR process were published simultaneously in 1990, arbitrarily primed PCR (AP-PCR; Welsh and McClelland, 1990) and random amplified polymorphic DNA (RAPD; Williams et al., 1990). The key innovation of these methods is the use of DNA amplification to generate genetic markers that require no prior knowledge of target DNA sequence. Both methods use a single oligonucleotide primer of arbitrary sequence. A discrete PCR product is produced when, at an appropriate annealing temperature, the single primer binds to sites on opposite strands of the genomic DNA that are within an amplifiable distance, generally less than 3,000 base pairs. The presence or absence of this specific PCR product is assumed to represent mutations in the primer-binding sites of the genomic DNA (but see Limitations below).

The RAPD protocol uses shorter primers (usually 10 bp) and a constant annealing temperature (generally 34–37°C). The resulting PCR products are generally resolved on 1.5–2.0% agarose gels and stained with ethidium bromide (EtBr); polyacrylamide and silver staining are sometimes used. The AP-PCR protocol uses longer primers (20 bp or greater), an initial low stringency annealing temperature (40°C) followed by high stringency annealing (60°C), and resolution is generally on polyacrylamide. Variations on the foregoing protocols include modifications in amplification profiles, the use of primers as short as five nucleotides (Caetano-Anollés et al., 1991; Caetano-Anollés and Gresshoff, 1994), the use of two arbitrary primers in a reaction (Jain et al., 1994), restriction digestion prior to amplification (Caetano-Anollés et al., 1993) and alternative methods of fragment resolution and staining.

The RAPD protocol and its variants have been used with minor modifications in hundreds of plant studies (Table 2.1; see also Chapters 15 and 16, as well as reviews in Waugh and Powell, 1992; Bachmann, 1994; Nybom, 1994; Smith and Williams, 1994; Sobral and Honeycutt, 1994; Whitkus et al., 1994; Powell et al., 1995c; Weising et al., 1995; Karp et al., 1996; Fritsch and Rieseberg, 1996; Rieseberg, 1996). The broad use of RAPD markers can be attributed to several factors: (1) no prior sequence information is needed; (2) small (approximately 25 nanogram per reaction) amounts of DNA can be used; (3) the method requires a minimal amount of laboratory supplies and equipment; (4) detection is nonradioactive; and (5) a large number of potential markers can be generated using readily available primers. All of these factors contribute to efficiency and speed of data gathering. The rapid pace of obtaining RAPD data is an important factor in their popularity (Nybom, 1994), making the acronym particularly appropriate. On the other hand, the use of the term "random" is unfortunate, for the primers and their amplification targets are more accurately described as "arbitrary." Nevertheless, the term RAPD is much more widely employed than AP-PCR or DAF (DNA amplification fingerprints; Caetano-Anollés et al., 1991).

Limitations

Ideally, RAPD polymorphisms (band presence/absence) result from nucleotide sequence mutations that prevent the amplification of a particular marker in some individuals. This is the generally unstated assumption used when bands are scored as independent, binary characters in a data matrix. However, many well-documented factors may invalidate this basic assumption. The basic limitations of RAPD markers can be categorized as: (1) deviations from the expectations of strict Mendelian inheritance (these may be caused by artifactual, nongenetic, variation; organellar bands; and epigenetic interactions); and (2) problems of homology assessment. If these factors are disregarded, the utility of RAPD markers is diminished (Hadrys et al., 1992; Clark and Lanigan, 1993; Lamboy, 1994a; Lynch and Milligan, 1994; Smith et al., 1994; Pillay and Kenny, 1995; Rieseberg, 1996).

Table 2.1. Representative RAPD studies that included a phenetic and/or cladistic estimation of relationships.

Taxon[a]	Taxonomic Scope and Sampling[b]	Primers Used[c]	Bands Scored[d]	Percent Poly-morphic[d]	Polym. bands/Primer[d]	Data Analysis[e]	Homology Testing[f]	Reference
Intergeneric Studies								
Brassicaceae: Brassica (6 spp.), Raphanus (1), Sinapis (1)	8 spp. 10 accs. (bulk)	17	284	ns	16.7	PCO (S_G)	No	Demeke et al., 1992
Brassicaceae: Brassica (6), Raphanus (1)	7 spp., 19 accs. (1 or bulk)	41	432	ops	10.5	UPGMA (S_J)	Southern 12/15 yes	Thormann et al., 1994
Fabaceae: Cajanus (9), Dunbaria (1), Flemingia (1), Rhynchosia (3)	14 spp. 15 accs. (ns)	16	247	ns	15.4	UPGMA (S_D)	No	Ratnaparkhe et al., 1995
Fabaceae: Cassia (3), Chamaecrista (12), Senna (13)	28 spp. 28 accs. (1)	6	ns	100%		PCO (S_D)	No	Whitty et al., 1994
Meliaceae: Cedrela (1), Khaya (3), Lovoa (1), Swietenia (3)	8 spp. 29 accs. (ns)	5	65	100%	13	PCO SL (S_J)	No	Chalmers et al., 1994
Poaceae: Brachypodium (7), Bromus (1), Festuca (1), Oryza (1), Triticum (1), Vulpia (1)	12 spp. 14 accs. (1 or bulk)	20	395	100%	19.8	MP	Southern unquantified	Catalán et al., 1995
Poaceae: Lolium (6), Festuca (8)	14 spp. and vars. 16 accs. (5)	3	ns			LS (S_D)	Southern 4/6 yes	Stammers et al., 1995
Intergeneric Means		15.4	284.6	100%	15.1			
Interspecific Studies								
Asteraceae: Senecio	5 spp. 17 pops. (1–3)	23	166	91%	6.6	NJ (S_J) MP	No	Purps and Kadereit, 1998
Azollaceae: Azolla	8 spp. and vars. 25 accs. (1)	22	486	ns	22.1	PCO UPGMA (S_D)	No	Van Coppenolle et al., 1993
Cupressaceae: Juniperus	44 spp. and vars. 44 accs. (1–5)	13	186	ns	14.3	PCO (S_G)	No	Adams and Demeke, 1993
Fabaceae: Arachis	9 spp. 26 accs. (1)	10	132	ois	13.2 ois	PCO UPGMA (S_D)	Southern unquantified	Hilu and Stalker, 1995
Fabaceae: Lens	6 spp. and vars. 36 accs. (ns)	10	50	90%	4.5	UPGMA $(1-S_{SM})^{1/2}$	No	Abo-elwafa et al., 1995
Fabaceae: Lens	6 spp. and vars. 54 accs. (1)	24	88	ops	3.7	PCO SL (S_D)	No	Sharma et al., 1995b
Fabaceae: Leucaena	24 spp. and ssps. 24 accs. (2)	19	319	99.7%	16.7	LS (S_J) MP (decay)	No	Harris, 1995

Table 2.1. Representative RAPD studies that included a phenetic and/or cladistic estimation of relationships, *continued*

Taxon[a]	Taxonomic Scope and Sampling[b]	Primers Used[c]	Bands Scored[d]	Percent Polymorphic[d]	Polym. bands/ Primer[d]	Data Analysis[e]	Homology Testing[f]	Reference
Interspecific Studies								
Fabaceae: *Lotus*	5 spp. 17 accs. (ns)	20	144	23%	1.65	MP	No	Campos et al., 1994
Fabaceae: *Medicago*	6 spp. 33 accs. (1)	10	178	100%	17.8	MP	No	Brummer et al., 1995
Liliaceae: *Allium*	5 spp. 11 accs. (ns)	16	91	ns	5.7	UPGMA $(1-S_J)$	Southern 7/7 yes	Wilkie et al., 1993
Liliaceae: *Asphodelus*	3 spp. 15 pops. (ns)	8	97	63%	7.6	UPGMA (S_{SM})	No	Díaz Lifante and Aguinagalde, 1996
Malvaceae: *Iliamna*	3 spp. 6 pops. (1–9)	2	35	89%	15.5	UPGMA (S_D) MP (Dollo)	No	Stewart and Porter, 1995
Poaceae: *Echinochloa*	8 spp. 38 accs. (1)	9	117	94%	12.2	UPGMA $(S_D$ and $S_{SM})$	No	Hilu, 1994
Poaceae: *Eleusine*	5 spp. and ssps. 22 accs. (1)	12	103	ops	8.6	PCO	No UPGMA (S_D)	Hilu, 1995
Poaceae: *Hordeum*	19 spp. and ssps. 26 accs. (2–3 bulk)	13	370	100%	28.5	UPGMA (S_J)	No	González and Ferrer, 1993
Poaceae: *Hordeum*	39 spp. and ssps. 40 accs. (ns)	12	76	100%	6.33	UPGMA (S_J)	Southern 2/2 yes	Marillia and Scoles, 1996
Poaceae: *Panicum*	4 spp.	11	294	99%	26.5	UPGMA (S_D)	No	M'Ribu and Hilu, 1994
Ranunculaceae: *Ranunculus*	5 spp. and vars. 13 pops. (4)	23	326	ns	14.2	UPGMA (S_J) MP	No	Van Buren et al., 1994
Rosaceae: *Malus*	17 spp. and vars. 18 accs. (1)	16	49	ops	3.1	UPGMA (S_J)	No	Dunemann et al., 1994
Rosaceae: *Rubus*	13 spp. 24 accs. (bulk)	10	372	100%	37.2	PCO UPGMA (S_D)	No	Graham and McNicol, 1995
Rubiaceae: *Coffea*	4 spp. and vars. 27 accs. (ns)	25	41	ois	1.64 ois	PCO SL (S_D)	Southern 1/1 yes	Orozco-Castillo et al., 1994
Salicaceae: *Populus*	12 spp. 32 clones	4	120	92%	27.5	UPGMA (S_D)	No	Castiglione et al., 1993
Interspecific Means		14.2	174.6	87.7%	13.4			

Table 2.1. Representative RAPD studies that included a phenetic and/or cladistic estimation of relationships, *continued*

Taxon[a]	Taxonomic Scope and Sampling[b]	Primers Used[c]	Bands Scored[d]	Percent Polymorphic[d]	Polym. bands/Primer[d]	Data Analysis[e]	Homology Testing[f]	Reference
Intraspecific Studies								
Apiaceae: *Apium graveolens*	23 cvs. (5–20 bulk)	28	309	9.3%	1.0	MP	3:1 segregation 5/5 yes	Yang and Quiros, 1993
Asteraceae: *Microseris bigelovii*	13 pops. (bulk-self)	22	194	77%	6.8	UPGMA (S_J) MP	No	Van Heusden and Bachmann, 1992c
Asteraceae: *Microseris elegans*	10 pops. (bulk-self)	17	134	83%	6.5	UPGMA (S_J) MP	No	Van Heusden and Bachmann, 1992b
Asteraceae: *Microseris pygmaea*	10 pops. (bulk-self)	24	208	56%	4.8	UPGMA (S_J) MP	No	Van Heusden and Bachmann, 1992a
Brassicaceae: *Brassica juncea*	23 accs. (ns)	34	595	84%	14.7	UPGMA (S_J)	No	Jain et al., 1994
Brassicaceae: *Capsella bursa-pastoris*	21 pops. (2)	7	180	ois	25.7 ois	NJ (ns)	No	Neuffer, 1996
Fabaceae: *Cajanus cajan*	10 cvs. (ns)	16	127	ns	7.9	UPGMA (S_D)	No	Ratnaparkhe et al., 1995
Fabaceae: *Gliricidia sepium*	10 accs. (5)	9	63	61%	7.0	UPGMA (S_D)	Southern 1/1 yes	Chalmers et al., 1992
Liliaceae: *Allium aaseae* (8), *A. simillimum* (6)	2 spp. 14 pops. (25)	12	65	76%	5.4	LS (Nei-78)	No	Smith and Pham, 1996
Myrtaceae: *Eucalyptus globulus*	4 subsps. 37 pops. (3–5)	10	162	92%	14.9	PCO (1-S_{SM}) UPGMA	No	Nesbitt et al., 1995
Oleaceae: *Olea europaea*	17 cvs. (1)	17	47	ops	2.8	UPGMA (S_D)	No	Fabbri et al., 1995
Pinaceae: *Abies alba* (3), *A. nebrodensis* (1)	2 spp. 4 pops. (14–35)	12	84	65%	4.6	UPGMA (Apuya et al.)	No	Vicario et al., 1995
Poaceae: *Avena sterilis*	24 accs. (bulk-self)	21	177	65%	5.5	PCO (1-S_J) UPGMA	No	Heun et al., 1994
Poaceae: *Buchloe dactyloides*	4 pops. (12)	7	98	100%	14.0	UPGMA (1-S_{SM})	No	Huff et al., 1993
Poaceae: *Hordeum spontaneum*	10 pops. (4–5)	10	152	24%	3.6	UPGMA (Nei-78)	No	Dawson et al., 1993
Poaceae: *Sorghum bicolor*	36 accs. (ns)	29	262	55%	5.0	UPGMA (1-S_D)	segregation-unquantified	Tao et al., 1993

Table 2.1. Representative RAPD studies that included a phenetic and/or cladistic estimation of relationships, *continued*

Taxon[a]	Taxonomic Scope and Sampling[b]	Primers Used[c]	Bands Scored[d]	Percent Polymorphic[c,d]	Polym. bands/Primer[d]	Data Analysis[e]	Homology Testing[f]	Reference
Intraspecific Studies								
Poaceae: *Triticum aestivum*	15 cvs. (bulk)	32	109	65%	2.2	UPGMA (S_J)	No	Joshi et al., 1993
Solanaceae: *Lycopersicon esculentum* (46), *L. cheesmanii* (2)	2 spp. 48 accs. (bulk)	24	215	37%	3.3	UPGMA (S_D) MP	3:1 segregation 7/7 yes	Williams and St. Clair, 1993
Solanaceae: *Solanum melongena*	52 cvs. (bulk)	22	130	55%	3.3	UPGMA (S_J)	No	Karihaloo et al., 1995
Sterculiaceae: *Theobroma cacao*	25 accs. (1)	9	75	80%	6.7	PCO NJ (S_D)	No	Russell et al., 1993
Sterculiaceae: *Theobroma cacao*	41 accs. (1)	24	105	30%	1.3	NJ (ns)	No	de la Cruz et al., 1995
Theaceae: *Camellia sinensis*	38 clones (1)	21	253	62%	7.5	SL (S_D)	No	Wachira et al., 1995
Intraspecific Means		18.5	170.2	62%	7.0			
Single Cultivar or Population								
Asteraceae: *Microseris douglasii* Cortina Ridge pop.	25 inbred lines	17	150	31%	2.7	MP (bootstrap)	No	Roelofs and Bachmann, 1995
Brassicaceae: *Brassica carinata* cv. 'Dodola'	6 individuals	13	69	8.7%	0.5	none	No	Demeke et al., 1992
Poaceae: *Spartina alterniflora* Willapa Bay pop.	9 clones	17	70	16%	0.65	UPGMA (ns)	No	Stiller and Denton, 1995
Poaceae: *Yushania niitakayamensis* "site 2"	51 clones	12	23	ops	1.9	UPGMA (S_D)	No	Hsiao and Rieseberg, 1994
Single Population Means		14.8	78	18.6%	1.44			

[a]For the intergeneric studies, the number of species sampled in a genus is given in parentheses after the genus name. For intraspecific studies with two closely related species, the number of populations sampled in each is given in parentheses.

[b]The number of taxa sampled (species, subspecies, and varieties) is given on the first line. The second line gives the total number of accessions (accs.), cultivars (cvs.), populations (pops.), clones, or individuals, followed by the number of individuals sampled (and whether these were bulked) in each of these units.

[c]Only primers that were assayed for all taxa are included.

[d]Abbreviations used include not stated (ns), only polymorphic bands scored (ops), and only informative bands scored (ois).

[e]Methods of data analysis include principal coordinates analysis (PCO) and phenograms constructed with the following methods: least squares (LS), neighbor joining (NJ), single linkage (SL), and UPGMA. For these phenetic methods, the (dis)similarity measure used is given in parentheses and include the Dice (S_D), Gower (S_G), Jaccard's (S_J), and simple matching (S_{SM}) coefficients, and genetic distances of Nei (1978) and Apuya et al. (1988). Cladistic methods include maximum (Wagner) parsimony (MP). Studies that used Dollo parsimony, the bootstrap, or the decay index are noted.

[f]If band homology was tested by Southern or segregation analysis, the fraction of bands that proved homologous is given.

Deviations from Mendelian Inheritance

Accurate scoring of RAPD markers depends on the reliable amplification of diagnostic fragments. The reliability or robustness of a particular marker can be influenced by several "environmental" factors. For example, variation in DNA purity, the concentration of magnesium chloride, and the primer to template ratio can change banding patterns (Ellsworth et al., 1993; Williams et al., 1993). Some RAPD markers may amplify despite primer-template mismatch (Venugopal et al., 1993). It is expected that such fragments will be particularly sensitive to changes in the foregoing parameters. Different annealing temperatures and different thermostable polymerases can also have dramatic effects on RAPD patterns (Williams et al., 1993). Rigorous standardization of the RAPD protocol is necessary to reduce such artifactual variation (Dowling et al., 1996). A noteworthy modification of the RAPD protocol using pairs of primers, increasing polymerase concentration, and slow ramping from the annealing to the extension temperature resulted in increased reproducibility across a range of human DNA concentrations and extraction methods (Benter et al., 1995).

Genetic factors can also impact the assumptions of binary scoring. When a RAPD polymorphism results from insertion/deletions in the amplified sequence, and both bands are resolved, codominant inheritance will result. In *Helianthus*, approximately 5% of RAPD loci display codominant inheritance (Rieseberg et al., 1993). A further complication with codominance is the common occurrence of heteroduplex bands in individuals heterozygous for a codominant locus (Ayliffe et al., 1994; Davis et al., 1995). If RAPD priming sites occur in precisely repetitive DNA, a single band will be obtained. However, degenerate repeats or multiple priming sites within a locus may produce several non-independent markers (Smith et al., 1994; Hilu and Stalker, 1995). Scoring of codominant alleles, heteroduplex bands, or repetitive markers as independent loci can result in an overestimate of relatedness (Smith et al., 1994).

Anomalous RAPD bands may also occur when competition among priming sites exists (Heun and Helentjaris, 1993). For example, a particular marker might amplify in one genetic background but not in another due to competition from unlinked sites, as demonstrated by Smith et al. (1994). Primer competition could also explain nonparental bands in segregating progeny, as observed in primates by Riedy et al. (1992). However, such "epistatic" interactions were not observed in *Zea mays* F_1 hybrids (Heun and Helentjaris, 1993). Other factors that may affect the robustness of RAPD amplifications include variation in nuclear DNA content and polyploidy (Bachmann, 1997). RAPD bands originating from parasitic and symbiotic organisms, rather than the organism of interest, are as yet undocumented, but the potential for their occurrence must also be considered (Newbury and Ford-Lloyd, 1993). Van Coppenolle et al. (1993) compared *Azolla* accessions with and without *Anabaena*. Although they expected to observe only a subset of the RAPD bands in the endosymbiont-free strains, this was not observed.

The expectation of Mendelian inheritance will not be met when RAPD markers originate from uniparentally inherited organellar genomes (Lorenz et al., 1994; Thormann et al., 1994; Aagard et al., 1995). In *Brassica*, Southern hybridization revealed that 0.7% of examined RAPD bands were of chloroplast origin and 4.9% were of mitochondrial origin (Thormann et al., 1994). In *Pseudotsuga menziesii*, 45% of scored RAPD bands were found to exhibit strict maternal inheritance, and thus were hypothesized to be of mitochondrial origin (Aagard et al., 1995). The large size of the mitochondrial genome in conifers may explain this result (Aagard et al., 1995). On the other hand, uniparental inheritance of RAPD markers was not found in *Zea mays* (Heun and Helentjaris, 1993). If organellar and nuclear RAPD markers are not distinguished, errors in calculations of genetic diversity and misinterpretations in studies of hybridization and introgression could result.

Problems of Homology Assessment

All pairwise comparisons of RAPD fragments begin with the assumption that co-migrating bands represent homologous loci. However, as in any study based on electrophoretic resolution, the assumption that equal length equals homology can be mistaken. More accurate resolution of fragment size (e.g., by using polyacrylamide gels and silver staining; Huff et al., 1993) can reduce such errors (Williams et al., 1993). Few RAPD studies in plants have used these alternatives to agarose gel electrophoresis and EtBr staining, perhaps due to the greater time and expense involved.

In RAPD studies, homology can be tested by Southern hybridization. A single band is excised from a gel and used as a hybridization probe (Williams et al., 1993). If the probe hybridizes to the equal length bands in other amplified samples, the assumption of length homology is correct. Although the approach is straightforward, it is also "extremely time-consuming" (Rieseberg, 1996). Among the studies cited in Table 2.1, only eight used this method, examining 1–8% of the scored bands. In several studies, bands of equal length have been found not to be homologous (Table 2.1; see also Pillay and Kenny, 1995).

An alternative test of homology is via restriction enzyme digestion of gel-isolated fragments (Fritsch and Rieseberg, 1992; Rieseberg, 1996). Fragments that generate the same restriction profile for two (or two of three) tested endonucleases are considered homologous. Although this provides only an approximate test of homology, it is much simpler than the alternatives and is preferable to the common practice of relying solely on fragment size as an indicator of homology.

The construction of a genomic map from segregating progeny may be used to test the homology of RAPD markers (Rieseberg, 1996). In practice, such segregation analysis is usually conducted as part of a genomic mapping study, and the discovery of nonhomologous bands is simply a by-product of map generation. Conducting segregation analysis solely for testing homology is usually impractical.

Genomic mapping in *Helianthus* revealed that approximately 13% of "homologous" (determined by Southern hybridization or restriction digestions) RAPD bands actually represent paralogous loci (various members of a gene family) and not orthologous genes (Rieseberg, 1996). Paralogous loci may result from amplifi-

cation of repetitive DNA sequences (Rieseberg, 1996) or homoeologs in polyploids (Jessup, 1993). Williams et al. (1993) and Thormann et al., (1994) found that 2 of 11 (18%) and 4 of 12 (25%) tested fragments hybridized to highly repeated sequences in *Glycine* and *Brassica,* respectively. The use of paralogous loci may bias estimates of phylogenetic relationships (Doyle, 1992).

RAPD fragments often vary in band intensity. In addition, methods of DNA detection that are more sensitive than EtBr have regularly demonstrated that many bands exist that are not observed in standard RAPD protocols (Williams et al., 1993; Bachmann, 1997). Band intensity differences could result from primer mismatch or relative sequence abundance (Devos and Gale, 1992; Williams et al., 1993). However, Thormann et al. (1994) found no correlation between copy number and band intensity. The fact that fainter bands are generally less robust in RAPD experiments (Ellsworth et al., 1993; Heun and Helentjaris, 1993) suggests that varying degrees of primer mismatch may account for many band intensity differences. Sequencing of RAPD markers has confirmed that loci flanked by perfect inverted repeats (priming sites) result in stronger amplification than loci with a 1–2-bp mismatch at one annealing site (Venugopal et al., 1993). Although most studies disregard differences in band intensity, scoring markers as binary (presence/absence) characters, Demeke et al. (1992) and Adams and Demeke (1993) used a 7-state scale of band intensity. However, because the source of band intensity differences is uncertain, the use of this character state scoring is not recommended.

Applications

Noteworthy applications of RAPD markers in plant systematics and evolutionary biology (see also Chapters 15 and 16) include the documentation of interspecific hybridization (e.g., Crawford et al., 1993; Smith et al., 1996) and introgression (e.g., Arnold et al., 1991; Dean and Arnold, 1996), identification of clones (e.g., Hsiao and Rieseberg, 1994; Stiller and Denton, 1995), development of markers linked to sex expression (e.g., Mulcahy et al., 1992; Hormaza et al., 1994) and B chromosome presence (e.g., Gourmet and Rayburn, 1996), estimation of breeding systems (e.g., Fritsch and Rieseberg, 1992), measurement of levels and distribution of genetic diversity (e.g., Gustafsson and Gustafsson, 1994, Smith and Pham, 1996), and description of patterns of intrapopulation (e.g., Roelofs and Bachmann, 1995) and intraspecific (e.g., Van Heusden and Bachmann, 1992a, 1992b, 1992c; Neuffer, 1996) differentiation. Additional applications in plant ecology (Bachmann, 1994) and conservation genetics (Fritsch and Rieseberg, 1996) have been recently reviewed. RAPD markers have also made feasible the generation of genomic maps in uncultivated species in order to address the mechanisms of plant speciation (e.g., Rieseberg et al., 1993, 1996; Bradshaw et al., 1995; Bachmann and Hombergen, 1996). The foregoing applications illustrate the potential for applying efficiently generated polymorphic DNA markers to a wide array of basic questions in plant systematics and evolutionary biology. However, the application in which RAPD markers have been most widely used in published literature is the estimation of phenetic and cladistic relationships (Table 2.1). Ironically, this is also the application for which RAPD markers have been considered unsuitable (Smith et al., 1994; Rieseberg, 1996; Backeljau et al., 1995; van de Zande and Bijlsma, 1995). This paradox is treated in the next section.

Table 2.1 contains representative RAPD marker studies that include a phenetic and/or cladistic algorithm to estimate relationships. Studies that used RAPD markers solely to identify taxa or individuals (clones), to document hybridization or introgression, for genetic mapping, for breeding system estimates, or to estimate levels and distribution of population genetic variation were not included. These selected studies illustrate the range of taxonomic levels at which RAPD markers have been applied in plants (Table 2.1). Relatively few studies have attempted to assess intergeneric relationships with RAPD markers, and most of these cases involve taxa that may be considered congeneric (e.g., Chalmers et al., 1994; Thormann et al., 1994; Whitty et al., 1994). As genetic distances increase, the probability of equating fragment size with homology decreases (van de Zande

and Bijlsma, 1995). In accord with this finding from *Drosophila,* Southern hybridization demonstrated that a putative RAPD locus was homologous in three species of *Brachypodium,* but not in *Festuca* (Catalán et al., 1995). Clearly, RAPD markers are of little utility in comparisons among plant genera. The same reservations should be extended to interspecific comparisons in large, highly diverse genera (e.g., Adams and Demeke, 1993; González and Ferrer, 1993; Graham and McNicol, 1995).

The number of primers included in these studies ranged from 2 to 41, generating between 23 and 595 polymorphic markers. An average of 10 polymorphic bands was obtained per primer. As expected, the average number of polymorphic bands per primer was lowest in the single population studies and higher in the intergeneric and interspecific studies (Table 2.1). Because different researchers used different criteria for band scoring, it is difficult to compare levels of observed polymorphism among studies. For example, Dunemann et al. (1994) recorded "only 'significant' RAPD loci." The criteria for "significance" were not reported. However, relatively few bands were scored, resulting in a reduced number of polymorphic bands per primer, compared to most interspecific studies. Likewise, the 4.5-fold difference in polymorphic bands per primer obtained in two studies of *Hordeum* (González and Ferrer, 1993; Marillia and Scoles, 1996, Table 2.1) clearly represent different criteria for scoring bands, and not biological differences. It is also difficult to make correlations between levels of RAPD marker polymorphism and life history traits, because most studies did not use population samples.

In a test of the reliability of band scoring, D. Huff et al. (pers. comm.) conducted RAPD analyses for two grass species (*Poa pratensis* and *Buchloe dactyloides*) in three independent laboratories. Although the same primers and DNA samples were used in each lab, the resulting data matrices were substantially different. These inconsistencies resulted from differences in the amplification and scoring of faint bands. Nevertheless, the independent phylogenetic analyses resulted in the same topologies for *Poa pratensis,* an apomictic species. In the outcrossing *Buchloe dactyloides,* a different topology was encountered in one laboratory, suggesting that more caution in scoring may be needed, particularly in outcrossing (polymorphic) taxa.

Analysis

In plants, most RAPD studies that estimate relationships have used calculations of pairwise distance and clustering approaches (Table 2.1) such as unweighted pair group method with arithmetic averaging (UPGMA) or neighbor joining (NJ). In general, the binary matrix of band presence/ absence was transformed to similarity (or distance) measures using the coefficients of Jaccard or Dice. Both measures exclude negative matches. Note that Equation 21 of Nei and Li (1979) and the DNA-fingerprint similarity of Lynch (1990) are equivalent to the similarity coefficient of Dice. Dice's coefficient gives more weight to matches than to mismatches (Sneath and Sokal, 1973), and has a direct biological meaning: It is an estimate of the expected portion of amplified fragments shared by two samples due to inheritance from a common ancestor (Nei and Li, 1979). Dice's coefficient is also less biased by the occurrence of a given level of artifactual bands (Lamboy, 1994a). A few studies have used the simple matching coefficient, which includes negative matches (shared absence of a band). Because band absence is more likely to be nonhomologous, the use of the simple matching coefficient is not recommended. Both the UPGMA and neighbor-joining methods can be influenced by the input order of operational taxonomic units (OTUs), resulting in alternative topologies or "ties" from the same data set (Backeljau et al., 1996). Ties are expected to be more common in binary-coded data such as RAPD markers, and may result in strikingly different topologies (Backeljau et al., 1996). If UPGMA or NJ algorithms are used, it is recommended that potential alternative topologies are evaluated.

As discussed previously, estimates of relationships from RAPD markers can be biased by the occurrence of false positives and false negatives (band presence or absence resulting from experimental error). Lamboy (1994b) has proposed estimating the probability of false positives and false negatives from replicate RAPD experiments. These values can be used to calculate corrected values of a similarity coefficient. Whereas the preferred approach is to optimize RAPD conditions to ensure consistent amplifications, the method of Lamboy is appropriate when this is not possible (e.g., when DNA quality cannot be controlled for).

Disadvantages of using the above phenetic approaches for phylogenetic estimation include (Swofford et al., 1996): (1) information is lost when characters are transformed to pairwise distances; (2) distance analysis does not facilitate the combination of different types of characters in a total evidence approach; (3) the influence of particular characters on the analysis cannot be directly examined; and (4) evaluation of suboptimal topologies is not feasible.

Although parsimony methods are the predominant approach to phylogenetic inference in the systematic literature, a minority of RAPD studies uses these methods (Table 2.1). Why? Concerns over assessments of band homology and independence are often cited (e.g., Smith et al., 1994). This objection applies equally to phenetic and cladistic analysis of any fragment-based data (Swofford et al., 1996). For RFLP analyses, Bremer (1991) has argued that by including "enough" fragment data, errors of homology assessment and nonindependence will be swamped. However, this "brute force" approach does not ensure an accurate phylogenetic estimate (Swofford et al., 1996). Studies are needed that quantify how inaccurate assessments of homology and independence bias phylogenetic analyses, as no source of comparative data is immune from these problems (Bachmann, 1995). In a study at the intraspecific level, Van Heusden and Bachmann (1992a, 1992b, 1992c) estimated that up to 10% of the RAPD markers might violate the assumptions of independence and size homology. Nevertheless, their phylogenetic analyses were consistent with other estimates of relationships.

Beyond the issues of homology and independence, Backeljau et al. (1995) have given three additional reasons for considering RAPD data inappropriate for parsimony analysis: (1) the higher G+C content of RAPD primers will result in the biased amplification of hypervariable parts of the genome; (2) appropriate models for

RAPD character state change do not exist; and (3) scoring the RAPD product as a binary character ignores potentially non-independent alleles. Each of these objections can be countered: (1) If rapidly mutating parts of the genome are indeed targeted (this has not been experimentally demonstrated) the higher level of polymorphism would provide potentially informative characters for taxa that cannot be distinguished by other molecular methods. (2) The absence of a model for character state change is not unique to RAPD data. The same objection could be applied to morphological data or nucleotide sequences from regions where the constraints on sequence evolution are not understood (e.g., nrDNA ITS). (3) Deficiencies of the independent allele model can be addressed by considering RAPD loci as diallelic (Lynch and Milligan, 1994). Although this makes the unrealistic assumption that "band absence" represents a single allele, it could potentially be used in conjunction with a step matrix that weighs RAPD band gains over losses.

The null hypothesis in a cladistic analysis of taxa connected by gene flow is the absence of structure in the resulting cladogram, due to apparent homoplasy resulting from recombination (Roelofs and Bachmann, 1997). Thus when a cladistic analysis of RAPD data results in well-resolved and well-supported topologies, barriers to gene flow can be inferred (e.g., Van Heusden and Bachmann, 1992a, 1992b, 1992c; Van Buren et al., 1994; Roelofs and Bachmann, 1995). Alternatively, an unresolved cladogram does not suggest that cladistic analysis is inappropriate, but rather provides evidence for gene flow among the examined taxa. To address this issue, the use of "population pools" has been promoted, to exclude intrapopulation RAPD variation from the analysis (Furman et al., 1997). An alternative approach is to conduct parsimony analysis of frequency data (Swofford et al., 1996). However, the performance of this method has not been tested with RAPD markers.

Despite the foregoing limitations, a large amount of RAPD data has been collected and analyzed (Table 2.1). These results must be evaluated by systematists interested in the phylogeny of the examined taxa. Rather than dismiss all phylogenetic analyses of RAPD data, it seems more constructive to differentiate between appropriate and inappropriate applications. When a large number of markers is included and the analysis is restricted to taxa in which band homology assessments can be confidently made, the results of RAPD analysis have been successfully incorporated into assessments of phylogenetic relationships. Studies that directly compare phylogenetic estimates from RAPD and other sources of data have demonstrated that RAPD markers can be a component of a multifaceted approach to phylogenetic reconstruction (Borowsky et al., 1995; Peakall et al., 1995; Roelofs and Bachmann, 1997).

AMPLIFIED FRAGMENT LENGTH POLYMORPHISMS

Description of the Method

(AFLP) Amplified fragment length polymorphism (AFLP) data consist of genetic markers generated by a combination of restriction enzyme digest of DNA, ligation of specific nucleotide sequences linked to the ends of the restriction fragments followed by two rounds of PCR amplification using labeled primers based on the linked sequences (Fig. 2.1; Vos et al., 1995). The AFLP technique has been patented (Vos et al., 1995), and licensed kits are available from various suppliers. Total DNA is first digested with two restriction enzymes (e.g., a 6-bp cutter such as *EcoRI* and a 4-bp cutter such as *MseI*) and oligonucleotide sequences are then ligated to the cut ends. In this example, the ligation mixture contains fragments with *EcoRI* sites on both ends, *MseI* sites on both ends, and fragments resulting from the double digest of *EcoRI* and *MseI*. The first round of amplification is a preselection for the subset of restriction fragments that have the linked primer sequences for the double digest (*EcoRI* site on one end and *MseI* site on the other end) plus a single specific nucleotide in the original fragment sequence. This reduces the pool of fragments significantly from the original mixture. The second round of amplification further selects for an additional subset of nucleotides in the preselected pool of fragments (typically three nucleotides) to reduce the final pool to a manageable number

a. Restriction digest and ligation of linkers.

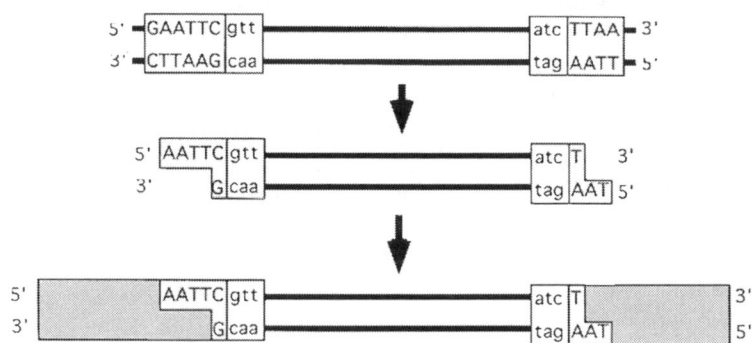

b. Preselective amplification using a single nucleotide extension sequence.

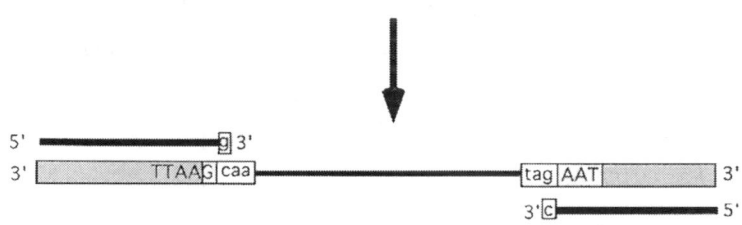

c. Selective amplification using additional nucleotide(s) extension sequence.

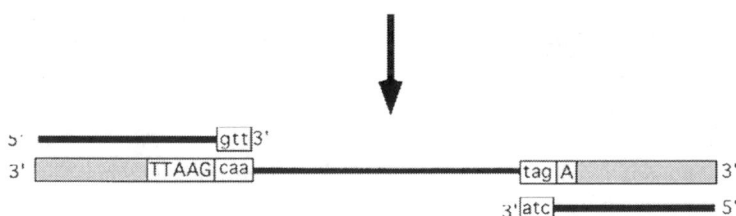

Figure 2.1. Diagrammatic representation of the AFLP technique showing the three steps involved in generating fragments. The lower case letters represent sequences found within the amplified fragment. The upper case letters illustrate the restriction enzyme recognition sequences. The gray boxes depict the linker oligonucleotide sequences ligated to the restriction fragments. *a.* Restriction digest with two endonucleases (*EcoRI* on left and *MseI* on right) followed by ligation of linker sequences. *b.* First round of amplification selects for a subset of the total fragments produced in the double digest using a single nucleotide extension into the restriction fragment. Primers are based on the linker sequences, which include the restriction enzyme recognition site. *c.* Further selection of a subset of the fragment pool using a three-nucleotide extension of the linker sequence for a second round of PCR amplification.

(approximately 50–100 fragments) and includes a radiolabeled primer. The amplified DNAs are then visualized via autoradiography after separation on denaturing polyacrylamide gels (Vos et al., 1995), generating a DNA fingerprint. Alter-

natively, fluorescently labeled fragments can be characterized in an automated sequencer using software designed for that purpose (see Perkin Elmer Website and catalog for additional information).

Applications

Most applications of AFLP have been for genome mapping and/or plant breeding (Ballvora et al. 1995; Becker et al. 1995; Meksem et al. 1995), but the technique has potential for populational studies where fine-scale monitoring of individual genotypes is important. For example, Travis et al. (1996) used AFLP to assess genetic diversity within and among all populations of the critically endangered taxon, *Astragalus cremnophylax* var. *cremnophylax*. Nine primer sets were used to generate 352 bands (220 variable) for 143 individuals sampled from three populations. Bands were scored as present or absent and F-statistics were generated using software developed for RAPD bands. Analysis of molecular variance (AMOVA; Excoffier et al., 1992) was conducted in combination with UPGMA clustering and principal coordinate analysis. (AMOVA is a phenetic approach of characterizing population structure, based on analysis of variance on the matrix of genetic distances among all individuals within and among populations. It is potentially useful in the analysis of dominant markers such as RAPD data (e.g., Huff et al., 1993), inter-simple sequence repeat (ISSR) (see below), and AFLPs.) From the results of the AFLP survey, Travis et al. (1996) were able to characterize the genetic diversity and population structure of *A. cremnophylax* var *cremnophylax* and suggest a management plan for the conservation of this rare plant.

Advantages and Limitations

Advantages of the technique include: (1) No prior sequence information is required for AFLP analysis; (2) a large number of polymorphic bands are produced; and (3) the technique is highly reproducible and standardized kits are available. Limitations involve the number of steps required to produce results. For example, most PCR-based applications involve only one round of amplification, whereas this technique requires two. Additional expenses and/or steps are the restriction enzyme digestion of total DNA and ligation of linkers requiring the purchase of enzymes and a licensed kit. In addition, to separate the radiolabeled fragments it is necessary to use polyacrylamide gel electrophoresis in a sequencing gel apparatus or to have access to an automated sequencer for nonradioactively labeled fragments. In contrast, most other PCR-based methods can be done using agarose gel electrophoresis and EtBr staining. It should be noted, however, that the sensitivity of the technique offers much to offset the few limitations outlined here.

PCR APPLICATIONS IN THE DETECTION OF STRUCTURAL MUTATIONS

Structural Rearrangements

Major structural rearrangements of the chloroplast genome have been successfully used as phylogenetic markers in many taxonomic groups (reviewed in Downie and Palmer, 1992; Olmstead and Palmer, 1994; see Chapter 1). Three basic types of chloroplast rearrangements are known: large inversions (1,000 bp), the loss of genes and introns, and the loss of the inverted repeat. Initial documentation of these rearrangements has been aided by the complete sequencing of several plastid genomes and detailed comparative mapping in species representing over 100 families (Downie and Palmer, 1992; Olmstead and Palmer, 1994). In addition, the taxonomic distributions of several specific rearrangements have been examined via extensive Southern hybridization surveys (Lavin et al., 1990; Downie et al., 1991; Downie and Palmer, 1992; Downie et al., 1994). Southern hybridization approaches have now been supplemented by, and in some cases replaced by, PCR-based documentation of rearrangements (Table 2.2).

The PCR approaches require the design of primers that flank the structural mutation. In the case of intron loss, primers are designed from conserved regions of the surrounding gene. Presence/ absence of the intron is determined by comparing the size of amplified products. The expected length of fragments with and without the intron is easily determined from published plastome sequences. Representative products can also be sequenced to confirm intron loss. The results can be further verified by hybridizing gene and intron specific probes to Southern blots of the PCR products (Wallace and Cota,

Table 2.2. PCR-based determination of structural mutations.

Taxon	Scope	Genome	Target	Confirmation[a]	Reference
Fabaceae	61 spp.	cp	Inverted repeat loss	Restriction sites	Liston, 1995
Angiosperms	54 families, 107 spp.	cp	*rpoC1* intron loss	Sequence	Downie et al., 1996
Angiosperms	70 families	cp	ORF2280 indels	None	Downie et al., 1997
Cactaceae	5 families, 38 spp.	cp	*rpoC1* intron loss	Southern	Wallace and Cota, 1996
Anemone	"large number" of spp.	cp	*rps12* intron loss	Southern	Hoot and Palmer, 1994
Monocots	22 families, 40 spp.	cp	3 inversions	Sequence	Doyle et al., 1992
Fabaceae	"several taxa" at least 4 losses	cp	*rpl2* intron loss	Southern	Doyle et al., 1995
Fabaceae	132 genera	cp	50 kb inversion	Southern	Doyle et al., 1996

[a]Indicates whether the results were confirmed with Southern blot analysis, nucleotide sequence, or restriction site studies.

1996), however this added step is generally unnecessary. The PCR approach has been successfully used to document the loss of the *rps12* intron in three species of *Anemone* (Hoot and Palmer, 1994); the loss of the *rpl2* intron in at least four separate clades of Fabaceae (Doyle et al., 1995); and the loss of the *rpoC1* intron in Cactaceae subfamily Cactoideae (Wallace and Cota, 1996), Poaceae and three related families (Katayama and Ogihara, 1996), four species of *Passiflora,* one species of *Medicago,* and two genera of Aizoaceae (Downie et al., 1996). Doyle et al. (1995) also used PCR to verify that the *trnI* intron was not lost in certain Fabaceae. PCR has also been used to document the deletion of the ORF2280 homolog in Poaceae and Joinvilleaceae (Hahn et al., 1995; Katayama and Ogihara, 1996).

To determine the presence or absence of plastid inversions with a PCR approach, primer pairs flanking the inversion endpoints have been used (Doyle et al., 1992; Doyle et al., 1996; Katayama and Ogihara, 1996). The basic strategy involves two diagnostic amplifications (Fig. 2.2). Inversion presence is determined by positive amplification with an "anchoring" primer A positioned outside of an inversion endpoint and a second primer B internal to the endpoint. A negative amplification with primer A and primer C positioned at the opposite side of the inversion confirms the rearrangement. In taxa lacking the inversion, primer pair A and C should result in a positive amplification, whereas primer pair A and B should not. A similar set of primers can be designed for the opposite end of the inversion, resulting in a total of four diagnostic pairs of primers. In a survey of three inversions known from the monocotyledons, this PCR strategy was very effective, but not without some difficulties (Doyle et al., 1992). For example, primers flanking one end of the large 28-kb inversion were considered unreliable because they resulted in faint amplifications (probably due to nonspecific priming) in taxa serving as negative controls. Other taxa gave negative results with the four diagnostic primer pairs for a particular inversion. Similar ambiguous results were obtained for many taxa surveyed for the presence of a 50-kb inversion in Fabaceae (Doyle et al., 1996), and it was necessary to return to a Southern hybridization approach that had been successfully used in an earlier study (Bruneau et al., 1990). Presence of the inversion in many taxa was subsequently

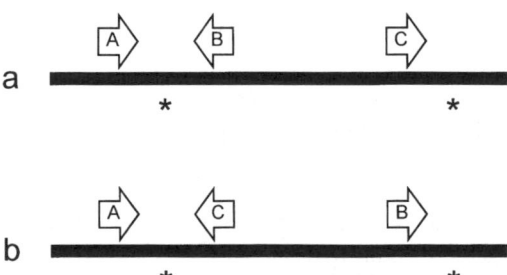

Figure 2.2.   The use of PCR to document inversions. Arrows represent PCR primers, asterisks represent inversion endpoints. *a.* Inversion present, primers, A and B amplify, whereas A and C do not. *b.* Inversion absent, primers A and C amplify, while A and B do not.

confirmed by PCR screening of a single endpoint. Fortunately, abundant DNA was available for most of the legume taxa, unlike the previously mentioned monocot survey, which made extensive use of herbarium material. The limited success of the PCR approach in this study may result from the fact that the region of the 50-kb inversion in the Fabaceae appears to be a hotspot for insertion deletion mutations and/ or nucleotide sequence divergence (Doyle et al., 1995).

A single large inverted repeat (IR) is a characteristic of most land plant plastid genomes; however its independent loss has been documented in at least five separate lineages (reviewed in Olmstead and Palmer, 1994). In an extensive Southern hybridization survey of IR absence in Fabaceae, Lavin et al. (1990) examined 95 species. The IR was absent in the temperate liana *Wisteria* (tribe Millettieae) and representatives from six "temperate herbaceous" tribes of subfamily Papilionoideae. Liston (1995) developed a PCR-based approach to extend the survey to 61 additional species. The taxonomic distribution of the IR loss was extended to additional members of the six temperate herbaceous tribes and to a subtropical genus, *Callerya* (Millettieae), including two lianas and a tree. IR presence was confirmed in additional genera from three other tribes. The PCR strategy was similar to that used in the inversion examples above, relying on a single "anchoring" primer outside of the IR and two diagnostic primers. The anchor and a primer located in the IR amplified when the IR was present, the anchor and a primer located in the small single copy region amplified when the IR was absent. The diagnostic PCR was carried out in a single reaction including template genomic DNA and the three primers. The two possible PCR products produced by the three primers differed in length. Furthermore each product contained unique gene sequences with conserved diagnostic restriction sites. Restriction digests of the PCR products confirmed their identity.

A critical factor to the success (or failure) of the foregoing PCR-based approaches is primer design. A primer should be based on a sequence that is conserved in as wide a range of taxa as possible, particularly at the terminal bases of the 3′ end. For many plastid genes, numerous sequences have been published. For other regions, the availability of complete plastome sequences from several taxa facilitates primer design. Further aspects of PCR primer design are discussed in Palumbi (1996). Although the primers designed for inversion endpoints or a copy of the IR are not informative outside of a particular taxonomic group, the primer pairs used to document intron loss are potentially valuable in all land plants.

Identical cpDNA structural rearrangements may occur independently in unrelated lineages (Downie et al., 1991, 1997), as well as within clades (Hoot and Palmer, 1994). As long as other sources of data are taken into consideration, these relatively rare events can be important phylogenetic markers at the intrageneric (e.g., Hoot and Palmer, 1994; Downie et al., 1996) and intrafamilial levels (e.g., Doyle et al., 1995; Downie et al., 1996; Wallace and Cota, 1996). The introns cited above, as well as several other introns (and genes) that are apparently lost in some land plant plastids (Downie and Palmer, 1992), remain excellent candidates for PCR-based surveys at the intrafamilial and intrageneric levels. The reacquisition of a cpDNA intron has not been confirmed. However, the loss of the *rpoC1* intron in the common ancestor of Poaceae, Restionaceae, Centrolepidaceae, and Anarthriaceae, and its presence in Joinvilleaceae, the sister family to the Poaceae (Katayama and Ogihara, 1996), could be explained by such an event.

Length Differences

To determine the prevalence of large length mutations in the chloroplast ORF2280 sequence, Downie et al. (1997) used PCR to survey representatives of 70 angiosperm families. Species from 19 of these families possessed deletions (relative to *Nicotiana tabacum*) greater than 100 bp. Although these results do not have immediate phylogenetic utility, they do identify taxa in which length variation in the ORF2280 homolog may be phylogenetically informative. Hahn et al. (1995) documented an apparently nested set of deletions in the ORF2280 homolog among the Commeliniflorae. The polarity of the length mutations was inferred by comparison to *rbcL* sequence phylogeny. Length variation in the cpDNA intergenic spacer between *trnL* and *trnF* has also been examined with PCR, and confirmed by sequencing studies to be phylogenetically informative within the Crassulaceae (van Ham et al., 1994).

Length variation in the nuclear rDNA internal transcribed spacer (ITS) region of 32 genera of nonflowering seed plants has also been documented in a PCR-based survey (Liston et al., 1996). Unlike the relatively length-constant angiosperm ITS region, gymnosperm ITS regions vary from 975 to 3,125 bp. The ITS region was PCR-amplified and restriction-site-mapped with 11 endonucleases. Restriction site mapping permitted a more accurate assessment of ITS region length and provided evidence for internal subrepeats in several taxa. In addition, conserved restriction sites allowed the determination of the relative size of the ITS-1 and 5.8S + ITS-2 regions, and identified regions of potential ITS-1 sequence conservation among investigated Cupressaceae.

PCR APPLICATIONS INVOLVING THE GENERATION OF MICROSATELLITE-BASED MARKERS

Systematic studies often use genetic markers to assay organismal relationships, biogeography, speciation, gene flow, genetic variability, and population structure (reviewed in P. Soltis et al., 1992a; Olmstead and Palmer, 1994). The effort and cost of using conventional molecular methods (nucleotide sequencing and restriction-site mapping of organellar or nuclear DNA) for population-level studies can be quite high in terms of time, technical training, facilities, and the amount of plant material needed.

Several classes of genetic markers generated from PCR amplification of DNA have recently been introduced that are appropriate for generating molecular markers for population-level studies. These include markers based on nuclear and chloroplast DNA sequences such as variable number tandem repeats (VNTRs), genetic loci, and noncoding genomic sequences.

Variable Number Tandem Repeats

VNTR markers include long repetitive sequences such as minisatellites (Scribner et al., 1994; Sharon et al., 1995; van Pijlen et al., 1995) and simple sequence repeats (SSR) or microsatellites (Litt and Luty, 1989; Edwards et al., 1991; Jacob et al., 1991). Both classes of VNTRs are distributed throughout the genome of eukaryotic organisms and both have been used for DNA fingerprinting (Rogstad et al., 1988; Nybom and Kraft, 1995), genome mapping, and cultivar identification. Minisatellite repeat motif sequences can be as long as 200 bp with allele sizes up to 50 kb (Jeffreys et al., 1985; Bruford and Wayne, 1993). PCR amplification has not been widely used for the development of genetic markers of minisatellite repeat motifs because of the length of the sequences between conserved flanking regions (Bruford and Wayne, 1993). In this review, we concentrate on the development of simple sequence repeat (SSR) genetic markers using PCR techniques.

SSR or DNA Microsatellites

Simple sequence repeat (SSR), simple tandem repeat (STR) and DNA microsatellite are synonyms for short (<6 bp) tandem nucleotide repeats scattered throughout the genome of eukaryotic organisms (Hamada and Kakunaga, 1982; Tautz and Renz, 1984; Tautz et al., 1986; Weising et al., 1989). The occurrence of several SSR motifs has been screened for many plant groups, including gymnosperms (Smith and Devey, 1994) and angiosperms (Condit and Hubbell, 1991; Poulsen et al., 1993; Wang et al., 1994; Baker et al., 1995). Estimates of the abundance of SSR motifs in plants range from one SSR every 7 kb to 1.5 Mb for plant genomes (Condit and Hubbell, 1991; Akkaya et al., 1992; Lagercrantz et al., 1993; Morgante and Olivieri, 1993; Lavi et al., 1994; Smith and Devey, 1994; Wang et al., 1994; Wu and Tanksley, 1993; Becker and Heun, 1995; Röder et al., 1995; Broun and Tanksley, 1996; Chase et al., 1996). The most common dinucleotide SSR motif in mammals is $(AC/TG)_n$ (Beckmann and Weber, 1992), whereas $(AC/TG)_n$, $(AT/TA)_n$, and $(AG/TC)_n$ are all common in plants (Condit and Hubbell, 1991; Akkaya et al., 1992; Lagercrantz et al., 1993; Morgante and Olivieri, 1993; Wu and Tanksley, 1993; Terauchi and Konuma, 1994; Smith and Devey, 1994; Becker and Heun,

1995; Depeiges et al., 1995; Dow et al., 1995; Röder et al., 1995; Broun and Tanksley, 1996; Chase et al., 1996). Common trinucleotide repeat motifs in plants are $(AAG/TTC)_n$, $(AAT/TTA)_n$, $(ATG/TAC)_n$, $(GGC/CCG)_n$ and $(GTG/CAC)_n$ (Akkaya et al., 1992; Morgante and Olivieri, 1993; Zhao and Kochert, 1993; Wang et al., 1994; Depeiges et al., 1995).

ST-SSR

Early, as well as some recent, studies employing SSR genetic markers used conventional molecular methods: digestion of genomic DNA, separation of fragments in agarose gel, transfer of DNA fragments to nylon membrane, hybridization of labeled microsatellite or microsatellite-like probes for visualization of a genetic fingerprint (Tautz and Renz, 1984; Weising et al., 1989; Rassmann et al., 1991; Poulsen et al., 1993; Nybom and Kraft, 1995). PCR technology enabled the design of sequence-tagged SSR (ST-SSR) primers for a particular organism, which are locus specific. ST-SSR markers are PCR-amplified microsatellite sequences where the primers have been designed from conserved flanking regions of the SSR motif. Usually, the flanking sequences differ sufficiently from one another and the PCR product sizes are distinct enough that multiple PCR reactions can be done in the same tube for multilocus analysis on the same gel (multiplex PCR; Queller et al., 1993; Saghai Maroof et al., 1994). ST-SSR loci have primarily been used as molecular markers in studies of animal genetics, human forensics, medicine, and plant breeding (reviewed in Bruford and Wayne, 1993). Only recently have ST-SSR markers been introduced into studies of population genetics and systematics of nondomesticated organisms (Amos et al., 1993; Hughes and Queller, 1993; Bowcock et al., 1994; Roy et al., 1994; reviewed in Schlötterer and Pemberton, 1994). In animals, there is considerable conservation of ST-SSR loci across large evolutionary distances (Moore et al., 1991; Schlötterer et al., 1991). For example, Moore et al. (1991) were successful in amplifying ST-SSR loci in sheep from bovine primers, and Schlötterer et al. (1991) found that ST-SSR primers developed for one whale genus could be

used to amplify ST-SSR loci in other whale genera. These results suggest the possibility of developing universal primers that would be useful across a wide spectrum of organisms. However, the few studies examining conservation of ST-SSR loci in plants are not encouraging (Zhao and Kochert, 1993; Röder et al., 1995; Brown et al., 1996; Whitton et al., 1997).

Most studies employing ST-SSR primers across wide phylogenetic distances in plants have revealed a lack of conservation in sequences flanking SSRs, although the SSR is usually conserved (Zhao and Kochert, 1993; Röder et al., 1995; Brown et al., 1996). As phylogenetic distance shortens, however, the likelihood of using ST-SSRs from one genus in another or from one species in another increases (Akkaya et al., 1992; Wu and Tanksley, 1993; Kijas et al., 1995; Whitton et al., 1997). Possibly, this phylogenetic pattern results from the nature of ST-SSR loci: Markers are selectively neutral and not subject to functional constraints (Chase et al., 1996) and mutations in the primer region may not be selected against. Support for this hypothesis comes from the discovery of null alleles, where the SSR motif is present, but no amplification product is obtained because of nonconserved flanking sequences (Callen et al., 1993). Most plant SSRs are in noncoding regions (Wang et al., 1994). For a universal primer to be developed from a ST-SSR locus, the SSR flanking sequence would have to be maintained under some selective constraint. For example, if the flanking sequences of a SSR were within gene sequences, the development of universal primers would be more likely than if the flanking sequences were in noncoding regions.

ST-SSR Primer Design

In most plant studies the development of ST-SSR primers involves the construction and screening of a genomic library, isolation and sequencing of DNA from positive clones, followed by primer design from the cloned sequence (Table 2.3). Strategies for SSR enrichment of clones and for ST-SSR development without bacterial transformation of ligated vectors have also been developed (Skinner, 1992; Grist et al., 1993; Kijas et al., 1994). Other

Table 2.3. Summary of studies utilizing plant simple sequence repeats (SSRs).

Taxon	No. of SSRs surveyed	Avg. no. of alleles /locus	No. primers with poly-morphisms (polymorphic bands/primer)	PCR product resolution[a]	Design parameters[b]	Reference
ST-SSR Studies						
Asteraceae: *Helianthus annuus* (18 hybrids, 13 lines)	2	1.5	—	B	gls	Brunel, 1994a
Asteraceae: *H. cusickii*	8	2.8	—	AE, Rn	gls, PCR-gls, db	Whitton et al., 1997
Asteraceae: *H. pumilis*	8	2.5	—	AE, rn	gls, PCR-gls, db	Whitton et al., 1997
Asteraceae: *Tagetes erecta*	7	2.9	—	AE, Rn	gls, PCR-gls, db	Whitton, et al., 1997
Brassicaceae: *Arabidopsis thaliana* (4 ecotypes)	36	1.8	—	Rp	gls, db	Depeiges et al., 1995
Brassicaceae: *Arabidopsis thaliana*	30	4.1	—	AE, Rp	gls, db	Bell and Ecker, 1994
Brassicaceae: *Brassica napus* (15 individuals)	4	1.5	—	Rp	gls	Lagercrantz et al., 1993
Brassicaceae: *Brassica oleracea* (15 individuals)	4	3.5	—	Rp	gls	Lagercrantz et al., 1993
Brassicaceae: *Brassica rapa* (15 individuals)	4	2.8	—	Rp	gls	Lagercrantz et al., 1993
Chenopodiaceae: *Beta vulgaris* (64 accessions)	4	9	—	nPE	gls	Mörchen et al., 1996
Dioscoreaceae: *Dioscorea tokoro* (23 individuals)	6	6.2	—	Rp	gls	Terauchi and Konuma, 1994
Fabaceae: *Glycine* spp. (43 accessions)	3	7	—	Rn	gls, db	Akkaya et al., 1992
Fabaceae: *Glycine* spp. (10 accessions)	2	7.5	—	AE	db, lit	Morgante and Olivieri, 1993
Fabaceae: *Glycine max* (47 accessions)	7	3.3	—	Rp	lit	Morgante et al., 1994
Fabaceae: *Glycine max* (62 accessions)	5	5.2	—	Rn???	db	Maughan et al., 1995
Fabaceae: *Glycine soja* (14 accessions)	7	5.1	—	Rp	lit	Morgante et al., 1994
Fabaceae: *Glycine soja* (32 accessions)	5	13.8	—	Rn???	db	Maughan et al., 1995
Fabaceae: *Glycine* (13 spp.; 141 accessions)	2 (cpDNA)	3.7	—	Rp	db	Powell et al., 1995a
Fabaceae: *Glycine* (91 accessions)	7	18.6	—	Rn	gls	Rongwen et al., 1995
Fabaceae: *Pithecellobium elegans* (52 individuals, 2 pops.)	6	5.3	—	Rn	gls	Chase et al., 1996
Fagaceae: *Quercus macrocarpa* (61 individuals)	3	14.3	—	Rn	gls	Dow et al., 1995
Lauraceae: *Persea americana* (12 cultivars)	2	9.5	—	Rp	gls	Lavi et al., 1994
Pinaceae: *Pinus* spp. (11 spp)	1 (cpDNA)	9	—	Rp	db	Powell et al., 1995b
Pinaceae: *Pinus radiata* (40 individuals)	2	6	—	nPE	gls	Smith and Devey, 1994
Pinaceae: *Pinus sylvestris*	2	ns	—	Rn	gls, db	Kostia et al., 1995

Table 2.3. Summary of studies utilizing plant simple sequence repeats (SSRs), *continued*

Taxon	No. of SSRs surveyed	Avg. no. of alleles /locus	No. primers with poly-morphisms (polymorphic bands/primer)	PCR product resolution[a]	Design parameters[b]	Reference
Poaceae: *Hordeum vulgare* ssp. *vulgare* (104 accessions)	4	11.8	—	Rn	db	Saghai Maroof et al., 1994
Poaceae: *Hordeum vulgare* ssp. *spontaneum* (103 accessions)	4	15.3	—	Rn	db	Saghai Maroof et al., 1994
Poaceae: *Hordeum vulgare* ssp. *vulgare* (10 cultivars)	15	2.1	—	AE	db	Becker and Heun, 1995
Poaceae: *Oryza* spp. (20 cultivars of *O. sativa;* 17 spp. of *Oryza*)	1	6	—	Rn	lit	Zhao and Kochert, 1993
Poaceae: *Oryza* spp. (14 accessions of *O. sativa;* 7 spp. of *Oryza*)	8	7.6	—	Rn	gls, db	Wu and Tanksley, 1993
Poaceae: *Oryza sativa* (238 accessions)	10	9.3	—	Rn	db, lit	Yang et al., 1994
Poaceae: *Sorghum biocolor* (17 inbred lines)	126	ns*	—	AE, nPE	gls, cb, lit	Brown et al., 1996
Poaceae: *Triticum aestivum* (18 inbred lines)	15	3.2	—	nPE	gls	Röder et al., 1995
Poaceae: *Triticum aestivum* (8 varieties)	2	4.5	—	AE, nPE	db	Devos et al., 1995
Poaceae: *Zea mays* (8 inbred lines)	6	3.7	—	AE	db	Senior and Heun, 1993
Rutaceae: *Citrus* spp. (12 spp.)	2	5.5	—	C, A	gls	Kijas et al., 1995
Scrophulariaceae: *Mimulus* (5 spp.; 10 populations)	6	16.7	—	nPE	gls	P. Awadalla, pers. comm.
Solanaceae: *Lycopersicon* spp.	9	1.8	—	Rn	gls	Broun and Tanksley, 1996
Vitaceae: *Vitis vinifera* (26 cultivars)	5	8.4	—	Rn	gls	Thomas and Scott, 1993

ISSR

Taxon	No. of SSRs surveyed	Avg. no. of alleles /locus	No. primers with poly-morphisms (polymorphic bands/primer)	PCR product resolution[a]	Design parameters[b]	Reference
Asteraceae: *Dendranthema grandiflora* (25 accessions)	4	—	ns*	Rn???	lit	Wolff et al., 1995
Asteraceae: *Lactuca* spp. (3 varieties)	23	—	4 (ns)	AE, RP	ns	Gupta et al., 1994
Brassicaceae: *Brassica napus* ssp. *oleifera* (20 cultivars)	14	—	5 (ns)	dPS	lit	Charters et al., 1996
Fabaceae: *Cicer* spp. (7 accessions, 3 spp.)	28	—	13 (ns)	AE	ns	Sharma et al., 1995a
Pinaceae: *Pinus* spp. (several pops.)	23	—	7 (ns)	AE, Rp	ns	Gupta et al., 1994
Pinaceae: *Pinus sylvestris*	3	—	3 (4.7)	Rn	lit	Kostia et al., 1995
Pinaceae: *Pseudotsuga menziesii* S31t20	22	—	19 (1.8)	AE	lit	Tsumura et al., 1996a
Pinaceae: *Pseudotsuga menziesii* S45t11	21	—	16 (1.2)	AE	lit	Tsumura et al., 1996a
Poaceae: *Eleusine* spp. (5 spp., 22 accessions)	6	—	ns*	Rn	ns	Salimath et al., 1995
Poaceae: *Hordeum vulgare* spp. *vulgare* (14 cultivars)	7	—	ns	nPS	ns	Sánchez de la Hoz et al., 1996

Table 2.3. Summary of studies utilizing plant simple sequence repeats (SSRs), *continued*

Taxon	No. of SSRs surveyed	Avg. no. of alleles /locus	No. primers with poly- morphisms (polymorphic bands/primer)	PCR product resolution[a]	Design parameters[b]	Reference
Poaceae: *Zea mays* (40 inbred lines)	23	—	7 (ns)	AE, Rp	ns	Gupta et al., 1994
Solanaceae: *Lycopersicon* spp. (3 varieties)	23	—	3 (ns)	AE, Rp	ns	Gupta et al., 1994
Taxodiaceae: *Cryptomeria japonica* Midori 5	24	—	13 (1.8)	AE	lit	Tsumura et al., 1996a
Vitaceae: *Vitus vinifera* (2 varieties)	23	—	9 (ns)	AE, Rp	ns	Gupta et al., 1994
RAMPs						
Poaceae: *Hordeum vulgare* ssp. *vulgare* (2 cultivars)	2 & 10[c]	—	9 (4.3)	Rp	ns	Becker and Heun, 1995
Poaceae: *Hordeum vulgare* ssp. *vulgare* (14 cultivars)	2 & 30	—	ns*	nPS	ns	Sánchez de la Hoz et al., 1996

Note: ST-SSR = sequence-tagged SSR; ISSR = inter-SSR; RAMPs = random amplified microsatellite polymorphism; ns = not stated; ns* = not explicitly stated, but high degree of polymorphism implied.
[a]Abbreviations used for PCR product resolution: B = denaturing polyacrylamide gel electrophoresis, blotting, hybridization to radioactive probe; C = agarose gel electrophoresis, Southern blot, and detection via chemiluminescence; A = fluorescent labeling of PCR amplicons and detection in denaturing polyacrylamide gel electrophoresis in automated sequencer; Rn = [$\alpha-^{32}$P] dCTP used in PCR, denaturing polyacrylamide gel electrophoresis and autoradiography; nPE = nondenaturing PAGE stained with ethidium bromide; Rp = radiolabeled primers used in PCR, with silver stain, nPS = nondenaturing PAGE with silver stain.
[b]Abbreviations used for design parameters: gls = genomic library screen, db = data base search.
[c]Number of microsatellite primers and number of RAPD or arbitrary primers.

strategies for development of ST-SSR loci include public database searches for sequences flanking microsatellite motifs from which primers can be designed (Table 2.3) and experimentation with primers designed for other organisms (Table 2.3). The database search strategy is feasible only for plant taxa that have a large amount of sequence data deposited in public databases such as GenBank. In nondomesticated plant taxa, a very limited number of appropriate sequences is available.

The PCR-assisted screen and design of ST-SSR primers outlined by Grist et al. (1993) offers a technically feasible method for plant systematists and evolutionary biologists to develop SSR loci without the culturing and blotting steps of genomic library screening (Sambrook et al., 1989). The method involves the construction of a genomic library through the digestion of total DNA with a frequently cutting enzyme such as *MboI* and ligation of the fragments into a cloning vector (see fig. 7 of Dowling et al., 1996 for a diagrammatic representation of the usual route to ST-SSR development). The ligation mixture is a library of genomic clones and is used in a series of PCR and sequencing reactions. Phase I is to find the optimal concentration of library template DNAs for PCR. A serial dilution of the genomic library mixture is made and aliquots from each dilution are used in an initial PCR amplification of SSR loci by using an Inter-SSR (ISSR) primer (see below in ISSR) and a primer specific to the cloning vector. The dilution mix that results in the clearest resolution of PCR products on an agarose gel becomes the library template DNA solution. The second step of phase I is to amplify all of the genomic cloned inserts of the library DNA by using the optimal concentration of the DNA templates (from step 1) and the vector primers flanking the cloned insert. These PCR products then become the working solution for isolating specific SSR loci. Phase II is to isolate specific SSR loci as follows: (1) the working solution is used for

PCR with an ISSR primer and one vector primer flanking the cloned DNA insert; (2) one or more PCR products are isolated and cleaned; (3) at least one PCR product is sequenced and a first primer is designed from the sequence flanking the SSR motif. Phase III is to develop the complementary flanking sequence of the SSR locus found in Phase II as follows: (1) the working solution is reamplified using the ST-SSR primer developed in Phase II and the vector primer flanking the other side of the cloned insert; (2) the PCR products are isolated and cleaned; (3) the PCR product is sequenced and the second ST-SSR primer is designed. After these steps, a ST-SSR has been developed and can be used for population-level studies.

Clearly, much effort is required to develop ST-SSR loci. However, the advantages of using these nuclear markers for plant systematic studies include: (1) markers are inherited in a codominant Mendelian fashion (Morgante and Olivieri, 1993; Thomas and Scott, 1993; Zhao and Kochert, 1993; Lavi et al., 1994; Morgante et al., 1994; Dow et al., 1995; Kijas et al., 1995; Maughan et al., 1995; Chase et al., 1996; Mörchen et al., 1996), which means that measures of genetic diversity can be calculated; (2) there is generally a much greater degree of allelic polymorphism of SSR loci than has been found for isozymes and RFLP patterns (Table 2.3: Goodfellow, 1992; Morgante and Olivieri 1993; Wu and Tanksley, 1993; Lavi et al., 1994; Morgante et al., 1994; Saghai Maroof et al., 1994; Terauchi and Konuma, 1994; Yang et al., 1994; Dow et al., 1995; Maughan et al., 1995; Röder et al., 1995; Rongwen et al., 1995; Chase et al., 1996); (3) PCR-based studies of ST-SSRs require small amounts of plant material; (4) ST-SSR primers are highly specific allowing simultaneous amplification of two or more loci in the same reaction (multiplex PCR; Saghai Maroof et al., 1994; Powell et al., 1995a; Brown et al., 1996) and (5) once ST-SSR primers have been developed, results can be quickly obtained.

ST-SSR Allelic Differences

ST-SSR alleles are defined by changes in the number of nucleotide repeat units and are resolved using several different methods (Table 2.3). The amount of allelic diversity observed varies with the resolution capacity of the particular DNA fragment separation matrix. For example, standard agarose electrophoresis and EtBr staining are inadequate for separation of fragments differing by a single dinucleotide repeat unit, but may be sufficient to separate fragments that differ by tens of base pairs. Resolution of PCR products improve dramatically with the use of low melt, sieving agaroses (AE of Table 2.3). Polyacrylamide gel electrophoresis (nondenaturing and denaturing) has also been used to resolve ST-SSR alleles either with EtBr staining or by the incorporation of radionucleotide labels in the PCR reaction and autoradiography (Table 2.3). Another factor affecting allelic diversity is the occurrence of null alleles that result from changes in the flanking sequences and interference with the primer binding site (Callen et al., 1993).

One of the major advantages of using ST-SSR markers is that genetic diversity estimates can be made using allele frequencies. The statistics employed in most plant studies to date (Morgante et al., 1994; Saghai Maroof et al., 1994; Terauchi and Konuma, 1994; Yang et al., 1994; Maughan et al., 1995; Rongwen et al., 1995) have been similar measures of genetic diversity or polymorphism information content (PIC) used with isozyme data (Nei, 1973; Anderson et al., 1993). It has been suggested, however, that measures of population structure and genetic diversity usually employed for isozyme data that are based on an infinite-alleles model can yield biased estimates except where divergence times are relatively short (Slatkin, 1995). Because SSR alleles differ by the number of repeat units in a particular motif, presumably arising through strand slippage replication errors (Levinson and Gutman, 1987; Schlötterer and Tautz, 1992), a step-wise mutation model has been proposed as the most appropriate model for estimating population subdivision structure and genetic distances (Valdes et al., 1993; Slatkin, 1995; Goldstein et al., 1995).

Applications

Most plant studies using ST-SSR markers have been for genomic fingerprinting of domesticated plants for cultivar identification, quantitative trait loci (QTL) mapping, and plant breeding (Table 2.3). Only a few studies have specifically targeted nondomesticated taxa and/or natural populations. Terauchi and Konuma (1994) surveyed six ST-SSR loci in 23 individuals of *Dioscorea tokoro,* a wild yam species of Japan. All six loci were polymorphic (Table 2.3). Terauchi and Konuma (1994) calculated observed and expected heterozygosity and Nei's (1973) genetic diversity statistic. The F values were similar for allozyme and ST-SSR results.

Two natural populations of the tree species *Quercus macrocarpa* (Dow et al., 1995) and *Pithecellobium elegans* (Chase et al. 1996) have also been surveyed for ST-SSR polymorphisms (Table 2.3). Dow et al. (1995) surveyed three ST-SSR loci in 61 adults in a population of *Q. macrocarpa* to assess the of these markers for determining parentage in a wind-pollinated temperate species. The high degree of polymorphism in *Q. macrocarpa* (Table 2.3) led to the conclusion that these markers would allow determination of parentage of seedlings/saplings in a population. For the tropical tree species, *P. elegans,* Chase et al. (1996) surveyed five ST-SSR loci in 52 individuals from two populations. Multilocus genotypes for all individuals in both populations were assigned based on the high degree of polymorphism in three of the five ST-SSR loci. The authors conclude that paternity analysis and gene flow among populations would be possible using the ST-SSR markers developed.

cpDNA ST-SSR

Most ST-SSR studies of plant DNAs have concentrated on nuclear markers where the repeat motifs are usually di-, tri-, and tetranucleotide sequences. Chloroplast DNA (cpDNA) is also a source of ST-SSR markers (Powell et al., 1995a, 1995b). Powell et al. (1995a) searched the database of all plants with complete chloroplast genome sequences and determined that the mononucleotide $(A/T)_n$ motif was the most abundant SSR for cpDNA. Furthermore, SSRs are distributed throughout the chloroplast genome, particularly in the large single-copy region, and SSR motifs are conserved

in the inverted repeat region suggesting an ancestral condition predating the divergence of chloroplast genomes (Powell et al., 1995a). Depending on the mode of inheritance of the chloroplast, cpDNA ST-SSR markers can be used for tracking gene flow of paternal (*Pinus*) or maternal parents (Powell et al., 1995b).

ISSR

Inter-SSR (ISSR) genetic markers were developed from the common SSR motifs present in eukaryotic organisms (Gupta et al., 1994; Zietkiewicz et al., 1994). The basic premise is that SSR loci are dispersed evenly throughout the genome (Hamada and Kakunaga, 1982; Tautz and Renz, 1984; Tautz et al., 1986; Tautz, 1989; Condit and Hubbell, 1991; Epplen et al., 1991) and the chances of "hitting" two SSRs with a common motif, oriented on opposing DNA strands, within an amplifiable distance of one another is high enough that single-primer amplifications should yield a high degree of polymorphic bands. The resulting PCR products are anonymous SSR loci.

ISSR PCR reactions use primers that are developed from within the SSRs themselves [e.g., $(GA)_{10}$, $(AGG)_6$] with or without a 5'- or 3'-anchor sequence [e.g., 5'-GG-$(CA)_6$-3', 5'-$(AGC)_4$-GC-3']. Incorporation of anchoring sequences purportedly eliminate strand-slippage artifacts. The resulting amplified fragments are resolved on agarose or polyacrylamide gels (Table 2.3).

ISSR markers are inherited as dominant or codominant genetic markers in a Mendelian fashion (Gupta et al., 1994; Tsumura et al., 1996a). They are interpreted as diallelic with alleles designated as "band present" or "band absent." Presumably, the absence of a band means that divergence has occurred at one or both of the primer sites. Other possibilities include the loss of a SSR site (one of the ISSR primer annealing sites) or a chromosomal structural rearrangement. The number of bands per lane resolved with ISSR primers range from 1 to 4 for *Pseudotsuga menziesii* and *Cryptomeria japonica* (Tsumura et al., 1996a), 3 to 10 for *Cicer* (Sharma et al., 1995a), 26 to 27 for *Dendranthema* (Wolff et al., 1995), and 15 to 66 for

Eleusine (Salimath et al., 1995). It is conceivable that SSR allelic differences could be interpreted as separate bands if the resolving power of the separation matrix is high enough. In practice, however, most studies using ISSRs use agarose gel electrophoresis and EtBr staining (Table 2.3).

Applications

Studies of plant SSR loci employing ISSR primers have been conducted to: (1) assess the abundance and/or inheritance of SSR loci in the genome (Gupta et al., 1994; Rogstad, 1994; Kostia et al., 1995), (2) facilitate cultivar identification (Charters et al., 1996; Wolff et al., 1995), and (3) gather preliminary data for genomic mapping projects (Tsumura et al., 1996a). All studies involving plant ISSRs to date have focused on cultivated species. In a study of four diploid and one polyploid species of *Eleusine*, ISSR markers were used to assess the allopolyploid origin of *E. coracana*, which has an AABB genome (Salimath et al., 1995). ISSR amplification of six primers yielded 15–66 bands per primer and 96% of the bands were polymorphic. ISSR markers confirm that the AA genome of *E. coracana* comes from *E. indica,* but none of the species assayed contributed the BB genome.

Studies using ISSR markers in natural populations are currently being conducted in both of the authors' laboratories. Preliminary analyses of a known hybrid complex of *Penstemon* (Wolfe and Elisens, 1993, 1994, 1995) have revealed ISSR markers that may be useful for assaying patterns of gene flow among species and populations (fig. 2.3; A. Wolfe et al., unpubl.). ISSR markers are also being used to study clonal diversity in *Festuca idahoensis,* a perennial bunch grass (W. Robinson et al., unpubl.).

RAMP

Random amplified microsatellite polymorphisms (RAMP) are genetic markers generated

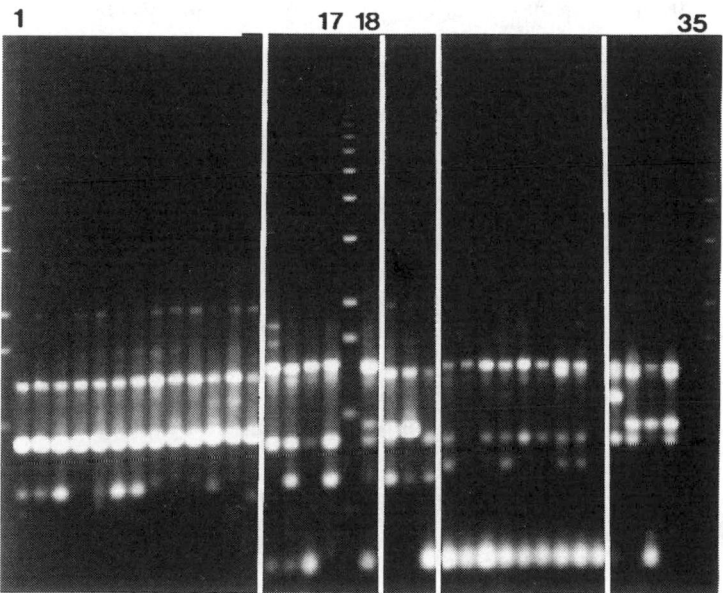

Figure 2.3.   ISSR banding patterns for primer AW3 in populations of *Penstemon centranthifolius* (lanes 1–4 = CE209, lanes 5–7 = CE308, lanes 8–10 = CE 309, lanes 11–13 = CE346), *P. clevelandii* (lanes 14–16 = CL313, lane 17 = CL206, lane 18 = CL198), the natural hybrid between *P. centranthifolius* and *P. spectabilis, P.* × *parishii* (lanes 19–21 = XP222), *P. spectabilis* (lanes 22–25 = SP223, lanes 26–27 = SP319, lanes 28–30 = SP323), and *P. grinnellii* (lane 31 = GR333, lanes 32–34 = GR299); lane 35 = negative control and size standards are a 1-kb ladder. Distinctive banding patterns are present in each of the species and bands from one or both parental profiles are present in all individuals of the hybrid. Population designations are the same as in Wolfe and Elisens, 1995.

from the PCR amplification of DNA using an ISSR primer and a RAPD primer in the same reaction (Wu et al., 1994). Three different band patterns may appear simultaneously in a RAMP reaction: ISSR bands, RAPD bands, and bands that are generated from the combination of the ISSR and RAPD primers. To identify bands that are generated from the ISSR primers, the ISSR primer is radioactively labeled prior to PCR and only the bands generated from the ISSR primer will appear on an autorad. Nonradioactive alternative methods include: (1) conducting a RAPD PCR and running the product alongside the RAMP to identify bands that are generated by the RAPD reaction (Sánchez de la Hoz et al., 1996); and (2) a double stringency PCR involving two phases with a high annealing temperature for the initial 15 rounds of PCR, which is essentially an ISSR reaction (reducing the pool of candidate templates), followed by 25 rounds at a reduced annealing temperature to bring in the RAMPs pattern (Matioli and deBrito, 1995).

RAMP is a useful tool for genetic mapping (Becker and Heun, 1995), but this technique has also been used to study genetic relationships among barley cultivars (Sánchez de la Hoz et al., 1996). Genetic relationships inferred from RAMP data are determined by band-sharing algorithms in the same fashion as RAPD data (Sánchez de la Hoz et al., 1996). RAMP markers are inherited in a dominant and codominant fashion. For example, Sánchez de la Hoz et al. (1996) reported that 35% of the primer combinations assayed in 24 doubled-haploid lines from a cross between two barley cultivars produced codominant markers. Wu et al. (1994) reported that 50% of the polymorphic loci found with RAMPs showed codominant alleles. Allelic differences at RAMP loci should be similar in principle to allelic differences at ST-SSR, ISSR, and RAPD loci.

OTHER NUCLEAR AND ORGANELLAR GENE MARKERS GENERATED BY PCR

In addition to RAPD, AP-PCR, AFLP, and PCR markers based on SSR, several other classes of nuclear and organellar markers are generated through the use of PCR. These include nuclear and organellar markers generated from known

genic sequences such as cleaved amplified polymorphic sequence (CAPS) and amplified sequence length polymorphisms (ASLP) (Maughan et al., 1995).

Cleaved Amplified Polymorphic Sequences and Mapped Restriction Site Polymorphisms

The primary sequence yields the maximum amount of information that can be extracted from a PCR product. However, in many cases, obtaining the complete nucleotide sequence is prohibitively expensive or simply unnecessary. The equipment, supplies, and training needed for DNA sequencing may be out of reach for a small laboratory. Likewise, the number of individuals sampled may be too large, or the length of the PCR product too long, to justify complete sequencing. If a diagnostic marker is needed, a cleaved amplified polymorphic sequence (CAPS) can provide an efficient and economical solution. In plants, this approach was first used in a study of hybridization and introgression among populations of the Louisiana irises (Arnold et al., 1991). Phylogenetic analysis can also be conducted on mapped restriction site polymorphisms (MRSP). Whereas PCR-MRSP has been much used for strain identification and phylogenetic analysis in microorganisms (Ralph and McClelland, 1994), its use in plants has been limited (Wolfe et al., 1997; Table 2.4). The advantages and limitations of this approach are discussed below.

The restriction digestion of a PCR product is a straightforward process. A pair of primers is used to amplify a particular sequence. Generally a relatively large amplification reaction is carried out (100 μl) to generate sufficient template for restriction digestion. The PCR product may be purified subsequent to verification of its size and intensity on an agarose gel. However, we have found that most restriction enzymes are sufficiently active in standard PCR reaction buffers. Most CAPS and PCR-MRSP studies have used agarose gel electrophoresis and EtBr staining. The increased availability of precast polyacrylamide gels is making polyacrylamide gel electrophoresis (PAGE) a more attractive option for obtaining better resolution. Likewise,

Table 2.4. PCR-MRSP studies.

Taxon (outgroups not listed)	Scope	PCR[a]	No. of Enzymes[b]	Total Length[c]	Total bp Sampled[d]	No. Var. Sites	No. Inf. Sites	Inf. Sites/ Var. Sites	Inf. Sites/ Frag. Size	Reference
Coniferales	24 gen, 45 spp.	frxC rbcL psbA psbD trnK 16S	8–12	8091	1109	222	188	0.847	0.023	Tsumura et al., 1995
Datiscaceae	6 gen, 14 spp	rbcLJ ORF106	22 (30)	3210 + 110	232	53	38	0.717	0.012	Rieseberg et al., 1992
Dipterocarpaceae	10 gen, 30 spp	rbcL rpoB petB psbA psbD atpH rpoC trnK 16S psaA petA	3–11	18,752	1126	141	82	0.581	0.004	Tsumura et al., 1996b
Fabaceae: *Astragalus*	14 spp.	rpoC	14 (23)	4100	615	38	9	0.236	0.002	Liston, 1992b
Fabaceae: Galegeae	21 gen, 51 spp	rpoC	19	4100+ 50	556	89	45	0.506	0.011	Liston and Wheeler, 1994
Fabaceae: *Lathyrus*	44 spp	rpoC psbA/ ndhF	31 (35) 28 (35)	5975 + 475	1573	227	109	0.480	0.018	Asmussen and Liston, 1998
Grossulariaceae: *Ribes*	33 spp	rbcLJ ORF106 rpoC	13 15	7300	ns	55	30	0.545	0.004	Messinger, 1994
Papaveraceae: Papaveroideae	9 gen, 42 spp	rpoC trnK trnTltrnF	8 21 2	8300	ns	190	121	0.637	0.015	Schwarzbach and Kadereit, 1995; Jork and Kadereit, 1995
Pinaceae: *Pinus*	13 spp	trnTltrnF	10 (26)	1455 + 80	143	19	11	0.579	0.008	Perez de la Rosa et al., 1995
Rosaceae: *Prunus*	9 spp	rbcLJ ORF106 trnTltrnF	20	5600 + 150	558	22	9	0.409	0.002	Badenes and Parfitt, 1995
Scrophulariaceae: Cheloneae	20 gen, 32 spp	trnK rbcL rps2	15–18 11	4598	845	138	86	0.623	0.019	Wolfe et al., 1997

[a]The gene amplified, or the locator of primers flanking a non-coding region, are given.

[b]The number of restriction enzymes used to detect polymorphisms. Some studies indicated that additional non-polymorphic enzymes were excluded, if so, the total number of enzymes surveyed is given in parentheses.

[c]The minimum observed length in base pairs is given, followed by the size of the longest observed length mutations, if any.

[a]Total base pairs sampled is the number of observed restrictions sites (invariant and polymorphic) multiplied by the number of base pairs in each recognition site.

new DNA-detection compounds (Morin and Smith, 1995) that are more sensitive than EtBr may facilitate CAPS and PCR-MRSP analysis by reducing the amount of DNA and restriction enzyme required in a reaction.

Because each PCR product encompasses a relatively short region of DNA, many of the "six-cutter" restriction enzymes used in traditional RFLP studies will not have recognition sites. Potentially useful enzymes can be chosen by identifying recognition sites in published nucleotide sequences. The polymorphisms detected in most studies have come primarily from enzymes with recognition sites that are four, rather than six, base pairs in length. A limitation of using "four-cutter" restriction enzymes is that fragments may be generated that are too small to be detected by agarose gel electrophoresis, complicating the interpretation of mutations. The recognition site of a four-cutter enzyme may be coincident with a six-cutter enzyme used in the same study, e.g., *DpnII* (5'-GATC-3') and *BamHI* (5'-GGATCC-3'). Such coincident enzymes can be used together as long as care is taken to score identical mutations only once. The location of mutations can be established by mapping polymorphic restriction sites with double or triple digests. The disadvantage of this approach is the requirement for additional PCR product.

Applications

To date, most PCR-MRSP studies in plants have amplified sequences from the chloroplast genome (Table 2.4). Population level analyses have generally utilized intron containing genes and/or intergenic regions (Taberlet et al., 1991; Demesure et al., 1995). Taxon-specific cpDNA markers have been developed in *Iris* (Arnold et al., 1991), *Actinidia* (Cipriani and Morgante, 1993), *Helianthus* (Rieseberg et al., 1994), *Pinus* (Boscherini et al., 1994), *Senecio* (Liston and Kadereit, 1995), and *Lemna* and *Spirodela* (Jordan et al., 1996). Infraspecific cpDNA polymorphisms have been obtained in *Datisca* (Liston et al, 1992), *Quercus* (Demesure et al., 1995), *Fagus* (Demesure et al., 1996), *Argania* (El Mousadik and Petit, 1996) and *Silene* (McCauley et al., 1996). Although infraspecific

cpDNA variation is widespread in plants (reviewed in Soltis et al., 1997), levels of polymorphism are relatively low. The amount of nucleotide sequence surveyed in a CAPS study is typically much less than a Southern blotting approach using heterologous probes representing the entire cpDNA genome. Thus, CAPS analysis of the *rbcL/* ORF106 and *rpoC1/rpoC2* regions failed to detect intraspecific cpDNA variation in *Tiarella;* one informative mutation was observed with a Southern hybridization approach (D. E. Soltis et al., 1992). This is not surprising, considering that approximately 3,600 bp of sequence were surveyed with Southern hybridization, which is substantially higher than even the most extensive PCR-MRSP study (Table 2.4). Due to low levels of cpDNA variation, other studies have also failed to detect inter- or intraspecific variation with CAPS (Boscherini et al., 1994, Vicario et al., 1995; Smith et al., 1996). The probability of detecting cpDNA polymorphisms can be increased by amplifying a large number of small cpDNA regions and using a single frequently cutting enzyme (El Mousadik and Petit, 1996) or using long-PCR to amplify a larger portion of the genome in a single step (V. Symonds and K. Schierenbeck, pers. comm.).

Taxon-specific markers have also been produced from restriction digests of mitochondrial genes (Rieseberg et al., 1994; Dawson et al., 1995; Wang et al., 1996) and anonymous nuclear loci (sequence-tagged sites) generated from clones in a genomic library (Dawson et al., 1996). Additional primers that amplify mitochondrial introns have also been developed (Demesure et al., 1995; Dumolin-Lepegue et al., 1997). Amplification and subsequent digestion of the rDNA ITS region with restriction enzymes has been little utilized in angiosperms (e.g., Swensen et al., 1995), but may have utility for the relatively long ITS regions of gymnosperms (Liston et al., 1996).

One of the more unusual studies using CAPS involved the use of RT-PCR (reverse transcriptase PCR) to examine expression of homologous genes in allopolyploid species of *Brassica* (Song and Osborn, 1994). Allopolyploids have multiple copies of genes that are inherited from each of the parental species. It has been demonstrated that gene silencing occurs in polyploids, so much so that in ancient polyploids only a diploid complement of genes is expressed (Haufler and Soltis, 1986; Haufler, 1987). The difficulties of determining gene silencing in polyploids using isozyme analysis, and the advantages of using a RNA-based assay were reviewed in Song and Osborn (1994). If there are distinct restriction-site patterns for genes inherited from parental species, it is possible to detect which of the gene copies has been inherited and/or silenced by using RT-PCR CAPS. In RT-PCR, reverse transcriptase uses RNA to make DNA (the template for PCR). The template DNA is then amplified and the resulting amplicons are digested as described above. Allopolyploids that express genes from all parental species will have additive patterns of RFLP profiles if the parental taxa differ in their RFLP profiles.

Phylogenetic analyses of MRSPs have relied on data from both noncoding regions and chloroplast genes (Table 2.4). The limiting factor in phylogenetic reconstruction using this method is the total number of nucleotides sampled (determined by the number of restriction sites observed per enzyme multiplied by the number of base pairs in each recognition site). Studies that sample too small a portion of the plastid genome generally cannot adequately resolve phylogenetic relationships. By including several genes with varying substitution rates, the number of nucleotides sampled can be increased while potentially resolving all levels of the tree (Wolfe et al., 1997). Including intergenic regions can provide increased levels of polymorphism. However, the presence of length mutations can complicate homology assessment. Restriction site mapping can be used to determine the position of length mutations, and these can be incorporated into a phylogenetic analysis.

The potential effectiveness of PCR-MRSP can be measured by the number of informative sites per fragment size (Table 2.4). Based on this measure, the method appears most suitable for resolving relationships at the intergeneric level (Table 2.4). Unless more nucleotides are sampled the method is not as effective at the interspecific level. An exception is an interspecific study of the genus *Lathyrus* (Asmussen and Lis-

ton, 1998). *Lathyrus* belongs to the clade of legumes lacking the inverted repeat, and this structural mutation is sometimes correlated with a higher rate of cpDNA evolution. The chloroplast genome of the genus appears to also have additional structural rearrangements (Asmussen and Liston, 1998), a factor that would further complicate a standard Southern hybridization study.

As an alternative to traditional Southern hybridization approaches, PCR-MRSP analysis has the attractive features of other PCR-based methods (reduced quantity and quality of DNA required, reduction in time and expense of a study, and elimination of radioisotopes). The method can also be extremely useful in conducting a feasibility study prior to a major sequencing effort. If low levels of sequence divergence are encountered for a particular region, it is unlikely that nucleotide sequencing will adequately resolve relationships.

Variations on a Theme: Generation of Other Genetic Markers Using PCR Technology

Several alternative methods for generating genetic markers either without or with prior sequence information have been developed in recent years. Markers developed without prior sequence information include: (1) amplification of particular restriction fragments after the development of primers through inverse PCR techniques (Ochman et al., 1988; Livneh et al. 1992) and (2) AFLP techniques (see foregoing text).

Markers developed from known DNA sequences include: (1) PCR amplification of repetitive DNA sequences other than those discussed above (Jarret et al., 1993), (2) amplification of specific gene sequences or genic regions (D'Ovidio et al., 1990; Welsh and McClelland, 1991; Williams et al., 1991; Maughan et al., 1995; Nishio et al., 1996; Strand et al., 1997); and (3) allele-specific polymerase chain reaction (ASPCR), allele-specific amplification (ASA), or allele-specific associated primers (ASAP), including amplification of sequenced RAPD bands (Shattuck-Eidens et al., 1991; Paran and Michelmore, 1993; Amato and Gatesy, 1994; Xu et al., 1995; Yu et al., 1995). In ASPCR, a PCR primer is designed to match a polymorphic nucleotide site at an allele's 3′ end. Under stringent PCR conditions, amplification will occur only for one of the variant sequences. This approach has been used to survey polymorphisms in animal mtDNA (Amato and Gatesy, 1994) and nuclear genes in maize (Shattuck-Eidens et al., 1991).

Most of the markers developed from known sequences have been for cultivar identification and/or plant breeding research. Of particular interest to plant systematists, however, is a set of universal primers developed by Strand et al. (1997) for several nuclear gene sequences, including alcohol dehydrogenase (*Adh*), aspartate aminotransaminase (*Aat*), calmodulin (*Cam*), chalcone isomerase (*Chi*), chalcone synthase (*Chs*), glyceraldehyde 3-phosphate dehydrogenase (*G3pdh*), phosphoglucose isomerase (*Pgi*), and triose phosphate isomerase (*Tpi*). Each set of primers developed from these gene sequences span an intron. Most primer pairs tried in taxa traversing seven orders of angiosperms produced multiple bands on agarose gels, which are presumably from members of the target low-copy number gene families. Of the primer pairs designed, only those for *Aat* and *Pgi* amplified a limited subset of the taxa sampled. Analysis of amplicons from these nuclear genes could include nucleotide sequencing, CAPS, SSCP and DGGE (see below), and heteroduplex analysis.

OTHER PCR APPLICATIONS

SSCP and DGGE

Single-strand conformation polymorphism (SSCP) and denaturing gradient gel electrophoresis (DGGE) are additional methods that can be used to distinguish among DNA sequences amplified by PCR. In SSCP (Orita et al., 1989) a DNA sequence is amplified by PCR, denatured, and run in a nondenaturing polyacrylamide gel, whereas in DGGE (Fischer and Lerman, 1983; Myers et al., 1985, 1987) the PCR-amplified fragments are loaded onto gels that have a linear gradient of denaturant. If fragments differ in their nucleotide sequences, they will migrate at different mobilities due to conformational differences (SSCP) or melting-domain differences (DGGE). Fragments are visu-

alized by EtBr or silver staining of the gels. Both methods have been used in the detection of sequence polymorphisms to facilitate molecular differentiation of cultivars or species (He et al., 1992; Dweikat et al., 1993, 1994; Boge et al., 1994; Brunel, 1994; Hiss et al., 1994; Widjojoatmodjo et al., 1994; Marklund et al., 1995; Watano et al., 1995), genome mapping (He et al., 1992), and tracking of particular genetic differences (To et al., 1993; Wehling et al., 1994; Procunier et al., 1995).

Applications

Although most applications of SSCP and DGGE have been in plant breeding, the development of these genetic markers for addressing questions in systematics and evolutionary biology has also been initiated. For example, Watano et al. (1995) used PCR and SSCP of a chloroplast intergenic spacer to examine purported hybrid speciation and introgression in pines, which have paternally inherited cpDNA. Diagnostic PCR-SSCP profiles were obtained for the putative parental taxa, *Pinus pumila* and *P. parviflora* var. *pentaphylla*, as well as the proposed hybrid derivative, *P. hakkodensis*. The hypothesized parental species had diagnostic SSCP profiles, whereas morphologically intermediate individuals had the SSCP profile of only one purported parental species. Watano et al. (1995) concluded that pollen-mediated gene flow is the evolutionary process that explains the distribution of the PCR-SSCP profiles.

Strand et al. (1996) surveyed a 525-bp noncoding region of the chloroplast genome using DGGE in 251 individuals from 18 populations (three taxa) of *Aquilegia*. Two alternative hypotheses regarding population structure were addressed: (1) whether the patterns of genetic variation were due to ongoing gene flow, and (2) whether the patterns were due to historical associations among populations. Five distinct cpDNA haplotypes were found and the distribution of these haplotypes among populations supported the latter hypothesis. In another study concentrating on introgression in plants, Leebens-Mack (1995) studied several hybrid swarm populations of *Baptisia leucophaea* and *B. sphaerocarpa*. DGGE was used to assay nucleotide variation in an *Adh* intron and patterns of interspecific gene flow between hybridizing populations. The degree of localized introgression at the *Adh* locus was found to vary among localities in the mosaic hybrid zone of these two *Baptisia* species.

RECOMMENDED APPLICATIONS OF PCR TECHNIQUES FOR PLANT SYSTEMATICS AND POPULATION BIOLOGY

The practical approach to PCR methodologies, as applied to plant systematics and population biology, would lead one to use the simplest and least expensive protocol that will yield the appropriate data to address the question of interest. Where genetic markers are needed to track gene flow or simple patterns of paternity, protocols for RAPD, ISSR, and RAMP are ideal because no prior sequence data are needed, a large number of markers can be generated, the supplies and equipment required is minimal, and these techniques do not require the use of radioisotopes. On the other hand, these techniques all have limitations (e.g., problems of homology assessment and sensitivity to minor modifications in experimental protocols).

More robust techniques (in terms of data interpretation and reproducibility) for addressing gene flow and paternity analysis include AFLP, ST-SSR, and PCR markers developed from known sequences and markers generated from CAPS. With the exception of AFLP, all of these techniques require knowledge of a sequence to be amplified. AFLP and ST-SSR loci have high degrees of polymorphism and are likely to exhibit additive patterns of Mendelian inheritance. For genetic diversity estimations, ST-SSR data is clearly superior. However, of all the techniques discussed in this chapter, development of ST-SSR primers requires the most laboratory expertise and has the longest lead time to obtaining results. AFLP and ST-SSR protocols also frequently employ radioisotope labeling and polyacrylamide gel electrophoresis and are therefore the most expensive protocols discussed here.

Many studies have used some of these methods (e.g., ISSR, RAPDs, PCR-MRSP) to estimate phylogenetic relationships. While ac-knowledging that nucleotide sequences and mapped chloroplast DNA restriction sites are the preferred data for inferring relationships, there is a place for alternative methods. Examples include the use of RAPD markers and related methods in the characterization of natural populations or large collections of cultivars. It is anticipated that sequencing costs will continue to drop. However, until that time many unstudied groups of plants could be economically and efficiently subjected to phylogenetic analysis with PCR-based methods other than sequencing and genomic analyses. PCR methods appropriate for determining whether a nucleotide sequence is sufficiently polymorphic to justify a full-scale sequencing effort include CAPS, SSCP, and DGGE. Of these, CAPS requires the least resources in terms of equipment, but the most time in generating sufficient information to make a determination (e.g., numerous restriction digests are needed). With SSCP and DGGE the information regarding polymorphism is obtained with one or two gels, but the equipment requirements and additional laboratory expertise needed are more involved than what is needed for CAPS analysis. In addition, the data from CAPS can be easily subjected to phylogenetic analysis, contrary to SSCP and DGGE results.

FINAL NOTE

PCR technology has clearly changed the way in which molecular data are gathered for applications ranging from nucleotide sequencing and phylogenetic reconstruction to population level studies. Many of the advantages and limitations of PCR applications have been discussed in the preceding sections, and they are also summarized in Table 2.5. Many PCR-based techniques have not yet made their way into the mainstream of research in plant systematics and evolutionary biology but nonetheless have great potential for studies of hybridization and introgression (ST-SSR, ISSR, RAMP, SSCP, DGGE, CAPS), population structure and genetic diversity (ST-SSR, SSCP, DGGE), conservation biology (ST-SSR, ISSR, RAMP, CAPS, AFLP), and phylogenetic reconstruction (PCR-MRSP). These PCR applications are in addition to the methodologies already established in plant systematics,

Table 2.5. Types of genetic markers commonly generated using PCR.

	RAPD	ST-SSR	ISSR	RAMP	AFLP	CAPS and PCR-MRSP
Principle	Arbitrary primers used for DNA amplification	Site-specific amplification of SSRs	Non-specific amplification of SSRs	Combination of RAPDs and ISSR	Amplification of fragment length polymorphisms	Restriction digest of PCR amplifications
Type of Polymorphism	Nucleotide substitutions indels	Changes in number of units in repeated motif	Nucleotide substitutions indels	Nucleotide substitutions indels	Nucleotide substitutions indels	Nucleotide substitutions indels
Inheritance	Dominant/ Codominant Mendelian and non-Mendelian	Codominant Mendelian (non-Mendelian for organellar)	Dominant/ Codominant Mendelian	Dominant/ Codominant Mendelian	Codominant Mendelian for nuclear	Codominant (nuclear) Dominant (organellar) Mendelian for nuclear
Prior sequence knowledge needed?	No	Yes	No	No	No	Yes
Radiolabeling needed?	No	Yes/No	No	No	Yes/No	No
Applications	Tracking gene flow, paternity analysis, cultivar identification, analysis of population structure, phylogenetic reconstruction, conservation biology, genetic mapping	Tracking gene flow, paternity analysis, cultivar identification, analysis of population structure, genetic diversity estimates, genetic mapping	Tracking gene flow, paternity analysis, cultivar identification, analysis of population structure, conservation biology, genetic mapping	Tracking gene flow, paternity analysis, cultivar identification, analysis of population structure, conservation biology, genetic mapping	Tracking gene flow, paternity analysis, cultivar identification, analysis of population structure, conservation biology, genetic mapping	Tracking gene flow, phylogenetic reconstruction, genetic mapping, preliminary screening for sequencing

Source: Modeled after Rafalski and Tingey, 1993.

such as nucleotide sequencing, and those summarized in Tables 2.1 and 2.2.

We can summarize the major advantages of PCR techniques compared to conventional molecular methods, as discussed in previous sections and in Wolfe et al. (1997), to include: (1) a smaller amount of DNA is required for PCR-based methods compared to other molecular techniques; (2) elimination of radioisotope use compared to conventional methods, at least for most PCR-based assays; (3) the ability to amplify DNA sequences from preserved tissues; (4) assessibility of the methodology for smaller labs in terms of equipment, facilities, and cost; (5) no prior sequence knowledge is required for many applications (e.g., RAPD, AFLP, and ISSR); (6) many genetic markers can be generated in a short time period; and (7) the ability to screen many genes simultaneously either for direct collection of data or as a feasibility study prior to nucleotide sequencing efforts.

LITERATURE CITED

Aagaard, J. E., S. S. Vollmer, F. C. Sorensen, and S. H. Strauss. 1995. Mitochondrial DNA products among RAPD profiles are frequent and strongly differentiated between races of Douglas-fir. Molecular Ecology 4:441–447.

Abo-elwafa, A., K. Murai, and T. Shimada. 1995. Intra- and inter-specific variations in Lens revealed by RAPD markers. Theoretical and Applied Genetics 90:335–340.

Abrol, S., and V. K. Chaudhary. 1993. Excess PCR primers inhibit DNA cleavage by some restriction endonucleases. BioTechniques 15:630–631.

Adams, R. P., and T. Demeke. 1993. Systematic relationships in Juniperus based on random amplified polymorphic DNAs (RAPDs). Taxon 42:553–571.

Akkaya, M. S., A. A. Bhagwat, and P. B. Cregan. 1992. Length polymorphisms of simple sequence repeat DNA in soybean. Genetics 132:1131–1193.

Amato, G., and J. Gatesy. 1994. PCR assays of variable nucleotide sites for identification of conservation units. In Molecular Ecology and Evolution: Approaches and Applications, eds. B. Schierwater, B. Streit, G. P. Wagner, and R. DeSalle, pp. 215–226. Birkhäuser Verlag, Basel, Switzerland.

Amos, B., C. Schlötterer, and D. Tautz. 1993. Social structure of pilot whales revealed by analytical DNA profiling. Science 260:670–672.

Anderson, J. A., G. A. Churchill, J. E. Autrique, S. D. Tanksley, and M. E. Sorrells. 1993. Optimizing parental selection for genetic linkage maps. Genome 36:181–186.

Apuya, N. R., B. L. Frazier, P. Keim, E. J. Roth, and K. G. Lark. 1988. Restriction fragment length polymorphisms as genetic markers in soybean Glycine max (L.) Merrill. Theoretical and Applied Genetics 75:889–901.

Arnold, M. L., C. M. Buckner, and J. J. Robinson. 1991. Pollen mediated introgression and hybrid speciation in Louisiana irises. Proceedings of the National Academy of Sciences U.S.A. 88:1398–1402.

Asmussen, C., and A. Liston. 1998. Chloroplast DNA characters, phylogeny, and classification of Lathryus (Fabaceae). American Journal of Botany, in press.

Ayliffe, M. A., G. J. Lawrence, J. G. Ellis, and A. J. Pryor. 1994. Heteroduplex molecules formed between allelic sequences cause nonparental RAPD bands. Nucleic Acids Research 22:1632–1636.

Bachmann, K. 1994. Molecular markers in plant ecology. New Phytologist 126:403–418.

Bachmann, K. 1995. Progress and pitfalls in systematics: cladistics, DNA and morphology. Acta Botanica Neerlandica 44:403–419.

Bachmann, K. 1997. Nuclear DNA markers in biosystematic research. Opera Botanica, in press.

Bachmann, K., and E. J. Hombergen. 1996. Mapping genes for phenotypic variation in Microseris (Lactuceae) with molecular markers. In Compositae: Systematics, Biology, Utilization, eds. D. J. N. Hind and P. D. S. Caligari, Proceedings of the International Compositae Conference, Kew, 1994, Vol. 2, Biology and Utilization, pp. 22–43. Royal Botanical Gardens, Kew, England.

Backeljau, T., L. De Bruyn, H. De Wolf, K. Jordaens, S. Van Dongen, R. Verhagen, and B. Winnepenninckx. 1995. Random amplified polymorphic DNA (RAPD) and parsimony methods. Cladistics 11:119–130.

Backeljau, T., L. De Bruyn, H. De Wolf, K. Jordaens, S. Van Dongen, and B. Winnepenninckx. 1996. Multiple UPGMA and neighbor-joining trees and the performance of some computer packages. Molecular Biology and Evolution 13:309–313.

Badenes, M. L., and D. E. Parfitt. 1995. Phylogenetic relationships of cultivated Prunus species from an analysis of chloroplast DNA variation. Theoretical and Applied Genetics 90:1035–1041.

Baker, R. J., J. L. Longmire, and R. A. Van Den Bussche. 1995. Organization of repetitive elements in the upland cotton genome (Gossypium hirsutum). Journal of Heredity 86:178–185.

Ballvora, A., J. Hesselbach, J. Niewöhner, D. Leister, F. Salamini, and C. Gebhardt. 1995. Marker enrichment and high-resolution map of the segment of potato chromosome VII harbouring the nematode resistance gene Gro1. Molecular and General Genetics 249:82–90.

Baum, D. A., and K. J. Sytsma. 1994. A phylogenetic analysis of Epilobium (Onagraceae) based on nuclear ribosomal DNA sequences. Systematic Botany 19:363–388.

Becker, J., and M. Heun. 1995. Barley microsatellites: allele variation and mapping. Plant Molecular Biology 27:835–845.

Becker, J., P. Vos, M. Kuiper, F. Salamini, and M. Heun. 1995. Combined mapping of AFLP and RFLP markers in barley. Molecular and General Genetics 249:65–73.

Beckmann, J. S., and J. L. Weber. 1992. Survey of human and rat microsatellites. Genomics 12:627–631.

Bell, C. J., and J. R. Ecker. 1994. Assignment of 30 microsatellite loci to the linkage map of *Arabidopsis*. Genomics 19:137-l44.

Benter, T., S. Papadopoulos, M. Pape, M. Manns, and H. Poliwoda. 1995. Optimization and reproducibility of random amplified polymorphic DNA in human. Analytical Biochemistry 230:92–100.

Bock, J. H., and J. L. Slightom. 1995. Fluorescence-based cycle sequencing with primers selected from a nonamer library. BioTechniques 19:60–62.

Boge, A., R. Gerstmeier, and R. Einspanier. 1994. Molecular polymorphism as a tool for differentiating ground beetles (*Carabus* species): application of ubiquitin PCR/SSCP analysis. Insect Molecular Biology 3:267–271.

Borowsky, R. L., M. McClelland, R. Cheng, and J. Welsh. 1995. Arbitrarily primed DNA fingerprinting for phylogenetic reconstruction in vertebrates: the *Xiphophorus* model. Molecular Biology and Evolution 12:1022–1032.

Boscherini, G., M. Morgante, P. Rossi, and G. G. Vendramin. 1994. Allozyme and chloroplast DNA variation in Italian and Greek populations of *Pinus leucodermis*. Heredity 73:284–290.

Bowcock, A. M., A. Ruiz-Linares, J. Tomfohrde, E. Minch, J. R. Kidd, and L. L. Cavalli-Sforza. 1994. High resolution of human evolutionary trees with polymorphic microsatellites. Nature 368:455–457.

Bradshaw, H. D., S. M. Wilbert, K. G. Otto, and D. W. Schemske. 1995. Genetic mapping of floral traits associated with reproductive isolation in monkeyflowers (*Mimulus*). Nature 376:762–765.

Bremer, B. 1991. Restriction data from chloroplast DNA for phylogenetic reconstruction: is there only one accurate way of scoring? Plant Systematics and Evolution 175:39–54.

Broun, P., and S. D. Tanksley. 1996. Characterization and genetic mapping of simple repeat sequences in the tomato genome. Molecular and General Genetics 250:39–49.

Brown, S. M., M. S. Hopkins, S. E. Mitchell, M. L. Senior, T. Y. Wang, R. R. Duncan, F. Gonzalez-Candelas, and S. Kresovich. 1996. Multiple methods for the identification of polymorphic simple sequence repeats (SSRs) in sorghum [*Sorghum bicolor* (L.) Moench.]. Theoretical and Applied Genetics 93:190–198.

Bruford, M. W., and R. K. Wayne. 1993. Microsatellites and their application to population genetic studies. Current Biology 3:939–943.

Brummer, E. C., J. H. Bouton, and G. Kochert. 1995. Analysis of annual *Medicago* species using RAPD markers. Genome 38:362–367.

Bruneau, A., J. J. Doyle, and J. D. Palmer. 1990. A chloroplast DNA inversion as a subtribal character in the Phaseoleae (Leguminosae). Systematic Botany 15:378–386.

Brunel, D. 1994a. A microsatellite marker in *Helianthus annuus* L. Plant Molecular Biology 24:397–400.

Brunel, D. 1994b. Denaturing gradient gel electrophoresis (DGGE) and direct sequencing of PCR amplified genomic DNA: Rapid and reliable identification of *Helianthus annuus* L. cultivars. Seed Science and Technology 22:185–194.

Caetano-Anollés, G. and P. M. Gresshoff. 1994. DNA amplification fingerprinting: a general tool with applications in breeding, identification and phylogenetic analysis of plants. **In** Molecular Ecology and Evolution: Approaches and Applications, eds. B. Schierwater, B. Streit, G. P. Wagner and R. DeSalle, pp. 17–31. Birkhäuser Verlag, Basel, Switzerland.

Caetano-Anollés, G., B. J. Bassam, and P. M. Gresshoff. 1991. DNA amplification fingerprinting using very short arbitrary oligonucleotide primers. Bio/Technology 9:553–557.

Caetano-Anollés, G., B. J. Bassam, and P. M. Gresshoff. 1993. Enhanced detection of polymorphic DNA by multiple arbitrary amplicon profiling of endonuclease digested DNA: identification of markers linked to the supernodulation locus in soybean. Molecular and General Genetics 241:57–64.

Callen, D. F., A. D. Thompson, Y. Shen, H. A. Phillips, R. I. Richards, J. C. Mulley, and G. R. Sutherland. 1993. Incidence and origin of 'null' alleles in the $(AC)_n$ microsatellite markers. American Journal of Human Genetics 52:922–927.

Campos, L. P., J. V. Raelson, and W. F. Grant. 1994. Genome relationships among *Lotus* species based on random amplified polymorphic DNA (RAPD). Theoretical and Applied Genetics 88:417–422.

Castiglione, S., G. Wang, G. Damiani, C. Bandi, S. Bisoffi, and F. Sala. 1993. RAPD fingerprints for identification and for taxonomic studies of elite poplar (*Populus* spp.) clones. Theoretical and Applied Genetics 87:54–59.

Catalán, P., Y. Shi, L. Armstrong, J. Draper, and C. A. Stace. 1995. Molecular phylogeny of the grass genus *Brachypodium* P. Beauv., based on RFLP and RAPD analysis. Botanical Journal of the Linnean Society 117:263–280.

Cha, R. S., and W. G. Thilly. 1995. Specificity, efficiency, and fidelity of PCR. **In** PCR Primer: A Laboratory Manual, eds. C. W. Dieffenbach and G. S. Dveksler, pp. 37–51. Cold Spring Harbor Laboratory Press, Plainview, New York.

Chalmers, K. J., R. Waugh, J. I. Sprent, A. J. Simons, and W. Powell. 1992. Detection of genetic variation between and within populations of *Gliricidia sepium* and *G. maculata* using RAPD markers. Heredity 69:465–472.

Chalmers, K. J., A. C. Newton, R. Waugh, J. Wilson, and W. Powell. 1994. Evaluation of the extent of genetic variation in mahoganies (Meliaceae) using RAPD markers. Theoretical and Applied Genetics 89:504–508.

Charters, Y. M., A. Robertson, M. J. Wilkinson, and G. Ramsay. 1996. PCR analysis of oilseed rape cultivars (*Brassica napus* L. ssp. *oleifera*) using 5'-anchored simple sequence repeat (SSR) primers. Theoretical and Applied Genetics 92:442–447.

Chase, M., R. Kesseli, and K. Bawa. 1996. Microsatellite markers for population and conservation genetics of tropical trees. American Journal of Botany 83:51–57.

Cheng, S., C. Fockler, W. Barnes, and R. Higuchi. 1994. Effective amplification of long targets from cloned inserts and human genomic DNA. Proceedings of the National Academy of Sciences U.S.A. **91**:5695–5699.

Cipriani, G., and M. Morgante. 1993. Evidence of chloroplast DNA variation in the genus *Actinidia* revealed by restriction analysis of PCR-amplified fragments. Journal of Genetics and Breeding **47**:319–326.

Clark, A. G., and M. S. Lanigan. 1993. Prospects for estimating nucleotide divergence with RAPDs. Molecular Biology and Evolution **10**:1096–1111.

Clegg, M. T., and M. L. Durbin. 1990. Molecular approaches to the study of plant biosystematics. Australian Systematic Botany **3**:1–8.

Cohen, S. N., A. C. Y. Chang, H. W. Boyer, and R. B. Helling. 1973. Construction of biologically functional bacterial plasmids in vitro. Proceedings of the National Academy of Sciences U.S.A. **70**:3240–3244.

Condit, R., and S. P. Hubbell. 1991. Abundance and DNA sequence of two-base repeat regions in tropical tree genomes. Genome **34**:66–71.

Coruzzi, G., R. Broglie, C. Edwards, and N.-H. Chua. 1984. Tissue-specific and light-regulated expression of a pea nuclear gene encoding the small subunit of ribulose-1,5-bisphosphate carboxylase. EMBO Journal **3**:1671–1679.

Crawford, D. J., S. Brauner, M. B. Cosner, and T. F. Stuessy. 1993. Use of RAPD markers to document the origin of the intergeneric hybrid × *Margyracaena skottsbergii* (Rosaceae) on the Juan Fernandez islands. American Journal of Botany **80**:89–92.

Davis, T. M., H. Yu, K. M. Haigis, and P. J. McGowan. 1995. Template mixing: a method for enhancing detection and interpretation of codominant RAPD markers. Theoretical and Applied Genetics **91**:582–588.

Dawson, I. K., K. J. Chalmers, R. Waugh, and W. Powell. 1993. Detection and analysis of genetic variation in *Hordeum spontaneum* populations from Israel using RAPD markers. Molecular Ecology **2**:151–159.

Dawson, I. K., A. J. Simons, R. Waugh, and W. Powell. 1995. Diversity and genetic differentiation among subpopulations of *Gliricidia sepium* revealed by PCR-based assays. Heredity **74**:10–18.

Dawson, I. K., A. J. Simons, R. Waugh, and W. Powell. 1996. Detection and pattern of interspecific hybridization between *Gliricidia sepium* and *G. maculata* in Meso-America revealed by PCR-based assays. Molecular Ecology **5**:89–98.

Dean, R., and J. Arnold. 1996. Cytonuclear disequilibria in hybrid zones using RAPD markers. Evolution **50**:1702–1705.

de la Cruz, M., R. Whitkus, A. Gómez-Pompa, and L. Mota-Bravo. 1995. Origins of cacao cultivation. Nature **375**:542–543.

Demeke, T., R. P. Adams, and R. Chibbar. 1992. Potential taxonomic use of random amplified polymorphic DNA (RAPD): a case study in *Brassica*. Theoretical and Applied Genetics **84**:990–994.

Demesure, B., N. Sodzi, and R. J. Petit. 1995. A set of universal primers for amplification of polymorphic non-coding regions of mitochondrial and chloroplast DNA in plants. Molecular Ecology **4**:129–131.

Demesure, B., B. Comps, and R. J. Petit. 1996. Chloroplast DNA phylogeography of the common beech (*Fagus sylvatica* L.) in Europe. Evolution **50**:2515–2520.

Depeiges, A., C. Goubely, A. Lenoir, S. Cocherel, G. Picard, M. Raynal, F. Grellet, and M. Delseny. 1995. Identification of the most represented repeated motifs in *Arabidopsis thaliana* microsatellite loci. Theoretical and Applied Genetics **91**:160–168.

Devos, K. M., and M. D. Gale. 1992. The use of random amplified polymorphic DNA markers in wheat. Theoretical and Applied Genetics **84**:567–572.

Devos, K. M., G. J. Bryan, A. J. Collins, P. Stephenson, and M. D. Gale. 1995. Application of two microsatellite sequences in wheat storage proteins as molecular markers. Theoretical and Applied Genetics **90**:247–252.

Díaz Lifante, Z., and I. Aguinagalde. 1996. The use of random amplified polymorphic DNA (RAPD) markers for the study of taxonomical relationships among species of *Asphodelus* sect. *Verinea* (Asphodelaceae). American Journal of Botany **83**:949–953.

Dieffenbach, C. W., and G. S. Dveksler. 1995. PCR Primer: A Laboratory Manual. Cold Spring Harbor Laboratory Press, Plainview, New York.

D'Ovidio, R., O. A. Tanzarella, and E. Porceddu. 1990. Rapid and efficient detection of genetic polymorphism in wheat through amplification by polymerase chain reaction. Plant Molecular Biology **15**:169–171.

Dow, B. D., M. V. Ashley, and H. F. Howe. 1995. Characterization of highly variable(GA/CT)n microsatellites in the bur oak, *Quercus macrocarpa*. Theoretical and Applied Genetics **91**:137–141.

Dowling, T. E., C. Moritz, J. D. Palmer, and L. H. Rieseberg. 1996. Nucleic Acids III: analysis of fragments and restriction sites. **In** Molecular Systematics, second edition, eds. D. M. Hillis, B. K. Mable, and C. Mortiz, pp. 249–320. Sinauer Associates, Sunderland, Massachusetts.

Downie, S. R., and J. D. Palmer. 1992. Use of chloroplast DNA rearrangements in reconstructing plant phylogeny. **In** Molecular Systematics of Plants, eds. P. S. Soltis, D.E. Soltis, and J. J. Doyle, pp. 14–35. Chapman & Hall, New York.

Downie, S. R., R. G. Olmstead, G. Zurawski, D. E. Soltis, P. S. Soltis, J. C. Wa, and J. D. Palmer. 1991. Six independent losses of the chloroplast DNA *rpl2* intron in dicotyledons. Evolution **45**:1245–1259.

Downie, S. R., D. S. Katz-Downie, K. H. Wolfe, P. J. Calie, and J. D. Palmer. 1994. Structure and evolution of the largest chloroplast gene (ORF2280): internal plasticity and multiple gene loss during angiosperm evolution. Current Genetics **25**:367–378.

Downie, S. R., E. Llanas, and D. S. Katz-Downie. 1996. Multiple independent losses of the *rpoC1* intron in angiosperm chloroplast DNAs. Systematic Botany **21**:135–151.

Downie, S. R., D. S. Katz-Downie, and K.-J. Cho. 1997. Relationships in the Caryophyllales as suggested by phylogenetic analyses of partial chloroplast DNA

ORF2280 homolog sequences. American Journal of Botany 84:253–273.

Doyle, J. J. 1992. Gene trees and species trees: molecular systematics as one-character taxonomy. Systematic Botany 17:144–163.

Doyle, J. J., J. I. Davis, R. J. Soreng, D. Garvin, and M. J. Anderson. 1992. Chloroplast DNA inversions and the origin of the grass family (Poaceae). Proceedings of the National Academy of Sciences U.S.A. 89:7722–7726.

Doyle, J. J., J. L. Doyle, and J. D. Palmer. 1995. Multiple independent losses of two genes and one intron from legume chloroplast genomes. Systematic Botany 20:272–294.

Doyle, J. J., J. L. Doyle, J. A. Ballenger, and J. D. Palmer. 1996. The distribution and phylogenetic significance of a 50-kb chloroplast DNA inversion in the flowering plant family Leguminosae. Molecular Phylogenetics and Evolution 5:429–438.

Dumolin-Lepegue, S., M.-H. Pemonge, and R. J. Petit. 1997. An enlarged set of consensus primers for the study of organelle DNA in plants. Molecular Ecology 6:393–397.

Dunemann, F., R. Kahnau, and H. Schmidt. 1994. Genetic relationships in Malus evaluated by RAPD 'fingerprinting' of cultivars and wild species. Plant Breeding 113:150–159.

Dweikat, I., S. Mackenzie, M. Levy, and H. Ohm. 1993. Pedigree assessment using RAPD-DGGE in cereal crop species. Theoretical and Applied Genetics 85:497–505.

Dweikat, I., H. Ohm, S. Mackenzie, F. Patterson, S. Cambron, and R. Ratcliffe. 1994. Association of a DNA marker with Hessian fly resistance gene H9 in wheat. Theoretical and Applied Genetics 89:964–968.

Edwards, A., A. Civitello, H. A. Hammond, and C. T. Caskey. 1991. DNA typing and genetic mapping with trimeric and tetrameric tandem repeats. American Journal of Human Genetics 49:746–756.

Ellsworth, D. L., K. D. Rittenhouse, and R. L. Honeycutt. 1993. Artifactual variation in randomly amplified polymorphic DNA banding patterns. BioTechniques 14:214–217.

El Mousadik, A., and R. J. Petit. 1996. Chloroplast DNA phylogeography of the argan tree of Morocco. Molecular Ecology 5:547–555.

Epplen, J. T., H. Ammer, C. Epplen, C. Kammerbauer, R. Mitreiter, L. Roewer, W. Schwaiger, V. Steimle, H. Zischler, E. Albert, A. Andreas, B. Beyermann, W. Meyer, J. Buitkamp, I. Nanda, M. Schmid, P. Nürnberg, S. D. J. Pena, H. Pöche, W. Sprecher, M. Schartl, K. Weising, and A. Yassouridis. 1991. Oligonucleotide fingerprinting using simple repeat motifs: a convenient, ubiquitously applicable method to detect hypervariability for multiple purposes. In DNA Fingerprinting: Approaches and Applications, eds. T. Burke, G. Dolf, A. J. Jeffreys, and R. Wolff, pp. 50–69. Birkhäuser Verlag, Basel, Switzerland.

Erlich, H. A., D. Gelfand, and J. J. Sninsky. 1991. Recent advances in the polymerase chain reaction. Science 252:1643–1651.

Estep, P. 1995. Long PCR reagents and guidelines. http://twod.med.harvard.edu/labgc/estep/longPCR protocol.html

Excoffier, L., P. E. Smouse, and J. M. Quattro. 1992. Analysis of molecular variance inferred from metric distances among DNA haplotypes: application to human mitochondrial DNA restriction data. Genetics 131:479–491.

Fabbri, A., J. I. Hormaza, and V. S. Polito. 1995. Random amplified polymorphic DNA analysis of olive (Olea europaea L.) cultivars. Journal American Society Horticultural Science 120:538–542.

Fischer, S. G., and L. S. Lerman. 1983. DNA fragments differing by single base-pair substitutions are separated in denaturing gradient gels: correspondence with melting theory. Proceedings of the National Academy of Sciences U.S.A. 80:1579–1583.

Foord, O. S., and E. A. Rose. 1995. Long-distance PCR. In PCR Primer: A Laboratory Manual, eds. C. W. Dieffenbach and G. S. Dveksler, pp. 63–77. Cold Spring Harbor Laboratory Press, Plainview, New York.

Fritsch, P., and L. H. Rieseberg. 1992. High outcrossing rates maintain male and hermaphrodite individuals in populations of the flowering plant Datisca glomerata. Nature 359:633–636.

Fritsch, P., and L. H. Rieseberg. 1996. The use of random amplified polymorphic DNA (RAPD) in conservation genetics. In Molecular Genetic Approaches in Conservation, eds. T.B. Smith and B. Wayne, pp. 54–73. Oxford University Press, New York.

Furman, B. J., D. Grattapaglia, W. S. Dvorak, and D. M. O'Malley. 1997. Analysis of genetic relationships of Central American and Mexican pines using RAPD markers that distinguish species. Molecular Ecology 6:321–331.

Goldstein, D. B., A. R. Linares, L. L. Cavalli-Sforza, and M. W. Feldman. 1995. An evaluation of genetic distances for use with microsatellite loci. Genetics 139:463–471.

Golenberg, E. M. 1991. Amplification and analysis of Miocene plant fossil DNA. Philosophical Transactions of the Royal Society B 333:419–427.

Golenberg, E. M., D. E. Giannasi, M. T. Clegg, C. J. Smiley, M. Durbin, D. Henderson, and G. Zurawski. 1990. Chloroplast DNA sequence from a Miocene Magnolia species. Nature 344:656–658.

González, J. M., and E. Ferrer. 1993. Random amplified polymorphic DNA analysis in Hordeum species. Genome 36:1029–1031.

Goodfellow, P. N. 1992. Variation is now the theme. Nature 359:777–778.

Gourmet, C., and A. L. Rayburn. 1996. Identification of RAPD markers associated with the presence of B chromosomes. Heredity 77:240–244.

Graham, J., and R. J. McNicol. 1995. An examination of the ability of RAPD markers to determine the relationships within and between Rubus species. Theoretical and Applied Genetics 90:1128–1132.

Grist, S. A., F. A. Firgaira, and A. A. Morley. 1993. Dinucleotide repeat polymorphisms isolated by the polymerase chain reaction. BioTechniques 15:304–309.

Gupta, M., Y.-S. Chyi, J. Romero-Severson, and J. L. Owen. 1994. Amplification of DNA markers from evolutionarily diverse genomes using single primers of simple-sequence repeats. Theoretical and Applied Genetics **89:**998–1006.

Gustafsson, L., and P. Gustafsson. 1994. Low genetic variation in Swedish populations of the rare species *Vicia pisiformis* (Fabaceae) revealed with RFLP (rDNA) and RAPD. Plant Systematics and Evolution **189:**133–148.

Hadrys, H., M. Balick, and B. Schierwater. 1992. Applications of random amplified polymorphic DNA (RAPD) in molecular ecology. Molecular Ecology **1:**55–63.

Hahn, W. J., T. J. Givnish, and K. J. Sytsma. 1995. Evolution of the monocot chloroplast inverted repeat: I. Evolution and phylogenetic implications of the ORF2280 deletion. **In** Monocotyledons: Systematics and Evolution, eds. P. J. Rudall, P. J. Cribb, D. F. Cutler, and C. J. Humphries, pp. 579–587. Royal Botanic Gardens, Kew, England.

Hamada, H., and T. Kakunaga. 1982. Potential Z-DNA forming sequences are highly dispersed in the human genome. Nature **298:**396–398.

Harris, S. A. 1995. Systematics and randomly amplified polymorphic DNA in the genus *Leucaena* (Leguminosae, Mimosoideae). Plant Systematics and Evolution **197:**195–208.

Haufler, C. H. 1987. Electrophoresis is modifying our concepts of evolution in homosporous pteridophytes. American Journal of Botany **74:**953–966.

Haufler, C. H., and D. E. Soltis. 1986. Genetic evidence suggests that homosporous ferns with high chromosome numbers are diploid. Proceedings of the National Academy of Sciences U.S.A. **83:**4389–4393.

He, S., H. Ohm, and S. Mackenzie. 1992. Detection of DNA sequence polymorphisms among wheat varieties. Theoretical and Applied Genetics **84:**573–578.

Heun, M., and T. Helentjaris. 1993. Inheritance of RAPDs in F1 hybrids of corn. Theoretical and Applied Genetics **85:**961–968.

Heun, M., J. P. Murphy, and T. D. Phillips. 1994. A comparison of RAPD and isozyme analyses for determining the genetic relationships among *Avena sterilis* L. accessions. Theoretical and Applied Genetics **87:**689–696.

Hilu, K. W. 1994. Evidence from RAPD markers in the evolution of *Echinochloa* millets (Poaceae). Plant Systematics and Evolution **189:**247–257.

Hilu, K. W., 1995. Evolution of finger millet: evidence from random amplified polymorphic DNA. Genome **38:**232–238.

Hilu, K. W., and H. T. Stalker. 1995. Genetic relationships between peanut and wild species of *Arachis* sect. *Arachis* (Fabaceae): evidence from RAPDs. Plant Systematics and Evolution **198:**167–178.

Hiratsuka, J., H. Shimada, R. Whittier, T. Ishibashi, M. Sakamoto, M. Mori, C. Kondo, Y. Honji, C. R. Sun, B. Y. Meng, Y.Q. Li, A. Kanno, Y. Nishizwa, A. Hirai, K. Shinozaki, and M. Sugiura. 1989. The complete sequence of the rice (*Oryza sativa*) chloroplast genome: intermolecular recombination between distinct tRNA genes accounts for a major plastid DNA inversion during the evolution of the cereals. Molecular and General Genetics **217:**185–194.

Hiss, R. H., D. E. Norris, C. H. Dietrich, R. F. Whitcombe, D. F. West, C. F. Bosio, S. Kambhampati, J. Piesman, M. F. Antolin, and W. C. I. Black. 1994. Molecular taxonomy using single-strand conformation polymorphism (SSCP) analysis of mitochondrial ribosomal DNA genes. Insect Molecular Biology **3:**171–182.

Hoot, S. B., and J. D. Palmer. 1994. Structural rearrangements, including parallel inversions, within the chloroplast genome of *Anemone* and related genera. Journal of Molecular Evolution **38:**274–281.

Hormaza, J. I., L. Dollo, and V. S. Polito. 1994. Identification of RAPD marker linked to sex determination in *Pistacia vera* using bulked segregant analysis. Theoretical and Applied Genetics **89:**9–13.

Hsiao, J. Y., and L. H. Rieseberg. 1994. Population genetic structure of *Yushania niitakayamensis* (Bambusoideae, Poaceae) in Taiwan. Molecular Ecology **3:**201–208.

Huff, D. R., R. Peakall, and P. E. Smouse. 1993. RAPD variation within and among natural populations of outcrossing buffalograss [*Buchloe dactyloides* (Nutt.) Engelm.]. Theoretical and Applied Genetics **86:**927–934.

Hughes, C. R., and D. C. Queller. 1993. Detection of highly polymorphic microsatellite loci in a species with little allozyme polymorphism. Molecular Ecology **2:**131–137.

Innis, M. A., and D. H. Gelfand. 1990. Optimization of PCRs. **In** PCR Protocols, eds. M. A. Innis, D. H. Gelfand, J. J. Sninsky, and T. J. White, pp. 3–12. Academic Press, San Diego.

Innis, M. A., D. H. Gelfand, and J. J. Sninsky, eds. 1995. PCR Strategies. Academic Press, San Diego.

Jacob, H. J., K. Lindpaintner, S. E. Lincoln, K. Kusumi, R. K. Bunker, Y.-P. Mao, D. Ganten, V. J. Dzau, and E. S. Lander. 1991. Genetic mapping of a gene causing hypertension in the stroke-prone spontaneously hypertensive rat. Cell **67:**213–224.

Jain, A., S. Bhatia, S. S. Banga, and S. Prakash. 1994. Potential use of random amplified polymorphic DNA (RAPD) technique to study the genetic diversity in Indian mustard (*Brassica juncea*) and its relationship to heterosis. Theoretical and Applied Genetics **88:** 116–122.

Jarret, R. L., D. R. Vuylsteke, N. J. Gawel, R. B. Pimentel, and L. J. Dunbar. 1993. Detecting genetic diversity in diploid bananas using PCR and primers from a highly repetitive DNA sequence. Euphytica **68:**69–76.

Jeffreys, A. J., V. Wilson, and S. L. Thein. 1985. Hypervariable 'minisatellite' regions in human DNA. Nature **314:**67–73.

Jessup, S. L. 1993. Reticulate evolution in *Gaudichaudia* (Malpighiaceae): phylogenetic and biogeographic analysis of molecular and morphological variation in a polyploid complex of neotropical vines. American Journal of Botany (suppl.) **80:**154.

Jordan, W. C., M. W. Courtney, and J. E. Neigel. 1996. Low levels of intraspecific genetic variation at a rapidly evolving chloroplast DNA locus in North American duckweeds (Lemnaceae). American Journal of Botany 83:430–439.

Jork, K. B., and J. W. Kadereit. 1995. Molecular phylogeny of the Old World representatives of Papaveraceae subfamily Papaveroideae with special emphasis on the genus *Meconopsis*. Plant Systematics and Evolution (suppl.) 9:171–180.

Joshi, C. P., and H. T. Nguyen. 1993. RAPD (random amplified polymorphic DNA) analysis based intervarietal genetic relationships among hexaploid wheats. Plant Science 93:95–103.

Karihaloo, J. L., S. Brauner, and L. D. Gottlieb. 1995. Random amplified polymorphic DNA variation in the eggplant, *Solanum melongena* L. (Solanaceae). Theoretical and Applied Genetics 90:767–770.

Karp, A., O. Seberg, and M. Buiatti. 1996. Molecular techniques in the assessment of botanical diversity. Annals of Botany 78:143–149.

Katayama, H., and Y. Ogihara. 1996. Phylogenetic affinities of the grasses to other monocots as revealed by molecular analysis of chloroplast DNA. Current Genetics 29:572–581.

Kijas, J. M. H., J. C. S. Fowler, C. A. Garbett, and M. R. Thomas. 1994. Enrichment of microsatellites from the citrus genome using biotinylated oligonucleotide sequences bound to streptavidin-coated magnetic particles. BioTechniques 16:656–662.

Kijas, J. M. H., J. C. S. Fowler, and M. R. Thomas. 1995. An evaluation of sequence tagged microsatellite site markers for genetic analysis with *Citrus* and related species. Genome 38:349–355.

Kleppe, K., E. Ohtsuka, R. Kleppe, I. Molineux, and H. G. Khorana. 1971. Studies on polynucleotides XCVI. Repair replication of short synthetic DNA's as catalyzed by DNA polymerases. Journal of Molecular Biology 56:341–361.

Kostia, S., S.-L. Varvio, P. Vakkari, and P. Pulkkinen. 1995. Microsatellite sequences in a conifer, *Pinus sylvestris*. Genome 38:1244–1248.

Lagercrantz, U., H. Ellegren, and L. Andersson. 1993. The abundance of various polymorphic microsatellite motifs differs between plants and vertebrates. Nucleic Acids Research 21:1111–1115.

Lamboy, W. F. 1994a. Computing genetic similarity coefficients from RAPD data: the effects of PCR artifacts. PCR Methods and Applications 4:31–37.

Lamboy, W. F. 1994b. Computing genetic similarity coefficients from RAPD data: Correcting for the effects of PCR artifacts caused by variation in experimental conditions. PCR Methods and Applications 4:38–43.

Lavi, U., M. Akkaya, A. Bhagwat, E. Lahav, and P. B. Cregan. 1994. Methodology of generation and characteristics of simple sequence repeat DNA markers in avocado (*Persea americana* M.). Euphytica 80:171–177.

Lavin, M., J. J. Doyle, and J. D. Palmer. 1990. Evolutionary significance of the loss of the chloroplast-DNA inverted repeat in the Leguminosae subfamily Papilionoideae. Evolution 44:390–402.

Leebens-Mack, J. 1995. An investigation of the consequence of natural hybridization between two east Texas *Bap-*

tisia (Fabaceae) species. Austin, Texas, University of Texas.

Levinson, G., and G. A. Gutman. 1987. Slipped-strand mispairing: a major mechanism for DNA sequence evolution. Molecular Biology and Evolution 4:203–221.

Liston, A. 1992a. Restriction site mapping of PCR-amplified products. Ancient DNA Newsletter 1:11.

Liston, A. 1992b. Variation in the chloroplast genes *rpoC1* and *rpoC2* of the genus *Astragalus* (Fabaceae): evidence from restriction site mapping of a PCR-amplified fragment. American Journal of Botany 79:953–961.

Liston, A. 1995. Use of the polymerase chain reaction to survey for the loss of the inverted repeat in the legume chloroplast genome. In Advances in Legume Systematics, Part 7, eds. M. D. Crisp and J. J. Doyle, pp. 31–40. Royal Botanic Gardens, Kew, England.

Liston, A., and J. W. Kadereit. 1995. Chloroplast DNA evidence for introgression and long distance dispersal in the desert annual *Senecio flavus* (Asteraceae). Plant Systematics and Evolution 197:33–41.

Liston, A., and J. A. Wheeler. 1994. The phylogenetic position of the genus *Astragalus* (Fabaceae): evidence from the chloroplast genes *rpoC1* and *rpoC2*. Biochemical Systematics and Ecology 22:377–388.

Liston, A., L. H. Rieseberg, and M. A. Hanson. 1992. Geographic partitioning of chloroplast DNA variation in the genus *Datisca* (Datiscaceae). Plant Systematics and Evolution 181:121–132.

Liston, A., W. A. Robinson, J. Oliphant, and E. R. Alvarez-Buylla. 1996. Length variation in the nuclear ribosomal internal transcribed spacer region of non-flowering seed plants. Systematic Botany 21:109–120.

Litt, M., and J. A. Luty. 1989. A hypervariable microsatellite revealed by in vitro amplification of a dinucleotide repeat within the cardiac muscle actin gene. American Journal of Human Genetics 44:397–401.

Livneh, O., E. Vardi, Y. Stram, O. Edelbaum, and I. Sela. 1992. The conversion of a RFLP assay into PCR for the determination of purity in a hybrid pepper cultivar. Euphytica 62:97–102.

Lorenz, M., A. Weihe, and T. Börner. 1994. DNA fragments of organellar origin in random amplified polymorphic DNA (RAPD) patterns of sugar beet (*Beta vulgaris* L.). Theoretical and Applied Genetics 88:775–779.

Lynch, M. 1990. The similarity index and DNA fingerprinting. Molecular Biology and Evolution 7:478–484.

Lynch, M., and B. G. Milligan. 1994. Analysis of population genetic structure with RAPD markers. Molecular Ecology 3:91–99.

Maier, R. M., K. Neckermann, G. L. Igloi, and H. Kossel. 1995. Complete sequence of the maize chloroplast genome: gene content, hotspots of divergence and fine tuning of genetic information by transcript editing. Journal of Molecular Biology 251:614–628.

Manhart, J. R., and J. D. Palmer. 1990. The gain of two chloroplast tRNA intron marks the green algal ancestors of land plants. Nature 345:268–270.

Marillia, E. F., and G. J. Scoles. 1996. The use of RAPD markers in *Hordeum* phylogeny. Genome 39:646–654.

Marklund, S., R. Chaudhary, L. Marklund, K. Sandberg, and L. Andersson. 1995. Extensive mtDNA diversity in horses revealed by PCR-SSCP analysis. Animal Genetics **26**:193–196.

Matioli, S. R., and R. A. DeBrito. 1995. Obtaining genetic markers by using double-stringency PCR with microsatellites and arbitrary primers. BioTechniques **19**:752–754.

Maughan, P. J., M. A. Saghai Maroof, and G. R. Buss. 1995. Microsatellite and amplified sequence length polymorphisms in cultivated and wild soybean. Genome **38**:715–723.

McBride, L. J., S. M. Koepf, R. A. Gibbs, W. Salser, P. E. Mayrand, M. W. Hunkapiller, and M. N. Kronick. 1989. Automated DNA sequencing methods involving polymerase chain reaction. Clinical Chemistry **35**:2196–2201.

McCauley, D. E., J. E. Stevens, P. A. Peroni, and J. A. Raveill. 1996. The spatial distribution of chloroplast DNA and allozyme polymorphisms within a population of *Silene alba* (Caryophyllaceae). American Journal of Botany **83**:727–731.

Meksem, K., D. Leister, J. Peleman, M. Zabeau, F. Salamini, and C. Gebhardt. 1995. A high-resolution map of the vicinity of the *R1* locus on chromosome V of potato based on RFLP and AFLP markers. Molecular and General Genetics **249**:74–81.

Messinger, W. 1994. Molecular systematic studies in the genus *Ribes* (Grossulariaceae). M.Sc. thesis. Oregon State University, Corvallis, Oregon.

Moore, S. S., L. L. Sargeant, T. J. King, J. S. Mattick, M. Georges, and D. J. S. Hetzel. 1991. The conservation of dinucleotide microsatellites among mammalian genomes allows the use of heterologous PCR primer pairs in closely related species. Genomics **10**:654–660.

Mörchen, M., J. Cuguen, G. Michaelis, C. Hänni, and P. Saumitou-Laprade. 1996. Abundance and length polymorphism of microsatellite repeats in *Beta vulgaris* L. Theoretical and Applied Genetics **92**:326–333.

Morgante, M., and A. M. Olivieri. 1993. PCR-amplified microsatellites as markers in plant genetics. Plant Journal **3**:175–182.

Morgante, M., A. Rafalski, P. Biddle, S. Tingey, and A. M. Olivieri. 1994. Genetic mapping and variability of seven soyean simple sequence repeat loci. Genome **37**:763–769.

Morin, P. A., and D. G. Smith. 1995. Non-radioactive detection of hypervariable simple sequence repeats in short polyacrylamide gels. BioTechniques **19**:223–228.

M'Ribu, H. K., and K. W. Hilu. 1994. Detection of interspecific and intraspecific variation in *Panicum* millets through random amplified polymorphic DNA. Theoretical and Applied Genetics **88**:412–416.

Mulcahy, D. L., N. F. Weeden, R. Kesseli and S. B. Carroll. 1992. DNA probes for the Y-chromosome of *Silene latifolia*, a dioecious angiosperm. Sexual Plant Reproduction **5**:86–88.

Mullis, K. B. 1994. PCR and scientific invention: the trial of Dupont vs. Cetus. **In** The Polymerase Chain Reaction, eds. K. B. Mullis, F. Ferre, and R. A. Gibbs, pp. 427–441. Birkhäuser, Boston, Massachusetts.

Mullis, K. B., and F. A. Faloona. 1987. Specific synthesis of DNA in vitro via a polymerase-catalyzed chain reaction. Methods in Enzymology **155**:335–350.

Mullis, K., F. Faloona, S. Scharf, R. Saiki, G. Horn, and H. Erlich. 1986. Specific enzymatic amplification of DNA in vitro: the polymerase chain reaction. Cold Spring Harbor Symposium Quntitative Biology **51**: 263–273.

Myers, R. M., S. G. Fischer, T. Maniatis, and L. S. Lerman. 1985. Modification of the melting properties of duplex DNA by attachment of GC-rich DNA sequence as determined by denaturing gradient gel electrophoresis. Nucleic Acids Research **13**:3111–3129.

Myers, R. M., T. Maniatis, and L. S. Lerman. 1987. Detection and localization of single base changes by denaturing gradient gel electrophoresis. Methods in Enzymology **155**:501–527.

Nei, M. 1973. Analysis of gene diversity in subdivided populations. Proceedings of the National Academy of Sciences U.S.A. **70**:3321–3323.

Nei, M. 1978. Estimation of average heterozygosity and genetic distance from a small number of individuals. Genetics **89**:583–590.

Nei, M., and W. Li. 1979. Mathematical model for studying genetic variation in terms of restriction endonucleases. Proceedings of the National Academy of Sciences U.S.A. **76**:5269–5273.

Nesbitt, K. A., B. M. Potts, R. E. Vaillancourt, A. K. West, and J. B. Reid. 1995. Partitioning and distribution of RAPD variation in a forest tree species, *Eucalyptus globulus* (Myrtaceae). Heredity **74**:628–637.

Neuffer, B. 1996. RAPD analyses in colonial and ancestral populations of *Capsella bursa-pastoris* (L.) Med. (Brassicaceae). Biochemical Systematics and Ecology **24**:393–403.

Newbury, H. J., and B. V. Ford-Lloyd. 1993. The use of RAPD for assessing variation in plants. Plant Growth Regulation **12**:43–51.

Nickrent, D. L. 1994. From field to film: rapid sequencing methods for field-collected plant species. BioTechniques **16**:470–475.

Nishio, T., M. Kusaba, M. Watanabe, and K. Hinata. 1996. Registration of *S* alleles in *Brassica campestris* L. by the restriction fragment sizes of *SLGs*. Theoretical and Applied Genetics **92**:388–394.

Nybom, H. 1994. DNA fingerprinting—a useful tool in fruit breeding. Euphytica **77**:59–64.

Nybom, H., and T. Kraft. 1995. Application of DNA fingerprinting to the taxonomy of European blackberry species. Electrophoresis **16**:1731–1735.

Ochman, H., A. S. Gerber, and L. H. Hartl. 1988. Genetic application of inverse polymerase chain reaction. Genetics **120**:621–623.

Ohyama, K., H. Fukuzawa, T. Kohchi, H. Shirai, T. Sano, S. Sano, K. Umesono, Y. Shiki, M. Takeuchi, Z. Chang, S. Aota, H. Inokuchi, and H. Ozeki. 1986. Chloroplast gene organization deduced from complete sequence of liverwort *Marchantia polymorpha* chloroplast DNA. Nature **322**:572–574.

Olmstead, R. G., and J. D. Palmer. 1994. Chloroplast DNA systematics: a review of methods and data analysis. American Journal of Botany 81:1205–1224.

Orita, M., Y. Suziki, T. Sekiya, and K. Hayashi. 1989. Rapid and sensitive detection of point mutations and DNA polymorphisms using the polymerase chain reaction. Genomics 5:874–879.

Orozco-Castillo, C., K. J. Chalmers, R. Waugh, and W. Powell. 1994. Detection of genetic diversity and selective gene introgression in coffee using RAPD markers. Theoretical and Applied Genetics 87:934–940.

Palumbi, S. R. 1996. Nucleic acids II: the polymerase chain reaction. In Molecular Systematics, second edition, eds. D. M. Hillis, B. K. Mable, and C. Mortiz, pp. 205–247. Sinauer Associates, Sunderland, Massachusetts.

Paran, I., and R. W. Michelmore. 1993. Development of reliable PCR-based markers linked to downy mildew resistance genes in lettuce. Theoretical and Applied Genetics 85:985–993.

Peakall, R., P. E. Smouse, and D. R. Huff. 1995. Evolutionary implications of allozyme and RAPD variation in diploid populations of dioecious buffalograss Buchloe dactyloides. Molecular Ecology 4:135–147.

Perez de la Rosa, J., S. A. Harris, and A. Farjon. 1995. Noncoding chloroplast DNA variation in Mexican pines. Theoretical and Applied Genetics 91:1101–1106.

Pillay, M., and S. T. Kenny. 1995. Anomalies in direct pairwise comparisons of RAPD fragments for genetic analysis. BioTechniques 19:694–698.

Poinar, H. N., M. Höss, J. L. Bada, and S. Pääbo. 1996. Amino acid racemization and the preservation of ancient DNA. Science 272:864–866.

Poulsen, G. B., G. Kahl, and K. Weising. 1993. Abundance and polymorphism of simple repetitive DNA sequences in Brassica napus L. Theoretical and Applied Genetics 85:994–1000.

Powell, W., M. Morgante, C. Andre, J. W. McNicol, G. C. Machray, J. J. Doyle, S. V. Tingey, and J. A. Rafalski. 1995a. Hypervariable microsatellites provide a general source of polymorphic DNA markers for the chloroplast genome. Current Biology 5:1023–1029.

Powell, W., M. Morgante, R. McDevitt, G. G. Vendramin, and J. A. Rafalski. 1995b. Polymorphic simple sequence repeat regions in chloroplast genomes: applications to the population genetics of pines. Proceedings of the National Academy of Sciences U.S.A. 92:7759–7763.

Powell, W., C. Orozco-Castillo, K. J. Chalmers, J. Provan, and R. Waugh. 1995c. Polymerase chain reaction-based assays for the characterization of plant genetic resources. Electrophoresis 16:1726–1730.

Procunier, J. D., T. F. Townley Smith, S. Fox, S. Prashar, M. Gray, W. K. Kim, E. Czarnecki, and P. L. Dyck. 1995. PCR-based RAPD/DGGE markers linked to rust resistance genes Lr29 and Lr25 in wheat (Triticum aestivum L.). Journal of Genetics and Breeding 49:87–91.

Purps, D. M. L., and J. W. Kadereit. 1998. RAPD evidence for a sister group relationship of the presumed progenitor-derivative species pair Senecio nebrodensis L. and S. viscosus L. (Asteraceae). Plant Systematics and Evolution, in press.

Queller, D. C., J. E. Strassmann, and C. R. Hughes. 1993. Microsatellites and kinship. Trends in Ecology and Evolution 8:285–288.

Rafalski, J. A., and S. V. Tingey. 1993. Genetic diagnostics in plant breeding: RAPDs, microsatellites and machines. Trends in Genetics 9:275–280.

Ralph, D., and M. McClelland. 1994. Mapped restriction site polymorphisms (MRSPS) in PCR products for rapid identification and classification of genetically distinct organisms. In PCR Technology: Current Innovations, eds. H. G. Griffin and A. M. Griffin, pp. 121–131. CRC Press, Boca Raton, Florida.

Rassmann, K., C. Schlötterer, and D. Tautz. 1991. Isolation of simple-sequence loci for use in polymerase chain reaction-based DNA fingerprinting. Electrophoresis 12:113–118.

Ratnaparkhe, M. B., V. S. Gupta, M. R. Ven Murthy, and P. K. Ranjekar. 1995. Genetic fingerprinting of pigeonpea [Cajanus cajan (L.) Millsp.] and its wild relatives using RAPD markers. Theoretical and Applied Genetics 91:893–898.

Riedy, M. F., W. J. Hamilton III, and C. F. Aquadro. 1992. Excess of non-parental bands in offspring from known primate pedigrees assayed using RAPD PCR. Nucleic Acids Research 20:918.

Rieseberg, L. H. 1996. Homology among RAPD fragments in interspecific comparisons. Molecular Ecology 5:99–105.

Rieseberg, L. H., M. A. Hanson, and C. T. Philbrick. 1992. Androdioecy is derived from dioecy in Datiscaceae: evidence from restriction site mapping of PCR-amplified chloroplast DNA fragments. Systematic Botany 17:324–336.

Rieseberg, L. H., H. Choi, R. Chan, and C. Spore. 1993. Genomic map of a diploid hybrid species. Heredity 70:285–293.

Rieseberg, L. H., C. Van Fossen, D. Arias and R. L. Carter. 1994. Cytoplasmic male sterility in sunflower: origin, inheritance, and frequency in natural populations. Journal of Heredity 85:233–238.

Rieseberg, L. H., B. Sinervo, C. R. Linder, M. C. Ungerer, and D. M. Arias. 1996. Role of gene interactions in hybrid speciation: evidence from ancient and experimental hybrids. Science 272:741–745.

Röder, M., J. Plaschke, S. U. König, A. Börner, M. E. Sorrells, S. D. Tanksley, and M. W. Ganal. 1995. Abundance, variability and chromosomal location of microsatellites in wheat. Molecular and General Genetics 246:327–333.

Roelofs, D., and K. Bachmann. 1995. Chloroplast and nuclear DNA variation among homozygous plants in a population of the autogamous annual Microseris douglasii (Asteraceae, Lactuceae). Plant Systematics and Evolution 196:185–194.

Roelofs, D. and K. Bachmann. 1997. Comparison of chloroplast and nuclear phylogeny in the autogamous annual Microseris douglasii (Asteraceae, Lactuceae). Plant Systematics and Evolution 204:49–63.

Rogers, S. O., and A. J. Bendich. 1985. Extraction of DNA from milligram amounts of fresh, herbarium, and mummified plant tissues. Plant Molecular Biology **5**:69–76.

Rogstad, S. H. 1994. Inheritance in turnip of variable-number tandem-repeat genetic markers revealed with synthetic repetitive DNA probes. Theoretical and Applied Genetics **89**:824–830.

Rogstad, S. H., J. C. Patton II, and B. A. Schaal. 1988. M13 repeat probe detects DNA minisatellite-like sequences in gymnosperms and angiosperms. Proceedings of the National Academy of Sciences U.S.A. **85**:9176–9178.

Rongwen, J., M. S. Akkaya, A. A. Bhagwat, U. Lavi, and P. B. Cregan. 1995. The use of microsatellite DNA markers for soybean genotype identification. Theoretical and Applied Genetics **90**:43–48.

Roy, M. S., E. Geffen, D. Smith, E. A. Ostrander, and R. K. Wayne. 1994. Patterns of differentiation and hybridization in North American wolflike canids, revealed by analysis of microsatellite loci. Molecular Biology and Evolution **11**:553–570.

Russell, J. R., F. Hosein, E. Johnson, R. Waugh, and W. Powell. 1993. Genetic differentiation of cocoa (*Theobroma cacao* L.) populations revealed by RAPD analysis. Molecular Ecology **2**:89–97.

Saghai Maroof, M. A., R. M. Biyashev, G. P. Yang, Q. Zhang, and R. W. Allard. 1994. Extraordinarily polymorphic microsatellite DNA in barley: Species diversity, chromosomal locations, and population dynamics. Proceedings of the National Academy of Sciences U.S.A. **91**:5466–5470.

Saiki, R. K., D. H. Gelfand, S. Stoffel, S. J. Scharf, R. Higuchi, G. T. Horn, K. B. Mullis, and H. A. Erlich. 1988. Primer-directed enzymatic amplification of DNA with thermostable DNA polymerase. Science **259**:487–491.

Salimath, S. S., A. C. de Oliveira, I. D. Godwin, and J. L. Bennetzen. 1995. Assessment of genome origins and genetic diversity in the genus *Eleusine* with DNA markers. Genome **38**:757–763.

Sambrook, J., E. F. Fritsch, and T. Maniatis. 1989. Molecular Cloning: A Laboratory Manual. Cold Spring Harbor Laboratory Press, Plainview, New York.

Sánchez de la Hoz, M. P., J. A. Dávila, Y. Loarce, and E. Ferrer. 1996. Simple sequence repeat primers used in polymerase chain reaction amplifications to study genetic diversity in barley. Genome **39**:112–117.

Sanger, F., S. Nicklen, and A. R. Coulson. 1977. DNA sequencing with chain-terminating inhibitors. Proceedings of the National Academy of Sciences U.S.A. **74**:5463–5467.

Savolainen, V., P. Cuénoud, R. Spichiger, M. D. P. Martinez, M. C. Crevecoeur, and J.-F. Manen. 1995. The use of herbarium specimens in DNA phylogenetics: evaluation and improvement. Plant Systematics and Evolution **197**:87–98.

Schlötterer, C., and J. Pemberton. 1994. The use of microsatellites for genetic analysis of natural populations. **In** Molecular Ecology and Evolution: Approaches and Applications, eds. B. Schierwater, B. Streit, G. P. Wagner, and R. DeSalle, pp. 203–214. Birkhäuser Verlag, Basel, Switzerland.

Schlötterer, C., and D. Tautz. 1992. Slippage synthesis of simple sequence DNA. Nucleic Acids Research **20**:211–215.

Schlötterer, C., B. Amos, and D. Tautz. 1991. Conservation of polymorphic simple sequence loci in cetacean species. Nature **354**:63–65.

Schwarzbach, A. E., and J. W. Kadereit. 1995. Rapid radiation of North American desert genera of the Papaveraceae: Evidence from restriction site mapping of PCR-amplified chloroplast DNA fragments. Plant Systematics and Evolution (suppl.) **9**:159–170.

Scribner, K. T., J. W. Arntzen, and T. Burke. 1994. Comparative analysis of intra- and interpopulation genetic diversity in *Bufu bufo,* using allozyme, single-locus microsatellite, minisatellite, and multilocus minisatellite data. Molecular Biology and Evolution **11**:737–748.

Senior, M. L., and M. Heun. 1993. Mapping maize microsatellites and polymerase chain reaction confirmation of the targeted repeats using a CT primer. Genome **36**:884–889.

Sharma, P. C., B. Hüttel, P. Winter, G. Kahl, R. C. Gardner, and K. Weising. 1995a. The potential of microsatellites for hybridization- and polymerase chain reaction-based DNA fingerprinting of chickpea. Electrophoresis **16**:1755–1761.

Sharma, S. K., I. K. Dawson, and R. Waugh. 1995b. Relationships among cultivated and wild lentils revealed by RAPD analysis. Theoretical and Applied Genetics **91**:647–654.

Sharon, D., A. Adato, S. Mhameed, U. Lavi, J. Hillel, M. Gomolka, C. Epplen, and J. T. Epplen. 1995. DNA fingerprints in plants using simple-sequence repeat and minisatellite probes. HortScience **30**:109–112.

Shattuck-Eidens, D. M., R. N. Bell, J. T. Mitchell, and V. C. Mcwhorter. 1991. Rapid detection of maize DNA sequence variation. Genetic Analysis, Techniques and Applications **8**:240–245.

Shinozaki, K., M. Ohme, M. Tanaka, T. Wakasugi, N. Hayashida, T. Matsubayashi, N. Zaita, J. Chunwongse, J. Obokata, K. Yamaguchi-Shinozaki, C. Ohto, K. Torazawa, B. Y. Meng, M. Sugita, H. Deno, T. Kamogashira, K. Yamada, J. Kusuda, F. Takaiwa, A. Kato, N. Tohdoh, H. Shimada, and M. Sugiura. 1986. The complete nucleotide sequence of the tobacco chloroplast genome: its gene organization and expression. EMBO Journal **5**:2043–2049.

Skinner, D. Z. 1992. A PCR-based method of identifying species-specific repeated DNAs. BioTechniques **13**:210–214.

Slatkin, M. 1995. A measure of population subdivision based on microsatellite allele frequencies. Genetics **139**:457–462.

Smith, D. N., and M. E. Devey. 1994. Occurrence and inheritance of microsatellites in *Pinus radiata*. Genome **37**:977–983.

Smith, J. F., and T. V. Pham. 1996. Genetic diversity of the narrow endemic *Allium aaseae* (Alliaceae). American Journal of Botany **83**:717–726.

Smith, J. F., C. C. Burke, and W. L. Wagner. 1996. Interspecific hybridization in natural populations of *Cyrtandra*

(Gesneriaceae) on the Hawaiian Islands: evidence from RAPD markers. Plant Systematics and Evolution **200:**61–77.

Smith, J. J., J. S. Scott-Craig, J. R. Leadbetter, G. L. Bush, D. L. Roberts, and D. W. Fulbright. 1994. Characterization of random amplified polymorphic DNA (RAPD) products from *Xanthomonas campestris* and some comments on the use of RAPD products in phylogenetic analysis. Molecular Phylogenetics and Evolution **3:**135–145.

Smith, J. S. C., and J. G. K. Williams. 1994. Arbitrary primer mediated fingerprinting in plants: Case studies in plant breeding, taxonomy and phylogeny. In Molecular Ecology and Evolution: Approaches and Applications, eds. B. Schierwater, B. Streit, G. P. Wagner, and R. DeSalle, pp. 5–15. Birkhäuser Verlag, Basel, Switzerland.

Sneath, P. H. A., and R. R. Sokal. 1973. Numerical Taxonomy. Freeman, San Francisco.

Sobral, B. W. S., and R. J. Honeycutt. 1994. Genetics, plants, and the polymerase chain reaction. In The Polymerase Chain Reaction, eds. K. B. Mullis, F. Ferre, and R. A. Gibbs, pp. 304–319. Birkhäuser, Boston, Massachusetts.

Soltis, D. E., P. S. Soltis, R. K. Kuzoff, and T. L. Tucker. 1992. Geographic structuring of chloroplast DNA genotypes in *Tiarella trifoliata* (Saxifragaceae). Plant Systematics and Evolution **181:**203–216.

Soltis, D. E., M. A. Gitzendanner, D. D. Strenge, and P. S. Soltis. 1997. Chloroplast DNA intraspecific phylogeography of plants from the Pacific Northwest of North America. Plant Systematics and Evolution **206:**353–373.

Soltis, P. S., D. E. Soltis, and J. J. Doyle, eds. 1992a. Molecular Systematics of Plants. Chapman & Hall, New York.

Soltis, P. S., D. E. Soltis, and C. J. Smiley. 1992b. An *rbc*L sequence from a Miocene *Taxodium* (bald cypress). Proceedings of the National Academy of Sciences U.S.A. **89:**449–451.

Song, K., and T. C. Osborn. 1994. A method for examining expression of homologous genes in plant polyploids. Plant Molecular Biology **26:**1065–1071.

Stammers, M., J. Harris, G. M. Evans, M. D. Hayward, and J. W. Forster. 1995. Use of random PCR (RAPD) technology to analyze phylogenetic relationships in the *Lolium/Festuca* complex. Heredity **74:**19–27.

Stewart, C. N., and D. M. Porter. 1995. RAPD profiling in biological conservation: An application to estimating clonal variation in rare and endangered *Iliamna* in Virginia. Biological Conservation **74:**135–142.

Stiller, J. W., and A. L. Denton. 1995. One hundred years of *Spartina alterniflora* (Poaceae) in Willapa Bay, Washington: random amplified polymorphic DNA analysis of an invasive population. Molecular Ecology **4:**355–363.

Strand, A. E., B. G. Milligan, and C. M. Pruitt. 1996. Are populations islands? Analysis of chloroplast DNA variation in *Aquilegia.* Evolution **50:**1822–1829.

Strand, A. E., J. Leebens-Mack, and B. G. Milligan. 1997. Nuclear DNA-based markers for plant evolutionary biology. Molecular Ecology **6:**113–118.

Swensen, S. M., G. J. Allan, M. Howe, W. J. Elisens, S. A. Junak, and L. H. Rieseberg. 1995. Genetic analysis of the endangered island endemic *Malacothamnus fasciculatus* (Nutt.)Greene var. *nesioticus* (Rob.) Kearn. (Malvaceae). Conservation Biology **9:**404–415.

Swerdlow, H., and R. Gesteland. 1990. Capillary gel electrophoresis for rapid high resolution DNA sequencing. Nucleic Acids Research **18:**1415–1420.

Swofford, D. L., G. J. Olsen, P. J. Waddell, and D. M. Hillis. 1996. Phylogenetic inference. In Molecular Systematics, second edition, eds. D. M. Hillis, B. K. Mable, and C. Mortiz, pp. 407–514. Sinauer Associates, Sunderland, Massachusetts.

Taberlet, P., L. Gielly, G. Pautou, and J. Bouvet. 1991. Universal primers for amplification of three non-coding regions of chloroplast DNA. Plant Molecular Biology Reporter **17:**1105–1109.

Tao, Y., J. M. Manners, M. M. Ludlow, and R. G. Henzell. 1993. DNA polymorphisms in grain sorghum (*Sorghum bicolor* (L.) Moench). Theoretical and Applied Genetics **86:**679–688.

Tautz, D. 1989. Hypervariability of simple-sequences as a general source for polymorphic DNA markers. Nucleic Acids Research **17:**6463–6471.

Tautz, D., and M. Renz. 1984. Simple sequences are ubiquitous repetitive components of eukaryotic genomes. Nucleic Acids Research **12:**4127–4138.

Tautz, D., M. Trick, and G. A. Dover. 1986. Cryptic simplicity in DNA is a major source of genetic variation. Nature **322:**652–656.

Terauchi, R., and A. Konuma. 1994. Microsatellite polymorphism in *Dioscorea tokoro,* a wild yam species. Genome **37:**794–801.

Thomas, M. R., and N. S. Scott. 1993. Microsatellite repeats in grapevine reveal DNA polymorphisms when analysed as sequence-tagged sites (STSs). Theoretical and Applied Genetics **86:**985–990.

Thormann, C. E., M. E. Ferreira, L. E. A. Camargo, J. G. Tivang, and T. C. Osborn. 1994. Comparison of RFLP and RAPD markers to estimating genetic relationships within and among cruciferous species. Theoretical and Applied Genetics **88:**973–980.

To, K.-Y., C.-I. Liu, S.-T. Liu, and Y.-S. Chang. 1993. Detection of point mutations in the chloroplast genome by single-stranded conformation polymorphism analysis. Plant Journal **3:**183–186.

Tracy, T. E., and L. S. Mulcahy. 1991. A simple method for direct automated sequencing of PCR fragments. BioTechniques **11:**74–75.

Travis, S. E., J. Maschinski, and P. Keim. 1996. An analysis of genetic variation in *Astragalus cremnophylax,* a critically-endangered plant, using AFLP markers. Molecular Ecology **5:**735–745.

Tsumura, Y., K. Yoshimura, N. Tomaru, and K. Ohba. 1995. Molecular phylogeny of conifers using RFLP analysis of PCR-amplified specific chloroplast genes. Theoretical and Applied Genetics **91:**1222–1236.

Tsumura, Y., K. Ohba, and S. H. Strauss. 1996a. Diversity and inheritance of inter-simple sequence repeat polymorphisms in Douglas-fir (*Pseudotsuga menziesii*) and sugi (*Cryptomeria japonica*). Theoretical and Applied Genetics **92**:40–45.

Tsumura, Y., T. Kawahara, R. Wickneswari, and K. Yoshimura. 1996b. Molecular phylogeny of Dipterocarpaceae in Southeast Asia using RFLP of PCR-amplified chloroplast genes. Theoretical and Applied Genetics **93**:22–29.

Valdes, A. M., M. Slatkin, and N. B. Freimer. 1993. Allele frequencies at microsatellite loci: the stepwise mutation model revisited. Genetics **133**:737–749.

Van Buren, R., K. T. Harper, W. R. Andersen, D. J. Stanton, S. Seyoum, and J. L. England. 1994. Evaluating the relationship of autumn buttercup (*Ranunculus acriformis* var. *aestivalis*) to some close congeners using random amplified polymorphic DNA. American Journal of Botany **81**:514–519.

Van Coppenolle, B., I. Watanabe, C. Van Hove, G. Second, N. Huang, and S. R. McCouch. 1993. Genetic diversity and phylogeny analysis of *Azolla* based on DNA amplification by arbitrary primers. Genome **36**:686–693.

van de Zande, L., and R. Bijlsma. 1995. Limitations of the RAPD technique in phylogeny reconstruction in *Drosophila*. Journal of Evolutionary Biology **8**:645–656.

van Ham, R. C. H. J., H. 't Hart, T. H. M. Mes, and J. M. Sandbrick. 1994. Molecular evolution of noncoding regions of the chloroplast genome in the Crassulaceae and related species. Current Genetics **25**:558–566.

Van Heusden, A. W., and K. Bachmann. 1992a. Genetic differentiation of *Microseris pygmaea* (Asteraceae, Lactuceae) studied with DNA amplification from arbitrary primers (RAPDs). Acta Botanica Neerlandica **41**:385–395.

Van Heusden, A. W., and K. Bachmann. 1992b. Genotype relationships in *Microseris elegans* (Asteraceae, Lactuceae) revealed by DNA amplification from arbitrary primers (RAPDS). Plant Systematics and Evolution **179**:221–233.

Van Heusden, A. W., and K. Bachmann. 1992c. Nuclear DNA polymorphisms among strains of *Microseris bigelovii* (Asteraceae: Lactuceae) amplified from arbitrary primers. Botanica Acta **105**:331–336.

van Pijlen, I. A., B. Amos, and T. Burke. 1995. Patterns of genetic variability at individual minisatellite loci in minke whale *Balaenoptera acutorostrata* populations from three different oceans. Molecular Biology and Evolution **12**:459–472.

Venugopal, G., S. Mohapatra, D. Salo, and S. Mohapatra. 1993. Multiple mismatch annealing: basis for random amplified DNA polymorphic fingerprinting. Biochemical and Biophysical Research Communications **197**:1382–1387.

Vicario, F., G. G. Vendramin, P. Rossi, P. Liò, and R. Giannini. 1995. Allozyme, chloroplast DNA and RAPD markers for determining genetic relationships between *Abies alba* and the relic population of *Abies nebrodensis*. Theoretical and Applied Genetics **90**:1012–1018.

Vos, P., R. Hogers, M. Bleeker, M. Reijans, T. van de Lee, M. Hornes, A. Frijiters, J. Pot, J. Peleman, M. Kuiper, and M. Zabeau. 1995. AFLP: a new technique for DNA fingerprinting. Nucleic Acids Research **23**:4407–4414.

Wachira, F. N., R. Waugh, C. A. Hackett, and W. Powell. 1995. Detection of genetic diversity in tea (*Camellia sinensis*) using RAPD markers. Genome **38**:201–210.

Wakasugi, T., J. Tsudzuki, S. Ito, K. Nakashima, T. Tsudzuki, and M. Sugiura. 1994. Loss of all *ndh* genes as determined by sequencing the entire chloroplast genome of the black pine *Pinus thunbergii*. Proceedings of the National Academy of Sciences U.S.A. **91**:9794–9798.

Wallace, R. S., and J. H. Cota. 1996. An intron loss in the chloroplast gene *rpo*C1 supports a monophyletic origin for the subfamily Cactoideae of the Cactaceae. Current Genetics **29**:275–281.

Wang, X.-R., A. E. Szmidt, and M.-Z. Lu. 1996. Genetic evidence for the presence of cytoplasmic DNA in pollen and megagametophytes and maternal inheritance of mitochondrial DNA in *Pinus*. Forest Genetics **3**:37–44.

Wang, Z., J. L. Weber, G. Zhong, and S. D. Tanksley. 1994. Survey of plant short tandem repeats. Theoretical and Applied Genetics **88**:1–6.

Watano, Y., M. Imazu, and T. Shimuzu. 1995. Chloroplast DNA typing by PCR-SSCP in the *Pinus pumila-P. parviflora* var. *pentaphylla* complex (Pinaceae). Journal of Plant Research **108**:493–499.

Waugh, R., and W. Powell. 1992. Using RAPD markers for crop improvement. Trends in Biotechnology **10**:186–191.

Wehling, P., B. Hackauf, and G. Wricke. 1994. Identification of S-locus linked PCR fragments in rye (*Secale cereale* L.) by denaturing gradient gel electrophoresis. Plant Journal **5**:891–893.

Weising, K., F. Weigand, A. J. Driesel, G. Kahl, H. Zischler, and J. T. Epplen. 1989. Polymorphic simple GATA/GACA repeats in plant genomes. Nucleic Acids Research **17**:10128.

Weising, K., H. Nybom, K. Wolff, and W. Meyer. 1995. DNA Fingerprinting in Plants and Fungi. CRC Press, Boca Raton, Florida.

Welsh, J., and M. McClelland. 1990. Fingerprinting genomes using PCR with arbitrary primers. Nucleic Acids Research **18**:7213–17218.

Welsh, J., and M. McClelland. 1991. Genomic fingerprints produced by PCR with consensus transfer RNA gene primers. Nucleic Acids Research **19**:861–866.

Whitkus, R., J. Doebley, and J. F. Wendel. 1994. Nuclear DNA markers in systematics and evolution. In DNA-based Markers in Plants, eds. R. L. Phillips and I. K. Vasil, pp. 116–141. Kluwer Academic Publishers, Netherlands.

Whitton, J., L. H. Rieseberg, and M. C. Ungerer. 1997. Microsatellite loci are not conserved across Asteraceae. Molecular Biology and Evolution **14**:204–209.

Whitty, P. W., W. Powell, and J. I. Sprent. 1994. Molecular separation of genera in Cassiinae (Leguminosae), and analysis of variation in the nodulating species of *Chamaecrista*. Molecular Ecology 3:507–515.

Widjojoatmodjo, M. N., A. C. Fluit, and J. Verhoef. 1994. Rapid identification of bacteria by PCR-single-strand conformation polymorphism. Journal of Clinical Microbiology **32**:3002–3007.

Wilkie, S. E., P. G. Isaac, and R. J. Slater. 1993. Random amplified polymorphic DNA (RAPD) markers for genetic analysis in *Allium*. Theoretical and Applied Genetics **86**:497–504.

Williams, C. E., and D. A. St. Clair. 1993. Phenetic relationships and levels of variability detected by restriction fragment length polymorphism and random amplified polymorphic DNA analysis of cultivated and wild accessions of *Lycopersicon esculentum*. Genome **36**:619–630.

Williams, J. G. K., A. R. Kubelik, K. J. Livak, J. A. Rafalski, and S. V. Tingey. 1990. DNA polymorphisms amplified by arbitrary primers are useful as genetic markers. Nucleic Acids Research **18**:6531–6535.

Williams, M. N. V., N. Pande, S. Nair, M. Mohan, and J. Bennett. 1991. Restriction fragment length polymorphism analysis of polymerase chain reaction products amplified from mapped loci of rice *Oryza sativa* L. genomic DNA. Theoretical and Applied Genetics **82**:489–498.

Williams, J. G. K., M. K. Hanafey, J. A. Rafalski, and S. V. Tingey. 1993. Genetic analysis using random amplified polymorphic DNA markers. Methods in Enzymology **218**:704–740.

Wolfe, A. D., and W. J. Elisens. 1993. Diploid hybrid speciation in *Penstemon* (Scrophulariaceae) revisited. American Journal of Botany **80**:1082–1094.

Wolfe, A. D., and W. J. Elisens. 1994. Nuclear ribosomal DNA restriction-site variation in *Penstemon* section *Peltanthera* (Scrophulariaceae): an evaluation of diploid hybrid speciation and evidence for introgression. American Journal of Botany **81**:1627–1635.

Wolfe, A. D., and W. J. Elisens. 1995. Evidence of chloroplast capture and pollen-mediated gene flow in *Penstemon* sect. *Peltanthera* (Scrophulariaceae). Systematic Botany **20**:395–412.

Wolfe, A. D., W. J. Elisens, L. E. Watson, and C. W. dePamphilis. 1997. Using restriction-site variation of PCR-amplified cpDNA genes for phylogenetic analysis: a case study of Cheloneae (Scrophulariaceae). American Journal of Botany **84**:555–564.

Wolfe, K. H., C. W. Morden, and J. D. Palmer. 1992. Function and evolution of a minimal plastid genome from a nonphotosynthetic parasitic plant. Proceedings of the National Academy Sciences U. S. A. **89**:10648–10652.

Wolff, K., E. Zietkiewicz, and H. Hofstra. 1995. Identification of chrysanthemum cultivars and stability of DNA fingerprint patterns. Theoretical and Applied Genetics **91**:439–447.

Wu, K.-S., and S. D. Tanksley. 1993. Abundance, polymorphism and genetic mapping of microsatellites in rice. Molecular and General Genetics **241**:225–235.

Wu, K., R. Jones, L. Danneberger, and P. A. Scolnik. 1994. Detection of microsatellite polymorphisms without cloning. Nucleic Acids Research **22**:3257–3258.

Xu, H., D. J. Wilson, S. Arulsekar, and A. T. Bakalinsky. 1995. Sequence-specific polymerase chain-reaction markers derived from randomly amplified polymorphic DNA markers for fingerprinting grape (*Vitis*) rootstocks. Journal American Society Horticultural Science **120**:714–720.

Yang, X., and C. Quiros. 1993. Identification and classification of celery cultivars with RAPD markers. Theoretical and Applied Genetics **86**:205–212.

Yang, G. P., M. A. Saghai Maroof, C. G. Xu, Q. Zhang, and R. M. Biyashev. 1994. Comparative analysis of microsatellite DNA polymorphism in landraces and cultivars of rice. Molecular and General Genetics **245**:187–194.

Yu, J., W. K. Gu, R. Provvidenti, and N. F. Weeden. 1995. Identifying and mapping two DNA markers linked to the gene conferring resistance to pea enation mosaic virus. Journal American Society Horticultural Science **120**:730–733.

Zhao, X., and G. Kochert. 1993. Phylogenetic distribution and genetic mapping of a GGCn microsatellite from rice *Oryza sativa* L. Plant Molecular Biology **21**:607–614.

Zietkiewicz, E., A. Rafalski, and D. Labuda. 1994. Genome fingerprinting by simple sequence repeat (SSR)-anchored polymerase chain reaction amplification. Genomics **20**:176–183.

Zurawski, G., B. Perrot, W. Bottomley, and P. R. Whitfeld. 1981. The structure of the gene for the large subunit of ribulose-1,5-bisphosphate carboxylase from spinach chloroplast DNA. Nucleic Acids Research **9**:3251–3270.

Zurawski, G., M. T. Clegg, and A. H. D. Brown. 1984. The nature of nucleotide sequence divergence between barley and maize chloroplast DNA. Genetics **106**:735–749.

3

Comparative Utility of Chloroplast DNA Restriction Site and DNA Sequence Data for Phylogenetic Studies in Plants

Robert K. Jansen, James L. Wee, and David Millie

Restriction enzyme comparisons of plant genomes have been used widely to address systematic and evolutionary questions during the past 15 years. Most studies have examined variation in chloroplast DNA (cpDNA) or nuclear ribosomal DNA (rDNA) to estimate phylogenetic relationships among species, genera, and in a few cases families of plants (e.g., Jorgensen and Cluster, 1988; Palmer et al., 1988; Olmstead and Palmer, 1994). The restriction site approach also has been used to examine intraspecific variation and genetic diversity among populations (e.g., Neale et al., 1986; Soltis et al., 1989, 1991a, 1997; Whittemore and Schaal, 1991; Fenster and Ritland, 1992; Kim et al., 1992a; Hong et al., 1993; Petit et al., 1993; Byrne and Moran, 1994; Dong and Wagner, 1994; McCauley, 1994; Mason-Gamer et al., 1995; Bain and Jansen, 1996). Restriction site and DNA sequence data have three characteristics that make them especially useful for phylogenetic analyses (Holsinger and Jansen, 1993). First, discrete character states can be scored unambiguously. Second, a large number of characters potentially can be obtained for each taxon. Third, both types of data provide valuable information on the extent and nature of divergence between sequences. None of these features are present in earlier DNA methods, such as DNA–DNA hybridization.

In spite of the widespread acceptance of the utility of restriction site comparisons for systematic and evolutionary studies of plants, there has been a recent decrease in the use of this approach and a very rapid increase in the use of DNA sequence data (Fig. 3.1). Clearly DNA sequencing is now the method preferred by most plant systematists (see Chapter 1). This can be attributed, at least in part, to the fact that more rapid and inexpensive DNA sequencing methods have been developed recently, including the use of polymerase chain reaction (PCR) to amplify rather than clone DNA to be sequenced,

We thank the following for providing copies of published or unpublished data sets: D. Baum, B. Bremer, S. Downie, T. Givnish, S. Graham, S. Hoot, T. Kellogg, R. Mason-Gamer, S. Okane, D. Olmstead, J. Panero, E. Schilling, D. Soltis, P. Soltis, K. Sytsma, L. Urbatsch, J. Wendel, and Q. Xiang. We also thank J. Bain, U. Boehle, J. Clement, K. Hansen, R. Mason-Gamer, J. Panero, and members of the systematics discussion group at The University of Texas for critical comments on an earlier version of the manuscript. Finally, one of us (RKJ) would like to thank all current and past members of his lab for the many stimulating discussions on the relative merits of cpDNA restriction site and DNA sequencing approaches. This work was supported by NSF grants to RKJ (DEB 9318278) and JLW (DEB 9408283).

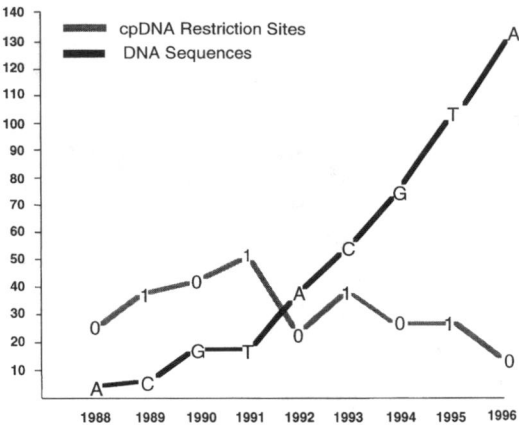

Figure 3.1. Comparison of numbers of restriction site and DNA sequencing papers presented at the annual meetings of the Systematics Section of the Botanical Society of America between 1988 and 1996. The number of papers was determined from the published abstracts for these meetings, which were published as a supplement to the *American Journal of Botany*.

and the increased availability of automated DNA sequencers.

A critical question is whether DNA sequence data are really more useful than restriction site data for phylogenetic comparisons. One study has demonstrated that restriction site data are more likely to produce the correct phylogeny than DNA sequences of four genes (Hillis et al., 1994). Much of this success of restriction sites can be attributed to the fact that in constructing trees it is preferable to use sites scattered throughout a genome rather than contiguous sites (Hillis et al., 1994). In this paper we explore the relative utility of DNA sequencing and cpDNA restriction site analysis for systematic and evolutionary studies in plants. Our contention is that restriction site analysis remains a very powerful technique that in certain cases is preferable to DNA sequences, and thus its use in plants should not be abandoned. The basis of this conclusion is a rigorous comparison of cpDNA restriction site and DNA sequence data for 33 groups of taxa. As noted in the section below (under Restriction Site Methods), the utility of the restriction site approach is dependent on how carefully the data are generated, interpreted, and analyzed. Thus, we first discuss in

some detail the two primary methods for performing comparative restriction site analyses.

RESTRICTION SITE METHODS

Southern Hybridization

The most widely used approach for restriction site comparisons is based on Southern hybridization methods, which involve digestion of DNAs with restriction endonucleases, size separation of DNA fragments via agarose gel electrophoresis, transfer of DNA fragments to a filter membrane, and hybridization of labeled probes to the filters. Several modifications have been developed to make this approach cheaper and faster (see Palmer et al., 1988 for a detailed description). These modifications include performing bidirectional transfer (Smith and Summers, 1980) of DNA fragments to generate two duplicate filters from a single gel, hybridization of many filters either in large buckets that accommodate up to 50 filters or in one of the commercially available hybridization ovens, the use of a modified gel apparatus that accommodates up to 40 samples, and the use of reusable nylon membrane which permits reprobing of filters as many as 20 times.

The Southern hybridization approach has been used in a wide diversity of plant groups and at a wide range of taxonomic levels, from population studies (e.g., Neale et al., 1986; McCauley, 1994; Dong and Wagner, 1994; Mason-Gamer et al., 1995; Bain and Jansen, 1996) to phylogenetic comparisons among genera (e.g., Soreng et al., 1990; Ranker et al., 1990; Bremer and Jansen, 1991; Jansen et al., 1991; Olmstead and Palmer, 1992; Sytsma and Smith, 1992; Johannson and Jansen, 1993; Soltis et al., 1993) and families (Downie and Palmer, 1992, 1994; Manos et al., 1993). The availability of clone banks (Palmer, 1986; Jansen and Palmer, 1987a; Chase and Palmer, 1989; Olmstead and Palmer, 1992) and complete sequences of the chloroplast genomes from a wide diversity of plants including a bryophyte (Ohyama et al., 1986), a conifer (Wakasugi et al., 1994), and three flowering plants (rice, Hiratsuka et al., 1989; tobacco, Shinozaki et al., 1986; beech drops, Wolfe et al., 1992) have provided an important resource for cpDNA restriction site comparisons. The complete sequences can assist in deciding which restriction enzymes are most appropriate based on the number of times they cut chloroplast genomes. At higher taxonomic levels enzymes that cut infrequently or with moderate frequency would be more appropriate, whereas at the population level, frequent-cutting enzymes, including those with 4-base pair (bp) recognition sequences, are often required to provide sufficient variation.

PCR Approaches

The second method uses PCR to amplify specific regions of DNA, followed by digestion of the resulting fragments with restriction enzymes and comparisons of fragment patterns by agarose gel electrophoresis (e.g., Liston, 1992; Rieseberg et al., 1992; Arnold, 1993; Cruzan and Arnold, 1993; Cruzan et al., 1993; McCauley, 1994). The PCR approach has the advantage of requiring very small quantities of plant material, such as DNA isolated from herbarium specimens or fossils (Golenberg et al., 1990; Soltis et al., 1992; Thomas and Paabo, 1993; Loockerman and Jansen, 1996). Furthermore, many specific chloroplast primers have been developed that are useful for a wide diversity of plant taxa (Arnold et al., 1991; Taberlet et al., 1991; Liston, 1992; Demesure et al., 1995). In spite of this advantage, restriction site analysis of PCR fragments is of limited phylogenetic utility because it examines such a small proportion of the DNA region or genome being studied and it detects so few characters. However, this approach is very useful for population-level studies or for examining hybridization and introgression (Rieseberg et al., 1992; Arnold, 1993; Cruzan and Arnold, 1993; Cruzan et al., 1993).

Importance of Mapping Restriction Sites

There are two primary methods of interpreting restriction enzyme data: they can be characterized by the presence or absence of either fragments or of sites. Restriction fragment differences can result from either nucleotide substitutions or insertions and deletions (indels).

Mapping sites enables the two types of changes to be distinguished from one another. If one is comparing very closely related taxa with a low level of sequence divergence (ca. <1.0 %), it is possible to interpret accurately restriction site changes by simply inspecting fragment patterns. At higher levels of divergence, however, it is essential to map both restriction site changes and indels to make an accurate determination of character homology for phylogenetic comparisons. Bremer (1991) suggested that mapping and aligning restriction sites may not be worth the time and effort, but recent experimental evidence from the T7 bacteriophage studies indicated that mapped sites generated the correct phylogeny whereas fragment data did not (Hillis et al., 1994). Thus, although restriction site mapping is clearly more time consuming than simply interpreting fragment data, it must be done in comparisons of genomes with higher levels of sequence divergence (ca. $>1.0\%$) to obtain correct gene trees.

COMPARISONS OF THE PHYLOGENETIC UTILITY OF CPDNA RESTRICTION SITES AND GENE SEQUENCES

Data Sets Examined

We gathered 33 pairs of data sets from 15 laboratories involved in molecular phylogenetic studies of plants (Table 3.1). Requests were made to all known plant molecular systematists in 1995 and all data sets received that included both cpDNA restriction site and DNA sequence data for the same set of taxa were included in the comparisons. The number of taxa ranged from eight to 39 with an average of 20 and the number of enzymes used ranged from eight to 34 with an average of 16.

Phylogenetic Comparisons

Phylogenetic analyses employed unweighted parsimony using PAUP (version 3.1.1, Swofford, 1993). All analyses used ACCTRAN, MULPARS, TREE BISECTION RECONNECTION (TBR) and 100 random addition sequences of taxa. The 100 bootstrap (Felsenstein

Table 3.1. List of data sets examined for phylogenetic utility of restriction site and DNA sequences.

Taxon	Number of Taxa	Number of Enzymes	DNA Markers	Reference
Argyranthemum (Asteraceae)	20	21	Rest. site, ITS[a]	Francisco-Ortega et al., 1995, 1996, 1997
Asteraceae	23	10	Rest. site, *rbcL*[b]	Jansen et al., 1991; Kim et al., 1992b
Asteraceae	23	10	Rest. site, *ndhF*[b]	Jansen et al., 1991; Kim and Jansen, 1995
Berberidaceae	10	10	Rest. site, *rbcL*[b]	Kim and Jansen, 1996, 1997
Brocchinia (Bromeliaceae)	29	30	Rest site, cpDNA Intergenic spacer	Givnish et al., 1997
Caryophyllales	19	10	Rest. site, ORF2280[b]	Downie and Palmer, 1994, Downie et al., 1997
Clarkia (Onagraceae)	14	18	Rest. site, ITS[a]	Sytsma et al., 1990; W. Hahn et al. unpubl.
Cornus (Cornaceae)	11	21	Rest. site, *matK*[b]	Xiang et al., 1996
Cornus (Cornaceae)	9	21	Rest. site, *rbcL*[b]	Xiang et al., 1993, 1996
Epilobium (Onagraceae)	10	10	Rest. site, ITS[a]	K. Sytsma and J. Smith, unpubl.; Baum et al., 1994
Galinsoginae (Asteraceae)	14	16	Rest. site, ITS[a]	J. Panero, unpubl.
Gossypium (Malvaceae)	39	25	Rest. site, ITS[a]	Wendel and Albert, 1992; Wendel et al., 1995
Helianthinae (Asteraceae)	24	16	Rest. site, ITS[a]	Schilling and Jansen, 1989; Schilling and Panero, 1996
Heuchera (Saxifragaceae)	31	22	Rest. site, ITS[a]	Soltis et al., 1991b; Soltis and Kuzoff, 1995
Krigia (Asteraceae)	16	34	Rest. site, ITS[a]	Kim et al., 1992c; Kim and Jansen, 1994
Lactuceae (Asteraceae)	17	17	Rest. site, *ndhF*[b]	Whitton et al., 1995; E. Korompei and R. Jansen, unpubl.
Lopezia (Onagraceae)	19	26	Rest. site, ITS[a]	S. O'Kane, unpubl.
Montanoa (Asteraceae)	23	20	Rest. site, ITS[a]	J. Panero et al., unpubl.
Onagraceae	8	10	Rest. site, *rbcL*[b]	K. Systma et al., unpubl.; D. Baum et al., unpubl.
Onagraceae	13	10	Rest. site, ITS[a]	K. Systma et al., unpubl.; Baum et al. 1994, unpubl.
Pontederiaceae	24	10	Rest. site, *rbcL*[b]	Kohn et al., 1996; S.S. Graham et al., unpubl.
Pontederiaceae	24	10	Rest. site, *ndhF*[b]	Kohn et al., 1996; S. Graham et al., unpubl.
Ranunculaceae	21	11	Rest. site, *rbcL*[b]	Johannson and Jansen, 1993; Hoot, 1995
Ranunculaceae	21	11	Rest. site, *atpB*[b]	Johannson and Jansen, 1993; Hoot, 1995
Rubiaceae	33	8	Rest. site, *rbcL*[b]	Bremer and Jansen, 1991; Bremer et al., 1995
Rudbeckinae (Asteraceae)	39	22	Rest. site, ITS[a]	Urbatsch and Jansen, 1995; L. Urbatsch et al., unpubl.
Saxifragaceae	20	23	Rest. site, *matK*[b]	Soltis et al., 1993, 1996; Johnson and Soltis, 1995
Saxifragaceae	20	23	Rest. site, *rbcL*[b]	Soltis et al., 1993, 1996; Johnson and Soltis, 1995
Solanaceae	18	10	Rest. site, *rbcL*[b]	Olmstead and Sweere, 1994
Solanaceae	18	10	Rest. site, *ndhF*[b]	Olmstead and Sweere, 1994
Triticeae (Poaceae)	15	14	Rest. site, 5S-long	Mason-Gamer and Kellogg, 1995; Kellogg and Appels, 1995
Triticeae (Poaceae)	22	14	Rest. site, 5S-short	Mason-Gamer and Kellogg, 1995; Kellogg and Appels, 1995
Triticeae (Poaceae)	25	14	Rest. site, ITS[a]	Mason-Gamer and Kellogg, 1995; Hsiao et al. 1995

[a]Indicates the 13 pairs of ITS sequences/cpDNA restriction site comparisons.
[b]Indicates the 17 pairs of chloroplast gene sequences/cpDNA restriction site comparisons (see Table 3.2).

1985, Sanderson, 1989, 1995; Hillis and Bull, 1993) replicates used the same options except that only a single random addition sequence was performed. Nine parameters were calculated for each of the 66 data sets, including: (1) number of variable positions, (2) number of shortest trees, (3) number of phylogenetically informative characters, (4) number of resolved nodes in the strict consensus tree, (5) number of base pairs examined, (6) consistency index (CI, excluding autapomorphies, Kluge and Farris, 1969), (7) retention index (RI, Farris, 1989), (8) the average bootstrap value of all resolved nodes on the strict consensus tree, and (9) the G_1 value (Hillis, 1991). These parameters were selected because they are used widely in cladistic parsimony analyses to measure homoplasy (Archie, 1996, Sanderson and Donoghue, 1996) and to assess confidence in a phylogenetic tree (Goloboff, 1991a, 1991b; Archie, 1996; Sanderson and Donoghue, 1996). Although it has been shown that the CI is greatly affected by the number of taxa (Sanderson and Donoghue, 1989, 1996), this effect was not a problem in our comparisons because the cpDNA restriction site and DNA sequence data sets for each of the 33 groups contained the same number of taxa. All nine parameters were calculated by PAUP except for the number of base pairs examined and average bootstrap value. For sequence data, the number of base pairs examined was simply the length of the aligned sequences. For restriction site data the value was calculated by multiplying the number of restriction sites examined by the number of base pairs in the recognition site for each enzyme. The number of restriction sites was provided either by the investigators (Table 3.1), or if it was not available, the number of times the enzymes cut the sequenced tobacco chloroplast genome was used. Calculating the number of restriction sites in this manner is likely to be precise for enzymes that have a 6-bp recognition site (6-bp enzymes). However, for those enzymes that have a 4-bp recognition site (4-bp enzymes) our estimates are certainly high because it is not possible to observe many of the smaller restriction fragments (<100 bp) on gel systems used in most restriction site comparisons. Fortunately, we only included one data set (*Argyranthemum*,

Francisco-Ortega et al., 1996) that used any enzymes with 4-bp recognition sites (only seven of the 21 enzymes examined). Thus, overall we believe that our estimates of the number of base pairs examined in the restriction site comparisons are accurate. Average bootstrap values were calculated by adding the values for all nodes on the strict consensus tree and dividing this number by the number of resolved nodes. The G_1 values were calculated using 100,000 random trees.

Statistical Analyses

Statistical means were compared for all nine parameters using a two sample Student's t-test assuming unequal variance among means (Snedecor and Cochran, 1980). The 66 data sets (33 cpDNA restriction site and 33 DNA sequence) were compared in three groups: (1) all 33 pairs; (2) the 13 pairs for which there were both cpDNA restriction site data and ITS sequences (indicated by *a* in Table 3.1); and (3) the 17 pairs for which there were both cpDNA restriction site data and sequences from a chloroplast gene (indicated by *b* in Table 3.1). For each comparison the mean, variance, t statistic (paired-two sample for means), and significance values were calculated.

Results and Discussion

Statistical comparisons of all 66 data sets indicated that cpDNA restriction site data are more useful for phylogenetic comparisons than DNA sequence data (Table 3.2). For eight of the nine parameters examined (all except the G_1 value, which was the same) the restriction site data had better values than the DNA sequence data and five of these were significant at the 0.05 level (see *a* in Table 3.2). Two values, the retention index and bootstrap average, were particularly notable because they indicated that restriction site data were significantly less homoplastic and provided stronger support for monophyletic groups. The other three parameters that were significantly better, number of variable and informative positions and number of base pairs examined, were not surprising because it was evident that the restriction site

Table 3.2. Statistical comparisons of restriction site and DNA sequence data sets (see Table 3.1 for list of data sets).

Parameter	Rest. Site (all 33 comparisons)	Sequence	Rest. Site (13 ITS data sets)	Sequence	Rest Site (17 cp gene sequences)	Sequence
Variable Positions	270.24	221.55	192.46	198.92	340.65	236.06
	26348.63	13737.00	13055.77	9594.58	26110.87	16767.06
	1.77		−0.29		2.95	
	0.04[a]		0.39		0.0047[a]	
Number of Shortest Trees	10.30	199.18	7.69	79.38	12.76	15.65
	152.47	835841.65	84.23	28117.59	225.69	424.74
	−1.19		−1.57		−0.53	
	0.12		0.07		0.30	
Number of Informative Positions	147.15	104.03	114.92	111.31	177.94	96
	8710.45	4028.84	5337.74	4945.88	9208.93	3503.75
	2.45		0.19		3.76	
	0.009[a]		0.42		0.0009[a]	
Number of Resolved Nodes	14.03	13.18	14.92	14.23	13.24	12.35
	28.97	38.15	26.24	58.19	28.57	29.87
	1.47		0.63		1.54	
	0.07		0.27		0.07	
Number of Base Pairs Examined	3955.15	1014.18	5276	605.38	3264.7	1372
	8152869.76	226461.22	14816401	8397.76	2653563.97	146227.63
	5.64		4.34		4.87	
	0.00000015[a]		0.0005[a]		0.00008[a]	
Consistency Index	0.67	0.64	0.77	0.67	0.56	0.59
	0.03	0.01	0.02	0.01	0.02	0.005
	1.23		3.06		−.96	
	0.11		0.005[a]		0.18	
Retention Index	0.77	0.68	0.88	0.73	0.66	0.63
	0.03	0.02	0.009	0.013	0.03	0.02
	4.13		6.44		1.02	
	0.0001[a]		0.000016[a]		0.16	
Average Bootstrap Value	79.09	71.70	79.15	74.31	78.05	72
	143.71	173.97	211.97	167.88	119.3	169.34
	2.87		1.38		1.66	
	0.003[a]		0.097		0.058	
G_1 Value	0.90	0.90	0.88	1.01	0.93	0.86
	0.11	0.20	0.13	0.44	0.12	0.05
	−0.04		−0.88		1.13	
	1.69		0.20		0.14	

Note: The following four values are listed under each of the nine parameters: mean (1st row), variance (2nd row), t statistic (3rd row), and significance value (4th row).
[a]Indicates values that were significantly different at the 0.05 level or below.

approach samples more sequence variation than DNA sequences of any single gene or DNA region.

We also partitioned the data sets into two categories. One of these compared the 13 data sets for which both cpDNA restriction site and ITS sequence data were available (indicated by *a* in Table 3.1). Several studies have demonstrated that levels of sequence divergence are higher in the ITS region than in the chloroplast genome (e.g., Baldwin, 1992; Sang et al., 1994, 1995;

Baldwin et al., 1995). This has prompted some to suggest that at lower taxonomic levels ITS sequences may be preferable to cpDNA restriction site comparisons because ITS sequences will provide more variation (Sang et al., 1994, 1995; Baldwin et al., 1995). Our comparisons do not support this contention because the cpDNA restriction site data performed better for eight of the nine parameters examined (Table 3.2). However, in this case, only three of the parameters, number of base pairs examined, CI, and RI, were significantly better at the 0.05 level. The significantly higher CI and RI values for cpDNA restriction site data indicated that there was less homoplasy in these data than in the ITS sequences. This could be due to the highly conserved nature of cpDNA and the uncertainty of aligning ITS sequences (see Chapter 4), especially in comparisons of more divergent taxa that contain many indels.

The second group of data sets compared the utility of restriction site data with sequences of several chloroplast genes, including *atpB, matK, ndhF, ORF2280,* and *rbcL.* The values for cpDNA restriction site data again were better for eight of the nine parameters examined (all except the CI). In this case, however, the differences were only significant at the 0.05 level for three parameters, number of variable and informative positions, and number of base pairs examined. Thus, although the cpDNA restriction site approach does generate significantly more characters than sequencing any single chloroplast gene, levels of homoplasy for the two types of data are very similar (Table 3.2). This may reflect the fact that most of the data sets compared were from studies at higher taxonomic levels (generic level and above) where there is more uncertainty about homology of mapped restriction sites due to size variation of the chloroplast genomes. Or, alternatively, because the genes included in the sequencing studies are from the same genome used for the restriction site analyses, the similar levels of homoplasy may simply reflect the rate of change of the chloroplast genome. Combining sequence data from two or more chloroplast genes may provide as much or more phylogenetic information as the restriction site approach. Two chloroplast genes were available for five of the data sets examined (Tables 3.1, 3.3). In four of the five cases even combining the sequences for two genes does not sample as much sequence diversity or generate as many characters as the restriction site approach. Thus, in these groups it would be necessary to sequence three chloroplast genes, each with a rate of change comparable to *rbcL,* to provide as much phylogenetic information as the restriction site data.

ADVANTAGES AND LIMITATIONS OF RESTRICTION SITE DATA

There are two primary advantages of restriction site data over DNA sequences for phylogenetic studies. First, given the use of a sufficient number of enzymes, the restriction site approach samples more sequence diversity than DNA sequencing. This fact is evident from the statistical

Table 3.3. Comparison of numbers of variable and informative characters for groups that have sequences of two chloroplast genes (see Table 3.1).

Group	cpDNA Restriction Site Data				Chloroplast Gene Sequences		
	bp examined	Variable	Informative		bp examined	Variable	Informative
Asteraceae	2,400	537	173	*rbcL*	1,458	265	124
				ndhF	2,238	521	232
				Total	3,696	786	356
Pontederiaceae[a]	2,136	213	104	*rbcL*	1,343	103	61
				ndhF	490	90	56
				Total	1,833	193	107
Ranunculaceae	3,204	537	369	*atpB*	1,427	288	133
				rbcL	1,488	406	145
				Total	2,915	694	278
Saxifragaceae	3,528	315	169	*matK*	1,087	304	99
				rbcL	1,440	128	44
				Total	2,527	432	133
Solanaceae	6,444	388	209	*rbcL*	1,408	181	63
				ndhF	2,086	471	100
				Total	3,494	652	163

[a]Only 490 bp of *ndhF* were sequenced in this group.

comparisons of the 33 groups (Tables 3.1 and 3.2), which showed that the restriction site analyses provide significantly more variation and informative changes than DNA sequences of any widely used genes. In fact, in all cases examined, it required sequences of two or three chloroplast genes to provide the equivalent amount of information as cpDNA restriction site comparisons (Table 3.3). Furthermore, the use of frequent-cutting enzymes with 4-bp recognition sites, higher concentration and longer agarose gels, and small hybridization probes has enabled the detection of sufficient variation to use this approach for population studies (e.g., Mason-Gamer et al., 1995). Second, two recent studies have suggested that using sites scattered throughout the genome is more likely to produce the whole genome (Cummings et al., 1995) or correct (Hillis et al., 1994) phylogeny than contiguous sites. According to Hillis et al. (1994) this may be the result of nonindependent evolution of some nucleotides within genes. Because restriction sites are scattered throughout the genome they are more likely to vary independently of one another.

There are four potential limitations of restriction site analyses. First, this approach generally requires large quantities of highly purified DNA, whereas DNA sequencing can be performed on very small quantities of DNA isolated from herbarium or even fossil remains. Restriction site analysis obviously requires the collection of bulk plant material, either fresh or preserved in silica gel (e.g., Chase and Hills, 1991). Sequencing of PCR product assures improved taxon sampling in many groups due to the ease of isolating small quantities of DNA from herbarium or cultivated material. Second, DNA sequence data from different laboratories are combined easily, whereas this is not the case for restriction site data. The combining of *rbcL* sequences from numerous labs (Chase, Soltis, and Olmstead et al., 1993) is an excellent example of this advantage. There have been some attempts to combine cpDNA restriction site data from different labs (e.g., Lane et al., 1996), but these are complicated by differences between labs in gel conditions (agarose concentration and length), estimation of fragment sizes, and mapping and aligning of sites. Third, intuitively one might

suspect that there may be more ambiguity about character homology in restriction site data than DNA sequences. This potential limitation is not supported, however, by our comparisons of the CI and RI values, which were generally higher for restriction site data than DNA sequences (Table 3.2). The amount of ambiguity in both types of data will depend on the level of sequence divergence and the frequency of indels. At the interspecific level there is usually very little doubt about alignment of restriction sites in cpDNA, especially if smaller hybridization probes ($<5-10$ kb) are used. However, as one encounters higher levels of sequence divergence and more indels, it becomes difficult to align sites reliably, which has resulted in the generalization that restriction site comparisons of the entire chloroplast genome generally are not useful at or above the family level. Furthermore, in some groups even comparisons among genera are not possible, especially if the chloroplast genomes have structural rearrangements, although these changes can provide powerful phylogenetic markers that have little or no homoplasy (e.g., Jansen and Palmer, 1987b; Raubeson and Jansen, 1992; Doyle et al., 1992; Knox et al., 1993; Hoot and Palmer, 1994). Character homology also may be a problem for sequence data, especially from noncoding regions (see Chapter 4). For example, DNA sequences of the ITS region of nuclear ribosomal DNA may be so divergent and have so many indels that it is not possible to make accurate alignments (e.g., Kim and Jansen, 1996). However, this is not a problem for chloroplast encoded protein genes, which have either no length variation (e.g., *rbcL*) or limited length variation that is easily aligned (e.g., *ndhF*). Fourth, restriction site data are not as amenable to molecular evolutionary comparisons as DNA sequences. Knowledge of the actual nucleotide sequence enables more accurate comparisons of the patterns and rates of sequence evolution.

COST COMPARISON OF RESTRICTION SITE ANALYSES AND DNA SEQUENCING

Cost may be a very important consideration in deciding whether to take a restriction site or

DNA sequencing approach. The relative costs of cpDNA restriction site analysis and DNA sequencing can vary a great deal from lab to lab; hence, it is very difficult to generalize. In the lab of one author (RKJ) both cpDNA restriction site and DNA sequence comparisons are routinely performed. Thus, we calculated the costs of these approaches based on a recent study involving 20 taxa in the genus *Argyranthemum* and related genera in the subtribe Chrysantheminae (Asteraceae, Anthemideae) (Francisco-Ortega et al., 1996, 1997). Our cpDNA restriction site analyses involved 21 enzymes, 14 with 6-bp recognition sites and 7 with 4-bp recognition sites. These 21 enzymes sampled approximately 3,190 restriction sites (978 from 6-bp and 2,112 from 4-bp enzymes, respectively) for a total of 14,716 bp (Francisco-Ortega et al., 1996). We sequenced the same 20 taxa for the ITS 1 and 2 regions of the nuclear ribosomal repeat, which had aligned sequences 503-bp long (Francisco-Ortega et al., 1997). We compared the cost of the

cpDNA restriction site analysis with both manual and automated sequencing (Table 3.4). The results show that although it is more expensive overall to perform the cpDNA restriction site comparisons, the cost per base pair, variable site, and informative site are far lower for the restriction site data. This comparison does not include labor, which also varies considerably among labs. The restriction site approach is clearly more labor intensive, especially in comparison with automated sequencing. Thus, if labor is included the difference in cost may be less.

CONCLUSIONS

Molecular phylogenetics has now moved into an era in which it is widely accepted that multiple markers should be used to reconstruct the evolutionary history of a group (see Chapters 1 and 2). Several phenomena, such as hybridization, lineage sorting of ancestral polymorphisms, and

Table 3.4. Costs of restriction site analysis and DNA sequencing.

cpDNA Restriction Site Analysis		Manual DNA Sequencing[c]		Automated DNA Sequencing	
Item	Cost	Item	Cost	Item	Cost
Restriction enzymes[a]	131.94	*Taq* polymerase	2.20	*Taq* polymerase	2.20
Agarose	44.40	Sequenase	22.00	Purification columns	22.00
Nylon membrane	236.22	Xray film	30.60	Sequencing charge[e]	600.00
32_P	354.00	35_S	69.20	Miscellaneous[d]	30.00
X-ray film	262.00	Dideoxy nucleotides	16.00		
Miscellaneous[b]	100.00	Miscellaneous[d]	30.00		
Total	1128.56	Total	170.00	Total	652.20
Number of bp examined per taxon: 14,716 × 20 taxa = 1,030,120 bp total. Thus, the cost per bp is 0.1 cents		Number of bp examined per taxon = 503 × 20 = 10,060 bp total. Thus, the cost per bp is 1.7 cents.		Number of bp examined per taxon = 503 × 20 = 10,060 bp total. Thus, the cost per bp is 6.5 cents.	
Cost per variable site[f]: $1128.56/122 = $9.25		Cost per variable site[f]: $170/41 = $4.15		Cost per variable site[f]: $652.20/41 = $15.90	
Cost per informative site[f]: $1128.56/72 = $15.67		Cost per informative site[f]: $170/7 = $24.29		Cost per informative site[f]: 652.20/7 = $93.17	

Note: The costs include supplies but not labor and are calculated for 20 taxa of the subtribe Chrysantheminae (Asteraceae, Anthemideae) examined in Francisco-Ortega et al. (1996, 1997).
[a]The cost of enzymes was calculated for 21 enzymes. For 6-bp enzymes 10 units of enzyme were used per digest, whereas 15 units were used for 4-bp enzymes. The enzymes used were provided in Francisco-Ortega et al. (1996).
[b]Includes all other components for agarose gel electrophoresis and filter hybridizations (i.e., nonradioactive nucleotides, buffers, DNA polymerase).
[c]Manual sequencing involved using the four ITS primers described in Kim and Jansen (1994) to sequence both strands of ITS 1 and 2.
[d]Includes primers, agarose, gene cleaning reagents, and buffers.
[e]The Automated DNA Sequencing Facility at the Molecular Biology Institute at the University of Texas charges $15 per primer/template reaction. Because it is usually possible to read up to 550 bp per primer, both strands of the ITS region can be sequenced with two primers.
[f]The number of variable and informative changes is from Francisco-Ortega et al. (1997).

mistaken orthology, can produce gene trees that are not equivalent to species trees (e.g., Pamilo and Nei, 1988; reviewed by Doyle, 1992; see Chapters 4 and 10). Unfortunately, there is a paucity of independent molecular markers available for phylogenetic studies in plants (see Chapter 1). Furthermore, the utility of some of these is restricted to a narrow range of taxonomic levels. The phylogenetic utility of one approach, restriction site analysis of cpDNA, has been demonstrated during the past 15 years. In spite of this usefulness, there has been a recent downward surge in the use of cpDNA restriction site analysis and a corresponding increase of DNA sequencing for phylogenetic studies of plants (Fig. 3.1). We believe strongly that the restriction site analysis of cpDNA should not be abandoned, and in fact, in certain applications it is preferable to DNA sequencing of any widely used genes or DNA regions. Having multiple gene trees is clearly important for phylogenetic studies of any plant group, and cpDNA restriction site data can generate an accurate and well-supported chloroplast genome tree. Our comparisons of 33 groups of plants from 15 molecular systematic labs demonstrated that cpDNA restriction site data generate as many, or more, phylogenetically informative characters as any single chloroplast gene sequence (Table 3.2), and in most cases more than any two genes combined (Table 3.3). Furthermore, these same comparisons indicated that restriction site data have less homoplasy than the DNA sequences, especially when comparing cpDNA restriction site data with ITS sequences. In general, we recommend that at higher taxonomic levels (generic level and above) it is preferable to sequence two or more chloroplast genes rather than conduct restriction site analyses of the entire chloroplast genome. At these higher taxonomic levels the uncertainty in aligning mapped sites, the amount of time required to construct detailed site maps, the benefits of being able to combine data from different labs readily, and the use of small quantities of DNA favor a sequencing approach. At lower taxonomic levels (interspecific and below), however, the cpDNA restriction site approach provides more reliable and a larger number of phylogenetic characters than any currently available gene or DNA region. Thus, to produce a chloroplast phylogeny it would be better to perform a restriction site analysis of the entire genome than to sequence multiple genes or even noncoding regions. To achieve the same benefits from DNA sequencing it would be necessary to sequence a large number of fragments scattered throughout the genome. Thus, we recommend that restriction site analyses continue to be used for phylogenetic studies in plants, especially at lower taxonomic levels where the amount of sequence divergence is low enough (usually <1%) that mapping sites is not necessary.

LITERATURE CITED

Archie, J. W. 1996. Measures of homoplasy. In Homoplasy: the Recurrence of Similarity in Evolution, eds. M. J. Sanderson and L. Hufford, pp. 153–188. Academic Press, San Diego.

Arnold, M. L. 1993. *Iris nelsonii* (Iridaceae): origin and genetic composition of a homoploid hybrid species. American Journal of Botany **80:**577–583.

Arnold, M. L., C. M. Buckner, and J. J. Robinson. 1991. Pollen-mediated introgression and hybrid speciation in Louisiana irises. Proceedings of the National Academy of Sciences U.S.A. **88:**1398–1402.

Bain, J. F., and R. K. Jansen. 1996. Numerous chloroplast DNA polymorphisms are shared among different populations and species in the aureoid *Senecio* (*Packera*) complex. Canadian Journal of Botany **74:**1719–1728.

Baldwin, B. G. 1992. Phylogenetic utility of the internal transcribed spacers of nuclear ribosomal DNA in plants: an example from the Compositae. Molecular Phylogenetics and Evolution **1:**3–16.

Baldwin, B. G., M. J. Sanderson, J. M. Porter, M. F. Wojciechowski, C. S. Campbell, and M. J. Donoghue. 1995. The ITS region of nuclear ribosomal DNA: a valuable source of evidence on angiosperm phylogeny. Annals of the Missouri Botanical Garden **82:**247–277.

Baum, D. A., K. J. Sytsma, and P. C. Hoch. 1994. A phylogenetic analysis of *Epilobium* based on nuclear ribosomal DNA sequences. Systematic Botany **19:**363–388.

Bremer, B. B. 1991. Restriction data from chloroplast DNA for phylogenetic reconstruction: is there only one accurate way of scoring? Plant Systematics and Evolution **175:**39–54.

Bremer, B. B., and R. K. Jansen. 1991. Comparative restriction site mapping of chloroplast DNA implies new phylogenetic relationships within Rubiaceae. American Journal Botany **78:**198–213.

Bremer, B., K. Andreasen, and D. Olsson. 1995 Subfamilial and tribal relationships in the Rubiaceae based on *rbcL* sequence data. Annals of the Missouri Botanical Garden **82:**383–397.

Byrne, M., and G. F. Moran. 1994. Population divergence in the chloroplast genome of *Eucalyptus nitens*. Heredity **73**:18–28.

Chase, M. W., and H. H. Hills. 1991. Silica gel: an ideal material for field preservation of leaf samples for DNA studies. Taxon **40**:215–220.

Chase, M. W., and J. D. Palmer. 1989. Chloroplast DNA systematics of lilioid monocots: resources, feasibility, and an example from the Orchidaceae. American Journal of Botany **76**:1720–1730.

Chase, M. W., D. E. Soltis, R. G. Olmstead, D. Morgan, D. H. Les, B. D. Mishler, M. R. Duvall, R. A. Price, H. G. Hills, Y.-L. Qiu, K. A. Kron, J. H. Rettig, E. Conti, J. D. Palmer, J. R. Manhart, K. J. Sytsma, H. J. Michaels, W. J. Kress, K. G. Karol, W. D. Clark, M. Hedrén, B. S. Gaut, R. K. Jansen, K.-J. Kim, C. F. Wimpee, J. F. Smith, G. R. Furnier, S. H. Strauss, Q.-Y. Xiang, G. M. Plunkett, P. S. Soltis, S. M. Swensen, S. E. Williams, P. A. Gadek, C. J. Quinn, L. E. Eguiarte, E. Golenberg, G. H. Learn, Jr., S. W. Graham, S. C. H. Barrett, S. Dayanandan, and V. A. Albert. 1993. Phylogenetics of seed plants: an analysis of nucleotide sequences from the plastid gene *rbc*L. Annals of the Missouri Botanical Garden **80**:528–580.

Cruzan, M. B., and M. L. Arnold. 1993. Ecological and genetic associations in an *Iris* hybrid zone. Evolution **47**:1432–1445.

Cruzan, M. L., M. L. Arnold, S. E. Carney, and K. R. Wollenberg. 1993. cpDNA inheritance in interspecific crosses and evolutionary inference in Louisiana irises. American Journal of Botany **80**:344–350.

Cummings, M. P., S. P. Otto, and J. Makeley. 1995. Sampling properties of DNA sequence data in phylogenetic analyses. Molecular Biology and Evolution **12**:814–822.

Demesure, B., N. Sodzi, and J. Petit. 1995. A set of universal primers for amplification of polymorphic non-coding regions of mitochondrial and chloroplast DNA in plants. Molecular Ecology **4**:129–131.

Dong, J., and D. B. Wagner. 1994. Paternally inherited chloroplast polymorphism in *Pinus:* estimation of diversity and population subdivision, and tests of disequilibrium with a maternally inherited mitochondrial polymorphism. Genetics **136**:1187–1194.

Downie, S. R., and J. D. Palmer. 1992. Restriction site mapping of the chloroplast DNA inverted repeat: a molecular phylogeny of the Asteridae. Annals of the Missouri Botanical Garden **79**:266–283.

Downie, S. R., and J. D. Palmer. 1994. A chloroplast DNA phylogeny of the Caryophyllales based on structural and inverted repeat restriction site variation. Systematic Botany **19**:236–252.

Downie, S. R., D. S. Katz-Downie, and K. Cho. 1997. Relationships in the Caryophyllales as suggested by phylogenetic analysis of partial chloroplast DNA ORF 2280 homolog sequences. American Journal of Botany **84**:253–273.

Doyle, J. J. 1992. Gene trees and species trees: molecular systematics as one-character taxonomy. Systematic Botany **17**:144–163.

Doyle, J. J., J. I. Davis, R. J. Soreng, D. Garvin, and M. J. Anderson. 1992. Chloroplast DNA inversions and the origin of the grass family (Poaceae). Proceedings of the National Academy of Sciences U.S.A. **89**:7722–7726.

Farris, J. S. 1989. The retention index and rescaled consistency index. Cladistics **5**:417–419.

Felsenstein, J. 1985. Confidence limits on phylogenies: an approach using the bootstrap. Evolution **46**:159–173.

Fenster, C. B., and K. Ritland. 1992. Chloroplast DNA and isozyme diversity in two *Mimulus* species (Scrophulariaceae) with contrasting mating systems. American Journal of Botany **79**:1440–1447.

Francisco-Ortega, J., R. K. Jansen, D. J. Crawford, and A. Santos-Guerra. 1995. Chloroplast DNA evidence for intergeneric relationships of the Macaronesian endemic genus *Argyranthemum* (Asteraceae). Systematic Botany **20**:413–422.

Francisco-Ortega, J., R. K. Jansen, and A. Santos-Guerra. 1996. Chloroplast DNA evidence of colonization, adaptive radiation, and hybridization in the evolution of the Macaronesian flora. Proceedings of the National Academy of Sciences U.S.A. **93**:4085–4090.

Francisco-Ortega, J., A. Santos-Guerra, A. Hines, and R. K. Jansen. 1997. Molecular evidence for a Mediterranean origin of *Argyranthemum* (Asteraceae). American Journal of Botany **84**:1595–1613.

Givnish, T. J., K. J. Sytsma, J. F. Smith, W. J. Hahn, D. H. Benzing, and E. M. Burkhardt. 1997. Molecular evolution and adaptive radiation atop tepuis in *Brocchinia* (Bromeliaceae: Pitcairnioideae). **In** Molecular Evolution and Adaptive Radiation, eds. T. J. Givnish and K. J. Sytsma. Cambridge University Press, pp. 407–431.

Golenberg, E. M., D. E. Giannasi, M. T. Clegg, C. J. Smiley, M. Durbin, D. Hendersen, and G. Zurawski. 1990. Chloroplast DNA sequence from a Miocene *Magnolia* species. Nature **344**:656–658.

Goloboff, P. 1991a. Homoplasy and the choice among cladograms. Cladistics **7**:215–232.

Goloboff, P. 1991b. Random data, homoplasy and information. Cladistics **7**:395–406.

Hillis, D. M. 1991. Discriminating between phylogenetic signal and random noise in DNA sequences. **In** Phylogenetic Analysis of DNA Sequences, eds. M. M. Miyamoto and J. Cracraft, pp. 278–294. Oxford University Press, New York.

Hillis, D. M., and J. J. Bull. 1993. An empirical test of bootstrapping as a method for assessing confidence in phylogenetic analysis. Systematic Biology **42**:182–192.

Hillis, D. M., J. P. Huelsenbeck, and C. W. Cunningham. 1994. Application and accuracy of molecular phylogenies. Science **264**:671–677.

Hiratsuka, J., H. Shimada, R. Whittier, T. Ishibashi, M. Sakamoto, M. Mori, C. Kondo, Y. Honji, C.-R. Sun, B.-Y. Meng, Y.-Q. Li, A. Kanno, Y. Nishizawa, A. Hirai, K. Shinozaki, and M. Sugiura. 1989. The complete sequence of the rice (*Oryza sativa*) chloroplast genome: intermolecular recombination between distinct tRNA genes accounts for a major plastid DNA inversion during the evolution of the cereals. Molecular and General Genetics **217**:185–194.

Holsinger, K. E., and R. K. Jansen. 1993. Phylogenetic analysis of restriction site data. Methods in Enzymology 224:439–455.

Hong, Y., V. D. Hipkins, and S. H. Strauss. 1993. Chloroplast DNA diversity among trees, populations, and species in the California closed-cone pines (*Pinus radiata, Pinus muricata,* and *Pinus attenuata*). Genetics 135:1187–1196.

Hoot, S. B. 1995. Phylogeny of the Ranunculaceae based on preliminary *atpB, rbcL,* and 18S nuclear ribosomal DNA sequence data. Plant Systematics and Evolution (suppl.) 9:241–251.

Hoot, S. B., and J. D. Palmer. 1994. Structural rearrangements, including parallel inversions, within the chloroplast genome of *Anemone* and related genera. Journal of Molecular Evolution 38:274–281.

Hsiao, C., N. J. Chatterton, K. H. Asay, and K. B. Jensen. 1995. Phylogenetic relationships of the monogenomic species of the wheat tribe, Triticeae (Poaceae), inferred from nuclear rDNA (internal transcribed spacer) sequences. Genome 38:211–223.

Jansen, R. K., and J. D. Palmer. 1987a. Chloroplast DNA from lettuce and *Barnadesia* (Asteraceae): structure, gene localization, and characterization of a large inversion. Current Genetics 11:553–564.

Jansen, R. K., and J. D. Palmer. 1987b. A chloroplast DNA inversion marks an ancient evolutionary split in the sunflower family (Asteraceae). Proceedings of the National Academy of Sciences U.S.A. 84:5818–5822.

Jansen, R. K., H. J. Michaels, and J. D. Palmer. 1991. Phylogeny and character evolution in the Asteraceae based on chloroplast DNA restriction site mapping. Systematic Botany 16:98–115.

Johansson, J. T., and R. K. Jansen. 1993. Chloroplast DNA variation and phylogeny of the Ranunculaceae. Plant Systematics and Evolution 187:29–49.

Johnson, L. A., and D. E. Soltis. 1995. Phylogenetic inference in Saxifragaceae sensu stricto and *Gilia* (Polemoniaceae) using *matK* sequences. Annals of the Missouri Botanical Garden 82:371–382.

Jorgensen, R. A., and P. D. Cluster. 1988. Modes and tempos in the evolution of nuclear ribosomal DNA: new characters for evolutionary studies and new markers for genetic and population studies. Annals of the Missouri Botanical Garden 75:1238–1247.

Kellogg, E. A., and R. Appels. 1995. Intraspecific and interspecific variation in 5S RNA genes are decoupled in diploid wheat relatives. Genetics 140:325–343.

Kim, K.-J., and R. K. Jansen. 1994. Comparisons of phylogenetic hypotheses among different data sets in dwarf dandelions (*Krigia*): additional information from internal transcribed spacer sequences of nuclear ribosomal DNA. Plant Systematics and Evolution 190:157–185.

Kim, K.-J., and R. K. Jansen. 1995. *ndhF* sequence evolution and the major clades in the sunflower family. Proceedings of the National Academy of Sciences U.S.A. 92:10379–10383.

Kim, K.-J., R. K. Jansen, and B. L. Turner. 1992a. Evolutionary implications of intraspecific chloroplast DNA variation in dwarf dandelions (*Krigia*-Asteraceae). American Journal of Botany 79:708–715.

Kim, K.-J., R. K. Jansen, R. S. Wallace, H. J. Michaels, and J. D. Palmer. 1992b. Phylogenetic implications of *rbcL* sequence variation in the Asteraceae. Annals of the Missouri Botanical Garden 79:428–445.

Kim, K.-J., B. L. Turner, and R. K. Jansen. 1992c. Phylogenetic and evolutionary implications of interspecific chloroplast DNA variation in dwarf dandelions (*Krigia*-Lactuceae-Asteraceae). Systematic Botany 17:449–469.

Kim, Y.-D., and R. K. Jansen. 1996. Phylogenetic implications of *rbcL* and ITS sequence variation in the Berberidaceae. Systematic Botany, 21:381–396.

Kim, Y.-D., and R. K. Jansen. 1997. Chloroplast DNA restriction site variation and phylogeny of the Berberidaceae. American Journal of Botany.

Kluge, A., and J. S. Farris. 1969. Quantitative phyletics and the evolution of anurans. Systematic Zoology 18:1–32.

Knox, E. B., S. R. Downie, and J. D. Palmer. 1993. Chloroplast genome rearrangements and the evolution of giant lobelias from herbaceous ancestors. Molecular Biology and Evolution 10:414–430.

Kohn, J. R., S. W. Graham, B. Morton, J. J. Doyle, and S. C. H. Barrett. 1996. Reconstruction of the evolution of reproductive characters in Pontederiaceae using phylogenetic evidence from chloroplast DNA restriction-site variation. Evolution 50:1454–1469.

Lane, M. A., D. R. Morgan, Y. Suh, R. K. Jansen, and B. B. Simpson. 1996. Relationships of North American genera of Astereae based on chloroplast DNA restriction site data. In Proceedings of Compositae Symposium, eds. N. Hind, H. Beentje, and G. Pope, pp. 49–77. Royal Botanic Gardens, Kew, England.

Liston, A. 1992. Variation in the chloroplast genes *rpoC1* and *rpoC2* of the genus *Astragalus* (Fabaceae): evidence from restriction site mapping of a PCR-amplified fragment. American Journal of Botany 79:953–961.

Loockerman, D., and R. Jansen. 1996. The use of herbarium material for molecular systematic studies. In Sampling the Green World, eds. T. F. Stuessy and S. Sohmer, pp. 205–220. Columbia University Press, New York.

Manos, P. S., K. C. Nixon, and J. J. Doyle. 1993. Cladistic analysis of restriction site variation within the chloroplast DNA inverted repeat region of selected Hamamelididae. Systematic Botany 18:551–562.

Mason-Gamer, R., and E. A. Kellogg. 1995. Chloroplast DNA analysis of the monogenomic Triticeae: phylogenetic implications of genome-specific markers. In Methods of Genome Analysis in Plants, ed. P. P. Jauhar, pp. 301–325. CRC Press, Boca Raton, Florida.

Mason-Gamer, R., K. E. Holsinger, and R. K. Jansen. 1995. Chloroplast DNA haplotype variation within and among populations of *Coreopsis grandiflora* (Asteraceae). Molecular Biology and Evolution 12:371–381.

McCauley, D. E. 1994. Contrasting the distribution of chloroplast DNA and allozyme polymorphism among local populations of *Silene alba:* implications for studies of gene flow in plants. Proceedings of the National Academy of Sciences U.S.A. **91:**8127–8131.

Neale, D. B., M. A. Saghai-Maroof, R. W. Allard, Q. Zhang, and R. A. Jorgensen. 1986. Chloroplast DNA diversity in populations of wild and cultivated barley. Genetics **120:**1105–1110.

Ohyama, K., H. Fukuzawa, T. Kohchi, H. Shirai, T. Sano, K. Umesono, Y. Shiki, M. Takeuchi, Z. Chang, S. Aota, H. Inokuchi, and H. Ozeki. 1986. Chloroplast gene organization deduced from complete sequence of liverwort (*Marchantia polymorpha*) chloroplast DNA. Nature **322:**572–574.

Olmstead, R. G., and J. D. Palmer. 1992. A chloroplast DNA phylogeny of the Solanaceae: subfamilial relationships and character evolution. Annals of the Missouri Botanical Garden **79:**346–360.

Olmstead, R. G., and J. D. Palmer. 1994. Chloroplast DNA systematics: a review of methods and data analysis. American Journal of Botany **81:**1205–1224.

Olmstead, R. G., and J. A. Sweere. 1994. Combining data in phylogenetic systematics: an empirical approach using three molecular data sets in the Solanaceae. Systematic Biology **43:**467–481.

Palmer, J. D. 1986. Isolation and structural analysis of chloroplast DNA. Methods in Enzymology **118:**167–186.

Palmer, J. D., R. K. Jansen, H. Michaels, J. Manhart, and M. Chase. 1988. Chloroplast DNA variation and plant phylogeny. Annals of the Missouri Botanical Garden **75:**1180–1206.

Pamilo, P., and M. Nei. 1988. Relationships between gene trees and species tree. Molecular Biology and Evolution **5:**568–583.

Petit, R. J., A. Kremer, and D. B. Wagner. 1993. Geographic structure of chloroplast DNA polymorphisms in European oaks. Theoretical and Applied Genetics **87:**122–128.

Ranker, T. A., D. E. Soltis, P. S. Soltis, and A. J. Gilmartin. 1990. Subfamilial phylogenetic relationships of the Bromeliaceae: evidence from chloroplast DNA restriction site variation. Systematic Botany **15:**425–434.

Raubeson, L. A., and R. K. Jansen. 1992. Chloroplast DNA evidence on the ancient evolutionary split in vascular land plants. Science **255:**1697–1699.

Rieseberg, L. H., M. A. Hanson, and C. T. Philbrick. 1992. Androdioecy is derived from dioecy in Datiscaceae: evidence from restriction site mapping of PCR-amplified chloroplast DNA fragments. Systematic Botany **17:**324–336.

Sanderson, M. J. 1989. Confidence limits on phylogenies: the bootstrap revisited. Cladistics **5:**113–129.

Sanderson, M. J. 1995. Objections to bootstrapping phylogenies: a critique. Systematic Biology **44:**299–320.

Sanderson, M. J., and M. J. Donoghue. 1989. Patterns of variation in levels of homoplasy. Evolution **43:**1781–1795.

Sanderson, M. J., and M. J. Donoghue. 1996. The relationship between homoplasy and confidence in a phylogenetic tree. **In** Homoplasy: The Recurrence of Similarity in Evolution, eds. M. J. Sanderson and L. Hufford, pp. 67–89. Academic Press, San Diego.

Sang, T., D. J. Crawford, S. Kim, and T. F. Stuessy. 1994. Radiation of the endemic genus *Dendroseris* (Asteraceae) on the Juan Fernandez Islands: evidence from sequences of the ITS regions of nuclear ribosomal DNA. American Journal of Botany **81:**1494–1501.

Sang, T., D. J. Crawford, T. F. Stuessy, and M. Silva O. 1995. ITS sequences and the phylogeny of the genus *Robinsonia* (Asteraceae). Systematic Botany **20:**55–64.

Schilling, E. E., and R. K. Jansen. 1989. Restriction fragment analysis of chloroplast DNA and the systematics of *Viguiera* and related genera (Asteraceae: Heliantinae). American Journal of Botany **76:**1769–1778.

Schilling, E. E., and J. L. Panero. 1996. Phylogenetic reticulation in subtribe Helianthinae. American Journal of Botany **83:**939–948.

Shinozaki, K., M. Ohme, M. Tanaka, T. Wakasugi, N. Hayashida, T. Matsubayashi, N. Zaita, J. Chunwongse, J. Obokata, K. Yamaguchi-Shinozaki, C. Ohto, K. Torazawa, B.-Y. Meng, M. Sugita, H. Deno, T. Kamogashira, K. Yamada, J. Kusuda, F. Takaiwa, A. Kato, N. Tohdoh, H. Shimada, and M. Sugiura. 1986. The complete nucleotide sequence of the tobacco chloroplast genome: its gene organization and expression. EMBO Journal **5:**2043–2049.

Smith, G. E., and M. D. Summers. 1980. The bidirectional transfer of DNA and RNA to nitrocellulose or diazobenzyloxymethyl paper. Annals of Biochemistry **109:**123–129.

Snedecor, G. W., and W. G. Cochran. 1980. Statistical Methods, Seventh edition. Iowa State University Press, Ames.

Soltis, D. E., and R. K. Kuzoff. 1995. Discordance between nuclear and chloroplast phylogenies in the *Heuchera* group (Saxifragaceae). Evolution **49:**727–742.

Soltis, D. E., P. S. Soltis, T. A. Ranker, and B. D. Ness. 1989. Chloroplast DNA variation in a wild plant, *Tolmiea menziesii*. Genetics **121:**819–826.

Soltis, D. E., M. S. Mayer, P. S. Soltis, and M. Edgerton. 1991a. Chloroplast-DNA variation in *Tellima grandiflora*. American Journal of Botany **78:**1379–1390.

Soltis, D. E., P. S. Soltis, T. G. Collier, and M. L. Edgerton. 1991b. Chloroplast DNA variation within and among genera of the *Heuchera* group (Saxifragaceae): evidence for chloroplast transfer and paraphyly. American Journal of Botany **78:**1091–1112.

Soltis, D. E., D. R. Morgan, A. Grable, P. S. Soltis, and R. Kuzoff. 1993. Molecular systematics of Saxifragaceae sensu stricto. American Journal of Botany **80:**1056–1081.

Soltis, D. E., R. K. Kuzoff, E. Conti, R. Gornall, and K. Ferguson. 1996. *matK* and *rbcL* gene sequencing data indicate that *Saxifraga* (Saxifragaceae) is polyphyletic. American Journal of Botany **83:**371–382.

Soltis, D. E., M. A. Gitzender, D. D. Strenge, and P. S. Soltis. 1997. Chloroplast DNA intraspecific phylogeography from the Pacific Northwest of North America. Plant Systematics and Evolution **206:**353–373.

Soltis, P. S., D. E. Soltis, and C. J. Smiley. 1992. An *rbcL* sequence from a Miocene *Taxodium* (bald cypress).

Proceedings of the National Academy of Sciences U.S.A. **89**:449–451.

Soreng, R. J., J. I. Davis, and J. J. Doyle. 1990. A phylogenetic analysis of chloroplast DNA restriction site variation in Poaceae subfam. Pooideae. Plant Systematics and Evolution **172**:83–97.

Swofford, D. L. 1993. PAUP: Phylogenetic Analysis Using Parsimony, version 3.1.1. Illinois Natural History Survey, Champaign.

Sytsma, K. J., and J. F. Smith. 1992. Molecular systematics of Onagraceae: examples from *Clarkia* and *Fuchsia*. In Molecular Systematics of Plants, eds. P. S. Soltis, D. E. Soltis, and J. J. Doyle, pp. 295–323. Chapman & Hall, New York.

Sytsma, K. J., J. F. Smith, and L. D. Gottlieb. 1990. Phylogenetics in *Clarkia* (Onagraceae): restriction site mapping of chloroplast DNA. Systematic Botany **15**:280–295.

Taberlet, P., L. Gielly, P. Guy and J. Bouvet. 1991. Universal primers for amplification of three non-coding regions of chloroplast DNA. Plant Molecular Biology **17**:1105–1109.

Thomas, W. K., and S. Paabo. 1993. DNA sequences from old tissue remains. Methods in Enzymology **224**:407–419.

Urbatsch, L., and R. K. Jansen. 1995. Phylogenetic affinities among and within coneflower genera (Asteraceae, Heliantheae), a chloroplast DNA analysis. Systematic Botany **20**:28–39.

Wakasugi, T., J. Tsudzuki, S. Ito, T. Nakashima, T. Tsudzuki, and M. Sugiura. 1994. Loss of all *ndh* genes as determined by sequencing the entire chloroplast genome of the black pine *Pinus thunbergii*. Proceedings of the National Academy of Sciences U.S.A. **91**:9794–9798.

Wendel, J. F., and V. A. Albert. 1992. Phylogenetics of the cotton genus (*Gossypium* L.): character-state weighted parsimony analysis of chloroplast DNA restriction site data and its systematic and biogeographic implications. Systematic Botany **17**:115–143.

Wendel, J. F., A. Schnabel, and T. Seelanan. 1995. An unusual ribosomal DNA sequence from *Gossypium gossypioides* reveals ancient, cryptic, intergenomic introgression. Molecular Phylogeny and Evolution **4**:298–313.

Whittemore, A. T., and B. A. Schaal. 1991. Interspecific gene flow in sympatric oaks. Proceedings of the National Academy of Sciences U.S.A. **76**:2540–2544.

Whitton, J., R. S. Wallace, and R. K. Jansen. 1995. Phylogenetic relationships and patterns of character change in the tribe Lactuceae (Asteraceae) based on chloroplast DNA restriction site variation. Canadian Journal of Botany **73**:1058–1073.

Wolfe, K. H., C. W. Morden, and J. D. Palmer. 1992. Function and evolution of a minimal plastic genome from a non photosynthetic plant. Proceedings of the National Academy of Sciences U.S.A. **89**:10648–10652.

Xiang, Q.-Y., D. E. Soltis, D. R. Morgan, and P. S. Soltis. 1993. Phylogenetic relationships of *Cornus* sensu lato and putative relatives inferred from *rbcL* sequence data. Annals of the Missouri Botanical Garden **80**:723–734.

Xiang, Q.-Y., S. J. Brunsfeld, D. E. Soltis, and P. S. Soltis. 1996. Phylogenetic relationships in *Cornus* based on chloroplast DNA restriction sites: implications for biogeography and character evolution. Systematic Botany **21**:515–534.

4

Homology in Molecular Phylogenetics: A Parsimony Perspective

Jeff J. Doyle and Jerrold I. Davis

Homology is central to systematics. That the very definition of homology is controversial (e.g., Hall, 1994) does not diminish the validity of this statement. The role of homology is as central for DNA sequences as for other sources of phylogenetic data; indeed, at least some homology issues described solely for gene sequences have parallels in nonmolecular data sets (Patterson, 1988; de Pinna, 1991; see also Chapter 12). Here we explore several aspects of the way in which homology permeates the theory and practice of systematics, with specific attention to the use of DNA sequences. Issues addressed include general definitions of homology at the level of genes, homology in alignment of nucleotide sequences, construction of gene trees, and inferences about species relationships made from such gene trees. The perspective is that of phylogenetic systematics— "cladistics"—in which parsimony plays a dominant role. One of our goals is to discuss the ways in which parsimony analysis is distinguished from other approaches to phylogeny reconstruction, such as distance or maximum likelihood estimation (ML), and to relate these differences to general concerns about character homology.

HOMOLOGY AND PHYLOGENY

Homology is similarity due to common descent, and a definition that is widely accepted among systematists is that homology is synonymous with synapomorphy (e.g., Patterson, 1982, p. 29). According to other authors (e.g., Roth, 1988), such a "phylogenetic" definition describes only one of several classes of homology, because common descent can refer to structural relationships (e.g., iterative homologies) as well as to taxic ones. Patterson (1982) discussed the distinction between taxic homology, which is concerned with hierarchical descent relationships (monophyly of groups) and transformational homology, which addresses issues of character change and may not imply taxic grouping. De Pinna (1991), however, noted that most transformational homologies contain a taxic component, and suggested that only character transformations that occur within an organism (producing serial homologues) lack taxic information.

Whether or not one partitions homology into different classes, it is clear that homology relationships exist at several organizational levels (e.g., Roth, 1991; Doyle, 1997). It is now widely

We thank our colleagues in the L. H. Bailey Hortorium for many discussions of systematic theory. We are grateful to Paul Lewis, Anne Bruneau, Doug Soltis, and Donovan Bailey for critical reading of the manuscript. This work was supported in part by various NSF grants.

accepted that a distinction can and should be made between trees that depict relationships among genes (gene trees) and trees that represent taxic relationships (species trees; e.g., Nei, 1987). Within the framework of phylogenetics, we will distinguish between homologies at these distinct organizational levels: *genic homologies* are synapomorphies in gene trees, whereas *taxic homologies* are synapomorphies that become fixed in organismal lineages. Both fit Patterson's (1982) definition of taxic homologies, in that they involve relationships among separate lineages in lineal descent systems (among lineages of gene copies in the case of genic homologies). Note, however, that from the perspective of organismal taxa, the gene tree for alleles of a single genetic locus can be viewed as a transformation series that links a series of states of a single character (Doyle, 1992). What are genic homologies from the perspective of the genetic locus are transformational homologies for the taxa in which the alleles occur. The situation is analogous, though at a lower organizational level, to biogeographic methods in which taxon trees become transformational hypotheses for geographic areas (e.g., Page, 1993).

A phylogenetic definition of homology has long been implicit in systematic practice, as taxonomists interested in discovering relationships among taxa formulated their intuitive hypotheses on the basis of similarities that were assumed to be due to common ancestry. Systematics texts (e.g., Simpson, 1961; Davis and Heywood, 1973) have been explicit: Characters used to postulate relationships must be homologous. To bring systematic theory to the stage of equating homology with synapomorphy required, therefore, only that two hierarchical levels of phylogenetic homology be distinguished from one another. Ancestral homologies variably retained by members of a group of taxa (plesiomorphies) are uninformative as to relationships among those taxa, whereas derived homologies (apomorphies) shared by two or more taxa within a group are evidence of shared common ancestry of that subset to the exclusion of other members of the group. Plesiomorphies are simply synapomorphies at a higher level of universality, and thus are homologies as well. It also is necessary to differentiate the organiza-

tional and hierarchical level at which homologies are being considered. Thus, a particular nucleotide can be a genic and a taxic homologue, or it can be a genic homologue but simultaneously a homoplasy in the taxon tree.

The long tradition of concern with character homology in systematics has recently come face to face with a new source of data, termed *molecular,* which in current usage is most commonly equated with DNA. The DNA–DNA hybridization method, in which whole genomes (or at least the single-copy components of whole genomes) are compared to produce a single estimate of relatedness between two species, has been claimed by its proponents (Sibley and Ahlquist, 1987) to eliminate any concern with distinguishing homology from analogy. This approach, however, has been criticized on numerous technical and theoretical grounds (e.g., Cracraft, 1987), and in any case it does not deal directly with homologies of individual characters. Any method that deals with discrete character data, molecular or otherwise, must necessarily be concerned with homology, and indeed homology (and homoplasy) have been key issues in discussions concerning the relative utility of DNA and nonmolecular data (e.g., Sanderson and Donoghue, 1989; Sytsma, 1990). "Homology" has been a particularly difficult term for molecular biologists, who have often confused homology with similarity, commonly using the term *percent homology* to denote the percentage of observed nucleotide matches between two sequences. Consequently, 50% identity of nucleotides or amino acids shared between two genes is often called "50% homology," whereas, as noted by Hillis (1994), the expression should mean (if anything) that the two genes share one structural domain but not a second one, as could occur in multifunctional genes due to exon shuffling (Gilbert, 1978). This confusion led Reeck et al. (1987) to clarify "homology" for the audience of the journal *Cell.* As they noted, similarity between two nucleotide sequences is a quantitative term. Homology, on the other hand, denotes shared common ancestry, and is thus qualitative: Two genes or gene regions either are or are not homologous. Degree of similarity, then, is not degree of homology, but it is a valid criterion for hypothesizing homology: If two genes share a level of nucleotide similarity greater than would be expected by chance alone, they may share that similarity because of common ancestry, and thus may be hypothesized to be homologous. Search algorithms (e.g., NCBI's BLAST search; Altschul et al., 1990) utilize similarity in this way, to measure the probability of matching an input sequence with sequences in nucleotide or protein databases by chance alone. A low probability of random match, obtained when two sequences are very similar, constitutes evidence that two sequences are homologous. It was this statistical quantification of similarity as a criterion for hypothesizing *gene* homology that led Patterson (1987, 1988) to conclude that the definition of homology in molecular biology is statistical. He did not deal with issues concerning homologies of individual *nucleotides* in genes already determined (or assumed) to be homologous. In any case, as Mindell (1991) has pointed out, statistical approaches may also be used in nonmolecular studies to quantify similarity and to hypothesize homology, so the concept is not unique to molecular systematics.

Homology and Gene Duplication

Clearly, all members of a multigene family are homologous; they are derived from a common ancestor by one or more gene duplications, and thus their overall levels of nucleotide similarity are due to common descent. A gene tree could be constructed for a multigene family by isolating all of its members from a single individual organism. This would provide a complete picture of the history of gene duplication in the gene family unless some members of the family had been lost from the genome of that organism. The gene family's anagenetic history would be represented by the nucleotide divergence (including length mutation) observed among the particular members of each gene subfamily sampled from that individual organism.

The organism from which these genes were sampled would be, of course, a member of a population—a species, a genus, and so on. In short, the organism is an element of taxic history, a history that is expected to be reticulate within the species and divergent above that level

(e.g., Hennig, 1966; Davis and Nixon, 1992; Doyle, 1995; Maddison, 1995). Variation generated by nucleotide substitution and recombination leads to allelic variation at each locus in the gene family. Partitioning of this allelic variation occurs during the speciation process (Nixon and Wheeler, 1992), after which additional anagenetic variation accumulates in each species. The depiction of taxic phylogeny has been termed a *species tree* by Nei (e.g., 1987) and others, regardless of the rank of the terminal taxa. We use the more general term *taxon trees* to refer to trees that depict descent relationships among groups or lineages of related organisms.

Thus, there are two independent dimensions, one genic, the other taxic, to the divergence of a gene family when it is sampled from more than one taxon (Fig. 4.1). A molecular phylogeny generated with the goal of identifying taxic relationships may misrepresent those relationships if it is interpreted without recognition of the distinction between genic and taxic differentiation. Such failures can arise from errors in sampling or from extinction of gene lineages (Fig. 4.2). Fitch (1970) introduced the terms *paralogy* and *orthology* to distinguish between genic and taxic homologies, respectively, in multigene families. Homology that arises by duplication of all or part of a gene is paralogy; homology that arises due to divergence of organismal lineages is orthology. The discovery of two or more distinct but homologous genetic loci coexisting in a single individual indicates that a gene duplication has occurred, and the loci are by definition paralogous. Orthology is more difficult to define empirically; the occurrence of only a single locus in two taxa does not guarantee that these loci are orthologous. It is possible that the loci are in fact paralogous, but extinction has eliminated one orthologue in each case (Fig. 4.2). Orthology can be defined only by reference to a particular kind of phylogenetic tree: a gene tree in which paralogous loci have been sampled from multiple organismal taxa.

When gene duplications occur simultaneously with taxic cladogenesis, homology relationships become more complex (Fig. 4.3). If only one member of a pair of sister species experiences a duplication, the two genes present in each of the individuals of that species are clearly paralogous with each other. The single gene present in individuals of the sister species, however, is equally related to either of these two paralogues. Thus the single gene is orthologous with both genes present in the clade with the duplication, despite the fact that these genes are paralogous with one another. Although the existing terminology is inadequate to cover this situation, the key point for phylogeny reconstruction is that if one descendant clade experiences a gene duplication and another clade does not, taxic relationships can be reconstructed equally well with the single gene that occurs in the clade that lacks the duplication and *either* of the genes that are present in the clade that bears the duplication. However, taxic relationships within the

Figures 4.1–4.3.   Gene family evolution. In all three figures and in Fig. 4.5, evolution is depicted in three dimensions: a genic dimension, a taxic dimension, and a time dimension. Diversification begins at the bottom of the box and proceeds upward through time, with cladogenic events occurring in either of the remaining two dimensions; gene duplication events are shown as white diamonds to distinguish these events from taxic cladogenesis. Taxa (e.g., species) are shown as numbers; genes are represented by letters. A two-dimensional view (i.e., viewing one face of the cube) reveals either the gene relationships or the taxon relationships, but not both. The full gene tree is obtained by rotating each of the subtrees through 90° and viewing the resulting tree from the genic perspective.

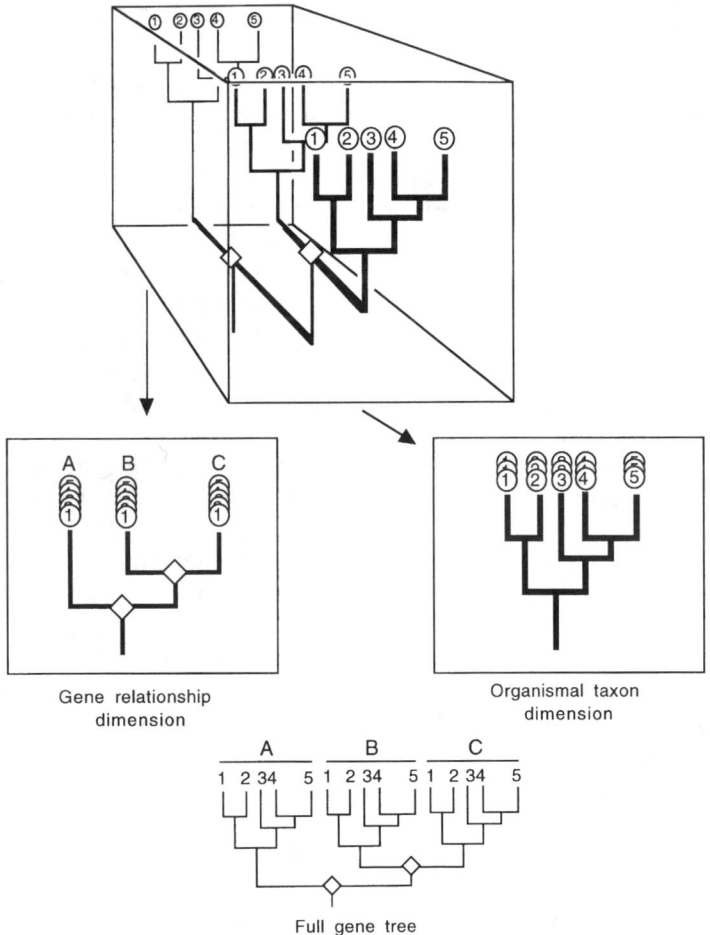

Figure 4.1.   A gene family in which two gene duplications precede all taxic cladogenesis events. Three paralogous genes (A, B, C) are evident in the genic dimension; five orthologous sequences are evident in the foreground of the taxic dimension (at locus C), with an additional set of five orthologous sequences at each of the two other paralogous loci. The total complexity of the family is evident in the full gene tree.

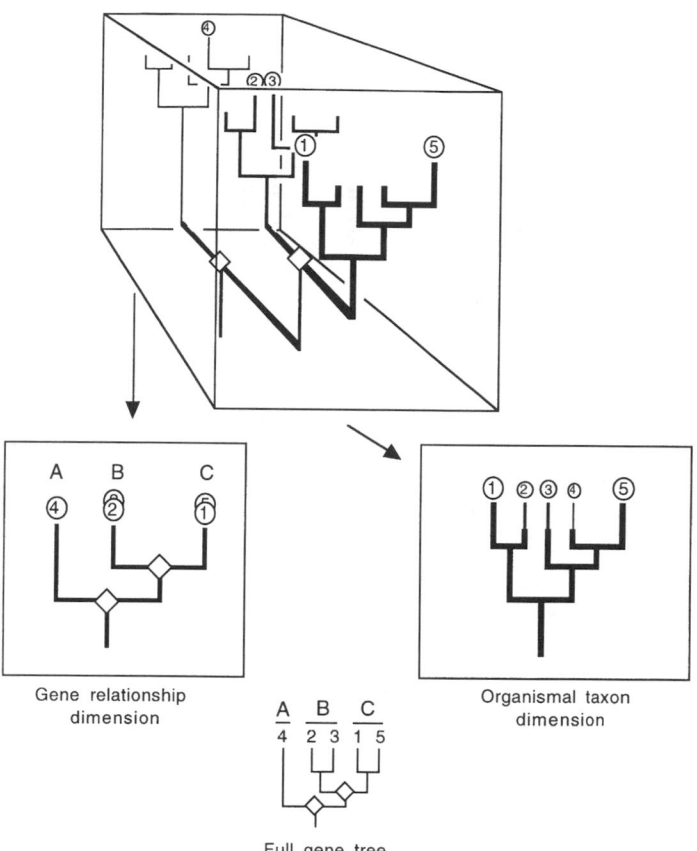

Figure 4.2. Gene lineage extinction (shown by truncated branches lacking taxon numbers at tips) results in the loss of some orthologous sequences at each paralogous locus, with each taxon retaining only one copy of the gene family. Note that gene relationships are unaffected—the genic complexity of the family remains, and phylogenetic reconstruction would correctly identify the points of gene duplication leading to paralogous genes. However, the full gene tree, though faithfully portraying the relationships among the sequences sampled, does not permit the accurate reconstruction of taxic relationships. If no ancillary criteria were available for establishing orthologies, the gene tree would be positively misleading as a taxon tree (e.g., taxa 1 and 5 would be hypothesized to be sister taxa). Note that poor sampling by the systematist could produce the same effect shown here for gene lineage extinction.

clade that bears the duplication can only be hypothesized from orthologous sequences.

Polyploidy adds more complexity. The haploid genome of a tetraploid is $2x$ rather than $1x$, so there are two copies of each locus in every gamete of a tetraploid. An allotetraploid is therefore equivalent to a diploid with two paralogous loci, but in this case the duplication event that generated the paralogues was the duplication of the entire genome during polyploidization, rather than the duplication of a single gene or chromosomal segment. In a genomic allotetraploid (Stebbins, 1947), independently segregating "homeologous" loci are clearly paralogous. In contrast, in a classic autotetraploid (one exhibiting tetrasomic inheritance), there are effectively four alleles at a single homologous locus in each individual, and it is impossible to distinguish orthologous and paralogous loci. In practical terms, the need to distinguish orthology from paralogy arises from the recognition that taxic phylogenies can be reconstructed accurately only using orthologues. Because both auto- and allopolyploids may often be of hybrid origin (e.g., Stebbins, 1947), a unique, divergently structured phylogeny cannot be reconstructed for such taxa (e.g., McDade, 1995). However, in the absence of recombination, the gene tree that includes sequences from

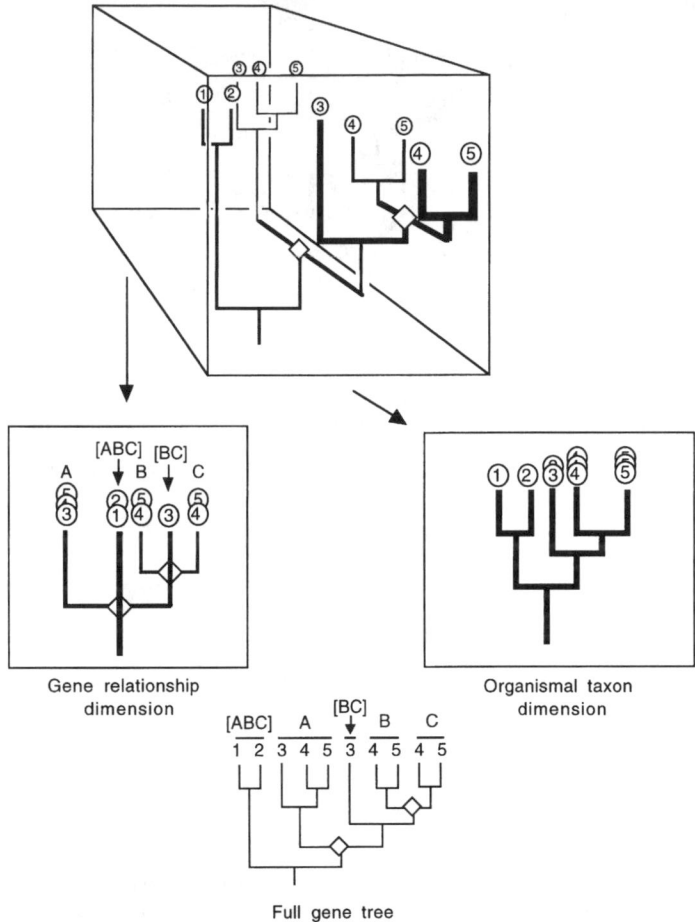

Gene relationship
dimension

Organismal taxon
dimension

Full gene tree

Figure 4.3.   The occurrence of gene duplications during taxic diversification produces complex homology relationships. As in the previous examples there are five organismal taxa and two gene duplications. The taxic dimension remains unchanged from that shown in Figs. 4.1 and 4.2. The genic dimension, in contrast, is quite different, and is reflected in the gene designations. Taxa 4 and 5 possess all three genes (A–C) and homology relationships are therefore clear when their genomes are compared: Genes with the same letter designation are orthologous, and genes with different letter designations are paralogous. Taxon 3 possesses only two of these genes. Its A locus is clearly orthologous with A loci in taxa 4 and 5, but its second locus (designated [BC]) is equally related to the B and C loci of taxa 4 and 5, as can be seen in the genic dimension view. In taxa 1 and 2, only a single gene exists, because both gene duplications took place after the divergence of their common ancestor from the common ancestor of taxa 3–5. This single gene is equally related to both genes in taxon 3 and to all three of the genes in taxa 4 and 5. The terms *paralogy* and *orthology* are insufficient to describe the full range of relationships among genes in different taxa; however, the definition of paralogy in any one taxon remains simple: When more than one locus exists, the two or more loci are paralogous. Reconstruction of taxic relationships is possible in this case regardless of whether the B or C gene is sampled in taxa 4 and 5. If additional taxic cladogenesis occurred in the lineages of these two taxa, however, it would be essential once again to sample only orthologous sequences in order to reconstruct taxic relationships accurately; the situation would be like that shown in Fig. 4. 1 for these taxa.

each of the different loci should track the contribution of the various different diploid genomes involved in the formation of the polyploid. Orthologous sequences, as defined for diploids, should be apparent in the polyploid as well, and therefore should provide a practical means for hypothesizing "phylogenies" of diploid genomes in diploid taxa and polyploid derivatives.

Inclusion of paralogous and xenologous (foreign sequences in a genome due to horizontal transfer; see Patterson, 1988) sequences may lead to incorrect inferences of taxic phylogenetic patterns. However, homologies at the level of the gene tree may still be correctly hypothesized: a gene tree constructed from paralogous sequences is a good gene tree, and is potentially an accurate representation of *genic* relationships. The homologies of each nucleotide in a sequence are unaffected by whether that sequence is an orthologue or a paralogue relative to other sequences. Similarly, the position of a particular sequence in a gene tree does not change even if the sequence moves from one taxon to another.

In contrast, recombination of various types, including allelic recombination and concerted evolution, presents additional challenges to concepts of homology at the gene tree level, where gene is defined in the standard molecular sense as a contiguous set of nucleotides (e.g., Rieger, 1991). Recombination is the gene-tree equivalent of hybridization in organismal taxa, a phenomenon long known to be troublesome for systematists (e.g., Funk, 1985; McDade, 1995; Rieseberg and Morefield, 1995). In a recombinant allele, different regions have different histories, each supported by different synapomorphies (Doyle, 1997; Fig. 4.4). In such a case, homoplasy does not result from failure to assess

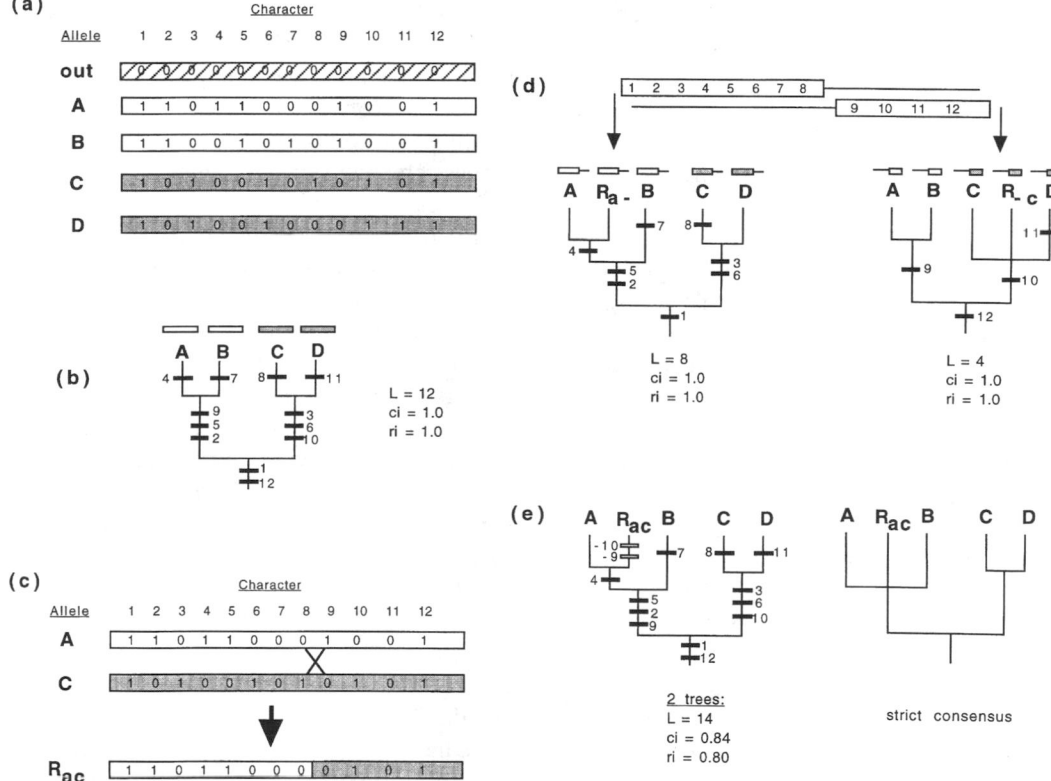

Figure 4.4.   Allelic recombination (from Doyle, 1997). *a.* Relationships among four ingroup alleles and one outgroup allele. States of 12 characters (binary, for simplicity) are shown from 5′ to 3′ along the length of each allele. The allele tree resolves two pairs of sister alleles and has no homoplasy. *b.* A reciprocal recombination event (crossover) between alleles A and C results in recombinant alleles, one of which (R_{ac}) combines character states from the 5′ end of allele A with character states from the 3′ end of allele C. *c.* Reconstruction of allele trees using either of the two regions delineated by the recombination event results in faithful recovery of historical relationships among segments of all alleles, including the separate sections of the recombinant. In neither case is there any homoplasy, only partitioning of the original set of characters. *d.* Reconstruction using all of the characters and treating each entire allele as a single historical unit introduces homoplasy due to conflict among characters having different histories in the recombinant allele. *e.* Two equally parsimonious trees are resolved, one of which is shown (the alternative tree unites alleles A and B); the strict consensus is less resolved. These trees require two additional transformations (steps), and this homoplasy comes about as a result of the reticulation produced by recombination—none of the homoplasy is due to parallelism.

homologies correctly, but rather from conflict among the topologies supported by different regions of the gene (Nixon and Wheeler, 1992). Hudson (1990) has stated that the proper analytical approach in such cases is to treat regions that differ in their historical patterns as separate taxa (separate genes, termed *coalescent genes* or *c-genes* by Doyle, 1995) in phylogenetic analyses. However, as Hudson noted, this separation will often be impractical, both because of the difficulty of recognizing regions with separate histories and because such regions could be as small as single nucleotides if recombination is extensive. Phylogeny reconstruction at recombining loci will be an increasingly important concern as nuclear genes become more widely used by systematists (see Chapter 1). Studies involving large numbers of ADH alleles of *Drosophila pseudoobscura* (Schaeffer and Miller, 1992; Doyle, 1995) exemplify this problem, with substantial amounts of homoplasy and low resolution of allele trees apparently attributable to recombination.

Standard recombination involves genetic exchange between alleles at a single locus. Concerted evolution (e.g., Zimmer et al., 1980), in contrast, is genetic exchange between two or more paralogous loci. This process usually is hypothesized to occur either by unequal crossing over (in tandemly repeated gene families; e.g., Smith, 1974) or gene conversion (in either tandem or dispersed families; e.g., Nagylaki and Petes, 1982; Ohta, 1984). Concerted evolution is best known for its homogenizing effect on gene families, as, for example, when it produces nearly uniform arrays of nuclear ribosomal gene (rDNA) sequences. This homogenization eliminates the phylogenetic distinction between paralogous and orthologous sequences. If all species in a large clade possess multiple copies of a gene, then the simplest explanation is that their common ancestor, and all subsequent ancestors within the clade, also possessed multiple paralogous copies. When these paralogous loci are transmitted to descendant species, the copies inherited by different descendants are orthologous, and it should be possible to reconstruct a conventional gene tree that depicts relationships of paralogy and orthology (e.g., Fig. 4.1). If, instead, all loci within each species were observed

to be closest relatives of each other in the gene tree (Fig. 4.5), and none, therefore, could be shown to be orthologous across species, two explanations would be possible. The first is that a gene amplification event occurred in each species following divergence of its genes from the most recent common ancestor shared with other species. This, however, would also require that prior to this amplification all but one copy of the gene family inherited from the common ancestor was lost in one or both species. Otherwise, if all existing genes were included in the phylogenetic analysis, some genes in different species would be shown to be orthologous.

The alternative explanation for the absence of clear orthologues in different taxa is concerted evolution (Fig. 4.5). Mutations that occur in one member of a multigene family are spread during concerted evolution to other members of the family, homogenizing the family and in the process creating synapomorphies that unite the members of the family in phylogenetic analyses (Fig. 4.6). Such shared mutations truly are genic homologues—the identity of nucleotides in erstwhile paralogues is due to descent from a common sequence. But that sequence occurred within the same genome rather than in an ancestral taxon, so these mutations are synapomorphies for the different members of the gene family within the individual plant, and ultimately (by gene exchange) within the species. To the extent that concerted evolution homogenizes loci efficiently and is continuous from generation to generation across speciation events, orthologues never become recognizable even though ancestors contribute paralogous loci to their descendants. The rate of homogenization and the rate of taxic cladogenesis together determine the topology of the multigene family tree.

Patterson (1988) termed the relationships among concertedly evolving gene family members "plerology." In a practical sense, if concerted evolution is complete, any member of the gene family can be used for reconstructing taxic phylogeny, making *complete* concerted evolution in general a useful attribute of a gene family. However, concerted evolution involving homeologous loci in allopolyploids can have the undesirable effect of making it impossible to trace the contribution of different diploid ances-

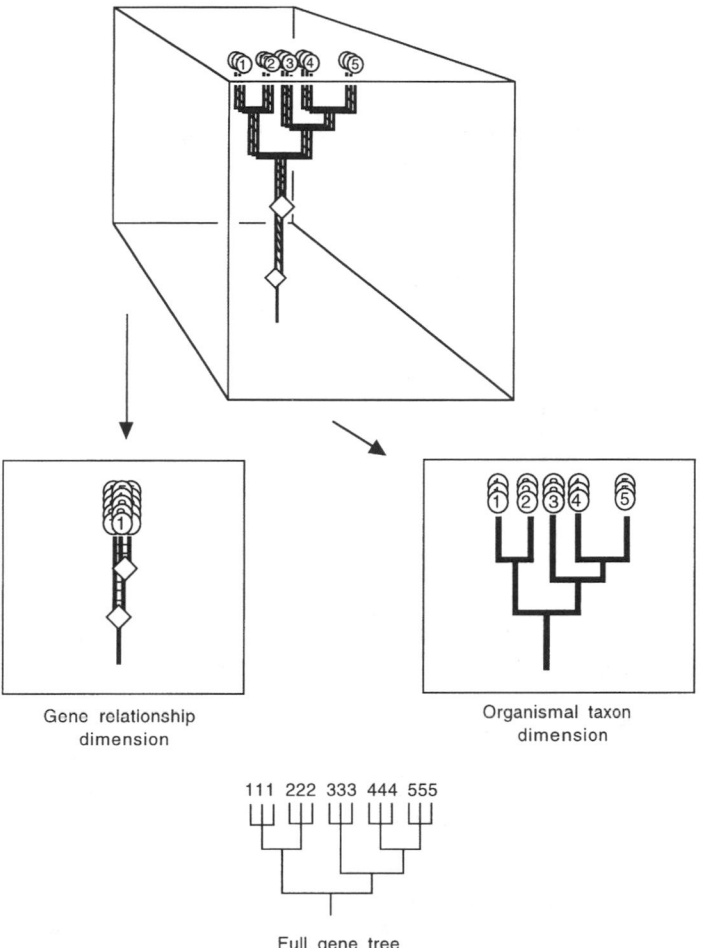

Gene relationship
dimension

Organismal taxon
dimension

111 222 333 444 555

Full gene tree

Figure 4.5.   Concerted evolution removes the distinction between orthologous and paralogous sequences. The taxic
dimension is unaffected by concerted evolution (cf. Fig. 4.1). Concerted evolutionary events (e.g., gene conversions) are
shown as lines linking paralogous gene clades (horizontal lines in the genic dimension and diagonal lines in the three-
dimensional perspective). The result in the genic dimension is that there is no evidence of differentiated paralogous loci,
despite the fact that two gene duplications have occurred (cf. Fig. 4.1). Sampling within any individual genome would still
reveal the existence of three genes, but these genes would be more similar to one another than to any genes in other taxa.
Thus, the full gene tree shows no evidence of paralogy/orthology relationships, and any of the three genes could be
sampled from any taxon to reconstruct an accurate taxon tree.

tors (Wendel et al., 1995). Intermediate levels of concerted evolution, in contrast, always present problems in the reconstruction of gene trees. Like allelic recombination, partial concerted evolution can produce "hybrid" sequences, in this case at coexisting paralogous loci rather than at a single locus; construction of accurate gene trees is compromised in such cases (Sanderson and Doyle, 1992). Two sets of conflicting homologies cause this problem and result in mixed signal homoplasy: synapomor-

phies arising from orthology relationships on the one hand and synapomorphies produced by concerted evolution on the other (Fig. 4.7; Sanderson and Doyle, 1992).

Homology and Characters

In addition to these new concerns and resulting additions to the lexicon of homology, molecular systematics has brought changes to the basic method of acquiring data. This method has until

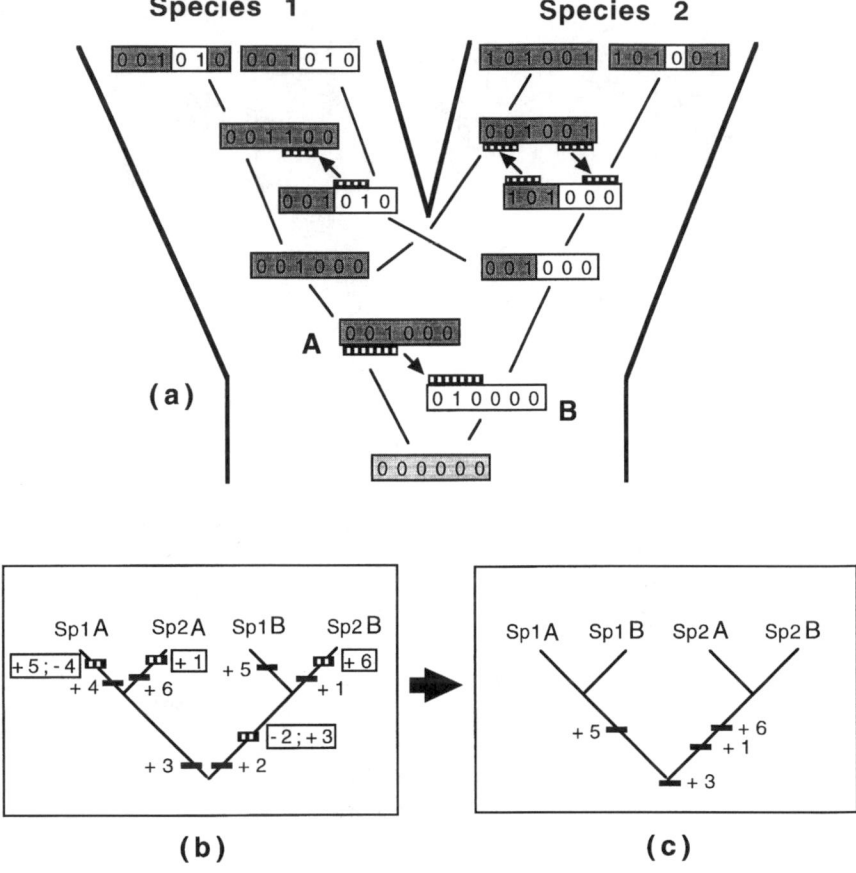

Figure 4.6. Effect of concerted evolution on character homologies in paralogous and orthologous genes in two sister taxa. *a.* In the common ancestor of taxa 1 and 2, an ancestral gene (light gray; binary characters labeled 5′ to 3′, left to right) undergoes a duplication, and the two products, A and B (white vs. dark gray boxes) accumulate apomorphies relative to one another (characters 3 and 2 in A and B, respectively). Genes in each lineage are connected by narrow lines. A gene conversion event (boxes with vertical black stripes, direction of conversion denoted by arrow) replaces the 5′ region of B with that of A, eliminating state 1 of character 2 in B and simultaneously replacing state 0 of character 3 with state 1. The two resulting genes (A and modified B) are transmitted to the two taxa that result from taxic cladogenesis, and each gene subsequently accumulates apomorphies separately (e.g., characters 6 and 5 in the B gene of taxa 1 and 2, respectively). These genes may also participate in gene conversion events (e.g., two conversion events shown between the A and B genes in taxon 2). In each taxon, the two paralogous genes within the taxon become identical. *b.* The same events are shown in a tree format, with character-state changes shown sequentially from the bottom; gene conversions (boxed, and indicated by vertically striped rectangles) can involve both the gain and the loss of a character in a single event. *c.* The single most parsimonious tree reconstructed for these four loci relative to an undiverged outgroup. Only four synapomorphies are observed, three of which resolve ingroup gene tree relationships. None of these homologies support the recognition of paralogous genes, despite the fact that the first event in the evolutionary history of these sequences was a gene duplication. Instead, all synapomorphies are the products of concerted evolution. These synapomorphic character states are true genic homologues, in each case being derived from a shared most recent common ancestral sequence.

recently remained very stable: systematists have *observed.* They have observed living plants or herbarium specimens representative of the taxa under study. They have noted the characteristics that distinguish these taxa, rejecting those that show plasticity or polymorphism within taxa,

choosing those that are stable. In this process, systematists have needed to define the characters and their states, an often difficult and controversial process (e.g., Stevens, 1991). Regardless of such problems, however, characters have been gathered one at a time, in each of the taxa

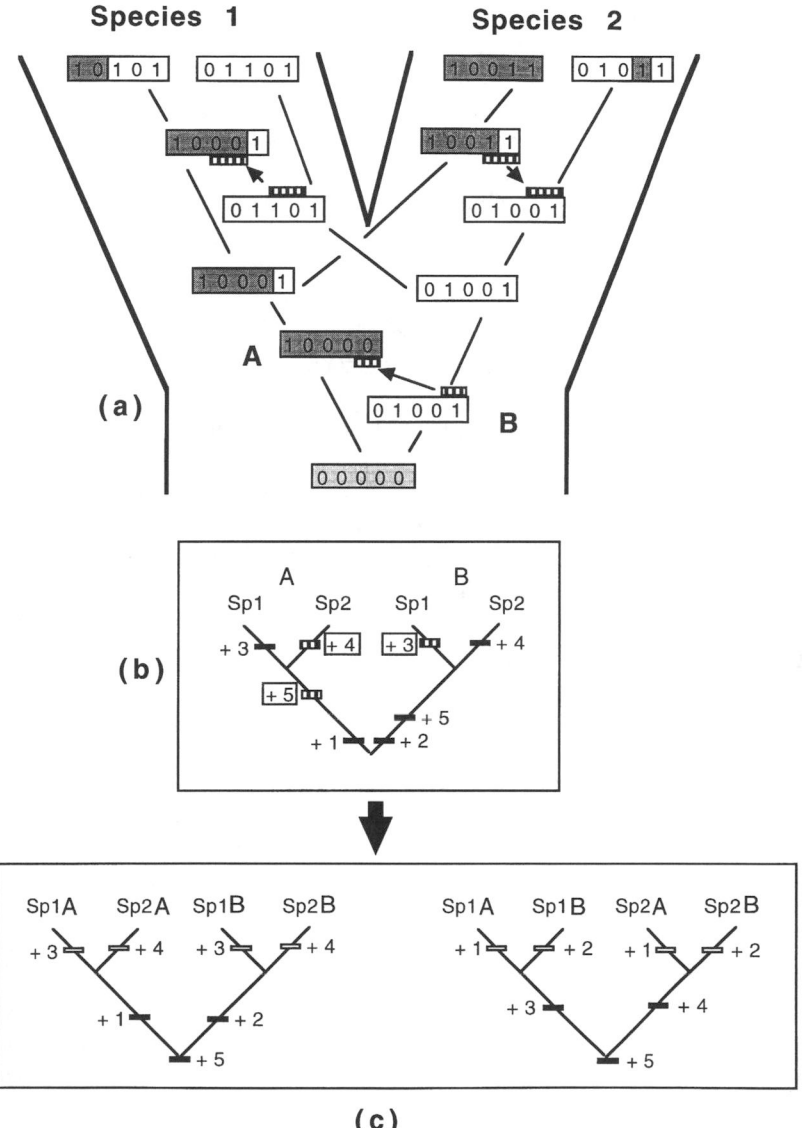

Figure 4.7. Partial concerted evolution produces homoplasy. *a, b.* Divergence and concerted evolution of paralogous loci in two taxa is depicted as in Fig. 4.6. *c.* Two equally parsimonious trees are resolved on which two conflicting classes of homologies (synapomorphies) exist. The first class comprises synapomorphies maintaining the distinction between paralogous gene copies (characters 1 and 2) that are not affected by concerted evolution; these support the left-hand tree. The second class comprises characters that arise in one gene copy in one taxon but are subsequently spread to the paralogous copy in that taxon (characters 3 and 4); these characters are also genic homologues, but track taxon histories. The conflict between these classes of characters is due to the different histories they track and is the only source of homoplasy in this example. In this case the homoplasy is balanced, and the strict consensus tree is completely unresolved; other resolutions (e.g., grouping Sp1A, Sp2A, and Sp2B) are possible with other distributions of mutations and gene conversions.

under study. The eventual character set (whether or not a data matrix was ever formally constructed) was therefore built one column at a time, with the homologies of the columns (characters) already thoroughly considered. It is clear at what levels hypotheses of homology are explicitly or implicitly invoked in this process: first, when choosing characters to score across taxa; second, when deciding which states of each character are shared by different taxa and so should be scored as identical. These two steps are not only logically separate but procedurally so as well: one chooses characters first, then decides on the scoring of states in the taxa at hand. Character 25 might be leaf pubescence, with the states in the taxa under study being glabrous, tomentose, and puberulent, while character 26 could be inflorescence type, with the states raceme, cyme, or panicle observed among the study group. Decisions to compare "inflorescences" or "pubescence" on "leaves" are often implicit statements of homology. Explicit statements of homology are made when it is claimed that the pubescence on the leaves of two or more different species is "tomentose" or that the inflorescences of some species are of the type called a *raceme*. The procedural separation between hypothesizing character versus character-state homologies ensures that there is little possibility of confusion between characters: The systematist would not assign the state raceme to character 25, which deals with leaf pubescence, or puberulent to character 26, which deals with inflorescence type.

Restriction site data are assessed in a similar manner (see Hillis, 1994, for further discussion of homology in restriction site and other classes of molecular data), with the homologies of individual sites hypothesized individually from observed fragment phenotypes. As a result, the addition of one new taxon to an existing restriction site data matrix requires a character-by-character homology assessment for the new taxon, a process that can often be arduous and time consuming (Olmstead and Palmer, 1994).

ALIGNMENT

With nucleotide sequence data, the situation is different from that described above. Data are gathered for many potential characters simultaneously, taxon by taxon. The character in a sequence data set is a particular position along the sequence. In the absence of length mutation, only the states A, G, C, or T will be available for each of the potentially thousands of characters of a DNA sequence, and all transformations must occur among these four nucleotide states. Moreover, in this simple case, alignment is trivial, with position always maintained: For example, position 25 will always be position 25 throughout the history of the sequence. As with the morphological example described previously, there will be no possibility of confusing characters 25 and 26. In the case where only base substitution is possible, the string of nucleotides derived from sequencing reactions is a row of characters in the data matrix for each of the taxa represented (assuming minimally that a common start point is found). Thus, in analyses of nucleotide sequences, potential characters are gathered row by row and not column by column.

Nucleotides of homologous genes are *potential* characters because if insertions/deletions can occur in the evolution of the sequences in question, then the data that are gathered are not necessarily rows of "characters" at all. The reason is that a character, in a phylogenetic sense, is a putative homologue, and prior to the alignment of length-variable sequences there is no hypothesis of homology for individual nucleotide positions, and thus no set of characters. Rather than establishing the characters to be sampled and then observing their states, the opposite is done: States are observed prior to character definition. To return to the difference in data gathering, not only is it true that nucleotide sequence data *are not* gathered column by column, they *cannot* be collected in this manner, because the only useful homology criterion for characters having identical ranges of states is position, and position is not established until sequences are aligned. Kjer (1995, p. 326) has stated quite bluntly that "Sequences alone, as submitted to GenBank, are no more meaningful than random lists of four letters since all phylogenetic conclusions are based on the alignment of these characters." Homology is equivalent to synapomorphy, which is a "phylogenetic conclusion," and therefore Kjer's comment refers as much to genic homologies at individual nucleotides as it does to the topology of the gene tree resulting from an analysis of aligned sequences.

The issue, then, becomes one of deciding how to insert spaces—gaps—in the alignment so as to recover the actual character homologies along columns of nucleotides. Failure to insert a gap correctly will produce inappropriate associations of states with characters. To return to the morphological example, if a species lacked leaves it would be necessary to score leaf pubescence character 25 as "missing." However, if nothing were placed in the cell for character 25 in that species (such as a "?"), all subsequent states would be shifted in the matrix, with the state of character 26 appearing in the column for character 25, the state of character 27 in column 26, and so on. The mistake would be fairly obvious if character and state names were used, because "leaf pubescence a cyme" would make no sense. The mistake would be much less apparent if codings such as 0 and 1 were used for character states, and anyone who has typed a large data matrix consisting only of these two states will be familiar both with the problem of insertion and deletion, and the difficulty of locating a mistake once it is made. The same situation applies for state-poor sequence data.

Clearly, were there no length mutation there would be no alignment problem: no matter how many substitutions had occurred, character homologies would always be assigned correctly. Of course, if two sequences differ by many substitutions it may be difficult to recognize them as being homologous, but this is a different (though related) issue. On the other hand, if there were no nucleotide substitution, there would also be little problem aligning sequences, because once a mismatch was observed, a gap would be added and extended until a perfect match was again observed. Nucleotide substitutions blur homologies within regions that have suffered length mutations; given the paucity of character states available for substitutional transformations, it is possible to insert gaps in different ways to hypothesize nucleotide homologies. Homology is hypothesized on the basis of position and similarity. But position of a given nucleotide in one sequence shifts relative to those in other sequences depending on where gaps are inserted, and similarity at any given position is a weak criterion because of the simplicity of nucleotide character states (e.g., one adenine looks like any other). Any difference between two homologous sequences can be explained by length mutation or nucleotide replacement but one can never be certain which process accounts for any particular case of observed variation.

Alignment is thus critical, because in establishing character homologies for a set of nucleotide sequences it simultaneously provides the provisional homology hypotheses for each of the character states (nucleotides) that will be used in the phylogenetic analysis. Analytical rigor at later stages will be wasted if the initial character homologies are incorrectly hypothesized. Despite this, alignment has received far less attention and discussion by systematists than have later analytical steps. Felsenstein (1988) pointed out how poorly understood this critical step was, and this observation was echoed by Hillis et al. (1996a) nearly a decade later. A comparison of the treatment of alignment with that of tree building methods in the first and second editions of *Molecular Systematics* (Hillis and Moritz, 1990; Hillis et al., 1996b) presents a striking contrast. Whereas discussion of the latter topic has expanded markedly in the second edition, and includes many methods that did not exist in 1990, discussion of alignment has increased only marginally. This is despite the fact that, in the intervening time period, DNA sequence data have come to dominate molecular systematics.

Several problems concerning the alignment of nucleotide sequences for phylogeny reconstruction continue to be debated. One controversy involves the approach used to perform alignments, specifically, whether (or how) alignment algorithms should be used. A second set of questions concerns regions that are difficult to align, or that vary depending on the parameters of the alignment algorithm used. Finally, the treatment of gaps as characters in phylogenetic analysis is logically dealt with under the heading of alignment, because gaps are only inferred during alignment and are not directly observed when sequence data are gathered. Each of these issues is related to the more general issue of homology either explicitly or implicitly.

Algorithms and Eyes: Statistics and Homology

Numerous computer programs are available for alignment of nucleotide or protein sequences (see Hillis et al., 1996a), with new programs and approaches appearing regularly (e.g., Hirosawa et al., 1995; Thompson, 1995; Notredame and Higgins, 1996). Two basic classes of alignment programs exist, local and global, and their names are accurate descriptions of their goals and general approaches. Local alignment programs do not attempt to align entire sequences; rather, they seek to identify regions of good match. These programs are primarily descendants of the method of Smith and Waterman (1981). Global alignment algorithms, in contrast, are used for aligning entire sequences, and thus are the algorithms primarily used in phylogeny reconstruction. Global multiple sequence alignments are in general based on the pairwise dynamic programming alignment method of Needleman and Wunsch (1970). The remaining discussion of algorithms deals primarily with global alignment.

Global alignment programs attempt to find an optimal alignment, one that maximizes or minimizes some overall score over entire sequences. For example, matches can be assigned a positive score and mismatches a negative score; the optimal alignment is the one that has the highest positive score. Any two sequences can be aligned so that there are no mismatches if gaps can be added wherever a mismatch would otherwise occur. Therefore, gaps must be penalized by being given a negative score that is higher than the cost of a mismatch. Alignment algorithms also commonly utilize a gap length penalty, such that longer gaps can receive a larger penalty than shorter ones. The gap-opening and gap-extending penalties together influence the number and length of the gaps that are hypothesized. Different alignment programs permit varying degrees of flexibility in penalizing gaps in an effort to capture the nuances of sequence evolution, such as the expectation that gaps in a protein-coding gene sequence should occur in multiples of three nucleotides to preserve reading frame.

Simultaneous alignment of multiple sequences, like exhaustive consideration of phylogenetic trees, becomes a difficult computational problem as the number of sequences increases: computational time is at least proportional to n^r, where n is the number of nucleotides and r is the number of sequences (e.g., Waterman et al., 1991). To overcome this problem, sequences can be aligned progressively in a pairwise manner, but the alignments will differ from one another depending on the order in which sequences are added, even when the same set of alignment parameters is used (e.g., Lake, 1991). A number of workers have suggested that more realistic alignments are produced when alignment order reflects descent relationships among the sequences being aligned. To quote Mindell (1991): "Use of phylogeny to inform alignments may, at first, appear circular because alignments themselves are used to infer phylogeny. However, this is an artifact of viewing alignment and phylogenetic analysis as independent procedures with different objectives when, in fact, they are mutually dependent procedures in assessment of evolutionary history based on nucleotide sequences." In some methods (e.g., Sankoff and Cedergren, 1983), the tree is provided by the user, which seems reasonable enough where the goal is to produce an alignment to study protein or gene evolution, but is clearly circular where the goal is phylogeny reconstruction. In other algorithms, such as those of Feng and Doolittle (1987, 1990) and the CLUSTAL programs (Higgins and Sharp, 1988, 1989; Higgins et al., 1992), a tree is constructed using distances computed from pairwise alignments, and this tree is used to determine input order. Hein (1989, 1990) and Konings et al. (1987) use the alignment produced as a result of such a distance-based method to calculate a new tree, which is used in a second alignment round, and continue this process through several iterations until the tree and alignment stabilize. In these methods the tree construction criterion, distance, is one that is rejected by many systematists. In contrast, the MALIGN package of Wheeler and Gladstein (1993, 1994) can be used in a similar, iterative fashion, but it uses parsimony under the assumption that the best alignment is that which produces the most parsimonious cladogram for a given set of gap, gap length, and substitution costs.

An overriding concern in the use of any alignment algorithm is the choice of parameters: different parameters produce different alignments, and these different homology hypotheses can support different topologies. Precisely what scores should be used to hypothesize nucleotide homologies is a matter of speculation, although efforts have been made to identify the range of biologically meaningful parameters (Vingron and Waterman, 1994). Choices of parameters, including even the decision to penalize gaps more heavily than mismatches, are clearly based on assumptions about the process of molecular evolution. Some criteria do exist for deciding what constitutes a reasonable alignment—one that is likely to identify homologies of nucleotides correctly. For example, the organization of functional protein-coding genes into codons offers important positional clues: Selection will most likely have preserved the reading frame, so the expectation is that length variation will most often occur in multiples of three, although not necessarily involving the precise excision or addition of a codon. Slipped-strand mispairing of short repeats is considered a major source of length variation (Levinson and Gutman, 1987), and this model suggests that repeat motifs should not be broken up in alignments (e.g., TATA–TA and not TAT–ATA). For ribosomal RNA sequences, comparisons across many taxa have revealed the existence of highly conserved sequences responsible for the secondary structure that is the critical functional property of rRNA (see Chapter 7). These regions represent useful landmarks for alignment (Kjer, 1995).

Some alignment programs provide considerable flexibility in an attempt to incorporate such assumptions into the choice of substitution and gap costs. In MALIGN (Wheeler and Gladstein, 1993, 1994), for example, gaps in multiples of three nucleotides can be penalized less than other gap lengths, and differential mismatch penalties can be invoked for transitions versus transversions or conservative versus nonconservative amino acid substitutions (e.g., using change matrices such as those of Dayhoff et al., 1978 or Koshi and Goldstein, 1995). Even such flexibility, however, seems unlikely to capture the complexity of molecular evolution. Kjer (1995, p. 314) concluded that even if a single best set of global gap and change costs were found for a given set of sequences, they could at most only be correct on average, and would therefore not be correct for any particular stretch of nucleotides being aligned: ". . . each region, and possibly each position within the molecule, has its own individual 'gap penalty.' " That the likelihood of mutational change varies across a

nucleotide sequence is widely appreciated. Genes, whether encoding proteins or structural RNAs, are usually composed of relatively conserved sequences separated by more variable regions. Not only are the overall levels of mutation likely to differ along a sequence, but so may the balance between substitution and length mutation. This can be seen trivially by noting that length mutation is often common in introns but not in adjacent exons, due to conservation of reading frame in exons. Different portions of a coding sequence are not expected to show such dramatic differences, in part because higher-level protein structure imposes constraints that are spread over the length of the sequence. However, constraints on length mutation are likely to vary from one segment to another. In addition to differences along the gene, there are also likely to be taxic differences in propensity for length mutation. For example, the evolutionary mode of ADH in *Drosophila* varies with species group, the gene evolving almost exclusively by substitution in the *D. pseudoobscura* group but showing considerable length mutation in *D. melanogaster* (Schaeffer et al., 1987).

McClure et al. (1994) tested a number of local and global protein alignment programs for their ability to identify and align highly conserved amino acid motifs; they found that no existing programs were uniformly successful, and that, surprisingly, global alignment programs outperformed the local alignment methods that were designed precisely for the task of identifying conserved features. If highly conserved features are not identified correctly, the situation is likely to be worse in regions where homologues are less readily hypothesized. Thus, Hillis et al. (1996a, p. 375) stated that "Visual inspection usually is necessary to ensure that the most reasonable alignment has been generated." The definition of what is reasonable, however, is likely to vary with the investigator, the sequence, and the group in question.

Difficult-to-Align Regions

In some gene regions, homologies are difficult to hypothesize by eye, and global algorithms produce quite different alignments with even relatively minor differences in gap and mismatch parameters. Swofford and Olson (1990) stated categorically that regions that cannot be aligned by eye should be eliminated from phylogenetic analyses, a view reiterated by Swofford et al. (1996). They justified the elimination of potential characters by reference to traditional systematic practice, in which the systematist chooses to include or exclude characters at the earliest stage of the analysis, often on the basis of the confidence with which homologies can be assessed a priori. Gatesy et al. (1993) explored the logical and practical consequences of culling (excluding) regions of doubtful character homology from sequence analyses. They conducted global alignments of crocodilian and insect 12S mitochondrial rDNA genes in which gap/change cost ratios were varied over a wide range of parameters, on the assumption that it is difficult if not impossible to know what constitutes "realistic" values for these parameters. This work produced alignments in which only a few regions were aligned identically in all cases, that is, where homology hypotheses were unambiguous. Culling all but such alignment-insensitive regions produced a data set with so few characters that it was virtually uninformative phylogenetically. They concluded that if realistic values cannot be suggested for parameters, then many characters will be alignment-variable; if all such characters are excluded, then the cost paid in phylogenetic resolution is likely to be unreasonably high. It could be argued, however, that resolution is highly suspect if it depends on transformations at characters whose homologies are likely to be incorrect.

These same workers subsequently proposed a solution to this dilemma in a method they termed *elision* (Wheeler et al., 1995; elision has some similarities to the ML-based method of Thorne and Kishino, 1992). Individual alignments were again produced under a wide range of parameters, but instead of culling variable regions, all of the different alignments were joined end to end to produce a single large data set which was then analyzed phylogenetically. In such an elided data matrix, any character that is alignment-invariable appears as an identical column in each of the separate data matrices, and therefore each transformation in that column is weighted by the number of different data matri-

ces being combined. For example, a single synapomorphic change appearing in such a column appears as six identically distributed synapomorphies if six different alignments are constructed. Transformations at characters sensitive to alignment parameters would receive less than this maximum amount of phylogenetic weight. Although elision thus has the benefit of weighting more heavily changes at those characters for which homologies are least ambiguous, Wheeler et al. (1995) noted that their method raises serious questions concerning the very definition of character homology in sequence analysis. In an elided matrix, various different homology assessments simultaneously coexist for a single set of observations (data). For example, for a given gene under one set of alignment parameters the 100th nucleotide in taxon A might be hypothesized to be homologous with the 97th nucleotide in taxon B, whereas with different parameters that same nucleotide might be hypothesized to be homologous with the 101st nucleotide in taxon B. In the elided matrix, constructed to provide a single overall set of homology hypotheses, both alternatives would coexist as different characters. There would seem to be no counterpart to this situation in a morphological analysis.

For systematists the usual goal of alignment is to produce provisional hypotheses of character homology for phylogeny reconstruction and not, for example, to analyze protein structure and function. Alignment-variable regions are therefore of concern primarily because of their potential effect on topologies produced from sequence data. It would seem reasonable, therefore, to explore the effect of different alignments (different homology hypotheses for some subset of characters) on phylogenetic analysis. One way to do so is to conduct phylogenetic analyses using different alignments and to report the sensitivity of topologies to alignment. In a recent study of histone H3 intron sequences in *Glycine* (Doyle et al., 1996), for example, variation in alignment affected only the robustness of support for clades, and not the overall topology of the most parsimonious tree. Soltis et al. (1996), in their study of rDNA ITS sequences of the *Boykinia* group (Saxifragaceae), found that the use of different alignment parameters also led to the resolution of different most parsimonious topologies. More rigorously, Wheeler (1995) used taxonomic and character congruence methods in a detailed "sensitivity analysis" to assess the robustness of various groups of arthropods to different alignments of 18S rDNA and polyubiquitin gene sequences. The particular parameters analyzed were gap/change and transition/transversion costs, but Wheeler (1995) pointed out that other parameters could also be analyzed similarly, and concluded (p. 331) that "Through the explicit examination of the parameters of analysis . . . coupled with the use of congruence based optimality, assumption-specific unstable conclusions can be avoided."

Gaps as Phylogenetic Characters

The gaps produced during the alignment of sequences represent hypothesized evolutionary events and are therefore potential phylogenetic characters. They may on first consideration appear different from nucleotide substitutions, in that gaps are only inferred, whereas nucleotides are observed. However, in reality there is little difference, because, as noted above, nucleotides themselves are inferred to be characters (homologous features) only after alignment. The process that produces gap characters (alignment) is therefore precisely the same as that which produces nucleotide characters. The use of gaps as characters may be difficult for methods of phylogenetic analysis that require a specific model of molecular evolution, insofar as the models commonly used in distance or ML calculations do not deal with insertions or deletions (but see Thorne and Kishino, 1992). This is no impediment for parsimony; indeed, Wheeler and Gladstein (1993) recommend using the same set of penalties for both substitutions and gaps in the alignment and cladogram construction phases of the phylogenetic study. Baldwin et al. (1995) list several other options for coding gaps and note some general difficulties; in practice the issue is, once again, the initial assessment of homology: gaps are produced by insertion or deletion events whose signatures may or may not be apparent by inspection of aligned sequences. A general concern is that it may be more difficult to hypothesize homologies for

length mutations than for nucleotide substitutions. Homologies will be difficult to hypothesize with confidence for complex and overlapping length mutations, such as could occur in hotspots for insertion or deletion (e.g., Palmer et al., 1985; Freudenstein and Doyle, 1994). For both the chloroplast genome and the rDNA ITS it appears that length mutations are no more homoplasious in general than are nucleotide substitutions (Olmstead and Palmer, 1994; Baldwin et al., 1995). Whether or not this is so is determined best by using insertions and deletions directly in parsimony analyses.

CHARACTER HOMOLOGY IN PHYLOGENETIC ANALYSIS: SYNAPOMORPHY VERSUS PROBABILITY

Homology Hypotheses, Extra Steps and Parsimony

Sequence alignment results in hypotheses of character homology. If the same alignment is used throughout a phylogenetic analysis, then character homologies remain unchanged; no further testing of those hypotheses is possible unless the alignment is adjusted. This is not true of hypotheses of character-state homology, which are also produced as a result of alignment. At each character (nucleotide sequence position), the distribution of character states is a set of hypotheses of homology (synapomorphies) that may disagree with homologies implied at other characters. As in any cladistic analysis, testing these homology hypotheses involves the reconstruction of phylogenies and the determination of the distribution of character-state changes on the resulting tree or trees. In this stage of the analysis conflicting hypotheses compete against one another, and a unified hypothesis is chosen in which some apparent homologies must be explained as homoplasies, the products of parallel substitution. For a change to be interpreted as parallel, it must occur in at least two different places on a tree. The parallel origin of a particular state produces an "extra step," obviously. More importantly, a single initial hypothesis of homology, derived from the initial observation that a particular nucleotide occurs in two or more taxa at a particular position in an aligned set of sequences, is replaced by at least one additional hypothesis. A parallelism in which the same nucleotide occurs at an identical position in two groups of unrelated taxa must be interpreted as two separate homologies (synapomorphies), one for each group, which are phylogenetically but not physically distinguishable. This explanation is special, produced only as a result of the phylogenetic analysis and after the observational fact—an ad hoc hypothesis. The logic of parsimony as applied to this problem was described years ago by Farris (1983):

That ad hoc hypotheses are to be avoided wherever possible in scientific investigation is, so far as I am aware, not seriously controversial. . . . Avoiding them is no less than essential to science itself. Science requires that choice among theories be decided by evidence, and the effect of an ad hoc hypothesis is precisely to dispose of an observation that otherwise would provide evidence against a theory. . . . The requirement that a hypothesis of kinship minimize ad hoc hypotheses of homoplasy is thus no more escapable than the general requirement that any theory should conform to observation; indeed, the one derives from the other.

Choice of parsimony as a method for phylogeny reconstruction should need no more justification than this philosophical one. However, if parsimony is defined solely as a method that accepts the tree topology that minimizes a value called *steps,* it can be viewed merely as one of several possible optimality criteria for tree selection (e.g., Swofford et al., 1996). This may be useful in explaining how a parsimony *procedure* works, and it relates such a procedure to other methods that choose a "best" tree based on some criterion (e.g., summed branch lengths or likelihood). But treating nucleotide character transformations merely as steps, an aggregate value to be minimized in an optimal tree, ignores the essential features of cladistic parsimony, which are its concern with individual character transformations (homologies) and its innate grounding in patterns of character change. This concern with character homology distinguishes parsimony (as used here and in Swofford et al., 1996, p. 426) from other methods. It has been stated that one should be careful to distinguish be-

tween parsimony and the computer programs used to search for parsimonious trees (Swofford and Olsen, 1990). It is at least equally important not to conflate cladistic parsimony as an approach to phylogeny reconstruction with the operation of counting steps on trees.

Contrasting Treatments of Homology

It is useful to contrast the cladistic parsimony approach with other methods currently used to produce phylogenetic hypotheses from DNA sequence data. Distance, ML, and parsimony analysis all begin with the assumption that identical positions in an aligned set of sequences are homologous. Cladistic methods treat characters individually, each character potentially providing (with multistate sequence data) one or more hypotheses of character-state homology (synapomorphies), from which relationships among sequences can be inferred. Distance methods, in contrast, use pairwise comparisons of entire sequences; the single homologous character with its states contributes only to this aggregate value. To the extent that one would be interested in the homologies of states at any individual character, it would be necessary to map (optimize) that character onto the distance tree topology. This would most likely use a parsimony criterion, presumably because the parsimony criterion is widely accepted in science, as noted by Farris (1983). Parsimonious optimization of nonmolecular characters (e.g., morphology, behavior) on distance trees is not uncommon in the scientific literature, and this suggests that even adherents of competing analytical methods accept the value of parsimony in studying character evolution.

ML is more directly connected to individual characters than is the distance approach. Indeed, every character, whether synapomorphic, autapomorphic, or constant, contributes directly to the likelihood estimate for a given tree, making the approach more "powerful" (in the statistical sense; Penny et al., 1992) than parsimony, in which only a subset of characters is informative. But, like distance methods, there is no concept of character-state homology comparable to that which exists in cladistic parsimony. To see this, consider Fig. 4.8, which has been modified from the example presented in the section entitled

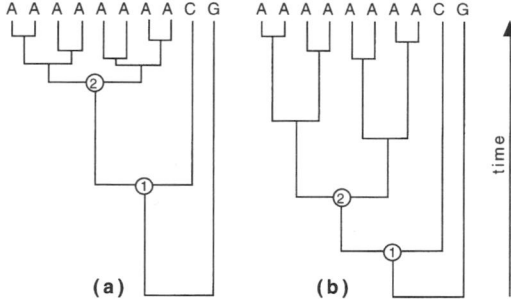

(a) **(b)**

time

Figure 4.8. Nucleotide homologies on gene trees (modeled on Swofford et al., 1996, fig. 9). The relationships among a group of 10 alleles are shown, with character states at a particular position given. Two ancestral nodes (1 and 2) are also shown. The two trees (a and b) are identical in topology but differ considerably in the relative lengths of their branches, as determined by character optimization for parsimony or estimated from the entire data set by ML. In both tree *a* and tree *b*, parsimonious optimization of character states results in an A being assigned to this position in ancestor 2, based on the observation of an A in all its descendants. The As found in all of the descendants of this ancestor are, consequently, homologous, and the A is synapomorphic for this group of alleles. In the maximum likelihood view, there is some probability that ancestor 2 possessed an A at this position, and that probability is lower in tree *b* than in tree *a*, because branches leading from ancestor 2 to the terminal taxa are longer in tree *b*.

"Differences in Perspective between Parsimony and Maximum Likelihood" of Swofford et al. (1996, pp. 428–429, fig. 9). The topologies in (a) and (b) are identical, but differ in relative branch lengths. Imagine that in both cases the topology is supported as the most parsimonious and also the most likely; all members of a clade that are descendants of ancestor 2 are observed to possess an A at a given sequence position. The cladistic interpretation of this observation is identical in both (a) and (b): The A is homologous in these taxa because each ancestor in the clade possessed this A and transmitted it to its descendants. For ML there is instead a probability statement contingent on the branch lengths derived from the character distribution the tree suggests. According to Swofford et al. (1996, p. 429):

from a maximum likelihood perspective, ancestor 2 would have an A in those histories (ancestral state reconstructions) having the highest probability of giving rise to the observed nucleotides. Although histories in which ancestor 2 had a C, G, or T would also

contribute to the overall probability of the specified tree having generated the observed data, if all branches in the subtree were very short, histories with an A at ancestor 2 would contribute the vast majority of the total probability. This is as close as maximum likelihood gets to saying "ancestor 2 had an A."

Thus, it is also as close as ML comes to saying that the A is homologous in the taxa that are observed to possess it, because if the ancestor did not possess an A, then the As found in at least some of its descendants cannot be homologous. With longer branches between ancestor 2 and the observed sequences (Fig. 4.8b), homology of the observed As would become more improbable under ML, because the fact that other characters have changed along the same branches suggests that change is also likely to have occurred at this site. The probability of identity by descent thus decreases even in the absence of any direct character conflict. ML provides a probability statement while cladistic parsimony offers a homology hypothesis; proponents of each view claim that their method is preferable.

Assumptions, Models, and Unobserved Substitutions

There has been much disagreement about the degree to which parsimony analysis does or does not invoke models and require assumptions (e.g., Farris, 1983; Goldman, 1990; Patterson, 1994; Penny et al., 1994; Sanderson, 1995). To a great extent, the debate focuses on what constitutes a model, or, as some have indicated, a *specific* model (Sanderson, 1995); that issue will not be discussed further here.

Computer simulations have shown that in most cases parsimony is statistically consistent (i.e., the method converges on the correct topology as data are added), but that, in agreement with theoretical predictions (Felsenstein, 1978; Steel et al., 1993), incorrect topologies are identified under some circumstances (e.g., Huelsenbeck, 1995). Using these results, it has been stated that parsimony "assumes" relatively equal rates of evolution or overall low rates of evolution (Felsenstein, 1988), or homogeneous nucleotide compositions (Steel et al., 1993). As Swofford et al. (1996, p. 426) put it, ". . . acceptance of an optimal tree under the parsimony criterion requires one to assume that conditions that can cause parsimony to estimate an incorrect tree are unlikely to have occurred." Goldman (1990) has claimed that because "parsimony" optimality solutions can be duplicated by particular choices of parameters in general maximum likelihood models, parsimony can simply be subsumed as one possible model under ML, and not a particularly realistic one at that. Many recommended alternatives to parsimony explicitly use models of molecular evolution in their search for optimal trees (see Chapter 5). It is not the models themselves that are the essential difference, however; indeed, weighting schemes used in parsimony are often based on the same evolutionary models as are used in other methods (e.g., transition bias, relative lability of third codon positions). Rather, the critical difference between parsimony and other methods, once again, is in the role played by individual observations—hypotheses of character-state homology—in the reconstruction of phylogenies. The primary focus of that difference in molecular systematics is unobserved substitutions.

As two homologous sequences diverge from their common ancestor, the number of observed nucleotide changes between them increases. When gene lineages diverge in a hierarchical pattern, these substitutions become the characters that describe the resulting phylogeny. Initially, when few changes have occurred, it is unlikely that a second substitution will occur at a position that has already experienced a substitution. All substitutions that have occurred, therefore, are likely to be observed. But as additional changes accumulate, there is an increasing chance that a substitution will occur at a site that has experienced a previous substitution. Such superimposed, unobserved substitutions (multiple hits) will obscure whatever changes had occurred previously at that site. For an informative character in a parsimony analysis, a later change to any of the three remaining states eliminates evidence of that synapomorphy. If there were a way to know that a further transformation had occurred, that is, that a transformation series 1 → 2 → 3 existed, this would not pose a problem, because state 3 would be recognized as a transformed state of 2, and taxa observed to have state 3 would be scored as having both apomor-

phic conditions for the character. Hypothesizing such transformation series is possible, though difficult, with morphological data, where characters are structurally complex and there are many dimensions to similarity (e.g., Mickevich and Lipscomb, 1991). But because of the simplicity of nucleotide characters, a second transformation in a DNA sequence at best obscures prior homologies by creating an autapomorphy. At worst, a superimposed substitution creates a false homology—a homoplasy—when the transformation is to a state that also occurs in an unrelated taxon. Reversion to a plesiomorphic condition can either produce the appearance of no change or create a homoplasious reversal.

False sister-group relationships are resolved by parsimony analysis when the relative number of homoplasies becomes great enough to swamp out the original phylogenetic signal that links taxa with their true relatives. Homoplasy from both first-hit parallelisms and superimposed changes increases as the amount of divergence increases, but in most circumstances synapomorphies are also expected to increase with divergence and cladogenesis. However, if only some sequences accumulate the preponderance of changes, or if sampling density is low and uneven, the problem of long-branch attraction can arise (e.g., Felsenstein, 1978). When such conditions are simulated, parsimony can become statistically inconsistent (converging more and more strongly on the wrong topology as data are added; e.g., Penny et al., 1992; Huelsenbeck, 1995).

In distance methods, equations such as the Jukes-Cantor (Jukes and Cantor, 1969) or Kimura (1980) two-parameter models "correct" for superimposed changes by providing a term that is added to the observed dissimilarities between pairs of aligned sequences. This term increases as the observed dissimilarity increases, by an amount determined by the values chosen for parameters in the particular model of molecular evolution invoked (e.g., Jukes and Cantor, Kimura, etc.; see Swofford et al., 1996 for general discussion). Thus, although the correction factor (and consequently the particular model invoked) is relatively unimportant at low levels of divergence, it becomes increasingly prominent as observed dissimilarity increases (e.g., Nei,

1987). The ML approach to phylogeny reconstruction also assumes that the greater the observed divergence between any two sequences, the greater the probability that other changes have gone unobserved (see Chapter 5). In ML, however, divergence is not measured by a single pairwise comparison of each aligned sequence with each other sequence as in distance methods. Rather, the probability of change for each individual nucleotide along a given branch of a particular topology depends on the overall length of that branch, as represented by all characters. The exact probability depends on the particular model of molecular evolution, but branch length remains a critical factor in the estimation procedure regardless of the model. Correction for unobserved substitutions is built into the basic working of the ML method, and this is the greatest difference between parsimony and ML, and is claimed as the primary advantage of the latter approach by its proponents (Swofford et al., 1996).

That superimposed substitutions occur as part of the molecular evolutionary process is irrefutable and is not questioned by adherents of cladistic methods. Homoplasy is expected, from this or other sources, and requires ad hoc hypotheses to explain; this is not unique to molecular data. But the key to the attitude of advocates of the cladistic method toward superimposed substitutions is that such substitutions are, quite literally, "unobserved." At the stage of hypothesizing character-state homologies, standard parsimony requires individual observations that, through alignment, constitute individual hypotheses that particular states were inherited from a common ancestor. Superimposed substitutions cannot be observed and therefore cannot meet this minimal requirement of parsimony. This difference between parsimony and ML is far more fundamental than acceptance or rejection of the use of evolutionary models to correct for superimposed substitutions. From the standpoint of the parsimony advocate, it is, quite simply, whether primacy is to be accorded to actual individual observations (data, evidence, homologies) or to assumptions about unobserved character-state changes. Of course, actual individual observations are also used to estimate likelihoods; indeed, all observations contribute to the likelihood score, including those that for

parsimony are considered phylogenetically uninformative (e.g., autapomorphies). But the observations are not discrete hypotheses of homology, as they are in cladistic parsimony.

In their comparison of parsimony and ML, Swofford et al. (1996, p. 426) criticize "some" (unnamed) parsimony adherents for believing that ". . . since the truth is unknowable, we should abandon the search for it and simply choose the most parsimonious solution for its own sake." This belief may be held by some persons; however, if many systematists choose parsimony it is not because they have abandoned the search for an unknowable truth, but because the sequences they have produced are the one thing that they *do* know. Each observed change at an aligned site is evidence that provides a hypothesis of relationship and that evidence, regardless of assumptions about unseen events, requires explanation.

Parsimony requires that observations of individual character transformation be respected, a position that precludes any direct use of presumed but unobserved changes. However, an effort can be made to ameliorate the problems of homoplasy, including homoplasy produced as a result of multiple hits, by giving less weight to characters that either are considered likely to be homoplasious or that actually are observed to be homoplasious on particular trees. A full discussion of weighting is beyond the scope of this chapter; suffice it to say here that various methods that can be applied to individual characters are consistent with the cladistic method. A priori weights can be assigned on the basis of presumed differences among categories of characters (e.g., particular codon positions or paired vs. unpaired nucleotides in rRNA) or character-state transformations (e.g., transitions vs. transversions) based on general presumptions about the molecular evolutionary process; the underlying models may be the same as those used in distance corrections or ML. An important and controversial topic is how to determine the appropriate values to be used (e.g., Wheeler, 1990; Albert and Mishler, 1992; Albert et al., 1993). Perhaps the most troubling aspect of a priori weighting is that a uniform set of weights must be applied to predetermined sets of characters across an entire data set (for 18S rDNA sequences see Chapter 7; for chloroplast genes see Chapter 6). As with alignment, it is unlikely that any one set of weights is correct for all characters of a particular class in any given data set. It may therefore be desirable to avoid, as much as possible, weighting methods that require assumptions about a process (molecular evolution) for which we have a great deal of information but relatively little specific, predictive knowledge. One alternative to a priori weighting is the successive approximations method of Farris (1969; see also Carpenter, 1988, 1994). Successive weighting is an iterative, a posteriori method in which an equally weighted parsimony analysis is first performed, followed by calculation of a weight for each character inversely proportional to its degree of homoplasy on the most parsimonious tree(s). A second analysis is conducted using these weights and new weights are calculated from the resulting trees; the process is continued until the weights and tree topologies stabilize (see Chapter 7 for an application of this approach in analyses of 18S rDNA sequences). Another alternative to a priori weighting is the implied weighting method of Goloboff (1993), which calculates weights during tree search, and thus is independent of tree topologies obtained from an initial unweighted analysis.

Unlike a priori weighting, successive and implied weighting methods do not make explicit assumptions about the evolutionary process. Application of either method to molecular data for protein-coding genes would presumably result in relative downweighting of many third positions, were such positions generally more "noisy" than other positions. However, those first or second positions that proved unreliable would also be downweighted, whereas those third positions that showed low homoplasy would not be downweighted. In both methods, it is the reliability of individual characters—observations, homology hypotheses—as determined against the backdrop of the entire data set that is used in the weighting, and no prior assumption is made that particular classes are less reliable than others. Neither method, however, accounts for differences in reliability of character-state transformations, because weights are applied to all classes of transformation at a given character. Methods do exist that use the principle of a priori weighting but permit character transformation weighting (e.g., Williams and Fitch, 1989, 1990; Fitch and Ye, 1991; Horovitz and Meyer, 1995).

Different Roles for Different Methods

It is useful to know the circumstances under which parsimony analysis may yield misleading results, and our understanding of long-branch attraction and similar problems has added a valuable cautionary note to molecular systematics. But in practice we do not know the conditions actually present during the evolution of the vast majority of taxa studied by systematists. In the absence of such knowledge, standard unweighted cladistic parsimony represents a baseline approach that accounts individually for each observation and that is consistent with the long tradition of systematic practice. It is noteworthy that after extensive experimentation with weighting strategies, this same conclusion is reached for both 18S rDNA (see Chapter 7) and chloroplast genes (see Chapter 6).

Alternatively, data can be explored using methods that permit the incorporation of explicit models of molecular evolution. The use of various methods, with different implicit or explicit assumptions, may be useful in identifying relationships that are sensitive to those assumptions. In the absence of detailed knowledge about the evolutionary parameters associated with any particular taxonomic group, it may be well to regard such "sensitive" topologies with even more than the usual degree of skepticism that should be exercised in the acceptance of any phylogenetic hypothesis.

FROM GENIC TO TAXIC HOMOLOGY

Synapomorphy and Pseudosynapomorphy

Various aspects of genic homology have been discussed in preceding sections. In the context of gene tree structure, the equation of homology with synapomorphy follows from the observation that alleles that lie on any specified branch of a gene tree share the historic presence of the descendant states of all homologous mutations (nucleotide replacements and structural changes) that occurred in their common ancestors. If intragenic recombination has not occurred, hierarchically structured gene trees can be constructed by phylogenetic analysis of observed allelic variation. The hypothesized homologues are those nucleotide substitutions or length mutations that are resolved as synapomorphies of two or more alleles, and, to the extent that the analysis is successful, these can be discriminated from homoplasies. Therefore, the gene tree, if it is reconstructed accurately, depicts each state change and thereby represents every similarity either as a genic homologue or a genic homoplasy. In this manner character differences among allelic variants are used as evidence in phylogenetic analyses, and these analyses in turn help to verify or contradict the initial hypotheses of homology, though every homologue or homoplasy inferred from such an analysis remains provisional and subject to further analysis.

A character may be consistent with a correct gene tree and still may not be an actual homologue. If alleles that are sister to one another in an allele tree experience parallel nucleotide substitutions, those changes will appear to be synapomorphies on a reconstructed gene tree, though they are nonhomologous, and correspondingly, are not synapomorphic. Although such parallel gains cannot in fact be observed, they are homoplasies that are consistent with phylogenetic relationships and are not true indicators of relationship. Characters of this sort are "shared" by two alleles due to parallel mutations in sister alleles; such characters that did not actually originate during a period of common ancestry of the two alleles can be called pseudosynapomorphies.

Gene Trees Versus Taxon Trees

The reconstruction of descent relationships among alleles (or loci) is only the first stage in the search for relationships among hierarchically related organismal groupings (taxa). With a shift in focus from gene trees to taxon trees comes the requirement for further clarification of the concept of homology as synapomorphy, necessary in large part because in molecular systematics taxic relationships are often inferred directly from gene trees. Operationally, this clarification involves the substitution of a taxon name for the name of a particular sequence (e.g., "*Ceratophyllum*" for "*rbcL* in *Ceratophyllum*"), the same process used when the names of areas are substituted for taxa in biogeographic studies (Doyle, 1992; Page, 1993). This substitution process assumes that gene tree synapomorphies are also synapomorphies for the taxon tree, and therefore that genic homology is synonymous with taxic homology. The assumption is clearly not valid if the gene tree and the species tree are incongruent, which under many circumstances is expected to be the case (see Chapter 10). In general terms, genic homology at a single locus is not itself taxic homology, but it contributes to an hypothesis of transformational homology (the gene tree) that may contain a taxic homology component and therefore may be relevant for reconstructing a taxon tree.

For a simple example of the potential for genic and taxic homology to be uncorrelated, consider a set of alleles (A, B, and C) at a locus for which the gene tree relationships and allele distributions among populations of two taxa are known. Alleles B and C are sister alleles, and allele A is sister to the BC allele pair. Allele C is fixed in one taxon, whereas both B and A occur in the other and are intermixed in all populations of that taxon. Although *alleles* B and C share the sorts of genic synapomorphies that have been discussed throughout this paper (nucleotide substitutions, insertions, etc.), *individuals* having allele B interbreed regularly with individuals having allele A and are more closely related to them than to individuals having allele C. Genic synapomorphy is not equivalent to taxic synapomorphy in this case.

Every statement of homology (synapomorphy) is specific to a particular level of descent hierarchy and biological organization, because different evolutionary processes are specific to different levels (see also Doyle, 1997). The genic apomorphies that distinguish alleles are produced by mutations; taxic apomorphies, in contrast, are produced by the apportionment and fixation of allelic polymorphisms among different taxic lineages. A taxic synapomorphy is a character that becomes fixed in an individual species and later is shared as a synapomorphy by all species that are descendants of that species (Nixon and Wheeler, 1992). Consider a gene tree (Fig. 4.9a) and a sequence of taxic divergence events (Fig. 4.9b). A taxon that is fixed for allele A becomes subdivided into two lineages, and, after the cladogenic event, a new allele, B, originates by mutation from allele A. Allele B then becomes fixed in one of the two descendant lineages, and this lineage subsequently undergoes a second cladogenic event. In the resulting sublineage that leads to taxon 3, a new allele, C, originates by mutation from B and becomes fixed in taxon 3. In this example, what were originally autapomorphic nucleotide characters of the ancestral allele B have become genic synapomorphies (i.e., homologues) of alleles B and C (Fig. 4.9a). Phylogenetic analysis of taxa 1, 2, and 3, using as a source of data the gene whose history was just described, and including appropriate outgroups, will correctly resolve 2

and 3 as sister taxa, and 1 as their sister group, if the gene tree is resolved accurately. In this case the history of genetic diversification is congruent with the history of taxic divergence, so the gene tree can be used to track the taxon tree, and genic homologues are also taxic homologues.

In this simple example both the appearance and the fixation of allele B occurred in the interval between cladogenic events, so there was no opportunity for retained polymorphisms to be apportioned into different taxa. However, when allelic polymorphism exists during a speciation event, different alleles or subsets of an allele pool eventually become apportioned among the various descendant species (Fig. 4.9c, d). When such lineage sorting occurs, a genic homologue may mark a nonmonophyletic group of taxa (e.g., Neigel and Avise, 1986; Nei, 1987; Pamilo and Nei, 1988; see Chapter 10), and thus genic synapomorphy and taxic synapomorphy are no longer congruent (Fig. 4.9a, c). Lineage sorting

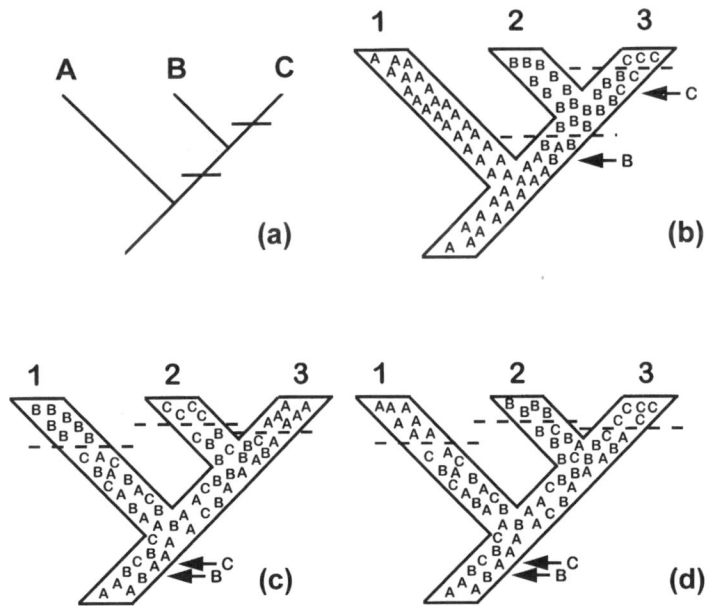

Figure 4.9.   Gene tree and three phylogenies for three taxa (1, 2, and 3) in which this gene occurs. *a.* Descent relationships among alleles A, B, and C. Alleles B and C share one synapomorphic nucleotide substitution, depicted by the lower horizontal bar; allele C has one autapomorphic substitution, depicted by the higher horizontal bar. *b–d.* Times of origin, presence, and fixation of alleles as three lineages diverge from one ancestor. Allele A is present and fixed in the ancestral taxon at the point of origin of each phylogeny; arrows depict times of origin of alleles B and C; dashed lines depict times of fixation of alleles in descendant taxa. *b.* Allele B originates and becomes fixed in an organismal lineage between lineage divergence events; allele C also originates and becomes fixed after a lineage divergence event. *c, d.* Alleles B and C originate prior to any lineage divergence events, but no allele is lost in any lineage until all three lineages have diverged.

does not always result in incongruence between the gene tree and the taxon tree (e.g., Nei, 1987, p. 289), but even so it distorts the relationship between genic and taxic homology. The fixation of sister alleles B and C in sister taxa 2 and 3 (Fig. 4.9d) permits the correct inference of taxic relationships from the gene tree (Fig. 4.9a). However, this is a case of independent fixation of sister alleles and is therefore a taxic pseudosynapomorphy. Thus even when genic and taxic homology apparently are congruent there may be historical differences between them.

Another process often recognized as a potential source of gene tree/taxon tree incongruence is hybridization. Hybridization usually is described in terms of a sequence of events that begins with one or more cladogenic episodes that generate a phylogeny in which the gene tree and the taxon tree are congruent (e.g., Fig. 4.8b). After this phylogenetic structure has developed, a hybridization event allows alleles to migrate between two or more of the descendant taxa; this secondary contact is the source of the incongruence (see discussion and references in Chapter 10). Whereas lineage sorting is a primary process that can lead to incongruence, hybridization is a secondary process. There is, however, an internal contradiction within this narrative, because it explicitly invokes gene flow between taxa while maintaining that the previously existing phylogenetic structure remains the correct tree for the taxa. However, descent relationships among the taxa have been altered by the hybridization event itself, and it is self-contradictory to state that two populations that have just exchanged genes are distantly related. Complex relationships among characters, including incongruence among gene trees, arise in such cases, particularly when introgressive hybridization occurs between well-differentiated taxa (e.g., Anderson, 1949; Arnold, 1994). Hybridization in any case violates the general requirement of phylogenetic analysis that there be a hierarchy to reconstruct (e.g., McDade, 1995; Rieseberg and Morefield, 1995), and as such it will not be considered further.

In cases of gene tree/taxon tree incongruence involving either lineage sorting or hybridization, the relevant unit of incongruence is not an individual nucleotide. Fixation during lineage sorting or transmission during hybridization is not of a single genic homologue (i.e., a single nucleotide) independent of all others, but rather of all nucleotide transformations at all characters of the allele in question. This raises both practical and theoretical concerns with character independence in the taxon tree (e.g., Doyle, 1992, 1995), the source of which is the fundamental difference between genic and taxic homology. Some proposed solutions to the problem treat genes explicitly as sources of transformational homologies for inferring taxic relationships (e.g., Doyle, 1992, 1996).

Multiple Gene Trees: Conflict and Homology

An important aspect of the gene tree/taxon tree problem lies in the fact that systematists seek to discover relationships among taxa, yet they do so with characters, including the nucleotide substitution characters from which gene trees are also constructed. Thus, incongruence between gene trees and taxon trees is recognized in theory as a profound problem, but in the absence of conflicting evidence regarding the structure of "the" taxon tree, incongruence cannot be observed. All that was usually available in the earliest days of plant molecular systematics was a single hypothesis of relationship, often derived from the plastid genome. Systematists frequently noted anecdotally that portions of resolved trees were inconsistent with prior notions of relationship, but were unable to demonstrate incongruence among character sets. With growing numbers of parallel data sets available for various sets of taxa, systematists now find themselves with enough information to be certain that a problem exists.

A growing literature has addressed the problems that arise in phylogenetic studies when multiple gene trees are available for the same sample of taxa and the trees are incongruent (e.g., Bull et al., 1993; Kluge and Wolf, 1993; de Queiroz et at., 1995; Miyamoto and Fitch, 1995; Nixon and Carpenter, 1996; see Chapter 11). If two gene trees are available for a set of taxa, and if the two trees are incongruent with each other, at least one of them is an inaccurate representation of relationships among the species. The reverse, however, is not true; if two gene trees are congruent in all respects they still may provide an inaccurate representation of taxon relationships. This could be the case, for example, if two genes showed the same pattern of lineage sorting, perhaps because they were genetically linked. Hence, systematists are concerned with the problem of incongruence between gene and taxon trees, but this problem is addressed indirectly by examining patterns of incongruence among data sets that represent different genes, and there can be no certainty that any of the individual sources of evidence provides an accurate assessment of relationships.

If the problem of intragenic recombination is set aside for the moment, incongruence between gene trees and taxon trees can be seen to arise from two general sources (see Chapter 10, this volume, for discussion of the many particular sources of such incongruence). First, a gene tree may be resolved incorrectly, which is to say that the gene tree that has been generated from the available observations is not an accurate depiction of descent relationships among the sampled alleles. Error in reconstruction of the gene tree can itself be attributed to any of several causes, including items discussed above, such as homoplasious nucleotide substitutions and misalignment during analysis; what unites these factors is that they contribute to incongruence between the actual descent relationships among alleles and the inferred descent relationships that are resolved by analysis. They are problems of genic homology. The second general source of gene tree/taxon tree incongruence includes the various factors such as lineage sorting that cause correctly resolved gene trees to be incongruent with relationships among taxa and with each other. These are problems of taxic homology. These two very different sources of error may co-occur in any particular analysis and may interact in pernicious ways. For example, the alleles that differ in occurrence among a group of closely related species may be differentiated from each other by small numbers of nucleotide substitutions. Thus there may be weak support for relationships on whatever gene trees are resolved; at the same time there would also be a strong possibility that lineage sorting has influenced the patterns of occurrence of these alleles

among these taxa due to their recency of divergence from a common ancestor.

If genic and taxic sources of incongruence are to be disentangled from one another, additional information is necessary, and this is usually provided by the sampling of one or more additional genes. Because no homology hypothesis can be assumed to be correct, the inclusion of another gene simultaneously introduces a second source of intragenic incongruence and a second potential source of incongruence between gene trees and taxon trees. Both of these phenomena may be manifested as incongruence among gene trees, because even incongruence among nucleotides within a particular genomic region (intragenic incongruence) may cause observed gene trees to be incongruent with one another, if it causes one or more gene trees to be resolved incorrectly. Because every separate single-gene analysis contains its own components of intragenic incongruence, observed incongruence among gene trees cannot be interpreted automatically as evidence that different genes have different descent histories. In short, any or all reconstructed gene trees may be incorrect as gene trees, so incongruence among them does not necessarily imply that taxic and genic homologues are incongruent, that is, that either hybridization or lineage sorting has occurred. At the same time, incongruence between two gene trees could be due to the genes having separate histories, so incongruence does not necessarily imply that genic homologies have been assessed incorrectly. DNA segments in tight linkage with one another are exceptions to this second statement. One of the reasons the plastid genome is considered to be so tractable for phylogeny reconstruction is that intragenomic recombination is regarded as an extremely unlikely possibility (e.g., Olmstead and Palmer, 1994). To the extent that this is true, incongruence within any set of DNA segments sampled from the chloroplast genome can be attributed to homoplasious mutations, whether those segments are from one or several structural genes. Conversely, all variation must be explainable by a single true genome history; genic and taxic homology must be identical if the "taxa" in question are chloroplast genomes, regardless of whether or not the chloroplast genomes track the histories of the plant taxa from which they have been sampled. This property of organellar genomes has been used to test the suitability of different animal mitochondrial genes for phylogeny reconstruction (Miyamoto et al., 1994).

On the other hand, if there has been intragenic recombination, some portion of the observed incongruence is attributable to incongruence among different regions of the gene, because the different regions have different descent relationships (see foregoing discussion). When two or more genes are examined and there has been intragenic recombination in any of them, it is possible that various regions within a single gene will have different descent histories, while some regions of some genes will have descent histories that are identical with only some regions of other genes.

Two general approaches to the summarization of incongruence have been developed: character incongruence analysis and taxon incongruence analysis (e.g., Mickevich and Farris, 1981; Miyamoto, 1985; Kluge, 1989; Kluge and Wolf, 1993; Miyamoto and Fitch, 1995). Character incongruence among portions of a data set can be measured and tested for statistical significance (e.g., Bull et al., 1993; Farris et al., 1994; Huelsenbeck and Bull, 1996; see Chapter 11). Data partitioning tests have at times been employed with the notion that it is meaningful to combine separate data sets only if statistically significant incongruence is not discovered (e.g., Bull et al., 1993). Incongruence among gene trees may suggest hybridization or lineage sorting as potential causes of intergenic incongruence, but any molecular evolutionary process that allows different phylogenetic signals to develop in different genes can account for incongruence just as effectively. Suppose, for example, that two taxa are resolved as sister groups by a particular gene, and that this result is inconsistent with the relationships that are favored by all other genes. Introgression or horizontal gene transfer might explain the signal that is observed in the gene with the incongruent signal (i.e., the two taxa are not sisters but they do bear actual sister alleles at this locus, as the result of a hybridization event). Alternatively, however, a past episode of selection on the locus in question is one of many possible alternative explanations for the incongruent signal, and these cannot be ruled out. When two or more genes support incongruent relationships, each may have a secondary signal that supports the correct relationships, and this signal may be hidden in each case by a primary signal that reflects idiosyncratic biases peculiar to each gene. In such cases, statistically significant levels of incongruence may exist among all pairwise combinations of genes, but simultaneous analysis of the entire data set might allow the secondary signals, which are congruent with each other, to emerge (Kluge and Wolf, 1993; Nixon and Carpenter, 1997). In such cases it would be inadvisable to avoid combining the data from the various genes solely because they provided conflicting signals.

CONCLUSIONS

The homology relationships that systematists hypothesize are translated into character scores in data matrices, but homology is never self-evident. In particular, sequence alignment is of particular concern for molecular systematists because it provides the provisional hypotheses of genic character homology upon which all subsequent analytical steps depend, yet it remains one of the most difficult and least understood stages in the construction of a gene tree. Even apart from problems at this fundamental stage of phylogenetic analysis, the reasons for conflict among homology hypotheses—character incongruence—are many, both at the genic level and when genic homologues are interpreted as taxic synapomorphies. When information from more than one gene is analyzed, the potential for intergenic incongruence exists, but with the additional data also comes the potential for countering the homoplasious elements of signal in any single gene. Parsimony analysis involves the subjection of individual hypotheses of homology to potential corroboration or contradiction by the weight of the discrete homology hypotheses provided by all available data (Ferris, 1983). Furthermore, character optimization (i.e., mapping of transformations) according to the criterion of parsimony allows each character to be evaluated against the best trees, and allows the investigator to make bold hypotheses concerning the history of each character. Parsimony analysis therefore isolates hypotheses of homology during tree construction and during subsequent evaluation, and it allows each of these hypotheses to be scrutinized individually. Cladistic parsimony, in thus allowing the investigator to deal explicitly with genic or taxic homology as synapomorphies of the gene or taxon tree, respectively, is consistent with the history of systematics as an observational and evidential science.

LITERATURE CITED

Albert, V. A., and B. D. Mishler. 1992. On the rationale and utility of weighting nucleotide sequence data. Cladistics **8**:73–83.

Albert, V. A., M. W. Chase, and B. D. Mishler. 1993. Character-state weighting for cladistic analysis of protein-coding DNA sequences. Annals of the Missouri Botanical Garden **80**:752–766.

Altschul, S. F., W. Gish, W. Miller, E. W. Myers, and D. J. Lipman. 1990. Basic local alignment tool. Journal of Molecular Biology **215**:403–410.

Anderson, E. 1949. Introgressive Hybridization. Wiley, New York.

Arnold, M. L. 1994. Natural hybridization and Louisiana irises: Defining a major factor in plant evolution. Bio-Science **44**:141–147.

Baldwin, B. G., M. J. Sanderson, J. M. Porter, M. F. Wojciechowski, C. S. Campbell, and M. J. Donoghue. 1995. The ITS region of nuclear ribosomal DNA: a valuable source of evidence on angiosperm phylogeny. Annals of the Missouri Botanical Garden **82**:257–277.

Bull, J. J., J. P. Huelsenbeck, C. W. Cunningham, D. L. Swofford, and P. J. Waddell. 1993. Partitioning and combining data in phylogenetic analysis. Systematic Biology **42**:384–387.

Carpenter, J. M. 1988. Choosing among multiple equally parsimonious cladograms. Cladistics **4**:291–296.

Carpenter, J. M. 1994. Successive weighting, reliability, and evidence. Cladistics **10**:215–220.

Cracraft, J. 1987. DNA hybridization and avian phylogenetics. Evolutionary Biology **21**:47–96.

Davis, J. I., and K. C. Nixon. 1992. Populations, genetic variation, and the delimitation of phylogenetic species. Systematic Biology **41**:421–435.

Davis, P. H., and V. H. Heywood. 1973. Principles of Angiosperm Taxonomy. Krieger, Huntington, New York.

Dayhoff, M. O., R. M. Schwartz, and B. C. Orcutt. 1978. A model of evolutionary change in proteins. Atlas of Protein Sequence and Structure (suppl. 3) **5**:345–352.

de Pinna, M. C. C. 1991. Concepts and tests of homology in the cladistic paradigm. Cladistics **7**:367–394.

de Queiroz, A., M. J. Donoghue, and J. Kim. 1995. Separate versus combined analysis of phylogenetic evidence. Annual Review of Ecology and Systematics **26**:657–681.

Doyle, J. J. 1992. Gene trees and species trees: molecular systematics as one-character taxonomy. Systematic Botany **17**:144–163.

Doyle, J. J. 1995. The irrelevance of allele tree topologies for species delimitation, and a non-topological alternative. Systematic Botany **20**:574–588.

Doyle, J. J. 1996. Homoplasy connections and disconnections: genes and species, molecules and morphology. **In** Homoplasy and the Evolutionary Process, eds. M. J. Sanderson and L. Hufford, pp. 37–66. Academic Press, New York.

Doyle, J. J. 1997. Trees within trees: genes and species, molecules and morphology. Systematic Biology **46**:537–553.

Doyle, J. J., V. Kanazin, and R. C. Shoemaker. 1996. Phylogenetic utility of histone H3 intron sequences in the perennial relatives of soybean (*Glycine:* Leguminosae). Molecular Phylogenetics and Evolution **6**:438–447.

Farris, J. S. 1969. A successive approximations approach to character weighting. Systematic Zoology **18**:374–385.

Farris, J. S. 1983. The logical basis of phylogenetic analysis. **In** Advances in Cladistics 2, eds. N. I. Platnick and V. A. Funk, pp. 7–36. Columbia University Press, New York.

Farris, J. S., M. Källersjö, A. G. Kluge, and C. Bult. 1994. Testing significance of incongruence. Cladistics **10**:315–319.

Felsenstein, J. 1978. Cases in which parsimony or compatibility methods will be positively misleading. Systematic Zoology **27**:401–410.

Felsenstein, J. 1988. Phylogenies from molecular sequences: inference and reliability. Annual Review of Genetics **22**:521–565.

Feng, D., and R. F. Doolittle. 1987. Progressive sequence alignment as a prerequisite to correct phylogenetic trees. Journal of Molecular Evolution **25**:351–360.

Feng, D., and R. F. Doolittle. 1990. Progressive alignment and phylogenetic tree construction of protein sequences. Methods in Enzymology **183**:375–387.

Fitch, W. M. 1970. Distinguishing homologous from analogous proteins. Systematic Zoology **19**:99–113.

Fitch, W. M., and J. Ye. 1991. Weighted parsimony: does it work? **In** Phylogenetic Analysis of DNA Sequences, eds. M. M. Miyamoto and J. Cracraft, pp. 147–154. Oxford Press, New York.

Freudenstein, J. V., and J. J. Doyle. 1994. Character transformation and evolution in *Corallorhiza* (Orchidaceae: Epidendroideae). I. Plastid DNA. American Journal of Botany **81**:1458–1467.

Funk, V. 1985. Phylogenetic patterns and hybridization. Annals of the Missouri Botanical Garden **72**:681–715.

Gatesy, J., R. DeSalle, and W. Wheeler. 1993. Alignment-ambiguous nucleotide sites and the exclusion of systematic data. Molecular Phylogenetics and Evolution **2**:152–157.

Gilbert, W. 1978. Why genes in pieces? Nature **271**:501.

Goldman, N. 1990. Maximum likelihood inference of phylogenetic trees, with special reference to a Poisson process model of DNA substitution and to parsimony analyses. Systematic Zoology **39:**345–361.

Goloboff, P. A. 1993. Estimating character weights during tree search. Cladistics **9:**83–91.

Hall, B. K., ed. 1994. Homology: The Hierarchical Basis of Comparative Biology. Academic Press, New York.

Hein, J. 1989. A new method that simultaneously aligns and reconstructs ancestral sequence for any number of homologous sequences, when a phylogeny is given. Molecular Biology and Evolution **6:**649–668.

Hein, J. 1990. Unified approach to alignment and phylogenies. Methods in Enzymology **183:**626–644.

Hennig, W. 1966. Phylogenetic Systematics. University of Illinois Press, Urbana.

Higgins, D. G., and P. M. Sharp. 1988. CLUSTAL: a package for performing multiple sequence alignment on a microcomputer. Gene **73:**237–244.

Higgins, D. G., and P. M. Sharp. 1989. Fast and sensitive multiple sequence alignments on a microcomputer. CABIOS **5:**151–153.

Higgins, D. G., A. J. Bleasby, and R. Fuchs. 1992. CLUSTAL V: Improved software for multiple sequence alignment. CABIOS **8:**189–191.

Hillis, D. M. 1994. Homology in molecular biology. In Homology: The Hierarchical Basis of Comparative Biology, ed. B. K. Hall, pp. 339–368. Academic Press, New York.

Hillis, D. M., and C. Moritz, eds. 1990. Molecular Systematics. Sinauer Associates, Sunderland, Massachusetts.

Hillis, D. M., B. K. Mable, A. Larson, S. K. Davis, and E. A. Zimmer. 1996a. Nucleic acids IV: sequencing and cloning. In Molecular Systematics, Second Edition, eds. D. M. Hillis, C. Moritz, and B. K. Mable, pp. 321–381. Sinauer Associates, Sunderland, Massachusetts.

Hillis, D. M., C. Moritz, and B. K. Mable, eds. 1996b. Molecular Systematics, Second Edition. Sinauer Associates, Sunderland, Massachusetts.

Hirosawa, M., Y. Totoki, M. Hoshida, and M. Ishikawa. 1995. Comprehensive study on iterative algorithms of multiple sequence alignment. CABIOS **11:**13–18.

Horovitz, I., and A. Meyer. 1995. Systematics of New World monkeys (Platyrrhini, Primates) based on 16S mitochondrial DNA sequences: a comparative analysis of different weighting methods in cladistic analysis. Molecular Phylogenetics and Evolution **4:**448–456.

Hudson, R. R. 1990. Gene genealogies and the coalescent process. Oxford Surveys in Evolutionary Biology **7:**1–44.

Huelsenbeck, J. P. 1995. Performance of phylogenetic methods in simulation. Systematic Biology **44:**17–48.

Huelsenbeck, J. P., and J. J. Bull. 1996. A likelihood ratio test to detect conflicting phylogenetic signal. Systematic Biology **45:**92–98.

Jukes, T. H., and C. R. Cantor. 1969. Evolution of protein molecules. In Mammalian Protein Metabolism, ed. H. N. Munro, pp. 21–132. Academic Press, New York.

Kimura, M. 1980. A simple method for estimating evolutionary rate of base substitutions through comparative studies of nucleotide sequences. Journal of Molecular Evolution **16:**111–120.

Kjer, K. M. 1995. Use of rRNA secondary structure in phylogenetic studies to identify homologous positions: an example of alignment and data presentation from the frogs. Molecular Phylogenetics and Evolution **4:**314–330.

Kluge, A. G. 1989. A concern for evidence and a phylogenetic hypothesis of relationships among *Epicrates* (Boidae, Serpentes). Systematic Zoology **38:**7–25.

Kluge, A. G., and A. J. Wolf. 1993. Cladistics: what's in a word? Cladistics **9:**183–199.

Konings, D. A. M., P. Hogeweg, and B. Hesper. 1987. Evolution of the primary and secondary structures of the E1a mRNAs of the adenovirus. Molecular Biology and Evolution **4:**300–314.

Koshi, J. M., and R. A. Goldstein. 1995. Context-dependent optimal substitution matrices. Protein Engineering **8:**641–645.

Lake, J. A. 1991. The order of sequence alignment can bias the selection of tree topology. Molecular Biology and Evolution **8:**378–385.

Levinson, G., and G. A. Gutman. 1987. Slipped-strand mispairing: a major mechanism for DNA sequence evolution. Molecular Biology and Evolution **4:**203–221.

Maddison, W. P. 1995. Phylogenetic histories within and among species. In Experimental and Molecular Approaches to Plant Biosystematics, eds. P. C. Hoch, A. G. Stevenson, and B. A. Schaal, pp. 273–287. Missouri Botanical Garden, St. Louis.

McClure, M. A., T. K. Vasi, and W. M. Fitch. 1994. Comparative analysis of multiple protein-sequence alignment methods. Molecular Biology and Evolution **11:**571–592.

McDade, L. A. 1995. Hybridization and phylogenetics. In Experimental and Molecular Approaches to Plant Biosystematics, eds. P. C. Hoch, A. G. Stevenson, and B. A. Schaal, pp. 305–331. Missouri Botanical Garden, St. Louis.

Mickevich, M. F., and J. S. Farris. 1981. The implications of congruence in *Menidia*. Systematic Zoology **30:**351–370.

Mickevich, M. F., and D. Lipscomb. 1991. Parsimony and the choice between different transformations for the same character set. Cladistics **7:**111–140.

Mindell, D. 1991. Aligning DNA sequences: homology and phylogenetic weighting. In Phylogenetic Analysis of DNA Sequences, eds. M. J. Miyamoto and J. Cracraft, pp. 73–89. Oxford University Press, New York.

Miyamoto, M. M. 1985. Consensus cladograms and general classifications. Cladistics **1:**186–189.

Miyamoto, M. M., and W. M. Fitch. 1995. Testing species phylogenies and phylogenetic methods with congruence. Systematic Biology **44:**64–76.

Miyamoto, M. M., M. W. Allard, R. M. Adkins, L. L. Janecek, and R. L. Honeycutt. 1994. A congruence test of reliability using linked mitochondrial DNA sequences. Systematic Biology **43:**236–249.

Nagylaki, T., and T. D. Petes. 1982. Intrachromosomal gene conversion and the maintenance of sequence homogeneity among repeated genes. Genetics **100:**315–337.

Needleman, S. B., and C. D. Wunsch. 1970. A general method applicable to the search for similarities in the amino acid sequences of two proteins. Journal of Molecular Biology **48**:443–453.

Nei, M. 1987. Molecular Evolutionary Genetics. Columbia University Press, New York.

Neigel, J. E., and J. C. Avise. 1986. Phylogenetic relationships of mitochondrial DNA under various demographic models of speciation. **In** Evolutionary Processes and Theory, eds. S. Karlin and E. Nevo, pp. 515–534. Academic Press, New York.

Nixon, K. C., and J. M. Carpenter. 1996. On simultaneous analysis. Cladistics **12**:221–241.

Nixon, K. C., and Q. D. Wheeler. 1992. Extinction and the origin of species. **In** Extinction and Phylogeny, eds. M. J. Novacek and Q. D. Wheeler, pp. 119–143. Columbia University Press, New York.

Notredame, C., and D. G. Higgins. 1996. SAGA: sequence alignment by genetic algorithm. Nucleic Acids Research **24**:1515–1524.

Ohta, T. 1984. Some models of gene conversion for treating the evolution of multigene families. Genetics **106**:517–528.

Olmstead, R. G., and J. D. Palmer. 1994. Chloroplast DNA systematics: a review of methods and data analysis. American Journal of Botany **81**:1205–1224.

Page, R. D. M. 1993. Genes, organisms, and areas: the problem of multiple lineages. Systematic Biology **42**:77–84.

Palmer, J. D., R. A. Jorgensen, and W. F. Thompson. 1985. Chloroplast DNA variation and evolution in *Pisum*: patterns of change and phylogenetic analysis. Genetics **109**:195–213.

Pamilo, P., and M. Nei. 1988. Relationships between gene trees and species trees. Molecular Biology and Evolution **5**:568–583.

Patterson, C. 1982. Morphological characters and homology. **In** Problems of Phylogenetic Reconstruction, eds. K. A. Joysey and A. E. Friday, pp. 21–74. Academic Press, London.

Patterson, C. 1987. Introduction. **In** Molecules and Morphology in Evolution: Conflict or Compromise? ed. C. Patterson, pp. 1–22. Cambridge University Press, Cambridge.

Patterson, C. 1988. Homology in classical and molecular biology. Molecular Biology and Evolution **5**:603–625.

Patterson, C. 1994. Null or minimal models. **In** Systematics Association Special Volume, Vol. 52, Models in Phylogeny Reconstruction, eds. R. W. Scotland, D. J. Siebert, and D. M. Williams, pp. 173–192. Oxford University Press, Oxford.

Penny, D., M. D. Hendy, and M. A. Steel. 1992. Progress with methods for constructing evolutionary trees. Trends in Ecology and Evolution **7**:73–79.

Penny, D., P. J. Lockhart, M. A. Steel, and M. D. Hendy. 1994. The role of models in reconstructing evolutionary trees. **In** Systematics Association Special Volume, Vol. 52, Models in Phylogeny Reconstruction, eds. R. W. Scotland, D. J. Siebert, and D. M. Williams, pp. 211–230. Oxford University Press, Oxford.

Reeck, G. R., C. de Häen, D. C. Teller, R. F. Doolittle, W. M. Fitch, R. E. Dickerson, P. Chambon, A. D. McLachlan, E. Margoliash, T. H. Jukes and E. Zuckerkandl. 1987. "Homology" in proteins and nucleic acids: a terminology muddle and a way out of it. Cell **50**:667.

Rieger, R. 1991. Glossary of Genetics: Classical and Molecular. Springer-Verlag, New York.

Rieseberg, L. H., and J. D. Morefield. 1995. Character expression, phylogenetic reconstruction, and the detection of reticulate evolution. **In** Experimental and Molecular Approaches to Plant Biosystematics, eds. P. C. Hoch, A. G. Stevenson, and B. A. Schaal, pp. 333–353. Missouri Botanical Garden, St. Louis.

Roth, V. L. 1988. The biological basis of homology. **In** Ontogeny and Systematics, ed. C. J. Humphries, pp. 1–26. Columbia University Press, New York.

Roth, V. L. 1991. Homology and hierarchies: problems solved and unresolved. Journal of Evolutionary Biology **4**:167–194.

Sanderson, M. J. 1995. Objections to bootstrapping phylogenies: a critique. Systematic Biology **44**:299–320.

Sanderson, M. J., and M. J. Donoghue. 1989. Patterns of variation in levels of homoplasy. Evolution **43**:1781–1795.

Sanderson, M. J., and J. J. Doyle. 1992. Reconstruction of organismal phylogenies from multigene families: paralogy, concerted evolution, and homoplasy. Systematic Biology **41**:4–17.

Sankoff, D. D., and R. J. Cedergren. 1983. Simultaneous comparison of three or more sequences related by a tree. **In** Time Warps, String Edits, and Macromolecules: The Theory and Practice of Sequence Comparison, eds. D. Sankoff and J. B. Kruskal, pp. 253–264. Addison-Wesley, Reading, Massachusetts.

Schaeffer, S. W., and E. L. Miller. 1992. Estimates of gene flow in *Drosophila pseudoobscura* determined from nucleotide sequence analysis of the alcohol dehydrogenase region. Genetics **132**:471–480.

Schaeffer, S. W., C. F. Aquadro, and W. W. Anderson. 1987. Restriction-map variation in the alcohol dehydrogenase region of *Drosophila pseudoobscura*. Molecular Biology and Evolution **4**:254–265.

Sibley, C. G., and J. E. Ahlquist. 1987. Avian phylogeny reconstructed from comparisons of the genetic material, DNA. **In** Molecules and Morphology in Evolution: Conflict or Compromise? ed. C. Patterson, pp. 95–121. Cambridge University Press, Cambridge.

Simpson, G. G. 1961. Principles of Animal Taxonomy. Columbia University Press, New York.

Smith, G. P. 1974. Unequal crossover and the evolution of multigene families. Cold Spring Harbor Symposium in Quantitative Biology **38**:507–513.

Smith, T. F., and M. S. Waterman. 1981. Identification of common molecular subsequences. Journal of Molecular Biology **147**:195–197.

Soltis, D. E., L. A. Johnson, and C. Looney. 1996. Discordance between ITS and chloroplast topologies in the *Boykinia* group (Saxifragaceae). Systematic Botany **21**:169–185.

Stebbins, G. L. 1947. Types of polyploids: their classification and significance. Advances in Genetics **1**:403–429.

Steel, M. A., P. J. Lockhart, and D. Penny. 1993. Confidence in evolutionary trees from biological sequence data. Nature **364**:440–442.

Stevens, P. F. 1991. Character states, morphological variation, and phylogenetic analysis: a review. Systematic Botany **16**:553–583.

Swofford, D. L., and G. J. Olsen. 1990. Phylogeny reconstruction. **In** Molecular Systematics, eds. D. M. Hillis and C. Moritz, pp. 411–501. Sinauer Associates, Sunderland, Massachusetts.

Swofford, D. L., G. J. Olsen, P. J. Waddell, and D. M. Hillis. 1996. Phylogenetic inference. **In** Molecular Systematics, Second Edition, eds. D. M. Hillis, C. Moritz, and B. K. Mable, pp. 407–514. Sinauer Associates, Sunderland, Massachusetts.

Sytsma, K. J. 1990. DNA and morphology: inference of plant phylogeny. Trends in Ecology and Evolution **5**:104–110.

Thompson, J. D. 1995. Introducing variable gap penalties to sequence alignment in linear space. CABIOS **11**:181–186.

Thorne, J. L. and H. Kishino. 1992. Freeing phylogeny from artifacts of alignment. Molecular Biology and Evolution **9**:1148–1162.

Vingron, M., and M. S. Waterman. 1994. Sequence alignment and penalty choice: review of concepts, case studies and implications. Journal of Molecular Biology **235**:1–12.

Waterman, M. S., J. Joyce, and M. Eggert. 1991. Computer alignment of sequences. **In** Phylogenetic Analysis of DNA Sequences, eds. M. J. Miyamoto and J. Cracraft, pp. 59–72. Oxford University Press, New York.

Wendel, J. F., A. Schnabel, and T. Seelanan. 1995. Bidirectional interlocus concerted evolution following allopolyploid speciation in cotton (*Gossypium*). Proceedings of the National Academy of Sciences U.S.A. **92**:280–284.

Wheeler, W. C. 1990. Combinatorial weights in phylogenetic analysis: a statistical parsimony procedure. Cladistics **6**:269–276.

Wheeler, W. C. 1995. Sequence alignment, parameter sensitivity, and the phylogenetic analysis of molecular data. Systematic Biology **44**:321–331.

Wheeler, W. C., and D. S. Gladstein. 1993. MALIGN: Program and Documentation, version 1.73. American Museum of Natural History, New York.

Wheeler, W. C., and D. S. Gladstein. 1994. MALIGN: a multiple sequence alignment program. Journal of Heredity **85**:417–418.

Wheeler, W. C., J. Gatesy, and R. DeSalle. 1995. Elision: a method for accommodating multiple molecular sequence alignments with alignment-ambiguous sites. Molecular Phylogenetics and Evolution **4**:1–9.

Williams, P. L., and W. M. Fitch. 1989. Finding the minimum change in a given tree. **In** The Hierarchy of Life, eds. B. Fernholm, K. Bremer, and H. Jörnvall, pp. 453–470. Elsevier, Amsterdam.

Williams, P. L., and W. M. Fitch. 1990. Phylogeny determination using dynamically weighted parsimony method. Methods in Enzymology **183**:615–626.

Zimmer, E. A., S. L. Martin, S. M. Beverley, Y. W. Kan, and A. C. Wilson. 1980. Rapid duplication and loss of genes coding for the a chain of hemoglobin. Proceedings of the National Academy of Sciences U.S.A. **77**:2158–2162.

5

Maximum Likelihood as an Alternative to Parsimony for Inferring Phylogeny Using Nucleotide Sequence Data

Paul O. Lewis

Much of what we know about the phylogenetic relationships among organisms has resulted from the application of parsimony methods to discrete character data. Huelsenbeck (1995) reported that 59% of phylogenetic analyses reported in the journal *Systematic Biology* over the past decade used parsimony rather than compatibility, invariants, distance, or likelihood methods. Parsimony will no doubt continue to be important for many years to come, especially in the analysis of large data sets involving many taxa, where the speed of parsimony methods will always allow them to conduct a much more thorough examination of the space of tree topologies than many other methods. Given the speed of parsimony and its demonstrated ability to produce reasonable phylogenetic hypotheses, are there any good reasons to entertain other methods or adopt criteria other than the principle of parsimony for choosing among competing tree topologies? If it could be shown that other optimality criteria produced trees that are consistently more accurate representations of the underlying phylogeny, most practicing systematists would probably abandon parsimony in favor of the better criterion. Unfortunately, be-cause we are reconstructing (inferring) past historical events, we have no way of determining *with certainty* the effectiveness of any criterion, including parsimony, at recovering the true phylogeny.

We do have ways of measuring the relative accuracy of phylogenetic optimality criteria under controlled conditions. We can perform computer simulations (Huelsenbeck, 1995; Huelsenbeck and Hillis, 1993; Kuhner and Felsenstein, 1994; Nei, 1991) and for some cases obtain analytical results (Felsenstein, 1978), but these depend on specifying both a detailed model tree and the rules for changing sequences along the branches of the tree. Such simulation experiments are nevertheless valuable, as they provide some means of measuring the efficacy of different criteria against a standard, known set of circumstances. The value of appraisals based on computer simulations depends directly on the validity of the model used to create the simulated data, although if a method behaves poorly for quite simplistic models of evolution, it is difficult to imagine it doing well under more realistic conditions (Huelsenbeck, 1995). A similar caution applies to inferences about the perfor-

I would like to thank those who helped make this a better chapter by providing stimulating conversation and much needed criticism, including Jeff Doyle, Louise Lewis, David Swofford, Jeff Thorne, and one anonymous reviewer.

mance of different optimality criteria based on artificial phylogenies constructed using experimental organisms such as bacteriophage in the laboratory (Bull et al., 1993; Hillis, 1995). Once again, the model tree problem arises, for there is no way to know whether the branching patterns imposed on bacteriophage lineages reflect branching patterns in real phylogenies, and substitution processes certainly differ from those in nature due to the necessity of using mutagens to speed up the process. These types of experiments do, however, uncover another side to the relative performance of various phylogeny algorithms, as they introduce complexities that are missing in the comparatively clean world of computer simulation.

Whereas much has been learned using computer simulations and bacteriophage experiments, we must admit that our preference for one criterion over others is, at its heart, a largely philosophical decision. Even if computer simulation studies overwhelmingly preferred one criterion over others, this may (in the worst case scenario) have no relevance at all to the question of which of these methods is actually the best at recovering the true phylogeny. It would simply reflect the fact that one method was better at solving the puzzle presented to it under the artificial conditions of the simulation model. It is important to keep in mind that no one ever knows for certain which method is best in an absolute sense, and therefore our judgment must fall back on other criteria.

I argue in the remainder of this chapter for a model-based (specifically maximum likelihood) approach to phylogenetic inference. These alternatives to parsimony are model-based because they depend on an explicit model of evolutionary change. Model-based approaches comprise distance methods, which involve a matrix of pairwise estimated evolutionary distances between taxa, and the method of maximum likelihood, which, like parsimony, makes use of the original discrete data matrix.

My argument for using model-based approaches is grounded in the ability to build into these models what we feel we already know about phylogenies and substitution processes. Model-based methods also allow testing of the underlying models used to make the phyloge-

netic inference (Goldman, 1993; Penny et al., 1994; Yang et al., 1994). It is not necessary to assume that a model is adequate; the adequacy of the model being used can be objectively measured. Although this takes considerably more time and computational effort, researchers should explore their data enough to avoid at least the most obvious pitfalls of the phylogenetic method they plan to use.

I appeal to the desire on the part of systematists to tailor their analyses to the particular details of the group and gene under study. That such a desire exists is apparent in the evolution of parsimony itself—the descendants of the ancestral version of parsimony (Camin and Sokal, 1965, Camin-Sokal parsimony) sport modifications that allow: (1) reversibility of character-state changes (Kluge and Farris, 1969, Wagner parsimony); (2) asymmetrical rates of change (Farris, 1977, Dollo parsimony); (3) lack of a natural ordering of character states (Fitch, 1971, Fitch parsimony); (4) differences in rates of change within and between certain classes of character states (Sankoff, 1975, weighted parsimony); and (5) differences across characters in levels of homoplasy (due, for example, to among-character differences in selective constraints or rates of evolution) (Farris, 1969, successive weighting). Modifications to standard, equal-weighted parsimony resulted from the recognition that not all data are alike, and that either adding or relaxing some of the constraints of standard parsimony were reasonable in view of the perceived differences in the nature of different data sets.

The primary aim of this chapter is to provide an intuitive explanation of the maximum likelihood (ML) approach to phylogenetic inference, highlighting some of the interesting insights and hypothesis tests that the ML approach allows. Because both ML and distance methods make use of many of the same evolutionary models, much of the discussion applies equally well to both approaches to phylogenetic inference. The concentration on maximum likelihood allows a more in-depth discussion of this method, which, being a discrete-character method, provides an alternative to parsimony both in terms of phylogeny reconstruction and estimation of ancestral character states. I first discuss models in general, with the goal of showing how making the model explicit can add flexibility to an analysis rather than making it more restrictive and assumption-laden (as is commonly assumed). The next major section describes specific substitution models currently used in phylogenetic inference. The chapter finishes with an examination of how likelihood methods are especially useful in framing and testing specific hypotheses. Enumerating all substitution models and tree-building methods is not an aim of this chapter; I instead concentrate on basic concepts. An attempt is made to use explanations that do not rely heavily on mathematical notation, however this has necessarily resulted in the omission of many details needed for a complete understanding of the methods presented. Some of these details have been placed in footnotes, but for those who find they wish to learn more, the reviews by Felsenstein (1988), Swofford et al. (1996), and Weir (1996, chap. 10) are recommended.

MODELS

A common misconception of model-based approaches is that they are restrictive; that is, a model forces the investigator to make many assumptions about the evolutionary process that do not need to be made with model-free approaches. This misconception is not the case, but to understand why, it is important to understand first what models are and how they are used. The goal when reconstructing phylogeny using nucleotide sequence data is to use circumstantial evidence (present-day sequences) to arrive at an educated guess (an *inference*) about the evolutionary history of a group of organisms or genes (Felsenstein, 1973b). We cannot, in general, observe any part of this history directly. Despite the deterministic nature of the algorithms used for phylogeny reconstruction, these algorithms simply provide rules for arriving at an *estimate* of the phylogeny, which may or may not be the same as the true underlying phylogeny. The quality of the estimate of phylogeny depends upon the quality of the assumptions being made in the process of arriving at the estimate. It is difficult, if not impossible, to arrive at an estimate of anything without making assump-

tions, although it is quite easy to fail to realize the specific assumptions being made. We often make assumptions in everyday life that go unrecognized unless they are violated. Simply crossing a street makes use of a model that has been in the process of fine tuning by experience since birth. This model provides, for instance, the very important probability of getting hit by a car while crossing the street based on such factors as the distance to the nearest moving car and the perceived speed at which that car is traveling. There are a number of assumptions that are made by this model of which we are usually only dimly aware. For example, our model assumes that we will not develop a serious leg cramp, drop our keys, or suffer any other major distraction while in the middle of the street!

All Statistical Inferences Are Based on a Model

Suppose someone told you that he or she had just flipped a coin 50 times, and all 50 times the coin turned up heads! You are likely to be very surprised at hearing this result, until it is revealed that the coin is a trick coin with heads on both sides. Once this fact is known, it is not surprising at all that each of the 50 tosses resulted in heads. Even though you were probably not aware of the fact, you were using two different models before and after the revelation that the coin was a trick coin. The first model involved the assumption that the coin used was fair and the tosses were independent of one another. The second model assumed that the coin was a two-headed trick coin. Under the fair-coin model, the data (50 tosses resulted in heads) are very improbable. The probability in this case is about one in 1,000 trillion, and you are appropriately very surprised. Under the second model, the probability of the data is 1.0, and you are not surprised at all. The point is that the level of surprise you feel when you observe the data is inversely related to the probability of the data under the currently assumed model.

The two models discussed are but two points along a continuum of conceivable models. It is possible, for example, to have a two-tailed coin in which the probability of heads (call it p) for any given toss is exactly 0.0. It is also possible to have a coin that is intermediate; in fact, many coins are undoubtedly unfair to some extent due to variation introduced in the manufacturing process and the asymmetry of design on the front versus the back. The continuum of conceivable models thus stretches all the way from $p = 0.0$ to $p = 1.0$. In such circumstances, it is usual to refer to this entire continuum not as an infinity of distinct models, but as a single model having the parameter p. This makes it clear that p is not something to which we can assign a value with absolute certainty. In either view, it is necessary to hypothesize some numerical value for p in order to obtain the probability of the data. When the parameters of a model are assigned actual values, the model is said to be *fully specified*. Whether one considers the coin flipping example to involve a comparison of two distinct models (fair coin vs. two-headed coin) or two different parameter settings of a single model ($p = 0.5$ vs. $p = 1.0$) is largely a matter of personal preference, but it is certainly true that these two alternatives represent distinct, fully specified models.

In model-based phylogeny reconstruction, the fully specified model includes all the information needed to calculate the probability of the data, including not only the assumption of character independence, but also the topology of the tree. One possible method for reconstructing phylogeny would be to consider a number of distinct, fully specified models, all differing only with respect to the tree topology (and associated branch lengths), and choose as best the one making us the least surprised when we look at the data. The problem with this approach is that sequence data are much more complicated than the data in the coin flipping example above. We were able to register the appropriate level of surprise in that case because the data and the competing models were simple. The reason computers must be used to estimate phylogenies from sequence data is that we are unable to formulate an appropriate response (i.e., degree of surprise) regardless of the model assumed because of the complexity of the data presented to us. The method of maximum likelihood uses as its optimality criterion the probability of the data given an assumed model of evolution. It works by essentially telling us how surprised we

should be were we able to comprehend the data fully. This method instructs us to accept the (fully specified) model making the data the least surprising.

Is There a Parsimony Model?

If the term *model* is defined rather loosely to encompass all the assumptions necessary to estimate some quantity of interest from the data, then all methods of reconstructing phylogeny can be said to use a model. Few would deny that even methods not explicitly model based, such as parsimony, assume at least that different sites (characters) are independent and that a bifurcating tree (acyclic graph) adequately depicts the underlying phylogeny. However, can parsimony be said to depend on a model in the statistical sense of the term? That is, can parsimony be viewed as a model-based approach, the only difference between parsimony and likelihood being the model used? If this were true, these two discrete character methods could be evaluated solely on the relative merits of the models used.

Models have been developed that, when used in a maximum likelihood context, produce results identical to those from standard parsimony. That is, if one selects the maximum likelihood tree under such a model, it will always be identical to the tree favored by standard, equally weighted parsimony. Although several such models exist (Farris, 1973, 1983; Felsenstein, 1973a, 1981b; Goldman, 1990; Penny et al., 1994), I have chosen to consider only one of them in detail. The model presented by Goldman (1990) simply assumes that the same amount of evolutionary change (on average) occurs on every branch of the phylogenetic tree. In order to see the relationship between likelihoods obtained assuming Goldman's model and tree lengths obtained by applying the parsimony criterion, it is useful to apply both to a simple, contrived data set (Table 5.1). It will be assumed for simplicity that there are only two character states, 0 and 1, but there is no difficulty generalizing to four states (Goldman, 1990).

There are only three possible unrooted tree topologies for four taxa, and Fig. 5.1 shows, for each of these three topologies, one possible most parsimonious mapping of character-state

Table 5.1.   Sample data set with four taxa (A, B, C, D) and four binary characters (1, 2, 3, 4).

	1	2	3	4
A	1	1	0	1
B	1	0	0	0
C	0	1	1	0
D	0	0	1	0

Table 5.2.   Comparison of likelihood to tree length for each possible combination of tree topology and ancestral character-state assignments.

Topology	Character	Steps	ln-Likelihood
Tree 1	1	1	−7.545
$\hat{p} = 0.25$	2	2	−8.644
	3	1	−7.545
	4	1	−7.545
	Total	5	−31.280
Tree 2	1	2	−9.521
$\hat{p} = 0.3$	2	1	−8.674
	3	2	−9.521
	4	1	−8.674
	Total	6	−36.390
Tree 3	1	2	−10.547
$\hat{p} = 0.35$	2	2	−10.547
	3	2	−10.547
	4	1	−9.928
	Total	7	−41.568

Note: The value of p maximizing the overall likelihood ($\hat{p}$) for each topology was used to compute the numerical value of the likelihood listed for each character individually. Because the likelihoods are such small numbers, the natural logarithm of the likelihood score is reported instead of the likelihood itself. This simply has the effect of scaling the reported values. Tree 1 requires the fewest steps and has the highest likelihood.

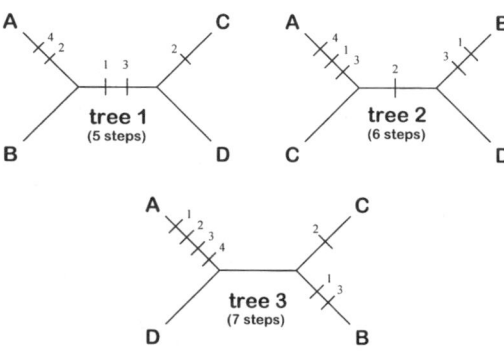

Figure 5.1.   The three possible unrooted tree topologies for four taxa (A, B, C, and D) showing one possible most parsimonious reconstruction of character-state transformations for each topology. The most parsimonious tree topology requires a total of 5 steps (tree 1).

changes. The most parsimonious topology is Tree 1, which requires only 5 steps compared to the 6 steps required by Tree 2 and the 7 steps required by Tree 3.

Under Goldman's model, Tree 1 also has the highest likelihood score of the three (Table 5.2). Consider the data for character 3 on the topology represented by Tree 1. Assume for the moment that the ancestral character states are known and are those depicted in Fig. 5.2. Goldman's model assumes that there is a probability p of a change in state along any branch of the tree, which implies (because there are only two possible states) that the probability of no change along any given branch is $1 - p$. The likelihood is defined as the probability of the data given the model (Edwards, 1972). The data in this case are the states of the four tip nodes for character 3; that is, taxon A and taxon B both have state 0, and taxon C and taxon D both have state 1. The model includes the following parameters: (1) the tree topology; (2) the parameter p; and (3) the states for the interior nodes of the tree

topology. Once actual values have been plugged in for each of these parameters, the model is fully specified and a likelihood can be computed.

The likelihood under this simple model equals the probability of starting with the state present at the root node (which is simply one-half; it is assumed that the root node had an equal chance of being in state 0 or state 1) times the product of probabilities corresponding to the events along each of the five branches (Fig. 5.2). One feature of this model is that it does not matter which node is considered to be the root; the likelihood is the same for each possible choice of root node. Each hypothesized combination of ancestral character states potentially results in a different likelihood. For character 3, there are four possible combinations of ancestral character states for Tree 1 (Fig. 5.3). The likelihood is highest (and the number of required steps fewest) for the character-state assignment shown in Fig. 5.3b.

Table 5.2 compares the maximum likelihood scores to the parsimony tree length for each pos-

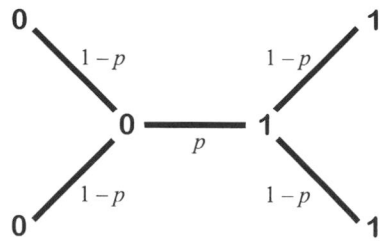

$$L = \tfrac{1}{2}(1-p)(1-p)(p)(1-p)(1-p)$$
$$= \tfrac{1}{2}p(1-p)^4$$
$$= \Pr(\text{state at root}) \times \Pr(\text{change}) \times [\Pr(\text{no change})]^4$$

Figure 5.2. Calculation of the likelihood under the Goldman parsimony model. The probability of the data (tip states) for character 3 (Table 5.1) and hypothesized ancestral-state assignments equals the probability of starting with state 0 in one of the two tips on the left side of the tree (the one-half term) times the probability of changing across the middle branch (the single p term) times the probability of not changing across the remaining four branches (the $1-p$ terms). The likelihood is the same regardless of which node is specified as the root of the tree, and the two possible states at the root (0 and 1) are considered to have equal prior probabilities.

sible combination of topology and ancestral character-state assignment. For this example, Tree 1 is best according to both the parsimony criterion and the criterion of maximum likelihood; it requires the fewest inferred character-state changes and also has the highest overall likelihood. It can be shown (Goldman, 1990) that there is a one-to-one correspondence between parsimony tree length (assuming standard, equal-weighted parsimony) and likelihood (assuming Goldman's model) as long as the probability of change (p) along each branch is less than the probability of no change ($1 - p$). Although not encountered in this example, if there were multiple most parsimonious trees, there would also be multiple maximum likelihood trees using the Goldman model because of this one-to-one correspondence.

It should be emphasized that the point of this exercise was to illustrate that it is possible, using a likelihood framework, to compare the parsimony and likelihood criteria directly, which was also the motivation behind the development of the model (Goldman, 1990). The primary value in considering Goldman's model is that it brings

a.

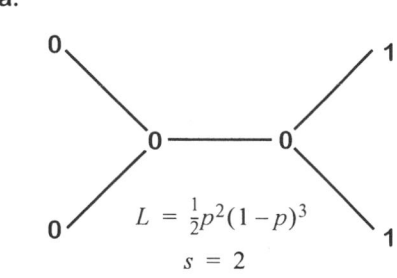

$$L = \tfrac{1}{2}p^2(1-p)^3$$
$$s = 2$$

b.

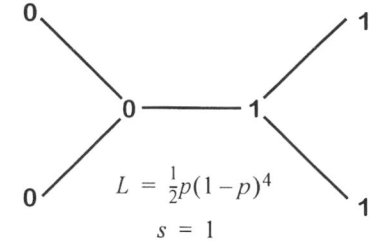

$$L = \tfrac{1}{2}p(1-p)^4$$
$$s = 1$$

c.

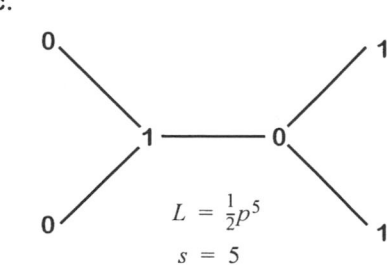

$$L = \tfrac{1}{2}p^5$$
$$s = 5$$

d.

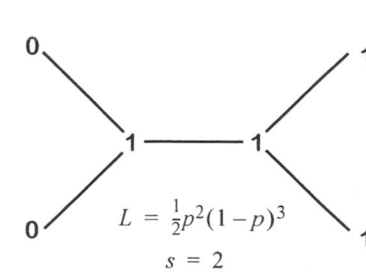

$$L = \tfrac{1}{2}p^2(1-p)^3$$
$$s = 2$$

Figure 5.3.   Likelihoods (L; according to the parsimony model of Goldman) and most parsimonious tree lengths (s) for the four possible combinations of ancestral character states for tree 1 (Fig. 5.1), character 3 (Table 5.1). Note that the most parsimonious combination of character-states occurs in b, which also has the highest likelihood score assuming only that change along individual branches is rare compared to no change (i.e., $p < 0.5$).

to light some potentially hidden assumptions of standard parsimony. For instance, the Goldman model assumes that the probability of character change is the same for all branches in a tree. This contrasts with models commonly used in conjunction with the maximum likelihood criterion, which allow each branch to have its own probability of change, thus allowing for variation in rates of evolution (or simply lineage ages) across the tree. A second difference between the Goldman model and the commonly used maximum likelihood phylogeny models is that (in the latter) the likelihood for a tree does not depend upon any particular combination of ancestral character-state assignments. To eliminate this dependence, the likelihood is normally summed over all possible combinations of ancestral character states. This could have been done in generating the likelihood values for Table 5.2, but the one-to-one correspondence between maximum parsimony and maximum likelihood would no longer be guaranteed; that is, the model could no longer be considered a "parsimony model."

Has this side-by-side comparison of parsimony and likelihood indeed allowed us any insight into hidden assumptions underlying the parsimony criterion? Under the Goldman model, the infamous long-branch attraction phenomenon (discussed below in *Consequences of Ignoring "Important" Factors*) may be viewed as the consequence of violating the model's central assumption that all branches (in the true tree) are equal in length.[1] Does this mean that parsimony *assumes* that all branches have the same length (i.e., the probability of change is the same for all branches)? Not necessarily. In order to make this jump, it is necessary to show that the Goldman model is the *only* model that produces a one-to-one correspondence between maximum likelihood trees and maximum parsimony trees. If it could be shown that another model exists that also produces this one-to-one correspondence but which does not assume that all branches have the same probability of

change, it would be hard to claim that parsimony assumes equal branch lengths. It turns out that at least one such alternative model has been found (Penny et al., 1994). This fact leaves open the question of whether a parsimony model can be said to exist. It appears that the parsimony criterion can be mapped to more than one distinct type of model when viewed in a likelihood framework.

A Possible Common Denominator

One feature that is present in all such parsimony models identified to date is that the number of parameters requiring estimation grows with the amount of data available, leading to potential instability in the inferences made (Felsenstein, 1978). In the example above, two extra parameters need to be estimated for every new character added to the data set. These two extra parameters are the states at the two interior nodes. It is not possible to obtain a parsimony score without assigning some state to these interior nodes. This is not the case with typical maximum likelihood phylogeny models, in which the likelihood is summed over all possible states at these interior nodes. This avoids "committing" to any one combination (or few combinations) of ancestral states. The same problem arises in other areas of biology. For example, one approach to assessing male fertility in plants is to determine the most likely male parent for each seed sampled in a population. The male fertility can be estimated for each male parent as the proportion of seeds sired by that parent (Meagher, 1986). Unfortunately, this approach tends to overestimate the reproductive success of males homozygous at marker loci (Devlin et al., 1988). This bias can be eliminated if the estimation method takes into consideration all possible male parents for each seed rather than the single most likely male parent (Devlin et al., 1988; Roeder et al., 1988).

The fact that the usual likelihood models do not commit to any particular combination of ancestral states for the purposes of reconstructing phylogeny does not mean that likelihood methods *cannot* commit to particular combinations of ancestral states if the goals of the investigation change. For example, once the investigator has

---

[1]It is interesting to note that parsimony can perform quite well even at very high rates of change so long as all branches are equal in length (Hillis et al., 1994; Huelsenbeck and Hillis, 1993), a behavior that is expected under Goldman's model.

estimated the phylogeny, attention may turn toward estimating ancestral states, in which case it becomes desirable to find the "most likely" states for each ancestral node for each character. The concern for bias in estimating the tree topology is now past, because the topology is fixed during this phase of the study. Likelihood provides alternatives for estimating ancestral character states, and these are discussed in The Perspective of Likelihood on Reconstruction of Ancestral States.

Models Can Add Flexibility Rather Than Restrictions

The development of variations on the parsimony theme reflect a desire to add flexibility of inference to parsimony. There is a certain satisfaction in making use of our hard-won knowledge of the evolution of DNA sequences in order to make inferences about that which we do not know, namely the phylogenetic history of the sequences in our sample. Model-based methods are often criticized as requiring too many assumptions, but in fact models can be quite restrictive *or* very flexible, depending on how many model parameters are estimated versus fixed at some specified value.

Adding parameters to a model increases the model's flexibility. For example, adding a transition/transversion rate ratio parameter allows the model to account for any bias that might exist in the relative rates of transitions compared to transversions (regardless of the underlying cause of such bias). If this parameter is fixed at some value, then the quality of the analysis will depend to some extent on the quality of this fixed parameter value. This may be quite valid in cases in which a parameter does not vary much among lineages or genes. The ratio of transitions to transversions is nearly always greater than 0.5, which is the value expected if there were no difference in underlying rates of transitions and transversions. Thus, specifying any value greater than 0.5 (up to a point, of course) should provide better performance than using a simpler model (e.g., the Jukes and Cantor, model [Jukes and Cantor, 1969]) that assumes a transition/transversion ratio of 0.5.

We need not fix this parameter at a specific value, however. It can be estimated from the data

along with the other parameters (such as branch lengths) in the model. This approach requires only that the existence of two classes of substitutions be recognized (e.g., transition-type vs. transversion-type), allowing for the possibility that there is a real substitution bias in nature. Note that this approach also allows for the possibility that *there is no such bias* (i.e., there is no real difference between the classes recognized in the model). In the absence of bias, the estimated parameter value will simply be close to 0.5.

The Best Models Capture Only the Essential Features of the Process

The best model is one that captures only those features of reality that are necessary to answer the question at hand; more detailed models simply cloud the picture. For example, maps depicting metro rail systems are notoriously skimpy with details. There are few, if any, details about the topography of the land or the city streets that are crossed, and even most of the actual curves in the rail line itself have been smoothed out into impressively straight lines. Two things that are preserved, however, are the order of the stations along the lines and the general direction in which the lines travel; this is, of course, exactly the information that commuters need. A lot of additional cartographic detail would only make it more difficult to figure out how many stations are between the commuter and his or her destination.

In terms of models used for statistical inference, the model should include only enough parameters to capture the most important aspects of the process being modeled (here, the substitution process). This philosophy is reflected in one of the most well-known measures used in comparing models, the Akaike information criterion (Akaike, 1974), which counterbalances the increase in the fit of the model to the data when additional parameters are added by penalties for these additional parameters. One effect of adding too many parameters in the context of phylogenetic inference is that the distinctions between alternative tree topologies become increasingly small (i.e., the picture becomes cloudy). The extreme case is the multinomial model (Navidi et al., 1991; Goldman, 1993), in which a separate parameter governs the probability of each possible pattern of nucleotides in the data. This model can be used to provide an upper bound on the likelihood (Goldman, 1993) because it will always provide a better fit to the data than models that assume a tree structure. It is useless, however, for purposes of phylogeny reconstruction because it provides no way to rank alternative trees: all trees have the same likelihood under this model!

Consequences of Ignoring "Important" Factors

Thus, flexibility comes with a cost in terms of the ability to discriminate among trees, and it pays to explore the data in order to discover the factors most important for the particular group, and gene, under study. How does one define "most important" in this case? The most important factors are those that, if not accounted for

by the model, lead us to select the incorrect tree topology. Recently there has there been a concerted effort to develop models incorporating those factors thought to be most important (Table 5.3).

In many cases, the pathological behavior resulting from failure to account for an important factor is *long-branch attraction,* a phenomenon in which lineages exhibiting a relatively large amount of evolutionary change (the long branches) are placed together in the estimated tree even if there are intervening branches in the true tree. Longer-branched lineages may share similarities due to independent substitutions to the same base from different ancestral bases. The extent of the long-branch attraction problem is directly related to the degree of heterogeneity in branch lengths and inversely related to the number of possible states. Estimating the branch lengths for each lineage can reduce or eliminate long-branch attraction by allowing the model to

Table 5.3.   Substitution models (and model variations) described since the classic model of Jukes and Cantor (1969), along with the molecular evolutionary factors they were designed to accommodate.

Evolutionary Factor	Described Model or Model Variation
Transition rate potentially differs from transversion rate	(Hasegawa et al., 1985; Kimura, 1980; Tamura and Nei, 1993)
Equilibrium base frequencies potentially not all 0.25	(Felsenstein, 1981a; Hasegawa et al., 1985)
Base frequencies and transition/transversion ratio potentially vary from lineage to lineage	(Yang and Roberts, 1995)
Substitution rates potentially vary from site to site	(Churchill et al., 1992; Reeves, 1992; Sidow and Speed, 1992; Yang, 1993; Yang, 1994b; Felsenstein and Churchill, 1995)
Nonindependence among sites due to codon structure	(Goldman and Yang, 1994; Muse and Gaut, 1994)
Nonindependence among sites due to stem-loop secondary structure (e.g., in rRNA genes)	(Muse, 1995; Tillier and Collins, 1995)
Nonindependence between neighboring sites	(Felsenstein and Churchill, 1995; Schöniger and von Haeseler, 1994)
Rate of A $\leftrightarrow$ G transitions potentially differs from rate of C $\leftrightarrow$ T transitions	(Tamura and Nei, 1993)
Rate of A $\leftrightarrow$ T, C $\leftrightarrow$ G transversions potentially differs from rate of A $\leftrightarrow$ C, C $\leftrightarrow$ G transversions	(Kimura, 1981)
Rates of A $\leftrightarrow$ C, A $\leftrightarrow$ G, A $\leftrightarrow$ T, C $\leftrightarrow$ G, C $\leftrightarrow$ T, and C $\leftrightarrow$ G substitutions all may differ	(Tavaré, 1986; see also Yang, 1994a)

Note: This list is hopefully representative, but by no means should be considered exhaustive. Several very general models (e.g., Barry and Hartigan, 1987; Lake, 1994; Steel, 1994) are not presented here because, while very useful for estimating evolutionary distances (see for example Lockhart et al., 1994), they are currently not able to be employed in a likelihood context.

explain these similarities some other way than by inferring a shared history (Gaut and Lewis, 1995; Huelsenbeck, 1995).

Thus, it becomes less surprising that taxa A and C (Fig. 5.4a) have the same state at several sites if we know that a large amount of change has taken place along the lineages leading to taxon A and taxon C. You might now ask how we know that these lineages have undergone a tremendous amount of change. This information comes, in part, from autapomorphous sites; that is, sites in which all taxa except one have the same state. If most autapomorphies are restricted to the lineages leading to taxon A and taxon C, this provides evidence for an elevated substitution rate in these two lineages. Thus, while autapomorphies are not considered "phylogenetically informative" in a parsimony analysis, they do provide information that in some cases may be critical to getting the right answer.

It is possible for long-branch attraction to occur even if individual branch lengths are estimated. Failing to account for the presence of site-to-site rate heterogeneity provides an example of a circumstance in which this situation can occur (Chang, 1996; Gaut and Lewis, 1995).

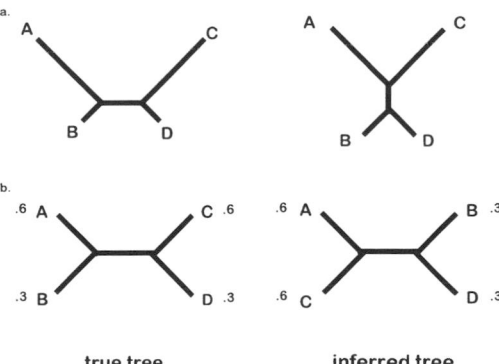

true tree   **inferred tree**

Figure 5.4. Illustration of long-branch attraction and base compositional attraction. a. A phylogeny having long branches leading to taxon A and taxon C can fool phylogenetic methods that do not address branch-length heterogeneity. The length of a branch is proportional to the amount of evolutionary change occurring along the lineage represented by the branch. b. phylogeny in which taxon A and taxon C have independently acquired high G+C contents (0.6). The inferred tree will place A and C together unless the model used allows for nucleotide composition to change across the tree.

Failing to account for nonindependence among sites within a sequence provides another example (Schöniger and von Haeseler, 1995).

A second type of pathological behavior could be termed *base compositional attraction*. Here again taxa share similarities not due to identity by descent, but the spurious similarities are due to similar G+C content in unrelated lineages. Like long-branch attraction, in this situation unrelated lineages are joined because of similarities that are not the result of simply having a shared history. One of the best examples is the misplacement of the *Euglena* chloroplast in a tree derived from 16S rDNA sequences of chloroplasts and cyanobacteria (Lockhart et al., 1994). Consider two unrelated lineages in which substitutions to A or to T for some reason no longer occurred. Over time, these two lineages would increase in G+C content (but not because their most recent common ancestor had a high G+C content). If the two lineages suffered substitutions at homologous sites, the chances that these substitutions would result in both lineages having a G or both having a C at these sites are considerably greater than if these lineages were allowed to use all four nucleotides in equal proportions (Fig. 5.4b).

Genes encoding proteins critical to cellular respiration and/or photosynthesis are all but guaranteed to have a well-developed codon structure, whereas adjacent sites within a pseudogene may be independent or nearly so. One way to deal with dependence among nucleotide sites within codons is by simply using the amino acid sequences rather than the underlying nucleotide sequence, but this often results in a paucity of information because much of the evolutionary change is in the form of synonymous substitutions. Recently models have been developed that make use of nucleotide sequence data and at the same time allow nonindependence of sites within codons (Goldman and Yang, 1994; Muse and Gaut, 1994). Another example of nonindependence among sites involves the stem-loop structure of ribosomal DNA. Variation in paired sites within stem structures is constrained compared to variation in sites present in loops (see Chapter 7). Models have recently been developed that explicitly account for this type of gene structure (Schöniger and von

Haeseler, 1994; Muse, 1995; Tillier and Collins, 1995).

What Are the Components of the Model?

Normally, when someone speaks of the model being used in a phylogenetic analysis, he or she is referring to the substitution model; however, the model always encompasses more than just the substitution process. To compute a likelihood, one must also specify such things as a particular tree topology and a particular set of branch lengths. Thus, the topology and branch lengths are components of the model in this broader sense. The substitution models in common use require a bit more explanation than these other model components because they arise from a consideration of the properties of stochastic processes. The next section is therefore devoted to an introduction to the basic features of substitution models.

SUBSTITUTION MODELS

The Parking Lot Metaphor

Imagine that it is the busiest time of the day in the parking lot of the local shopping mall. If you keep your eye on a single parking spot for a long enough time, you will see what is termed a *substitution event:* The car currently parked in that spot will leave and be "instantly" replaced by another car (instantly because this is the busiest time of the day and there is tremendous competition for parking spaces). Let us further suppose that the cars in this town only come in four colors: red, green, blue, and yellow. Furthermore, assume that the four car colors are equally frequent, both in town and at the parking lot of the mall. When a substitution event occurs, it is important to note that we are just as likely to see a red car being replaced by another red car as we are of seeing a red car being replaced by a car of a different color. Let us reserve the term *color-changing event* for the case of a substitution event that results in a car of a different color. Thus, color changes are less common than substitutions, because some proportion of substitutions are by cars of the same color as the previous occupant. Now, suppose it is known that, on average, substitution events occur at a rate of one substitution every 1,800 seconds (one every half hour). How many substitutions would be expected, on average, if one parking space were watched for three hours? There should be an average of

$$\left(\frac{1 \text{ substitution}}{1800 \text{ seconds}}\right)(10800 \text{ seconds})$$
$$= 6 \text{ substitutions} \qquad (5.1)$$

The number 6 is the product of the substitution rate, which will be symbolized μ, and the time t. The product μt is thus the expected number of substitutions per parking space over the time period represented by t.

I have already made the distinction above between ordinary substitutions and color-changing substitutions. Suppose we were interested in the expected number of color changes rather than the expected number of substitutions. In working out simple Mendelian genetics problems, a device called the Punnett square is used to determine the expected numbers of offspring of different genotypes given the proportions of the different haplotypes produced by the parents. We can use a similar approach to find the expected numbers of different types of substitutions occurring in the parking lot. Figure 5.5a shows all possible combinations of the color of the car being replaced and the color of the replacing car. For simplicity, these colors will be referred to as the "before" and "after" colors.

To obtain the expected number of color-changing substitutions, we simply add together all the relevant squares. Given that 12 of the 16 squares represent changes in color, the expected number of color changes is simply $12(\frac{1}{16}\mu t) = \frac{3}{4}\mu t$. Because the expected number of substitutions per parking space was 6 in a 3-hour period, the expected number of color changes per parking space is thus 4.5.

Note that the same principles would apply if the different colors were not equally frequent; in such case the one-fourth terms on the margins of the square in Fig. 5.5 would be replaced with the correct color frequencies. For example, suppose the relative frequencies of the four colors of automobile were $\pi_r = \frac{1}{10}$ (red), $\pi_g = \frac{1}{5}$ (green), π_b

a.

after

		red $\frac{1}{4}$	green $\frac{1}{4}$	blue $\frac{1}{4}$	yellow $\frac{1}{4}$
	red $\frac{1}{4}$	$\frac{1}{16}\mu t$	$\frac{1}{16}\mu t$	$\frac{1}{16}\mu t$	$\frac{1}{16}\mu t$
before	green $\frac{1}{4}$	$\frac{1}{16}\mu t$	$\frac{1}{16}\mu t$	$\frac{1}{16}\mu t$	$\frac{1}{16}\mu t$
	blue $\frac{1}{4}$	$\frac{1}{16}\mu t$	$\frac{1}{16}\mu t$	$\frac{1}{16}\mu t$	$\frac{1}{16}\mu t$
	yellow $\frac{1}{4}$	$\frac{1}{16}\mu t$	$\frac{1}{16}\mu t$	$\frac{1}{16}\mu t$	$\frac{1}{16}\mu t$

b.

after

		red π_r	green π_g	blue π_b	yellow π_y
	red π_r	$\pi_r^2 \mu t$	$\pi_r \pi_g \mu t$	$\pi_r \pi_b \mu t$	$\pi_r \pi_y \mu t$
before	green π_g	$\pi_g \pi_r \mu t$	$\pi_g^2 \mu t$	$\pi_g \pi_b \mu t$	$\pi_g \pi_y \mu t$
	blue π_b	$\pi_b \pi_r \mu t$	$\pi_b \pi_g \mu t$	$\pi_b^2 \mu t$	$\pi_b \pi_y \mu t$
	yellow π_y	$\pi_y \pi_r \mu t$	$\pi_y \pi_g \mu t$	$\pi_y \pi_b \mu t$	$\pi_y^2 \mu t$

Figure 5.5.   Punnett square for determining the proportion of the total expected number of replacements (μt) corresponding to each different combination of before and after colors. *a.* Assuming each car color occurs in equal frequency. *b.* Allowing each car color to be present in a different frequency, with the frequency of red cars being π_r, the frequency of green cars being π_g, the frequency of blue cars being π_b, and the frequency of yellow cars being π_y.

$= \frac{3}{10}$ (blue), and $\pi_y = \frac{2}{5}$ (yellow). Adding up the contributions of the 12 squares representing color-changing substitutions in this case (see Fig. 5.5b) results in the formula below, which makes use of the fact that the sum over all squares is μt.

$$(1 - \pi_r^2 - \pi_g^2 - \pi_b^2 - \pi_y^2)\mu t \quad (5.2)$$

Plugging in the actual values of the relative frequencies produces the result that for every 6 substitutions expected, $(1 - \frac{1}{100} - \frac{1}{25} - \frac{9}{100} - \frac{4}{25})(6) = 4.2$ replacements should be expected. Note that this value is close to, but not equal to, the value 4.5 obtained for the case of equal color frequencies. Differences in values for specific squares are more noticeable. For example, $(\frac{1}{16})(6) = 0.375$ substitutions are ex-

pected to occur between blue and yellow cars in the case of equal frequencies, but a much higher number $(\frac{3}{10})(\frac{2}{5})(6) = 0.72$ is expected in the case of the unequal base frequencies used in the foregoing example.

Evolutionary Distances

We are now in good position to make the jump from parking lots to DNA, where nucleotide sites are analogous to parking spaces and cars of four different colors analogous to the four bases. Letting the rate of nucleotide substitution be μ, the expected number of nucleotide-changing substitutions between two sequences separated by an amount of time t is

$$d = (1 - \pi_A^2 - \pi_C^2 - \pi_G^2 - \pi_T^2)\mu t \quad (5.3)$$

If all four bases are equal in frequency, then $\pi_A = \pi_C = \pi_G = \pi_T = 0.25$ and Eq. (5.3) reduces to just

$$d = \frac{3}{4}\,\mu t \qquad (5.4)$$

These formulas are directly analogous to the formulas obtained previously for replacements of cars in parking spaces. The quantity d (the expected number of nucleotide-changing substitutions) is often referred to as the evolutionary distance between the two sequences. The concept of an evolutionary distance is central to model-based methods for inferring phylogeny. In both distance and maximum likelihood methods, the lengths of a tree's branches are usually measured in terms of evolutionary distances. Distance methods work by finding the tree topology and branch lengths that together best fit a matrix of pairwise evolutionary distances. Maximum likelihood methods work by finding the particular combination of tree topology and branch lengths that maximizes the probability of the observed data.

The evolutionary distance in Eq. (5.4) is called the Jukes-Cantor, or JC, distance, because it is based on the substitution model proposed by Jukes and Cantor (1969). This model assumes that the four bases are present in equal frequencies and there is a single rate of substitution that applies to all possible types of transformations. These same assumptions were made in the parking lot analogy: All four colors of car are present in equal frequency and the substitution rate did not depend on the car color. The more general distance formula given by Eq. (5.3) allows for differences in frequency between the four bases. This distance formula corresponds to the Felsenstein (1981a) model, and will be referred to as the F81 distance. The F81 distance formula collapses to the JC distance formula when the four base frequencies are equal. The F81 model is a good example of how flexibility can be added to a model by the addition of parameters without sacrificing generality. The F81 distance formula with its extra frequency parameters works just as well as the JC distance for cases in which the base frequencies are equal, but has the benefit of also working in cases of unequal base frequencies.

Estimating an Evolutionary Distance

A small problem presents itself if a formula like Eq. (5.3) is used to estimate an evolutionary distance from real sequence data. The problem is that such formulas require that the quantity μt be estimated. Unfortunately, obtaining the exact value of μt requires continuous observation. For parking lots, this value could be obtained by recording every substitution that occurred. In the case of sequences, we never have that luxury: We have to work with two sequences that have been diverging for some time from a common ancestral sequence, and unfortunately no one was around to monitor both lineages continuously for substitution events.

The time can be defined in one of two ways (Fig. 5.6), but in this chapter t always refers to the total amount of time along the path from one sequence to the other (Fig. 5.6a). Most substitution models assume that the substitution process has no memory. That is, it is assumed that an adenine nucleotide at a particular site has a certain probability of changing at any given instant to a cytosine, regardless of what nucleotide was present at that site in the recent or distant past. Such models are termed *time reversible,* as likelihoods computed using them are the same regardless of where the root of the tree is assumed

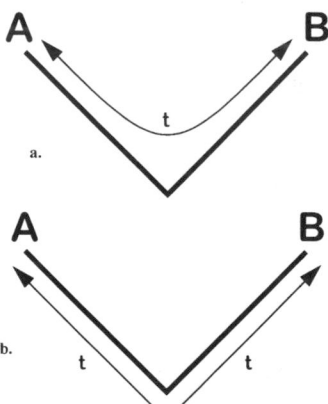

Figure 5.6.   Two ways to measure the amount of time that has elapsed between two sequences. *a.* Some workers define time *t* to be the total amount of time along the path from sequence A to sequence B. *b.* Others define time *t* to be just the amount of time between either sequence and the common ancestor. In the latter case, the total time along the path from A to B is 2*t*.

to lie. All that is visible today are the endpoints of this time line; the substitutions that occurred along the path between the two sequences were not observed and hence could not be counted. This situation is analogous to conducting two surveys in our parking lot. Suppose that, during each survey (which were spaced, say, 1 hour apart) the color of car in every one of the 1,000 spaces in the entire lot was recorded. After the second survey, it would be possible to fill in a four-by-four matrix (see Fig. 5.7a) showing the proportion of each type of color change (or non-change) that occurred over the 2-hour period in the whole parking lot.

Note that, because no observer was present for a full hour, this divergence matrix does *not* represent a Punnett-square-style breakdown of the total number of substitutions that occurred,

because undoubtedly many of these substitutions were missed. Instead, the elements of this matrix record only the effect of the last substitution to occur before the second survey was taken (if indeed any substitutions occurred). Even though there is less information in these survey data than could be obtained by continuous observation, there is still some basis for estimating μt and thus obtaining a value that can be plugged into Eq. (5.2).

For any specified value for μt, a prediction can be made concerning the expected values of the 16 cells in the divergence matrix. The link between μt and these expected values is provided by the theory of stochastic processes. A stochastic event is one that may occur at a predictable rate but is not able to be predicted exactly. An example of a typical stochastic event is

a.

	after			
before	red	green	blue	yellow
red	0.083	0.061	0.053	0.058
green	0.050	0.088	0.051	0.062
blue	0.054	0.042	0.085	0.05
yellow	0.049	0.062	0.049	0.103

b.

	after			
before	red	green	blue	yellow
red	$\frac{1}{4}[1-P(t)]+\frac{1}{16}P(t)$	$\frac{1}{16}P(t)$	$\frac{1}{16}P(t)$	$\frac{1}{16}P(t)$
green	$\frac{1}{16}P(t)$	$\frac{1}{4}[1-P(t)]+\frac{1}{16}P(t)$	$\frac{1}{16}P(t)$	$\frac{1}{16}P(t)$
blue	$\frac{1}{16}P(t)$	$\frac{1}{16}P(t)$	$\frac{1}{4}[1-P(t)]+\frac{1}{16}P(t)$	$\frac{1}{16}P(t)$
yellow	$\frac{1}{16}P(t)$	$\frac{1}{16}P(t)$	$\frac{1}{16}P(t)$	$\frac{1}{4}[1-P(t)]+\frac{1}{16}P(t)$

Figure 5.7.   Observed and expected divergence matrices for the parking lot example. *a.* Four-by-four divergence matrix showing the proportions of observed substitutions of each different type for the 1-hour period between surveys. *b.* Four-by-four expected divergence matrix in terms of the probability $P(t)$ that at least one substitution occurred during the time period t that elapsed between surveys.

the failure of a light bulb. While it is not possible to know exactly how long any given light bulb will burn before failing, it is possible to describe the process statistically. The graph in Fig. 5.8 shows the relationship between time (horizontal axis) and the probability of failure of a particular type of light bulb (vertical axis). As time goes on, it becomes increasingly probable that the burning bulb will fail. Because the probability of failure cannot surpass 1.0, while time can continue to infinity, such plots must necessarily flatten out and asymptotically approach 1.0. Substitution events are analogous stochastic processes: It is impossible to specify when the next substitution will occur, but the probability of seeing at least one substitution event after waiting a specified amount of time will follow a trajectory similar to that of the burning light bulb with respect to time. This probability of seeing at least one substitution event as a function of time will is referred to hereafter as $P(t)$. With substitution events, as with light bulb failure events, time goes on indefinitely whereas the probability $P(t)$ is constrained to be less than 1.0. An interesting and useful fact for both burning bulbs and substitution events is that the shape of the relationship between time and $P(t)$ is determined completely by the slope of a line tangent to the curve at time zero (Fig. 5.8)!

Every distinct curve relating time to $P(t)$ has associated with it a unique value for this slope. This slope is called the *instantaneous rate* and equals the quantity μ that forms an integral part of the formula for estimating an evolutionary distance. Thus, the relationship depicted in Fig. 5.8 provides the means for estimating μt from survey data (parking lot example) or sequences sampled only at the endpoints of a timeline extending from one species to another passing through their common ancestor.

In order to see how this estimation might be accomplished, compare the divergence matrix resulting from comparison of the before and after surveys of the parking lot (Fig. 5.7a) to the expected value of this matrix (assuming equal car color frequencies), which is given in terms of the probability of at least one substitution $P(t)$ (Fig. 5.7b). As an example, the element corresponding to "red before, blue after," has an expected value equal to

Pr(red car before) $\times$ Pr(at least one subst.)
$\times$ Pr(last subst. results in blue car) $= \frac{1}{4}P(t)\frac{1}{4}$
$= \frac{1}{16}P(t)$

Thus, the expected values of the off-diagonal elements of the divergence matrix are determined using the Punnett-square approach,

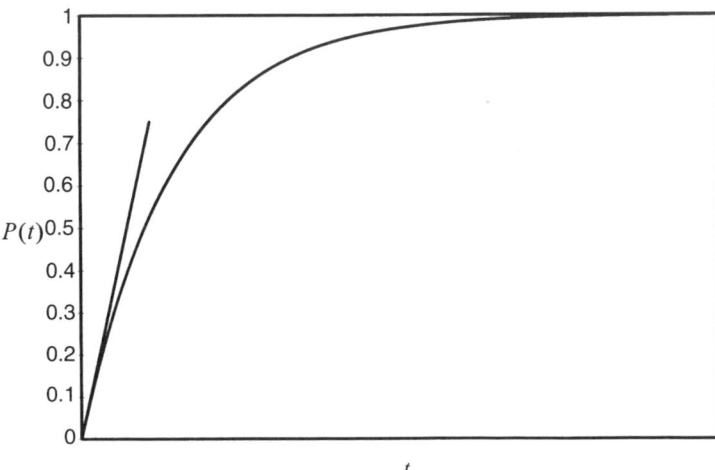

Figure 5.8. The curved line represents the relationship of time t to the probability of light bulb failure (or at least one substitution). The straight line tangent to the curve depicts the slope of the curve at the first instant of time (instantaneous rate of change). After an infinite amount of time has elapsed, the curve becomes flat (zero slope), and the probability of failure (or at least one substitution) is 1.0.

divvying up the probability of at least one substitution in a way similar to the divvying up of the total number of substitutions (μt) in Fig. 5.5. The expected values of the diagonal elements are different, however, in that for these elements the possibility that no substitutions occurred must be considered. The expected value for the element "red before, red after" is therefore computed as follows:

Pr(red car before) $\times$ Pr(at least one subst.)
$\times$ Pr(last subst. results in red car)
$+$ Pr(red car before) $\times$ Pr(no subst.)
$= \frac{1}{4}P(t)\frac{1}{4} + \frac{1}{4}[1 - P(t)]$

The first term above, $\frac{1}{16}P(t)$, is identical to the expected value of each off-diagonal element. The second term, $\frac{1}{4}[1 - P(t)]$, is what accounts for the possibility that no substitutions occurred. Estimation of μt can be accomplished by simply varying the value of μt until the expected divergence matrix approximates as closely as possible the observed divergence matrix.[2] Increasing the value of the product μt leads to higher values of $P(t)$, which in turn lead to increased expected values for the off-diagonal elements and correspondingly decreased values for the diagonal elements. Decreasing μt has the opposite effect. Once the optimal value for μt has been obtained, the number of color-changing substitutions can be estimated using the same formula as before, $\frac{3}{4}\mu t$. Even though the parking lot metaphor has been used to describe the principles underlying this estimation, the mechanics are identical for estimating evolutionary distances under the JC model.

In the two substitution models we have considered thus far (JC and F81), a single instantaneous rate is assumed to apply to all 16 possible transformations between bases. In many cases, the assumption of a single rate is not appropriate. Consider two primate mitochondrial DNA sequences, one human and one from chimpanzee (Hayasaka et al., 1988), each of length 896 nucleotides. A matrix showing the number of observed differences falling into each of the 16 possible substitution categories is shown in Fig. 5.9. It is instructive to assume a particular substitution model and use it to predict the number of differences for each of the 16 categories. Using the JC model forces us to assume that all bases are equally frequent. Using the technique described above wherein μt is adjusted until the expected values in the divergence matrix approximate as closely as possible the observed values, the best estimate[3] of μt turns out to be 0.126752. Note that, even using this optimal value for μt, the predicted numbers are noticeably different from the numbers actually observed (Fig. 5.10a). The most conspicuous disagreement is in the number of sites in which the human and chimpanzee sequences both have the base G. The observed value for this cell is 82, whereas the JC model predicts the value to be 202.7! One problem in using the JC model in this case is its inflexibility with respect to inequalities in base frequency. The empirical base frequencies estimated from these two sequences are $\pi_A = 0.307$, $\pi_C = 0.328$, $\pi_G = 0.104$, and $\pi_T = 0.261$. This much variation in base frequencies makes the use of the JC model questionable.

A better fit can be obtained using the F81 model, which is identical to the JC model except

		Pan			
		A	C	G	T
Homo	A	261	1	9	2
	C	1	267	0	29
	G	14	0	82	0
	T	1	23	0	206

Figure 5.9. Matrix showing the numbers of each type of transformation between a human mitochondrial DNA sequence and the homologous sequence in chimpanzee. Data from Hayasaka et al. (1988).

[2]While a plot of the relationship between μt and $P(t)$ was presented in Fig. 5.8, the actual functional form of this relationship has not been presented. This function is $P(t) = 1 - e^{-\mu t}$ for this example. The term $e^{-\mu t}$ is the Poisson probability of observing *no* substitution events over the time period t given that the substitution rate is μ.

[3]The free computer program MLCalc, written by the author to accompany this chapter, can be obtained using the directions supplied in the section entitled Software at the end of this chapter. This program, among other things, allows the user to estimate parameters for this example by trial-and-error or using the built-in numerical optimization algorithm.

a.

		Pan			
		A	C	G	T
Homo	A	202.7	7.1	7.1	7.1
	C	7.1	202.7	7.1	7.1
	G	7.1	7.1	202.7	7.1
	T	7.1	7.1	7.1	202.7

b.

		Pan			
		A	C	G	T
Homo	A	241.7	11.7	3.7	9.4
	C	11.7	269.9	4.1	10.5
	G	3.7	4.1	83.3	3.3
	T	9.4	10.5	3.3	215.7

Figure 5.10. Expected numbers of differences between human and chimpanzee mitochondrial DNA sequences. *a.* Jukes-Cantor model. *b.* F81 model. Note how allowing for unequal base frequencies (F81 model) greatly improves the predicted values along the main diagonal, but still does not produce a completely satisfactory explanation of the off-diagonal elements: There are many more observed transition-type differences than either model predicts. The expected values in both matrices shown here have been rounded off to one decimal place for clarity.

that it allows the base frequencies to differ. The total number of substitutions per site according to the F81 model is estimated to be $\mu t = 0.13252$, and when this is partitioned into each of the 16 cells according to the base frequencies, the matrix shown in Fig. 5.10b is obtained. Note the improvement in the prediction for the GG cell: The F81 model predicts a value of 83.3, which is quite close to the observed value of 82. There are still some discrepancies between observed and expected values for the off-diagonal elements, however. For example, in the observed matrix, the values corresponding to transition-type differences (AG, CT, GA, TC) are much higher than values corresponding to transversion-type differences. It is obvious that just accounting for unequal base frequencies has not captured this aspect of the real sequence data, as such big differences are not evident in the matrix of expected values corresponding to the F81 model (Fig. 5.10b).

To account for this transition/transversion bias adequately in the data, we can add one more parameter to our model to provide for the possibility that transition-type substitutions occur at a rate that is different than the transversion rate. Whereas in the JC and F81 models, all substitutions were treated as if they occurred at the single rate μ, transitions are now treated as if they occurred at a rate equal to $\kappa\mu$, where the parameter κ is called the *transition/transversion rate ratio*. If κ equals 1.0, then the new model becomes exactly equal to the F81 model, because the rate of transition-type substitutions is in this case identical to the rate of transversion-type substitutions. If κ is greater than zero, however, then the rate of transition-type substitutions is κ times greater than the rate of transversion-type substitutions. If κ is a fractional quantity (i.e., between 0.0 and 1.0), then the rate of transitions will be less than the rate of transversions. Using this model, known as the HKY model after Hasegawa et al. (1985), it is possible to estimate the total number of substitutions per site μt and the transition/transversion rate ratio κ. These estimates are $\mu t = 0.011582$ and $\kappa = 34.365593$, indicating the presence of a much larger instantaneous rate of transition-type substitutions over transversion-type substitutions. Using the HKY model to predict the values of each cell in the matrix of expected differences, the result portrayed in Fig. 5.11 is obtained. Now the greater number of transversion differences seen in the real data is reflected in the matrix of values predicted by the HKY model.

		Pan			
		A	C	G	T
Homo	A	263.1	1.0	11.5	0.8
	C	1.0	261.0	0.4	30.3
	G	11.5	0.4	81.7	0.3
	T	0.8	30.3	0.3	201.6

Figure 5.11. Expected differences in each possible category based on the HKY model, which allows base frequencies to differ and the transition rate to differ from the transversion rate.

The foregoing exercise is intended to show how more complex models involving larger numbers of parameters add the flexibility required for making sensible estimates of quantities of interest. Estimates of the evolutionary distance between the human and chimpanzee sequences based on the JC, F81, and HKY models are 0.0951, 0.0953, and 0.0991, respectively. Because of the much better fit[4] of predicted to observed differences, the estimate 0.0991 (based on the HKY model) would be preferable to the estimates based on the other two models. This example also illustrates the value of exploring the data to determine those evolutionary factors that have contributed significantly to the present-day sequences. Allowing base frequencies to differ had little impact on the estimated distance (0.0953 vs. 0.0951) compared to allowing for transition/transversion rate bias (0.0991 vs. 0.0953). Many factors other than variation in base frequency and transition/transversion bias might, of course, also affect sequences, and a summary of some of the factors that have been incorporated into substitution models is provided below. For a more in-depth consideration of substitution models, I recommend Tavaré (1986).

USING SUBSTITUTION MODELS TO INFER TREES

How is a data matrix translated into an inference of phylogeny? There are two general model-based approaches to this problem: distance matrix and maximum likelihood methods. Both approaches technically require an examination of every possible tree topology connecting the sequences under investigation. The computational difficulties of performing an exhaustive search such as this are well known, and

---

[4]The fit can be tested by means of a chi-square goodness-of-fit test. The chi-square test statistics, associated degrees of freedom (d.f.), and significance level (P) for the three models are: (1) 267.284786 (JC model, 14 d.f., $P = 0.000000$); (2) 133.552885 (F81 model, 11 d.f., $P = 0.000000$); and (3) 6.096059 (HKY model, 10 d.f., $P = 0.807129$). The significant results for the JC and F81 models are somewhat suspect because several of the expected counts are less than 1.0; however, the HKY model clearly fits these data quite well.

for problems of any size (e.g., more than about 11 sequences) we must settle for an approximate method, using a heuristic search that avoids looking at tree topologies that are obviously bad. As these strategies are the same for both model-based and parsimony methods, they are not discussed further here, and the interested reader is referred to the extensive description of the different heuristic search strategies in Swofford et al. (1996). Just as with parsimony, each tree topology investigated during a search is compared to the best tree found thus far by means of an optimality criterion. The optimality score is some measure of the "goodness" of a particular tree topology, and most of the differences of opinion on methods for reconstructing trees boil down to what optimality criterion is believed to be the best general measure of tree "goodness." The following section describes the optimality criterion used by the method of maximum likelihood. As discussed in the introduction, emphasis is placed here on maximum likelihood methods because they, like parsimony (and unlike distance methods), make use of the discrete taxon-by-character data matrix directly.

Maximum Likelihood Phylogenies

The method of maximum likelihood (ML) is one of the most widely used and most general methods of estimation available (Kiefer, 1987, p. 65). It is used for a vast array of different applications, of which phylogenetic inference occupies but a minuscule corner. Its generality lies in the fact that the same approach may be used in every application, the only difference being the probability model employed. If a mathematical expression, however complicated, can be obtained for the probability of the observed data, then ML estimation can be performed. It was mentioned earlier that computing the probability of the data allows one to quantify the degree of surprise that should be felt when the data are considered. In situations where the data are simple, such as in the coin tossing example, these probabilities can be adequately approximated by simple intuition. As the data become more complex, however, intuition becomes a poor guide and, in fact, does not help at all in the case of a large matrix of sequences. The likelihood of a specific tree[5] is proportional to the probability of the data assuming that tree (and a particular substitution model).

It is important to remember that the likelihood of a tree refers to the probability of the data given the tree, not the other way around (Felsenstein, 1981b). It is not correct to think of a likelihood as being the probability of a particular tree being the correct tree, because probabilities refer to the relative frequencies of particular outcomes of a random process (Weir, 1996, p. 80). While a tree and its associated branch lengths may indeed be thought of as outcomes of a random process (as the past history and evolutionary direction of any group is contingent upon multitudes of stochastic events), we are a very long way from being able to describe adequately the stochastic processes underlying the generation of phylogenies in nature (Felsenstein, 1973a, 1981b). We are much better, however, at modeling the processes that produce sequences from trees. Likelihoods allow us to choose among different trees based on the predictions those trees imply. Some tree topologies make the observed sequence data much more likely to have arisen than other tree topologies. Assuming the ML tree topology makes us (or at least should make us) less surprised when looking upon our sequence data than does any other tree topology.

Because there are two components to a tree, the topology and branch lengths, it is possible for the correct topology to have a lower likelihood than one of the alternative topologies if the branch lengths have not been estimated correctly. Under the ML criterion, the branch lengths are adjusted until the probability of the data (i.e., the likelihood) is greatest. Estimates thus obtained are called maximum likelihood estimates (MLEs). For each tree topology investigated during a search, care must be taken to ensure that the branch lengths used in computing the likelihood of the tree are the MLEs. Failure to do this can easily result in an incorrect inference for the topology, and more than one simulation study comparing ML to other phylogeny reconstruction methods has been misleading because the algorithms used were faulty in this respect (Hasegawa et al., 1991).

Finding the MLEs of the branch lengths is a multidimensional optimization problem that can be solved by a variety of standard numerical methods. Perhaps the best way to achieve an understanding of maximum likelihood estimation in the context of phylogenetic inference is by considering the simplest of all trees: a single branch connecting two terminal taxa. There are two sequences involved and a single branch length to estimate. While there is only one tree for two terminal taxa, delving into this example serves two purposes: It illustrates the relationship between the length of a branch and the probability of the data, and also shows how maximum likelihood estimates of evolutionary distances are obtained, for ML distances are simply MLEs of the single branch length connecting two taxa. Our simple example involves two sequences of length five nucleotides (see Fig. 5.12). There is a single difference between the sequence at one end of the branch and its counterpart at the other end of the branch. Imagine three different lengths for the branch connecting the two sequences. The first branch has zero length (Fig. 5.12a). Because branch length is measured as the evolutionary distance (expected number of base-changing substitutions per site), assuming a branch length of zero would result in a very low likelihood score. This is because it would be extremely surprising to find a difference at even a single nucleotide site given that there was zero chance of any substitutions occurring! At the other end of the scale, assuming that the branch is infinitely long (Fig. 5.12c) also leads to a low likelihood, because if the true branch length had been infinitely long we would expect many more sites to be different. Said another way, assuming there have been an infinite number of substitutions at each site, it is very surprising that only one site showed a difference. Thus, there must be some intermediate (maximum-likelihood) branch length that is "just right" (Fig. 5.12b). Assuming that particular maximumlikelihood length makes the fact that we observed only one difference between the sequences least surprising. Figure 5.13 plots the likelihood as a function of the length of the

---

[5]The term *tree* will be taken hereafter to include the topology and branch lengths.

a. ACCGT ACCGC

b. ACCGT ——————— ACCGC

c. ACCGT ══════════════╱╱═══════════════ ACCGC

Figure 5.12. A simple example illustrating the estimation of an evolutionary distance by maximum likelihood. Three different putative branch lengths are shown. *a*. Branch having length zero. *b*. Branch having an intermediate length. *c*. Branch having an infinite length. Note that the only difference between the sequences at either end of the branch in each case is that a T (on the left) has changed into a C (on the right).

single branch connecting the two sequences under the JC model of substitution.[6]

_____

[6]The likelihood is the probability of a sequence ACCGT changing to ACCGC over an amount of time t at a substitution rate μ. The branch length provides the product μt needed to compute this probability. As we saw earlier, the branch length for the JC model is just $v = \frac{3}{4}\mu t$, and thus the quantity μt can be obtained for any specified branch length v by inverting this relationship: $\mu t = \frac{4}{3}v$. The first four sites (ACCG) have not changed (or have changed and then reverted back), whereas at least one nucleotide-changing substitution was observed at the fifth site, resulting in the difference T $\leftrightarrow$ C. Using the JC model, the probability of observing no change is $\frac{1}{16}P(t) + \frac{1}{4}[1 - P(t)] = \frac{1}{16}[1 - e^{-\mu t}] + \frac{1}{4}[e^{-\mu t}] = \frac{1}{16}[1 + 3e^{-\mu t}] = \frac{1}{16}[1 + 3e^{-\frac{4}{3}v}]$, whereas the probability of observing a change is $\frac{1}{16}P(t) = \frac{1}{16}[1 - e^{-\mu t}] = \frac{1}{16}[1 - e^{-\frac{4}{3}v}]$. Thus, the likelihood function plotted in Fig. 5.13 can be written as a function of the branch length (v) as follows:

$$L(v) = \left\{ \frac{1}{16} [1 + 3e^{-\frac{4}{3}v}] \right\} \left\{ \frac{1}{16} [1 - e^{-\frac{4}{3}v}] \right\}.$$

Finding the MLE of the single branch length in the example above is similar to turning the dial on a radio to locate the exact frequency being broadcast by a particular radio station. Imagine that if you turn the dial in a clockwise direction, the branch length becomes larger; turning the dial in a counterclockwise direction results in a shorter branch. Suppose that you can watch a gauge that plots the likelihood as a function of the position of the dial (similar to the graph in Fig. 5.13). Starting at a branch length of zero and turning clockwise, the likelihood rises dramatically. At one point, however, the curve tops out (near 0.23) and then begins to fall. This point at which the likelihood is greatest is the maximum likelihood estimate of the branch length.

Finding the likelihood of a tree with many branches is entirely analogous to the situation just described, except that now we have a dial to

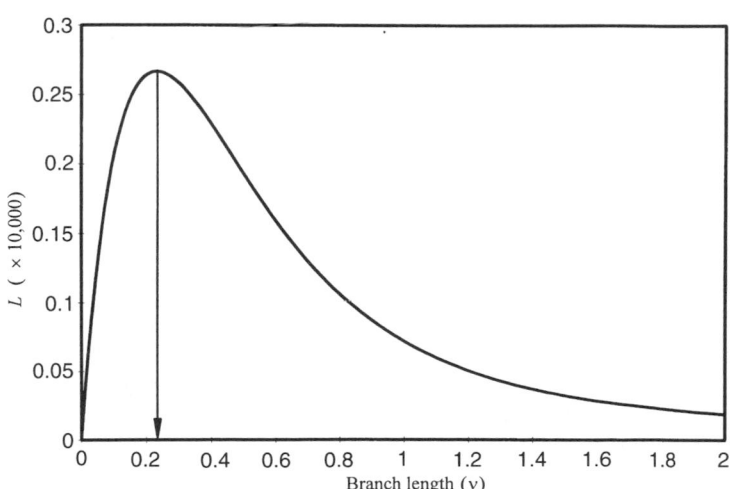

Figure 5.13. Plot of the likelihood as a function of branch length for the data in Fig. 5.12 (i.e., one difference observed out of five nucleotide sites). The vertical line is located at a branch length of 0.232616, which is the maximum likelihood estimate (MLE). For the case of a tree with one branch, the branch length is identical to the MLE of the evolutionary distance between the two taxa.

turn for each branch in the tree. You can imagine turning each dial clockwise and counterclockwise until the gauge is at its highest point. When this maximum likelihood point is reached, turning any of the dials will result in a drop in the likelihood. The likelihood at this point is the score that is used to compare the present tree topology with all other tree topologies visited during the search. When a computer program such as DNAML in the PHYLIP package (Felsenstein, 1993) is used to perform a search for the maximum likelihood tree, numerical algorithms within the program metaphorically turn the dials for each tree topology visited during a search (see Felsenstein, 1981b). It is largely this process of fine tuning branch lengths for each tree that accounts for the difference in speed between likelihood and parsimony tree searches.

In the case of the JC substitution model, the branch lengths themselves are the only parameters that need to be estimated in order to obtain the likelihood of a tree topology. To account for transition/transversion bias, however, one might wish to use a model that allows the instantaneous rate of transition-type substitutions to be greater than the rate for transversion-type substitutions by a factor κ (as was discussed in the section on substitution models). Adding an additional parameter (i.e., κ) to the model just means that there is one more dial to turn when maximizing the likelihood for each topology visited. Just as for branch lengths, there will be a point at which turning any of the dials, including the one for κ, will result in a lower likelihood. The value of the likelihood at this point is the score used to compare the present tree topology to all other topologies. It is certainly possible to tape down the dial for κ so that it cannot be turned, thus fixing the value of the κ parameter for all comparisons, but there is no rule that says that parameters such as κ *must* be fixed in this way. A rather widespread myth concerning ML inference is that somehow the investigator must know the proper values to plug into the substitution model. The conclusion commonly drawn is that ML inference forces one to make many more assumptions than is necessary with other methods, such as parsimony. In reality, one can allow any of the parameters of the model to be estimated

jointly with all other parameters, adding flexibility rather than restrictions to the analysis. There is, however, an advantage in terms of speed to fixing parameter values (taping down dials, so to speak). The fewer dials there are to adjust, the more quickly one can arrive at the maximum value of the likelihood for any given tree topology.

Although I have plotted the likelihood itself in the graph of Fig. 5.13, the score that is given by most computer programs that perform maximum likelihood inference is the natural logarithm of the likelihood. Because the likelihood is a probability, its value must lie between 0.0 and 1.0. For even small numbers of taxa and moderate amounts of sequence data, the likelihoods of even the best trees are quite small numbers. For example, the maximum likelihood tree for a study involving 25 taxa and a gene of length 1,400 nucleotides might easily have a likelihood that is on the order of 1×10^{-4000}. To make the estimation possible, the likelihood is often represented internally (and also reported) as the natural log of the likelihood, or the ln-likelihood for short. The graph in Fig. 5.14 shows the relationship between a number and its natural logarithm. Note that any number between 0.0 and 1.0 (such as a probability) has a natural logarithm less than zero. Thus, ln-likelihoods are always negative numbers, and there is a monotonic relationship between the likelihood and its natural logarithm. Because of this monotonic relationship, estimates of parameters made by maximizing the ln-likelihood will be

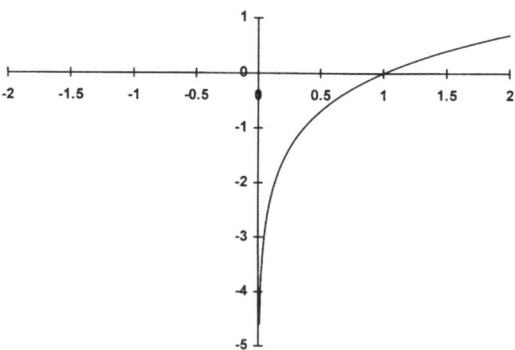

Figure 5.14. Relationship between a number (horizontal axis) and its natural logarithm (vertical axis).

identical to estimates made by maximizing the likelihood.

For those who wish to delve further into the details of maximum likelihood phylogeny inference and understand more about how likelihoods are computed from sequence data, I recommend Felsenstein (1981a, 1983) and Swofford et al. (1996).

Pros and Cons of Maximum Likelihood Compared to Parsimony

One primary advantage of the parsimony approach is speed. Finding the MLEs of all branches in the tree along with any additional parameters specified in the substitution model is a lengthy process, and thus ML inference will always be very slow compared to a parsimony approach. Why, given the choice of using a fast parsimony method that is able to evaluate many more trees in a given amount of time, would anyone choose to estimate trees using maximum likelihood?

There are two distinct advantages that all model-based methods, including maximum likelihood, appear to have over all parsimony-based methods. First, distance and likelihood methods explicitly allow for the possibility of heterogeneity of branch lengths in the true tree. Although parsimony methods do not *explicitly* assume anything in particular about branch lengths, they may *implicitly* assume that branch lengths are all equal in length. An assumption of homogeneous branch lengths would be expected to lead to problems with long-branch attraction. Parsimony methods do appear to have more difficulty with long-branch attraction than do model-based approaches, but the reasons for this are far from clear (Kim, 1996). Second, the optimality scores produced by model-based approaches do not depend on particular combinations of character states at the internal nodes of the tree. Parsimony tree lengths, on the other hand, are based on one or a few maximally parsimonious character-state assignments at the interior nodes of the tree. This dependency means, in effect, that these ancestral character states are parameters of the parsimony model that need to be estimated in order to obtain the optimality score. If this is the case, and it is admittedly difficult to prove anything about the assumptions made by the parsimony model, then the number of parameters needing to be estimated grows as fast (or faster) than the data, a condition known to lead to inconsistent estimation. Model-based methods such as distance and likelihood obtain their optimality scores by considering every possibility at these ancestral states, thereby eliminating the ancestral states as parameters of the model.

In addition to these advantages, model-based approaches allow for data exploration. The method of maximum likelihood provides a nonarbitrary means for estimating parameters of the model from the data. Allowing likelihood to estimate the transition/transversion rate ratio along with branch lengths and tree topology provides a natural weighting for transformations between these two different recognized classes of substitutions. While it is possible to supply weights for transformations in parsimony analyses, it is far from clear how to arrive at the correct weights in a nonarbitrary fashion. Furthermore, if two different substitution models are used and result in different estimated tree topologies, then it is possible to examine the assumptions made by each of the substitution models and ascertain why the two models produced different estimates of the phylogeny. This very important advantage comes simply from making the model explicit and thus enumerating all the assumptions being made in order to arrive at the inference.

HYPOTHESIS TESTING AND DATA EXPLORATION

This final section introduces several models (or additions to existing models) that are useful for phylogenetic inference, data exploration, or hypothesis testing (or some combination of these). The purpose is not to describe each of these in detail, but to alert interested readers of their existence and usefulness.

Testing the Assumption of a Molecular Clock

Ordinarily, the length of each branch in a tree is considered independent of the lengths of all other branches in the tree. That is, it is considered possible for each distinct lineage to have its own substitution rate. Long branches are long either because the lineage was in existence a longer time or because it experienced higher substitution rates than more slowly evolving lineages (or both). It is possible, however, to place constraints on branch lengths that allow certain hypotheses about substitution rates to be tested. One of the most straightforward tests is of the hypothesis that the substitution rate has remained constant over an entire tree. If the substitution rate is constant, then long branches are long only because they have been in existence for a longer period of time. This leads to the expectation that the two paths leading backward in time from each of two taxa to their common ancestor must be the same total length. For example, in the tree in Fig. 5.15a, the molecular clock hypothesis forces the length of the branch t^*_1 and the length of the branch t_2 to add up exactly to the length of the branch t_1. Because the lengths of some of the branches in the "constrained" tree (i.e., the one in which the substitution rate is assumed constant) may be obtained by subtraction, there are fewer branch lengths that need to be estimated in order to compute the likelihood of the tree. In fact, for a tree with n tip taxa, there are $n - 1$ branch lengths to estimate in the constrained tree compared to $2n - 3$ in the unconstrained tree. This is an example of a situation that is perfectly suited for a type of statistical test known as a *likelihood ratio test*. A likelihood ratio test may be applied when there are two models, one of which is created by placing constraints on some of the parameters in the other model. In this case, the parameters being constrained are branch lengths. To conduct the test (Felsenstein, 1983), one must maximize the likelihood for both the constrained and the unconstrained models. The likelihood L_0 for the constrained model will necessarily be the same or lower than the likelihood L_1 for the unconstrained, because placing constraints on a model prevents it from explaining the data any better than the unconstrained version. If L_0 is much less than L_1, we must reject the idea of a molecular clock because assuming a molecular clock does not allow us to explain the data nearly as well as an hypothesis that relaxes this assumption. In other words, we are (read "should be") much more surprised when looking at the data if it is assumed that there has been a molecular clock in operation than if variation in substitution rate across lineages is allowed. If L_0 is only slightly less than L_1, however, there is a point at which the two alternative models must be considered statistically indistinguishable in their ability to explain the data. To find this point, we can make use of the fact that the quantity $-2(\ln L_0 - \ln L_1)$ is approximately distributed as a chi-square random variable with v degrees of freedom, where v equals the difference in the number of free parameters in the two models. In this case, $v = n - 2$ because there are $n - 2$ fewer branch lengths needing to be estimated in the constrained model compared to the unconstrained model. Thus, given the likelihoods L_0 and L_1, the significance level for the test can be obtained by consulting a standard table of chi-square probabilities. The test is referred to as a likelihood ratio test be-

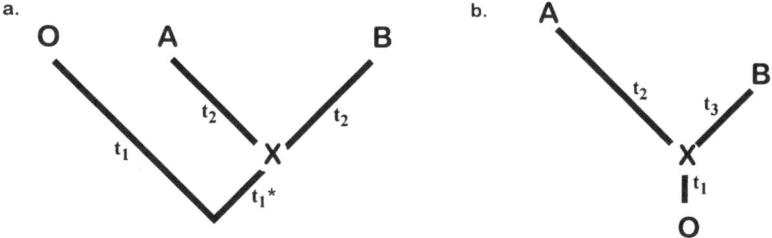

Figure 5.15. Two trees, one of which (left) has branch lengths conforming to the constraints of a strict molecular clock (substitution rate constant), the other (right) having branch lengths that are unconstrained (each branch can have its own substitution rate and time). In the constrained tree, there are only two independent branch lengths, t_1 and t_2: The branch length labeled t^*_1 can be obtained by subtraction (i.e., $t_1 - t_2$) and hence is not independent. The constraints imposed on branch lengths by the molecular clock assumption cause different rootings of the tree to have different likelihoods. The unconstrained tree on the right has three independent branch lengths and all possible roofings result in the same likelihood.

cause the test statistic given above is numerically equivalent to the quantity $-2\ln(L_0/L_1)$, the term *ratio* coming from the fact that this expression involves the quotient of two likelihoods.

Relative Rate Tests

Likelihood ratios can also be used to perform relative rate tests. Consider again Fig. 5.15. Suppose it was necessary to test whether the lineage leading from X to A was evolving at the same substitution rate as the lineage leading from X to B. If it is assumed that both lineages have been evolving at the same rate, then a likelihood L_0 can be computed in which the same branch length (t_2 in Fig. 5.15a) is applied to both lineages. The alternative hypothesis is that the lineages have not been evolving at the same rate, and the likelihood L_1 computed under this unconstrained model allows each branch to have its own length (t_2 vs. t_3 in Fig. 5.15b). Conducting this likelihood ratio test (Muse and Weir, 1992) involves computing these two likelihoods, using them to construct the likelihood ratio test statistic $-2(\ln L_0 - \ln L_1)$, and comparing this value to a table of chi-square probabilities for one degree of freedom. The single degree of freedom for this test comes from the fact that the unconstrained model estimates one more parameter (the branch length t_3) than its constrained counterpart. In this case, the branch leading from X to the outgroup O forms a single additional parameter that is estimated in both constrained and unconstrained models.

The first time likelihood ratio tests are encountered, it is difficult to see why L_0 must always be less than or equal to L_1. Because there is no rule that the branch lengths t_2 and t_3 in the unconstrained model cannot be equal to each other, this means that it is always possible for L_1 to have exactly the same value as L_0. Thus, for any given value attained by L_0, L_1 can either match or exceed it.

Accounting for Variation in Rates Among Sites

Substitution rates not only vary from lineage to lineage, but also from site to site within a sequence. Considerable attention has been invested recently in ways to allow for such site-to-site rate heterogeneity within substitution models. The most straightforward approach is to treat each site as if it had its own, independent rate. Such a substitution model has k rate parameters if the sequence of interest has k sites. This approach has the benefit of not making any assumptions about the nature of the variation in rates among sites, but has the disadvantage of requiring that many more parameters be added to the model. Adding this many more parameters to the substitution model means a considerably higher expenditure of computing effort in the calculation of likelihoods, and thus other approaches to modeling site-to-site rate heterogeneity have been investigated. One alternative is to assume that the distribution of rates across sites takes the shape of a well-defined distribution. The distribution of choice in this regard has been the gamma distribution (Jin and Nei, 1990; Tamura and Nei, 1993; Yang, 1993, 1994b, 1995), which has several desirable properties as a distribution of relative rates. The gamma density[7] function can take on two qualitatively different shapes, depending on the value of its "shape" parameter, which ranges from zero to infinity. For values of the shape parameter less than or equal to 1.0, the gamma density has the shape illustrated by the curved line in Fig. 5.16a. This shape means that most sites have low substitution rates (with a relative rate less than 1.0) and a very few sites have quite large rates. The closer the shape parameter gets to zero, the larger this disparity in relative rates becomes. Figure 5.16b illustrates a gamma density with a shape parameter equal to 10. As the shape parameter increases, the gamma density becomes more bell shaped, until, at very large values of the shape parameter, the entire area is compressed into a single spike at the value 1.0. In the limit (the shape parameter is at infinity), the single spike at 1.0 implies that all sites have the same rate (they all have a relative rate of 1.0).

What advantage is there to assuming that rates follow a gamma distribution? One advantage is that, by adding just one more parameter

---

[7]The term *density function* is used when describing the distribution of continuous random variables. It is the area under the curve that has meaning as a probability.

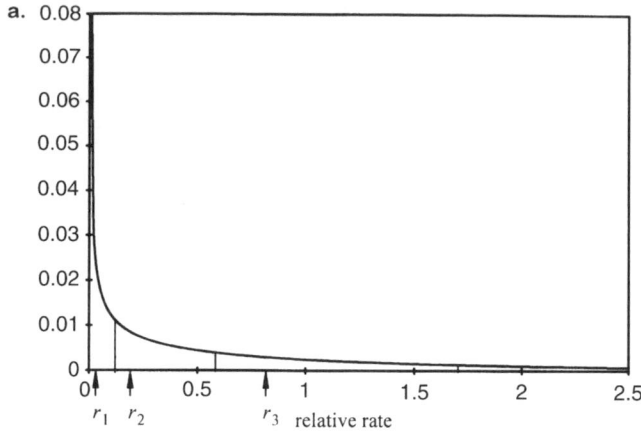

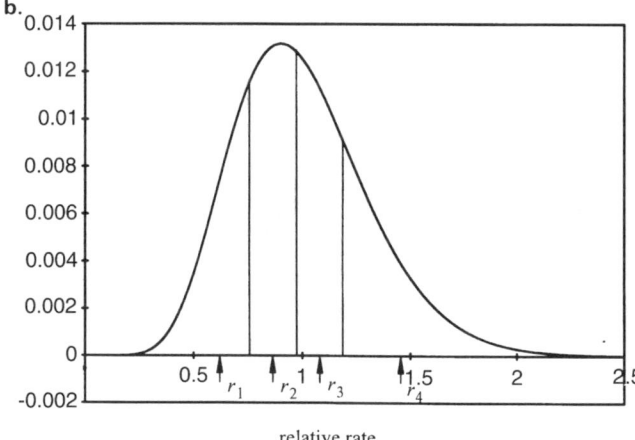

relative rate

Figure 5.16.   Plot of the gamma probability density function. *a.* The shape parameter of the gamma distribution is equal to 0.5. *b.* The shape parameter of the gamma distribution is equal to 10. The three vertical lines in each case partition the distribution into four regions having equal areas under the curve. The means of each of these four regions (indicated by arrows and the symbols r_1, r_2, r_3, and r_4) are used as relative rates in modeling site-to-site rate heterogeneity. The mean rate r_4 for the gamma density function having shape 0.5 was not plotted, as it falls just off the right side of the graph.

to our substitution model—the gamma shape parameter—we can allow anything from complete homogeneity of rates (shape parameter equal to infinity) to quite extreme rate heterogeneity (shape parameter near zero). It is also quite easy to modify most common distance measures to allow for gamma-distributed relative rates. While it is not practical to incorporate the gamma distribution directly when computing likelihoods of trees (Yang, 1993), it is quite easy to incorporate a discrete version of the gamma distribution into the likelihood calculations. Figure 5.16 shows how the continuous gamma distribution may be divided into four sections, each

cutting off an equal proportion of the total area under the curve. The means of each of these sections are used as relative rates in computing the likelihood. The sections are referred to as rate categories, and the means of these sections (labeled r_1, r_2, r_3, and r_4) are referred to as category means. Just as likelihood scores take into account all possible states at each of the internal nodes of the tree, it is possible to take into account the possibility that a given site has a rate equal, in turn, to each of the category means. This frees the likelihood from dependence on a particular choice of rate at each site. If phylogeny reconstruction is no longer the goal (i.e., a satisfactory estimate of the phylogeny is already in hand), interest may turn to estimating rates at individual sites. In this case, it is desirable to find the most likely rate at each site or at specific sites of interest. As the number of categories is increased, the discrete gamma distribution becomes increasingly similar, in its effect on the likelihood score, to a continuous gamma distribution.

A second approach to modeling site-to-site rate heterogeneity is simply to assume that all sites fall into two categories: those having a rate of zero and those having some nonzero rate. This approach also involves only estimating one parameter related to rate heterogeneity, which is in this case the probability that any given site has a rate of zero. This two-category model has been used by several investigators, including Reeves (1992), Sidow and Speed (1992), and Churchill et al. (1992).

A third way to model site-to-site rate heterogeneity was proposed by Felsenstein and Churchill (1995) and implemented in Felsenstein's PHYLIP computer package (Felsenstein, 1993). The primary advantage of Felsenstein and Churchill's hidden Markov chain approach is allowance for correlation between the rates at adjacent nucleotide sites. It is also more general in that it does not specify a particular distribution (e.g., gamma) for the relative rates. This generality comes at a cost, however, because there are two parameters (the relative rate itself and the probability that that rate applies to any given site) to estimate for each relative rate category proposed[8] in addition to the correlation parameter.

Which of the above methods of handling site-to-site rate heterogeneity is preferred depends on the goal of the study. If the primary goal is to infer phylogeny, then the parameters involved with modeling rate heterogeneity are included because they add realism to the model and stability to the inference of the parameters of interest; these parameters are of no particular interest in and of themselves for this type of study. For phylogenetic reconstruction, therefore, it is preferable to keep the number of parameters involved with modeling rate heterogeneity to a minimum, for purposes of speed, if for no other reason. The gamma distribution is ideal in this case, because it adds only one more parameter. On the other hand, if the primary goal of the study is to infer something about the nature of the variation in rates across sites, then it is worth adding more flexibility to the model, such as allowing correlations between relative rates at adjacent sites (Felsenstein and Churchill, 1995) or even allowing each site to have its own rate, as this makes the fewest assumptions about the nature of the rate heterogeneity. This approach has been especially useful for identifying quickly and slowly evolving regions of the ribosomal RNA genes because there is a tremendous amount of data that can be applied to this question.

Accounting for Independence among Nucleotide Sites

Among the problems that have been recognized longest in the area of phylogeny reconstruction is the problem posed by nonindependence of characters. Nearly all phylogeny methods assume that all characters (nucleotide sites in the case of sequence data) have evolved independently of all other characters, even though researchers have known from the outset that this assumption is violated by almost all data sets (an exception being some pseudogenes). There are two special cases of nonindependence in sequence data that have proven amenable to modeling. The first is the nonindependence among the three nucleotide sites of a codon, the second being the nonindependence among the sites forming stem structures in RNA-coding genes. Two models have been recently proposed for dealing with codon structure (Goldman and Yang, 1994; Muse and Gaut, 1994). These codon models allow nucleotide sequence data to be used, with its richer supply of variation, while still treating codons as the independent units rather than individual nucleotide sites. This treatment makes possible relative rate tests that can separate the effects of synonymous substitutions from nonsynonymous substitutions (Muse and Gaut, 1994; Muse, 1996).

Models have also been developed for dealing with secondary structure in RNA genes. In one

---

[8]Except for the last category, since the rate probabilities must add to one and the expected overall rate must be unity.

model (Muse, 1995; Tillier and Collins, 1995), sites that do not participate in stem structures are treated as one class of sites, while the remaining stem sites are treated as a second class. A standard substitution model is applied to the non-stem sites, while the substitution model applied to the stem sites contains a parameter that specifies the degree of correlation in the evolution of sites that oppose each other in a stem structure. Although many rules of thumb have been supplied for dealing with this problem using parsimony (such as weighting changes in stem sites by a factor of one-half), without using a model it is difficult to know whether these rules of thumb have any basis in reality. With a model-based approach, it is possible to estimate this correlation parameter and obtain a reasonable idea of the magnitude of the nonindependence.

Testing for Homogeneity among Disparate Data Sets

Another example of the utility of a model-based approach involves testing whether data derived from two genes were generated by the same underlying phylogeny. Cases in which gene trees conflict might arise as the result of lineage sorting or horizontal transfer events. Once again, a likelihood ratio test proves quite useful (Fig. 5.17). The unconstrained model allows the data for each of two (or more) genes to be explained using potentially different tree topologies, whereas the constrained likelihood is computed by forcing the topology to be the same for all genes. This can be accomplished by analyzing each of the data sets separately to obtain the unconstrained likelihood ($\ln L_0$ is simply the sum of the ln-likelihoods resulting from each of the separate analyses), and then finding the single maximum likelihood tree for the combined data, which provides the constrained likelihood. The main difficulty with this test comes in determining the number of degrees of freedom. It would appear that the number of degrees of freedom for this test should be $2n-3$, where n is the number of terminal taxa, because this is the number of branches in one tree, and the unconstrained model requires us to estimate one entire tree more than the constrained model. Apparently, the situation is not this simple, and Huelsenbeck and Bull (1996) suggest determining the significance of the likelihood ratio statistic by simulation. This is accomplished by simulating a number of independent data sets according to the constrained model and constructing the likelihood ratio statistic for each of these simulated data sets. Computer simulation is being used in this case to simulate the distribution of the test statistic under the null hypothesis that the genes were produced by a common true gene tree. This application of computer simulation is known as parametric bootstrapping (Felsenstein, 1988) and is now receiving much attention (Hillis et al., 1996) as a general method for testing specific hypotheses in which the underlying null distribution does not fit easily into any of the standard continuous distributions (i.e., normal, gamma, beta, etc.). If the test statistic from the original data falls inside the right-hand 5% tail of the simulated distribution, it is declared statistically significant at the 5% level. Determining significance in this way, while perfectly valid, is quite time consuming, and more work is needed to see if a computationally tractable means of determining significance can be found.

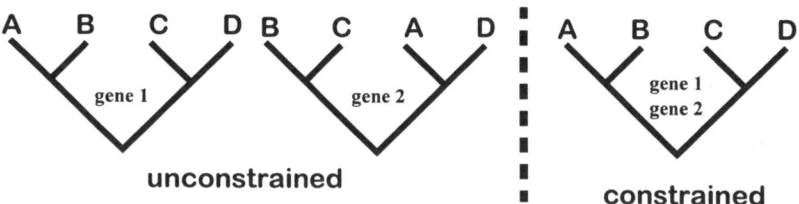

Figure 5.17.   Likelihood ratio test for a common gene tree. The unconstrained model allows (but does not require) the topology to be different for the two genes, whereas the constrained model forces the topology used for both genes to be the same. A likelihood ratio test statistic is constructed to decide whether forcing both genes to have the same tree topology results in a significantly worse fit to the data.

Evolutionary Distances Free from Assumptions about Alignment

Another long-standing problem faced by molecular systematists is that of alignment (see also Chapter 4). Phylogeny algorithms assume that the data are aligned correctly, even while alignment itself is widely recognized as a problem equal in magnitude to phylogeny reconstruction. One solution to this problem is to construct evolutionary distances that represent a weighted average over all possible alignments. Some alignments will be better than others, and thus it is imperative that the weights used in the average reflect the quality of the alignment associated with that weight. Thorne et al. (1992) accomplished this by developing a substitution model that incorporates insertion and deletion events. Although this model is too computationally intensive at present to use for purposes of reconstructing trees with multiple sequences, it can be effectively used in estimating pairwise distances that are then submitted to a distance method for phylogenetic inference. The Thorne and coworkers model is used to compute a likelihood for each possible alignment of the pair of sequences being considered. The best feature of this approach is that alignment is effectively removed as a consideration—it has been integrated out of the distance estimate—which is a highly desirable feature for data sets containing many highly diverged sequences.

Testing Whether Two Tree Topologies Are Significantly Different

Kishino and Hasegawa (1989) introduced a computationally efficient method of testing whether two tree topologies differ to a statistically significant degree. This test involves likelihoods but is not a likelihood ratio test because there is no difference between two tree topologies in the number of parameters estimated (one tree cannot be considered to be derived from the other tree by placing constraints on any parameters). The Kishino and Hasegawa test makes use of the fact that the sum of a collection of random variables is approximately normally distributed, even if the components of the sum are not themselves normally distributed; this result is known as the Central Limit theorem. As might be ex-

pected, this approximation increases in accuracy with the number of elements in the sum. In the case of nucleotide sequences, the elements making up the sum are the ln-likelihoods of individual sites. The total ln-likelihood of a tree is the sum of the ln-likelihoods for individual sites, much as the parsimony tree length is the sum of the tree lengths for each site. The Central Limit theorem states that even though we have no knowledge of how ln-likelihoods for individual sites are distributed, the sum of a collection of site ln-likelihoods will approach a normal distribution given that there are enough site ln-likelihoods in the sum. Kishino and Hasegawa show how the variation among site ln-likelihoods can be used to estimate the variance of their sum in the same manner that variation in heights of individuals may be used to obtain an estimate of the variance of the mean height. The resulting variance estimate may then be used to decide whether the difference in ln-likelihood between the two trees being compared is statistically significant. It should be pointed out, however, that Yang (1994c) has questioned the validity of this test on the basis of simulations that revealed that the distribution of ln-likelihood differences may in some cases be highly skewed.

The Perspective of Likelihood on Reconstruction of Ancestral States

The fact that both the maximum likelihood and parsimony criteria operate on a matrix of discrete character states (as opposed to a matrix of pairwise distances) means that both provide some means of inferring ancestral character states. These two optimality criteria differ, however, in their relative perspectives on the reconstruction of ancestral character states. Once again, we run into the problem of not knowing exactly what is assumed by the parsimony model (or the possibility that parsimony analyses can be viewed as embodying more than one model), which makes it difficult to characterize the parsimony perspective on inferring character states. In any case, maximum likelihood will sometimes infer an ancestral state that is unfavorable to parsimony, and vice versa. Figure 5.18 illustrates a situation where this difference in perspective is evident.

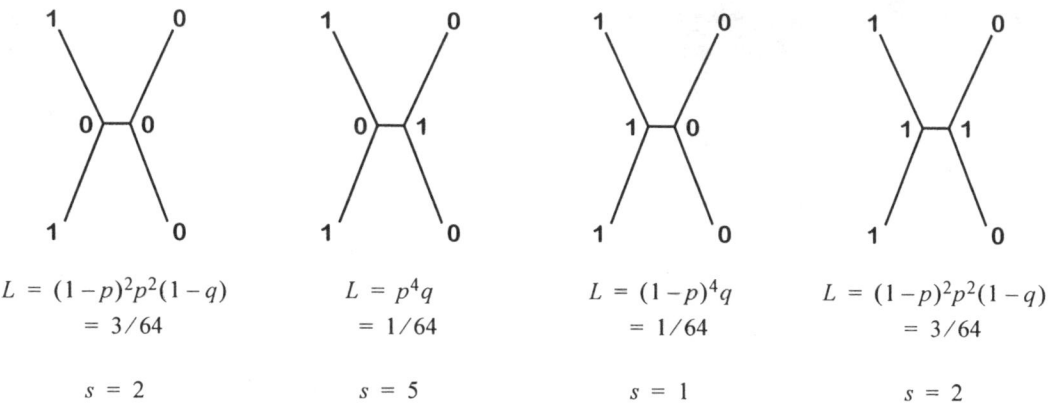

$$L = (1-p)^2 p^2 (1-q)$$
$$= 3/64$$

$$s = 2$$

$$L = p^4 q$$
$$= 1/64$$

$$s = 5$$

$$L = (1-p)^4 q$$
$$= 1/64$$

$$s = 1$$

$$L = (1-p)^2 p^2 (1-q)$$
$$= 3/64$$

$$s = 2$$

Figure 5.18. Illustration of a special case in which maximum likelihood and maximum parsimony differ in their inference of the ancestral state assigned to the internal nodes of a four-taxon tree. For each of the four trees, the longer, terminal branches have a length such that the probability of a change across the branch is p. For the shorter, internal branch, the length is such that the probability of a change is q. For the purposes of computing the numerical likelihood of each of these trees, which differ only in the assignment of states to the internal nodes, it is necessary to assign values to p and q. For this example, the values assigned were $p = 0.5$ and $q = 0.25$; however, likelihood will prefer the same assignments as long as $p \leq \frac{1}{2}$ and $q < \dfrac{p^2}{p^2 + (1-p)^2}$. Regardless of the values assigned to p and q, parsimony will prefer the third assignment because of the fact that this assignment of states to the internal nodes requires only a single hypothesized character-state change.

In this four-taxon example, the likelihood and parsimony scores have been calculated for each possible reconstruction of ancestral states for one binary character. It is assumed, for purposes of illustration, that there are only two branch lengths in the true tree. The four branches having the longer of these two branch lengths lead to the four tips of the tree, leaving the relatively shorter branch connecting the two interior nodes. It is also assumed that the branch lengths are such that the probability of observing a change across any one of the four terminal branches is p, whereas the probability of observing a change across the internal branch is q. For any values of p and q, parsimony prefers the ancestral-state reconstruction (1, 0), because this reconstruction requires only one step. Letting $p = \frac{1}{2}$ and $q = \frac{1}{4}$ leads to likelihood favoring either the ancestral-state reconstruction (0, 0) or (1, 1) over the reconstruction favored by parsimony. In fact, it can be shown that whenever

$$p < \tfrac{1}{2} \text{ and } q < \dfrac{p^2}{p^2 + (1-p)^2},$$ likelihood will favor this alternative reconstruction. What is the reason behind this difference in the preferences of parsimony and likelihood? From the likeli-

hood perspective, as the internal branch becomes shorter and shorter, it becomes increasingly improbable that a change occurred across the branch. Thus, when the length of the central branch becomes small enough, likelihood begins favoring an ancestral-state reconstruction that does not require a change across this short branch. Parsimony prefers the same ancestral-state configuration regardless of the lengths of branches. It is important to note at this point that likelihood uses information from all the data in order to estimate branch lengths. There is not enough information in the data shown (from one site) to allow accurate estimation of branch lengths. As more data are considered, however, likelihood estimates of branch lengths will become increasingly more accurate, and thus the reconstruction of ancestral states for individual sites will become correspondingly more accurate.

Because maximum likelihood makes use of more information than parsimony, it has the capability to decide among alternative character-state assignments that parsimony considers equivalent. Inferring character states on a specific tree by maximum likelihood is not costly;

the cost of likelihood methods (in terms of computing time) is felt most during heuristic searches when the likelihood scores for thousands of separate tree topologies must be computed. Thus, given that a reasonable tree is in hand, it is worthwhile to explore the maximum likelihood ancestral-state reconstructions, especially in cases where there is a large cost (monetarily or otherwise) to getting the answer wrong. An example would be experiments involving synthesis of putatively ancestral genes for purposes of inferring ancestral gene function or ancestral efficiency of enzymes (e.g., Chandrasekharan et al., 1996).

There are two different approaches to inferring ancestral states using the maximum likelihood criterion. In *joint* ancestral-state reconstruction, the goal is to find a single combination of states for all ancestral nodes for which the likelihood score is greatest. This can be accomplished using the method of Fig. 5.3 in which all possible combinations of ancestral states are considered.

An alternative is to focus attention on one ancestral node and infer its state only. In this *marginal* method for ancestral-state reconstruction, the likelihood score is computed by summing over all possible character-state assignments at all internal nodes in the tree *except* the one being inferred. The marginal method allows more of the data to be applied to the estimation of one particular ancestral state of interest. The joint approach is preferred when there is no reason to focus attention on one particular ancestral node.

SUMMARY

This chapter focuses on the method of maximum likelihood as an alternative to parsimony for phylogenetic inference. Because the method of maximum likelihood is both model-based and (like parsimony) makes direct use of the discrete character data, it allows estimation of quantities of interest to molecular systematists such as the transition/transversion rate ratio, nucleotide frequencies, ancestral character states, and the degree of variation in rates across sites and among lineages. In addition to estimation, likelihood ratio tests provide a statistically sound framework for testing many types of hypotheses that arise in molecular systematic studies. Examples include: 1) testing the molecular clock hypothesis;

2) relative rate tests; 3) testing an assumption of homogeneity of substitution rate across sites; 4) testing specific hypotheses concerning the nature of dependence among sites; and 5) testing the hypothesis of homogeneity among gene trees reconstructed from different sequence data sets. Finally, model-based methods such as maximum likelihood provide an alternative to parsimony for phylogeny reconstruction itself, providing a perspective that differs from parsimony in significant ways and makes full use of previous gains in the field of molecular evolution.

SOFTWARE

The best general source of information about software for performing the analyses described herein is the well-annotated World Wide Web site maintained by Joe Felsenstein. This site may be approached from the PHYLIP web site, which is located at *http://evolution.genetics. washington.edu/phylip.html*. Also, consult Swofford et al. (1996) for a hardcopy listing of many of these sources.

In addition, a Microsoft Windows™ application has been written to accompany this chapter. This program, called *MLCalc* for *M*aximum *L*ikelihood *C*alculator, illustrates four substitution models commonly used in obtaining evolutionary distances or in maximum likelihood phylogeny reconstruction. It allows the user to enter a four-by-four divergence matrix made up of counts of differences (off-diagonal elements) and similarities (diagonal elements), and interactively manipulate parameter values in the currently selected model in an attempt to maximize the likelihood. MLCalc is a free program that can be obtained via the World Wide Web from *http://biology.unm.edu/~lewisp/mlcalc.html*. There are two versions, a 16-bit version for Windows 3.1x and a 32-bit version for Windows 95/NT.

LITERATURE CITED

Akaike, H. 1974. A new look at the statistical model identification. Institute of Electrical and Electronics Engineers Transactions on Automatic Control **19**:716–723.

Barry, D., and J. A. Hartigan. 1987. Asynchronous distance between homologous DNA sequences. Biometrics **43**:261–276.

Bull, J. J., C. W. Cunningham, I. J. Molineux, M. R. Badgett, and D. M. Hillis. 1993. Experimental molecular evolution of bacteriophage T7. Evolution **47:**993–1007.

Camin, J. H., and R. R. Sokal. 1965. A method for deducing branching sequences in phylogeny. Evolution **19:**311–326.

Chandrasekharan, U. M., S. Sanker, M. J. Glynias, S. S. Karnik, and A. Husain. 1996. Angiotensin II-forming activity in a reconstructed ancestral chymase. Science **271:**502.

Chang, J. 1996. Inconsistency of evolutionary tree topology reconstruction methods when substitution rates vary across characters. Mathematical Biosciences **134:**189–215.

Churchill, G. A., A. von Haeseler, and W. C. Navidi. 1992. Sample size for a phylogenetic inference. Molecular Biology and Evolution **9:**753–769.

Devlin, B., K. Roeder, and N. C. Ellstrand. 1988. Fractional paternity assignment: theoretical development and comparison to other methods. Theoretical and Applied Genetics **76:**369–380.

Edwards, A. W. F. 1972. Likelihood. Oxford University Press, Oxford.

Farris, J. S. 1969. A successive approximations approach to character weighting. Systematic Zoology **18:**374:385.

Farris, J. S. 1973. On the use of the parsimony criterion for inferring evolutionary trees. Systematic Biology **22:**250–256.

Farris, J. S. 1977. Phylogenetic analysis under Dollo's law. Systematic Zoology **26:**77–88.

Farris, J. S. 1983. The logical basis of phylogenetic analysis. **In** Advances in Cladistics, eds. N. I. Platnick and V. A. Funk, pp. 7–36. Columbia University Press, New York.

Felsenstein, J. 1973a. Maximum likelihood and minimum-steps methods for estimating evolutionary trees from data on discrete characters. Systematic Zoology **22:**240–249.

Felsenstein, J. 1973b. Maximum-likelihood estimation of evolutionary trees from continuous characters. American Journal of Human Genetics **25:**471–492.

Felsenstein, J. 1978. Cases in which parsimony or compatibility methods will be positively misleading. Systematic Zoology **27:**401–410.

Felsenstein, J. 1981a. Evolutionary trees from DNA sequences: a maximum likelihood approach. Journal of Molecular Evolution **17:**368–376.

Felsenstein, J. 1981b. A likelihood approach to character weighting and what it tells us about parsimony and compatibility. Biological Journal of the Linnean Society **16:**183–196.

Felsenstein, J. 1983. Statistical inference of phylogenies. Journal of the Royal Statistical Society A **146:**246–272.

Felsenstein, J. 1988. Phylogenies from molecular sequences: inference and reliability. Annual Review of Genetics **22:**521–565.

Felsenstein, J. 1993. PHYLIP (Phylogeny Inference Package), version 3.5c. University of Washington, Seattle.

Felsenstein, J., and G. A. Churchill. 1995. A hidden Markov chain approach to variation among sites in rate of evolution. Molecular Biology and Evolution **13:**93–104.

Fitch, W. M. 1971. Toward defining the course of evolution: minimal change for a specific tree topology. Systematic Zoology **20:**406–416.

Gaut, B., and P. O. Lewis. 1995. Success of maximum likelihood phylogeny inference in the four-taxon case. Molecular Biology and Evolution **12:**152–162.

Goldman, N. 1990. Maximum likelihood inference of phylogenetic trees, with special reference to a Poisson process model of DNA substitution and to parsimony analyses. Systematic Zoology **39:**345–361.

Goldman, N. 1993. Statistical tests of models of nucleotide substitution. Journal of Molecular Evolution **36:**182–198.

Goldman, N., and Z. Yang. 1994. A codon-based model of nucleotide substitution for protein-coding DNA sequences. Molecular Biology and Evolution **11:**725–736.

Hasegawa, M., H. Kishino, and T. Yano. 1985. Dating of the human-ape splitting by a molecular clock of mitochondrial DNA. Journal of Molecular Evolution **21:**160:174.

Hasegawa, M., H. Kishino, and N. Saitou. 1991. On the maximum likelihood method in molecular phylogenetics. Journal of Molecular Evolution **32:**443–445.

Hayasaka, K., T. Gojobori, and S. Horai. 1988. Molecular phylogeny and evolution of primate mitochondrial DNA. Molecular Biology and Evolution **5:**626–644.

Hillis, D. M. 1995. Approaches for assessing phylogenetic accuracy. Systematic Biology **44:**3–16.

Hillis, D. M., J. P. Huelsenbeck, and D.L. Swofford. 1994. Hobgoblin of phylogenetics? Nature **369:**363–364.

Hillis, D. M., B. K. Mable, and C. Moritz. 1996. Applications of molecular systematics: the state of the field and a look to the future. **In** Molecular Systematics, Second Edition, eds. D. M. Hillis, C. Moritz, and B. K. Mable, pp. 515–543. Sinauer Associates, Sunderland, Massachusetts.

Huelsenbeck, J. P. 1995. Performance of phylogenetic methods in simulation. Systematic Biology **44:**17–48.

Huelsenbeck, J. P., and D. M. Hillis. 1993. Success of phylogenetic methods in the four-taxon case. Systematic Biology **42:**247–264.

Huelsenbeck, J. P., and J. J. Bull. 1996. A likelihood ratio test to detect conflicting phylogenetic signal. Systematic Biology **45:**92–98.

Jin, L., and M. Nei. 1990. Limitations of the evolutionary parsimony method of phylogenetic analysis. Molecular Biology and Evolution **7:**82–102.

Jukes, T. H., and C. R. Cantor. 1969. Evolution of protein molecules. **In** Mammalian Protein Metabolism, ed. H. N. Munro, pp. 21–132. Academic Press, New York.

Kiefer, J. C. 1987. Introduction to statistical inference. Springer-Verlag, Berlin.

Kim, J. 1996. General inconsistency conditions for maximum parsimony: effects of branch lengths and increasing numbers of taxa. Systematic Biology **45:**363–374.

Kimura, M. 1980. A simple method for estimating evolutionary rate of base substitutions through comparative studies of nucleotide sequences. Journal of Molecular Evolution **16:**111–120.

Kimura, M. 1981. Estimation of evolutionary distances between homologous nucleotide sequences. Proceedings of the National Academy of Sciences U.S.A. **78**:454–458.

Kishino, H., and M. Hasegawa. 1989. Evaluation of the maximum likelihood estimate of the evolutionary tree topologies from DNA sequence data, and the branching order in hominoidea. Journal of Molecular Evolution **29**:170–179.

Kluge, A. G., and J. S. Farris. 1969. Quantitative phyletics and the evolution of anurans. Systematic Zoology **18**:1–32.

Kuhner, M. K., and J. Felsenstein. 1994. A simulation comparison of phylogeny algorithms under equal and unequal evolutionary rates. Molecular Biology and Evolution **11**:459–468.

Lake, J. A. 1994. Reconstructing evolutionary trees from DNA and protein sequences: paralinear distances. Proceedings of the National Academy of Sciences U.S.A. **91**:1455–1459.

Lockhart, P. J., M. A. Steel, M. D. Hendy, and D. Penny. 1994. Recovering evolutionary trees under a more realistic model of sequence evolution. Molecular Biology and Evolution **11**:605–612.

Meagher, T. R. 1986. Analysis of paternity within a natural population of *Chamaelirium luteum*. I. Identification of most-likely male parents. American Naturalist **128**:199–215.

Muse, S. V. 1995. Evolutionary analyses of DNA sequences subject to constraints on secondary structure. Genetics **139**:1429–1439.

Muse, S. V. 1996. Estimating synonymous and nonsynonymous substitution rates. Molecular Biology and Evolution **13**:105–114.

Muse, S. V., and B. S. Gaut. 1994. A likelihood approach for comparing synonymous and nonsynonymous substitution rates, with application to the chloroplast genome. Molecular Biology and Evolution **11**:715–724.

Muse, S. V., and B. S. Weir. 1992. Testing for equality of evolutionary rates. Genetics **132**:269–276.

Navidi, W. C., G. A. Churchill, and A. von Haeseler. 1991. Methods for inferring phylogenies from nucleic acid sequence data by using maximum likelihood and linear invariants. Molecular Biology and Evolution **8**:128–143.

Nei, M. 1991. Relative efficiencies of different treemaking methods for molecular data. **In** Phylogenetic Analysis of DNA Sequences, eds. M. M. Miyamoto and J. Craycraft, pp. 90–128. Oxford University Press, New York.

Penny, D., P. J. Lockhart, M. A. Steel, and M. D. Hendy. 1994. The role of models in reconstructing evolutionary trees. **In** Models in Phylogeny Reconstruction, eds. R. W. Scotland, D. J. Siebert, and D. M. Williams, pp. 211–230. Clarendon Press, Oxford.

Reeves, J. H. 1992. Heterogeneity in the substitution process of amino acid sites of proteins coded for by mitochondrial DNA. Journal of Molecular Evolution **35**:17–31.

Roeder, K., B. Devlin, and B. G. Lindsay. 1988. Application of maximum likelihood methods to population genetic data for the estimation of individual fertilities. Biometrics **45**:363–379.

Sankoff, D. 1975. Minimal mutation trees of sequences. SIAM Journal of Applied Mathematics **28**:35–42.

Schöniger, M., and A. von Haeseler. 1994. A stochastic model for the evolution of autocorrelated DNA sequences. Molecular Phylogenetics and Evolution **3**:240–247.

Schöniger, M., and A. von Haeseler. 1995. Performance of the maximum likelihood, neighbor joining, and maximum parsimony methods when sequence sites are not independent. Systematic Biology **44**:533–547.

Sidow, A., and T. P. Speed. 1992. Estimating the fraction of invariable codons with a capture-recapture method. Journal of Molecular Evolution **35**:253–260.

Steel, M.A. 1994. Recovering a tree from the leaf colourations it generates under a Markov model. Applied Mathematics Letters **7**:19–23.

Swofford, D. L., G. J. Olsen, P. J. Waddell, and D. M. Hillis. 1996. Phylogenetic inference. **In** Molecular Systematics, Second Edition, eds. D. M. Hillis, C. Moritz, and B. K. Mable, pp. 407–514. Sinauer Associates, Sunderland, Massachusetts.

Tamura, K., and M. Nei. 1993. Estimation of the number of nucleotide substitutions in the control region of mitochondrial DNA in humans and chimpanzees. Molecular Biology and Evolution **10**:512–526.

Tavaré, S. 1986. Some probabilistic and statistical problems on the analysis of DNA sequences. Lectures in Mathematics of the Life Sciences **17**:57–86.

Thorne, J. L., H. Kishino, and J. Felsenstein. 1992. Inching toward reality: an improved likelihood model of sequence evolution. Journal of Molecular Evolution **34**:3–16.

Tillier, E. R. M., and R. A. Collins. 1995. Neighbor joining and maximum likelihood with RNA sequences: addressing the interdependence of sites. Molecular Biology and Evolution **12**:7–15.

Weir, B. S. 1996. Genetic Data Analysis II. Sinauer Associates, Sunderland, Massachusetts.

Yang, Z. 1993. Maximum-likelihood estimation of phylogeny from DNA sequences when substitution rates differ over sites. Molecular Biology and Evolution **10**:1396–1401.

Yang, Z. 1994a. Estimating the pattern of nucleotide substitution. Journal of Molecular Evolution **39**:105–111.

Yang, Z. 1994b. Maximum likelihood phylogenetic estimation from DNA sequences with variable rates over sites: approximate methods. Journal of Molecular Evolution **39**:306–314.

Yang, Z. 1994c. Statistical properties of the maximum likelihood method of phylogenetic estimation and comparison with distance matrix methods. Systematic Biology **43**:329–342.

Yang, Z. 1995. A space-time process model for the evolution of DNA sequences. Genetics **139**:993–1005.

Yang, Z., and D. Roberts. 1995. On the use of nucleic acid sequences to infer early branchings in the tree of life. Molecular Biology and Evolution **12**:451–458.

Yang, Z., N. Goldman, and A. Friday. 1994. Comparison of models for nucleotide substitution used in maximum-likelihood phylogenetic estimation. Journal of Molecular Evolution **11**:316–324.

6

Patterns of Sequence Evolution and Implications for Parsimony Analysis of Chloroplast DNA

Richard G. Olmstead, Patrick A. Reeves, and Alan C. Yen

One of the advantages frequently cited for the use of DNA sequence data in phylogeny reconstruction is that explicit models of character evolution, based on our understanding of sequence evolution, may be incorporated into the analysis. In this way the method for recovering the pattern of evolutionary divergence can be tied more closely to the processes that produce the variation (Clegg and Zurawski, 1992; Simon et al., 1994; Moritz and Hillis, 1996). Within the context of parsimony analysis, it has been assumed that the reliability of information for phylogeny reconstruction can be inferred from estimates of substitution rate for different sites, with the most slowly evolving sites being the most reliable and, therefore, most heavily weighted (Farris, 1969, 1977; Felsenstein 1981a; Swofford et al. 1996). However, parsimony analyses based on chloroplast DNA (cpDNA) typically are conducted with equal weight given to all changes (e.g., Kim et al., 1992; Olmstead et al., 1993; Price and Palmer, 1993; Hoot et al., 1995; Johnson and Soltis, 1995; Gadek et al., 1996). The main reasons for this may be historical, stemming from a reluctance to weight characters in morphological cladistic analyses, or operational, due to the difficulty of implementing the more complicated analyses necessary for incorporating weights in phylogenetic analysis. In addition, there is a paucity of statistical analyses of cpDNA sequence substitution patterns (Clegg, 1993), which results in an inability to provide a firm basis for establishing weights.

Chloroplast DNA sequences are likely to provide abundant data for plant phylogenetic study for many years to come, so an examination of substitution patterns is warranted at this time. In this chapter, we examine patterns of cpDNA sequence evolution at two commonly used chloroplast genes for three taxonomic groups to examine the suitability of weighting methods and to explore prospective methods for future parsimony analyses of cpDNA sequence data. The following introduction is not intended to review the literature on evolutionary rates, phylogenetic methods for sequence analysis, or weighting. Useful introductions to these subjects may be found in Simon et al. (1994), Swofford et al. (1996), and Yang (1996b).

Simulation studies (Huelsenbeck and Hillis, 1993; Huelsenbeck, 1995) and studies of experimental phylogenies (Hillis et al., 1994) indicate that differentially weighted parsimony often outperforms equally weighted parsimony in

We thank Jeff Doyle, Sean Graham, Paul Lewis, Doug Soltis, and Jonathan Wendel for helpful comments on the manuscript, Anna Goebel for discussions of patterns of sequence evolution, and D. Swofford for permission to use the test version of PAUP*. This work was supported by NSF Grants BSR-9107827 and DEB-9509804.

phylogeny reconstruction over a broad range of conditions (e.g., varying rates among lineages or sites). Furthermore, maximum likelihood (ML) methods may outperform even differentially weighted parsimony (Huelsenbeck and Hillis, 1993; Hillis et al., 1994; Kuhner and Felsenstein, 1994; Tateno et al., 1994; Huelsenbeck, 1995). The extent to which weights assist parsimony in recovering correct trees under a variety of conditions will depend on the "fit" of the weighting scheme to the actual processes giving rise to sequence variation. While the application of ML approaches (see Chapter 5) may one day supplant the use of weighted parsimony as the primary means of conducting phylogenetic analyses (when it becomes more tractable to use this method), an accurate understanding of the patterns of sequence change would improve the implementation of both methods. At present it is much more feasible to apply parsimony methods to large sets of sequence data (Swofford et al., 1996; Yang, 1996b), so the decision regarding how to weight parsimony analyses best will continue to confront molecular systematists.

Early efforts to model sequence evolution assumed that all sites had an equal probability of substitution, that change between any two nucleotides was equiprobable, and that the processes responsible for change were stochastic (Jukes and Cantor, 1969; Fitch, 1971). A corollary of this was the prediction that a comparison of sequences that had diverged under this model would exhibit a Poisson distribution of change for sites in the sequence, with most sites exhibiting little or no change and a few sites showing many changes (Bishop and Friday, 1985, Goldman, 1990). The assumptions of the Jukes and Cantor model are essentially equivalent to those of equally weighted parsimony (Yang, 1996a). Felsenstein (1981a) has shown that when the amount of change is small the Jukes and Cantor model assumptions approximate reality, but when change is greater this model may lead to erroneous results (Felsenstein, 1978; Hillis et al., 1994). Examples of refinements to this model take into account inferred differences in numbers of transitions or transversions relative to the random expectation (Kimura, 1980), differences in base composition (Felsenstein, 1981b), inferred differences in rates of change by codon position for protein coding genes (Fitch and Markowitz, 1970; Fitch, 1986; Albert et al., 1993), and among-site rate heterogeneity (Jin and Nei, 1990; Yang, 1993). The first two refinements affect the assumption of equiprobable change among nucleotides and the last two contradict the assumption of equiprobable change among sites. Inferring differences in rate among codon positions divides the data into classes, within which the assumption of equiprobable change is maintained (e.g., Albert et al., 1993), but between which differences are presumed to exist. Differences among sites within any class are still expected as a result of the stochastic nature of the underlying processes.

Estimating substitution rates commonly is done by one of two methods: (1) pairwise comparisons, with correction for multiple substitutions between sequences, or (2) parsimony optimization using a "best" tree (see Swofford et al., 1996). The pairwise comparison method generally has been used in association with distance based methods of phylogeny construction (e.g., Fitch and Margoliash, 1967), but also has been used with parsimony methods for estimation of substitution rates or transition-transversion ratios. When using corrected pairwise comparisons, the correction itself is based on assumptions of underlying evolutionary processes and one has no way of knowing if the correction over- or underestimates change. The parsimony optimization method provides a correction for multiple substitutions by making explicit hypotheses of multiple changes on the tree for homoplastic characters, but always will be either a correct estimate of change, if small, or an underestimate of change, if great (Swofford et al., 1996; Yang, 1996b). Likewise, if two classes of changes (e.g., transversions and transitions) are unequal in frequency, the effect on the estimate of the one with the higher rate is expected to be greater. The estimate provided by parsimony is conservative in that one knows which direction the error, if any, will be. In the following analyses, we take a tree-based approach using the parsimony criterion to estimate change in cpDNA sequences.

In examining the estimates of change inferred over the "accepted" trees for each data set, we apply statistical tests to address questions concerning the differences in rate among putative classes of sites (e.g., those defined by gene or codon position) and use correlation analysis to explore site-to-site variability in substitution rates in an effort to account better for variation in probability of change by site for cpDNA sequence analysis. Although this study is exploratory, it may provide useful guidelines for future phylogenetic studies of these and other chloroplast genes (see Vawter and Brown, 1993, for similar exploration of small subunit rRNA gene sequences).

METHODS

Data

The experimental design of this study incorporated data from three different taxonomic groups or lineages, families Solanaceae, Lamiaceae, and Scrophulariaceae. For each group, data from two chloroplast genes, rbcL, and ndhF were examined, all of which have been the subject of previously published phylogenetic analyses (Solanaceae: Olmstead and Sweere, 1994; Olmstead et al., 1998; Lamiaceae: Wagstaff and Olmstead, 1997; Wagstaff et al., 1998; Scrophulariaceae: Olmstead and Reeves, 1995). These three taxonomic groups exhibit a range of cpDNA sequence divergence. Sequences of Solanaceae (represented by 17 species) are the least diverged, Lamiaceae (33 species) are intermediate in divergence, and Scrophulariaceae (29 species), are the most divergent. These three groups are independent, except that three species of Lamiaceae are included in the Scrophulariaceae data set. Scrophulariaceae are a polyphyletic assemblage (Olmstead and Reeves, 1995), so sampling includes members of several other families. To minimize the problem of estimating change on long branches connecting the outgroups to the ingroups in each data set, the outgroups used in the published phylogenetic results were removed from the data sets, but the inferred topology of the ingroup was kept the same. In sum, for most of the results reported here, there will be six data sets, three taxonomic groups for each of two gene sequences.

The aligned sequence length for the *rbcL* data sets was 1,402 nucleotides (nt) with no gaps in any alignment. In each data set the first aligned site corresponded to position 27 in tobacco, which is the first site following the PCR primer, and the last site is position 1,428, the point where unambiguous alignment ceases. For *ndhF,* the first aligned position corresponded to position 24 in tobacco, which is the first site following the PCR primer, and the last position corresponded to position 2,109 in tobacco, the last position before the PCR primer near the 3′ end of the gene (Olmstead and Sweere, 1994). Because *ndhF* is variable in length, for the purposes of comparing substitution patterns and correlations between data sets, alignments maintaining putative positional homology across all three taxonomic groups were used. Gaps were removed so that only those positions for which data were available for all data sets were used (a total of 2,086 nt).

Inferred Character Change

To study patterns of sequence evolution, inference must be made regarding explicit numbers of changes per site. In this study, the parsimony criterion was used to optimize changes at each position in the gene sequences over a selected tree (Figs. 6.1–6.3) for each set of taxa, using the computer program MacClade vers. 3.01 (Maddison and Maddison, 1992). This provides a minimum estimate of change for each site for each data set. Trees were selected on the basis of the best available inference from previously published studies (Solanaceae: Olmstead and Sweere, 1994; Olmstead et al., 1998; Lamiaceae; Wagstaff and Olmstead, 1997; Wagstaff et al., 1998; Scrophulariaceae; Olmstead and Reeves, 1995). In each case the tree was constructed primarily using a combined analysis of *rbcL* and *ndhF* sequence data and an equally weighted parsimony approach. However, for Solanaceae, an extensive cpDNA restriction site data set was analyzed in combination with the sequence data (Olmstead and Sweere, 1994) and larger sequence and restriction site data sets have been analyzed that corroborate these results (Olmstead et al., 1998). For Lamiaceae, cpDNA restriction site and morphological studies are avail-

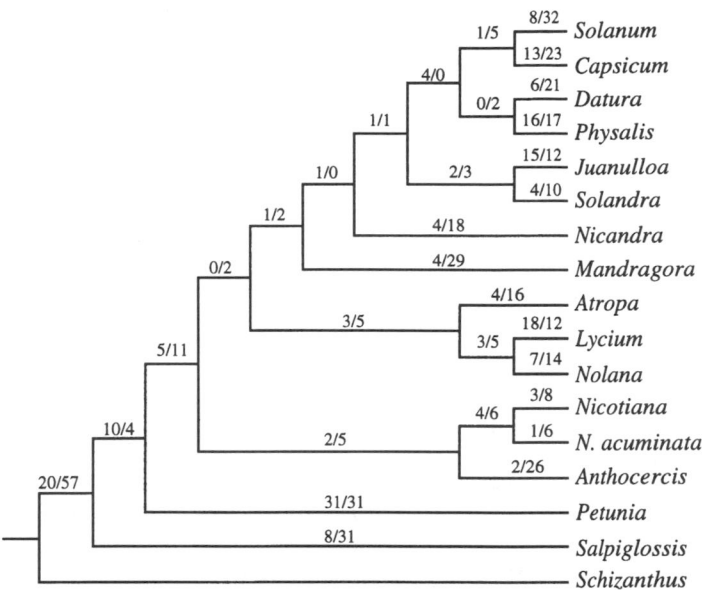

Figure 6.1.   Tree depicting relationships among representatives of Solanaceae. Branch lengths are inferred by ACCTRAN optimization for *rbcL*/*ndhF*.

able that corroborate portions of the tree, where these analyses overlap. For comparison, a maximum likelihood tree was obtained for each taxonomic group using the combined data from both genes and using an approximate gamma distribution (discrete gamma; Yang, 1994) of rates (with four rate classes) to accommodate among site rate heterogeneity (Yang, 1996b; implemented with test version 4.0d49 of PAUP* kindly provided by D. Swofford). Transition/transversion ratio was not fixed and a gamma distribution of 0.5 was used. A gamma distribution allows the likelihood function to be optimized independently for each site. Each of these trees then was subject to the Kishino-Hasegawa test (Kishino and Hasegawa, 1989) for tree congruence with the corresponding parsimony tree.

Sequence Analyses

A series of analyses was conducted to examine the patterns of variation in sequence evolution in cpDNA. For all analyses, the minimum inferred number of substitutions at each site provided the primary data. To examine the inferred change, positions in the sequences of each data set were partitioned into a hierarchical array of three classes in the following order of increasing inclusiveness: (1) phylogenetically (cladistically) informative sites; (2) all variable sites; and (3) all sites in the sequence. The first two partitions contain different sites in different taxonomic groups. Because both genes are protein-coding, a second partitioning of the data was examined in which the first, second, and third codon positions define the classes.

Using the minimum number of inferred changes as an estimate of substitution rate for each site within each data set, we applied tests to ask the following questions. (1) Does the pattern of change among sites conform to a Poisson distribution? This was assessed using a chi-square goodness-of-fit test comparing the expected with the observed numbers of sites exhibiting different amounts of change. (2) Is there a significant difference in mean substitution rate between the two genes, *rbcL* and *ndhF*, within each taxonomic group? (3) Is there a significant difference in mean substitution rate among codon positions within each gene? A Poisson model is used for the last two tests, because the amongsite distribution was found to be similar to a Poisson model. There may not always be a simple model for data sets with rate heterogeneity.

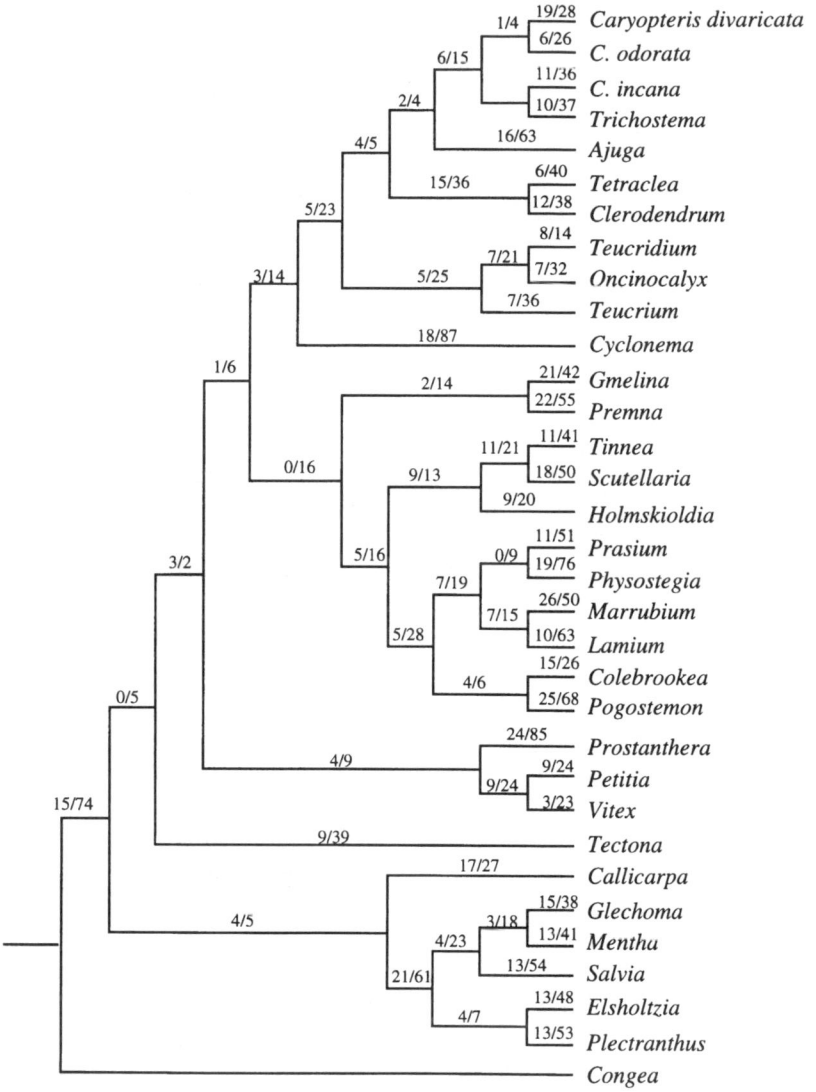

Figure 6.2.   Tree depicting relationships among representatives of Lamiaceae. Branch lengths are inferred by ACCTRAN optimization for *rbcL*/*ndhF*.

Even the gamma distribution model, which assumes a Poisson process at each site, but allows for rate heterogeneity between sites (Yang, 1993, 1996b), may be rejected for some sequence data sets (Sullivan et al., 1995). These tests using the Poisson model should be considered approximate tests made using a reasonable model.

For each of these questions, we partitioned the data into the hierarchical classes, enabling us to determine if there was a difference between the class of characters that is recognized in a parsimony analysis (informative sites only) and those classes (including all variable sites and all sites) that are considered in phylogenetic analyses using distance or maximum likelihood methods. A sequential Bonferroni correction to the experiment-wise error level was made because of multiple tests (Rice, 1989). This correction guards against the random occurrence of a significant test result when many tests are conducted and may result in some tests being non-

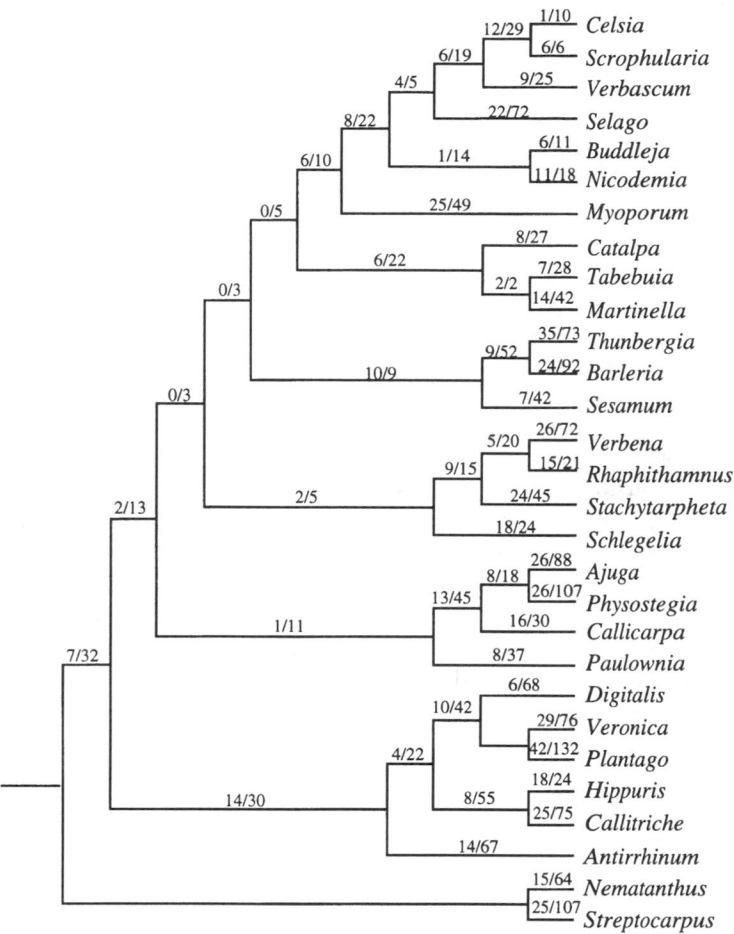

Figure 6.3.   Tree depicting relationships among representatives of Scrophulariaceae and some related families. Branch lengths are inferred by ACCTRAN optimization for *rbcL/ndhF.*

significant that would be significant if tested alone. In addition to estimates of substitution rate, the mean consistency index, or CI (Kluge and Farris, 1969), was calculated for each codon position for the informative sites to compare levels of homoplasy for each codon position (Kim et al., 1992).

If all sites have equal substitution probabilities and substitutions are stochastic, then one would predict that the inferred frequency of change at one site in one group would not be correlated with the inferred frequency of change at the homologous site in another taxonomic group. To test this, we used Spearman's rank correlation test for all pairwise combinations of the three taxonomic groups for each sequence. A

test of the informative sites common to pairs of taxonomic groups was conducted to see if the most consistently informative sites among the three groups showed correlated amounts of change. Separate correlation analyses also were conducted for each codon position. In addition, the test was conducted on a subset of the sites where sites that exhibited no change in any of the three groups were excluded. This test was done to reflect the concern that some proportion of sites is not free to vary at all (Hasegawa et al., 1985; Palumbi, 1989). Removal of invariant sites provided a conservative test for the remaining sites, because we eliminated the largest possible cohort of invariant sites, even though many of these sites are known to vary in other studies (*rbcL:* Chase, Soltis, Olmstead et al., 1993; *ndhF:* R. Olmstead and R. Jansen, unpubl.).

Estimates of transition to transversion ratios (ts/tv) were calculated in two ways. First, numbers of unambiguous transitions and transversions were inferred over each tree using the parsimony criterion and the computer program MacClade. Because this method is likely to underestimate transitions relative to transversions, particularly among deep divisions and long branches (Sullivan et al., 1996; Yang, 1996a), and because many transformations cannot be interpreted unambiguously in this manner, a second estimate was made by counting only changes inferred to occur on terminal branches of sister taxa. This counting should estimate ts/tv more accurately, because fewer hidden changes are likely to exist among closely related taxa near the tips of trees. However, the number of transformations will be smaller when examining only terminal sister branches, so statistical power may be reduced. The transition-transversion ratio was calculated for the most highly variable sites (defined as those exhibiting inferred numbers of changes above the midpoint of the range of variation exhibited) in each data set and this figure was compared to the overall estimates. This comparison was designed to detect whether there was a different ts/tv bias in the most highly variable (and, therefore, most homoplastic) sites. Maximum likelihood methods may allow for a more accurate estimate of ts/tv ratio, particularly when substitution rates are high, but are not considered in this paper.

Base composition for each data set and inferred proportions of transformations between bases in a representative data set (Scrophulariaceae *rbcL*) were calculated using MacClade. The expected number of base transformations, given the observed base compositions, and the ratio of inferred to expected numbers for the transformation from base i to base j, was calculated as follows, where P_i = proportion of base i; P_j = proportion of base j; T = total number of inferred transformations for a given tree (tree length); and $t_{i \to j}$ = inferred number (by parsimony criterion) of transformations from base i to j:

$$P_i \frac{P_j}{1 - P_i} = \text{Expected proportion of transformations from } i \text{ to } j \quad (6.1)$$

$$T \times P_i \frac{P_j}{1 - P_i} = \text{Expected number of transformations from } i \text{ to } j \quad (6.2)$$

$$\frac{t_{i \to j}}{T\,[(P_i)(P_j) / (1 - P_i)]} = \text{Ratio of inferred to expected transformations from } i \text{ to } j \text{ scaled to base composition} \quad (6.3)$$

RESULTS

The accepted trees used for the character optimizations for both genes are depicted in Figs. 6.1–6.3 with branch lengths inferred by an accelerated transformation (ACCTRAN) optimization criterion. The root node in each case indicates the inferred point of attachment of the outgroups to the ingroup. For all three taxonomic groups the ML trees obtained (not shown) using the approximate gamma distribution method (Swofford et al., 1996; Yang, 1996b) differed topologically in minor ways, but were not significantly different (as indicated by the Kishino-Hasegawa test: Kishino and Hasegawa, 1989) from the parsimony trees used for comparisons in the analyses reported here.

A value representing the mean substitution rate for each gene in each data set and for each codon position in each gene was estimated using the inferred number of changes on each tree. No effort was made to calculate absolute substitution rates, so comparisons across taxonomic groups are not valid due to different ages for the three groups. However, the estimated values are valid for comparisons between genes within a taxonomic group and among codon positions within a gene and taxonomic group.

The distributions of numbers of changes at each site in all of the data sets are left-skewed and Poisson-like. They consist of a large number of sites that exhibit no change and a small number of sites that exhibit much change (see Fig. 6.4 for an example). These distributions also hold for all hierarchical subsets of the data and for codon position subsets. However, a goodness-of-fit test of the observed distributions to expected distributions under a Poisson model (Sokal and Rohlf, 1981), for all sites in each

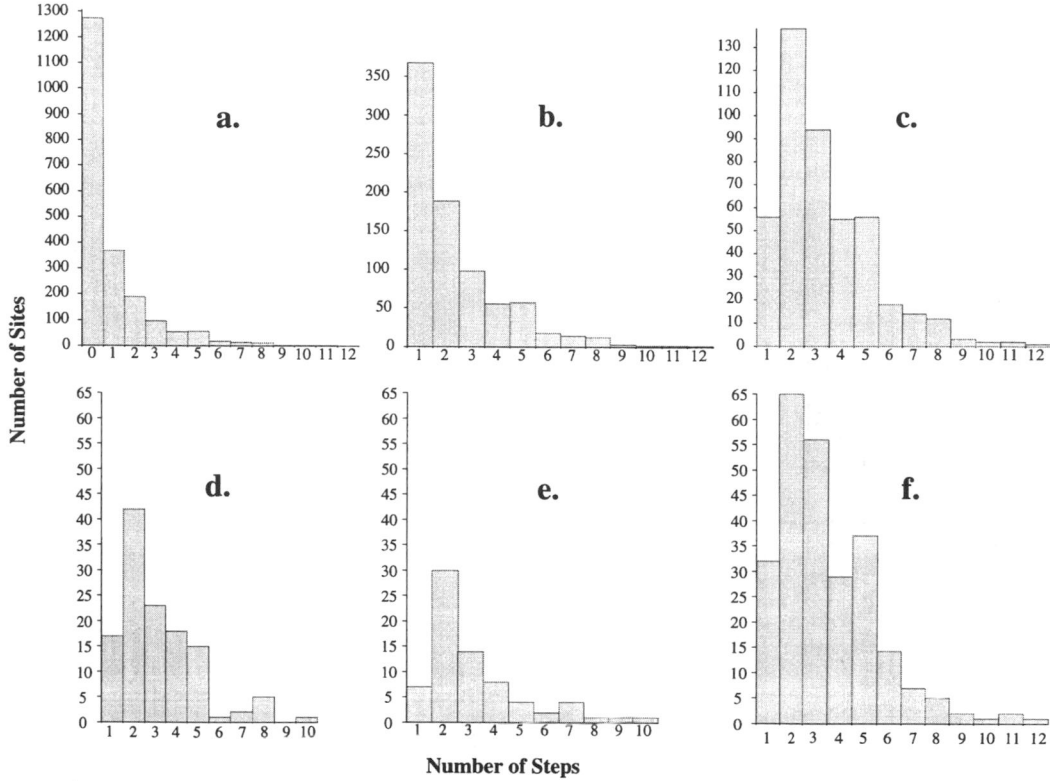

Figure 6.4.   Distribution of the number of nucleotide sites exhibiting inferred amounts of change for *ndhF* in Lamiaceae. A. All sites, *b.* variable sites only, *c.* informative sites only, *d.* informative first codon positions, *e.* informative second positions, *f.* informative third positions. The vertical scale expands threefold from a to b and from b to c and twofold c to d, e, and f. Horizontal axis remains constant for all graphs.

gene together or for particular codon positions within each gene, showed that the observed distributions have an excess in the invariant class and in the high value classes and thus differ significantly from a Poisson distribution (Table 6.1). The same test on informative sites indicated that some data sets do not differ significantly from the Poisson model. Specifically, neither of the genes in Solanaceae, nor *rbcL* in Scrophulariaceae were significantly different from a Poisson distribution for any codon position, or for all informative sites combined. The other three data sets (both *rbcL* and *ndhF* in Lamiaceae and *ndhF* in Scrophulariaceae) differ significantly from a Poisson distribution when all informative sites were considered, but in 4 of 9 cases the informative sites in individual codon positions do not differ significantly from a Poisson distribution (Table 6.1). The smaller sample sizes of the informative sites subsets may make it more difficult to reject a Poisson distribution in this test.

Even though the distributions do not conform exactly to a Poisson model, their close approximation to this distribution provides the basis for tests of rate differences. Deviation from the Poisson may have an impact on some comparisons that are near the limit of statistical power, but should not impact most comparisons (Sneath [1986] suggests that a binomial model might be better for data sets with invariant sites removed, but that the difference from a Poisson distribution is not great). Use of an approximate model serves the useful heuristic purpose of examining the difference between mean rates. Mean substitution rates for *rbcL* and *ndhF* in all three taxonomic groups were significantly different when all sites were included, with *ndhF* having the

Table 6.1. Distribution of inferred change for all sites for both *rbcL* and *ndhF* in the Solanaceae, Lamiaceae, and Scrophulariaceae data sets. The p-value indicates whether the observed distribution differs from a Poisson distribution. n, number of sites; ns, not significant; *, $p < 0.05$; **, $p < 0.01$; ***, $p < 0.001$.

Data Set	Data Type	Codon Position	0	1	2	3	4	5	6	7	8	9	10	11	12	13	n	Mean	p
Solanaceae *rbcL*	All sites	1	434	17	5	9	2										467	0.133	***
		2	448	13	3	1	1	1									467	0.066	***
		3	388	60	14	4	2										468	0.231	***
		combined	1270	90	22	14	5	1									1402	0.143	***
	Informative sites	1		4	5	9	2										20	2.450	ns
		2		4	3	1	1	1									10	2.200	ns
		3		7	9	3	2										21	2.000	ns
		combined		15	17	13	5	1									51	2.216	ns
Lamiaceae *rbcL*	All sites	1	406	31	7	5	5	1	3	5	1	2				1	467	0.379	***
		2	433	17	11	3	1	2									467	0.133	***
		3	287	93	49	20	6	4	3	1	3	1		1			468	0.778	***
		combined	1126	141	67	28	12	7	6	6	4	3		1		1	1402	0.430	***
	Informative sites	1		3	5	5	5	1	3	5	1	2				1	31	4.677	ns
		2		1	8	3	1	2									15	2.667	ns
		3		19	43	20	6	4	3	1	3	1		1			101	2.752	***
		combined		23	56	28	12	7	6	6	4	3		1		1	147	3.150	***
Scroph *rbcL*	All sites	1	408	24	15	6	2	3	2	2	1	1	1		1		467	0.353	***
		2	424	33	1	5		3	1								467	0.152	***
		3	250	118	48	24	12	9	4	2	1						468	0.908	***
		combined	1082	175	64	35	14	15	7	6	2	1	1		1		1402	0.471	***
	Informative sites	1		2	12	6	2	3	2	2	1	1	1		1		34	4.029	ns
		2		2		5		3	1								11	3.455	ns
		3		14	36	23	12	9	4	2	1						101	2.911	ns
		combined		18	48	34	14	15	7	6	2	1	1		1		146	3.212	ns

Table 6.1. *Continued*

Data Set	Data Type	Codon Position	Number of Steps														n	Mean	p
			0	1	2	3	4	5	6	7	8	9	10	11	12	13			
Solanaceae *ndhF*	All sites	1	618	66	8		2		1								695	0.138	***
		2	639	47	6	2	1										695	0.099	***
		3	512	139	33	7	3	2									696	0.356	***
		combined	1769	252	47	9	6	2	1								2086	0.198	***
	Informative sites	1		7	6		2		1								16	2.063	ns
		2		1	3	2	1										7	2.429	ns
		3		13	22	7	3	2									47	2.128	ns
		combined		21	31	9	6	2	1								70	2.143	ns
Lamiaceae *ndhF*	All sites	1	453	115	60	25	18	15	1	2	5		1				695	0.758	***
		2	532	86	42	15	8	4	2	4	1	1					695	0.466	***
		3	290	164	86	59	28	37	14	7	5	1	1	2	1	1	696	1.501	***
		combined	1275	365	188	99	54	56	17	13	11	2	2	2	1	1	2086	0.909	***
	Informative sites	1		17	42	23	18	15	1	2	5		1				124	3.121	***
		2		7	30	14	8	4	2	4	1	1					71	3.070	*
		3		32	65	56	28	37	14	7	5	1	1	2	1	1	250	3.448	*
		combined		56	137	93	54	56	17	13	11	2	2	2	1	1	445	3.297	***
Scroph *ndhF*	All sites	1	456	111	60	31	16	9	8	2	2						695	0.737	***
		2	516	88	39	14	19	11	4	2	2						695	0.565	***
		3	269	161	88	67	34	29	23	11	9	3	1	1			696	1.658	***
		combined	1241	360	187	112	69	49	35	15	13	3	1	1			2086	0.987	***
	Informative sites	1		10	35	30	16	9	8	2	2						112	3.196	ns
		2		7	23	14	19	11	4	2	2						82	3.415	ns
		3		21	63	64	34	29	23	11	9	3	1	1			259	3.687	**
		combined		38	121	108	69	49	35	15	13	3	1	1			453	3.517	***

higher rate in all comparisons (Fig. 6.5, Table 6.1). When only informative sites were considered, no significant difference was found between the two genes in any of the three taxonomic groups (Fig. 6.5). Similarly, a comparison of substitution rates among codon positions within each gene showed a significant difference ($p < 0.001$ for all comparisons), with second positions being lowest and third positions highest in both genes for all three taxonomic groups (Fig. 6.6, Table 6.1). However, when only informative sites were considered, most comparisons (14 of 18) were not significant. The only significant differences observed among codon positions suggest that *rbcL* first codon positions in Lamiaceae are more variable than second ($p < 0.01$) or third positions ($p < 0.001$), that *rbcL* first positions in Scrophulariaceae are more variable than third positions ($p < 0.05$), and that *ndhF* third codon positions in Lamiaceae are more variable than first positions ($p < 0.01$).

An example of the distribution of among-site variation for amount of change in one data set is illustrated for the gene *ndhF* in Lamiaceae (Fig. 6.4) and is typical of all six data sets examined. With all sites considered, the distribution is Poisson-like (Fig. 6.4a), though it is not a perfect fit to a Poisson model because of an excess of invariant sites and sites with three or more changes. When uninformative sites are removed, a skewed, unimodal distribution is obtained (Fig. 6.4c). When informative sites are partitioned by codon position (Fig. 6.4d–f), similarly shaped distributions are found.

Mean CI values for informative sites partitioned by codon position indicate no clear tendency with respect to degree of homoplasy relative to codon position (Table 6.2). Homoplasy never was lowest for second positions and mean CI was highest for third positions in four of six cases.

The test for correlated change at homologous positions among taxonomic groups found highly significant correlations ($p < 0.001$) for all comparisons among taxonomic groups for each gene sequence when all sites were considered together and when the codon position partitions were considered separately (Figs. 6.7–6.8, Table

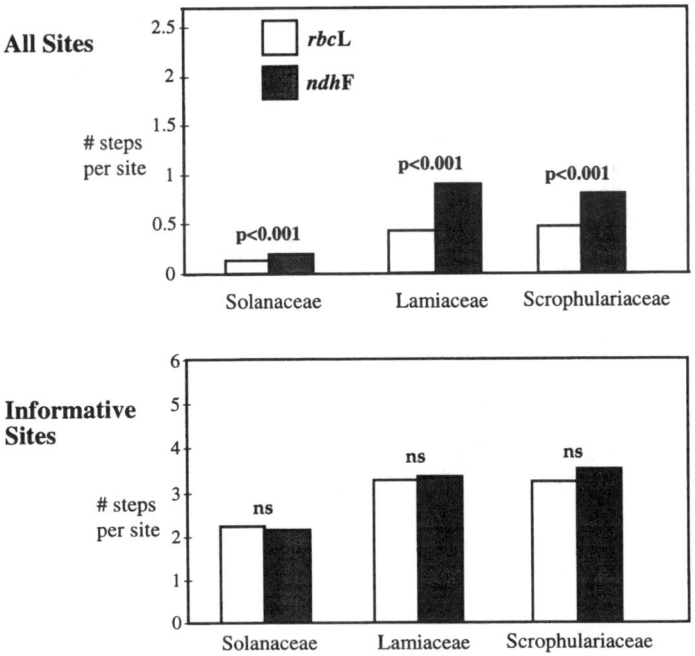

Figure 6.5.   Mean number of inferred changes per site for *rbcL* and *ndhF* for all sites and for informative sites only. Results of statistical tests for difference in mean rates using a Poisson model are indicated above the bars for each pair of genes in each family.

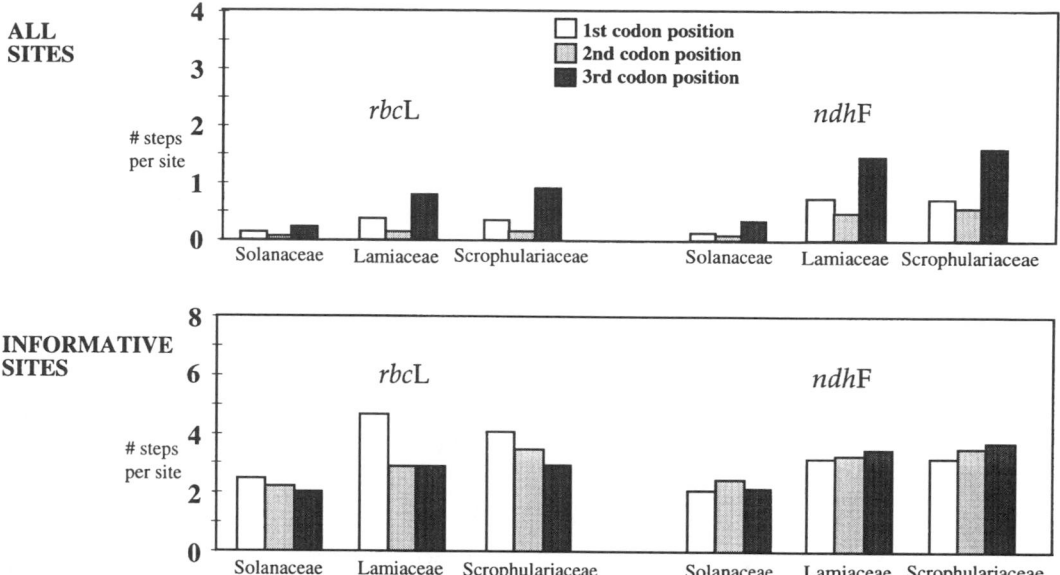

Figure 6.6.   Mean number of inferred changes per site by codon position for *rbcL* and *ndhF* for all sites and for informative sites only.

6.3). When sites that are invariant in all three taxonomic groups were removed, significant correlations were found for all three comparisons for *ndhF*, but only for the Lamiaceae/Scrophulariaceae comparison for *rbcL* (Table 6.3). For each gene the greatest number of invariant sites were at second positions and the fewest at third positions. Comparisons of the codon partitions with the invariant sites removed showed the fewest significant correlations for second position partitions (only *ndhF* for the Lamiaceae/Scrophulariaceae comparison) and significant correlations were most frequent for third position partitions (Table 6.3). Correlations for sets of sites that are informative in both groups compared were positive in all comparisons, but were significant only in the comparison of Lamiaceae and Scrophulariaceae for both genes. The other two *rbcL* comparisons (Solanaceae-Lamiaceae and Solanaceae-Scrophulariaceae) had significant correlations ($p < 0.05$) that disappeared after Bonferroni adjustment for multiple comparisons.

Transition transversion ratios were higher in all data sets than the random expectation of 0.5 if all changes are equally likely, and were consistently higher in *rbcL* than in *ndhF* (Table

6.4). Estimates based on changes inferred over the whole tree (reported in Table 6.4 as minimum and maximum estimates based on different optimization criteria, or as estimates based only on unambiguously optimized changes) were consistently higher than those based on comparison of terminal sister groups. Base composition in each gene shows an excess of A and T relative to C and G (Table 6.4), but with a higher A+T content in *ndhF* than *rbcL* and remarkably little variation in A+T content between taxonomic groups for each gene (Table 6.4). Comparing inferred numbers of each base transformation type with calculated expected numbers for each transformation type for the Scrophulariaceae *rbcL* data (Table 6.5)

Table 6.2. Consistency index for informative sites.

	rbcL			*ndhF*		
	Solan-aceae	Lami-aceae	Scrophu-lariaceae	Solan-aceae	Lami-aceae	Scrophu-lariaceae
all sites	0.500	0.380	0.420	0.586	0.466	0.477
1st pos.	0.442	0.276	0.314	0.576	0.487	0.516
2nd pos.	0.455	0.425	0.368	0.529	0.447	0.450
3rd pos.	0.591	0.428	0.476	0.598	0.461	0.470

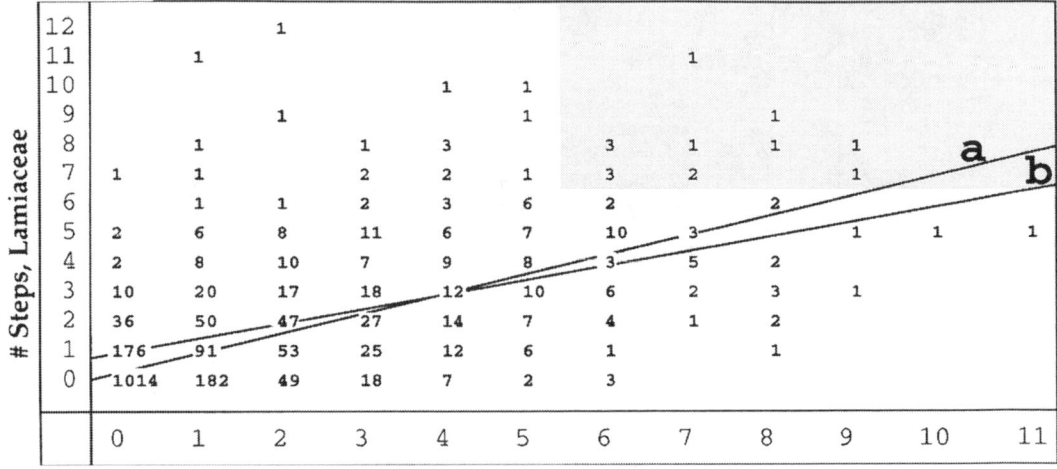

Figure 6.7. Plot of correlation study of change in *ndhF* sequences in Lamiaceae versus Scrophulariaceae. Correlation line a includes all sites (r = 0.624); correlation line b (r = 0.414) excludes sites that are invariant in all three groups. Both are significant (*p* < 0.001). Regions in each distribution that are most variable (above the midpoint in the range of values exhibited for inferred numbers of change) are shaded.

indicates that all transversions are underrepresented and all transitions are overrepresented. However, there is a threefold range in the proportion of inferred to expected numbers for the different transversion types and a twofold range for the different transition types.

DISCUSSION

As a starting point, we took the best available trees for three taxonomic groups for which we have comparable data from each of two chloroplast genes, *rbcL* and *ndhF*. We proceeded to

Table 6.3. Correlation coefficients for comparison of change in *rbcL* and *ndhF* sequences in pairwise comparisons of taxonomic groups. Spearman's rank order correlation results: ns, not significant;*, p < 0.05**, p < 0.01;*** p < 0.001.

Data Type	Gene	Data Set	Codon Position 1		Codon Position 2		Codon Position 3		Overall	
			Solanaceae	Labiatae	Solanaceae	Labiatae	Solanaceae	Labiatae	Solanaceae	Labiatae
All Sites	*rbcL*	Lamiaceae	0.422,***		0.414,***		0.215,***		0.345,***	
		Scrophulariaceae	0.476,***	0.559,***	0.431,***	0.462,***	0.194,***	0.451,***	0.345,***	0.531,***
	ndhF	Lamiaceae	0.249,***		0.232,***		0.269,***		0.300,***	
		Scrophulariaceae	0.259,***	0.568,***	0.268,***	0.567,***	0.327,***	0.628,***	0.336,***	0.624,***
Sites Variable in	*rbcL*	Lamiaceae	0.243, ns		0.201, ns		0.040, ns		0.114, ns	
at Least One		Scrophulariaceae	0.299, ns	0.338,*	0.142, ns	−0.022, ns	0.000, ns	0.220, **	0.093, ns	0.217, ***
Group	*ndhF*	Lamiaceae	0.028, ns		−0.015, ns		0.124, ns		0.090, *	
		Scrophulariaceae	0.027, ns	0.331,***	0.010, ns	0.278,***	0.498,***	0.192,***	0.133,***	0.414,***
Intersection of	*rbcL*	Lamiaceae							0.547, ns	
Informative Sites		Scrophulariaceae							0.432, ns	0.445,***
	ndhF	Lamiaceae							0.106, ns	
		Scrophulariaceae							0.161, ns	0.470,***

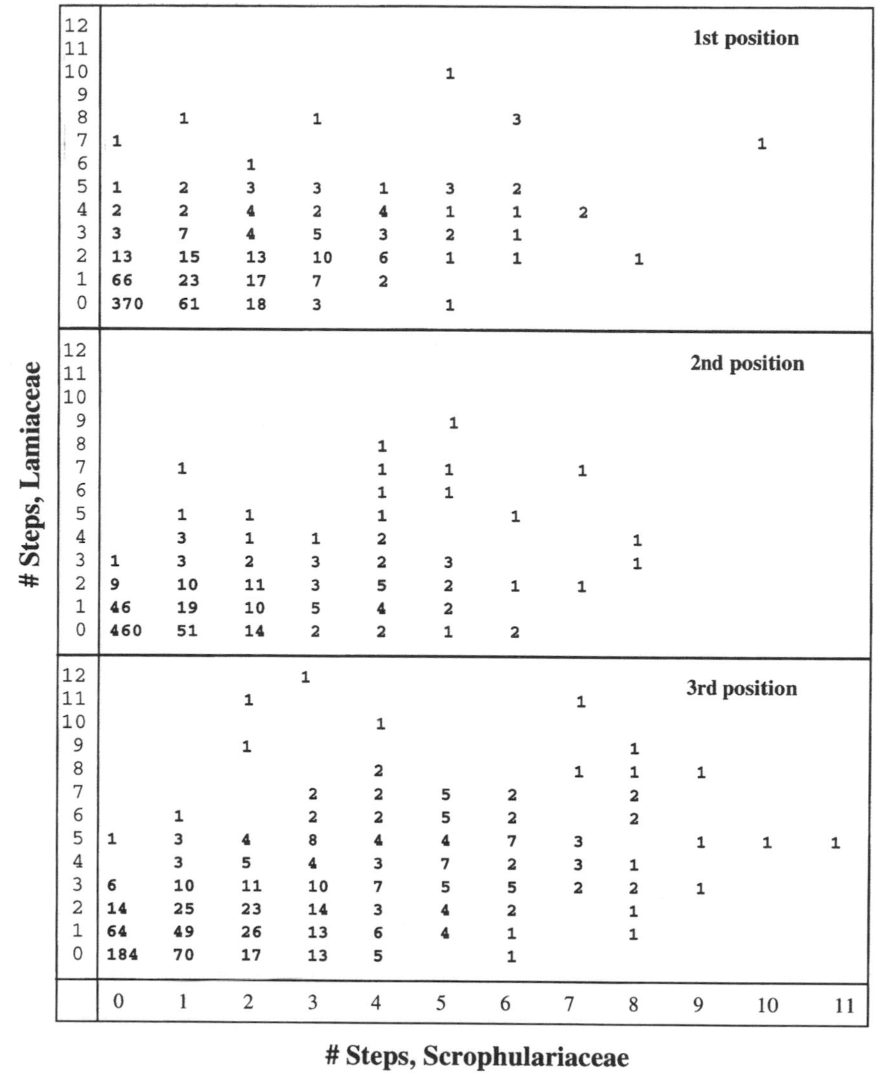

 # Steps, Lamiaceae

1st position

	0	1	2	3	4	5	6	7	8	9	10	11
12												
11												
10						1						
9												
8		1		1			3					
7	1									1		
6			1									
5	1	2	3	3	1	3	2					
4	2	2	4	2	4	1	1	2				
3	3	7	4	5	3	2	1					
2	13	15	13	10	6	1	1		1			
1	66	23	17	7	2							
0	370	61	18	3		1						

2nd position

	0	1	2	3	4	5	6	7	8	9	10	11
12												
11												
10												
9						1						
8					1							
7		1			1	1		1				
6					1	1						
5		1	1		1		1					
4		3	1	1	2				1			
3	1	3	2	3	2	3			1			
2	9	10	11	3	5	2	1	1				
1	46	19	10	5	4	2						
0	460	51	14	2	2	1	2					

3rd position

	0	1	2	3	4	5	6	7	8	9	10	11
12				1								
11			1					1				
10					1							
9			1						1			
8					2			1	1	1		
7				2	2	5	2		2			
6		1		2	2	5	2		2			
5	1	3	4	8	4	4	7	3		1	1	1
4		3	5	4	3	7	2	3	1			
3	6	10	11	10	7	5	5	2	2	1		
2	14	25	23	14	3	4	2	1				
1	64	49	26	13	6	4	1	1				
0	184	70	17	13	5		1					

 # Steps, Scrophulariaceae

Figure 6.8.   Plots of correlation study of change in *ndhF* sequences in Lamiaceae versus Scrophulariaceae for codon position classes. Correlation coefficients: $r = 0.331$, 0.278, and 0.192, for first, second, and third positions, respectively (all $p < 0.001$).

draw inferences regarding historical patterns of change in those sequences in those groups. Although there are several sources of error in this approach (see below), it provides a valuable approximation of sequence evolution and is a refinement over previous pioneering efforts (e.g., Albert et al., 1993; 1994; Kim and Jansen, 1995), which used a parsimony approach to estimate substitution probabilities in cpDNA sequences. First, to assess substitution probabilities accurately using parsimony and a tree-based inference approach, a group of relatively closely related sequences is needed. Second, two or more independent data sets are needed to be confident that the results are consistent among lineages. Finally, data from two or more genes are necessary to see whether patterns inferred for one gene (e.g., *rbcL*) are consistent with other chloroplast genes.

Sources of error that may affect the tabulations and tests include: (1) incorrectly inferred trees; (2) random error due to sampling of sequence or

Table 6.4. Base composition and transition-transversion ratio.

	Base Composition					Transition/Transversion ratio		
						whole tree		terminal sister group
	A	C	G	T	A+T	range	unambiguous changes	
Solanaceae *rbcL*	0.272	0.190	0.248	0.290	0.562	1.11	1.18	1.07
Lamiaceae *rbcL*	0.274	0.192	0.246	0.288	0.562	1.70–1.83	1.88	1.63
Scrophlariaceae *rbcL*	0.273	0.192	0.246	0.289	0.562	1.64–1.75	1.71	1.65
Solanaceae *ndhF*	0.278	0.152	0.184	0.386	0.664	0.96–1.00	1.02	0.85
Lamiaceae *ndhF*	0.274	0.156	0.180	0.390	0.664	1.42–1.57	1.57	1.36
Scrophulariaceae *ndhF*	0.308	0.146	0.231	0.314	0.622	1.16–1.34	1.32	1.23

Table 6.5. Effect of base composition on expected transformation probabilities.

Transformation type	Expected proportion	Expected number	Inferred number[1]	Ratio: inferred/expected
Transitions				
a → g	0.092	60.81	83.15	1.367
c → t	0.069	45.61	122.72	2.691
g → a	0.089	58.83	113.99	1.938
t → c	0.078	51.56	98.00	1.901
Transversion				
a → c	0.072	47.59	32.40	0.681
a → t	0.109	72.05	21.41	0.297
c → a	0.065	42.97	33.29	0.775
c → g	0.058	38.34	32.28	0.842
g → c	0.063	41.64	31.90	0.766
g → t	0.094	62.13	26.88	0.433
t → a	0.111	73.37	18.46	0.252.
t → g	0.100	66.10	46.55	0.704

[1]Calculated by MacClade as the average number considering alternative optimizations for ambiguous transformations.

taxa (sampling error); (3) underestimation of substitution probabilities by the parsimony criterion (see foregoing discussion); and (4) the "best" trees are based on equally weighted parsimony analysis of the data being analyzed, thereby creating the possibility of bias. Because we are concerned about the tree of the plastome sequences only in this analysis, any source of error that may exist due to the chloroplast tree not being a true reflection of the organismal tree is irrelevant (Doyle, 1992; Miyamoto et al., 1994).

We suggest that bias resulting from these sources of error is minimal. There is no way we can know the correct tree for any group, but the sampling of sequence data for each data set was large relative to most cpDNA systematic studies published to date (ca. 3,500 bp of sequence per taxon). For two of the taxonomic groups examined, other studies of cpDNA restriction sites (Solanaceae: Olmstead and Sweere, 1994; Labiatae: Wagstaff et al., 1995) or morphology (Labiatae: Cantino, 1992) provide estimates of phylogeny that corroborate results obtained for these data or contribute to the results when analyzed in combination with these data. We examined change over relatively closely related species, so that the parsimony criterion could reconstruct character change better than if more distantly related sequences were used. In addition to the trees used for the sequence analyses reported here, we conducted maximum likelihood analyses on each data set using an approximate gamma distribution for substitution probabilities. In each case the ML trees differed slightly from the parsimony trees, but were not significantly different using the Kishino-Hasegawa (1989) test of tree congruence. In all cases the only differences between the parsimony and ML trees fell in areas where few sites in these sequences provide information. This finding suggests that even if our accepted tree is not entirely correct, acceptable alternate tree topologies based on a different criterion would not affect patterns of inferred sequence change significantly (Swofford et al., 1996).

The distribution of change by site often has been assumed to follow a Poisson model (Jukes and Cantor, 1969; Goldman, 1990; Albert et al., 1993). However, in this study, the distribution of change (Fig. 6.4, Table 6.1), when entire sequences were considered, did not entirely fit a Poisson distribution for either gene in any of the three taxonomic groups, thereby suggesting among-site rate heterogeneity. A comparison of observed to expected numbers of sites in each class of number of inferred changes (Table 6.1) showed an excess of invariant sites and high rate sites. Sullivan et al. (1995) found similar distributions for *cytb* and 12S mitochondrial sequences in a group of rodents. They tested the complete *cytb* and complete 12S sequences for fit to a negative binomial distribution (indicating a gamma distribution of rates) and also found that there was a significant deviation from the expectation for that model. We did not test a gamma distribution model, but suspect by comparison of our distributions with theirs that it may not apply. However, when informative sites alone were considered, the Poisson model could not be rejected for either gene in Solanaceae (Table 6.1) or for *rbcL* in Scrophulariaceae and could not be rejected for several of the codon position subsets in the other three data sets, suggesting that the excess of invariant sites may be primarily responsible for the deviation from Poisson for all sites in Solanaceae, the least diverged set of sequences. It is possible that with greater sequence divergence (as in Lamiaceae or Scrophulariaceae *ndhF* data sets) a greater deviation from the Poisson model becomes apparent in sequence data as multiple hits accumulate at the most variable sites.

Whether the distribution of change for a particular lineage fits a particular distribution model (e.g., Poisson) does not reveal what the cause of the observed variance might be. For example, conformation to a Poisson model does not require equal rates at all sites, although such an underlying process may exist. Considering that the Poisson model does not explain the observed changes over the entire sequences and is unable to explain all of the changes observed in the cladistically informative subset of sites, it is important to look for patterns that might offer insight into the underlying substitution process.

We first tested whether the two genes have significantly different underlying rates of substitution. When all sites were considered, the answer was yes, *ndhF* evolves at a significantly faster rate than *rbcL*. However, when only informative sites were considered, the answer was no. There was no significant rate difference between the two genes for those sites that would be useful in a parsimony analysis, suggesting that the primary reason for the overall rate difference between the genes is the greater proportion of sites that are invariant in *rbcL*. This issue has significance in the debate over combining data from different sources in phylogenetic analysis (Bull et al., 1993; Chippindale and Weins, 1994; de Queiroz et al., 1995). Bull et al. (1993) argue that combining data from two DNA regions with significantly different rates may reduce the accuracy of phylogenetic analysis relative to the better of two separate analyses. However, Sullivan (1996) points out that this conclusion relies on a model in which all sites in each region have the same underlying substitution rate. If rate heterogeneity exists within each region, so that the means are different, but substantial overlap exists in the distribution of rates from site to site, then the conclusion of Bull et al. (1993) is not valid. The fact that *rbcL*, a relatively slowly evolving gene, and *ndhF*, a relatively rapidly evolving gene, have similar rate distributions suggests that most genes in the chloroplast genome probably are appropriate for combining, even for methods that consider the entire sequence (e.g., maximum likelihood). This similarity may be true both for coding and noncoding regions.

Another implication of these results is that whole-gene rate may be an unimportant factor when considering the utility of these genes for parsimony analysis. There was a weak trend toward a greater increase in rate among informative sites in *ndhF* than in *rbcL* going from the least diverged (Solanaceae) to the most divergent (Scrophulariaceae) data sets (Fig. 6.5, Table 6.1). Alignment of gaps, on the other hand, which tend to be clustered in discrete regions in the *ndhF* sequences (Olmstead and Reeves, 1995; Scotland et al., 1995), is more likely to cause problems for inference of deep phylogenies than rate differences, possibly indicating that other genes with fewer alignment problems, but higher overall rates than *rbcL* (e.g., *matK*, Johnson and Soltis, 1995; see Chapter 1) may be useful at deeper levels than commonly thought. This finding also suggests that the primary consideration when choosing chloroplast genes for parsimony analysis should be the number of informative characters per kilobase, rather than overall substitution rate, as long as there are minimal alignment problems. At this time, there are insufficient examples of broad taxonomic sampling from genes of varying rates to evaluate this suggestion. A detailed study of *ndhF* in the Asteridae with a substantial number of outgroups, including some basal angiosperms (R. Jansen and R. Olmstead, unpubl.), finds results that are highly congruent with *rbcL* results, but with greater resolution, both within the Asteridae (Olmstead et al., 1993) and among its outgroups (Chase, Soltis, Olmstead et al., 1993; Rice et al., 1997). The overlapping distribution of rates and nonsignificant difference in mean rates at informative sites in two genes with significantly different overall rates, suggests that the axiom that one must find a match of gene with proper rate for a particular phylogenetic problem is overstated. We suggest that most genes in the chloroplast genome will be informative over a much larger range of phylogenetic divergence than currently conceived, perhaps encompassing most questions in angiosperm phylogeny from the level of genus upward.

For entire sequences, codon positions represent classes with significantly different rates. However, as with the comparison of genes, the primary difference between codon positions in terms of the distribution of changes was the number of invariant sites. The amount of change at informative sites at the three codon positions within each gene was not significantly different in most cases (Table 6.1). Among sites that were informative, the distribution of sites exhibiting various amounts of change was remarkably similar in all three codon positions for both genes and in all three taxonomic groups (Fig. 6.4, Table 6.1), indicating that homoplasy is as likely to occur at second positions as at first or third positions among informative sites, as was reported previously for *rbcL* in Asteraceae (Kim et al., 1992). Consistency indices for informative sites indicate that all codon positions are roughly comparable and that second codon position sites never had the highest CI as a group in any gene for any taxonomic group. Comparable results were obtained for *rbcL* and *ndhF* in the Asteraceae (Kim et al., 1992; Kim and Jansen, 1995). The distribution of changes at informative sites suggests that weighting by codon position in cpDNA is a bad idea.

Given that we cannot reject the Poisson model for some informative subsets of the data, what evidence do we have that the observed among-site differences in amounts of change represent heterogeneous rates caused by some underlying processes, rather than stochastic variation acting on sites all having the same underlying rate? Correlation analysis of amounts of change at sites in independent groups provided the strongest evidence that rate heterogeneity is real in these sequences (Figs. 6.7–6.8, Table 6.3). Highly significant comparisons were found for all three pairwise comparisons among taxonomic groups, for both genes, and for each codon position within each gene. This means that the amount of change observed over a tree at a particular site in one taxonomic group is likely to be a good predictor of change at that site in an independent group. Our test of the effect of removing some portion of sites that are invariant and, therefore, not free to change (Hasegawa et al., 1985) was a conservative one, insofar as we removed all sites that did not vary in any of our three taxonomic groups. We know from other studies that many of these sites may be variable (e.g., Chase, Soltis, Olmstead et al. 1993, for *rbcL* and R. Jansen and R. Olmstead, unpubl., for *ndhF*). Even with such a dramatic removal of a large class of sites at one end of the correlation distribution, significant positive correlations were found for most comparisons and all but one of the nonsignificant correlations had positive r-values. A comparison of intersecting sets of informative sites among these taxonomic groups also revealed significant positive correlations for some comparisons and positive r-values for all comparisons (Table 6.3), indicating that variation among informative sites is nonrandom. Despite the positive correlation, there was a large amount of scatter around the correlation line (Figs. 6.7–6.8), which frustrates efforts to establish universal weights based explicitly on the amounts of inferred change at individual sites.

Transition/transversion ratios for the three *rbcL* and *ndhF* data sets averaged 1.45 and 1.15, respectively, as estimated by the terminal sister group method (Table 6.4). The consistently higher ts/tv estimates based on whole-tree inferences, relative to the sister group method, suggest that the differences are not due to small sample sizes in the terminal sister group comparisons, but rather are due to underestimation of transversions over the whole tree. Only counting unambiguously optimized changes is expected to overestimate ts/tv, because many ambiguous changes involve situations in trees where three descendant branches have different states. In these cases the alternate optimizations will be either one transition and one transversion or two transversions. The most highly variable sites in each data set (i.e., those that exhibited inferred numbers of changes above the midpoint of the range of variation) showed substantially lower ts/tv ratios in Solanaceae and Labiatae for both genes (data not shown), but were more or less equal to the overall rate for both genes in Scrophulariaceae, the taxonomic group with the greatest phylogenetic diversity. Estimates of ts/tv for other cpDNA studies (calculated in various ways) include the following: *rbcL*—0.97 in Asteraceae (Kim and Jansen, 1995), 1.25 in Lardizabalaceae (Hoot et al., 1995), 1.41 in Saxifragaceae (Johnson and Soltis, 1995), 1.51 in Onagraceae (Conti et al., 1993), 2.27 in Dipsacales (Donoghue et al., 1992); *ndhF*—0.99 in Asteraceae (Kim and Jansen, 1995); atpB—2.20 in Lardizabalaceae (Hoot et al., 1995); *matK*—0.94 in Saxifragaceae (Johnson and Soltis, 1995).

Base composition was remarkably uniform between taxonomic groups for each gene, yet differed between genes (Table 6.4), suggesting a constraint at the level of the gene or the gene product. Among taxa within each data set and for each gene, base frequencies varied by about 0.5% to 2.5% and sequences differed at as many as 12% of their positions (data not shown), yet A + T content for *rbcL* for all three taxonomic groups is 56.2% and for *ndhF* ranges from 62.2 to 66.4% (Table 6.4; compared with 56.3% and 67.6%, respectively in Asteraceae, Kim and Jansen, 1995).

Most ts/tv weighting schemes (e.g., Huelsenbeck and Hillis, 1993) treat the various types of base transformations (e.g., A to C, A to G) as two classes (transitions and transversions), within which the various types are assumed to occur with equal probability, but between which rates differ. For all data sets, A ↔ T transversions were the lowest-frequency character-state transformations, despite the fact that A and T are the bases at highest frequency. For example, in Scrophulariaceae only 6.0% of inferred changes in *rbcL* are between A and T; this figure represents only 27% of the expected number, given the observed base composition and assuming that all transformations are equally likely (Table 6.5). In contrast, C ↔ G transversions accounted for 9.7% of all observed transformations in the same data set. This number is 80% of the random expectation given the base composition of the gene. A threefold difference exists in the substitution probability among the various types of transversions (Table 6.5). Likewise, there was a substantial difference in the observed versus expected number of changes for the two different classes of transitions (A ↔ G = 165%, C ↔ T = 227% of random expectations based on base composition) In this case, taking into account base composition, all transversions exhibited fewer than expected numbers of changes and all transitions exhibited more than expected, but the rates for the transformation types within each class varies considerably. The implication of this observation for parsimony analysis is that simple weighting of transitions and transversions, even with base composition taken into account, does not reflect real substitutional processes (see Knight and Mindell, 1993, for a method of weighting based on unequal occurrence of classes of transformations). Weighting all such transformations equally may not be the best solution, but imposing differential weights based on classes of transformations also has problems. For example, examination of highly homoplastic sites (results not shown) indicates that there are some such sites in which all changes are transversions. Such sites would be up-weighted erroneously in a simple ts/tv weighting system in a parsimony analysis. Maximum likelihood methods, which are better able to accommodate variation in substitution rates, without necessarily imposing an explicit "weight" on individual transformation types, may be better suited than parsimony to handle this sort of variation (Swofford et al., 1996).

What do these results suggest about the validity of previous suggestions for weighting of protein coding genes? Miyamoto et al. (1994) suggested that deleting third positions improved the congruence between trees resulting from independent analysis of mammalian mitochondrial DNA (mtDNA) genes. This approach has been used for human immunodeficiency viruses (HIV) (Mindell et al., 1995), plant nuclear genes (e.g., glutamine synthase; Doyle, 1991) and cpDNA (e.g., Conti et al., 1993). While the much higher synonymous substitution rate in animal mtDNA (Wolfe et al., 1987) and HIV (Mindell et al., 1995) than in cpDNA may make this approach necessary, there seems to be no justification for this in cpDNA. Huelsenbeck and Hillis (1993) conclude from simulation studies that weighted parsimony (corrected for ts/tv ratio) outperforms equally weighted parsimony. However, in their simulations the substitution probabilities did not include the transition and transversion rate heterogeneity observed here (e.g., they treated all transversion types as equally likely and both transition types as equally likely) and they used ts/tv rate ratios of 5:1 and 10:1, which are much higher than that inferred for cpDNA (examples cited previously and calculated here). Also, they used a much higher base substitution rate than has been found in virtually all cpDNA sequence studies, thereby assuring many more multiple hits at sites on longer branches, making it harder for equally weighted parsimony to account for all changes.

Combining weights from estimates of ts/tv and codon position substitution probabilities calculated over entire sequences (e.g., Albert et al., 1993) compounds the problems identified with either one of these approaches.

Few would dispute that the goal in parsimony analysis is the accurate reconstruction of phylogeny. In the concern over applying a realistic model of sequence change to parsimony analysis, the risk exists that increase in resolution (i.e., finding fewer most parsimonious trees or a more fully resolved consensus tree) will be interpreted uncritically to mean an increase in accuracy, especially when no corroborating evidence is available. As a general rule, differential weighting increases resolution, because a larger number of optimal solutions is possible when all changes are equal (Olmstead and Palmer, 1992; Kron and Chase, 1993; Rodman et al., 1993; Albert et al., 1993; Smith and Sytsma, 1994; Wiegrefe et al., 1994). A method that increases resolution (e.g., weighted parsimony, successive weighting) is only beneficial as long as the model underlying it provides an accurate description of how the characters evolve. A model that is erroneous increases the chance that the correct tree is excluded as resolution is increased. A more general model increases the chance that the true tree is within the tree space accepted by the parsimony criterion. Assigning equal weights in a parsimony analysis invokes an evolutionary model (Swofford et al., 1996), but it is more general because it is less discriminating in the trees that it rejects. In many cases this approach is preferable to a more explicit model even when we have information on character evolution. An analysis that discriminates more finely among trees, but which is based on an erroneous model, is likely to increase resolution, but also more likely to exclude the correct tree. The pattern that emerges from the analyses that are reported here suggests that there is a complicated series of processes that constrain the evolution of sites in heterogeneous ways. These processes are overlain with stochastic variation arising from mutation and substitution. The implication for parsimony analysis is that weights based on models of evolution, in which characters or transformation types (ts or tv) are assigned to classes that are assumed to differ from each other, yet to be homogeneous internally, may increase resolution at the expense of accuracy.

So, how can we apply what we have learned from the study of these two genes in these three taxonomic groups to future parsimony analyses? Here we present one simple suggestion and evidence for its efficacy. While correlations identified in Table 6.3 are real, the variation around the correlation line (Fig. 6.7) probably is too great to permit an explicit weighting scheme based on the correlation observed between different taxonomic groups. However, correlations between data sets may provide sufficient information to justify an objective criterion for the rejection of characters for subsequent analysis.

A series of parsimony analyses was carried out in which the most variable sites were deleted and the results compared with those from the analysis of an entire gene sequence. Then both are compared with the results from the accepted tree (Table 6.6). These were done for *rbcL* and *ndhF* sequence data sets in Solanaceae and the *rbcL* data set in Lamiaceae, the taxonomic groups for which the best-corroborated hypotheses of phylogeny exist, based on these and other data (*ndhF* alone yields a nearly fully resolved accepted tree for Lamiaceae, so is not considered here). Two approaches were taken. In one, an external criterion was applied, in which the sites that were most variable (arbitrarily defined as those changing more frequently than the midpoint change class, see foregoing discussion) in two of the groups were removed from the third data set under consideration. For *rbcL* in Lamiaceae and Scrophulariaceae this number is ≥ 7 changes per site, for Solanaceae ≥ 3 changes per site. This external criterion was applied in two levels of stringency, removing only those sites that meet the most variable criterion in both groups and a relaxed version in which sites are excluded if they are in the most variable category in one taxonomic group and no more than two steps less variable (one step for Solanaceae) in the other group. In the other approach, an internal criterion was applied, in which the most variable sites identified in each data set were removed from subsequent analysis of the same data set (e.g., the sites identified as most variable in Solanaceae by inference from the accepted tree were removed from the Solanaceae

analysis). The results of these analyses were compared with the analysis of complete *rbcL* sequences using uniform weighting and with a successive approximations weighting approach (Farris, 1969) for their ability to recover clades on the accepted tree. Successive approximations is an a posteriori weighting approach in which an iterative series of analyses are run with character weights based on their consistency and retention indices in the previous iteration starting with equal weights.

The reanalysis of *rbcL* and *ndhF* data in Solanaceae and *rbcL* data in Lamiaceae after removal of the highly variable sites resulted in trees that often were less well resolved, but which had a higher proportion of clades congruent with the accepted trees (Table 6.6). For example, analysis of the complete Solanaceae *rbcL* data set resulted in 67 trees with nine clades resolved in the strict consensus tree. Of these nine clades, only three were congruent with clades represented in the accepted tree (33% correct). By deleting three characters that were identified by the most stringent external criterion (among the most highly variable in both of the two other groups), only two of which were informative in Solanaceae, the reanalysis resulted in 695 trees with six clades resolved in the strict consensus, four of which were congruent with the accepted tree (67% correct). Application of the internal criterion required the removal of 20 characters and resulted in 15 trees with seven clades, six of which were congruent with the accepted tree (86% correct). In Lamiaceae, analysis of complete *rbcL* sequences resulted in 210 trees with 21 resolved clades, of which 14 were congruent with the accepted tree (67% correct). By removal of the nine sites identified by the relaxed external criterion, 204 trees were recovered and 21 clades were resolved, 18 of which were congruent with clades in the accepted tree (86% correct). The internal criterion removed 15 sites and resulted in 416 trees with 22 clades, of which 19 were congruent with the accepted tree (86% correct). For all three data sets examined, the internal criterion resulted in a greater number of clades congruent with the accepted tree and a smaller number of incongruent clades than did the external criterion. The successive approximations approach always resulted in greater resolution (i.e., more

Table 6.6 Character delation results.

		Solanaceae rbcL	Solanaceae ndhF	Lamiaceae rbcL
accepted clades		14	14	30
all sites	#characters	51	70	147
	#trees	67	3	210
	congruent	3	9	14
	incongruent	6	3	7
external	#characters removed	3/9[1]	14/29	3/9
criteria	#trees	695/64	17/24	1440/204
	congruent	4/4	8/8	14/18
	incongruent	2/2	1/1	4/3
internal	#characters removed	20	9	15
criteria	#trees	15	12	416
	congruent	6	9	19
	incongruent	1	0	3
successive	#trees	37	2	2
approximation	congruent	3	9	20
	incongruent	7	4	9

[1]More/less stringent removal criterion. Some of the characters removed based on external criteria are not informative in the group from which they are removed.

resolved clades), but consistently found a much greater number of incongruent clades than either of the character removal criteria (Table 6.6). This argues against the use of successive approximations in weighting cpDNA data.

We have tried to take an empirical approach to understanding the patterns of evolution in chloroplast gene sequences by examining two genes in three taxonomic groups. The two genes are viewed conventionally to represent slow (*rbcL*) and fast (*ndhF*) evolving genes and have been suggested to have optimum utilities at different phylogenetic levels (Olmstead and Palmer, 1994; see also Chapter 1), so the similarity between these genes in terms of their characteristics in a parsimony context is somewhat unexpected. The detailed examination of more chloroplast genes (e.g., *atpB* and *matK*) will provide valuable additional information about chloroplast gene evolution and how best to analyze cpDNA sequence data for phylogenetic inference.

In conclusion, we have tried to compile some important take-home messages from our studies of the patterns of sequence variation and phylogeny:

1. Patterns of variation in cpDNA gene sequences are likely to reflect a combination of interacting processes; simple evolutionary models that treat classes of characters (e.g., codon positions) and transformations (e.g., transitions and transversions) as homogeneous are unlikely to represent them accurately and may be as likely to mislead as to inform.

2. For parsimony analysis, knowledge of individual positions in particular sequences may be very important for selection of characters. Through study of data sets from independent groups, predictions may be made concerning the phylogenetic utility of individual sites that will be useful for other taxonomic groups.

3. Variable sites with low rates of change are best for parsimony analysis. Therefore selection of sequences should optimize for maximum number of sites while keeping the number of sites with a very high rate of evolution low. For some questions such as deep radiations, it may be that the most important consideration is avoiding high rate sites, even if this requires obtaining more sequence to get enough informative characters. It may be that regions with low synonymous substitution rates, such as the chloroplast inverted repeat (Wolfe et al., 1987) will be best for such questions.

4. In some cases, it may be preferable to *not* use all of the data gathered. Systematists always are selective in the data they gather; for example, by selecting particular genes for comparison. If we have reason to believe that particular regions or sites within genes are not reliable indicators of relationship at the level in which we are working, then it is permissible to delete them.

5. Accuracy is more important than resolution. A slight loss of resolution to achieve a greater proportion of clades that reflect accurate phylogenetic inference is permissible, even desirable. While it may seem counterintuitive, removal of even a small number of highly homoplastic sites is likely to decrease resolution, because those homoplastic sites may provide inaccurate resolution in places where the taxa in question otherwise would be unresolved.

6. External criteria for choice of data is logical and avoids circular reasoning, but evidence from external data sources may not be readily available. Internal criteria (i.e., degree of homoplasy on most parsimonious or other "accepted" tree) may yield results that are as good or better than those from external criteria. The best corroborated trees will provide the best inference for an internal criterion for character selection. Using more than one tree derived from different sources of data or different methods of analysis may provide some amelioration of the bias that may result from the use of a single tree.

7. Given that loss of resolution may result from the deletion of the most homoplastic characters, it may be necessary to obtain a larger quantity of information. We believe that many previously published phylogenetic studies contain inaccurately resolved clades, because of insufficient data. While this may be minimized by the elimination of a relatively small number of characters, more accurate and highly resolved trees are possible by the addition of more data.

LITERATURE CITED

Albert, V. A., M. W. Chase, and B. D. Mishler. 1993. Weighting for cladistic analysis of protein-coding DNA sequences. Annals of the Missouri Botanical Garden **80:**752–766.

Albert, V. A., A. Backlund, K. Bremer, M. W. Chase, J. R. Manhart, B. D. Mishler, and K. C. Nixon. 1994. Functional constraints and *rbcL* evidence for land plant phylogeny. Annals of the Missouri Botanical Garden **81:**534–567.

Bishop, M. J., and A. E. Friday. 1985. Evolutionary trees from nucleic acid and protein sequences. Proceedings of the Royal Society London B **226:**271–302.

Bull, J. J., J. P. Huelsenbeck, C. W. Cunningham, D. L. Swofford, and P. J. Waddell. 1993. Partitioning and combining data in phylogenetic analysis. Systematic Biology **42:**384–397.

Cantino, P. D. 1992. Evidence for a polyphyletic origin of the Labiatae. Annals of the Missouri Botanical Garden **79:**361–379.

Chase, M. W., D. E. Soltis, R. G. Olmstead, D. Morgan, D. H. Les, B. D. Mishler, M. R. Duvall, R. Price, H. G. Hills, Y.-L. Qui, K. A. Kron, J. H. Rettig, E. Conti, J. D. Palmer, J. R. Manhart, K. J. Sytsma, H. J. Michaels, W. J. Kress, K. G. Karol, W. D. Clark, M. Hedren, B. S. Gaut, R. K. Jansen, K.-J. Kim, C. F. Wimpee, J. F. Smith, G. R. Furnier, S. H. Strauss, Q.-Y. Xiang, G. M. Plunkett, P. S. Soltis, S. E. Williams, P. A. Gadek, C. J. Quinn, L. E. Eguiarte, E. Golenberg, G. H. Learn, S. Graham, S. C. H. Barrett, S. Dayanandan, and V. A. Albert. 1993. Phylogenetics of seed plants: an analysis of nucleotide sequences from the plastic gene *rbcL*. Annals of the Missouri Botanical Garden **80:**528–580.

Chippindale, P. T., and J. J. Wiens. 1994. Weighting, partitioning and combining characters in phylogenetic analysis. Systematic Biology **43:**278–287.

Clegg, M. T. 1993. Chloroplast gene sequences and the study of plant evolution. Proceedings of the National Academy of Science U.S.A. **90:**363–367.

Clegg, M. T., and G. Zurawski. 1992. Chloroplast DNA and the study of plant phylogeny: present status and future prospects. **In** Molecular Systematics of Plants, eds. P. S. Soltis, D. E. Soltis, and J. J. Doyle, pp. 1–13. Chapman & Hall, New York.

Conti, E., A. Fischback, and K. J. Sytsma. 1993. Tribal relationships in Onagraceae: implications from *rbcL* data. Annals of the Missouri Botanical Garden **80:**672–685.

de Queiroz, A., M. J. Donoghue, and J. Kim. 1995. Separate versus combined analysis of phylogenetic evidence. Annual Review of Ecology and Systematics **26:**657–681.

Donoghue, M. J., R. G. Olmstead, J. Smith, J. D. Palmer, and K. J. Sytsma. 1992. Phylogenetic relationships of Dipsacales based on *rbcL* sequence data. Annals of the Missouri Botanical Garden **79:**333–345.

Doyle, J. J. 1991. Evolution of higher plant glutamine synthetase genes: regulatory specificity as a criterion for predicting orthology. Molecular Biology and Evolution **8:**366–377.

Doyle, J. J. 1992. Gene trees and species trees: molecular systematics as one-character taxonomy. Systematic Botany **17:**144–163.

Farris, J. S. 1969. A successive approximations approach to character weighting. Systematic Zoology **18:**374–385.

Farris, J. S. 1977. Phylogenetic analysis under Dollos's law. Systematic Zoology **26**:77–88.

Felsenstein, J. 1978. Cases in which parsimony or compatibility methods will be positively misleading. Systematic Zoology **27**:401–410.

Felsenstein, J. 1981a. A likelihood approach to character weighting and what it tells us about parsimony and compatibility. Biological Journal of the Linnean Society **16**:183–196.

Felsenstein, J. 1981b. Evolutionary trees from DNA sequences: a maximum likelihood approach. Journal of Molecular Evolution **17**:368–376.

Fitch, W. M. 1971. Towards defining the course of evolution: minimal change for a specific tree topology. Systematic Zoology **20**:406–416.

Fitch, W. M. 1986. The estimate of total nucleotide substitutions from pairwise differences is biased. Philosophical Transactions of the Royal Society of London B **312**:317–324.

Fitch, W. M., and E. Margoliash. 1967. Construction of phylogenetic trees. Science **155**:279–284.

Fitch, W. M., and E. Markowitz. 1970. An improved method for determining codon variability in a gene and its application to the rate of fixation of mutations in evolution. Biochemical Genetics **4**:579–593.

Gadek, P. A., E. S. Fernando, C. J. Quinn, S. B. Hoot, T. M. Terrazas, C. Sheahan, and M. W. Chase. 1996. Sapindales: molecular delimitation and infraordinal groups. American Journal of Botany **83**:802–811.

Goldman, N. 1990. Maximum likelihood inference of phylogenetic trees, with special reference to a Poisson process model of DNA substitution and to parsimony analyses. Systematic Zoology **39**:345–361.

Hasegawa, M., H. Kishino, and T. Yano. 1985. Dating the human-ape splitting by a molecular clock of mitochondrial DNA. Journal of Molecular Evolution **22**:160–174.

Hillis, D. M., J. P. Huelsenbeck, and C. W Cunningham. 1994. Application and accuracy of molecular phylogenies. Science **264**:671–677.

Hoot, S. B., A. Culham, and P. R. Crane. 1995. The utility of *atpB* gene sequences in resolving phylogenetic relationships: comparison with *rbcL* and 18S ribosomal DNA sequences in the Lardizabalaceae. Annals of the Missouri Botanical Garden **82**:194–207.

Huelsenbeck, J. P. 1995. The performance of phylogenetic methods in simulation. Systematic Biology **44**:17–48.

Huelsenbeck, J. P., and D. M. Hillis. 1993. Success of phylogenetic methods in the four-taxon case. Systematic Biology **42**:247–264.

Jin, L., and M. Nei. 1990. Limitations of the evolutionary parsimony method of phylogenetic analysis. Molecular Biology Evolution **7**:82–102.

Johnson, L. A., and D. E. Soltis. 1995. Phylogenetic inference in Saxifragaceae sensu stricto and Gilia (Polemoniaceae) using *matK* sequences. Annals of the Missouri Botanical Garden **82**:149–175.

Jukes, T. H., and C. R. Cantor. 1969. Evolution of protein molecules. **In** Mammalian Protein Metabolism, ed. H. W. Munro, pp. 21–120. Academic Press, New York.

Kim, K.-J., and R. K. Jansen. 1995. *ndhF* sequence evolution and the major clades in the sunflower family. Proceedings of the National Academy of Sciences U.S.A. **92**:10379–10383.

Kim, K.-J., R. K. Jansen, R. S. Wallace, H. J. Michaels, and J. D. Palmer. 1992. Phylogenetic implications of *rbcL* sequence variation in the Asteraceae. Annals of the Missouri Botanical Garden **79**:428–445.

Kimura, M. 1980. A simple method for estimating evolutionary rate of base substitutions through comparative studies of nucleotide sequences. Journal of Molecular Evolution **16**:111–120.

Kishino, H., and M. Hasegawa. 1989. Evaluation of the maximum likelihood estimate of the evolutionary tree topologies from DNA sequence data, and the branching order in Hominoidea. Journal of Molecular Evolution **29**:170–179.

Kluge, A. R., and J. S. Farris. 1969. Quantitative phyletics and the evolution of anurans. Systematic Zoology **18**:1–32.

Knight, A., and D. P. Mindell. 1993. Substitution bias, weighting of DNA sequence evolution, and the phylogenetic position of Fea's viper. Systematic Biology **42**:18–31.

Kron, K. A., and M. W. Chase. 1993. Systematics of the Ericaceae, Empetraceae, Epacridaceae and related taxa based upon *rbcL* sequence data. Annals of the Missouri Botanical Garden **80**:735–741.

Kuhner, M. K., and J. Felsenstein. 1994. A simulation comparison of phylogeny algorithms under equal and unequal evolutionary rates. Molecular Biology and Evolution **11**:459–468.

Maddison, W. P., and D. R. Maddison. 1992. MacClade, Analysis of Phylogeny and Character Evolution, version 3.01. Sinauer Associates, Sunderland, Massachusetts.

Mindell, D. P., J. W. Shultz, and P. W. Ewald. 1995. The AIDS pandemic is new, but is HIV new? Systematic Biology **44**:77–92.

Miyamoto, M. M., M. W. Allard, R. M. Adkins, L. L. Lanecek, and R. L. Honeycutt. 1994. A congruence test of reliability using linked mitochondrial DNA sequences. Systematic Biology **43**:236–249.

Moritz, C., and D. M. Hillis. 1996. Molecular systematics: context and controversies. **In** Molecular Systematics, second edition, eds. D. M. Hillis, C. Moritz, and B. K. Mable, pp. 1–13. Sinauer Associates, Sunderland, Massachusetts.

Olmstead, R. G., and J. D. Palmer. 1992. A chloroplast DNA phylogeny of the Solanaceae: subfamilial relationships and character evolution. Annals of the Missouri Botanical Garden **79**:346–360.

Olmstead, R. G., and J. D. Palmer. 1994. Chloroplast DNA systematics: a review of methods and data analysis. American Journal of Botany **81**:1205–1224.

Olmstead, R. G., and P. A. Reeves. 1995. Polyphyletic origin of the Scrophulariaceae: evidence from *rbcL* and *ndhF* sequences. Annals of the Missouri Botanical Garden **82**:176–193.

Olmstead, R. G., and J. A. Sweere. 1994. Combining data in phylogenetic systematics: an empirical approach using three molecular data sets in the Solanaceae. Systematic Biology **43**:467–481.

Olmstead, R. G., B. Bremer, K. Scott, and J. D. Palmer. 1993. A parsimony analysis of the Asteridae sensu lato based on *rbcL* sequences. Annals of the Missouri Botanical Garden **80**:700–722.

Olmstead, R. G., J. A. Sweere, R. E. Spangler, L. Bohs, and J. D. Palmer. 1998. Phylogeny and provisional classification of the Solanaceae based on chloroplast DNA. **In** Solanaceae IV, Advances in Biology and Utilization, eds M. Nee, D. E. Symon, and J. G. Hawkes. Royal Botanic Gardens, Kew, England. in press.

Palumbi, S. R. 1989. Rates of molecular evolution and the function of nucleotide positions free to vary. Journal of Molecular Evolution **29**:180–187.

Price, R. A., and J. D. Palmer. 1993. Phylogenetic relationships of the Geraniaceae and Geraniales from *rbcL* sequence comparisons. Annals of the Missouri Botanical Garden **80**:661–671.

Rice, W. R. 1989. Analyzing tables of statistical tests. Evolution **43**:223–25.

Rice, K. A., M. J. Donoghue, and R. G. Olmstead. 1997. Reanalysis of a large *rbcL* dataset: implications for future phylogenetic studies. Systematic Biology, 46:554–563.

Rodman, J., R. A. Price, K. Karol, E. Conti, K. J. Sytsma, and J. D. Palmer. 1993. Nucleotide sequences of the *rbcL* gene indicate monophyly of mustard oil plants. Annals of the Missouri Botanical Garden **80**:686–699.

Scotland, R. W., J. S. Sweere, P. A. Reeves, and R. G. Olmstead. 1995. Higher level systematics of Acanthaceae determined by chloroplast DNA sequences. American Journal of Botany **82**:266–275.

Simon, C., F. Frati, A. Beckenbach, B. Crespi, H. Liu, and P. Flook. 1994. Evolution, weighting, and phylogenetic utility of mitochondrial gene sequences and a compilation of conserved polymerase chain reaction primers. Annals of the Entomological Society of America **87**:651–701.

Smith, J. F., and K. J. Sytsma. 1994. Evolution in the Andean epiphytic genus *Columnea* (Gesneriaceae). II. Chloroplast DNA restriction site variation. Systematic Botany **19**:317–336.

Sneath, P. H. A. 1986. Estimating uncertainty in evolutionary trees from Manhattan-distance triads. Systematic Zoology **35**:470–488.

Sokal, R. R., and F. J. Rohlf. 1981. Biometry, second edition. W. H. Freeman & Company, San Francisco.

Sullivan, J. 1996. Combining data with different distributions of among-site rate variation. Systematic Biology **45**:375–379.

Sullivan, J., K. E. Holsinger, and C. Simon. 1995. Among-site rate variation and phylogenetic analysis of 12S rRNA in sigmodontine rodents. Molecular Biology and Evolution **12**:988–1001.

Sullivan, J., K. E. Holsinger, and C. Simon. 1996. The effect of topology on estimates of among-site rate variation. Journal of Molecular Evolution **42**:308–312.

Swofford, D. L., G. J. Olsen, P. J. Waddell, and D. Hillis. 1996. Phylogenetic inference. **In** Molecular Systematics, second edition, eds. D. M. Hillis, C. Moritz, and B. K. Mable, pp. 407–514. Sinauer Associates, Sunderland, Massachusetts.

Tateno, Y., N. Takezaki, and M. Nei. 1994. Relative efficiencies of the maximum-likelihood, neighbor-joining, and maximum-parsimony methods when substitution rate varies with site. Molecular Biology and Evolution **11**:261–277.

Vawter, L., and W. M. Brown. 1993. Rates and patterns of base change in the small subunit ribosomal RNA gene. Genetics **134**:597–608.

Wagstaff, S. J., and R. G. Olmstead. 1997. Phylogeny of the Labiatae and Verbenaceae inferred from *rbcL* sequences. Systematic Botany, 22:165–177.

Wagstaff, S. J., R. G. Olmstead, and P. D. Cantino. 1995. Parsimony analysis of chloroplast DNA restriction site variation in subfamily Nepetoideae (Labiatae). American Journal of Botany **82**:886–892.

Wagstaff, S. J., P. A. Reeves, L. Hickerson, R. E. Spangler, and R. G. Olmstead. 1998. Phylogeny and character evolution in Labiatae s. l. inferred from cpDNA sequences. Plant Systematics and Evolution, in press.

Wiegrefe, S. J., K. J. Sytsma, and R. P. Guries. 1994. Phylogeny of Elms (*Ulmus*, Ulmaceae): molecular evidence for a sectional classification. Systematic Botany **19**:590–612.

Wolfe, K. H., W.-H. Li, and P. M. Sharp. 1987. Rates of nucleotide substitution vary greatly among plant mitochondrial, chloroplast, and nuclear DNAs. Proceedings of the National Academy of Sciences U.S.A. **84**:9054–9058.

Yang, Z. 1993. Maximum likelihood estimation of phylogeny from DNA sequences when substitution rates differ over sites. Molecular Biology and Evolution **10**:1396–1401.

Yang, Z. 1994. Maximum likelihood phylogenetic estimation from DNA sequences with variable rates over sites: approximate methods. Journal of Molecular Evolution **39**:306–314.

Yang, Z. 1996a. Phylogenetic analysis using parsimony and likelihood methods. Journal of Molecular Evolution **4**:294–307.

Yang, Z. 1996b. Among-site rate variation and its impact on phylogenetic analyses. Trends in Ecology and Evolution **11**:367–372.

7

Molecular Evolution of 18S rDNA in Angiosperms: Implications for Character Weighting in Phylogenetic Analysis

Pamela S. Soltis and Douglas E. Soltis

Ribosomal DNAs (rDNAs) have inherent appeal for phylogeny estimation because they are present in all prokaryotic and eukaryotic organisms and thus have the potential to provide characters common to all of life. In fact, rDNA (or rRNA) sequences have been used to infer relationships in animals (e.g., Sogin et al., 1986; Field et al., 1988; Wainright et al., 1993; Wada and Satoh, 1994), protozoa (Schlegel et al., 1991; Van de Peer et al., 1996b), algae (e.g., Buchheim et al., 1990; Huss and Sogin, 1990; Kantz et al., 1990; Buchheim and Chapman, 1991; Chapman and Buchheim, 1991; Hendriks et al., 1991; Bakker et al., 1994; Ragan et al., 1994; Olsen et al., 1994; Medlin et al., 1996), fungi (e.g., Forster et al., 1990; Swann and Taylor, 1993; Hinkle et al., 1994; Berbee, 1996), lichens (e.g., Gargas et al., 1995), bryophytes (e.g., Waters et al., 1992; Mishler et al., 1994; Capesius, 1995; Kranz et al., 1995; Hedderson et al., 1996), pteridophytes (e.g., Kranz and Huss, 1996), gymnosperms (e.g., Chaw et al., 1993, 1995, 1997), and even among the deepest branches of life (e.g., Pace et al., 1986; Wolters and Erdmann, 1986; Olsen, 1987; Woese, 1987;

Embley et al., 1994; Bhattacharya and Medlin, 1995). Despite this wide use, relatively few studies have focused on patterns of rDNA evolution, particularly in plants; most have concentrated instead on the phylogenetic patterns suggested by the rDNA sequences.

The utility of rDNAs for estimating angiosperm phylogeny has been debated extensively during the past several years. Attempts to reconstruct plant phylogeny have used 5S rRNA sequences (Hori et al., 1985; Hori and Osawa, 1987) and portions of the 18S and 26S rRNAs (and rDNAs) (e.g., Hamby and Zimmer, 1988, 1992; Zimmer et al., 1989; Nickrent and Franchina, 1990). However, some of the criticisms directed at phylogenetic analyses of 5S rRNA sequences (e.g., Erdmann et al., 1985; Hori et al., 1985; Hendriks et al., 1986; Wolters and Erdmann, 1986; Hori and Osawa, 1987) apply equally well to other ribosomal DNAs (see discussions by Bremer et al., 1987; Mishler et al., 1988; Steele et al., 1988, 1991; Halanych, 1991). Although the 18S–26S rDNAs are much longer than 5S rDNA (the aligned sequences in Hori et al. [1985] are only 120 bp) and appear to

We thank Dan Nickrent for critical reading of the manuscript and many good suggestions for improvement. This work was supported in part by NSF grant DEB 9307000 to DES and by a grant from the Mellon Foundation to E. A. Zimmer, PSS, and DES.

evolve more slowly, these larger rDNAs also are potentially plagued by substitution biases, problems in homology assessment, and compensatory changes dictated by the secondary structure of the rRNA molecule. Furthermore, based on a fairly small number of angiosperm 18S rDNA sequences, several additional problems with 18S rDNAs have been noted. Among these were the perceptions that (1) 18S rDNA sequences are highly prone to insertion and deletion, (2) these indels make 18S rDNA sequences difficult to align, and (3) the rate of nucleotide substitution of 18S rDNA is too slow for use in phylogenetic reconstruction of the angiosperms. However, based on a data set of over 200 angiosperm 18S rDNA sequences (Soltis, Soltis, Nickrent et al., 1997), these perceptions are largely incorrect. Instead, indels are neither as common nor as problematic for alignment as they have been perceived; alignment of sequences is therefore straightforward; and the rate of evolution of 18S rDNA is roughly one-third to one-half that of *rbcL,* providing enough variable nucleotide positions to identify major clades, even if there is little or no resolution within some clades.

As described below, many of the past criticisms of 18S rDNA for phylogeny reconstruction are unfounded; however, other aspects of 18S rDNA evolution have potentially important implications for phylogeny estimation (e.g., Rothschild et al., 1986; Mishler et al., 1988; Steele et al., 1988, 1991; Wheeler and Honeycutt, 1988; Mindell and Honeycutt, 1990; Dixon and Hillis, 1993). We will explore (1) patterns of insertion and deletion, and the implications of indels for alignment; (2) stems, loops, and compensatory change; (3) conserved and variable domains; (4) transition/transversion ratios and other substitution biases in angiosperm 18S rDNA; and (5) approaches to character and character-state weighting for 18S rDNA that incorporate these aspects of molecular evolution.

Because studies of molecular evolution and phylogeny reconstruction are, or at least should be, intimately intertwined (cf. Gutell, 1993, 1996), the patterns of molecular evolution of angiosperm 18S rDNA described in the following discussion are based on estimates of angiosperm phylogeny derived from analysis of 223 an-

giosperm 18S rDNA sequences (Soltis, Soltis, Nickrent et al., 1997). Although heuristic searches failed to swap to completion and shorter trees likely exist for the data set, the trees obtained share many features with the large *rbcL* tree for seed plants (Chase, Soltis, Olmstead et al., 1993), suggesting that many of the clades recovered truly reflect angiosperm evolutionary history. The analyses of molecular evolution described in the following discussion are based on character-state distributions over Tree 1, arbitrarily selected from the 5,294 minimal-length trees recovered (3,923 steps) (Soltis, Soltis, Nickrent et al., 1997). Plots of sequence variability and estimates of substitution biases were generated using MacClade 3.03 (Maddison and Maddison, 1992); structural analyses are based on the model of 18S rDNA secondary structure for *Glycine max* developed by D. Nickrent (Soltis, Soltis, Nickrent et al., 1997). The aligned sequences for 223 angiosperms plus five Gnetales used in Soltis, Soltis, Nickrent et al. (1997) are available at http://www.wsu.edu: 8080/~soltilab/.

MOLECULAR EVOLUTION OF 18S RDNA IN ANGIOSPERMS

Frequency and Distribution of Indels

Angiosperm 18S rDNA is more highly constrained in length and has far fewer indels than suggested in previous studies (e.g., Nickrent and Franchina, 1990; Hamby and Zimmer, 1992), as inferred from the aligned sequences of 223 angiosperms. We attribute the large number of small indels reported previously for some sequences to technical difficulties in sequencing rRNA or rDNA. In fact, resequencing the 18S rDNA of some species (using cycle sequencing techniques and automated gel runs) produced cleaner sequences with fewer indels than the previously published versions (see Soltis, Soltis, Nickrent et al., 1997). When the "false" gaps deriving from sequencing errors are eliminated from the data set, most of the remaining indels are only one or two nucleotides in length. Furthermore, most of these indels are localized in specific regions of the rRNA secondary structure model (Fig. 7.1). The termini of helices

18S rRNA Secondary Structure

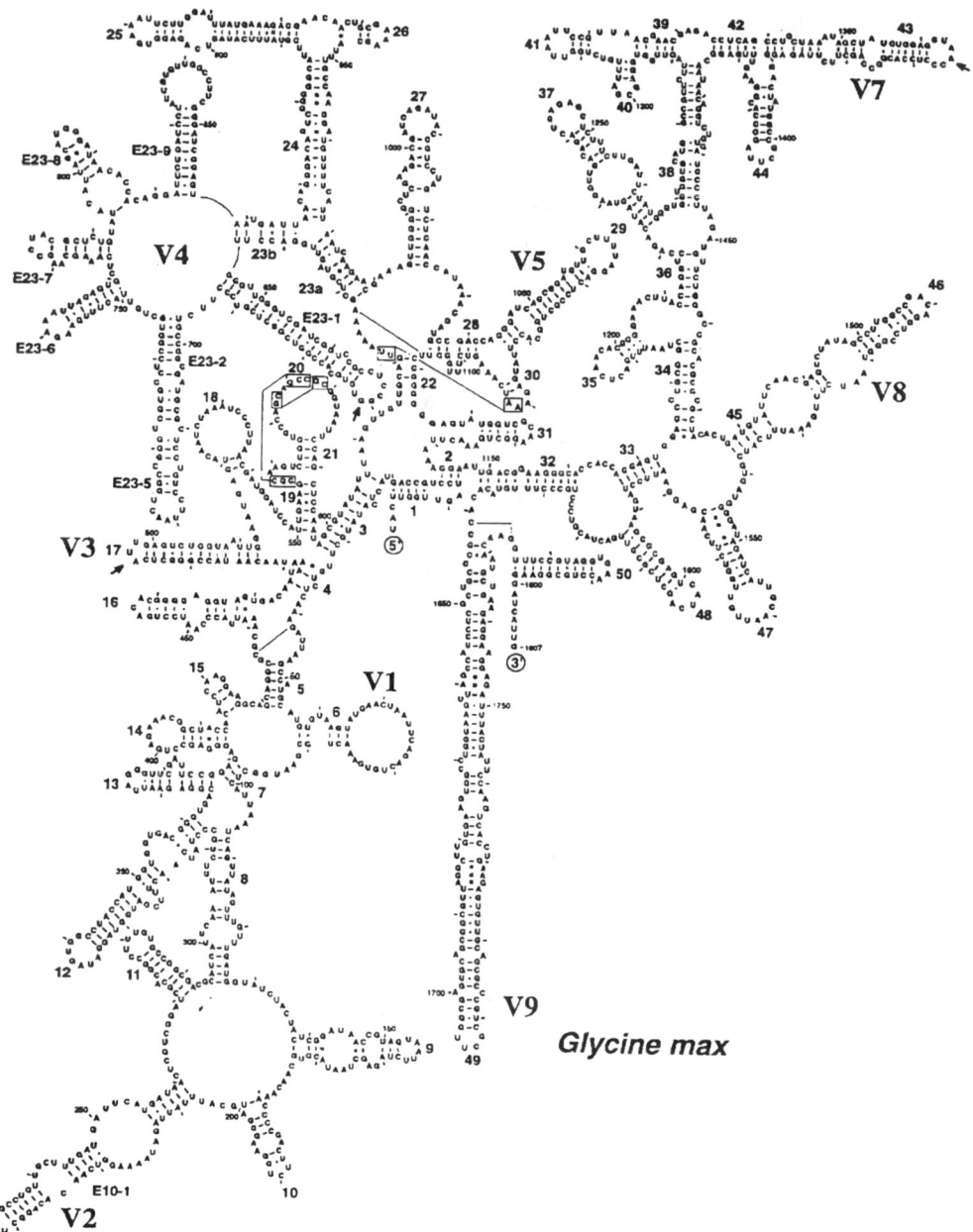

Figure 7.1. Secondary structure of 18S rRNA for *Glycine max* (D. Nickrent, in Soltis, Soltis, Nickrent et al., 1997).

E10-1, 17, E23-1, and 43 are particularly prone to insertion and deletion and are far less constrained in length than the remainder of the sequence. The small size and highly localized distribution of indels in angiosperm 18S rDNA make sequence alignment, and thus homology assessment for nucleotide characters, straightforward, except in the indel-prone helices.

Some indels may even be parsimony-informative, serving as synapomorphies for clades. For example, a deletion of a single nucleotide at position 1,568 on the aligned sequence unites all higher eudicots. This position is present in Gnetales, magnoliids, monocots, Platanaceae, Proteaceae, Trochodendraceae, Tetracentraceae, and ranunculids, but is absent from the rosid clade and Asteridae s. l. This deletion would therefore appear as a synapomorphy for the rosid/Asteridae s. l. clade (Fig. 7.2). A second such parsimony-informative indel is an insertion of one nucleotide at position 1,406 that unites the saxifragoids. This position is present only in Saxifragaceae s. s. and related rosids (i.e., some Saxifragaceae s. l., Crassulaceae, Haloragaceae) plus Paeoniaceae, Hamamelidaceae, Cercidiphyllaceae, and Daphniphyllaceae and would serve as an additional synapomorphy for the saxifragoid clade that is already strongly supported by nucleotide sequence data (Soltis, Soltis, Nickrent et al., 1997).

Stems, Loops, and Compensatory Change

The propensity for rRNAs to undergo base pairing across portions of their sequences results in characteristic secondary structures related to the

18S rDNA Phylogeny

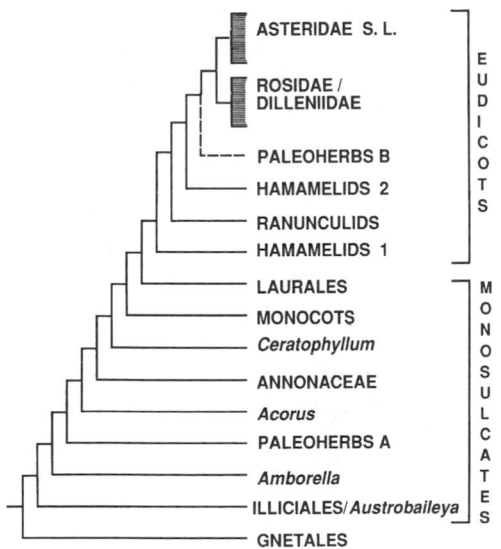

Figure 7.2. Summary of angiosperm phylogeny inferred from parsimony analysis of 18S rDNA sequences (from Soltis, Soltis, Nickrent et al., 1997).

function of the rRNA in the ribosome (Fig. 7.1). Base pairing occurs in rRNA not only between Watson-Crick (canonical) nucleotide pairs (A-U and G-C) but also between other pairs of nucleotides. Of these noncanonical pairs, G-U is the most common and is thought to maintain stem structure (Gutell et al., 1994). Mispairing during DNA replication generates noncanonical pairs of nucleotides, such as G-U, G-A, A-C, C-C, and G-G, in rRNA secondary structure. Although previous hypotheses of the mutational process invoked the incorporation of disfavored tautomeric forms of the four bases (such as enol or imino forms) during replication (Watson and Crick, 1953; Topal and Fresco, 1976), it appears instead that standard forms are misincorporated directly into a growing DNA strand. Both purine–purine and pyrimidine–pyrimidine pairs are formed, but selective constraints, including the pressure to maintain Watson-Crick structure in the growing strand, limit the rates of mispairing. The most frequent mispairs that occur during replication are G-T, G-A, and A-C (Echols and Goodman, 1991), producing a transition bias even at the point of mutation.

Using the model of 18S rRNA secondary structure for *Glycine max* (developed by D. Nickrent, given in Soltis, Soltis, Nickrent et al., 1997) and incorporating the 43 additional positions required by the aligned sequences, approximately 71% of angiosperm 18S rDNA lies in stem regions, broadly defined as those bases that typically participate in base pairing, with the remainder occurring in loops (i.e., those regions that do not undergo base pairing). Although an oversimplification of the pairing process, inferences of base pairing were restricted to canonical and G-U pairs, following analyses by Wheeler and Honeycutt (1988) and Dixon and Hillis (1993). In contrast, in an analysis that identified regions of animal 18S rDNA as stems, loops, "other" single-stranded regions that interact with proteins, and "unknown," only 43% of the 18S rDNA sequence was inferred to represent stem structure (Vawter and Brown, 1993).

Both theoretical and empirical studies have suggested that different evolutionary processes operate on stem versus loop regions of rDNA, with stems subject to compensatory change (cosubstitution sensu Wheeler and Honeycutt,

1988). Compensatory change is here defined as a nucleotide substitution that maintains or restores base pairing (i.e., stem structure); such cosubstitution renders changes in paired bases nonindependent and violates the underlying assumption of parsimony analyses that all characters are independent. Again, because pairing can also occur between noncanonical bases, this definition of compensatory change is a somewhat simplistic view of nucleotide substitution in rDNA. However, because noncanonical pairings (other than G-U) are relatively rare and play an uncertain role in maintaining stem structure, substitutions that generate or disrupt these pairs are not considered in our assessment of compensatory change (see also Wheeler and Honeycutt, 1988; Dixon and Hillis, 1993).

Compensatory changes to correct for disruptions of base pairing are not necessarily instantaneous (Kraus et al., 1992). Compensation may occur by complementary substitution (G-C to G-A to U-A) or by reversal (G-C to G-A to G-C), and the time required for these changes may span several nodes of a phylogenetic tree (see fig. 1 of Gatesy et al., 1994). Within these time constraints, several types of compensatory change may be identified (cf. Dixon and Hillis, 1993), depending on whether a single substitution or two substitutions occur and whether or not uracil is involved. Single compensatory changes are those that maintain or restore base pairing with a change in only one of the interacting nucleotides. Typical examples are those in which a substitution changes a noncomplementary base pair to a complementary pair, restoring base pairing (e.g., G-A to G-C). In addition, single substitutions involving uracil may also be compensatory because uracil can pair not only with adenine but also with guanine; therefore, changes that create, maintain, or restore G-U pairs also contribute to stem structure and should be considered compensatory (e.g., G-C to G-U; A-U to G-U). Double compensatory changes are those that involve two nucleotide changes and change one complementary base pair to another (e.g., G-C to A-U).

How frequent are compensatory changes in rDNA? Estimates vary, from essentially complete compensation in stem regions of animal 5S rDNA (e.g., Wheeler and Honeycutt, 1988) to only 20% of the amount of compensation possible in animal 28S rDNA (e.g., Dixon and Hillis, 1993). Not only are the patterns and frequency of compensatory change of interest for discerning differences in the evolutionary processes of stem and loop regions, but they are also important for evaluating the usefulness of these regions as sources of phylogenetic information. If compensatory change is common in stems, then stem regions contain numerous nonindependent characters and should be given lower weights than loops to reduce the effects of constrained substitutions and of doubly scoring the same changes. In fact, some models and a few empirical analyses have gone so far as to suggest that stems may not contain interpretable historical information and should therefore be eliminated from phylogenetic analyses. However, estimates from different rDNAs provide different views of the importance of compensatory change in rDNA evolution and consequently carry different messages about the relative weighting of stems and loops in phylogenetic analyses.

Discrepancies about the frequency of compensatory change may in part be due to estimations derived from different rRNA genes. The highest estimates of compensatory change are for 5S rDNA (both animal [Wheeler and Honeycutt, 1988] and plant [Steele et al.'s 1991 analysis of Hori et al.'s 1985 data]); much lower estimates have been reported for larger rDNAs, such as the 28S rDNA (Dixon and Hillis, 1993). Stem regions of different rDNAs seem to exhibit different levels of compensatory change, but some of the apparent differences may also be artifactual, resulting from optimization of states over phylogenetic trees that unevenly sample the taxonomic spectrum of interest. For example, a change from a G-C pair to A-U would be viewed as double compensatory in sister taxa A and B (Fig. 7.3A); however, with greater sampling of this clade to include taxa C, D, and E, it becomes apparent that G-C changed first to G-A and then back to G-C prior to the changes that produced the A-U pair (Fig. 7.3B). In this example, the frequency of compensatory change would be considered high in Fig. 7.3A, with all substitutions being instantaneously compensatory, and lower in Fig. 7.3B, with only one of three (or two of four, depending on the scoring

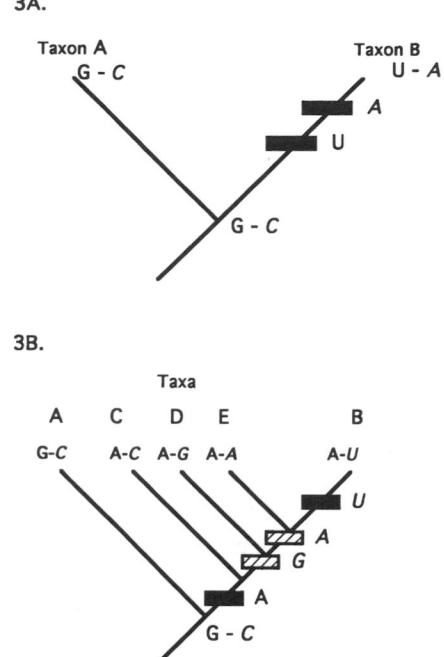

3A.

Taxon A
G - C

Taxon B
U - A

A

U

G - C

3B.

Taxa

A   C   D   E   B

G-C A-C A-G A-A   A-U

U

A

G

A

G - C

Figure 7.3. Nucleotide substitutions at paired bases, showing that estimates of levels of compensatory change may vary as taxon sampling changes. Paired nucleotides that form stem structure are indicated as G-C. Substitutions are given for each position: letters in italics represent changes at one position, and Roman letters represent changes at the paired position. *3A.* A G-C pair apparently undergoes the following changes to produce the divergent bases present in Taxa A and B. Along the lineage to Taxon B, G changes to U, followed by C changing to A. In this cladogram, the compensatory change from C to A appears to follow immediately the change from G to U, implying strong selection for compensation. *3B.* Lower levels of compensatory change would be inferred from this cladogram, where taxon sampling has been increased to include Taxa C, D, and E. Here, compensation is delayed. In this scenario, given an ancestral condition of G-C, G changes to A, and C changes to G and then to A prior to the change to U that restores Watson-Crick base pairing in Taxon B. The two changes inferred in 3A are shown as black bars in 3B; the two additional changes in 3B are shown as hatched bars. In 3A, 50% of the substitutions are compensatory (the C to U change); in 3B, only 25% are compensatory (the A to U change), and the compensation is delayed rather than instantaneous or nearly so. Differences in taxon sampling may therefore affect estimates of compensatory change.

of double compensatory changes) being compensatory. Therefore, estimates of the frequency of compensatory change are at least somewhat

specific to the data set at hand, making it difficult to set a priori weights for stem versus loop regions to reflect levels of compensatory change. However, it may still be possible to infer relative levels of compensatory change in different rDNAs and for different taxa and to provide very general guidelines for weighting stem and loop bases in phylogenetic analyses. Such topological effects on estimates of rates and patterns of nucleotide substitution (e.g., Sullivan et al., 1996) have contributed to the argument that estimation procedures should be independent of topologies (see review by Wakeley, 1996; Van de Peer et al., 1993, 1996a, 1996b).

How much of the phylogenetic signal present in stem regions is either constrained or doubly represented because of compensatory change, and what weights, if any, should be applied to counteract the effects of cosubstitution? In a preliminary survey (Soltis, Soltis, Nickrent et al., 1997) of the frequency of compensatory change in angiosperm 18S rDNA, 216 variable stem positions were examined for evidence of compensation. Of these 216 substitutions, 73% are compensatory, maintaining or restoring Watson-Crick or G-U base pairing; 27% destroyed such base pairing. Most of the compensatory changes (46% of the 216 substitutions) are single substitutions involving uracil, with smaller numbers of other single substitutions and double compensatory changes (8% and 19% of the 216 substitutions, respectively). Given these values, the evolution of 18S rDNA seems to be constrained by the secondary structure of the rRNA. Following the rationale of Wheeler and Honeycutt (1988), the high frequency of compensatory change might warrant severe downweighting of stems relative to loops. However, relatively few substitutions in angiosperm 18S rDNA are double compensatory and thus cause the same substitution to be represented twice in a data matrix. Perhaps, therefore, stem changes should be downweighted according to the frequency of double compensatory changes rather than all compensatory changes. For the angiosperm 18S rDNA data set, this would mean downweighting stems in phylogenetic analyses by perhaps 10% rather than 35%. However, weighting all stem bases less than all loop bases disregards other patterns of molecular evolution that may be operating in specific regions of the gene. That is, some stem bases may not undergo compensatory change, and some loop regions may be so highly variable that positions become saturated with substitutions; furthermore, some loop regions are highly constrained in sequence (e.g., Woese et al., 1990; Gutell, 1993). Assigning weights to stems and loops without regard for such changes would not adequately reflect patterns of evolution. Instead, we suggest that specific changes be mapped onto the secondary structure model and that only substitutions identified as double compensatory be weighted by 0.5. Attention to substitution patterns at individual positions, rather than broad regions, is more likely to produce an effective weighting scheme. Wakeley (1996) reviews methods of estimating transition bias that correct for multiple substitutions (e.g., Yang and Kumar, 1996) and incorporate heterogeneity of evolutionary rate among sites (e.g., Yang, 1993, 1994). These considerations apply equally to estimating other aspects of molecular evolution. A similar call for position-specific weighting, rather than by codon-position category, has also been made for *rbcL* sequence analysis (see Chapter 6). Molecular characters should be evaluated with the same rigor that is applied to the interpretation of morphological characters.

Conserved and Variable Domains of 18S rDNA

A serious criticism of 5S rRNAs/rDNAs for phylogeny estimation across green plants is the high level of homoplasy in the data (e.g., Bremer et al, 1987). Extensive homoplasy may mask phylogenetic signal, particularly if the signal is weak. Therefore, assessment of levels and patterns of variation (and ultimately of homoplasy) is necessary if the effects of homoplasious characters are to be minimized in phylogenetic analyses.

Angiosperm 18S rDNA has been described as a mosaic of conserved and variable regions (e.g., Nickrent and Soltis, 1995). Highly conserved segments of rDNA, with few or no substitutions, are interspersed with highly variable regions that undergo dozens of substitutions per site. Analysis of the large angiosperm 18S rDNA data set using Tree 1 of Soltis, Soltis, Nickrent et al. (1997) and the CHART option of MacClade showed a mosaic similar to that reported previously (Fig. 7.4). In fact, the pattern shown in Fig. 7.4 is even more striking than that presented by Nickrent and Soltis (1995; see their fig. 4); our analysis, using the same *x*-axis of 4-bp intervals, found longer stretches of sequence where no substitutions occurred, even though the number

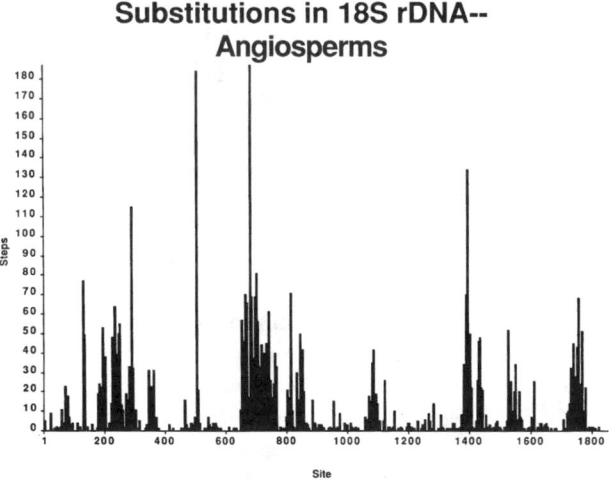

Substitutions in 18S rDNA-- Angiosperms

Figure 7.4. Patterns of nucleotide substitution in 18S rDNA across angiosperms, as estimated using the CHART option of MacClade 3.03 and Tree 1 from the phylogenetic analysis of angiosperms based on 18S rDNA sequences (Soltis, Soltis, Nickrent et al., 1997). The *x*-axis (Site) consists of 4-bp intervals rather than single nucleotide sites, for consistency with previous analyses of 18S rDNA evolution (Nickrent and Soltis, 1995). The number of steps on Tree 1 for each 4-bp interval is plotted on the *y*-axis.

of taxa included in our analysis is roughly four times greater (228) than that analyzed previously (59; Nickrent and Soltis, 1995). We attribute the larger number of invariant positions (or 4-bp intervals) observed in the larger study to the analysis of more accurate sequences, with fewer errors and fewer gaps, than were available previously.

The most conserved regions of angiosperm 18S rDNA are generally 50 or so nucleotides in length (Table 7.1) and tend to occur in stem regions of the secondary structure model (Fig. 7.5), although some small loops are also included in these conserved segments. The highly variable regions are much shorter, typically only 5–10 nucleotides in length (Table 7.1), and correspond to terminal helices (Fig. 7.5). Three of the most variable regions, with 130 to over 180 substitutions per 4-bp interval, also correspond to those helices prone to length variation, and alignment problems in the regions of helices at positions 290–300, 680–688, 505–510, and 1,390–1,396 may contribute to the apparently high levels of sequence variability. Other variable segments, however, are not prone to insertion and deletion, so the estimates of sequence variability in these regions likely reflect real nucleotide substitution alone and not the confounding effects of misaligned sequences.

The pattern of conserved and variable domains (Fig. 7.4) is based on substitution patterns across the angiosperms, and it is unclear from this analysis where the sources of variation lie: Are the highly variable regions the result of comparison across a broad evolutionary range of species, or do the variable regions evolve more quickly even within specific clades? To address this question, we relied on the 18S rDNA consensus tree (Soltis, Soltis, Nickrent et al., 1997) and examined patterns of substitution within seven clades (Fig. 7.2): the monocots, ranunculids, saxifragoids, glucosinolates, Caryophyllidae s. l., Asteridae s. s., and Asteridae s. l. These clades were selected because they are present in all of the shortest trees obtained in the phylogenetic analyses (although shorter trees likely exist). The taxonomic composition of each clade analyzed is given in Table 7.2. Data for each clade were charted on the topology for that clade in Tree 1 (Soltis, Soltis, Nickrent et al., 1997) using MacClade. Similar patterns of conserved

and variable domains were observed within each clade (Fig. 7.6), even though the number of species included in these clades varied substantially (Table 7.2). Therefore, these variable regions must evolve quite rapidly under few selective constraints, even within small, well-defined groups like the glucosinolate clade and the Caryophyllidae s. l., each of which is represented in the analysis by only nine species.

Ribosomal DNAs of other organisms are also typically mosaics of highly conserved and highly variable regions (Van de Peer et al., 1996a; but see Vawter and Brown, 1993, on animal 18S rDNA). For example, the 16S rDNA of prokaryotes contains nine highly variable regions (Van de Peer et al., 1996a), at least three of which (helices P10, P23-1, and P43) appear to correspond to hypervariable regions of angiosperm 18S rDNA. In contrast, helix P27 of the prokaryotic 16S rDNA is highly conserved, with eight of nine bases conserved across 500 sequences (Van de Peer et al., 1996a). This region appears to be involved in the association of the ribosomal subunits (Tapprich and Hill, 1986) and in the initiation of protein synthesis (Tapprich et al., 1989). The corresponding helix E27 on the angiosperm 18S secondary structure model also appears to be highly conserved. Functional constraints throughout most of the rDNA sequence are poorly understood but are thought to maintain regions important for the structure and function of the rRNA, particularly in rRNA-protein interactions or as functional sites (e.g., Raue et al., 1988; Powers and Noller, 1990; Brimacombe, 1995).

Table 7.1. Highly conserved and variable regions of the 18S rRNA gene in angiosperms.

Conserved Regions	Variable Regions
390–450	120–125
580–630	290–300
980–1,050	505–510
1,180–1,230	680–688
1,340–1,389	800–810
1,650–1,700	1,390–1,396
	1,720–1,730

Note: Nucleotide positions correspond to the alignment for 228 species of angiosperms and five Gnetales (Soltis, Soltis, Nickrent et al., 1997); this alignment is available at http://www.wsu .edu:8080/~soltilab/.

Conserved & Variable Domains

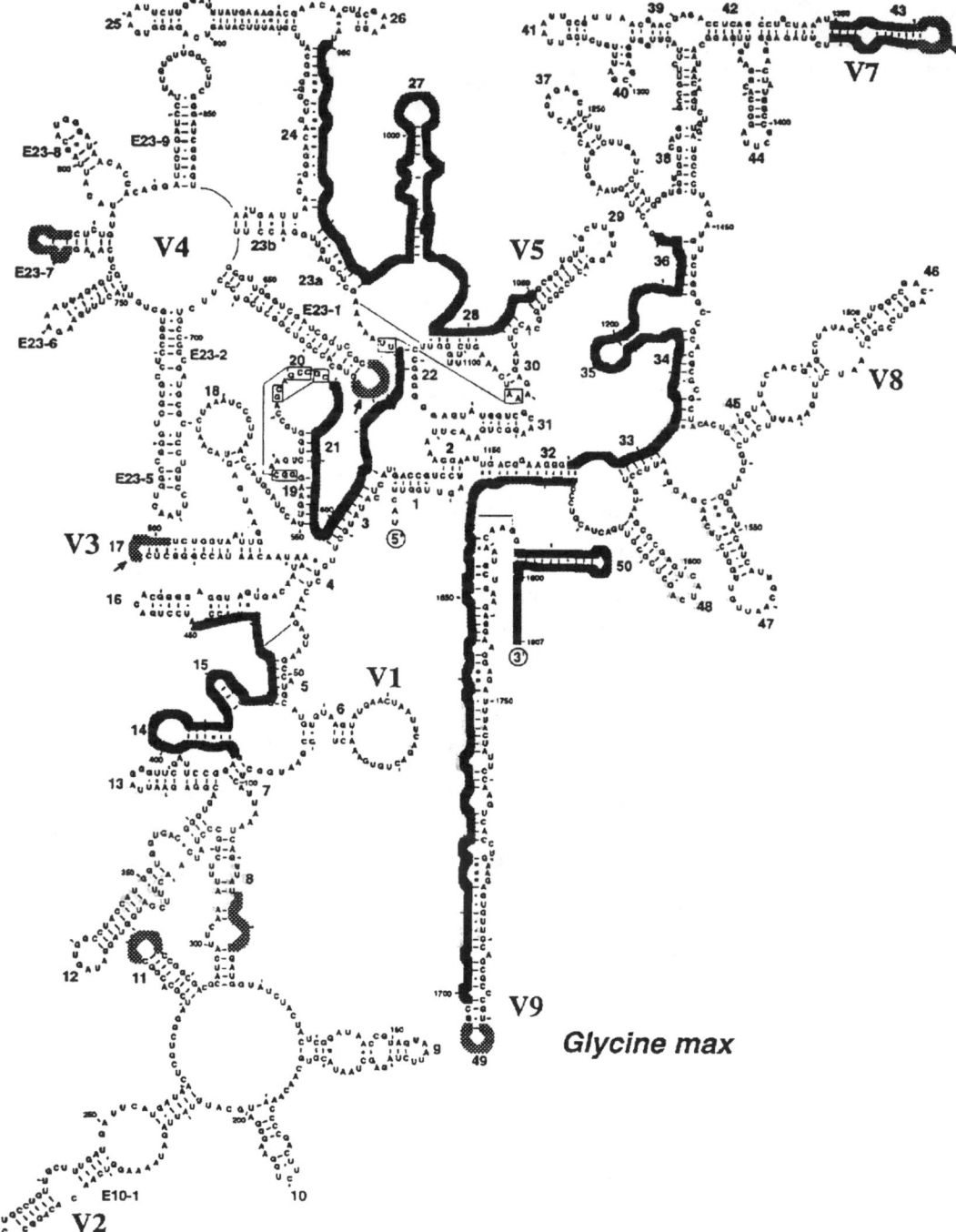

Figure 7.5. The secondary structure of 18S rDNA, showing the locations of the highly conserved and variable regions listed in Table 7.1. Conserved regions are shown in black, highly variable regions in gray.

Table 7.2. Taxonomic composition of clades (from Soltis, Soltis, Nickrent et al., 1997) analyzed for patterns of molecular evolution.

Clade	Genera in Clade	Clade	Genera in Clade
Monocots	*Tacca*	Saxifragoids, cont.	*Daphniphyllum*
	Bowiea		*Cercidiphyllum*
	Chlorophyton		*Liquidambar*
	Allium		*Altingia*
	Sagittaria	Glucosinolates	*Floerkea*
	Hippeastrum		*Koeberlinia*
	Eucharis		*Arabidopsis*
	Gladiolus		*Brassica*
	Isophysis		*Cleome*
	Xanthorrhoea		*Batis*
	Veitchia		*Moringa*
	Cyanella		*Carica*
	Helmholtzia		*Tropaeolum*
	Maranta	Caryophyllidae s. l.	*Drosera*
	Zingiber		*Nepenthes*
	Costus		*Plumbago*
	Canna		*Cocoloba*
	Heliconia		*Mollugo*
	Musa		*Tetragonia*
	Glomeropitcairnia		*Phytolacca*
	Zea		*Spinacia*
	Oryza		*Mirabilis*
	Cyperus	Asteridae s.s.	*Garrya*
	Sparganium		*Pittosporum*
	Elasis		*Hedera*
	Colchicum		*Lomatium*
	Calla		*Escallonia*
	Oncidium		*Eremosyne*
Ranunculids	*Akebia*		*Corokia*
	Sargentodoxa		*Symphoricarpos*
	Ranunculus		*Dipsacus*
	Coptis		*Lonicera*
	Xanthorhiza		*Roussea*
	Menispermum		*Campanula*
	Tinospora		*Lobelia*
	Caulophyllum		*Tragopogon*
	Podophyllum		*Tagetes*
	Euptelea		*Eucommia*
	Dicentra		*Aucuba*
	Hypecoum		*Helwingia*
Saxifragoids	*Tetracarpaea*		*Phyllonoma*
	Haloragis		*Olea*
	Penthorum		*Pachystachys*
	Paeonia		*Byblis*
	Itea		*Lamium*
	Pterostemon		*Buddleja*
	Crassula		*Parmentiera*
	Sedum		*Pedicularis*
	Dudleya		*Orthocarpus*
	Kalenchoe		*Linaria*
	Heuchera		*Ipomoea*
	Boykinia		*Cuscuta*
	Saxifraga		*Aeschynanthus*
	Ribes		*Phacelia*

Table 7.2. *Continued*

Clade	Genera in Clade	Clade	Genera in Clade
Asteridae s.s., cont.	*Bourreria*	Asteridae s. l., cont.	*Acanthogilia*
	Tabernaemontana		*Gilia*
	Mitchella		*Sarracenia*
	Brunfelsia		*Clethra*
	Montinia		*Arctostaphylos*
	Vahlia		*Vaccinium*
	Berzelia		*Styrax*
Asteridae s. l.	Asteridae s. s.		*Cyrilla*
	Caryophyllidae s. l.		*Monotropa*
	Camptotheca		*Pyrola*
	Manilkara		*Actinidia*
	Fouquieria		*Diospyros*
	Diapensia		*Camellia*
	Galax		*Symplocus*
	Impatiens		*Philadelphus*
	Primula		*Hydrangea*
	Cobaea		

Note: Species included are given in Soltis, Soltis, Nickrent et al. (1997), along with sources of material and voucher information.

Transition/Transversion Biases

Transitions are purine–purine (A ↔ G) or pyrimidine–pyrimidine (C ↔ T) substitutions. Transversions are purine–pyrimidine or pyrimidine–purine changes (A ↔ C, A ↔ T, G ↔ C, G ↔ T). Because there are eight possible transversions and only four possible transitions, transversions are expected to occur twice as frequently as transitions, assuming equal probabilities of all nucleotide substitutions. However, observed changes in DNA instead reveal an excess of transitions over transversions. This transition bias was observed in the first comparisons of DNA sequences (Vogel and Rorhborn, 1966; Fitch, 1967) and is considered a "striking and consistently observed feature of DNA-sequence change" (Wakeley, 1996, p. 162).

A.

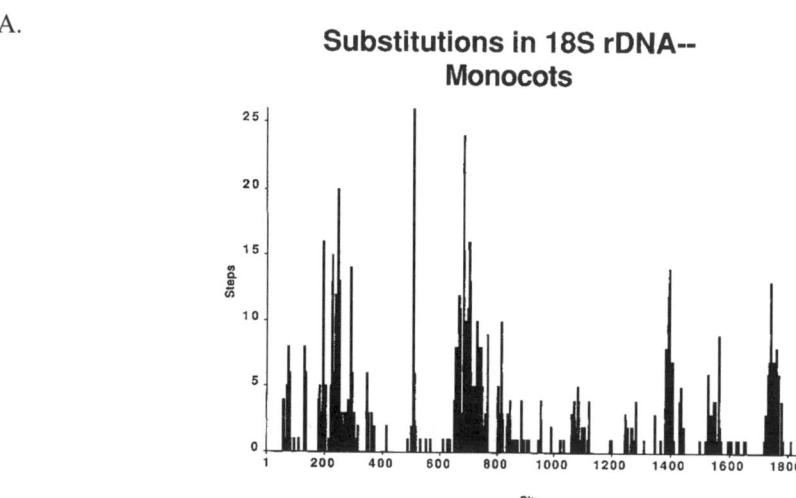

Figure 7.6. Patterns of nucleotide substitution in 18S rDNA in specific clades of angiosperms, as described for Fig. 7.4. *A.* Monocots. *B.* Ranunculids. *C.* Saxifragoids. *D.* Glucosinolates. *E.* Caryophyllidae s. l. *F.* Asteridae s. s. *G.* Asteridae s. l.

B.

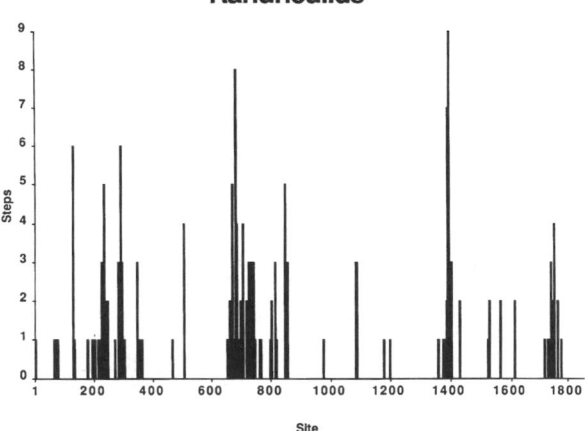

**Substitutions in 18S rDNA--
Ranunculids**

C.

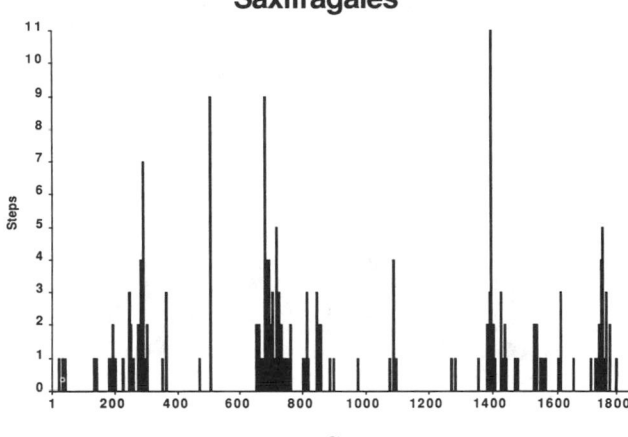

**Substitutions in 18S rDNA--
Saxifragales**

D.

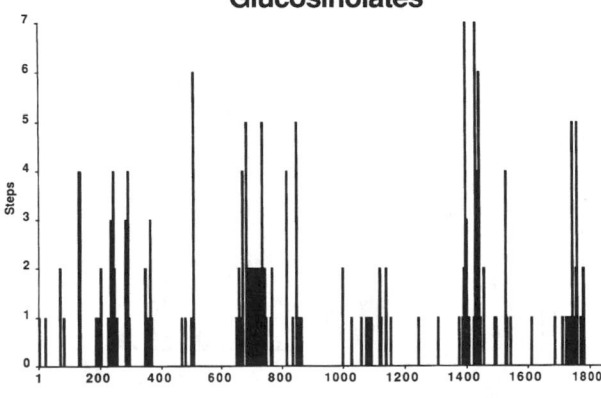

**Substitutions in 18S rDNA--
Glucosinolates**

E.

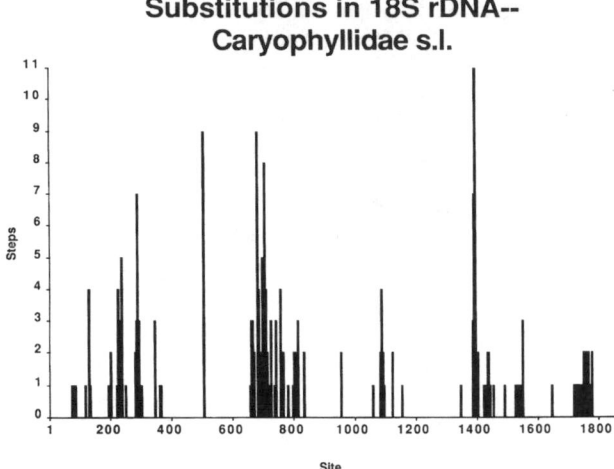

F.

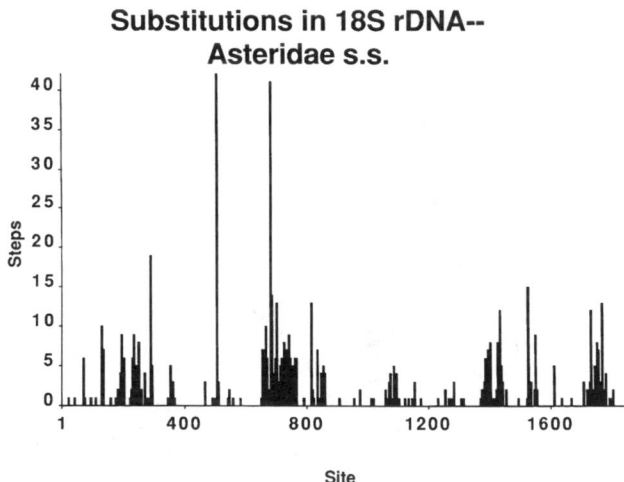

G.

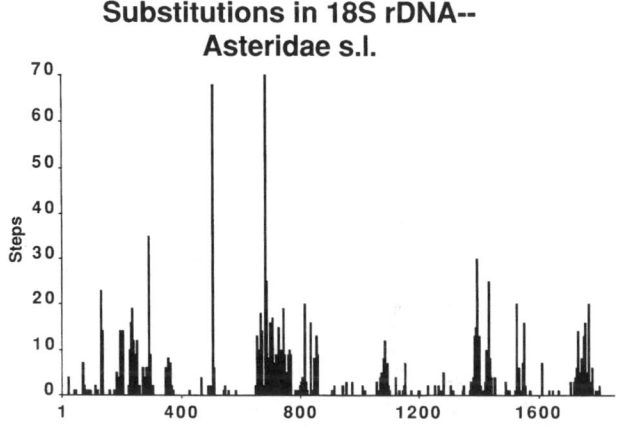

Substitution biases were evident when nucleotide changes were plotted over Tree 1 of the angiosperm phylogeny (Soltis, Soltis, Nickrent et al., 1997). With the data set for 223 angiosperms and five Gnetales and Tree 1, we used MacClade to CHART transitions and transversions, analyzing only unambiguous changes. The transition/transversion ratio of 1.9 is similar to that of 2.0 reported for 18S rDNA for a smaller group of species (Nickrent and Soltis, 1995). The most frequent single type of nucleotide substitution in 18S rDNA across the angiosperms is the C to T transition; A to G transitions were relatively rare (Fig. 7.7). Similar patterns of nucleotide substitution were observed within specific clades: monocots, ranunculids, saxifragoids, glucosinolates, Caryophyllidae s. l., Asteridae s. s., and Asteridae s. l. Transition/transversion ratios of 1.7–2.1 were observed in all clades except the Caryophyllidae s. l. (ts/tv of 1.4), values similar to that obtained for the angiosperms as a whole (Table 7.3). Frequencies of C to T and A to G transitions within clades were also similar to those observed across the angiosperms (Fig. 7.7). The C to T transition is also the most common substitution in animal 18S rDNA (Vawter and Brown, 1993). In contrast, G to A transitions are also common in stems (but not loops) of animal 18S rDNA and are infrequent in angiosperm 18S rDNA. A transition bias in rDNA stem regions may help maintain secondary structure (Vawter and Brown, 1993; Wakeley, 1996).

Transition biases, while present in nuclear (e.g., Gojobori et al., 1982; Li et al., 1984), chloroplast (e.g., Curtis and Clegg, 1984; see Chapter 6), and animal mitochondrial genomes (e.g., Brown et al., 1982) and in both genes and pseudogenes, vary in their magnitude. However, transition biases in rDNA have not been reported for most groups of organisms. Although a transition bias has been assumed for rDNAs (e.g., Mishler et al., 1988; Steele et al., 1988, 1991), such a bias is lacking in most rDNA data sets (see Vawter and Brown, 1993). In animal 18S rDNA, no consistent transition bias was detected (Vawter and Brown, 1993). Instead, the relative rates of the various transitions and transversions vary substantially among different regions of the rRNA molecule. Vawter and Brown (1993) conclude that an assumption of transition bias is inappropriate for phylogenetic studies of rDNA sequences. The transition biases reported here for angiosperm 18S rDNA support similar earlier observations on a smaller data set (Nickrent and Soltis, 1995) and suggest that, for angiosperm 18S rDNA, the assumption of a transition bias is indeed valid.

APPLYING WEIGHTS TO 18S RDNA IN PHYLOGENETIC ANALYSES

We explored the topological effects of various weighting schemes, based on the patterns of 18S

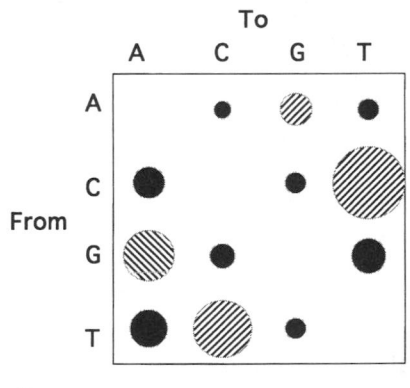

Transitions
Transversions

Figure 7.7. Summary of classes of nucleotide substitutions in 18S rDNA across the angiosperms, as calculated using the CHART option of MacClade 3.03, Tree 1 of the phylogenetic analysis of angiosperms based on 18S rDNA sequences (Soltis et al., 1997), and only unambiguous changes.

Table 7.3. Transition/transversion ratios in 18S rDNA across all angiosperms and in the seven clades listed in Table 7.2.

Clade	Transition/Transversion Ratio
Angiosperms	1.9:1
Monocots	2.1:1
Ranunculids	1.9:1
Saxifragoids	1.7:1
Glucosinolates	1.7:1
Caryophyllidae s. l.	1.4:1
Asteridae s. s.	2.0:1
Asteridae s. l.	2.0:1

Note: Ratios were calculated using the CHART option of MacClade 3.03, Tree 1 from the phylogenetic analysis of angiosperms based on 18S rDNA (Soltis, Soltis, Nickrent et al., 1997), and only unambiguous changes.

rDNA evolution observed across the angiosperms. In these analyses, we did not use the entire angiosperm 18S data set because the multiple analyses we planned to conduct could not be performed adequately within a reasonable time frame, given the large size of the data set (P. S. Soltis and D. E. Soltis, 1996; Rice et al., 1997). The heuristic analyses performed on the angiosperm 18S data set required over two years of computer time (Soltis, Soltis, Nickrent et al., 1997), and it is likely that the shortest trees were not even recovered. Given that we wanted to experiment with five or more weighting schemes, analyses of the entire data set would have required nearly five years of computer time, and we still would not have been satisfied that differences in topology were due to alternative weightings rather than to the vagaries of heuristic searches of large data sets. Instead, we constructed a data set that corresponds to the saxifragoid clade of Soltis, Soltis, Nickrent et al. (1997) and conducted a series of weighted analyses on this data set. The saxifragoid clade, comprising 18 genera (Table 7.2), is the most strongly supported clade (of more than five taxa) recovered in the 18S rDNA analysis (Soltis, Soltis, Nickrent et al., 1997), with a jackknife value of 68%. It was therefore an obvious clade for further study. Outgroups were *Trochodendron, Tetracentron, Platanus,* and *Sabia.* We opted against study of the larger clades, such as Asteridae s. l., because their large size (72 genera for Asteridae s. l.; Table 7.2) precluded the timely completion of multiple analyses.

We used the aligned sequences of 1,850 nucleotides (Soltis et al., 1997), excluding the 131 nucleotide positions described in that analysis. Using PAUP* 4.0d42, 49, and 52 (courtesy of D. Swofford), we conducted a series of heuristic searches with 100 replicates using random taxon addition and tree bisection-reconnection (TBR) branch swapping, saving all most parsimonious trees. In the initial analysis of the saxifragoid data set, all characters (i.e., nucleotide positions) had equal weights. Subsequent analyses invoked the following weighting schemes: (1) a stepmatrix weighting transversions twice as much as transitions, based on the approximate 2:1 transition/transversion bias detected across the angiosperms and even within the saxifragoid clade

(Table 7.3); (2) loops only, stem bases excluded (i.e., weighted to 0); (3) stems only, loop bases excluded (i.e., weighted to 0); and (4) loop bases weighted twice as much as stems, following a model of complete compensation in stem regions. Bootstrap analyses (Felsenstein, 1985) using 100 replicates and the Fast Bootstrap option of PAUP* 4.0, in which stepwise addition but no branch swapping is performed, were conducted to assess support for the inferred relationships.

We also applied the successive approximations approach (Farris, 1969) to character weighting, in which characters are assigned weights based on their fit to an initial tree (or trees). These weights are then used in a subsequent search, and new weights are assigned based on the fit of the characters to the new tree(s). This iterative process is repeated until either the weights do not change between successive analyses or the topology does not change. The purpose of a successive approximations analysis is to assign greater weight to those characters that are "cladistically reliable" (sensu Farris, 1969), that is, that are correlated because of phylogenetic history rather than chance. These cladistically reliable characters are those with little or no homoplasy; thus, character weighting by successive approximations will assign greater weights to those characters with little or no homoplasy. This analysis was implemented using PAUP* 4.0; weights were assigned using the rescaled consistency index (RCI) as a measure of homoplasy, and values were rounded rather than truncated. Multiple topologies (i.e., the shortest trees of the initial heuristic searches) were considered, and the fit of the characters to the multiple trees was assessed using both the RCI value giving the best fit and the mean RCI calculated across the multiple trees. These two approaches produced identical results with the saxifragoid data set.

Analyses of Saxifragoid Clade

Heuristic searches of the saxifragoid clade, applying equal weights to all nucleotide positions, recovered three minimal-length trees of 204 steps (consistency index [CI] = 0.775; retention index [RI] = 0.725; Fig. 7.8). A posteriori character weighting using the successive approxima-

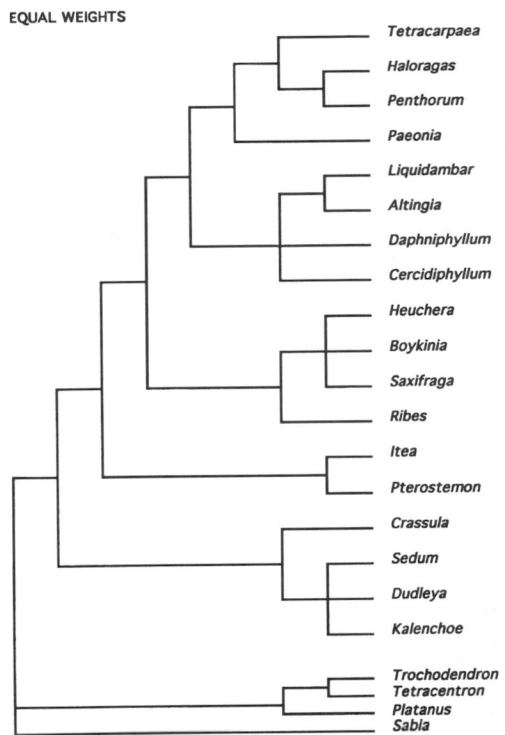

EQUAL WEIGHTS

Tetracarpaea
Haloragas
Penthorum
Paeonia
Liquidambar
Altingia
Daphniphyllum
Cercidiphyllum
Heuchera
Boykinia
Saxifraga
Ribes
Itea
Pterostemon
Crassula
Sedum
Dudleya
Kalenchoe
Trochodendron
Tetracentron
Platanus
Sabia

Figure 7.8. Strict consensus of the three most parsimonious trees recovered in phylogenetic analyses of the saxifragoid clade using equal weighting of all nucleotide positions.

imal-length trees of 273 steps. Included in these 15 trees are the three topologies obtained by equal weighting. The effect of transition/transversion character-state weighting in this data set is to increase the number of most parsimonious trees and consequently reduce the resolution of the strict consensus tree (Fig. 7.9).

Weighting Stems and Loops

Separate analyses were conducted using loop bases only, stem bases only, and loop bases weighted 2:1 over stem bases. Nucleotide positions were designated as stem bases if they participated in base pairing (or appeared to have the potential to pair) in the 18S rRNA secondary structure model for *Glycine max* (D. Nickrent, in Soltis, Soltis, Nickrent et al., 1997). Bases were designated as loops if they were not involved in base pairing in the *G. max* rRNA model. In addition, loops were arbitrarily defined as being four bases or more in length; both hairpin and internal loops (terminology sensu Chastain and

tions approach (Farris, 1969) recovered the same three trees. This suggests, at least at this rather focused level within the angiosperms, that no differential weighting between generally conservative versus variable domains is needed or warranted. However, at a broader scale, such weighting might be necessary, as levels of homoplasy in variable domains would be expected to increase. Within the saxifragoid clade, the same basic subclades were recovered as in previous analyses (e.g., D. E. Soltis and P. S. Soltis, 1997; Soltis, Soltis, Nickrent et al., 1997), but the relationships among the subclades differed slightly. However, this set of trees will serve as a reference point for examining the effects of various weighting schemes for 18S rDNA characters.

Weighting Transitions and Transversions

Using a stepmatrix to weight transitions half as much as transversions, we recovered 15 min-

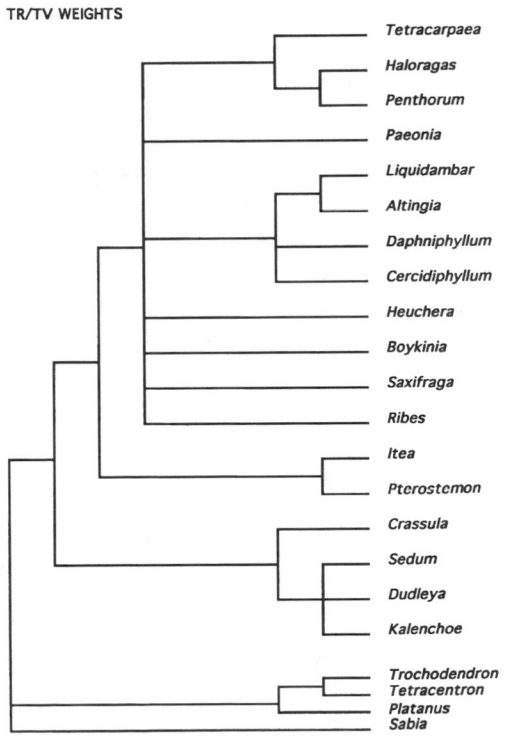

TR/TV WEIGHTS

Tetracarpaea
Haloragas
Penthorum
Paeonia
Liquidambar
Altingia
Daphniphyllum
Cercidiphyllum
Heuchera
Boykinia
Saxifraga
Ribes
Itea
Pterostemon
Crassula
Sedum
Dudleya
Kalenchoe
Trochodendron
Tetracentron
Platanus
Sabia

Figure 7.9. Strict consensus of the 15 most parsimonious trees recovered in phylogenetic analyses of the saxifragoid clade weighting transitions half as much as transversions.

Tinoco, 1991) were included. Hairpin loops of four nucleotides (i.e., tetra-loops) are a common feature of both small subunit (SSU) and large subunit (LSU) rRNAs, and make up over 50% of all hairpin loops in 16S rRNA (Woese et al., 1990; Gutell, 1993). Thus, most loop regions are four or more bases in length, and our designation of loops largely reflects patterns of base pairing in the rRNA. Several small regions, one to three nucleotides in length, were unpaired (forming bulges or small internal loops; Chastain and Tinoco, 1991) but were omitted from the analysis. Exceptions were positions 55–56 of the aligned sequence (between helices 5 and 6 of domain V1) and position 586 in helix 20 (domain V3); these positions were obvious components of loop structures and were analyzed as such even though their adjacent positions were paired (either in secondary structure or tertiary folding) so that they were not part of a string of four or more unpaired nucleotides.

Analysis of "loops only" identified 528 positions as loop bases. Of these, 40 positions were removed because they are among the 131 positions that were excluded from all analyses because they fall in hypervariable regions or at the beginning or the end of the 18S gene. Of the 488 remaining loop characters, 438 were constant in the saxifragoid clade and outgroups. Only 50 characters (10%) were variable, and 16 (3.3% of all characters, 32% of variable characters) were parsimony-informative (as calculated by PAUP* 4.0; see discussion by Rice et al. [1997] on calculations of the number of parsimony-informative characters). Phylogenetic analyses of the "loops only" data recovered five shortest trees, each of 90 steps (CI = 0.700; RI = 0.719). As might be predicted given the small number of (variable) loop characters, the strict consensus of these five trees is not well resolved. However, this lack of resolution seems not to be a function of low resolution in each of the minimal trees; instead, it is due to conflict among the shortest trees. The shortest trees portray relationships among saxifragoids that differ, in some cases substantially, from those suggested both by analyses using equal weights (Fig. 7.8) and previous studies using 18S rDNA (e.g., D. E. Soltis and P. S. Soltis, 1997; Soltis, Soltis, Nickrent et al., 1997) and *rbcL* (Chase, Soltis,

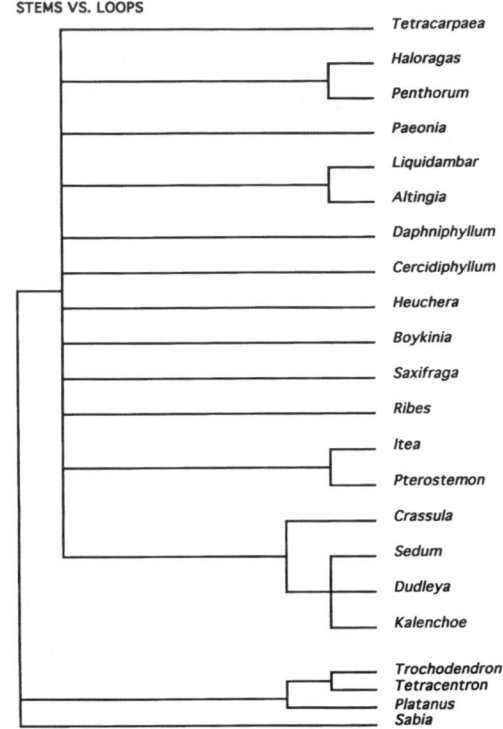

STEMS VS. LOOPS

Figure 7.10. Strict consensus of the 51 most parsimonious trees recovered in phylogenetic analyses of the saxifragoid clade weighting loop regions twice as much as stems.

Olmstead et al., 1993; Morgan and Soltis, 1993) sequences. Bootstrap support for most of the relationships is quite low; only *Itea-Pterostemon* (84%), *Liquidambar-Altingia* (82%), and saxifragoids (relative to outgroups; 82%) received bootstrap support over 50%. Loop bases evolve more rapidly, at least in length and perhaps in primary sequence (but see below), than do stems.

Exclusion of the 528 loop bases produced a data set of 1,322 stem bases, of which 91 were removed because they fell among the beginning, ending, or hypervariable portions of the sequence. Of the remaining 1,231 characters included in the analyses of the saxifragoids, 1,061 were invariant; 170 characters (14%) were variable, and 44 (3.5% of all characters, 26% of variable characters) were parsimony-informative. The percentages of variable and parsimony-informative characters are similar for stem and loop bases. The heuristic searches of the "stems only" data set found 866 trees, each of 151 steps (CI = 0.781; RI = 0.723). The strict consensus of these 866 trees identifies only clades corresponding to Crassulaceae, Saxifragaceae s. s., and Hamamelidaceae; these clades, plus *Itea-Pterostemon* and saxifragoids relative to the outgroups, are the only ones that received bootstrap support over 50%. Analyses of neither the stem bases only nor of the loop bases only recovered the *Paeonia-Tetracarpaea-Haloragas-Penthorum* clade found in the analysis of all 18S characters, equally weighted, and in other studies.

Analyses weighting stems half as much as loops recovered 51 shortest trees, each of 256 steps, distributed on three islands. The topologies of some of the shortest trees are identical to those found in the searches where equal weights were applied, but the strict consensus (Fig. 7.10) retains only Crassulaceae, Hamamelidaceae, *Itea-Pterostemon,* and *Haloragas-Penthorum,* although much greater structure is preserved in over 80% of the shortest trees. With the exceptions of *Itea-Pterostemon* (95%), Crassulaceae (95%), Hamamelidaceae (93%), and the saxifragoid clade (96%), bootstrap values were low. The sister relationship between *Haloragas* and *Penthorum* received only 64% bootstrap support, and the clade of *Sedum-Dudleya-Kalenchoe* has a bootstrap value of only 59%.

Effects of Weighting

The general effect of the weighting schemes attempted was to decrease the resolution of the strict consensus trees, either by reducing the number of characters available for analysis or by recovering additional conflicting trees. The weighted analyses may provide more conservative estimates of relationships, based on the consensus trees. However, relationships supported by high bootstrap values did not change markedly among any of the analyses, regardless of how characters and character states were weighted. Analysis of the saxifragoid clade leads us to conclude that most efforts to weight stem versus loops and transitions versus transversions, and even conserved versus variable domains, are probably not worth the effort or extra computer time required to conduct some of the analyses.

IMPLICATIONS FOR PHYLOGENY ESTIMATION USING 18S RDNA

The more highly conserved length of angiosperm 18S rDNA makes alignment of sequences much more straightforward than previously reported. Across the angiosperms, or a large subset of angiosperms, manual alignment may be difficult in those few, short regions prone to length variation; regions of uncertain alignment should obviously be omitted from phylogenetic analysis. However, these indel-prone regions appear quite stable in length within more focused groups, permitting confident alignments over smaller taxonomic distances and providing additional stretches of sequence, some of which may be appropriately variable, for phylogeny estimation.

The high frequency of compensatory change in angiosperm 18S rDNA suggests that some downweighting of stem positions may be warranted. However, our application of various weighting schemes to stem and loop bases had very little effect on the resultant topologies. Relationships that were supported by high bootstrap values in any analysis were likewise well supported in other analyses using different weights. Furthermore, patterns of compensatory change are complex, and weights based on a simple dichotomy between stem and loop positions do not adequately reflect underlying patterns of molecular evolution of 18S rDNA. If differential weighting of stems and loops is to be applied, we recommend instead that weight be assigned to specific positions, or perhaps regions, that tend to undergo double compensatory changes, if such regions can be identified. A potential danger with this approach is that a posteriori weighting may reinforce the initial topologies recovered. However, the double effect of changes in nonindependent characters, if no weights are applied, may in some cases (although not the ones we examined) bias the outcome of a phylogenetic analysis. We hope that through detailed analysis of substitution patterns for large data sets, such as the angiosperm 18S rDNA matrix analyzed here, generalities will emerge that can guide future analyses of both larger and smaller groups of species.

Differential weights might also be applied to transition and transversion character-state changes to reflect the bias observed in the data. However, incorporating small transition/transversion ratios, either as character-state weights or in models, seems to have had little effect on the resultant topologies (but see Van de Peer et al., 1993, 1996b; Wakeley, 1996; Sullivan et al., 1996, for a different perspective). Perhaps more meaningful, and appropriate, weights might reflect the frequency of each type of substitution. The most common substitution across the angiosperms was the C-to-T transition, and the least common was the A-to-G transition. Changes from C to A, G to T, and T to A were more common than several others (Fig. 7.7). The frequencies of these changes might also be incorporated into a character-state weighting system for 18S rDNA.

The mosaic structure of 18S rDNA, comprising both highly conserved and highly variable domains, also has implications for phylogeny reconstruction. Most important is the observation that some regions may evolve too quickly to retain historical signal across broad evolutionary distances. These highly variable regions, even when alignable, should perhaps be downweighted relative to more conserved regions in phylogenetic analyses of all angiosperms or large groups of angiosperms. The successive approximations approach to character weighting (Farris, 1969), which attempts to minimize the effects of "cladistically unreliable" characters, may be useful for applying relative weights based on homoplasy in an rDNA data set. However, this approach applied to the saxifragoid clade yielded topologies identical to those recovered in analyses using equal weights. Furthermore, the more variable regions may be valuable sources of phylogenetic information in smaller groups of species.

Although there may be sound theoretical reasons for applying differential weights to characters and character-state changes in 18S rDNA, the empirical effect of these weighting schemes in the cases that we explored was minimal. If this finding proves to be generally true, we recommend that complex weighting be abandoned to shorten run times and permit more thorough exploration of tree space in a given amount of time.

CONCLUSIONS

Ribosomal DNAs offer a potentially large set of molecular characters for phylogeny reconstruction, but several aspects of rDNA evolution may have important implications for phylogenetic analysis. We investigated patterns of molecular evolution in the 18S rDNA of flowering plants using a data set of 223 angiosperm sequences and phylogenetic trees estimated from these sequences. We explored (1) patterns of insertion and deletion, (2) patterns of stem and loop formation in the secondary structure of the rRNA and levels of compensatory change in stem regions, (3) the distribution of conserved and variable domains, (4) substitution biases, including transitions versus transversions, and (5) the effects of various weighting schemes based on the patterns of molecular evolution observed. We found that indels are largely localized to a few helical regions and cause few alignment problems, even across the angiosperms. Compensatory changes in stem regions were inferred, with double compensatory changes occurring at a frequency of approximately 20%. The angiosperm 18S rDNA is a mosaic of highly conserved and highly variable regions. These conserved and variable domains were observed not only across the angiosperms, but also within specific clades, suggesting that some regions may be under limited selective constraints and thus evolve quite rapidly. A transition/transversion ratio of approximately 2:1 was observed across the angiosperms and within several specific clades.

Weighting schemes to incorporate these patterns of molecular evolution into phylogenetic analyses were attempted for the saxifragoid clade. In general, weighting had little effect on the resultant topologies or their bootstrap support; relationships that were strongly supported in analyses with equally weighted characters and character states were maintained in analyses using differential weighting. This general result was also obtained in studies of weighting schemes for chloroplast genes (see Chapter 6).

LITERATURE CITED

Bakker, F. T., J. L. Olsen, W. T. Stam, and C. Van Den Hoek. 1994. The *Cladophora* complex (Chlorophyta): new views based on 18S rRNA gene sequences. Molecular Phylogenetics and Evolution 3:365–382.

Berbee, M. L. 1996. Loculoascomycete origins and evolution of filamentous ascomycete morphology based on 18S rRNA gene sequence data. Molecular Biology and Evolution 13:462–470.

Bhattacharya, D., and L. Medlin. 1995. The phylogeny of plastids: A review based on comparisons of small-subunit ribosomal RNA coding regions. Journal of Phycology 31:489–498.

Bremer, K., C. J. Humphries, B. D. Mishler, and S. P. Churchill. 1987. On cladistic relationships in green plants. Taxon 36:339–349.

Brimacombe, R. 1995. The structure of ribosomal RNA: a three-dimensional jigsaw puzzle. European Journal of Biochemistry 230:365–383.

Brown, W. M., E. M. Prager, A. Wang, and A. C. Wilson. 1982. Mitochondrial DNA sequences of primates: the tempo and mode of evolution. Journal of Molecular Evolution 18:225–239.

Buchheim, M. A., and R. L. Chapman. 1991. Phylogeny of the colonial green flagellates: a study of 18S and 26S rRNA sequence data. BioSystems 25:85–100.

Buchheim, M. A., M. Turmel, E. A. Zimmer, and R. L. Chapman. 1990. Phylogeny of *Chlamydomonas* (Chlorophyta) based on cladistic analysis of nuclear 18S rRNA sequence data. Journal of Phycology 26:689–699.

Capesius, I. 1995. A molecular phylogeny of bryophytes based on the nuclear encoded 18S rRNA genes. Journal of Plant Physiology 146:59–63.

Chapman, R. L., and M. Buchheim. 1991. Ribosomal RNA gene sequences: analysis and significance in the phylogeny and taxonomy of green algae. C.R.C. Critical Reviews in Plant Science 10:343–368.

Chase, M. W., D. E. Soltis, R. G. Olmstead, D. Morgan, D. H. Les, B. D. Mishler, M. R. Duvall, R. A. Price, H. G. Hills, Y.-L. Qiu, K. A. Kron, J. H. Rettig, E. Conti, J. D. Palmer, J. R. Manhart, K. J. Sytsma, H. J. Michaels, W. J. Kress, K. G. Karol, W. D. Clark, M. Hedren, B. S. Gaut, R. K. Jansen, K.-J. Kim, C. F. Wimpee, J. F. Smith, G. R. Furnier, S. H. Strauss, Q.-Y. Xiang, G. M. Plunkett, P. S. Soltis, S. M. Swensen, S. E. Williams, P. A. Gadek, C. J. Quinn, L. E. Eguiarte, E. Golenberg, G. H. Learn, Jr., S. W. Graham, S. C. H. Barrett, S. Dayanandan, and V. A. Albert. 1993. Phylogenetics of seed plants: an analysis of nucleotide sequences from the plastid gene *rbcL*. Annals of the Missouri Botanical Garden 80:528–580.

Chastain, M., and I. Tinoco, Jr. 1991. Structural elements in RNA. Progress in Nucleic Acid Research and Molecular Biology 41:131–177.

Chaw, S.-M., H. Long, B.-S. Wang, A. Zharkikh, and W.-H. Li. 1993. The phylogenetic position of Taxaceae based on 18S rRNA sequences. Journal of Molecular Evolution 37:624–630.

Chaw, S.-M., H.-M. Sung, H. Long, A. Zharkikh, and W.-H. Li. 1995. The phylogenetic positions of the conifer genera *Amentotaxus, Phyllocladus,* and *Nageia* inferred from 18S rRNA sequences. Journal of Molecular Evolution 41:224–230.

Chaw, S.-M., A. Zharkikh, H.-M. Sung, T.-C. Lau, and W.-H. Li. 1997. Molecular phylogeny of gymnosperms and seed plant evolution: analysis of 18S rRNA sequences. Journal of Molecular Evolution 14:56–68.

Curtis, S. E., and M. T. Clegg. 1984. Molecular evolution of chloroplast DNA sequences. Molecular Biology and Evolution 1:291–301.

Dixon, M. T., and D. M. Hillis. 1993. Ribosomal RNA secondary structure—compensatory mutations and implications for phylogenetic analysis. Molecular Biology and Evolution 10:256–267.

Echols, H., and M. F. Goodman. 1991. Fidelity mechanisms in DNA replication. Annual Review of Biochemistry 60:477–511.

Embley, M. T., R. P. Hirt, and D. M. Williams. 1994. Biodiversity at the molecular level: the domains, kingdoms and phyla of life. Philosophical Transactions of the Royal Society of London B 345:21–31.

Erdmann, V. A., J. Wolters, E. Huysmans, and R. de Wachter. 1985. Collection of published 5S, 5.8S and 4.5S ribosomal RNA sequences. Nucleic Acids Research 13:r105–r153.

Farris, J. S. 1969. A successive approximations approach to character weighting. Systematic Zoology 18:374–385.

Felsenstein, J. 1985. Confidence limits on phylogenies: an approach using the bootstrap. Evolution 39:783–791.

Field, K. G., G. J. Olsen, D. J. Lane, S. J. Giovannoni, M. T. Ghiselin, E. C. Raff, N. R. Pace, and R. A. Raff. 1988. Molecular phylogeny of the animal kingdom. Science 239:748–753.

Fitch, W. M. 1967. Evidence suggesting a non-random character to nucleotide replacements in naturally occurring mutations. Journal of Molecular Biology 26:499–507.

Forster, H., M. D. Coffey, H. Elwood, and M. L. Sogin. 1990. Sequence analysis of the small subunit ribosomal RNAs of three zoosporic fungi and implications for fungal evolution. Mycologia 82:306–312.

Gargas, A., P. T. DePriest, M. Grube, and A. Tehler. 1995. Multiple origins of lichen symbioses in fungi suggested by SSU rDNA phylogeny. Science 268:1492–1495.

Gatesy, J., C. Hayashi, R. DeSalle, and E. Vrba. 1994. Rate limits for mispairing and compensatory change: the mitochondrial ribosomal DNA of antelopes. Evolution 48:188–196.

Gojobori, T., W.-H. Li, and D. Graur. 1982. Patterns of nucleotide substitution in pseudogenes and functional genes. Journal of Molecular Evolution 18:360–369.

Gutell, R. R. 1993. Comparative studies of RNA: inferring higher-order structure from patterns of sequence variation. Current Opinions in Structural Biology 3:313–322.

Gutell, R. R. 1996. Comparative sequence analysis and the structure of 16S and 23S rRNA. **In** Ribosomal RNA:

Structure, Evolution, Processing, and Function in Protein Biosynthesis, eds. R. A. Zimmermann and A. E. Dahlberg, pp. 111–128. CRC Press, Boca Raton, Florida.

Gutell, R. R., N. Larson, and C. R. Woese. 1994. Lessons from an evolving rRNA: 16S and 23S rRNA structures from a comparative perspective. Microbiological Review **58:**10–26.

Halanych, K. M. 1991. 5S ribosomal RNA sequences inappropriate for phylogenetic reconstruction. Molecular Biology and Evolution **8:**149–153.

Hamby, R. K., and E. A. Zimmer. 1988. Ribosomal RNA sequences for inferring phylogeny within the grass family (Poaceae). Plant Systematics and Evolution **160:**29–37.

Hamby, R. K., and E. A. Zimmer. 1992. Ribosomal RNA as a phylogenetic tool in plant systematics. **In** Molecular Systematics of Plants, eds. P. S. Soltis, D. E. Soltis, and J. J. Doyle, pp. 50–91. Chapman & Hall, New York.

Hedderson, T. A., R. L. Chapman, and W. L. Rootes. 1996. Phylogenetic relationships of bryophytes inferred from nuclear-encoded rRNA gene sequences. Plant Systematics and Evolution **200:**213–224.

Hendriks, L., E. Huysmans, A. Vandenberghe, and R. deWachter. 1986. Primary structures of the 5S ribosomal RNAs of 11 arthropods and applicability of 5S RNA to the study of metazoan evolution. Journal of Molecular Evolution **24:**103–109.

Hendriks, L., R. De Baere, Y. Van de Peer, J. Neefs, A. Goris, and R. De Wachter. 1991. The evolutionary position of the rhodophyte *Porphyra umbilicalis* and the basidiomycete *Leucosporidium scottii* among other eukaryotes as deduced from complete sequences of small ribosomal subunit RNA. Journal of Molecular Evolution **32:**167–177.

Hinkle, G., J. K. Wetterer, T. R. Schultz, and M. L. Sogin. 1994. Phylogeny of the attine ant fungi based on analysis of small subunit ribosomal RNA gene sequences. Science **266:**1695–1697.

Hori, H., and S. Osawa. 1987. Origin and evolution of organisms as deduced from 5S ribosomal RNA sequences. Molecular Biology and Evolution **4:**445–472.

Hori, H., B.-L. Kim, and S. Osawa. 1985. Evolution of green plants as deduced from 5S rRNA sequences. Proceedings of the National Academy of Sciences U.S.A. **82:**820–823.

Huss, V. A. R., and M. L. Sogin. 1990. Phylogenetic position of some *Chlorella* species within the Chlorococcales based upon complete small-subunit ribosomal RNA sequences. Journal of Molecular Evolution **31:**432–442.

Kantz, T. S., E. C. Theriot, E. A. Zimmer, and R. L. Chapman. 1990. The Pleurastrophyceae and Micromonadophyceae: cladistic analysis of nuclear rRNA sequence data. Journal of Phycology **26:**711–721.

Kranz, H. D., and V. A. R. Huss. 1996. Molecular evolution of pteridophytes and their relationship to seed plants: evidence from complete 18S rRNA gene sequences. Plant Systematics and Evolution **202:**1–11.

Kranz, H. D., D. Miks, M.-L. Siegler, I. Capesius, Ch. W. Sensen, and V. A. R. Huss. 1995. The origin of land plants: phylogenetic relationships between Charophytes, Bryophytes, and vascular plants inferred from complete small subunit ribosomal RNA gene sequences. Journal of Molecular Evolution **41:**74–84.

Kraus, F., L. Jarecki, M. Miyamoto, S. Tanhauser, and P. Laipis. 1992. Mispairing and compensational changes during the evolution of mitochondrial ribosomal RNA. Molecular Biology and Evolution **9:**770–774.

Li, W.-H., C.-I. Wu, and C.-C. Luo. 1984. Nonrandomness of point mutations reflected in nucleotide substitutions in pseudogenes and its evolutionary implications. Journal of Molecular Evolution **21:**58–71.

Maddison, W. P., and D. R. Maddison. 1992. MacClade, Analysis of Phylogeny and Character Evolution, version 3.03. Sinauer Associates, Sunderland, Massachusetts.

Medlin, L. K., W. H. C. F. Kooistra, R. Gersonde, and U. Wellbrock. 1996. Evolution of the diatoms (Bacillariophyta). II. Nuclear-encoded small-subunit rRNA sequence comparisons confirm a paraphyletic origin for the centric diatoms. Molecular Biology and Evolution **13:**67–75.

Mindell, D. P., and R. L. Honeycutt. 1990. Ribosomal RNA in vertebrates: evolution and phylogenetic implications. Annual Review of Ecology and Systematics **21:**541–566.

Mishler, B. D., K. Bremer, C. J. Humphries, and S. P. Churchill. 1988. The use of nucleic acid sequence data in phylogenetic reconstruction. Taxon **37:**391–395.

Mishler, B. D., L. A. Lewis, M. A. Buchheim, K. S. Renzaglia, D. J. Garbary, C. F. Delwiche, F. W. Zechman, T. S. Kantz, and R. L. Chapman. 1994. Phylogenetic relationships of the "green algae" and "bryophytes." Annals of the Missouri Botanical Garden **81:**451–483.

Morgan, D. R., and D. E. Soltis. 1993. Phylogenetic relationships among members of Saxifragaceae sensu lato based on *rbcL* sequence data. Annals of the Missouri Botanical Garden **80:**631–660.

Nickrent, D. L., and C. R. Franchina. 1990. Phylogenetic relationships of the Santalales and relatives. Journal of Molecular Evolution **31:**294–301.

Nickrent, D. L., and D. E. Soltis. 1995. A comparison of angiosperm phylogenies from nuclear 18S rDNA and *rbcL* sequences. Annals of the Missouri Botanical Garden **82:**208–234.

Olsen, G. J. 1987. Earliest phylogenetic branchings: comparing rRNA-based evolutionary trees inferred with various techniques. Cold Spring Harbor Symposium on Quantitative Biology **52:**825–837.

Olsen, J. L., W. T. Stam, S. Berger, and D. Menzel. 1994. 18S rDNA and evolution in the Dasycladales (Chlorophyta): modern living fossils. Journal of Phycology **30:**729–744.

Pace, N. R., G. Olsen, and C. R. Woese. 1986. Ribosomal RNA phylogeny and the primary lines of evolutionary descent. Cell **45:**325–326.

Powers, T., and H. F. Noller. 1990. Dominant lethal mutations in a conserved loop in 16S rRNA. Proceedings of the National Academy of Sciences U.S.A. **87:** 1042–1046.

Ragan, M. A., C. J. Bird, E. L. Rice, R. R. Gutell, C. A. Murphy, and R. K. Singh. 1994. A molecular phylogeny of the marine red algae (Rhodophyta) based on nuclear small-subunit rRNA gene. Proceedings of the National Academy of Sciences U.S.A. **91:**7276–7280.

Raue, H. A., J. Klootwijk, and W. Musters. 1988. Evolutionary conservation of structure and function of high molecular weight ribosomal RNA. Progress in Biophysics and Molecular Biology **51:**77–129.

Rice, K. A., M. J. Donoghue, and R. G. Olmstead. 1997. A reanalysis of the large *rbcL* data set: implications for future phylogenetic studies. Systematic Biology **46:**554–563.

Rothschild, L. J., M. A. Ragan, A. W. Coleman, P. Heywood, and S. A. Gerbi. 1986. Are rRNA sequence comparisons the Rosetta stone of phylogenetics? Cell **47:**640.

Schlegel, M., H. J. Elwood, and M. L. Sogin. 1991. Molecular evolution in hypotrichous ciliates: sequence of the small subunit ribosomal RNA genes from *Onychodromus quadricornutus* and *Oxytricha granulifera* (Oxytrichidae, Hypotrichida, Ciliophora). Journal of Molecular Evolution **32:**64–69.

Sogin, M. L., H. J. Elwood, and J. H. Gunderson. 1986. Evolutionary diversity of eukaryotic small-subunit rRNA genes. Proceedings of the National Academy of Sciences U.S.A. **83:**1383–1387.

Soltis, D. E., and P. S. Soltis. 1997. Phylogenetic relationships among Saxifragaceae sensu lato: a comparison of topologies based in 18S rDNA and *rbcL* sequences. American Journal of Botany **84:**504–522.

Soltis, D. E., P. S. Soltis, D. L. Nickrent, L. A. Johnson, W. J. Hahn, S. B. Hoot, J. A. Sweere, R. K. Kuzoff, K. A. Kron, M. W. Chase, S. M. Swensen, E. A. Zimmer, S.-M. Chaw, L. J. Gillespie, W. J. Kress, and K. J. Sytsma. 1997. Angiosperm phylogeny inferred from 18S ribosomal DNA sequences. Annals of the Missouri Botanical Garden **84:**1–49.

Soltis, P. S., and D. E. Soltis. 1996. Phylogenetic analysis of large molecular data sets. Boletin de la Sociedad Botanica Mexicana **59:**99–114.

Steele, K. P., K. E. Holsinger, R. K. Jansen, and D. W. Taylor. 1988. Phylogenetic relationships in green plants—a comment on the use of 5S ribosomal RNA sequences by Bremer et al. Taxon **37:**135–138.

Steele, K. P., K. E. Holsinger, R. K. Jansen, and D. W. Taylor. 1991. Assessing the reliability of 5S rRNA sequence data for phylogenetic analysis in green plants. Molecular Biology and Evolution **8:**240–248.

Sullivan, J., K. E. Holsinger, and C. Simon. 1996. The effect of topology on estimates of among-site rate variation. Journal of Molecular Biology **42:**308–312.

Swann, E. C., and J. W. Taylor. 1993. Higher taxa of Basidiomycetes: an 18S rRNA perspective. Mycologia **85:**923–936.

Tapprich, W. E., and W. E. Hill. 1986. Involvement of bases 787–795 of *Escherichia coli* 16S ribosomal RNA in ribosomal subunit association. Proceedings of the National Academy of Sciences U.S.A. **83:**556–560.

Tapprich, W. E., D. J. Goss, and A. E. Dahlberg. 1989. Mutation at position 791 in *Escherichia coli* 16S ribosomal RNA affects processes involved in the initiation of protein synthesis. Proceedings of the National Academy of Sciences U.S.A. **86:**4927–4931.

Topal, M. D., and J. R. Fresco. 1976. Complementary base pairing and the origin of substitution mutations. Nature **263:**285–289.

Van de Peer, Y., J.-M. Neefs, P. De Rijk, and R. De Wachter. 1993. Reconstructing evolution from eukaryotic small-ribosomal-subunit RNA sequences: calibration of the molecular clock. Journal of Molecular Evolution **37:**221–232.

Van de Peer, Y., S. Chapelle, and R. De Wachter. 1996a. A quantitative map of nucleotide substitution rates in bacterial rRNA. Nucleic Acids Research **24:**3381–3391.

Van de Peer, Y., G. Van der Auwera, and R. De Wachter. 1996b. The evolution of stramenopiles and alveolates as derived by "substitution rate calibration" of small ribosomal subunit RNA. Journal of Molecular Evolution **42:**201–210.

Vawter, L., and W. M. Brown. 1993. Rates and patterns of base change in the small subunit ribosomal RNA gene. Genetics **134:**597–608.

Vogel, F., and G. Rohrborn. 1966. Amino-acid substitutions in haemoglobins and the mutation process. Nature **210:**116–117.

Wada, H., and N. Satoh. 1994. Details of the evolutionary history from invertebrates to vertebrates, as deduced from the sequences of 18S rDNA. Proceedings of the National Academy of Sciences U.S.A. **91:**1801–1804.

Wainright, P. O., G. Hinkle, M. L. Sogin, and S. K. Stickel. 1993. Monophyletic origins of the Metazoa: an evolutionary link with fungi. Science **260:**340–342.

Wakeley, J. 1996. The excess of transitions among nucleotide substitutions: new methods of estimating transition bias underscore its significance. Trends in Ecology and Evolution **11:**158–163.

Waters, D. A., M. A. Buchheim, R. A. Dewey, and R. L. Chapman. 1992. Preliminary inferences of the phylogeny of bryophytes from nuclear-encoded ribosomal RNA sequences. American Journal of Botany **79:** 459–466.

Watson, J. D., and F. H. C. Crick. 1953. A structure for deoxyribose nucleic acid. Nature **171:**737–738.

Wheeler, W. C., and R. L. Honeycutt. 1988. Paired sequence difference in ribosomal RNAs: evolutionary and phylogenetic implications. Molecular Biology and Evolution **8:**90–96.

Woese, C. R. 1987. Bacterial evolution. Microbiological Review **51:**221–271.

Woese, C. R., S. Winker, and R. R. Gutell. 1990. Architecture of ribosomal RNA: constraints on the sequence of "tetra-loops." Proceedings of the National Academy of Sciences U.S.A. **87:**8467–8471.

Wolters, J., and V. A. Erdmann. 1986. Cladistic analysis of 5S rRNA and 16S rRNA secondary and primary structure—the evolution of eukaryotes and their relation to Archaebacteria. Journal of Molecular Evolution **24:**152–166.

Yang, Z. 1993. Maximum likelihood estimation of phylogeny from DNA sequences when substitution rates differ over sites. Molecular Biology and Evolution **10:**1396–1401.

Yang, Z. 1994. Estimating the pattern of nucleotide substitution. Journal of Molecular Evolution **39:**105–111.

Yang, Z., and S. Kumar. 1996. Approximate methods for estimating the pattern of nucleotide substitution and the variation of substitution rates among sites. Molecular Biology and Evolution **13:**650–659.

Zimmer, E. A., R. K. Hamby, M. L. Arnold, D. A. LeBlanc, and E. C. Theriot. 1989. Ribosomal RNA phylogenies and flowering plant evolution. **In** The Hierarchy of Life, eds. B. Fernholm, K. Bremer, and J. Jörnvall, pp. 205–214. Elsevier Science Publishers, Amsterdam.

8

Molecular Phylogenetic and Evolutionary Studies of Parasitic Plants

Daniel L. Nickrent, R. Joel Duff, Alison E. Colwell, Andrea D. Wolfe, Nelson D. Young, Kim E. Steiner, and Claude W. dePamphilis

The parasitic nutritional mode is a frequently evolved adaptation in animals (Price, 1980), as well as in flowering plants (Kuijt, 1969). Heterotrophic angiosperms can be classified as either *mycotrophs* or as *haustorial parasites*. The former derive nutrients via a symbiotic relationship with mycorrhizal fungi. Haustorial parasites, in contrast, directly penetrate host tissues via a modified root called a haustorium and thereby obtain water and nutrients. Although such categories are often a matter of semantics, we use the term *parasite* in a strict sense to refer to haustorial parasites. Angiosperm parasites are restricted to the dicot subclasses Magnoliidae, Rosidae, and Asteridae; have evolved approximately 11 times; and represent approximately 22 families, 265 genera, and 4,000 species, that is, about 1% of all angiosperms (Fig. 8.1). Owing to their unique adaptations, parasitic plants have long been the focus of anatomical, morphological, biochemical, systematic, and ecological research (Kuijt, 1969; Press and Graves, 1995). For the vast majority of parasitic plants, negative effects upon the host are difficult to detect, yet others (e.g., *Striga, Orobanche*) are serious weeds of economically important crops (Kuijt, 1969; Musselman, 1980; Eplee, 1981; Stewart and Press, 1990; Press and Graves, 1995).

The degree of nutritional dependence on the host varies among haustorial parasites. Hemiparasites are photosynthetic during at least one phase of their life cycle and derive mainly water and dissolved minerals from their hosts. Obligate hemiparasites require a host plant to complete their life cycles whereas facultative hemiparasites do not. Hemiparasites can be found in Laurales (*Cassytha*), Polygalales (*Krameria*), and all families of Santalales. In Solanales (*Cuscuta*) and Scrophulariales, some species are chlorophyllous hemiparasites whereas other species are achlorophyllous holoparasites. Holoparasites represent the most extreme manifestation of the parasitic mode because they lack

This work was supported by grants from the National Science Foundation (DEB 94-07984 to DLN, DEB 91-20258 to CWD, and BIR 93-03630 to ADW), the Special Research Program of the Office of Research Development Administration, SIUC and the University Research Council of Vanderbilt University. Thanks go to C. Augspurger, W. Barthlott, J. Beaman, D. E. Bran, S. Carlquist, W. Forstreuter, J. Leebens-Mack, A. Markey, C. Marticorena, D. McCauley, S. Medbury, M. Melampy, Willem Meijer, B. Molloy, L. Musselman, R. Narayana, M. Nees, J. Paxton, S. Sargent, B. Swalla, W. Takeuchi, and M. Wetherwax for helpful discussions and/or for contributing plant material. The manuscript was improved by the critical comments of M. Bowe and an anonymous reviewer.

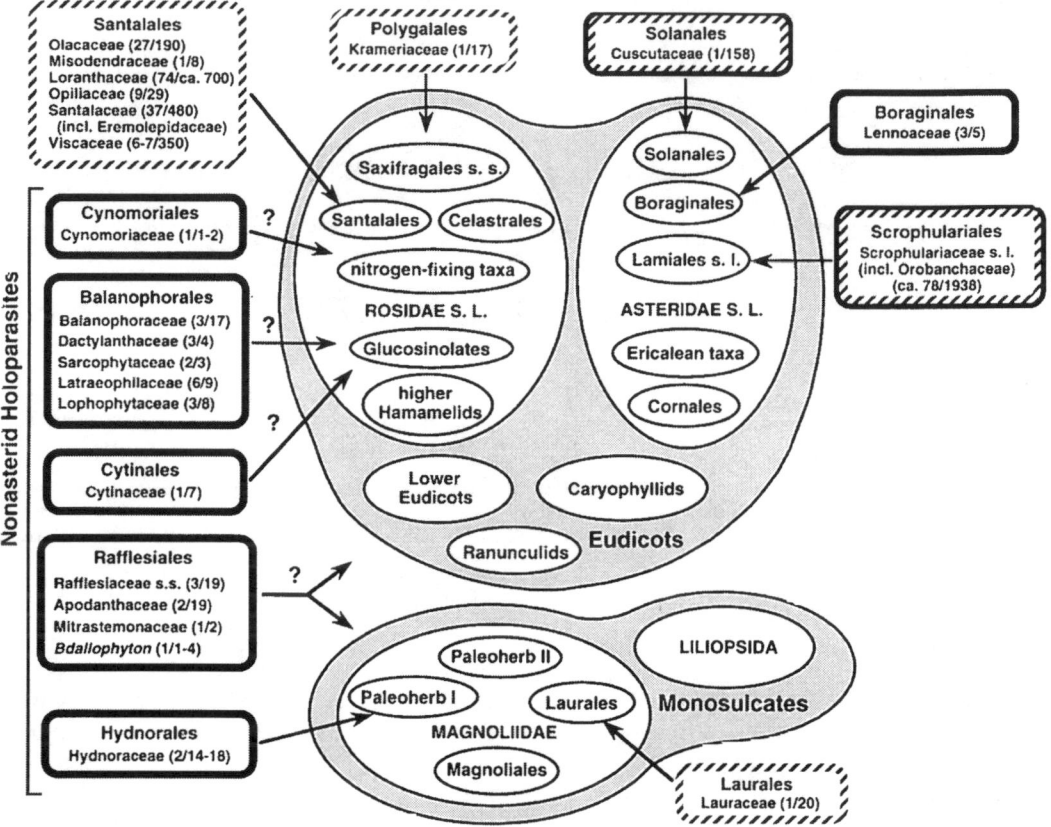

Figure 8.1. The distribution of haustorial parasitism among angiosperms. This generalized diagram incorporates information from global molecular phylogenetic studies using *rbcL* (Chase, Soltis, Olmstead et al. 1993) and nuclear 18S rDNA (Soltis, Soltis, Nickrent et al., 1997). No attempt was made to show all taxa, only to indicate groups that were supported by both studies. Hemiparasitic angiosperms are enclosed within dashed borders and holoparasites by black borders. Both trophic modes occur in Scrophulariaceae s. l. and *Cuscuta*. Arrows that touch a group indicate that strong evidence exists for the placement of that parasitic taxon within the group. Uncertain affinities are indicated by arrows with question marks. The familial classification of the nonasterid holoparasites is modified from Tahktajan (1987); however the placement of these orders is not concordant with his superordinal classification. The number of genera and species is indicated in parentheses following each family name. For Scrophulariaceae s. l., only the parasitic members are tabulated.

photosynthesis and must rely upon the host for both water and inorganic and organic nutrients. Six groups (orders or families—Fig. 8.1) are represented entirely by holoparasites: Balanophorales, Cynomoriaceae, Cytinaceae, Hydnoraceae, Lennoaceae, and Rafflesiales.

The relationships shown in Figure 8.1 are based upon those of Takhtajan (1987) as well as results of recent molecular analyses. For this paper, Santalales are considered, in a strict sense, to include Olacaceae, Misodendraceae, Loranthaceae, Opiliaceae, Santalaceae, and Viscaceae. This composition differs from that of Cronquist

(1988) by excluding Balanophoraceae, Medusandraceae, and Dipentodontaceae. Morphological, cytological, and molecular evidence all point toward the separation of Cynomoriaceae from Balanophoraceae and of Cytinaceae from Rafflesiaceae (Takhtajan, 1987; Nickrent and Duff, 1996; Pazy et al., 1996). Traditional classifications have often allied these holoparasites with Santalales; however, considerable variation can be seen in alternate classifications and such an affinity is not apparent from molecular investigations (see below). Given this, the term *nonasterid holoparasites* will be used to distinguish these plants from holoparasites in Asteridae such as those found in Scrophulariales, Boraginales, and Solanales. The nonasterid holoparasites are Balanophorales, Cynomoriales, Cytinales, Hydnorales, and Rafflesiales (Fig. 8.1). Results of molecular phylogenetic studies indicate that the nonasterid holoparasites are not closely related to each other (Nickrent and Duff, 1996), hence the term is applied for reference purposes only.

Additional parasitic plants can be found in Scrophulariaceae, a large family (over 250 genera) that is broadly and at present inexactly defined within Scrophulariales/Lamiales (Burtt, 1965; Cronquist, 1981; Thorne, 1992; Olmstead and Reeves, 1995). Although most species are completely autotrophic, the members of two tribes (Buchnereae, Pediculareae; [Pennell, 1935]) display a wide range of parasitic modes from fully photosynthetic, facultative hemiparasites, to nonphotosynthetic holoparasites. Orobanchaceae are a group of nonphotosynthetic holoparasites closely related to the holoparasitic Scrophulariaceae. A continuum of morphological and physiological traits unites the two, as do several "transitional genera" (*Harveya, Hyobanche,* and *Lathraea*) that have been classified alternatively in one family or the other (Boeshore, 1920; Kuijt, 1969; Minkin and Eshbaugh, 1989). Orobanchaceae are alternatively included within Scrophulariaceae (Takhtajan, 1987; Thorne, 1992) or recognized, by tradition, at the family level (Cronquist, 1981). Regardless of rank, most workers are in agreement that Orobanchaceae are derived from within the parasitic Scrophulariaceae. Numerous lines of evidence support this conclusion, including the shared presence of several morphological characters (Boeshore, 1920; Weber, 1980), pollen features, and a derived chloroplast DNA restriction site loss present in both groups (C. dePamphilis, unpubl.). For these reasons, this chapter considers Scrophulariaceae s. l. to include Orobanchaceae (Fig. 8.1).

In this chapter we discuss the results of macromolecular studies of the groups shown in Fig. 8.1. Most attention is, by necessity, directed toward four groups for which DNA sequence data are available, Scrophulariaceae s. l., *Cuscuta,* Santalales, and the nonasterid holoparasites. Our goals are to demonstrate the utility of these molecular markers in documenting phylogenetic relationships (at the genus level and above) and to show how parasitic plants represent unique models that can be used to study molecular evolutionary and genetic processes such as the structure, function, and evolution of plant genomes. For example, the continuum of trophic modes in Scrophulariaceae s. l. from nonparasitic to hemiparasitic to holoparasitic makes this group ideal for investigating questions concerning the evolution of parasitism and the molecular changes that accompany adaptation to a heterotrophic existence. Holoparasitic plants that show increased rates of molecular evolution pose particular problems for phylogenetic analysis but at the same time provide intriguing subjects for studying genome reorganizations that accompany the loss of photosynthesis.

Problems with the Classification of Parasitic Plants

The placement of many parasitic plants within the global angiosperm phylogeny is not disputed. For example, despite questions about familial boundaries and interfamilial relationships within Scrophulariales, it is clear that Scrophulariaceae are allied with other sympetalous dicots of Asteridae s. l. Such is not the case, however, for the nonasterid holoparasites whose higher-level classification still remains problematic. Two processes that occur during the evolution of advanced parasitism are reduction (and/or extreme modification) of morphological features and convergence. The first may involve loss of leaves, chlorophyll, perianth parts, or even ovular integuments. Loss of features confounds phylogenetic analysis unless clear transformation series are apparent. Convergences in parasitic plants are rampant because similar features have evolved in unrelated groups. For example, the squamate habit (i.e., with scale leaves) has evolved independently in aerial parasites of Viscaceae, Santalaceae, and Misodendraceae. One of the most striking examples of convergence involves *Cuscuta* (Asteridae) and *Cassytha* (Magnoliidae), two twining, yellow to orange parasites that are frequently confused based on a superficial examination of morphology. With reference to the internal haustorial tissues (endophytes) of Rafflesiaceae and Viscaceae, Cronquist (1981, p. 698) states ". . . parallelism in a number of features is in itself some indication of relationship, to be considered along with other evidence." Because parallelism involves similarities due to analogy *and* homology, and given that the common ancestry of the two families has not been established, this situation should best be described as convergence. A more serious problem is that this statement implies that all characters, whether they evolved via parallelism or convergence, should receive equal consideration when constructing a phylogeny.

The classifications of parasitic plants are often plagued by past misconceptions or overemphasis on a few conspicuous characters. The parasitic habit itself, known to have had multiple origins, was used by Cronquist (1981) to link Balanophoraceae, Hydnoraceae, and Rafflesiaceae to Santalales. It is not our contention that molecular data are more immune to the effects of convergence, parallelism, and reversal (i.e., homoplasy) than morphological characters, only that homologous DNA sequences provide additional, independent genetic characters that can be rigorously analyzed to test phylogenetic hypotheses. Furthermore, molecular data can be used to explore the dynamics of molecular evolution in parasitic plants by examining the genetic structure and biochemical processes underlying evolutionary change.

THE PLASTOME OF PARASITIC PLANTS

The chloroplast genomes (cpDNA) of photosynthetic plants are circular molecules ranging in size from about 120 to 217 kilobase pairs (kb) (Downie and Palmer, 1992). These genomes typically have a large (about 25 kb) inverted repeat that contains rRNA-encoding and other genes, and separates the large and small single-copy regions of the molecule. A typical chloroplast genome, such as the completely sequenced cpDNA of *Nicotiana* (Shimada and Sugiura, 1991), consists of about 112 genes and potentially functional open reading frames (ORFs). Although a substantial number of the genes (29,

approximately 25%) encode proteins for photosynthetic carbon fixation and electron transport, about half (60, 54%) encode proteins or RNAs involved in gene expression (transcription, translation, and related processes). An additional 11 genes (10%; the *ndh* genes) encode putative "chlororespiratory" proteins, based on their sequence similarity to mitochondrial *ndh* genes (Shimada and Sugiura, 1991). Ten or more large ORFs (and many smaller ones) also represent potentially functional protein genes, but their functions are largely unknown. Among the many hundreds of species whose cpDNAs have been characterized using restriction endonuclease mapping (Downie and Palmer, 1992), most retain a very similar gene content and gene order. Molecular systematic research on photosynthetic plants has extensively used phylogenetically informative variation in the presence and absence of restriction endonuclease sites, as well as a very large and rapidly growing comparative database of DNA sequences for the plastid genes *rbcL, matK,* and *ndhF,* and smaller databases for other plastid genes (see Chapter 1). Rare structural mutations such as inversions and insertions or deletions of genes or introns provide additional characters of special phylogenetic significance (reviewed by Downie and Palmer, 1992; see also Chapter 1).

Because many of the parasitic plants discussed herein are nonphotosynthetic, the more general term *plastid* will be used instead of *chloroplast* and ptDNA (plastid DNA or plastome) will be used instead of cpDNA (chloroplast DNA). Given that a large fraction of the plastome is devoted to photosynthetic function and to the expression of photosynthetic and other genes, parasitic plants provide a unique opportunity to determine the extent to which the evolutionary conservation of ptDNA is related to photosynthetic ability. In this regard, parasitic plants may be considered to be natural genetic mutants for the dissection of ptDNA function and evolutionary processes (dePamphilis, 1995, and references therein). At the same time, development of a phylogenetic framework to interpret significant variation among parasitic and nonparasitic lineages is complicated by the absence of many plastid genes from at least some parasitic lineages and by the acceleration of evolutionary rate for many other plastid and nuclear genes (see Rate Variation among Genomes and Lineages).

Epifagus

Epifagus virginiana (beechdrops, Scrophulariaceae s. l.) is a holoparasite native to eastern North America whose sole host is *Fagus grandifolia* (American beech). Although entirely lacking chlorophyll and photosynthetic ability, *Epifagus* retains plastids (Walsh et al., 1980) and ptDNA (dePamphilis and Palmer, 1989). Detailed mapping (dePamphilis and Palmer, 1990) and complete sequencing of the *Epifagus* plastome (Wolfe et al., 1992c) revealed a greatly reduced genome of only 70,028 bp, containing

just 42 intact genes. Surprisingly, although the *Epifagus* ptDNA had sustained a large number of deletions (and some small insertions as well) relative to another asterid, *Nicotiana,* the two genomes are almost entirely colinear (dePamphilis and Palmer, 1990). Complete sequencing revealed only one small inversion in *trnL* of the small single-copy region of *Epifagus* relative to *Nicotiana.* Furthermore, the parasite retains a nearly full-sized inverted repeat of 22,735 bp that separates the greatly reduced large and small single-copy regions of the genome.

Comparison of the ptDNA gene content of *Epifagus* with that of *Nicotiana* revealed a highly selective pattern of gene deletion in the holoparasitic plant (Table 8.1). All plastid-encoded photosynthetic genes have been deleted

Table 8.1. Plastid genes in *Epifagus* (42 total[a]) compared to *Nicotiana* (112 total).

	Gene Present in *Epifagus*	Deleted or Pseudogene (ψ) in *Epifagus*
Photosynthesis		
Photosystem I		*psaA, B, C, I, J*
Photosystem II		ψ*psbA,* ψ*B, C, D, E, F, psbH, I, J, K, L, M*
Cytochrome b/f		*petA, B, D, G*
ATP synthase		ψ*atpA,* ψ*B, E, F, H, I*
Calvin cycle		ψ*rbcL*
Chlororespiration		*ndhA,* ψ*B, C, D, E, F, G, ndhH, I, J, K*
Gene Expression		
rRNA	16S, 23S, 4.5S, 5S	
Ribosomal protein	*rps2, 3, 4, 7, 8, 11, rps14, 18, 19, rpl2, 16, 20, 33, 36*	*rps15, 16,* ψ*14, rpl22,* ψ*23, 32*
Transfer RNA	*trnD*$_{GUC}$*, E*$_{UUC}$*, F*$_{GAA}$*, trnH*$_{GUG}$*, I*$_{CAU}$*, L*$_{CAA}$*, trnL*$_{UAG}$*, M*$_{CAU}$*, N*$_{GUU}$*, trnP*$_{UGG}$*, Q*$_{UUG}$*, R*$_{ACG}$*, trnS*$_{GCU}$*, S*$_{UGA}$*, W*$_{CCA}$*, trnY*$_{GUA}$*, trnfM*$_{CAU}$	ψ*trnA*$_{UGC}$*,* ψ*C*$_{GCA}$*, G*$_{GCC}$*, trnG*$_{UCC}$*,* ψ*I*$_{GAU}$*, K*$_{UUU}$*, trnL*$_{UAA}$*,* ψ*R*$_{UCU}$*,* ψ*S*$_{GGA}$*, trnT*$_{GGU}$*, T*$_{UGU}$*, V*$_{GAC}$*, trnV*$_{UAC}$
RNA polymerase		ψ*rpoA, B, C1, C2*
Maturase	*matK*	
Initiation factor	*infA*	
Other protein genes	*clpP, accD,* ORF1738[b], ORF2216[b]	ORF29, 31, 34, 62, 168, ORF184, 229, 313

[a]Data based on gene mapping (dePamphilis and Palmer, 1990) and complete sequence (Wolfe et al., 1992c). All genes listed are present in *Nicotiana* chloroplast DNA except *infA,* which is a pseudogene in that plant (Wolfe et al., 1992c).

[b]ORF1738 and ORF2216 in *Epifagus* are homologs of *Nicotiana* ORF1901 and ORF2280, respectively (Wolfe et al., 1992c).

from *Epifagus* or remain as small fragments or pseudogenes. Similarly, the 11 *ndh* genes have been lost. A majority of genes of unknown function (ORFs 168, 184, 229, and 313, plus a host of smaller ones) has been completely lost from *Epifagus*. In contrast, 38 (or over 90%) of the retained genes represent components of the plastid machinery for gene expression. All of the four ribosomal RNA genes are retained, as are intact copies of 13 of 19 plastid-encoded ribosomal protein genes and 17 of 30 plastid tRNA genes. The implications of these results are that ptDNA can evolve very rapidly under certain predictable conditions, and that DNA deletions, both large and small, are a dominant mode of structural evolution in that molecule. This is a striking example of the role natural selection plays in shaping genome structure and suggests that rapid structural alteration is possible when photosynthetic constraints are altered.

These results also have important implications for our understanding of the function of ptDNA. RNA polymerase subunit genes (*rpoA, B, C1,* and *C2*), which encode the central enzymes of transcription, are present in tobacco but absent in *Epifagus*. If plastid-encoded RNA polymerase is solely responsible for ptDNA transcription, we would expect the beechdrops ptDNA to be unexpressed. To the contrary, despite the loss of all RNA polymerase genes, the plastome of *Epifagus* is clearly transcribed and is probably also translated (Wolfe et al., 1992c). The evidence comes from several sources, including direct observation of transcripts of plastid rRNA genes (dePamphilis and Palmer, 1990), ribosomal protein genes (Ems et al., 1995), and additional plastid ORFs (Ems et al., 1995). *Epifagus* plastid RNAs are subject to appropriate intron splicing and even experience RNA editing, as observed in other plastid RNAs (Ems et al., 1995). Furthermore, the highly specific pattern of gene retention and bias of synonymous over nonsynonymous base substitutions, particularly in large genes subject to significant levels of sequence divergence (Wolfe et al., 1992b, 1992c; dePamphilis et al., 1997) is clearly indicative of genes that have evolved under functional constraint.

That gene expression in plastids is linked to nuclear-encoded genes is well documented (Mayfield et al., 1995). Expression of plastid genes in the absence of plastid-encoded RNA polymerase would imply that another RNA polymerase, presumably of nuclear origin, is involved in the transcription of beechdrops ptDNA (Morden et al., 1991; Ems et al., 1995). It is interesting to note also that virtually all of the proposed eubacterial-type -35 and -10 promoter regions have either diverged or been deleted in *Epifagus* (Wolfe et al., 1992a, 1992c), consistent with the possibility that these genes now use different promoters, possibly ones capable of interacting with an RNA polymerase of nonplastid origin. Similarly, the loss of 13 of the 30 normally plastid-encoded tRNA genes would suggest that plastid translation, to the extent that it occurs, must also require the import of some tRNAs from outside the plastid (Wolfe et al., 1992c). Finally, the retention of just four protein-coding genes not involved in gene expression (*clpP, accD,* ORF1738, and ORF2216) in the plastome of *Epifagus* suggests that one or more of these genes encodes a protein required for a nonbioenergetic (i.e., not involved in photosynthetic or chlororespiratory) function in the *Epifagus* plastid and may be the raison d'être for the retention of a functional ptDNA (Wolfe et al., 1992c; dePamphilis, 1995).

A related study investigated the phylogenetic distribution of one of the lost tRNA genes, tRNA-cys, in 11 parasitic and nonparasitic plants related to *Epifagus* (Taylor et al., 1991). The polymerase chain reaction was used to amplify fragments that would contain the tRNA-cys locus from each plant. The fragments were then cloned and sequenced to reveal that all of the photosynthetic plants had a normal tRNA-cys gene, whereas all of the nonphotosynthetic plants had lost the gene. This result suggests that the tRNA-cys gene was lost at about the same time as photosynthesis was lost and that the two events may be causally related.

Conopholis

Concurrent with the early work on *Epifagus,* the plastome of another orobanchaceous holoparasite, *Conopholis americana* (squawroot), was being examined at the molecular level. Using heterologous probes, Wimpee et al. (1991) documented the modification or absence of many photosynthetic genes. Later, the rDNA operon was cloned and sequenced, thereby demonstrating that the 16S/23S spacer lacked tRNA genes and contained substantial deletions (Wimpee et al., 1992a). Experiments using the polymerase chain reaction and hybridizations failed to detect the lost tRNA genes elsewhere in the squawroot genome (Wimpee et al., 1992b). Colwell (1994) conducted restriction site mapping of the plastome of squawroot, documenting its size as 43 kb, making it the smallest ptDNA molecule then observed in plants. This genome is only slightly smaller than that of *Epifagus* if the length of one inverted repeat is removed from the latter genus (70 kb − 22.7 kb = 47.3 kb).

Cuscuta

Other molecular genetic investigations have focused on *Cuscuta,* a genus sometimes placed in its own family (Cuscutaceae) but widely recognized to share a common ancestor with nonparasitic Convolvulaceae. *Cuscuta,* like Scrophulariaceae, includes hemiparasitic species (e.g., *C. reflexa*) as well as holoparasitic species (e.g., *C. europaea*) that lack thylakoids, chlorophyll, RUBISCO (ribulose bisphosphate carboxylase/oxygenase) and light-dependent CO_2 fixation, but (strangely) retain *rbcL* (Machado and Zetsche, 1990). Although the plastome of *Cuscuta* is yet to be sequenced completely, significant progress is being made. A 6-kb portion of ptDNA from *C. reflexa* that includes *petG, trnV, trnM, atpE, atpB,* and *rbcL* was sequenced (Haberhausen et al., 1992; Haberhausen and Zetsche, 1992). Further work on this species (Bömmer et al., 1993) resulted in sequences for 16S rDNA, *psbA, trnH,* ORF740, ORF77, *trnL,* and ORF55. Later, a 9-kb portion of ptDNA from this same species that included a large portion of inverted repeat A spanning a segment from *trnA* to *trnH* was cloned and sequenced (Haberhausen and Zetsche, 1994) Although some short sequences were identical to those of *Nicotiana* (e.g., *trnI*), many deletions were observed. For example, *rpl2* and *rpl23* were both deleted, and ORF2280 was reduced to only 740 bp. These results show that, like *Epifagus, C. reflexa* has experienced major deletions in the plastome. The complete loss of ribosomal protein genes such as *rpl2* invites questions about how the translational apparatus of the plastid functions and its relationship to the nuclear and mitochondrial genomes. The plastid genome of the holoparasite *C. europaea* has undergone additional deletions in comparison to *C. reflexa,* such as the loss of the cis-spliced intron of *rps12* (Freyer et al., 1995). Despite this alteration, correctly processed transcripts and ribosomal proteins are produced, thus providing evidence that the translational apparatus of the plastid may remain functional following significant structural alterations.

Do the Nonasterid Holoparasites Retain a Plastome?

As described in the foregoing discussion, the plastomes of *Epifagus, Conopholis,* and *Cuscuta* have experienced extensive losses of photosynthetic genes, yet all have retained intact and functional ribosomal cistrons. It was therefore reasoned that if any portion of the plastome remained in nonasterid holoparasites, it would likely be the ribosomal cistron. The first evidence that rDNA was present in these holoparasites was obtained when plastid 16S rDNA from *Cytinus* (Cytinaceae) was amplified and sequenced (Nickrent et al., 1995; Nickrent and Duff, 1996). In pairwise comparisons to other angiosperms, this 16S sequence contained approximately three times more base substitutions, yet this rRNA retained all major secondary structural features (because most changes were compensatory), thus suggesting functionality. Since these initial studies, 16S rDNA sequences have been obtained from all nonasterid holoparasite lineages (Nickrent et al., 1997a). These represent the most diverse 16S rRNA sequences documented among angiosperms and contain several structural features unknown among land plants.

Southern blots using both homologous and heterologous probes developed for five plastid genes (16S rDNA and four ribosomal protein genes) resulted in positive hybridizations to digested total genomic DNA from *Cytinus* (Nickrent et al., 1997b). These preliminary data suggest the *Cytinus* genome is approximately 20 kb in size, thus the smallest yet documented for angiosperms. Hybridizations using more derived holoparasites such as *Rafflesia* and *Balanophora* were negative, thus suggesting their plastome is either absent or even more extensively modified than that of *Cytinus*. Recently, sequences of PCR-amplified products derived from the plastid 16S–23S spacer have been obtained from *Cytinus* and *Cynomorium* (D. Nickrent and R. Duff, unpubl.), further suggesting the retention of a plastome in these plants.

MOLECULAR PHYLOGENETIC STUDIES OF SCROPHULARIACEAE S. L.

Systematic and Phylogenetic Problems

Until recently, most opinions on the circumscription of Scrophulariaceae could be traced to concepts from the nineteenth century, such as the recognition of three subfamilies (Pseudosolaneae, Antirrhinoideae, and Rhinanthoideae) and a distinct Orobanchaceae. Rhinanthoideae were traditionally considered to include the parasitic tribes Buchnereae and Euphrasieae, as well as the nonparasitic tribes Veronicae and Digitaleae. Anatomical evidence was used to refute a presumed relationship between Rhinanthoideae and the slightly zygomorphic Pseudosolaneae and Solanaceae (Thieret, 1967). Other family-level issues concern the presumed "edges" of the family, including whether to recognize the Selagineae and/or the Globulariaceae as formal taxa within Scrophulariaceae. A general trend has been the recognition of various small tribes (including Angeloneae, Russelieae, Collinsieae, Melospermeae, and many others) and especially the splitting of three large and heterogeneous tribes, the Digitaleae (Hong, 1984), Gratioleae, and Cheloneae (Thieret, 1967) into smaller, presumably monophyletic tribes. Attempts to analyze cladistically morphological variation across the entire family have met with little success, leading An-Ming (1990) to speculate about the possible paraphyly, or even polyphyly, of the family as presently defined.

Although a close relationship between Orobanchaceae and parasitic Scrophulariaceae seems assured, genera such as *Harveya*, *Hy-*

obanche, and *Lathraea*, have been alternatively placed in either family (Kuijt, 1969). Does the evolutionary series nonparasite·hemiparasite· holoparasite explain the pathway to all holoparasites? Are the transitional genera and Orobanchaceae monophyletic or are they independent lineages that have converged on a nonphotosynthetic lifestyle? If the latter is true, how many times has photosynthesis been lost? The availability of numerous extant species displaying all modes of parasitism makes these questions tractable via systematic analyses.

The molecular phylogeny of parasitic Scrophulariaceae has been examined using sequences from three plastid genes: *rps2*, *matK*, and *rbcL*. The first two are relatively rapidly evolving plastid genes that encode nonphotosynthetic proteins (see Chapter 1) and that we have found are present in all parasitic Scrophulariaceae regardless of photosynthetic ability. Although the photosynthetic gene *rbcL* is absent from some holoparasites (see Phylogeny of Scrophulariaceae s. l. using *rbcL*), it is present in most taxa. Phylogenetic analyses have revealed much about the molecular evolution of this key plastid gene in relation to photosynthesis.

Phylogeny of Scrophulariaceae s. l. Using rps2

Phylogenetic analysis of plastid ribosomal protein gene *rps2* was conducted for 55 species of Scrophulariaceae and related families (Fig. 8.2), including 32 species of parasites (dePamphilis et al., 1997). Nonparasites were selected to reflect the patterns shown following analysis of a larger data set (58 genera of nonparasitic Scrophulariaceae and relevant outgroup taxa; C. dePamphilis unpubl.). Parsimony analysis of *rps2* identifies a single clade (bootstrap value of 53%) containing a broad sampling of parasitic Scrophulariaceae s. l. With the exception of *Lathraea*, the "transitional" genera *Harveya* and *Hyobanche* are present on the *rps2* consensus tree as a clade (88% bootstrap support) of hemiparasitic Scrophulariaceae; without these genera, the remaining orobanchaceous taxa are weakly supported as monophyletic. These data indicate that hemiparasites gave rise to holoparasites a minimum of five separate times (dePamphilis et al.,

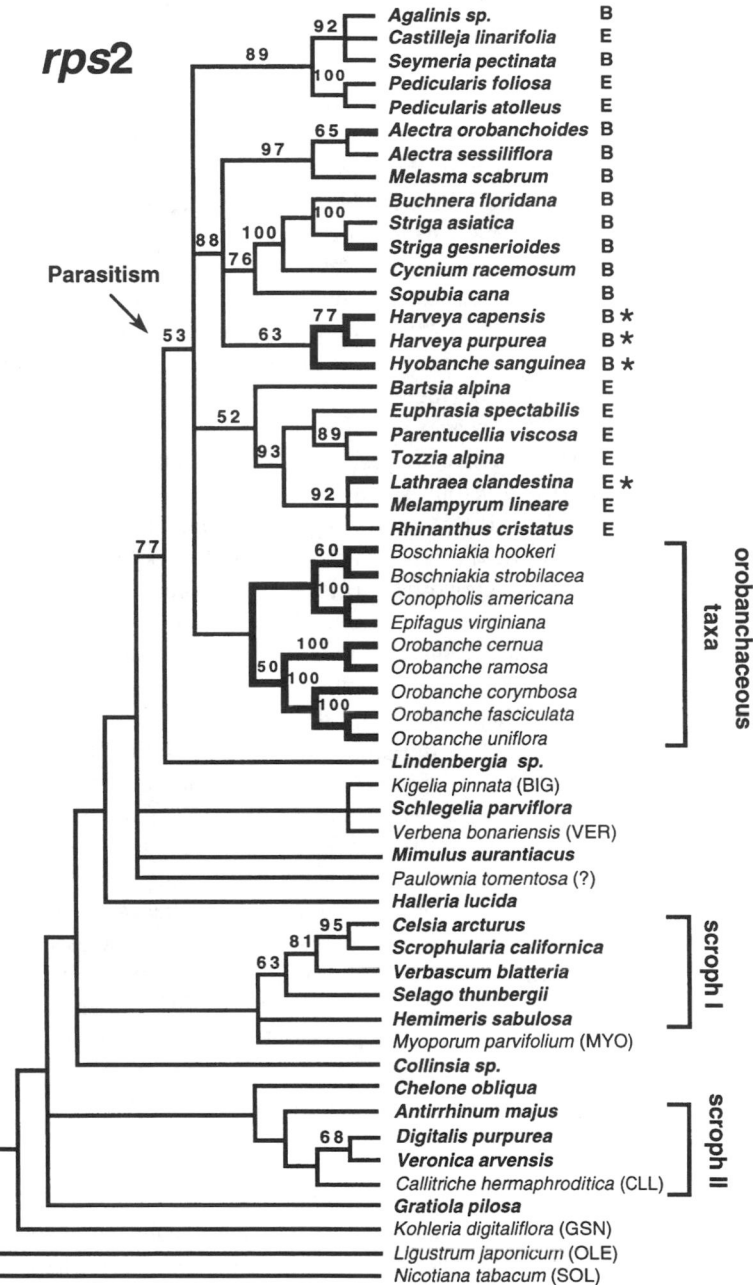

Figure 8.2. Phylogenetic reconstruction of Scrophulariaceae and outgroups based on maximum parsimony analysis of plastid *rps2* gene sequences. The tree was constructed using 174 phylogenetically informative base substitutions. Strict consensus of 78 trees at 660 steps (Consistency Index [CI] without autapomorphies = 0.492; Retention Index [RI] = 0.657). Bootstrap values shown above branches; nodes without values were supported in less than 50% of the replications. The arrow indicates a parasite clade including genera traditionally assigned to Orobanchaceae. Nonphotosynthetic holoparasites are indicated by thick lines. Taxa in bold font are traditionally placed in Scrophulariaceae. The traditional parasitic tribes Buchnereae and Euphrasieae are indicated by B and E, respectively. Taxa followed by an asterisk (*) are "transitional" genera classified alternatively in Scrophulariaceae or Orobanchaceae. Scroph I and II refers to clades identified in Olmstead and Reeves (1995). Abbreviations for outgroups and families in Scrophulariales: BIG = Bignoniaceae; CLL = Callitrichaceae, GSN = Gesneriaceae; MYO = Myoporaceae; OLE = Oleaceae; SOL = Solanaceae; VRB = Verbenaceae. The "?" for *Paulownia* indicates its higher-level classification is uncertain.

1997). Seven of the nine genera classified as members of Euphrasieae occur in a single clade; the remaining two genera (*Castilleja* and *Pedicularis*) occur in a clade with two genera of Buchnereae (*Agalinis* and *Seymeria*).

In contrast to the apparent monophyly of the parasitic taxa, the nonparasitic Scrophulariaceae are only weakly resolved with *rps2* and appear as a complex paraphyletic assemblage with representatives of other families in Scrophulariales such as Bignoniaceae, Verbenaceae, and Myoporaceae. Low bootstrap values suggest that the power of this short gene to resolve phylogenetic relationships in the nonparasitic lineages is low. Yet, two of the nonparasitic lineages recovered correspond to the scroph I and scroph II clades identified by Olmstead and Reeves (1995) in a phylogenetic analysis using *rbcL* and *ndhF* sequences. Other genera not sampled by Olmstead and Reeves (*Gratiola, Collinsia, Halleria, Mimulus,* and *Lindenbergia* plus the entire parasitic lineage) are not components of these clades. These results suggest that at least three and possibly additional major lineages may represent what has now clearly been identified as a polyphyletic Scrophulariaceae (Olmstead and Reeves, 1995). Neither *Veronica* nor *Digitalis,* genera usually included within the rhinanthoid subfamily, are close relatives of the parasitic plants. In contrast, *Lindenbergia,* a nonparasitic genus originally classified in the Gratioleae (Wettstein, 1897), occurs as sister to all other parasitic Scrophulariaceae.

Phylogeny of Scrophulariaceae s. I. Using rbcL

Studies employing *rbcL* sequencing for phylogenetic reconstruction in Scrophulariaceae have been, until recently, restricted to nonparasitic plants (Olmstead and Reeves, 1995; Wolfe and dePamphilis, 1995, 1997; dePamphilis et al., 1997). Recent studies by Wolfe and dePamphilis (1995, 1997) have revealed accelerated synonymous and nonsynonymous substitution rates for *rbcL* in several lineages of derived members of Scrophulariaceae. In comparison to nonparasitic plants of Scrophulariales, the evolutionary rate of *rbcL* in those parasites that retain this gene is faster, thus enabling more resolved phylogenetic

reconstructions. The *rbcL* locus from 30 additional taxa representing 20 genera of Scrophulariaceae has been sequenced (Wolfe and dePamphilis 1995, 1997). Phylogenetic reconstruction based on the new sequences and those from Olmstead and Reeves (1995) reveals that Scrophulariaceae, as traditionally circumscribed, are not monophyletic (Fig. 8.3). Although Olmstead and Reeves (1995), arrived at the same conclusion based on a combined analysis of *rbcL* and *ndhF* sequences, specific relationships between major clades differ between the two studies. The "scroph I" and "scroph II" clades were distinct in Olmstead and Reeves (1995) but in the present analysis they occur in one clade with representatives of other Scrophulariales.

One of the major results of the phylogenetic analysis based on both *rps2* (Fig. 8.2) and *rbcL* (Fig. 8.3) sequences is that there is a single origin of the parasitic habit in Scrophulariaceae. Taxa traditionally placed in Orobanchaceae are intercalated with hemi- and holoparasitic genera of the Scrophulariaceae, hence Orobanchaceae (as traditionally defined) are paraphyletic. The genus *Orobanche* is also not monophyletic according to the *rbcL* strict consensus tree (Fig. 8.3). The *rps2* results suggested at least five independent losses of photosynthesis. A strict interpretation of the *rbcL* tree indicates seven losses of photosynthesis, assuming that it is unlikely that the nonphotosynthetic ancestors of *Orobanche, Boschniakia,* and others gave rise to photosynthetic hemiparasites. Some of these relationships, such as lack of monophyly for *Orobanche* or a distant relationship between *Harveya* and *Hyobanche,* are not in accord with *rps2* or 18S rDNA data (see next section); hence further examination of the phylogenetic utility of *rbcL* in these plants is in order.

In those holoparasites where *rbcL* no longer produces a functional product, it is considered a pseudogene. In most cases of pseudogene formation, a functional gene is retained in addition to a degenerated duplicated copy (e.g., β-globin genes [Li, 1983]); however, in Scrophulariaceae no duplication has occurred, and only the primary *rbcL* pseudogene remains. These data indicate that primary pseudogene formation occurred independently three times (Fig. 8.3). The most surprising discovery from this *rbcL* phylogenetic analysis is

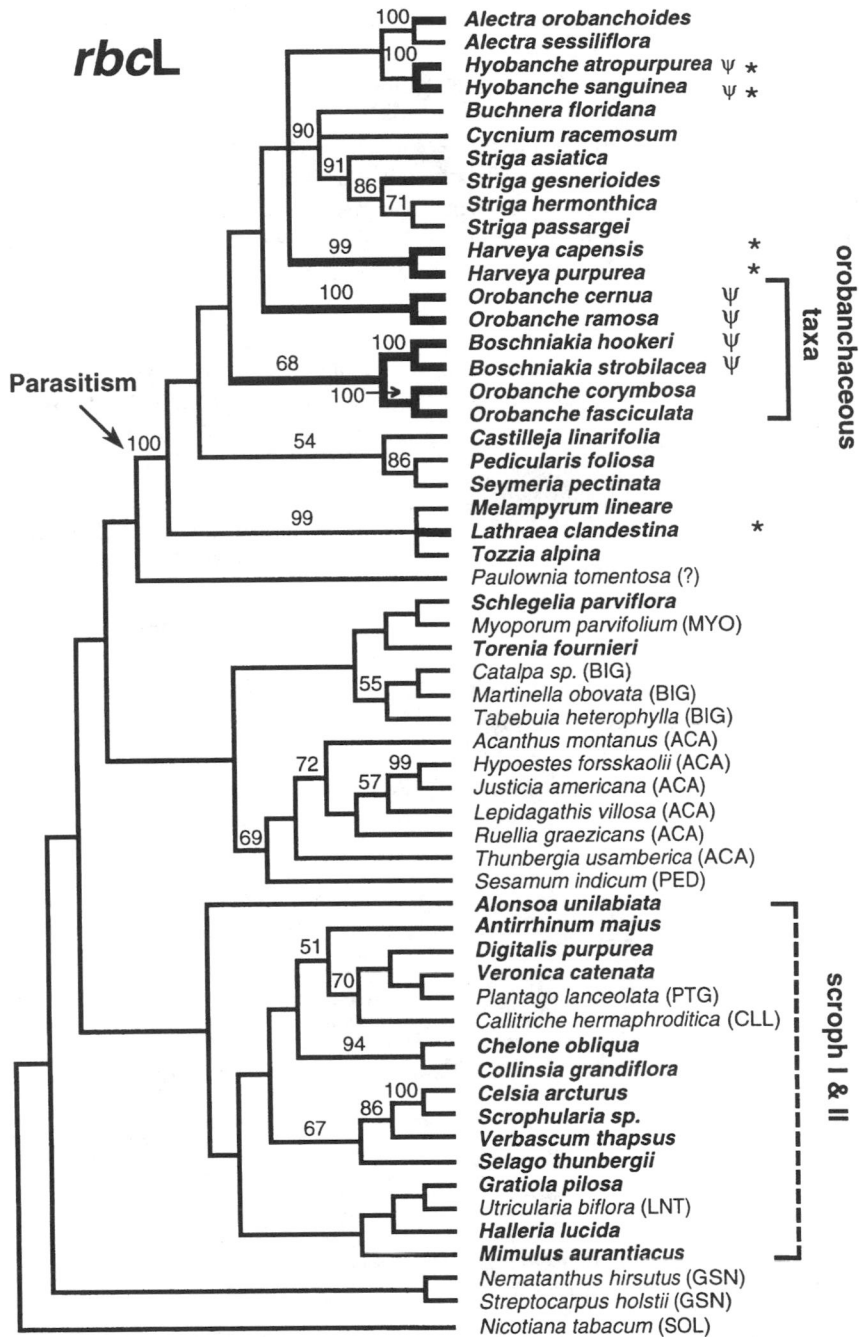

Figure 8.3. Phylogenetic reconstruction of Scrophulariaceae based on maximum parsimony analysis of *rbcL* nucleotide sequences. Taxa with primary pseudogenes are indicated as ψ (see Fig. 8.6 and text). Strict consensus of 12 most-parsimonious trees with 1,482 steps (CI = 0.550, RI = 0.521). The arrow indicates a parasite clade including genera traditionally assigned to Orobanchaceae. Nonphotosynthetic holoparasites are indicated by thick lines. Taxa in bold font are traditionally placed in Scrophulariaceae. The clade identified as scroph I and II refers to taxa described in Olmstead and Reeves (1995), but here also includes representatives of other families. Taxa followed by an asterisk (*) are "transitional" genera classified alternatively in Scrophulariaceae or Orobanchaceae. Abbreviations for families in Scrophulariales in addition to those in Fig. 8.2: ACA = Acanthaceae, GLB = Globulariaceae, LNT = Lentibulariaceae, PED = Pedaliaceae, PTG = Plantaginaceae.

that many holoparasitic Scrophulariaceae (e.g., *Alectra, Harveya, Lathraea, Orobanche,* and *Striga*) retain an open reading frame for the gene (Fig. 8.3). The independence of structural mutations among and/or within genera for the pseudogene sequences (see Structural Analysis of RUBISCO) suggests that disruption of the gene occurred independently following the loss of photosynthesis in *Boschniakia, Conopholis, Hyobanche, Epifagus,* and *Orobanche* (dePamphilis and Palmer, 1990; Wolfe et al., 1992c; Colwell, 1994; Wolfe and dePamphilis, 1997). Combined, these data suggest that a *rbcL* ORF may have been retained in these holoparasitic lineages after the loss of photosynthesis and that perhaps the gene was functional at some point following the adaptation to heterotrophy.

Phylogeny of Scrophulariaceae s. l. Using Nuclear 18S rDNA

A general review of the organization of nuclear 18S rDNA and its use in phylogenetic studies of plants can be found in Chapter 1. Phylogenetic relationships among hemi- and holoparasitic Scrophulariaceae using 18S rDNA sequences

were determined by Colwell (1994). That study examined nine parasitic genera; subsequent sequencing has increased to 18 the number of 18S rDNA sequences of Scrophulariaceae s. l. now available for analysis (Nickrent and Duff, 1996). Using *Glycine, Lycopersicon,* and *Ipomoea* as outgroups, a clade is retrieved (Fig. 8.4) containing the nonparasitic *Linaria* and *Chionophila;* this clade is sister to the remaining Scrophulariaceae s. l. These results based on 18S rDNA sequences indicate, as did *rps2* and *rbcL,* that parasitism arose just once in this group. Although sampling was limited compared to analyses of the chloroplast genes, a number of relationships are concordant, such as clades containing: (1) *Pedicularis, Orthocarpus,* and *Castilleja;* (2) *Harveya* and *Hyobanche;* and (3) *Epifagus* and *Conopholis.* Unlike the results from the chloroplast genes, 18S rDNA sequence analysis resulted in a strong association (91% bootstrap support) between the "transitional" genus *Lathraea* and an *Orobanche* clade that is composed of only North American taxa. The rDNA tree also indicates (as does *rps2*) that this clade is not closely related to the North American genera *Epifagus, Conopholis,* and *Boschniakia,* making Orobanchaceae (in the tra-

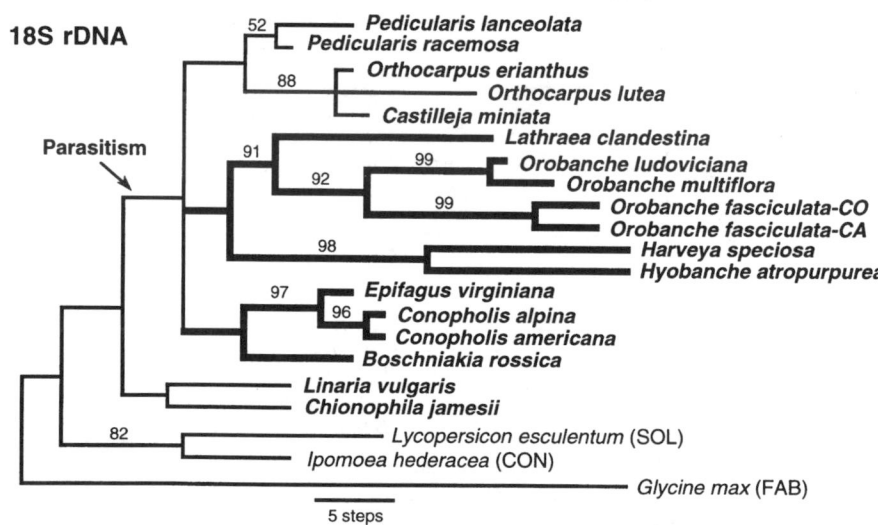

Figure 8.4. Parsimony analysis of nuclear 18S rDNA sequences from Scrophulariaceae s. l. Strict consensus phylogram of four trees at 246 steps (CI without autapomorphies = 0.553; RI = 0.643. Bootstrap values (from 100 replications) shown above branches; nodes without values were supported in less than 50% of the replications. The arrow indicates a parasite clade including genera traditionally assigned to Orobanchaceae. Nonphotosynthetic holoparasites are indicated by thick lines. *Linaria* and *Chionophila* are nonparasitic Scrophulariaceae. *Lycopersicon* (Solanaceae), *Ipomoea* (Convolvulaceae) and *Glycine* (Fabaceae) were included as outgroups.

ditional sense) paraphyletic. Although nuclear 18S rDNA sequences usually provide insufficient data at the rank of family and below, increased rates of nucleotide substitution, especially in orobanchaceous and transitional genera, have allowed greater resolution of some relationships. From the same taxa sampled for *rps2* and *rbcL*, 18S rDNA sequences should be obtained, thereby allowing direct comparisons of resulting trees and analyses of combined data sets.

MOLECULAR PHYLOGENETIC STUDIES OF SANTALALES AND NONASTERID HOLOPARASITES

Systematic and Phylogenetic Problems

Traditional classifications place Santalales in Rosidae, often near Celastrales (Cronquist, 1981; Takhtajan, 1987). In the system of Cronquist (1981), Santalales comprise ten families (including Balanophoraceae) and this order is allied with Rafflesiales (Rafflesiaceae and Hydnoraceae). The phylogenetic system proposed by Takhtajan (1987) is similar, except for the placement of the nonasterid holoparasites (Fig. 8.1). Takhtajan (1987) places Rafflesiaceae, Hydnoraceae, and Balanophoraceae (minus Cynomoriaceae) in subclass Magnoliidae, not Rosidae. For Santalales, concepts of interfamilial relationships date to the nineteenth century. A noncladistic phylogeny of the order was proposed by Fagerlind (1948); however, this system was overly influenced by the presence of *Allium* type embryo sacs and reductions seen in the gynoecium, thereby resulting in unlikely relationships such as the derivation of Balanophoraceae from Viscaceae. As with others before and after him, Fagerlind's system failed to deal with convergent features in unrelated groups.

At least three systems for interfamilial relationships in Santalales have been proposed that differ mainly in their circumscription of Eremolepidaceae and Santalaceae. The first places Santalaceae as sister to Viscaceae and Eremolepidaceae as sister to Loranthaceae (Kuijt, 1968, 1969). The second also places Santalaceae as sister to Viscaceae, but does not consider Eremolepidaceae as distinct from the former family (Wiens and Barlow, 1971). The third hypothesis (Bhandari and Vohra, 1983) allies Loranthaceae with Viscaceae (the latter including Eremolepidaceae). All three of these systems include Opiliaceae as a part of Olacaceae; the latter family being the least derived in the order because it contains both parasitic and nonparasitic members. These differing hypotheses have provided the impetus for molecular phylogenetic investigations (Nickrent and Franchina, 1990).

Phylogeny of Santalales Using Nuclear 18S rDNA and rbcL

At present, 62 nuclear 18S rDNA (Nickrent and Duff, 1996) and 37 *rbcL* sequences exist for representatives of six families of Santalales (Fig. 8.1). For Viscaceae and Misodendraceae, all genera have been sampled. Phylogenetic analyses for the same suite of 37 taxa using 18S rDNA and *rbcL* sequence data, separately and in combination, allow the utilities of these molecules to be compared directly (Table 8.2). The

Table 8.2. Comparison of the results of phylogenetic analyses of Santalales using 18S rDNA, *rbcL*, and 18S rDNA/*rbcL* combined.[a]

Tree Characteristic	18S rDNA	*rbcL*	18S + *rbcL*
Length of shortest tree (steps)	1047	1134	2212
Number of shortest trees	15	8	11
Number of characters	1741	1410	3151
Number of variable characters	433	468	901
Number of potentially informative characters	226	275	501
Consistency Index/C.I. minus uninformative sites	0.5/0.39	0.54/0.44	0.51/0.40
Homoplasy Index/H.I. minus uninformative sites	0.5/0.61	0.46/0.56	0.48/0.60
Retention Index	0.57	0.61	0.58
Rescaled Consistency Index	0.29	0.33	0.30

[a]Sequences for the same suite of 38 taxa were used in a heuristic search using PAUP.

number of variable characters and the number of phylogenetically informative characters for the two molecules are similar, as are the consistency indices. The analysis of the 37-taxon 18S rDNA data set yields trees generally concordant with those obtained following analysis of the 62-taxon matrix. Trees summarizing family-level relationships derived from the 18S rDNA and *rbcL* analyses are generally quite similar and differ mainly in the placement of Opiliaceae (Fig. 8.5). Given the overall concordance between the topologies, the data sets were combined and analyzed together. The topology of the combined-analysis phylogram is congruent with that obtained from the separate 18S rDNA data set, including the position of Opiliaceae; hence it will be used in the following discussion of relationships.

In agreement with traditional phylogenetic systems, Olacaceae represent the least derived family in the order. Despite limited generic sampling, neither 18S nor *rbcL* phylograms indicate that the family is monophyletic (Fig. 8.5). In the combined 18S/*rbcL* analysis, *Schoepfia* is sister to *Misodendrum,* and this clade is sister to Loranthaceae. Anatomical evidence supporting the distinctiveness of *Schoepfia* from other Olacaceae include its possession of aliform-confluent parenchyma (Sleumer, 1984) and tracheid/vessel features (Reed, 1955). The latter study also noted similarities in the pollen of *Schoepfia* with the more derived family Santalaceae. Taken as a whole, it is worth considering possible phylogenetic relationships between *Schoepfia* and genera of Loranthaceae that are root parasites. Biogeographic information and the long branch connecting *Schoepfia* with *Misodendrum* suggest the latter genus may represent a relictual taxon that evolved on the Gondwanan landmass along with the aerial parasites of Loranthaceae. The three genera of Loranthaceae used for the combined analysis form a clade (100% bootstrap) that is sister to the *Misodendrum/Schoepfia* clade. Analysis of a larger 18S data set (23 of the 74 genera, data not shown) resolves two clades, one composed of New World and the other of Old World mistletoes (Nickrent and Duff, 1996). A more rapidly evolving gene is required to provide sufficient phylogenetic signal to address relationships among all genera of this family. The combined analysis indicates that Opiliaceae form a well-supported (100% bootstrap) clade that is sister to the remaining members of the order. In contrast to traditional classifications, sequence data indicate that Opiliaceae are distinct from and evolutionarily more derived than Olacaceae. Using *rbcL* data alone, Opiliaceae are intercalated between a group of New World and Old World Santalaceae (the latter including Eremolepidaceae).

Analyses of 18S rDNA and *rbcL* sequences separately and in combination show that Santalaceae are not monophyletic but a grade that culminates in Viscaceae (Fig. 8.5). Two groups of Santalaceae are resolved representing mainly New World genera (*Acanthosyris, Buckley, Jodina,* and *Pyrularia*) and Old World genera (*Exocarpos, Osyris,* and *Santalum*) plus Eremolepidaceae (*Antidaphne, Eubrachion,* and *Lepidoceras*). These latter three genera are not monophyletic using either molecular data set, thus supporting the concept of Wiens and Barlow (1971) who used karyological and morphological evidence to suggest that Eremolepidaceae are aerial parasites of Santalaceae. Viscaceae are the most derived family of the order, and bootstrap support for the clade is high using either 18S or *rbcL* sequences. Molecular analyses indicate that Viscaceae is distinct from Loranthaceae, as was previously shown by data derived from biogeography, floral morphology, and cytology (Barlow, 1964; Barlow and Wiens, 1971; Wiens and Barlow, 1971; Barlow, 1983). Strongly supported intergeneric relationships are obtained only for *Phoradendron/Dendrophthora* and *Korthalsella/Ginalloa*. Despite complete generic sampling in the family, 18S and *rbcL* sequence data, and accelerated substitution rates, relationships among all genera are not fully resolved, suggesting that the evolution of the viscaceous genera may represent a "hard polytomy" (Maddison, 1989), that is, a true rapid radiation.

The foregoing discussion demonstrates that analyses of a nuclear and plastid gene are concordant in resolving relationships within Santalales. In addition, hypotheses proposed by the three traditional classification systems can now be evaluated in light of new evidence. For example, it is apparent that aerial parasitism has evolved in Santalales in at least four separate

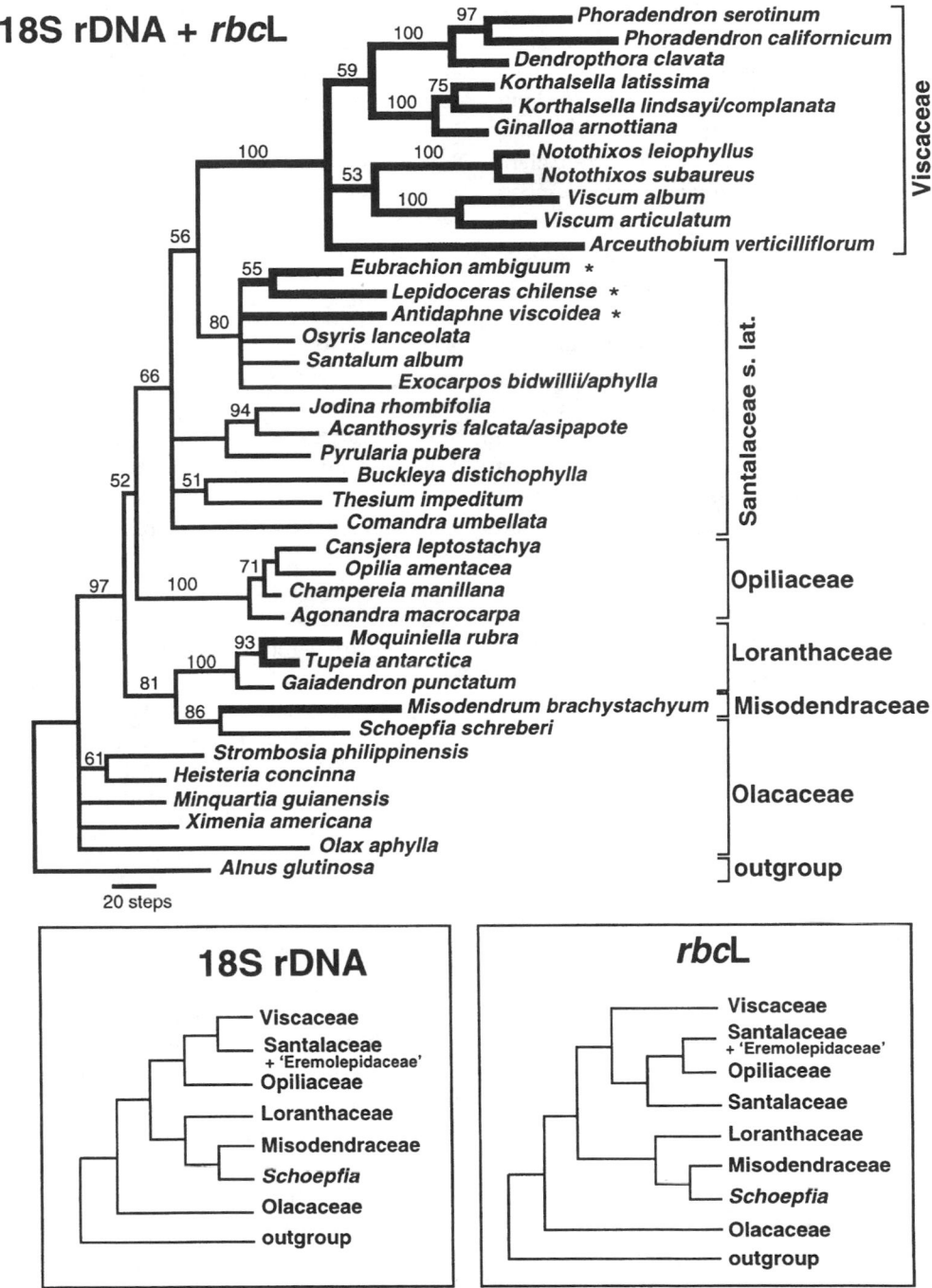

Figure 8.5. Strict consensus phylogram of 11 trees of length 2,212 derived from a parsimony analysis of nuclear 18S rDNA and *rbcL* sequences combined from 37 taxa of Santalales and *Alnus* (outgroup). For taxa with two specific epithets, 18S rDNA was sequenced from the first species and *rbcL* from the second. The phylograms enclosed as insets indicate the topologies of trees derived from analyses of the 18S rDNA and *rbcL* sequences separately. Numbers above the nodes indicate bootstrap values (from 100 replications). Nodes without values were supported in less than 50% of the replications. See Table 8.2 for additional details pertaining to this analysis. The genera indicated by an asterisk have traditionally been classified in Eremolepidaceae but are here considered part of Santalaceae s. l. The multiple origins of aerial parasitism are indicated by taxa with thick lines.

lineages (Misodendraceae, Loranthaceae, "Eremolepidaceae," and Viscaceae; thick lines on Fig. 8.5). Aside from such phylogenetic implications, what molecular evolutionary insights might be gained by comparing Santalales to other parasitic plants? In contrast to Scrophulariaceae, the holoparasitic trophic mode has apparently never evolved among the diverse lineages of root and stem parasitic Santalales. Although Olacaceae have been proposed as the evolutionary origination point for Balanophoraceae (Cronquist, 1981), such an association is not supported by nuclear 18S rDNA sequence analyses (see next section). If such a relationship is real, then no intermediate (less derived) taxa exist that might provide a link between the two families. It is noteworthy that, like Balanophoraceae, base substitution rates in 18S rDNA are increased among Viscaceae in general and *Arceuthobium* in particular. For the latter genus, base substitution rates are also increased for *rbcL*. Significantly, only about 30% of the carbon needed for growth is fixed in shoot tissue of *Arceuthobium* (Hull and Leonard, 1964). Because these parasites are obtaining a large portion of their photosynthate from the host, can we expect the eventual evolution of holoparasitism? It appears that such a trophic mode is not compatible with the aerially parasitic habit. Seedlings of *Arceuthobium* (Tocher et al., 1984) and those of other mistletoes are actively photosynthetic and must exist autotrophically prior to attachment to the host branch. Such is not the case for root holoparasites that germinate in response to chemical stimulants (carried in a moist rhizosphere) and attach to closely proximal host roots (Press and Graves, 1995). Mistletoe seeds do not undergo dormancy, are not adapted to long-term storage, and germinate in the absence of chemical cues; therefore, attachment to the host must be rapid. These features impose critical evolutionary constraints upon mistletoes that do not exist for root holoparasites.

Phylogenetic Studies of the Nonasterid Holoparasites Using Nuclear 18S rDNA Sequences

As discussed previously, traditional means of classifying the nonasterid holoparasites have met with difficulty owing to the extreme reduction and/or modification of morphological structures that have accompanied the evolution of these lineages. Although sequences of *rbcL* have proved extremely valuable in elucidating relationships throughout the angiosperms (see Chapter 17), this gene does not amplify for the nonasterid holoparasites using standard PCR procedures (D. Nickrent, unpubl.). Because nuclear 18S rDNA sequences have proved useful for addressing phylogenetic relationships in other parasitic plants, data from this molecular marker were obtained for representatives of all nonasterid holoparasites (Fig. 8.1). Relative rate tests showed that 18S sequences in these holoparasites were evolving, on average, 3.5 times faster than nonparasitic and most hemiparasitic plants (Nickrent and Starr, 1994), thus complicating their use in phylogenetic analyses. When these divergent sequences were included in analyses with nonparasites, the resulting topologies frequently showed aberrant relationships attributed to "long-branch attractions" (Felsenstein, 1978). Similar results were seen using parsimony, neighbor-joining, and maximum likelihood methods.

Long-branch artifacts are sometimes the result of incomplete sampling where intermediate lineages that could serve to "break up" a long branch have been omitted (Chase, Soltis, Olmstead et al., 1993). Recently, increased taxon density for nuclear 18S rDNA has been achieved such that, at present, more than 400 complete nuclear 18S sequences exist for angiosperms (Nickrent and Soltis, 1995; Soltis, Soltis, Nickrent et al., 1997). A 223-taxon data set (that included a greater sampling of monosulcate taxa than previous studies) allowed a reexamination of the placement of the nonasterid holoparasites. Sequences from the following holoparasites were added to the alignment: Balanophorales (*Balanophora fungosa, Corynaea crassa, Helosis cayennensis, Ombrophytum subterraneum, Rhopalocnemis phalloides,* and *Scybalium jaimaicense*), Cynomoriales (*Cynomorium coccineum*), Cytinales (*Cytinus ruber*), Rafflesiales (*Rafflesia keithii* and *Rhizanthes zippelii*), and Hydnorales (*Hydnora africana* and *Prosopanche americana*). Given the size of this data set, only heuristic search strategies could be used and branch swapping did not go to completion.

The tree resulting from this analysis (not shown) retained the major topological features found in the more extensive analysis reported in Soltis, Soltis, Nickrent et al. (1997). None of the nonasterid holoparasites were associated with Santalales. As seen in previous analyses, Rafflesiales and Balanophorales exhibited long-branch attractions, for example *Rafflesia* and *Rhizanthes* were intercalated in the middle of Balanophorales—an unlikely relationship. This composite clade was placed with taxa allied with Saxifragales, an equally unlikely relationship. This analysis showed that *Cytinus* and *Rafflesia* were only distantly related, thus supporting the segregation of Cytinaceae as a distinct family of Rosidae s. l. (Takhtajan et al., 1985; Takhtajan, 1987). The disposition of the remaining families of Rafflesiales has been hampered by the lack of nuclear 18S rDNA sequences from Apodanthaceae, Mitrastemonaceae, and *Bdallophyton*. 18S rDNA sequences of representatives of Balanophorales show that Cynomoriaceae differ markedly (more than 100 steps) from other members of the order, thus supporting the segregation of this family as proposed by Tahktajan (1987) and Thorne (1992).

In large- and small-scale analyses, Hydnoraceae are consistently placed as sister to the "paleoherb I" clade (sensu Donoghue and Doyle, 1989), that is, Aristolochiaceae, Chloranthaceae, Lactoridaceae, Piperaceae, all members of Magnoliidae. In contrast to Balanophorales and Rafflesiales, 18S rDNA rate increases for Hydnoraceae are less than half that seen in other holoparasites such as *Rafflesia* and *Balanophora;* hence, long-branch attraction appears to be less problematic. Furthermore, when taxa such as *Rafflesia* and *Balanophora* are included in the same analysis as Hydnoraceae, the association of the latter family with the paleoherbs is retained. The relationship between Hydnoraceae and Aristolochiaceae is concordant with the concepts proposed by Solms-Laubach (1894), Harms (1935), and more recently Cocucci (1983). The results with Hydnoraceae demonstrate the utility of molecular phylogenetic methods in providing independent evidence useful in resolving long-standing problems in the higher-level classification of these unusual holoparasites.

MOLECULAR EVOLUTIONARY STUDIES OF PARASITIC PLANTS

Molecular evolutionary processes occurring in parasitic plants represent some of the most extreme and dynamic known among all angiosperms. As evidenced by advances made from studying *Epifagus, Conopholis,* and *Cuscuta,* these plants can serve as important genetic models for facilitating the characterization of genome evolution. It is therefore worthwhile to compare and contrast genetic processes found in evolutionarily distinct groups in an attempt to uncover common themes, mechanisms, and evolutionary trends. As shown below for RUBISCO, fine-scale examination of the types of mutations that have occurred at the molecular level help illuminate the progression of change among diverse members of Scrophulariaceae.

Structural Analysis of RUBISCO Large Subunit Protein in Scrophulariaceae

The structure and function of the RUBISCO large subunit has been explored in great depth (Roy and Nierzwicki-Bauer, 1991; van der Vies et al., 1992; Schreuder et al., 1993; Adam, 1995; Kellogg and Juliano, 1997 and references therein). A comparative analysis of the amino acid substitutions of 499 flowering plants (Kellogg and Juliano, 1997) reveals a conservative mode of evolution of the protein in that there is a high degree of toggling among biochemically similar amino acids along the length of the polypeptide. The important structural motifs are particularly conserved in most flowering plants. To examine the effect of relaxed functional constraints on the evolution of the RUBISCO large subunit in parasitic Scrophulariaceae, several approaches were used, including: (1) an analysis of the synonymous and nonsynonymous substitution rates (Table 8.3); (2) mapping amino acid substitutions onto the strict consensus tree using MacClade 3.0 (Maddison and Maddison, 1992); (3) a PAM250 analysis (Dayhoff et al., 1978); and (4) assessment of substitutions in important structural motifs (Table 8.4); Wolfe and dePamphilis, 1997).

Average synonymous (K_s) and nonsynonymous (K_n) substitution rates and K_s/K_n for *rbcL*

Table 8.3. Average substitution rates of *rbcL* sequences for parasitic Scrophulariaceae and average PAM250 scores for RUBISCO large subunit polypeptide.

Parasitic Plant Category	No. Species	$K_s{}^a$	$K_N{}^a$	K_s/K_N	PAM250[b]
Nonparasitic	16	0.2022	0.0281	7.19	1113
Photosynthetic hemiparasitic	7	0.2227	0.0223	9.98	1124
Nonphotosynthetic[c] hemiparasitic	11	0.2463	0.0270	9.12	1112
Holoparasitic[d]	6	0.2482	0.0450	5.52	902

[a]Synonymous (K_s) and nonsynonymous (K_N) substitution rates based on Jukes-Cantor corrections determined using the computer program MEGA (Kumar et al., 1993).
[b]*Nicotiana tabacum* as a reference protein.
[c] Hemiparasitic lineages with holoparasitic members and holoparasitic *rbcL* ORFs (see Fig. 8.3).
[d] *rbcL* pseudogene sequences only.

from different categories of parasitic Scrophulariaceae are listed in Table 8.3. As heterotrophy increases in Scrophulariaceae, K_s also increases. In contrast, K_n does not appear to be affected greatly until the selective constraint is removed and pseudogenes are formed. There is a bias toward nonsynonymous substitutions in parasitic plants with *rbcL* open reading frames, as evidenced by the higher K_s/K_n ratios compared to nonparasitic Scrophulariaceae. However, the average K_s/K_n for *rbcL* pseudogene sequences is much lower, suggesting a relatively higher rate of nonsynonymous substitutions after selection is no longer a factor.

The *rbcL* nucleotide sequences were translated to inferred protein sequences. Deletions were maintained in the pseudogene sequences, but insertions were removed to retain the accuracy of the amino acid alignment. The step changes along the length of the polypeptide were plotted against the strict consensus tree based on *rbcL* sequences using MacClade. The primary finding of this analysis is the apparent loss of conserved sequences along the RUBISCO large subunit polypeptide with increasing heterotrophy. For nonparasitic Schrophulariaceae, there are long sequences in which no amino acid replacements are observed. This

Table 8.4. PAM250 analysis of important *rbcL* structural motifs for representative Scrophulariaceae.[a]

Taxon	No. of Events	α	β	Intradimer	Between Subunits	Dimer–dimer	Active Site	Other	PAM250 Difference
Holoparasites									
Orobanche ramosa	20	6	1	3	2	2	0	9	103
Boschniakia hookeri	13	4	2	0	3	0	1	4	64
Boschniakia strobilacea	10	2	2	0	2	0	1	3	57
Hyobanche sanguinea	7	3	1	1	1	1	0	3	41
Hyobanche atropurpurea	6	4	1	1	0	0	1	1	25
Other Holoparasites									
Lathraea clandestina	6	1	0	1	0	1	0	4	32
Orobanche corymbosa	2	0	1	0	0	0	0	1	5
Orobanche fasciculata	1	0	0	0	0	0	0	1	2
Hemiparasites									
Alectra sessiliflora	1	0	0	0	0	0	0	1	6
Striga hermonthica	1	0	1	0	0	0	0	0	8
Nonparasites									
Alonsoa unilabiata	1	0	1	0	0	0	0	0	3
Chelone obliqua	1	1	0	0	0	0	0	0	5

[a]See text for further explanation and interpretation.

same pattern was found by Kellogg and Juliano (1997) for 499 *rbcL* sequences from the Chase, Soltis, Olmstead et al. (1993) study. Kellogg and Juliano (1997) noted that important structural domains were conserved across a wide sampling of flowering plants and that there were additional conserved sequences of no known function. Hemi- and holoparasitic plants of Scrophulariaceae have fewer conserved regions across the length of the RUBISCO large subunit polypeptide. However, it is important to note that many biochemically similar amino acid replacements (e.g., replacing leucine with isoleucine) do not affect the structure and/or function of the protein and are, therefore, neutral replacements. Plotting the amino acid replacements against the strict consensus tree gives a first indication of molecular evolution of the protein, but does not elucidate the effect of molecular evolution on the structure and/or function of the protein.

To assess the effect of amino acid replacements on the structure and function of the RUBISCO large subunit polypeptide, Wolfe and dePamphilis (1995, 1997) conducted a PAM250 analysis (Dayhoff et al., 1978) on the translated protein sequences for selected taxa from all categories of parasitic plants in Schrophulariaceae. Dayhoff et al. (1978), using empirical data for many proteins (hemoglobins, fibrinopeptides, cytochrome c), constructed a table (the PAM250 matrix) of amino acid substitution probabilities; in other words, a table that shows the probabilities of the substitution of one amino acid for another. Amino acids that have similar biochemical properties are more likely to be substituted with one another than with biochemically dissimilar amino acids, and this is reflected in the values given in the PAM250 matrix. These substitution probabilities are measured in terms of PAM (percent accepted mutations) with one unit representing one amino acid substitution per 100 residues.

Using the reference protein from the outgroup, *Nicotiana tabacum,* PAM250 scores were calculated over the entire length of the polypeptide and from the critical structural motifs (Tables 8.3, 8.4) by assigning a PAM250 value for each amino acid of *Nicotiana* and comparing the PAM250 score for each amino acid substitution in the other taxa. The total PAM250 scores for the entire polypeptide and for each critical structural motif were compared between the reference protein and each taxon. Although the amount of amino acid replacement increases with increasing heterotrophy, the overall effect of these replacements on the protein structure as inferred from the PAM250 analysis is minimal. The only PAM250 scores significantly different from the reference protein are pseudogene sequences. The translated protein sequences from the holoparasite *rbcL* ORFs are nearly identical to those of photosynthetic plants in terms of their inferred structure. Hence, despite increased synonymous and nonsynonymous substitutions in nonphotosynthetic hemiparasites, the functionality of RUBISCO does not appear to be negatively impacted, which implies that RUBISCO is functional in parasitic plants even after the loss of photosynthesis (see next section).

Plastome Reduction in Holoparasitic Plants

For those holoparasites whose plastomes have been more fully characterized at the sequence level, the loss of photosynthesis is accompanied by the loss or modification of photosynthetic genes, a process that reduces the overall genome size. Empirical evidence for genome reduction can be seen in Scrophulariaceae s. l., *Cuscuta,* and the nonasterid holoparasites. The genes responsible for plastid gene expression (rDNA, ribosomal proteins) are most often retained, as evidenced by data on more extreme holoparasites such as *Cytinus.* In *Epifagus,* plastid-encoded tRNA genes appear more labile as evidenced by their loss and apparent replacement by nuclear-encoded tRNAs (Wolfe et al., 1992b). Can the reductional trend be taken to the extreme where the plastid and/or its plastome is lost entirely? Based on ultrastructural studies of holoparasite plastids, Dodge and Lawes (1974) concluded that plastids are retained as storage organs or for other functions. Because the plastome of *Epifagus* contains only four genes that are not involved in gene expression (dePamphilis and Palmer, 1990; Wolfe et al., 1992c), any one may serve some indispensable function in the nonphotosynthetic plastid, thus explaining the retention of the entire plastome.

Studies of the *rbcL* locus in holoparasitic Scrophulariaceae (Fig. 8.6) are especially informative in that they show a range in the degree of structural modifications experienced by different lineages, thereby implying that such changes occur independently and result in different degrees of pseudogene formation. It is of interest that the distribution of *rbcL* pseudogenes on the phylogeny of Scrophulariaceae (Fig. 8.3) is different among Old and New World species of *Orobanche* (the Old World species *Orobanche cernua* and *O. ramosa* have pseudogenes, whereas the New World *O. corymbosa* and *O. fasciculata* do not). It is therefore important to determine not just the presence of *rbcL* genes but also whether they are expressed in holoparasites.

It is clear from recent studies of parasitic Scrophulariaceae and Cuscutaceae (Machado and Zetsche, 1990; Haberhausen et al., 1992; Delavault et al., 1995; Wolfe and dePamphilis, 1997) that *rbcL* ORFs are maintained in many independent lineages of parasitic plants. In two holoparasites, *Lathraea* (Bricaud et al., 1986; Thalouarn et al., 1989; Thalouarn et al., 1991; Delavault et al., 1995, 1996) and *Cuscuta* (Press

et al., 1986, 1991; Machado and Zetsche, 1990; Haberhausen and Zetsche, 1992), the *rbcL* locus is minimally expressed in terms of gene transcription and/or protein activity. No *rbcL* expression was detected in several species of *Orobanche* (Thalouarn et al., 1994); however, such studies have not been conducted on species with intact *rbcL* ORFs (e.g., *O. corymbosa* or *O. fasciculata*). An examination of the 5′ and 3′ untranslated regions (UTRs) of the *rbcL* locus in *Orobanche* (Wolfe and dePamphilis, 1997) revealed that the promoter and/or ribosome binding sites are maintained in *O. corymbosa* and *O. fasciculata,* but the 5′ UTRs of the pseudogene sequences have structural mutations or nucleotide substitutions. The 3′ UTRs of all species of *Orobanche* examined to date (Wolfe and dePamphilis, 1997) have intact stem-loop structures but have major deletions upstream of the palindromic sequence. Phylogenetic analyses, gene expression studies, and sequence analyses of *rbcL* all suggest that the retention of *rbcL* ORFs in *O. corymbosa, O. fasciculata,* and in other lineages of holoparasitic Scrophulariaceae can be explained by the recency of their transition to holoparasitism.

Wolfe and dePamphilis (1997) presented three hypotheses to explain the retention of *rbcL*

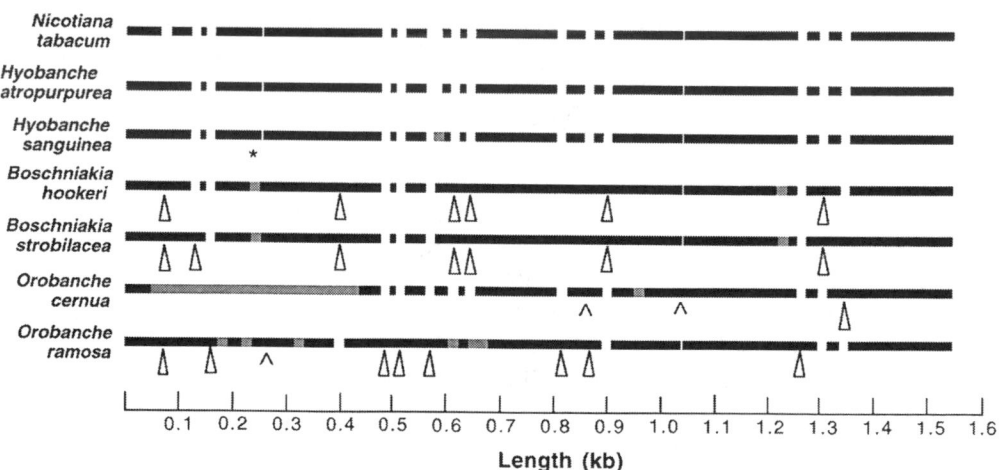

Figure 8.6. Structural maps of *rbcL* pseudogenes from orobanchaceous holoparasites compared with *Nicotiana tabacum.* Solid black lines indicate sequence not affected by structural changes in most other taxa. Blank regions demarcate insertions from other taxa. Open triangles represent multinucleotide insertions, ^ = single nucleotide insertions. Gray lines are multinucleotide deletions, * = single nucleotide deletion. Note that the *rbcL* locus of *Hyobanche atropurpurea* lacks structural mutations, but is a pseudogene, presumably because of a premature stop codon. *Conopholis* and *Epifagus rbcL* loci are not shown (missing in *Conopholis,* unalignable fragment in *Epifagus*).

in holoparasitic plants of Scrophulariaceae and Cuscutaceae: (1) the plant still requires the RUBISCO protein for low levels of autotrophic carbon fixation; (2) the oxygenase activity may function in glycolate metabolism; and (3) its maintenance or loss is a result of stochastic events. For the latter, the presence or absence of the gene in closely related members of a lineage may simply reflect the fact that insufficient time has passed for the accumulation of deleterious mutations. Whereas the carboxylase activity of RUBISCO fixes carbon dioxide into reduced carbon (which is subsequently metabolized by the plant), the oxygenase activity is a competitive reaction involved in glycine and serine biosynthesis (glyoxylate metabolism). Holoparasitic plants receive reduced carbon from their host, eliminating the need for RUBISCO activity (Press et al., 1991); however, some species maintain very low levels of photosynthetic pigment production and photosynthesis, or perform photosynthesis only during a discrete phase of the life cycle such as seedling development or host attachment. This may explain the retention of a minimally expressed *rbcL* in the stem parasite, *Cuscuta reflexa* (Haberhausen et al., 1992). It is unclear how important the glycolate pathway is for amino acid biosynthesis in photosynthetic or parasitic plants. Press et al. (1986) reported a decrease in photorespiration with increasing heterotrophy. Although unlikely, it is possible that the oxygenase activity of RUBISCO is maintained in parasitic plants until the adaptation to heterotrophy is complete, and acquisition of host amino acids reduces the necessity for photorespiration. Glycolate metabolism has also been suggested as a possible cause for the retention of *rbcL* in the plastome of the nonphotosynthetic euglenoid *Astasia longa* which has experienced deletions of nearly all other photosynthetic genes (Gockel et al., 1994).

To determine whether stochastic factors were the sole reason for maintenance of *rbcL* ORFs in *O. corymbosa* and *O. fasciculata,* Wolfe and dePamphilis (1997) developed a probability model using the following assumptions: (1) *O. corymbosa* and *O. fasciculata* had a nonphotosynthetic ancestor, and all *rbcL* sequence divergence occurred without photosynthetic constraint; (2) the probability of random mutations generating a stop codon is 4/63. Under these assumptions, and if all mutations accumulating in the *rbcL* sequences are independent, then, if the sequences are pseudogenes, the probability of retaining an ORF without a stop codon is: $P = (59/63)^n$, where n is the number of nucleotide substitution differences between them. Using this probability model, Wolfe and dePamphilis (1997) calculated a probability of 2.79%; therefore it is unlikely that chance alone explains the maintenance of the *rbcL* ORF in *O. corymbosa* and *O. fasciculata.*

As suggested by the differential occurrence of pseudogenes in different species of *Orobanche,* the short time scale for evolutionary diversification may explain differences in the plastome structure of *Epifagus* and *Conopholis,* shown to be sister taxa by nuclear 18S rDNA, plastid 16S rDNA, and *rps2* data. The 16S/23S rDNA spacer of *Epifagus* contains two nearly full-length tRNA pseudogenes whereas these genes are essentially absent in the spacer of *Conopholis,* thus suggesting a recent loss. Moreover, both 18S rDNA and *rps2* data indicate that *Boschniakia* is sister to the *Epifagus/Conopholis* clade, hence the ancestor of the latter two taxa was also a holoparasite (whose plastome must have contained inverted repeats). Therefore, the loss of one copy of the inverted repeat in *Conopholis* took place after its divergence from *Epifagus,* likely less than 5 myr ago (Muller, 1981). The case of the malarial *Plasmodium* (an apicomplexan) contrasts with such rapid plastome reorganizations. Apicomplexans are thought to have evolved from photosynthetic chromophytes, hence the malarial plastome has been retained for possibly 800 myr or more (Escalante and Ayala, 1995). As with holoparasitic plants, the plastome of *Plasmodium* has a compact organization and retains a specific suite of genes involved in gene expression *and* inverted repeats. These two examples demonstrate the complexity of selectional factors involved in either maintaining or deleting major structural features of the plastome. Although the selective environments of malarial parasites and plants are clearly quite different, both have followed parallel pathways in shaping the overall plastome structure, and both suggest that this genome may contribute essential metabolic products required by the parasite (Wilson et al., 1996).

RATE ACCELERATIONS

One of the fundamental questions in evolutionary biology centers around the concept of the molecular clock, specifically the application of a strict molecular clock that infers equal rates of nucleotide substitution for organisms with equal generation times (the generation-time effect). As has been shown for *rbcL* in monocots (Bousquet et al., 1992; Gaut et al., 1992), *rps2* in parasitic Scrophulariaceae (dePamphilis et al., 1997), and 18S rDNA in Santalales (Nickrent and Starr, 1994), a strict molecular clock cannot be universally applied, at least for these loci and taxa. As discussed below in Rate Variation among Genomes and Lineages, rate accelerations occur in genes from all three subcellular genomes in parasitic plants, hence these organisms provide ideal models for studying molecular evolutionary questions such as the effect of rate heterogeneity on phylogeny reconstruction.

Nuclear 18S rRNA Genes

The first documentation of rate acceleration in nuclear 18S rDNA in plants was the study by Nickrent and Starr (1994) wherein relative rates tests were conducted using one hemiparasite (*Arceuthobium*) and four holoparasites (*Balanophora, Prosopanche, Rafflesia,* and *Rhizanthes*). These tests showed that 18S rDNA sequences in these holoparasites were evolving, on average, at least three times faster than nonparasitic and most hemiparasitic plants. Rate heterogeneity of this magnitude has not been detected among over 200 angiosperm sequences that have been examined (D. Nickrent, unpubl.); however, statistically significant rate increases have been measured in some Scrophulariaceae, *Cuscuta, Pholisma* (Lennoaceae), and mycoheterotrophic angiosperms (A. Colwell and D. Nickrent, unpubl.). These four nonasterid holoparasites and the dwarf mistletoe represent four distinct orders, thus accelerated substitution rates have occurred independently in an ancestor of each of these lineages. In Santalales, accelerated rates are not universally seen, as many other nonparasites and

hemiparasites show no indication of acceleration. These data suggest that substitution rate acceleration of nuclear rDNA occurs only in lineages where photosynthesis is absent or diminished, thereby requiring an advanced state of nutritional dependence upon the host. It is not presently understood why these changes in life history are correlated with changes at nuclear rDNA loci, although relaxation of selectional constraints on rRNA structure and function, small effective population size, and molecular drive have been proposed (Nickrent and Starr, 1994).

Plastid Genes: rps2 and rbcL

Phylogenetic comparisons of *Epifagus* to *Nicotiana* and more distant outgroups revealed longer branches leading to *Epifagus* for virtually all of the retained protein-coding genes and plastid-encoded rRNA genes (Morden et al., 1991; Wolfe et al., 1992b, 1992c). Increased numbers of substitutions were observed at both synonymous and nonsynonymous sites between the parasitic and nonparasitic lineages. These studies provided the first indication that the plastome of *Epifagus* may be subject to increased rates of molecular evolution, even at functionally constrained genes. As shown on the *rps2* phylogeny (dePamphilis et al., 1997), the branch lengths for both *Epifagus* and *Conopholis* are considerably longer than those of other Scrophulariaceae, and formal relative rates tests show that both have increased evolutionary rates relative to nonparasitic and most hemiparasitic lineages (dePamphilis et al., 1997). Also evolving at an accelerated rate are the ptDNAs of *Orobanche ramosa* (nonsynonymous only), the *Buchnera-Striga-Cycnium* clade (nonsynonymous sites only), and *Euphrasia* (synonymous sites only). These complex patterns of rate acceleration indicate not only that rate asymmetries in this group are not uniquely tied to the loss of photosynthesis, but that they may have more than one distinct underlying cause, resulting in rate accelerations at synonymous and nonsynonymous sites.

For *rbcL,* synonymous and nonsynonymous substitution rates in nonparasitic and hemiparasitic (photosynthetic) Scrophulariaceae are very similar to those of all other flowering plants examined to date (Wolfe and dePamphilis, 1995,

1997). This similarity is probably the result of the strong functional constraints imposed on the protein to maintain important structural motifs such as the alpha-beta barrels forming the active site, the active site itself, intradimer and dimer–dimer interactions, and intersubunit interactions (Kellogg and Juliano, 1997). Parasitic plants that have lost photosynthesis are no longer under selective constraints to maintain the structure and function of proteins involved in photosynthetic reactions. This has become evident in examination of holoparasitic Scrophulariaceae. For example, the entire *rbcL* locus is missing in *Conopholis* (Colwell, 1994), and only a remnant of the gene remains in *Epifagus* (dePamphilis and Palmer, 1990; Wolfe et al., 1992c).

Plastid Genes: 16S rDNA

Plastid-encoded 16S rDNA sequences have seldom been used in phylogenetic studies of angiosperms given their overall sequence conservation (see Chapter 1). The 16S rDNA sequences from five nonasterid holoparasites were included in a phylogenetic analysis with other land plants, algae, and cyanobacteria to show their affinity to angiosperms and graphically demonstrate rate increases. The topology of the strict consensus phylogram (Fig. 8.7) retains the major features obtained by others using different methods of phylogenetic reconstruction such as distance-based methods and neighbor joining (summarized by Nelissen et al., 1995). Despite their extremely long branches, the holoparasite 16S rDNA sequences are most closely related to other angiosperm plastid 16S rDNAs. This position is significant because previous analyses using nuclear 18S rDNA sequences of extreme holoparasites resulted in the migration of these long-branch taxa to the base of the tree (Nickrent and Starr, 1994). Similar perturbations of the position of divergent taxa on mitochondrial

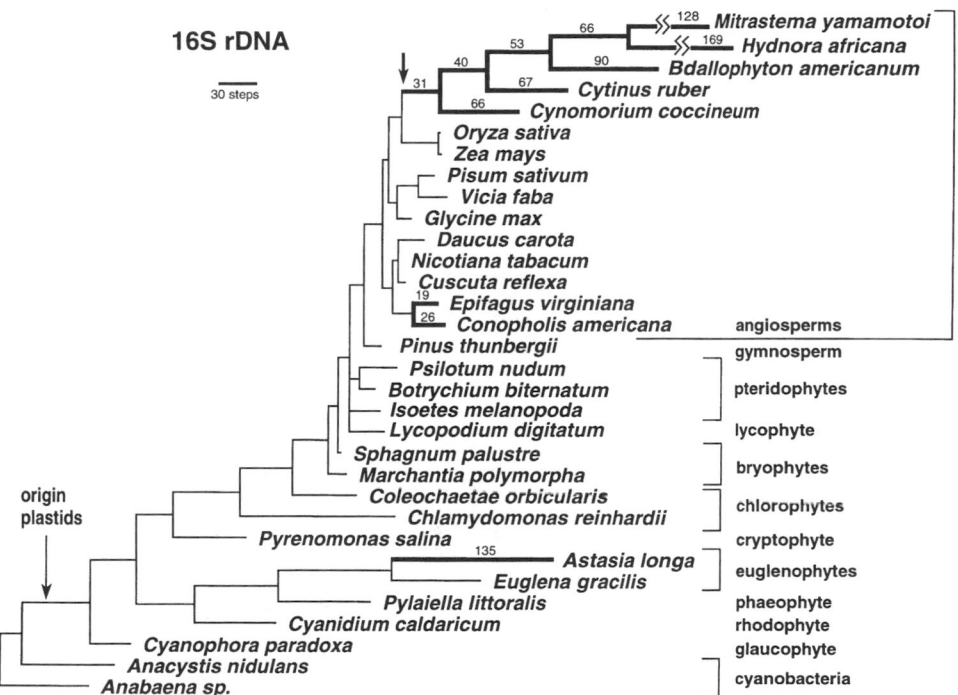

Figure 8.7. Strict consensus phylogram of four minimum-length trees of 2,576 steps derived from a parsimony analysis of 30 plastid-encoded 16S rDNA sequences and two cyanobacterial outgroup taxa (*Anacystis* and *Anabaena*). Numbers above the branches of the nonphotosynthetic taxa indicate number of substitutions (steps) and are shown in bold. The nonasterid holoparasite clade (arrow) is present within the angiosperms; however, intergeneric relationships are influenced by long-branch attraction artifacts.

16S rRNA trees has been observed (Olsen, 1987). For this character-based analysis (parsimony) of the holoparasites, the types of changes, not simply the number of changes, apparently determined the placement of the clade. On the other hand, interrelationships *among* the nonasterid holoparasites are likely artifactual because topological positions simply reflect an increasing number of substitutions.

Most angiosperm 16S rDNA sequences differ by 2–3% when compared with tobacco, whereas the nonasterid holoparasites show an increasingly greater number of mutations: *Cynomorium* (7.3%), *Cytinus* (8.0%), *Bdallophyton* (12.7%), *Mitrastema* (14.9%), *Hydnora* (19.4%), *Pilostyles* (30.4%), and *Corynaea* (35.9%) (Nickrent et al., 1997a). The high sequence variation suggests that these 16S rDNA sequences may possibly be pseudogenes. As with *Cytinus,* however, these rRNAs can be folded into secondary structures and most retain the typical complement of 50 helices (Nickrent et al., 1997a). The single exception is the 16S rRNA of *Pilostyles,* likely the most unusual 16S rRNA structure yet documented; it not only lacks four helices (9, 10, 11, and 37), but also contains large insertions

on helices 6, 23-1, 29, and 48. Further molecular and biochemical work is needed to determine whether these genes are expressed and whether the rRNA is functional.

To quantify rate heterogeneity among various 16S rDNA sequences, formal relative rates tests were conducted. The parametric test described by Wu and Li (1985) as implemented for nuclear 18S rDNA sequences in holoparasites by Nickrent and Starr (1994) was used. In the absence of information on actual divergence times, this method employs a variance estimation to determine whether substitution rates (K) between two lineages differ. Because the time of divergence of the holoparasites relative to monocots and dicots is uncertain, *Marchantia* was chosen as the reference. Use of the phylogenetically closer *Pinus* as the reference did not change the results. The magnitude of rate increase among the nonasterid holoparasites matches the overall number of substitutional differences described above with *Cynomorium* at one extreme and *Hydnora* at the other (Fig. 8.8). The highly divergent sequences of *Pilostyles* and *Corynaea* would show K values even greater than *Hydnora* and are therefore not shown. The

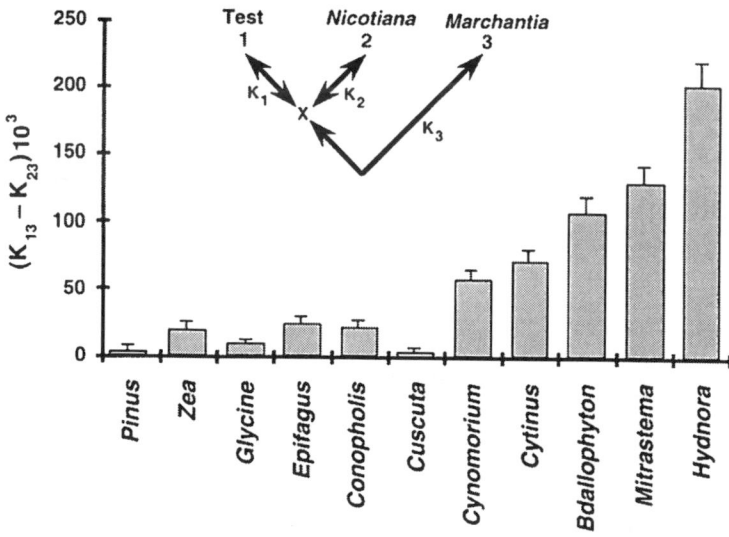

Figure 8.8. Histogram showing results of relative rate tests using plastid 16S rDNA sequences. The 11 land plants indicated on the abscissa are the test organisms (taxon 1), *Nicotiana* was taxon 2, and *Marchantia* was the reference or outgroup (taxon 3). The three-taxon tree uses K_1, the number of nucleotide substitutions per site for taxon 1, K_2 for taxon 2, and K_3 for taxon 3. The difference in nucleotide substitutions per site ($K_{13} - K_{23}$) is multiplied by 1,000 for graphical purposes. Standard error values are included above the bar for each taxon. The fastest rates are observed in the five nonasterid holoparasites.

16S rDNA sequences of *Epifagus* and *Conopholis* exhibited slightly accelerated rates relative to most angiosperms (in agreement with Wolfe et al., 1992a), but not of the magnitude seen in the nonasterid holoparasites.

Mitochondrial rDNA

Increased nucleotide substitution rates among both nuclear 18S rDNA and plastid-encoded 16S rDNA of holoparasites raises the question "what is happening in the mitochondrial genome?" The plant mitochondrial genome has been largely ignored as a source of data for phylogenetic and population genetic studies of plants (see Chapter 1), partly because the rate of synonymous nucleotide substitution in this genome is only one-third to one-quarter that of chloroplast DNA, despite the high frequency of genetic recombination (Palmer, 1990, 1992).

Small subunit sequences of mitochondrial rDNA (here designated 19S rDNA) exhibit especially low levels of divergence relative to their chloroplast and nuclear counterparts in angiosperms. The six published photosynthetic angiosperm sequences (Table 8.5) exhibit less than 1.3% sequence divergence. This value was determined from the "core" of the molecule, that is, excluding two variable (and unalignable) regions (helices 6 and 43). Six new mitochondrial 19S rDNA sequences have been generated (Duff

and Nickrent, submitted) that include four holoparasites (*Cytinus, Hydnora, Corenaea, Epifagus*), and *Nicotiana.* These sequences were aligned with six published sequences and compared with *Glycine* (Table 8.5). As with the 18S and 16S rDNA data, *Cytinus* exhibited the smallest degree of sequence divergence (2.3%) followed by *Cornaea* (2.8%) and *Hydnora* (3.1%). The moderate divergence of the *Epifagus* sequence (1.5%) is comparable to divergence levels of its plastid 16S rDNA.

Formal relative rates tests using mitochondrial 19S rDNA (similar to those performed on the 18S and 16S rDNA sequences) showed that only the nonasterid holoparasites had significantly increased substitution rates (Duff and Nickrent, 1997). Borderline rate increase was seen for *Epifagus.* In contrast, rates of sequence evolution for a second mitochondrial gene, *coxI*, are not elevated for *Epifagus* or for *Cytinus* relative to nonparasitic relatives (C. dePamphilis et al., unpubl.). Unlike the 18S and 16S rDNA sequences of *Cytinus, Hydnora,* and *Corynaea,* which exhibit base composition biases, the mitochondrial sequences of these three genera show no such bias. In addition to sequence divergence, these sequences are characterized by a higher frequency of indels than sequences of nonparasitic plants. Models constructed from these sequences indicate that the majority of mutations do not disrupt the higher-order structure, thus providing

Table 8.5. Angiosperm mitochondrial 19S rDNA sequences compared with *Glycine.*

Species	GenBank Number	Substitutions[a]	Indels[b] (bp)	Helix 6 (bp)	Helix 43 (bp)
Secale cereale	Z14049	18	0	151	366
Triticum aestivum	Z14078	18	0	151	344
Zea mays	X60794	20	1(14)	144	350
Oenothera berteriana	X61277	10	3(3)	106	338
Glycine max	M16859	—	—	102	427
Lepidoceras chilense	U82641	9	1(1)	103	364
Lupinus luteus	Z11512	11	1(1)	103	461
Nicotiana tabacum	U82638	15	0	102	335
Cytinus ruber[c]	U82639	33	4(10)	99	330
Epifagus virginiana[c]	U82642	22	2(2)	87	not sequenced
Hydnora africana[c]	U82637	41	3(4)	89	449
Corynaea crassa[c]	U82636	40	3(39)	75	271

[a]Substitutions from comparisons excluding helix 6 and 43.

[b]Frequency and size of indels were determined by sequence alignment aided by secondary structure analysis based on a model for *Zea* (Gutell, 1993).

[c]Holoparasitic taxa.

strong indirect evidence that these genes are functional.

The presence of RNA editing in plant mitochondria has recently received increased attention (Arts and Benne, 1996; Bowe and dePamphilis, 1996; Sper-Whitis et al., 1996). Although RNA editing is recognized as a significant process in many mitochondrial genes, the holoparasite 19S rDNA sequences reported here are unlikely to be significantly affected as determined by reverse-transcriptase PCR experiments (R. Duff and D. Nickrent, unpubl.). Despite the numerous reports of editing among protein-coding genes in plant mitochondria, no cases of editing of mitochondrial 195 rRNA are known (Schuster et al., 1991). Neither the pattern of substitutions nor the presence of indels follows that seen in reported cases of RNA editing. For these reasons, we do not feel editing can be used to explain the observed pattern of substitutions in mitochondrial 19S rDNA.

RATE VARIATION AMONG GENOMES AND LINEAGES

The patterns of nucleotide substitution across lineages and organellar gene loci provide a framework for addressing the underlying causes of rate variation. Relatively few studies exist in which rate comparisons are made between different organellar genes across lineages. An exception is recent work showing that synonymous (but not nonsynonymous) substitution rates are slower for palms than grasses for both

nuclear-encoded *Adh* and plastid-encoded *rbcL* (Gaut et al., 1996). Three basic models of rate variation can be defined: (1) organismal, (2) genome-specific, and (3) gene-specific. The first model requires that DNA replication rate for each subcellular genome be correlated; hence if one locus is accelerated, all loci will be similarly affected. In the absence of selection, comparisons of molecular phylogenies using (say) two different organellar loci would show each taxon to have the same relative branch lengths on the respective trees. In the genome-specific model, there is no assumption that a correlation exists between DNA replication and/or substitution rates between organellar genomes. Such a model could result from differences in DNA replication and/or repair mechanisms in the distinct organellar genomes. Here, comparisons of molecular phylogenies derived from the different organellar loci would not be expected to show the same distribution of rate variation among taxa. Finally, the gene-specific model implies that each subcellular genome, and possibly individual genes within each genome, are constrained by their own selectional environment. This model would be manifest as molecular phylogenies with patterns of rate variation that are not correlated between gene loci.

Sequences of parasitic plant genes derived from all three subcellular genomes can be used to address the above models and to examine the association between rate changes and loss of photosynthesis. As shown in Table 8.6, the patterns of rate variation are complex; some ex-

Table 8.6. Rate comparisons for parasitic plants across genomes.

Taxon	Nuclear 18S rDNA	Plastid			Mitochondrial 19S rDNA
		16S rDNA	*rbcL*	*rps2*	
Nonasterid holoparasites	++[a]	++	absent	?	++
Santalales (minus Viscaceae)	−	−	−	?	−
Arceuthobium	++	+	+	?	?
Epifagus	+/−	+	ψ	+	+/−
Conopholis	+/−	+	absent	+	?
Hyobanche	+	−	ψ	−	?
Cuscuta	+[b]	−[c]	+[d]	?	−

[a]Molecular evolutionary rates: ++ = very fast, + = fast, +/− = borderline fast, − = not accelerated in relative rate tests, ψ = pseudogene.
[b]From *Cuscuta gronovii* (A. Colwell and D. Nickrent, unpubl.).
[c]From *Cuscuta reflexa* (Haberhausen and Zetsche, 1994).
[d]Gene from *Cuscuta reflexa* present but with low transcription levels (Haberhausen et al., 1992).

amples appear to support one model, whereas others do not. Consider first the nonasterid holoparasites. Formal relative rate tests, performed using sequences from nuclear, plastid, and mitochondrial small subunit rDNA, revealed that all three genomes are accelerated relative to autotrophic plants. These results support the organismal model of rate variation.

In other instances, the available data do not support the organismal model. Considering a particular gene locus, different substitution rates can be observed among different parasites of the same lineage. For 18S rDNA and *rbcL,* rates are elevated for Viscaceae, but not for other families of Santalales (Nickrent and Soltis, 1995; Nickrent and Duff, 1996), demonstrating that rate acceleration is not a general evolutionary feature of the order but has occurred only following the divergence of Viscaceae (see Fig. 8.5). Similarly, rate acceleration for *rps2* can be seen in *Epifagus, Conopholis,* and *Orobanche,* but not in other holoparasitic Scrophulariaceae such as *Hyobanche* (dePamphilis et al., 1997). Rate elevations may occur differentially among genes within a single genome. For example, *Arceuthobium* shows elevated rates for *rbcL,* but not for 16S rDNA, which indicates that elevated substitution rates can exist in genes of fully functional chloroplasts, but that conservative loci may lag behind more variable ones. In Scrophulariaceae, the pattern of rate increase (or pseudogene formation) for *rbcL* does not parallel that of 16S rDNA, again, with the latter showing more conservative rates of change. Finally, differences in substitution rates may be observed between different subcellular genomes within the same species. For *Arceuthobium, Hyobanche,* and *Cuscuta,* rates for nuclear but not plastid small subunit rDNA are elevated. These data do not tend to support either the organismal or the genome-specific model of rate variation, thus leaving only the gene-specific model as an explanation.

The relationship between the loss of photosynthesis and rate accelerations in nuclear and plastid genes can be interpreted using the data presented in Table 8.6. As shown by *Arceuthobium,* not all plants with rate increases (for the nuclear or plastid genomes) are nonphotosynthetic. Conversely, not all plants that lose photosynthesis show increased rates, as evidenced by *rps2* in *Alectra orobanchoides* or *Hyobanche.* Further support for the latter concept is seen following relative rate tests of nuclear rDNA in ericaceous mycotrophs (e.g., *Monotropa*) or nonphotosynthetic orchids (e.g., *Corallorhiza*) where no rate increases were observed (A. Colwell and D. Nickrent, unpubl.). Results from *rps2* analyses of Scrophulariaceae indicate that rate increases can occur in hemiparasites prior to the loss of photosynthesis (dePamphilis et al., 1997), thus providing evidence supporting the recency and rapidity of this trophic change.

The relationship between rate accelerations in plastomes and the nuclear genome is not clear at present. One might predict that changes in the plastome occur first and that the comparatively more conservative nuclear ribosomal loci follow. That the two events are linked is evidenced by the paucity of cases, at least in parasitic plants, where plastome genes are clearly accelerated but nuclear rDNA genes are not. Such correlations suggest a greater degree of signal transduction from the plastid to the nucleus than has previously been suspected (Susek and Chory, 1992).

CONCLUSIONS

Parasitic plants represent some of the most unusual organisms on earth, and each independent lineage represents a natural genetic experiment whose organelles and genes have evolved under relaxed functional constraints. Although *Cuscuta* is amenable to culture and direct experimental manipulation, very few holoparasitic parasitic plants have been established as model systems for biochemical and genetic research. Certainly, attempts to culture holoparasites would be hindered by lack of knowledge of their life histories. To date, studies of Santalales, nonasterid holoparasites, Schrophulariaceae s. l., and *Cuscuta* have yielded a wealth of new insights into molecular phylogenetic and molecular evolutionary processes. Despite this progress, a number of questions about the evolutionary process still remain, which will hopefully stimulate continued study of these fascinating plants.

LITERATURE CITED

Adam, Z. 1995. A mutation in the small subunit of ribulose-1,5-biphosphate carboxylase/oxygenase that reduces the rate of its incorporation into holoenzyme. Photosynthesis Research **43**:143–147.

An-Ming, L. 1990. A preliminary cladistic study of the families of the superorder Lamiiflorae. Botanical Journal of the Linnean Society **103**:39–57.

Arts, G. J., and R. Benne. 1996. Mechanism and evolution of RNA editing in kinetoplastida. Biochimica Biophysica Acta **1307**:39–54.

Barlow, B. A. 1964. Classification of the Loranthaceae and Viscaceae. Proceedings of the Linnaean Society of New South Wales **89**:268–272.

Barlow, B. A. 1983. Biogeography of Loranthaceae and Viscaceae. In The Biology of Mistletoes, eds. M. Calder and P. Bernhardt, pp. 19–45. Academic Press, New York.

Barlow, B. A., and D. Wiens. 1971. The cytogeography of the loranthaceous mistletoes. Taxon **20**:291–312.

Bhandari, N. N., and S. C. A. Vohra. 1983. Embryology and affinities of Viscaceae. In The Biology of Mistletoes, eds., M. Calder and P. Bernhardt, pp. 69–86. Academic Press, New York.

Boeshore, I. 1920. The morphological continuity of Scrophulariaceae and Orobanchaceae. Contributions of the Botanical Laboratory of the Morris Arboretum **5**:139–177.

Bömmer, D., G. Haberhausen, and K. Zetsche. 1993. A large deletion in the plastid DNA of the holoparasitic flowering plant *Cuscuta reflexa* concerning two ribosomal proteins (*rpl2, rpl23*), one transfer RNA (*trnI*) and an ORF 2280 homologue. Current Genetics **24**:171–176.

Bousquet, J., S. H. Strauss, A. H. Doerksen, and R. A. Price. 1992. Extensive variation in evolutionary rate of *rbcL* gene sequences among seed plants. Proceedings of the National Academy of Sciences U.S.A. **89**:7844–7848.

Bowe, L. M., and C. W. dePamphilis. 1996. Effects of RNA editing and gene processing on phylogenetic reconstruction. Molecular Biology and Evolution **13**:1159–1166.

Bricaud, C. H., P. Thalouarn, and S. Renaudin. 1986. Ribulose 1,5-bisphosphate carboxylase activity in the holoparasite *Lathraea clandestina* L. Journal of Plant Physiology **125**:367–370.

Burtt, B. L. 1965. The transfer of *Cyrtandromoea* from Gesneriaceae to Scrophulariaceae, with notes on the classification of that family. Bulletin of the Botanical Survey **7**:73–88.

Chase, M. W., D. E. Soltis, R. G. Olmstead, D. Morgan, D. H. Les, B. D. Mishler, M. R. Duvall, R. A. Price, H. G. Hills, Y.-L. Qiu, K. A. Kron, J. H. Rettig, E. Conti, J. D. Palmer, J. R. Manhart, K. J. Sytsma, H. J. Michaels, W. J. Kress, K. G. Karol, W. D. Clark, M. Hedrén, B. S. Gaut, R. K. Jansen, K.-J. Kim, C. F. Wimpee, J. F. Smith, G. R. Furnier, S. H. Strauss, Q.-Y. Xiang, G. M. Plunkett, P. S. Soltis, S. M. Swensen, S. E. Williams, P. A. Gadek, C. J. Quinn, L. E. Eguiarte, E. Golenberg, G. H. Learn, Jr., S. W. Graham, S. C. H. Barrett, S.

Dayanandan, and V. A. Albert. 1993. Phylogenetics of seed plants: an analysis of nucleotide sequences from the plastid gene *rbcL*. Annals of the Missouri Botanical Garden **80**:528–580.

Cocucci, A. E. 1983. New evidence from embryology in angiosperm classification. Nordic Journal of Botany **3**:67–73.

Colwell, A. E. 1994. Genome evolution in a non-photosynthetic plant, *Conopholis americana*. Ph.D. dissertation, Washington University, St. Louis, Missouri.

Cronquist, A. 1981. An Integrated System of Classification of Flowering Plants. Columbia University Press, New York.

Cronquist, A. 1988. The Evolution and Classification of Flowering Plants. New York Botanical Garden, New York.

Dayhoff, M. O., R. M. Schwartz, and B. C. Orcutt. 1978. A model of evolutionary change in proteins. In Atlas of Protein Sequence and Structure, ed. M. O. Dayhoff, pp. 345–352. National Biomedical Research Foundation, Silver Springs, Maryland.

Delavault, P., V. Sakanyan, and P. Thalouarn. 1995. Divergent evolution of two plastid genes, *rbcL* and *atpB*, in a non-photosynthetic parasitic plant. Plant Molecular Biology **29**:1071–1079.

Delavault, P. M., N. M. Russo, N. A. Lusson, and P. A. Thalouarn. 1996. Organization of the reduced plastid genome of *Lathraea clandestina*, an achlorophyllous parasitic plant. Physiologia Plantarum **96**: 674–682.

dePamphilis, C. W. 1995. Genes and genomes. In Parasitic Plants, eds. M. C. Press and J. D. Graves, pp. 176–205. Chapman & Hall, London.

dePamphilis, C. W., and J. D. Palmer. 1989. Evolution and function of plastid DNA: a review with special reference to nonphotosynthetic plants. In Physiology, Biochemistry, and Genetics of Nongreen Plastids, eds. C. D. Boyer, J. C. Shannon, and R. C. Hardison, pp. 182–202. American Society of Plant Physiologists, Rockville, Maryland.

dePamphilis, C. W., and J. D. Palmer. 1990. Loss of photosynthetic and chlororespiratory genes from the plastid genome of a parasitic flowering plant. Nature (London) **348**:337–339.

dePamphilis, C. W., N. D. Young, and A. D. Wolfe. 1997. Evolution of plastid gene *rps2* in a lineage of hemiparasitic and holoparasitic plants: many losses of photosynthesis and complex patterns of rate variation. Proceedings of the National Academy of Sciences U.S.A. **94**:7367–7372.

Dodge, J. D. and G. B. Lawes. 1974. Plastid ultrastructure in some parasitic and semi-parasitic plants. Cytobiology **9**:1–9.

Donoghue, M. J., and J. A. Doyle, 1989. Phylogenetic analysis of angiosperms and the relationships of Hamamelidae. In Evolution, Systematics, and Fossil History of the Hamamelidae, eds. P. R. Crane and S. Blackmore, pp. 17–45. Clarendon Press, Oxford.

Downie, S. R., and J. D. Palmer. 1992. Use of chloroplast DNA rearrangements in reconstructing plant phylogeny. In Molecular Plant Systematics, eds. P. S.

Soltis, D. E. Soltis, and J. J. Doyle, pp. 14–35. Chapman & Hall, New York.

Duff, R. J., and D. L. Nickrent. 1997. Characterization of mitochondrial small-subunit ribosomal RNAs from holoparasitic plants. Journal of Molecular Evolution **45**:631–639.

Ems, S. C., C. W. Morden, C. K. Dixon, K. H. Wolfe, C. W. dePamphilis, and J. D. Palmer. 1995. Transcription, splicing and editing of plastid RNAs in the nonphotosynthetic plant *Epifagus virginiana.* Plant Molecular Biology **29**:721–733.

Eplee, R. E. 1981. Striga's status as a plant parasite in the United States. Plant Disease **56**:951–954.

Escalante, A. A., and F. J. Ayala. 1995. Evolutionary origin of *Plasmodium* and other Apicomplexa based on rRNA genes. Proceedings of the National Academy of Sciences U.S.A. **92**:5793–5797.

Fagerlind, F. 1948. Beitrage zür Kenntnis der Gynaceummorphologie und Phylogenie der Santalales-Familien. Svensk Botanisk Tidskrift **42**:195–229.

Felsenstein, J. 1978. Cases in which parsimony or compatibility will be positively misleading. Systematic Zoology **27**:401–410.

Freyer, R., K. Neckermann, R. M. Maier, and H. Kössel. 1995. Structural and functional analysis of plastid genomes from parasitic plants: loss of an intron within the genus *Cuscuta.* Current Genetics **27**:580–586.

Gaut, B., S. Muse, W. Clark, and M. Clegg. 1992. Relative rates of nucleotide substitution at the *rbcL* locus of monocotyledonous plants. Journal of Molecular Evolution **35**:292–303.

Gaut, B. S., B. R. Morton, B. C. McCaig, and M. T. Clegg. 1996. Substitution rate comparisons between grasses and palms: synonymous rate differences at the nuclear gene *Adh* parallel rate differences at the plastid gene *rbcL.* Proceedings of the National Academy of Sciences U.S.A. **93**:10274–10279.

Gockel, G., W. Hachtel, S. Baier, C. Fliss, and M. Henke. 1994. Genes for components of the chloroplast translational apparatus are conserved in the reduced 73-kb plastid DNA of the nonphotosynthetic euglenoid flagellate *Astasia longa.* Current Genetics **26:** 256–262.

Gutell, R. R. 1993. Collection of small subunit (16S and 16S-like) ribosomal RNA structures. Nucleic Acids Research **21**:3051–3054.

Haberhausen, G., and K. Zetsche. 1992. Nucleotide sequences of the *rbcL* gene and the intergenic promoter region between the divergently transcribed *rbcL* and *atpB* genes in *Ipomoea purpurea* (L.). Plant Molecular Biology **18**:823–825.

Haberhausen, G., and K. Zetsche. 1994. Functional loss of all *ndh* genes in an otherwise relatively unaltered plastid genome of the holoparasitic flowering plant *Cuscuta reflexa.* Plant Molecular Biology **24**:217–222.

Haberhausen, G. K., Valentin, and K. Zetsche. 1992. Organization and sequence of photosynthetic genes from the plastid genome of the holoparasitic flowering plant *Cuscuta reflexa.* Molecular and General Genetics **232**:154–161.

Harms, H. 1935. Hydnoraceae. In Die Natürlichen Pflanzenfamilien, eds. A. Engler and H. Harms, pp. 282–295. W. Engelmann, Leipzig.

Hong, D.-Y. 1984. Taxonomy and evolution of the Veroniceae (Scrophulariaceae) with special reference to palynology. Opera Botanica **75**:1–60.

Hull, R. J., and O. A. Leonard. 1964. Physiological aspects of parasitism in mistletoes (*Arceuthobium* and *Phoradendron*). 2. The photosynthetic capacity of mistletoes. Plant Physiology **39**:1008–1017.

Kellogg, E. A., and N. D. Juliano. 1997. The structure and function of RuBisCo and their implications for systematic studies. American Journal of Botany **84**:413–428.

Kuijt, J. 1968. Mutual affinities of Santalalean families. Brittonia **20**:136–147.

Kuijt, J. 1969. The Biology of Parasitic Flowering Plants. University of California Press, Berkeley.

Kumar, S., K. Tamura, and M. Nei. 1993. MEGA: Molecular Evolutionary Genetics Analysis, version 1.01. The Pennsylvania State University, University Park.

Li, W.-H. 1983. Evolution of duplicate genes and pseudogenes. In Evolution of Genes and Proteins, eds. M. Nei and R. K. Koehn, pp. 14–37. Sinauer Associates, Sunderland, Massachusetts.

Machado, M. A., and K. Zetsche. 1990. A structural, functional and molecular analysis of plastids of the holoparasites *Cuscuta reflexa* and *Cuscuta europaea.* Planta **181**:91–96.

Maddison, W. P. 1989. Reconstructing character evolution on polytomous cladograms. Cladistics **5**:365–377.

Maddison, W. P., and D. R. Maddison. 1992. MacClade, Analysis of Phylogeny and Character Evolution, version 3.01. Sinauer Associates, Sunderland, Massachusetts.

Mayfield, S. P., C. B. Yohn, A. Cohen, and A. Danon. 1995. Regulation of chloroplast gene expression. Annual Review of Plant Physiology and Plant Molecular Biology **46**:147–166.

Minkin, J. P., and W. H. Eshbaugh. 1989. Pollen morphology of the Orobanchaceae and rhinanthoid Scrophulariaceae. Grana **28**:1–18.

Morden, C. W., K. H. Wolfe, C. W. dePamphilis, and J. D. Palmer. 1991. Plastid translation and transcription genes in a nonphotosynthetic plant: intact, missing and pseudo genes. European Molecular Biology Organization Journal **10**:3281–3288.

Muller, J. 1981. Fossil pollen records of extant angiosperms. Botanical Review **47**:1–142.

Musselman, L. J. 1980. The biology of *Striga, Orobanche,* and other root-parasitic weeds. Annual Review of Phytopathology **18**:463–489.

Nelissen, B., Y. Van de Peer, A. Wilmotte, and R. DeWachter. 1995. An early origin of plastids within the cyanobacterial divergence is suggested by evolutionary trees based on complete 16S rRNA sequences. Molecular Biology and Evolution **12**:1166–1173.

Nickrent, D. L., and R. J. Duff. 1996. Molecular studies of parasitic plants using ribosomal RNA. In Advances in Parasitic Plant Research, eds. M. T. Moreno, J. I. Cubero, D. Berner, D. Joel, L. J. Musselman, and

C. Parker, pp. 28–52. Junta de Andalucia, Dirección General de Investigación Agraria, Cordoba, Spain.

Nickrent, D. L., and C. R. Franchina. 1990. Phylogenetic relationships of the Santalales and relatives. Journal of Molecular Evolution 31:294–301.

Nickrent, D. L., and D. E. Soltis. 1995. A comparison of angiosperm phylogenies based upon complete 18S rDNA and *rbcL* sequences. Annals of the Missouri Botanical Garden 82:208–234.

Nickrent, D. L., and E. M. Starr. 1994. High rates of nucleotide substitution in nuclear small-subunit (18S) rDNA from holoparasitic flowering plants. Journal of Molecular Evolution 39:62–70.

Nickrent, D. L., Y. Ouyang, and C. W. dePamphilis. 1995. Presence of plastid genes in representatives of the holoparasitic families Balanophoraceae, Hydnoraceae, and Rafflesiaceae. American Journal of Botany (suppl.) 82:75.

Nickrent, D. L., R. J. Duff, and D. A. M. Konings. 1997a. Structural analyses of plastid-encoded 16S rRNAs in holoparasitic angiosperms. Plant Molecular Biology 74:731–743.

Nickrent, D. L., Y. Ouyang, R. J. Duff, and C. W. dePamphilis. 1997b. Do nonasterid holoparasitic flowering plants have plastid genomes? Plant Molecular Biology 34:717–729.

Olmstead, R. G., and P. A. Reeves. 1995. Evidence for the polyphyly of the Scrophulariaceae based on chloroplast *rbcL* and *ndhF* sequences. Annals of the Missouri Botanical Garden 82:176–193.

Olsen, G. J. 1987. Earliest phylogenetic branchings: comparing rRNA-based evolutionary trees inferred with various techniques. Cold Spring Harbor Symposium on Quantitative Biology 52:825–837.

Palmer, J. D. 1990. Contrasting modes and tempos of genome evolution in land plant organelles. Trends in Genetics 6:115–120.

Palmer, J. D. 1992. Mitochondrial DNA in plant systematics: applications and limitations. In Molecular Plant Systematics, eds. P. S. Soltis, D. E. Soltis, and J. J. Doyle, pp. 36–49. Chapman & Hall, New York.

Pazy, B., U. Plitmann, and O. Cohen. 1996. Bimodal karyotype in *Cynomorium coccineum* L. and its systematic implications. Journal of the Linnaean Society, London 120:279–281.

Pennell, F. W. 1935. The Scrophulariaceae of Eastern Temperate North America. Academy of Natural Sciences, Philadelphia.

Press, M. C., and J. D. Graves. 1995. Parasitic Plants. Chapman & Hall, London.

Press, M. C., N. Shah, and G. R. Stewart. 1986. The parasitic habit: trends in metabolic reductionism. In Biology and Control of *Orobanche*, ed. S. J. ter Borg, pp. 96–106. LH/VPO, Wageningen, Netherlands.

Press, M. C., S. Smith, and G. R. Stewart. 1991. Carbon acquisition and assimilation in parasitic plants. Functional Ecology 5:278–283.

Price, P. 1980. Evolutionary Biology of Parasites. Monographs in Population Biology No. 15. Princeton University Press, Princeton, New Jersey.

Reed, C. F. 1955. The comparative morphology of the Olacaceae, Opiliaceae, and Octoknemaceae. Memorias da Sociedade Broteriana 10:29–79.

Roy, H., and S. A. Nierzwicki-Bauer. 1991. RuBisCO: genes, structure, assembly, and evolution. In The Photosynthetic Apparatus: Molecular Biology and Operation, eds. L. Bogorad and I. K. Vasil, pp. 347–364. Academic Press, San Diego.

Schreuder, H. A., S. Knight, P. M. G. Curmi, I. Andersson, D. Cascio, C.-I. Brändén, and D. Eisenberg. 1993. Formation of the active site of ribulose-1,5-bisphosphate carboxylase/oxygenase by a disorder-order transition from the unactivated to the activated form. Proceedings of the National Academy of Sciences U.S.A. 90:9968–9972.

Schuster, W., R. Ternes, V. Knoop, R. Hiesel, B. Wissinger, and A. Brennicke. 1991. Distribution of RNA editing sites in *Oenothera* mitochondrial mRNAs and rRNAs. Current Genetics 20:397–404.

Shimada, H., and M. Sugiura. 1991. Fine structural features of the chloroplast genome: comparison of the sequenced chloroplast genomes. Nucleic Acids Research 19:983–995.

Sleumer, H. O. 1984. Flora Neotropica. Olacaceae Monograph No 38. New York Botanical Garden, New York.

Solms-Laubach, H. 1894. Hydnoraceae. In Die Natürlichen Planzenfamilien, Part III, eds. A. Engler and K. Prantl, pp. 282–285. W. Engelmann, Leipzig.

Soltis, D. E., P. S. Soltis, D. L. Nickrent, L. A. Johnson, W. J. Hahn, S. B. Hoot, J. A. Sweere, R. K. Kuzoff, K. A. Kron, M. W. Chase, S. M. Swensen, E. A. Zimmer, S.-M. Chaw, L. J. Gillespie, W. J. Kress, and K. J. Sytsma. 1997. Angiosperm phylogeny inferred from 18S ribosomal DNA sequences. Annals of the Missouri Botanical Garden 84:1–49.

Sper-Whitis, G. L., J. L. Moody, and J. C. Vaughn. 1996. Universality of mitochondrial RNA editing in cytochrome-c oxidase subunit I (*coxI*) among the land plants. Biochimica Biophysica Acta 1307:301–308.

Stewart, G. R., and M. C. Press. 1990. The physiology and biochemistry of parasitic angiosperms. Annual Review of Plant Physiology and Plant Molecular Biology 41:127–151.

Susek, R. E., and J. Chory. 1992. A tale of two genomes: role of a chloroplast signal in coordinating nuclear and plastid genome expression. Australian Journal of Plant Physiology 19:387–399.

Takhtajan, A. L. 1987. Sistema magnoliofitov [in Russian]. Nauka, Leningrad.

Takhtajan, A. L., N. R. Meyer, and V. N. Kosenko. 1985. Morfologiya pyl'tsy i klassifikatsiya semeystva Rafflesiaceae s. 1. Botanicheskii Zhurnal 70:153–162.

Taylor, G. W., K. H. Wolfe, C. W. Morden, C. W. dePamphilis, and J. D. Palmer. 1991. Lack of a functional plastid tRNA[cys] gene is associated with loss of photosynthesis in a lineage of parasitic plants. Current Genetics 20:515–518.

Thalouarn, P., M.-C. Arnaud, and S. Renaudin. 1989. Evidence of ribulose-bisphosphate carboxylase in the Scrophulariaceae holoparasite *Lathraea clandestina* L.

Comptes Rendu Academy of Science Paris **309**: 275–280.

Thalouarn, P., C. Theodet, and S. Renaudin. 1991. Evidence of plastid and nuclear genes for the large and small subunits of Rubisco in the Scrophulariaceae holoparasite *Lathraea clandestina* L. Comparison with the autotroph *Digitalis purpurea* L. and hemiparasite *Melampyrum pratense* L. Comptes Rendu Academy of Science Paris **312**:1–6.

Thalouarn, P., C. Theodet, N. Russo, and P. Delavault. 1994. The reduced plastid genome of a nonphotosynthetic angiosperm *Orobanche hederae* has retained the *rbcL* gene. Plant Physiology and Biochemistry **32**:233–242.

Thieret, J. W. 1967. Supraspecific classification in the Scrophulariceae: a review. Sida 387–106.

Thorne, R. F. 1992. An updated phylogenetic classification of the flowering plants. Aliso **13**:365–389.

Tocher, R. D., S. W. Gustafson, and D. M. Knutson. 1984. Water metabolism and seedling photosynthesis in dwarf mistletoes. **In** Biology of Dwarf Mistletoes: Proceedings of the Symposium, eds. F. G. Hawksworth and R. F. Scharpf, pp. 62–69. USDA, Rocky Mt. Forest & Range Experimental Station, Ft. Collins, Colorado.

van der Vies, S. M., P. V. Viitanen, A. A. Gatenby, G. H. Lorimer, and R. Jaenicke. 1992. Conformational states of ribulosebisphosphate carboxylase and their interaction with chaperonin 60. Biochemistry **31**:3635–3644.

Walsh, M. A., E. A. Rechel and T. M. Popovich. 1980. Observations of plastid fine structure in the holoparasitic angiosperm *Epifagus virginiana*. American Journal of Botany **67**:833–837.

Weber, H. C. 1980. Evolution of parasitism in Scrophulariceae and Orobanchaceae. Plant Systematics and Evolution **136**:217–232.

Wettstein, R. 1897. Scrophulariceae. **In** Die Natürlichen Pflanzenfamilien Nachtrage I–IV, eds. A. Engler and K. Prantl, pp. 293–299. W Engelmann, Leipzig.

Wiens, D., and B. A. Barlow. 1971. The cytogeography and relationships of the viscaceous and eremolepidaceous mistletoes. Taxon **20**:313–332.

Wilson, R. J. M., P. W. Denny, P. R. Preiser, K. Rangachari, K. Roberts, A. Roy, A. Whyte, M. Strath, D. J. Moore, P. W. Moore, and D. H. Williamson. 1996. Complete gene map of the plastid-like DNA of the malaria parasite *Plasmodium falciparum*. Journal of Molecular Biology **261**:155–172.

Wimpee, C. F., R. L. Wrobel, and D. K. Garvin. 1991. A divergent plastid genome in *Conopholis americana,* an achlorophyllous parasitic plant. Plant Molecular Biology **17**:161–166.

Wimpee, C. F., R. Morgan, and R. L. Wrobel. 1992a. An aberrant plastid ribosomal RNA gene cluster in the root parasite *Conopholis americana*. Plant Molecular Biology **18**:275–285.

Wimpee, C. F., R. Morgan, and R. L. Wronel. 1992b. Loss of transfer RNA genes from the plastid 16S–23S ribosomal RNA gene spacer in a parasitic plant. Current Genetics **21**:417–422.

Wolfe, A. D., and C. W. dePamphilis. 1995. Systematic implications of relaxed functional constraints on the RUBISCO large subunit in parasitic plants of the Scrophulariaceae and Orobanchaceae. American Journal of Botany (suppl.) **82**:6.

Wolfe, A. D., and C. W. dePamphilis. 1997. Alternate paths of evolution for the photosynthetic gene *rbcL* in four nonphotosynthetic species of *Orobanche*. Plant Molecular Biology **33**:965–977.

Wolfe, K. H., D. S. Katz-Downie, C. W. Morden, and J. D. Palmer. 1992a. Evolution of the plastid ribosomal RNA operon in a nongreen parasitic plant: accelerated sequence evolution, altered promoter structure, and tRNA pseudogenes. Plant Molecular Biology **18**:1037–1048.

Wolfe, K. H., C. W. Morden, S. C. Ems, and J. D. Palmer. 1992b. Rapid evolution of the plastid translational apparatus in a nonphotosynthetic plant: loss or accelerated sequence evolution of tRNA and ribosomal protein genes. Journal of Molecular Evolution **35**: 304–317.

Wolfe, K. H., C. W. Morden, and J. D. Palmer. 1992c. Function and evolution of a minimal plastid genome from a nonphotosynthetic parasitic plant. Proceedings of the National Academy of Sciences U.S.A. **89**: 10648–10652.

Wu, C.-I., and W.-H. Li. 1985. Evidence for higher rates of nucleotide substitution in rodents than in man. Proceedings of the National Academy of Sciences U.S.A. **82**:1741–1745.

9

Estimating Rate and Time in Molecular Phylogenies: Beyond the Molecular Clock?

Michael J. Sanderson

The study of rates of character evolution has been a cornerstone of evolutionary biology since the pioneering work of Simpson (1944). It has occupied a similar position in molecular evolutionary studies since Zuckerkandl and Pauling's (1962, 1965) proposal of the molecular clock. There is a fascinating contrast between these two works, however. Simpson used information about time, from the fossil record, to draw inferences about rates and modes of evolution. His main conclusion was that such rates are highly variable. Although also using information from fossils, Zuckerkandl and Pauling came to just the opposite conclusion about rates of protein evolution. They then argued that if proteins evolved at a roughly constant rate, a study of rates and modes of evolution could be used to say something about timing of events in evolutionary history. Both these ideas about the tempo of character evolution have achieved nearly the status of null hypotheses in their respective disciplines. Although Simpson clearly inferred that some morphological rates have been nearly linear, or "clock-like" over at least moderate periods of time (e.g., Simpson, 1944, pp. 203–204), few paleontologists or morphologists give credence to the notion of morphological clocks. And although there is indisputable evidence that many genes and proteins do not evolve at a con-

stant rate through time (Britten, 1986; Avise, 1994), molecular rate constancy continues to be viewed as a reasonable model even across vast reaches of the tree of life (Wray et al., 1996).

One obvious reason for this difference is that there was no neutral theory that predicted rate constancy at the morphological level, and thus never any process-based justification for using morphological divergence to date events in history. The pervasiveness of the null model of molecular rate constancy, however, led to many reconstructions of timings of key evolutionary events, often without even a passing attempt to test for the existence of a clock, even in cases where the neutral theory might not be expected to hold, such as in nonsynonymous substitution rates. A classic example is Ramshaw et al.'s (1972) reconstruction of an angiosperm age of 300–400 Ma based on amino acid replacement rates in cytochrome *c,* clearly at odds with the Early Cretaceous origin strongly suggested by the fossil record (Doyle and Donoghue, 1993).

Despite this contrast there are parallels between the development of the study of morphological and molecular rates of evolution. Just as the fossil record led Simpson to conclude that rates could vary over time and across taxa, evidence left in genes and proteins eventually led molecular evolutionists to conclude the same

---

I thank Steve Nadler, the editors and their students, and an anonymous reviewer for comments on the manuscript.

thing about rates at the molecular level. It was also clear from the outset that rates varied among regions of genes and proteins subject to constraints imposed by the genetic code, protein structure or function, expression, and so on. This evidence continues to play a key role in testing theories of evolution at the molecular level, including neutralist versus selectionist models (Kimura, 1983; Gillespie, 1991). Tests of rate constancy across lineages became possible even in the absence of a fossil record using relative rate tests (Sarich and Wilson, 1967), and considerable heterogeneity across lineages was discovered (Britten, 1986; Avise, 1994). Notable cases included the higher rate in rodents versus primates (Wu and Li, 1985), grasses versus palms (Wilson et al., 1990; Gaut et al., 1992), and annual versus perennial angiosperms (Savard et al., 1994).

The literature on rates has been reviewed with respect to the neutral theory (Gillespie, 1986, 1991) and with respect to phylogenetic inference and the many potential biases that phylogenetic practice must contend with (Mindell and Thacker, 1996). Reviews of the empirical literature have been infrequent and not comprehensive (Britten, 1986; Wolfe et al., 1987; Gaut et al., 1993; Avise, 1994). Estimation of divergence times in particular has been reviewed briefly by Springer (1995), and Hillis et al. (1996), who both emphasize calibration issues. However, there has not been a comprehensive review of the methodology for assessing rate variation across lineages and for estimating divergence times. Many of these methods have been developed within the last 5 years. This paper reviews methods aimed at estimating lineage specific rates of evolution and/or divergence times. It does not consider the equally rich and biologically interesting set of issues associated with estimating rates across different entities *within the same lineage:* that is, rates in different genes, different sites, synonymous versus nonsynonymous rates, and so forth. Some of these issues can be easily built into the models of molecular evolution used to study lineage variation in rate (e.g., third position versus first or second positions in codons), but others raise technical problems that are daunting in their own right, such as a truly correct inclusion of synonymous

versus nonsynonymous rate parameters in models of protein coding sequences (Li et al., 1985; Goldman and Yang, 1994; Muse and Gaut, 1994).

Some methods of estimating evolutionary rates and divergence times require a phylogeny; others do not. If a phylogeny is needed for a particular method, it will be assumed that an algorithm is available to provide it (e.g., Swofford et al., 1996). The error introduced into the estimation of rates or divergence times because of inaccurate phylogenetic estimation is not considered further.

DEFINITIONS OF TERMS

A *rate* is a change in some quantity in an interval of time. An *absolute rate* specifies the time scale, such as numbers of substitutions per million years, whereas a *relative rate* merely compares two quantities observed during the same time interval, such as the inferred number of substitutions in two sister lineages known to be the same age. In general, a rate can be thought of as a property of some deterministic or random process that proceeds in time. Thus it might be an average taken over many discrete time intervals, or it might be a parameter of some very detailed model of evolution that prescribes the instantaneous rate of change of DNA sequences. In either case it is usually something that is estimated from data via a statistical procedure for extracting signal from noise.

The various meanings of branch length have caused endless confusion. *Branch length* refers to the amount of character-state change occurring along a branch. This may be an integer in the case of discrete characters, or a real number in the case of genetic distances that are continuous valued. In trees reconstructed with the aid of a probabilistic model, branch lengths usually refer to the expected (i.e., mean) number of changes occurring as an outcome of some probability of change over some duration. Less commonly in the empirical literature but not infrequently in theoretical papers (Cavender, 1978), branch length refers simply to the *probability* of states differing at the endpoints of a branch. As such it is a real number between 0 and 1. This meaning will not be used further. Length should not be confused with duration. *Branch duration* refers to the amount of *time* associated with a branch in a tree.

Two other important terms are *dissimilarity, d,* the fraction of observed differences between sequences, and *distance, D,* the inferred number of substitutions that occurred along the lineages separating those two sequences. It is possible to estimate D from d, by using standard statistical methods such as maximum likelihood, in the context of a particular model of evolution. One way to think of a pairwise distance is as a "correction" function, Φ, that maps sequence dissimilarity onto numbers of substitutions (Steel et al., 1996). That is $D = \Phi(d)$. The phrase "correction for multiple hits" refers to the behavior of the model-based correction functions, which attempt to account for homoplasy that adds to distance but is not observed in dissimilarity.

EVOLUTIONARY MODELS

An elaborate theory has developed around the estimation of molecular rates of evolution, because understanding the details of molecular variation has allowed construction of a diversity of models of molecular evolution (Rodriguez et al., 1990). Unlike the study of morphological rates of evolution (Stanley, 1979; Walker, 1985), where models of character evolution are much rarer and estimation of rates is often a straightforward process of sampling data through time in the fossil record (but see Gittleman et al., 1996), much of rate estimation in molecular studies focuses on estimation of the rate *parameters* of the underlying model of molecular evolution, in which the data are only available at the endpoints of the process.

The diversity of stochastic models of evolutionary change in molecular characters has been reviewed thoroughly elsewhere (Rodriguez et al., 1990; Gillespie, 1991; Swofford et al., 1996). Models for change at the molecular level exist for restriction sites, nucleotide and amino acid sequences, allozymes, insertion-deletion events, and so on, but the key issues can be understood by reference to nucleotide substitution and that will be the focus in this paper. Tests of the validity of these models are discussed in Box 9.1. The foundation of all models of sequence evolution is the Poisson process, perhaps the fundamental stochastic process describing discrete changes in continuous time. The connection with the familiar Poisson distribution is that the probability of observing b substitutions (a branch length of b), in any branch of temporal duration, T, follows a Poisson distribution with mean, λT, where λ is the substitution rate.

All analyses of molecular evolutionary rates implicitly make use of a fundamental equation describing sequence evolution over time. This equation connects the rates of evolution to the probabilities of observing differences as a function of time. The probability of observing a change from nucleotide base i to j after a period of time T along a branch, $P_{ij}(T)$, is given by

$$\mathbf{P}(T) = \exp(\mathbf{R}T), \qquad (9.1)$$

where $\mathbf{R}$ is the matrix of instantaneous rates between different nucleotides, and $\mathbf{P} = \{P_{ij}(T)\}$. The form of $\mathbf{R}$ depends on the kind of assumptions made regarding the process, such as how many parameters are involved, and how they are constrained. For example, the Jukes-Cantor model has only a single unknown parameter in the matrix. A much fuller exposition of this equation is given by Swofford et al. (1996), along with a discussion of the various special cases of models embodied by the $\mathbf{R}$ matrix.

Although statistical inferences that rely on the models described above are aimed at estimating the unknown parameters in $\mathbf{R}$, it is often enough to estimate an overall rate of substitution per site, K, which is the rate averaged over all the possible kinds of transformations from base i to j. The overall rate of substitution per site per unit time is

$$K = \sum_i \pi_i \left(\sum_{j \neq i} R_{ij} \right) = -\sum_i \pi_i R_{ii}, \qquad (9.2)$$

where π_i is the equilibrium frequency of base i, and the simplification occurs because the sum of any row in $\mathbf{R}$ is zero (Cox and Miller, 1977). This equation relates the overall substitution rate to the individual parameters in $\mathbf{R}$, which is of use if the model is complex and several parameters in $\mathbf{R}$ are estimated. However, K can also be estimated more directly (see section on estimation under rate constancy).

BOX 9.1.   TESTS OF THE MODELS

Any model-based rate estimation procedure rests on the validity of the parametric model, and some attempt should be made to test that model against more general models (Gillespie and Langley, 1979; Navidi et al., 1991; Penny et al., 1992; Goldman, 1993; Yang et al., 1994). Cases in which this is done are surprisingly rare in the literature, but the few exceptions are instructive. Gillespie and Langley reanalyzed a model of protein sequence evolution that had been found to be non-rate-constant assuming a Poisson process model. A more general, but still rate-constant episodic model was found to fit the data adequately. Goldman (1993) also found cases in which conclusions based on one model were overturned upon the discovery of a significantly better model. The conclusion was doubly illustrative because both a rate-constant and non-rate-constant model were found to be "adequate" when tested against a very general unconstrained model, but the non-constant model was still significantly better than the rate-constant model. Hence, a general caveat is to not merely test for rate constancy but to test the test.

   Few general conclusions are available about the adequacy of models of molecular evolution, but it is important to review briefly the extensive work of Gillespie (1991), which calls into question the utility of the Poisson model, and the response by Goldman (1994), which defends it. Gillespie examined the quantity $R(T) = V(T)/m(T)$, the variance to mean ratio of substitutions in lineages as a function of divergence time, in nucleotide sequence data. This ratio, also known as the index of dispersion, is 1.0 for a Poisson process and greater than 1.0 for many other stochastic processes, including so-called episodic ones in which bursts of change are clustered in time (Gillespie, 1991; Takahata, 1991). Gillespie's survey (1989, 1991) indicated that in real sequences $R(T)$ varied from less than 1.0 to 40, with a mean of about 7.0 for amino acid replacements, suggesting a significant departure from a Poisson process. Goldman (1994) has recently responded to this work by arguing that values of $R(T)$ have been inflated by several factors including: (1) deviation of the true phylogeny from an assumed star phylogeny, (2) inappropriate choice of Poisson model (i.e., the form of the **R** matrix), (3) inappropriate selection of weighting factors aimed at removing lineage specific rate variation in Gillespie's (1989) analysis, and (4) nonindependence of paired comparisons (see also Bulmer, 1989). This inflation may be enough to account for observed indices of dispersion even if a Poisson model does hold. On the other hand, Goldman admits a weakness in his analysis is the lack of consideration of the sequences' degeneracy classes, which was an integral element of Gillespie's analysis. Clearly much more work is needed in this important area.

Making matters more complicated is a generalization of the preceding evolutionary model to allow rate variation across lineages. One way to do this is to multiply the rate matrix, **R** by a constant, c_k associated with branch k (note, this constant is *not* the branch length as defined above). This is the gentlest way to allow rates to vary across the tree because **R** itself is not affected. Each component of **R** retains the same relative relationship to each another, but the effective instantaneous rates are actually determined by the product c_k**R.** The model would be even more complicated if the relationship of the rates within the **R** matrix also changed among lineages. The former case has $N + r$ free parameters, where N is the number of branches and r is the number of free parameters in **R**. In the latter case, there are Nr parameters.

   The difficulty in estimating rates in this now very general model should be apparent. For a given branch, k, the relevant probability of observing sequence differences is now

$$\mathbf{P} = \exp(c_k \mathbf{R} T_k). \qquad (9.3)$$

For example, if we knew how to calculate this expression for a given **R**, c_k, and T_k, we could calculate the probability of a given kind of change along this branch and, multiplying this by the length of the sequence, L, could calculate the expected number of sites that differ between ancestor and descendant in the specified way (e.g., from site i to j, or transitions, transversions, etc.). Working backward from **P**, at best we can hope to estimate one or more (depending on how complex **R** is) products of these three unknown quantities—unless further assumptions are made.

A final generalization is to vary the matrix **R** across *sites* in a sequence. This can be modeled by selecting the rate for a given site according to some specified probability distribution which itself may be characterized by some parameter to be estimated. There have been two basic approaches which differ only in whether the rate distribution is discrete or continuous. A common discrete example is one in which a site is allowed to be either invariable (rate of zero) or have some positive rate (Palumbi, 1989). For continuous distributions the usual choice is the gamma distribution, because it models a broad class of distributions, including normal, exponential, and constant, depending on the two parameters of the gamma distribution (one parameter if only the shape of the distribution is at issue). These parameters can be set ahead of time or estimated from the data, most commonly by maximum likelihood or by fitting the moments of the inferred distribution of substitutions along branches to the expected distribution. If rates vary across sites according to a gamma distribution, the distribution of changes along branches should follow a negative binomial distribution (Yang, 1996). An interesting variation on the discrete approach uses a hidden Markov model (Felsenstein and Churchill, 1996) which lets a stochastic process choose the rates at a particular site as it marches down the sequence. This allows for rates at neighboring sites to be autocorrelated as one might expect because of functional and structural constraints. Because Yang (1996) provides a thorough review on the important subject of rate variation, it is not considered in detail in this chapter. Its relevance, however, is underscored by the point that ignoring rate variation leads to estimates of divergence times that are biased to an extent that

depends on the distance between the sequences (Yang, 1996).

TESTS OF RATE CONSTANCY

Most available methods for reconstructing rates and divergence times require explicit or implicit assumption of rate constancy. Such tests should always be performed prior to attempting to draw inferences about those rates or times. Of course, these tests also have implications in testing theories of molecular evolution, such as the molecular clock, but these broader process issues are not considered further here.

Relative Rate Tests

The simplest and most widely used tests of rate constancy compare two lineages descended from a common ancestor, such that branch i and j are sister lineages, versus a more distantly related outgroup taxon. Sarich and Wilson (1967) pointed out that the relative distance between the outgroup and each ingroup is expected to be the same under rate constancy. The only complications that hinder application of this insight relate to (1) how to choose and estimate an appropriate measure of distance, and (2) how to establish the statistical significance of differences in relative distance. See Box 9.2.

Most of these tests use distances derived from pairwise sequence dissimilarities as measures of divergence. Wu and Li's (1985) test is used more commonly than any other. It uses a Kimura two-parameter model of sequence evolution to estimate the three pairwise distances in a three-taxon statement, $(OG, (A, B))$, in which OG is the outgroup and A and B are the two ingroup taxa. Let $D(x, y)$ be the distance between taxa x and y. The difference in distance from outgroup to each ingroup is $\Delta = D(OG, A) - D(OG, B)$. The test uses the same model to predict the variance of Δ, under the assumption of rate constancy. A standard two-tailed z-test is then applied to

$$z = \frac{\Delta}{\sqrt{\text{var}(\Delta)}} \qquad (9.4)$$

which is expected to be normally distributed with mean zero and variance one. An extreme

BOX 9.2. RELATIVE RATE TESTS OF RATE CONSTANCY

The most recent common ancestor of two taxa provides a useful temporal landmark for comparison of rates in two lineages. If rates are constant, the accumulated changes from the common ancestor of A and B (Fig. A) are expected to be the same except for differences due to sampling error (finite sequences). The same is true for the accumulated changes from the outgroup taxon to A and B. Relative rate tests construct a statistic, Δ, the difference in some measure of distance between the outgroup and each of the two ingroup taxa, which is expected to be zero under rate constancy.

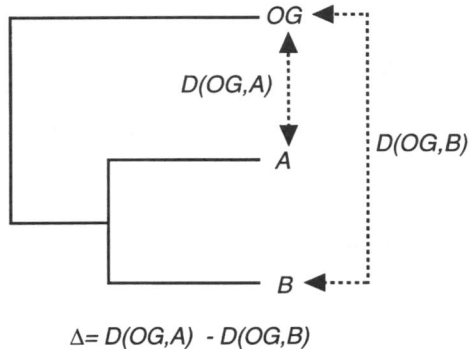

$$\Delta = D(OG,A) - D(OG,B)$$

Fig. A. Schematic phylogeny showing comparisons used in relative rate tests. $D(A,B)$ refers to the distance between taxa A and B. OG is the outgroup and Δ is the statistic tested against the expectation of zero when rates are constant.

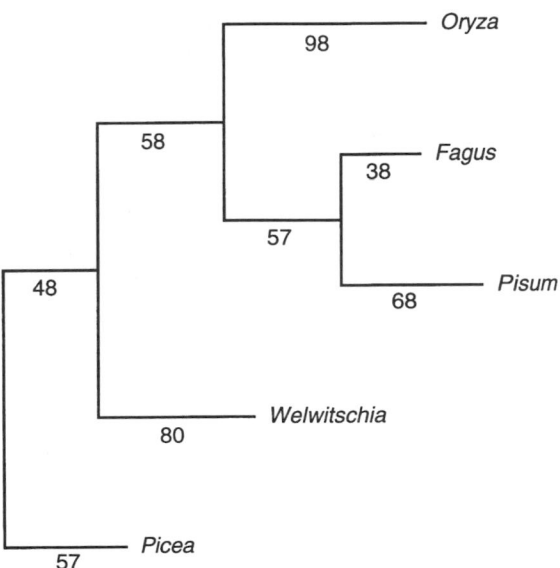

Fig. B. Example data set and phylogeny used to illustrate various rate tests. Sequence data were obtained for the chloroplast *rbcL* gene for five species of seed plants (data provided in electronic form by M. J. Donoghue; see Chase, Soltis, Olmstead et al., 1993). The topology is based on a synthesis of hypotheses about seed plant relationships. Branch lengths were estimated using the default optimization algorithm in PAUP. Note that to estimate the first two ingroup branch lengths, it was necessary to include an additional outgroup, *Lycopodium*. Without this, PAUP's algorithms have no way to distribute relative lengths between the branch leading to *Picea* and that subtending the remaining seed plants.

Various relative rate tests use different measures of distance, D, and assess the statistical significance of the deviation of Δ from zero in different ways. Wu and Li's (1985) method accounts for differences in transitions and transversions and attempts to correct for multiple hits. They provide analytical expressions for the variance of Δ which are used to construct an "exact" z-score. Alternatively, bootstrapping can be used to resample from the original sequences, estimate the variance of Δ, and determine a nonparametric z-score. For the *rbcL* data and tree in Figure B, table A reports all possible relative rate tests using both these methods for estimating a significant departure from rate constancy. Note the close correspondence between analytical and bootstrap results. Table B reports the Tajima relative rate tests. In that test, a simple χ^2 statistic (see Eq. 9.5) is tested with one degree of freedom. The measure of distance used is much simpler and does not keep track of transitions versus transversions or correct for multiple hits. Nonetheless, results are quite similar. Although relative rate tests are significant using one method and not using another in a few cases, in these cases, the test values are near the corresponding significance cutoff points for the tests.

Table A. Relative rate tests (method of Wu and Li, 1985)

(Outgroup	(Ingroup A	IngroupB))	z("exact")[a]	z(bootstrap)[b]
(PICEA	(ORYZA	WELWITSCHI))	+2.0317*	+2.0767*
(WELWITSCHI	(ORYZA	FAGUS))	+2.4615*	+2.5351*
(WELWITSCHI	(ORYZA	PISUM))	+1.4581	+1.4821
(PICEA	(ORYZA	FAGUS))	+2.3334*	+2.1517*
(PICEA	(ORYZA	PISUM))	+0.2304	+0.2579
(ORYZA	(FAGUS	PISUM))	−3.661***	−3.3924***
(PICEA	(FAGUS	WELWITSCHI))	−0.0037	+0.0739
(PICEA	(PISUM	WELWITSCHI))	+1.9063	+1.8589
(WELWITSCHI	(FAGUS	PISUM))	−0.9722	−0.9753
(PICEA	(FAGUS	PISUM))	−2.3971*	−2.7942**

[a]Positive z-score means higher rate in first ingroup. Significance levels: * = $P < 0.05$; ** = $P < 0.01$; *** = $P < 0.001$.
[b]Bootstrap estimates based on 500 replicates.

Table B. Relative rate tests (method of Tajima, 1993)

(Outgroup	(Ingroup A | Ingroup B))	χ^{2a}
(PICEA	(ORYZA | WELWITSCHI))	4.6875*
(WELWITSCHI	(ORYZA | FAGUS))	3.0303
(WELWITSCHI	(ORYZA | PISUM))	2.2989
(PICEA	(ORYZA | FAGUS))	3.5588
(PICEA	(ORYZA | PISUM))	0.0514
(ORYZA	(FAGUS | PISUM))	6.8571**
(PICEA	(FAGUS | WELWITSCHI))	0.0000
(PICEA	(PISUM | WELWITSCHI))	3.9836*
(WELWITSCHI	(FAGUS | PISUM))	0.2941
(PICEA	(FAGUS | PISUM))	5.9024*

[a]Significance levels: * = $P < 0.05$; ** = $P < 0.01$; *** = $P < 0.001$.

All analyses were performed using the author's program r8s. Source code available by anonymous ftp to "loco.ucdavis.edu."

value on either tail causes a rejection of rate constancy. Several approximations are used to obtain the rather complicated expressions for the variance described in Wu and Li (1985; see also Muse and Weir, 1992), the most important of which is probably the assumption that the distances are sufficiently high that z in Eq. 9.4 is normally distributed. Wu and Li (1985) suggest that about 20 inferred substitutions are sufficient.

Steel et al. (1996) have suggested a much simpler test that uses an uncorrected dissimilarity as a distance. This ignores the transition/transversion biases in sequence evolution and counts all substitutions equally. Their z-test can be constructed using a much simpler equation for the variance. In general, any measure of pairwise distance can be used, relying on as simple or complex a model of sequence evolution as desired, as long as an estimate of the variance of Δ can be obtained.

Tajima (1993) describes a slight variant of the relative rate test based on χ^2. Let m_A be the number of sites in which (1) ingroup A has one nucleotide, and (2) ingroup B and the outgroup share the same nucleotide, but one that differs from that of ingroup A. Define m_B analogously. Clearly $m_A - m_B$ is expected to be zero under rate constancy, and Tajima showed that the quantity,

$$\frac{(m_A - m_B)^2}{(m_A + m_B)} \tag{9.5}$$

is distributed approximately as χ^2 with one degree of freedom (Box 9.2). Despite the absence of the details of the substitution process that are found in the Wu and Li test, Tajima provides simulation evidence to suggest that the power of his method is about the same.

Any tree with reconstructed branch lengths can also provide the raw data for relative rate tests. Most commonly these data will be parsimony reconstructions of branch lengths, although some authors have worried about parsimony's bias toward underestimating length (Nei, 1987). The most conservative test is a simple binomial test of the proportion of length allocated to the lineage from the common ancestor of the ingroups to one ingroup against the expected frequency of 50% (Mindell and Honeycutt, 1990).

Relative rate tests based on statistical likelihood (Felsenstein, 1988) have been described by Muse and Weir (1992). In a three-taxon tree, the likelihood of a specific model assuming rate constancy is compared against the likelihood of the same model with rate constancy removed. As sample size increases it is known that a test based on the likelihood ratio, specifically,

$$G = -2\ln\left(\frac{L^0}{L^1}\right) \tag{9.6a}$$

or, equivalently,

$$G = -2(\ln L^0 - \ln L^1), \tag{9.6b}$$

converges to the χ^2 distribution, with degrees of freedom equal to the difference in the number of free parameters, when the null hypothesis is true. Here L^0 is the maximum likelihood value under the null (rate constant) hypothesis, and L^1 is the maximum likelihood value under the alternative (rate variable) hypothesis. Muse and Gaut (1994) showed that this χ^2 approximation worked well for sequences longer than 250 nt. The power of the likelihood ratio test versus variance-based distance tests was usually similar, but the likelihood test was occasionally much more powerful at detecting differences in rate concentrated in either transitions or transversions in one lineage.

Relative rate tests have been generalized to test for rate constancy between sister clades in which divergences among and between all the taxa are considered. This is done by generalizing Δ (see foregoing discussion) to be the mean of all pairwise differences in distance. In other words, one examines

$$\Delta^* = \frac{1}{nm} \sum_{i=1}^{n} \sum_{j=1}^{m} \Delta(i, j) \tag{9.7}$$

relative to the variance in Δ^*, as in Eq. 9.4, where $\Delta(i, j)$ is the difference in distance from an outgroup to ingroups i and j. Once again the difficulty is in obtaining good estimates of the variance of Δ^* so that confidence limits on the test can be implemented. Li and Bousquet (1992) and Steel et al. (1996) have discussed these methods further.

The nonparametric bootstrap may provide a general method of putting confidence limits on tests based on Δ or Δ^* in complex models or cases in which the assumptions underlying the analytical expressions for the variance are not met (Dopazo, 1994). The bootstrap resamples the original sequence data many times (sampling with replacement among the columns of characters in the aligned sequences), constructing Δ_i for the ith replicate. Examination of the distribution of the Δ_i allows construction of confidence intervals on the observed Δ directly, or in cases in which this distribution is nearly normal, the variance of Δ can be estimated and simply plugged into Eq. 9.4 to perform a z-test.

Variance to Mean Ratios

Under a Poisson process with a constant rate, the variance in the number of substitutions in lineages of equal duration is expected to equal the mean, which suggests a test of rate constancy that can be applied to any set of lineages of the same duration, such as the extant lineages descended from a common ancestor. More lineages are better, and therefore this test has been applied to so-called star phylogenies in which several lineages are thought to have originated more or less contemporaneously. This test is therefore restricted to nodes that are "hard polytomies," that is, multiple speciation events, rather than "soft polytomies" that merely represent lack of information or too much character conflict to resolve the relationships.

Consider N sequences and let D_{ij} be the pairwise distance between them. This may be any of the available model-based transformations of pairwise dissimilarity, or it may even be uncorrected dissimilarity. Let $\bar{D}$ be the mean. Kimura (1983) proposed as an estimator of the variance to mean ratio

$$\hat{r} = \frac{2\sum_{i<j}(D_{ij} - \bar{D})^2}{(n - 1)(n - 2)\bar{D}} \qquad (9.8)$$

where

$$\bar{D} = \frac{\sum_{i<j}D_{ij}}{(1/2)\,n(n - 1)} \qquad (9.9)$$

He used uncorrected distances (raw dissimilarities). For corrected distances, the above estimator is not expected to have a mean of one. Bulmer (1989) modified Eq. 9.8 to account for multiple

hits and also for the variance/covariance structure of the distances, and his estimator should have a mean of one and be distributed as χ^2. Goldman (1994) confirmed that both the expected value and the null distribution of Bulmer's corrected estimator matched predictions in a series of computer simulations, but Kimura's did not. He also showed that the assumption of a star phylogeny is key. If some taxa are actually more closely related to each other than they are to others, the variance/mean ratio will be inflated even when rates are constant.

Distance or Branch Length Tests across a Tree

It is possible to combine relative rate tests to give some sense of rate constancy over an entire tree (Takezaki et al., 1995). There are two closely related methods that rely on general linear models. Pick one outgroup and for every ingroup node construct a difference-in-distance statistic (as in Eq. 9.4), Δ_i. Let Δ be the vector of individual Δ_i values, and let $\mathbf{V}$ be the variance/covariance matrix of these, such that $V_{ij} = \text{cov}(\Delta_i, \Delta_j)$. Then the statistic,

$$U = \Delta^t \mathbf{V}^{-1} \Delta, \qquad (9.10)$$

is distributed approximately as χ^2 with $n - 1$ degrees of freedom, where n is the number of nodes tested. The components of $\mathbf{V}$ are given in Takezaki et al. (1995). The quantity U is dependent on the individual relative rate differences combined over all nodes but weighted by the covariance between branch lengths that stems from shared phylogenetic history.

A more direct method is based on the weighted comparison of all distances from the root to the tips. The vector of distances between the root and each tip, $\mathbf{b}$, is compared to the average distance, $\bar{\mathbf{b}}$, and then a new statistic, whose expectation is zero can be examined (Takezaki et al., 1995; Uyenoyama, 1995). Here the relevant statistic is

$$U = (\mathbf{b} - \bar{\mathbf{b}})^T \mathbf{V}^{-1}(\mathbf{b} - \bar{\mathbf{b}}) \qquad (9.11)$$

This is also distributed as χ^2 with $n - 1$ degrees of freedom, where n is the number of sequences.

Again, the weights depend on the phylogenetic covariance structure.

Other direct tests that take account of information about an entire tree have been available for some time. A distance-based F-test was suggested by Felsenstein (1984, 1988). In a tree constructed according to a distance method such as Fitch-Margoliash, the criterion that is being optimized is a deviation of branch lengths from the ideal under additivity (that is, that the sum of intervening branch lengths adds up to the observed pairwise difference). The goodness of fit of the tree to this model is based on the residual error variance. Comparison of this goodness of fit under a clock, U_c versus without a clock U_{nc} can be made with an F-test based on

$$F = \frac{(U_c - U_{nc}) / (n - 2)}{U_{nc} / [n(n - 1) / 2 - (2n - 3)]}$$

(9.12)

with d.f. of $n - 2$ and $n(n - 1)/2 - (2n - 3)$. Felsenstein (1988, 1993) pointed out that the assumptions for this quantity actually to follow the F distribution are probably not met if the tree and/or branch lengths are constructed by ordinary or weighted least squares methods, because of nonindependence of the branch lengths, but generalized least squares could be used to estimate branch lengths instead (Takezaki et al., 1995).

A likelihood approach using branch lengths to test for rate constancy across a tree is due to Langley and Fitch (1974). Their method assumes that branch lengths are known, perhaps from a parsimony reconstruction. The likelihood of these data is calculated as a function of the unknown branching times and a single unknown rate of evolution assumed to be constant across the tree. The model of sequence evolution assumes that all substitutions occur according to a single Poisson process. Numerical algorithms are used to maximize this likelihood and estimate the unknown parameters. Finally, the maximum likelihood rate estimate is used to generate expected branch lengths, and a χ^2 test is used to assess whether the observed branch lengths deviate from the expectation under the rate constant model. Thus, we obtain $\hat{K}$ and $\hat{t}_k$, as ML estimates, from which we can construct $\hat{T}_k$, the es-

timated duration for a particular branch. On branch k, the expected length is just $\hat{K}\hat{T}_k$, so the overall χ^2 test is constructed as

$$\sum_k \frac{(O_k - E_k)^2}{E_k} = \sum_k \frac{(O_k - \hat{K}\hat{T}_k)^2}{\hat{K}\hat{T}_k}$$

(9.13)

There are $m - k - 1$ degrees of freedom, where m is the number of branches and k is the number of internal nodes whose times are estimated (the last degree of freedom is for the overall rate, $\hat{K}$). Nei (1987) points out that use of parsimony-inferred branch lengths underestimates true branch lengths, and it may be preferable to estimate branch lengths using maximum likelihood, although there is no hard evidence that this is any less biased or has lower mean error. See Box 9.3 for an example.

Character-Based Global Tests

Felsenstein (1988, 1993) proposed a likelihood ratio analog of this method in which a rate constant model is tested against the model described earlier in Eq. 9.3, in which every branch of a tree can have its rate matrix multiplied by a free parameter. The likelihood of a given tree is first calculated assuming rate constancy. Then the likelihood is calculated without it, and the likelihood ratio test described in Eq. 9.6 is used assuming it converges on χ^2. Any model of sequence evolution can be used as long as it differs only with respect to rate constancy in the two hypotheses (it would be difficult to interpret the results otherwise). The degrees of freedom are given by the difference in the number of free parameters in the two models. In the rate constant model, one overall rate and k branching times are estimated, where k is the number of branch points in the tree (excluding the root), whereas in the unconstrained model m branch parameters are estimated, where m is the number of branches (Felsenstein, 1993). The degrees of freedom used in the χ^2 approximation to the likelihood ratio test is thus d.f. $= m - (k + 1)$. In a completely dichotomous tree, $k = N - 2$, and $m = 2N - 3$, so the number of degrees of freedom is $N - 2$. Each model may have additional parameters that are estimated, but so long as these are shared, this will not affect the degrees of freedom. This test has been used rarely (Bruns and

BOX 9.3. GLOBAL TREE-BASED TESTS OF RATE CONSTANCY

Because it is problematic to combine the statistical conclusions from multiple relative rate tests that share nodes in common, tests that simultaneously consider an entire tree are an attractive alternative. They test the global hypothesis of rate constancy. If even a single branch deviates from rate constancy by enough, it will cause rejection of the rate-constant null hypothesis. This lack of discrimination may be disadvantageous in some applications, but judicious combination of these tests with more phylogenetically restricted relative rate tests may be illuminating.

Langley and Fitch (1974) developed an especially straightforward global test that uses branch lengths. As an example, the test was run on the tree and branch lengths shown in Figure B of Box 9.2. The likelihood optimization indicated a (log) maximum likelihood value of -87.6 and an estimated (constant) rate of 138.8 nucleotide substitutions per unit time (where a unit of time is the time from the root to the tips of the tree). The algorithm also estimates the three unknown internal node times to be at 78.3%, 55.6%, and 29.7% of the time back from tip to root. Using this information and the estimated rate, expected branch lengths could be calculated. Comparison of these to the "observed" lengths in Figure B of Box 9.2 using the χ^2 test of Eq. 9.5, results in a χ^2 value of 124.3. The degrees of freedom are four because there are eight pieces of data and four parameters to estimate. This is an extremely significant deviation from the null hypothesis of rate constancy. This test is implemented in the author's 'r8s' program. Source code available by anonymous ftp to "loco.ucdavis.edu."

Felsenstein's (1988, 1993) method uses the raw sequence data directly. The *rbcL* sequence data used to estimate branch lengths in Figure B of Box 9.2 and the tree were used to test for rate constancy using this method. The method rests on a comparison of the maximum likelihood values of two different models of evolution on the same tree. One model assumes rate constancy; the other does not; PHYLIP's programs DNAMLK and DNAML, respectively, were used (as described in the manual). For the tree shown in Box 9.2, the log maximum likelihood under a clock is -4322.2, whereas the log maximum likelihood without assuming a clock was -4282.6. The quantity $-2 (\log L_{clock} - \log L_{no\ clock})$ should be distributed as with $N - 2$ degrees of freedom, where N is the number of taxa in this tree. This value is 79.4 and d.f. are 3. This is also an extremely significant deviation from rate constancy. Both these global rate tests reject rate constancy, and the level of rejection is much higher than any single relative rate test suggested in Box 9.2. This is presumably because the probabilities combine to reject (strongly) clocklike evolution over the entire phylogeny.

Szaro, 1992; Nadler, 1992; Hasegawa et al., 1993). See Box 9.3 for an example.

Pitfalls in Application of Tests of Rate Constancy

Despite their widespread use, problems with relative rate tests have been frequently discussed (Fitch, 1976; Gingerich, 1986; Springer, 1995; Mindell and Thacker, 1996). An obvious problem is that a relative rate test is local; it tests for rate constancy only in a region of a phylogeny covered by three taxa. As such, it averages over sublineages, either implicitly in ordinary relative rate tests, or explicitly in pooled relative rate tests that examine sister clades.

More to the point, these tests do not test rate constancy from a branch earlier in time to one later in time. They only test rate differences between lineages descended from a common point in time. Thus, great caution must be exercised in attempting to make statements about rates increasing or decreasing based on relative rate tests. For example, in a four-taxon tree (OG,(A,(B,C))), it is easy to imagine cases in which every three-taxon relative rate test says there is rate constancy, and yet there is a shift in rate in the branch subtending (B,C). All that is required is a parallel shift in rate in lineage A. Relative rate tests simply cannot detect such departures from rate constancy. Perhaps it could be argued that parallel rate shifts are unlikely, but this sort of "homoplasy" may be very reasonable if rates are affected by traits such as generation time that in turn undergo homoplasy.

Multiple relative rate tests are often performed to assess rate constancy over an entire phylogeny. For example, Gaut et al. (1992) examined all 595 possible relative rate tests in their study of rates of *rbcL* evolution in 35 monocots, using one outgroup. This practice raises two statistical issues. Under the best case, in which no taxa are duplicated in these multiple tests, one must still correct the significance levels to account for so-called multiple tests problems. If one performs enough tests looking to reject rate constancy at the α level, roughly α of those tests will indeed cause rejection even if the null hypothesis is true. To counteract this, significance levels must be made more stringent.

Moreover, sets of relative rates tests are usually phylogenetically nonindependent because some of the tests use the same taxa. If 50 tests are performed and 49/50 fail to reject rate constancy, even under fairly strict multiple test criteria one would ordinarily be tempted to accept the null hypothesis. However, this assumes independence of tests. Unfortunately, the dependence introduced by using the same taxa affects the size of the critical α region in unpredictable ways. This problem cannot be corrected by any simple procedure. This may explain why most workers do not pursue the matter, preferring to present a table of multiple tests and proceed in an exploratory manner, rather than in a hypothesis-testing framework one (e.g., Gaut et al., 1992; Nickrent and Starr, 1994).

Treewide character-based likelihood tests, on the other hand, may be overly sensitive to local variations in rate. They have been used less frequently than one might expect despite the existence of a clear description of their use in the PHYLIP documentation (Felsenstein, 1993). It may be that most data sets deviate significantly from a clock in *some* part of the tree, which can lead to a rejection of the strong null hypothesis of a clock over the *entire* tree. Using this test, Baldwin and Sanderson (in prep.) could not reject rate constancy in nrDNA ITS sequences for an assemblage of Hawaiian silverswords, but once enough of the California relatives were included in the analysis, deviations from rate constancy emerged. This is almost inevitably true if enough taxa are examined.

An important statistical problem is that the use of the χ^2 approximation in likelihood ratio tests may not be sufficiently accurate in practice. It is a large sample approximation, and Goldman (1993; see also Takezaki et al., 1995) suggests that it requires at least a few sites in every configuration of nucleotides among taxa, which will almost never occur except in the smallest data sets, because for T taxa there are 4^T possible configurations of nucleotides among taxa for a single site. Goldman (1993) suggests Monte-Carlo simulation of the likelihood ratio as a general solution, and finds that a mean-corrected χ^2 can often be substituted for χ^2.

A conservative procedure—that is, one that is unlikely to find rate constancy erroneously—is to use the global likelihood ratio test on an entire phylogeny. If rate constancy is found under these stringent conditions it is probably a strong result. If rate constancy across the tree is rejected then more localized tests should be done. This testing might include a combination of relative rate tests and global likelihood ratio tests on subclades of the tree (S. Muse, pers. comm.). These tests might uncover patterns in which there are a small number of subclades in which rate constancy (at different rates) occurs. If this procedure does not work, it may be necessary to invoke methods that do not require rate constancy at all.

ESTIMATION OF RATES OF EVOLUTION AND/OR TIMES OF DIVERGENCE UNDER RATE CONSTANCY

Statement of General Problem

The estimation of rate and time are inextricably linked and it is cumbersome to try to discuss them separately. All methods that estimate rates of evolution across a tree also simultaneously estimate the unknown divergence times, and vice versa. In any tree of m internal unknown node times, we must know at least one of those times to estimate the other $m - 1$ times. This is the issue of calibration (see Box 9.4). If relative rates or times are sufficient, then times can be measured relative to the (arbitrary) time of the root node of the tree, and rates can be scaled relative to the largest or smallest rate estimated. In general, we can write all the unknown parameters as the set $\theta = \{\mathbf{R}, K(\mathbf{R}), t_1, \ldots, t_{m-2}\}$, where $\mathbf{R}$ is the rate matrix of the substitution process, $K(\mathbf{R})$ is the overall substitution rate (see Eq. 9.2), and the t_i are the unknown internal node times. The general problem is to estimate all of these parameters.

BOX 9.4. CALIBRATION USING FOSSIL EVIDENCE

To obtain absolute rate estimates by using Eq. 9.14 (or indeed any of the methods described herein), at least one date must be known. This is usually obtained by reference to the fossil record or geological history. An example of the latter might be the date for the breakup of a continent (Hillis et al., 1996). The problem with attempts to assign a date to a node is that fossil and geological evidence can constrain such dates only within limits. The existence of a fossil can provide a minimum age for a nearby node but not a maximum age. A maximum age can be postulated only in those rare instances in which the fossil record is so complete that absence of fossil evidence for a lineage can be treated as true absence rather than mere failure of discovery. Likewise a vicariance event such as a continent breaking up also assigns only a minimum age for a splitting of a lineage.

Doyle and Donoghue (1993) have characterized the kind of time constraints that can be inferred from a fossil in conjunction with phylogenetic information. Suppose we wish to use a particular branching point, x, that defines a clade in a phylogeny as a calibration point for the estimation of rates (Fig. A). What sort of evidence constrains the timing of that event? A fossil can be used if enough is known about its phylogenetic relationships. In order to postulate a minimum age for node x, the age of a fossil from a subclade of x must be known, such as fossil 2. This taxon must therefore not merely belong to clade x and possess the apomorphy a that defines x, it must also belong to a clade y that is nested within x and possess those additional synapomorphies (such as b) not found in x but found in y. It is easy to assume erroneously that any fossil that shares the synapomorphies of x, such as fossil 1, will provide a minimum age of clade x. However, fossil 1 might be the sister group of x rather than be nested within it.

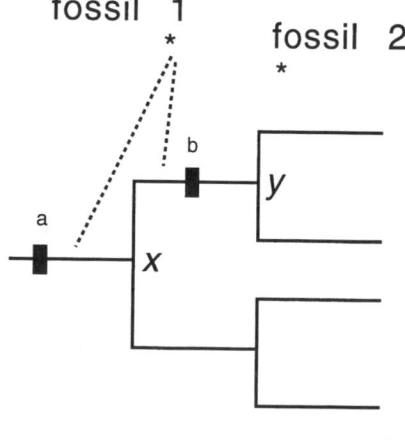

Fig. A

The Two-Species Case

Although it is possible to estimate the rate parameters in **R,** or the overall substitution rate, K, over an entire phylogeny, until very recently it has been much more common to approach the problem piecemeal using pairwise sequence comparisons singly or perhaps in small sets of comparisons. Indeed this technique has completely dominated empirical investigations of molecular evolutionary rates and is consistently cited in standard texts (Nei, 1987; Li and Graur, 1991). For two species that share a common ancestor of age t, and which have undergone an overall rate of substitution per site per unit time K, the distance that accumulates, or the mean number of substitutions per site D, that will have accumulated along the two lineages separating them will be

$$D = 2Kt \qquad (9.14)$$

So that one natural estimator of substitution rate is just

$$\hat{K} = \frac{D}{2t} \qquad (9.15)$$

Again, note that the pairwise distance, D, is not a rate. Without an accurate estimate of the divergence time nothing further can be said about rate (see Box 9.4). Note also that there is an implicit assumption that the rate is the same in both lineages descended from the divergence (splitting) event. Another way of looking at this method is that it effectively estimates one branch "length" between the two tips of a tree and pulls the branch down to its common ancestor, making two branches of equal length out of it. See Box 9.5.

Divergence times can be estimated from pairwise distances by noting that under rate constancy, K, the instantaneous substitution rate per site per time is constant, so D and t are linearly related by Eq. 9.14. By using one calibration point to estimate K, one can use another value of D to infer a different point in time. If several calibration points are available, a regression of D on t will provide an estimate of K that can then be used to predict times using standard regression methods (Box 9.5).

Tree-Based Methods Relying on Branch Length Inferences

The preceding method can be seen as a special case of a general procedure of estimating branch lengths and branch durations across an entire tree and then combining this information to obtain an estimate of rate. One method already described that uses branch length estimates directly is that of Langley and Fitch (1974). Another method is a simple generalization of the method described previously for a pair of species. In a tree in which branch lengths are available, a rate may be estimated for any pair of lineages even if they are not sister groups. Rather than using the corrected pairwise distance between them, one could sum the branch lengths (obtained by any method) and use this as an estimate of distance. As long as a relative rate test indicated constancy of rate, then Eq. 9.14 could be used by calibrating relative to the root node of the most recent common ancestor of the two taxa.

Distance methods for estimating branch lengths can also be constrained over and above the additivity assumption. If the further assumption of *ultrametricity* is assumed, meaning the total distance between root and every tip is the same, then the branching points on the resulting tree will correspond to estimates of internal node times. For example, Felsenstein (1993), in his Fitch-Margoliash program, KITSCH, constrains his additive least squares calculations by forcing ultrametricity. UPGMA might also be used. It implicitly assumes rate constancy as it builds a tree and therefore can be used to estimate times if a rate constancy holds. UPGMA has been shown to estimate branch lengths in such a way that (1) ultrametricity holds, and (2) the least square deviations from an additive tree are minimized (Farris, 1969; Chakraborty, 1977). In either case it is desirable to estimate the branch lengths on a tree reconstructed using a more accurate phylogenetic inference algorithm than these.

Tree-Based Generalized Distance Methods

Any method that combines information from pairs of taxa that share some part of their histories in common has to account for the fact that

BOX 9.5   ESTIMATING K AND DIVERGENCE TIMES USING REGRESSION TECHNIQUES

Given the expected linear relationship between time and (corrected) genetic distance shown by Eq. 9.14, linear regression methods can be used to provide estimates of the overall substitution rate and particular divergence times (see also Hillis et al., 1996). Because of the theoretical expectation that distance should be zero at time zero, the regression must be forced through the origin (Fig. A). Moreover, because the variance of genetic distance should increase with time under standard Poisson substitution models, weighted linear regression is most appropriate (Hillis et al., 1996). A data point typically consists of the genetic distance between a pair of taxa for which fossil information is available to provide an age estimate. Multiple data points represent multiple pairs of taxa. The slope of the line is an estimate of $2K$, and confidence intervals on that slope can be estimated from the variation in the data. A test of rate constancy could be performed by testing the linear regression model against a polynomial alternative. Estimates of divergence times at nodes without fossil evidence could be made by prediction from the regression equation. Note that the errors of such predictions are generally much higher than the errors implied by the confidence intervals on the slope (Fig. A; see also Hillis et al., 1996).

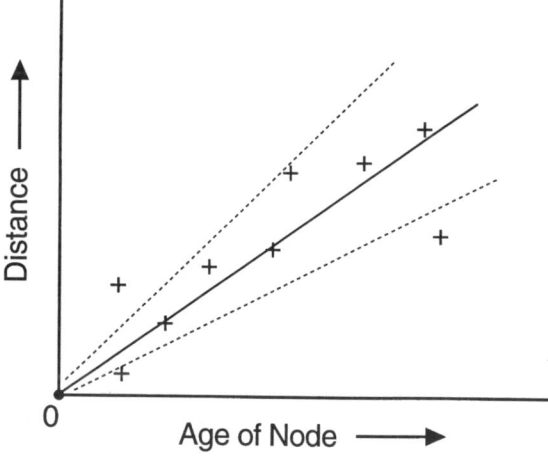

Fig. A. Schematic diagram of regression analysis used to estimate rates of substitution, K, and divergence times. Crosses represent data points from pairwise distances between taxa that span nodes that can be dated using fossil evidence. Solid line is the estimated weighted linear regression slope, which is an estimate of the quantity $2K$, and dotted lines are $1 - \alpha$ confidence intervals on the slope.

Despite the appeal of this approach, which combines data from potentially many pairwise comparisons (e.g., Wray et al. [1996] used over 1,700 calibration points for α-hemoglobin dating of metazoa), it is worth summarizing some of the possible sources of error that may arise. Some of these are accounted for in the scatter around the regression line. Others are not. (1) All else being optimal, there will still always be variance in the genetic distances owing to the stochastic substitution process; (2) rates of substitution may not be constant after all, even if there is not enough data (statistical power) to reject constancy; (3) age estimates of nodes using fossils will be off because of inaccuracies in obtaining absolute dates from geologic data (which adds variance) and because of the intrinsic bias toward using minimum ages (which adds statistical bias); (4) the degrees of freedom in any tests of the regression will be incorrect if some of the calibration points overlap phylogenetically and are nonindependent (Wray et al. [1996], for example, reduced the degrees of freedom value of over a thousand to under 50 because of phylogenetic nonindependence); and (5) if phylogenetic relationships are poorly understood the fossil dates may actually be assigned to a node in the tree different from the one representing the common ancestor of the two taxa in the pairwise divergences.

these distances are correlated with one another. Unless nonindependence is incorporated into significance tests, the tests will be biased. Hasegawa et al. (1985) estimate rate parameters and times by combining information from all pairwise comparisons for a set of taxa in conjunction with an estimate of the variance-covariance matrix of these distances. Let d_{ij} be the pairwise dissimilarity between sequences i and j for a set of s sequences. We can write as a vector all of the pairwise values of d_{ij}: $\mathbf{d}(\theta) = (d_{11}, d_{12}, \ldots d_{1s}, d_{23} \ldots d_{2s}, d_{(s-1)s})$. The probability distribution depends on the model of evolution and its associated parameters, including rates and times, $\theta = \{\mathbf{R}, K(\mathbf{R}), t_1, \ldots, t_{m-2}\}$. Because each of these dissimilarities is the sum of probabilities acting at a large number of sites, the distribution of $\mathbf{d}$ is multivariate normal with a mean of $\overline{\mathbf{D}}(\theta)$ and a variance-covariance matrix $\Omega(\theta)$. Each of these can be calculated or estimated from the data (Hasagawa et al., 1985). A likelihood solution would entail maximizing the equation for the multivariate normal distribution of $\mathbf{d}(\theta)$ and finding the θ that produced this maximum:

$$L(\theta) = C(\theta) \exp(-(1/2)[\mathbf{d}(\theta) - \overline{\mathbf{d}}(\theta))^T \Omega^{-1}(\theta)(\mathbf{d}(\theta) - \overline{\mathbf{d}}(\theta))]), \quad (9.16)$$

all of which is a very nasty nonlinear function of θ, the quantity to be optimized. Hasegawa et al. (1985) actually do a simpler optimization, which entails finding the value of θ that maximizes the term *within* the exponentiation. This is a generalized least squares approximation to the maximum likelihood estimator.

Tree-Based, Character-Based Maximum Likelihood Methods

Under the assumption of rate constancy, θ can be estimated from the sequence data directly using maximum likelihood over a given phylogeny (e.g., by enforcing the molecular clock in the likelihood options in PAUP* 4.0), at least for the general six-parameter reversible model or special cases thereof (including almost all the familiar models). The likelihood as a function of θ is then a much more complex function of the data matrix, determined by recursively calculating conditional likelihoods on a site-by-site basis and then combining them across sites (Felsenstein, 1981; Swofford et al., 1996).

Given a calibration point somewhere in the tree (Box 9.4), the parameter estimates in $\hat{\theta}$ can be scaled to obtain a set of absolute rate and time estimates. It cannot be emphasized too strongly, however, that the rate estimates for $\mathbf{R}$ obtained in the *absence* of the assumption of rate constancy in a phylogenetic maximum likelihood analysis (e.g., in PAUP* 4.0) cannot be translated into actual rates of evolution, because they are scaled by branch length constants that are in turn confounded by time (Eq. 9.3).

Errors in Estimated Times

Errors that can affect the accuracy of estimated times arise from several sources. Calibration of the clock is subject to error, the stochastic processes of substitution have intrinsic noise, and finally, even if rate constancy is not rejected, there may be real variation in rate that is below the power of test of the method to detect.

The error due to disagreement among multiple calibration points can be estimated as the residual error term in the linear regression of distance on time (Hillis et al., 1996; for an example, see Wray et al., 1996). This can be quite large owing to the difficulty of correlating a fossil to a node in a tree (see previous discussion), but it still is only one component of error that can affect estimates of divergence times.

Another, often large, source of error is due to the noise in the Poisson substitution process itself. It can be characterized on theoretical grounds, because it produces a variance in inferred distances, D. In other words, if the same process were run repeatedly over the same course of time, there would be variation in the actual number of substitutions that occurred. For many models of sequence evolution, this variance has been derived exactly (e.g., Kumar et al., 1993). For example, for the Jukes-Cantor model, the variance in D is

$$V(D) = \frac{d(1 - d)}{(1 - 4d / 3)^2 L} \qquad (9.17)$$

where L is the length of the sequence and d is the pairwise dissimilarity. Steel et al. (1996) suggest that the values $\{D_u = D + \sigma, D_l = D - \sigma\}$ (or whatever significance level is desirable) can be inserted in Eq. 9.15 for a given estimate of K to predict the confidence interval on time $[t_u, t_l]$. However, this interval is wider if K itself is being estimated. Steel et al. (1996) also demonstrated that it is possible to narrow these confidence intervals by sampling multiple sequences in each clade descended from the branching event of interest. The variance of the mean pairwise distance is smaller than for individual distances and hence the error on the inferred time is also smaller.

Hasegawa et al.'s (1985) tree-based distance methods estimate the variances in the unknown time parameters via the curvature of the likelihood surface (the so-called Fisher information; Lindgren, 1976). Although this method has been briefly discussed by Felsenstein (1981, 1988) it has not been used to assign confidence intervals to divergence times inferred over an entire tree for character-based likelihood methods except in Yang's rate program PAML (Yang, 1995). Bootstrapping may once again provide a practical answer. This could be implemented by (1) generating a fixed tree topology (by whatever means), and (2) generating a series of bootstrap replicates of data sets using PHYLIP's bootstrap program, and then running these through the likelihood algorithms available with either PHYLIP or PAUP* 4.0 constrained by a clock and by the tree. The variation in estimated times for a given branch could

be used to provide an assessment of the confidence intervals on that time.

Coalescent Theory Methods

Coalescent theory (reviewed in Ewens, 1990; Hudson, 1990) is aimed at describing the tree-like history of samples of sequences taken from a population. Its role as a theory has been twofold. First, it has provided an alternative framework for the derivation of many familiar estimators for population genetic parameters. Second, it has been used to develop improved estimators of these parameters that take into account the phylogeny of the sequences. In practice this has proven to be computationally and mathematically difficult (closed analytical expressions are exceedingly rare), but it is clear that such estimators are often better than the traditional nonphylogenetic ones. For example, Felsenstein (1992) showed that standard estimators of effective population size are less efficient than their phylogenetic alternatives.

Coalescent estimates for the age of a sample of sequences have been derived. The standard result is that the mean age back to the most recent common ancestor (MRCA) of a set of randomly sampled sequences is

$$E(T_{MRCA}) = 4N_e(1 - 1 / n) \qquad (9.18)$$

where n is the number of sequences sampled, assumed to be much less than N_e, the effective population size. This result does not rest on information contained in the estimated gene phylogeny for the sample. Griffiths and Tavare (1994) derive an improved coalescent-based estimation procedure for the age of the sample that does. They first calculate the posterior probability of observing the age, given the data, and then derive the conditional expectation of that age given the data. For a sample of mitochondrial sequences from a human population, they showed that this phylogenetic estimator tends to be 40% lower (more recent) than the estimator in Eq. 9.18.

Note that the same kind of Bayesian approach can be taken in the absence of detailed information about the phylogeny. The recent discussion of the age of the human Y chromosome based on a completely invariant sample of sequences has revolved around the calculation of such conditional expectations (Dorit et al., 1995; Donnelly et al., 1996; Fu and Li, 1996; Weiss and von Haeseler, 1996). Fu (1996) has examined these methods in great detail. One important conclusion is that the information contained in the number of segregating sites regarding the age of the sample is dependent on θ, so that sequences with high mutation rates and long lengths are best.

ESTIMATION OF RATES OF EVOLUTION AND/OR TIMES OF DIVERGENCE IN THE ABSENCE OF RATE CONSTANCY

The frequent discovery of violations of rate constancy in relative rate comparisons has led to many ad hoc approaches to estimating divergence times even under those conditions. For example, Li and Tanimura (1987) selectively pruned away lineages that deviated from rate constancy in a series of relative rate tests in primates and then reconstructed divergence times using remaining lineages under the assumption that they were clocklike. Takezaki et al. (1995) formalized this procedure by using a nested series of relative rate tests and combining information from multiple sequences within clades, to eliminate lineages that deviated significantly from rate constancy. They also used the deviations of the total branch length from root to tip from the mean length to prune lineages. They referred to these pruned trees as "linearized" trees, not because they look untreelike, but rather because they make the resultant times linear with distance.

Model Selection Methods

An important general approach was developed in a series of papers beginning with Hasegawa et al. (1989) and Kishino and Hasegawa (1990), which generalized earlier work that had used maximum likelihood techniques to estimate branch times and rates using pairwise dissimilarities between tips and specific models of sequence evolution. The generalization allowed different rate parameters in different parts of the tree. For example, in a four-taxon primate tree,

they allowed three transition and three transversion rates. Maximum likelihood was then used to simultaneously estimate the three different pairs of rates and the divergence times.

The key question that arises in this approach relates to "model selection" (Linhart and Zucchini, 1986; see Box 9.6). Models differ in how they distribute rate parameters among branches, and even in very small trees there are many possible models. It is generally not possible to test them all, so how should some subset be selected a priori?

One way is to use prior biological or evolutionary hypotheses about where rates are likely to change. In a thorough application of model selection methods, Uyenoyama (1995) estimated rates and divergence times in phylogenies of self-incompatibility alleles in plants. Especially in the case of distinct functional classes of sequences, it appears to be reasonable to postulate that functional constraints may alter lineage-specific rates. Uyenoyama used a prior model based on assigning different lineage rates to different functional classes of alleles to show that the age of sporophytic self-incompatibility alleles in *Brassica* is several times older than the genus itself. As with Kishino and Hasegawa's work, the supposition is that there may be "locally" constant rates in part of a tree, which may vary at a larger phylogenetic scale. This paper also uses generalized-distance methods for testing rate constancy as described above.

The next two methods hold some promise for allowing lineage-specific rate variation without forcing the specification of a prior model for the exact assignment of rate parameters on the phylogeny. What this requires is some kind of assumption of rate autocorrelation through time (Gillespie, 1991, p. 140), which might come about if rates of evolution had at least some heritable component to them, even an environmentally induced one.

Hidden Markov Models

Hidden Markov models have been used to study sequence analysis (Hughey and Krogh, 1996), sequence alignment (Mitchison and Durbin, 1995), and to model rate variation across sites in a sequence (Felsenstein and Churchill, 1996).

BOX 9.6. MODEL SELECTION METHODS WHEN RATES VARY ACROSS THE TREE

Mathematically, the number of possible models for j parameters distributed on m branches is given by the so-called Stirling numbers of the second kind (Bogart, 1990), $S(m,j)$. For example, there are 301 three-parameter models in a tree of seven branches. What is worse, the number of possible models is much larger, because one could construct models with as many as m parameters on m branches. The number of all possible models is equal to the number of partitions of a set of m objects, which is given by the Bell numbers, $B_m = \sum_{n=0}^{m} S(m,n)$ (Bogart, 1990). Figure A shows that even in the very simplest of cases, in which there are only two taxa and three branches, there are easily five different models to be considered. For seven branches there are 877 different models, and for 20 branches nearly 10^{14} models. Except for the smallest trees, this approach has to be guided by prior assumptions about where rates are likely to change on the tree.

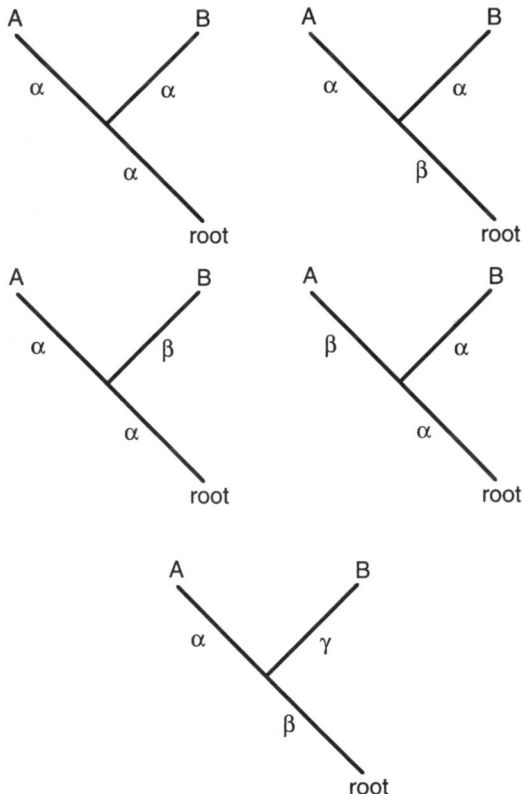

Fig. A. The five possible rate models for two taxa and three branches. Rate parameters are given by α, β, and γ. Models in this example can have as few as one parameter or as many as three.

Assuming the number of models is somehow restricted in number, Kishino and Hasegawa (1990) suggest the use of the Akaike information criterion (AIC) to choose among them. Because many of the models have the same number of free parameters (consider all the three parameter models in a tree with 10 branches, for example), a standard likelihood ratio test of the two models is inappropriate. Such tests require that one model be "nested" within the other—that it be a specialization of the other with fewer parameters. The AIC is a measure that balances the likelihood of the model with its complexity by minimizing

$$AIC = 2m - 2L$$

where m is the number of parameters in the model and L is its maximum log likelihood value. Kishino and Hasegawa (1990) used this approach to select among a number of reasonable alternative models of rate variation in the primate case.

Briefly, a hidden Markov is a model in which the rate is itself determined by a stochastic process of change. For example, the rate might be modeled to be in one of two alternative states, "fast" and "slow," with a given probability of changing between these states. This probability of change of rates is the autocorrelation of rate mentioned previously. The process of substitution is then modeled by first determining what rate state the process is found in along a particular branch, and second by determining the number of substitutions according to the selected rate.

Henze and Sanderson (in prep.) have examined this kind of model in the context of data consisting of inferred branch lengths, a Poisson substitution process, and a fixed rate-transition matrix. The magnitudes of the fast and slow rates can be estimated by maximum likelihood at the same time as the other parameters, such as divergence times, are estimated. As with estimation problems in hidden Markov models generally (Rabiner, 1989), the shape of the likelihood surface is complex and multiple optima are common. Evidently these correspond to discrete alternative "reconstructions" of the ancestral rate states.

Nonparametric Rate Smoothing

Another, less model-dependent, approach is derived by analogy to nonparametric regression techniques. Briefly, the idea is to smooth local transformations in rate as it changes over the tree. This induces an autocorrelation of rate but does not fix the autocorrelation at any one value, unlike the hidden Markov method

described above. One simple procedure for doing this (Sanderson, 1997) is to use as a local estimate of rate,

$$\hat{r}_k = \frac{b_k}{T_k} \qquad (9.19)$$

where b_k is the branch length of branch k (subtending node k) and T_k is the duration of the branch between node k and its ancestor (i.e., $t_{anc(k)} - t_k$), and then minimize the ancestral-descendant changes in this rate given by

$$(|\hat{r}_k - \hat{r}_{\text{left-desc}(k)}|)^p + (|\hat{r}_k - \hat{r}_{\text{right-desc}(k)}|)^p \qquad (9.20)$$

where p is an exponent such as 2 (which would correspond to a least squares minimization). Terms like this one are then summed over the whole tree, and standard optimization techniques are used to find the set of unknown times. Local rates can then be estimated either directly by substituting the estimated divergence times into Eq. 9.19 or by averaging over larger pieces of the tree. Note that this is a very different kind of least squares minimizations than those described above. First, it is a minimization of a change in *rate,* not distance, and second, it is a minimization of a change from ancestral lineage to its immediate descendant lineage, rather than merely the sum of deviations from some treewide mean deviation or local expected deviation. This approach is subject to local fluctuations in rate, so that a larger window for defining local rate might be appropriate for additional smoothing.

Finally, it is worth pointing out that simulation studies (Sanderson, 1997) suggest that it is difficult to estimate divergence times accurately in the absence of rate constancy, even though it is certainly possible to improve upon available methods. We may ultimately be forced to come to terms with the possibility that the quantity of data needed in such cases is much greater than currently available for most key divergence time questions.

LITERATURE CITED

Avise, J. C. 1994. Molecular Markers, Natural History and Evolution. Chapman & Hall, New York.

Bogart, K. P. 1990. Introductory Combinatorics, second edition. Harcourt Brace Jovanovich, San Diego.

Britten, R. J. 1986. Rates of DNA sequence evolution differ between taxonomic groups. Science 231:1393–1398.

Bruns, T. D., and T. M. Szaro. 1992. Rate and mode differences between nuclear and mitochondrial small-subunit rRNA genes in mushrooms. Molecular Biology and Evolution 9:836–855.

Bulmer, M. 1989. Estimating the variability of substitution rates. Genetics 123:615–619.

Cavender, J. A. 1978. Taxonomy with confidence. Mathematical Biosciences 40:271–280.

Chakraborty, R. 1977. Estimation of time of divergence from phylogenetic studies. Canadian Journal of Genetics and Cytology 19:217–223.

Chase, M. W., D. E. Soltis, R. G. Olmstead, D. Morgan, D. H. Les, B. D. Mishler, M. R. Duvall, R. A. Price, H. G. Hills, Y.-L. Qiu, K. A. Kron, J. H. Rettig, E. Conti, J. D. Palmer, J. R. Manhart, K. J. Sytsma, H. J. Michaels, W. J. Kress, K. G. Karol, W. D. Clark, M. Hedren, B. S. Gaut, R. K. Jansen, K.-J. Kim, C. F. Wimpee, J. F. Smith, G. R. Furnier, S. H. Strauss, Q.-Y. Xiang, G. M. Plunkett, P. S. Soltis, S. M. Swensen, S. E. Williams, P. A. Gadek, C. J. Quinn, L. E. Eguiarte, E. Golenberg, G. H. Learn, Jr., S. W. Graham, S. C. H. Barrett, S. Dayanandan, and V. A. Albert. 1993. Phylogenetics of seed plants: an analysis of nucleotide sequences from the plastid gene rbcL. Annals of the Missouri Botanical Garden 80:528–580.

Cox, D. R., and H. D. Miller. 1977. The Theory of Stochastic Processes. Chapman & Hall, London.

Donnelly, P., S. Tavare, D. J. Balding, and R. C. Griffiths. 1996. Estimating the age of the common ancestor of men from the ZFY intron. Science 272:1357–1359.

Dopazo, J. 1994. Estimating errors and confidence intervals for branch lengths in phylogenetic trees by a bootstrap approach. Journal of Molecular Evolution 38:300–304.

Dorit, R. L., H. Akashi, and W. Gilbert. 1995. Absence of polymorphism at the ZFY locus on the human Y chromosome. Science 268:1183–1185.

Doyle, J. A., and M. J. Donoghue. 1993. Phylogenies and angiosperm diversification. Paleobiology 19:141–167.

Ewens, W. J. 1990. Population genetics theory—the past and future. In Mathematical and Statistical Developments of Evolutionary Theory, ed. S. Lessard, pp. 177–227. Kluwer Academic Publishers, Dordrecht, The Netherlands.

Farris, J. S. 1969. On the cophenetic correlation coefficient. Systematic Zoology 18:279–285.

Felsenstein, J. 1981. Evolutionary trees from DNA sequences: a maximum likelihood approach. Journal of Molecular Evolution 17:368–376.

Felsenstein, J. 1984. Distance methods for inferring phylogenies: a justification. Evolution 38:16–24.

Felsenstein, J. 1988. Phylogenies from molecular sequences: inference and reliability. Annual Review of Genetics 22:521–565.

Felsenstein, J. 1992. Estimating effective population size from samples of sequences: inefficiency of pairwise and segregating sites as compared to phylogenetic estimates. Genetics Research 56:139–147.

Felsenstein, J. 1993. PHYLIP: Phylogenetic Inference Package. University of Washington, Seattle.

Felsenstein, J., and G. A. Churchill. 1996. A hidden markov model approach to variation among sites in rate of evolution. Molecular Biology and Evolution 13:93–104.

Fitch, W. M. 1976. Molecular evolutionary clocks. In Molecular Evolution, ed. F. J. Ayala, pp. 160–178. Sinauer Associates, Sunderland, Massachusetts.

Fu, Y.-X. 1996. Estimating the age of the common ancestor of a DNA sample using the number of segregating sites. Genetics 144:829–838.

Fu, Y.-X., and W.-H. Li. 1996. Estimating the age of the common ancestor of men from the ZFY intron. Science 272:1356–1357.

Gaut, B., S. V. Muse, W. D. Clark, and M. T. Clegg. 1992. Relative rates of nucleotide substitution at the rbcL locus in monocotyledonous plants. Journal of Molecular Evolution 35:292–303.

Gaut, B., S. V. Muse, and M. T. Clegg. 1993. Relative rates of nucleotide substitution in the chloroplast genome. Molecular Phylogenetics and Evolution 2:89–96.

Gillespie, J. H. 1986. Rates of molecular evolution. Annual Review of Ecology and Systematics 17:637–665.

Gillespie, J. H. 1989. Lineage effects and the index of dispersion of molecular evolution. Molecular Biology and Evolution 6:636–647.

Gillespie, J. H. 1991. The Causes of Molecular Evolution. Oxford University Press, New York.

Gillespie, J. H., and C. H. Langley. 1979. Are evolutionary rates really variable? Journal of Molecular Evolution 13: 27–34.

Gingerich, P. D. 1986. Temporal scaling of molecular evolution in primates and other mammals. Molecular Biology and Evolution 3:205–221.

Gittleman, J. L., C. G. Anderson, M. Kot, and H.-K. Luh. 1996. Comparative tests of evolutionary lability and rates using molecular phylogenies. In New Uses for New Phylogenies, eds. P. H. Harvey, A. J. L. Brown, J. Maynard Smith, and S. Nee, pp. 289–307. Oxford University Press, Oxford.

Goldman, N. 1993. Statistical tests of models of DNA substitution. Journal of Molecular Evolution 36:182–198.

Goldman, N. 1994. Variance to mean ratio, R(t), for Poisson process on phylogenetic trees. Molecular Phylogenetics and Evolution 3:230–239.

Goldman, N., and Z. Yang. 1994. A codon-based model of nucleotide substitution for protein-coding DNA sequences. Molecular Biology and Evolution **11**: 725–736.

Griffiths, R. C., and S. Tavare. 1994. Ancestral inference in population genetics. Statistical Science **9**:307–319.

Hasegawa, M., H. Kishino, and T. Yano. 1985. Dating of the human-ape splitting by a molecular clock of mitochondrial DNA. Journal of Molecular Evolution **22**: 160–174.

Hasegawa, M., H. Kishino, and T. Yano. 1989. Estimation of branching dates among primates by molecular clocks of nuclear DNA which slowed down in Hominoidea. Journal of Human Evolution **18**:461–476.

Hasegawa, M., A. Di Rienzo, T. D. Kocher, and A. C. Wilson. 1993. Toward a more accurate time scale for the human mitochondrial DNA tree. Journal of Molecular Evolution **37**:347–354.

Hillis, D., B. K. Mable, and C. Moritz. 1996. Applications of molecular systematics: the state of the field and a look to the future. In Molecular Systematics, second edition, eds. D. M. Hillis, C. Moritz, and B. K. Mable, pp. 515–543. Sinauer Associates, Sunderland, Massachusetts.

Hudson, R. R. 1990. Gene genealogies and the coalescent process. Oxford Surveys in Evolutionary Biology **7**:1–44.

Hughey, R., and A. Krogh. 1996. Hidden markov models for sequence analysis: extension and analysis of the basic method. Computer Applications in the Biosciences **12**:95–107.

Kimura, M. 1983. The Neutral Theory of Molecular Evolution. Cambridge University Press, Cambridge.

Kishino, H., and M. Hasegawa. 1990. Converting distance to time: application to human evolution. Methods in Enzymology **183**:550–570.

Kumar, S., K. Tamura, and M. Nei. 1993. MEGA: Molecular Evolutionary Genetic Analysis, version 1.0. Pennsylvania State University, University Park.

Langley, C. H., and W. Fitch. 1974. An estimation of the constancy of the rate of molecular evolution. Journal of Molecular Evolution **3**:161–177.

Li, P., and J. Bousquet. 1992. Relative rate test for nucleotide substitutions between lineages. Molecular Biology and Evolution **9**:1185–1189.

Li, W.-H., and D. Graur. 1991. Fundamentals of Molecular Evolution. Sinauer Associates, Sunderland, Massachusetts.

Li, W.-H., and M. Tanimura. 1987. The molecular clock runs more slowly in man than in apes and monkeys. Nature **326**:93–96

Li, W.-H., C.-I. Wu, and C.-C. Luo. 1985. A new method for estimating synonymous and nonsynonymous rates of nucleotide substitution considering the relative likelihood of nucleotide and codon changes. Molecular Biology and Evolution **2**:150–174.

Lindgren, B. W. 1976. Statistical Theory, third edition. Macmillan, New York.

Linhart, H., and W. Zucchini. 1986. Model Selection. Wiley, New York.

Mindell, D. P., and R. L. Honeycutt, 1990. Ribosomal RNA evolution in vertebrates: evolution and phylogenetic applications. Annual Review of Ecology and Systematics **21**:541–566.

Mindell, D. P., and C. E. Thacker. 1996. Rates of molecular evolution: phylogenetic issues and applications. Annual Review of Ecology and Systematics **27**:279–303.

Mitchison, G., and R. Durbin. 1995. Tree-based maximal likelihood substitution matrices and hidden Markov models. Journal of Molecular Evolution **41**: 1139–1151.

Muse, S. V., and B. S. Gaut. 1994. A likelihood approach for comparing synonymous and nonsynonymous nucleotide substitution rates, with application to the chloroplast genome. Molecular Biology and Evolution **11**:715–724.

Muse, S. V., and B. S. Weir. 1992. Testing for equality of evolutionary rates. Genetics **132**:269–276.

Nadler, S. A. 1992. Phlogeny of some Ascaridoid nematodes, inferred from comparison of 18S and 28S rRNA sequences. Molecular Biology and Evolution **9**:932–944.

Navidi, W. C., G. A. Churchill, and A. von Haeseler. 1991. Methods for inferring phylogenies from nucleic acid sequence data by using maximum likelihood and linear invariants. Molecular Biology and Evolution **3**:418–426.

Nei, M. 1987. Molecular Evolutionary Genetics. Columbia University Press, New York.

Nickrent, D., and E. M. Starr. 1994. High rates of nucleotide substitution in nuclear small-subunit (18S) rDNA from holoparasitic flowering plants. Journal of Molecular Evolution **39**:62–70.

Palumbi, S. R. 1989. Rates of molecular evolution and the function of nucleotide positions free to vary. Journal of Molecular Evolution **29**:180–187.

Penny, D., M. D. Hendy, and M. A. Steel. 1992. Progress with methods for constructing evolutionary trees. Trends in Ecology and Evolution **7**:73–79.

Rabiner, L. R. 1989. A tutorial on hidden markov models and selected applications in speech recognition. Proceedings of the Institute of Electrical and Electronics Engineers **77**:257–286.

Ramshaw, J. A. M., D. L. Richardson, B. T. Meatyard, R. H. Brown, M. Richardson, E. W. Thompson, and D. Boulter. 1972. The time of origin of the flowering plants determined by using amino acid sequence data of cytochrome c. New Phytologist **71**:773–779.

Rodriguez, F., J. L. Oliver, A. Marin, and J. R. Medina. 1990. The general stochastic model of nucleotide substitution. Journal of Theoretical Biology **142**:485–501.

Sanderson, M. J. 1997. A nonparametric approach to estimating divergence times in the absence of rate constancy. Molecular Biology and Evolution **14**:1218–1231.

Sarich, V., and A. C. Wilson. 1967. Rates of albumin evolution in primates. Proceedings of the National Academy of Sciences U.S.A. **58**:142–148.

Savard, L., P. Li, S. H. Strauss, M. W. Chase, M. Michaud, and J. Bousquet. 1994. Chloroplast and nuclear gene sequences indicate Late Pennsylvanian time for the last common ancestor of extant seed plants. Proceedings of the National Academy of Sciences U.S.A. **91**: 5163–5167.

Simpson, G. G. 1944 [reprinted 1984]. Tempo and Mode in Evolution. Columbia University Press, New York.

Springer, M. 1995. Molecular clocks and the incompleteness of the fossil record. Journal of Molecular Evolution **41**:531–538.

Stanley, S. M. 1979. Macroevolution. W. H. Freeman & Company, San Francisco.

Steel, M., A. Cooper, and D. Penny. 1996. Estimating the time to divergence for groups of taxa. Systematic Biology **45**:127–134.

Swofford. D. L., G. K. Olsen, P. J. Waddell, and D. M Hillis. 1996. Phylogeny reconstruction. **In** Molecular Systematics, second edition, eds. D. M. Hillis, C. Moritz, and B. K. Mable, pp. 407–514. Sinauer Associates, Sunderland, Massachusetts.

Tajima, F. 1993. Simple methods for testing the molecular evolutionary clock hypothesis. Genetics **135**: 599–607.

Takahata, N. 1991. Statistical models of the overdispersed molecular clock. Theoretical and Population Biology **39**:329–344.

Takezaki, N., A. Rzhetsky, and M. Nei. 1995. Phylogenetic test of the molecular clock and linearized trees. Molecular Biology and Evolution **12**:823–833.

Uyenoyama, M. K. 1995. A generalized least-squares estimate for the origin of sporophytic self-incompatibility. Genetics **139**:975–992.

Walker, T. D. 1985. Diversification functions and the rate of taxonomic evolution. **In** Phanerozoic Diversity Patterns, ed. J. W. Valentine, pp. 311–334. Princeton University Press, Princeton, New Jersey.

Weiss, G., and A. von Haeseler. 1996. Estimating the age of the common ancestor of men from the *ZFY* intron. Science **272**:1359–1360.

Wilson, M. A., B. Gaut, and M. T. Clegg. 1990. Chloroplast DNA evolves slowly in the Palm family (Arecaceae). Molecular Biology and Evolution **7**:303–314.

Wolfe, K. H., W.-H. Li, and P. M. Sharp. 1987. Rates of nucleotide substitution vary greatly among plant mitochondria, chloroplast, and nuclear DNAs. Proceedings of the National Academy of Sciences U.S.A. **84**:9054–9058.

Wray, G. A., J. S. Levinton, and L. H. Shapiro. 1996. Molecular evidence for deep Precambrian divergences among metazoan phyla. Science **274**:568–573.

Wu, C.-I., and W-H. Li. 1985. Evidence for higher rates of nucleotide substitution in rodents than in man. Proceedings of the National Academy of Sciences U.S.A. **82**:1741–1745.

Yang, Z. 1995. Phylogenetic analysis by maximum likelihood (PAML), version 1.1. Institute of Molecular Genetics, Pennsylvania State University, University Park.

Yang, Z. 1996. Among-site rate variation and its impact on phylogenetic analyses. Trends in Ecology and Evolution **11**:367–372.

Yang, Z., N. Goldman, and A. Friday. 1994. Comparison of models for nucleotide substitution used in maximum-likelihood phylogenetic estimation. Molecular Biology and Evolution **11**:316–324.

Zuckerlandl, E., and L. Pauling. 1962. Molecular disease, evolution, and genetic heterogeneity. **In** Horizons in Biochemistry, eds. M. Kasha and B. Pullman, pp. 189–225. Academic Press, New York.

Zuckerkandl, E., and L. Pauling. 1965. Evolutionary divergence and convergence. **In** Evolving Genes and Proteins, eds. V. Bryson and H. J. Vogel, pp. 97–166. Academic Press, New York.

10

Phylogenetic Incongruence: Window into Genome History and Molecular Evolution

Jonathan F. Wendel and Jeff J. Doyle

The field of systematic biology has been revitalized and transformed during the last few decades by the confluence of phylogenetic thinking with ready access to the tools of molecular biology. Indeed, the title of this volume and the fact that it is already in its second edition offers ample testimony to the impact that molecular approaches have had on efforts to reconstruct the phylogenetic history of plants. Concomitant with the proliferation of molecular tools has been a growing awareness that reliance on a single data set may often result in insufficient phylogenetic resolution or misleading inferences. Accordingly, it is an increasingly widespread practice to apply multiple data sets to a common group of taxa. One of the consequences of analyzing multiple data sets is that the phylogenies inferred may differ from each other in one or more details. This phylogenetic incongruence is not rare; to the contrary, it is almost the rule rather than the exception, being evident to varying degrees.

Given the prevalence of phylogenetic incongruence, the question naturally arises as to whether two or more independent data sets should only be analyzed separately or whether they should be combined into a global analysis. This question stems in part from the recognition that different character sets may have different underlying evolutionary histories and are therefore expected, in many instances, to lead to different reconstructions of the sampled taxa. Despite considerable past and present discussion regarding optimal treatment of multiple data sets, no clear consensus has emerged as to the most appropriate course of action (Miyamoto, 1985; Hillis, 1987; Kluge, 1989; Barrett et al., 1991; Doyle, 1992; Bull et al., 1993; Eernisse and Kluge, 1993; de Queiroz et al., 1995; Miyamoto and Fitch, 1995). The reader is referred to these sources for explication of arguments on both sides of the issue.

Complicating this issue is the realization that not all incongruence is equal in magnitude and that topological differences between competing phylogenies may have a number of different causes, including some that are artifactual. Consequently, there have been efforts to measure incongruence and evaluate whether it is "real" and hence potentially biologically significant, or spurious, due to insufficient character evidence, excessive homoplasy, or some other

We thank our many graduate students and colleagues with whom we have shared innumerable illuminating conversations. We also thank R. Cronn, R. Small, and L. Clark for comments on the manuscript and T. Seelanan for assistance with the figures. Much of the authors' research has been sponsored by the National Science Foundation, whose support we gratefully acknowledge.

cause (Mickevich and Farris, 1981; Templeton, 1983; Faith, 1991; Swofford, 1991; Rodrigo et al., 1993; Lutzoni and Vilgalys, 1994; Farris et al., 1995; Mason-Gamer and Kellogg, 1996; Kellogg et al., 1996; Sites et al., 1996; Lyons-Weiler et al., 1996, 1997; Graham et al., 1997; Seelanan et al., 1997). The question of whether different data sets should be combined into a single global analysis thus becomes intertwined with the issue of assessing the reality of topological differences. Because this latter topic is discussed in detail in Chapter 11, it is not expanded upon here.

Surrounding much of the discussion of the treatment of multiple data sets has been an implicit or explicit assumption that phylogenetic incongruence is inherently undesirable. That is, incongruence of topologies derived from different data sets is often viewed as an unfortunate but unavoidable side effect of phylogenetic analysis, and as such it represents an impediment to achieving phylogenetic understanding. Against this backdrop, perhaps it is not surprising that a considerable portion of the literature on the treatment of multiple data sets has focused on *reconciliation* of alternative estimates of phylogeny or on assessing which of two competing resolutions is better supported by the available evidence.

In this chapter, we present a different perspective on phylogenetic incongruence. The unifying theme of the chapter is that rather than representing an undesirable outcome or a problem that requires a solution, phylogenetic incongruence is touted as an important observation that often reflects something interesting in the biology of the group under study, and accordingly, its appearance may alert us to one or more evolutionary processes that would not have been suspected in the absence of incongruence. To this extent, phylogenetic incongruence may be *desirable,* as it often illuminates previously poorly understood evolutionary phenomena. In this chapter we enumerate and discuss the various processes that underlie phylogenetic discord and attempt to assess their relative importance as causative agents. In expanding upon previous treatments of this subject (Sytsma, 1990; Doyle, 1992; Kadereit, 1994; de Queiroz et al., 1995; Brower et al., 1996), our intention is to provide

an introduction to the relevant issues and to the causes of phylogenetic incongruence, as well as to offer a single-source entry point to the recent literature. We focus on plants, although literature from other groups is cited where appropriate or particularly illustrative.

CAUSES OF PHYLOGENETIC INCONGRUENCE

Before delving into the causes of phylogenetic incongruence, it is necessary to address the question of whether the evidence for it, in any particular case, is substantial enough to warrant a conclusion that it actually exists. Phrased another way, it is important to evaluate whether the conflict is "significant" and therefore possibly reflective of different evolutionary histories for two or more data sources, or "insignificant," meaning that it does not hold up to inspection and standard measures of statistical evaluation (Mickevich and Farris, 1981; Templeton, 1983; Faith, 1991; Swofford, 1991; Rodrigo et al., 1993; Farris et al., 1995; Mason-Gamer and Kellogg, 1996; Huelsenbeck et al., 1996; Kellogg et al., 1996; Sites et al., 1996; Seelanan et al., 1997; see Chapter 11). If the conflict is judged to be insignificant, using appropriate criteria, the possibility remains that the phylogenetic incongruence *may* reflect one or more underlying biological processes that are differentially affecting distinct data sets, but it is also possible that the incongruence has a more mundane genesis. Examples of the latter would include the many cases where particular clades are weakly supported and where alternative resolutions are only slightly less parsimonious. Indeed, this type of "soft incongruence" (Seelanan et al., 1997) is actually *expected* to occur as long as weakly supported nodes exist in trees arising from different data sources (e.g., Kim and Jansen, 1994; Olmstead and Sweere, 1994; Hoot et al., 1995; Kellogg et al., 1996; Seelanan et al., 1997). In *Krigia,* for example, the expected and cladistically supported (by morphology and cpDNA) sister-species relationship between *Krigia biflora* and *K. montana* was not obtained in a tree based on internal transcribed spacer (ITS) data; this incongruence, however, disappeared in trees one step longer than the shortest.

Similarly, in an analysis of the Solanaceae based on three different molecular data sets, Olmstead and Sweere (1994) only observed disagreement in the Solanoideae, where character support was minimal, suggesting that the lack of complete congruence among gene trees reflects an absence of sufficient signal rather than fundamentally different evolutionary histories.

The foregoing discussion and examples are intended to underscore an important notion, namely, that *not all incongruence is created equal*. Minor points of topological disparity, often reflecting no special biological process or differential history, are commonly observed, and in fact are expected in many situations. In the context of the present chapter, this type of conflict is not especially interesting in that it cannot be assumed to result from an evolutionary process that differentially affects data sets. More interesting are examples of "hard incongruence" (Seelanan et al., 1997), where alternative resolutions in two or more trees derived from distinct data sets are statistically supported as incongruent. In these instances one might justifiably suspect that the discord was generated by an evolutionary process that differentially affected the various sources of data, so it becomes important to ask what this process or processes are.

This form of phylogenetic incongruence provides the basis for the remainder of this chapter. As shown in Table 10.1, the evolutionary processes potentially responsible are many and varied. For purposes of presentation, these have been divided into three broad categories, reflecting technical causes, processes that operate at the whole organism level, and gene or genome-level processes. To a certain extent, this classification is subjective, in that several of the listed phenomena might reasonably be placed elsewhere in the table. Phylogenetic sorting of ancestral polymorphisms (lineage sorting), for example, might be conceived of as either an organism-level or genome-level process, depending on context and one's perspective. Notwithstanding this arbitrary aspect of the listing, each category of phenomena and its subdivisions are discussed in turn.

A final comment concerns the "phenotype" in question, namely, differing phylogenetic resolutions of two or more taxa based on two or more

Table 10.1. Phenomena that may lead to conflicting phylogenetic hypotheses.

Technical causes	Insufficient data
	Gene choice
	Sequencing error
	Taxon sampling
Organism-level processes	Convergent or rapid morphological evolution
	Rapid diversification
	Hybridization/Introgression
	Lineage sorting
	Horizontal transfer
Gene and genome-level processes	Intragenic recombination
	Orthology/paralogy conflation
	Interlocus interactions and concerted evolution
	Rate heterogeneity among taxa
	Rate heterogeneity among sites
	Base compositional bias
	RNA editing
	Nonindependence of sites

data sets. In some if not most cases, more than one of the myriad phenomena listed in Table 10.1 might conceivably have generated the discord of interest. This fact alone serves to underscore an important cautionary note about phylogenetic incongruence, specifically, that even when the discord is striking and strongly supported, its genesis will probably not be evident from phylogenetic analysis alone, and may not be evident even when all available evidence is considered. Thus, inferring the cause is often problematic.

Technical Causes

Phylogenetic resolutions based on two or more data sets may differ due to experimental or technical reasons rather than differing underlying histories. The examples mentioned above, where weakly supported clades differ from alternative resolutions by only one to several steps, are a case in point, with the technical cause being "insufficient data." Other experimental variables that may play a role in generating discordant trees include appropriateness of the molecular tool, quality of the data, and density of taxon sampling.

Gene Choice   One of the fundamental principles of molecular systematics is that the rate of molecular evolution for a particular sequence should be optimized, to the extent possible, to the scale of divergence in the taxa whose phylogenetic relationships are being explored. If the gene evolves too slowly, signal will be insufficient and hence little phylogenetic resolution may be expected. Alternatively, if the rate of evolution is too high relative to the scale of taxon divergence, phylogenetic signal may be obscured by homoplasy. Less than optimal choices at the outset of a study may lead to incongruence among the resulting gene trees, either from lack of sufficient signal in one or more of the data sets (hence, a spurious resolution) or from excessive sequence evolution, evidenced as long branches and the associated problem of homoplasy masquerading as synapomorphy ("long-branch attraction"; Felsenstein, 1978). These issues of gene choice and its relationship to taxonomic rank (or more precisely, scale of divergence, as not all families and genera, for example, are equivalently aged or evolve at the same rate) are discussed in greater depth in Chapter 1. The essential point here is that topological incongruence among gene trees may have an experimental basis.

A related technical or experimental issue concerns the length of DNA sequences employed for the construction of gene trees. Just as substitution rates need to be sufficient to provide phylogenetic signal, enough nucleotides need to be sampled to overcome the sampling error inherent in data representing single stretches of DNA (Cummings et al., 1995; de Quieroz et al., 1995; Hillis, 1996). Thus, sequence lengths need to be considered in addition to evolutionary rates in the design of phylogenetic projects. The total length of sequence needed is difficult to predict a priori; it depends on many factors, including taxon sampling density, the amount of and level of heterogeneity in molecular evolutionary rates, the degree of divergence among taxa within the ingroup, and the amount and distribution of homoplasy. The simulation by Hillis (1996) is instructive in this regard. Using a model phylogeny of 228 angiosperms based on 18S ribosomal DNA sequences (generated by D. Soltis, P. Soltis, D. Nickrent, and colleagues; see Soltis, Soltis, Nickrent et al., 1997), Hillis explored the relationship between accuracy of phylogenetic inference and number of nu-

cleotides in the data set. He found that over 90% of the model tree was accurately reconstructed using sequences approximately the length of *rbcL*, with an asymptotic approach to over 99% correct using 5,000 base pairs.

Sequencing Error   In addition to gene choice and sequence length, the accuracy of the sequence data collected may influence the topologies obtained (Clark and Whittam, 1992), and therefore possibly lead to discordant gene trees. Sequencing error is probably of most concern in studies where sequence divergence, and hence the number of potentially informative sites, is very small, although it is conceivable that it also influences resolution among closely related terminals in studies involving more divergent taxa. Although most investigators are alert to this potential problem, it is often impractical to verify the accuracy of each and every apomorphy scored during an initial round of sequencing. Moreover, in studies of recently diverged taxa, where sequencing errors are likely to have their biggest impact, a common solution to the problem of insufficient sequence evolution is to employ nongenic sequences such as introns or spacers. These regions, while often providing the desired higher rates of sequence evolution, may also lead to higher error rates in scoring, because of the absence of the continual alignment checks present in genes (e.g., codons), and because of their inherently higher indel frequencies (Golenberg et al., 1993; Morton and Clegg, 1993; Gielly and Taberlet, 1994; Johnson and Soltis, 1994, 1995; van Ham et al., 1994; Baldwin et al., 1995; Kelchner and Wendel, 1996). Indels may complicate sequence alignments and hence homology assessments (see Chapter 4), potentially leading to alignment errors. Sequencing error may also result from sampling nuclear loci that are heterozygous, particularly when sequences are generated from PCR pools rather than cloned products. Finally, sequencing error may arise from amplification of contaminating DNAs, as a number of recent studies have documented (e.g., Olmstead and Palmer, 1994; Liston et al., 1996; Zhang et al., 1997).

Taxon Sampling   A final and related technical issue concerns taxon sampling, which, if insufficiently dense, has long been recognized as a cause of long-branch attraction (Felsenstein, 1978) and hence, potentially, phylogenetic incongruence. This problem may or may not be avoidable, depending on the study group in question and its historical pattern of speciation and extinction. In some cases increasing taxon sampling density may lead to improved phylogenetic accuracy (Wheeler, 1992), although this is not assured (Kim, 1996).

In addition to the issue of taxon sampling *density,* taxon *identification* is also relevant, in that misidentified samples in one study may lead to an erroneous phylogenetic inference and hence incongruence with a second study or data set. In many molecular phylogenetic studies, one or more taxa may be obtained as DNA extracts, or samples for DNA extraction may be secured as fresh or dried leaves from collectors, colleagues, or botanical gardens. The potential for mislabeling or misidentification underscores the necessity of using appropriately vouchered materials in molecular phylogenetic work.

Organism-level Processes

Perhaps more interesting than cases of incongruence resulting from technical causes are those where the discord reflects different underlying evolutionary histories for the data sets in question. As listed in Table 10.1 and discussed subsequently, a variety of processes may be involved. Most of these entail some variant on a common theme, whereby one or more genes has experienced a perturbation in its history relative to other genes. Because the genes will have experienced different histories, it is expected that this will be reflected in reconstructions of these histories.

A simple example illustrates an important point. Imagine a single allele that somehow is transferred across a species barrier, perhaps through hybridization and subsequent introgression, from a donor taxon to a recipient species, where it becomes fixed. If this is the sole introgressed allele, a gene tree based on this particular gene will likely be incongruent with the "species tree" and may also conflict with trees based on other, nonintrogressant genes. Notice that in the latter case, that involving a comparison of gene trees, *neither* gene tree is more correct than the other, given the important caveats that sufficient data have been accurately gathered and that there have been no other perturbing forces. Phrased alternatively, phylogenetic hypotheses based on sequences from one DNA segment may be in conflict with hypotheses generated from a different DNA segment *even when both gene trees are correct in all details.* This, in fact, is the expectation in many cases, particularly at the species level (Davis and Nixon, 1992; Doyle, 1992, 1995; Baum and Shaw, 1995; Maddison, 1995, 1996). In the example given, each gene tree faithfully reproduces the history of diversification of the alleles sampled in the various taxa studied. The fact that one allele was interspecifically introgressed across a species barrier does not invalidate its gene tree. A pitfall, however, arises from a tempting interpretive extrapolation, namely, that a gene tree faithfully reproduces the history of organismal divergence. This may or may not be true, as this example underscores.

So gene trees may be incongruent with other gene trees and/or with species trees. The challenge then, is not only to infer the phylogeny of the organisms under study, but also to interpret the evolutionary history of the data sources used. These twin objectives often involve reciprocal illumination, whereby phylogenetic analysis leads to the recognition of incongruence, which informs subsequent experiments or facilitates revised interpretations. The latter, in turn, provide a more complete portrait of organismal history.

Convergent or Rapid Morphological Evolution
The relative merits of morphological and molecular data in phylogeny reconstruction have been discussed at length (e.g., Hillis, 1987; Donoghue and Sanderson, 1992; Patterson et al., 1993; Kadereit, 1994). A point on which there is near universal agreement is that morphological evolution may differ fundamentally from molecular evolution in that single genetic changes often underlie dramatic morphological transformations (Gottlieb, 1984; Doebley, 1993). Sytsma (1990) and Kadereit (1994) highlight these transformations as a root cause of phylogenetic incongruence, suggesting that they are

typically unaccompanied by similar levels of molecular divergence. Essentially this is a form of incongruence arising from rate differences, where the rates in question are morphological on the one hand and molecular on the other. To the extent that this phenomenon causes incongruence, it may serve to identify cases where genes with major influences on morphogenesis have had profound evolutionary effects.

In addition to evolutionary rate differences and timing of morphological transformations, evolutionary convergence in morphology may cause incongruence between trees inferred from molecular and morphological data, as noted by Sytsma (1990). In some cases this convergence may be striking, as with the repeated evolution of carnivory in plants (Albert et al., 1992).

Issues concerning convergent morphological evolution and the evolutionarily sporadic nature of morphological change are already widely appreciated and need not be elaborated here. Perhaps it is worthwhile, though, to draw a parallel with molecular evolution, which is often viewed as being more stochastically regular and free from the type of convergence envisioned for morphological traits. Thus, there is no evidence that major convergence events have taken place in *rbcL* evolution, for example, at least to the extent that sequences from disparate lineages would have become phylogenetically linked via this process. It seems likely that this freedom from wholesale convergence will generally be true for molecular data, or at least for DNA sequence data, although there may be exceptions.

On the other hand, it is equally clear that molecular tools are not completely free from convergence. There are now many examples, at a variety of taxonomic levels, of convergent molecular changes in characters that at one time might have been presupposed to be unique. These include the repeated losses of cpDNA genes and introns (Downie et al., 1991, 1994; Doyle et al., 1995), convergent inversions of minute as well as large pieces of cpDNA (Downie and Palmer, 1994; Hoot and Palmer, 1994; Kelchner and Wendel, 1996), and the remarkable discovery of "intron homing" in rDNA of fungi (Hibbett, 1996), where a particular group I intron is reported to have been precisely inserted at an identical nucleotide position

in disparate groups of homobasidiomycetes. These and other examples bear witness to the potential for molecular convergence, and to the extent that these labile characters influence gene tree topologies, they may also be a source of phylogenetic incongruence. Given this potential lability, individual structural mutations may prove most informative when evaluated in the context of phylogenies inferred using other data sources.

Rapid Diversification   If organismal divergence events are temporally compressed relative to the scale of molecular evolution, phylogenetically inferred internodes on gene trees may be short and difficult to resolve with confidence. In these cases, the relevant clades tend to be weakly supported, often decaying in trees only one to several steps longer than the most parsimonious trees. These same clades are often not recovered in subsequent analyses using additional molecular markers. This short internode or short interior branch phenomenon may be a common cause of misleading phylogenetic inference as well as phylogenetic incongruence.

The short internode problem is a relative concept, meaning that it is dependent on the scale of divergence and the data type being used. It is, however, applicable to all levels of organismal divergence, and examples of phylogenetic incongruence attributable to this cause abound at a variety of taxonomic levels in both plants and animals (e.g., Olmstead and Sweere, 1994; Fehrer, 1996; Lara et al., 1996; Baldwin, 1997). At the family level, for example, several short branches are apparent near the base of the Poaceae in trees based on cpDNA characters, regardless of the data set, and alternative molecular data sets often yield conflicting resolutions of these major clades. Clark et al. (1995) and Mathews and Sharrock (1996) found a BOP clade (Bambusoid, Oryzoid, Pooid) using *ndhF* and phytochrome data, respectively, whereas alternative resolutions were obtained using *rbcL* (Barker et al., 1995; Duvall and Morton, 1996) and cpDNA restriction site (Davis and Soreng, 1993) data. Similarly, some cladistic relationships in the Asteraceae (Kim et al., 1992; Kim and Jansen, 1995), Lardizabalaceae (Hoot et al., 1995), and Saxifragaceae (Morgan and Soltis,

1993; Soltis et al., 1993, 1996; Johnson and Soltis, 1994, 1995; Soltis et al., 1996) are unstable vis-à-vis source of molecular data. At the tribal and generic levels in the Gossypieae (Malvaceae) and in *Gossypium* itself, Seelanan et al. (1997) attributed varying resolutions, as inferred from ITS, *ndhF,* and cpDNA restriction site data, to the short interior branch phenomenon.

In these and many other examples, the absolute meaning of "short interior branch" is dependent on the scale of divergence. Thus, a short interior branch in, for example, a phylogenetic study of families or orders using a slowly evolving molecule such as *rbcL* will have a rather different temporal interpretation than a similar branch in a phylogeny of closely related species using a fast-evolving sequence such as a nuclear intron. In both cases, short refers to the relative paucity, and hence presumable instability, of support, but in the former case the actual temporal spacing of divergence events may be an order of magnitude or more greater than in the latter case. Both types of phenomena are commonly referred to in the literature as "rapid radiations," so the expression is a relative one. The pattern underlying the interpretation, though, namely a star phylogeny, is the same regardless of scale.

Regardless of the issue of absolute temporal scale, the many examples of unstable short interior branches highlight an evolutionary phenomenon that may be fairly common. In many cases, it may be that this form of phylogenetic incongruence is revealing with respect to the history of the group under study. Thus, rather than focusing exclusively on resolution or reconciliation of the conflict between data sets, it might be worthwhile to acknowledge the potential significance of the incongruence and seek its genesis. Star phylogenies might, for example, alert us to a history of fragmentation of a formerly widespread population system into a series of geographically isolated populations, or to a relatively rapid diversification following a novel ecological or morphological adaptation (see fig. 16, p. 176, Soltis and Soltis, 1995), such as radiation of allopolyploids following their formation (e.g., Wendel, 1989). In practice, alternative resolutions of short interior nodes represent a form of soft incongruence, as defined in the introduction to this chapter, and

in these cases the conflict might disappear with additional data. Nevertheless, the perspective offered here is that the incongruence is important in its own right, in that it focuses our attention on the possibility of a history of rapid radiation, whatever that means ecologically or temporally for the group in question.

Hybridization and Introgression   One of the most significant and conspicuous insights to emerge from molecular systematic investigations in the last decade is that hybridization and introgression are even more widespread in plant populations than earlier suspected on morphological grounds (Anderson, 1949; Heiser, 1973; Rieseberg and Wendel, 1993; Rieseberg, 1995). In nearly all cases, the initial evidence for the phenomenon has consisted of an unexpected phylogenetic result, usually between cpDNA restriction sites and morphology but also among various other combinations of organellar and nuclear markers. Many of these cases apparently represent ancient introgression episodes, events that were wholly unsuspected prior to their serendipitous discovery using molecular markers. A testimonial to the prevalence of cryptic hybridization and introgression is that examples of the process have accumulated quickly in the literature (e.g., Palmer et al., 1983, 1985; Doebley, 1989; Smith and Sytsma, 1990; Rieseberg et al., 1990; Rieseberg, 1991; Wendel et al., 1991). In their 1991 review, Rieseberg and Soltis listed 37 examples of cpDNA introgression, and within the span of just 5 years this number had grown to over 100 (Rieseberg et al., 1996). Mechanisms that underlie interspecific introgression of organellar genomes have been reviewed (Rieseberg et al., 1996), as has the potential evolutionary significance of the process (Rieseberg and Soltis, 1991; Arnold, 1992; Rieseberg and Wendel, 1993; Rieseberg, 1995; Rieseberg et al., 1996). Irrespective of these important questions of mechanism and significance, cpDNA introgression evidently is a common process, and hence it is necessary to consider its phylogenetic implications.

cpDNA introgression is diagrammatically illustrated in Fig. 10.1. If plastome transfer is unaccompanied by nuclear introgression, an inferred cpDNA tree will likely differ from gene

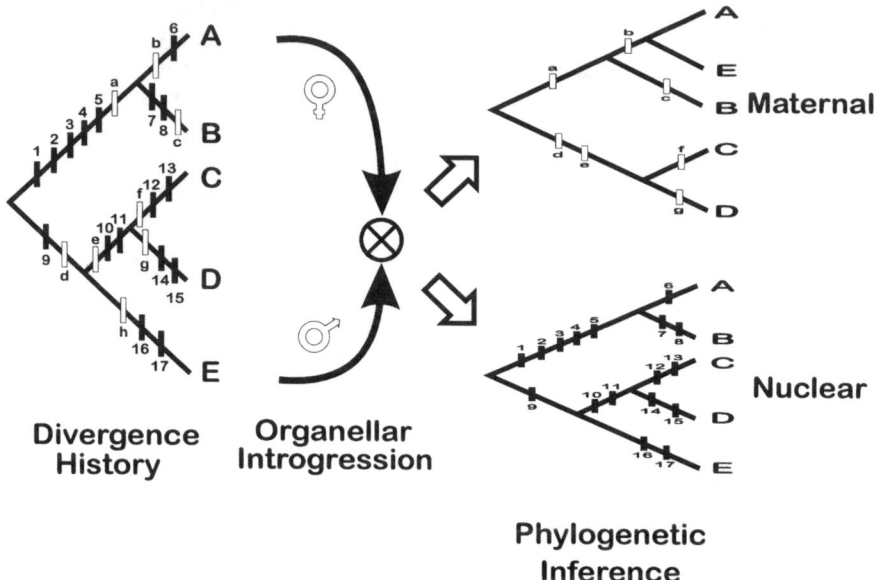

Figure 10.1. Introgression as a cause of phylogenetic incongruence. A divergence history for five taxa (A through E) is illustrated on the left. During divergence, mutations occur in both nuclear (solid bars) and organellar (open bars) genes. Hybridization between Taxon A, as female, and Taxon E, as the pollen parent, followed by organellar introgression into the paternal lineage, leads to cytoplasmic replacement in Taxon E, which acquires an A organellar complement. Phylogenies based on maternal (top right) and nuclear (bottom right) genes will therefore be incongruent. For simplicity, there is no homoplasy in either data set and no molecular evolution subsequent to hybridization.

trees based on nuclear sequences and from phylogenies inferred from morphological data. Hence, phylogenetic incongruence is the expected result. Given the apparently high frequency of cytoplasmic introgression, the phenomenon is likely to be among the most common causes of phylogenetic discord. An important point about cpDNA introgression is that because the plastome is nonrecombinant, it is inherited as a unit, and so all cpDNA sequences and/or restriction sites are expected to be transmitted together as a single haplotype, regardless of how many nucleotides or restriction sites are scored (Doyle, 1992). Consequently, phylogenetic incongruence is not expected among gene trees based on two or more different cpDNA data sets.

Among the several generalizations to emerge with respect to introgression is that cytoplasmic gene flow occurs at an apparently higher frequency than nuclear introgression. Moreover, patterns of organellar and nuclear introgression are typically asymmetrical. That is, cytoplasmic gene flow is frequently observed without evidence of nuclear introgression, whereas nuclear introgression without concomitant cytoplasmic introgression has rarely been demonstrated. Rieseberg et al. (1996) present a useful analysis of why patterns of nuclear and cytoplasmic gene flow may differ in plant populations and include a discussion of some of the potentially important ecological and genetic factors. Especially uncommon have been examples of introgression of nuclear genes between plant populations for which there is no ongoing evidence of hybridization (Talbert et al., 1990; Wendel et al., 1995b). Presumably, this scarcity of evidence does not reflect the actual absence of ancient nuclear introgression, *nor necessarily* a quantitative difference in its frequency relative to cytoplasmic introgression. Instead, it probably reflects both methodological and conceptual factors, including the difficulty of detecting introgressant loci in large, complex plant genomes, and problems in demonstrating that candidate markers are truly introgressant, as opposed to phenotypically convergent, symplesiomorphic, or a consequence of sorting of ancestral polymorphisms (see Lineage Sorting, below). As nuclear molecular tools become more widely applied, examples of nuclear introgression will continue to accumulate (Rieseberg et al., 1991, 1996; Brubaker et al., 1993; Wendel et al., 1995b).

From a phylogenetic reconstruction standpoint, nuclear introgression may differ from cytoplasmic introgression in significant ways, although both may lead to phylogenetic incongruence. For a single, nonrecombining introgressant nuclear gene, the effects will be similar to those of plastome introgression, in that gene trees based on this sequence will reflect the evolutionary history of that particular gene rather than the taxa from which they arose. If, however, a reconstruction is based on a number of independent nuclear markers, such as restriction fragment length polymorphism (RFLP) or allozyme loci for which individual gene trees are not first reconstructed, the resolution observed should depend on many factors, including the proportion of loci that are introgressed, their linkage relationships in the genome, and the antiquity of the introgression event. First generation hybrids are expected to resolve in cladistically basal positions (McDade, 1990, 1992), and it is likely that introgressants will be found to resolve at every level in a hierarchy, depending on the factors listed previously as well as the breeding history of the introgressed population.

Given these considerations, it should be apparent why interpretation of introgression may not be straightforward, except in those cases where evidence for cytoplasmic introgression is convincing. That is, the phenotype of incongruence may be self-evident, but its genesis may be elusive. Further complicating the analysis is the issue of antiquity of the putative reticulation event, both in an absolute sense and relative to the scale of cladogenesis and extinction. The more ancient the reticulation, the more difficult will be its detection, particularly for nuclear markers, which tend to be more polymorphic than cytoplasmic makers and which recombine. On the other hand, there are continuing reports of reasonably ancient reticulations first evidenced by "aberrant" plastomes (Doebley, 1989;

Wendel et al., 1991; Soltis and Kuzoff, 1995), demonstrating that the effects of reticulation may remain phylogenetically detectable for perhaps millions of years.

One important implication of these examples, and others, is that hybridization and introgression may influence interpretations of relationships not only at the very tips of phylogenetic trees, which is where interspecific progeny are most likely possible, but also at higher levels of divergence among species that no longer are able to form fertile hybrids. In this respect it is noteworthy that many of the reports of cytoplasmic introgression involve ancient hybrids among lineages that presently cannot be crossed, but apparently once did (Palmer et al., 1983; Doebley, 1989; Wendel et al., 1991; Soltis and Kuzoff, 1995). A particularly striking example of this phenomenon is offered by *Gossypium gossypioides,* a species from Oaxaca, Mexico (Wendel et al., 1995b). Evidence indicates that the nuclear genome of this species is heavily introgressed with sequences from taxa that presently exist in Africa and Asia. This is an example of introgression between taxa whose modern derivatives are not only incompatible, but have geographic ranges in different hemispheres, and hence, have no opportunity for sexual contact.

Perhaps it is in this arena of hybridization and introgression that phylogenetic incongruence has had its greatest impact in systematic biology. Despite the many uncertainties of interpretation given any single set of conflicting cladograms, the incongruence itself has inspired a search for explanations, which has led to a renewed interest in the possibility that hybridization is a potent force in evolution (Arnold, 1992; Rieseberg, 1995).

Lineage Sorting Genetic polymorphism is a conspicuous characteristic of plant populations. This polymorphism has been extensively documented by allozyme analysis (Hamrick and Godt, 1989; Soltis and Soltis, 1989), and more recently has been confirmed and quantified by DNA sequencing (e.g., Gaut and Clegg, 1993b; Hanson et al., 1996; Innan et al., 1996). Thus, it comes as no surprise that individuals may be heterozygous at a particular nuclear gene used for phylogeny reconstruction, or that variation exists for a locus within a population, or that multiple alleles exist within a particular species. These multiple alleles are related to each other by a complex history involving myriad factors, such as mutational processes, interaction with other alleles via recombination or gene conversion, and a whole series of population level features that may have influenced genetic transmission. Genes within species therefore have a history, which may be reconstructed by phylogenetic analysis to yield an intraspecific gene tree (Doyle, 1992, 1995, 1996; Maddison, 1995, 1996).

It is of interest to consider the nature of gene trees in the context of naturally occurring polymorphism levels and to extend the gene tree concept to interspecific comparisons. Consider, for example, the case where ancestral alleles are polymorphic and a particular polymorphism is maintained through one or more speciation events. The daughter lineages arising from this speciation event each have two alleles, and these alleles are shared across the species boundary. Accordingly, each of the two alleles may be referred to as being older than the species to which they belong. Extending this process of organismal divergence through time, and introducing allele extinction through normal stochastic or selective means and allele genesis through mutation it is possible to envision how gene trees might fail to reflect the phylogeny of the species that house them.

Lineage sorting, or phylogenetic sorting, refers to a special subset of gene tree—species tree relationships (Neigel and Avise, 1986; Pamilo and Nei, 1988). Specifically, if a polymorphism transcends one or more organismal divergence events, and if the polymorphism fails to survive later speciation events, the polymorphism is "sorted" into its component alleles. A simple example of this process is illustrated in Fig. 10.2. Notice that the sine qua non of lineage sorting is the existence of an ancestral polymorphism, which may exist for any nuclear or organellar sequence, or, for that matter, for a morphological trait (Roth, 1991). As long as there is polymorphism, the possibility exists that it may survive speciation and thereby contribute to a future lineage sorting event.

With respect to the subject of the present chapter, the significance of lineage sorting is

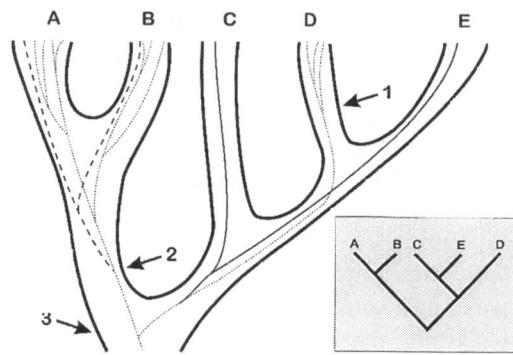

Figure 10.2. Lineage sorting, coalescence, and phylogenetic incongruence. Illustrated are the historical relationships among alleles at a single locus in five hypothetical species, A through E. Alleles are continually created through mutational processes and lost via selection and drift, but for simplicity, only alleles that have survived to the present are depicted. Alleles within a species may be monophyletic, as exemplified in this example by the three alleles in species D; these alleles share a most recent common ancestor, or *coalesce*, at the point marked by arrow number 1. In other cases, alleles within a species are not monophyletic, as shown by species A and B, each of which possess both dotted and dashed alleles; coalescence of these alleles occurs at the point marked by arrow number 2, antedating the divergence of species A and B. Coalescence of all alleles in the figure (including solid alleles) occurs at arrow number 3. Differential survival of alleles in alternative lineages will lead to instances where historical relationships among alleles (= gene tree) are not the same as those of the species that house them (= species tree). In the example shown, dotted alleles survive to the present in species D but solid alleles do not, whereas the opposite allelic survivorship pattern occurs in species C and E. Hence, sampling of a single allele from each taxon will lead to recovery of the gene tree shown in the inset, which is incongruent with the history of organismal divergence. The process by which the allelic polymorphism present in the common ancestor of species C, D, and E was lost is termed *lineage sorting*.

that it may be a cause of phylogenetic incongruence. This is explained most easily by referring to Figs. 10.2 and 10.3, which show how lineage sorting may lead to discordant gene trees and species trees, and by extension, between two or more gene trees. One important manifestation of lineage sorting is that alleles *from different* species may be more closely related to each other than are alleles *within the same* species (Figs. 10.2, 10.3). Consequently, and because the process by definition involves alleles that are older than their species, alleles at any given locus within a particular species are not necessar-

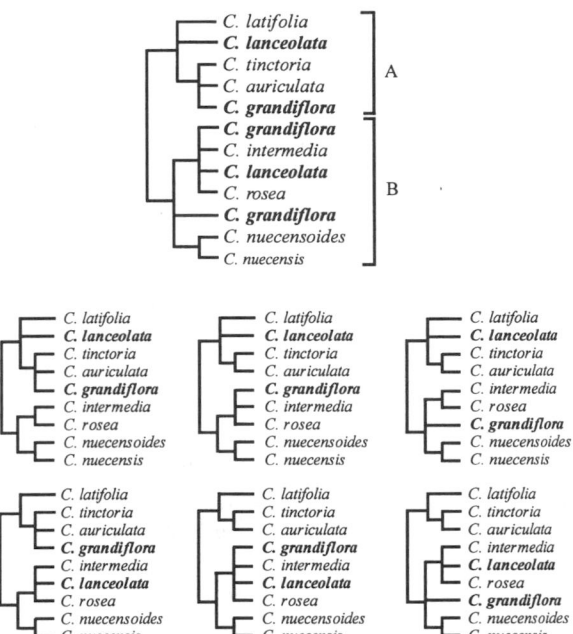

Figure 10.3. The effect of sampling or lineage sorting on a real data set. *Coreopsis lanceolata* and *C. grandiflora* are each polymorphic for cpDNA haplotypes, whose relationships to one another and to cpDNA haplotypes from other *Coreopsis* species are shown in the top cladogram (after Mason-Gamer et al., 1995). Haplotypes fall into two primary clades, designated A and B. If only one individual was sampled from each of the two species *C. lanceolata* and *C. grandiflora,* or if there had been additional haplotype lineage extinction, any one of the six topologically different hypotheses of species relationships (below) might have been obtained.

ily expected to be monophyletic. To this extent, phylogenetic incongruence among gene trees is an *expected* outcome of lineage sorting at lower taxonomic ranks, the divergence levels for which lineage sorting processes are usually thought to operate.

Lineage sorting may affect both organellar and nuclear genes, but because of the generally slow rate of cpDNA evolution and population genetic considerations, there are relatively few examples of cpDNA polymorphisms that transcend species boundaries. One example is offered by *Coreopsis,* in which two plastome classes (A and B) found in a wide survey of *Coreopsis grandiflora* may have been phylogenetically sorted into various populations of this species and other *Coreopsis* taxa (Mason-Gamer et al., 1995). Additional examples include wild potatoes (Castillo and Spooner, 1997) and populations of two species of *Phacelia* (Levy et al., 1996).

More common are empirical examples of nuclear genes where alleles transcend species boundaries, and this number is expected to grow rapidly in the coming years as additional surveys are undertaken of multiple alleles within species (see, for example, Cronn et al., 1996 for 5S rDNA polymorphisms shared among closely related *Gossypium* species). At present, some of the best examples in plants come from maize and its relatives, where alleles that are more ancient than their species have been described for *Adh* (Gaut and Clegg, 1993a; Goloubinoff et al., 1993), *c1* (Hanson et al., 1996), and the internal transcribed spacer region of the major rDNA repeat (Buckler and Holtsford, 1996a). In each of these examples, gene tree topologies are consistent with the operation of lineage sorting, although the authors correctly take pains to discuss the alternative possibility of interspecific gene flow. This uncertainty as to cause deserves emphasis here: The incongruence footprint left by lineage sorting is often identical to that expected from other processes, such as introgression (see Assessing the Cause).

The fact that alleles may transcend species boundaries has a number of important implications. First, because alleles at each locus have their own history, and because there may be considerable discord among their gene trees, we are challenged to come to terms with what is meant by the expression "species tree." This rich subject is perhaps beyond the scope of this chapter, but it has been thoughtfully discussed elsewhere (see in particular Baum and Shaw, 1995; Doyle, 1995, 1996; Maddison, 1995, 1996).

A second consequence of the realization that alleles have complex histories is the emergence and development of an entire discipline within population genetics known as coalescence or coalescent theory (Hudson and Kaplan, 1988; Ewens, 1990; Hudson, 1990; Slatkin, 1991). Alleles at any locus are subject to the full spectrum of internal (e.g., mutation, concerted evolution) and external forces (e.g., breeding history, selection, drift). Effects of these various influences are expected to differ for alleles within a species as well as for alleles from different species. At any locus, a given pair of alleles may share a relatively recent common ancestor, in which case coalescence is said to be relatively fast, or alternatively, the most recent common ancestor may antedate several speciation events, in which case coalescence is relatively ancient.

One result from coalescent theory that is particularly relevant to phylogenetic issues is that effective population size has a large effect on coalescence times: Alleles in small populations reach fixation or are lost more rapidly than alleles from large populations, and hence the time to coalescence is less in small than in large populations. Accordingly, coalescence of organellar genes may be faster than nuclear genes (Moore, 1995), all else being equal (see, however Hoelzer, 1997), because organellar genes often have lower effective population sizes than nuclear genes (Birky et al., 1983; see, however, Chesser and Baker, 1996). This population genetical process, combined with the generality that infraspecific polymorphism levels are typically low for plastome genes, accounts for the relative rarity of lineage sorting for cpDNA-based gene trees (see, however, Mason-Gamer et al., 1995). It also accounts for the expectation that in general, lineage sorting is only expected to be a potential source of dis-

cord at lower taxonomic ranks, levels in the hierarchy that antedate coalescence. Moore (1995) models coalescence for neutral nuclear and organellar genes and shows that for a three-species portion of a phylogeny, when the probability of congruence between an organellar tree and a species tree is 0.95, the probability of congruence for a nuclear tree is 0.62. Moreover, 16 independent nuclear trees are required to generate as much confidence of congruence as obtained with a single organellar tree.

It is tempting to conclude from this discussion that phylogenetic hypotheses derived from nuclear genes will be wholly unreliable at lower taxonomic ranks, and indeed, the foregoing examples for *Zea* illustrate the potential for discord attributable to lineage sorting and coalescence issues. These problems may dissolve, however, for species with small effective population sizes, such as inbreeding species or those whose populations have few individuals. Unfortunately, this information rarely exists in the detail necessary to assess accurately the likelihood that lineage sorting is a factor in allele topologies (see Doyle, 1995 for additional discussion). At present, we must simply be alert to the possibility that it is a factor, especially in phylogenetic analyses at the generic level and below.

A final comment concerns the possibility that lineage sorting may affect tree topologies at higher levels of divergence. This prospect emerges from polymorphisms that are maintained by natural selection through many speciation events (e.g., Gaur et al., 1992). Polymorphisms maintained by balancing selection will have longer coalescence times than neutral loci, and thus may contribute to phylogenetic incongruence at higher taxonomic ranks. Perhaps the most compelling example of this process in plants concerns the self-incompatibility (S) locus in Solanaceae, where high molecular diversity is ensured by enforced outcrossing (Rivers et al., 1993). Phylogenetic and phenetic analyses of alleles at this locus have demonstrated the existence of alleles from different genera that are grouped together in clades (Ioerger et al., 1990; Clark and Kao, 1991; Richman et al., 1996). Some of these allelic polymorphisms have survived what must be an enormous number of speciation events, as coalescence times for alleles shared between *Petunia* and

Solanum appear to precede the divergence of these two genera, which may have last shared a common ancestor over 30 million years ago.

This example from the S locus in the Solanaceae demonstrates that while lineage sorting is generally of importance to population and species-level studies, it may occasionally be significant at higher levels of divergence. In this respect, lineage sorting is similar to introgression of nuclear or cytoplasmic genes, in that these processes also are expected to influence phylogenetic inference mostly at lower taxonomic ranks.

Horizontal Transfer   Among the more unexpected conclusions drawn from molecular phylogenetic studies over the past decade has been that genes occasionally traverse species boundaries through nonsexual means. This process, termed horizontal or lateral transfer (Amábile-Cuevas and Chicurel, 1993; Kidwell, 1993; Syvanen, 1994; "xenology" or "paraxenology" of Patterson, 1988), has long been suspected to play a role in the evolution of bacteria (Lan and Reeves, 1996), but in recent years an increasing number of putative examples have been reported from eukaryotes (Houck et al., 1991; Clark et al., 1994; Robertson and Lampe, 1995; Delwiche and Palmer, 1996; Hibbett, 1996; Moens et al., 1996). From a phylogenetic standpoint, horizontal transfer generates discordant trees exactly like the processes of hybridization and introgression, although perhaps only for a single gene and potentially across great phylogenetic distances.

One of the more compelling and well-known examples involves "P element" transposons in *Drosophila* (Houck et al., 1991; Clark et al., 1994). A wealth of phylogenetic and phenetic evidence supports the proposal that these elements first made their way into the *D. melanogaster* lineage via horizontal transfer from a taxon in the *D. willistoni* clade. This evidence includes the near sequence identity of elements extracted from *D. melanogaster* populations that possess the element, the high sequence similarity between elements from the *D. melanogaster* and *D. willistoni* lineages, and peculiarities of their distribution in *D. melanogaster* populations and laboratory stocks. A remarkable and unique feature of this example is that both a likely vector and an eco-

logical context for interspecific DNA transfer have been identified, namely, a mite (*Proctolaelaps regalis*) that feeds on *Drosophila* eggs and larvae. Houck et al. (1991) demonstrated that the mite was capable of carrying P elements subsequent to feeding on fruit fly strains that possessed the transposon, thereby providing a rare glimpse into vector-mediated horizontal transfer.

For most other putative examples of horizontal transfer, the evidence is not as compelling, although there are several recent and interesting exceptions (Delwiche and Palmer, 1996; Hibbett, 1996; Moens et al., 1996). Delwiche and Palmer's study is especially noteworthy, as their report involves the widely utilized gene *rbcL,* for which more sequences exist in plants than any other gene. From phylogenetic analyses of bacterial and plant *rbcL* sequences, Delwiche and Palmer (1996) inferred approximately six horizontal transfer events involving *rbcL* and various combinations of proteobacteria, cyanobacteria, and plastids (see Chapter 13).

These and other examples demonstrate that this non-Mendelian process almost certainly occurs in nature, although its extent and evolutionary significance remain unclear. In addition to the process being relatively rare and difficult to catch in the act, the initial evidence for it in any particular study usually consists of the simple observation of a noteworthy discord between a gene tree and the expected organismal relationships. As is clear from the present chapter, many other evolutionary processes may result in the same observation of phylogenetic incongruence, and these alternatives are not always readily excluded as formal possibilities (Cummings, 1994). Thus, the patterns reported by Delwiche and Palmer (1996) may have resulted, as the authors note, from a complex history of gene duplication and loss, with differential survival in various lineages resulting in "apparent" horizontal transfer. Similarly, the putative horizontal transfer of group I introns into the rDNA of mushroom-forming fungi may also have resulted from a highly homoplasious pattern of intron loss in clades lacking the putative insertion (Hibbett, 1996). Inferences of lateral transfer of members of replicative families, such as P elements and retrotransposons (VanderWiel et al., 1993) are additionally complicated by their inherent history of duplication and differential survival of orthologous genes (see also section Orthology/Paralogy Conflation).

At present, it seems clear that horizontal transfer occurs, although we do not yet understand its frequency or significance. Given the many other causes of phylogenetic incongruence, it seems unwise, however, to invoke it on this basis alone, without due consideration of the other possibilities and without additional supporting evidence (Cummings, 1994). This evidence may be difficult to garner, particularly if the putative transfer event is ancient, but some of the alternatives may still be addressed using arguments based on parsimony (cf. Delwiche and Palmer, 1996; Hibbett, 1996).

Gene and Genome-level Processes

Most of the evolutionary processes included under this heading cause phylogenetic incongruence because of some manifestation of nonindependence. This applies to both intra- and intergenic recombination, various forms of concerted evolution, evolutionary rate heterogeneity among sites or taxa, and most obviously, when nucleotide sites in a molecule evolve in a directly dependent fashion, such as stemmed bases in ribosomal RNAs. Despite the diversity of underlying molecular mechanisms encompassed by these phenomena, they exhibit the common thread of perturbing relationships among terminals in gene trees through the effects of interaction (among alleles or genes) or nonindependence (among sites), with the consequence that the resulting gene tree is incongruent with other gene trees. Our understanding of these processes is still incomplete, and at present empirical demonstrations of their effects on phylogenetic inference remain reasonably small in number. As a result, the relative significance of each phenomenon to systematists is not known.

Intragenic Recombination An evolutionary history that involves reticulation is not accurately depicted by a strictly bifurcating tree. Trees are, of course, the products of cladistic analysis, and hence the reality of hybridization poses problems for the ideal of phylogeny reconstruction. This problem applies equally to whole organisms, using morphological data sets (Funk, 1985; McDade, 1990, 1992, 1995) and to genes, using sequence data from a sampling of alleles. In the latter case, intragenic (or interallelic) recombination (or gene conversion) leads to the evolution of composite molecules that possess characteristics of both parental alleles. When this occurs, alleles will no longer have arisen exclusively from normal mutational processes, and their relationships may no longer be depicted in a strictly hierarchical fashion. Forcing a treelike topology onto a sampling of alleles that originated from both nonrecombinant and recombinant processes will result in a reconstruction that is, at least in part, erroneous. As a consequence, the resulting gene tree may be incongruent with phylogenetic estimates obtained using other sources of data.

That interallelic recombination occurs is already evident, despite the fact that there are still relatively few studies where multiple alleles of a locus have been sequenced from individual plant taxa. Perhaps the best example concerns alcohol dehydrogenase in maize (Gaut and Clegg, 1993a; Hanson et al., 1996), where recombination was documented among alleles at *Adh1* (nine recombinations among six alleles sampled) and *Adh2* (five recombinants in 12 alleles sampled). Hanson et al. (1996) also suggest that there have been three recombination events among 27 sequences at the anthocyanin regulatory locus, *c1*. Innan et al. (1996) described the evolutionary history of 17 sequences of *Arabidopsis Adh* as having included four intragenic recombination events.

Detecting recombination in a sampling of alleles is not always straightforward, and the reliability of an inference is likely to depend on several factors, including the level of confidence that the alleles involved are from the same locus as opposed to a paralogous locus, the relative antiquity of the recombination event, the amount of sequence divergence between the two alleles involved, and the distribution of differences along the length of the gene. Thus, ancient recombination between two similar alleles is likely to remain undetected, whereas a single recent gene conversion or reciprocal recombination between two highly dissimilar alleles will probably be readily apparent. In the latter case, this may be manifested as a "chimeric" allele, where the 5′

and 3' portions of the allele (or smaller regions) appear to have different phylogenetic histories and phenetic relationships with other alleles sampled (e.g., Innan et al., 1996). Despite these potential difficulties in detection, several tests for recombination have been developed (Sawyer, 1989; Fitch and Goodman, 1991; Hein, 1993; Templeton and Sing, 1993; Jakobsen and East-eal, 1996) and computer programs that assist in the analysis are available (Hein, 1993; Jakobsen and Easteal, 1996; Hey and Wakeley, 1997).

In some cases, recombination may be so extensive that it will remain undetected as such except under extraordinary circumstances. A case in point concerns the rDNA ITS from *Gossypium gossypioides,* which available evidence suggests was derived from recombination between two highly divergent ITS sequences that were brought together in a common nucleus as a result of inter-

specific hybridization (Wendel et al., 1995b). The ITS sequence in question is not a simple chimeric product, but instead consists of a mosaic of nucleotides contributed by both parental clades distributed in an apparently random fashion along the length of the sequence. In a follow-up study, phylogenetic data were gathered that suggest that the 5S rDNA locus in *G. gossypioides* was similarly recombined (Cronn et al., 1996).

Recombination, like hybridization, causes homoplasy when included in a phylogenetic analysis. Consequently it is not surprising that recombinant alleles behave cladistically like morphological hybrids (Funk, 1985; McDade, 1990, 1992, 1995) in that they resolve in relatively basal positions in clades occupied by one parent or the other. This prediction is fulfilled by the recombinant rDNA alleles from *Gossypium gossypioides,* as shown in Fig. 10.4. The three

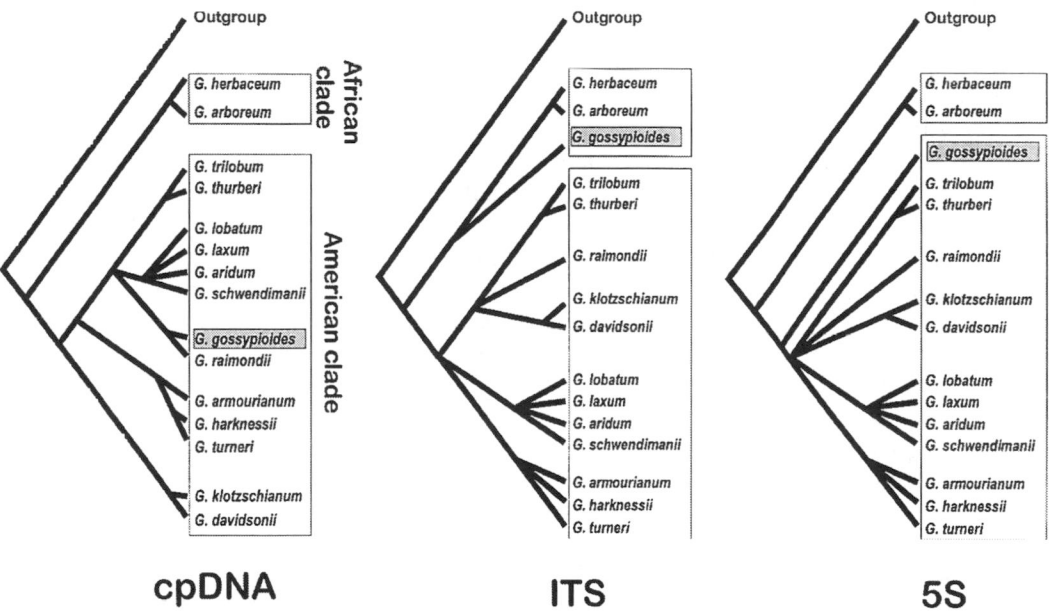

cpDNA   **ITS**   **5S**

Figure 10.4. Effect of recombination on phylogenetic resolution. Shown on the left are relationships among American and African diploid species of *Gossypium,* as inferred from cpDNA restriction site variation (Wendel and Albert, 1992; Seelanan et al., 1997). The primary division is into strongly supported clades consisting of diploids from different hemispheres. Within the American clade, *G. gossypioides* appears as the sister taxon of *G. raimondii,* which is the conventional placement supported by most other genetic and cytogenetic data (Wendel et al., 1995b). Evidence suggests, though, that the history of *G. gossypioides* includes an episode of hybridization and introgression with a genome like that presently found in African diploid species. This hypotheses was initially invoked to account for phylogenetic trees that are incongruent with the conventional tree shown on the left. Specifically, in trees based on both ITS (center) and 5S (right) ribosomal DNA sequences (redrawn from Wendel et al., 1995b and Cronn et al., 1996, respectively) *G. gossypioides* occupies phylogenetically basal positions, although in different primary clades in each case. This basal resolution is consistent with the highly recombined nature of the ribosomal repeats in *G. gossypioides,* and illustrates that the phylogenetic effects of molecular recombination are similar to those of morphological hybridization (Funk, 1985; McDade, 1990, 1992, 1995).

panels illustrate phylogenetic observations from a nonrecombinant molecule (cpDNA, left), which cladistically resolves in a terminal "non-recombinant" position, and from the two recombinant rDNA genes (ITS, center; 5S, right), which resolve basally, although in different parental clades in each case.

Recombination may be difficult to distinguish from other causes of phylogenetic discord, such as undetected and unexpected interspecific introgression. In some cases, an inference of recombination may be justified by the pattern of molecular variation, such as when an obvious chimeric allele is observed. In other cases, the source of the conflict may not be so readily apparent. With respect to the peculiar rDNA sequences of *Gossypium gossypioides,* both processes appear to have been involved. Specifically, an organismal history of hybridization and introgression is suggested to have brought "native" and "alien" rDNA repeats into contact, whereafter they extensively recombined. The inference of recombination, as opposed to excessive homoplasy or aberrant molecular evolution, is based on extensive supporting evidence, including comparative morphology, cytogenetics, and independent molecular evidence from other nuclear and organellar markers. In the absence of this additional evidence, the genesis of the aberrant rDNAs would have been more difficult to infer.

Orthology/Paralogy Conflation Sequences from the plastid genome exist in only a single copy, except for genes located in the inverted repeat. Hence, a particular cpDNA gene from different individuals or species is likely to be related by direct descent, with no intervening history of gene duplication and extinction of related genes. This simple form of homology is termed *orthology,* and two genes are said to be *orthologous* if their relationship originated from organismal cladogenesis (see Chapter 4). Orthologous sequences are appropriate for phylogenetic analysis, in that their history may reveal organismal divergence events, assuming the absence of the other potentially confounding influences described in this chapter.

Genic relationships may take forms other than orthology, however. Consider the case, for example, where a gene becomes duplicated (yielding locus A and locus B) and each copy of the gene is transmitted to two daughter lineages (1 and 2) following an organismal divergence event. In this case there are two sets of orthologous relationships, one involving locus A from both lineages (A1 vs. A2) and the other involving locus B from both lineages (B1 vs. B2). In addition, however, there are relationships among non-orthologous genes from different species (A1 vs. B2 and A2 vs. B1) as well as between the duplicated copies within each species (A1 vs. B1, A2 vs. B2). In this example the duplicated sequences within and between lineages exhibit *paralogy,* and the genic relationships are termed *paralogous* (Fig. 10.5). Paralogy is simpler to define than orthology, in that a rigorous criterion exists: If two loci occur in the same individual genome, then they must be paralogous. Orthology, however, describes a genetic relationship that is inextricably tied to a phylogenetic concept involving two or more taxa (see Chapter 4).

When paralogous genes are unwittingly included in phylogenetic analysis to the exclusion of appropriate orthologous comparisons, the resulting gene tree will confound organismal divergence events with a tracking of the history of duplication. Hence, erroneous assessments of orthology and paralogy will lead to phylogenetic incongruence, as will sampling an unintended mixture of orthologous and paralogous sequences in a study (Doyle, 1992; Sanderson and Doyle, 1992; VanderWiel et al., 1993).

The trivial example shown in Fig. 10.5 represents an oversimplification of the complexity of genic relationships from the nuclear genomes of nearly all plants. In actuality, genes that exist in only a single copy are rare relative to those that exist in several to many copies. Even in the remarkably streamlined genome of *Arabidopsis thaliana,* a minimum of 15% of "single-copy" nuclear sequences are duplicated (Kowalski et al., 1994; McGrath et al., 1993). Most nuclear genes from most plants exist in several to many copies, these reflecting a complicated history of sequence duplication, through polyploidization (Masterson, 1994), retrotransposition (Vander-Wiel et al., 1993; Wessler et al., 1995; Kumar, 1996), or other mechanisms, as well as processes of sequence loss, via genic and/or chromosomal deletion (e.g., Dubcovsky and

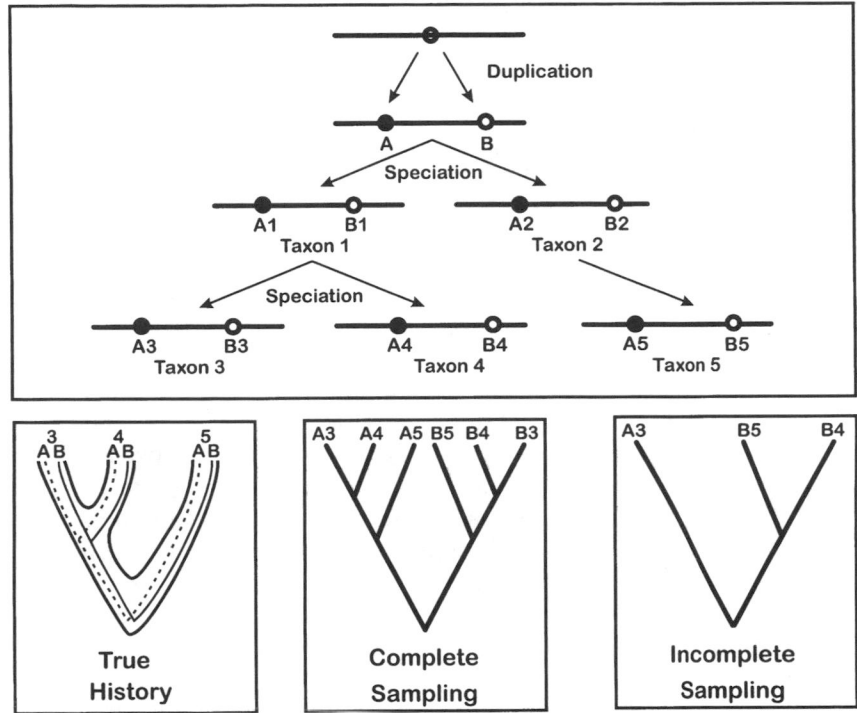

Figure 10.5. Orthology, paralogy, and phylogenetic incongruence. Top: Gene duplication yields two loci, A and B, which are each transmitted into daughter lineages (Taxon 1 and Taxon 2) following an organismal divergence event. For illustrative purposes only, the duplication is shown as occurring on the same chromosome. Loci A1 and A2 become distinct from one another as a consequence of organismal divergence, as do loci B1 and B2; hence, each of these locus pairs represents a set of *orthologous* genes or *orthologues*. Nonorthologous genes are termed *paralogues*. *Paralogous* genes arise from gene duplication; the term applies to relationships among genes from different species (A1 vs. B2 and A2 vs. B1) and to duplicated copies within species (A1 vs. B1; A2 vs. B2). Following additional divergence events, the two sets of orthologues (A and B) are transmitted into three taxa (3, 4, and 5). Each of these three organismal lineages thus contains two gene copies, A and B, whose phylogenetic relationships are shown in the bottom left panel. If all six copies are sampled, two gene trees will be recovered, each faithfully reproducing the history of organismal divergence (bottom center). If sampling is incomplete, however, or if one or more genes fails to survive to the present, a gene tree may be obtained that is incongruent with the true organismal history (bottom right).

Dvořák, 1995; Gottlieb and Ford, 1996) and pseudogene formation (Buckler and Holtsford, 1996b; Gottlieb and Ford, 1996; Seberg et al., 1996). Given this history of gene duplication and loss, homology for any given gene system from two or more species *may* include orthologous comparisons, but will almost certainly include a diverse array of paralogous relationships. Duplications responsible for the latter may have been ancient or more recent. Hence paralogy is a relative concept (VanderWiel et al., 1993), as two genes may be more or less paralogous than two other genes, depending on the relative recency of their shared common ancestry.

The foregoing discussion highlights why sequence similarity per se is an unreliable indicator of orthology. Given that orthologous comparisons are the only ones potentially useful for phylogeny reconstruction, it is essential that orthology be established prior to a phylogenetic study. Orthologous relationships can be hypothesized with varying degrees of confidence from using criteria such as tissue specificity (Doyle, 1991), expression patterns (Doyle, 1994a), Southern blot analysis (Matthews and Sharrock, 1996), and perhaps most convincingly, comparative mapping (Bonierbale et al., 1988; Whitkus et al., 1992; Kowalski et al., 1994; Pereira et al.,

1994; Brubaker et al., 1997). For most plants, however, this information is likely to be difficult to generate. In some cases, it may be possible to amplify apparent orthologues selectively using PCR, but in the absence of other data, the assumption of strict orthology may not be justified. Reasonable evidence for orthology may often be obtained, though, through a combination of Southern blot analysis, PCR specificity, and perhaps most critically, phylogenetic analysis, particularly at lower taxonomic ranks, where duplication and divergence histories are less complex.

Many recent empirical presentations of gene trees illustrate how phylogenetic incongruence may result from orthology/paralogy conflation (VanderWiel et al., 1993; Kolukisaoglu et al., 1995; Mathews et al., 1995; Mathews and Sharrock, 1996; Seberg et al., 1996). Studies of phytochrome genes (Kolukisaoglu et al., 1995; Mathews et al., 1995; Mathews and Sharrock, 1996) demonstrate nicely some of the complexities in the history of this gene family and show that different orthology and paralogy issues apply to various levels in the taxonomic hierarchy. VanderWiel et al. (1993) show how incongruence with organismal relationships is the *expected* outcome for phylogenetic analyses of multigene families such as retrotransposable elements, due to the complex history of duplicative transposition combined with incomplete sampling.

A somewhat different type of confounded orthology/paralogy relationship may arise in cases involving organellar genes where there has been duplicative gene transfer to the nucleus during organellar evolution (Nugent and Palmer, 1991; Gantt et al., 1991). If nuclear copies of the duplicated gene are sampled from only a portion of the taxa included in a study, gene trees may show unexpected relationships. To the extent that the underlying phenomenon may be diagnosed, the inferred phylogeny may therefore provide critical information on the timing of the gene transfer event.

A final comment concerns the intriguing possibility that a "species tree" may be estimated from gene trees even in the presence of confounded orthology/paralogy relationships. Guigó et al. (1996) present a method that recon-ciles conflicting gene trees by seeking a species tree that minimizes the total number of gene duplications and losses. This interesting approach may facilitate insight into both gene-family and organismal histories.

Interlocus Interactions and Concerted Evolution As previously discussed, evolutionary histories that involve reticulation are not accurately depicted by bifurcating trees, regardless of whether the reticulation involves an organismic-level process (see Hybridization and Introgression, above) or a process at the genic level (see Intragenic Recombination). In both cases, characters from two parental lineages are mixed, leading to phylogenetic reconstructions that have a high likelihood of being erroneous, and as a consequence, incongruent with estimates obtained using other data sources. In this section, we discuss other common processes in genic evolution that may be construed to be forms of reticulation, that is, when different genes interact. These interactions are multifarious and complex, but they all lead to nonindependent character evolution, through the recombining, blending, or altering of different DNA sequences.

Intergenic recombination may influence gene tree interpretation at any level in the taxonomic hierarchy. In a phylogenetic reconstruction of serine protease genes from humans, for example, different trees were obtained from different domains of the proteins; these were interpreted as reflecting ancient mosaicism, or "exon shuffling" between formerly independent gene regions (Ikeo et al., 1995). This example is only one of many of exon shuffling, mostly from vertebrates (Stone and Schwartz, 1990; Patthy, 1991; Doolittle, 1995) but including at least a couple of examples from higher plants (Domon and Steinmetz, 1994; Schena and Davis, 1994). To the degree that proteins have been assembled by this process, their gene sequences will be reticulate, which has obvious ramifications for phylogenetic analysis.

The extent to which exon shuffling needs to be considered by plant systematists is not known, but it is likely that the phenomenon applies primarily to the deepest levels in plant phylogeny. Other, better-known forms of genic interaction apply to more recent divergences, particularly between duplicated genes and among members of multigene families. Given that most sequences in plant genomes have a long history of duplication and divergence (see previous section, Orthology/Paralogy Conflation), the potential for intergenic interaction applies to many if not most nuclear sequences. It is also likely that the relevant genetic processes influence phylogeny reconstruction at a wide variety of taxonomic levels, making this a general concern.

One form of nonindependence among genes arises from intergenic recombination or gene conversion. These two mechanisms are related and are usually difficult to distinguish, although gene conversion should result in a higher degree of sequence homogenization than classical reciprocal recombination, which merely redistributes variation. Examples of gene conversion may include alcohol dehydrogenase alleles from *Gossypium,* where different regions of two alleles demonstrate remarkably different frequencies of indels and nucleotide substitutions (Millar and Dennis, 1996). In *Pisum,* duplicated copies of a glutamine synthetase gene are over 99% identical at the nucleotide level, except in the middle portion of the genes. Walker et al. (1995) attribute high sequence similarity in the distal regions to the homogenizing influence of gene conversion, and interpret the reduced similarity in the middle part of the genes as representing unconverted portions of sequence. Similarly, paralogous copies of *PgiC* in some species of *Clarkia* may have experienced gene conversion (Gottlieb and Ford, 1996).

These examples have several implications. First, in each case, the chimeric or recombinant nature of the genes involved was instrumental in revealing an important aspect of their evolutionary history, namely, interaction with other, similar sequences. Thus, the reticulation that confounds phylogeny reconstruction yields insights into molecular evolutionary process. Second, there are still relatively few molecular systematic investigations using nuclear genes (other than rDNA, but see below), yet given even this small number, phylogenetically significant gene conversion and recombination events have been detected, suggesting that the process may be common, as has been shown in yeast (Goldman and Lichten, 1996), and hence of consequence for the practicing systematist. Third, it may be that most gene conversion or intergenic recombination events will remain undetected, because the ability to detect chimeric sequences should decrease with increasing antiquity of the intergenic interaction. Moreover, repeated interlocus interactions over time may yield mosaic sequences whose composite histories defy disclosure.

Additional complexities are involved in the consideration of larger multigene families and repetitive DNAs, like ribosomal RNA loci, which are especially subject to the homogenizing forces collectively referred to under the umbrella heading of "concerted evolution" (reviewed in Arnheim, 1983; Elder and Turner, 1995). Phylogenetic implications of concerted evolution have been discussed based on simulations as well as empirical examples (e.g., Hillis and Dixon, 1991; Sanderson and Doyle, 1992; Baldwin et al., 1995; Wendel et al., 1995a). Effects of concerted evolution may range from complete homogenization of multigene families to occasional gene conversion events that affect only a small portion of the members of a given multigene family. Meagher et al. (1989), for example, showed that gene conversion has played a prominent role in the evolution of RUBISCO small subunit genes (*rbcS*) in plants, whereby sequences from single lineages are more similar to each other than would be expected under a scenario of independent evolution. Concerted evolution is incomplete, however, as evidenced by the considerable *rbcS* diversity that is retained within species. A similar process of partial and intermittent homogenization has been invoked for actin evolution in angiosperms (Moniz de Sá and Drouin, 1996). This latter study demonstrates that gene conversion events have occurred at various times during evolution of the actin gene family, and that these may be detected at phylogenetic depths ranging from the rank of class (Liliopsida) to genus (*Zea*).

The mechanisms by which these occasional gene conversion events take place, and the forces responsible for their sporadic nature, are incompletely understood. From a phylogenetic perspective, though, these examples of partial homogenization pose real problems for interpretation of both organismal and gene-family history (Sanderson and Doyle, 1992). A reasonably accurate gene tree is a minimum requirement for inferring a species tree, and this may be obtained both in cases where concerted evolutionary forces are absent and when they are strong enough to maintain relative infraspecific homogeneity. Partial homogenization, however, generates recombinant sequences that are neither orthologous nor paralogous. Hence, no history of bifurcation among genes or organisms may be recovered. Consequently, nuclear genes that are subject to *incomplete* concerted evolution are expected to be the least useful for reconstruction of organismal relationships.

A variant on this theme of incomplete concerted evolution involves cases where the degree of homogenization varies among taxa. This variation has been shown in legumes, for example, where both leghemoglobin (Doyle, 1994b) and 7S seed storage protein genes (Doyle et al., 1992) appear to evolve concertedly in some but not all legume taxa. This process may lead to cases where orthologues of particular genes in a taxon experiencing weak concerted evolutionary forces may simply no longer exist in related taxa where the genes have been homogenized by stronger concerted evolutionary processes.

Concerted evolutionary forces are often stronger for tandemly repeated sequences, such as ribosomal genes, than they are for dispersed members of multigene families spread across several or more chromosomal loci, at least in part because one major mechanism, unequal crossing over, is only possible for tandemly arranged families. Ribosomal DNA genes are organized into arrays consisting of hundreds to thousands of identical to nearly identical repeats, indicative of their concerted mode of evolution. This feature of rDNA, in addition to their high copy number, has been responsible for the rapid growth and utilization of these sequences in phylogeny reconstruction (Hillis and Dixon, 1991; Baldwin et al., 1995; see Chapter 7).

Numerous phylogenetic studies in plants have been conducted using the 18S–26S rDNA repeat, allowing several insights into the various possibilities for molecular evolution of repeated sequences; each of these has a differing relevance to phylogenetic inference and phylogenetic incongruence. One possibility is that ho-

mogenization of repeats may be incomplete, and indeed, considerable sequence variation may exist among repeats within species. Buckler and Holtsford (1996b), for example, in an analysis of ITS sequences, demonstrate both intra-array sequence diversity and the presence of rDNA pseudogenes. Because of incomplete homogenization, interspecific orthology is thus not assured. Similarly, inter-array homogenization may not occur, even in circumstances where rDNA repeats within arrays are reasonably homogeneous. This results in the long-term maintenance of more than one rDNA repeat type (e.g., Suh et al., 1993; Kim and Jansen, 1994; Sang et al., 1995; Waters and Schaal, 1996), which may in some cases actually preserve paralogous relationships. An additional evolutionary outcome, especially in cases where divergent repeats are united in a common nucleus as a result of hybridization or polyploidy, is homogenization of repeats within (intralocus concerted evolution) and among (interlocus concerted evolution) arrays (Wendel et al., 1995a; Brochmann et al., 1996). Wendel et al. (1995a) showed that this process might occur bidirectionally in allopolyploids derived from a common progenitor bearing divergent rDNA arrays, such that different derivative taxa are fixed for alternative repeat types. This situation is formally equivalent to the random loss of different paralogous copies of duplicated genes in different lineages, a process that also has been demonstrated for rRNA genes (Dubcovsky and Dvorák, 1995; Danna et al., 1996).

These examples bear witness to the many possibilities for intergenic interactions in the evolution of multigene families, and accordingly, illustrate the complexities that need to be considered if they are to be used in phylogeny reconstruction, especially at the level of genus and below. There is irony in the fact that much of what has been learned about the evolutionary behavior of these multigene families was first revealed by the discordant gene trees they produce, once again underscoring the enhanced understanding of process that emerged from an initial observation of phylogenetic incongruence.

Rate Heterogeneity among Taxa A common observation in molecular systematic studies is evolutionary rate heterogeneity among the sequences sampled (Wu and Li, 1985; Britten, 1986). This phenomenon is widely acknowledged (Bousquet et al., 1992; Gaut et al., 1992, 1993), and is evidenced either intuitively by the observation of dramatic branch length variation in phylogenetic reconstructions, or by statistical tests of rate homogeneity (Wu and Li, 1985; Tajima, 1993; Takezaki et al., 1995; Gaut et al., 1996, 1997; see Chapter 9). Substitution rate heterogeneity may have a number of different underlying causes, but in some cases it is partially accounted for by variation in generation times, with which rates are often inversely correlated (Wu and Li, 1985; Li et al., 1987; Ohta, 1993; Gaut et al., 1996, 1997).

Regardless of the causative mechanism, rate heterogeneity is of interest from a phylogenetic perspective. In particular, when some branches are long and others are short, the long branches may experience long-branch attraction (Felsenstein, 1978), due to parallel substitutions in two lineages becoming mistakenly identified as actual synapomorphy. This phenomenon has been much discussed and is appreciated as a cause of misleading phylogenetic inference, and hence as a cause of phylogenetic incongruence.

A striking example both of rate heterogeneity and its influence on phylogeny reconstruction is offered by 18S rDNA sequence variation in parasitic and nonparasitic angiosperms (Nickrent and Starr, 1994; see Chapter 8). Relative rate tests showed that substitution rates were greatly accelerated in holoparasitic taxa: When compared with *Glycine,* the mean number of substitutions per site (K) for five autotrophic angiosperms was 0.036, whereas for four holoparasites the value was 3.5 times higher (K = 0.126). Nickrent and Starr (1994) further discuss how the accelerated substitution rates in the parasitic angiosperms confound attempts to deduce their phylogenetic placement, presumably due to long-branch attraction.

It is of interest to note that long-branch attraction has at least two fundamentally different causes. One, rate heterogeneity, is a molecular evolutionary phenomenon and thus it might be modeled in such a way that it is accommodated in phylogenetic reconstruction methods. The other cause relates to taxon sampling density, as

discussed earlier under Taxon Sampling. In this case it may be possible to "break up" long branches by a strategic increase in sampling (Wheeler, 1992; see, however, Kim, 1996), although the long branches may also arise from natural divergence and extinction patterns that are beyond investigator control. Distinguishing whether long branches reflect molecular evolutionary phenomena or sampling phenomena may not always be straightforward, although in most cases tests of rate heterogeneity (Wu and Li, 1985; Tajima, 1993; Takezaki et al., 1995; Gaut et al., 1996, 1997; see Chapter 9) should prove diagnostic.

Rate Heterogeneity among Sites   One of the more obvious aspects of DNA sequence data for protein-encoding genes is that the amount of variation detected varies by codon position. This is for the most part a straightforward consequence of the degeneracy of the genetic code by codon position in conjunction with the operation of selective contraints. Given that 70% of the possible nucleotide changes in the third position are silent, whereas nearly all substitutions in the first and second codon positions result in amino acid replacements, it is not surprising that in most molecular systematic studies greater variation is detected at the third position than in the other two codon positions. This is an example of amongsite rate heterogeneity, but it is not the only one. Substitution rates among sites may vary systematically in non-protein-encoding sequences just as they do among different regions or domains of protein-encoding genes (see Chapter 6).

When among-site rate variation exists, evolutionary change may be concentrated in a minority of nucleotide positions, while many other sites experience few to no substitutions. In many cases, this circumstance poses few problems vis-à-vis phylogeny estimation, particularly when substitution rates at the variable sites are low enough that evolutionary history is not obscured by multiple hits. This characterization may apply, for example, to most published studies at the family level using *rbcL,* where a preponderance of changes are in third codon positions. In other instances, third positions may be too homoplasious, while other sites provide insufficient information. Consequently, sequences that are characterized by severe rate variation tend to be less informative on a per nucleotide basis than those with a more even distribution of substitutions among sites.

A good example of this rate distributional phenomenon is provided by Steele and Vilgalys (1994) in their comparison of *matK* and *rbcL* sequence evolution. They showed that the former sequence evolves at twice the rate of the latter, but in addition, that substitutions are more evenly distributed among codon positions. In a comparison of tobacco and rice *rbcL* genes, for example, 79% of differences are at third positions, whereas for *matK,* this value is much lower (50%). Thus, the likelihood of multiple substitutions that obscure phylogenetic information is greater in the case of the more rate-heterogeneous (*rbcL*) gene. Similar results have been reported for other taxonomic groups; in a comparison of 25 saxifragaceous taxa, for example, substitutions are far more evenly distributed among codon positions for *matK* than for *rbcL* (Johnson and Soltis, 1995).

Considerable attention has recently been given to among-site rate variation and its potential phylogenetic consequences (Kuhner and Felsenstein, 1994; Wakeley, 1994; Sullivan et al., 1995; Sullivan, 1996; Yang, 1996; see Chapter 6). These consequences include phylogenetic incongruence, as exemplified by the study of Sullivan et al. (1995) on murid rodents. In this case, cytochrome *b* gene sequences, but not 12S rRNA gene sequences, yielded a phylogeny that is well corroborated by morphology and other molecular data sets. Sullivan et al. (1995) attributed the misleading signal in the latter gene to extreme among-site rate variation.

Because of the potential for among-site rate variation to yield misleading phylogenetic signal and hence phylogenetic discord, approaches that accommodate among-site rate variation in phylogenetic analysis have been developed. These often involve likelihood and distance methods, as recently reviewed by Yang (1996; see Chapter 5).

Base Compositional Biases In most DNA sequences the four different nucleotide bases do not occur in equal proportions. This may be true for sequences as a whole, for different partitions such as stemmed and looped portions of rRNA molecules, and, in the case of protein-encoding genes, for different codon positions (e.g., Vawter and Brown, 1993; Martin, 1995; Hershkovitz and Zimmer, 1996). These compositional biases may be associated with asymmetries in nucleotide transformational probabilities (Albert and Mishler, 1992; Albert et al., 1993; Vawter and Brown, 1993; Collins et al., 1994b), which may systematically distort patterns of character-state reconstruction. In response to recognition of these biases, a number of corrective measures have been developed for inclusion in phylogenetic reconstruction methods (Albert and Mishler, 1992; Albert et al., 1993; Knight and Mindell, 1993; Collins et al., 1994a; Lockhart et al., 1994; Gaultier and Gouy, 1995).

Compositional bias, and the related phenomenon of transformational bias, may lead to misleading phylogenetic inferences, as noted in several studies (Lockhart et al., 1994; Martin, 1995; Steel et al., 1995; see Chapter 5). In particular, taxa typically become spuriously linked by virtue of sharing a compositional bias, such as a high G+C content. A case in point is the study by Lockhart et al. (1994), who showed that the apparent support for a bird/mammal clade in vertebrate 18S rRNA sequences is due to independently acquired compositional similarities, and that this support evaporates when compositional biases are accounted for. Once the data are transformed (using LogDet transformation) there is actually strong support for a bird/crocodilian relationship using the same data. Similarly, inferred phylogenetic relationships in sharks are distorted by compositional variation in cytochrome *b* sequences (Martin, 1995). Because compositional biases may lead to erroneous reconstructions, as these examples show, they may also be a source of phylogenetic incongruence with other gene trees or trees derived from morphological data sets.

RNA Editing In this ubiquitous process, mRNA sequences, primarily from organellar genes, are modified by the editing of particular bases, usually by conversion of C to U, but also by U to C substitutions, at least in plant mitochondria (Gray and Covello, 1993; Araya et al., 1994; Hiesel et al., 1994; Malek et al., 1996;

Yoshinaga et al., 1996). As a result of these nucleotide changes, the genetic information in the transcript may differ from that of the gene, leading to a translation product with one or more amino acids that would not have been predicted by the DNA sequence. Explanations for the existence of RNA editing include preservation of highly conserved amino acids (e.g., Hirose et al., 1994), restoration of function to DNA sequences that have suffered mutations that create premature stop codons (Yoshinaga et al., 1996), and compensation for T to C transitional drift (Malek et al., 1996).

That this process may be a potential source of phylogenetic incongruence was recently highlighted in phylogenetic analyses of plant mitochondrial *cox* genes (Bowe and dePamphilis, 1996; C. dePamphilis, pers. comm.). As discussed by Bowe and dePamphilis (1996), genomic sequences that undergo RNA editing are appropriate to include in phylogenetic analysis, because the process operates at the transcriptional level and thus should not affect historical information stored in the DNA sequences. Inclusion in an analysis of both DNA sequences and mRNA sequences that have been edited, however, may lead to conflict, as mRNA sequences will be spuriously linked by apparent synapomorphies of edited bases at positions affected by RNA editing. Because mRNAs (via cDNA synthesis) are sometimes incorporated into phylogenetic studies, especially when sequences are down-loaded from databases, this may be a practical consideration. RNA editing is also a concern, however, due to a peculiar form of orthology and paralogy conflation. Specifically, Bowe and dePamphilis (1996) show that phylogenetic incongruence may arise from inclusion of a mixture of unprocessed DNA sequences with processed genes ("processed paralogs" in their terminology) resulting from genomic integration of reverse-transcribed, previously edited RNAs. This result reinforces the necessity, invoked above, of documenting orthology during phylogenetic analysis. In addition, this example underscores the insights into molecular evolutionary history that may be revealed through phylogenetic discord. In this respect RNA editing may be distinguished from other potential sources of paralogy by careful analyses of potentially edited sites (Nugent and Palmer, 1991; Bowe and dePamphilis, 1996).

Nonindependence of Sites   One of the assumptions of phylogenetic analysis of nucleotide sequences is that each position is independent of other positions, at least in models where characters (= positions) are equally weighted (see, however, Albert et al., 1993, for an example of character weighting). While the condition of complete independence among sites is probably never met in a strict sense, in many instances it may not be too seriously violated. For some types of sequences, however, structural considerations alone predict that the assumption of independence is unwarranted. Perhaps best known in this respect are genes that encode ribosomal RNAs, which are single-stranded but have a secondary structure that includes stemmed regions containing Watson-Crick base pairs. Because rRNA is divided into domains where bases are either paired or unpaired, it is likely that different evolutionary constraints operate in each case.

Stemmed bases are subject to selection for compensatory mutations so that base-pairing is maintained, although phylogenetic evidence indicates that an unpaired base may persist for some time following a mutation, even if it does ultimately become compensated (Gatesy et al., 1994). Because mutations typically become compensated, some thought has been given to differential weighting of paired and unpaired nucleotide positions for purposes of phylogenetic analysis (see Chapter 7). Dixon and Hillis (1993) studied 28S RNA genes from selected vertebrates and recommend reducing the weight accorded stem characters by 20% relative to loop characters, whereas Springer et al. (1995), in a study of 12S rRNA gene sequences from mammals, suggest a more extreme weighting of stemmed positions (circa 40% downweighting). Though these weighting recommendations differ quantitatively, they underscore the possibility of nonindependence of nucleotide positions in sequence data. Importantly, Dixon and Hillis (1993) further show how alternative weighting can modify the phylogenetic resolution obtained (see, however, Chapter 7), raising the possibility that failure to account for nonindependence among sites may lead to discordance with other phylogenetic estimates.

ASSESSING THE CAUSE

In this chapter we have drawn attention repeatedly to the enhanced understanding of evolutionary process that has derived from observations of discordant phylogenies. Given that many phenomena may lead to the same phylogenetic "phenotype," that is, incongruence, the question naturally arises as to how to discriminate among the many possible causes when one is faced with a particular unexpected or conflicting resolution. Often, there may be insufficient information to render a confident diagnosis, and the cause of incongruence simply remains unknown. Rodman et al. (1993), for example, report discordant resolutions for the Limnanthaceae in a comparison of trees based on morphological and molecular data, and note that the explanation for incongruence is obscure. Unexplained phylogenetic discord, as in this example, is prevalent, and it is also disquieting in that it hampers robust phylogenetic conclusions for the taxa involved, without an apparent rationalization that might yield insight into process.

This difficulty of inferring the underlying phenomenon when different causes result in similar phylogenetic observations is a common and vexing problem. Take lineage sorting and introgression, for example, which often leave the same incongruence footprint. Mason-Gamer et al. (1995) note the difficulty of distinguishing these causes in their analysis of the distribution of cpDNA haplotypes in *Coreopsis*. Hanson et al. (1996) and Buckler and Holtsford (1996a) note a similar difficulty in distinguishing between the two alternatives for shared alleles at nuclear loci (*c1* and ITS, respectively) in *Zea*.

Given an observation of incongruence and in the absence of other information, it is unlikely that its cause will be confidently diagnosed in many cases. An escape from the impasse is often provided by taking advantage of ancillary information and other data sources. Geographic distribution data, for example, may be instrumental in favoring introgression over lineage sorting by virtue of putative introgressant alleles occurring only in areas of sympatry (Rieseberg and Wendel, 1993). Alternatively, it may be possible to distinguish between lineage sorting and introgression on the basis of the distribution of alleles at other loci and by the occurrence patterns of alleles among individuals. Although both introgression and lineage sorting may affect loci widely distributed in the genome, in some cases these two processes may be distinguished, particularly if most putatively introgressed alleles co-occur in the same individuals while most other individuals appear pure (Brubaker and Wendel, 1994; Brubaker et al., 1993). Under a scenario of lineage sorting, the marker alleles in question might be expected to be more randomly distributed among individuals. Quantitative arguments may also facilitate diagnosis of cause, as for example in many of the cases of plastome introgression (Rieseberg and Soltis, 1991; Reiseberg and Wendel, 1993; Rieseberg, 1995; Rieseberg et al., 1996). To attribute the alien cpDNA to ancestral retention may invoke the long-term maintenance of a polymorphism level that is deemed improbable, and hence the alternative of hybridization and introgression may be preferred. A clever, complementary approach was recently suggested by Hanson et al. (1996), who, in noting the presence of *c1* alleles shared among *Zea* taxa, used postulated divergence times and mutation rates in simulations of sequence evolution, with the aim of addressing the likelihood that alleles could have survived since speciation without accumulating differences. The authors take pains to point out that their method does not preclude the possibility of introgression, merely that introgression need not be invoked to account for alleles shared among taxa.

Additional quantitative arguments may derive from parsimony itself. When confronted with a possible case of trans-specific, nonsexual gene flow (horizontal transfer), it is necessary to consider the formal alternative of differential duplications and losses of genes, leading to a mixed sampling of orthologous and paralogous sequences (Fig. 10.5; see also fig. 4 in VanderWiel et al., 1993). Resolution often relies on parsimony criteria, where, for example, the number of homoplasious gene gains and/or losses that would need to be invoked is deemed too high (Delwiche and Palmer, 1996; Hibbett, 1996).

These examples, and many others cited in this chapter, indicate that a holistic approach often will prove helpful in assessing the cause of phylogenetic incongruence. It is difficult to generalize about recommended approaches or guidelines, however, because each instance of incongruence has peculiarities unique to the taxa involved and their histories. When a diversity of evidence is harnessed, though, the discord often dissolves into understanding.

CONCLUSIONS

Our intention in preparing this chapter was to draw attention to the many biological processes that lead to discordant phylogenetic hypotheses by providing brief explanations of the various phenomena and providing pointers to relevant literature. In addition, though, we hoped to emphasize the many insights into evolutionary process that were stimulated by initial observations of incongruence. These insights have been many and varied, and have led to an enhanced appreciation of phenomena that operate at the organismal as well as molecular levels. A partial list would include issues surrounding the terms *gene tree* and *species tree;* the phenomena of lineage sorting and introgression; the potential for reticulation and/or concerted evolution among alleles, genes, and among organisms; the possibility of myriad forms of hybrid or recombinational speciation as well as rapid radiation; and the potential significance of horizontal gene transfer in evolution. Each of these processes confounds phylogeny reconstruction while offering insight into some aspect of the biology of the organisms under study. This is an impressive list, and it includes phenomena about which little was known prior to molecular systematic investigations that led to an alerting and informing incongruence. Accordingly, one might justifiably argue that an enhanced understanding of the evolutionary process is the most important achievement of molecular systematics.

Perhaps it is of more than just historical interest to note the interplay between the tools used to infer phylogeny and our appreciation of the appropriate use of those tools. In our effort to infer phylogeny, incongruence has emphasized the need to understand the evolutionary history of the data sources used. These interconnected objectives have often involved reciprocal illumination, whereby phylogenetic analysis using a particular molecular tool has led to incongruence, which has led to increased knowledge of the evolutionary forces that impact the use of that particular tool in phylogeny reconstruction. A case in point would be ribosomal DNA, where phylogenetic analysis led to recognition of the possibility of interlocus concerted evolution, which has obvious relevance to orthology and paralogy concerns. Similarly, phylogenetic use of rDNA sequences has enhanced our understanding of character and character-state weighting issues, which led to refinements in phylogenetic methodology.

So we return to the premise of the chapter: Phylogenetic incongruence is not merely an evil to be reconciled, but is instead a glimmer of light. When it shines with sufficient brightness, we are offered a window through which evolutionary process may be seen more clearly.

LITERATURE CITED

Albert, V. A., and B. D. Mishler. 1992. On the rationale and utility of weighting nucleotide sequence data. Cladistics **8**:73–83.

Albert, V. A., S. E. Williams, and M. W. Chase. 1992. Carnivorous plants: phylogeny and structural evolution. Science **257**:1491–1495.

Albert, V. A., M. W. Chase, and B. D. Mishler. 1993. Character-state weighting for cladistic analysis of protein-coding DNA sequences. Annals of the Missouri Botanical Garden **80**:752–766.

Amábile-Cuevas, C. F., and M. E. Chicurel. 1993. Horizontal gene transfer: gene flow from parent to child is the basis of heredity. American Scientist **81**:332–341.

Anderson, E. 1949. Introgressive Hybridization. Wiley, New York.

Araya, A., D. Begu, and S. Litvak. 1994. RNA editing in plants. Physiologia Plantarum **91**:543–550.

Arnheim, N. D. 1983. Concerted evolution of multigene families. **In** Evolution of Genes and Proteins, eds. M. Nei. and R. K. Koehn, pp. 38–61. Sinauer Associates, Sunderland, Massachusetts.

Arnold, M. L. 1992. Natural hybridization as an evolutionary process. Annual Review of Ecology and Systematics **23**:237–261.

Baldwin, B. G. 1997. Adaptive radiation of the Hawaiian silversword alliance: congruence and conflict of phylogenetic evidence from molecular and non-molecular investigations. **In** Molecular Evolution and Adaptive Radiation, eds. T. J. Givnish and K. J. Systma, pp. 103–128. Cambridge University Press, Cambridge.

Baldwin, B. G., C. S. Campbell, J. M. Porter, M. J. Sanderson, M. F. Wojciechowski, and M. J. Donoghue. 1995. The ITS region of nuclear ribosomal DNA: a valuable source of evidence on angiosperm phylogeny. Annals of the Missouri Botanical Garden 82:247–277.

Barker, N. P., H. P. Linder, and E. H. Harley. 1995. Polyphyly of Arundinoideae (Poaceae): evidence from rbcL sequence data. Systematic Botany 20:423–435.

Barrett, M., M. J. Donoghue, and E. Sober, 1991. Against consensus. Systematic Zoology 40:486–493.

Baum, D. A., and K. L. Shaw. 1995. Genealogical perspectives on the species problem. In Experimental and Molecular Approaches to Plant Biosystematics, Monographs in Systematic Biology No. 53, eds. P. C. Hoch and A. G. Stevenson, pp. 289–303. Missouri Botanical Garden, St. Louis.

Birky, C. W., T. Maruyama, and P. Fuerst. 1983. An approach to population and evolutionary genetic theory for genes in mitochondria and chloroplasts and some results. Genetics 103:513–527.

Bonierbale, M. W., R. L. Plaisted, and S. D. Tanksley. 1988. RFLP maps based on a common set of clones reveal modes of chromosomal evolution in potato and tomato. Genetics 120:1095–1103.

Bousquet, J., S. H. Strauss, A. H. Doerksen, and R. A. Price. 1992. Extensive variation in evolutionary rate of rbcL gene sequences among seed plants. Proceedings of the National Academy of Sciences U.S.A. 89:7844–7848.

Bowe, L. M., and C. W. dePamphilis. 1996. Effects of RNA editing and gene processing on phylogenetic reconstruction. Molecular Biology and Evolution 13:1159–1166.

Britten, R. J. 1986. Rates of DNA sequence evolution differ between taxonomic groups. Science 231:1393–1398.

Brochmann, C., T. Nilsson, and T. M. Gabrielsen. 1996. A classic example of postglacial allopolyploid speciation re-examined using RAPD markers and nucleotide sequences: Saxifraga osloensis (Saxifragaceae). Symbolae Botanicae Upsaliensis 31:75–89.

Brower, A. V. Z., R. DeSalle, and A. Vogler. 1996. Gene trees, species trees, and systematics: a cladistic perspective. Annual Review of Ecology and Systematics 27:423–450.

Brubaker, C. L., and J. F. Wendel. 1994. Reevaluating the origin of domesticated cotton (Gossypium hirsutum: Malvaceae) using nuclear restriction fragment length polymorphisms (RFLPs). American Journal of Botany 81:1309–1326.

Brubaker, C. L., J. A. Koontz, and J. F. Wendel. 1993. Bidirectional cytoplasmic and nuclear introgression in the New World cottons, Gossypium barbadense and G. hirsutum. American Journal of Botany 80:1203–1208.

Brubaker, C. L., A. H. Paterson, and J. F. Wendel. 1997. Chromosome structural evolution in allotetraploid cotton and its diploid progenitors. Genetics, in press.

Buckler, E. S., and T. P. Holtsford. 1996a. Zea systematics: ribosomal ITS evidence. Molecular Biology and Evolution 13:612–622.

Buckler, E. S., and T. P. Holtsford. 1996b. Zea ribosomal repeat evolution and substitution patterns. Molecular Biology and Evolution 13:623–632.

Bull, J. J., J. P. Huelsenbeck, C. W. Cunningham, D. L. Swofford, and P. J. Waddell. 1993. Partitioning and combining data in phylogenetic analysis. Systematic Biology 42:384–397.

Castillo, R. O., and D. M. Spooner. 1997. Phylogenetic relationships of wild potatoes, Solanum Series Conicibaccata (Sect. Petota). Systematic Botany 22:45–83.

Chesser, R. K., and R. J. Baker. 1996. Effective sizes and dynamics of uniparentally and diparentally inherited genes. Genetics 144:1225–1235.

Clark, A. G., and T.-H. Kao. 1991. Excess nonsynonymous substitution at shared polymorphic sites among self-incompatibility alleles of Solanaceae. Proceedings of the National Academy of Sciences U.S.A. 88:9823–9827.

Clark, A. G., and T. S. Whittam. 1992. Sequencing errors and molecular evolutionary analysis. Molecular Biology and Evolution 9:744–752.

Clark, J. B., W. P. Maddison, and M. G. Kidwell. 1994. Phylogenetic analysis supports horizontal transfer of P transposable elements. Molecular Biology and Evolution 11:40–50.

Clark, L. G., W. Zhang, and J. F. Wendel. 1995. A phylogeny of the grass family based on ndhF sequence data. Systematic Botany 20:436–460.

Collins, T. M., F. Kraus, and G. Estabrook. 1994a. Compositional effects and weighting of nucleotide sequences for phylogenetic analysis. Systematic Biology 43:449–459.

Collins, T. M., P. H. Wimberger, and G. J. P. Naylor. 1994b. Compositional bias, character-state bias and character-state reconstruction using parsimony. Systematic Biology 43:482–496.

Cronn, R. C., X. Zhao, A. H. Paterson, and J. F. Wendel. 1996. Polymorphism and concerted evolution in a multigene family: 5S ribosomal DNA in diploid and allopolyploid cottons. Journal of Molecular Evolution 42:685–705.

Cummings, M. P. 1994. Transmission patterns of eukaryotic transposable elements: arguments for and against horizontal transfer. Trends in Ecology and Evolution 9:141–145.

Cummings, M. P., S. P. Otto, and J. Wakeley. 1995. Sampling properties of DNA sequence data in phylogenetic analysis. Molecular Biology and Evolution 12:814–822.

Danna, K. J., R. Workman, V. Coryell, and P. Keim. 1996. 5S rRNA genes in tribe Phaseoleae: array size, number, and dynamics. Genome 39:445–455.

Davis, J. I., and K. C. Nixon. 1992. Populations, genetic variation, and the delimitation of phylogenetic species. Systematic Biology 41:421–435.

Davis, J. I., and R. J. Soreng. 1993. Phylogenetic structure of the grass family (Poaceae) as inferred from chloroplast DNA restriction site variation. American Journal of Botany 80:1444–1454.

Delwiche, C. F., and J. D. Palmer. 1996. Rampant horizontal transfer and duplication of rubisco genes in eubacteria and plastids. Molecular Biology and Evolution 13:873–882.

de Queiroz, A., M. J. Donoghue, and J. Kim. 1995. Separate versus combined analysis of phylogenetic evidence.

Annual Review of Ecology and Systematics **26**: 657–681.

Dixon, M. T., and D. M. Hillis. 1993. Ribosomal RNA secondary structure: compensatory mutations and implications for phylogenetic analysis. Molecular Biology and Evolution **10**:256–267.

Doebley, J. 1989. Molecular evidence for a missing wild relative of maize and the introgression of its chloroplast genome into *Zea perennis*. Evolution **43**:1555–1559.

Doebley, J. 1993. Genetics, development and plant evolution. Current Opinion in Genetics and Development **3**:865–872.

Domon, C., and A. Steinmetz. 1994. Exon shuffling in anther-specific genes from sunflower. Molecular and General Genetics **244**:312–317.

Donoghue, M. J., and M. J. Sanderson. 1992. The suitability of molecular and morphological evidence in reconstructing plant phylogeny. **In** Molecular Systematics of Plants, eds. P. S. Soltis, D. E. Soltis, and J. J. Doyle, pp. 340–368. Chapman & Hall, New York.

Doolittle, R. F. 1995. The multiplicity of domains in proteins. Annual Review of Biochemistry **64**:287–314.

Downie, S. R., and J. D. Palmer. 1994. A chloroplast DNA phylogeny of the Caryophyllates based on structural and inverted repeat restriction site variation. Systematic Botany **19**:236–252.

Downie, S. R., R. G. Olmstead, G. Zurawski, D. E. Soltis, P. S. Soltis, J. C. Watson, and J. D. Palmer. 1991. Six independent losses of the chloroplast DNA *rpl2* intron in dicotyledons: molecular and phylogenetic implications. Evolution **45**:1245–1259.

Downie, S. R., D. S. Katz-Downie, K. H. Wolfe, P. J. Calie, and J. D. Palmer. 1994. Structure and evolution of the largest chloroplast gene (ORF2280): internal plasticity and multiple gene losses during angiosperm evolution. Current Genetics **25**:367–378.

Doyle, J. J. 1991. Evolution of higher-plant glutamine synthetase genes: tissue specificity as a criterion for predicting orthology. Molecular Biology and Evolution **8**:366–377.

Doyle, J. J. 1992. Gene trees and species trees: molecular systematics as one-character taxonomy. Systematic Botany **17**:144–163.

Doyle, J. J. 1994a. Evolution of a plant homeotic multigene family: toward connecting molecular systematics and molecular developmental genetics. Systematic Biology **43**:307–328.

Doyle, J. J. 1994b. Phylogeny of the legume family: an approach to understanding the origins of nodulation. Annual Review of Ecology and Systematics **25**:325–349.

Doyle, J. J. 1995. The irrelevance of allele tree topologies for species delimitation and a non-topological alternative. Systematic Botany **20**:574–588.

Doyle, J. J. 1996. Homoplasy connections and disconnections: genes and species, molecules and morphology. **In** Homoplasy and the Evolutionary Process, eds. M. J. Sanderson and L. Hufford, pp. 37–66. Academic Press, New York.

Doyle, J. J., M. Lavin, and A. Bruneau. 1992. Contributions of molecular data to Papilionoid legume systematics.

In Molecular Systematics of Plants, eds. P. S. Soltis, D. E. Soltis and J. J. Doyle, pp. 223–251. Chapman & Hall, New York.

Doyle, J. J., J. L. Doyle, and J. D. Palmer. 1995. Multiple independent losses of two genes and one intron from legume chloroplast genomes. Systematic Botany **20**:272–294.

Dubcovsky, J., and J. Dvořák. 1995. Ribosomal RNA multigene loci: nomads of the Triticeae genomes. Genetics **140**:1367–1377.

Duvall, M. R., and B. R. Morton. 1996. Molecular phylogenetics of Poaceae: an expanded analysis of *rbcL* sequence data. Molecular Phylogenetics and Evolution **5**:352–358.

Eernisse, D. J., and A. G. Kluge. 1993. Taxonomic congruence versus total evidence and amniote phylogeny inferred from fossils, molecules and morphology. Molecular Biology and Evolution **10**:1170–1195.

Elder J. F., and B. J. Turner. 1995. Concerted evolution of repetitive DNA sequences in eukaryotes. Quarterly Review of Biology **70**:297–320.

Ewens, W. 1990. Population genetics theory—the past and the future. **In** Mathematical and Statistical Developments of Evolutionary Theory, ed. S. Lessard, pp. 177–227. Kluwer, New York.

Faith, D. P. 1991. Cladistic permutation tests for monophyly and nonmonophyly. Systematic Zoology **40**:366–375.

Farris, J. S., M. Källersjö, A. G. Kluge, and C. Bult. 1995. Testing significance of incongruence. Cladistics **10**:315–319.

Fehrer, J. 1996. Conflicting character distribution within different data sets on cardueline finches: artifact or history? Molecular Biology and Evolution **13**:7–20.

Felsenstein, J. 1978. Cases in which parsimony or compatibility methods will be positively misleading. Systematic Zoology **27**:401–410.

Fitch, D. H. A., and M. Goodman. 1991. Phylogenetic scanning: a computer-assisted algorithm for mapping gene conversions and other recombinational events. Computer Applications in the Biosciences **7**:207–216.

Funk, V. A. 1985. Phylogenetic patterns and hybridization. Annals of the Missouri Botanical Garden **72**:681–715.

Gantt, J. S., S. L. Baldauf, P. J. Calie, N. F. Weeden, and J. D. Palmer. 1991. Transfer of *rpl22* to the nucleus greatly preceded its loss from the chloroplast and involved the gain of an intron. EMBO Journal **10**:3073–3078.

Gatesy, J., C. Hayashi, R. Desalle, and E. Vrba. 1994. Rate limits for mispairing and compensatory change: the mitochondrial ribosomal DNA of antelopes. Evolution **48**:188–196.

Galtier, N., and M. Gouy. 1995. Inferring phylogenies from DNA sequences of unequal base compositions. Proceedings of the National Academy of Sciences U.S.A. **92**:11317–11321.

Gaur, L. K., Hughes, E. R. Heise, and J. Gutknecht. 1992. Maintenance of *DQB1* polymorphisms in primates. Molecular Biology and Evolution **9**:599–609.

Gaut, B. S., and M. T. Clegg. 1993a. Molecular evolution of the *Adh1* locus in the genus *Zea*. Proceedings of the National Academy of Sciences U.S.A. **90**:5095–5099.

Gaut, B. S., and M. T. Clegg. 1993b, Nucleotide polymorphism in the *Adh1* locus of pearl millet (*Pennisetum glaucum*) (Poaceae). Genetics **135**:1091–1097.

Gaut, B. S., S. V. Muse, W. D. Clark, and M. T. Clegg. 1992. Relative rates of nucleotide substitution at the *rbcL* locus of monocotyledonous plants. Journal of Molecular Evolution **35**:292–303.

Gaut, B. S., S. V. Muse, and M. T. Clegg. 1993. Relative rates of nucleotide substitution in the chloroplast genome. Molecular Phylogenetics and Evolution **2**:89–96.

Gaut, B. S., B. R. Morton, B. C. McCaig, and M. T. Clegg. 1996. Substitution rate comparisons between grasses and palms: synonymous rate differences at the nuclear gene *Adh* parallel rate differences at the plastid gene *rbcL*. Proceedings of the National Academy of Sciences U.S.A. **93**:10274–10279.

Gaut, B. S., L. G. Clark, J. F. Wendel, and S. V. Muse. 1997. Comparisons of the molecular evolutionary process at *rbcL* and *ndhF* in the grass family (Poaceae). Molecular Biology and Evolution **14**:769–777.

Gielly, L., and P. Taberlet. 1994. The use of chloroplast DNA to resolve plant phylogenies: non-coding versus *rbcL* sequences. Molecular Biology and Evolution **11**:769–777.

Goldman, A. S. H., and M. Lichten. 1996. The efficiency of meiotic recombination between dispersed sequences in *Saccharomyces cerevisiae* depends upon their chromosomal location. Genetics **144**:43–55.

Golenberg, E. M., M. T. Clegg, M. L. Durbin, J. F. Doebley, and D. P. Ma. 1993. Evolution of a non-coding region of the chloroplast genome. Molecular Phylogenetics and Evolution **2**:52–64.

Goloubinoff, P., S. Pääbo, and A. C. Wilson. 1993. Evolution of maize inferred from sequence diversity of an *Adh2* gene segment from archaeological specimens. Proceedings of the National Academy of Sciences U.S.A. **90**:1997–2001.

Gottlieb, L. D. 1984. Genetics and morphological evolution in plants. American Naturalist **123**:681–709.

Gottlieb, L. D., and V. S. Ford. 1996. Phylogenetic relationships among the sections of *Clarkia* (Onagraceae) inferred from the nucleotide sequence of *PgiC*. Systematic Botany **21**:45–62.

Graham, S. W., J. R. Kohn, B. R. Morton, J. E. Eckenwalder, and S. C. H. Barrett. 1997. Phylogenetic congruence and discordance among one morphological and three molecular data-sets from Pontederiaceae. Systematic Biology, in press.

Gray, M. W., and P. S. Covello. 1993. RNA editing in plant mitochondria and chloroplasts. FASEB Journal **7**:64–71.

Guigó, R., I. Muchnik, and T. F. Smith. 1996. Reconstruction of ancient molecular phylogeny. Molecular Phylogenetics and Evolution **6**:189–213.

Hamrick, J. L., and M. J. W. Godt. 1989. Allozyme diversity in plant species. **In** Plant Population Genetics, Breeding and Genetic Resources, eds. A. H. D. Brown, M. T. Clegg, A. L. Kahler, and B. S. Weir, pp. 43–63. Sinauer Associates, Sunderland, Massachusetts.

Hanson, M. A., B. S. Gaut, A. O. Stec, S. I. Fuerstenberg, M. M. Goodman, E. H. Coe, and J. F. Doebley. 1996. Evolution of anthocyanin biosynthesis in maize kernels: the role of regulatory and enzymatic loci. Genetics **143**:1395–1407.

Hein, J. 1993. A heuristic method to reconstruct the history of sequences subject to recombination. Journal of Molecular Evolution **36**:396–405.

Heiser, C. B. 1973. Introgression re-examined. Botanical Review **39**:347–366.

Hershkovitz, M. A., and E. A. Zimmer. 1996. Conservation patterns in angiosperm rDNA ITS2 sequences. Nucleic Acids Research **24**:2857–2867.

Hey, J., and J. Wakeley. 1997. A coalescent estimator of the population recombination. Genetics **145**:833–846.

Hibbett, D. S. 1996. Phylogenetic evidence for horizontal transmission of group I introns in the nuclear ribosomal DNA of mushroom-forming fungi. Molecular Biology and Evolution **13**:903–917.

Hiesel, R., B. Combettes, and A. Brennicke. 1994. Evidence for RNA editing in mitochondria of all major groups of land plants except the Bryophyta. Proceedings of the National Academy of Sciences U.S.A. **91**:629–633.

Hillis, D. M. 1987. Molecular versus morphological approaches to systematics. Annual Review of Ecology and Systematics **18**:23–42.

Hillis, D. M. 1996. Inferring complex phylogenies. Nature **383**:130–131.

Hillis, D. M., and M. T. Dixon. 1991. Ribosomal DNA: molecular evolution and phylogenetic inference. Quarterly Review of Biology **66**:411–453.

Hirose, T., T. Wakasugi, M. Sugiura, and H. Kössel. 1994. RNA editing of tobacco *petB* mRNAs occurs both in chloroplasts and non-photosynthetic proplastids. Plant Molecular Biology **26**:509–513.

Hoelzer, G. A. 1997. Inferring phylogenies from mtDNA variation: mitochondrial gene-trees versus nuclear gene-trees revisited. Evolution **51**:622–626.

Hoot S. B., and J. D. Palmer. 1994. Structural rearrangements, including parallel inversions, within the chloroplast genome of *Anemone* and related genera. Journal of Molecular Evolution **38**:274–281.

Hoot, S. B., A. Culham, and P. R. Crane. 1995. The utility of *atpB* gene sequences in resolving phylogenetic relationships: comparison with *rbcL* and 18S ribosomal DNA sequences in the Lardizabalaceae. Annals of the Missouri Botanical Garden **82**:194–207.

Houck, M. A., J. B. Clark, K. R. Peterson, and M. G. Kidwell. 1991. Possible horizontal transfer of *Drosophila* genes by the mite *Proctolaelaps regalis*. Science **253**:1125–1129.

Hudson, R. R. 1990. Gene genealogies and the coalescent process. Oxford Surveys in Evolutionary Biology **7**:1–41.

Hudson, R. R., and N. L. Kaplan. 1988. The coalescent process in models with selection and recombination. Genetics **120**:831–840.

Huelsenbeck, J. P., J. J. Bull, and C. W. Cunningham. 1996. Combining data in phylogenetic analysis. Trends in Ecology and Evolution **11**:152–158.

Ikeo, K., K. Takahashi, and T. Gojobori. 1995. Different evolutionary histories of kringle and protease domains in serine proteases: a typical example of domain evolution. Journal of Molecular Evolution **40**:331–336.

Innan, H., F. Tajima, R. Terauchi, and N. T. Miyashita. 1996. Intragenic recombination in the *Adh* locus of the wild plant *Arabidopsis thaliana*. Genetics **143**:1761–1770.

Ioerger, T. R., A. G. Clark, and T.-H. Kao. 1990. Polymorphism at the self-incompatibility locus in Solanaceae predates specification. Proceedings of the National Academy of Sciences U.S.A. **87**:9732–9735.

Jakobsen, I. B., and S. Easteal. 1996. A program for calculating and displaying compatibility matrices as an aid in determining reticulate evolution in molecular sequences. Computer Applications in the Biosciences **12**:291–295.

Johnson, L. A., and D. E. Soltis. 1994. *matK* DNA sequences and phylogenetic reconstruction in Saxifragaceae *s. str.* Systematic Botany **19**:143–156.

Johnson, L. A., and D. E. Soltis. 1995. Phylogenetic inference in Saxifragaceae *sensu stricto* and *Gilia* (Polemoniaceae) using *matK* sequences. Annals of the Missouri Botanical Garden **82**:149–175.

Kadereit, J. W. 1994. Molecules and morphology, phylogenetics and genetics. Botanica Acta **107**:369–373.

Kelchner, S. A., and J. F. Wendel. 1996. Hairpins create minute inversions in non-coding regions of chloroplast DNA. Current Genetics **30**:259–262.

Kellogg, E. A., R. Appels, and R. Mason-Gamer. 1996. When genes tell different stories: incongruent gene trees for diploid genera of Triticeae (Gramineae). Systematic Botany **21**:321–347.

Kidwell, M. G. 1993. Lateral transfer in natural populations of eukaryotes. Annual Review of Genetics **27**:235–256.

Kim, J. 1996. General inconsistency conditions for maximum parsimony: effects of branch lengths and increasing numbers of taxa. Systematic Biology **45**:363–374.

Kim, K.-J., and R. K. Jansen. 1994. Comparisons of phylogenetic hypotheses among different data sets in dwarf dandelions (*Krigia,* Asteraceae): additional information from internal transcribed spacer sequences of nuclear ribosomal DNA. Plant Systematics and Evolution **190**:157–185.

Kim, K.-J., and R. K. Jansen. 1995. *ndhF* sequence evolution and the major clades in the sunflower family. Proceedings of the National Academy of Sciences U.S.A. **92**:10379–10383.

Kim, K.-J., R. K. Jansen, R. S. Wallace, H. J. Michaels, and J. D. Palmer. 1992. Phylogenetic implications of *rbcL* sequence variation in the Asteraceae. Annals of the Missouri Botanical Garden **79**:428–445.

Kluge, A. G. 1989. A concern for evidence and a phylogenetic hypothesis of relationships among *Epicrates* (Boidea, Serpentes). Systematic Zoology **38**:7–25.

Knight, A., and D. P. Mindell. 1993. Substitution bias, weighting of DNA sequence evolution and the phylogenetic position of Fea's viper. Systematic Biology **42**:18–31.

Kolukisaoglu, H. Ü., S. Marx, C. Wiegmann, S. Hanelt, and H. A. W. Schneider-Poetsch. 1995. Divergence of the phytochrome gene family predates angiosperm evolution and suggests that *Selaginella* and *Equisetum* arose prior to *Psilotum*. Journal of Molecular Evolution **41**:329–337.

Kowalski, S. P., T.-H. Lan, K. A. Feldmann, and A. H. Paterson. 1994. Comparative mapping of *Arabidopsis thaliana* and *Brassica oleracea* chromosomes reveals islands of conserved organization. Genetics **138**:499–510.

Kuhner, M. K., and J. Felsenstein. 1994. A simulation of phylogeny algorithms under equal and unequal evolutionary rates. Molecular Biology and Evolution **11**:459–468.

Kumar, A. 1996. The adventures of the Tyl-*copia* group of retrotransposons in plants. Trends in Genetics **12**:41–43.

Lan, R., and P. R. Reeves. 1996. Gene transfer is a major factor in bacterial evolution. Molecular Biology and Evolution **13**:47–55.

Lara, M. C., J. L. Patton, and M. N. F. da Silva. 1996. The simultaneous diversification of South American echimyid rodents (Hystricognathi) based on complete cytochrome b sequences. Molecular Phylogenetics and Evolution **5**:403–413.

Levy, F., J. Antonovics, J. E. Boynton, and N. W. Gillham. 1996. A population genetic analysis of chloroplast DNA in *Phacelia*. Heredity **76**:143–155.

Li, W.-H., M. Tanimura, and P. M. Sharp. 1987. An evaluation of the molecular clock hypothesis using mammalian DNA sequences. Journal of Molecular Evolution **25**:330–342.

Liston, A., W. A. Robinson, J. M. Oliphant, and E. R. Alvarez-Buylla. 1996. Length variation in the nuclear ribosomal DNA internal transcribed spacer region of nonflowering seed plants. Systematic Botany **21**:109–120.

Lockhart, P. J., M. A. Steel, M. D. Hendy, and D. Penny. 1994. Recovering evolutionary trees under a more realistic model of sequence evolution. Molecular Biology and Evolution **11**:605–612.

Lutzoni, F., and R. Vilgalys. 1994. Integration of morphological and molecular data sets in estimating fungal phylogenies. Canadian Journal of Botany **73**:S649–S659.

Lyons-Weiler, J., G. A. Hoelzer, and R. J. Tausch. 1996. Relative apparent synapomorphy analysis (RASA) I: the statistical measurement of phylogenetic signal. Molecular Biology and Evolution **13**:749–757.

Lyons-Weiler, J., G. A. Hoelzer, and R. J. Tausch. 1997. Relative apparent synapomorphy analysis (RASA) II: optimal outgroup analysis. Systematic Biology, in press.

Maddison, W. P. 1995. Phylogenetic histories within and among species. In Experimental and Molecular Approaches to Plant Biosystematics, Monographs in Systematic Biology No. 53, eds. P. C. Hoch and A. G. Stevenson, pp. 273–287. Missouri Botanical Garden, St. Louis.

Maddison, W. P. 1996. Molecular approaches and the growth of phylogenetic biology. **In** Molecular Zoology: Approaches, Strategies and Protocols, eds. J. D. Ferraris and S. R. Palumbi, pp. 47–63. Wiley, New York.

Malek, O., K. Lättig, R. Hiesel, A. Brennicke, and V. Knoop. 1996. RNA editing in bryophytes and a molecular phylogeny of land plants. EMBO Journal **15**:1403–1411.

Martin, A. P. 1995. Mitochondrial DNA sequence evolution in sharks: rates, patterns and phylogenetic inferences. Molecular Biology and Evolution **12**:1114–1123.

Mason-Gamer, R. J., and E. A. Kellogg. 1996. Testing for phylogenetic conflict among molecular data sets in the tribe Triticeae (Gramineae). Systematic Biology **45**:524–545.

Mason-Gamer, R. J., K. E. Holsinger, and R. K. Jansen. 1995. Chloroplast DNA haplotype variation within and among populations of *Coreopsis grandiflora* (Asteraceae). Molecular Biology and Evolution **12**:371–381.

Masterson, J. 1994. Stomatal size in fossil plants: evidence for polyploidy in majority of angiosperms. Science **264**:421–424.

Mathews, S., and R. A. Sharrock. 1996. The phytochrome gene family in grasses (Poaceae): a phylogeny and evidence that grasses have a subset of the loci found in dicot angiosperms. Molecular Biology and Evolution **13**:1141–1150.

Mathews, S., M. Lavin, and R. A. Sharrock. 1995. Evolution of the phytochrome gene family and its utility for phylogenetic analyses of angiosperms. Annals of the Missouri Botanical Garden **82**:296–321.

McDade, L. A. 1990. Hybrids and phylogenetic systematics I. Patterns of character expression in hybrids and their implications for cladistic analysis. Evolution **44**:1685–1700.

McDade, L. A. 1992. Hybrids and phylogenetic systematics II. The impact of hybrids on cladistic analysis. Evolution **46**:1329–1346.

McDade, L. A. 1995. Hybridization and phylogenetics. In Experimental and Molecular Approaches to Plant Biosystematics, Monographs in Systematic Botany No. 53, eds. P. C. Hoch and A. G. Stevenson, pp. 305–331. Missouri Botanical Garden, St. Louis.

McGrath, J. M., M. M. Jansco, and E. Pichersky. 1993. Duplicate sequences with a similarity to expressed genes in the genome of *Arabidopsis thaliana*. Theoretical and Applied Genetics **86**:880–888.

Meagher, R. B., S. Berry-Lowe, and K. Rice. 1989. Molecular evolution of the small subunit of ribulose bisphosphate carboxylase: nucleotide substitution and gene conversion. Genetics **123**:845–863.

Mickevich, M. F., and J. S. Farris. 1981. The implications of congruence in *Menidia*. Systematic Zoology **30**:351–370.

Millar, A. A., and E. S. Dennis. 1996. The alcohol dehydrogenase genes of cotton. Plant Molecular Biology **31**:897–904.

Miyamoto, M. M., 1985. Consensus cladograms and general classifications. Cladistics **1**:186–189.

Miyamoto, M. M., and W. M. Fitch. 1995. Testing species phylogenies and phylogenetic methods with congruence. Systematic Biology **44**:64–76.

Moens, L., J. Vanfleteren, Y. Van de Peer, K. Peeters, O. Kapp, J. Czeluzniak, M. Goodman, M. Blaxter, and S. Vinogradov. 1996. Globins in nonvertebrate species: dispersal by horizontal gene transfer and evolution of the structure-function relationships. Molecular Biology and Evolution **13**:324–333.

Moniz de Sá, M., and G. Drouin. 1996. Phylogeny and substitution rates of angiosperm actin genes. Molecular Biology and Evolution **13**:1198–1212.

Moore, W. S. 1995. Inferring phylogenies from mtDNA variation: mitochondrial-gene trees versus nuclear-gene trees. Evolution **49**:718–726.

Morgan, D. R., and D. E. Soltis. 1993. Phylogenetic relationships among members of the Saxifragaceae s.1. based on *rbcL* sequence data. Annals of the Missouri Botanical Garden **80**:631–660.

Morton, B. R., and M. T. Clegg. 1993. A chloroplast DNA mutational hotspot and gene conversion in a non-coding region near *rbcL* in the grass family (Poaceae). Current Genetics **24**:357–365.

Neigel, J. E., and J. C. Avise. 1986. Phylogenetic relationships of mitochondrial DNA under various demographic models of speciation. In Evolutionary Processes and Theory, eds. E. Nevo and S. Karlin, pp. 515–534. Academic Press, New York.

Nickrent, D. L., and E. M. Starr. 1994. High rates of nucleotide substitution in nuclear small-subunit (18S) rDNA from holoparasitic flowering plants. Journal of Molecular Evolution **39**:62–70.

Nugent, J. M., and J. D. Palmer. 1991. RNA-mediated transfer of the gene *coxII* from the mitochondrion to the nucleus during flowering plant evolution. Cell **66**:473–481.

Ohta, T. 1993. An examination of the generation-time effect on molecular evolution. Proceedings of the National Academy of Sciences U.S.A. **90**:10676–10680.

Olmstead, R. G., and J. D. Palmer, 1994. Chloroplast DNA systematics: a review of methods and data analysis. American Journal of Botany **81**:1205–1224.

Olmstead, R. G., and J. A. Sweere. 1994. Combining data in phylogenetic systematics: an empirical approach using three molecular data sets in the Solanaceae. Systematic Biology **43**:467–481.

Palmer, J. D., C. R. Shields, D. B. Cohen, and T. J. Orten. 1983. Chloroplast DNA evolution and the origin of amphidiploid *Brassica* species. Theoretical and Applied Genetics **65**:181–189.

Palmer, J. D., R. A. Jorgenson, and W. F. Thompson. 1985. Chloroplast DNA variation and evolution in *Pisum*: patterns of change and phylogenetic analysis. Genetics **109**:195–213.

Pamilo, P., and M. Nei. 1988. Relationships between gene trees and species trees. Molecular Biology and Evolution **5**:568–583.

Patterson, C. 1988. Homology in classical and molecular biology. Molecular Biology and Evolution **5**:603–625.

Patterson, C., D. M. Williams, and C. J. Humphries. 1993. Congruence between molecular and morphological phylogenies. Annual Review of Ecology and Systematics **24**:153–188.

Patthy, L. 1991. Modular exchange principles in proteins. Current Opinion in Structural Biology **1**:351–361.

Pereira, M. G., M. Lee, P. Bramel-Cox, W. Woodman, J. Doebley, and R. Whitkus. 1994. Construction of an

RFLP map in sorghum and comparative mapping in maize. Genome **37**:236–243.

Richman, A. D., M. K. Uyenoyama, and J. R. Kohn. 1996. Allelic diversity and gene genealogy at the self-incompatibility locus in the Solanaceae. Science **273**:1212–1216.

Rieseberg, L. H. 1991. Homoploid reticulate evolution in *Helianthus* (Asteraceae): evidence from ribosomal genes. American Journal of Botany **78**:1218–1237.

Rieseberg, L. H. 1995. The role of hybridization in evolution: old wine in new skins. American Journal of Botany **82**:944–953.

Rieseberg, L. H., and D. E. Soltis. 1991. Phylogenetic consequences of cytoplasmic gene flow in plants. Evolutionary Trends in Plants **5**:65–84.

Rieseberg, L. H., and J. F. Wendel. 1993. Introgression and its consequences. In Hybrid Zones and the Evolutionary Process, ed. R. Harrison, pp. 70–109. Oxford University Press, New York.

Rieseberg, L. H., S. Beckstrom-Sternberg, and K. Doan. 1990. *Helianthus annuus* ssp. *texanus* has chloroplast DNA and nuclear ribosomal RNA genes of *Helianthus debiolis ssp. cucumerifolius*. Proceedings of the National Academy of Sciences U.S.A. **87**:593–597.

Rieseberg, L. H., H. C. Choi, and D. Ham. 1991. Differential cytoplasmic versus nuclear introgression in *Helianthus*. Journal of Heredity **82**:489–493.

Rieseberg, L. H., J. Whitton, and C. R. Linder. 1996. Molecular marker incongruence in plant hybrid zones and phylogenetic trees. Acta Botanica Neerlandica **45**:243–262.

Rivers, B. A., R. Bernatzky, S. J. Robinson, and W. Jahnen-Dechent. 1993. Molecular diversity at the self-incompatibility locus is a salient feature in natural populations of wild tomato (*Lycopersicon peruvianum*). Molecular and General Genetics **238**:419–427.

Robertson, H. M., and D. J. Lampe. 1995. Recent horizontal transfer of a *mariner* transposable element among and between Diptera and Neuroptera. Molecular Biology and Evolution **12**:850–862.

Rodman, J., R. A. Price, K. Karol, E. Conti, K. J. Systma, and J. D. Palmer. 1993. Nucleotide sequences of the *rbcL* gene indicate monophyly of mustard oil plants. Annals of the Missouri Botanical Garden **80**:686–699.

Rodrigo, A. G., M. Kelley-Borges, P. R. Bergquist, and P. L. Bergquist. 1993. A randomization test of the null hypothesis that two cladograms are sample estimates of a parametric phylogenetic tree. New Zealand Journal of Botany **31**:257–268.

Roth, V. L. 1991. Homology and hierarchies: problems solved and unresolved. Journal of Evolutionary Biology **4**:167–194.

Sanderson, M. J., and Doyle, J. J. 1992. Reconstruction of organismal and gene phylogenies from data on multigene families: concerted evolution, homoplasy, and confidence. Systematic Biology **41**:4–17.

Sang, T., D. J. Crawford, and T. F. Stuessy. 1995. Documentation of reticulate evolution in peonies (*Paeonia*) using internal transcribed spacer sequences of nuclear ribosomal DNA: implications for biogeography and concerted evolution. Proceedings of the National Academy of Sciences U.S.A. **92**:6813–6817.

Sawyer, S. 1989. Statistical tests for detecting gene conversion. Molecular Biology and Evolution **6**:526–538.

Schena, M., and R. W. Davis. 1994. Structure of homeobox-leucine zipper genes suggests a model for the evolution of gene families. Proceedings of the National Academy of Sciences U.S.A. **91**:8393–8397.

Seberg, O., G. Petersen, and C. Baden. 1996. The phylogeny of *Psathyrostachys* Nevski (Triticeae, Poaceae)—are we able to see the wood for the trees? In Proceedings of the 2nd International Triticeae Symposium, eds. R. R.-C. Wang, K. B. Jensen, and C. Jaussi, pp. 247–253. Utah State University, Logan.

Seelanan, T., A. Schnabel, and J. F. Wendel. 1997. Congruence and consensus in the cotton tribe. Systematic Botany **22**:259–290.

Sites, J. W., S. K. Davis, T. Guerra, J. B. Iverson, and H. L. Snell. 1996. Character congruence and phylogenetic signal in molecular and morphological data sets: a case study in the living iguanas (Squamata, Iguanidae). Molecular Biology and Evolution **13**:1087–1105.

Slatkin, M. 1991. Inbreeding coefficients and coalescence times. Genetical Research of Cambridge **58**:167–175.

Smith, R. L., and K. J. Sytsma. 1990. Evolution of *Populus nigra* (sect. *Aegiros*): introgressive hybridization and the chloroplast contribution of *Populus alba* (sect. *Populus*). American Journal of Botany **77**:1176–1187.

Soltis, D. E., and R. K. Kuzoff. 1995. Discordance between nuclear and chloroplast phylogenies in the *Heuchera* group (Saxifragaceae). Evolution **49**:727–742.

Soltis, D. E., and P. S. Soltis (eds.). 1989. Isozymes in Plant Biology. Dioscorides Press, Portland Oregon.

Soltis, D. R. Morgan, A. Grable, P. S. Soltis, and R. Kuzoff. 1993. Molecular systematics of Saxifragaceae *sensu stricto*. American Journal of Botany **80**:1056–1081.

Soltis, D. E., L. A. Johnson, and C. Looney. 1996. Discordance between ITS and chloroplast topologies in the *Boykinia* group (Saxifragaceae). Systematic Botany **21**:169–185.

Soltis, D. E., P. S. Soltis, D. L. Nickrent, L. A. Johnson, W. J. Hahn, S. B. Hoot, J. A. Sweere, R. K. Kuzoff, K. A. Kron, M. W. Chase, S. M. Swensen, E. A. Zimmer, S.-M. Chaw, L. J. Gillespie, W. J. Kress, and K. J. Sytsma. 1997. Angiosperm phylogeny inferred from 18S ribosomal DNA sequences. Annals of the Missouri Botanical Garden **84**:1–49.

Soltis, P. S., and D. E. Soltis. 1995. Plant molecular systematics: inferences of phylogeny and evolutionary processes. Evolutionary Biology **28**:139–194.

Springer, M., L. J. Hollar, and A. Burk. 1995. Compensatory substitutions and the evolution of the mitochondrial 12S rRNA gene in mammals. Molecular Biology and Evolution **12**:1138–1150.

Steel, M., P. J. Lockhart, and D. Penny. 1995. A frequency-dependent significance test for parsimony. Molecular Phylogenetics and Evolution **4**:64–71.

Steele, K. P., and R. Vilgalys. 1994. Phylogenetic analyses of Polemoniaceae using nucleotide sequences of the plastid gene *matK*. Systematic Botany **19**:126–142.

Stone, E. M., and R. J. Schwartz (eds.). 1990. Intervening Sequences in Evolution and Development. Oxford University Press, New York.

Suh, Y., L. B. Thien, H. E. Reeve, and E. A. Zimmer. 1993. Molecular evolution and phylogenetic implications of internal transcribed spacer sequences of ribosomal DNA in Winteraceae. American Journal of Botany **80:**1042–1055.

Sullivan, J. 1996. Combining data with different distributions of among-site rate variation. Systematic Biology **45:**375–380.

Sullivan, J., K. E. Holsinger, and C. Simon. 1995. Among-site rate variation and phylogenetic analysis of 12S rRNA in Sigmodontine rodents. Molecular Biology and Evolution **12:**988–1001.

Swofford, D. L. 1991. When are phylogeny estimates from molecular and morphological data incongruent? **In** Phylogenetic Analysis of DNA Sequences, eds. M. M. Miyamoto and J. Cracraft, pp. 295–333. Oxford University Press, New York.

Sytsma, K. J. 1990. DNA and morphology: inference of plant phylogeny. Trends in Ecology and Evolution **5:**104–110.

Syvanen, M. 1994. Horizontal gene transfer: evidence and possible consequences. Annual Review of Genetics **28:**237–261.

Tajima, F. 1993. Simple methods for testing the molecular evolutionary clock hypothesis. Genetics **135:**599–607.

Takezaki, N., A. Rzhetsky, and M. Nei. 1995. Phylogenetic test of the molecular clock and linearized trees. Molecular Biology and Evolution **12:**823–833.

Talbert, L. E., J. F. Doebley, S. Larson, and V. L. Chandler. 1990. *Tripsacum andersonii* is a natural hybrid involving *Zea* and *Tripsacum:* molecular evidence. American Journal of Botany **77:**722–726.

Templeton, A. R. 1983. Phylogenetic inference from restriction endonuclease cleavage site maps with particular reference to the evolution of humans and the apes. Evolution **37:**221–244.

Templeton, A. R., and C. F. Sing. 1993. A cladistic analysis of phenotypic associations with haplotypes inferred from restriction endonuclease mapping. IV. Nested analysis with cladogram uncertainty and recombination. Genetics **134:**659–669.

VanderWiel, P. S., D. F. Voytas, and J. F. Wendel. 1993. *Copia*-like retrotransposable element evolution in diploid and polyploid cotton (**Gossypium** L.). Journal of Molecular Evolution **36:**429–447.

van Ham, R. C. H. J., H. T'Hart, T. H. M. Mes, and J. M. Sandbrink. 1994. Molecular evolution of non-coding regions of the chloroplast genome in the Crassulaceae and related species. Current Genetics **25:**558–566.

Vawter, L., and W. M. Brown. 1993. Rates and patterns of base change in the small subunit ribosomal RNA gene. Genetics **134:**597–608.

Wakeley, J. 1994. Substitution-rate variation among sites and the estimation of transition bias. Molecular Biology and Evolution **11:**426–442.

Walker, E. L., N. F., Weeden, C. B. Taylor, P. Green, and G. M. Coruzzi. 1995. Molecular evolution of duplicate copies of genes encoding cytosolic glutamine synthetase in *Pisum sativum.* Plant Molecular Biology **29:**1111–1125.

Waters, E. R., and B. A. Schaal. 1996. Biased gene conversion is not occurring among rDNA repeats in the *Brassica* triangle. Genome **39:**150–154.

Wendel, J. F. 1989. New World tetraploid cottons contain Old World cytoplasm. Proceedings of the National Academy of Sciences U.S.A. **86:**4132–4136.

Wendel, J. F., and V. A. Albert. 1992. Phylogenetics of the cotton genus (*Gossypium* L.): character-state weighted parsimony analysis of chloroplast DNA restriction site data and its systematic and biogeographic implications. Systematic Botany **17:**115–143.

Wendel, J. F., J. McD. Stewart, and J. H. Rettig., 1991. Molecular evidence for homoploid reticulate evolution in Australian species of *Gossypium.* Evolution **45:** 694–711.

Wendel, J. F., A. Schnabel, and T. Seelanan. 1995a. Bidirectional interlocus concerted evolution following allopolyploid speciation in cotton (*Gossypium*). Proceedings of the National Academy of Sciences U.S.A. **92:**280–284.

Wendel, J. F., A. Schnabel, and T. Seelanan. 1995b. An unusual ribosomal DNA sequence from *Gossypium gossypioides* reveals ancient, cryptic, intergenomic introgression. Molecular Phylogenetics and Evolution **4:**298–313.

Wessler, S. R., T. E. Bureau, and S. E. White. 1995. LTR-retrotransposons and MITEs: important players in the evolution of plant genomes. Current Opinion in Genetics and Development **5:**814–821.

Wheeler, W. C. 1992. Extinction, sampling, and molecular phylogenetics. **In** Extinction and Phylogeny, eds. M. J. Novacek and Q. D. Wheeler, pp. 205–215. Columbia University Press, New York.

Whitkus, R., J. Doebley, and M. Lee. 1992. Comparative genome mapping of sorghum and maize. Genetics **132:**1119–1130.

Wu, C.-I., and W-H. Li. 1985. Evidence for higher rates of nucleotide substitution in rodents than in man. Proceedings of the National Academy of Sciences U.S.A. **82:**1741–1745.

Yang, Z. 1996. Among-site rate variation and its impact on phylogenetic analyses. Trends in Ecology and Evolution **11:**367–372.

Yoshinaga, K., H. Iinuma, T. Masuzawa, and K. Ueda. 1996. Extensive RNA editing of U to C in addition to C to U substitution in the *rbcL* transcripts of hornwort chloroplasts and the origin of RNA editing in green plants. Nucleic Acids Research **24:**1008–1014.

Zhang, W., J. F. Wendel, and L. G. Clark. 1997. Bamboozled again! Inadvertent isolation of fungal rDNA sequences from bamboos (Poaceae: Bambusoideae). Molecular Phylogenetics and Evolution **8:** 205–217.

11

Assessing Congruence: Empirical Examples from Molecular Data

Leigh A. Johnson and Douglas E. Soltis

Numerous DNA regions representing the nuclear and both organellar genomes are now available for comparative sequencing in plants (see Chapter 1); in addition, morphological and chemical data can also be obtained for phylogenetic analyses. With such a diversity of potential data sets available and the relative ease with which DNA sequences can be obtained, the acquisition of multiple data sets for the same suite of taxa is straightforward. As a result, the number of groups for which multiple data sets is available is increasing rapidly. Although it is readily apparent that multiple data sets are needed for estimating phylogenetic relationships reliably, it is also recognized that different genes may, in fact, possess different branching histories (see Chapter 10). Consequently, incorporating multiple data sets into phylogenetic studies is not a casual undertaking. Essential tasks in the analysis of multiple data sets include assessing congruence between different phylogenetic trees and data sets, and ascertaining whether multiple data sets should be combined into a single data matrix prior to phylogenetic reconstruction.

Three alternatives have been proposed for handling multiple data sets in phylogenetic analyses: the *combined* approach, the *consensus* approach, and the *conditional combination* approach (reviewed in Huelsenbeck et al., 1996). Kluge (1989) and Kluge and Wolf (1993) advocate the first alternative, suggesting that all available data should be combined into a single matrix before phylogenetic analyses. Because the phylogenetic information from all characters is considered simultaneously, and conflict between individual characters can be assessed (e.g., Larson, 1994), this method has also been referred to as the *total evidence* or *character congruence* approach. In contrast, Miyamoto and Fitch (1995) argue that multiple data sets (representing natural partitions) should be analyzed separately and the different phylogenetic estimates compared. This method seeks similarities between independent analyses for phylogenetic corroboration and is also referred to as *taxonomic congruence. Conditional data combination* is essentially a hybrid of the first two approaches; it involves combining data except when significant heterogeneity exists between

We thank the molecular lab discussion groups at Rancho Santa Ana Botanic Garden and the Department of Botany at Washington State University for comments and criticisms; J. Wendel for a preprint of a manuscript and for helpful suggestions on an earlier draft of this chapter; R. Mason-Gamer and E. Kellogg for a preprint of their manuscript, and D. Swofford for permission to use prerelease versions of PAUP* 4.0. This work was supported by NSF grant 9307000 to DES and NSF grant 9321788 to DES and LAJ. NSF grant DEB 9509121 to J. M. Porter provided support to LAJ during much of the writing.

data sets and the heterogeneity is either attributable to different branching histories, or cannot be compensated for by revising the method of phylogenetic reconstruction (e.g., Bull et al., 1993; de Queiroz, 1993; Lutzoni and Vilgalys, 1995; de Queiroz et al., 1995).

There has been considerable debate regarding the advantages and limitations of the combined, consensus, and conditional combination approaches (c.f., Mickevich and Farris, 1981; Kluge, 1989; Swofford, 1991; Bull et al., 1993; Eernisse and Kluge, 1993; Jones et al., 1993; Lanyon, 1993; Rodrigo et al., 1993; Farris et al., 1995; Lutzoni and Vilgalys, 1995; Miyamoto and Fitch, 1995; Humphries, 1995; Huelsenbeck and Bull, 1996; reviewed in de Queiroz et al., 1995 and Huelsenbeck et al., 1996; thorough reviews are provided by de Queiroz et al., 1995 and Huelsenbeck et al., 1996). We therefore highlight only the key arguments.

As with any inferential method, phylogenetic reconstruction is subject to sampling error. Given finite sample sizes and the stochastic nature of molecular evolution (cf. Lanyon, 1988), a data set may, by chance, fail to contain character-state distributions that reflect the actual phylogenetic relationships. Alternatively, due to homoplasy, data sets may contain character-state distributions that support "erroneous" groupings of taxa. Systematists that favor the combined approach argue that combined analyses minimize sampling error and maximize the "explanatory power" of the data. That is, when data sets differ only due to sampling error, combining data sets increases the number of characters, which may facilitate the retrieval of, or increase the support for, "true" clades. However, the combined approach can also yield estimates of relationships that are spurious or even strongly contradicted by one or more of the data sets when there are major discrepancies between the data matrices. Unfortunately, the a priori combination of data without separate analyses precludes the discovery of the latter result.

In contrast, conducting separate analyses ensures that data sets that strongly contradict each other (due to different evolutionary processes or histories) will each provide a phylogenetic estimate that reveals the contrasting relationships. However, the decreased number of characters in analyses of separate data sets also increases the opportunity for stochastic errors, which may result in different topologies even when the characters in different data sets are homogeneous. By itself, the separate analysis approach does not distinguish between cases in which sampling error is responsible for different trees from different data sets and those instances in which the different topologies reflect different evolutionary mechanisms or branching histories in the data themselves. In the first instance, combining data sets should be advantageous and improve the estimate of phylogeny; in the second, combining data sets may be inappropriate.

The conditional combination approach attempts to reduce mistakes in estimating phylogeny by preventing the combination of heterogeneous data sets and ensuring the combination of homogeneous data sets. This approach employs statistical tests of a general null hypothesis that multiple data sets are homogeneous. However, because any statistical test has associated error, one might erroneously reject the null hypothesis. In this case, data sets that could be combined would not be. Of greater practical concern is the fact that statistical evidence of data set heterogeneity by itself is a weak argument for not combining data sets. Measurable heterogeneity may simply reflect stochastic variation or different evolutionary processes (e.g., Sullivan, 1996). Distinguishing between instances in which clades are weakly supported and alternative resolutions are only slightly less parsimonious ("soft" incongruence; Seelanen et al., 1997) and those instances in which alternative resolutions in trees from different data sets are strongly supported as incongruent ("hard" incongruence) is an important part of the conditional combination methodology. Substantive reasons for not combining data include: (1) evidence that heterogeneity reflects different branching histories between data sets, and (2) extreme heterogeneity caused by differences in evolutionary mechanisms that cannot be explained or compensated for by revising the method of phylogenetic reconstruction (de Queiroz et al., 1995; Huelsenbeck et al., 1996). Thus, determining when to and when not to combine data sets cannot always be made on the basis of statistical tests alone. Familiarity with possible biases in the data and a sound knowledge of the biology of the species under investigation are also needed for interpreting the meaning of significant heterogeneity.

This chapter complements Chapter 10, which examines the *sources* of incongruence. We illustrate the application of the conditional combination approach and apply an array of methods to multiple data sets from Polemoniaceae and Saxifragaceae sensu stricto (s. s.). Our primary objectives are to illustrate the use of several tests for quantifying congruence between trees and data sets, and assessing when incongruence is "significant"; we also provide guidance in determining when data sets may, or may not, be combined. The results of these tests are also used to identify possible sources of incongruence (see Chapter 10). We provide a series of boxes that contain the background information needed to conduct the tests employed; we encourage those interested in applying these tests to peruse the boxes and the references therein.

CAVEATS AND CLARIFICATIONS

We use the terms *congruence* and *incongruence* as general descriptors of agreement between trees, characters, or data sets. Two trees are congruent if they are identical (a strict definition) or if one tree is a more resolved version of the second tree (a loose definition). Similarly, congruent characters are those that specify compatible relationships (i.e., show no homoplasy). Because virtually all empirical data sets contain homoplasy, these data sets all contain some characters that are incongruent. When sets of characters show significantly lower incongruence within data sets and higher incongruence between data sets, the data sets are heterogeneous. Conversely, when sets of characters are no more congruent within data sets than between data sets, the data sets are homogeneous. When two trees are compared, the level of incongruence between topologies depends on the number or relative location of discordant relationships. "Significant" incongruence, however, is independent of the level of incongruence between topologies; significance is associated instead with a statistical probability and depends on the strength of support for the discordant relationships specified by the underlying data. Consequently, the task of assessing congruence for the purpose of determining the suitability of combining data sets rests on the characters themselves rather than simply on the topologies that the characters specify. Accordingly, determining whether data sets are significantly homogeneous or heterogeneous is a separate task from measuring the level of congruence between phylogenetic trees recovered from different data sets. Both types of congruence assessments can be useful, however, and in this chapter we assess both topological congruence and character congruence (also termed *data set congruence* or *character-set homogeneity*).

We use two and three clades from Polemoniaceae and Saxifragaceae s. s., respectively, as examples. These clades were chosen because multiple data sets representing chloroplast and nuclear genes are available, and also because our familiarity with the families and the data (c.f., Soltis, 1987; Soltis et al., 1991, 1993, 1996a, 1996b; Johnson and Soltis, 1994, 1995; Soltis and Kuzoff, 1995; Johnson, 1996; Johnson et al., 1996) enables us to address potential sources of incongruence (see Chapter 10). In addition, each clade exhibits different levels or kinds of incongruence and thus ultimately requires a different analytical solution. The systematic implications of these studies are not discussed here. Numerous other plant groups for which multiple data sets are available could similarly serve as examples of a rigorous assessment of congruence. Several published studies, in fact, provide quantitative assessments of congruence within plants and employ several of the approaches discussed here (e.g., Kim and Jansen, 1994; Smith and Sytsma, 1994; Mason-Gamer and Kellogg; 1996, Seelanan et al., 1997; see also Lutzoni and Vilgalys, 1995 for an example involving fungi).

We approach this topic from a parsimony perspective and use standard analytical techniques outlined elsewhere for recovering shortest trees, estimating bootstrap confidence levels, and determining decay values (e.g., Felsenstein, 1985a; Bremer, 1988, 1994; see also Maddison, 1991; Hillis and Bull, 1993; Johnson et al., 1996; Soltis et al., 1996b). These analyses were conducted using the Nexus standard programs

MacClade 3.05 (Maddison and Maddison, 1992), PAUP* 4.0 (Swofford, 1997), and COMPONENT 2.0 (Page, 1993). Consequently, our discussion is biased toward these programs although other software packages may perform many of the same tasks.

We restrict our study to parsimony-based methods; assuming the conditions hold under which parsimony is a consistent estimator of phylogeny (Huelsenbeck and Hillis, 1993; see also Chapter 4), these methods are appropriate and effective (to varying degrees) at describing levels of incongruence and indicating heterogeneity when it exists. We recognize that techniques for assessing data set homogeneity with maximum likelihood (ML) also have great utility (see Chapter 5). In particular, by incorporating evolutionary models of sequence evolution that are uniquely determined for each data set, ML methods can help distinguish when heterogeneity is attributable to stochastic variation or to different evolutionary processes acting at the DNA level (e.g., Huelsenbeck and Bull, 1996; Sullivan, 1996). Importantly, however, ML methods are not inherently better than parsimony-based methods at determining when data sets should be combined (cf. Sullivan, 1996). As with parsimony-based methods, a knowledge of both the biology of the species and the underlying data is important for interpreting significant heterogeneity and distinguishing between soft and hard incongruence.

All of our examples involve comparisons between molecular data sets (either cpDNA restriction sites or DNA sequences) that are taxonomically equivalent (in one instance, known sister species were substituted). The approach and methods we employ, however, work equally well for comparisons that include morphological or other types of data amendable to standard parsimony analysis. These approaches can also be extended to three or more data sets (see Example 5: Saxifragaceae s. s.) with some increase in complexity. Although several quantitative measures of congruence can be calculated for data sets with unequal taxon sampling (Bledsoe and Raikow, 1990; Miyamoto, 1996) and methods have been suggested for dealing with missing data (e.g., Baum and Ragan, 1993; Lanyon, 1993; Wiens and Reeder, 1995) the presence or absence of individual taxa can have a great effect on relationships when homoplasy exists (Lanyon, 1985). Careful consideration of relationships and possible biases are needed under these circumstances, and we do not address this issue further; we simply note that congruence and sources of heterogeneity can be explored much more thoroughly when taxonomically equivalent data sets are obtained.

METHODOLOGICAL FRAMEWORK

A practical framework for exploring congruence is provided by the conditional combination methodology, and we follow this approach for several reasons. (1) In the course of assessing data set heterogeneity, both separate and combined analyses are conducted; the results of both analyses are thus available for scrutiny. (2) It is clear that gene trees may not represent species trees (e.g., Doyle, 1992; see Chapter 10), and many unambiguous cases of discordance due to chloroplast capture have been documented (reviewed in Rieseberg and Soltis, 1991 and Rieseberg, 1995); thus, the argument that data partitions are artificial constructs (Kluge and Wolf, 1993) is difficult to accept. (3) When data sets are homogeneous, combined analyses may increase support or resolution in areas that are individually weakly supported or poorly resolved; because separate analyses were also conducted, the corroboration that exists from independent analyses is still present. (4) The conditional combination of data is in harmony with the multifaceted approach to assessing congruence advocated by Swofford (1991) and the conceptual framework advocated by de Queiroz et al. (1995). These approaches encourage, as we do, a thorough examination of the data.

Bull et al. (1993) present a flow chart that outlines the conditional combination approach; we expand on their theme to provide a more comprehensive model (Fig. 11.1). Data are first compiled, individual analyses conducted, and various aspects of topological and character incongruence are measured to gain insights into the level and distribution of any observed differences. Tests for homogeneity (which require either independent analyses of all data sets or both independent and combined analyses) are then

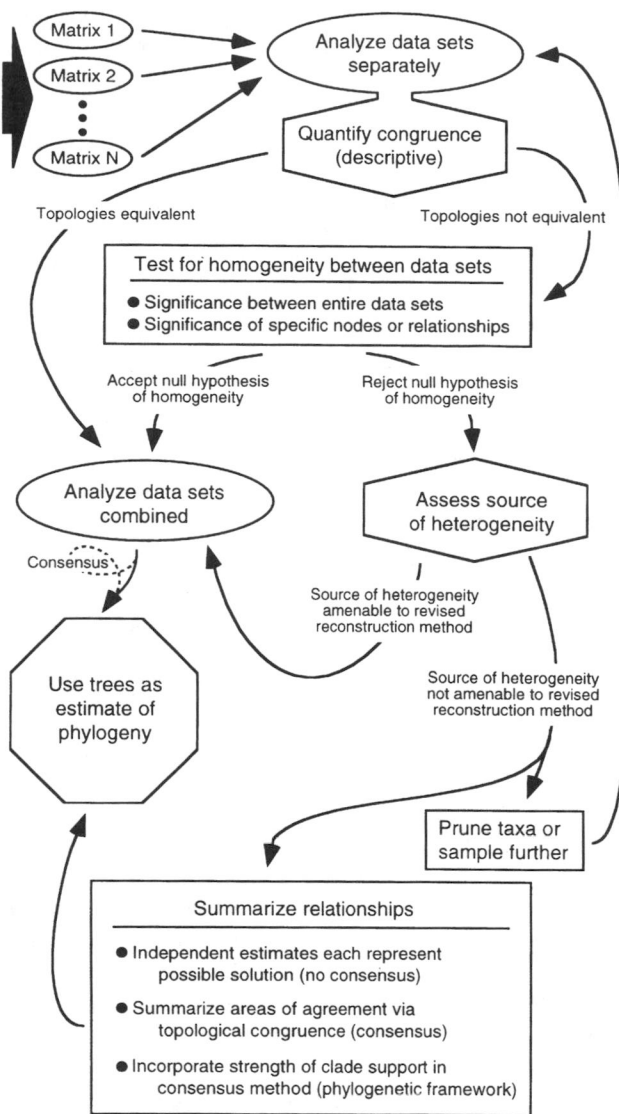

Figure 11.1. Diagram outlining the conditional combination approach to analyzing multiple data sets (modified from Bull et al., 1993).

conducted to assess the significance of the observed levels of incongruence. In the rare instance in which no topological incongruence is observed, combined analyses can be conducted directly given that the combined topology will necessarily match the separate topologies. Considering character congruence, if the null hypothesis of data set homogeneity cannot be rejected, combined analyses are conducted and the resulting tree or trees are used as the estimate of phylogeny. If, however, the null hypotheses is rejected, the source of heterogeneity is investigated. Some sources of heterogeneity, such as unequal rates of sequence evolution, may be adequately addressed with differential weighting schemes; in these instances combined analyses are still appropriate (Bull et al., 1993; Chippendale and Wiens, 1994; see also Sullivan, 1996). Other sources of heterogeneity may be difficult to identify, or may be explained by biological processes such as lineage sorting or introgression. In these latter cases, simple weighting schemes are inappropriate and combined analyses are not recommended (Huelsenbeck et al., 1994). If, however, the heterogeneity can be isolated to one or a few species, removing the offending species may lead to homogeneity between data sets; combined analyses of the reduced data sets are then appropriate (e.g., Rodrigo et al., 1993; Seelanan et al., 1997). Although elaboration is beyond the scope of this chapter, several options exist for summarizing the phylogeny of a group from the trees generated by independent analyses in the final step of the conditional combination methodology (Fig. 11.1; see Swofford, 1991; Baum and Ragan, 1993; Lanyon, 1993; Rodrigo et al., 1993; de Queiroz et al., 1995; Miyamoto and Fitch, 1995; Ronquist, 1996).

QUANTITATIVE MEASURES OF INCONGRUENCE

An observation used in support of the separate analysis (never combine) and the conditional combination approach is that phylogenetic hypotheses inferred from different data sets, though often similar, are seldom identical (Penny and Hendy, 1985a; Rodrigo et al., 1993). This observation leads to three interrelated issues. (1) How different are two phylogenetic trees? (2) How much conflicting phylogenetic information exists between two data sets? (3) How does an investigator distinguish among the various sources of conflict and identify those suggesting that data sets should not be combined? Many indices and measures have been developed to address these three questions (reviewed in Swofford, 1991; Larson, 1994; de Queiroz et al., 1995; Mason-Gamer and Kellogg, 1996). Rather than illustrate all available congruence techniques, we selected 11 metrics and tests that appear to be as good as, or better than, competing methods for addressing the questions just posed. We selected methods that are easily applied, appropriate, insightful, and that represent a variety of approaches used by others in the recent literature. By employing more than one index or test to address each question, we are able to compare and contrast

the information each method provides. Additionally, the procedures we use are generally straightforward to apply and are either directly implemented in current computer software (e.g., PAUP* 4.0 or COMPONENT 2.0), or can be computed by manipulating output produced by these programs. Summaries of each of the 11 methods we chose are presented in a series of boxes (see Boxes 11.1–11.8) to assist those wishing to use these congruence techniques in their own work.

Topological Congruence Indices

How different are two phylogenetic trees? Qualitative comparisons of topological congruence are common in the literature (e.g., Johnson and Soltis, 1994; Olmstead and Sweere, 1994), yet such comparisons usually fail to describe adequately the magnitude of observed similarities and differences. Quantitative measures, in contrast, provide discrete numerical values that are more satisfying from an objective perspective. Interpreting a single numerical index can be difficult, however, particularly in the absence of the trees from which the number was derived (Swofford, 1991). The most revealing course of action involves quantitative measures that are interpreted via qualitative assessments of the distribution of discordant relationships (Penny and Hendy, 1985b).

There are three groups of quantitative indices for assessing topological congruence; indices based on: (1) tree interconversion, (2) subtree similarities, and (3) shared clades identified through consensus procedures. For comparison, we selected six topological indices that represent these three groups (see Boxes 11.1–11.4): (1) the partition metric (PM); (2) the explicitly agree (EA) quartet/triplet similarity metric; (3) the greatest agreement subtree metric (D_1); (4) Colless's consensus fork index (CI_C); (5) Rohlf's modified consensus information index (CI_1); and (6) the index of topological identity (TI). The partition metric measures the number of branch contractions and decontractions required to change one tree into another (Box 11.1). The explicitly agree (Box 11.2) and greatest agreement subtree metrics (Box 11.3), in contrast, quantify similarities that exist in branching patterns

BOX 11.1. CONGRUENCE METRICS BASED ON TREE INTERCONVERSION: THE PARTITION METRIC

The partition metric (PM; Robinson and Foulds, 1979, 1981) is one of the earliest applied measures for assessing incongruence between trees (e.g. Penny et al., 1982; Penny and Hendy, 1985b). Given two trees, PM can be defined as the minimum number of branch contractions and decontractions needed to transform the first tree into the second tree. Alternatively, PM can be defined as the number of internal branches summed over both trees that, when divided, define two clades on one tree that cannot be defined by dividing any equivalent branch on the other tree.

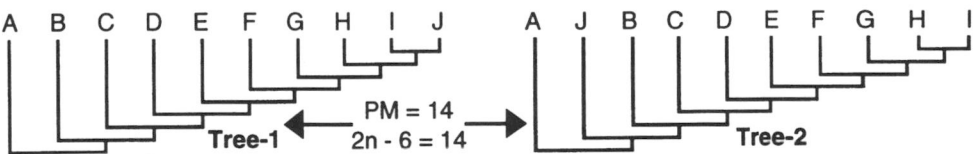

PM = 4 (four unique partitions, four steps to interconversion)

For N taxa, there are $N - 3$ internal branches and $2(N - 3)$ unique partitions for two trees; thus, the maximum value that the partition metric can obtain, $2N - 6$, is dependent on the number of taxa. The straightforward definition of the partition metric and its ease of calculation have made PM one of the more popular measures for assessing distance between trees in phylogenetic studies, especially when comparing observed distances to distances between random trees (Box 11.9) to test the null hypothesis of chance similarity (e.g., Penny et al., 1982; Rodrigo et al., 1993). However, the partition metric has also been criticized because, analogous to a strict consensus tree, the differential placement of a single taxon on two pectinate (maximally asymmetrical) trees will result in the greatest possible distance, as illustrated below.

Penny and Hendy (1985b) addressed this concern and suggest that valuable insights can be gained by pruning taxa individually at first, and then by pairs, and finally by larger subsets with PM calculated for each new tree. The influence of individual taxa, as well as groups of taxa, on the observed distance can thereby be assessed. Citing Robinson and Foulds (1981), they also suggest that PM can be weighted by counting the lengths of the branches that are different. Decay values (Bremer, 1988, 1994) may be more appropriate than observed lengths for weighting, but the merit of weighting PM by any means has yet to be determined.

The partition metric is calculated by COMPONENT 2.0 and by PAUP* 4.0 (the latter program uses the name "symmetrical difference"; use the command TREEDIST METRIC = SYMDIFF). PM can be determined for both rooted and unrooted trees with or without polytomies. Exact probability distributions have been worked out for up to 16 taxa (Hendy et al., 1984). For more than 16 taxa, probabilities based on comparisons with random pairs of trees can be determined when necessary (Box 11.9).

BOX 11.2. CONGRUENCE INDICES BASED ON TOPOLOGY SUBSTRUCTURE: QUARTET AND TRIPLET SIMILARITY MEASURES

An unrooted topology for five or more taxa can be described as a series of four-taxon statements that indicate relationships between unrooted quartets of taxa (Estabrook et al., 1985; Estabrook, 1992). Rooted three-taxon statements that indicate relationships between triplets of taxa provide an analogous measure for rooted topologies (Page, 1993; Critchlow et. al., 1996). Each quartet or triplet uniquely resolves relationships in one of four ways:

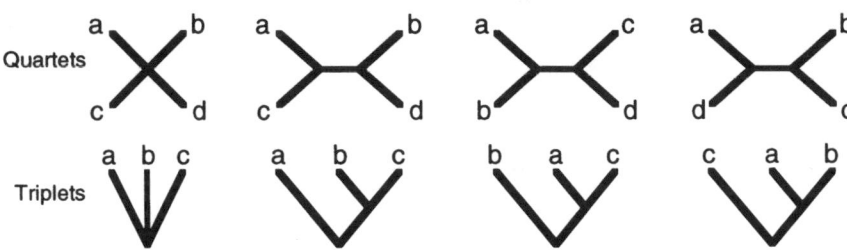

When two trees are compared, there are five possible outcomes between pairs of quartets or triplets:

s resolved and identical
d resolved and different
$r1$ resolved in tree one but not in tree two
$r2$ resolved in tree two but not in tree one
u unresolved in both tree one and tree two

The total number of quartets or triplets, $Q, = s + d + r1 + r2 + u$.

Using these outcomes, several congruence indices have been proposed (Estabrook et al., 1985; Day, 1986; Page, 1993):

Index	*Similarity*	*Dissimilarity*
Do not conflict (DC)	$(s + r1 + r2 + u) \div Q$	$d \div Q$
Explicitly agree (EA)	$s \div Q$	$(d + r1 + r2 + u) \div Q$
Strict joint assertions (SJA)	$s \div (s + d)$	$d \div (s + d)$
Symmetric distance (SD)	$2s \div (2d + 2s + r1 + r2)$	$(2d + r1 + r2)$ $\div (2d + 2s + r1 + r2)$

These measures are computed by COMPONENT 2.0.

One advantage of quartet/triplet indices over the partition metric (Box 11.1) is that the former are less sensitive to the placement of individual taxa. Whereas the partition metric can give a maximum dissimilarity value for the rearrangement of a single taxon on pectinate trees (Box 11.1), the EA similarity metric indicates that the same two 10 taxon trees illustrated in Box 11.1 are 60% similar. On the other hand, quartets/triplets are sensitive to the location of branch rearrangements. Single-branch discrepancies among terminal taxa affect quartet/triplet measures much less than do single rearrangements among internal branches. Consequently, as with other indices, examination of areas of discordance among topologies is necessary for interpreting the meaning of a particular value.

continued

continued

There are reasons for preferring quartets over triplets, and vice versa, in a particular study. If information regarding the root of the topologies being compared is unavailable or congruence between topologies exclusive of directionality imposed by the root is desired, quartets are more appropriate than triplets. However, for N taxa there are $[(N(N - 1)(N - 2)(N - 3)) \div 24]$ quartets compared to $[(N(N - 1)(N - 2)) \div 6]$ triplets (Page, 1993); for 70 taxa, this results in a 16-fold increase in computational burden for quartets over triplets (916,895 quartets verses 54,740 triplets). Thus, for computational reasons, triplets may be preferred, particularly for large data sets.

BOX 11.3. CONGRUENCE MEASURES BASED ON TOPOLOGY SUBSTRUCTURE: AGREEMENT SUBTREE

Agreement subtrees depict relationships among groups of taxa that branch identically in two or more trees (Gordon, 1980; Finden and Gordon, 1985; Kubicka et al., 1995). As a measure of similarity between trees, the *greatest agreement subtree* (=*largest common pruned tree*, LCPT) is most useful, because it requires the removal of the fewest taxa in order to transform discordant topologies into concordant topologies.

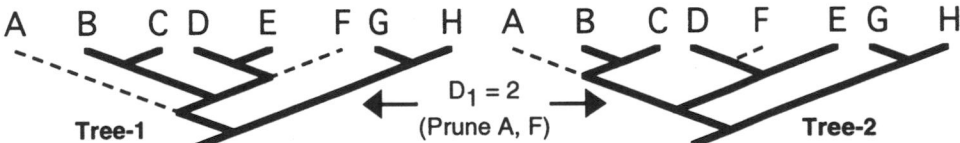

Given two or more trees, there may be more than one LCPT. COMPONENT 2.0 will find one of the possible LCPTs, and will show this tree as well as the number of taxa pruned to form this tree. The input trees must be fully binary, however, and if more than two input trees exist, LCPTs are computed only in pairwise comparisons (i.e., COMPONENT 2.0 will not find a LCPT that summarizes three or more trees). PAUP* 4.0 is more flexible than COMPONENT 2.0 and will compute a LCPT that summarizes any reasonable number of input trees, using the AGREE command. Additionally, with PAUP* 4.0 the input trees may contain polytomies, and an option to find all LCPTs exists. PAUP* 4.0 also provides two distance measures (using TREEDIST METRIC = AGD1 or AGR) that are derived from the greatest agreement subtrees. Agree D_1 (abbreviated D_1 in this text) equals the number of taxa pruned to form the LCPT, whereas agree D is the number of taxa pruned to form the LCPT weighted by distances between taxa. Normalized dissimilarity values for agree D and agree D_1 are computed by PAUP* 4.0 in the histogram portion of the output.

The agree D_1 similarity measure is less sensitive to the placement of discordant taxa than is the partition metric. The 10 taxon pectinate tree example given in Box 11.1 requires the pruning of a single taxon (taxon 10) to form the LCPT, which results in a normalized similarity D_1 of 0.86 compared to the 0.00 similarity value given by the normalized partition metric.

among smaller subsets of taxa. CI_C, CI_1, and TI (Box 11.4) quantify resolved clades either on consensus trees or those shared between the original trees. Note that CI_C and CI_1 are topological *consensus* indices and are not to be confused with the *consistency index* (CI or C; Kluge and Farris, 1969) that is commonly used as a measure of homoplasy.

A uniform convention for reporting these metrics does not exist. For example, PM is typically reported as the number of branch contraction/decontractions required to convert trees, and D_1 is reported as the number of taxa that require pruning to make two trees identical. PM and D_1 are thus whole numbers that increase in value as trees decrease in similarity. EA, in contrast, is usually reported as a normalized value (percentage of maximum possible score). Although EA can be computed as either a similarity or dissimilarity metric, reporting EA as a dissimilarity metric is

---

BOX 11.4. CONGRUENCE MEASURES BASED ON CONSENSUS METHODS: CI_C, CI_1, AND TI

Several indices have been described that quantify agreement between trees as summarized on a consensus tree (reviewed by Rohlf, 1982; Mickevich and Platnick, 1989; Swofford, 1991). Two of these, Colless's (1980) consensus fork index (CI_C) and Rohlf's (1982) consensus information index (CI_1), are discussed here. A third index also considered, the index of topological identity (TI; Bledsoe and Raikow, 1990), is similar to these consensus indices, but is computed from the original trees themselves rather than from a consensus tree.

Colless's consensus fork index is both easy to calculate and easy to interpret. $CI_C = B_i \div (B_t - R)$, where B_i is the number of internal branches (=number of resolved clades), B_t is the number of terminal branches, and R is either two or three depending on whether the tree is rooted with an outgroup or a basal polytomy, respectively. Thus, CI_C is the proportion of possible clades that is resolved on the consensus tree. Though straightforward to calculate, CI_C has been criticized because trees that resolve only a few branches will receive the same CI_C value regardless of the distribution of the branches:

$CI_C = 0.444$
$CI_1 = 0.333$
Tree-1　　　**Tree-2**
$CI_C = 0.444$
$CI_1 = 0.578$

Although both of the trees illustrated resolve three branches, the second tree implies synapomorphies that distinguish among a greater number of taxa and thus conveys a greater degree of phylogenetic information.

CI_1 is one of several consensus indices that attempts to improve upon CI_C by incorporating information regarding the numbers of taxa that appear in clades. $CI_1 = \Sigma N_i \div [\Sigma N_i + \Sigma \Delta_i]$ where N_i is the number of taxa minus one in each resolved clade, and Δ_i is an adjustment factor for the number of unresolved clades at each node. Thus, CI_1 is equal to one when the consensus tree is fully bifurcating (because Δ_i will equal zero) and equal to zero when the consensus tree is a bush (because N_i will equal zero). The adjustment for unresolved clades, Δ_i, is calculated by:

$$\Delta_i = \sum_{a=2}^{f_i-1}\left[\sum_{b=1}^{a} n_b - 1\right]$$

continued

continued

where n_b is the size, in decreasing order, of the subsets participating in the polytomy and f_i is the number of subsets in the polytomy at node i (Rohlf, 1982).

In terms of classifications and information content, both CI_C and CI_I possess the desirable property that they equal zero when the consensus tree is completely unresolved. However, as with the partition metric (Box 11.1), the position of a single taxon can result in a consensus bush with some consensus procedures. Additionally, two identical trees with at least one polytomy cannot achieve a perfect score even though they are in complete agreement. Although this is arguably reasonable in some applications (see Rohlf, 1982; Mickevich and Platnick, 1989; Swofford, 1991), it may be misleading when trying to assess congruence between data sets.

Bledsoe and Raikow (1990) used the index of topological identity (TI), which strictly measures the number of shared clades between two (or more) trees relative to the number of clades resolved in tree one summed with the number of clades resolved on tree two:

$$TI = \frac{kx - k}{\left(\sum_{i=1}^{k} y\right) - k}$$

where k is the number of topologies being compared, x is the number of clades identical in all topologies, and y is the number of resolved clades on topology i (note this formula is corrected from the original given in Bledsoe and Raikow [1990] which neglected to subtract k from the numerator, their computations, however, used the correct formula).

TI is analogous to a consensus measure based on a strict consensus tree, but measures congruence regardless of resolution. That is, TI equals one when two trees being compared are identical regardless of the number of polytomies they contain. Bledsoe and Raikow (1990) describe the application of TI to more than two trees and to trees with unequal taxonomic sampling. They also discuss a sister index, the index of topological compatibility (TC) that is analogous to the semistrict (combinable component) consensus measure.

In practice, Both CI_C and CI_I are easily obtained; both indices are calculated by PAUP* 4.0 using the INDICES = YES option of the CONTREE command (CI_C = normalized component information and CI_I = Rohlf's CI(1)). TI, in contrast, must be calculated by hand using the formula above. TI calculations can be facilitated by PAUP* 4.0, however, given that x equals the consensus fork plus one (nonnormalized CI_C + 1) calculated from the strict consensus tree and y equals the consensus fork plus one measured on each topology individually. In the special case of two fully bifurcating trees, TI = CI_C calculated from the strict consensus of those two trees (this holds true even if the binary trees are rooted with a basal polytomy for display purposes in PAUP* 4.0).

---

more common (e.g., COMPONENT 2.0; see Box 11.2). CI_C, CI_I, and TI are also reported as normalized values, but invariably as similarity metrics (e.g., PAUP* 4.0; see Box 11.4). To provide uniformity and increase the ease with which these metrics can be compared, we normalized all of the topological metrics in terms of *similarity*. All normalized values reported here for PM, EA, D_1, CI_C, CI_I, and TI range from zero, least similar, to one, most similar.

Character Congruence Indices

How much conflicting phylogenetic information exists between two data sets? The six indices introduced provide insights into *topological discordance* between trees. They do not, however, address the degree of *character incongruence* that leads to the topological discordance. We use two additional indices to quantify character incongruence: the incongruence metric of

Mickevich and Farris (1981; I_{MF}) and the incongruence metric of Miyamoto (I_M, in Swofford, 1991). These indices summarize support for discordant relationships while providing a quantitative index of incongruence between data matrices (Box 11.5). For both indices, incongruence is measured as the proportion of extra homoplasy required to explain an alternative topology, relative to the homoplasy that exists on shortest trees for each data set (Box 11.5). Thus, these indices range from zero when the shortest tree recovered from each data set is identical (regardless of how much homoplasy is observed within each data set), to one when no homoplasy is observed in either data set and the topologies of the shortest trees from each data set are unique (regardless of how topologically different the shortest trees are). Consequently, the amount of homoplasy within each data set (which corresponds to character conflict within a data set) is an important contributor to character conflict between data sets as measured by these incongruence indices.

BOX 11.5. INCONGRUENCE INDICES OF MICKEVICH AND FARRIS (1981; I_{MF}) AND MIYAMOTO (IN SWOFFORD, 1991; I_M)

Two similar indices, I_{MF} and I_M, measure incongruence in terms of the extra homoplasy required to explain specific alternative topologies. Both indices share a common mathematical relationship:

$$\text{incongruence}_{\text{Between data sets}} = \text{incongruence}_{\text{Total}} - \Sigma \text{ incongruence}_{\text{Within each data set}}$$

Both I_{MF} and I_M measure incongruence within data sets identically; they differ, however, in how the total incongruence is quantified and thus provide different estimates of the incongruence that exists between data sets:

- I_{MF} measures total incongruence as the number of homoplasious steps required by each individual data set to explain the shortest trees recovered from analysis of the combined data.
- I_M measures total incongruence by summing the number of homoplasious steps required to map data set {a} on the shortest trees recovered from data set {b} with the number of homoplasious steps required to map data set {b} on the shortest trees recovered from data set {a}.

The following procedures outline the computation of I_{MF} and I_M when congruence is being assessed between two data sets; the calculations are generalizable for more than two data sets.

$$I_{MF}$$

1. Construct two taxonomically equivalent data sets, {a} and {b}, using identical taxon labels for both data sets.
2. Compose a third, combined data set, {ab}, that simply includes the matrices for both data sets {a} and {b} in the Nexus data block.
3. Find and save the most parsimonious trees for all three data sets.
4. Record the values defined below from the LENFIT command with the CI option activated in PAUP* 4.0:

continued

continued

Len A{a} = length of most parsimonious trees, A, (recovered from data set {a}) as measured on data set {a}.

minL {a} = minimum tree length for data set {a} in the absence of homoplasy.

Len B{b} and minL {b} are recorded for data set B analogously to the corresponding values for data set {a}.

Len AB{ab} = length of most parsimonious trees, AB, (recovered from combined data set {ab}) as measured on data set {ab}.

minL {ab} = minL {a} + minL {b}.

5. Calculate I_{MF} using the formula:

$I_{MF} = (i_{Total} - i_{Within}) \div (i_{Total})$

where $\quad i_{Total} = $ Len AB{ab} $-$ minL {ab}

$\quad\quad i_{Within} = ($Len A{a} $-$ minL {a}$) + ($Len B{b} $-$ minL {b}$)$

$$I_M$$

1. Construct two taxonomically equivalent data sets, {a} and {b}, using identical taxon labels for both data sets.
2. Find and save the most parsimonious trees for both data sets.
3. Record the values defined below from the LENFIT command with the CI option activated in PAUP* 4.0:

   Len A{a}, minL {a}, Len B{b} and minL {b} are identical to the same values defined above for I_{MF}.

   Len A{b} = Shortest length of most parsimonious trees, A, recovered from data set {a} when measured on data set {b}. [Hint: execute data set {b}, then GETTREES A, from data set {a} and measure lengths LENFIT/CI; the shortest tree length will provide the most conservative estimate of incongruence].

   Len B{a} is defined analogous to the above value for Len A{b}.
4. Calculate I_M using the formula:

   $I_M = (i_{Total} - i_{Within}) \div (i_{Total})$

   where $\quad i_{Total} = ($Len A{b} $-$ minL {b}$) + ($Len B{a} $-$ minL{a}$)$

   $\quad\quad i_{Within} = ($Len A{a} $-$ minL {a}$) + ($Len B{b} $-$ minL {b}$)$

Scrutiny of the formulas above reveals a relative relationship between I_{MF} and I_M values: If the topology of trees A = B = AB, then $I_{MF} = I_M = 0$; otherwise $I_{MF} < I_M$.

Significance Tests for Heterogeneity

How does an investigator distinguish among the various sources of conflict and identify those that suggest that data sets should not be combined? The character-congruence indices I_M and I_{MF} provide measures of character conflict between data sets, but it is not clear how large these values need to be to indicate that significant conflict exists. Furthermore, given that the magnitudes of I_M and I_{MF} depend as much on

levels of homoplasy within data sets as they do on the levels of homoplasy between data sets, the magnitude of I_M or I_{MF} alone is insufficient for determining significant heterogeneity. Several tests have been developed, however, that assess the general null hypothesis of homogeneity between data sets (see Bull et al., 1993; Rodrigo et al., 1993; Farris et al., 1995; Huelsenbeck and Bull, 1996). We apply two of these tests: the homogeneity test of Farris et al. (1995; HT_F) and the homogeneity test of Rodrigo et al. (1993;

HT$_R$); the names and acronyms for both tests are ours.[1] HT$_F$ (Box 11.6) tests the null hypothesis that characters are randomly distributed between data sets with respect to the phylogenetic information they contain, whereas HT$_R$ (Box 11.7) tests the null hypothesis that the sampling of characters contained in each data set represents the same universe of characters.

We also apply the significantly less parsimonious test (Templeton, 1983; SLP$_T$; acronym 11.5). The test Farris et al. described (our HT$_F$) can be used with any metric including, but not restricted to, their ILD (=Swofford's I$_{MF}$). Similarly, SLP$_T$ (Box 11.8) is used as a general acronym for the test proposed by Templeton (1983) and expanded by Felsenstein (1985b). Recent publications often use the acronym WSR for Templeton's test even though a simple sign test is used rather than the Wilcoxon signed-rank test (we use WSR to refer specifically to the Wilcoxon signed-rank test).

---

[1]We use a consistent set of acronyms throughout this paper that we define explicitly the first time they are used. When an unambiguous acronym has been widely used in the literature, we also use that acronym in this chapter. Ambiguous or misleading acronyms have been avoided. Examples: Our HT$_F$ (Box 11.6) equates to the ILD test of Mason-Gamer and Kellogg (1996) and others. Farris et al. (1995), however, defined ILD as equivalent to I$_{MF}$, a widely used acronym coined earlier by Swofford (1991) that we also use here (Box

---

BOX 11.6. "CHARACTER-BASED" TEST FOR DATA SET HOMOGENEITY (HT$_F$)

Farris et al. (1995) discuss a test analogous to the Mann-Whitney U test (Sokal and Rohlf, 1995a) for assessing the significance of consensus indices. Two data sets of size M and N are considered as one possible partition of a greater (combined) data set composed of $M + N$ characters. All possible partitions of the combined data into sets of size M and N are considered equally likely, and all distances between alternative partitions form the null distribution against which the distance between the original data sets are compared (i.e., the null hypothesis of data set homogeneity is considered equivalent to the null hypothesis that characters are randomly distributed between data sets with respect to the phylogenetic information they contain). Because the number of possible partitions can be enormous for large numbers of characters (# partitions = $[M + N]! \div M!N!$), the null distribution is determined from an appropriate number of random partitions. If a number, S, of the W random partitions provides a distance value less than the distance between the original data sets, then the type 1 error rate, $\alpha = 1 - [S \div (W + 1)]$. In other words, significance at the 5% level is determined when 95% or more of all partitions provide a distance value that is less than or equal to the distance between the original data sets. This test is extendable to more than two data sets.

The test described above can be used to assess the significance of any of several incongruence indices against the null hypothesis of homogeneity in the distribution of phylogenetic information between the data sets. The original implementation of the test and suggestion by Farris et al. (1995), however, is to use I$_{MF}$ (Box 11.5) as the distance measure, which is appropriate given that I$_{MF}$ assesses incongruence between data sets in terms of the alternative of combining the data (Box 11.5). PAUP* 4.0 implements this test (the Partition Homogeneity Test) with I$_{MF}$ as the distance measure.

As detailed by Farris et al. (1995), the computation of I$_{MF}$ is not necessary for either the original partition or each random partition in conducting the significance test. Consequently, the output from PAUP* 4.0 shows only the value of [Len A{a} + Len B{b}] (see Box 11.5 for definitions of these terms) for the original and random partitions, and the P value indicating the level of significance. When a random [Len A{a} + Len B{b}] value is larger than the original value, the corresponding I$_{MF}$ value will be smaller for that random partition. The interpretation of the significance test thus follows: The only way to decrease the partitioning of incongruent characters between data sets (if 95% or more of the random partitions give a larger [Len A{a} + Len B{b}] value) is to destroy the partitioning that existed between the original data sets.

BOX 11.7. "TOPOLOGY-BASED" TEST
FOR DATA SET HOMOGENEITY (HT$_R$)

Rodrigo et al. (1993) present a protocol that has been modified to provide a conditional combination test of homogeneity (e.g., Lutzoni and Vilgalys, 1995). Following the diagram below, this procedure first tests the null hypothesis that the observed trees are no more similar than would be expected due to chance (see Box 11.9). Although Rodrigo et al. used the partition metric (Box 11.1) to measure distance, other indices, such as those based on quartet similarities (Box 11.2) could also be used. Rodrigo et al. (1993) consider cladograms to represent topological estimates of the "true" phylogenetic tree; cladograms may vary from the phylogenetic tree for a number of reasons, including sampling error, given that cladograms are constructed from only a sample of the universe of data. If the null hypothesis of random similarity is rejected, the spectrum of possible trees for a given data set is estimated by saving the trees generated by bootstrapping (1,000 replications is recommended; cf. Lutzoni and Vilgalys, 1995). If identical topologies are recovered by the different data sets during the bootstrapping procedure, then evidence exists that the two data sets are estimating the same phylogenetic tree. The next step is to test whether random processes acting on the data sets could generate trees with the observed distance even though they estimate the same phylogeny. This is accomplished by comparing the observed distance between data sets to the distribution of distances observed between pairs of trees generated from different bootstrap replications of the same data set (one distribution per data set). If the observed distance is no greater than 95% of the bootstrapped distances, the null hypothesis of sampling error cannot be rejected. At this point, Rodrigo et al. suggest a method for combining the information from both data sets that takes into account the variability of the cladistic structure. If the null hypothesis of sampling error is rejected, a method is also proposed for pruning the data sets to remove potentially confounding information and the process is repeated. In the context of conditional combination of data, however, the choice could also be made to analyze the data sets combined or separately, respectively.

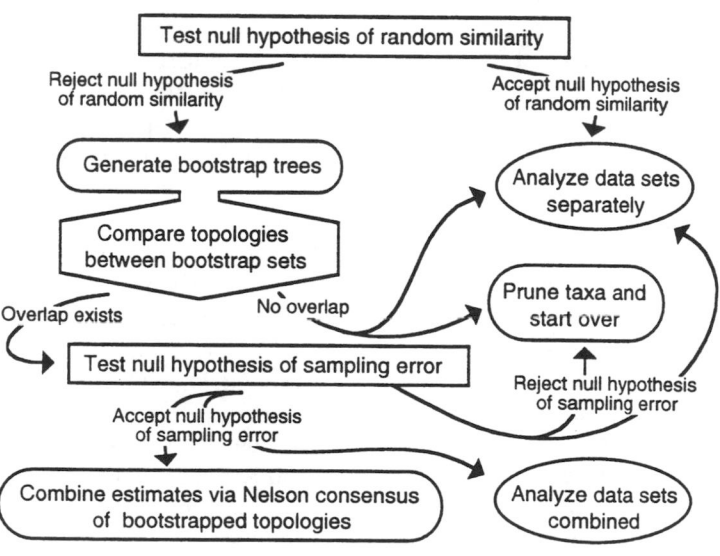

ours), that uses the sign test or Wilcoxon signed-rank tests (Siegel, 1956; Sokal and Rohlf, 1995a). This test can be used to assess the significance of topological differences between entire trees, or between individual discordant branches (Box 11.8). SLP_T tests whether character reconstructions for a given data set are significantly more parsimonious on one topology than on an alternative topology. SLP_T can be particularly valuable for identifying significant

incongruence attributable to individual taxa or clades.

All of the metrics and tests just described have strengths and weaknesses (see Boxes 11.1–11.8) that are not necessarily apparent when only formula and technical descriptions are available. In the following examples, we apply these measures to empirical data and discuss aspects of the data that influence the values obtained, what the measures can and cannot tell us about similari-

---

BOX 11.8 TEMPLETON'S (1983) SIGNIFICANTLY LESS PARSIMONIOUS TEST (SLP_T) FOR CHARACTER-STATE RECONSTRUCTIONS ON COMPETING TOPOLOGIES

Templeton (1983) suggests using the Wilcoxon signed-rank test (WSR; Siegel, 1956; Wilcoxon et al., 1970; Sokal and Rohlf, 1995a), to assess whether data provide significantly less support for a specified alternative topology compared to a most parsimonious topology. We designate this test SLP_T (significantly less parsimonious test of Templeton) for ease of reference. In data sets with homoplasy, some characters require a greater number of steps on an alternative topology compared to the most parsimonious tree, whereas other characters may require fewer steps. When the increased number of steps required by characters on the second topology is not significantly different from the decreased number of steps allowed by other characters, the topologies are considered statistically equally well supported by the data set tested. Larson (1994) succinctly outlines this method, which has since been used in the context of a test for data-set homogeneity (Mason-Gamer and Kellogg, 1996; Miyamoto, 1996; Seelanan et al., 1997).

SLP_T is easily implemented using MacClade 3.0 (see Larson, 1994) and is automated in PAUP* 4.0 (using the PSCORE/NONPARA command or by using the nonparametric pairwise comparison option in the tree scores/parsimony pull down menu). When each character that varies increases or decreases the same number of steps, the sign test (Siegel, 1956; Sokal and Rohlf, 1995a) is more appropriate than WSR. PAUP* 4.0 simultaneously calculates probabilities for both the WSR and sign tests; for both tests, at least six characters must vary to show significance (Siegel, 1956; Sokal and Rohlf, 1995a, 1995b).

SLP_T assesses whether character reconstructions for a given data set are significantly more parsimonious on one topology (usually the most parsimonious topology for that data set) than on an alternative topology. Specifically, the null hypothesis being tested is that the increase in steps shown by some characters on the alternative topology is not significantly greater than the decrease in steps shown by other characters. Thus, multiple data sets themselves are not tested directly for homogeneity with SLP_T; rather, most parsimonious topologies from one data set are tested to see if they represent statistically supported, suboptimal topologies for a second data set. When statistical suboptimality exists in at least one direction (i.e., topology Y is shown to be suboptimal by data set X even though topology X is not shown to be suboptimal by data set Y), homogeneity may be inferred.

As a test for character support for suboptimal topologies, SLP_T is versatile and provides valuable insights into the data (see Mason-Gamer and Kellogg, 1996). Individual characters can be scrutinized for their support of alternative topologies, and phylogenetic signal for relationships not shown on the most parsimonious trees can be revealed. Furthermore, any topol-

continued

continued

ogy can be tested to see if it is a suboptimal topology for a given data set. This allows specific areas of topological incongruence to be tested, from entire trees to individual branches. Thus, a topology from data set Y that differs in five branch arrangements from topology X can be broken down into five topologies that each differ in only a single branch arrangement from topology X; the significance of each branch difference with respect to data set X can then be tested individually. This allows heterogeneity to be pinpointed to specific locations in two conflicting topologies.

When multiple most parsimonious trees are recovered by a given data set, the most conservative approach is to conduct all pairwise SLP_T tests and choose the outcome that shows the greatest probability (i.e., least significance; see Miyamoto, 1996). A correction for multiple tests, such as the sequential Bonferroni method (Rice, 1989) should be employed in this circumstance. The sequential Bonferroni is conducted by ordering all probability values from lowest to highest. The desired probability for type-1 errors (e.g., 0.05) is then divided by the number (N) of probability values to give the tablewide experimental error rate. If the lowest probability is less than the data set wide experimental error rate, then that sample is deemed significant and is removed from the list. This process is repeated by subtracting one from N for each repetition (so that the divisor is always the number of samples still on the list) until a nonsignificant result is obtained, at which point all remaining samples are deemed nonsignificant. An alternative to conducting all pairwise comparisons is to use only the tree that requires the fewest additional net steps when measured on the alternative topology.

Constraint topologies, formed by using either the strict consensus tree or a strict consensus tree with all nodes with bootstrap percentages below 70% collapsed, can also be employed to make the test a more conservative estimator of well-supported differences. If the constraint is formed based on trees from data set X, then it is used as a topological constraint to search for shortest trees from data set Y that are compatible with this constraint (i.e., using the ENFORCE command in Paup* 4.0). The tree(s) recovered from this search is then used as the alternative topology for assessing the significance of data set X relationships on data set Y. Although this approach is conservative, relationships in the unconstrained portions of the recovered trees may be resolved in ways that are unique compared to the sets of most parsimonious trees from either data set. SLP_T tests using trees with such unique relationships test whether any trees compatible with a specified constraint topology are significantly less parsimonious than the most parsimonious trees for a given data set, rather than specifically whether the shortest trees from one data set are significantly less parsimonious than the shortest trees from the second data set. Constraint based trees that resolve only single discordant nodes can similarly be used to test the significance of taxonomic partitions that vary between data sets.

ties between trees and data sets, and features of data that suggest sufficient heterogeneity to warrant that data sets not be combined.

EMPIRICAL EXAMPLES

Example 1: Temperate Polemoniaceae

The process outlined in Figure 11.1 is first illustrated with chloroplast (*matK*) and nuclear (ITS) DNA sequences from 26 species representing the "temperate" clade of Polemoniaceae (Johnson et al., 1996). We use these two data sets to illustrate how quantitative congruence indices and tests of homogeneity are applied (following Boxes 11.1–11.8). The 26 species were selected from an analysis of 70 species representing the entire Polemoniaceae (Johnson, 1996); the branching pattern for these 26 species is identical in the shortest trees retrieved in analyses of the 70-taxon matrices for both *matK* and ITS. Hence, we expect a priori that the recovered trees will be highly concordant. In fact, an identical topology should be recovered by both data sets if the 44 pruned taxa are not influencing the relationships among these 26 species (e.g., in the absence of homoplasy; cf. Lanyon, 1985). This "expected" topology should match the branching pattern that exists among the 26 species in the larger 70-taxon analysis.

Separate parsimony analyses of the *matK* and ITS data sets recover multiple shortest trees (e.g., Fig. 11.2; Appendix 11.1). One of the shortest *matK* trees is identical to the "expected" topology, but the recovery of three additional topologies indicates that the narrower sampling of taxa does, in fact, alter the distribution of homoplasy. The redistribution of homoplasy in the ITS data set is even more profound; 20 most parsimonious trees, each different from both the "expected" and *matK* topologies, are recovered. Clearly, *matK* and ITS sequences from these 26 species do not convey identical phylogenetic information (compare Figs. 11.3A, B).

Partition Metric The first topological congruence measure we applied is the partition metric (PM; Box 11.1). Pairwise similarity values were computed between all trees, both within and between the *matK* and ITS data sets. The level of incongruence between trees within data sets provides a background of values against which discordance between trees from the different data sets can be compared. Because trees within data sets will likely show differing levels of similarity to trees from opposing data sets, pairwise comparisons also permit the identification of those trees that are most similar between data sets. For example, the greatest similarity between the *matK* and ITS trees for Polemoniaceae as measured with the partition metric is between trees *matK*-1 and ITS-4 (Fig. 11.2). The similarity between these trees is within the range of similarity values observed just within the ITS data set (Appendix 11.2). The trees *matK*-1 and ITS-4 (Fig. 11.2) differ in the placement of taxa 12 and 14, and the discrepancy involves only three branches giving a PM of 6 (normalized = 0.87). In con-

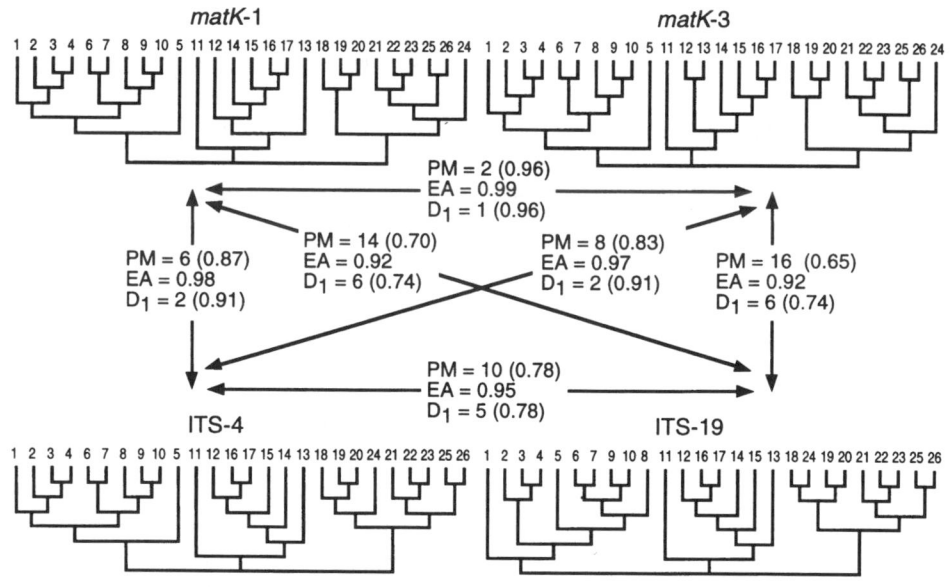

Figure 11.2. Representative shortest trees obtained from parsimony analysis of *matK* and ITS data for 26 taxa from Example 1 (Polemoniaceae). Numbers representing taxa are associated with numbers listed to the right of the taxon names in Figure 11.3C. Values for the topological congruence indices PM (partition metric), EA (explicitly agree metric) and D_1 (agreement subtree metric) are given for all pairwise comparisons as indicated by arrows. Normalized similarity values for PM and D_1 are given in parentheses.

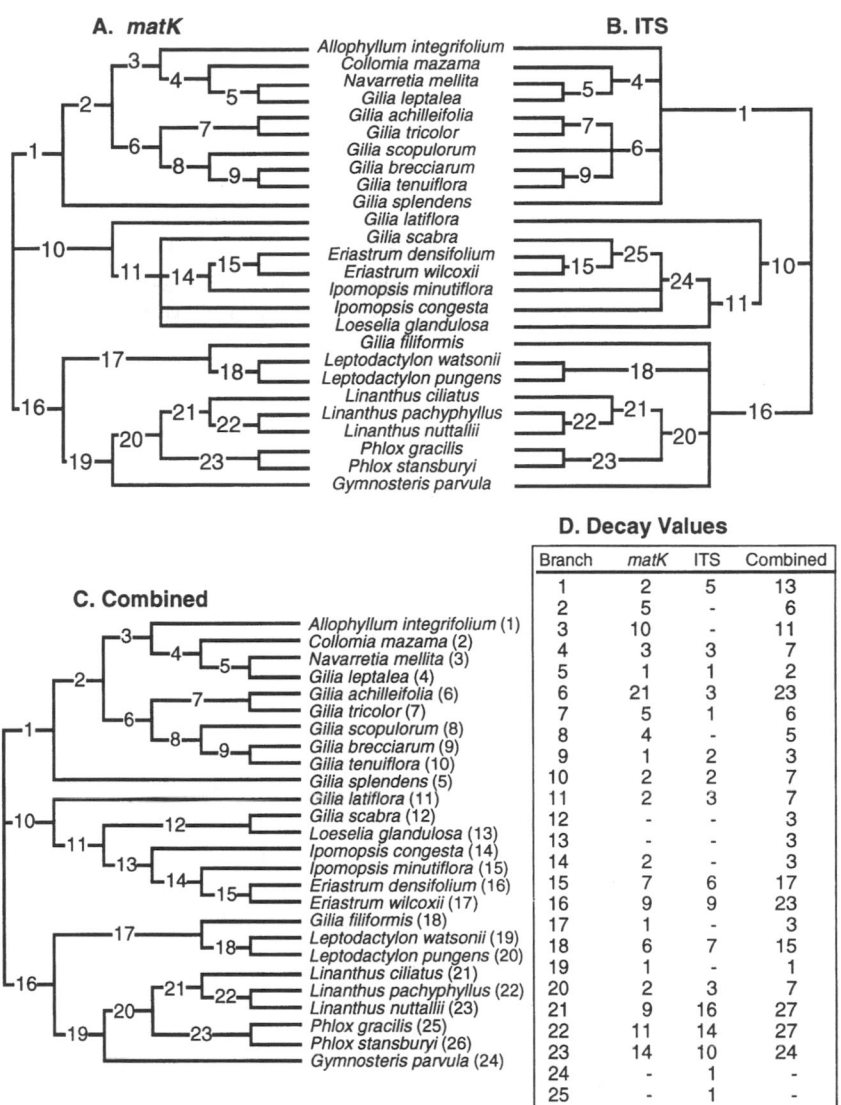

Figure 11.3. Phylogenetic trees for Example 1 (Polemoniaceae) with decay values for all clades. Numbers on branches identify identical clades among trees. *A.* Strict consensus of four shortest *matK* trees. *B.* Strict consensus of 20 shortest ITS trees. *C.* Single shortest tree obtained from combining *matK* and ITS data. Numbers to right of taxon names correspond to numbers shown in place of taxon names in Figure 11.2. *D.* Table of decay values for each tree; branch labels correspond to numbered branches on trees.

trast, trees ITS-4 and ITS-19 differ in the placement of taxa 1, 5, 8, 14 (or 15), and 18 (or 24) and the discrepancy involves five branches, giving a PM of 10 (normalized = 0.78). However, that some ITS trees are more similar to some *matK* trees than to other ITS trees (in terms of PM) does not in itself suggest the two data sets are congruent.

Penny et al. (1982) were among the first to use the partition metric and test whether greater similarity exists between the observed trees than would be expected by chance. With the 26-taxon Polemoniaceae data set, comparisons with random trees (see Box 11.9 for calculation of random trees) indicate that the least similar *matK* and ITS trees are still above the range of

BOX 11.9. DISTRIBUTIONS OF RANDOM TREES

The probability of observing a given value of any tree comparison metric is finite and dependent on the number of taxa included in the study. Depending on the metric used, whether the trees are rooted or unrooted, and whether the trees are fully binary may also effect the probability. For small numbers of taxa, exact probabilities have been determined for indices such as the partition metric (Box 11.1; Hendy et al., 1984). Probability distributions for larger numbers of taxa, and for other metrics, can be determined by applying the metric to pairs of trees drawn at random from an appropriate tree distribution (e.g., Simberloff, 1987; Slowinski, 1990; Steel and Penny, 1993).

Several random tree distributions have been proposed, each with different assumptions depending on the constraints placed on the kinds of trees included in the distribution (cf. Page, 1991; Maddison and Slatkin, 1991; Steel and Penny, 1993). Common distributions (following Steel and Penny, 1993) include:

1. D_{all}: equal probability of all trees, both binary and nonbinary.
2. D_{bin}: equal probability of all binary trees.
3. D_{tip}: binary trees generated from a Markovian model of branching; favors balanced or symmetrical trees over asymmetric or pectinate trees.

MacClade 3.0 and COMPONENT 2.0 will generate trees from D_{bin} and D_{tip}.

Steel and Penny (1993) discuss a fourth distribution, D_r, that we used throughout this chapter. Trees from D_r are obtained by randomly permuting character states within characters (cf. Archie, 1989) and obtaining a tree from the randomized data via parsimony. The D_r distribution is appropriate for testing whether trees recovered from different data sets are more similar than trees generated from data sets that contain no common hierarchical information; thus, D_r is well suited for testing the null hypothesis of random or chance similarity. Additionally, trees drawn from D_r tend to favor topologies similar in shape to those actually observed, and a fraction of the branches are often unresolved—an expected feature of "real" topologies as the number of taxa increases (Day, 1986). Thus, D_r often provides a more conservative and realistic model of chance similarity.

The shuffle option of MacClade will produce random permutations of a data set, but a considerable amount of time is required to generate a large number of permutations. Below we offer a quick solution based on just five permutations of each data set; distributions provided by this quick method and the more exact method (where one tree is generated per permutation) show no distinguishable differences based on exploratory analyses we conducted:

1. Open data set in MacClade and remove all invariant characters.
2. Permute character states within characters using the SHUFFLE option; save shuffled output and repeat to generate five randomly permuted data sets.
3. Open the five shuffled data sets in PAUP and merge all matrices into one data file using the INTERLEAVE option; save "merged" data file.
4. Execute "merged" data file, conduct jackknife analysis with 80% character deletion, quickswap, save trees to file; perform 1,000 replications.
5. Repeat for second data set, compare trees between tree files with COMPONENT or PAUP.

similarity observed between pairs of random trees (Appendix 11.2). Thus, despite differences, the *matK* trees are more similar to the ITS trees than expected merely from chance. This is not surprising, especially among large data sets with presumed shared ancestry, but testing the null hypothesis of chance similarity becomes increasingly important with small data sets because the probability of chance similarity increases as the number of taxa decreases (Rodrigo et al., 1993). Demonstrating topological similarity greater than chance expectations is insufficient, however, for determining that the character-state distributions in the data sets are homogeneous.

Explicitly Agree Metric   The second measure we used is the explicitly agree similarity metric (EA; Box 11.2). The 26 species analyzed represent three well-supported lineages of temperate Polemoniaceae (Johnson et al., 1996; Porter, 1997), but no species outside of these lineages were included because it was desirable to compare relationships irrespective of the location of the root (cf. Penny and Hendy, 1985a; trees were rooted for display purposes in Figs. 11.2–11.3 with a basal polytomy leading to the three primary lineages of temperate Polemoniaceae). Because an outgroup was not specified, we used quartets, rather than triplets, to determine EA similarity values (the pros and cons of triplets vs. quartets are provided in Box 11.2). EA indicates greater similarity between the *matK* and ITS trees than does PM (Appendix 11.2), in part because of the greater discriminating power of quartets and triplets over the partition metric (Steel and Penny, 1993). That is, there are 14,590 possible quartets for 26 taxa and only 44 possible partitions. Therefore, minor differences in the placement of taxa, such as the rearrangement of taxa 12 and 14 between trees *matK*-1 and ITS-4 (Fig. 11.2) involve a smaller proportion of the total quartet similarity value than the proportion of the total PM value (Appendix 11.2). As with PM, EA values between *matK* and ITS trees are above the range of similarity values for random trees; however, the distribution of EA values from the random trees is bell shaped (rather than leptokurtic as is the PM curve), and reaches higher similarity scores

than the distribution of PM values (Fig. 11.4; Appendix 11.2). In this example, both PM and EA values provide similar indications of topological congruence between data sets because the trees are not highly asymmetrical (see Box 11.2) and most discrepancies involve rearrangements among neighboring taxa; thus, factors associated with tree structure that can lower similarity values are not at play in this comparison.

Agreement Subtrees   The greatest agreement subtree metric (D_1; Box 11.3) applied to the Polemoniaceae data provides a view of similarity between the *matK* and ITS trees that is different from, yet complementary to, PM and EA. Similar to PM, a single incremental increase in the distance between trees lowers the normalized D_1 similarity score more than it does EA similarity values. D_1 is insensitive to the location of a single taxon discrepancy, however, and thus provides a more telling index of similarity than may be afforded by PM. Whereas both D_1 and EA values are based on similarities in topological substructure, D_1 values are easier to interpret. For example, the 97.6% EA similarity observed between Polemoniaceae trees *matK*-1 and ITS-4 (Fig. 11.2) results from differences in 350 of 14,590 quartets, but it is not immediately obvious which quartets these are (although they likely involve taxa 12 and 14). The 91.3% D_1 similarity observed between the same two trees, in contrast, results from the greatest agreement subtree being formed by pruning taxa 12 and 14. The pruning of two taxa is measured directly by D_1, and the effect of such removal is readily seen on the greatest agreement subtree. Considering the Polemoniaceae data, D_1 values between *matK* and ITS trees are again above the range reached by the distribution of random trees (Appendix, 11.2), and, like EA, these values are bell shaped and reach higher similarity scores than do values for PM (Fig. 11.4).

Consensus Tree Indices   Colless's consensus fork index (CI_C), Rohlf's consensus information index (CI_1), and the index of topological compatibility (TI) are related to one another (see Box 11.4) and, consequently, are discussed together here. Each of these indices is sensitive to

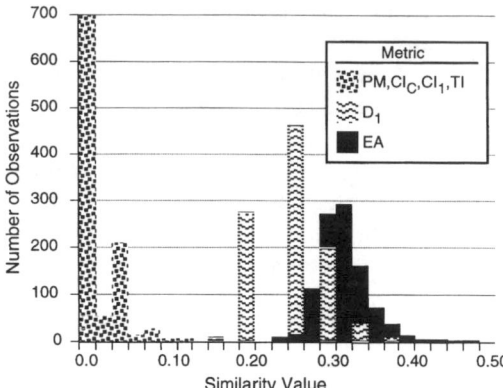

Figure 11.4. Distribution of topological similarity values from 1,000 pairs of random trees for Example 1 (Polemoniaceae).

different aspects of tree similarity (Box 11.4). Thus, although CI_C and CI_1 values are similar, and CI_C = TI in this example using Polemoniaceae, such is not always the case. The similarity in CI_C and CI_1 values suggests that polytomies formed on pairwise consensus trees are relatively minor with respect to the number of phylogenetic statements that can be made from each consensus tree. Because these measures are all affected by the decrease in resolution observed when two trees are compared or combined via some consensus measure, the choice of consensus measure affects the magnitude of the index. Throughout this chapter, we used strict consensus trees in the calculation of CI_C and CI_1 (TI is, by definition, based on a strict consensus comparison). However, semistrict (combinable component; Bremer, 1990) consensus trees may be more desirable in circumstances in which polytomies exist.

Because CI_C was calculated from strict consensus trees and each of the *matK* and ITS trees for Polemoniaceae is fully resolved, CI_C = TI in these comparisons. Additionally, when normalized (as was done in this study), the partition metric can be defined as the proportion of shared clades between two trees, which is equivalent to the proportion of resolved clades when a strict consensus tree is formed from those two trees. Thus, CI_C = TI = normalized PM in this example (compare values, Appendix 11.2). In other circumstances (see additional examples

below), CI_C can be slightly higher than the normalized PM because declaring an outgroup automatically resolves one branch on the consensus tree. Conversely, CI_C can be lower than the normalized PM if one or both trees have polytomies. The identity of PM and CI_C in this example gives rise to identical distributions of similarity values between random pairs of trees (Appendix 11.2). Again, however, as with PM, EA, and D_1, the evidence of greater similarity than chance expectations is insufficient for declaring the data sets homogeneous.

Character Congruence   The six topological indices calculated above indicate that, although identical *matK* and ITS trees were not recovered, at least some of the trees in each data set are very similar. Trees *matK*-1 and ITS-4 (Fig. 11.2) are the most similar trees between the two data sets, differing only in the placement of two taxa (D_1 = 2) that are attached to the tree in close proximity (PM = 6). These measures do not, however, address the strength of support exhibited by the data themselves for the discordant relationships. To assess *character incongruence,* we applied I_M and I_{MF} (Box 11.5), the data set homogeneity tests HT_F and HT_R (Boxes 11.6–11.7), and the test for significant departure from parsimony, SLP_T (Box 11.8).

As with most data sets, the *matK* and ITS data sets for Polemoniaceae contain homoplasies in the distribution of character states on the most parsimonious trees. Ninety-one homoplasious steps are required to explain the *matK* data on the most parsimonious trees recovered from analyses of the *matK* data (Appendix 11.1). Similarly, 224 homoplasious steps are required to explain the ITS data on the most parsimonious trees recovered from analyses of the ITS data (Appendix 11.1). To explain any tree other than the most parsimonious trees associated with a given data set, additional steps, or homoplasies, are required. For example, mapping the ITS data on tree *matK*-3 (Fig. 11.2) requires 572 steps, or two additional homoplasies, compared to the shortest ITS trees, whereas mapping the *matK* data on tree ITS-4 requires 439 steps or eight additional homoplasies compared to the shortest *matK* trees. These additional steps are minimally based on the decay values required to disrupt any monophyletic groups recognized on the trees recovered by a given data set that are not recognized on the trees recovered by the alternative data set (Farris et al., 1995). That is, the length of a tree from data set X when measured on data set Y will be minimally as long as the length of the most parsimonious trees from data set Y plus the largest decay value associated with any clades unique to data set Y with respect to the tree from data set X.

Based on the logic described above, I_M measures incongruence between data sets in terms of the number of extra steps required to explain the trees recovered by the *opposing data sets* (Box 11.5). We used the lengths of the trees requiring the fewest additional steps (i.e., *matK*-3 and ITS-4) in calculating I_M, although average values across all recovered topologies could also be used (Swofford, 1991). Compared to the total homoplasy observed in the *matK* and ITS data sets (91 + 224 = 315 steps, see above and Appendix 11.1), the sum of the number of extra steps required to explain the trees *matK*-3 and ITS-4 is quite small (2 + 8 = 10 steps). Accordingly, the incongruence between data sets as measured by I_M is modest: $I_M = \{[(315 + 10) - 315] \div (315 + 10)\} \times 100 = 3.08\%$. This value provides our first quantitative indication that not only are the topologies recovered from analyses of *matK* and ITS data similar, but that the degree of support within each data set for the discordant relationships is also relatively weak.

Whereas I_M measures incongruence relative to topologies supplied by the opposing matrix, I_{MF} assesses incongruence between data sets relative to the topology resulting from analysis of the *combined data sets* (Box 11.5). For the combined Polemoniaceae data, a single most parsimonious tree is recovered (Fig. 11.3C) that is equivalent to the "expected" tree retrieved from analyses of the larger 70-taxon data sets from which these 26 taxa were chosen (see above in Example 1). The topology for the combined data sets requires 317 homoplasious steps, compared to the 315 steps summed from the independent analyses of the *matK* and ITS data (Appendix 11.1). Thus, $I_{MF} = [(317 - 315) \div 317] \times 100 = 0.63\%$. In this example, the topology resulting from the combined data sets (Fig. 11.3C) is identical to one of the four shortest *matK* trees

(*matK*-3; Fig. 11.2). Consequently, there are no extra steps of homoplasy required by the *matK* data set to explain the topology based on the combined data sets. The ITS data require only two additional steps to explain the topology of the combined data sets. Note that I_{MF} provides a smaller estimate of incongruence between data sets than does I_M, because the distance to the topology of the combined data sets (as measured by summing additional homoplasies) will always be less than the sum of the shortest distance to the topologies recovered by the opposing data sets when the topologies generated from the two data sets are unique.

Homogeneity Test HT$_F$ The low I_{MF} value for Polemoniaceae suggests homogeneity between the *matK* and ITS data sets, but it is desirable to associate this I_{MF} value with a *probability* to determine the significance of the result. Farris et al. (1995) suggest such a test (HT$_F$; Box 11.6) based on random repartitioning of the combined data. HT$_F$ estimates a data-set-dependent probability level for I_{MF} values for testing the null hypothesis that characters are effectively random in distribution between data sets with respect to their phylogenetic content. To reject the null hypothesis, we set the type I error rate at 5%, as is common practice (i.e., when $P < 0.05$, we reject the null hypothesis and conclude that the data sets are significantly heterogeneous). The application of HT$_F$ to Polemoniaceae resulted in 64 of 499 random partitions of the combined data with I_{MF} values smaller than the 0.63% I_{MF} value observed among the original partition defined by the two DNA regions. The probability (P) of observing an I_{MF} of 0.63% or smaller, then, is $1 - [64 \div (499 + 1)] = 0.872$ (Box 11.6). Because this value of P is greater than 0.05, we conclude that the data sets are homogeneous. This test suggests that, although the *matK* and ITS data sets do not contain completely congruent phylogenetic information, the amount of incongruence between the data sets is trivial and that analysis of a combined data set is appropriate.

Homogeneity Test HT$_R$ HT$_F$ can be considered a "character-based" test in that incongruence between data sets is measured in terms of additional steps of character evolution needed to explain other topologies. An alternative test for data set homogeneity (Rodrigo et al., 1993; Box 11.7; HT_R) is "topology based" in that data sets are considered heterogeneous when the sets of all topologies ("spectrum" in Rodrigo et al., 1993) recovered from bootstrapping each data set do not overlap. The distinction between a character-based and topology-based test is fuzzy, however, because both HT_F and HT_R rely on the phylogenetic distribution of characters states in the data matrices; both tests use alternative topologies for assessing incongruence, and both tests account for the strength of discordant relationships in assessing the homogeneity of data.

Following Box 11.7, HT_R was applied to Polemoniaceae using EA as the measure of topological similarity. As discussed above, EA values between the *matK* and ITS trees are above the range of EA values for random trees (Appendix 11.2), so we immediately can reject the null hypothesis of random similarity. We next bootstrapped the *matK* and ITS data sets (1,000 replications with no maxtree limit) and saved all of the trees generated by each replication. For the *matK* data, 9,046 bootstrapped trees were recovered; these represent the spectrum of topologies that can be derived from *matK* simply due to sampling error. For the ITS data, 6,037 bootstrapped trees were recovered. Comparing these *matK* and ITS trees (using PAUP* 4.0 command GETTREES/MODE = 2) reveals overlap in the tree spectra because 196 trees representing 27 unique topologies are shared between the two data sets. The overlap in bootstrapped topologies for *matK* and ITS is the critical step in this test because it indicates that the discordant relationships are *not* strongly supported. Although bootstrap values themselves are not compared (it may, however, be useful to do so in some cases; see Mason-Gamer and Kellogg, 1996), if overlap in bootstrapped topologies occurs, discordance is *not* strongly supported by the data sets. For example, a discordant clade showing 100% bootstrap support by one data set is not resolved differently through character resampling from that data set; thus, this discordance is strongly supported by the data. Even a discordant clade receiving low bootstrap support (e.g., 60%) by one data matrix may contribute to the lack of overlap in boot-

strapped topologies if the taxa are never resolved in that same arrangement in the bootstrapped trees from the second data set. Thus, this method is sensitive to both the strength of support for discordant relationships by the tree's own data set and the lack of support by that data set for discordant relationships recovered by the other data set.

Returning to the Polemoniaceae example, the overlap in bootstrapped topologies suggests that the *matK* and ITS data sets are estimating the same phylogeny and differ only due to character sampling error. To test the hypothesis of sampling error directly, EA similarity values were measured between 1,000 pairs of trees derived from 2,000 bootstrap replications of the *matK* data; this measurement was repeated for the ITS data (this can be done quickly either by selecting the FASTBOOT option or turning MULPARS off in PAUP). The distribution of EA values from these bootstrapped trees provides a null distribution of values possible within each data set when sampling error can explain the different topologies. Comparison of the range of EA values initially calculated between *matK* and ITS trees (Appendix 11.2) with the distributions from bootstrapped trees provides probability estimates that range from 18% to 94%. Probability values below 5% are needed to reject the hypothesis of sampling error. Consequently, as with HT_F, HT_R suggests that we proceed with an analysis of a combined *matK* + ITS data set.

SLP_T The final test we applied to Polemoniaceae is the significantly less parsimonious test (SLP_T) proposed by Templeton (1983; Box 11.8; see also Larson, 1994), that uses the sign test or Wilcoxon signed-rank test (Siegel, 1956; Sokal and Rohlf, 1995a). SLP_T assesses whether characters in a data set statistically support specific topological arrangements not found in the set of shortest trees recovered for that data set. Following the procedure outlined in Box 11.8, we first used the ITS data set to test whether any of the four shortest *matK* trees is significantly less parsimonious than the 20 shortest ITS trees. All 80 pairwise tree comparisons between data sets were conducted. Probability values ranged from 0.37 to 0.82, indicating that the ITS data statistically support the *matK* topologies (e.g., Appendix 11.3). Correcting for multiple comparisons using the sequential Bonferroni technique (Rice, 1989; Miyamoto, 1996) only makes these values more conservative and does not alter the interpretation. The *matK* data were next used to test whether any of the 20 ITS trees were significantly less parsimonious than the four *matK* trees. The 80 probability values (ranging from <0.001 to 0.04, Appendix 11.3), all indicate, in contrast, that the ITS trees are significantly less parsimonious than the *matK* trees when measured with the *matK* data. Correcting for multiple comparisons with the sequential Bonferroni technique, however, found that trees ITS-4 and ITS-14 were not significantly less parsimonious than any of the *matK* trees. These results suggest that the ITS data contain substantial support for relationships proposed by the *matK* data, but the *matK* data provide only minimal support for relationships proposed by the ITS data. Because tree *matK*-3 (Fig. 11.2) is equivalent to the tree recovered by the combined data, it is clear that both data sets statistically support the topology recovered by the combined data, and analysis of a combined data set is appropriate.

Combined Analysis   Quantitative measures of congruence reveal that the *matK* and ITS data sets examined in this example are highly congruent. We therefore proceed with analysis of a combined *matK* + ITS data set. We have already shown that a single tree is recovered from such an analysis; this tree has the topology "expected" based on relationships recovered in the larger (70-taxon) analysis (Fig. 11.3C). In this example, discordance between data sets can confidently be attributed to sampling error. The combination of the ITS and *matK* data restricted the ambiguity that caused multiple most-parsimonious resolutions in both data sets, and a single tree with the expected topology was recovered (Fig. 11.3C). In addition to fewer phylogenetic hypotheses (which in itself is not a reason to favor analyses of combined data sets), increased support is provided for most clades through the synergism of the two combined data sets (see also Doyle et al., 1994; Olmstead and Sweere, 1994; Soltis et al., 1996b; Sullivan, 1996). For example, of the 23 clades resolved on the combined tree, 17 show larger decay values

than the sum of the decay values for the corresponding branches in the separate analyses (Fig. 11.3D; compare Figs. 11.3A–11.3C). Thus, combining homogeneous data sets increases the number of characters and decreases the effects of sampling error on estimating phylogeny.

Example 2: Phlox and Related Genera

Phlox, Linanthus, Leptodactylon, Gymnosteris, and several species of *Gilia* are shown by *matK* and ITS sequences to form a strongly supported clade within temperate Polemoniaceae (Johnson et al., 1996; Porter, 1997). Sequences of *matK* and ITS from 14 species representing this clade, as well as the outgroup *Gilia splendens,* are used to illustrate additional aspects of the conditional combination approach.

Following the outline in Figure 11.1, separate analyses of the *matK* and ITS matrices each recovered a single, yet different, phylogenetic tree (Figs. 11.5A, B; Appendix 11.1). Comparisons of these trees show that both DNA regions recognize two large clades: (1) the *Gilia filiformis* clade, with *Gilia filiformis* as sister to *Leptodactylon* and two species of *Linanthus;* and (2) the *Gymnosteris* clade, with *Gymnosteris* as sister to *Phlox* and four species of *Linanthus* (Figs. 11.5A, B).

In the *Gilia filiformis* clade, relationships among *Linanthus* and *Leptodactylon* are resolved differently by the two DNA regions, and it is difficult to assess any similarities by visual comparison. In the *Gymnosteris* clade, relationships are resolved identically with the exception that *matK* places *Linanthus breviculus* as sister to *L. pachyphyllus,* whereas ITS sequences show *L. nuttallii* as sister to *L. pachyphyllus.* For all six quantitative topological measures (Appendix 11.2), observed levels of similarity between the *matK* and ITS trees are greater than chance expectations. The EA similarity metric (Box 11.2) provides the greatest measure of similarity (above 97%), whereas CI_C provides the lowest estimate (69%). An odd number of branch contractions/decontractions is necessary to convert the *matK* tree into the ITS tree (i.e., PM = 7) because the *matK* tree has a polytomy whereas the ITS tree does not. Likewise (in contrast to example 1), PM ≠ CI_C ≠ TI because the

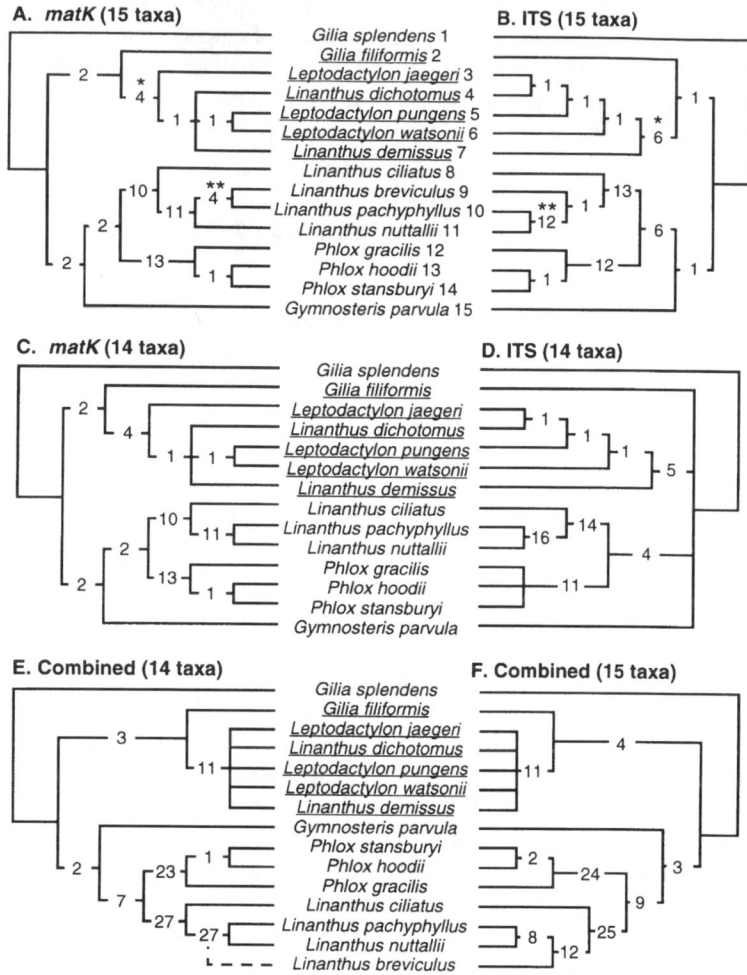

Figure 11.5. Phylogenetic trees for Example 2 (*Phlox* and related genera). Numbers on branches are decay values for each branch. Taxon names for taxa belonging to the *Gilia filiformis* clade are underlined, whereas taxon names for taxa belonging to the *Gymnosteris* clade are plain, as is the outgroup taxon, *Gilia splendens*. * and ** represent nodes that were switched between trees to test the significance of the relationships shown using SLP$_T$. A. Single shortest *matK* tree. B. Single shortest ITS tree. C. Single shortest *matK* tree after pruning *Linanthus breviculus*. D. Strict consensus of three shortest ITS trees after pruning *Linanthus breviculus*. E. Strict consensus of four shortest trees obtained from combining *matK* and ITS sequences after pruning *Linanthus breviculus*, with *L. breviculus* subsequently grafted back onto the tree in a position that reflects agreement between ITS sequences and morphology. F. Strict consensus of four shortest trees obtained from combining *matK* and ITS sequences before pruning *Linanthus breviculus*.

matK tree has a polytomy and both trees are rooted by the outgroup (no outgroup was used in example 1). The D$_1$ metric indicates that the removal of three taxa will result in a greatest agreement subtree. Scrutiny of the most parsimonious trees (Figs. 11.5A, B) shows that the removal of either *Linanthus breviculus, L. pachyphyllus,* or *L. nuttallii* will result in identical relationships among the taxa in the *Gymnos-*

teris clade. The other taxa requiring pruning to form the greatest agreement subtree are *Leptodactylon jaegeri* and *Linanthus dichotomus* from the *Gilia filiformis* clade (identifying these as the taxa to be pruned was facilitated by the AGREE option in PAUP* 4.0).

Although all of the quantitative measures above indicate greater than random similarity in the trees recovered by *matK* and ITS sequences

(Appendix 11.2), the application of HT_F (Box 11.6), HT_R (Box 11.7), and SLP_T (Box 11.8) suggests that the *matK* and ITS data sets are significantly heterogeneous. With HT_F, the probability of observing the 5.31% I_{MF} value is significant at the 5% level (P = 0.034; Appendix 11.1). With HT_R, 1,000 bootstrap replications recovered 5,577 trees from the ITS data set and 4,676 trees from the *matK* data set, yet no trees were shared between these sets of bootstrap topologies.

To localize the source of heterogeneity, SLP_T was applied to specific differences between the shortest trees recovered from each data set (Figs. 11.5A, B). First, the ITS data set was used to test whether the relationships revealed in the *matK* tree were significantly less parsimonious than the relationships revealed in the ITS tree. The topological differences in the *Gilia filiformis* clade were tested together by changing the relationships in this clade on the ITS tree (Fig 11.5B, marked with *) to reflect the relationships found in the *matK* tree (Fig. 11.5A, marked with *). Three characters required a total increase of four steps on the modified tree, but an additional three characters each decreased a single step, giving an overall insignificant increase in steps (P = 0.82; Appendix 11.3). The topological difference involving *Linanthus breviculus, L. pachyphyllus,* and *L. nuttallii* (Figs. 11.5A, B, marked with **) was similarly tested. A total of 13 characters each required a single step increase by the ITS data set to support the *matK* arrangement for these three taxa, whereas a single character decreased one step. This result is highly significant (P = 0.002; Appendix 11.3) and pinpoints the source of heterogeneity between the ITS data set and the *matK* topology to the differences in relationships among *Linanthus breviculus, L. pachyphyllus,* and *L. nuttallii* (compare Figs. 11.5A, B).

Significance for either of these topological differences could not be determined with the *matK* data because at least six differences (six characters that vary in steps) must exist to show significance at the $\alpha = 0.05$ level with SLP_T (Box 11.8). It is noteworthy, however, that two characters increase in one step each to map the *matK* data on the relationships in the *Gilia filiformis* clade revealed by ITS, and four characters increase one step each to map the *matK* data

on the relationships among *Linanthus breviculus, L. pachyphyllus,* and *L. nuttallii* suggested by ITS (Appendix 11.3). The *matK* data thus agree with the ITS data and indicate that the cause of heterogeneity is due to differences in relationships among *L. breviculus, L. pachyphyllus,* and *L. nuttallii.*

This same conclusion is suggested by comparing decay values (Fig. 11.5). Branches indicating the very different topologies within the *Gilia filiformis* clade are all weakly supported by both data sets, but the single discordance in the *Gymnosteris* clade is strongly supported by both data sets. Thus, the incongruence observed in the *G. filiformis* clade appears to be the result of sampling error. Consequently, removing *Leptodactylon jaegeri* or *Linanthus dichotomus* from the data set and repeating the analyses does not remove the significance of the heterogeneity (at the $\alpha = 0.05$ level) for HT_F, or result in overlap in bootstrap topologies with HT_R (Table 11.1). In contrast, the discordance observed in the *Gymnosteris* clade does appear to reflect different branching histories in the two DNA regions (different evolutionary rates alone seem unlikely to produce the magnitude of character incongruence observed). Removing either *Linanthus nuttallii, L. pachyphyllus,* or *L. breviculus* results in nonsignificant heterogeneity with both HT_F and HT_R (Table 11.1).

Thus, analysis of a combined *matK* + ITS data set is appropriate if either *L. nuttallii, L. pachyphyllus,* or *L. breviculus* is first removed from the data matrices (Rodrigo et al., 1993; Box 11.7). Pruning *L. pachyphyllus* results in the lowest measure of incongruence between data sets (see I_{MF} values; Table 11.1), but pruning *L. breviculus* results in the largest probability value for I_{MF} (based on the application of HT_F), as well as the largest overlap in shared trees from bootstrap replications using HT_R (Table 11.1). An objective argument could thus be made for pruning either of these taxa. Ultimately, we relied on morphology and pruned *L. breviculus* because a possible cause of the observed incongruence involves chloroplast capture. That is, *L. nuttallii* and *L. pachyphyllus* are morphologically similar perennials that were considered conspecific until 20 years ago (Patterson, 1977), whereas *L. breviculus* and

Table 11.1. Results of tests of homogeneity for *Phlox* and related genera (example 2) before and after pruning individual taxa. Probabilities for the observed I_{MF} values were derived using HT_F (Box 11.6) with 499 random partitions. The number of trees recovered from 1000 bootstrap replications, and the number of unique topologies among those trees, are given for the *matK* data set, the ITS data set, and the intersection of the tree files from both data sets following HT_R (Box 11.7).

Data set	HT_F I_{MF}/Probability	HT_R Number of trees/Number of unique topologies		
		matK	ITS	intersection
All 15 taxa	5.31/0.034	4676/917	5577/2114	0/0
14 taxa (-*Leptodactylon jaegeri*)	3.81/0.046	3348/880	4374/621	0/0
14 taxa (-*Linanthus dichotomus*)	5.77/0.018	4470/1118	4176/863	0/0
14 taxa (-*Linanthus nuttallii*)	1.94/0.790	5047/2059	5892/1275	1/1
14 taxa (-*Linanthus pachyphyllus*)	1.89/0.790	5513/2129	4596/857	6/6
14 taxa (-*Linanthus breviculus*)	2.15/0.902	5624/1187	4526/1330	454/10

L. ciliatus are annuals. Thus, general agreement exists between morphology and relationships within the *Gymnosteris* clade as reconstructed by the ITS data, but not between morphology and the *matK* data.

The analysis of a combined data set of 14 taxa recovered four most parsimonious trees (Appendix 11.1; Fig. 11.5E) that resolve relationships within the *Gymnosteris* clade identically. However, in the *Gilia filiformis* clade, the weakly supported discordance in relationships among the *Linanthus* and *Leptodactylon* species seen in the separate analyses (Figs. 11.5C, D) still exists; hence, the strict consensus tree shows a polytomy among these five species (Fig. 11.5E). Combining data in this case, then, neither results in one data set swamping the other nor provides sufficient information to overcome the effects of sampling error. Furthermore, because *L. breviculus* was pruned before the analysis of the combined data sets was conducted, we must either graft this species back onto the tree, or be satisfied with a tree that does not include it. If grafting is employed (see Rodrigo et al., 1993), placing *L. breviculus* as sister to *L. pachyphyllus-L. nuttallii* (Fig. 11.5E) may be the most appropriate solution when both molecular and morphological data are considered. Although the tree constructed by grafting *L. breviculus* as sister to *L. pachyphyllus* and *L. nuttallii* is identical to the tree recovered by the analysis of the combined, unpruned data (Fig. 11.5F), such agreement is neither expected nor an indication that conduct-

ing a combined analysis a priori will always lead to reliable phylogenetic hypotheses.

Example 3: Boykinia Group

The *Boykinia* group (Saxifragaceae s. s.) is a strongly supported clade comprising six genera, most of which contain fewer than five species (species number given in parentheses): *Bolandra* (3), *Boykinia* (7), *Jepsonia* (3), *Suksdorfia* (2), *Sullivantia* (3), and *Telesonix* (2) (Gornall and Bohm, 1985; Soltis, 1987; Soltis et al., 1993, 1996a). Qualitative assessments of congruence among trees have been conducted for the *Boykinia* group using data sets with unequal taxon sampling (Soltis et al., 1996a). Here, we assess congruence quantitatively using ITS sequences and cpDNA restriction site data for 13 taxa representing all six genera of the *Boykinia* group (with *Tanakaea radicans* and *Leptarrhena pyrolifolia* as outgroups).

Considerable length variation attributed to multiple insertion/deletion events (indels) exists in the ITS sequences of the *Boykinia* group (Soltis et al., 1996a). Consequently, alignment of these sequences was problematic, and 18 alignments were generated using CLUSTALW (Thompson et al., 1994) by employing a variety of gap-open and gap-extension penalties (see Soltis et al., 1996a for details). Because parsimony analyses of all 18 alignments recovered similar topologies, we arbitrarily selected one of the alignments to use as our primary ITS data set for the congruence analyses. However, tests for homogeneity were conducted between the cpDNA restriction site data set and all 18 ITS alignments to assess the impact of ambiguous alignment.

Analyses of the *Boykinia* group recovered one and two most parsimonious trees, respectively, from the cpDNA restriction site and ITS sequence data sets (Appendix 11.1; Fig. 11.6). These trees differ in the placement of several taxa. cpDNA restriction sites place *Sullivantia* as sister to a clade comprising *Suksdorfia, Bolan-*

dra, and *Boykinia,* whereas ITS sequences place *Sullivantia* as sister to *Telesonix* and *Jepsonia* (Fig. 11.6). Other discordant relationships involve the placement of *Suksdorfia ranunculifolia, Boykinia aconitifolia,* and the two accessions of *Boykinia major* (compare Figs. 11.6A, B). Individually, each of these discrepancies is minor (i.e., the alternative relationship suggested by the opposing data set usually involves just one or two branches); however, the combination of all discordant relationships results in substantially

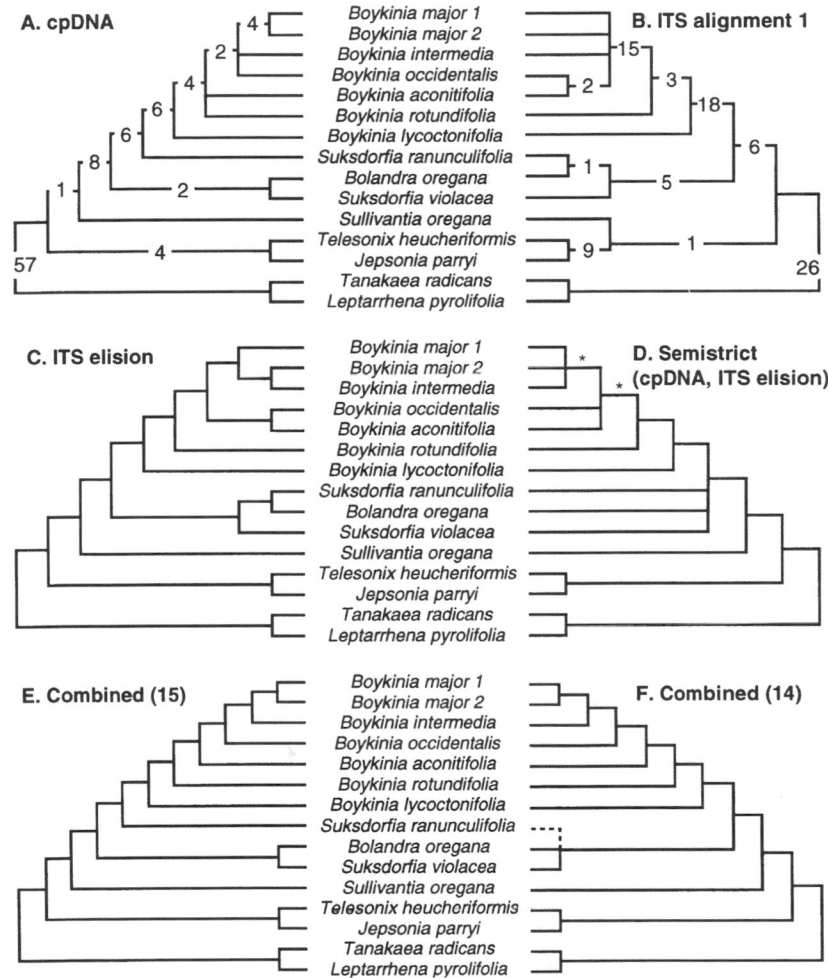

Figure 11.6. Phylogenetic trees for Example 3 (the *Boykinia* group). Numbers on branches for trees A and B are decay values. *A.* Single shortest cpDNA restriction site tree. *B.* Strict consensus of two shortest ITS trees. *C.* Single shortest ITS tree after eliding all 18 ITS sequence alignments (see text). *D.* Semistrict consensus of trees A and C; branches not found on strict consensus tree are indicated by *. *E.* Single shortest tree obtained by combining elided ITS and cpDNA data sets (see text). *F.* Single shortest tree obtained by combining elided ITS and cpDNA data sets after pruning *Suksdorfia ranunculifolia. S. ranunculifolia* was subsequently grafted back in a position that reflects agreement between ITS sequences and morphology.

lower similarity indices compared to examples 1 and 2 involving Polemoniaceae. The D_1 metric indicates that a third of the taxa must be pruned to form the greatest agreement subtree and the normalized PM = 0.5, indicating that half of all partitions are unique to either the ITS or cpDNA restriction site trees (Appendix 11.2). The consensus indices CI_C, CI_1, and TI are correspondingly low, with a CI_1 value of 0.38, indicating a substantial decrease in the number of phylogenetic statements that can be made from the strict consensus tree. Only the EA values ($\approx$90%; Appendix 11.2) suggest that the individual discrepancies are relatively minor and that considerable agreement exists in the underlying pattern of relationships. Nevertheless, all of these metrics, except D_1, provide values that are above the range of similarity observed among random trees (Appendix 11.2). The lowest D_1 value observed between the cpDNA restriction site tree and the ITS trees (58%), is equivalent to the upper limit of values observed among random trees (but still greater than 95% of the random D_1 values).

Comparison of the decay values supporting the discordant branches (Fig. 11.6) on the ITS and cpDNA trees suggests that the number of extra steps required to map each tree on the opposing data set is *not* trivial. SLP_T confirms this: Mapping the ITS data on the cpDNA tree requires an increase in 39 steps (from 38 characters) and allows a decrease in only nine steps (from nine characters) compared to one of the ITS trees ($P < 0.001$; Appendix 11.3). Similarly, the ITS tree requires an increase in 18 steps (from 18 characters) and allows a decrease in only two steps (from two characters) using the cpDNA data ($P < 0.001$; Appendix 11.3). These large departures from the most parsimonious solution for each data set are associated with an I_M value of 21.7%. I_{MF} is much lower, however, measuring only 4.6%. This large disparity is readily explained based on the different manner in which I_M and I_{MF} are calculated (Box 11.5). That is, calculation of I_{MF} involves comparison of the topologies from the individual data sets to the topology retrieved when the data sets are combined (Figs. 11.6A, B, E). The combined data set topology is one of the possible fully bifurcating resolutions of the cpDNA tree that contains two polytomies due to zero length branches. Consequently, the contribution to I_{MF} by the cpDNA data set is decreased by 16 homoplasies that are required in the I_M calculation, but not in the calculation of I_{MF}, to explain the ITS topology. An additional 22 homoplasies are eliminated from the ITS data set in the calculation of I_{MF}, in part because the fully resolved tree from the combined data sets agrees with the ITS data in recognizing a clade comprising *Boykinia intermedia*, *B. occidentalis*, *B. aconitifolia*, *B. major*-1, and *B. major*-2. The cpDNA polytomies remove *Boykinia aconitifolia* from this clade, and the decreased resolution is costly in the calculation of I_M. This same rationale explains why SLP_T values decrease when the topology from the analysis of the combined data sets is used in place of the shortest trees from the opposing data set (Appendix 11.3).

Although the 4.6% I_{MF} value is low, HT_F indicates significant heterogeneity between data sets at the 5% level ($P = 0.032$). Additionally, significant heterogeneity is indicated by HT_R given that no shared trees were recovered by the 1,000 bootstrap replicates conducted for each data set. Because ambiguity exists in the alignment of the ITS sequences, it is possible that the discordance between data sets is simply the result of inaccurate homology assignments. This explanation by itself, however, does not account for all of the heterogeneity. First, application of HT_R and HT_F using each of the 18 ITS alignments results in significant heterogeneity in all instances with HT_R (data not shown), and in all but one instance with HT_F (Table 11.2). Second, it is noteworthy that all of the ITS sequence alignments place both species of *Suksdorfia* in the same lineage with *Bolandra* (*Suksdorfia* is monophyletic in analyses of some alignments, contra Fig. 11.6B), a relationship supported by morphology and chemistry. In contrast, cpDNA data place *Suksdorfia ranunculifolia* as sister to *Boykinia*, rather than with *S. violacea* and *Bolandra* (Fig. 11.6A; see also Soltis et al., 1996a). This relationship is strongly supported by the cpDNA data set, with nine characters increasing one step each, and no characters decreasing in steps, when relationships among these species are rearranged to reflect the ITS topology ($P = 0.004$ using SLP_T; Appendix 11.3). Thus, the cpDNA data strongly reject placing the two *Suksdorfia*

Table 11.2. Results from HT_F (Box 11.6) applied to cpDNA restriction site data and 18 ITS alignments from the *Boykinia* group (example 3). Results are given for comparisons both before and after pruning discordant taxa. Pruned taxa are listed as column headings. I_{MF} values are given only for the original, unpruned data set. Probabilities for the observed I_{MF} values are based on 499 random partitions of the combined data. "*" indicates significant heterogeneity at the 5% level.

Alignment	Before pruning I_{MF}/Probability	After pruning (Probability)			
		(*S. ranunculifolia*)	(*B. aconitifolia*)	(*B. major*-1)	(*S. oregana*)
ITS alignment-1	4.60/0.032*	0.326	0.054	0.044*	0.002*
ITS alignment-2	7.38/0.002*	0.060	0.004*	0.018*	0.002*
ITS alignment-3	6.88/0.002*	0.018*	0.002*	0.014*	0.002*
ITS alignment-4	5.92/0.006*	0.158	0.014*	0.012*	0.002*
ITS alignment-5	5.43/0.004*	0.104	0.010*	0.006*	0.002*
ITS alignment-6	4.71/0.016*	0.522	0.014*	0.018*	0.002*
ITS alignment-7	4.59/0.014*	0.558	0.018*	0.006*	0.002*
ITS alignment-8	5.96/0.006*	0.070	0.026*	0.092	0.004*
ITS alignment-9	4.76/0.018*	0.076	0.062	0.138	0.004*
ITS alignment-10	4.81/0.008*	0.208	0.016*	0.028*	0.004*
ITS alignment-11	4.17/0.034*	0.410	0.024*	0.020*	0.002*
ITS alignment-12	4.08/0.030	0.476	0.038*	0.016*	0.002*
ITS alignment-13	7.24/0.002*	0.050	0.002*	0.008*	0.002*
ITS alignment-14	5.95/0.016*	0.114	0.048*	0.050	0.002*
ITS alignment-15	4.09/0.062	0.324	0.056	0.056	0.004*
ITS alignment-16	5.00/0.008*	0.158	0.008*	0.022*	0.002*
ITS alignment-17	5.32/0.012*	0.436	0.014*	0.008*	0.002*
ITS alignment-18	5.18/0.012*	0.456	0.010*	0.010*	0.002*

species in the same lineage with *Bolandra* as suggested by the ITS data. Despite a decay value of five (Fig. 11.6B), however, the ITS data do not statistically reject the cpDNA topology ($P = 0.332$ using SLP_T; Appendix 11.3). This is true for all 18 alignments, with SLP_T probabilities ranging from 0.076 to 0.63.

The different placement of *Suksdorfia ranunculifolia* in the ITS and cpDNA trees is the primary source of incongruence in these data (Appendix 11.3). When evidence from morphology and the biology of the group is taken into consideration, this discordance is likely attributable to chloroplast capture (Soltis et al., 1996a). The impact of pruning *S. ranunculifolia* and repeating HT_F is dramatic, resulting in nonsignificant heterogeneity with 17 of the 18 ITS alignments (Table 11.2). The discordant relationships involving *Boykinia aconitifolia* and the two populations of *B. major*, on the other hand, are each only well supported by one of the data sets (ITS sequences and cpDNA restriction sites, respectively; Fig. 11.6), but in both cases statistical significance is not provided by SLP_T (Appendix 11.3). The significance level for the observed

I_{MF} actually increases for 13 alignments when either *B. aconitifolia* or *B. major*-1 are pruned (Table 11.2; *B. major*-1 was chosen arbitrarily over *B. major*-2 for pruning). Similarly, the discordant relationship involving *Sullivantia oregana* is weakly supported by both data sets (Appendix 11.3), and pruning this species increases the significance of heterogeneity for all alignments (Table 11.2).

This example illustrates the effect that different plausible alignments can have on assessments of data set homogeneity. Each of the 18 ITS alignments considered provides different views of both the significance of heterogeneity between cpDNA restriction sites and ITS sequences and the degree with which particular taxa are responsible for the heterogeneity. Consequently, different conclusions regarding the appropriateness of combining data could be reached if only one alignment were considered.

Seventeen of the 18 ITS alignments show probability levels from HT_F that indicate the ITS and cpDNA data sets are not significantly heterogeneous if *Suksdorfia ranunculifolia* is first pruned (Table 11.2). Analysis of a combined data set could thus be conducted after removing *S. ranunculifolia,* but it is not obvious how best to proceed with a combined analysis that incorporates the ambiguity that exists between the individual ITS alignments. Furthermore, there may be sound biological reasons for *not* combining the data (e.g., chloroplast capture). For illustrative purposes, we proceed with both separate and combined analyses. In both cases, we use the "elision" method (Wheeler et al., 1995) and combine all ITS alignments into a single, grand data set (all 18 ITS alignments were placed interleaved in the matrix block of a Nexus file). Phylogenetic analysis of this "elided" data set resulted in one most parsimonious tree (Fig. 11.6C) nearly identical to that obtained with a single ITS alignment (compare Fig. 11.6B and C). A semistrict consensus of this ITS tree and the shortest cpDNA tree (Fig. 11.6A) was also constructed (Fig. 11.6D); this tree is highly resolved showing considerable agreement between the elided ITS and cpDNA topologies.

For the analysis of the combined ITS and cpDNA data sets, the cpDNA data set was also replicated 18 times to provide equal weighting to the 18 elided ITS alignments. Analysis of this combined data set produced a single tree that places *Suksdorfia ranunculifolia* as sister to *Boykinia* (Fig. 11.6E). This shows the influence of the cpDNA characters in the placement of this taxon despite contrary evidence from ITS sequences. Pruning *S. ranunculifolia* from this tree results in a topology that is identical to that recovered from analysis of combined elided ITS and cpDNA data sets from which *S. ranunculifolia* has been removed (Fig. 11.6F). For illustrative purposes, we grafted *S. ranunculifolia* back onto this tree as part of a polytomy with *Bolandra* and *S. violacea,* a position that reflects evidence from ITS sequences, as well as ancillary evidence from morphology and chemistry (Fig. 11.6F).

Example 4: Heuchera Group

The *Heuchera* group is another well-supported monophyletic group of genera in Saxifragaceae s. s. (e.g., Soltis et al., 1993; Soltis and Kuzoff, 1995); five of the nine genera (*Bensoniella,* *Conimitella, Elmera, Tellima,* and *Tolmiea*) are monotypic; *Tiarella* and *Lithophragma,* comprise three and nine species, respectively, whereas the largest genera, *Mitella* and *Heuchera,* consist of approximately 20 and 50 species, respectively. Both ITS sequences and cpDNA restriction sites are available (Soltis and Kuzoff, 1995) for the same 30 accessions representing all nine genera in the group; these collections also represent well the species-level diversity within genera. Soltis and Kuzoff (1995) explored the congruence of the cpDNA and ITS data sets from a qualitative standpoint. Here we conduct a quantitative assessment of congruence in the *Heuchera* group using *Darmera* as an outgroup.

Parsimony analyses of the *Heuchera* group recovered one and 66 shortest trees, respectively, from the cpDNA restriction site and ITS sequence data sets (Appendix 11.1). The large number of ITS trees reflects homoplasy among characters attributable to sources other than alignment ambiguities. In contrast to the *Boykinina* group (Example 3) alignment of ITS sequences for the *Heuchera* group was not problematic and a single alignment was determined (Soltis and Kuzoff, 1995).

Comparison of trees, both within the ITS data set and between the cpDNA and ITS data sets, provides a more extreme view of incongruence than that exhibited by the foregoing Examples 1, 2, and 3. Substantial differences exist just within the set of ITS topologies. The EA metric again provides the highest estimate of similarity between trees, but even this metric records only 80% similarity between some pairs of ITS trees (Appendix 11.2). PM is as low as 66%, and as many as 13 taxa must be removed to form the greatest agreement subtree (Appendix 11.2). Comparing the single cpDNA tree to the ITS trees reveals varying degrees of similarity, but even the most similar pairwise comparison reflects considerable topological dissimilarity. The highest normalized PM is 0.32, the highest D_1 is 0.36 and requires pruning 18 taxa, and the highest strict consensus indices indicate very little resolution with values that range from 7% to 21% similarity (Appendix 11.2). The EA metric indicates up to 48% of the triplets agree between the cpDNA and ITS trees, a value markedly lower than any of the EA values observed in the other examples provided in this chapter.

It is reasonable to ask whether these very low similarity values are any greater than chance expectations. Previously, this exercise has been a formality, with similarity greater than all random values observed with every metric in all three examples above (except D_I in the *Boykinia* group). For the *Heuchera* group, however, only PM shows greater similarity than the entire range of random values. CI_C, CI_I and TI show greater similarity than 95% of the random values, but because these indices and PM can show maximum discordance due the placement of a single taxon (e.g., Box 11.1), random trees tend toward maximum dissimilarity and rejecting the null hypothesis of chance similarity is relatively easy. EA and D_I values between the cpDNA and ITS trees, in contrast, all fall within the range of similarity scores exhibited by random trees. Furthermore, at least some pairwise comparisons are no more similar than 95% of the values obtained from random trees. Thus, it is appropriate to question whether the cpDNA and ITS trees are actually estimating the same phylogeny.

Detecting similarity values between trees that are greater than chance expectations is expected (Rodrigo et al., 1993; Swofford, 1991). Thus, finding that similarities between trees are only marginally greater than chance expectations is very unexpected; such results are particularly surprising with a sizable data set such as that used for the *Heuchera* group. How do we interpret the cpDNA and ITS trees for the *Heuchera* group if they are not estimating the same phylogenetic history? The answer rests with the biology of this group of plants. The *Heuchera* group provides an important model for the analysis of chloroplast capture and its impact on phylogenetic reconstruction because hybridization is well documented within and even between genera. Intergeneric hybrids involving six of the nine genera have been reported. Based on comparisons of cpDNA restriction sites with ITS sequences and allozyme data, more than five examples of chloroplast capture have been documented for the *Heuchera* group, several of which involve species from different genera (Soltis et al., 1991; Soltis and Kuzoff, 1995). The strong likelihood that ancient events of

chloroplast capture occurred between lineages during the early diversification of the *Heuchera* group even further complicates the use of cpDNA data for phylogenetic reconstruction among these taxa (Soltis and Kuzoff, 1995). Thus, the discordance between ITS and cpDNA restriction site data sets for the *Heuchera* group reflects fundamental differences in the genealogies of the chloroplast and nuclear DNA regions and is likely attributable to hybridization and subsequent chloroplast capture.

Whereas substantial discordance between cpDNA and ITS trees is evident by simple visual inspection (Figs. 11.7A, B), quantitative measures emphasize the severity of the underlying incongruence between the data sets themselves. For example, I_M and I_{MF} values (67.8% and 32.5%, respectively) reveal high levels of character incongruence, and both HT_F and HT_R indicate significant heterogeneity between the ITS and cpDNA data sets. SLP_T shows that both data sets strongly reject the opposing topologies (Appendix 11.3). To test the significance of individual discordant nodes, constraint trees with a single branch resolved (Box 11.8) were constructed for five discordant branches from each data set (branches a1–a5 in Fig. 11.7A and b1–b5 in Fig. 11.7B). SLP_T applied to these 10 branches provides further evidence that incongruence in the *Heuchera* group is well supported and likely reflects extreme differences in the branching histories of the two DNA regions. Significant SLP_T results are observed for seven of the 10 individual branch comparisons (Appendix 11.3). A similar example of extreme discordance between nuclear and chloroplast topologies exists within Triticeae (Mason-Gamer and Kellogg, 1996).

Because the number of discordant taxa is so large in the *Heuchera* group (the greatest agreement subtrees of all 67 most parsimonious trees from both data sets combined contain only nine taxa) and many of the discordant nodes are well-supported, no attempt was made to assess heterogeneity after pruning taxa. Rather, we retain the strict consensus of the ITS topologies and the cpDNA topology as independent gene phylogenies that, when compared, provide the clearest insights into the evolutionary processes involved in the diversification of the *Heuchera* group.

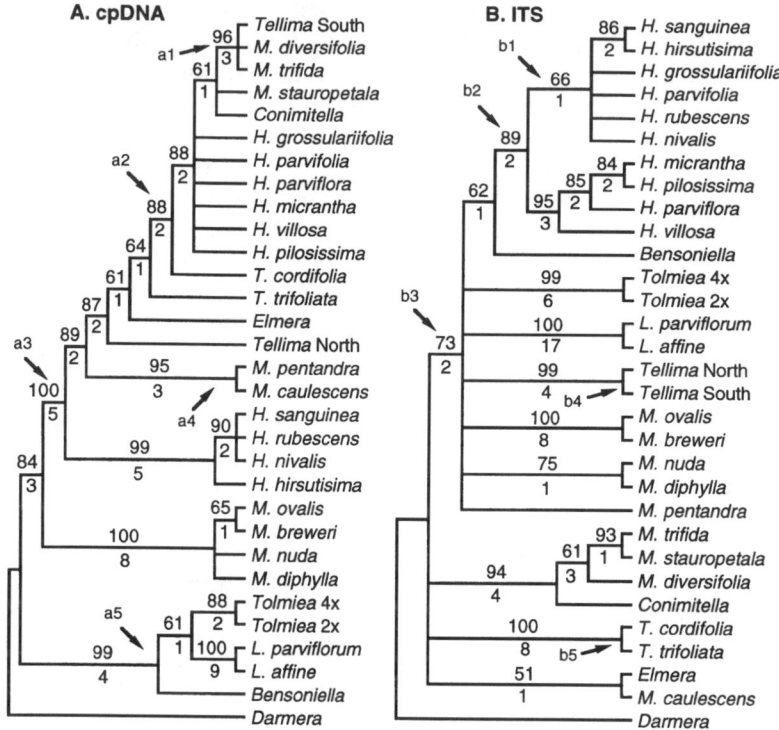

Figure 11.7. Phylogenetic trees for Example 4 (the *Heuchera* group). Bootstrap percentages from 500 bootstrap replications are shown above branches. Decay values are shown below branches. Arrows with alphanumeric designations (a1–a5 and b1–b5) indicate nodes resolved on single-node constraint trees for SLP$_T$ tests (Appendix 11.3). *A.* Single shortest cpDNA restriction sites tree. *B.* Strict consensus of 66 shortest ITS trees.

Example 5: Saxifragaceae s. s.

Saxifragaceae s. s. comprise approximately 30 genera of herbaceous perennials. *Saxifraga* (≈400 species), *Heuchera* (≈50 species), and *Mitella* (≈20 species) are the three largest genera. Over half of the remaining genera include only one to three species (Spongberg, 1972; Soltis et al., 1993). Recent molecular (Soltis et al., 1991, 1993, 1996a, 1996b; Johnson and Soltis, 1994, 1995) and biosystematic studies (e.g., Wells, 1984; Gornall and Bohm, 1985; Soltis, 1987) reveal several well-supported clades of genera within Saxifragaceae s. s. that have been given informal names (e.g., the *Boykinia* and *Heuchera* groups discussed previously). Multiple data sets derived from the chloroplast genome, as well as nuclear ITS sequences, exist for representative taxa from throughout Saxifragaceae s. s. Here, we assess congruence among some of these data at the fam-

ily level using *matK, rbcL,* and ITS sequences for 36 species representing 21 genera of Saxifragaceae s. s. Two genera from outside the family (*Pterostemon* and *Itea*) are included as outgroups. The *matK* and *rbcL* sequences were obtained from identical species and have been reported (Soltis et al., 1993; Johnson and Soltis, 1994) and even combined (Johnson and Soltis, 1995; Soltis et al., 1996b) in previous studies. ITS sequences are also available for 34 of the 38 species for which *matK* and *rbcL* sequences have been gathered (Soltis and Kuzoff, 1995; Soltis et al, 1996a; D. Soltis et al., unpubl.). For four species (all in *Saxifraga;* labels S1–S4 in Figs. 11.8 and 11.9), ITS sequences from sister species were substituted for the species sequenced for *matK* and *rbcL;* based on the congruence analyses conducted below, such substitution did not substantially increase discordance.

The *matK* and *rbcL* sequences are free from alignment ambiguities; the few indels that exist in

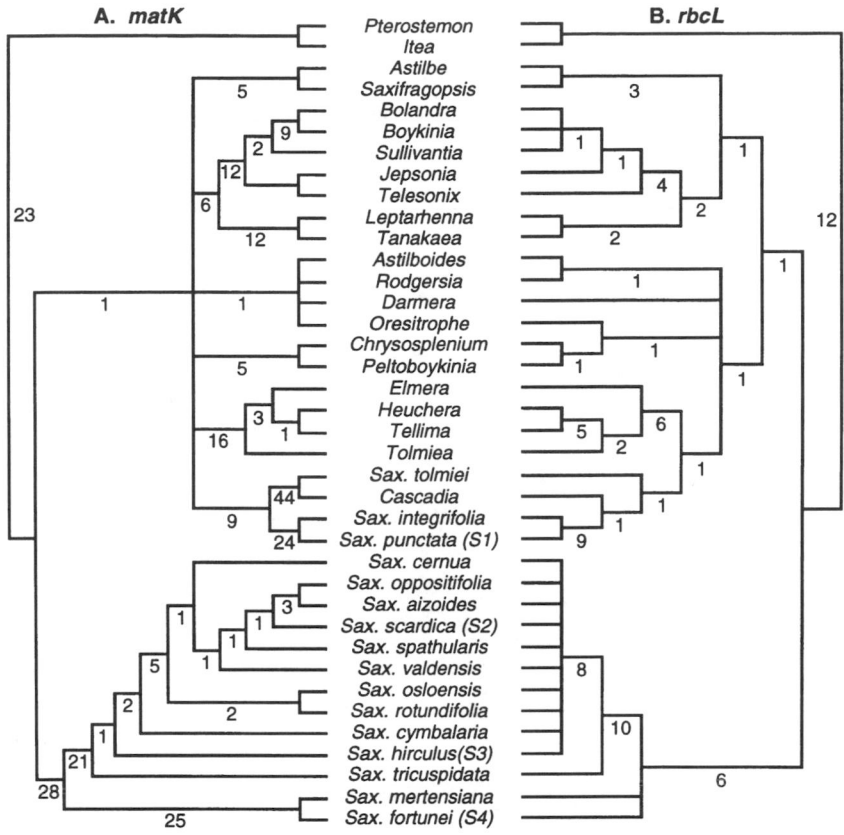

Figure 11.8. Phylogenetic trees for Example 5 (Saxifragaceae s. s.); *matK* and *rbcL*. Decay values are shown below branches. *A*. Strict consensus of 60 shortest *matK* trees. *B*. Strict consensus of 92 shortest *rbcL* trees.

matK are easily positioned (Johnson and Soltis, 1994, 1995; Soltis et al., 1996b). Although ITS sequences are usually easily aligned both within genera and within most well-supported groups of genera (e.g., within the *Heuchera* and *Darmera* groups, and within *Saxifraga*), alignment across Saxifragaceae s. s. is problematic. Following the general approach used for the *Boykinia* group (Soltis et al., 1996a), we experimented extensively with different gap-open and gap extension penalties using CLUSTALW (Thompson et al., 1994) and then manually adjusted these alignments. Several reasonable alignments were produced for the family; we arbitrarily chose one to use here (problems associated with multiple alignments were addressed in Example 3). Furthermore, the several alignments all recover identical or highly similar trees.

The *matK, rbcL,* and ITS sequences were compiled into three data sets. However, because

the chloroplast genome is generally considered free from recombination and the genes therein share a common phylogenetic history, a strong argument could be made to combine the *matK* and *rbcL* data sets a priori. Conversely, one could argue that ITS sequences are composed of two unique DNA regions (i.e., ITS-1 and ITS-2 regions) that should first be subjected to homogeneity tests before being combined into a single matrix. Thus, depending on your view, these data represent two, three, or four separate data sets. When more than two data sets are compared, assessments of congruence and data set heterogeneity become more complex. The topological congruence metrics D_1, CI_C, CI_1, and TI, the incongruence indices I_M and I_{MF}, and the homogeneity test HT_F can be readily extended to three or more trees or data sets simultaneously with available software (e.g., PAUP* 4.0). However, the results may be less informative and

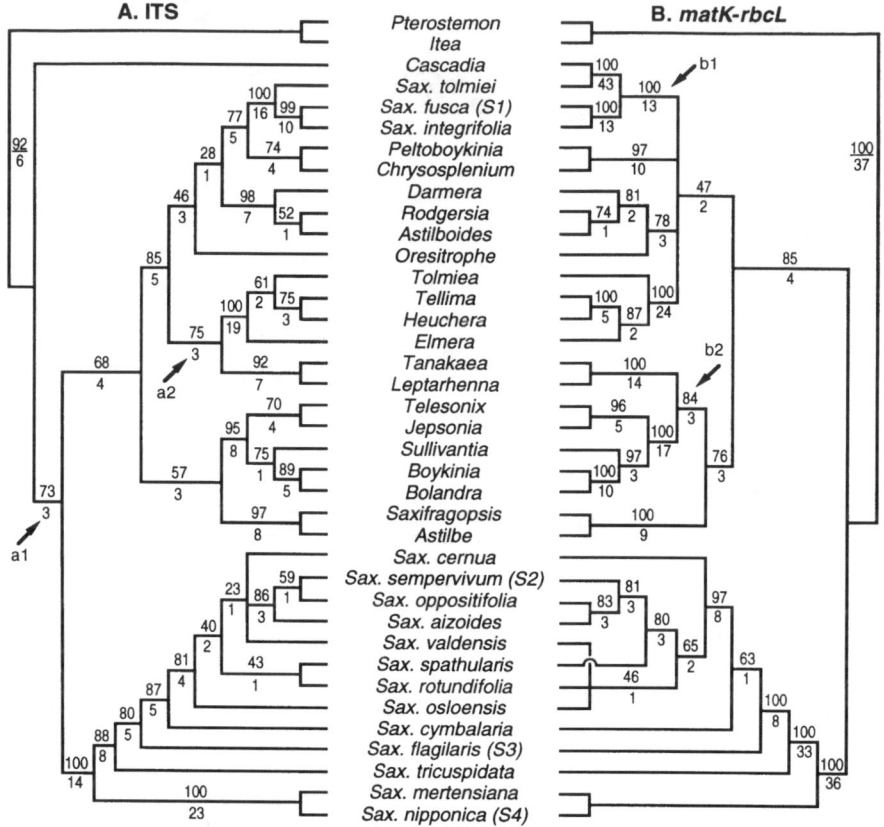

Figure 11.9. Phylogenetic trees for Example 5 (Saxifragaceae s. s.); ITS and *matK-rbcL*. Bootstrap percentages from 500 bootstrap replications are shown above branches. Decay values are shown below branches. Arrows with alphanumeric designations (a1–a2 and b1–b2) indicate nodes resolved on single-node constraint trees for SLP_T tests (Appendix 11.3). A. Strict consensus of two shortest ITS trees. B. Strict consensus of four shortest trees obtained from combined *matK* and *rbcL* data.

more difficult to interpret than for standard pairwise comparison. For example, if HT_F indicates significant heterogeneity when applied to three data sets simultaneously, as it does in this example ($P = 0.002$), it is not clear whether the three data sets are heterogeneous with respect to each other or if two of the data sets are homogeneous and heterogeneous with respect to the third data set. Thus, pairwise comparisons involving all possible combinations of the data sets are better indicators of congruence.

Although we agree that multiple cpDNA data sets for the same suite of species can be combined a priori, for the sake of demonstration we initially considered the *matK, rbcL,* and ITS (combined ITS-1 and ITS-2) data sets to represent three independent matrices. Parsimony analyses recovered 60 shortest *matK* trees, 92

shortest *rbcL* trees, and two shortest ITS trees (Appendix 11.4; Fig. 11.8). To simplify our assessment of congruence, we used only PM, EA, and D_1 to quantify topological congruence (CI_C, CI_1, and TI all provide estimates of similarity between trees that are highly comparable, though not always identical, to PM; compare values in Appendix 11.2). As with all of the previous examples, EA values are higher than the PM or D_1 values. The lowest EA value, between two *matK* trees, is above 90%, whereas the lowest PM and D_1 values, between two *rbcL* trees, is 69% (Appendix 11.5). When comparisons are made between data sets, different metrics suggest different conclusions: the least similar trees using PM occur in the *rbcL*-ITS comparison; the least similar trees using EA occur in the *matK*-ITS comparison; and the least similar trees using D_1 actually occur in the *matK-rbcL* comparison (Appendix 11.5). Overall, the greatest topological similarity is found between pairs of *matK* and *rbcL* trees; the *rbcL* and ITS trees show the lowest maximum similarity of the three pairwise comparisons (Appendix 11.5).

Pairwise comparisons of character incongruence between the *matK, rbcL,* and ITS matrices for Saxifragaceae s. s. contrast with the measures of topological congruence described above. Whereas the greatest topological similarity occurs between some *matK* and *rbcL* trees, the *matK* and *rbcL* data sets show the greatest amount of character incongruence as measured by both I_M and I_{MF} (Appendix 11.5). Likewise, the greatest topological discordance occurs between some *rbcL* and ITS trees, yet the *rbcL* and ITS matrices exhibit the lowest character incongruence based on I_M and I_{MF} values (Appendix 11.5).

The level of homoplasy within data sets also affects the I_M and I_{MF} incongruence indices. The number of extra homoplasious steps required to explain the combined topologies in the calculation of I_{MF} is 18 steps for *matK-rbcL,* 19 steps for *rbcL*-ITS, and 52 steps for *matK*-ITS. These extra steps are associated with I_{MF} values of 4.0%, 1.6%, and 3.9%, respectively (Appendix 11.5). Although the *matK-rbcL* and *rbcL*-ITS comparisons differ only by one extra step to explain their respective combined topologies, their I_{MF} values differ by a factor of 2.5 because the amount of homoplasy within the ITS data is substantially greater than the amount of homoplasy within the *matK* data set (Appendix 11.4). These results suggest that caution be exercised in comparing I_{MF} and I_M values between different data sets for the same taxa, because increased homoplasy within a data set will generally decrease its relative contribution to the measurement of character incongruence between data sets. These results also help explain the probability for each I_{MF} value provided by HT_F. The level of significance is associated with the number of extra steps required to explain the combined topology, rather than the proportion of extra steps as measured by I_{MF} (compare extra steps, above, with probability values in Appendix 11.5).

HT_F indicates that significant heterogeneity exists in all three pairwise comparisons between the *matK, rbcL,* and ITS data sets (Appendix

11.5). SLP_T provides additional insights into the nature of this incongruence. The ITS data strongly reject the topologies recovered by the *matK* and *rbcL* data sets, the *matK* data strongly reject the topologies recovered by the *rbcL* and ITS data sets, and the *rbcL* data strongly reject the topologies recovered by the ITS data set, but do not statistically reject at least some of the topologies recovered by the *matK* data set ($P = 0.051$; Appendix 11.3). Additionally, both chloroplast data sets statistically support the topologies recovered when the *matK* and *rbcL* sequences are combined into a single data matrix prior to phylogenetic analysis (data not shown). Consequently, although HT_F suggests the *matK* and *rbcL* data sets are significantly heterogeneous at the 5% level (Appendix 11.5), there is no compelling reason not to combine these data.

Rather than indicating that these cpDNA data sets should not be combined, this example illustrates well the sensitivity of HT_F to differences in character-state distributions between data sets with identical branching histories. Combining these two cpDNA data sets reveals some relationships unique to the combined data set, increased support for some relationships, and a better overall estimate of phylogeny (Fig. 11.9). Similar results have been obtained in other studies involving multiple cpDNA data sets (e.g., Johnson and Soltis, 1994; Olmstead and Sweere, 1994; Soltis et al. 1996a), as well as multiple mtDNA data sets in animals (e.g., Cummings et al., 1995; Sullivan, 1996). These examples suggest that evidence for the unique relationships is present in each data set, but is not recovered in individual analyses due to sampling error. In addition, these studies emphasize that a knowledge of the data sets and organisms is essential to handling multiple data sets; better phylogenetic estimates may be obtained in some instances by combining data sets even if the general null hypothesis of data set homogeneity is rejected (Sullivan, 1996; J. Farris, pers. comm.; D. Swofford, pers. comm.).

Parsimony analysis of the combined *matK* and *rbcL* data sets yielded four shortest trees (Fig. 11.9; Appendix 11.4). Comparison of these trees with the ITS trees shows discordance in some minor rearrangements among more terminal branches (e.g., *Saxifraga valdensis*) and a few major rearrangements involving several branches (e.g., *Cascadia*). Whereas the PM value is largely influenced by the discordant placement of *Cascadia*, D_1 values indicate that 11 or 12 additional taxa are also resolved differently by the two data sets. Although the I_{MF} and I_M values are relatively small, the I_{MF} value is determined to be highly significant by HT_F. This significance is expected based on the results of the pairwise comparisons discussed previously, given that larger sets must show significance if their smaller components showed significance (Sokal and Rohlf, 1981, p. 437). Although we could again chose to overlook the significance of HT_F between the combined *matK-rbcL* data and the ITS data and proceed with a combined analysis of all three data sets, we keep these matrices separate because we have no reason to assume the chloroplast and nuclear data sets are specifying the same pattern of relationships and vary only due to sampling error. In fact, SLP_T rejects the opposing topologies for both data sets, with the discordant placement of *Cascadia* strongly supported (Fig. 11.9; Appendix 11.3) by both nuclear and chloroplast data when single branch constraint trees are used in the SLP_T test. Not all individual discordances are significantly supported, however, as evidenced by *P* values greater than 0.05 for the discordant placement of *Tanakaea* and *Leptarrhena* (Fig. 11.9; Appendix 11.3).

We therefore end our analysis of Saxifragaceae s. s. with the conclusion *not* to combine the ITS and two chloroplast data sets. Some of the discordance involving minor rearrangements could represent sampling error. Another source of incongruence includes alignment ambiguity of ITS across Saxifragaceae s. s.. Particularly intriguing are taxa that appear to be the major sources of "hard" incongruence (e.g., *Cascadia*). One possible avenue for additional study would be to prune these taxa from the data sets and reanalyze the separate and combined data sets (see Examples 2 and 3).

DISCUSSION

The foregoing examples illustrate many facets of assessing congruence between trees and data matrices. We have employed a general framework for exploring congruence based on the conditional combination methodology (Fig. 11.1) that includes assessments of topological concordance, character incongruence, statistical tests of homogeneity, and, finally, evaluation of all of these results in an effort to localize and distinguish the cause of incongruence. Here we summarize the important aspects of this framework.

Topological Congruence

Quantitative assessments of topological congruence employ any of several metrics that describe different aspects of tree similarity (Swofford, 1991). The six indices we used include metrics based on tree interconversion (PM; Box 11.1), subtree similarities (EA and D_1; Boxes 11.2, 11.3), and shared clades between trees (CI_C, CI_1, and TI; Box 11.4). In practice, we found PM, EA, and D_1 to have the greatest utility.

CI_C, CI_1, and TI measure different aspects of consensus tree resolution that can potentially give rise to very different values for these metrics; however, our results show that similarity values are not only highly comparable for these three metrics, but they are often similar or identical to PM values (Appendix 11.2). The superiority of PM over CI_C, CI_1, and TI is due to the interpretation of PM as both a measure of the number of inequivalent taxonomic partitions between two trees, and as the number of branch contractions-decontractions required to convert one tree into a second tree. Both interpretations are easily understood, and the first interpretation can be particularly useful given that it quantifies the total number of SLP_T tests needed to assess the significance of every discordant branch found on two trees being compared. However, because only one, or a few, branch differences can result in minimal similarity using PM, random trees often show maximum dissimilarity (Fig. 11.4). This makes PM a less conservative metric in tests designed to assess whether two topologies show significant departure from random similarity (Box 11.9).

In contrast to PM, EA and D_1 are less sensitive to the placement of individual taxa and both provide more conservative distributions of similarity between random trees (Fig. 11.4). Because

EA is based on identity in branching pattern among either triplets (rooted) or quartets (unrooted) of taxa, this metric is useful for quantifying underlying similarities in relationships that are not always apparent at the whole tree level. Accordingly, EA provides the most generous estimate of similarity between even random trees, and we favor its use for testing departure from random similarity. As noted by Steel and Penny (1993), EA may be additionally valuable when the trees being compared are constructed by methods such as quartet puzzling (Strimmer and von Haeseler, 1996). The straightforward interpretation of D_1 as the minimum number of taxa that must be pruned to make two trees identical gives this metric utility in identifying potentially problematic taxa that may be contributing to significant incongruence between data matrices. Because D_1, like PM, is implemented in PAUP* 4.0, whereas EA is implemented in COMPONENT 2.0, we expect future studies will favor the use of PM and D_1 due to the broader distribution of PAUP (nonetheless, COMPONENT is a versatile application that we highly recommend). In our estimation, PM, EA, and D_1 are equally "good" metrics; each provides useful information that is not quantified by the other two indices.

Assessments of topological congruence, whether quantitative or qualitative, are insufficient by themselves for determining the appropriateness of combining data. Consequently, several studies have simply noted that topological discordance exists in the course of assessing data set heterogeneity, rather than quantifying levels of topological incongruence with metrics (e.g., Huelsenbeck and Bull, 1996; Mason-Gamer and Kellogg, 1996; Sullivan, 1996; Seelanan et al., 1997). Nonetheless, we advocate the use of topological indices because their application, followed by careful scrutiny of the trees, is an important means of identifying problematic taxa and sources of incongruence. Topological congruence indices are also useful for quantifying similarities between trees within a single data set given that both separate and combined analyses often recover multiple most-parsimonious trees. Applied in this context, these indices provide means other than consensus trees for assessing similarities

among trees, and they can help identify taxa that are particularly sensitive to the distribution of homoplasy in these data sets.

Character Incongruence

The character incongruence indices employed in this chapter, I_M and I_{MF} (Box 11.5), summarize the strength of support for discordant relationships across entire topologies; they provide an initial assessment of character incongruence between data sets. The primary difference between I_M and I_{MF} is that I_M measures incongruence in relation to the sum of the additional homoplasy required to map each data set on a shortest tree from the opposing data set, whereas I_{MF} measures incongruence in relation to the sum of the additional homoplasy required to map each data set on the topologies recovered from an analysis of both data sets combined. Rather than one index being superior to the other, the two indices are complementary and contrasting their values (i.e., subtracting I_{MF} from I_M) provides an estimate of the decrease in incongruence between data sets that occurs when data sets are combined, relative to that present in the individual data sets. Both metrics measure incongruence between data sets relative to the magnitude of homoplasy within each data set. That is, maximum I_M and I_{MF} values are observed when individual data sets contain no homoplasy, regardless of the magnitude of homoplasy between data sets or the degree of topological discordance. Because the degree of incongruence within data sets can vary greatly, it is necessary to consider the magnitude of homoplasy within data sets when comparing or considering the meaning of an I_M or I_{MF} value (e.g., see Example 5). It is also important to consider the distribution of the discordant relationships giving rise to an observed I_M or I_{MF} value. A single discordant branch with high decay values for two trees can result in an I_M or I_{MF} value equivalent to that observed between two different trees with many, weakly supported discordant branches. The former case may result in significant heterogeneity (by HT_F, HT_R, or SLP_T) and, if the heterogeneity is attributed to different branching histories, pruning the offending taxon may be appropriate before proceeding with a combined analysis. In the latter case, heterogeneity may be insignificant and simply reflect sampling error or excessive stochastic variation; a combined analysis may thus be appropriate without any modifications to the matrix.

Assessments of character incongruence via the application of I_M and I_{MF} are also insufficient by themselves for determining the appropriateness of combining data; the tests for heterogeneity discussed below can all be employed without having first quantified character incongruence with these two indices. Nonetheless, I_M and I_{MF} provide a starting point for investigating the significance of character incongruence. We gained important insights into the nature and extent of homoplasy within and between data sets by attempting to explain the observed levels of character incongruence measured by these indices, in much the same way that the topological indices provided insights into similarities between trees and the distribution of discordant taxa. Thus, not only do I_M and I_{MF} quantify an aspect of incongruence between data sets that is of interest in and of itself, these metrics ultimately contribute toward the decision-making process of whether data sets should be combined by familiarizing an investigator with pertinent aspects of his or her data.

Significance Tests for Heterogeneity

We have employed three statistical methods. HT_F (Box 11.6), HT_R (Box 11.7), and SLP_T (Box 11.8), that have been advocated by others as methods for assessing the general null hypothesis of data set homogeneity (e.g., Rodrigo et al., 1993; Farris et al., 1995; Lutzoni and Vilgalys, 1995; Mason-Gamer and Kellogg, 1996; Miyamoto, 1996; Seelanan et al., 1997). We found HT_F to be the easiest test to apply, and SLP_T to have the greatest versatility and utility for distinguishing between well-supported and weakly supported incongruencies.

HT_F The implementation of HT_F in PAUP* 4.0 is straightforward, and we recommend its use as a general test for heterogeneity. The interaction of data from both matrices through random repartitioning of the data (Box 11.6) provides a conceptually simple approach to assessing ho-

mogeneity. However, because this test summarizes heterogeneity across data sets, a significant result can occur from the combination of many weakly supported discordant relationships or a single, strongly supported discordant relationship (see previous I_{MF} discussion). Consequently, when heterogeneity is suggested, the SLP_T test is recommended for assessing the source of heterogeneity more specifically (see also Mason-Gamer and Kellogg, 1996). When applying HT_F, a large number of replications (≥ 500) is required to obtain meaningful and repeatable probability values. This is particularly important when the observed probability value is near the level of significance determined before conducting the test. Because the number of possible partitions increases with the number of characters in each data set, a large number of replications is increasingly important for larger data sets even though it may be tempting to use even fewer replications to reduce computational time.

HT_R Although HT_R is not incorporated as a specific test into any software, the steps required to conduct this test (Box 11.7) are available in PAUP* 4.0. HT_R assesses, through the use of nonparametric bootstrapping, the possibility that sampling error underlies topological differences between data sets. Parametric bootstrapping (cf. Hillis, 1995) could alternatively be used, and may, in fact, provide a superior test when evolutionary parameters that accurately describe an entire data set can be estimated reliably from the data. In practice, we found the use of HT_R to be most effective with smaller data sets, when few discordant relationships were present, and when relatively few trees were recovered from each bootstrap replication (see also Mason-Gamer and Kellogg, 1996). When a large number of trees exists per bootstrap replicate during the search for overlapping tree spectra, researchers may find it useful to generate the bootstrap data sets using the SEQBOOT component of PHYLIP (Felsenstein, 1993), and recover trees with PAUP* 4.0 via a HEURISTIC search. Then, if, for example, 25,000+ trees are found and only 5,000 are swapped on, the run can be aborted and all 25,000 trees saved for comparison with bootstrapped trees from the other data set; aborting a run when bootstrapping in PAUP* 4.0 causes all trees for that run to be lost. Another strategy for data sets that recover numerous trees per replication involves conducting a bootstrap analysis with a large number of replications (say 1,000) and saving only 100 or 200 trees per replication in combination with multiple RANDOM addition HEURISTIC searches to more fully explore tree space. Bootstrap values from both data sets are then compared for all discordant branches. This enables the degree of support from each data set for discordant relationships on both sets of discordant trees to be assessed. If one of the data matrices never recovers a particular clade that is recovered by the second data matrix, the researcher can be fairly confident that a shared topology is unlikely to be generated by bootstrapping. Although HT_R provides a conceptually interesting approach for assessing homogeneity, its unwieldy application with large or less decisive data matrices decreases its general utility.

SLP_T   We found SLP_T (Box 11.8) to be the most versatile and insightful of the three statistical tests for heterogeneity that we employed (see also Mason-Gamer and Kellogg, 1996; Seelanan et al., 1997). Although this test is the least direct in assessing the general null hypothesis that two data sets are homogeneous, it provides the most information on the nature, distribution, and statistical significance of character incongruence. SLP_T distinguishes whether one, both, or neither data set rejects specified alternative topologies, and the alternative topologies can range from shortest trees from the opposing data set (to test incongruence on a data-set-wide basis), to shortest trees that are compatible with a constraint tree that resolves only a single node (to test the significance of a given partition). Single node constraint trees that represent nodes present on the strict consensus tree for a given data set can also be used to recover trees from heuristic searches of that data set that fail to satisfy the constraint topology (PAUP* 4.0 commands ENFORCE CONVERSE). The recovered trees can then be used to test the significance of a large decay value for a specific branch, which extends the versatility of SLP_T beyond assessing incongruence between data sets.

An additional advantage of SLP_T is that the number of characters that either increase or decrease in steps (i.e., that reject or support the alternative topology, respectively) is provided along with a list of which characters vary. Thus, the data can be closely scrutinized to determine possible causes for the incongruence. Because the sign and Wilcoxon sign-ranked tests (Box 11.8) are conservative (i.e., large differences are required to show significant heterogeneity), some relationships that seem well-supported (see Example 2; Fig. 11.5A) are not deemed significant with SLP_T because at least six characters must vary to show significance with these tests (Siegel, 1956).

The methodology for SLP_T is easily applied, but we encountered multiple shortest trees in most analyses using both constraint-based and conventional tree searches. Multiple trees require the application of the sequential Bonferroni method (Box 11.8), or selecting a single topology by some criterion, such as the tree that requires the fewest additional steps on the opposing data set (Miyamoto, 1996). Additionally, we found it necessary to consider carefully what is being tested with a particular alternative topology, given that constraint-based topologies often possess novel relationships in unconstrained portions of the tree (Box 11.8). Furthermore, fully resolved alternative topologies that contain many weakly supported nodes may unfairly increase the assessment of incongruence because many of the discordant nodes may be only weakly supported by the data set that produced them (Mason-Gamer and Kellogg, 1996). In such situations, a tree that resolves only branches with bootstrap percentages above 70% has been recommended for use as a topological constraint tree (Mason-Gamer and Kellogg, 1996). In spite of these caveats, SLP_T is the best method of those we employed for distinguishing between weakly supported and strongly supported incongruent nodes. Accordingly, close scrutiny of topologies on a node-by-node basis with SLP_T can be helpful for assessing the meaningfulness of significant heterogeneity evidenced at the whole data set level. We found this process, in conjunction with a knowledge of the organisms involved, to be a valuable means of identifying likely causes of heterogeneity and, ultimately, in determining whether data should be combined.

Additional Considerations Whereas we agree that statistical methods provide information necessary for determining whether data sets should be combined, we also foresee the potential misuse of these tests if they are applied in the absence of relevant biological information, or without an understanding of what these tests do, and do not, reveal about the appropriateness of combining data. For example, each of these tests for heterogeneity is subject to statistical error, and the sensitivities to various aspects of phylogenetic character distributions are not fully understood. Although we followed the common practice of setting the type-1 error rate to 0.05 (that is, the probability of rejecting the null hypothesis when it is true is 5%), the actual incidence of heterogeneity in phylogenetic data may be much lower than this value, in which case many homogeneous data sets would be incorrectly deemed heterogeneous (Huelsenbeck et al., 1996).

Recognizing which feature of heterogeneity each test assesses is also important. HT_F, HT_R, and SLP_T have been employed as tests of the *general* null hypothesis that two data sets are homogeneous. More precisely, however, HT_F tests the null hypothesis that characters are randomly distributed between data sets with respect to the phylogenetic information they contain; HT_R tests the specific null hypothesis that the sampling of character-state distributions contained in each data set represents the same universe of characters; and SLP_T tests the null hypothesis that character-state reconstructions for a given data set are not significantly more parsimonious on one topology than on an alternative topology. These specific null hypotheses are based on the traditional statistical tests from which they are derived. That is, HT_F is based on the Mann-Whitney U test (Farris et al., 1995), HT_R uses a number of statistical methods, primary of which is nonparametric bootstrapping (Rodrigo et al., 1993), and SLP_T employs either the sign test or the Wilcoxon signed-rank test (Templeton, 1983; Felsenstein, 1985b).

Understanding the specific null hypothesis addressed by each of these tests, and the assumptions built into the methods of statistic inference employed (cf. Sokal and Rohlf, 1995a), is essential, both for interpreting statistically significant heterogeneity, and for recognizing conditions or aspects of the data that contribute to statistical significance. Evidence of significant heterogeneity using HT_F indicates that the distribution of the phylogenetic information contained in characters is not randomly distributed between two data matrices, but it does not distinguish among the many potential sources of incongruence. Some sources, such as rate variation among characters, may still be effectively dealt with in a combined analysis either with or without modifying the method of reconstruction (e.g., character weighting; Bull et al., 1993; Chippendale and Wiens, 1994; Sullivan, 1996). For example, because different nucleotide sites within genes evolve at different rates, characterizing entire data sets as "rapidly evolving" or "slowly evolving" may be inappropriate; combining statistically heterogenous data with among-site rate variation without weighting can still improve phylogenetic estimates (Sullivan, 1996). Other sources of heterogeneity, such as different branching histories for the two data sets, require only separate analyses unless specific problem taxa are identified and removed from the data because an implicit assumption of combined analyses is that both data sets reflect a common phylogenetic history (de Queiroz et al., 1995). Statistical tests can only indirectly assess the appropriateness of combining data (Huelsenbeck et al., 1996; Sullivan, 1996) because they do not distinguish the cause of the heterogeneity. Systematists must carefully explore their data and familiarize themselves with the biology of the species involved to determine the source of heterogeneity when it is deemed significant by these tests.

The Conditional Combination Methodology in Practice

The examples used here demonstrate how the conditional combination methodology can be used as a framework for assessing congruence. We have applied a broad view of congruence that includes descriptive assessments of both topological and character incongruence, rather than simply addressing the appropriateness of combining data with statistical tests. This strategy (Fig. 11.1) included conducting separate analyses followed by quantitative (use of PM, EA, D_1, CI_C, CI_1, and TI) and qualitative (identification of discordant taxa) assessments of topological congruence. Levels of character incongruence were then quantified using I_M and I_{MF}, and the strength of support for discordant relationships was assessed via the statistical tests HT_F, HT_R, and SLP_T. The insights and information gained from this process were then used to decide whether combined or separate analyses were warranted.

Additional facets of the conditional combination methodology explored here include pruning taxa that can be identified as the primary source of heterogeneity between data sets, and assessing congruence among more than two data sets. A single taxon was identified in both example 2 (*Phlox* and related genera) and example 3 (*Boykinia* group) that appears to have an introgressed chloroplast genome (Johnson, 1996; Soltis et al., 1996a). Pruning the well-supported discordant taxon resulted in nonsignificant heterogeneity with both data sets, even though topological concordance among trees recovered from the pruned data sets was not achieved. This underscores the important point that some discordances are "hard," and reflect real differences between the data sets that need to be addressed before data sets are combined, whereas other discordances are "soft," and may simply reflect sampling error, subtleties of evolutionary history (such as short nodes) or stochastic variation (Seelanan et al., 1997). Because ancillary evidence from morphology and chemistry exists that supports the nuclear DNA-based topologies for both the *Phlox* and *Boykinia* groups (Fig. 11.5; Johnson, 1996; Soltis et al., 1996a), we also grafted the pruned taxon back onto the tree in a location that reflects the agreement between the nuclear DNA topology and other lines of evidence. This approach was suggested by Rodrigo et al. (1993) and is in harmony with the "phylogenetic framework" methodology proposed by Lanyon (1993). Assessing congruence among more than two data sets was illustrated with Example 5 (Saxifragaceae s. s.). We found it most useful to employ pairwise comparisons

between the various data sets because this distinguishes which data sets are pairwise incongruent, and which are homogeneous. When pairs of data sets are deemed combinable, repeating the incongruence assessments between the combined and uncombined data sets (as we did between the combined *matK-rbcL* data set and the ITS data set for Saxifragaceae s. s.) is advisable because novel relationships may be recovered from combined analyses, and character support for various relationships is likely to change.

Although application of the quantitative indices and statistical tests is straightforward, we reiterate that the determination of when combined analyses are preferred over separate analyses cannot be based simply on the result of any single index or currently available test of homogeneity (de Queiroz et al., 1995; Huelsenbeck et al., 1996; Sullivan, 1996). Rather, a sound understanding of the data and the organisms involved is imperative in conjunction with appropriate metrics. This point is well illustrated by Example 3 involving the *Boykinia* group. All 18 ITS alignments used with the *Boykinia* group place *Suksdorfia ranunculifolia* in the lineage with *S. violacea* and *Bolandra*. This relationship is supported by morphology, but is not recovered by analyses of cpDNA restriction sites. One of the 18 ITS alignments fails to identify this discordance as significant, however, by failing to reject the null hypothesis of data set homogeneity. A knowledge of the difficulties involved in aligning the ITS sequences (see also Chapter 4) and a knowledge of the biology of these species, including the propensity for hybridization, was important for properly interpreting the source of incongruence. A reasonable scenario suggests that due to hybridization/introgression, a foreign chloroplast genome is present in *Suksdorfia ranunculifolia:* this results in different branching histories between cpDNA restriction sites and ITS sequences. However, because ITS alignment ambiguities also contribute to levels of character incongruence, correctly diagnosing the source of heterogeneity would not have occurred without a knowledge of the species and the data. Similarly, as evidenced with the two cpDNA data sets in example 5 (*matK* and *rbcL* sequences for Saxifragaceae s. s.), HT_F can be sensitive to relatively small differences in data sets that necessarily share a common branching history. Combining the cpDNA data sets despite significant heterogeneity in this example improved the phylogenetic estimate.

CONCLUDING REMARKS

Assessing congruence in phylogenetic studies is becoming an increasingly prominent and challenging topic. Our approach to this subject, which includes assessments of topological congruence, character congruence, and statistical tests of homogeneity, recognizes that the topic of congruence encompasses more than merely addressing the appropriateness of combining data. The emphasis in current literature, however, is appropriately focused on this important subject (cf. de Queiroz et al., 1995; Lutzoni and Vilgalys, 1995; Huelsenbeck et al., 1996; Mason-Gamer and Kellogg, 1996; Miyamoto, 1996; Sullivan, 1996; Seelanan et al., 1997). In this respect, the statistical techniques we employed, particularly HT_F and SLP_T, have general utility in that they are easily implemented with available software and can be applied to any data amenable to phylogenetic analyses using parsimony. SLP_T is especially versatile in its ability to distinguish between "soft" and "hard" incongruence on a node-by-node basis; SLP_T has broad applicability and appears as good as any available technique for pinpointing discordant relationships that merit closer scrutiny (see also Larson, 1994; Mason-Gamer and Kellogg, 1996; Miyamoto, 1996; Seelanan et al., 1997).

With the enormous interest in determining how best to handle multiple data sets, future prospects for increasing our understanding of congruence and developing more sensitive methods for identifying and describing discordances appear bright. Methods not covered in this chapter, such as parametric bootstrapping (Hillis, 1995) applied in either a parsimony or likelihood framework, can offer insights into congruence by using evolutionary parameters derived from one data set to assess the probability of generating topologies or specific nodes reflected in the second data set. Similarly, maximum likelihood approaches are apt to discern heterogeneity resulting from evolutionary processes at the DNA sequence level. Present methods use a likelihood ratio test to assess the significance of some characteristic of the data, which is usually the tree topology. For example, one can test whether a single data set has significantly different likelihoods when measured on alternative trees recovered from each data set (e.g., Lutzoni and Vilgalys, 1995), or whether a single topology better fits all of the data sets than do distinct topologies for each data set (Huelsenbeck and Bull, 1996). Because evolutionary parameters can vary between data sets, the probability that data sets represent discrete partitions with different parameters rather than a single partition with mixed evolutionary parameters can also be assessed (Sullivan, 1996). However, likelihood methods are subject to the same errors as parsimony methods, and significant heterogeneity may be suggested even when combined analyses improve the phylogenetic estimate (Sullivan, 1996). Although this only reemphasizes that statistical tests may not provide a definitive answer as to whether it is appropriate to combine data sets, this should not be viewed pessimistically with a "why bother" attitude. Rather, congruence tests represent a vehicle for exploring data; they provide the information necessary to make informed, justifiable decisions regarding how best to handle multiple data matrices. Application of these tests facilitates the discrimination between "hard" and "soft" incongruence and ultimately permits a better understanding of the underlying causes of incongruity.

LITERATURE CITED

Archie, J. W. 1989. A randomization test for phylogenetic information in systematic data. Systematic Zoology 38:239–252.

Baum, B. R., and M. A. Ragan. 1993. Reply to A. G. Rodrigo's "A comment on Baum's method for combining phylogenetic trees." Taxon 42:637–640.

Bledsoe, A. H., and R. J. Raikow. 1990. A quantitative assessment of congruence between molecular and nonmolecular estimates of phylogeny. Journal of Molecular Evolution 30:247–259.

Bremer, K. 1988. The limits of amino acid sequence data in angiosperm phylogenetic reconstruction. Evolution 42:795–803.

Bremer, K. 1990. Combinable component consensus. Cladistics 6:369–372.

Bremer, K. 1994. Branch support and tree stability. Cladistics **10**:295–304.

Bull, J. J., J. P. Huelsenbeck, C. W. Cunningham, D. L. Swofford, and P. J. Waddell. 1993. Partitioning and combining data in phylogenetic analysis. Systematic Biology **42**:384–397.

Chippindale, P. T., and J. J. Wiens. 1994. Weighting, partitioning, and combining characters in phylogenetic analysis. Systematic Biology **43**:278–287.

Colless, D. H. 1980. Congruence between morphometric and allozyme data for *Menidia* species: a reappraisal. Systematic Zoology **29**:288–299.

Critchlow, D. E., D. K. Pearl, and C. Qian. 1996. The triples distance for rooted bifurcating phylogenetic trees. Systematic Biology **45**:323–334.

Cummings, M. P., S. P. Otto, and J. Wakeley. 1995. Sampling properties of DNA sequence data in phylogenetic analysis. Molecular Biology and Evolution **12**:814–822.

Day, W. H. E. 1986. Analysis of quartet dissimilarity measures between undirected phylogenetic trees. Systematic Zoology **35**:325–333.

de Queiroz, A. 1993. For consensus (sometimes). Systematic Biology **42**:368–372.

de Queiroz, A., M. J. Donoghue, and J. Kim. 1995. Separate versus combined analysis of phylogenetic evidence. Annual Review of Ecology and Systematics **26**:657–681.

Doyle, J. J. 1992. Gene trees and species trees: molecular systematics as one-character taxonomy. Systematic Botany **17**:144–163.

Doyle, J. A., M. J. Donoghue, and E. A. Zimmer. 1994. Integration of morphological and rRNA data on the origin of angiosperms. Annals of the Missouri Botanical Garden **81**:419–450.

Eernisse, D. J., and A. G. Kluge. 1993. Taxonomic congruence versus total evidence, and the phylogeny of amniotes inferred from fossils, molecules and morphology. Molecular Biology and Evolution **10**:1170–1195.

Estabrook, G. F. 1992. Evaluating undirected positional congruence of individual taxa between two estimates of the phylogenetic tree for a group of taxa. Systematic Biology **41**:172–177.

Estabrook, G. F., F. R. McMorris, and C. A. Meacham. 1985. Comparison of undirected phylogenetic trees based on subtrees of four evolutionary units. Systematic Zoology **34**:193–200.

Farris, J. S., M. Källersjö, A. G. Kluge, and C. Bult. 1995. Testing significance of incongruence. Cladistics **10**:315–319.

Felsenstein, J. 1993. PHYLIP, version 3.5c. University of Washington, Seattle.

Felsenstein, J. 1985a. Confidence limits on phylogenies: an approach using the bootstrap. Evolution **39**:783–791.

Felsenstein, J. 1985b. Confidence limits on phylogenies with a molecular clock. Systematic Zoology **34**:152–161.

Finden, C. R., and A. D. Gordon. 1985. Obtaining common pruned trees. Journal of Classification **2**:255–276.

Gordon, A. D. 1980. On the assessment and comparison of classifications. In Analyse de Donnees et Informatique, ed. R. Tomassone, pp. 149–160. INRIA, Le Chesnay, France.

Gornall, R. J., and B. A. Bohm. 1985. A monograph of *Boykinia, Peltoboykinia, Bolandra,* and *Suksdorfia* (Saxifragaceae). Botanical Journal of the Linnean Society **90**:1–71.

Hendy, M. D., C. H. C. Little, and D. Penny. 1984. Comparing trees with pendant vertices labelled. SIAM Journal of Applied Mathematics **44**:1054–1065.

Hillis, D. M. 1995. Approaches for assessing phylogenetic accuracy. Systematic Biology **44**:3–16.

Hillis, D. M., and J. J. Bull. 1993. An empirical test of bootstrapping as a method for assessing confidence in phylogenetic analysis. Systematic Biology **42**:182–192.

Huelsenbeck, J. P., and J. J. Bull. 1996. A likelihood ratio test for detection of conflicting phylogenetic signal. Systematic Biology **45**:92–98.

Huelsenbeck, J. P., and D. M. Hillis. 1993. Success of phylogenetic methods in the four-taxon case. Systematic Biology **42**:247-264.

Huelsenbeck, J. P., D. L. Swofford, C. W. Cunningham, J. J. Bull, and P. J. Waddell. 1994. Is character weighting a panacea for the problem of data heterogeneity in phylogenetic systematics? Systematic Biology **43**:288–291.

Huelsenbeck, J. P., J. J. Bull, and C. W. Cunningham. 1996. Combining data in phylogenetic analysis. Trends in Ecology and Evolution **11**:152–158.

Humphries, C. 1995. Biodiversity and phylogeny: XIII international meeting of the Willi Hennig Society (meeting review). Cladistics **11**:385–398.

Johnson, L. A. 1996. A molecular approach to resolving phylogenetic relationships in Polemoniaceae. Ph.D. dissertation, Washington State University, Pullman.

Johnson, L. A., and D. E. Soltis. 1994. *matK* DNA sequences and phylogenetic reconstruction in Saxifragaceae s. str. Systematic Botany **19**:143–156.

Johnson, L. A., and D. E. Soltis. 1995. Phylogenetic inference in Saxifragaceae sensu stricto and *Gilia* (Polemoniaceae) using *matK* sequences. Annals of the Missouri Botanical Garden **82**:149–175.

Johnson, L. A., J. S. Schultz, D. E. Soltis, and P. S. Soltis. 1996. Monophyly and generic relationships of Polemoniaceae based on *matK* sequences. American Journal of Botany **83**:1207–1224.

Jones, T. R., A. G. Kluge, and A. J. Wolf. 1993. When theories and methodologies clash: a phylogenetic reanalysis of the North American ambystomatid salamanders (Caudata: Ambystomatidae). Systematic Biology **42**: 92–102.

Kim, K.-J., and R. K. Jansen. 1994. Comparisons of phylogenetic hypotheses among different data sets in dwarf dandelions (*Krigia*): additional information from internal transcribed spacer sequences of nuclear ribosomal DNA. Plant Systematics and Evolution **190**:157–185.

Kluge, A. G. 1989. A concern for evidence and a phylogenetic hypothesis of relationships among *Epicrates* (Boidae, Serpentes). Systematic Zoology **38**:7–25.

Kluge, A. G., and J. S. Farris. 1969. Quantitative phyletics and the evolution of anurans. Systematic Zoology **18**:1–32.

Kluge, A. G., and A. J. Wolf. 1993. Cladistics: what's in a word? Cladistics **9**:183–199.

Kubicka, E., G. Kubicki, and F. R. McMorris. 1995. An algorithm to find agreement subtrees. Journal of Classification **12**:91–101.

Lanyon, S. M. 1985. Detecting internal inconsistencies in distance data. Systematic Zoology **34**:397–403.

Lanyon, S. M. 1988. The stochastic mode of molecular evolution: what consequences for systematic investigations? Auk **105**:565–573.

Lanyon, S. M. 1993. Phylogenetic frameworks: towards a firmer foundation for the comparative approach. Biological Journal of the Linnaean Society **49**:45–61.

Larson, A. 1994. The comparison of morphological and molecular data in phylogenetic systematics. **In** Molecular Ecology and Evolution: Approaches and Applications, eds. B. Schierwater, B. Streit, G. P. Wagner, and R. DeSalle, pp. 371–390. Birkhaüser Verlag, Basel, Switzerland.

Lutzoni, F., and R. Vilgalys. 1995. Integration of morphological and molecular data sets in estimating fungal phylogenies. Canadian Journal of Botany (suppl.) **73**:S649–S659.

Maddison, D. R. 1991. The discovery and importance of multiple islands of most-parsimonious trees. Systematic Zoology **40**:315–328.

Maddison, W. P., and D. R. Maddison. 1992. MacClade, version 3.05. Sinauer Associates, Sunderland, Massachusetts.

Maddison, W. P., and M. Slatkin. 1991. Null models for the number of evolutionary steps in a character on a phylogenetic tree. Evolution **45**:1184–1197.

Mason-Gamer, R. J., and E. A. Kellogg. 1996. Testing for phylogenetic conflict among molecular data sets in the tribe Triticeae (Gramineae). Systematic Biology **45**:522–543.

Mickevich, M. F., and J. S. Farris. 1981. The implications of congruence in *Menidia*. Systematic Zoology **30**:351–370.

Mickevich, M. F., and N. I. Platnick. 1989. On the information content of congruence in *Menidia*. Systematic Zoology **30**:351–370.

Miyamoto, M. M. 1996. A congruence study of molecular and morphological data for Eutherian mammals. Molecular Phylogenetics and Evolution 6:373–390.

Miyamoto, M. M., and W. M. Fitch. 1995. Testing species phylogenies and phylogenetic methods with congruence. Systematic Biology **44**:64–76.

Olmstead, R. G., and J. A. Sweere. 1994. Combining data in phylogenetic systematics: an empirical approach using three molecular data sets in the Solanaceae. Systematic Biology **43**:467–481.

Page, R. D. M. 1991. Random dendrograms and null hypotheses in cladistic biogeography. Systematic Zoology **40**:54–62.

Page, R. D. M. 1993. COMPONENT 2.0. The Natural History Museum, London.

Patterson, R. 1977. A revision of *Linanthus* sect. *Siphonella* (Polemoniaceae). Madroño **24**:36–48.

Penny, D., and M. D. Hendy. 1985a. Testing methods of evolutionary tree construction. Cladistics **1**:266–278.

Penny, D., and M. D. Hendy. 1985b. The use of tree comparison metrics. Systematic Zoology **34**:75–82.

Penny, D., L. R. Foulds, and M. D. Hendy. 1982. Testing the theory of evolution by comparing phylogenetic trees constructed from five different protein sequences. Nature **297**:197–200.

Porter, J. M. 1997. Phylogeny of Polemoniaceae based on nuclear ribosomal transcribed spacer DNA sequences. Aliso **15**:57–77.

Rice, W. R. 1989. Analyzing tables of statistical tests. Evolution **43**:223–225.

Rieseberg, L. H. 1995. The role of hybridization in evolution: old wine in new skins. American Journal of Botany **78**:1218–1237.

Rieseberg, L. H., and D. E. Soltis. 1991. Phylogenetic consequences of cytoplasmic gene flow in plants. Evolutionary Trends in Plants **5**:65–84.

Robinson, D. F., and L. R. Foulds. 1979. Comparisons on weighted labelled trees. **In** Lecture Notes in Mathematics, Volume 748, pp. 119–126. Springer-Verlag, Berlin.

Robinson, D. F., and L. R. Foulds. 1981. Comparison of phylogenetic trees. Mathematical Biosciences **53**:131–147.

Rodrigo, A. G., M. Kelly-Borges, P. R. Bergquist, and P. L. Bergquist. 1993. A randomisation test of the null hypothesis that two cladograms are sample estimates of a parametric phylogenetic tree. New Zealand Journal of Botany **31**:257–268.

Rohlf, F. J. 1982. Consensus indices for comparing classifications. Mathematical Biosciences **59**:131–144.

Ronquist, F. 1996. Matrix representation of trees, redundancy, and weighting. Systematic Biology **45**:247–253.

Seelanan, T., H. Schnabel, and J. F. Wendel. 1997. Congruence and consensus in the cotton tribe. Systematic Botany **22**:259–290.

Siegel, S. 1956. Nonparametric Statistics for the Behavioral Sciences. McGraw-Hill, New York.

Simberloff, D. 1987. Calculating probabilities that cladograms match: a method of biogeographical inference. Systematic Zoology **36**:175–195.

Slowinski, J. B. 1990. Probabilities of n-trees under two models: a demonstration that asymmetrical interior nodes are not improbable. Systematic Zoology **39**:89–94.

Smith, J. F., and K. J. Sytsma. 1994. Molecules and morphology: congruence of data in *Columnea* (Gesneriaceae). Plant Systematics and Evolution **193**:37–52.

Sokal, R. R., and F. J. Rohlf. 1981. Biometry, second edition, W. H. Freeman & Co., New York.

Sokal, R. R., and F. J. Rohlf. 1995a. Biometry, third edition, W. H. Freeman & Co., New York.

Sokal, R. R., and F. J. Rohlf. 1995b. Statistical Tables, third edition, W. H. Freeman & Co., New York.

Soltis, D. E. 1987. Karyotypes and relationships among *Bolandra, Boykinia, Peltoboykinia,* and *Suksdorfia* (Saxifragaceae: Saxifrageae). Systematic Botany **12**:14–20.

Soltis, D. E., and R. K. Kuzoff. 1995. Discordance between molecular and chloroplast phylogenies in the *Heuchera* group (Saxifragaceae). Evolution **49**:727–742.

Soltis, D. E., P. S. Soltis, T. G. Collier, and M. L. Edgerton. 1991. Chloroplast DNA variation within and among genera of the *Heuchera* group (Saxifragaceae): evidence for chloroplast transfer and paraphyly. American Journal of Botany **78**:1091–1112.

Soltis, D. E., D. R. Morgan, A. Grable, P. S. Soltis, and R. K. Kuzoff. 1993. Molecular systematics of Saxifragaceae sensu stricto. American Journal of Botany **80**:1056–1081.

Soltis, D. E., L. A. Johnson, and C. Looney. 1996a. Discordance between ITS and chloroplast topologies in the *Boykinia* group (Saxifragaceae). Systematic Botany **21**:169–185.

Soltis, D. E., R. K. Kuzoff, E. Conti, R. Gornall, and K. Ferguson. 1996b. *matK* and *rbcL* gene sequence data indicate that *Saxifraga* (Saxifragaceae) is polyphyletic. American Journal of Botany **83**:371–382.

Spongberg, S. A. 1972. The genera of Saxifragaceae in the southeastern United States. Journal of the Arnold Arboretum **53**:409–498.

Steel, M. A., and D. Penny. 1993. Distributions of tree comparison metrics—some new results. Systematic Biology **42**:126–141.

Strimmer, K, and A. von Haeseler. 1996. Quartet puzzling: a maximum likelihood method for reconstructing tree topologies. Molecular Biology and Evolution **13**:964–969.

Sullivan, J. 1996. Combining data with different distributions of among-site rate variation. Systematic Biology **45**:375–380.

Swofford, D. L. 1991. When are phylogeny estimates from molecular and morphological data incongruent? **In** Phylogenetic Analysis of DNA Sequences, eds. M. M. Miyamoto and J. Cracraft, pp. 295–333. Oxford University Press, New York.

Swofford, D. L. 1997. PAUP*: Phylogenetic Analyses Using Parsimony, version 4.0. Sinauer Associates, Sunderland, Massachusetts, in press.

Templeton, A. R. 1983. Phylogenetic inference from restriction endonuclease cleavage site maps with particular reference to the evolution of humans and the apes. Evolution **37**:221–244.

Thompson, J. D., D. G. Higgins, and T. J. Gibson. 1994. CLUSTALW: improving the sensitivity of progressive multiple sequence alignment through sequence weighting, position-specific gap penalties and weight matrix choice. Nucleic Acids Research **22**:4673–4680.

Wells, E. F. 1984. A revision of the genus *Heuchera* (Saxifragaceae) in eastern North America. Systematic Botany Monographs **3**:45–121.

Wheeler, W. C., J. C. Gatesy, and R. Desalle. 1995. Elision: a method for accommodating multiple molecular sequence alignments with alignment-ambiguous sites. Molecular Phylogenetics and Evolution **4**:1–9.

Wiens, J. J., and T. W. Reeder. 1995. Combining data sets with different numbers of taxa for phylogenetic analysis. Systematic Biology **44**:548–558.

Wilcoxon, F., S. K. Katti, and R. A. Wilcox. 1970. Critical values and probability levels for the Wilcoxon rank sum test and the Wilcoxon signed rank test. In Selected Tables in Mathematical Statistics, Vol. 1, eds. H. L. Harter and D. B. Owen, pp. 171–259. Markham Publishing, Chicago.

Appendix 11.1. Comparison of sequence variation and character-state reconstructions on most parsimonious trees from separate and combined analyses of data sets from examples 1–4. cpDNA refers to the chloroplast data set (= *matK* for examples 1 and 2; cpDNA restriction sites for examples 3 and 4).

EXAMPLE Comparison	cpDNA	ITS	Combined
EXAMPLE 1: Polemoniaceae			
Number of potentially informative characters	158	159	317
Number of most parsimonious trees	4	20	1
Minimum tree length in absence of homoplasy	340	346	686
Observed tree length	431	570	1003
Tree lengths measured with cpDNA data	431	439–462	431
Tree lengths measured with ITS data	572–576	570	572
Tree lengths measured with combined data	1003–1007	1009–1032	1003
EXAMPLE 2: *Phlox* and related genera; 15 Species			
Number of potentially informative characters	68	102	170
Number of most parsimonious trees	1	1	4
Minimum tree length in absence of homoplasy	153	212	365
Observed tree length	176	296	478
Tree lengths measured with cpDNA data	176	182	180–182
Tree lengths measured with ITS data	312	296	296–298
Tree lengths measured with combined data	488	478	478
EXAMPLE 2: *Phlox* and related genera; 14 Species (*Linanthus breviculus* pruned)			
Number of potentially informative characters	64	96	160
Number of most parsimonious trees	1	3	4
Minimum tree length in absence of homoplasy	150	206	356
Observed tree length	172	275	449
Tree lengths measured with cpDNA data	172	174–178	172–174
Tree lengths measured with ITS data	279	275	275–277
Tree lengths measured with combined data	451	449–453	449
EXAMPLE 3: *Boykinia* group.			
Number of potentially informative characters	108	189	297
Number of most parsimonious trees	1	2	1
Minimum tree length in absence of homoplasy	132	399	531
Observed tree length	151	546	705
Tree lengths measured with cpDNA data	151	167	151
Tree lengths measured with ITS data	576	546	554
Tree lengths measured with combined data	727	713	705
EXAMPLE 4: *Heuchera* group.			
Number of potentially informative characters	62	89	151
Number of most parsimonious trees	1	66	6
Minimum tree length in absence of homoplasy	103	245	348
Observed tree length	113	337	499
Tree lengths measured with cpDNA data	113	217–233	144–151
Tree lengths measured with ITS data	448	337	348–355
Tree lengths measured with combined data	561	554–670	499

Appendix 11.2. Similarity values for six topological congruence indices within and between data sets for examples 1–4. PM = partition metric, EA = explicitly agree metric, D_1 = greatest agreement subtree metric, CI_C = Colless' consensus fork index, CI_1 = Rohlf's consensus information index, and TI = index of topological identity. Normalized values are given in parentheses for metrics commonly reported as numberical values. "Random" values were gathered from comparisons of 1000 pairs of trees selected from D_r (Box 11.9).

EXAMPLE Comparison	PM	EA	D_1	CI_C	CI_1	TI	I_M	I_{MF}	HT_F
EXAMPLE 1: Polemoniaceae									
Within *matK*									
Median	3 (0.94)	0.992	1 (0.96)	0.94	0.93	0.94			
Range	2–4 (0.91–0.96)	0.986–0.995	1–2 (0.91–0.96)	0.91–0.96	0.91–0.95	0.91–0.96			
Within ITS									
Median	6 (0.87)	0.976	2 (0.91)	0.87	0.84	0.87			
Range	1–10 (0.78–0.96)	0.932–0.997	1–5 (0.78–0.96)	0.78–0.96	0.66–0.97	0.78–0.96			
Between *matK* and ITS									
Median	12 (0.74)	0.940	4 (0.83)	0.74	0.62	0.74	3.08%	0.63%	0.872
Range	6–16 (0.65–0.87)	0.917–0.976	1–6 (0.74–0.96)	0.65–0.87	0.54–0.85	0.65–0.87			
Random									
Median	46 (0.00)	0.326	17 (0.26)	0.00	0.00	0.00			
Range	40–46 (0.0–0.13)	0.256–0.482	14–19 (0.17–0.39)	0.00–0.13	0.00–0.09	0.00–0.13			
EXAMPLE 2: *Phlox* and related genera; 15 Species									
Between *matK* and ITS	7 (0.71)	0.978	3 (0.75)	0.69	0.86	0.72	17.05%	5.31%	0.034
Random									
Median	24 (0.00)	0.457	8 (0.33)	0.08	0.14	0.08			
Range	18–24 (0.00–0.25)	0.306–0.726	5–10 (0.17–0.58)	0.08–0.31	0.14–0.30	0.08–0.31			
EXAMPLE 2: *Phlox* and related genera; 14 Species (*Linanthus breviculus* pruned)									
Within ITS									
Median	4 (0.82)	0.887	1 (0.91)	0.83	0.61	0.83			
Range	2–6 (0.73–0.91)	0.885–0.997	1–2 (0.81–0.91)	0.75–0.92	0.59–0.98	0.75–0.92			
Between *matK* and ITS									
Median	7 (0.68)	0.973	2 (0.82)	0.67	0.84	0.70	6.19%	2.15%	0.902
Range	5–9 (0.59–0.77)	0.863–0.975	1–3 (0.73–0.91)	0.58–0.75	0.50–0.86	0.61–0.78			
Random									
Median	22 (0.00)	0.467	8 (0.27)	0.08	0.02	0.08			
Range	16–22 (0.00–0.36)	0.321–0.731	5–9 (0.18–0.55)	0.08–0.33	0.02–0.42	0.08–0.33\			

Appendix 11.2. Similarity values for six topological congruence indices within and between data sets for examples 1–4. PM = partition metric, EA = explicitly agree metric, D₁ = greatest agreement subtree metric, CI_C = Colless' consensus fork index, CI₁ = Rohlf's consensus information index, and TI = index of topological identity. Normalized values are given in parentheses for metrics commonly reported as numberical values. "Random" values were gathered from comparisons of 1000 pairs of trees selected from D_R.

EXAMPLE Comparison	PM	EA	D₁	CI_C	CI₁	TI	I_M	I_MF	HT_F
EXAMPLE 3: *Boykinia* group									
Within ITS	4 (0.83)	0.987	2 (0.83)	0.85	0.90	0.85			
Between cpDNA and ITS	12 (0.50)	0.898–0.907	4–5 (0.58–0.67)	0.46	0.38	0.50	21.70%	4.60%	0.032
Random									
Median	24 (0.00)	0.547	7 (0.42)	0.08	0.14	0.08			
Range	18–24 (0.00–0.25)	0.290–0.708	5–9 (0.25–0.58)	0.08–0.31	0.14–0.31	0.08–0.31			
EXAMPLE 4: *Heuchera* group									
Within ITS									
Median	7 (0.87)	0.963	5 (0.82)	0.86	0.76	0.87			
Range	1–19 (0.66–0.98)	0.804–0.999	1–13 (0.54–0.96)	0.66–0.97	0.40–0.98	0.65–0.98			
Between *matK* and ITS									
Median	41 (0.27)	0.468	19 (0.32)	0.14	0.07	0.17	67.22%	32.45%	0.002
Range	38–41 (0.27–0.32)	0.403–0.479	18–20 (0.29–0.36)	0.14–0.17	0.07–0.08	0.14–0.21			
Median	56 (0.00)	0.387	21 (0.25)	0.03	0.07	0.03			
Range	48–56 (0.0–0.13)	0.248–0.564	17–24 (0.14–0.39)	0.03–0.14	0.07–0.14	0.03–0.14			
Upper 5%	52 (0.07)	0.478	22 (0.29)	0.069	0.069	0.069			

Appendix 11.3. Summary SLP$_T$ results for examples 1–5. Probability values greater than 0.05 indicate that the alternative topology is not significantly less parsimonious than at least one of the shortest trees from the data set being tested. Non-significant results that could not be explicitly tested because fewer than six characters vary are denoted with "†"

EXAMPLE Data set	Alternative topology	Number of steps Increase	Decrease	Net	Probability
EXAMPLE 1: Polemoniaceae					
matK	ITS-4	9	1	8	0.04
ITS	*matK*-3	12	7	5	0.37
EXAMPLE 2: *Phlox* and related genera					
matK	ITS	6	0	6	0.032
	Gilia filiformis clade	2	0	2	†
	Gymnosteris clade	4	0	4	†
ITS	*matK*	18	2	16	0.002
	Gilia filiformis clade	5	1	4	†
	Gymnosteris clade	13	1	12	0.002
EXAMPLE 3: *Boykinia* group					
cpDNA	ITS	18	2	16	0.002
	Combined	0	0	0	†
	Sullivantia	3	2	1	†
	Suksdorfia	9	0	9	0.004
	Boykina aconitifolia	2	0	2	†
	Boykinia major	4	0	4	†
ITS	cpDNA	39	9	30	<0.001
	Combined	20	12	8	0.08
	Sullivantia	5	4	1	1.00
	Suksdorfia	11	6	5	0.332
	Boykina aconitifolia	7	2	5	0.454
	Boykinia major	3	1	2	†
EXAMPLE 4: *Heuchera* group					
cpDNA	ITS	104	0	104	<0.001
	b1 (Fig. 11-7)	10	0	10	0.002
	b2 (Fig. 11-7)	10	0	10	0.002
	b3 (Fig. 11-7)	26	1	25	<0.001
	b4 (Fig. 11-7)	10	0	10	0.002
	b5 (Fig. 11-7)	2	0	2	†
ITS	cpDNA	84	1	83	<0.001
	a1 (Fig. 11-7)	24	2	22	<0.001
	a2 (Fig. 11-7)	38	4	34	<0.001
	a3 (Fig. 11-7)	13	6	7	0.211
	a4 (Fig. 11-7)	13	2	11	0.013
	a5 (Fig. 11-7)	11	4	7	0.146
EXAMPLE 5: Saxifragaceae s.s.					
matK	*rbcL*	75	2	73	<0.001
matK	ITS	81	0	81	<0.001
rbcL	*matK*	17	5	12	0.051
rbcL	ITS	22	2	20	<0.001
ITS	*matK*	98	24	74	<0.001
ITS	*rbcL*	131	27	104	<0.001
ITS	*matK*+*rbcL*	104	29	75	<0.001
	b1 (Fig. 11-9)	50	25	25	0.015
	b2 (Fig. 11-9)	25	15	10	0.155
matK+*rbcL*	ITS	96	5	91	<0.001
	a1 (Fig. 11-9)	60	5	55	<0.001
	a2 (Fig. 11-9)	12	4	8	0.076

Appendix 11.4. Comparison of sequence variation and character-state reconstructions for 38 species representing Saxifragaceae s.s. and two outgroups (example 5).

Comparison	*matK*	*rbcL*	[*matK+rbcL*]	ITS	[*matK+rbcL*+ITS]
Number of potentially informative characters	290	124	414	395	809
Number of most parsimonious trees	60	92	4	2	1
Minimum tree length in absence of homoplasy	632	267	899	983	1882
Observed tree length	897	431	1338	1993	3386
Tree lengths measured with *matK* data	897	970–1016	899–900	978	903
Tree lengths measured with *rbcL* data	443–451	431	438–439	451	443
Tree lengths measured with [*matK+rbcL*] data	1340–1348	1401–1447	1338	1429	1346
Tree lengths measured with ITS data	2067–2090	2097–2165	2073–2077	1993	2040
Tree lengths measured with combined data	3409–3438	3498–3612	3411–3415	3422	3386

Appendix 11.5. Example 5, Saxifragaceae s.s. Comparison of similarity values within and between data sets for six metrics: PM = partition metric, EA = explicitly agree quartet metric, D_1 = Greatest agreement subtree metric. Normalized values are given in parentheses for metrics commonly reported as numerical values. "Random" values were gathered from comparisons of 1000 pairs of trees selected from D_r (Box 5)

Comparison	PM	EA	D_1	I_M	I_{MF}	Prob. (HT$_F$)
Within *matK*						
Median	6 (0.91)	0.947	6 (0.83)			
Range	1–12 (0.83–0.99)	0.904–0.999	1–10 (0.71–0.97)			
Within *rbcL*						
Median	12 (0.83)	0.991	5 (0.86)			
Range	1–22 (0.69–0.99)	0.981–0.999	1–11 (0.69–0.97)			
Within ITS	2 (0.97)	0.999	1 (0.97)			
Within [*matK* + *rbcL*]						
Median	4 (0.94)	0.981	6 (0.83)			
Range	2–6 (0.91–0.97)	0.981–0.999	1–7 (0.80–0.97)			
Between *matK* and *rbcL*				16.54%	4.03%	0.02
Median	31 (0.56)	0.949	14 (.60)			
Range	24–40 (0.43–0.66)	0.892–0.988	9–20 (0.43–0.74)			
Between *matK* and ITS				10.84%	3.92%	0.002
Median	31 (0.56)	0.883	13 (0.63)			
Range	29–34 (0.51–0.59)	0.854–0.908	9–17 (0.51–0.74)			
Between *rbcL* and ITS				9.55%	1.59%	0.032
Median	38 (0.46)	0.890	15 (0.571)			
Range	34–42 (0.40–0.51)	0.886–0.894	13–17 (0.51–0.63)			
Between [*matK* + *rbcL*] and ITS				10.56%	3.66%	0.002
Median	30 (0.57)	0.895	12.5 (0.643)			
Range	30–30 (0.57–0.57)	0.891–0.898	12–13 (0.629–0.657)			
Random						
Median	70 (0.00)	0.379	28 (0.20)			
Range	64–70 (0.0–0.09)	0.266–0.565	23–28 (0.2–0.34)			

12

Ontogenetic Systematics, Molecular Developmental Genetics, and the Angiosperm Petal

Victor A. Albert, Mats H. G. Gustafsson, and Laura Di Laurenzio

> *Anyone who pays a little attention to the growth of plants will readily observe that certain of their external members are sometimes transformed, so that they assume—either wholly or in some lesser degree—the form of the members nearest in the series ... If we note that it is in this way possible for the plant to take a step backwards and thus to reverse the order of growth, we shall obtain so much the more insight into Nature's regular procedure; and we shall make the acquaintance of the laws of transmutation, according to which she produces one part from another, and sets before us the most varied forms through modification of a single organ.*
>
> (J. W. von Goethe, 1790; translation by Arber, 1946, p. 90)

Phylogenetic systematics discovers pattern; ontogenetic systematics uncovers processes behind pattern. Plant diversity, as recognized in the field, herbarium, or library, stems from the diversity of plant form. How an organism develops determines its phenotype, and therefore differences among ontogenies are what generate diversity. The molecular basis for these differences is of fundamental importance to plant systematics, yet the topic remains poorly understood.

To date, comparative molecular studies in systematic botany have been concerned with the discovery of phylogenetic pattern (e.g., Sytsma and Gottlieb, 1986; Chase, Soltis, Olmstead et

VAA thanks Piero Delprete, Michael Donoghue, Paula Elomaa, Peter Endress, Peter Engström, Chuck Gasser, Yrjö Helariutta, Tom Lammers, Mika Kotilainen, Merja Mehto, Scott Mori, Lena Struwe, Karolina Tandre, Teemu Teeri, and Deyue Yu for helpful discussions and/or fruitful collaborations. We thank Peter Crane, Jeff Doyle, Susana Magallon, Ole Seberg, and an anonymous reviewer for comments on an earlier version of this paper. We also thank Mark Chase, Piero Delprete, Michael Donoghue, Peter Engström, Lena Struwe, Peter Stevens, Mats Svensson, Karolina Tandre, and the Teeri Lab (Institute of Biotechnology, Helsinki) for communicating unpublished observations.

Research on several of the systems described is supported by the Lewis B. and Dorothy Cullman Foundation. Research on the *Gerbera* MADS box gene system has been supported by grants from the Swedish Natural Science Research Council (to VAA) and the Academy of Finland (to Teemu H. Teeri). MHGG is supported by a postdoctoral fellowship from the Swedish Natural Science Research Council. Preparation of this paper was supported by the Cullman Foundation. The Bassett Maguire microscopic slide collection (NY) was examined for our studies of Clusiaceae.

al., 1993). Phylogenetic trees, among other important things, permit new, monophyletic classifications (e.g., Spooner et al., 1993; Struwe et al., 1994) as well as hypotheses of character evolution (e.g., Olmstead et al., 1993; Weller et al., 1995). The processes that generate character-state distributions will not be directly observable through tree topologies, but correlations among characters, possibly of a causal nature, may be testable (Maddison, 1990; cf. Werdelin and Tullberg, 1995). This approach to studying character evolution is pedagogically useful but biologically incomplete. It is one thing to test whether traits such as fleshy fruits and dioecy are correlated on a phylogenetic tree (Donoghue, 1989); it is another to pursue the common or distinct molecular mechanisms for dry versus fleshy fruits and monoecy versus dioecy among different lineages of plants.

Over the past decade, investigations of model plant systems such as *Arabidopsis thaliana* have uncovered an array of genes that control aspects of root, shoot, leaf, and flower morphogenesis (e.g., Coen and Meyerowitz, 1991; Mayer et al., 1991; Vollbrecht et al., 1991; Bowman, 1994; Di Laurenzio et al., 1996). To be useful as tools for ontogenetic systematics, these genes and their activities must be related through a comparative framework. Molecular and other data are now capable of providing phylogenetic links for a great many taxa, including *Arabidopsis* to the Brassicaceae, Capparales, mustard-oil producing clade, eudicots, and angiosperms as a whole (Chase, Soltis, Olmstead et al., 1993; Rodman et al., 1993, Price et al., 1994). Determining the "homology" of gene activity among species is also technically feasible through RNA detection (e.g., Ingram et al., 1995) and transgenics methods (e.g., complementation studies; Irish and Yamamoto, 1995), but terminology to describe such research is inadequate.

Homology has at least three distinct connotations in ontogenetic systematics. First is *historical homology*. A morphological trait such as a nectar spur may or may not be historically correlated through a single origin on a phylogenetic tree. For example, *Impatiens* (Balsaminaceae) and *Utricularia* (Lentibulariaceae) are only distantly related (Chase, Soltis, Olmstead et al., 1993), and therefore their nectar spurs are not

historically homologous. Even if historical homology could be demonstrated, a complication arises if nectar spurs of different taxa derive from different perianth whorls; *Impatiens* spurs are in the sepal whorl, whereas *Utricularia* spurs are in the petal whorl (Mabberley, 1987). Therefore, these nectar spurs do not share *positional homology*. However, nectar spurs of *Impatiens* may be truly petaloid in the genetic sense, that is, through a transference of function of the same petal-determining genes involved in the nectar spur development of *Utricularia*. This condition would indicate *process homology*. These concepts of homology are readily extended using Fitch's (1970) terminology (Fig. 12.1; cf. Patterson, 1988; De Pinna, 1991, pp. 388–390). *Historical orthology, positional orthology,* and *process orthology* all describe phenomena of the type above that are traceable (respectively) to single lineages of organisms, to single lineages of cells, organ primordia or whorls of primordia, or to developmental control by orthologous genes. In contrast, *historical paralogy, positional paralogy,* and *process paralogy* refer to phenomena that involve (respectively) different monophyletic groups of organisms, different cell lineages, organ primordia, or whorls, or differential developmental control by paralogous genes.

Here we discuss prospects for ontogenetic systematics research in plant diversity studies using, as an example, the development and evolution of angiosperm petals.

DETERMINATION AND DEVELOPMENT OF PETALS

Structural and Functional Morphology

Petals are determined by a discrete molecular developmental program that distinguishes them, in normal cases, from the other organs of the flower. Before discussing these mechanisms, it is useful to review the major differences between petals and sepals, which together comprise the perianth of most flowers (Endress, 1994). We restrict this review to eudicots (sensu Doyle and Hotton, 1991), which comprise approximately 75% of all angiosperm species (Drinnan et al., 1994) and all of the molecular developmental examples discussed in this chapter.

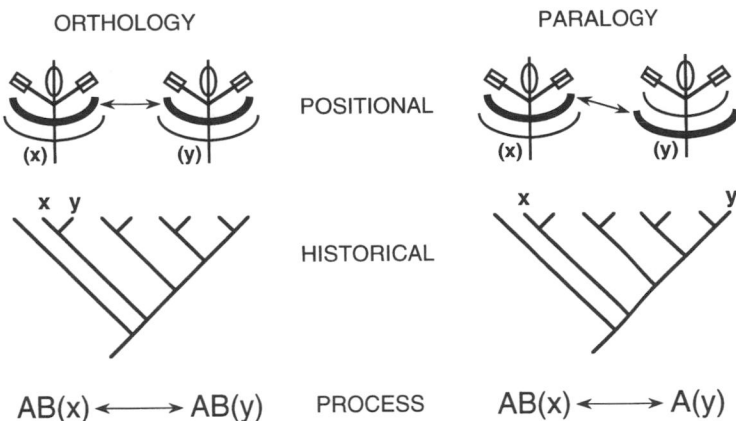

Figure 12.1. Homology concepts in ontogenetic systematics studies. A homology statement can be made at the positional, historical, or process levels. Use of the terms orthology and paralogy (Fitch, 1970) provides a more specific comparison (see Patterson, 1988; De Pinna, 1991). Positional orthology and paralogy refers to comparisons among structures by position, for example, when comparing the perianth whorls of angiosperm flowers. If sepals and petals are considered serially homologous, then intertaxon comparisons (arrows) of petals to petals versus petals to sepals reveal *positional orthology* versus *positional paralogy,* respectively. In turn, if the taxa in question (x and y) are analyzed for their phylogenetic relationship to one another, monophyly would suggest *historical orthology* of the perianth parts being compared, and polyphyly might indicate their *historical paralogy.* Finally, if developmental processes are shared among the relevant organ systems (AB = petals in the ABC model for floral organ determination; Coen and Meyerowitz, 1991), then these structures may be termed process orthologous. A comparison between petals and sepals (right) may demonstrate *process paralogy* with respect to B or A+B but *process orthology* with respect to A only.

Sepals comprise the first whorl of the flower, and petals the second (see Endress, 1994, for amplification on the synopsis below). Petals may function in the protection of the inner whorls of sex organs, as display features for attracting animal pollinators, and for the secretion of nectar rewards. Sepals also perform a protective function, although primarily in the earliest stages of flower ontogeny. Normal flower development usually leads to sepals being vastly exceeded in their dimensions by petals, and hence the loss of their early protective role (e.g., in most sympetalous angiosperms). In other cases, sepals may continue to enlarge and persist through fruit development, often as a dispersal aid.

Petals, like a single whorl of stamens, may initiate nearly simultaneously among eudicots, whereas sepals often have a more pronounced spiral phyllotaxis. Petals, like stamens, often display a developmental lag whereby rapid differential growth occurs near the time of anthesis. Sepals, as just described, are usually arrested in their growth at an early stage. The vascular systems of petals (and stamens) often lack the scle-

renchymatous cells characteristic of sepals and foliar leaves. Vascular patterning in petals is often derived from a single basal trace (as in stamens) that bifurcates and ramifies toward the sometimes bifid petal apex. In contrast, sepals often have three basal traces and acuminate tips. Additionally, the bases of petals are often narrow compared to their apices whereas the bases of sepals are often as broad as the entire organ.

Petals are often brightly colored with anthocyanins, carotenoids, or both, whereas sepals are characteristically green. Petals usually lack stomates except when they are very thick. Sepals may have stomates on both surfaces, and their indument of trichomes frequently matches that of foliar leaves. Petals sometimes show this similarity, but usually then on their abaxial surfaces. Petals usually have a simple parenchyma with extensive vacuolation, whereas sepals often have palisade parenchyma layers similar to those of foliage leaves.

To generalize, petals show similarities to stamens whereas sepals are similar to foliage leaves. Stamens, like petals, are often optical attractants, and some are capable of producing nectar. Of course, organs in the sepal whorl with petaloid features may also serve as optical attractants and secrete nectar (e.g., the sepalar spur of *Impatiens*), but are these organs process orthologous with normal sepals or only positionally orthologous?

Molecular Control of Organ Determination

MADS Box Genes
The identities of flower organs are controlled by a limited set of genes. Many of these so-called homeotic genes are members of the same eukaryotic gene superfamily, known as the MADS box genes (see the review by Theißen et al., 1996). Many MADS box genes have a determination function analogous to the homeobox pattern formation genes of animals. Both gene superfamilies encode transcription factors, proteins that regulate the expression of other genes. MADS box genes are known from animals and fungi, but unlike those of plants, most of these genes are involved in regulating cellular differentiation. Likewise, many homeobox genes of plants are members of signal transduction pathways (Mattsson et al., 1992; Carabelli et al., 1993; Sessa et al., 1993; Söderman et al., 1994), although there are exceptions (e.g., Vollbrecht et al., 1991; Reiser et al., 1995). MADS and homeobox genes do interact in yeast (Herschbach et al., 1994; Mead et al., 1996), so there is reason to believe that they may form regulatory relationships during plant development.

MADS box genes encode proteins with two distinctive domains; the MADS domain (Schwarz-Sommer et al., 1990), conserved across eukaryotic life, and the K domain (Ma et al., 1991), which is known only from plants. The MADS domain may be a site of DNA binding (Nurrish and Treisman, 1995) and both the MADS and K domains may function in protein dimerization (Davies and Schwarz-Sommer, 1994; Shore and Sharrocks, 1995).

Plant MADS box genes fall into different functional classes that strongly correlate with membership in particular gene lineages (J. J. Doyle, 1994; Purugganan et al., 1995; Tandre et al., 1995). These lineages, or orthologues, appear to be descended from a common ancestral gene through a number of duplication events. An an-

cient derivation for several of these orthologues is supported by including both angiosperm and Norway spruce MADS box genes in phylogenetic analysis (Tandre et al., 1995). The spruce sequences intercalate within previously recognized angiosperm gene lineages rather than forming a group of their own. Therefore, several MADS box orthologues are older than the divergence of conifers and angiosperms, which occurred over 300 million years ago.

Gene duplication is a common theme in the diversification of gene function (e.g., Helariutta et al., 1996). MADS box genes have various known and putative activities, ranging from root, embryo, and vegetative development, prepatterning of the floral meristem, determination of the four flower whorls, to ovule development (Rounsley et al., 1995). The ancestral function of plant MADS box genes remains unknown. MADS box genes being isolated from Lycopodiopsida (M. Svensson, pers. comm.; Svensson and Engström, 1997), the sister group of all other vascular plants (Raubeson and Jansen, 1992), may help resolve this issue.

The ABC Model
Genetic studies of *Arabidopsis* and *Antirrhinum* mutants led to a model for flower organ determination that included the combinatorial action of three functions (Coen and Meyerowitz, 1991). Two of these functions are now known to be encoded solely by MADS box genes, the third by MADS box and other, unrelated genes.[1]

The A function was seen to be expressed in sepals and petals. By itself, expression of A would lead to sepals. Combined with B, A would lead to petals. The *Arabidopsis* MADS box gene *APETALA1* (*AP1*; Mandel et al., 1992) and possibly also its orthologues (e.g., *SQUAMOSA* of *Antirrhinum*; Huijser et al., 1992) are A-function genes. The other principal A-function gene is *APETALA2* (*AP2*; Jofuku et al., 1994), a putative

---

[1]As MADS box genes have become the most useful molecular markers for studies of perianth identity, we have restricted our discussion to these genes and do not discuss other factors acting in flower organ development (for more general reviews of the genetic control of reproductive development, see Okamuro et al., 1993; Coen and Nugent, 1994; Weigel and Meyerowitz, 1994; Ma, 1994; Theißen and Saedler, 1995; Bradley et al., 1996; Weigel and Clark, 1996).

transcription factor that is part of a unique gene and protein family. It was later found that only *AP1* expression is truly restricted to the first and second floral whorls (Mandel et al., 1992); *AP2* expresses in carpels and the outer integuments of ovules (Jofuku et al. 1994; Modrusan et al., 1994), although expression patterns differ among different *AP2* alleles (C. S. Gasser, pers. comm.). *AP1* orthologues reside in a complex, larger clade that includes genes with diverse vegetative, inflorescence, and floral expression patterns (see Purugganan et al., 1995; Theißen et al., 1996).

The B function expresses principally in petals and stamens, but also in the outer integuments of ovules (Jack et al., 1994). All B-function genes identified so far are MADS box genes. B-function genes form a single lineage with at least two orthologues, one corresponding to the *Arabidopsis* gene *APETALA3* (*AP3;* Jack et al., 1992), the other to *PISTALLATA* (*PI*). *Antirrhinum* orthologues of these genes are *DEFICIENS* (*DEF;* Sommer et al., 1990) and *GLOBOSA* (*GLO;* Tröbner et al., 1992), respectively. Protein products of these genes may form AP3/PI and DEF/GLO heterodimers (Tröbner et al., 1992; Goto and Meyerowitz, 1994) that regulate further expression of both gene pairs (B orthologue dynamics differ in some species, e.g., *Petunia;* van der Krol et al., 1993; see B-Function MADS Box Genes: other Petal Functions). Factors that prevent B-function gene expression in the sepal whorl remain unknown.

The C function is expressed in the reproductive organs. C expression alone leads to carpels. Combined with B, C leads to stamens. All known C-function genes are MADS box genes related to the *Arabidopsis* gene *AGAMOUS* (*AG;* Yanofsky et al., 1990). The *Antirrhinum* orthologue is *PLENA* (*PLE;* Bradley et al., 1993). Some plants (e.g., *Petunia;* Tsuchimoto et al., 1993) have two *AG* orthologues. Some phylogenetic analyses suggest that the C- and B-function orthologues may be sister lineages, descended from a single gene duplication event (J. J. Doyle, 1994; Tandre et al., 1995).

To summarize, in its simplest form, the ABC model specifies that A = sepals, A + B = petals, B + C = stamens, and C = carpels. Recently, ovules have been recognized as a fifth organ category, and ovule-specific D-function MADS

box genes have been identified (Colombo et al., 1995). Further complexity beyond the simple ABC model is indicated by "ectopic" expression of A- and B-function genes in developing ovules (e.g., Jack et al., 1994; Modrusan et al., 1994).

An additional complication is that organ identity may be conferred in part by a concentration effect. For example, cosupression transgenics for the *Petunia* MADS box genes *PMADS1* and *FBP2* show different degrees of organ identity transformation depending upon the number of gene copies inserted (van der Krol et al., 1993; Angenent et al., 1994). Transcription factors such as MADS domain proteins act as on/off switches in the regulation of other genes, but this activity is modulated by the intracellular concentrations of these factors. Thus, greater or lesser transcription and/or post-translational modification (e.g., phosphorylation, dimerization, or nuclear localization) of A- or B-function genes could lead to aB or Ab petals (lower case indicating weak function, upper case, strong function) with different phenotypes, the latter possibly sepaloid. Ectopic expression of C in the petal whorl or A in the staminal whorl could likewise give rise to petal-stamen intermediates. Even though the C and A functions may share a negative regulatory relationship (in *Arabidopsis,* AP2 is a negative regulator of *AG* in the outer floral whorls, and AG is a negative regulator of *AP1* in the inner whorls; see Theißen and Saedler, 1995, p. 633), PLE (a C-function protein) and SQUA (a potential A-function protein) are nevertheless known to interact within some cells of *Antirrhinum* (Davies et al., 1996).

Serial Homology:
Are Petals Sterile Stamens?

It is a very probable idea that the colour and scent of the petals are to be attributed to the presence in them of the male fertilising substance. Probably this substance occurs in them in a state in which it is not yet sufficiently isolated, but mixed and diluted with other juices.

(J. W. von Goethe, 1790; translation by Arber, 1946, p. 99)

Goethe was the first to state clearly the special relationship of the organs of flowers. Foliage leaves, sepals, petals, stamens, and carpels were envisaged to be different transformations of the same basic appendicular structure. This theory of "serial homology" requires (1) process orthology of basic developmental regulation, (2) positional paralogy via phyllotactic differentiation, and (3) process paralogy of identity-determining mechanisms. Process orthology of foliar and flower organs is suggested by mutant analysis. When the A, B, and C functions of *Arabidopsis* are eliminated by mutagenesis, leafy structures form in place of floral organs (Bowman et al., 1991). Positional paralogy of appendicular organs is self-evident, but it is not so obvious how domains of altered organ identity are maintained. Establishment of the four floral whorls probably occurs before the ABC determination system begins to operate (Day et al., 1995), and when whorl identity is conferred, this is not influenced by the identities of previously initiated whorls. Process paralogy results from the differential, overlapping expression of A-, B-, and C-function genes during organogenesis.

Central to Goethe's theory of flowers—and supported by the ABC model—was that proximal whorls were each others' closest relatives. As such, he likened petals to stamens and described "intermediate organs" such as nectaries and coronas as "gradual transitions between the petals and the stamens" (Arber, 1946, pp. 100–101). These and other intermediate organs can be observed in many angiosperm families. For example, the stamens of many Magnoliales (e.g., *Austrobaileya;* Endress, 1994) are flat and petaloid. In some taxa of Nymphaeaceae, laminate perianth organs actually grade morphologically into stamens (Heywood, 1985). Based on such cases of similarity, Ĉelakovský (1896/1900) and others (e.g., Hiepko, 1965) have advanced the view that angiosperm petals are evolutionarily derived from stamens, and sepals from foliar bracts. Petals would then be sterile stamens modified for optical attraction of insect pollinators. Many petals are extremely similar to stamens in their morphology and ontogeny (e.g., in Ranunculales; Hiepko, 1965; Kosuge, 1994). Correspondingly, the perianth of Magnoliales and Laurales would represent sepals because of their more bractlike morphological characteristics (see foregoing discussion). Takhtajan (1991) propounded a variant

of this model with two separate (paralogous) origins for petals, from bracts in the Magnoliales and Illiciales (bracteopetals), and from stamens in Ranunculales, Cactaceae, Rosaceae, etc. (andropetals).

However, both models confound petaloidy (potential process orthology) with nominal typology, historical orthology/paralogy, or both. The name *sepal* applies well only when whorl identity can be clearly defined, and such is definitely not the case with many Magnoliales and Laurales, where floral organs may be borne in spiral cycles (Endress, 1994). The name *petal* can bear the same applicability problem, and as pointed out by Endress (1994), organs in the petal whorl may have the structural and developmental features usually associated with sepals! Therefore, an organ terminology based principally on process orthology/paralogy could be adopted for consistency; this would always distinguish petals (AB) from sepals (A), stamens (BC), and staminodes (ostensibly ABC), regardless of historical or positional orthology/paralogy. Use of A-, B-, and C-function gene and protein markers would seem to provide more objective criteria for organ identity diagnoses than phenotypic characteristics alone. In terms of gene expression, petals certainly are like stamens—but they are also like sepals—and so Goethe's insistence that serial homology must be bidirectional is clearly correct: "For we can just as well say that a stamen is a contracted petal, as we can say of a petal that it is a stamen in a state of expansion" (Arber, 1946, p. 115).

B-Function MADS Box Genes: Other Petal Functions

Expression of B-function MADS box genes provides the decisive identity difference between sepals versus petals and stamens versus carpels. In performing these organ determination functions, B-function genes may have played a major role in morphological and functional evolution of the angiosperm perianth and sexual systems. Other roles of B-function genes, in petal differentiation pathways, may also have been evolutionarily influential.

Molecular analysis of petal development in *Petunia* demonstrates that B-function genes control key aspects of angiosperm sympetaly. As noted above, gene products of B-function orthologues in *Petunia* show altered regulatory relationships with respect to *Arabidopsis* and *Antirrhinum*. There are two *PI/GLO* orthologues (*FBP1* and *PMADS2;* Angenent et al., 1993; van der Krol et al., 1993) and one *AP3/DEF* orthologue (*PMADS1*). Unlike the situation with *Antirrhinum DEF* and *GLO, FBP1* and *PMADS2* are regulated by PMADS1 in the petal whorl but not stamen whorl (van der Krol et al., 1993). *Antirrhinum* mutants that lack *DEF* expression have sepals in place of petals and carpels in place of stamens (Sommer et al., 1990), whereas *Petunia* mutants that lack *PMADS1* show the petal-to-sepal transformation while retaining normal stamens (van der Krol et al., 1993; van der Krol and Chua, 1993). Thus, *PMADS1* is a petal-specific B-function gene.

Elimination of *PMADS1* function produces another interesting effect, the abolishment of the petal/stamen adnation characteristic of sympetaly in the Asteridae (van der Krol et al., 1993). Congenital union of the corolla members is, however, retained in these mutants. Therefore, *PMADS1* is not required for congenital fusion of petals, but it is required for the congenital fusion of petals with stamens. Growth under the zone of petal and stamen attachment, which is lacking when *PMADS1* is not expressed (van der Krol et al., 1993), is what normally produces a stamen-corolla tube (Erbar, 1991). Growth past this point of attachment, which would normally lead to further tube and limb development and filament elongation, is not disrupted when *PMADS1* is not expressed (van der Krol et al., 1993).

Additional insight on this *PMADS1* function comes from transgenics that overexpress the gene in all parts of the flower (van der Krol and Chua, 1993; Halfter et al., 1994). As expected, ectopic expression of *PMADS1* in wild-type *Petunia* plants caused a partial organ identity transformation from sepal to petal. Ectopic expression of *PMADS1* in mutant plants lacking *PMADS1* showed partial rescue of the normal petal phenotype that was dependent upon the amount of *PMADS1* expression. Another feature of *PMADS1* overexpression in wild-type plants is that extra petal tissue forms on corolla tubes at

the points of petal–petal fusion, which may indicate that *PMADS1* expression is required for lateral petal growth before petal fusion, and that when *PMADS1* is overexpressed, this growth may outpace fusion competence. Another possible interpretation is that overexpressed *PMADS1* could delay the onset of petal fusion, which could normally be a signal for the inhibition of further lateral growth. Regardless, timing of petal–petal fusion and outgrowth of petal tissue at these boundaries are related developmental phenomena.

B-function MADS box genes are also implicated in other petal differentiation pathways. Anthocyanin pigmentation is often used as a marker for petal development (e.g., Helariutta et al., 1993). The production of anthocyanins in petals is finely regulated both spatially and temporally (e.g., Helariutta et al., 1995). Many of the regulatory factors in anthocyanin biosynthesis are known (Martin and Gerats, 1993). Transcription factors of the basic helix-loop-helix (bHLH) type have been shown to regulate the expression of chalcone synthase (CHS), which catalyzes the first committed step of the anthocyanin pathway (Jackson et al., 1992). Studies of B-function mutants that display sectors of green sepaloid and pigmented petaloid tissue suggest that anthocyanin regulation is linked to organ identity and therefore to B-function genes (see Martin and Gerats, 1993, pp. 1261–1262). Late-revertant petaloid sectors on transposon-induced *DEF* mutant sepals demonstrate that anthocyanogenesis can be induced when the B function is restored (Schwarz-Sommer et al., 1992). It is also known that *CHS* expression is down-regulated in transgenic plants lacking expression of certain B-function genes (Angenent et al., 1993). It has even been shown that a putative MADS box consensus binding site exists in the *CHS* promoter of *Antirrhinum* (Martin et al., 1991). It is more plausible, however, that B-function genes regulate anthocyanin biosynthesis further upstream, through control of bHLH factors known to regulate *CHS* (Martin and Gerats, 1993).

In summary, we can identify the following petal developmental modalities that are under the control of B-function MADS box genes: (1) petal and stamen identity, (2) congenital fusion of petals and stamens in asterid eudicots, (3) lateral growth of petals and the timing of petal fusion, and (4) petaloid organ differentiation, including anthocyanogenesis. The evolutionary impact of the first and last modalities are discussed later.

Differences in congenital fusion of petals and stamens are important features in some groups of angiosperms (Heywood, 1985). Among the orders of basal Asteridae, Ebenales have characteristic sympetalous corollas with adnate stamens, whereas Ericales have sympetalous corollas usually with free stamens. In *Ilex* (Aquifoliaceae), another basal asterid, the petals are characteristically fused at their bases but the stamens may be either adnate to the corolla or free. This phenomenon is not restricted to lower Asteridae, as many Goodeniaceae and Campanulaceae (Asterales) have sympetalous corollas with free stamens. There are also many examples of corolla tubes without stamen fusion in the Liliales/Asparagales. Lateral petal growth may be an important variable in some families where corolla tubes form late in flower ontogeny via postgenital fusion of previously free petals, for example, in Rutaceae (*Correa*), Oxalidaceae (some *Oxalis* species), Escalloniaceae (*Polyosma*), Pittosporaceae (*Pittosporum*), and some Rubiaceae (Endress, 1994; M. H. G. Gustafsson, pers. obs.). Other taxa, congenitally sympetalous, show distinct outgrowths of tissue at petal–petal boundaries, for example, *Catesbaea* and the *Portlandia* complex (Rubiaceae; Delprete, 1996a) and *Gentiana* (Gentianaceae; Adams, 1996).

B-Function MADS Box Genes: Are Outer Integuments Orthologous with Petals?

Although the C-function alone determines the identity of the carpel whorl, B-function gene expression is still present in developing ovules (Jack et al., 1994). Ovule development may be regulated independently of carpel development, and by other MADS box genes (Angenent et al., 1995). Overexpression of the *Petunia* MADS box gene *FBP11* leads to ectopic ovule formation on sepals and sometimes on petals (Colombo et al., 1995). These mutant ovules, which appear at the expense of trichomes, are anatomically normal. In mutant plants lacking

FBP11 function, carpeloid tissue forms in place of ovules (Angenent et al., 1995). Ovule induction appears not to be linked with the specific action of any other MADS box gene (Colombo et al., 1995). *FBP11* normally expresses in the developing gynoecium before ovule initiation, later settling into the developing ovules themselves (Angenent et al., 1995).

Experiments such as those with *FBP11* (now called a D-function MADS box gene; Colombo et al., 1995) and *BEL1* (a homeobox gene required for integument development; Reiser et al., 1995) have underscored the unique organ-identity status of the angiosperm ovule (see also Baker et al., 1997). For the present purpose, we shall concern ourselves with the possible role of B-function MADS box genes in the petaloid development of the outer integument. Research with *Arabidopsis* has uncovered late *AP3* expression during ovule development (Jack et al., 1994). *PI* does not limit ovule-specific *AP3* expression as it does in the petal and stamen whorls. Elimination of *BEL1* expression leads to ectopic *AP3* expression in the ovule and the disruption of normal integument development (Modrusan et al., 1994).

The A-function gene *AP2* (more specifically, its *AP2-5* and *AP2-6* alleles; Modrusan et al., 1994) and the *AP2*-like gene *AINTEGUMENTA* (*ANT;* Elliot et al., 1996; Klucher et al., 1996) are also required for ovule initiation. Mutant plants without *AP2* expression form carpeloid structures in place of normal ovules (the mutant *bel1* phenotype is similar but stronger; Modrusan et al., 1994). *AP2* and *AP3* (and even *PI;* C. S. Gasser, pers. comm.) normally express in the outer integuments, which renders these structures AB with respect to the ABC model and therefore process orthologous with angiosperm petals. Of course, this orthology is relevant only to these functions because ovules require the expression of several genes not involved with petal development (e.g., *FBP11* and *BEL1;* see Baker et al., 1997).

Historically speaking it is likely that the genetic complexity of outer integuments (ABD+ *BEL1* et al.) evolved before the simpler petal condition (AB). After all, ovules (and probably outer integuments; see Stewart and Rothwell, 1993, pp. 458–459) preceded petals in seed plant phylogeny by over a hundred million years. The petal may merely have been the beneficiary of transference of gene function. Most fossil and extant clades of seed plants have free ovules, but because the angiosperm condition involves ovules borne in laminate structures (carpels), it is natural to assume that male or sterile serial homologues of carpels would retain part of the ovule developmental program. The fact that ovules form on sepals and petals of plants overexpressing *FBP11,* but not on leaves, stems, or roots (Colombo et al., 1995), indicates sufficient genetic preconditioning of cells among the perianth whorls, and therefore their serial homology with carpels.

If each ovule (or, more generally, each sporangium) is considered to terminate a branch, a consequence of the above interpretation is that carpels, stamens, sepals, and petals are all basically branched structures. The carpel would represent a laminate version of a fertile truss of ovule-bearing branches, the stamen a reduced (tetrasporangiate) branch, and the perianth would represent sterile, laminate branching systems (see Stewart and Rothwell, 1993).

The genetic similarity (process orthology) between petals and outer integuments leads to a logical comparison between "true" flowers and ovules as "sub-flowers." This is not as far-fetched an idea as it may seem, considering the emerging evidence linking the developmental regulation of inflorescences and flowers in Asteraceae. A flower may simply be an inflorescence by another name, with branching order (i.e., paralogy of ovule position) providing the most fundamental difference. The evolutionary history of MADS box gene control over reproductive determination in seed plants may therefore resemble a set of Chinese boxes or a babushka doll, with an accumulation of internested developmental programs among complex, branching structures (as in angiosperm flowers and inflorescences) descending from simple, stalked sporangia (as in the sporophytes of mosses).

EVOLUTION OF THE BIPARTITE PERIANTH

Distinguishing sepals from petals is not always straightforward. Whereas most eudicots (e.g., *Arabidopsis*) have the clear sepal/petal distinction described above, most magnoliids do not. Many magnoliids (e.g., *Magnolia*) bear numerous, petaloid organs in a spiral arrangement. This condition, along with a basically strobilar gynoecial morphology, has been considered primitive among angiosperms by some workers (Bessey, 1915; Cronquist, 1981). Therefore, it seems possible that sepals evolved as a later development in angiosperms.

Reproductive structures of the extinct seed plant clade Bennettitales, a possible near relative to angiosperms (Crane, 1985; Doyle and Donoghue, 1986, 1992; Albert et al., 1994; Doyle et al., 1994; Nixon et al., 1994), superficially resemble *Magnolia* flowers, namely with a central, female strobilus, male organs, and shielding, perianthlike outer structures (see Stewart and Rothwell, 1993, pp. 350–365; J. A. Doyle, 1994). Some of the earliest angiosperm fossils do show this strobilar morphology, e.g., *Archaeanthus* of the Cretaceous-Cenomanian (Dilcher and Crane, 1984; cf. Crane et al., 1994), but most—and the oldest—are tiny flowers, sometimes unisexual, and usually with a simple perianth (Crane et al., 1994; Friis et al., 1994).

The evolution of sepal/petal distinction is most logically correlated with the appearance of a whorled, undifferentiated, bipartite perianth, and fossils with bipartite perianth are known from at least the mid-Early Cretaceous (Drinnan et al., 1991; Crane et al., 1994; Crepet and Nixon, 1994; Friis et al., 1994). Both bipartite and spiral perianths may have functioned in the attraction of insect pollinators. However, insects may have been attracted to early angiosperms for reasons other than mutualistic, and features to protect developing floral buds from predation probably appeared at the same time as insect pollination (on the latter, see Crepet et al., 1991; Crepet and Nixon, 1994). Spiral flowers sometimes have tough, greenish outer cycles that protect the buds (e.g., in *Nymphaea*) or even differentiated sepal-like organs that enclose the inner perianth (e.g., in *Illicium*) (Heywood, 1985). These organs are difficult to characterize as definitively sepal or petal (or even bracteal in origin). The first whorl of a differentiated bipartite perianth, however, is much more readily categorized. We speculate that bipartite perianths, with their basic whorled structure, probably evolved from early angiosperms with simple, whorled perianths rather than from spiral flowers, which likely represent a separate radiation from simple flowers. The outer whorl of the bipartite flower may have become rapidly fixed in its morphological (and developmental regulatory) characteristics because of functional constraints imposed by insect predation. Correspondingly, the necessary insect-attractant features of the second perianth whorl would have been equally canalized, which would have favored a clear developmental segregation (and morphological differentiation) between the whorls. By this argument, "true" sepals and their molecular developmental controls are restricted to flowers with bipartite perianths.

Phylogenetic Timing of Sepal/Petal Distinction

Charting the evolutionary appearance of "true" sepals requires some sort of working definition of what a sepal is. Morphological and anatomical differences between sepals and petals were examined previously, but as alluded to there, problems arise if all outer, greenish, protective organs are considered sepals. The useful distinction between basically spiral and bipartite flowers, which likely represent at least two separate clades, is obfuscated. Thus, it seems logical to consider the evolution of a true bipartite perianth as coincident with the evolution of sepals as they are expressed in *Arabidopsis* and *Antirrhinum,* the model systems from which our knowledge of sepal developmental regulation derives (Coen and Meyerowitz, 1991).

Estimating the phylogenetic timing of appearance of true sepals requires mapping alternative perianth states on a tree that includes both the basically spiral magnoliids and the basically bipartite eudicots. The phylogenetic hypothesis for angiosperm interrelationships presented by Chase, Soltis, Olmstead et al. (1993) provides an encompassing backbone for such an optimization experiment. Based on nucleotide variation among plastid *rbcL* sequences of 499 seed plant taxa, Search II of Chase, Soltis, Olmstead et al. (1993) is the broadest phylogenetic treatment of angiosperms yet published. We opti-

mized the following six perianth states on a reduced version of the Search II strict consensus tree: 0 = perianth absent, 1 = perianth simple, 2a = perianth spiral, 2b = perianth spiral with separate, sepal-like organs, 3 = perianth bipartite, and 4 = perianth multipartite. In order to recognize the difficulty in distinguishing between the two spiral perianth states (2a, 2b), we merged them into a single spiral perianth state (2) for a second optimization experiment.

In congruence with evidence from the fossil record, both optimizations result in simple perianth being resolved as the basalmost state for flowering plants (Fig. 12.2). Spiral perianth is unambiguously derived from simple perianth, and therefore taxa like *Archaeanthus* and *Magnolia* display a derived morphology (cf. Drinnan et al., 1994). Separation of the spiral perianth states illustrates two alternative scenarios for early evolution of the magnoliid flower. The simple perianth state optimizes much deeper into magnoliid angiosperms when both spiral perianth states are included, because the different spiral perianth states fail to mark monophyletic groups of taxa; instead, the character optimization indicates (outside the eudicots) up to two independent derivations of spiral perianth with no sepal-like organs and up to five independent origins of spiral perianth with separate, sepaloid organs. When the spiral perianth states are combined, a single origin of the state is predicted, marking most of the magnoliid angiosperms (Fig. 12.3). Monimiaceae, Lauraceae, and Chloranthaceae no longer optimize with simple perianth, and the possible character state connection to simple-flowered Myristicaceae is no longer available. This ambiguity may be due to several factors, from limitations in character coding to underrepresentation of magnoliid diversity because of extinction.

In contrast, the phylogenetic origins of bipartite perianth are unambiguous in either optimization experiment (Figs. 12.2 and 12.3). Four to five independent derivations are indicated: (1) in the paleoherb magnoliid *Saruma,* (2) in the monocot lineage, (3) in some Monimiaceae and Lauraceae, which are sister taxa, and (5) at the very base of the eudicot lineage. Our analysis suggests that the bipartite condition of eudicots is derived from a simple perianth, possibly by whorl duplication. This could hold also for *Saruma,* but Monimiaceae/Lauraceae and monocots have spiral-perianthed ancestors in our optimization that joined all spiral states (Fig. 12.3). Most importantly for the present argument, neither magnoliids nor monocots need be considered to have "true" sepals in the eudicot sense because they acquired their bipartite perianths independently.

Our optimizations suggest that the origins of true sepals may be found near the base of the eudicots. Despite being optimized as primitively bipartite, basal eudicot taxa are highly variable in perianth morphology. Several basal eudicot taxa (e.g., Buxaceae, Tetracentraceae, Gunneraceae, and some Lardizabalaceae) have undifferentiated perianth whorls, either both greenish or both petaloid (Heywood, 1985; Mabberley, 1987; Drinnan et al., 1994). Others are highly variable in their floral phyllotaxis (e.g., Berberidaceae/Ranunculaceae) whereas some are regularly multipartite (e.g., Papaveraceae, which usually have two bractlike "sepals" and two decussate petal whorls). Some basal eudicots have simple perianths (e.g., Proteaceae, some Lardizabalaceae, and Berberidaceae/Ranunculaceae) or no perianth at all (e.g., Eupteleaceae, Trochodendraceae). Distinctly greenish outer perianths with petaloid inner perianths do occur in multipartite Papaveraceae, some Berberidaceae/Ranunculaceae, some Menispermaceae, Platanaceae, and Sabiaceae. Some taxa of the Caryophyllales clade (e.g., Amaranthaceae, Chenopodiaceae, Nyctaginaceae, and Phytolaccaceae) show secondary reversal to a simple perianth. The characteristic sepaloid/petaloid bipartite condition of *Arabidopsis* may only have become fixed in the rosid/asterid clade, some five nodes away from the base of the eudicots on our tree. Of course, notable reversals to simple or absent perianth (e.g., in Hamamelidales, Urticales, Fagales, and Aucubaceae-Garryaceae-Eucommiaceae) have occurred within the rosids and asterids, but these appear in highly derived positions as the exception rather than the rule.

To summarize, the overall picture of true sepal evolution appears to be one of (1) innovation and preadaptation via development of a bipartite perianth early in eudicot evolution, followed by (2) morphological and functional diversification, as represented by perianth variability among basal eudicot clades, (3) canalization of true sepal/petal distinction, possibly in response to animal predation pressure, and (4) diversification of sepal/petal configurations through interactions with pollinating animals. The earliest diagnosable eudicot pollen is of Barremian-Cretaceous age (ca. 126 myr BP), and several basal eudicot lineages (e.g., Trochodendrales, Platanaceae, Buxaceae, and perhaps also Nelumbonaceae and Proteaceae) are recognizable by the Albian/Turonian-Cretaceous (ca. 97–90 myr BP; cf. Drinnan et al., 1994). Rosid and asterid eudicots of diverse morphologies (e.g., possible Hamamelidaceae and sympetalous Ericales) may date from the Turonian-Cretaceous as well (Crepet et al., 1992; Nixon and Crepet, 1993). Therefore, we suggest that the fixation and diversification of true sepals and petals occurred relatively early in angiosperm phylogeny.

Sepals and Developmental Intercalation

If we accept the argument that sepals are evolutionarily derived relative to petals (excluding those of putative staminal origin, e.g., Ranunculaceae; Hiepko, 1965; Kosuge, 1994), then how can the appearance of a new whorl between the flower and inflorescence be explained? Might sepals be derived from bracts (Ĉelakovský, 1896/1900), or, do sepals represent the restriction of petaloid developmental programs from the outer whorl of a bipartite, petaloid flower? Either scenario produces the net effect of developmental intercalation, whereby the sepal differentiation program is nonterminally added to the relatively linear sequence of phyllotactic organogenesis.

The organs of angiosperm flowers form a highly organized reproductive axis, and the outer whorl is no exception. Such organization, initiated by the action of several floral meristem genes (see Theißen and Saedler, 1995; Weigel and Clark, 1996), would not seem to lend itself to positional displacement by vegetative organ developmental programs. However, the evolution of added perianth structures (i.e., increased organization external to the normal four whorls) has occurred a number of times among eudicots,

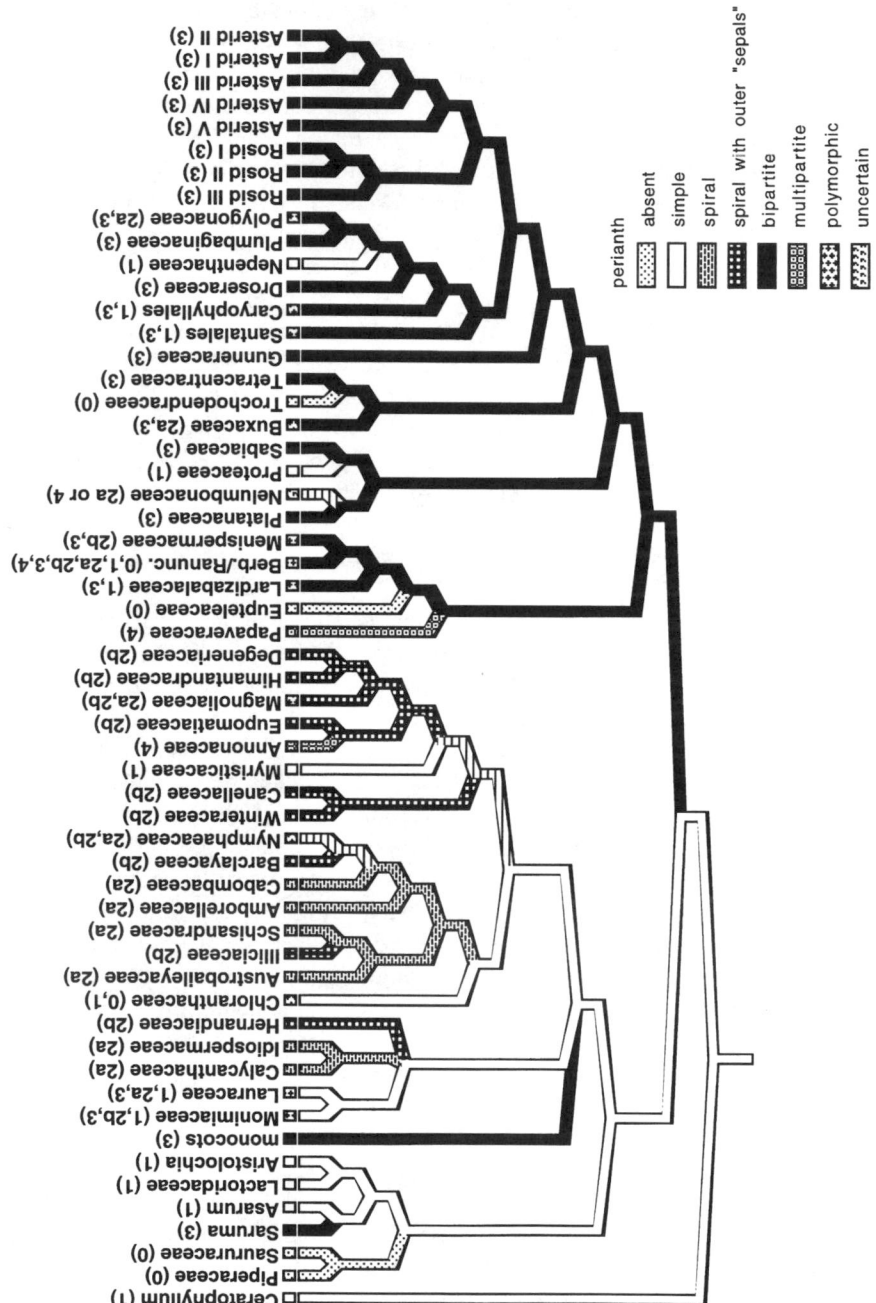

Figures 12.2 and 12.3. Optimization analyses of perianth evolution among angiosperms. Character states pertaining to perianth whorl number and phyllotaxis were optimized (using MacClade; Maddison and Maddison, 1992) onto a reduced version of the strict consensus tree for Search II of Chase, Soltis, Olmstead et al. (1993). The following character coding was considered (states are indicated after each taxon name): 0 = perianth absent, 1 = simple (one-whorl) perianth, 2a = spiral perianth, 2b = spiral perianth with outer, sepaloid organs, 3 = bipartite (two-whorled) perianth, and 4 = multipartite (multiple-whorled) perianth. Figure 12.2 shows the optimization of all states, whereas Figure 12.3 shows the result when states 2a and 2b are combined into a single spiral perianth state. Figures 12.2 and 12.3 show identical implications for the evolution of the bipartite perianth of eudicots: a single, basal origin of the bipartite condition followed by reversals to simple or absent perianth in several lineages. This optimization occurs even though the perianth of basal eudicots is structurally diverse

360

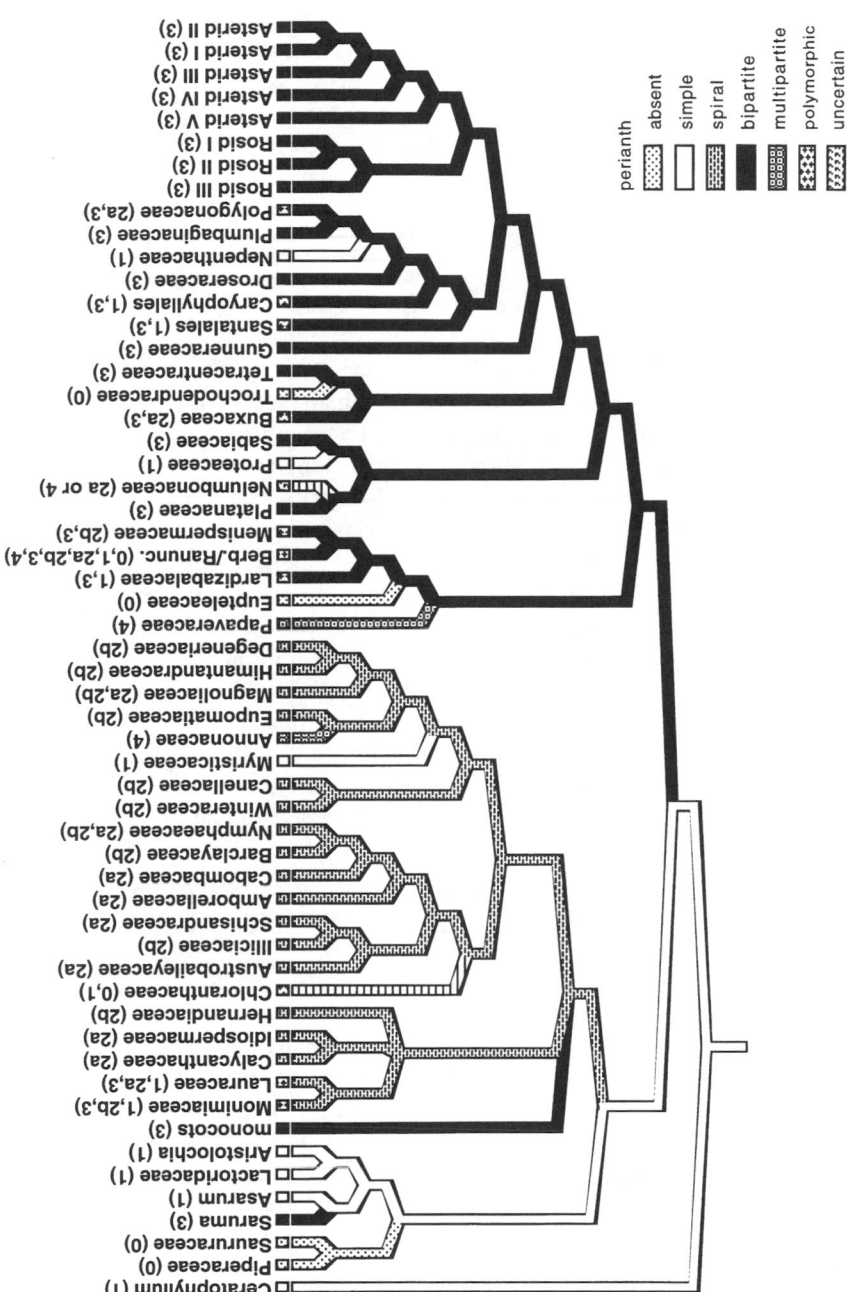

(Drinnan et al., 1994) and frequently polymorphic. Figures 12.2 and 12.3 do show different implications for the early evolution of flowers. In Figure 12.2, simple perianth optimizes as the basal state and extends well into the magnoliids, appearing basalmost in all major lineages and therefore ancestral to a polyphyletic spiral perianth condition. In Figure 12.3, simple perianth is similarly basalmost for angiosperms, but spiral perianth is derived monophyletically near the base of the tree. Our hypothesis that simple perianth represents the plesiomorphic angiosperm condition is largely dependent upon the "anchoring" position of *Ceratophyllum*, which is somewhat contentious (e.g., Donoghue, 1994). Indeed, trees based on 18S rDNA sequence data (Soltis, Soltis, Nickrent et al., 1997) do not place *Ceratophyllum* at the root, but rather spiral-flowered taxa instead, leading to a different conclusion about the ancestral perianth state. Sources for character state codings of the taxa included in Figures 12.2 and 12.3 include Heywood (1985), Mabberley (1987), Drinnan et al. (1994), and Endress (1994). States representing larger clades (e.g., Caryophyllales, Rosids I–III, Asterids I–V) were determined by separate optimizations on individual branches of the complete strict consensus tree of Search II (Chase, Soltis, Olmstead et al., 1993).

361

namely the integration of bracts into flowers of *Macrocarpaea* (Gentianaceae), *Thunbergia* (Acanthaceae), *Malope* (Malvaceae), and *Potentilla* (Rosaceae) (Ewan, 1948; Heywood, 1985; Mabberley, 1987; Endress, 1994). In the variable flowers of *Macrocarpaea bracteata,* for example, leaflike bracts may be partially fused to the sepals, whereas other sepals can be totally leaflike (L. Struwe and V. A. Albert, pers. obs.). These clear cases of vegetative integration into a preorganized floral axis imply that new features can indeed intercalate between vegetative and floral developmental programs, originating from the former and attaching to the base of the latter. Moreover, there may be vegetative involvement in some basal eudicot flowers (e.g., Buxaceae, Myrothamnaceae, and Berberidaceae), where bracts grade into floral organs in such a way that inflorescence/flower distinction is difficult (Drinnan et al., 1994).

Molecular developmental evidence, however, combined with our hypothesized plesiomorphy of bipartite perianth among eudicots, supports the alternative, exclusion-of-petaloidy hypothesis. Analysis of B-function MADS box gene expression in *Rumex* (Polygonaceae; Ainsworth et al., 1995) suggests that a secondary restriction of petaloidy (and an expansion of sepaloidy) accounts for its two sepaloid perianth whorls. The activities of two *DEF* orthologues were found to be limited to the staminal whorl only, which would render both perianth whorls A only with respect to organ identity. Since the last common ancestor of Polygonaceae, Plumbaginaceae, Nepenthaceae, and Droseraceae probably had petals (Figs. 12.2 and 12.3), the loss of petals in *Rumex* is likely secondary. Changes in the pattern of B-function gene expression, as suggested above for *Rumex,* are not unusual; recall the restriction of *PMADS1* to the petal whorl of *Petunia* (van der Krol et al., 1993). This lability in expression could account for a number of morphological differences among angiosperms, both proximal and distal in evolutionary terms. Proximally, B-function MADS box genes might control sex determination in some plants (i.e., B function on = male, B function off = female; cf. Dellaporta and Calderon-Urrea, 1993), and if so, this could have occurred many times in parallel (e.g., in *Rumex,* Begoniaceae, Clusiaceae, and

Aucubaceae; Heywood, 1985). More distally, it is possible that an ancient exclusion of the B function from the outer petaloid whorl of early eudicots—process orthologous but historically and positionally paralogous to the condition in *Rumex*—is the evolutionary basis of sepal/petal distinction. The sepaloid perianths of some basal eudicot taxa (e.g., Buxaceae) could imply that analogous restrictions in B-function gene expression occurred also at intermediate phylogenetic levels, that is, after the evolution of a petaloid, bipartite perianth but before the stabilization of discrete sepal and petal whorls.

To summarize, our proposed sequence for the evolution of sepals is (1) simple perianth, to (2) bipartite perianth, to (3) sepal/petal distinction. Derivation of sepals via a duplicated whorl of petaloid organs seems the most plausible based on character optimization and molecular developmental support. A "duplicated" petal whorl could have been the consequence of A-function expansion into a staminal whorl; the same mechanism could account for staminally derived petals in Ranunculaceae. The result could then have been developmental intercalation of a petaloid whorl (and not directly the addition of a bractlike one) followed by restriction of the B function from the outermost whorl. The opposite process, reexpansion of the B function into the sepal whorl, may have occurred several times in parallel among eudicots.

ONTOGENETIC SYSTEMATICS OF PETALOIDY

Much of angiosperm diversity, at least in terms of plant/pollinator interactions, is in their petals. Petals differ in many parameters, several of which can be listed here: thickness, texture, color, vascularization, connation, adnation to stamens, nectar secretion, presence/absence of a nectar cavity or spur, indumentum, and cell elongation. All of these parameters can vary substantially within flowering plant lineages. Among the sympetalous asterids, families may often be identified on rather uniform vegetative and floral characteristics although individual genera or species may be strikingly diverse in floral features related to pollination syndromes. In Gentianaceae, the elongate, red-yellow, tubular corollas of *Lagenanthus* (Ewan, 1952), probably hummingbird pollinated, contrast strongly with the green-purple, nectar-spurred petals of *Halenia* (Pringle, 1995), which are likely bee or butterfly pollinated. In Goodeniaceae, which are not closely related to Gentianaceae (Olmstead et al., 1993), parallel variation occurs in *Lechenaultia tubiflora* and *Goodenia macroplectra,* respectively (Carolin et al., 1992). No matter how remarkable the range of petal diversity may be among higher eudicots, determining organ identity is only rarely a problem. That is, the eudicot floral bauplan is rather fixed. When organ identity is at issue, however, taxa often deviate strikingly from this groundplan. Petaloid structures may appear in the position of stamens, and sepaloid organs may replace petals. Such modification, although phenotypically drastic, may not be so extreme in terms of the developmental mechanisms that control it. Ontogenetic systematic research seeks to describe these mechanisms across a diversity of relevant taxa and their phylogenetic contexts. Here, we describe several systems that highlight different aspects of altered organ identity and petaloidy among eudicots.

Petaloid Stamens and Synorganization

The transference of petaloid character to stamens may induce diversification through the production of novel functionalities. The sharing of features in organs of "hybrid" origin may permit novel and specialized relationships with pollinating animals. In the example of the Lecythidaceae (see following section), synorganization of stamens into a partly laminate, partly staminodial structure yields a pollinator-orientation device analogous to the personate flowers of *Antirrhinum* (Scophulariaceae), in which the bilabiate corolla is tightly appressed to prohibit simple entry to the inner whorls of the flower. In Lecythidaceae, fertile pollen and gynoecium are protected not by petals but by a petaloid outgrowth of the androecium. Synorganization of stamens into a ring in some *Clusia* species (Clusiaceae; see section on *Clusia*) may have opened up new possibilities for the presentation of pollen and resin, which may be an attractant for insect pollinators as well as a medium of pollen transport.

The Lecythidaceae Hood The Lecythidaceae are a medium-sized family of tropical trees related to the Ericales/Ebenales (Asterid III of Chase, Soltis, Olmstead et al., 1993; Morton et al., 1996). The family is well known because of the genus *Bertholletia,* which produces Brazil nuts (Mori and Prance, 1990a). The Lecythidaceae had previously been referred to the Myrtales because of their massively proliferated stamens (up to 500–1,200 per flower). These stamens characteristically initiate on a large ring primordium (Endress, 1994). Taxa such as *Barringtonia* and *Gustavia* have actinomorphic androecia, whereas others such as *Bertholletia, Corythophora, Couratari, Couroupita, Eschweilera,* and *Lecythis* have highly zygomorphic androecia (Prance and Mori, 1979; Mori and Prance, 1990b). In the zygomorphic-flowered genera, a hood develops as an outgrowth from one side of the androecial ring primordium, which originally resembles the same, unbroken, doughnut-shaped mound characteristic of the actinomorphic-flowered taxa. After initiation, the hood, composed of multiple, congenitally fused organs of apparent staminal origin, undergoes a petaloid differentiation, curving over the inner, fertile androecial ring and the gynoecium (Endress, 1994, pp. 244–246). Stamen development in the hood is variously elaborated in different taxa (Mori and Prance, 1990b). The hoods of some species bear only staminodial appendages without anthers (e.g., *Eschweilera* species), whereas those of others (e.g., *Couroupita* and *Corythophora* species) have complete stamens that produce fodder pollen (a reward for insect pollinators). In the latter taxa, fodder pollen (borne as tetrads in *Couroupita*) and fertile pollen are produced in different regions of the hood (Mori et al., 1980). Laminate, petaloid regions of the hood are variously elaborated among species (Mori and Prance, 1990b). In *Lecythis,* the hood curves over once and can be covered with nectar-producing staminodes (as in *L. holcogyne*) or even can be almost laminate (as in *L. confertiflora*). Hoods of *Eschweilera* species coil inwards, whereas those of *Couratari* have a double coil with nectar-producing staminodes developed in the inner loop. To summarize, the lecythidaceous hood appears to vary in its petaloid versus staminal nature from its proximal part to its distal part as well as between different taxa.

Specific hood configurations in zygomorphic Lecythidaceae probably correlate closely with pollination syndrome (Mori and Prance, 1990b). The hood serves many animal attractant functions associated with petals, and a few associated with stamens. Hoods may be highly colorful, like the petals, and those with antherless, staminodial masses produce nectar rewards (also in *Couratari,* which has anthers). Additionally, the variously curved hood lamina serves to orient the pollinating animal in many cases. The presentation of fodder pollen rewards in taxa with staminodial anthers is perhaps the only hood functionality that is not basically petaloid in nature.

The question then arises as to how the hood, staminal in origin by positional orthology, attained process orthology with petals. There are two key parameters that must be considered: Variation in petaloidy within the hood, and its zygomorphic derivation from the staminal ring primordium. Nonsymmetrical transference of the A function into the stamen ring primordium could give rise to ABC-genotype staminodia. The congenital fusion of these staminodia into a laminate structure could be controlled by the same mechanisms that determine sympetaly (see previous discussion). Variability of petal and stamen identity along the hood could correlate with varying A-function gene activity. Staminodia without anthers could have the highest A-function activity, followed by tetrad-bearing fodder staminodes, monad-bearing fodder staminodes, and fertile stamens of the inner ring. Gene expression studies via in situ hybridization with MADS box gene probes might reveal different patterns of A-function expansion in the hoods of different species of Lecythidaceae.

Petaloid Staminodes in Clusia (Clusiaceae)

The Clusiaceae (or Guttiferae) are a large family of rosid eudicots related to Malpighiaceae and Euphorbiaceae (M. Chase, pers. comm.). The family includes important fruit crops such as the mangosteen (*Garcinia*) and mammey apple (*Mammea*). Most species are tropical, woody plants, although the largely herbaceous St. John's wort (*Hypericum*) is widespread. The genus *Clusia* comprises approximately 200 mostly dioecious species native to the Neotropics (P. Stevens, pers. comm.). The number and phyllotaxis of floral parts is highly variable in *Clusia* (Engler, 1925). Unisexuality in *Clusia* flowers is often incomplete, and variously elaborated staminodes and pistillodes may be present in plants of opposite sexes. Staminodes of some species of *Clusia* may be petaloid, and petal/sepal distinction is not always marked, as in the petaloid calyx of *Clusia cochlanthera* (Maguire, 1977). In other words, floral organ categories in *Clusia* are somewhat labile, more so in this genus than in most eudicot families. In the small segregate genus *Quapoya,* the central androecium of male flowers is similar in structure to the gynoecium of female flowers (Maguire, 1972). The stamens (equal in number to staminodes plus carpels of female flowers) are radially organized like standard *Clusia* carpels, and the anthers are positioned similarly to the stigmatic lobes of gynoecia. No trace of female parts is visible in male flowers, which suggests that a one-for-one homeotic transformation determines floral sexuality in this genus.

Not all organ identity changes in Clusiaceae are equally drastic. The outer, fertile androecium of male *Clusia palmicida* flowers resembles the androecium of *Gustavia* (Lecythidaceae); one to several rows of elongate anthers (staminodes in female flowers) are borne on a petaloid ring (M. H. G. Gustafsson, pers. obs.). Other species have connate staminodial rings, such as in *Clusia minor* female flowers. Comparative anatomy of staminodes and petals in *Clusia minor* reveals very similar epidermal and parenchymatic features that distinguish both from sepals.

Staminodial diversity in *Clusia* may be correlated with a high degree of pollinator specificity. Both stamens and staminodes are known to secrete resin, which may serve as a reward for visiting pollinators in many species (see Armbruster, 1984; Bittrich and Amaral, 1996). In section *Euclusia,* for example, the production of resin by the massive central stamens or staminodes of male flowers often resembles resin secretion by lateral staminodes surrounding a massive gynoecium, as in female flowers. *Clusia palmicida* falls into this category (Armbruster,

1984), and the massive, gynoecially structured androecia of *Quapoya* may act similarly. In other groups (e.g., *Clusia* sect. *Cochlanthera* subsect. *Orthoneura;* Maguire, 1977), the resin-secreting staminodia are in the same positions (lateral) in male and female flowers. Thus, the structural lability of *Clusia* flowers may serve generally to promote extreme pollinator fidelity. As with Lecythidaceae, formation of laminate, petaloid staminodial structures may be correlated with expansion of the realm of A-function gene activity.

Petaloid Sepals

As in the case of petaloid stamens (above), transference of petaloid features to the sepal whorl may promote diversification of plant/pollinator interactions. Deviations from floral groundplans can provide flexible alternatives to highly constrained floral morphologies. In some species of *Clermontia* (Campanulaceae), sepals are entirely replaced by an additional whorl of valvate petals. These extra petals may contribute to both floral protection and floral display. The sepal whorl of *Impatiens* (Balsaminaceae), which includes its nectar spur, also takes on petaloid character, but to varying degrees in different species.

Clermontia (Campanulaceae) *Clermontia* comprises 22 species of lobelioid Campanulaceae endemic to the Hawaiian Islands (Lammers, 1991). Most species are confined to single islands in the chain, and all appear to be colonizers of disturbed habitats. Some *Clermontia* species are known to have been pollinated by endemic honeycreepers, and all are bird dispersed. *Clermontia* flowers are strongly protandrous, and they are capable of self-fertilization if pollen is not removed by a vector (Cory, 1984). Strikingly, flowers of 15 *Clermontia* species appear to have two whorls of petals, the outer occurring in place of the sepals (Lammers, 1991). Scanning electron micrographic and anatomical studies show that this sepal-to-petal transformation is complete for some species (V. A. Albert, L. Struwe, and L. Di Laurenzio, pers. obs.; Di Laurenzio et al., 1997). Phylogenetic analysis of 5S rDNA nontranscribed spacer sequences sug-

gests that the double-corolla trait arose early in *Clermontia* evolution, but that little subsequent genetic differentiation has occurred. The phenotypic diversity (including flowers with rotate, bilabiate, and tubular corollas; Lammers, 1991) and genetic similarity observed among double-corolla *Clermontia* species suggests that morphological radiation can follow rapid fixation of a new floral groundplan. This process would be enhanced by inbreeding (accompanying pollinator extinction or disturbance) and founder effect in island environments with rapidly evolving ecological niches. In keeping with this hypothesis, the *Clermontia* clade is most diverse and apomorphic on the geologically young islands of Maui and Hawaii, and most depauperate and plesiomorphic on the oldest high island, Kauai (Lammers, 1991, 1995).

As sepal/petal organ identity is affected in double-corolla clermontias, MADS box gene expression patterns are likely to be altered. Specifically, ectopic B-function MADS box gene activity is expected in the sepal whorl (Halfter et al., 1994; Krizek and Meyerowitz, 1996). In *Arabidopsis,* overexpression of *PI* results in partial conversion of sepals to petals without other organ identity changes (Krizek and Meyerowitz, 1996). Petalization of the first whorl is more complete in transgenic plants overexpressing both *PI* and *AP3,* but such flowers also have stamens replaced by carpels (Krizek and Meyerowitz, 1996). Thus, the double-corolla mutation in *Clermontia* may involve overexpression of the native *PI* orthologue or possibly an unknown regulatory gene that normally excludes both B-function genes from expressing in the sepal whorl.

Impatiens (Balsaminaceae) *Impatiens* is a large and widespread genus of the family Balsaminaceae. Molecular data have placed Balsaminaceae in the thealean alliance at the base of the Asteridae, most closely related to Marcgraviaceae and Tetrameristaceae (Morton et al., 1996). *Impatiens* flowers are often highly zygomorphic (Grey-Wilson, 1980; Heywood, 1985). There are three members in the outer floral whorl, one of which bears a nectar spur. This outer whorl, traditionally referred to as sepals, is variously differentiated. The spur-bearing sepal may be small and green, or large, pigmented, and petaloid. The lateral sepals may similarly be small and sepaloid or large and petaloid. The spur itself is variously elaborated, from long, thin, and green to highly inflated and pigmented. The inner perianth whorl of five members is always petaloid, with the four lower petals fused in pairs.

Transference of petaloid features to the sepal whorl of *Impatiens* always includes adaxially positioned, pollinator-related nectar production, an otherwise rare feature of sepals (but typical, e.g., of Malvales; Endress, 1994). Abaxial nectar production by sepals, which is more common, is probably related to extrafloral nectar production (e.g., attraction of ants and other nonpollinating insects). *Impatiens* sepals, when petaloid, tend to have the same color and texture as the petals. Therefore, if an extension of B-function gene expression is imagined for *Impatiens* sepals (cf. *Clermontia*, previous section), then this ectopic activity is likely to be as variable in extent as petaloidy is among the different species. Variation within the *Impatiens* groundplan, spread out among species native to North America, Eurasia, Southeast Asia, and Africa, may represent an innovation-based diversification analogous to but much older than that seen within the double-corolla species of *Clermontia*.

Petaloidy and Radial Heteromorphy

In some angiosperms, petaloidy varies according to a flower's position in an inflorescence. In Asteraceae, partitioning of flower types in an inflorescence can be very complex, and it is here that the genetic basis for floral heteromorphy is becoming best known. Other angiosperms show simpler peripheral-central inflorescence partitions, from sterile, petaloid flowers in peripheral inflorescence domains of *Viburnum*, petaloid sepals in peripheral *Hydrangea* flowers, single, petaloid calyx lobes in peripheral flowers of some Rubiaceae species, to petaloid, nectariferous, bractlike organs central to regular flowers of *Marcgravia*. In each case, radial heteromorphy also represents a partitioning of functions, for example, in the attraction of animal pollinators. Radially heteromorphic inflorescences, with outer attractive and inner sexual partitions,

may function like a single flower from the pollinator's standpoint. A flowerlike inflorescence can permit a greater variety of plant/pollinator interactions since each organ of each flower (e.g., the corolla) can vary independently, as is often seen in Asteraceae. Thus, radial heteromorphy can promote diversification by enhancing opportunities for pollinator specificity.

Gerbera (Asteraceae)

In the compressed inflorescences (capitula) of Asteraceae there are often concentric fields of different flower types. It has long been thought that a radial gradient effect at either the chemical or phyllotactic level might be responsible for heteromorphy in Asteraceae (Bachmann, 1983; Mauthe et al., 1984).

The heteromorphic capitula of Asteraceae are superficially similar to single flowers. Sexuality among flowers in an inflorescence is often differentially expressed (see Harris, 1995). For example, *Gerbera*, like the majority of Asteraceae, has female outer florets with greatly expanded corollas and bisexual inner florets with comparatively minute corollas (Helariutta et al., 1993). Numerous, sepallike involucral bracts surround the capitulum; true sepals are replaced by pappus bristles. Studies of MADS box gene expression and function in *Gerbera* have revealed that the entire capitulum is much more than a simple analog of a single flower. Perturbation of C-function MADS box gene expression reveals the C function to be strongest in the capitulum center (M. Kotilainen, D. Yu, P. Elomaa, M. Mehto, Y. Helariutta, V. A. Albert, T. H. Teeri, unpubl.; Albert et al., 1997; Kotilainen et al., 1997). The A function, which is mutually antagonistic with the C function (but differentially so through *AP1* versus *AP2;* see Coen and Meyerowitz, 1991; Theißen and Saedler, 1995), may be correspondingly weakest in the center but strongest toward the periphery, and different B-function genes show different mid-radial effects. Each function may be manifest as a gradient of gene expression, protein concentration, or differential protein dimerization. For example, antisense down-regulation of the *Gerbera* orthologue of *GLO* (a B-function MADS box gene) leads to pappoid transformation of the corolla in marginal and central flowers, but not in intermediate "trans" flowers (Yu et al., 1997; V. Albert et al., unpubl.). Therefore, pappus may be considered positionally as well as process orthologous with sepals. The special nature of the trans radial zone in *Gerbera* is reiterated in the antisense phenotype of one *DEF* orthologue transformant; here, sympetaly is disrupted, but only in the trans flowers (Yu et al., 1997; V. Albert et al., unpubl.). With the possible radial organization A (external) → B → C (internal), inflorescences of *Gerbera* (and other angiosperms) may share more developmental regulatory similarities with single flowers than previously suggested. Of course, individual flowers have their own ABC regulation of organ identity, utilizing the same genes that affect inflorescence development. Molecular developmental studies of other radially heteromorphic systems, including where petaloidy varies, should help elucidate this phenomenon more generally.

Viburnum (Adoxaceae) and Hydrangea (Hydrangeaceae)

Viburnum (Adoxaceae) is a northern temperate genus of Asteridae (Dipsacales; Chase, Soltis, Olmstead et al., 1993; Donoghue et al., 1992; Judd et al., 1994) with approximately 150 species (Mabberley, 1987). Some species, for example *Viburnum opulus,* bear highly expanded, petaloid flowers in the peripheral parts of their highly branched, umbel-like inflorescences. The third and fourth whorls of the outer flowers are suppressed, rendering them sterile. The central flowers, by contrast, are small and hermaphroditic, with a relatively inconspicuous corolla.

Hydrangea (Hydrangeaceae) is often confused with *Viburnum.* There are about 23 species distributed in East Asia and North America (Mabberley, 1987). *Hydrangea* also belongs to the Asteridae, close to Cornaceae and Nyssaceae (Chase, Soltis, Olmstead et al., 1993; Morgan and Soltis, 1993). Unlike the situation in *Viburnum,* in some *Hydrangea* species it is the calyx of peripheral flowers of the umbellike inflorescences that is expanded and petaloid. Interestingly, these petaloid calyces become more sepaloid with age, beginning white-pink/blue and later turning greenish in color.

Homomorphic "snowball" mutants of both *Viburnum* and *Hydrangea* are known, in which all flowers of an inflorescence show the features of the petaloid, peripheral flowers. In *Viburnum,* the snowball phenotype has probably evolved three times independently (M. Donoghue, pers. comm.). Shifts between heteromorphy and homomorphy, although exceptional in *Viburnum* and *Hydrangea* individuals, may underlie major diversifications in other families such as Asteraceae, where many taxa, including whole tribes (e.g., Cardueae and Lactuceae), are almost completely homomorphic (see Bremer, 1994).

The sterile, peripheral flowers of *Viburnum* may be related to strengthened A and B function (and correspondingly weakened C function) effects, whereas the sepals of peripheral flowers of *Hydrangea* may have ectopic expression of B-function genes (at least during early stages of development). Radial inflorescence effects may be caused by the same general mechanism described for *Gerbera* (previous section), but this would then have to occur within the inflorescence meristem itself to be retained in branched as opposed to synorganized inflorescences. Study of naturally occurring and snowball variants could help elucidate the molecular basis for radial heteromorphy in these taxa.

Calycophylls of Rubiaceae

Calycophylls (sometimes called pterophylls when occurring in fruit) are extended, petaloid calyx lobes characteristic of approximately 23 genera of Rubiaceae (Gentianales-Asteridae), which are spread polyphyletically over many tribes (Claßen-Bockhoff, 1996; Delprete, 1996b; P. G. Delprete, pers. comm.). Calycophylls, like the petaloid flowers of *Viburnum* and *Hydrangea,* usually appear in flowers at the peripheral margins of basically cymose inflorescences. Most often, only a single calyx lobe is altered, usually the abaxial one, but there are some mutants (e.g., in *Mussaenda*) and naturally occurring taxa (e.g., *Cruckshanksia lithiophila, Carphalea cloiselii*) in which each calyx lobe is modified (Claßen-Bockhoff, 1996). Calycophylls have many characteristics of petals, including papillate epidermis (which produces a velvety effect with saturated, warm colors) and internal anatomy with a simple mesophyll and large intercellular spaces (Endress, 1994; Claßen-Bockhoff, 1996). Most calycophylls are petiolate, but they do not have leaflike vascularization. The constricted bases of calycophylls are evocative of petals, which, unlike sepals with their broad attachment points, have narrow or ligulate attachment points. Calycophylls may be white, yellow, or conspicuously colored with anthocyanins. Sometimes, extrafloral inflorescence bracts are expanded and colored in a similar manner.

Calycophylls, like the petaloid calyx lobes of *Hydrangea, Impatiens,* and *Clermontia,* may reflect a transference of B-function gene expression to the outer whorl. In normal cases where this is restricted to the abaxial calyx lobes of peripheral flowers, whole-inflorescence radial effects may be governing the distribution of B-function activity.

Where inflorescence bracts show petaloid similarities to calycophylls, these may be process orthologous to the petaloid bracts of *Davidia, Cornus,* and *Euphorbia* species. Overexpression of both B-function orthologues in *Arabidopsis, AP3* and *PI,* results in partial transformation of upper leaves to petals (Krizek and Meyerowitz, 1996), and naturally occurring ectopic expression of these genes might be expected to yield similar petaloid leaf or bract phenotypes.

Marcgravia bracts (Marcgraviaceae)

Marcgravia is a medium-sized genus of approximately 45 species native to the Neotropics (Mabberley, 1987). The family Marcgraviaceae is related to the Ericales/Ebenales alliance of Asteridae, most particularly to Tetrameristaceae and Balsaminaceae (Morton et al., 1996). Like Balsaminaceae, Marcgraviaceae are known for their nectar spurs, which are extrafloral in this case (Bailey, 1922; Heywood, 1985). The pitcher-like, nectariferous bracts of *Marcgravia* are particularly well developed. Each bract appears to be longitudinally adnate to the pedicel of a single flower, which is usually abortive. The bracts are fleshy and petaloid, having the color (often red or reddish) and consistency of the calyptrate corollas and ripe fruits. The enlarged bracts arise in the central parts of the characteristically pendulous, umbellike inflorescences, surrounded by fertile flowers with only bracteoles on their pedicels. The switch between normal and bracteose flowers is very sharp, with intermediate structures being exceptional. Some other genera of Marcgraviaceae have pitcher-like bracts on the pedicels of fertile flowers (e.g., *Norantea* and *Souroubea*) but these are smaller than in *Marcgravia* and are not centrally located in the inflorescence. Production of nectar rewards, plesiomorphic in the family, followed by the evolution of pronounced radial heteromorphy, could have been the primary innovations leading to the *Marcgravia* condition.

Transference of A- and B-function gene expression to the *Marcgravia* bracts could give rise to the petaloid features of the central nectaries (see discussion of petaloid bracts, above). Radial heteromorphy, however, runs opposite to that of *Viburnum,* where petaloid flowers are peripheral rather than central. Although the A function may generally be stronger at the periphery of inflorescences (e.g., in Asteraceae; see above), this radial difference could be reversed in *Marcgravia.*

CONCLUSIONS: MACROMUTATION, PETALOIDY, AND DIVERSIFICATION

Petaloidy is a feature of the angiosperm perianth that may in fact be unique to angiosperms. The outer integument of Gnetales could be positionally orthologous with petals, and perhaps even historically and process orthologous as well (J. A. Doyle, 1994). MADS box gene lineages corresponding to the A, B, and C functions may have been present in ancestral seed plants, but so far, only the C function certainly predated the conifer/angiosperm common ancestor (Tandre et al., 1995; P. Engström and K. Tandre, pers. comm; Engström and Tandre, 1997). Regardless, what could be the structural equivalent (positional orthologue) of petals outside of angiosperms plus Gnetales? Cycads, ginkgo, and conifers do not have a simple equivalent to perianth. What they and the Gnetales do share is separate male and female reproductive axes, borne either on different plants or on different parts of the same plant. Male sporangia-bearing structures in angiosperms are clearly mechanistically related (process orthologous) to petals through shared expression of B-function genes. Petaloidy in angiosperms may therefore have evolved secondarily to a more primary innovation: B-function mediated sex determination in seed plants.

Simple changes in gene regulation may have macromorphological effects, but it is probably during the aftermath of such effects that radiations occur. The coexpression of the A and B function in angiosperm petals may represent the central innovation that permitted massive diversification based on morphological evolution of petals accompanied by plant/pollinator coevolution. The evolution of sex determination in seed plants, possibly the precursor of petals, resulted in an earlier radiation of its own. Molecular developmental processes, such as B-function gene activity, may remain essentially the same as organisms evolve around them, building developmental pathway on top of developmental pathway. By this we mean that a gene lineage may evolve by a chance duplication, serve one function initially, and have this function co-opted to other functions repeatedly over time.

In angiosperms, B-function MADS box genes express normally in the outer integument of ovules, in stamens, in petals, and possibly also in some sepals (as defined by positional orthology) and foliar bracts. Additionally, the B function in Asteraceae appears to differ in strength as a function of radial position in the capitulum. It would probably be "florocentric" to consider all of these B-orthologous functions to derive evolutionarily from the organ systems we know best: stamens and petals. After all, it is possible that flowers are nothing more than axes of sporangium-bearing branches (and their sterile serial homologs), and that the B-function had its origins closer to sex determination than is usually recognized. Just as some Lecythidaceae have extra, protective, staminodial structures surrounding their fertile reproductive structures, so might a sporangium benefit from its own protective integument. We would argue that this capacity, possibly innate in angiosperms, may have simply been reinvented in Lecythidaceae, just as petals themselves may likewise have been reinvented by some basal eudicots (e.g., Ranunculaceae). The sterile, protective role of the B-function gene lineage may be much more ancient, perhaps as old as the seed.

Transference of the position of morphogene expression may be a common mode of ontogenetic evolution in plants. Mechanisms leading to such changes may include gene overexpression, complete or partial loss-of-function gene mutation, and altered gene regulation. Little is known, however, about how these mechanisms may have impacted naturally occurring systems. For the present, the best we can do is to refer to transference of gene function as merely a phenomenon. Thus, in the eudicot flower, transference of gene function may have yielded petaloidy in the stamen or sepal whorls, effecting a macromorphological change. In some *Clermontia* species, for example, it appears that small, green sepals have undergone a homeotic change to elongate petals that are frequently colored. There is in fact a diversity of double-corolla species with various flower shapes and color combinations. However, homeotic innovation itself is not likely to be the diversity generator; rather, it is probably the new-found freedom from structural constraints that opens avenues for novel ontogenies, morphologies, and adaptive interactions with the environment. We expect that diversification will come to be seen as a patchy (i.e., punctuated) process, appearing as waves of change based upon new themes. An appropriate synthesis of phylogenetic, molecular developmental, and morphological/anatomical methods is required in order to describe diversity properly as resulting from both cladogenesis and its causal processes.

LITERATURE CITED

Adams, L. G. 1996. Gentianaceae. In Flora of Australia vol. 28, ed. A. Wilson, pp. 72–103. CSIRO, Australia.

Ainsworth, C., S. Crossley, V. Buchanan-Wollaston, M. Thangavelu and J. Parker. 1995. Male and female flowers of the dioecious plant sorrel show different patterns of MADS box gene expression. Plant Cell 7:1583–1598.

Albert, V. A., A. Backlund, K. Bremer, M. W. Chase, J. R. Manhart, B. D. Mishler, and K. C. Nixon. 1994. Functional constraints and *rbcL* evidence for land plant phylogeny. Annals of the Missouri Botanical Garden 81:534–567.

Albert, V. A., M. Kotilainen, P. Elomaa, J. W. Grimes, Y. Helariutta, D. Yu, and T. H. Teeri. 1997. Inflorescence development in Asteraceae—a radial morphogenetic gradient in *Gerbera*. In Evolution of Plant Development, Keystone Symposium B1, Taos, New Mexico, p. 22 [abstract].

Angenent, G. C., J. Franken, M. Busscher, L. Colombo, and A. J. van Tunen. 1993. Petal and stamen formation in petunia is regulated by the homeotic gene *fbp1*. Plant Journal **3**:101–112.

Angenent, G. C., J. Franken, M. Busscher, D. Weiss, and A. J. van Tunen. 1994. Co-suppression of the petunia homeotic gene *fbp2* affects the identity of the generative meristem. Plant Journal **5**:33–44.

Angenent, G. C., J. Franken, M. Busscher, A. van Dijken, J. L. van Went, H. J. M. Dons, and A. J. van Tunen. 1995. A novel class of MADS box genes is involved in ovule development in petunia. Plant Cell **7**:1569–1582.

Arber, A. 1946. Goethe's botany. Chronica Botanica **10**:63–126.

Armbruster, W. S. 1984. The role of resin in angiosperm pollination: ecological and chemical considerations. American Journal of Botany **71**:1149–1160.

Bachmann, K. 1983. Evolutionary genetics and the genetic control of morphogenesis in flowering plants. Evolutionary Biology **16**:157–208.

Bailey, I. W. 1922. The pollination of *Marcgravia:* a classical case of ornithophily? American Journal of Botany **9**:370–384.

Baker, S. C., K. Robinson-Beers, J. M. Villanueva, J. C. Gaiser, and C. S. Gasser. 1997. Interactions among genes regulating ovule development in *Arabidopsis thaliana.* Genetics **145**:1109–1124.

Bessey, C. E. 1915. The phylogenetic taxonomy of flowering plants. Annals of the Missouri Botanical Garden **2**:109–164.

Bittrich, V., and M. C. E. Amaral. 1996. Flower morphology and pollination biology of some *Clusia* species from the Gran Sabana (Venezuela). Kew Bulletin **51**:681–694.

Bowman, J. L. 1994. *Arabidopsis:* An Atlas of Development. Springer-Verlag, New York.

Bowman, J. L., D. R. Smyth, and E. M. Meyerowitz. 1991. Genetic interaction among floral homeotic genes of *Arabidopsis.* Development **112**:1–20.

Bradley, D., R. Carpenter, H. Sommer, N. Hartley, and E. S. Coen. 1993. Complementary floral homeotic phenotypes result from opposite orientations of a transposon at the *PLENA* locus of *Antirrhinum.* Cell **72**:85–95.

Bradley, D., R. Carpenter, L. Copsey, C. Vincent, S. Rothstein, and E. S. Coen. 1996. Control of inflorescence architecture in *Anthirrhinum.* Nature **379**:791–797.

Bremer, K. 1994. Asteraceae: Cladistics and Classification. Timber Press, Portland, Oregon.

Carabelli, M., G. Sessa, S. Baima, G. Morelli, and I. Ruberti. 1993. The *Arabidopsis Athb-2* and *-4* genes are strongly induced by far-red-rich light. Plant Journal **4**:469–479.

Carolin, R. C., M. T. M. Rajput, and D. Morrison. 1992. Goodeniaceae. **In** Flora of Australia **35,** ed. A. S. George, pp. 4–328. Australian Government Publishing Service, Canberra.

Čelakovský, L. J. 1896/1900. Ueber den phylogenetischen Entwicklungsgang der Blüthe und über den Ursprung der Blumenkrone 1/2. Sitzungsberichte der Königlichen Bömischen Gesellschaft der Wissenschaften in Prag. Mathematisch-naturwissenschaftliche Classe 1896(**40**):1–91, 1900(**3**):1–221.

Chase, M. W., D. E. Soltis, R. G. Olmstead, D. Morgan, D. H. Les, B. D. Mishler, M. R. Duvall, R. A. Price, H. G. Hills, Y.-L. Qiu, K. A. Kron, J. H. Rettig, E. Conti, J. D. Palmer, J. R. Manhart, K. J. Sytsma, H. J. Michaels, W. J. Kress, K. G. Karol, W. D. Clark, M. Hedrén, B. S. Gaut, R. K. Jansen, K.-J. Kim, C. F. Wimpee, J. F. Smith, G. R. Furnier, S. H. Strauss, Q.-Y. Xiang, G. M. Plunkett, P. S. Soltis, S. M. Swensen, S. E. Williams, P. A. Gadek, C. J. Quinn, L. E. Eguiarte, E. Golenberg, G. H. Learn, Jr., S. W. Graham, S. C. H. Barrett, S. Dayanandan, and V. A. Albert. 1993. Phylogenetics of seed plants: an analysis of nucleotide sequences from the plastid gene *rbc*L. Annals of the Missouri Botanical Garden **80**:528–580.

Claßen-Bockhoff, R. 1996. A survey of flower-like inflorescences in the Rubiaceae. Opera Botanica Belgica **7**:329–367.

Coen, E. S., and E. M. Meyerowitz. 1991. The war of the whorls: genetic interactions controlling flower development. Nature **353**:31–37.

Coen, E. S., and J. M. Nugent. 1994. Evolution of flowers and inflorescences. Development **107**:107–116.

Colombo, L., J. Franken, E. Koetje, J. van Went, H. J. M. Dons, G. C. Angenent, and A. J. van Tunen. 1995. The petunia MADS box gene *FBP11* determines ovule identity. Plant Cell **7**:1859–1868.

Cory, C. 1984. Pollination biology of two species of Hawaiian Lobelioideae (*Clermontia kakeana* and *Cyanea angustifolia*) and their presumed coevolved relationship with native honeycreepers (Drepanidae). M.S. thesis. California State University, Fullerton.

Crane, P. R. 1985. Phylogenetic analysis of seed plants and the origin of angiosperms. Annals of the Missouri Botanical Garden **72**:716–793.

Crane, P. R., E. M. Friis, and K. R. Pedersen. 1994. Paleobotany evidence on the early radiation of magnoliid angiosperms. Plant Systematics and Evolution (suppl.) **8**:51–72.

Crepet, W. L., and K. C. Nixon. 1994. Flowers of Turonian *Magnoliidae* and their implications. Plant Systematics and Evolution (suppl.) **8**:73–91.

Crepet, W. L., E. M. Friis, and K. C. Nixon. 1991. Fossil evidence for the evolution of biotic pollination. Philosophical Transactions of the Royal Society of London B **333**:187–195.

Crepet, W. L., K. C. Nixon, E. M. Friis, and J. V. Freudenstein. 1992. Oldest fossil flowers of hamamelidaceous affinity, from the late Cretaceous of New Jersey. Proceedings of the National Academy of Sciences U.S.A. **89**:8986–8989.

Cronquist, A. 1981. An Integrated System of Classification of Flowering Plants. Columbia University Press, New York.

Davies, B., and Z. Schwarz-Sommer. 1994. Control of floral organ identity by homeotic MADS box transcription factors. **In** Results and Problems in Cell Differentiation, ed. L. Nover, pp. 235–258. Springer-Verlag, Berlin.

Davies, B., M. Egea-Cortines, E. de Andrade Silva, H. Saedler, and H. Sommer. 1996. Multiple interactions amongst floral homeotic MADS box proteins. EMBO Journal **15**:4330–4343.

Day, C. D., B. F. Galgoci, and V. F. Irish. 1995. Genetic ablation of petal and stamen primordia to elucidate cell interactions during floral development. Development **121**:2887–2895.

Dellaporta, S. L., and A. Calderon-Urrea. 1993. Sex determination in flowering plants. Plant Cell **5**:1241–1251.

Delprete, P. G. 1996a. Evaluation of the tribes Chiococceae, Condamineeae and Catesbaeeae (Rubiaceae) based on morphological characters. Opera Botanica Belgica **7**:165–192.

Delprete, P. G. 1996b. Notes on calycophyllous Rubiaceae: Part 1. Morphological comparisons of the genera *Chimarrhis, Bathysa* and *Calycophyllum,* with new combinations and a new species, *Chimarrhis gentryana.* Brittonia **48**:35–44.

De Pinna, M. C. C. 1991. Concepts and tests of homology in the cladistic paradigm. Cladistics **7**:367–394.

Di Laurenzio, L., J. Wysocka-Diller, J. E. Malamy, L. Pysh, Y. Helariutta, G. Freshour, M. G. Hahn, K. A. Feldmann, and P. N. Benfey. 1996. The *SCARECROW* gene regulates an asymmetric cell division that is essential for generating the radial organization of the *Arabidopsis* root. Cell **86**:423–433.

Di Laurenzio, L., L. Struwe, A. S. Pepper, D. Kizirian, and V. A. Albert. 1997. Gene expression analysis of sepal identity in *Clermontia* (Lobelioideae: Campanulaceae): homeosis and floral diversification in the Hawaiian archipelago. In Evolution of Plant Development, Keystone Symposium B1, Taos, New Mexico, p. 24 [abstract].

Dilcher, D. L., and P. R. Crane. 1984. *Archaeanthus:* an early angiosperm from the Cenomanian of the western interior of North America. Annals of the Missouri Botanical Garden **71**:351–383.

Donoghue, M. J. 1989. Phylogenies and the analysis of evolutionary sequences, with examples from seed plants. Evolution **43**:1137–1156.

Donoghue, M. J. 1994. Progress and prospects in reconstructing plant phylogeny. Annals of the Missouri Botanical Garden **81**:405–418.

Donoghue, M. J., R. G. Olmstead, J. F. Smith, and J. D. Palmer. 1992. Phylogenetic relationships of Dipsacales based on *rbcL* sequences. Annals of the Missouri Botanical Garden **79**:333–345.

Doyle, J. A. 1994. Origin of the angiosperm flower: a phylogenetic perspective. Plant Systematics and Evolution (suppl.) **8**:7–29.

Doyle, J. A., and M. J. Donoghue. 1986. Seed plant phylogeny and the origin of angiosperms: an experimental cladistic approach. Botanical Review **52**:321–431.

Doyle, J. A., and M. J. Donoghue. 1992. Fossils and seed plant phylogeny reanalyzed. Brittonia **44**:89–106.

Doyle, J. A., and C. L. Hotton. 1991. Diversification of early angiosperm pollen in a cladistic context. In Pollen and Spores: Patterns of Diversification, eds. S. Blackmore and S. H. Barnes, pp. 169–195. Clarendon Press, Oxford.

Doyle, J. A., M. J. Donoghue, and E. A. Zimmer. 1994. Integration of morphological and ribosomal RNA data on the origin of angiosperms. Annals of the Missouri Botanical Garden **81**:419–450.

Doyle, J. J. 1994. Evolution of a plant homeotic multigene family: toward connecting molecular systematics and molecular developmental genetics. Systematic Biology **43**:307–328.

Drinnan, A. N., P. R. Crane, E. M. Friis, and K. R. Pedersen. 1991. Angiosperm flowers and tricolpate pollen of buxaceous affinity from the Potomac Group (mid-Cretaceous) of eastern North America. American Journal of Botany **78**:153–176.

Drinnan, A. N., P. R. Crane, and S. B. Hoot. 1994. Patterns of floral evolution in the early diversification of nonmagnoliid dicotyledons (eudicots). Plant Systematics and Evolution (suppl.) **8**:93–122.

Elliott, R. C., A. S. Betzner, E. Huttner, M. P. Oakes, W. Q. J. Tucker, D. Gerentes, P. Perez, and D. R. Smyth. 1996. *AINTEGUMENTA,* an *APETALA2*-like gene of *Arabidopsis* with pleiotropic roles in ovule development and floral organ growth. Plant Cell **8**:155–168.

Endress, P. K. 1994. Diversity and Evolutionary Biology of Tropical Flowers. Cambridge University Press, Cambridge.

Engler, A. 1925. Guttiferae. In Die Natürlichen Pflanzenfamilien, second edition, Vol. 21, eds. A. Engler and G. Prantl, pp. 154–237. Engelmann, Leipzig.

Engström, P., and K. Tandre. 1997. Conservation of transcription factor function between angiosperms and conifers. In Evolution of Plant Development, Keystone Symposium B1, Taos, New Mexico, p. 6 [abstract].

Erbar, C. 1991. Sympetaly—a systematic character? Botanische Jahrbücher für Systematik, Pflanzengeschichte und Pflanzengeographie **112**:417–451.

Ewan, J. 1948. A revision of *Macrocarpaea,* a neotropical genus of shrubby gentians. Contributions to the United States National Herbarium **29**:209–251.

Ewan, J. 1952. A review of the Neotropical lisianthoid genus *Lagenanthus* (Gentianaceae). Mutisia **4**:1–5.

Fitch, W. M. 1970. Distinguishing homologous from analogous proteins. Systematic Zoology **19**:99–113.

Friis, E. M., K. R. Pedersen, and P. R. Crane. 1994. Angiosperm floral structures from the Early Cretaceous of Portugal. Plant Systematics and Evolution (suppl.) **8**:31–49.

Goto, K., and E. M. Meyerowitz. 1994. Function and regulation of the *Arabidopsis* floral homeotic gene *PISTILLATA.* Genes and Development **8**:1548–1560.

Grey-Wilson, C. 1980. *Impatiens* of Africa. Morphology, Pollination and Pollinators, Ecology, Phytogeography, Hybridisation, Keys and Systematic Treatment of All African Species. Balkema, Rotterdam.

Halfter, U., N. Ali, J. Stockhaus, L. Ren, and N. H. Chua. 1994. Ectopic expression of a single homeotic gene, the petunia gene *GREEN PETAL,* is sufficient to convert sepals to petaloid organs. EMBO Journal **13**:1443–1449.

Harris, E. M. 1995. Inflorescence and floral ontogeny in Asteraceae: a synthesis of historical and current concepts. Botanical Review **61**:93–278.

Helariutta, Y., P. Elomaa, M. Kotilainen, P. Seppänen, and T. H. Teeri. 1993. Cloning of cDNA coding for dihydroflavonol-4-reductase (DFR) and characterization of *DFR* expression in the corollas of *Gerbera hybrida* var. Regina (Compositae). Plant Molecular Biology **22**:183–193.

Helariutta, Y., M. Kotilainen, P. Elomaa, and T. H. Teeri. 1995. *Gerbera hybrida* (Asteraceae) imposes regulation at several anatomical levels during inflorescence development on the gene for dihydroflavonol-4-reductase. Plant Molecular Biology **28**:935–941.

Helariutta, Y., M. Kotilainen, P. Elomaa, N. Kalkkinen, K. Bremer, T. H. Teeri, and V. A. Albert. 1996. Duplication and functional divergence in the chalcone synthase gene family of Asteraceae: evolution with substrate change and catalytic simplification. Proceedings of the National Academy of Sciences U.S.A. **93**:9033–9038.

Herschbach, B. M., M. B. Arnaud, and A. D. Johnson. 1994. Transcriptional repression directed by the yeast alpha 2 protein in vitro. Nature **370**:309–311.

Heywood, V. H. (ed.) 1985. Flowering Plants of the World. Croom Helm, London and Sydney.

Hiepko, P. 1965. Vergleichend-morphologische und entwicklungs-geschichtliche Untersuchungen über das Perianth bei den Polycarpicae. Botanische Jahrbücher für Systematik, Pfanzengeschichte und Pflanzengeographie **84**:359–508.

Huijser, P., J. Klein, W.-E. Lönnig, H. Meijer, H. Saedler, and H. Sommer. 1992. Bracteomania, an inflorescence anomaly, is caused by the loss of function of the MADS box gene *SQUAMOSA* in *Antirrhinum majus*. EMBO Journal **11**:1239–1249.

Ingram, G. C., J. Goodrich, M. D. Wilkinson, R. Simon, G. W. Haughn, and E. S. Coen. 1995. Parallels between *UNUSUAL FLOWER ORGANS* and *FIMBRIATA* genes controlling flower development in *Arabidopsis* and *Antirrhinum*. Plant Cell **7**:1501–1510.

Irish V. F., and Y. T. Yamamoto. 1995. Conservation of floral homeotic gene function between *Arabidopsis* and *Antirrhinum*. Plant Cell **7**:1635–1644.

Jack, T., L. L. Brockman, and E. M. Meyerowitz. 1992. The homeotic gene *APETALA3* of *Arabidopsis thaliana* encodes a MADS-box and is expressed in petals and stamens. Cell **68**:683–687.

Jack, T., G. L. Fox, and E. M. Meyerowitz. 1994. Arabidopsis homeotic gene *APETALA3* ectopic expression: transcriptional and posttranscriptional regulation determine floral organ identity. Cell **76**:703–716.

Jackson, D., K. Roberts, and C. Martin. 1992. Temporal and spatial control of expression of anthocyanin biosynthetic genes in developing flowers of *Antirrhinum majus*. Plant Journal **2**:425–434.

Jofuku, K. D., B. G. den Boer, M. van Montagu, and J. K. Okamuro. 1994. Control of *Arabidopsis* flower and seed development by the homeotic gene *APETALA2*. Plant Cell **6**:1211–1225.

Judd, W. S., R. W. Sanders, and M. J. Donoghue. 1994. Angiosperm family pairs: preliminary phylogenetic analyses. Harvard Papers in Botany **5**:1–51.

Klucher, K. J., H. Chow, L. Reiser, and R. L. Fischer. 1996. The *AINTEGUMENTA* gene of *Arabidopsis* required for ovule and female gametophyte development is related to the floral homeotic gene *APETALA2*. Plant Cell **8**:137–153.

Kosuge, K. 1994. Petal evolution in Ranunculaceae. Plant Systematics and Evolution (suppl.) **8**:185–191.

Kotilainen, M., D. Yu, P. Elomaa, V. A. Albert, Y. Helariutta, M. Mehto, and T. H. Teeri. 1997. Inflorescence development in Asteraceae—a novel MADS box gene *RCD1* is differentially required for C-function in marginal versus central florets. In Evolution of Plant Development, Keystone Symposium B1, Taos, New Mexico, p. 25 [abstract].

Krizek, B. A., and E. M. Meyerowitz. 1996. The *Arabidopsis* homeotic genes *APETALA3* and *PISTILLATA* are sufficient to provide B class organ identity function. Development **122**:11–22.

Lammers, T. G. 1991. Systematics of *Clermontia* (Campanulaceae-Lobelioideae). Systematic Botany Monographs **32**:1–97.

Lammers, T. G. 1995. Patterns of speciation and biogeography in *Clermontia* (Campanulaceae, Lobelioideae). In Hawaiian Biogeography: Evolution on a Hot Spot Archipelago, eds. W. L. Wagner and V. Funk, pp. 338–362. Smithsonian Institution Press, Washington, D.C.

Ma, H. 1994. The unfolding drama of flower development: recent results from genetic and molecular analyses. Genes and Development **8**:745–756.

Ma, H., M. F. Yanofsky, and E. M. Meyerowitz. 1991. *AGL1–AGL6*, an *Arabidopsis* gene family with similarity to floral homeotic and transcription factor genes. Genes and Development **5**:484–495.

Mabberley, D. J. 1987. The Plant Book. Cambridge University Press, Cambridge.

Maddison, W. P. 1990. A method for testing the correlated evolution of two binary characters: are gains or losses concentrated on certain branches of phylogenetic tree? Evolution **44**:539–557.

Maddison, W. P., and D. R. Maddison. 1992. MacClade: Interactive Analysis of Phylogeny and Character Evolution, version 3.0. Sinauer Associates, Sunderland, Massachusetts.

Maguire, B. 1972. Clusiaceae—*Quapoya* and *Renggeria*. In The Botany of the Guyana Highland, Part IX, Memoirs of the New York Botanical Garden Vol. 23, eds. B. Maguire and collaborators, pp. 192–196. New York Botanical Garden, New York.

Maguire, B. 1977. A revision of *Clusia* L. section *Cochlanthera* (Choisy) Engler. Caldasia **11**:129–146.

Mandel, M. A., C. Gustafson-Brown, B. Savidge, and M. F. Yanofsky. 1992. Molecular characterization of the *Arabidopsis* floral homeotic gene *APETALA1*. Nature **360**:273–277.

Martin C., and T. Gerats. 1993. Control of pigment biosynthesis genes during petal development. Plant Cell **5**:1253–1264.

Martin, C., A. Prescott, S. Mackay, J. Bartlett, and E. Vrijlandt. 1991. Control of anthocyanin biosynthesis in flowers of *Antirrhinum majus*. Plant Journal **1**:37–49.

Mattsson, J., E. Söderman, M. Svenson, C. Borkird, and P. Engström. 1992. A new homeobox-leucine zipper gene from *Arabidopsis thaliana.* Plant Molecular Biology **18:**1019–1022.

Mauthe, S., K. Bachmann, K. L. Chambers, and H. J. Price. 1984. Independent responses of two fruit characters to developmental regulation in *Microseris douglasii* (Asteraceae, Lactuceae). Experientia **40:**1280–1281.

Mayer, U., R. A. Torres-Ruiz, T. Berleth, S. Miséra, and G. Jürgens. 1991. Mutations affecting body organization in the *Arabidopsis* embryo. Nature **353:**402–407.

Mead, J., H. Zhong, T. B. Acton, and A. K. Vershon. 1996. The yeast alpha2 and Mcm1 proteins interact through a region similar to a motif found in homeodomain proteins of higher eukaryotes. Molecular and Cell Biology **16:**2135–2143.

Modrusan, Z., L. Reiser, K. A. Feldmann, R. L. Fischer, and G. W. Haughn. 1994. Homeotic transformation of ovules into carpel-like structures in *Arabidopsis.* Plant Cell **6:**333–349.

Morgan, D. R., and D. E. Soltis. 1993. Phylogenetic relationships among members of Saxifragaceae sensu lato based on *rbc*L sequence data. Annals of the Missouri Botanical Garden **80:**631–660.

Mori, S. A., and G. T. Prance. 1990a. Taxonomy, ecology, and economic botany of the Brazil nut (*Bertholletia excelsa* Humb. & Bonpl.: Lecythidaceae). Advances in Economic Botany **8:**130–150.

Mori, S. A., and G. T. Prance. 1990b. Lecythidaceae, part II: the zygomorphic-flowered New World genera (*Couroupita, Corythophora, Bertholletia, Couratari, Eschweilera,* & *Lecythis*). With a study of secondary xylem of neotropical Lecythidaceae by Carl de Zeeuw. FloraNeotropica Monographs **21:**1–376.

Mori, S. A., J. E. Orchard, and G. T. Prance. 1980. Intrafloral pollen differentiation in the New World Lecythidaceae, subfamily Lecythidoideae. Science **209:** 400–403.

Morton, C. M., M. W. Chase, K. A. Kron, and S. A. Swensen. 1996. A molecular evaluation of the monophyly of the order Ebenales based upon *rbc*L sequence data. Systematic Botany **21:**567–586.

Nixon, K. C., and W. L. Crepet. 1993. Late Cretaceous flowers of ericalean affinity. American Journal of Botany **80:**616–623.

Nixon, K. C., W. L. Crepet, D. M. Stevenson, and E. M. Friis. 1994. A reevaluation of seed plant phylogeny. Annals of the Missouri Botanical Garden **81:**484–533.

Nurrish, S. J., and R. Treisman. 1995. DNA binding specificity determinants in MADS-box transcription factors. Molecular and Cell Biology **15:**4076–4085.

Okamuro, J. K., B. G. W. den Boer, and K. D. Jofuku. 1993. Regulation of *Arabidopsis* flower development. Plant Cell **5:**1183–1193.

Olmstead, R. G., B. Bremer, K. M. Scott, and J. D. Palmer. 1993. A parsimony analysis of the Asteridae sensu lato based on *rbc*L sequences. Annals of the Missouri Botanical Garden **80:**700–722.

Patterson, C. 1988. Homology in classical and molecular biology. Molecular Biology and Evolution **5:**603–625.

Prance, G. T., and S. A. Mori. 1979. The actinomorphic-flowered New World Lecythidaceae. Lecythidaceae, part I. Flora Neotropica Monographs **21:**1–270.

Price, R. A., I. A. Al-Shehbaz, and J. D. Palmer. 1994. Systematic relationships of *Arabidopsis:* a molecular and morphological perspective. In *Arabidopsis,* eds. C. Somerville and E. M. Meyerowitz. Cold Spring Harbor Press, New York.

Pringle, J. S. 1995. Gentianaceae. In Flora of Ecuador, vol. 159A, eds. G. Harling and L. Andersson, pp. 1–131. Department of Systematic Botany, Göteborg University, Göteborg.

Purugganan, M. D., S. D. Rounsley, R. J. Schmidt, and M. F. Yanofsky. 1995. Molecular evolution of flower development: diversification of the plant MADS-box regulatory gene family. Genetics **140:**345–356.

Raubeson, L. A., and R. K. Jansen. 1992. Chloroplast DNA evidence on the ancient evolutionary split in vascular land plants. Science **255:**1697–1699.

Reiser, L., Z. Modrusan, L. Margossian, A. Samach, N. Ohad, G. W. Haughn, and R. L. Fischer. 1995. The *BELL1* gene encodes a homeodomain protein involved in pattern formation in the *Arabidopsis* ovule primordium. Cell **83:**735–742.

Rodman, J. E., R. A. Price, K. Karol, E. Conti, K. Sytsma, and J. D. Palmer. 1993. Nucleotide sequences of the *rbc*L gene indicate monophyly of mustard oil plants. Annals of the Missouri Botanical Garden **80:**686–699.

Rounsley, S. D., G. S. Ditta, and M. F. Yanofsky. 1995. Diverse roles for MADS box genes in *Arabidopsis* development. Plant Cell **7:**1259–1269.

Schwarz-Sommer, Z., P. Huijser, W. Nacken, H. Saedler, and H. Sommer. 1990. Genetic control of flower development: homeotic genes in *Antirrhinum majus.* Science **250:**931–936.

Schwarz-Sommer, Z., I. Hue, P. Huijser, P. J. Flor, R. Hansen, F. Tetens, W. E. Lönnig, H. Saedler, and H. Sommer. 1992. Characterization of the *Antirrhinum* floral homeotic MADS-box gene *DEFICIENS:* evidence for DNA binding and autoregulation of its persistent expression throughout flower development. EMBO Journal **11:**251–263.

Sessa, G., G. Morelli, and I. Ruberti. 1993. The Athb-1 and -2 HD-Zip domains homodimerize forming complexes of different DNA binding specificities. EMBO Journal **12:**3507–3517.

Shore, P., and A. D. Sharrocks. 1995. The MADS-box family of transcription factors. European Journal of Biochemistry **229:**1–13.

Söderman, E., J. Mattsson, M. Svenson, C. Borkird, and P. Engström. 1994. Expression pattern of novel genes encoding homeodomain leucine-zipper proteins in *Arabidopsis thaliana.* Plant Molecular Biology **26:**145–154.

Soltis, D. E., P. S. Soltis, D. L. Nickrent, L. A. Johnson, W. J. Hahn, S. B. Hoot, J. A. Sweere, R. K. Kuzoff, K. A. Kron, M. W. Chase, S. M. Swensen, E. A. Zimmer, S.-M. Chaw, L. J. Gillespie, W. J. Kress, and K. J. Sytsma. 1997. Angiosperm phylogeny inferred from 18S ribosomal DNA sequences. Annals of the Missouri Botanical Garden **84:**1–49.

Sommer, H., J.-P. Beltrán, P. Huijser, H. Pape, W.-E. Lönnig, H. Saedler, and Z. Schwarz-Sommer. 1990. *Deficiens,* a homeotic gene involved in the control of flower morphogenesis in *Antirrhinum majus:* the protein shows homology to transcription factors. EMBO Journal 9:605–613.

Spooner, D. M., G. J. Anderson, and R. K. Jansen. 1993. Chloroplast DNA evidence for the interrelationships of tomatoes, potatoes and pepinos (Solanaceae). American Journal of Botany 80:676–688.

Stewart, W. N., and G. W. Rothwell. 1993. Paleobotany and the Evolution of Plants, second edition. Cambridge University Press, Cambridge.

Struwe, L., V. A. Albert, and B. Bremer. 1994. Cladistics and family level classification of the Gentianales. Cladistics 10:175–206.

Svensson, M., and P. Engström. 1997. The *L421* gene isolated from *Lycopodium annotinum* L. strobilus cDNA library is a distant relative to seed plant MADS-box genes. In Evolution of Plant Development, Keystone Symposium B1, Taos, New Mexico, p. 27 [abstract].

Sytsma, K. J., and L. D. Gottlieb. 1986. Chloroplast DNA evidence for the origin of the genus *Heterogaura* from a species of *Clarkia* (Onagraceae). Proceedings of the National Academy of Sciences U.S.A. 83:5554–5557.

Takhtajan, A. L. 1991. Evolutionary Trends in Flowering Plants. Columbia University Press, New York.

Tandre, K., V. A. Albert, A. Sundås, and P. Engström. 1995. Conifer homologues to genes that control floral development in angiosperms. Plant Molecular Biology 27:69–78.

Theißen, G., and H. Saedler. 1995. MADS-box genes in plant ontogeny and phylogeny: Haeckel's "biogenetic law" revisited. Current Opinion in Genetics and Development 5:628–639.

Theißen, G., J. T. Kim, and H. Saedler. 1996. Classification and phylogeny of the MADS-box multigene family suggest defined roles of MADS-box gene subfamilies in the morphological evolution of eukaryotes. Journal of Molecular Evolution 43:484–516.

Tröbner, W., L. Ramirez, P. Motte, I. Hue, P. Huijser, W. Lönnig, H. Saedler, H. Sommer, and Z. Schwarz-Sommer. 1992. *GLOBOSA:* a homeotic gene which interacts with *DEFICIENS* in the control of *Antirrhinum* floral organogenesis. EMBO Journal 11:4693–4704.

Tsuchimoto, S., A. R. van der Krol, and N.-H. Chua. 1993. Ectopic expression of *pMADS3* in transgenic petunia phenocopies the petunia *blind* mutant. Plant Cell 5:843–853.

van der Krol, A. R., and N.-H. Chua. 1993. Flower development in petunia. Plant Cell 5:1195–1203.

van der Krol, A. R., A. Brunelle, S. Tsuchimoto, and N.-H. Chua. 1993. Functional analysis of petunia floral homeotic MADS box gene *pMADS1*. Genes and Development 7:1214–1228.

Vollbrecht, E., B. Veit, N. Sinha, and S. Hake. 1991. The developmental gene *KNOTTED-1* is a member of a maize homeobox gene family. Nature 350:241–243.

Weigel, D., and S. E. Clark. 1996. Sizing up the floral meristem. Plant Physiology 112:5–10.

Weigel, D., and E. M. Meyerowitz. 1994. The ABCs of floral homeotic genes. Cell 78:203–209.

Weller, S. G., M. J. Donoghue, and D. Charlesworth. 1995. The evolution of self-incompatibility in flowering plants: a phylogenetic approach. In Experimental and Molecular Approaches to Plant Biosystematics, eds. P. C. Hoch and A. G. Stephenson, pp. 355–382. The Missouri Botanical Garden, St. Louis.

Werdelin, L., and B. S. Tullberg. 1995. A comparison of the two methods to study correlated discrete characters on phylogenetic trees. Cladistics 11:265–277.

Yanofsky, M. F., H. Ma, J. L. Bowman, G. N. Drew, K. A. Feldmann, and E. M. Meyerowitz. 1990. The protein encoded by the *Arabidopsis* homeotic gene *AGAMOUS* resembles transcription factors. Nature 346: 35–39.

Yu, D., M. Kotilainen, P. Elomaa, M. Mehto, Y. Helariutta, V. A. Albert, and T. H. Teeri. 1997. Inflorescence development in Asteraceae: B-function MADS-box genes are required for congenital fusion in corolla and stamen, and the *DEFICIENS* ortholog has differential radial effects within capitulum. In Evolution of Plant Development, Keystone Symposium B1, Taos, New Mexico, p. 28 [abstract].

13

The Origin and Evolution of Plastids and Their Genomes

Jeffrey D. Palmer and Charles F. Delwiche

Plastids, the eukaryotic organelles responsible for photosynthesis and other biochemical tasks, are semiautonomous endosymbionts derived from previously free-living cyanobacteria (Gray, 1992; Douglas, 1994; Loiseaux-de Göer, 1994; Bhattacharya and Medlin, 1995). In this chapter we discuss, with an emphasis on molecular phylogenetic evidence, our current understanding of the origin and diversification of these organelles, provide an overview of plastid diversity in the context of endosymbiosis, and discuss the evolution of plastid genomes from that of a cyanobacterium. We place particular emphasis on the evidence from different cellular genomes—nuclear, plastid, and mitochondrial—that bears on the fundamental question of whether all plastids are derived from a single endosymbiotic event, or from two or more independent events. Although the bulk of the data support a monophyletic origin of plastids, some evidence, particularly from the nuclear genome, suggests at least two independent endosymbiotic events.

The phylogeny of plastids is complicated by the fact that a number of algal lineages, and apparently at least one nonphotosynthetic protist lineage, have acquired plastids indirectly, via acquisition of a *eukaryotic* endosymbiont, itself already equipped with plastids, giving rise to a "russian-doll" cell (Fig. 13.1; Gibbs, 1993; Whatley, 1993a, 1993b; Cavalier-Smith, 1995; McFadden and Gilson, 1995). Such secondary symbioses have given rise to the plastids of a variety of algae. In fact, only three algal lineages have plastids that are thought to be the direct products of primary endosymbioses: the green algae, the red algae, and a relatively obscure group, the glaucocystophytes. All other algal lineages seem, with various levels of confidence, to have acquired their plastids via secondary, or in some cases tertiary, symbiosis. Because of this reticulation, reconstruction of the history of plastid evolution requires careful attention to the distinction between primary and secondary plastids.

In addition to covering current information on plastid phylogeny, this chapter also reviews three other topics in plastid evolution of broad interest. First, we consider recent discoveries that have implications concerning the pigment composition of the cyanobacterial ancestor of plastids. Second, the phylogeny of the CO_2-fixing enzyme RUBISCO (ribulose-1,5-bisphosphate carboxylase/oxygenase) differs from that of all other known plastid genes and seems to reflect a history of relatively frequent lateral gene transfer and gene duplication. Finally, we review in some detail the evolution of plastid

We thank Chris Parkinson for generous help with the manuscript; Debashish Bhattacharya, Gertraud Burger, Mike Gray, Ben Hall, Franz Lang, Geoff McFadden, Charley O'Kelly, Diana Stein, John Stiller, Richard Triemer, and Seán Turner for generously supplying unpublished results; the NSF (grant DEB-9318594 to J.D.P.) for financial support; and the editors and publisher of this book for accepting this chapter so very late.

Primary Endosymbiosis

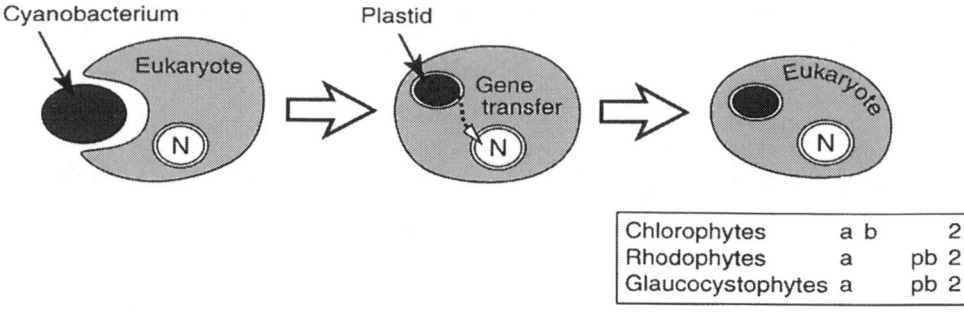

Chlorophytes	a b		2
Rhodophytes	a	pb	2
Glaucocystophytes	a	pb	2

Secondary Endosymbiosis

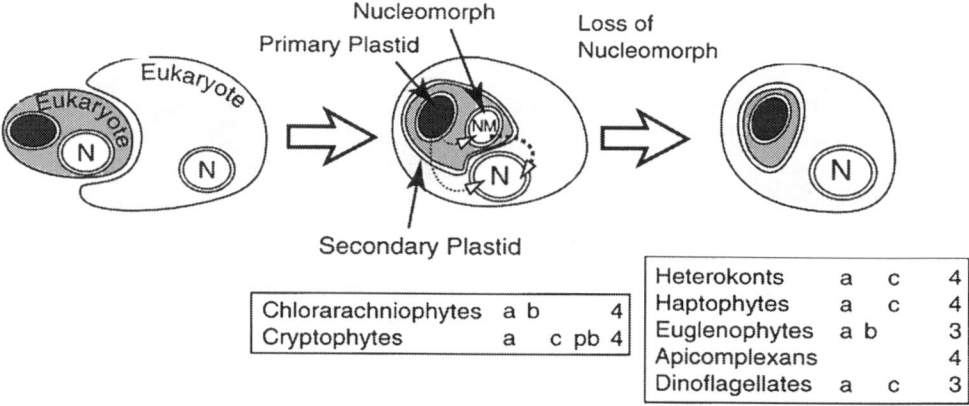

Chlorarachniophytes	a b		4
Cryptophytes	a	c pb	4

Heterokonts	a	c	4
Haptophytes	a	c	4
Euglenophytes	a b		3
Apicomplexans			4
Dinoflagellates	a	c	3

Figure 13.1. The origin of plastids by primary endosymbiosis (top) and their spread by secondary endosymbiosis (bottom). Groups of eukaryotes are indicated in boxes below the plastid/cell type they possess. The *a, b, c,* and/or pb shown in columns next to each group name indicate the presence of chlorophylls *a, b, c,* and/or phycobilins, respectively, in that group (chlorophyll *c* is also present in at least one green alga [Wilhelm, 1987]). The numbers in the boxes indicate the number of membranes surrounding each plastid type. Note that whereas some nucleomorph-lacking secondary plastids do have four bounding membranes (as shown), others have only three (not shown). In addition, some secondary plastids have a direct continuity (not shown) between the outermost plastid membrane (which sometimes contains ribosomes on its cytoplasmic side) and the outer nuclear membrane (Gibbs, 1993; Whatley, 1993a, 1993b).

genomes in the context of their cyanobacterial ancestry, with emphasis on how they have diverged from cyanobacterial genomes. This divergence is most pronounced in terms of genome size and gene content: A massive transfer of symbiont gene function to the nucleus of the host cell has occurred not only in the evolution of primary plastids, but also repeatedly during secondary symbiosis. Nonetheless, plastid genomes of diverse algal and plant lineages show surprising variation in number and kind of genes still resident. Other major ways in which plastid genomes have diverged from those of

cyanobacteria include their content of introns and repeated sequences.

AT LEAST THREE LINEAGES OF PRIMARY PLASTIDS

The most familiar and widely studied plastids are those of green algae and their land plant descendants. These plastids are pigmented by chlorophylls *a* and *b,* with carotenoids as accessory pigments; lutein or neoxanthin is generally the major xanthophyll. They are surrounded by two membranes and have stacked thylakoids.

Essentially all information indicates that green algal plastids are primary plastids, descended directly from free-living cyanobacterial ancestors (but see Stiller and Hall, 1997). Taxa with secondary plastids thought to be derived from green algal ancestors include the chlorarachniophytes, the euglenophytes, some dinoflagellates, and, apparently, the apicomplexans (Fig. 13.1). Throughout this chapter we refer to the monophyletic group composed of green algae and land plants as "green algae" or "chlorophytes," and use the term *chlorophytes s. l.* (sensu lato) to refer to all taxa with plastids, both primary and secondary, that derive from this lineage.

The red algae constitute a second lineage with plastids that are almost certainly primary in origin. Primary pigments in red algae are chlorophyll *a* and phycobilins, light-harvesting proteins that are organized into hemispherical antennae called phycobilisomes (similar structures are also found in cyanobacteria). Carotenoids are also present, with zeaxanthin typically being the major xanthophyll (Tan et al., 1997). Taxa with secondary plastids thought to be of red algal origin include the cryptophytes, the heterokonts (also known as chromophytes), the haptophytes, and peridinin-containing dinoflagellates (Fig. 13.1). Among these secondary plastids, the cryptophytes contain chlorophyll *c* in addition to chlorophyll *a* and phycobilins (although unlike red plastids, the phycobilins are not arranged in phycobilisomes, but rather are found in the thylakoid lumens), while the other secondary plastids have chlorophylls *a* and *c,* but no phycobilins (Fig. 13.1). We use the term *rhodophytes s. l.* to refer to red algae and their secondary symbiotic derivatives.

A small and relatively little-known group, the glaucocystophytes, contains a third lineage of probable primary plastids. Like the red algae, the glaucocystophytes contain chlorophyll *a* and phycobilins. Remarkably, the plastids of glaucocystophytes have a peptidoglycan cell wall located between their two plastid membranes (Kies and Kremer, 1990). This structure confounded researchers for many years, because it was unclear whether the photosynthetic structure was a plastid or a cyanobacterium. This confusion led to the use of the term *cyanelle* (after *Cyanophora paradoxa*) for the plastid of glaucocystophytes. However, the cell wall of the glaucocystophyte plastid is now regarded as simply an ancestral feature that has been retained in this plastid lineage. In all other particulars the glaucocystophyte plastid is a true plastid; it is an obligate endosymbiont that shares the marked reduction in genome size and gene content that is characteristic of plastids (Löffelhardt and Bohnert, 1994; Stirewalt et al., 1995; Bhattacharya and Schmidt, 1997). No algae are known to have acquired secondary plastids from glaucocystophytes, but one candidate is the poorly understood *Paulinella chromatophora* (see next paragraph and Bhattacharya et al., [1995a] and Bhattacharya and Medlin, [1995]).

The following sections of this chapter deal exclusively with these three groups of primary plastids, and with the above-mentioned groups of secondary plastids of green and red algal origin, as sufficient molecular data exist for these groups only to address questions relating to the origin and evolution of their plastids and plastid genomes. We should point out in passing, however, that there are several poorly characterized photosynthetic eukaryotes whose photosynthetic "organelles" may have arisen by still further primary or secondary endosymbioses. These include not only *Paulinella chromatophora* (Bhattacharya et al., 1995a; Bhattacharya and Medlin, 1995), but also *Psalteriomonas lanterna* (Hackstein, 1995) and *Dinophysis* spp. (Schnepf, 1993). The only one of these for which any molecular data are available is *P. chromatophora,* whose nuclear 18S rRNA sequence places it, with high bootstrap support, in a nested grouping with two nonphotosynthetic lineages (Bhattacharya et al., 1995a; Bhattacharya and Medlin, 1995). This observation, and the fact that *Paulinella ovalis* lacks any detectable vestige of a photosynthetic organelle but is otherwise morphologically identical to *P. chromatophora,* point to an independent origin of the *P. chromatophora* plastid (or "cyanelle," because it is bounded by a peptidoglycan cell wall) from that of any other plastids. What is unclear, however, is whether this was an independent primary endosymbiosis or a secondary endosymbiosis from, most likely, a glaucocystophyte (Bhattacharya and Medlin, 1995; Bhattacharya et al., 1995a).

ARE PRIMARY PLASTIDS MONOPHYLETIC OR POLYPHYLETIC IN ORIGIN?

Monophyly of the Groups in Question

Did the known groups of primary plastids arise by one or more than one cyanobacterial endosymbioses? Because all available data strongly indicate that each of the three groups of primary plastid-containing organisms—green algae, red algae, and glaucocystophytes—is itself monophyletic, this question boils down to "Were there one, two, or three primary endosymbioses?" Gene trees supporting monophyly of these three groups come from plastid and nuclear SSU rRNAs, the only molecules sampled multiply from all three groups (Cavalier-Smith, 1993a; Bhattacharya et al., 1995b; Bhattacharya and Medlin, 1995; Helmchen et al., 1995; Ragan and Gutell, 1995; Van de Peer et al., 1996); a number of molecules (plastid *rbcL,* nuclear actin, and a concatenation of multiple mitochondrial protein genes) sampled from multiple green and red algae but not glaucocystophytes (Delwiche and Palmer, 1996; Bhattacharya and Weber, 1997; Lang et al., 1997; Leblanc et al., 1997); and several other molecules sampled from green algae only (e.g., Keeling and Doolittle, 1996). An especially strong marker of red algal monophyly is the presence of a RUBISCO operon of proteobacterial origin, as opposed to the cyanobacterial RUBISCO operon of chlorophytes s. l. and glaucocystophytes (see section on RUBISCO Phylogeny and Horizontal Gene Transfer).

Plastid Evidence

Six plastid genes have been sequenced in multiple cyanobacteria and in at least one member of each of the three primary lineages of plastids. Phylogenetic analyses of these six data sets all support a monophyletic origin of plastids, albeit with varying degrees of confidence and subject to invoking a lateral gene transfer event for one of the genes. 16S rRNA provides the strongest evidence for plastid monophyly, both because cyanobacterial sampling is best for this molecule (11–15 complete cyanobacterial 16S se-

quences have been analyzed, with up to 89 partial sequences reported) and because bootstrap support for monophyly is highest (e.g., Bhattacharya and Medlin, 1995; Nelissen et al., 1995; Turner, 1997). Next best sampled among cyanobacteria (a total of 16 taxa) is *rpoC1,* which gives moderately high support for plastid monophyly. However, these analyses are problematic with respect to plastid sampling, as *rpoC1* sequences from all three groups of primary plastids have not yet been included in the same analysis (in one study, *Cyanophora* and four green algae were examined [Palenik and Haselkorn, 1992], while in another study, *Cyanophora* and two red algae were analyzed [Palenik and Swift, 1996]). The *psbA* gene is moderately well sampled among cyanobacteria (13 sequences) and also provides moderately strong support for plastid monophyly (Hess et al., 1995). The *tufA* gene is only modestly sampled among cyanobacteria (seven) and gives weak to moderate evidence for plastid monophyly (Fig. 13.2; Delwiche et al., 1995; Köhler et al., 1997). The *atpB* gene is poorly sampled for cyanobacteria (four) and gives moderate support for plastid monophyly (Douglas and Murphy, 1994). Lastly, the *rbcL* gene strongly supports monophyly of green algae and *Cyanophora.* However, only six cyanobacteria have been sampled, and red algae are problematic because their *rbcL* gene is not orthologous to those of other plastids and cyanobacteria, but is instead of proteobacterial origin as a result, most likely, of lateral gene transfer (see section on RUBISCO Phylogeny and Horizontal Gene Transfer).

Thus, setting aside this *rbcL* transfer, all[1] six relevant plastid genes examined to date point toward a monophyletic, cyanobacterial origin of

---

[1]The only apparently conflicting data for genes of plastid origin (but one which was anciently transferred to the nucleus) concerns HSP60, which in one analysis groups red algae and cyanobacteria to the exclusion of green algae (Viale and Arakaki, 1994). However, further sampling of cyanobacterial sequences (Viale et al., 1994) leads to the conclusion (Delwiche et al., 1995; Delwiche and Palmer, 1997) that the green and red algal HSP60s are probably paralogous (resulting from an ancient gene duplication within cyanobacteria), in which case this gene is at present silent on the question of plastid monophyly.

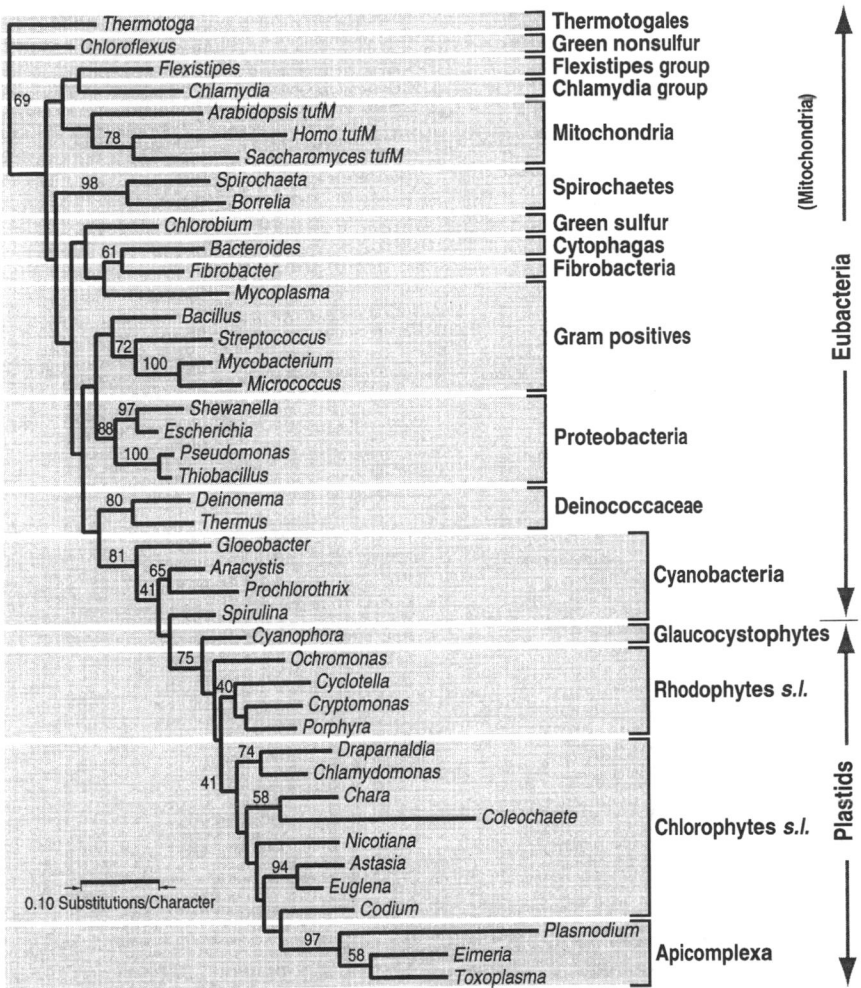

Figure 13.2. A *tufA* phylogeny showing monophyly of plastids and a green algal origin of the apicomplexan plastid. The tree shown had the highest likelihood in a maximum likelihood analysis of first and second position nucleotides (Ln likelihood = −11614 with an F84 model using empirical base frequencies and T = 0.8). Branch lengths are proportional to the number of inferred substitutions (see scale bar). Only those bootstrap values ≥40% are shown. Figure is modified from Fig. 3 of Köhler et al. (1997), which presents a tree showing monophyly of plastids relative to seven cyanobacterial sequences.

all primary plastids. We must, however, reemphasize the two major limitations of these analyses: the generally weak bootstrap support for plastid monophyly and the lack of extensive sampling among cyanobacteria. With present sampling, it is impossible to rule out two or even three independent endosymbiotic events involving closely related cyanobacteria.

Several other plastid traits support, or at least are consistent with, a monophyletic origin of all primary plastids. First, and perhaps most signif-

icantly, transit peptides from all three plastid types are functionally interchangeable; for example, a red algal transit peptide can direct import of a cytoplasmically synthesized protein into a land plant plastid (Apt et al., 1993; Jakowitsch et al., 1996). Second, the extensive overlap of plastid gene content among diverse eukaryotes, viewed against a backdrop of extreme reduction in all cases from cyanobacterial gene content, supports the notion of plastid monophyly (Kowallik, 1994). This argument is most compelling for glaucocystophytes and red algae, for example all but eight of the 191 plastid genes found in *Cyanophora paradoxa* are also present among the 251 genes of *Porphyra purpurea* (Reith, 1995). Third, all three primary plastids possess two gene clusters (*psbB/N/H* and *atp/rps/rpo*) that are not found in cyanobacteria (Reith and Munholland, 1993a), although again the force of this conclusion is limited by poor sampling of cyanobacteria. Fourth, the antenna complex light-harvesting proteins of photosystem I are related immunologically between green and red algae, but not between these groups and cyanobacteria; comparable data for glaucocystophytes are lacking (Wolfe et al., 1994). Fifth, all three lineages of primary plastids usually contain a large, rRNA-encoding, inverted repeat in their genomes; while this repeat structure could be a synapomorphy for plastids, it could also be a sympleisiomorphy common to cyanobacterial genomes as well or, conversely, of parallel origin within plastids (see Repeated Sequences), and thus the phylogenetic significance of this pattern is dubious (Palmer, 1985, 1991; Loiseaux-de Goër, 1994; Reith, 1995). Finally, the nuclear genes from red algae and land plants that encode plastid glyceraldehyde-3-phosphate dehydrogenase (GAPDH) possess an intron of potential homology, but the problematic alignment of the very divergent transit peptides housing this intron precludes any strong conclusions (Liaud et al., 1993; Zhou and Ragan, 1994).

Mitochondrial Evidence

With one major caveat, mitochondrial DNA data strongly support the plastid-derived evidence for a monophyletic origin of primary plastids, whereas nuclear DNA data are largely unresolved with respect to this issue. The caveat is that no mitochondrial data are available for any glaucocystophytes. In contrast, the complete sequencing of mitochondrial genomes from several red algae and green algae has allowed the construction of phylogenetic trees using sequences from between one and five protein genes (the latter analyzed in combination). In all of these trees, a sister-group relationship of red algal and green algal mitochondria is recovered

with high bootstrap support (see especially Lang et al., 1997, which has the best sampling, but also see Boyen et al., 1994; Leblanc et al., 1997; Paquin et al., 1997). Note that mitochondrial rRNA genes are so unreliable as to be no longer used in global studies of eukaryotic phylogeny, owing to their highly unequal, often very rapid rates of evolution in different groups (Gray, 1995; Gray and Spencer, 1996; Leblanc et al., 1997).

Nuclear Evidence

Phylogenetic analyses that include all three groups of primary plastid-containing organisms are available for two nuclear genes (18S rRNA and actin), while seven other nuclear genes address the question of red and green algal relationships. The numerous 18S analyses that include the latter two groups consistently place them as part of an unresolved "crown group radiation" that includes many of the major groups of eukaryotes (e.g., Cavalier-Smith, 1993a; Sogin, 1994, 1996; Bhattacharya and Medlin, 1995; Bhattacharya et al., 1995a, 1995b; Nelissen et al., 1995; Ragan and Gutell, 1995; Pawlowski et al., 1996; Van de Peer et al., 1996). The comparatively few 18S analyses that also include glaucocystophytes place them within this tight knot of unresolved lineages too, although the glaucocystophyes also tend to group, with modest support, with the host nuclear lineage of cryptophytes (Bhattacharya and Medlin, 1995; Bhattacharya et al., 1995b). The few published analyses of 28S and 5S rRNA sequences include only short sequence lengths and fail to resolve the placement of red and green algae relative to other eukaryotes (see Perasso et al., 1989; Ragan and Gutell, 1995; Delwiche and Palmer, 1997).

Among nuclear-encoded proteins, actin consistently recovers, albeit with modest bootstrap support, a specific grouping of red and green algae, and less consistently groups these algae with the only glaucocystophyte yet examined (Bhattacharya and Weber, 1997). Four other proteins either fail to resolve the relationships of red algae, green algae, and various other eukaryotes (β-tubulin and triosephosphate isomerase; Liaud et al., 1995; Keeling and Doolittle, 1997) or group red and green algae only weakly (cytosolic GAPDH and EF-1alpha; Liaud et al., 1994; Zhou and Ragan, 1995; Liu et al., 1996). Alone among the nine nuclear genes examined, the gene for the largest subunit of RNA polymerase II (RPB1) provides high bootstrap support against any simple theory of plastid monophyly, as green algae group with three non-plastid-containing lineages (animals, fungi, and *Acanthamoeba*) to the exclusion of red algae, although taxon sampling in this study is relatively poor (Stiller and Hall, 1997).

Overview of the Evidence

Overall, then, an impressive body of plastid data, from six different genes and various idiosyncratic characters, provides consistent support for a monophyletic origin of the plastids of green algae, red algae, and glaucocystophytes, the strength of this conclusion being limited largely by the degree of cyanobacterial sampling for any one data set. Extensive mitochondrial DNA data also strongly support the monophyly of green algae and red algae; unfortunately, no such data are available for glaucocystophytes. With the notable exception of the *RPB1* gene, which weighs in against plastid monophyly, the available nuclear data are either unresolved on this issue or point weakly to monophyly. On balance, we regard the total weight of evidence from the three genomes as providing clear support for the notion that primary plastids are monophyletic, arising from a single cyanobacterial endosymbiosis. At the same time, there is a clear need for substantially more data, from nuclear genes in general and from mitochondrial genes of glaucocystophytes, although our prediction is that these data will ultimately build an unassailable case for plastid monophyly.

Reconciling a Potential Conflict between Plastid and Nuclear Data

What if additional data fail, contrary to this prediction, to support the case for plastid monophyly? In particular, what if Stiller and Hall (1997) are correct, as suggested by their *RPB1* analyses, in arguing that green algae and red algae are not sister groups at the nuclear level?

Three main hypotheses have been advanced to reconcile a fundamental conflict between plastid gene trees that support monophyly of primary plastids and nuclear trees that do not (for recent review, see Stiller and Hall, 1997). First, as argued most forcefully by Cavalier-Smith (1993b), primary plastids may indeed have arisen only once, relatively early in eukaryotic evolution, but were subsequently lost in many descendant lineages. This hypothesis seems unlikely, both because it posits complete plastid loss (as opposed to loss of photosynthesis, a relatively common event) in so many lineages, and because it seems improbable that all traces of cyanobacterial genes would have vanished from such well-studied, even completely sequenced, nuclear genomes as those of fungi and animals. Second, the various groups of primary plastids may instead have arisen from independent endosymbioses, but these involved such closely related cyanobacteria that the currently poor sampling of cyanobacteria fails to resolve the plastid progenitors and thus results in a false picture of plastid monophyly.

A third hypothesis, apparently novel to Stiller and Hall (1997), is that red algal (or green algal) plastids are actually secondary plastids, but are not recognized as such because they have lost *both* extrabounding membranes regarded as signatures of secondary symbiosis. This idea is attractive both because membrane loss is already strongly suspected for those secondary plastids with three membranes (see Fig. 13.1 and below) and because this is the only hypothesis that allows one to reconcile the evidence from all three genomes. That is, while the first two hypotheses can reconcile the nuclear evidence for nonsisterhood of red and green algae with the strong evidence for monophyly of their plastids, these hypotheses are incompatible with the strong evidence for sisterhood of their mitochondria (instead, they predict congruent nuclear and mitochondrial phylogenies). But, as Stiller and Hall (1997) imply, the third hypothesis allows for congruent plastid and mitochondrial phylogenies by postulating that red algae might have acquired not only their plastids secondarily from a green alga but also their mitochondria. However, the concept of secondary mitochondria poses three specific problems (for elaboration, see Delwiche and Palmer, 1997), involving (1) wholesale transfer of genes of mitochondrial origin from the endocytosed cell to the host nucleus; (2) incompatibilities between mitochondrially and nuclearly encoded proteins (of different evolutionary origin) that must interact in the same multisubunit complex; and (3) the lack of evidence (see foregoing discussion) for a progenitor/derivative relationship among reds and greens as predicted by the membrane-loss, secondary-symbiosis hypothesis of Stiller and Hall (1997). Furthermore, these three problems only worsen when one adds glaucocystophytes to the picture. Thus, if a bona fide conflict between plastid and nuclear data concerning plastid monophyly were to arise (which we think is unlikely), and if the mitochondrial data remain in agreement with the plastid evidence (which seems likely), then *none* of the extant hypotheses suffices to reconcile all three lines of evidence.

Further Work on Plastid and Cyanobacterial Phylogeny

We have already mentioned the need for more nuclear and mitochondrial data in order to achieve comprehensive and robust phylogenies that allow one to test for congruence among all three genetic compartments, as predicted if plastids arose monophyletically. On the plastid side, whole-genome sequences are already available (see Table 13.2 and section on Evolution of Plastid Genomes section), or soon will be, from many diverse plastid types. (The only major group of algae for which we currently lack any plastid gene sequences are the dinoflagellates [see below], but assuming their plastids are indeed of secondary origin [Fig. 13.1], they are irrelevant to the issues in this section). However, cyanobacteria are relatively poorly studied from a comparative molecular standpoint, as only a single cyanobacterial genome has been sequenced (Kaneko et al., 1996), and only one gene (16S rRNA) has been widely sequenced within the group (Turner, 1997). Therefore, those plastid genes that appear most promising for elucidating deep phylogeny, for example, *tufA, atpB, rpoC1,* and 23S rRNA (but not *rbcL,* owing to its frequent lateral transfer, see section

RUBISCO Phylogeny and Horizontal Gene Transfer), should be sequenced from dozens of cyanobacteria that represent the diversity of the group as defined by current 16S rRNA analyses. In combination with the available 16S data, these enlarged data sets should produce a much more highly resolved picture of cyanobacterial phylogeny than we currently have (Turner, 1997). This, in turn, is crucial to settling, as definitively as possible, whether plastids arose from one or more than one lineage of cyanobacteria. The better the sampling and phylogenetic resolution of cyanobacteria, the better our chances of distinguishing true monophyly from cryptic polyphyly, that is, endosymbioses involving two or more *closely related* cyanobacterial lineages (see foregoing discussion). In addition, a firm resolution of cyanobacterial phylogeny should, at long last, allow identification of the specific cyanobacterial lineage(s) that gave rise to plastids. Such identification should stimulate intensive study of this progenitor group, to better our understanding of both those properties that plastids have faithfully retained from their specific cyanobacterial progenitors, and also the many ways in which plastid biology has diverged in response to the organelle's highly integrated partnership with its host cell.

Finally, there is good reason to think that a current paradox regarding the age of plastids and cyanobacteria should be resolved once we have a much better understanding of cyanobacterial relationships and of the place(s) of plastids among cyanobacteria. The fossil and biogeochemical records are commonly interpreted to suggest that plastids are probably a much younger group (between 1 and 2 billion years old) than are cyanobacteria, which seem to be at least 2.8 and quite likely 3.5 billion years of age, and which were well diversified 2 billion years ago (Knoll, 1992, 1994; Knoll and Golubic, 1992; Schopf, 1993, 1994, 1996). In contrast, however, most 16S rRNA phylogenies place plastids close to the base of cyanobacteria (Nelissen et al., 1995; Bhattacharya and Medlin, 1995; Turner, 1997), suggesting that the two groups are not very different in age. One solution to this paradox is a very different interpretation of the fossil record (e.g., Runnegar, 1992). However, an alternative explanation should also be considered—that plastids are placed artefactually deeply within cyanobacteria by 16S rRNA, most likely because rRNA evolution is relatively fast in plastids (Giovannoni et al., 1988). Comprehensive sampling and well-resolved phylogenies for several molecules are needed to distinguish between these alternative explanations.

TAXA WITH SECONDARY PLASTIDS

The Recognition of Secondary Plastids

It is now clear that many algal taxa gained plastids via acquisition of a *eukaryotic* endosymbiont, itself already equipped with plastids, and hence have secondary plastids. Ultrastructural studies of algae revealed that some taxa have more membranes surrounding the plastid than the two found in primary plastids (Fig. 13.1). In heterokonts (which include brown algae, diatoms, chrysophytes, xanthophytes, and other algae) and haptophytes (also known as prymnesiophytes), there are four membranes, two of which correspond to the dual membranes of the primary plastid, and two of which form an envelope that is continuous with the nuclear envelope and binds ribosomes. Because of their similarity to the endoplasmic reticulum, the outer two envelopes are often referred to as the chloroplast endoplasmic reticulum (CER). In cryptomonads and chlorarachniophytes the CER is still more complex, and includes a nucleomorph, a small DNA-containing structure interpreted as a reduced eukaryotic nucleus (Gibbs, 1962; Greenwood et al., 1977; Hibberd and Norris, 1984; Ludwig and Gibbs, 1989). The presence of a nucleomorph within the periplastidal space located between the inner and outer pair of plastid envelope membranes (Fig. 13.1) suggested that the CER and nucleomorph may be remnants of a eukaryotic endosymbiont, with the inner membrane of the CER corresponding to the plasma membrane of the endosymbiont, and the outer to the host food vacuole. In this case the nucleomorph, CER, and primary plastid together would constitute a secondary plastid. No evidence of a mitochondrion has been found in any secondary plastid. This in turn led to speculation that the CER of heterokonts and

haptophytes may similarly have arisen via acquisition of a eukaryotic endosymbiont. In these cases the nucleomorph is thought to have undergone reduction to the point of extinction, such that the only ultrastructural evidence of the secondary nature of plastids in these taxa is the additional set of membranes surrounding the plastid. This argument was further extended to include the plastids of euglenophytes and dinoflagellates, which lack a CER and are typically surrounded by three membranes (Gibbs, 1978, 1981).

Thus, the interpretation that many taxa have secondary plastids was based largely on membrane counting in ultrastructural studies. Remarkably, these early ideas—first widely dismissed as overly speculative—have borne up well under further investigation (reviewed in Whatley, 1993a, 1993b; McFadden and Gilson, 1995; Gilson and McFadden, 1997; McFadden et al., 1997a). The two membranes surrounding the plastids of red and green algae and glaucocystophytes are now thought to represent the base number for primary plastids, and additional membranes completely surrounding the plastid provide evidence for secondary origin. The concept of secondary symbiosis also helps explain taxa such as the euglenophytes, whose plastid is green alga-like but whose host cell much more closely resembles that of another protist group, the kinetoplastids. Recently, molecular phylogenetic analyses have permitted relatively detailed reconstruction of the evolution of these organisms by comparison of the genes and genomes of the chloroplast, nucleomorph, and host nucleus.

Taxa with Nucleomorphs

Nucleomorphs are found in two unrelated groups of algae, the cryptomonads and chlorarachniophytes. The secondary origin of plastids in both of these groups is strongly supported by molecular phylogenetic analyses, which place the nuclear and nucleomorph genomes of both groups in different clades (McFadden et al., 1995; Van de Peer et al., 1996; Cavalier-Smith et al., 1996). Placement of nuclear and nucleomorph genes from the same organism in different clades supports the identity of the nucleomorph as the reduced nucleus of a eukaryotic endosymbiont, while placement of cryptomonad and chlorarachniophyte genes in different clades confirms the independent acquisition of secondary plastids in both groups. This observation is consistent with differences in host cell structure and plastid pigmentation. Cryptomonads are biflagellate unicells with chlorophylls a and c_2 and phycobilins, suggesting the endosymbiont was a red alga, an inference supported by phylogenetic analysis of both plastid and nucleomorph genes (Fig. 13.3; Douglas, 1992; McFadden et al., 1995; Van de Peer et al., 1996; Cavalier-Smith et al., 1996). By contrast, chlorarachniophytes form an amoeboid reticulum, are uniflagellate in their motile stage, and have plastids pigmented with chlorophylls a and b, suggesting that the endosymbiont was a green alga, an inference also supported by phylogenetic analyses (Douglas, 1992; McFadden et al., 1995; Van de Peer et al., 1996; Cavalier-Smith et al., 1996).

Although they are almost certainly the products of independent endosymbiotic events, the nucleomorph genomes of cryptomonads and chlorarachniophytes show substantial convergence. Both have very small (380–660 kb) genomes organized in three chromosomes (Gilson and McFadden, 1996, 1997; McFadden et al., 1997b). This is a striking example of the recurring theme of reduction of endosymbiont genomes in both size and gene content (see Evolution of Plastid Genomes). Most of the genes still residing in the nucleomorphs of both groups function merely in genetic housekeeping (transcription, translation, splicing), that is, most genes of plastid origin have been transferred a second time, to the host cell nucleus (see previous references). The nucleomorph of chlorarachniophyes is fascinating in strictly molecular terms as a marvel of genetic compaction: Intergenic spacers are unusually short (averaging 65 nt in length), some genes overlap and others are cotranscribed, and the numerous spliceosomal introns are the tiniest known and are remarkably invariant in length (most are 19 nt, range is 18–20 nt; Douglas et al., 1992; McFadden et al., 1995; Van de Peer et al., 1996; Cavalier-Smith et al., 1997). Although secondary plastids with nucleomorphs were first identified in cryptomonads, recent work on the chlorarach-

niophytes has greatly improved understanding of this previously obscure group, which is now perhaps the best documented case of acquisition of a plastid via secondary endosymbiosis (McFadden and Gilson, 1995; Gilson and McFadden, 1996, 1997; McFadden et al., 1997b).

Taxa with a CER, but No Nucleomorph

Four membranes surround the plastids of heterokonts and haptophytes, but no nucleomorph or comparable extranuclear genome has been identified in these taxa. The strikingly parallel reduction in genome size and gene content of nucleomorphs in cryptomonads and chlorarachniophytes further supports the idea that in heterokonts and haptophytes the eukaryotic endosymbiont nucleus has been reduced to the point of extinction (Gibbs, 1993; Palmer and Delwiche, 1996). As with other reduced plastid genomes, the genes may have been transferred to the host nuclear genome, substituted by a host nuclear gene, or lost entirely (see Genome Size and Gene Content). Phylogenetic analyses of plastid rRNA genes place heterokont and haptophyte plastids with those of red algae, consistent with the belief that these are secondary plastids derived from a red alga (Medlin et al., 1995; Bhattacharya and Medlin, 1995). But perhaps the strongest evidence that heterokonts and haptophytes, like cryptomonads, have plastids derived from red algae is the surprising presence (discussed later in more detail) of a proteobacterial RUBISCO operon in each of these groups rather than the cyanobacterial RUBISCO found in glaucocystophytes and chlorophytes s. l. (Fig. 13.3; Fujiwara et al., 1993; Delwiche and Palmer, 1996; Medlin et al., 1997). Although the chloroplasts of heterokonts and haptophytes are relatively similar in pigmentation and ultrastructure, differences in host cell ultrastructure suggest that they may have acquired secondary plastids independently. Phylogenetic analyses of both plastid (16S rRNA and $rbcL$) and nuclear (18S rRNA and actin) sequences support this conclusion, placing heterokonts and haptophytes within different groups of nonphotosynthetic eukaryotes (Medlin et al., 1997; Leipe et al., 1994; Bhattacharya and Weber, 1997).

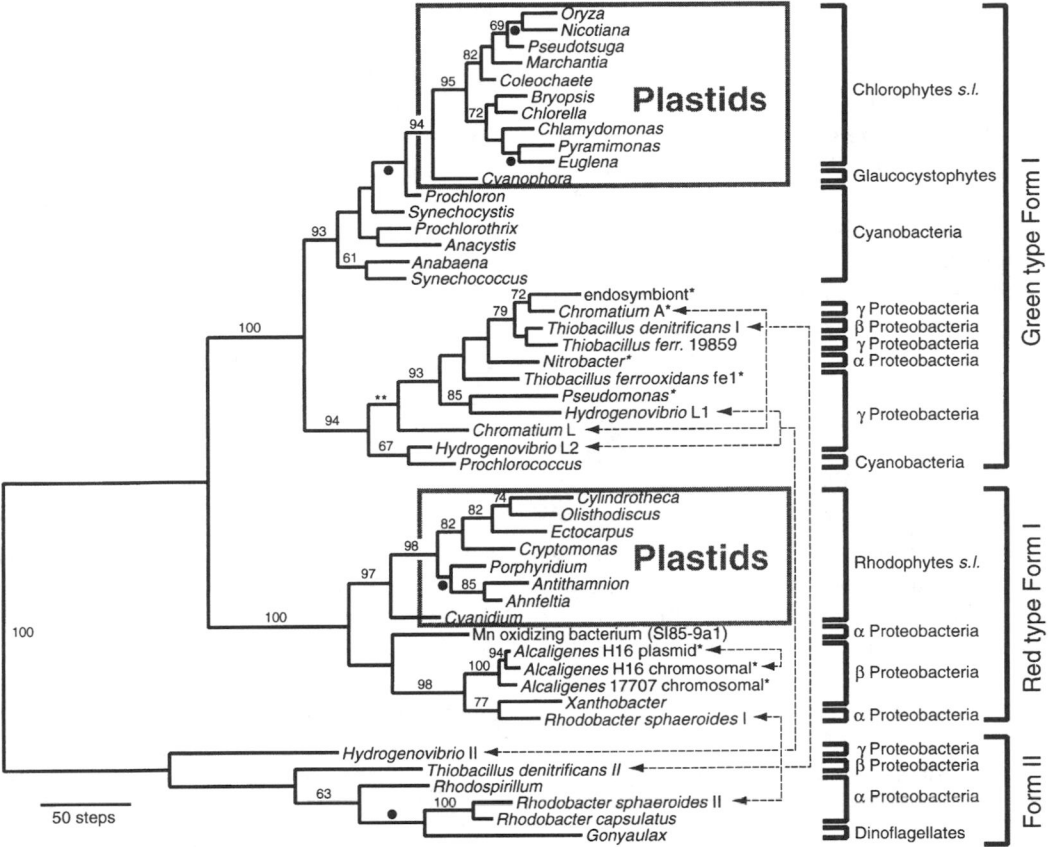

Figure 13.3. An *rbcL* phylogeny illustrating the highly discordant relationship between *rbcL* and organismal evolution for eubacteria and algae (modified from Fig. 2 of Delwiche and Palmer, 1996). Shown is one of 24 shortest trees from a maximum parsimony analysis of *rbcL* amino acid sequences from all available eubacteria and from representative algae. Organismal classification based on 16S rRNA and other evidence (see Delwiche and Palmer, 1996) is indicated to the right of the taxon names. Only those bootstrap values >60% are shown, and branches not present in the strict consensus tree (of the 24 shortest trees) are indicated by a bullet (•). Branch lengths are proportional to the number of changes assigned to each branch (see scale bar). Double-arrow-headed, dashed lines connect multiple *rbcL* sequences determined from the same taxon. *rRNA sequences were not available for this particular strain of the so-named taxon, and therefore the proteobacterial subgroup may be suspect. **Bootstrap analysis placed *Chromatium* L. in a clade with *Hydrogenovibrio* L2 and *Prochlorococcus* with 60% bootstrap support, but this topology did not occur among the shortest trees.

Taxa with Secondary Plastids, but No CER or Nucleomorph

Euglenophytes—The Third Membrane

A further state of reduction seems to have occurred in taxa with more than two membranes surrounding the plastid, but no evidence of a nucleomorph or CER per se. The euglenophytes have three membranes surrounding their green-pigmented chloroplasts. The extra membrane, in combination with the peculiar combination of plastid and host cell, led Gibbs (1978) to propose that the euglenophyte plastid is secondary in origin, but that the nucleomorph and outer membrane of the secondary plastid have been completely lost. The presence of only three membranes (rather than four) surrounding a secondary plastid is somewhat surprising. Complete loss of a secondary plastid membrane would be expected to be difficult, given the need to target nuclear-encoded peptides into the plastid. This view is supported by the uniform retention of both membranes of the CER throughout the many representatives of the four groups (cryptomonads, chlorarachniophytes, hetero-konts, and haptophytes) in which these structures are found.

However, despite the apparent improbability of loss of one of the CER membranes, molecular phylogenetic and additional ultrastructural studies have supported the secondary origin of euglenophyte plastids, and have given rise to the current view that the host cell in euglenophytes was a relatively derived member of a larger clade that includes kinetoplastids and euglenoids (Triemer and Farmer, 1991; Sogin, 1994; R. Triemer, pers. comm.). Phylogenetic analyses of the GAPDH gene also support a secondary origin of the euglenophyte plastid, although the frequent duplication and loss of this gene complicate interpretation of the data (Henze et al., 1995).

It is not clear what factors differ in euglenophytes from lineages that have retained four membranes surrounding the plastid. Assuming that the feeding mechanism of the ancestral cell required formation of a food vacuole during ingestion of the eukaryotic endosymbiont, then one of the two membranes corresponding to the food vacuole and endosymbiont cell membrane must have been lost (but see the discussion of myzocytosis in the next section). The lack of connections to the nuclear envelope and absence of ribosomes on the chloroplast envelope (Gibbs, 1978) argue that if one membrane was lost, it was probably the outer envelope.

Dinoflagellates—Plastid Diversity Run Amok

Dinoflagellates are unicellular flagellates that include both autotrophic and heterotrophic (or mixotrophic) taxa (Elbrächter, 1991). The photosynthetic dinoflagellates display a bewildering diversity of plastid types, probably because of many independent secondary endosymbioses. The most familiar dinoflagellate plastid is the peridinin-type plastid, which is surrounded by three membranes and pigmented with chlorophylls a and c_2 with peridinin, diadinoxanthin, and β-carotene as accessory pigments (Dodge, 1989; Schnepf, 1993; Whatley, 1993a). This kind of plastid is found only within the orders Gonyaulacales (Peridiniales) and Prorocentrales, but these account for the majority of photosynthetic dinoflagellates.

As with euglenophytes, it was proposed early on that peridinin-type dinoflagellate plastids may be of secondary origin (Gibbs, 1978). Like euglenophytes, dinoflagellates seem to have a mismatch between their plastid and host cell. Although their plastids are qualitatively similar to those of heterokonts and haptophytes, with chlorophylls a and c, and thylakoids stacked into groups of three, their host cell is dramatically different from the host cell in those groups. On the basis of both ultrastructural studies and phylogenetic analyses of nuclear genes, the dinoflagellate host cell may be placed with confidence among the alveolates, a group that includes the ciliates and apicomplexan parasites (Cavalier-Smith, 1993a; Sogin, 1994, 1996; Van de Peer et al., 1996). The incongruence between host cell and plastid, combined with the presence of three bounding membranes, suggests a secondary origin for the peridinin-type plastid. This hypothesis is further bolstered by the widespread occurrence of phagotrophy in dinoflagellates. Only about half of all dinoflagellates are pigmented, and many of those with plastids supplement their diet with the occasional predatory snack (Elbrächter, 1991; Raven, 1997). Such predation provides a ready mechanism for endosymbiont acquisition, and in some cases it is difficult to be certain whether a plastid is a permanent, inherited endosymbiont or a more short-lived "kleptochloroplast" (Fields and Rhodes, 1991; Schnepf, 1993; Raven, 1997). Unfortunately, no molecular information is yet available from the peridinin-type plastid genome, and until such information is available, all inferences about the phylogenetic history of these organelles must be based on ultrastructure and biochemistry alone.

The peridinin-type dinoflagellate plastid is also remarkable in the nature of its carbon-fixing enzyme and in the location of the gene for this enzyme. The peridinin-containing dinoflagellates *Gonyaulax* and *Symbiodinium* lack the form I RUBISCO (with eight large and eight small subunits) found in all other plastids and cyanobacteria. Instead, these dinoflagellates rely on a form II RUBISCO (with two large subunits only), a form known elsewhere only from proteobacteria (Morse et al., 1995; Whitney et al., 1995; Rowan et al., 1996). Form II RUBISCO is thought to be more sensitive to oxygen than form I RUBISCO (Tabita, 1995), and its presence in oxygenic phototrophs such as dinoflagellates is therefore unexpected. The nuclear location of the gene is similarly remarkable. The only other documented case of a nuclear-encoded RUBISCO is a genetically engineered species of *Nicotiana* (Kanevski and Maliga, 1994). At present one can only speculate on the origin of the form II RUBISCO gene in dinoflagellates. Because the mitochondrion is of proteobacterial descent (Gray, 1992, 1995),

one potential mechanism would involve horizontal transfer of a proteobacterial form II RUBISCO from the ancestral mitochondrion to the nuclear genome, followed by targeting of the form II RUBISCO to the plastid, with eventual loss of the plastid's original form I gene (Fig. 13.4; Morse et al., 1995; Palmer, 1995; Rowan et al., 1996). Another possibility is that the dinoflagellate nucleus acquired its form II *rbcL* gene directly by lateral transfer from a proteobacterium (Fig. 13.4).

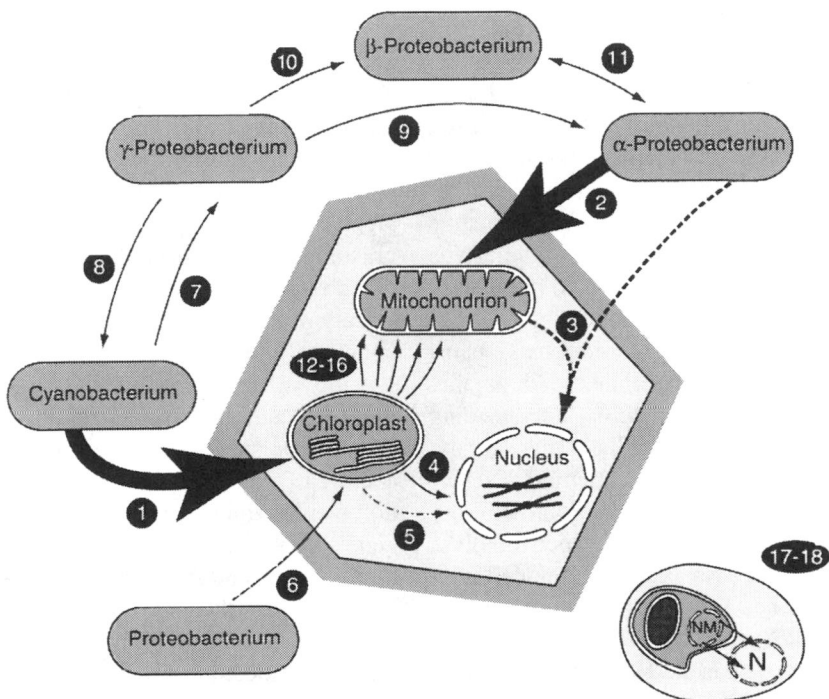

Figure 13.4. (Main figure): Widespread transfer of RUBISCO genes between organisms and between genetic compartments within a eukaryotic cell (modified and updated from Fig. 1 of Palmer, 1995). The thick arrows indicate the transfers of RUBISCO genes from free-living bacteria to incipient organelles that were part of the endosymbiotic origin of plastids (circled #1) and mitochondria (2) from cyanobacteria and alpha-proteobacteria, respectively. The former event most likely delivered form I RUBISCO genes only, while the latter delivered either form I, form II, both, or neither. The split, dashed arrow (3) indicates the transfer of a form II *rbcL* gene from either the mitochondrion or a proteobacterium to the nucleus of peridinin-containing dinoflagellates (see text and Morse et al., 1995; Palmer, 1995; Rowan et al., 1996). The remaining, thin arrows indicate evolutionary or engineered transfers of one or both form I RUBISCO genes, including: (4) *rbcS* transfer from the chloroplast to nucleus in a common ancestor of chlorophytes s. l.; (5) the engineered transfer of *rbcL* to the nucleus in *Nicotiana* (Kanevski and Maliga, 1994); (6) transfer of the entire *rbcL/rbcS* operon from some undetermined group of proteobacteria to the plastid of the common ancestor of rhodophytes s. l. (also see Fig.13.3 and Delwiche and Palmer, 1996); (7–11) five independent transfers of the RUBISCO operon tentatively inferred between eubacteria (also see Fig. 13.3 and Delwiche and Palmer, 1996); and (12–16) five independent, duplicative transfers of chloroplast *rbcL* to the mitochondrion within angiosperm evolution, twice in the Poaceae, and once each in the Brassicaceae, Cucurbitaceae, and Convolvulaceae (Lonsdale et al., 1983; Nugent and Palmer, 1988; Nakazono and Hirai, 1993; Olmstead and Palmer, 1994; M. P. Cummings, J. M. Nugent, R. G. Olmstead, and J. D. Palmer, unpubl. data). (Inset at lower right): Independent transfers (17–18) of *rbcS* from the nucleomorph to the nucleus in euglenophytes and chlorarachniophytes (Chan et al., 1990; G. McFadden, pers. comm.).

The two orders of dinoflagellates with peridinin-type plastids (the Gonyaulacales and Prorocentrales) are not thought to be sister taxa, and this hypothesis has been interpreted as evidence that these plastids are the ancestral type for the dinoflagellates. This would imply that many nonphotosynthetic dinoflagellates are descended from photosynthetic ancestors (Van den Hoek et al., 1995). If this is the case, then some exant taxa had peridinin-type plastids, lost them, and then later reacquired other types of plastids. One attractive feature of this hypothesis is that it would explain the apparent ease with which dinoflagellates acquire new endosymbionts; the possession by the nuclear genome of some elements necessary for the maintenance of the endosymbiont might facilitate capture. However, all nonphotosynthetic organisms known to be derived from a photosynthetic ancestor still retain some remnants of a plastid. An alternative explanation for the distribution of peridinin-type plastids in the dinoflagellates would involve independent acquisitions of secondary plastids in the two orders. Because heterotrophy remains widespread in the dinoflagellates, such parallel acquisition seems reasonable. The feeding mechanism of some dinoflagellates may also explain the presence of three membranes surrounding the secondary plastid. Some dinoflagellates feed by myzocytosis, which involves penetration of the prey cell, with the cell contents being sucked into a food vacuole without retention of the prey plasmalemma (Elbrächter, 1991; Van den Hoek et al., 1995). If the contents of such a feeding vacuole were to persist, it would ultimately give rise to a triple-membraned secondary plastid.

Although peridinin-type plastids account for the bulk of dinoflagellate plastids, a great diversity of other plastids have been documented, all apparently of secondary (or tertiary) origin. Several dinoflagellates have plastids that contain the carotenoid fucoxanthin, which is characteristic of heterokonts and haptophytes. In several of these, a eukaryotic nucleus has been identified in the secondary endosymbiont (Tomas and Cox, 1973; Kite and Dodge, 1985). In one such case, *Peridininum foliaceum,* phylogenetic analysis of the RUBISCO operon has placed the plastid solidly among those of diatoms, a het-

erokont group (Chesnick et al., 1996). Because the plastids of heterokonts are themselves secondary plastids, this is a *tertiary* endosymbiosis. It is not clear whether the endosymbiotic diatom nucleus in this organism has undergone reduction analogous to that seen in the nucleomorphs of cryptomonads and chlorarachniophytes. Another case of tertiary endosymbiosis in dinoflagellates can be found in *Gymnodinium acidotum,* which may be found in nature with a cryptomonad endosymbiont (Wilcox and Wedemayer, 1985). In this case, however, it is not clear whether the endosymbiont is a permanent, heritable resident or a short-term guest (Farmer and Roberts, 1990; Fields and Rhodes, 1991).

Finally, a fascinating situation is found in *Lepidodinium,* a green-pigmented dinoflagellate that lacks an endosymbiont nucleus and whose plastids resemble those of prasinophyte green algae. What is striking is that, in addition to the prasinophyte-like plastids, *Lepidodinium* displays complex scales on its outer surface that closely resemble prasinophyte scales (Watanabe et al., 1990). Other dinoflagellates form scales, but the structural complexity of the *Lepidodinium* scales and their striking similarity to those of prasinophytes suggest that they were acquired from the endosymbiont. This acquisition would require transfer of the scale-forming genes from the endosymbiont nucleus to the host nucleus (and successful expression of these genes) prior to the loss of the endosymbiont nucleus. Verification of this hypothesis will have to await identification and study of the scale-forming genes in prasinophytes and *Lepidodinium.* A trivial example of a green endosymbiont is found in some populations of *Noctiluca,* where an unreduced prasinophyte may be found swimming in the host vacuole (Sweeney, 1976).

Apicomplexan Parasites—Plastids without Photosynthesis The most recently identified secondary plastids are those found in apicomplexan parasites. Phylum Apicomplexa consists entirely of obligate parasites of animals and includes a number of important pathogens such as *Plasmodium,* the cause of malaria, and *Toxoplasma;* there are no known photosynthetic apicomplexans (Vivier and Desportes, 1990). Ultrastructural studies and phylogenetic analyses of nuclear genes place the apicomplexans along with the dinoflagellates and ciliates in the Alveolata (Cavalier-Smith, 1993a; Sogin, 1994, 1996; Van de Peer et al., 1996). Ultrastructural studies had identified a membrane-bound "spherical body" in apicomplexans, but the identity of this organelle as a plastid was unsuspected until the identification of two extrachromosomal genomes in *Plasmodium.* One of these extrachromosomal genomes is a 35-kb circular genome, which was at first confused with a mitochondrial genome, but is now known to be a plastid genome (Palmer, 1992; Feagin, 1994; Wilson et al., 1996; McFadden et al., 1997b; Wilson and Williamson, 1997). The second extrachomosomal genome, now known to be mitochondrial, is a linear genome of tandemly repeated 6-kb elements. As with most organellar genomes, the apicomplexan extrachromosomal genomes are uniparentally inherited (Vaidya et al., 1993). The 35-kb genome of *Plasmodium* has been sequenced and resembles those of plastids in the presence of an rRNA-encoding inverted repeat, genes encoding a eubacterial RNA polymerase, elongation factor Tu (*tufA*), and ORF470, and a group I intron in the UAA anticodon of the gene for tRNA-Leu (Preiser et al., 1995; Wilson et al., 1996; McFadden et al., 1997b; Köhler et al., 1997). None of these elements is at all common in mitochondria.

Phylogenetic analyses of genes from the apicomplexan 35-kb genome have been complicated by poor taxon sampling and the highly divergent and AT-rich nature of the genome (e.g., Egea and Lang-Unnasch, 1995). Recent analyses of the gene *tufA* from the 35-kb genome of three apicomplexans, *Plasmodium, Toxoplasma,* and *Eimeria,* place this organelle unequivocally among plastids, and, curiously, consistently among the green plastids (Fig. 13.2; Köhler et al., 1997). These data strongly support the belief that the 35-kb genome is plastidic in origin and, combined with the non-green-algal nature of the apicomplexan nucleus (Cavalier-Smith, 1993a; Sogin, 1994, 1996; Van de Peer et al., 1996), suggest that the apicomplexan plastid is a secondary plastid derived from green algae. Because apicomplexans are closely related to dinoflagellates at the nuclear level, it has been speculated that the apicomplexan plastid might be related to those of dinoflagellates (Palmer, 1992). However, even if this were the case (i.e., contrary to the *tufA* evidence [Fig. 13.2]), the diversity of plastid types within the dinoflagellates makes it impossible to predict what kind of plastid would be expected in apicomplexans without a finely detailed phylogeny of dinoflagellates.

In situ hybridization has been used to identify the subcellular location of the 35-kb genome in *Toxoplasma* (McFadden et al., 1996, 1997b; Köhler et al., 1997). The genome is located in a spherical, membrane-bound organelle that lies near the nucleus and golgi apparatus, toward the apical side of the nucleus. Originally given various names, including "spherical body" and "Hohlzylinder," this structure can now be recognized as a degenerate plastid. The number of membranes surrounding this plastid is a subject of debate. McFadden et al. (1996) report having found sections in which the *Toxoplasma* plastid is surrounded by only two membranes and interpret the additional membranes often seen as part of the golgi apparatus. This differs from the interpretation of Köhler et al. (1997), who present micrographs with good membrane preservation that clearly show four membranes surrounding the plastid and report having followed these membranes in serial sections through the plastids of several cells. The number of membranes surrounding the apicomplexan plastid needs to be firmly established, because the number of membranes provides important clues concerning the putative secondary origin of the plastid (from green algae). Careful serial sectioning and three-dimensional reconstruction of the plastid are necessary to eliminate the possibility that pores or evaginations of the plastid envelope or golgi apparatus create the appearance of a different number of membranes surrounding the plastid than are actually present. The apparent secondary origin of the plastid and its apparent derivation from green algae could also be confirmed by identification of (green) algal genes in the apicomplexan nuclear genome.

Other Organisms with Long-term Plastid Associations

The acquisition of plastids via endosymbiosis should not be viewed as a static historical event.

Phagotrophy continues to place photosynthetic organelles inside of nonphotosynthetic hosts, and some fraction of these might be expected to give rise eventually to a permanent and stable symbiotic relationship (Raven, 1997). A number of organisms not traditionally classified among the algae have close associations with phototrophs. Some phagotrophic ciliates retain photosynthetic prey for long periods, and at least one, *Mesodinium rubrum,* has what may be a permanent and heritable cryptomonad endosymbiont (Lindholm, 1985; Van den Hoek, 1995). A number of marine invertebrates have close symbiotic associations with various algae, often dinoflagellates (zooxanthellae), and in some cases these algae are endosymbiotic and inherited. Although this is a very close association, many zooxanthellae differ from secondary plastids in that they are capable of survival outside of their host (Van den Hoek, 1995). Recent studies have shown that the zooxanthellae are much more diverse than previously believed, and the various relationships between host and symbiont remain poorly understood (Rowan and Powers, 1991). Finally, some associations of plastids with nonphotosynthetic hosts may prove to be more complex than had previously been believed; the kleptochloroplasts of the marine sea slug *Elysia chlorotica,* which are captured from the heterokont *Vaucheria litorea,* continue to synthesize proteins after capture and may also import proteins from the animal's cytosol (Pierce et al., 1996; Mujer et al., 1996).

PIGMENTATION IN THE ANCESTRAL PLASTID

Pigmentation, and consequently the kind of plastid, has been a critical feature in identification and description of algae since ancient times. With the realization that plastids are endosymbiotic organelles, three hypotheses have emerged concerning the origin of pigment diversity among plastids. In the first of these hypotheses, the majority of diversification of photosynthetic pigments is assumed to have occurred among the ancestors of plastids while they were still free-living microorganisms, prior to their establishment as endosymbiotic organelles. According to this polyphyletic hypothesis, algal lineages with different pigmentations acquired their plastids by ingestion of variously colored microorganisms (Raven, 1970). Thus, while the red algae would have acquired cyanobacterial endosymbionts (both have chlorophyll *a* and phycobilins), the green algae would have acquired their plastids from hypothetical "green" bacteria pigmented with chlorophylls *a* and *b,* and the heterokonts theirs from hypothetical "golden" bacteria with chlorophylls *a* and *c.*

This hypothesis found widespread acceptance with the identification of the prochlorophytes, bacteria that perform oxygenic photosynthesis with green alga-like pigmentation (Bullerjahn and Post, 1993). However, mounting evidence makes this hypothesis increasingly unlikely. First, as discussed previously, there is considerable evidence that all plastids are monophyletic, which is inherently incompatible with the idea that variously pigmented endosymbionts gave rise to the different plastid lineages. Second, in phylogenetic analyses of all four of those genes (16S rRNA, *rpoC, psbA,* and *rbcL*) that have been sampled from more than one of the prochlorophytes (*Prochloron, Prochlorococcus,* and *Prochlorothrix*), these taxa do not form a monophyletic group and are scattered throughout the cyanobacteria (Palenik and Haselkorn, 1992; Urbach et al., 1992; Palenik and Swift, 1996; Delwiche and Palmer, 1996; Turner, 1997). Thus, despite the superficial similarity of their pigmentation, the prochlorophytes are not a monophyletic group, but rather are simply cyanobacteria that have independently acquired a chlorophyte-like pigmentation. Furthermore, there is scant evidence to indicate that any of the prochlorophytes is the sister to plastids (Palenik and Haselkorn, 1992; Urbach et al., 1992; Turner 1997). Current phylogenetic evidence indicates that the three prochlorophyte lineages and green algae have independently converged on (or retained an ancestral) green pigmentation. Consistent with this view, there are several noteworthy differences among the photosynthetic apparati of prochlorophytes and green plastids (Bullerjahn and Post, 1993). Most notably, the chlorophyll *a/b* light-harvesting complexes (LHCs) of the three prochlorophyte lineages have been shown to be related to one another and to two other classes of cyanobacterial LHC, but are distantly related to the functionally similar LHC proteins of plastids (La Roche et al., 1996).

Two other models for the evolution of plastid pigmentation are more compatible with current phylogenetic and biochemical information. Under the acquisitive hypothesis, the pigment diversity of plastids developed after their endosymbiotic association with the host cell. In this theory, the cyanobacterial progenitor of plastids would have had "typical" cyanobacterial pigmentation, that is, chlorophyll *a* and phycobilins. In this case, plastids with green (chlorophylls *a* and *b*) and brown (chlorophylls *a* and *c*) pigments would have independently lost phycobilins and acquired chlorophyll-based light antennae. A reductive model, first developed by Bryant (1992), proposes that the cyanobacterial ancestor of plastids had many, perhaps all, of the photosynthetic pigments now associated with different lineages. According to this model, differences in pigmentation among lineages represent independent losses of different pigments.

Two lines of evidence now suggest that plastid pigment diversity may have arisen by a combination of reduction and acquisition. First, the pigment composition of many taxa is more complex than originally realized. Both plastids and cyanobacteria have been identified that have chlorophylls *b* and *c* in addition to chlorophyll *a*. A chlorophyll *c*-like pigment in addition to the standard green algal pigments has been reported in the unicellular green alga *Mantoniella squamata* (Wilhelm, 1987; Schmitt et al., 1994), and a similar complement of chlorophylls has been detected in the prochlorophyte *Prochloron* (Larkum et al., 1994). Furthermore, phycobilins in combination with chlorophylls *a* and *b* have been identified in the prochlorophyte *Prochlorococcus marinus* (Hess et al., 1996). Second, study of the LHC proteins associated with antenna pigments in plastids and cyanobacteria has begun to clarify the underlying differences between various pigment systems. In red algae, photosystem II is associated with a phycobilisome, which functions as a light antenna, while photosystem I relies on LHC proteins that bind chlorophyll *a*, zeaxanthin, and β-carotene. These LHC proteins show immunological reac-

tivity with the LHCs of chlorophytes, but not with those of cyanobacteria (Wolfe et al., 1994; Tan et al., 1997). This indicates that homologs of these LHC proteins were present in the common ancestor of red and green plastids and suggests that this ancestor had two kinds of LHC, one with a phycobilisome antenna, and one with a chlorophyll/carotenoid antenna. This hypothesis is supported by the discovery in cyanobacteria with "typical" pigmentation of proteins that are similar to the chlorophyll *a/b* binding proteins of chlorophytes (Miroshnichenko Dolganov et al., 1995). None of these observations alone is convincing, but together they illustrate the complex combinations of pigments and proteins that occur in nature and bolster the idea that much of the diversity of modern plastids represents reduction and modification of the photosynthetic apparatus of the ancestral plastid.

RUBISCO PHYLOGENY AND HORIZONTAL GENE TRANSFER

Oxygenic photosynthesis is known in nature only among cyanobacteria and plastids. Other bacteria are able to capture light and fix carbon, but none have linked photosystems, and none evolve oxygen, which provides strong biochemical evidence that all plastids are cyanobacterial in origin. This conclusion has been further strengthened by phylogenetic analyses of several genes that place all plastids among the cyanobacteria (see foregoing). One pair of genes tells a different story, however. In rhodophytes s. l., including most taxa with secondary plastids thought to be derived from red algae (i.e., heterokonts, haptophytes, and cryptomonads, but not dinoflagellates; see foregoing discussion), the RUBISCO-encoding genes *rbcL* and *rbcS* are much more closely related to those of certain proteobacteria than they are to those of cyanobacteria (Boczar et al., 1989; Valentin and Zetsche, 1989; Douglas, 1994; Delwiche and Palmer, 1996). Because overwhelming evidence from biochemistry, ultrastructure, and phylogenetic analysis of other genes unequivocally places all plastids among the cyanobacteria, most authors have attributed the peculiar RUBISCO phylogeny to a single horizontal gene transfer from a proteobacterium to an ancestral red alga (see previous references and Morden et al., 1992). An alternative explanation proposed by Martin et al. (1992) is that the genes are paralogs, the products of an ancient gene duplication, with differential loss of one or the other copy from extant plastid and bacterial lineages. This hypothesis would explain the RUBISCO phylogeny but would require many independent losses of one of the two copies of the RUBISCO operon, and if this is correct, it is surprising that no extant taxon retains both copies of the operon from the predicted phylogenetic lineages.

A single horizontal gene transfer cannot, however, account for the RUBISCO phylogeny. Although the RUBISCO genes of rhodophytes s. l. resemble those of *some* proteobacteria, other proteobacteria with RUBISCO genes are much more closely related to those of cyanobacteria and plastids (Delwiche and Palmer, 1996; see also Watson and Tabita, 1996). When rooted by form II *rbcL* genes, form I *rbcL* genes cluster into two clearly distinct groups, one composed of chlorophytes s. l., glaucocystophytes, cyanobacteria, and some proteobacteria (green type in Fig. 13.3, which corresponds to types IA and IB of Tabita, 1995), the other composed of rhodophytes s. l. and other proteobacteria (red type/types IC and ID). Within these groups *rbcL* sequences are relatively similar (70% amino acid identity or greater), while between groups there is much greater divergence (less than 60% identity; Delwiche and Palmer, 1996). The *rbcL* phylogeny (Fig. 13.3) shows many incongruencies with organismal phylogeny as it is understood from other sources of information. If horizontal gene transfer is responsible for the RUBISCO phylogeny, then several (roughly six) transfers must have occurred (Fig. 13.4), five involving only proteobacteria or proteobacteria and cyanobacteria, and the sixth being the putative transfer of a proteobacterial operon to the plastid genome of a red algal ancestor.

Although horizontal gene transfer seems to be the most likely explanation for the split between green- and red-type RUBISCOS, other features of RUBISCO phylogeny make it clear that these genes have undergone a number of duplication and transfer events (Delwiche and Palmer, 1996). The chaos in RUBISCO phylogeny among eubacteria and major plastid groups is ironic given the widespread use of *rbcL* as a phylogenetic molecule in plants and algae (e.g., Clegg, 1993; Freshwater et al., 1994; Manhart, 1994; Sytsma and Hahn, 1994; see Chapters 17 and 18). While there is at present no reason to believe that lower-level studies (i.e., those within chlorophytes or rhodophytes) have been distorted by horizontal transfer of RUBISCO genes, the frequency with which RUBISCO seems to have undergone gene transfer suggests that this key photosynthetic enzyme is functionally interchangeable across many diverse taxa. Therefore, those systematists using this molecule for phylogenetic purposes should be alert to the possibility that it has undergone additional, as yet undocumented, transfers. A related complication to keep in mind is the relatively frequent (at least in angiosperms) duplication of *rbcL* genes in the mitochondrion (see legend to Fig. 13.4), which potentially can lead to the "wrong" gene being isolated and analyzed in studies of plastid phylogeny (Olmstead and Palmer, 1994). Lest we throw out the baby with the bathwater, however, it should be noted that within plants at least (but less so for bacteria), congruence seems to be the norm among molecular phylogenies based on different genes. Genetic inheritance is primarily through descent, and horizontal gene transfer remains the exception.

EVOLUTION OF PLASTID GENOMES

Our emphasis here is on the ways in which plastid DNAs (ptDNAs) have diverged from their cyanobacterial heritage. Plastid genome evolution has been highly reductive with respect to size and gene content, but largely acquisitive, sometimes highly so, in the case of intron content and repeated sequence content. However, most features of plastid gene organization and expression are still very much cyanobacterial-like and therefore are treated only briefly.

Genome Size and Gene Content

Like cyanobacterial genomes (Bancroft et al., 1989; Chen and Widger, 1993; Kaneko et al., 1996), all wild-type ptDNAs are circular in organization. The great majority of ptDNAs are between 100 and 200 kb in size (Table 13.1;

Table 13.1. Size and repeat content of sequenced plastid genome.[a]

Organism	Genome Size (bp)				Percent Repeated
	Total	LSC[b]	SSC[b]	Repeat[c]	
Rhodophytes s. l.					
Porphyra purpurea	191,028	147,768	33,620	4,820	5.0
Odontella sinesis	119,704	123,570	38,908	7,725	12.9
Glaucocystophytes					
Cyanophora paradoxa	135,599	94,946	18,083	11,285	16.6
Chlorophytes s. l.					
Euglena gracilis	143,170	125,343[d]	0[d]	5,918[d]	13.7
Marchantia polymorpha	121,024	81,095	19,813	10,058	16.6
Pinus thunbergii	119,707	65,696	53,021	495	0.8
Oryza sativa	134,525	80,592	12,334	20,799	30.9
Zea mays	140,387	82,355	12,536	22,748	32.4
Nicotiana tabacum	155,952	86,689	18,575	25,340	32.5
Epifagus virginiana[e]	70,028	19,799	4,579	22,735	64.9
Plasmodium falciparum[e]	34,682	~24,100[f]	~80[f]	~5,250[f]	30.3

[a]References for the complete sequences are: *Porphyra,* Reith and Munholland (1995); *Odontella,* Kowallik et al. (1995); *Cyanophora,* Stirewalt et al. (1995); *Euglena,* Hallick et al. (1993), Thompson et al. (1995); *Marchantia,* Ohyama et al. (1986); *Pinus,* Wakasugi et al. (1994); *Oryza,* Hiratsuka et al. (1989); *Zea,* Maier et al. (1995); *Nicotiana,* Shinozaki et al. (1986) and Olmstead et al. (1993); *Epifagus,* Wolfe et al. (1992); *Plasmodium,* Wilson et al. (1996) and Wilson and Williamson (1997).

[b]Large single-copy region (LSC) and small single-copy region (SSC).

[c]Only the largest repeat in each genome is given (all other repeats in these genomes are <1 kb in size; except see footnote *d*). All repeats are inverted duplications and are apparently identical within a genome except those of *Porphyra* (a direct repeat, with 41 of 4,280 positions differing in the two copies; Reith and Munholland, 1993b, 1995) and *Euglena* (footnote *d*).

[d]*Euglena* cpDNA contains 2.9 copies of a tandemly repeated 5,918-bp-long rRNA operon (Hallick et al., 1993). These repeats differ by a few substitutions and insertions/deletions. In addition, a fourth, partial operon (encoding a complete 16S rRNA gene) is located upstream of the first intact operon-repeat. In sum, there are 19.6 kb of repeated rDNA sequence.

[e]Nonphotosynthetic organisms.

[f]Has either no SSC or else a tiny one of only "tens of nucleotides," the uncertainty reflecting failure to sequence through the entire *Hind*III fragment located at the center of the palindrome (Wilson and Williamson, 1997).

Palmer, 1991; Loiseaux-de Goër, 1994; and Reith, 1995), or about 20–30 times smaller than cyanobacterial genomes (Herdman et al., 1979). The few ptDNAs that have been studied from permanently nonphotosynthetic plants, algae, and other protists are even smaller (35–73 kb), reflecting the loss of most or all photosynthetic genes (Table 13.2; also see Gockel et al., 1994, and below). The only ptDNAs from photosynthetic organisms that fall outside the 100–200 kb size range are those of certain green algae, which range from 89 kb in *Codium fragile* to 400 kb and possibly much larger in several other poorly characterized taxa (see above reviews).

Cyanobacterial genomes, like plastid genomes, have a compact, gene-dense organization. For example, the one completely sequenced genome, from *Synechocystis* PCC 6803, is 3,573 kb in size and contains roughly 3,200 genes that occupy nearly 90% of the genome (Kaneko et al., 1996). Thus, the massive reduction in plastid genome size discussed above means that plastid genomes have lost over 90% of the genes possessed by their founding genome. This massive loss of genes is a consequence of the highly integrated and interdependent relationship established between plastid and host cell during their 1–2 billion year endosymbiotic marriage.

Three processes underlie these many gene losses. First, many genes are no longer needed in the context of such a permanent symbiotic relationship and have been lost entirely from the cell with no functional replacement. For example, all genes involved in biosynthesis of the cyanobacterial cell wall have presumably been lost from rhodophytes s. l. and chlorophytes s. l., but retained in glaucocystophytes (see At Least Three Lineages of Primary Plastids). Second,

Table 13.2.   Gene content of sequenced plastid genome.[a]

Organism	Number of Genes[b]					Gene Density (#/kb)
	Total	Genet.	Photo.	Misc.	ORF	
Rhodophytes s. l.						
Porphyra purpurea	251	103	53	30	64	1.31
Odontella sinesis	165	77	41	10	37	1.45
Glaucocystophytes						
Cyanophora paradoxa	191	87	48	20	36	1.41
Chlorophytes s. l.						
Euglena gracilis	97	56	27	1	13	0.68
Marchantia polymorpha	120	62	42	7	9	0.99
Pinus thunbergii	108	62	30	5	11	0.90
Oryza sativa	110	61	41	1	7	0.82
Zea mays	110	61	41	1	7	0.78
Nicotiana tabacum	113	61	41	2	9	0.78
Epifagus virginiana[c]	42	37	0	2	2	0.60
Plasmodium falciparum[c]	57	48	0	2	7	1.64

[a]See Table 13.1 for references for the complete genome sequences.
[b]Duplicated genes are counted only once. Genet. stands for genetic genes, which are involved in DNA replication, transcription, RNA processing, and translation. Photo. stands for photosynthetic genes, which include the chlororespiratory NADH dehydrogenase genes in addition to standard photosynthetic genes. Misc. stands for miscellaneous genes, which are involved in the biosynthesis of amino acids, fatty acids, chlorophyll, carotenoids, phycobilins, pyrimidines, NAD^+, thiamine, and the peptidoglycan cell wall of the *Cyanophora* plastid; protein transport, processing, and assembly/chaperoning; the assimilation of nitrogen and sulfur; and the specification of thioredoxin and pyruvate dehydrogenase. ORF stands for open reading frames of unknown function, which are highly conserved and therefore almost certainly represent functional genes.
[c]Nonphotosynthetic organisms.

many still-essential genes have been transferred to the nucleus (Palmer, 1991; Loiseaux-de Goër, 1994; Löffelhardt and Bohnert, 1994; Reith, 1995). These gene transfers occur by a transitional stage of gene duplication in both plastid and nucleus, with the plastid gene being lost only after the nuclear gene becomes fully functional, that is, properly expressed, with the resulting protein correctly targeted to the plastid (Baldauf et al., 1990; Gantt et al., 1991; Nugent and Palmer, 1991). Third, many instances are now known of gene substitution, whereby an essential plastid gene is lost and its function supplied by a nuclear gene, usually one of strictly eukaryotic history, whose product becomes newly targeted to the plastid (Palmer, 1991). Most gene substitutions involve gene duplication, with one gene supplying the cytosol and the other the plastid (e.g., Tingey et al., 1988; Henze et al., 1994; Schmidt et al., 1995). In some cases, transfer and substitution events can be coupled, leading in the case of 3-phosphoglycerate kinase to a gene of plastid/cyanobacte-

rial ancestry taking residence in the nucleus, followed by a duplication event that led to this eubacterial gene family now supplying both plastid and cytosol with this glycolytic enzyme (Brinkman and Martin, 1996).

For many years following the landmark sequencing in 1986 of the tobacco and *Marchantia* plastid genomes (Ohyama et al., 1986; Shinozaki et al., 1986), it was thought that all plastid genomes had largely the same set of just over 100 genes. However, the recent sequencing of three ptDNAs from rhodophytes and glaucocystophytes has shattered this view, as they contain many more genes (165–251; Table 13.2). At the same time, even the most gene-rich plastid genome known, from the red alga *Porphyra* (Reith and Munholland, 1993a, 1995), contains only 8% of the genes found in the cyanobacterium *Synechocystis* PCC 6803 (Kaneko et al., 1996). Plastids are metabolically diverse organelles (Emes and Tobin, 1993) thought to contain over 1,000 different proteins, and thus roughly 80% of plastid proteins are now encoded by nuclear genes in *Porphyra,* and over 90% in chlorophytes (Table 13.2).

Most of the genes found in the smaller plastid genomes are also present in the larger ones, for example, all but 8 of 191 genes in *Cyanophora* ptDNA are present among the 251 genes in *Porphyra* ptDNA, as are all but 20 of 120 genes in *Marchantia.* This observation both supports the idea of plastid monophyly (see earlier section), and also suggests that the last common ancestor of all extant plastids had as few as 300 genes. If so, then roughly 90% of cyanobacterial genes were lost relatively rapidly, by one process or another, in the presumably short period between the onset of endosymbiosis and the emergence of extant groups of algae. Still, the range in gene content (97–251 genes) observed among the nine sequenced plastid genomes from photosynthetic eukaryotes shows that there has nonetheless been considerable loss, transfer, and substitution of plastid gene function during more recent times. Indeed, a number of these events have been characterized within the evolution of specific families of flowering plants (Gantt et al., 1991; Downie et al., 1994; Doyle et al., 1995).

The sequenced genomes from photosynthetic plastids all contain the same core set of about 80 genes, whose products function almost exclusively in just two processes: gene expression (primarily translation) and photosynthesis. Moreover, almost all of the plastid genes found in land plants and *Euglena gracilis* are geared toward these two processes. In the extreme case, the rice and maize plastid genomes may encode but a single protein (a protease subunit) involved in functions other than photosynthesis and gene expression (Hiratsuka et al., 1989; Maier et al., 1995).

The additional plastid genes present in rhodophytes and glaucocystophytes include not only more genes for these two processes (except for photosynthesis in the case of *Odontella;* Table 13.2), but also a surprising number and diversity of genes that contribute to other aspects of plastid metabolic and biochemical diversity. For example, the *Porphyra* genome (Reith and Munholland, 1993a, 1995) encodes at least 30 enzymes and factors (and probably many more, considering how many unidentified ORFs it has;

Table 13.2) involved in (1) biosynthesis of amino acids, fatty acids, chlorophyll, carotenoids, phycobilins, pyrimidines, and thiamine; (2) assimilation of nitrogen; (3) protein transport, processing, and assembly/chaperoning; and (4) the specification of thioredoxin and pyruvate dehydrogenase. Two additional biosynthetic functions—for NAD^+ and the peptidoglycan cell wall of its "cyanelle"/plastid—are in part encoded by *Cyanophora* ptDNA (Stirewalt et al., 1995). A far greater diversity of molecules involved in gene expression is also encoded by ptDNA in nonchlorophytes compared to chlorophytes, including tRNA synthetases, translation initiation factors, transcription regulators, a DNA helicase, RNAse E, and the RNA component of RNAse P (Kowallik et al., 1995; Reith and Munholland, 1995; Stirewalt et al., 1995).

Only one major set of genes is present in chlorophytic ptDNAs but absent in others, and these 11 NADH dehydrogenase (*ndh*) genes are enigmatic in two respects. First, although thought to be involved in photosynthetic metabolism, their specific biochemical functions are poorly understood. Indeed, it came as a complete surprise when these *ndh* genes, homologs of which play a well-characterized role in mitochondrial respiration (complex I) and are found in many mitochondrial genomes, were first discovered in ptDNA (Ohyama et al., 1986; Shinozaki et al., 1986). Second, the *ndh* genes have a puzzlingly relictual distribution in ptDNA, being present only in land plants (i.e., not only are they absent from rhodophytes s. l. and glaucocystophytes, but also from *Euglena* and *Chlamydomonas* [Boudreau et al., 1994]). This implies the repeated and pervasive loss of the entire set of *ndh* genes in all algae (charophytes, the green algal group from which land plants are derived, are important to examine here). Wholesale losses of plastid *ndh* genes have also been reported in two groups of photosynthetic land plants, *Pinus* (Wakasugi et al., 1994) and *Cuscuta* (Haberhausen and Zetsche, 1994), and may reflect a marginal selective value of the NADH dehydrogenase complex in plastids. This has unfortunate implications with respect to the phylogenetic utility of the plastid *ndhF* gene, a recently popular choice of plant molecular systematists (e.g., Olmstead and Sweere, 1994; Kim and Jansen, 1995; Neyland and Urbatsch, 1996), as this gene may be a pseudogene or even missing in many lineages of land plants. Indeed, Neyland and Urbatsch (1996) found *ndhF* to be a pseudogene in several lineages of orchids, and much of the gene is absent from the plastid genomes of the legume *Hebestigma* (unpubl. data) and the fern *Polystichum acrostichoides* (B. Georgieva and D. B. Stein, pers. comm.).

Because so much of the genome of all examined chlorophytes s. l. is geared toward photosynthesis and because streamlining pressures to maintain a compact genome are strong in ptDNA, virtually all that remains in the very reduced ptDNAs of their permanently nonphotosynthetic derivatives (e.g., the beechtree parasite *Epifagus virginiana* [Wolfe et al., 1992], the malarial parasite *Plasmodium falciparum* [Wilson et al., 1996], and the colorless euglenophyte *Astasia longa* [Gockel et al., 1994]) are genes "merely" involved in gene expression itself. Hence, it should come as no surprise if a plastid entirely devoid of a genome were to be found in one of the many unexamined lineages of nonphotosynthetic chlorophytes. If so, this would be analogous to the many lineages of anaerobic mitochondria (usually called hydrogenosomes) that have lost their genomes (Palmer, 1997a). In contrast, because their plastid gene content is so rich, it would be very surprising to find a nonphotosynthetic rhodophyte or glaucocystophyte whose plastid has lost its genome. Just as the repeated loss of *ndhF* poses a problem for its use in phylogenetic studies, so too does the likely repeated loss of virtually *all* photosynthetic genes in the many lineages of nonphotosynthetic plants and algae pose a problem for this entire category of genes, including the popular *rbcL* and *atpB* genes (see Chapters 1 and 8).

Although many hundreds of functional gene relocations (via transfer or substitution) have occurred during the evolution of primary plastids from their cyanobacterial ancestor(s), this is just the tip of the iceberg of repeated, massive gene relocation represented by secondary plastid evolution. Four of the molecularly characterized lineages of secondary plastids (heterokonts, haptophytes, euglenophytes, apicomplexans) lack any trace of a nucleomorph genome, while the nucle-omorphs of chlorarachniophytes and cryptophytes are tiny and appear to encode very few plastid proteins (Gilson and McFadden, 1996, 1997; Palmer and Delwiche, 1996; McFadden et al., 1997a). Thus, in all six independent cases of secondary symbiosis, most plastid proteins are now encoded by the host nucleus, and so genes for hundreds of plastid proteins have—six times over—been relocated a second time, from nucle-omorph to host nucleus. All this without even considering the many secondary and even tertiary symbioses within dinoflagellates!

Repeated Sequences

A much higher fraction of the genome consists of repeated sequences in ptDNAs (usually 10–35%, as much as 70%; Table 13.1; also see Palmer, 1991; Loiseaux de Goër, 1994; and Reith, 1995 for reviews on this and most other aspects of this section) than in cyanobacterial genomes (<2%; Bancroft et al., 1989; Chen and Widger, 1993; Kaneko et al., 1996). In the one sequenced cyanobacterial genome, most of the repeated DNA consists of 99 short transposase-like ORFs that cluster into six families of repeats, and together these compose but 1.5% of the genome (Kaneko et al., 1996). Transposases are unknown in ptDNAs, most of which also lack any families of short dispersed repeats. Instead, most of the repeated DNA in plastids consists of a single large repeated element that defines and dominates the architecture of the genome. In particular, most ptDNAs have a large duplication whose two copies always encode the rRNA operon and usually are identical to one another and present in inverted orientation (Table 13.1). The separation of the two copies of this inverted repeat (IR) by a significant amount of DNA defines the quadripartite organization of most ptDNAs, that is, two large repeats separated by small single-copy and large single-copy regions, each of significant size. In a range of diverse algae, the IR is typically 5–10 kb in size (but 20–40 kb in *Chlamydomonas*), whereas in angiosperms it is usually 20–30 kb in size. Overall, variation in IR size (0–76 kb) accounts for most of the total size variation (120–217 kb) known among ptDNAs of photosynthetic land plants.

While some differences in IR size reflect internal insertions and deletions (e.g., Downie et al., 1994), many reflect shifts in position of the IR junction, unaccompanied by internal change (reviewed in Goulding et al., 1996). Most of these shifts are thought to occur by a gene conversion-like process involving short tracts of DNA, whereby the repeat either expands into previously single-copy regions or else contracts; more rarely, large "ebbs and flows" of the IR can occur by the recombinational repair of double-strand breaks (Goulding et al., 1996). Several other interesting recombinational properties are associated with the large IRs of plastid genomes, including (1) "flip-flop" recombination that maintains a 50:50 mixture of two different genomic configurations, that is, inversion isomers differing only in the relative orientation of their single-copy regions; (2) copy-correctional recombination that maintains absolute identity between the two repeats within a genome; and (3) a potential inhibition of the recombination responsible for evolutionary inversions (reviewed in Palmer, 1991).

Because a rRNA-encoding IR is a common feature of ptDNAs (present widely in all three groups of primary plastids and their secondary derivatives) and is also present in at least certain cyanobacteria ([Kaneko et al., 1996]; duplicated rRNA operons of undetermined orientation are present in the other two mapped cyanobacterial genomes [Bancroft et al., 1989; Chen and Widger, 1993]), such an arrangement may have been present in the common ancestor of plastids. If so, then it must have been lost many times, as a number of lineages lacking an IR are known in both rhodophytes s. l. and chlorophytes s. l. Given this distribution, the possibility that this structure was independently evolved on many occasions during plastid evolution is also quite tenable, particularly given two other considerations. First, aside from the presence of rRNA genes, the repeats are otherwise highly variable, that is, in size, overall gene content, spacing of the two repeats, and location and orientation of the rRNA operon relative to the small single-copy region, which suggests that all IRs may not be homologous. Second, largely specific duplications of rRNA genes have been observed in multiple independent lineages of mitochondrial

DNAs, and rRNA genes are almost always multicopy in bacteria and nuclear genomes, all of which suggest that pressures to duplicate these genes in plastids may be high. Reith and Munholland (1993b) have even raised a third scenario, that the directly oriented rRNA-repeats of the red alga *Porphyra purpurea* may reflect the ancestral plastid situation, although this seems unlikely given that direct rRNA repeats are such a rare condition among plastid genomes and that inverted rRNA repeats have now been described in cyanobacteria. In any event, land plants, the only group whose sampling of ptDNA structure is dense enough to permit compelling inferences regarding the gain and loss of the IR, show a clear pattern: A large IR was unquestionably present in the common ancestor of all land plants and has since been lost (entirely or nearly so) only rarely, four times among the hundreds of diverse taxa examined.

Intron Content

Intron content is another way in which ptDNAs are generally thought to have changed radically from the cyanobacterial situation. However, there are two important distinctions between the evolutionary patterns observed for intron content and those described previously for gene content and genome size. In the latter cases, plastid evolution has been uniformly divergent in all plastid lineages and reductively so; the substantial differences in gene content noted among major lineages pale against the ≥ 10-fold overall reduction in gene number compared to cyanobacteria. With the recent surge of complete and partial genome sequences from rhodophytes s. l. and the glaucocystophyte *Cyanophora paradoxa,* it is now clear that these two groups have, without known exception, retained the cyanobacterial state of very few introns. In contrast, the ptDNAs of chlorophytes s. l. have in general acquired significant numbers of introns; indeed, three major groups of chlorophytes have independently acquired entirely different sets of introns and also show quite different patterns of within-group intron evolution (Table 13.3 and the following discussion).

Only two introns, both of which are self-splicing group I introns in tRNA genes, have

Table 13.3. Intron content of sequenced plastid genomes.

Organism	Number of Introns				Intron Density (#/kb)
	Total	Group I	Group II	Group III	
Rhodophytes s. l.					
Porphyra purpurea	0	0	0	0	0.00
Odontella sinesis	0	0	0	0	0.00
Glaucocystophytes					
Cyanophora paradoxa	1	1	0	0	0.01
Chlorophytes s. l.					
Chlamydomonas (all species)[a]	≥23	≥21	≥2	≥0	n.a.
Euglena gracilis	155	0	88	67	1.08
Marchantia polymorpha	19	1	18	0	0.16
Pinus thunbergii	16	1	15	0	0.13
Oryza sativa	18	1	17	0	0.13
Zea mays	18	1	17	0	0.13
Nicotiana tabacum	21	1	20	0	0.14
Epifagus virginiana[b]	6	0	6	0	0.09
Plasmodium falciparum[b]	1	1	0	0	0.03

Note: n.a. indicates that data are not available.
[a]All genomes listed are completely sequenced (for references see Table 13.1), except for that of *Chlamydomonas,* whose numbers are a compilation for the partly examined genomes of several species of *Chlamydomonas* (see text).
[b]Nonphotosynthetic organisms.

been characterized in cyanobacteria (Kuhsel et al., 1990; Xu et al., 1990; Biniszkiewicz et al., 1994). One of these two introns, in the gene for tRNA-Leu(UAA), was clearly transmitted vertically from cyanobacteria to plastids, as it is widespread in both groups; in fact, by phylogenetic criteria it is the oldest known intron (Kuhsel et al., 1990; Xu et al., 1990). This tRNA-Leu intron is the only intron present in the *Cyanophora* plastid genome (Stirewalt et al., 1995) and is also present in diverse rhodophytes s. l. (Kuhsel et al., 1990). This intron has been lost in other rhodophyte genomes (Kuhsel et al., 1990), including the completely sequenced genomes of the red alga *Porphyra* (Reith and Munholland, 1995) and the diatom *Odontella* (Kowallik et al., 1995), both of which entirely lack introns (Table 13.3). A few to many plastid genes have been sequenced in a number of other rhodophytes s. l., and only one other intron has been discovered, a degenerate, group II-like intron in the phycoerythrin gene *rpeB* of the red alga *Rhodella violacea* (Bernard et al., 1992). Thus, rhodophytes s. l. and *Cyanophora* inherited as few as a single intron from their cyanobacterial ancestor, and this is the only in-

tron still resident in some, perhaps many, of these algae.

All examined land plants also retain this ancient tRNA-Leu group I intron, as does the apicomplexan parasite *Plasmodium falciparum.* However, whereas this is the only intron present in *Plasmodium* ptDNA, thus echoing the *Cyanophora* situation, each of the other three major lineages of chlorophytes s. l. for which significant data exist has independently experienced major intron invasions (Table 13.3; reviewed in Palmer, 1991). This independence is complete, for none of the many introns present in ptDNAs of either land plants, euglenophytes, or *Chlamydomonas* is present in either of the other two groups.

Intron content is a rather stable feature of land plant ptDNAs. The 21 introns present in tobacco ptDNA, all of which are group II introns except for the ancient tRNA-Leu intron, represent the entire set of known land plant introns. Of these, 19 are present in the liverwort *Marchantia* and thus can be traced back to the common ancestor of land plants, one is absent from *Marchantia,* and one is of undefined status because the gene in which it resides is absent from *Marchantia* ptDNA. Three of these 20 group II introns are also present in one or more charophytic green algae (Manhart and Palmer, 1990; Lew and Manhart, 1993); the others have not yet been surveyed for in charophytes. Within land plants, intron loss is rare. Among the six sequenced genomes, only five cases of intron loss are known: three in the common ancestor of rice and maize, and two in *Pinus* (the other three "missing" introns reflect gene loss rather than intron loss, as do all 15 "missing" introns in the parasite *Epifagus;* Table 13.3).

In two large-scale surveys, the *rpl2* intron has been shown to be present in most of nearly 800 angiosperms examined, yet 10 independent losses were also inferred, including four losses among the ~400 legumes examined (Downie et al., 1991; Doyle et al., 1995). However, two other introns are present universally among the same set of legumes (Doyle et al., 1995). More data are needed to evaluate whether certain introns are particularly susceptible to loss, which is generally thought to occur by recombination of a cDNA molecule made from a spliced mRNA with the native gene.

Intron evolution has followed very different patterns in *Chlamydomonas,* an enormous, highly paraphyletic genus of chlorophycean green algae thought to be as old as land plants. First, all but two of the 23 introns found variously in different chlamydomonad ptDNAs are group I introns, while the other two are group II's; land plants show precisely the opposite pattern. Second, intron content is much more variable among chlamydomonads than among land plants. Only three chlamydomonad ptDNAs have been even moderately well characterized (and none completely): Two distant relatives, *C. reinhardtii* and *C. eugametos,* share only two of the total of 16 different introns found in the six intron-containing genes sequenced in common, while even two closely related taxa (e.g., with identical 23S rRNA genes), *C. eugametos* and *C. moewusii,* share only six of 13 introns found in the same six genes (Turmel et al., 1993a, 1993b, 1995a).

That intron content is highly variable within chlamydomonad ptDNAs, and that these differences reflect frequent, often recent, loss and gain of (group I) introns, is nicely illustrated by

an in-depth study of the 23S rRNA gene from 17 taxa (Turmel et al., 1993a). All told, no fewer than 12 different group I introns were found in this gene, and in any one gene, intron number varied from 0 to 6. The distribution of only one of the introns could be explained by a single gain without subsequent loss, and only two distributions could be explained by one gain with one loss. The other nine introns have highly sporadic distributions that imply many gains and/or losses. Remarkably, two of these rRNA introns have close homologs (inserted at identical sites in rRNA genes) in the mitochondrial genome of the amoeba *Acanthamoeba castellanii,* and are therefore hypothesized to have been transmitted laterally between the plastid of a *Chlamydomonas*-type organism and the mitochondrion of an *Acanthamoeba*-like organism (Turmel et al., 1995b).

The most dramatic proliferation, both quantitatively and qualitatively, of introns in plastid genomes has occurred in euglenophytes. The completely sequenced genome of *Euglena gracilis* contains at least 155 introns, several times as many as in any other examined organellar genome (Hallick et al., 1993; Thompson et al., 1995). These introns represent some 40% of the *E. gracilis* plastid genome. As in *Chlamydomonas,* but unlike land plants as a group, intron content within the genus *Euglena* is highly variable. In particular, there seems to have been a recent proliferation of introns in the *E. gracilis* lineage, with other members of the genus containing relatively few or no introns in the genes surveyed (Thompson et al., 1995, 1997). The *rbcL* gene nicely illustrates this extraordinary proliferation: No introns have been found in any of the thousands of *rbcL* genes sequenced from land plants, only a single alga (the green alga *Codium fragile;* Manhart and VonderHaar, 1991) other than the genus *Euglena* has even a single *rbcL* intron, and five basal *Euglena* species have only 0–2 *rbcL* introns. In contrast, *E. gracilis* and its close relatives have nine different *rbcL* introns (Thompson et al., 1995).

Introns are so numerous in the *E. gracilis* plastid genome that many instances of introns-within-introns (twintrons) are known. Twintrons are unique to euglenophyte ptDNAs, and have been shown by both functional and phylogenetic criteria to represent the insertion of one intron within another, more ancient intron (Thompson et al., 1997). Another unique feature of these genomes is a novel class of introns, called group III introns (Copertino and Hallick, 1993). These are thought to be highly degenerate group II introns that have lost the capacity for self-splicing and that are instead dependent on *trans*-acting small RNAs derived from a fragmented group II intron. If so, then this condition may represent a case of parallelism relative to the much-speculated evolutionary derivation of the nuclear spliceosome from group II introns (Copertino and Hallick, 1993).

Gene Organization and Expression

In contrast to the many structural and sequence-level differences discussed in the preceding sections, the fundamental organization and expression of genes has been largely conserved between plastids and their cyanobacterial ancestors. This conservation distinguishes plastids from mitochondria, which, with the prominent exception of the protist *Reclinomonas americana* (Lang et al., 1997; Palmer, 1997b), show remarkable deviation, often lineage-specific, from their eubacterial progenitors in many of these same properties. Because the emphasis here is on the many ways in which plastids resemble cyanobacteria, it will suffice to list these properties, and except where otherwise noted, this brief overview is adapted entirely from Gillham's recent book on organelle molecular biology (Gillham, 1994).

Most of the genes in plastid genomes are grouped into contranscribed clusters ("operons" in the loose sense of the word) that closely resemble those of cyanobacteria and other eubacteria in terms of gene order and content (also see Palmer, 1991; Douglas, 1994; Loiseaux-de Goër, 1994). In contrast, eubacterial operon structure has completely or partly disintegrated in most mitochondrial genomes. The only significant disintegration of operon structure known for any ptDNAs is restricted to the genus *Chlamydomonas,* and even this is less extreme than in many mitochondrial DNAs. Certain operon organizations can be viewed as synapomorphies grouping cyanobacteria together with

plastids, whereas others more tentatively group all plastids to the exclusion of cyanobacteria and thus support the notion of plastid monophyly (see previous discussion). Douglas (1994) provides a particularly nice overview of the evolution of operon organization in plastids and cyanobacteria.

Transcription in plastids generally resembles that in cyanobacteria. All examined ptDNAs contain most of the genes for the classic, multi-subunit RNA polymerase of eubacteria, and share with all examined cyanobacteria the fission of the *rpoC* gene into two separate genes (*rpoC1* and *rpoC2*). In contrast, virtually all mitochondrial genetic systems have lost their eubacterially-inherited RNA polymerase and replaced it with a single-subunit phage T7-like polymerase of uncertain derivation. Consistent with the presence of eubacterial polymerase genes in plastid genomes, many plastid genes also contain classic eubacterial-like -35/-10 promoter elements. In angiosperms, however, additional classes of plastid promoters are also present, and a second, nuclear-encoded polymerase (probably of the phage T7 class) is thought to be active in plastid transcription as well.

Translation in plastids is even more cyanobacterial-like than is transcription. Plastid and eubacterial ribosomes are sensitive to the same spectrum of antibiotics, are the same size, contain similar numbers of proteins, virtually all of which are homologs, and possess rRNAs of highly similar size, secondary structure, and primary sequence. Many mitochondrial lineages deviate significantly in virtually all of these aspects except for antibiotic sensitivity. All plastid tRNAs have the conventional cloverleaf structure (many mitochondrial tRNAs do not), and plastid genomes encode a largely self-sufficient set of tRNAs with little additional wobble needed to read all codons (many mitochondria import most or all of their tRNAs or use greatly expanded wobble). Virtually all plastid systems are thought to still use Shine-Dalgarno pairing between the 5′ end of mRNAs and the 3′ end of 16S rRNA, whereas mitochondria do not. Finally, all plastids use the "universal" genetic code, whereas many mitochondrial systems famously use a wide variety of deviant codes.

The one area in which gene expression differs significantly between (at least some) plastids and cyanobacteria is posttranscriptional processing. We have already discussed at length the large number of diverse introns acquired by the ptDNAs of most chlorophytes s. l. Precise RNA-level splicing of these introns is obviously required to produce functional gene products, and thus chlorophytes require much more extensive splicing than do cyanobacteria (and also rhodophytes s. l. and glaucocystophytes). Land plant plastids carry out very low levels of RNA editing (mostly C → U substitutions), another derived feature, although the magnitude of this process is far lower than in the many mitochondrial systems that have independently elaborated diverse kinds of RNA editing. Except for *Euglena gracilis,* which lacks any detectable RNA editing, algal systems are unexplored in terms of RNA editing. If rhodophytes and glaucocystophytes turn out to lack this process, then it will become increasingly appropriate to view their plastids as generally primitive (many more and diverse genes, no derived introns, no editing) compared to those of chlorophytes.

CONCLUSIONS

An impressive amount of evidence from both plastids and mitochondria supports the idea that the three well-characterized groups of primary plastids (those of chlorophytes, rhodophytes, and glaucocystophytes) are monophyletic in origin, that is, arose from the same cyanobacterial endosymbiosis. Nonetheless, much more information is still needed, from nuclear genes in general and mitochondrial genes of glaucocystophytes, before the monophyletic theory can be considered proven. In addition, much work is needed on cyanobacteria in order to identify the specific ancestor(s) of plastids and the nature of the photopigments and associated proteins inherited from that ancestor.

As time progresses, it becomes increasingly obvious that secondary symbiosis has played a major role in plastid evolution, as many of the major groups of algae (and even one phylum of parasitic protists) are now thought to have acquired their plastids in this manner. We can only wonder how many more examples of secondary, and even tertiary, symbiosis remain to be discovered; the diverse plastids of dinoflagellates are a gold mine waiting to be prospected.

The evolution of plastids and their genomes is largely one of lateral transfer, that is, horizontal evolution. The primary endosymbiosis that gave rise to plastids and the many subsequent secondary endosymbioses that have spread them across diverse phyla of eukaryotes are all examples of horizontal evolution—reticulation—in the extreme: Two organisms that were the products of hundreds of millions, sometimes billions, of years of independent vertical evolution merged and entered into a permanent symbiotic relationship of the most intimate kind. The genetic intermingling of host and endocytosed cells is a consequence of the massive lateral transfer—one at a time—of many hundreds of plastid genes to the host nucleus, accompanied by the equally massive loss of essential plastid genes and their replacement by functional substitutes already present in the nucleus. This massive flow of genetic information to the nucleus has happened many times over, initially from plastid to the nucleus in the case of primary plastids, and then, at least six times separately, from the nucleomorph to the nucleus in the case of the various groups of secondary plastid-containing organisms. Why the nucleus has repeatedly captured most of plastid gene function is unclear; equally unclear is why there is such diversity in numbers and kinds of genes still resident in the ptDNAs of extant algae and plants.

The RUBISCO genes are the epitome of lateral gene transfer: Not only do they show a history of these various kinds of symbiotic-specific transfers, but they also display an astonishing number and variety of *trans*-organismal and *trans*-compartmental transfers unrelated to symbiosis per se. Finally, the many introns present in most chlorophytic ptDNAs are examples of horizontal evolution in reverse, that is, the capture by plastid genomes of self-splicing classes of mobile genetic elements, accompanied by the spread of these introns within these genomes. The *Euglena gracilis* plastid genome best illustrates the impact of lateral transfer on plastid evolution: Descended from a self-sufficient genome of perhaps 3,000 genes and a handful of introns, its 97 genes are now outnumbered by introns.

Note added in proof: The first complete sequence of a green algal plastid genome was reported while this chapter was being edited and typeset (see T. Wakasugi, T. Nagai, M. Kapoor, M. Sugita, M. Ito, S. Ito, J. Tsudzuki, K. Nakashima, T. Tsudzuki, Y. Suzuki, A. Hamada, T. Ohta, A. Inamura, K. Yoshinaga, and M. Sugiura. 1997. Complete nucleotide sequence of the chloroplast genome from the green alga *Chlorella vulgaris:* the existence of genes possibly involved in chloroplast division. Proceedings of the National Academy of Sciences U.S.A. **94:**5967–5972). The *Chlorella vulgaris* plastid genome is 150,613 bp in size, lacks a large inverted repeat typical of most ptDNAs, and contains three introns (all group I). It contains at least 111 genes, similar to land plant genomes (Table 13.2), although it lacks the 11 NADH dehydrogenase genes found in most land plant genomes (but absent from all examined algal genomes; see text) and instead contains 10 genes absent from land plant genomes. Most of these extra genes are, however, present in one or more nonchlorophyte genomes. Two notable exceptions are *minC* and *minD,* which are homologous to bacterial genes involved in cell division, and which are unique to the *Chlorella* genome among all examined plastid genomes.

LITERATURE CITED

Apt, K. E., N. E. Hoffman, and A. R. Grossman. 1993. The gamma subunit of R-phycoerythrin and its possible mode of transport into the plastids of red algae. Journal of Biological Chemistry **268:**16206–16215.

Baldauf, S., J. Manhart, and J. D. Palmer. 1990. Different fates of the chloroplast *tufA* gene following its transfer to the nucleus in green algae. Proceedings of the National Academy of Sciences U.S.A. **87:**5317–5321.

Bancroft, I., C. P. Wolk, and E. V. Oren. 1989. Physical and genetic maps of the genome of the heterocyst-forming cyanobacterium *Anabaena* sp. strain PCC 7120. Journal of Bacteriology **171:**5940–5948.

Bernard, C., J. C. Thomas, D. Mazel, A. Mousseau, A. M. Castets, N. Tandeau de Marsac, and J. P. Dubacq. 1992. Characterization of the genes encoding phycoerythrin in the red alga *Rhodella violacea:* Evidence for a splitting of the *rpeB* gene by an intron. Proceedings of the National Academy of Sciences U.S.A. **89:**9564–9568.

Bhattacharya, D., and L. Medlin. 1995. The phylogeny of plastids: a review based on comparisons of small-subunit ribosomal RNA coding regions. Journal of Phycology **31:**489–498.

Bhattacharya, D., and H. A. Schmidt. 1997. Division Glaucocystophyta. *In* The Origins of Algae and Their Plastids, ed. D. Bhattacharya, pp. 139–148. Springer-Verlag, Vienna.

Bhattacharya, D., and K. Weber. 1997. The actin gene of the glaucocystophyte *Cyanophora paradoxa:* analysis of the coding region and introns and an actin phylogeny of eukaryotes. Current Genetics **31:**439–446.

Bhattacharya, D., T. Helmchen, and M. Melkonian. 1995a. Molecular evolutionary analyses of nuclear-encoded small subunit ribosomal RNA identify an independent rhizopod lineage containing the Euglyphina and the Chlorarachniophyta. Journal of Eukaryotic Microbiology **42:**65–69.

Bhattacharya, D., T. Helmchen, C. Bibeau, and M. Melkonian. 1995b. Comparisons of nuclear-encoded small-subunit ribosomal RNAs reveal the evolutionary position of the Glaucocystophyta. Molecular Biology and Evolution **12:**415–420.

Biniszkiewicz, D., E. Cesnaviciene, and D. A. Shub. 1994. Self-splicing group I intron in cyanobacterial initiator methionine tRNA: evidence for lateral transfer of introns in bacteria. EMBO Journal **13:**4629–4635.

Boczar, B. A., T. P. Delaney, and R. A. Cattolico. 1989. Gene for the ribulose-1,5-bisphosphate carboxylase small subunit protein is similar to that of a chemoautotrophic bacterium. Proceedings of the National Academy of Sciences U.S.A. **86:**4996–4999.

Boudreau, E., C. Otis, and M. Turmel. 1994. Conserved gene clusters in the highly rearranged chloroplast genomes of *Chlamydomonas moewusii* and *Chlamydomonas reinhardtii.* Plant Molecular Biology **24:** 585–602.

Boyen, C., C. Leblanc, G. Bonnard, J. M. Grienenberger, and B. Kloareg. 1994. Nucleotide sequence of the *cox3* gene from *Chondrus crispus:* evidence that UGA encodes tryptophan and evolutionary implications. Nucleic Acids Research **22:**1400–1403.

Brinkmann, H., and W. Martin. 1996. Higher-plant chloroplast and cytosolic 3-phosphoglycerate kinases: a case of endosymbiotic gene replacement. Plant Molecular Biology **30:**65–75.

Bryant, D. A. 1992. Puzzles of chloroplast ancestry. Current Biology **2:**240–242.

Bullerjahn, G. S., and A. F. Post. 1993. The prochlorophytes: are they more than just chlorophyll a/b-containing cyanobacteria? Critical Review in Microbiology **19:** 43–59.

Cavalier-Smith, T. 1993a. Kingdom Protozoa and its 18 phyla. Microbiology Reviews **47:**953–994.

Cavalier-Smith, T. 1993b. The origin, losses and gains of chloroplasts. *In* Origins of Plastids, ed. R. Lewin, pp. 291–348. Chapman & Hall, New York.

Cavalier-Smith, T. 1995. Membrane heredity, symbiogenesis, and the multiple origins of algae. *In* Biodiversity and Evolution, eds. R. Arai, M. Kato, and Y. Doi, pp. 75–114. National Science Museum, Tokyo.

Cavalier-Smith, T., J. A. Couch, K. E. Thorsteinsen, P. Gilson, J. A. Deane, D. R. A. Hill, and G. I. McFadden. 1996. Cryptomonad nuclear and nucleomorph 18S rRNA phylogeny. European Journal of Phycology **31**:315–328.

Chan, R. L., M. Keller, J. Canaday, J.-H. Weil, and P. Imbault. 1990. Eight small subunits of *Euglena* ribulose 1-5 bisphosphate carboxylase/oxygenase are translated from a large mRNA as a polyprotein. EMBO Journal **9**:333–338.

Chen, X., and W. R. Widger. 1993. Physical genome map of the unicellular cyanobacterium *Synechococcus* sp. Strain PCC 7002. Journal of Bacteriology **175**:5106–5116.

Chesnick, J. M., C. W. Morden, and A. M. Schmeig. 1996. Identity of the endosymbiont of *Peridinium foliaceum* (Pyrrophyta): analysis of the *rbcLS* operon. Journal of Phycology **32**:850–857.

Clegg, M. T. 1993. Chloroplast gene sequences and the study of plant evolution. Proceedings of the National Academy of Sciences U.S.A. **90**:363–367.

Copertino, D. W., and R. B. Hallick. 1993. Group II and group III introns of twintrons: potential relationships with nuclear pre-mRNA introns. Trends in Biochemical Sciences **18**:467–471.

Delwiche, C. F., and J. D. Palmer. 1996. Rampant horizontal gene transfer and duplication of rubisco genes in eubacteria and plastids. Molecular Biology and Evolution **13**:873–882.

Delwiche, C. F., and J. D. Palmer. 1997. The origin of plastids and their spread via secondary symbiosis. **In** The Origins of Algae and Their Plastids, ed. D. Bhattacharya, pp. 53–86. Springer-Verlag, Vienna.

Delwiche, C. F., M. Kunsel, and J. D. Palmer. 1995. Phylogenetic analysis of *tufA* sequences indicates a cyanobacterial origin of all plastids. Molecular Phylogenetics and Evolution **4**:110–128.

Dodge, J. D. 1989. Phylogenetic relationships of dinoflagellates and their plastids. **In** The Chromophyte Algae: Problems and Perspectives, eds. J. C. Green, B. S. C. Leadbeater, and W. L. Diver, pp. 207–227. Clarendon Press, Oxford.

Douglas, S. E. 1992. Eukaryote-eukaryote endosymbioses: insights from studies of a cryptomonad alga. BioSystems **8**:57–68.

Douglas, S. E. 1994. Chloroplast origins and evolution. **In** The Molecular Biology of Cyanobacteria, ed. D. A. Bryant, pp. 91–118. Kluwer Academic Publishers, Amsterdam.

Douglas, S. E., and C. A. Murphy. 1994. Structural, transcriptional, and phylogenetic analyses of the *atpB* gene cluster from the plastid of *Cryptomonas sp.* (Cryptophyaceae). Journal of Phycology **30**:329–340.

Downie, S. R., R. G. Olmstead, G. Zurawski, D. E. Soltis, P. S. Soltis, J. C. Watson, and J. D. Palmer. 1991. Six independent losses of the chloroplast DNA *rpl2* intron in dicotyledons: molecular and phylogenetic implications. Evolution **45**:1245–1259.

Downie, S. R., D. S. Katz-Downie, K. H. Wolfe, P. J. Calie, and J. D. Palmer. 1994. Structure and evolution of the largest chloroplast gene (ORF2280): internal plasticity and multiple gene loss during angiosperm evolution. Current Genetics **25**:367–378.

Doyle, J. J., J. L. Doyle, and J. D. Palmer. 1995. Multiple independent losses of two genes and one intron from legume chloroplast genomes. Systematic Botany **20**:272–294.

Egea, N., and N. Lang-Unnasch. 1995. Phylogeny of the large extrachromosomal DNA of organisms in the Phylum Apicomplexa. Journal of Eukaryotic Microbiology **42**:679–684.

Elbrächter, M. 1991. Food uptake mechanisms in phagotrophic dinoflagellates and classification. **In** The Biology of Free Living Heterotrophic Flagellates, Systematics Association Special Volume No. 45, eds. D. J. Patterson and J. Larsen, pp. 303–312. Clarendon Press, Oxford.

Emes, M. J., and A. K. Tobin. 1993. Control of metabolism and development in higher plant plastids. International Review of Cytology **145**:149–216.

Farmer, M. A., and K. R. Roberts. 1990. Organelle loss in the endosymbiont of *Gymnodinium acidotum* (Dinophyceae). Protoplasma **153**:178–185.

Feagin, J. E. 1994. The extrachromosomal DNAs of apicomplexan parasites. Annual Review of Microbiology **48**:81–104.

Fields, S. D., and R. G. Rhodes. 1991. Ingestion and retention of *Chroomonas* spp. (Cryptophyceae) by *Gymnodinium acidotum* (Dinophyceae). Journal of Phycology **27**:525–529.

Freshwater, D. W., S. Fredericq, B. S. Butler, M. Hommersand, and M. W. Chase. 1994. A gene phylogeny of the red algae (Rhodophyta) based on plastid *rbcL*. Proceedings of the National Academy of Sciences U.S.A. **91**:7281–7285.

Fujiwara, S., H. Iwahashi, J. Someya, S. Nishikawa, and N. Minaka. 1993. Structure and contranscription of the plastid-encoded *rbcL* and *rbcS* genes of *Pleurochrysis carterae* (Prymnesiophyta). Journal of Phycology **29**:347–355.

Gantt, J. S., S. Baldauf, P. Calie, N. Weeden, and J. D. Palmer. 1991. Transfer of *rpl22* to the nucleus greatly preceded its loss from the chloroplast and involved the gain of an intron. EMBO Journal **10**:3073–3078.

Gibbs, S. P. 1962. Nuclear envelope-chloroplast relationships in algae. Journal of Cell Biology **14**:433–444.

Gibbs, S. P. 1978. The chloroplasts of *Euglena* may have evolved from symbiotic green algae. Canadian Journal of Botany **56**:2883–2889.

Gibbs, S. P. 1981. The chloroplast endoplasmic reticulum: structure, function, and evolutionary significance. International Review of Cytology **72**:49–99.

Gibbs, S. P. 1993. The evolution of algal chloroplasts. **In** Origins of Plastids, ed. R. A. Lewin, pp. 107–121. Chapman & Hall, New York.

Gillham, N. W. 1994. Organelle Genes and Genomes. Oxford University Press, New York.

Gilson, P. R., and G. I. McFadden. 1996. The miniaturized nuclear genome of a eukaryotic endosymbiont contains genes that overlap, genes that are cotranscribed, and the smallest known spliceosomal introns. Proceedings of the National Academy of Sciences U.S.A. **93**:7737–7742.

Gilson, P. R., and G. I. McFadden. 1997. Good things in small packages: the tiny genomes of chlorarachniophyte endosymbionts. Bioessays 19:167–173.

Giovannoni, S., S. Turner, G. Olsen, S. Barns, D. Lane, and N. Pace. 1988. Evolutionary relationships among cyanobacteria and green chloroplasts. Journal of Bacteriology 170:3584–3592.

Gockel, G., W. Hachtel, S. Baier, C. Fliss, and M. Henke. 1994. Genes for components of the chloroplast translational apparatus are conserved in the reduced 73-kb plastid DNA of the nonphotosynthetic euglenoid flagellate Astasia longa. Current Genetics 26:256–262.

Goulding, S. E., R. G. Olmstead, C. W. Morden, and K. H. Wolfe. 1996. Ebb and flow of the chloroplast inverted repeat. Molecular and General Genetics 252:195–206.

Gray, M. W. 1992. The endosymbiont hypothesis revisited. International Review of Cytology 141:233–357.

Gray, M. W. 1995. Mitochondrial evolution. In The Molecular Biology of Plant Mitochondria, eds. C. S. Levings III and I. K. Vasil, pp. 635–659. Kluwer, Dordrecht, The Netherlands.

Gray, M. W., and D. F. Spencer. 1996. Organellar evolution. In Evolution of Microbial Life, eds. D. McL. Roberts, P. Sharp, G. Alderson, and M. Collins, pp. 109–126. Cambridge University Press, Cambridge.

Greenwood, A. D., H. B. Griffiths, and U. J. Santore. 1977. Chloroplasts and cell components in Cryptophyceae. British Phycological Journal 12:119.

Haberhausen, G., and K. Zetsche. 1994. Functional loss of all ndh genes in an otherwise relatively unaltered plastid genome of the holoparasitic flowering plant Cuscuta reflexa. Plant Molecular Biology 24:217–222.

Hackstein, J. H. P. 1995. A photosynthetic ancestry for all eukaryotes? Trends in Ecology and Evolution 10:247.

Hallick, R., L. Hong, R. Drager, M. Favreau, A. Monfor, B. Orsat, A. Spielmann, and E. Stutz. 1993. Complete sequence of Euglena gracilis chloroplast DNA. Nucleic Acids Research 21:3537–3544.

Helmchen, T. A., D. Bhattacharya, and M. Melkonian. 1995. Analyses of ribosomal RNA sequences from glaucocystophyte cyanelles provide new insights into the evolutionary relationships of plastids. Journal of Molecular Evolution 41:203–210.

Henze, K., C. Schnarrenberger, J. Kellermann, and W. Martin. 1994. Chloroplast and cytosolic triosephosphate isomerase from spinach: purification, microsequencing and cDNA sequence of the chloroplast enzyme. Plant Molecular Biology 26:1961–1973.

Henze, K., A. Badr, M. Wettern, R. Cerff, and W. Martin. 1995. A nuclear gene of eubacterial origin in Euglena gracilis reflects cryptic endosymbioses during protist evolution. Proceedings of the National Academy of Sciences U.S.A. 92:9122–9126.

Herdman, M., M. Janvier, R. Rippka, and R. Y. Stanier. 1979. Genome size of cyanobacteria. Journal of General Microbiology 111:73–85.

Hess, W. R., A. Weihe, S. Loiseaux-de Goër, F. Partensky, and D. Vaulot. 1995. Characterization of the single psbA gene of Prochlorococcus marinus CCMP 1375 (Prochlorophyta). Plant Molecular Biology 27:1189–1196.

Hess, W. R., F. Partensky, G. W. M. van der Staay, J. M. Garcia-Fernandez, T. Börner, and D. Vaulot. 1996. Coexistence of phycoerythrin and a chlorophyll a/b antenna in a marine prokaryote. Proceedings of the National Academy of Sciences U.S.A. 93:1126–1130.

Hibberd, D. J., and R. E. Norris. 1984. Cytology and ultrastructure of Chlorarachnion reptans (Chlorarachniophyta Divisio Nova, Chlorarachniophyceae Classis Nova). Journal of Phycology 20:310–330.

Hiratsuka, J., H. Shimada, R. Whittier, T. Ishibashi, M. Sakamoto, M. Mori, C. Kondo, Y. Honji, C.-R. Sun, B.-Y. Meng, Y.-Q. Li, A. Kanno, Y. Nishizawa, A. Hirai, K. Shinozaki, and M. Sugiura. 1989. The complete sequence of the rice (Oryza sativa) chloroplast genome: intermolecular recombination between distinct tRNA genes accounts for a major plastid DNA inversion during the evolution of the cereals. Molecular and General Genetics 217:185–194.

Jakowitsch, J., C. Neumann-Spallart, Y. Ma, J. Steiner, H. E. A. Schenk, H. J. Bohnert, and W. Löffelhardt. 1996. In vitro import of pre-ferredoxin-NADP+-oxidoreductase from Cyanophora paradoxa into cyanelles and into pea chloroplasts. FEBS Letters 381:153–155.

Kaneko, T., S. Sato, H. Kotani, A. Tanaka, E. Asamizu, Y. Nakamura, N. Miyajima, M. Hirosawa, M. Sugiura, S. Sasamoto, T. Kimura, T. Hosouchi, A. Matsuno, A. Muraki, N. Kakazaki, K. Naruo, S. Okumura, S. Shimpo, C. Takeuchi, T. Wada, A. Watanabe, M. Yamada, M. Yasudaand, and S. Tabata. 1996. Sequence analysis of the genome of the unicellular cyanobacterium Synechocystis sp. strain PCC6803. II. Sequence determination of the entire genome and assignment of potential protein-coding regions. DNA Research 3:109–136.

Kanevski, I., and P. Maliga. 1994. Relocation of the plastid rbcL gene to the nucleus yields functional ribulose-1,5-bisphosphate carboxylase in tobacco chloroplasts. Proceedings of the National Academy of Sciences U.S.A. 91:1969–1973.

Keeling, P. J., and W. F. Doolittle. 1996. Alpha-tubulin from early-diverging eukaryotic lineages and the evolution of the tubulin family. Molecular Biology and Evolution 13:1297–1305.

Keeling, P. J., and F. Doolittle. 1997. Evidence that eukaryotic triosephosphate isomerase is of alpha-proteobacterial origin. Proceedings of the National Academy of Sciences U.S.A. 94:1270–1275.

Kies, L., and B. P. Kremer. 1990. Phylum Glaucocystophyta. In Handbook of Protoctista, eds. L. Margulis, J. O. Corliss, M. Melkonian, and D. J. Chapman, pp. 152–166. Jones and Bartlett, Boston.

Kim, K.-J., and R. K. Jansen. 1995. ndhF sequence evolution and the major clades in the sunflower family. Proceedings of the National Academy of Sciences U.S.A. 92:10379–10383.

Kite, G. C., and J. D. Dodge. 1985. Structural organization of plastid DNA in two anomalously pigmented dinoflagellates. Journal of Phycology 21:50–56.

Knoll, A. 1992. The early evolution of eukaryotes: a geological perspective. Science **256:**622–627.

Knoll, A. 1994. Proterozoic and early Cambrian protists: evidence for accelerating evolutionary tempo. Proceedings of the National Academy of Sciences U.S.A. **91:**6743–6750.

Knoll, A., and S. Golubic. 1992. Proterozoic and living cyanobacteria. **In** Early Organic Evolution: Implications for Mineral and Energy Resources, eds. M. Schidlowski, S. Golubic, M. M. Kimberley, D. M. McKirdy, and P. A. Trudinger, pp. 450–462. Springer-Verlag, New York.

Köhler, S., C. F. Delwiche, P. W. Denny, L. G. Tilney, P. Webster, R. J. M. Wilson, J. D. Palmer, and D. S. Roos. 1997. A plastid of probable green algal origin in apicomplexan parasites. Science **275:**1485–1489.

Kowallik, K. V. 1994. From endosymbionts to chloroplasts: evidence for a single prokaryotic/eukaryotic endocytobiosis. Endocytobiosis Cell Research **10:**137–149.

Kowallik, K. V., B. Stoebe, I. Schaffran, P. Kroth-Panic, and U. Freier. 1995. The chloroplast genome of a chlorophyll a +c-containing alga, *Odontella sinensis*. Plant Molecular Biology Reporter **13:**336–342.

Kuhsel, M. G., R. Strickland, and J. D. Palmer. 1990. An ancient group I intron shared by eubacteria and chloroplasts. Science **250:**1570–1573.

Lang, B. F., G. Burger, C. J. O'Kelly, and M. W. Gray. 1997. Organelle genome megasequencing program. http://megasun.bch.umontreal.ca/maps/globaltree.gif

Larkum, A. W. D., C. Scaramuzzi, F. C. Cox, R. G. Hiller, and A. G. Turner. 1994. Light-harvesting chlorophyll c-like pigment in *Prochloron*. Proceedings of the National Academy of Sciences U.S.A. **91:**679–683.

La Roche, J., G. W. M. van der Staay, F. Partensky, A. Ducret, A. Aebersold, R. Li, S. S. Golden, R. G. Hiller, P. M. Wrench, A. W. D. Larkum, and B. R. Green. 1996. Independent evolution of the prochlorophyte and green plant chlorophyll a/b light-harvesting proteins. Proceedings of the National Academy of Sciences U.S.A. **93:**15244–15248.

Leblanc, C., O. Richard, B. Kloareg, S. Viehmann, K. Zetsche, and C. Boyen. 1997. Origin and evolution of mitochondria: what have we learnt from red algae? Current Genetics **31:**193–207.

Leipe, D. D., P. O. Wainwright, J. H. Gunderson, D. Porter, D. J. Patterson, F. Valois, S. Himmerich, and M. L. Sogin. 1994. The stramenophiles from a molecular perspective: 16S-like rRNA sequences from *Labyrinthuloides minuta* and *Cafeteria roenbergensis*. Phycologia **33:**369–377.

Lew, K. A., and J. R. Manhart. 1993. The *rps12* gene in *Spirogyra maxima* (Chlorophyta) and its evolutionary significance. Journal of Phycology **29:**500–505.

Liaud, M.-F., C. Valentin, U. Brandt, F.-Y. Bouget, B. Kloareg, and R. Cerff. 1993. The GAPDH gene system of the red alga *Chondrus crispus:* promotor structures, intron/exon organization, genomic complexity and differential expression of genes. Plant Molecular Biology **23:**981–994.

Liaud, M.-F., C. Valentin, W. Martin, F.-Y. Bouget, B. Kloareg, and R. Cerff. 1994. The evolutionary origin of red algae as deduced from the nuclear genes encoding cytosolic and chloroplast glyceraldehyde-3-phosphate dehydrogenases from *Chondrus crispus*. Journal of Molecular Evolution **38:**319–327.

Liaud, M.-F., U. Brandt, and R. Cerff. 1995. The marine red alga *Chondrus crispus* has a highly divergent β-tubulin gene with a characteristic 5' intron: functional and evolutionary implications. Plant Molecular Biology **28:**313–325.

Lindholm, T. 1985. *Mesodinium rubrum*—a unique photosynthetic ciliate. Advances in Aquatic Microbiology **3:**1–48.

Liu, Q., S. Baldauf, and M. Reith. 1996. Elongation factor 1δ genes of the red alga *Porphyra purpurea* include a novel, developmentally specialized variant. Plant Molecular Biology **31:**77–85.

Löffelhardt, W., and H. J. Bohnert. 1994. Molecular biology of cyanelles. **In** The Molecular Biology of the Cyanobacteria, ed. D. A. Bryant, pp. 65–89. Kluwer Academic Publishers, The Netherlands.

Loiseaux-de Göer, S. 1994. Plastid lineages. **In** Progress in Phycological Research 10, eds. F. E. Round and D. J. Chapman, pp. 137–177. Biopress, Ltd., London.

Lonsdale, D. M., T. P. Hodge, C. J. Howe, and D. B. Stern. 1983. Maize mitochondrial DNA contains a sequence homologous to the ribulose-1,5-bisphosphate carboxylase large subunit gene of chloroplast DNA. Cell **34:**1007–1014.

Ludwig, M., and S. P. Gibbs. 1989. Evidence that the nucleomorphs of *Chlorarachnion reptans* (Chlorarachniophyceae) are vestigial nuclei: morphology, division, and DNA-DAPI fluorescence. Journal of Phycology **25:**385–394.

Maier, R. M., K. Neckermann, G. L. Igloi, and H. Kössel. 1995. Complete sequence of the maize chloroplast genome: gene content, hotspots of divergence and fine tuning of genetic information by transcript editing. Journal of Molecular Biology **251:**614–628.

Manhart, J. R. 1994. Phylogenetic analysis of green plant *rbcL* sequences. Molecular Phylogenetics and Evolution **3:**114–127.

Manhart, J. R., and J. D. Palmer. 1990. The gain of two chloroplast tRNA introns marks the green algal ancestors of land plants. Nature **345:**268–270.

Manhart, J. R., and R. A. VonderHaar. 1991. Intron revealed by nucleotide sequence of large subunit of ribulose-1,5-bisphosphate carboxylase/oxygenase from *Codium fragile* (Chlorophyta): phylogenetic analysis. Journal of Phycology **27:**613–617.

Martin, W. S., C. C. Somerville, and S. Loiseaux-de Göer. 1992. Molecular phylogenies of plastid origins and algal evolution. Journal of Molecular Evolution **35:**385–404.

McFadden, G. I., and P. Gilson. 1995. Something borrowed, something green: lateral transfer of chloroplasts by secondary endosymbiosis. Trends in Ecology and Evolution **10:**12–17.

McFadden, G. I., P. R. Gilson, and R. F. Waller. 1995. Molecular phylogeny of chlorarachniophytes based on plastid rRNA and *rbcL* sequences. Archives für Protistenkund **145**:231–239.

McFadden, G. I., M. E. Reith, J. Munholland, and N. Lang-Unnasch. 1996. Plastid in human parasites. Nature **381**:482.

McFadden, G. I., P. R. Gilson, and C. J. B. Hofmann. 1997a. Division Chlorarachniophyta. **In** The Origins of Algae and Their Plastids, ed. D. Bhattacharya, pp. 175–186. Springer-Verlag, Vienna.

McFadden, G. I., R. F. Waller, M. E. Reith, and N. Lang-Unnasch. 1997b. Plastids in apicomplexan parasites. **In** The Origins of Algae and Their Plastids, ed. D. Bhattacharya, pp. 261–287. Springer-Verlag, Vienna.

Medlin, L. K., A. Cooper, C. Hill, S. Wrieden, and U. Wellbrock. 1995. Phylogenetic position of the Chromista plastids based on small subunit rRNA coding regions. Current Genetics **28**:560–565.

Medlin, L. K., W. H. C. F. Kooistra, D. Potter, G. W. Saunders, and R. A. Anderson. 1997. Phylogenetic relationships of the 'golden algae' (haptophytes, heterokont chromophytes) and their plastids. **In** The Origins of Algae and Their Plastids, ed. D. Bhattacharya, pp. 187–220. Springer-Verlag, Vienna.

Miroshnichenko Dolganov, N. A., D. Bhaya, and A. R. Grossman. 1995. Cyanobacterial protein with similarity to the chlorophyll a/b binding proteins of higher plants: evolution and regulation. Proceedings of the National Academy of Sciences U.S.A. **92**:636–640.

Morden, C. W., C. F. Delwiche, M. Kuhsel, and J. D. Palmer. 1992. Gene phylogenies and the endosymbiotic origin of plastids. BioSystems **28**:75–90.

Morse, D., P. Salois, P. Markovic, and J. W. Hastings. 1995. A nuclear-encoded form II RuBisCO in dinoflagellates. Science **268**:1622–1624.

Mujer, C. V., D. L. Andrews, J. R. Manhart, S. K. Pierce, and M. E. Rumpho. 1996. Chloroplast genes are expressed during intracellular symbiotic association of *Vaucheria litorea* plastids with the sea slug *Elysia chlorotica*. Proceedings of the National Academy of Sciences U.S.A. **93**:12333–12338.

Nakazono, M., and A. Hirai. 1993. Identification of the entire set of transfered chloroplast DNA sequences in the mitochondrial genome of rice. Molecular and General Genetics **236**:341–346.

Nelissen, B., Y. Van de Peer, A. Wilmotte, and R. De Wachter. 1995. An early origin of plastids within the cyanobacterial divergence is suggested by evolutionary trees based on complete 16S rRNA sequences. Molecular Biology and Evolution **6**:1166–1173.

Neyland, R., and L. Urbatsch. 1996. Phylogeny of subfamily Epidendroideae (Orchidaceae) inferred from *ndhF* chloroplast gene sequences. American Journal of Botany **83**:1195–1206.

Nugent, J. M., and J. D. Palmer. 1988. Location, identity, amount and serial entry of chloroplast DNA sequences in crucifer mitochondrial DNAs. Current Genetics **14**:501–509.

Nugent, J. M., and J. D. Palmer. 1991. RNA-mediated transfer of the gene *coxII* from the mitochondrion to the nucleus during flowering plant evolution. Cell **66**:473–481.

Ohyama, K., H. Fukuzawa, T. Kohchi, H. Shirai, T. Sano, S. Sano, K. Umesono, Y. Shiki, M. Takeuchi, Z. Chang, S.-I. Aota, H. Inokuchi, and H. Ozeki. 1986. Chloroplast gene organization deduced from complete sequence of liverwort *Marchantia polymorpha* chloroplast DNA. Nature **322**:572–574.

Olmstead, R. G., and J. D. Palmer. 1994. Chloroplast DNA systematics: a review of methods and data analysis. American Journal of Botany **81**:1205–1224.

Olmstead, R. G., and J. A. Sweere. 1994. Combining data in phylogenetic systematics: an empirical approach using three molecular data sets in the Solanaceae. Systematic Biology **43**:467–481.

Olmstead, R. G., J. A. Sweere, and K. H. Wolfe. 1993. Ninety extra nucleotides in *ndhF* gene of tobacco chloroplast DNA: a summary of revisions to the 1986 genome sequence. Plant Molecular Biology **22**:1191–1193.

Palenik, B., and R. Haselkorn. 1992. Multiple evolutionary origins of prochlorophytes, the chlorophyll *b*-containing prokaryotes. Nature **355**:265–267.

Palenik, B., and H. Swift. 1996. Cyanobacterial evolution and prochlorophyte diversity as seen in DNA-dependent RNA polymerase gene sequences. Journal of Phycology **32**:638–646.

Palmer, J. D. 1985. Comparative organization of chloroplast genomes. Annual Review of Genetics **19**:325–354.

Palmer, J. D. 1991. Plastid chromosomes: structure and evolution. **In** The Molecular Biology of Plastids, eds. L. Bogorad and I. K. Vasil, pp. 5–53. Academic Press, San Diego.

Palmer, J. D. 1992. Green ancestry of malarial parasites. Current Biology **2**:318–320.

Palmer, J. D. 1995. Rubisco rules fall; gene transfer triumphs. Bioessays **17**:1005–1008.

Palmer, J. D. 1997a. Organelle genomes: going, going, gone! Science **275**:790–791.

Palmer, J. D. 1997b. The mitochondrion that time forgot. Nature **387**:454–455.

Palmer, J. D., and C. Delwiche. 1996. Second-hand chloroplasts and the case of the disappearing nucleus. Proceedings of the National Academy of Sciences U.S.A. **93**:7432–7435.

Paquin, B., M. J. Laforest, L. Forget, I. Roewer, A. Wang, J. Longcore, and B. F. Lang. 1997. The fungal mitochondrial genome project: evolution of fungal mitochondrial genomes and their gene expression. Current Genetics **31**:380–395.

Pawlowski, J., I. Bolivar, J. Fahrni, T. Cavalier-Smith, and M. Gouy. 1996. Early origin of Foraminifera suggested by SSU rRNA gene sequences. Molecular Biology and Evolution **13**:445–450.

Perasso, R., A. Baroin, L. Qu, J. P. Bachellerie, and A. Adoutte. 1989. Origin of the algae. Nature **339**:142–144.

Pierce, S. K., R. W. Biron, and M. E. Rumpho. 1996. Endosymbiotic chloroplasts in molluscan cells contain proteins synthesized after plastid capture. Journal of Experimental Biology 199:2323–2330.

Preiser, P., D. H. Williamson, and R. J. M. Wilson. 1995. tRNA genes transcribed from the plastid-like DNA of *Plasmodium falciparum*. Nucleic Acids Research 23:4329–4336.

Ragan, M. A., and R. R. Gutell. 1995. Are red algae plants? Botanical Journal of the Linnean Society 118:81–105.

Raven, J. A. 1997. Phagotrophy in prototrophs. Limnology and Oceanography 42:198–205.

Raven, P. H. 1970. A multiple origin for plastids and mitochondria. Science 169:641–646.

Reith, M. 1995. Molecular biology of rhodophyte and chromophyte plastids. Annual Review of Plant Physiology and Plant Molecular Biology 46:549–575.

Reith, M., and J. Munholland. 1993a. A high-resolution gene map of the chloroplast genome of the red alga *Porphyra purpurea*. Plant Cell 5:465–475.

Reith, M., and J. Munholland. 1993b. The ribosomal RNA repeats are non-identical and directly oriented in the chloroplast genome of the red alga *Porphyra purpurea*. Current Genetics 24:443–450.

Reith, M., and J. Munholland. 1995. Complete nucleotide sequence of the *Porphyra purpurea* chloroplast genome. Plant Molecular Biology Reporter 13:333–335.

Rowan, R., and D. A. Powers. 1991. A molecular genetic classification of zooxanthellae and the evolution of animal-algal symbioses. Science 251:1348–1351.

Rowan, R., S. M. Whitney, A. Fowler, and D. Yellowlees. 1996. Rubisco in marine symbiotic dinoflagellates: form II enzymes in eukaryotic oxygenic phototrophs, encoded by a nuclear multi-gene family. Plant Cell 8:539–553.

Runnegar, B. 1992. The tree of life. In The Proterozoic Biosphere. A Multidisciplinary Study, eds. J. W. Schopf and C. Klein, C., pp., 471–475. Cambridge University Press, New York.

Schmidt, M., I. Dvensen, and J. Feierabend. 1995. Analysis of the primary structure of the chloroplast isozyme of triosephosphate isomerase from rye leaves by protein and cDNA sequencing indicates a eukaryotic origin of its gene. Biochimica and Biophysica Acta 1261:257–264.

Schmitt, A., G. Frank, P. James, W. Staudenmann, H. Zuber, and C. Wilhelm. 1994. Polypeptide sequence of the chlorophyll a/b/c-binding protein of the prasinophycean alga *Mantoniella squamata*. Photosynthesis Research 40:269–277.

Schnepf, E. 1993. From prey via endosymbiont to plastids: comparative studies in dinoflagellates. In Origins of Plastids, ed. R. A. Lewin, pp. 53–76. Chapman & Hall, New York.

Schopf, J. W. 1993. Microfossils of the early archean apex chert: new evidence of the antiquity of life. Science 260:640–646.

Schopf, J. W. 1994. Disparate rates, differing fates: tempo and mode of evolution changed from the Precambrian to the Phanerozoic. Proceedings of the National Academy of Sciences U.S.A. 91:6735–6742.

Schopf, J. W. 1996. Are the oldest fossils cyanobacteria? In Evolution of Microbial Life, eds. D. M. Roberts, P. Sharp, G. Alderson, and M.A. Collins, pp. 23–61. Cambridge University Press, Cambridge.

Shinozaki, K., M. Ohme, M. Tanaka, T. Wakasugi, N. Hayashida, T. Matsubayashi, N. Zaita, J. Chunwongse, J. Obokata, K. Yamaguchi-Shinozaki, C. Ohto, K. Torazawa, B. Y. Meng, J. Kusuda, F. Takaiwa, A. Kato, N. Tohdoh, H. Shimada, and M. Sugiura. 1986. The complete nucleotide sequence of the tobacco chloroplast genome: its gene organization and expression. EMBO Journal 5:2043–2049.

Sogin, M. 1994. The origin of eukaryotes and evolution into major kingdoms. In Early Life on Earth, Nobel Symposium No. 84, ed. S. Bengston, pp. 181–194. Columbia University Press, New York.

Sogin, M. 1996. Problems with molecular diversity in the Eukarya. In Evolution of Microbial Life, eds. D. M. Roberts, P. Sharp, G. Alderson, and M. A. Collins, pp. 167–184. Cambridge University Press, Cambridge.

Stiller, J. W., and B. D. Hall. 1997. The origin of red algae: implications for plastid evolution. Proceedings of the National Academy of Sciences U.S.A. 94:4520–4525.

Stirewalt, V., C. Michalowski, W. Löffelhardt, H. Bohnert, and D. Bryant. 1995. Nucleotide sequence of the cyanelle genome from *Cyanophora paradoxa*. Plant Molecular Biology Reporter 13:327–332.

Sweeney, B. M. 1976. *Pedinomonas noctilucae* (Prasinophyceae), the flagellate symbiotic in *Noctiluca* in southeast Asia. Journal of Phycology 12:460–464.

Sytsma, K. J., and W. J. Hahn. 1994. Molecular systematics: 1991–1993. Progress in Botany 55:307–333.

Tabita, F. R. 1995. The biochemistry and metabolic regulation of carbon metabolism and CO_2 fixation in purple bacteria. In Anoxygenic Photosynthetic Bacteria, eds. R. E. Blankenship, M. T. Madigan, and C. E. Bauer, pp. 885–914. Kluwer Academic Publishers, Amsterdam.

Tan, S., F. X. Cunningham, and E. Gantt. 1997. *LhcaR1* of the red alga *Porphyridium cruentum* encodes a polypeptide of the LHC1 complex with seven potential chlorophyll a-binding residues that are conserved in most LHCs. Plant Molecular Biology 33:157–167.

Thompson, M. D., D. W. Copertino, E. Thompson, M. R. Favreau, and R. B. Hallick. 1995. Evidence for the late origin of introns in chloroplast genes from an evolutionary analysis of the genus *Euglena*. Nucleic Acids Research 23:4745–4752.

Thompson, M. D., L. Zhang, L. Hong, and R. B. Hallick. 1997. Two new group-II twintrons in the *Euglena gracilis* chloroplast are absent in basally branching *Euglena* species. Current Genetics 31:89–95.

Tingey, S. V., F.-Y. Tsai, J. W. Edwards, E. L. Walker, and G. M. Coruzzi. 1988. Chloroplast and cytoplasmic glutamine synthetase are encoded by homologous nuclear genes which are differentially expressed *in vivo*. Journal of Biology Chemistry 263:9651–9657.

Tomas, R. N., and E. R. Cox. 1973. Observations on the symbiosis of *Peridinium balticum* and its intracellular alga I. Ultrastructure. Journal of Phycology 9:304–323.

Triemer, R. E., and M. A. Farmer. 1991. An ultrastructural comparison of the mitotic apparatus, feeding apparatus, flagellar apparatus and cytoskeleton in euglenoids and kinetoplastids. Protoplasma **164:**91–104.

Turmel, M., R. R. Gutell, J.-P. Mercier, C. Otis, and C. Lemieux. 1993a. Analysis of the chloroplast large subunit ribosomal RNA gene from 17 *Chlamydomonas* taxa. Three internal transcribed spacers and 12 group I intron insertion sites. Journal of Molecular Biology **232:**446–467.

Turmel, M., J.-P. Mercier, and M.-J. Côté. 1993b. Group I introns interrupt the chloroplast *psaB* and *psbC* and the mitochondrial *rrnL* gene in *Chlamydomonas.* Nucleic Acids Research **21:**5242–5250.

Turmel, M., Y. Choquet, M. Goldschmidt-Clermont, J.-D. Rochaix, C. Otis, and C. Lemieux. 1995a. The *trans*-spliced intron 1 in the *psaA* gene of the *Chlamydomonas* chloroplast: a comparative analysis. Current Genetics **27:**270–279.

Turmel, M., V. Côté, C. Otis, J.-P. Mercier, M. W. Gray, K. M. Lonergan, and C. Lemieux. 1995b. Evolutionary transfer of ORF-containing group I introns between different subcellular compartments (chloroplast and mitochondrion). Molecular Biology and Evolution **12:**533–545.

Turner, S. 1997. Molecular systematics of oxygenic photosynthetic bacteria. **In** The Origin of Algae and Their Plastids, ed. D. Bhattacharya, pp. 13–52. Springer-Verlag, Vienna.

Urbach, E., D. L. Robertson, and S. W. Chisholm. 1992. Multiple evolutionary origins of prochlorophytes within the cyanobacterial radiation. Nature **355:**267–269.

Vaidya, A. B., M. Morrisey, C. V. Plowe, D. C. Kaslow, and T. E. Wellems. 1993. Unidirectional dominance of cytoplasmic inheritance in two genetic crosses of *Plasmodium falciparum.* Molecular and Cellular Biology **13:**7349–7357.

Valentin, K., and K. Zetsche. 1989. The genes of both subunits of ribulose-1,5-bisphosphate carboxylase constitute an operon on the plastome of a red alga. Current Genetics **16:**203–209.

Van den Hoek, C., D. G. Mann, and H. M. Jahns. 1995. Algae: An Introduction to Phycology. Cambridge University Press, Cambridge.

Van de Peer, Y., S. A. Rensing, U.-G. Maier, and R. De Wachter. 1996. Substitution rate calibration of small subunit ribosomal RNA identifies chlorarachniophyte endosymbionts as remnants of green algae. Proceedings of the National Academy of Sciences U.S.A. **93:**7732–7736.

Viale, A. M., and A. K. Arakaki. 1994. The chaperone connection to the origins of the eukaryotic organelles. FEBS Letters **341:**146–151.

Viale, A. M., A. K. Arakaki, F. C. Soncini, and R. G. Ferreyra. 1994. Evolutionary relationships among eubacterial groups as inferred from GroEL (chaperonin) sequence comparisons. International Journal of Systematic Bacteriology **44:**527–533.

Vivier, E., and I. Desportes. 1990. Phylum Apicomplexa. **In** Handbook of Protoctista, eds. L. Margulis, J. O. Corliss, M. Melkonian, and D. J. Chapman, pp. 549–573. Jones and Bartlett, Boston.

Wakasugi, T., J. Tsudzuki, S. Ito, K. Nakashima, T. Sudzuki, and M. Sugiura. 1994. Loss of all *ndh* genes as determined by sequencing the entire chloroplast genome of the black pine *Pinus thunbergii.* Proceedings of the National Academy of Sciences U.S.A. **91:**9794–9798.

Watanabe, M. M., S. Suda, I. Inouye, T. Sawaguchi, and M. Chihara. 1990. Lepidodinium viride gen. et sp. nov. (Gymnodiniales, Dinophyta), a green dinoflagellate with a chlorophyll a- and b-containing endosymbiont. Journal of Phycology **26:**741–751.

Watson, G. M. F., and F. R. Tabita. 1996. Regulation, unique gene organization, and unusual primary structure of carbon fixation genes from a marine phycoerythrin-containing cyanobacterium. Plant Molecular Biology **32:**1103–1115.

Whatley, J. M. 1993a. Chloroplast ultrastructure. **In** Ultrastructure of Microalgae, ed. T. Berner, pp. 135–204. CRC Press, Boca Raton, Florida.

Whatley, J. M. 1993b. Membranes and plastid origins. **In** Origins of Plastids, ed. R. A. Lewin, pp. 77–106. Chapman & Hall, New York.

Whitney, S. M., D. C. Shaw, and D. Yellowlees. 1995. Evidence that some dinoflagellates contain a ribulose-1,5-bisphosphate carboxylase/oxygenase related to that of the a-proteobacteria. Proceedings of the Royal Society London B **259:**271–275.

Wilcox, L. W., and G. J. Wedemayer. 1985. Dinoflagellate with blue-green chloroplasts derived from an endosymbiotic eukaryote. Science **227:**192–227.

Wilhelm, C. 1987. The existence of chlorophyll *c* in the Chl *b*-containing light-harvesting complex of the green alga *Mantoniella squamata* (Prasinophyceae). Botanica Acta **101:**7–10.

Wilson, R. J. M., and D. H. Williamson. 1997. Extrachromosomal DNA in the apicomplexa. Microbiology and Molecular Biology Reviews **61:**1–16.

Wilson, R. J. M., P. W. Denny, P. R. Preiser, K. Rangachari, K. Roberts, A. Roy, A. Whyte, M. Strath, D. J. Moore, P. W. Moore, and D. H. Williamson. 1996. Complete gene map of the plastid-like DNA of the malaria parasite Plasmodium falciparum. Journal of Molecular Biology **261:**155–172.

Wolfe, G. R., F. X. Cunningham, D. Durnford, B. R. Green, and E. Gantt. 1994. Evidence for a common origin of chloroplasts with light-harvesting complexes of different pigmentation. Nature **367:**566–568.

Wolfe, K. H., C. W. Morden, and J. D. Palmer. 1992. Function and evolution of a minimal plastid genome from a nonphotosynthetic parasitic plant. Proceedings of the National Academy of Sciences U.S.A. **89:**10648–10652.

Xu, M. Q., S. D. Kathe, H. Goodrich-Blair, S. A. Nierzwicki-Bauer, and D. A. Shub. 1990. Bacterial origin of a chloroplast intron: conserved self-splicing group I introns in cyanobacteria. Science **250:**1566–1570.

Zhou, Y. H., and M. A. Ragan. 1994. Cloning and characterization of the nuclear gene encoding plastid glyceraldehyde-3-phosphate dehydrogenase from the marine red alga *Gracilaria verrucosa*. Current Genetics **26:**79–86.

Zhou, Y. H., and M. A. Ragan. 1995. The nuclear gene and cDNAs encoding cytosolic glyceraldehyde-3-phosphate dehydrogenase from the marine red alga *Gracilaria verrucosa:* cloning, characterization and phylogenetic analysis. Current Genetics **28:**324–332.

14

Molecular Phylogenetic Insights on the Origin and Evolution of Oceanic Island Plants

Bruce G. Baldwin, Daniel J. Crawford, Javier Francisco-Ortega,
Seung-Chul Kim, Tao Sang, and Tod F. Stuessy

The floras and faunas of oceanic islands have fascinated evolutionary biologists since the time of Darwin (1859) and Wallace (1880). Questions that first arose about island life are among those still being addressed today: How did terrestrial organisms come to exist on remote islands? What ancestors gave rise to island lifeforms? Why are so many insular species unique and different from those of other islands and continental settings? For Darwin, such questions about organisms of the Galápagos archipelago were helpful in formulating his ideas about evolutionary change.

More recently, works such as the books of Carlquist (1965, 1974, 1980) have focused on a wide diversity of biological and evolutionary questions about island endemics and have provided a framework for modern studies. As elucidated by Carlquist (1965), island plants can serve as model systems for studying plant evolution, in part because islands are natural laboratories: small, isolated, and relatively simple systems compared to most continental situations. Also, the age of an island can be determined by potassium-argon dating (e.g., Stuessy et al., 1984; Clague and Dalrymple, 1987), a method

that indirectly places a maximum age limit on any endemic plants that evolved in situ. The site of origin of insular endemic lineages may, in turn, be identifiable from biogeographic interpretations based on results of phylogenetic analyses, as seen in the following discussion.

Available data suggest that lineages of island plants are often younger than closely related continental groups of similar taxonomic rank (Carlquist, 1995), although relative duration of insular and continental plant lineages has only recently been tested from a phylogenetic perspective. The apparent youth of insular taxa is beneficial for studying diversification; the more recent the divergence, the higher the probability that species differences were causally associated with lineage branching or speciation (Templeton, 1981). Another general attribute of island plants that allows for detailed evolutionary study is interfertility between species of endemic lineages, despite often striking morphological and ecological differences between the taxa (e.g., Gillett and Lim, 1970; Humphries, 1976a; Carr, 1985; Lowrey, 1986). Hybridization studies have provided valuable cytogenetic evidence for studies of island plant evolution

We thank Jeff Doyle, Doug Soltis, and John Strother for helpful reviews of an earlier version of the paper. This work was supported in part by the U.S. National Science Foundation (DEB-9458237 to BGB, and BSR-8906988, DEB-9500499, and DEB-9521017 to DJC, TFS, and S.-C. K.) and the Ministerio de Educacion y Ciencia, Spain (PF92 42044506 to J. F.-O.).

(e.g., Carr and Kyhos, 1981, 1986; Rabakonan-drianina and Carr, 1981) and offer potential for dissecting the genetic bases of phenotypic differences between taxa, information that may allow more precise resolution of changes associated with diversification.

CHALLENGES TO EVOLUTIONARY STUDY OF ISLAND PLANTS

Outgroup Choice

Some of the most fascinating attributes of island plants may also pose analytical challenges for evolutionary studies. Determination of the closest outgroup of island lineages may prove difficult if divergence between insular and continental relatives has been too extensive to allow accurate assessment of character homology or phylogenetic estimation (see Felsenstein, 1978; Wheeler, 1990; Smith, 1994). Excessive divergence between insular species and noninsular close relatives, however, has rarely been problematical in phylogenetic analyses of oceanic island plants. A more fundamental problem can be uncertainty about the breadth and depth of outgroup sampling necessary to ensure discovery of the closest relatives (e.g., the sister group) of island lineages. Precise outgroup information may be crucial for resolving ancestral characteristics of island lineages and the sequence of evolutionary and biogeographic changes that occurred in the insular setting.

Circumscribing the Ingroup

Accurate identification of the species membership of island plant lineages is critical to reconstruction of relationships, character evolution, and biogeographic history, but may be difficult to achieve without extensive sampling and investigation. Morphological and ecological divergence between members of insular plant lineages was sufficient to obscure monophyly of some groups until in-depth morphological (e.g., Keck, 1936; Carlquist, 1959, in Hawaiian Madiinae) or molecular studies (e.g., Givnish et al., 1996, in Hawaiian lobelioids; S.-C. Kim et al., 1996b, in Macaronesian Sonchus and relatives) were undertaken with broad species sampling.

Conversely, one group of island plants regarded as monophyletic on the basis of nonphylogenetic morphological evidence, the Hawaiian endemic genus *Lipochaeta,* proved to be polyphyletic upon cytogenetic investigation (Rabakonandrianina and Carr, 1981).

Finding Useful Characters for Phylogenetic Analysis

Lack of variation among insular taxa in molecular characters from conventional sources, that is, chloroplast DNA (cpDNA) and nuclear ribosomal DNA (rDNA), has been an obstacle to achieving phylogenetic resolution within various groups of island plants, for example, Hawaiian *Bidens, Lipochaeta,* and Hawaiian *Tetramolopium* (see Baldwin, 1997). The pace of diversification in some insular groups has apparently been too rapid for fixation of enough synapomorphic mutations to allow robust phylogenetic reconstruction using a limited number of cpDNA or rDNA characters. New molecular phylogenetic markers need to be identified in plant gene regions that evolve rapidly and in ways that preserve evidence of site homology and lineage diversification (see Chapter 1). In contrast to excessive similarity of island species at the molecular level, divergence within island plant lineages in morphological and ecological characters may appear to be too extensive or to involve characteristics suspected of being too prone to homoplasy (e.g., in response to selection) to permit accurate resolution of phylogenetic relationships (Givnish et al., 1995).

At present, too few well-resolved, robustly supported morphological and molecular trees are available from the same island plant lineages to make general conclusions about the relative value of the two types of data for reconstructing relationships in insular groups. The possibility that parallel evolution of morphological characteristics is especially prevalent and phylogenetically misleading in island plants is not evident from published studies. Phylogenetic analyses of combined morphological and molecular data in Macaronesian *Aeonium* (Mes and 'T Hart, 1996) and in Hawaiian *Alsinidendron/Schiedea* (Sakai et al., 1997; Soltis et al., 1997) offered better resolution and support than analyses of either type of data independently, in keeping with predictions for simultaneous analysis of congruent sets of characters (see Barrett et al., 1991; de Queiroz et al., 1995). Inasmuch as morphological evidence from nonphylogenetic investigations has been the primary guide to choice of taxa for molecular phylogenetic studies, it must be concluded that both lines of evidence have contributed greatly to our recent progress in discerning relationships among island plants.

Extinction: Causes and Consequences

Extinction and fragmentation of indigenous plant populations on islands pose increasing challenges for systematists interested in reconstructing the evolutionary and biogeographic histories of insular groups. Human-caused habitat destruction and introduction of nonnative species have led to rapid loss of indigenous plant diversity in the floras of oceanic islands, with extinction rates exceeding those of continental areas of similar size (World Conservation Monitoring Centre, 1992; Smith et al., 1993). Susceptibility of island plants to human-caused and natural extinction may be particularly high in groups with narrow geographic distributions, small population sizes, limited dispersal ability, or restricted pollination systems—attributes that are common in species-rich insular lineages, for example, in Hawaiian *Cyanea/Rollandia,* which has lost 14 of 63 recorded species in historic times (Givnish et al., 1995). Even in insular groups that have lost minimal species diversity to recent extinction, extirpation and decimation of populations may have sufficiently altered genetic variation or ecological interactions to confound interpretation of systematic data.

Geological and climatic dynamism of oceanic islands may promote high rates of extinction in the absence of human-related factors. Erosion and subsidence of volcanic islands must ultimately result in extinction of terrestrial species that fail to disperse elsewhere. An age estimate of 15 to 20 Ma for the most recent common ancestor of Hawaiian lobelioids (Givnish et al., 1996) suggests that the group once occupied islands west of Kauai that are now small islets or atolls devoid of lobelioids. Phylogenetic and

biogeographic patterns inferred from rDNA internal transcribed spacer (ITS) data in the Hawaiian silversword alliance led Baldwin and Robichaux (1995) to conclude that the true silversword and greensword lineage (*Argyroxiphium*) may have experienced prehistoric extinction on the older islands of the modern Hawaiian chain, a possible result of high-elevation habitat loss via island erosion and subsidence.

Ancient and recent loss of species diversity creates a sampling problem that can reduce accuracy of hypotheses of phylogeny, character evolution, and timing of diversification (see Donoghue et al., 1989; Huelsenbeck, 1991). The effect of unequal branch lengths on accuracy of phylogenetic estimates is expected to be minimal, however, within groups with low genetic divergence between taxa (Huelsenbeck and Hillis, 1993), as is typical of island plants. A more serious potential problem is a bias toward extinction of ancient lineages, for example, those on older islands of a hot spot archipelago. Inability to sample extinct, basally divergent insular lineages prevents subdivision of the potentially long phylogeny branch between the apparent root node of an insular lineage and the noninsular sister group. Failure to break up the long branch between ingroup and outgroup species may reduce the potential under global parsimony for accurate root placement in the insular phylogeny (Felsenstein, 1978; Wheeler, 1990; see Swofford et al., 1996).

Even if accurate phylogenetic reconstruction of the insular ingroup is achieved, inability to identify the true root node of the island plant clade, or the island group plus sister group clade, may lead to error in reconstructions of historical biogeography and character evolution (see Sanderson, 1996). Loss of basally divergent insular groups is particularly likely in hot spot archipelagos such as the Hawaiian Islands, wherein island turnover may lead to extinction of old sublineages that are unsuccessful in dispersal to new islands. Extinction biased against old lineages may explain in part the long phylogeny branch that unites members of the Hawaiian silversword alliance in the ITS tree (B. G. Baldwin and M. J. Sanderson, in prep.; see Baldwin, 1997).

Despite the obstacles to molecular phylogenetic study of oceanic island plants, DNA data have proved invaluable for reconstructing historical patterns of insular plant evolution. In this chapter, we discuss how molecular phylogenetic data can be applied to the study of origin and evolution of plants endemic to oceanic islands. We also provide examples of how molecular evidence has helped to elucidate biogeographic and ecological history, morphological character evolution, and processes and rates of diversification in island plant lineages.

IDENTIFYING THE CLOSEST RELATIVES OF ISLAND PLANTS

Rigorous phylogenetic methodology, most often applied to molecular data, has provided more precise resolution of the closest relatives of distinctive insular plant lineages than was achieved before the use of such approaches in systematics. Well-known examples of island plant groups that have been positioned phylogenetically with the aid of molecular data include the silversword alliance in the Hawaiian Islands (Baldwin et al., 1991; Baldwin, 1992), *Dendroseris* in the Juan Fernandez Islands (Whitton et al., 1995; S.-C. Kim et al., 1996a, 1996b), and *Scalesia* in the Galápagos archipelago (Schilling et al., 1994); other examples are given in Table 14.1. Understanding of extrainsular relationships of island plant groups gained through molecular phylogenetic studies has offered insights into the site of origin, dispersal mode, and morphological characteristics of founders of island plant radiations.

Uncertainty about the identity of the closest living relatives of the endemic Hawaiian silversword alliance (28 species in *Argyroxiphium, Dubautia,* and *Wilkesia;* Asteraceae) was a longstanding problem (see Keck, 1936; Fosberg, 1948) because the plants are so divergent from each other and from other members of Asteraceae in gross morphological characteristics (see Carr, 1985). Anatomical studies by Carlquist (1959) revealed fine-scale vegetative and reproductive characteristics shared exclusively by the silversword alliance and the western American tarweed subtribe Madiinae (Heliantheae sensu lato). On the basis of the anatomical data, Carlquist (1959) made a convincing argument for an American origin of the

silversword alliance from a tarweed ancestor. Results from phylogenetic analyses of cpDNA restriction-site data (Baldwin et al., 1991) and sequences from the ITS region of rDNA (Baldwin, 1992, 1996; Baldwin and Robichaux, 1995) reinforced and refined Carlquist's (1959) hypothesis by demonstrating that the silversword alliance was derived from within a paraphyletic group of species in *Madia* and *Raillardiopsis* that is nearly restricted to the California Floristic Province (Fig. 14.1). Reconstructions of morphological and dispersal characteristics of the most recent common ancestor of the silversword alliance based on the molecular trees suggest that the ancestor was an herbaceous, radiate plant with pappose disk fruits and other characteristics conducive to external bird dispersal, in accord with Carlquist's predictions.

Relationships of *Dendroseris* (Asteraceae-Lactuceae), the largest endemic genus in the Juan Fernandez Islands (Sanders et al., 1987), have been a matter of considerable speculation, with various taxa in Lactuceae suggested as closest relatives (Stebbins, 1953, 1977; Jeffrey, 1966; Carlquist, 1967; Sanders et al., 1987). A broad-scale phylogenetic analysis of Lactuceae using cpDNA restriction-site variation included three species of *Dendroseris* (Whitton et al., 1995), which constituted a strongly supported clade with several species of *Sonchus* and closely related genera (Fig. 14.2). In a subsequent study by S.-C. Kim et al. (1996a), a phylogenetic analysis of ITS sequences from nearly all recognized genera of subtribe Sonchinae suggests that *Dendroseris* may be sister to a clade that includes two widespread species of *Sonchus* subg. *Sonchus* and two genera endemic to New Zealand (Fig. 14.3). Although the sister group relationship is not strongly supported, it does suggest that the closest relatives of *Dendroseris* are also from the Pacific rather than the Old World (where most members of subtribe Sonchinae occur), in keeping with predictions based on dispersal-likelihood considerations.

Morphological and phytochemical data on *Argyranthemum* (Asteraceae-Anthemideae), the largest endemic genus in Macaronesia (i.e., the Azores, Canary, Cape Verde, Madeira, and Selvagens Islands), indicated a close relationship to the three continental genera *Chrysanthemum,*

Table 14.1. Examples of genera or groups of genera that have been examined at the molecular level to resolve the relationships of island endemics.

Family/Genus	Island or Archipelago	Reference
Apiaceae		
Sanicula	Hawaiian	Vargas et al., 1998
Asteraceae		
Argyranthemum*	Macaronesian	Francisco-Ortega et al., 1995, 1996a, 1997a, 1997b
Argyroxiphium,* Dubautia,*	Hawaiian	Baldwin et al., 1990, 1991; Baldwin, 1992, 1996, 1997;
and *Wilkesia**		Baldwin and Robichaux, 1995
Babcockia,* Lactucosonchus,*	Macaronesian	Whitton et al., 1995; S.-C. Kim et al., 1996a, 1996b
Prenanthes, Sonchus,		
Sventenia,* and *Taeckholmia**		
Cheirolophus	Macaronesian	Garnatje et al., 1996
Dendroseris*	Juan Fernandez	Whitton et al., 1995; S.-C. Kim et al., 1996c
Hesperomannia*	Hawaiian	H.-G. Kim et al., 1996
Pericallis*	Canarian	Knox and Palmer, 1995; Kadereit and Jeffrey, 1996
Reichardia	Canarian	S.-C. Kim et al., 1996a, 1996b
Robinsonia	Juan Fernandez	Crawford et al., 1993; Sang et al., 1995b
Scalesia*	Galápagos	Schilling et al., 1994
Senecio	Canarian	Lowe and Abbott, 1996
Tetramolopium	Hawaiian, Cook Islands	Chan, 1994; Chan et al., 1994; Okada et al., 1994
Boraginaceae		
Echium	Macaronesian	Boehle et al., 1996
Brassicaceae		
Brassica	Canarian	Warwick and Black, 1993
Crambe (sect. *Dendrocrambe*)	Macaronesian	Francisco-Ortega et al., 1996b
Erucastrum	Canarian	Warwick and Black, 1993
Sinapidendron*	Madeira	Warwick and Black, 1993
Campanulaceae	Hawaiian	Givnish et al. 1994, 1995, 1996
Brighamia,* Cyanea,*		
Clermontia,*Delissea,* Lobelia,		
Rollandia,* Trematolobelia*		
Caryophyllaceae		
Alsinidendron*, Schiedea*	Hawaiian	Sakai et al., 1997; Soltis et al., 1997
Chenopodiaceae		
Beta	Canarian	Santoni and Berville, 1992
Crassulaceae		
Aeonium	Macaronesian	Mes, 1995; Mes and 'T Hart, 1996; Mes et al., 1996
Aichryson*	Macaronesian	Mes, 1995; Mes et al., 1996
Monanthes*	Canarian	Mes, 1995; Mes and 'T Hart, 1994; Mes et al., 1996, 1997
Fabaceae		
Adenocarpus	Macaronesian	Kaess and Wink, 1995
Chamaecytisus	Macaronesian	Badr et al., 1994; Kaess and Wink, 1995
Spartocytisus*	Macaronesian	Kaess and Wink, 1995
Teline	Macaronesian	Kaess and Wink, 1995
Geraniaceae		
Geranium	Hawaiian	Pax et al., 1997
Malvaceae		
Gossypium	Galápagos	Wendel and Percival, 1990
Gossypium	Hawaiian	Dejoode and Wendel, 1992
Lavatera	Canarian	Ray, 1995; Fuertes-Aguilar et al., 1996
Papaveraceae		
Argemone	Hawaiian	Schwartzbach and Kadereit, 1996
Pinaceae		
Pinus	Canarian	Strauss and Doerksen, 1990; Krupin et al., 1996
Poaceae		
Lolium	Canarian	Charmet et al., 1997
Rubiaceae		
Psychotria	Hawaiian	Nepokroeff and Sytsma, 1996

Table 14.1. continued

Family/Genus	Island or Archipelago	Reference
Scrophulariaceae		
Isoplexis	Canarian	Carvalho and Culham, 1997
Solanaceae		
Solanum	Canarian	Olmstead and Palmer, 1997
Violaceae		
Viola	Hawaiian	Ballard et al., 1996

Note: Genera denoted by an asterisk are endemic; others include island endemic species, or groups of species, plus continental taxa.

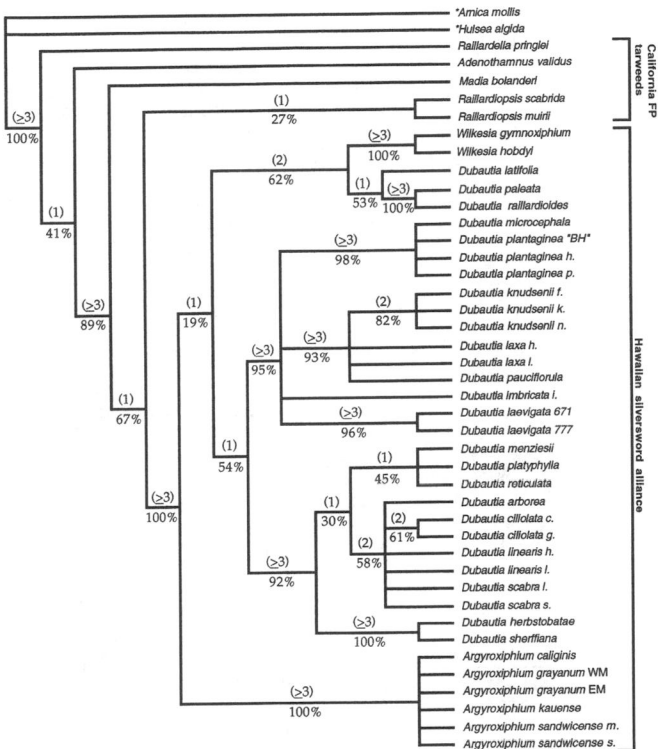

Figure 14.1. A phylogenetic hypothesis for the Hawaiian silversword alliance (Asteraceae-Madiinae); the strict consensus tree of eight equally most parsimonious trees based on ITS sequence variation from the Hawaiian endemic genera *Argyroxiphium, Dubautia,* and *Wilkesia* and related California tarweeds. Numbers above the branches are decay index values. Percentages below branches are bootstrap values from 100 replicate analyses. Reprinted from Baldwin and Robichaux (1995) with permission of the authors.

Ismelia, and *Heteranthemis* from northern Africa, southern Iberia, and the Mediterranean region (Humphries, 1976a, 1976b, 1979). On the basis of a cladistic study of morphological characters, Bremer and Humphries (1993) suggested that the three aforementioned continental genera are the closest relatives of *Argyranthemum.* Bremer and Humphries treated all four genera as the exclusive representatives of Chrysantheminae (Anthemideae). Others have suggested that the closest relative of *Argyranthemum* is the South African genus *Cymbopappus* (Hutchinson, 1917; Bramwell, 1976; Takhtajan, 1986) or the Japanese genus *Nipponanthemum* (Kitamura, 1978). Results from a phylogenetic analysis of cpDNA restriction-site mutations in the four genera of Chrysantheminae sensu Bremer and Humphries (1993) and representative genera from Leucantheminae, Anthemidinae, and Achilleinae provide strong

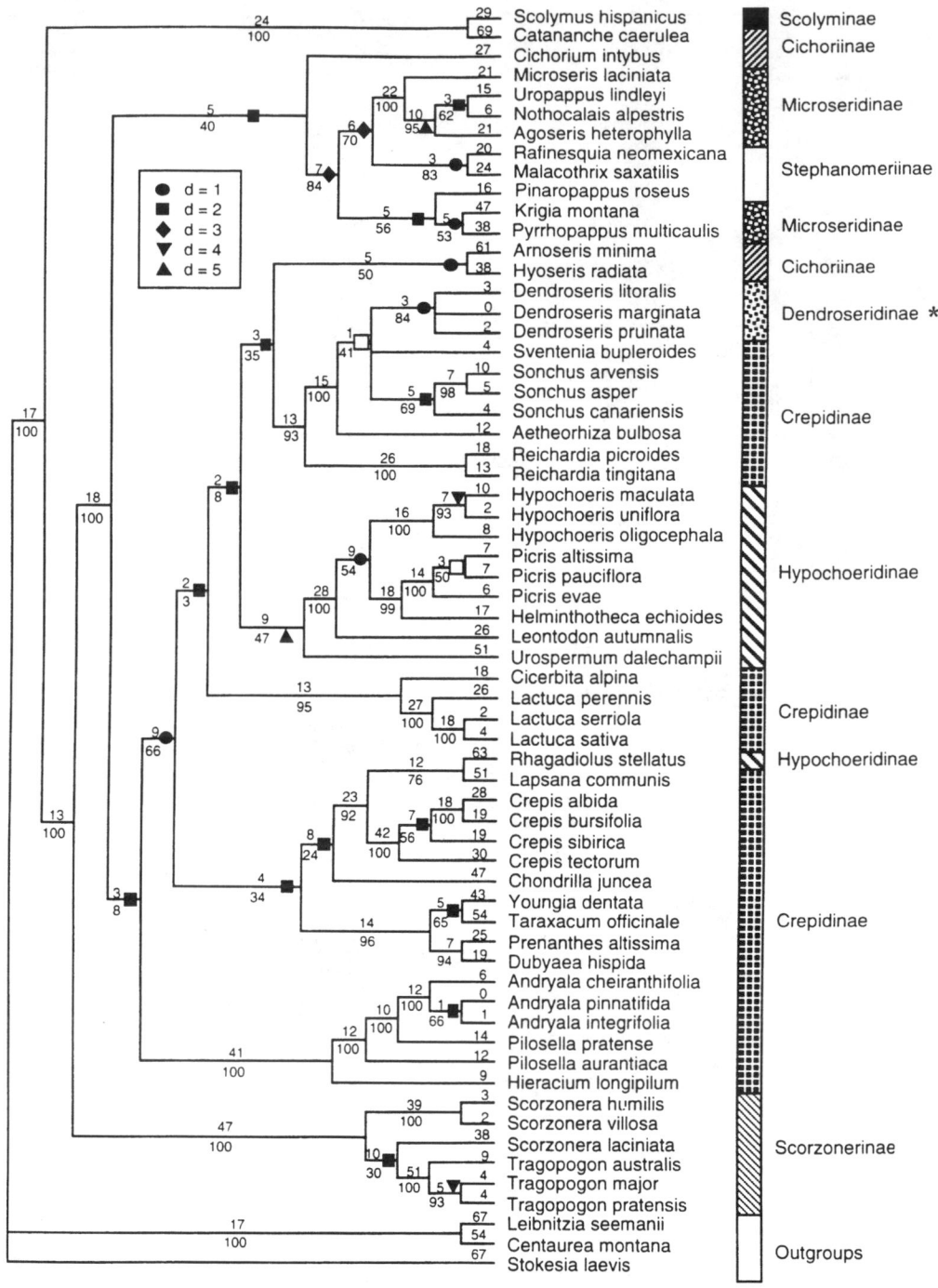

Figure 14.2. Phylogenetic hypothesis for Lactuceae (Asteraceae); one of eight maximally parsimonious trees for Lactuceae from analysis of cpDNA restriction-site variation. Nodes that collapse in the strict consensus tree are indicated by open boxes. Decay indices are represented by solid symbols. The number of character-state changes is given above each branch, and the bootstrap values are shown below. The three species of *Dendroseris* included in the analysis fall within a strongly supported clade (15 synapomorphic mutations, bootstrap value of 100) with species of *Sonchus, Sventenia,* and *Aetheorhiza.* Modified from Whitton et al. (1995) and used with permission from the *Canadian Journal of Botany.*

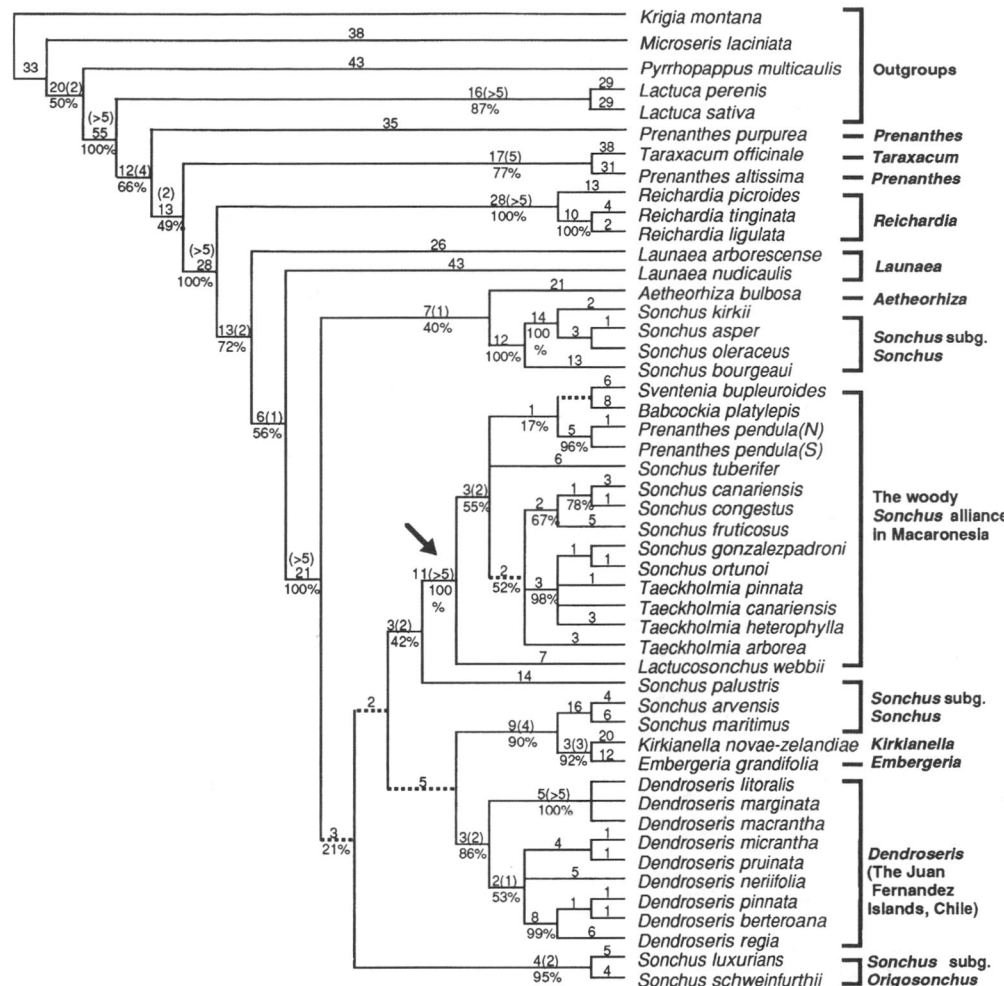

Figure 14.3. Phylogenetic hypothesis for Sonchinae (Asteraceae-Lactuceae); one of 144 maximally parsimonious trees from analysis of ITS sequence variation. Dashed lines designate branches that collapse in the strict consensus tree. Numbers above lines indicate how many characters support a branch; those in parentheses are decay index values. Percentages below branches are bootstrap values. *Dendroseris* is sister to a clade containing two species from subg. *Sonchus* and the genera *Embergeria* and *Kirkianella* from the South Pacific. The arrow designates the clade comprising the woody *Sonchus* alliance from Macaronesia (= the Azores, Canary, Cape Verde, Madeira, and Selvagens Islands). Reproduced from S.-C. Kim et al. (1996b), with permission from the *Proceedings of the National Academy of Sciences U.S.A.*

support for a clade including *Argyranthemum* and the three other genera of Chrysantheminae sensu Bremer and Humphries (Francisco-Ortega et al., 1995) (Fig. 14.4).

An additional phylogenetic investigation of ITS sequences with wider taxonomic sampling, from 32 genera and eight subtribes of Anthemideae, showed that *Argyranthemum* is not closely related to the South African and Japanese genera (Francisco-Ortega et al., 1997a) (Fig.

14.5). Analyses of both molecular data sets (Figs. 14.4, 14.5) demonstrate that the closest continental relatives of *Argyranthemum* reside within the genera *Chrysanthemum, Heteranthemis,* and *Ismelia,* all three of which occur naturally in the nearest continental source areas for the Macaronesian Islands.

An excellent example of how molecular data can lead to a complete rethinking of the extra-insular relationships of island plant lineages

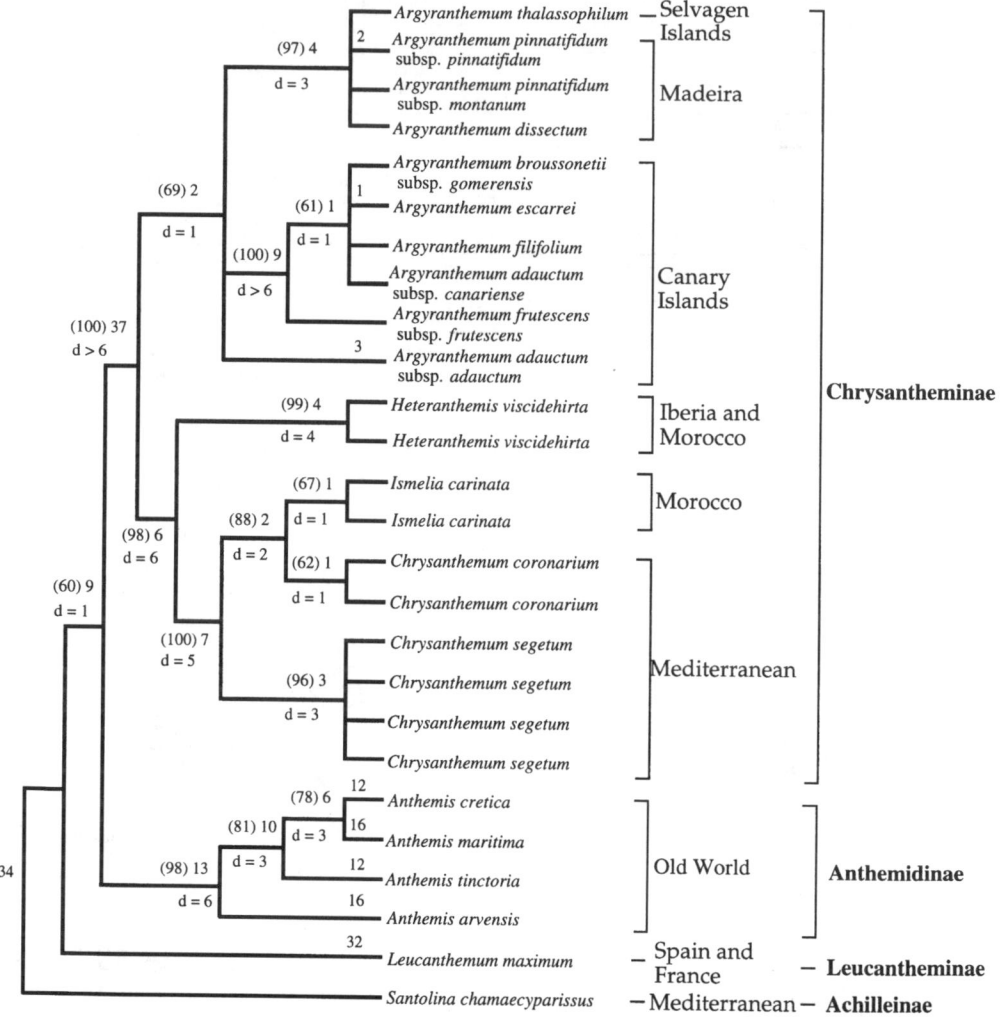

Figure 14.4. Phylogenetic hypothesis for Chrysantheminae (Asteraceae-Anthemideae); the single most parsimonious tree resulting from analysis of cpDNA restriction site mutations from the four genera constituting Chrysantheminae. The number of restriction-site changes is shown above each node or along each lineage. Bootstrap percentages are shown in parentheses above each node. Decay index values (= d) appear below each node. Note the strong support for the clade that includes the four genera traditionally assigned to Chrysantheminae. Reprinted from Francisco-Ortega et al. (1995) with permission of the American Society of Plant Taxonomists.

involves the endemic Hawaiian genus *Hesperomannia* (Asteraceae). *Hesperomannia* had long been treated as a member of Mutisieae, but results of phylogenetic analysis of *ndhF* sequence data showed that the genus belongs in Vernonieae (H.-G. Kim et al., 1996). An extensive sample of taxa of Vernonieae in the analysis allowed the conclusion that *Hesperomannia* is most closely related to African species of *Vernonia,* a surprising result in light of the im-

mense distance separating Africa and the Hawaiian Islands. Subsequent morphological studies of style and pollen characters confirmed that the Hawaiian genus is a member of Vernonieae (H.-G. Kim et al., 1996).

The precision of molecular methods for identifying the closest relatives of island plants depends on a variety of factors including taxon sampling, rates of molecular evolution, and timing of diversification. The potential for inadequate

Figure 14.5. Phylogenetic hypothesis for Anthemideae (Asteraceae); the single most parsimonious tree from analysis of ITS sequences of 32 genera of Anthemideae. Number of nucleotide changes is given above each branch and bootstrap values (from 100 replicates) are shown in bold print below branches. The arrow indicates a clade comprising taxa with a 17-bp deletion in ITS2. Symbols at tips of terminal branches denote generic distributions: closed circles = Mediterranean; open circles = mostly Mediterranean; open square = predominantly European and Eastern Asian; gray square = primarily European and Eastern Asian; and triangles = South African. Note the distant relationship between *Argyranthemum* and either *Cymbopappus* or *Nipponanthemum*, two of the genera outside Chrysantheminae that have been considered closely related to *Argyranthemum*. Reproduced from Francisco-Ortega et al. (1997a) with permission of the *American Journal of Botany*.

sampling of continental taxa (i.e., omission of sister species of an island lineage) depends, of course, on the degree of uncertainty about relationships of potential outgroups to one another and to the insular lineage in question. For example, a broad survey of Lactuceae using cpDNA restriction sites (Whitton et al., 1995) was necessary to find the closest relatives of *Dendroseris* because there was no strong evidence placing the genus with any particular subgroup in the tribe. By contrast, sampling of continental taxa for phylogenetic analysis of silversword alliance ancestry (Baldwin et al., 1991; Baldwin, 1996) was simplified by the availability of convincing anatomical evidence (Carlquist, 1959) for placement of the Hawaiian lineage in Madiinae, a much smaller group to sample than Lactuceae.

TESTING THE MONOPHYLY OF ENDEMIC ISLAND PLANT GROUPS

Identification of a monophyletic island plant group sets the stage for meaningful studies and interpretations of insular evolution in the lineage. Extensive morphological and ecological divergence in insular plant lineages may obscure close relationships unless phenotypic and molecular characters are carefully studied. In the following discussion, we provide examples of how molecular data have aided in testing the monophyly of island plant groups composed of morphologically and ecologically disparate species (also see Table 14.1).

Molecular data from cpDNA (Baldwin et al., 1990, 1991) and the ITS region (Baldwin and Robichaux, 1995) strongly reinforced the hypothesis of monophyly of the Hawaiian silversword alliance (Fig. 14.1), a group that is poorly diagnosed by shared morphological characteristics (e.g., expanded style-branch apices) in spite of numerous morphological features shared exclusively with the superficially dissimilar, continental members of Madiinae (Carlquist, 1959). Biosystematic evidence consistent with monophyly of the silversword alliance was obtained earlier by Carr and Kyhos (Carr, 1985; Carr and Kyhos, 1986), who demonstrated that hybrids of at least partial fertility could be produced in any intergeneric or interspecific crossing combination attempted in the group.

An example of how molecular phylogenetic data can lead to rethinking of the circumscription of an insular lineage is from work on the woody *Sonchus* alliance (Asteraceae) of Macaronesia, which has been understood to include subg. *Dendrosonchus* and four or five other genera, depending on the taxonomic treatment (S.-C. Kim et al., 1996a) (Fig. 14.6). Not surprisingly, the molecular data placed the monotypic genus *Sventenia* in the *Sonchus*-alliance clade (naturally occurring hybrids between *Sventenia* and species of subg. *Dendrosonchus* are known). Surprisingly, however, *Prenanthes pendula* and the monotypic *Lactucosonchus webbii* were also placed in the strongly supported island clade (Fig. 14.6). These results illustrate the importance of careful sampling of island plants for phylogenetic study, without overreliance on existing taxonomic concepts; placement of taxa in different genera or tribes should not be taken a priori as strong evidence that they arose from more than one insular introduction.

Givnish et al. (1996) used cpDNA spacer sequences to test whether *Cyanea* and several other genera of Hawaiian lobelioids form a monophyletic assemblage. The results place *Brighamia, Clermontia, Delissea, Lobelia* sect. *Galeatella, L.* sect. *Revolutella, Rollandia,* and *Trematolobelia* in a strongly supported endemic-Hawaiian clade with *Cyanea,* exclusive of non-Hawaiian species in subfamily Lobelioideae. Evidence for monophyly of Hawaiian lobelioids contradicts the traditional view that the species descended from at least three introductions of ancestral taxa to the Hawaiian Islands (Fosberg, 1948).

Not all molecular phylogenetic studies of island plants have supported monophyly of groups of seemingly closely related species, for example, congeners. Ray (1995) conducted an extensive phylogenetic study of *Lavatera* and *Malva* (Malvaceae) based on morphology and ITS sequences that included both species of *Lavatera* endemic to the Canary Islands. Results of the morphological analysis indicate that the two endemics, *L. acerifolia* and *L. phoenicea,* descended from separate introductions, with each species nested within different large clades. The ITS phylogeny is also consistent with polyphyly of *Lavatera* from the Canary Islands (Fig. 14.7).

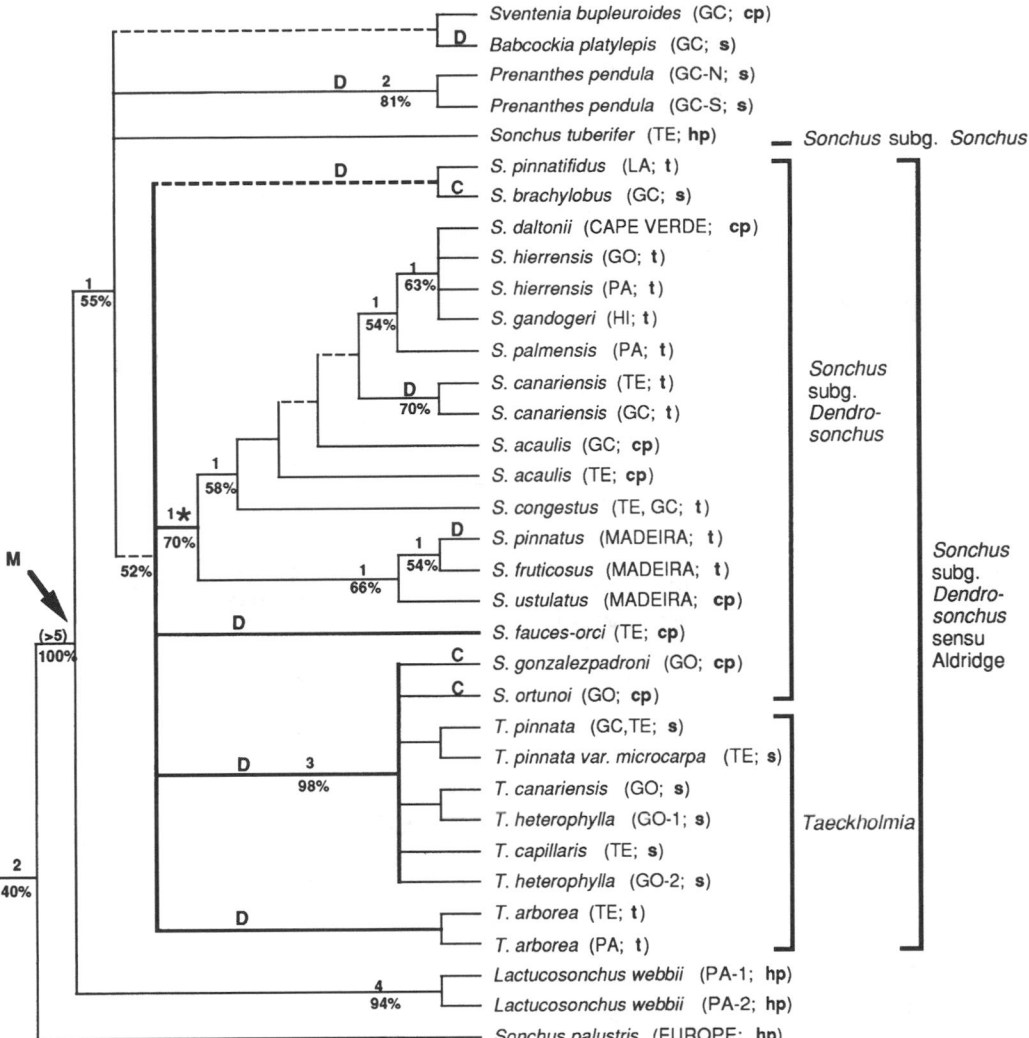

Figure 14.6. Phylogenetic hypothesis for the woody *Sonchus* alliance (Asteraceae-Lactuceae) of Macaronesia; one of 8,123 maximally parsimonious trees from analysis of ITS sequence variation. Dashed lines indicate branches that collapse in the strict consensus tree. Numbers above lines indicate how many characters support a branch. Percentages below branches are bootstrap values. The arrow designates the clade comprising the woody *Sonchus* alliance; the number in parentheses is the decay index for the clade. Abbreviations for the islands in the Canary archipelago are: F, Fuerteventura; GC, Gran Canaria; GO, La Gomera; HI, El Hierro; LA, Lanzarote; PA, La Palma; TE, Tenerife. Habit abbreviations given in bold print include: hp, herbaceous perennial; cp, caudex perennial; s, rosette subshrub; t, rosette tree-shrub. Habitat abbreviations given along the branches are: M, mesic; D, dry; C, coastal. Reproduced from S.-C. Kim et al. (1996b), with permission from the *Proceedings of the National Academy of Sciences U.S.A.*

CHOICE OF MOLECULAR REGIONS FOR PRODUCING PHYLOGENIES OF INSULAR ENDEMICS

The evident youth of most island plant lineages dictates that rapidly evolving DNA sequences must be used to achieve well-resolved phylogenies. cpDNA restriction-site mutations and variable ITS sequence characters have been the most successfully used molecular markers in phylogenetic studies of island plants, as in similar studies of congeneric angiosperms in general (see Chapter 1).

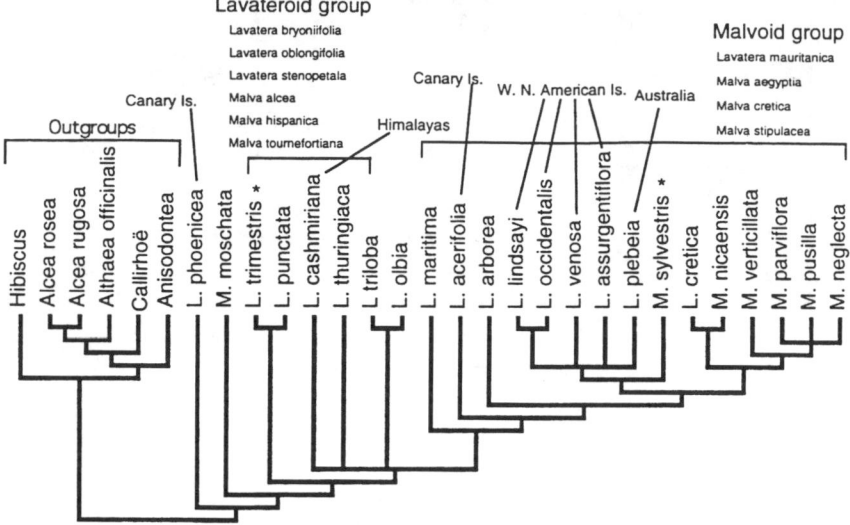

Figure 14.7. Phylogenetic hypothesis for species of *Lavatera* and *Malva* (Malvaceae); a strict consensus tree of six maximally parsimonious trees from analysis of ITS sequence variation. The two Canary Island endemics are indicated. The species listed under Lavateroid group and Malvoid group were not sampled in the ITS study, but were suspected to belong to the clades immediately below based on fruit morphology. Reproduced from Ray (1995) with permission from *Plant Systematics and Evolution.*

As mentioned previously, cpDNA (Baldwin et al., 1990; Crawford et al., 1993; Givnish et al, 1995; Francisco-Ortega et al., 1996a) and ITS (Sang et al., 1994; Baldwin and Robichaux, 1995; Mes, 1995; S.-C. Kim et al., 1996b; Francisco-Ortega et al., 1997a) data often provide too few informative characters for full resolution of relationships in island plant groups (Figs. 14.1, 14.4, 14.6, 14.8–14.10). One exception is the study of Sang et al. (1995b) wherein analysis of ITS sequences produced a fully resolved phylogeny for *Robinsonia* (Asteraceae), a genus of six species endemic to the Juan Fernandez Islands (Fig. 14.11). Attempts to use sequences from noncoding spacer regions of cpDNA to resolve relationships in island plants (Mes, 1995; S.-C. Kim et al., 1996c) have met with limited success because of low levels of variation. Extensive sampling of cpDNA restriction sites has been used successfully to reconstruct relationships in *Argyranthemum* despite high genetic similarities between species in cpDNA-spacer and ITS sequences (Francisco-Ortega et al., 1996a, 1997a). Rapidly evolving domains of floral homeotic genes in the MADS-box regulatory gene family are currently being assessed for

phylogenetic utility in the Hawaiian silversword alliance (M. D. Purugganan, B. G. Baldwin and colleagues, in prep.).

Lack of phylogenetic resolution among the major, basally divergent clades in island plant groups (Figs. 14.4, 14.6, 14.9, 14.11) is a common result that may reflect rapid early diversification of lineages, unstable rates of molecular evolution, or homoplasy effects. Typically, lack of synapomorphic characteristics rather than homoplasy appears to be responsible for poor resolution (e.g., in *Dendroseris,* Sang et al., 1994); few or no mutations apparently became fixed in the common ancestors of the early radiating lineages. A phylogeny based on a molecule that has evolved at a constant rate within the island lineage of interest is particularly valuable for assessing timing of diversification events. In the absence of such data, comparison of gene trees for a particular group based on unlinked molecular datasets (e.g., from nuclear and organellar DNAs) may allow discrimination between the hypotheses of rapid radiation or rate inconstancy of molecular evolution (see Baldwin, 1997). Acquisition of multiple molecular data sets is also desirable for improving phylogenetic resolution

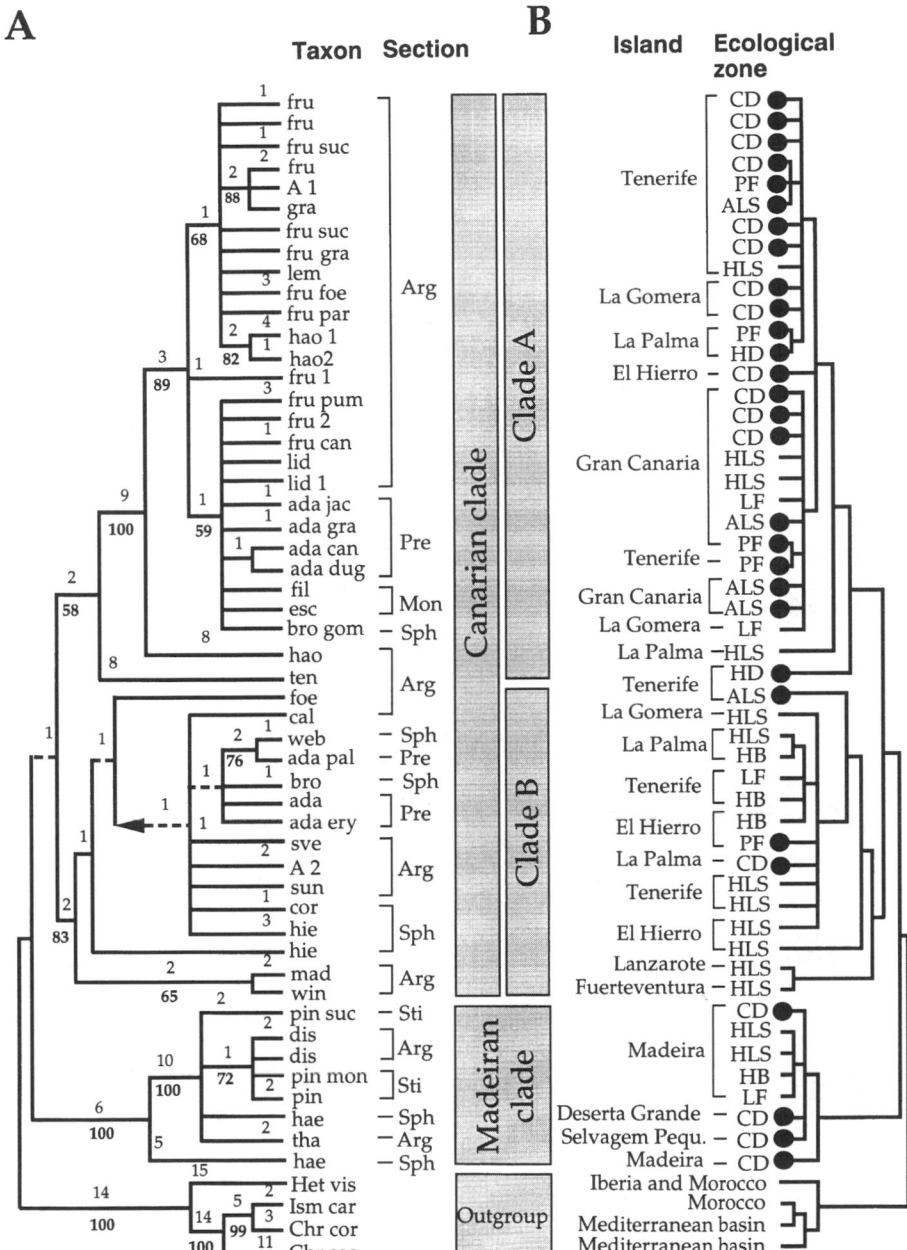

Figure 14.8. Phylogenetic hypothesis for *Argyranthemum* (Asteraceae-Anthemideae) in Macaronesia; one of two minimum-length trees from weighted parsimony analysis of cpDNA restriction-site variation. Dashed lines indicate branches that collapse in the strict consensus tree. Dashed line with arrow is a branch that collapses in the second weighted parsimony tree. The number of changes is given above each branch, and the bootstrap values from 100 replicates are given below the branches. See Francisco-Ortega et al. (1996a) for name abbreviations of all species. Second name for any species refers to the variety; numbers following names designate different populations of the same taxa. Sectional abbreviations are: Arg, *Argyranthemum;* Mon, *Monoptera;* Pre, *Preauxia;* Sph, *Sphenisimelia;* Sti, *Stigmatotheca.* Abbreviations for ecological zones are: CD, coastal desert; ALS, arid lowland scrub; HLS, humid lowland scrub; LF, level forest; HB, heath belt; PF, pine forest; HD, high altitude desert. *A.* Distribution of taxa within and among the taxonomic sections. *B.* Distribution of taxa among islands and ecological habitats. Solid circles indicate ecological habitats not affected by the trade winds; remaining branches are influenced by trade winds. Taken from Francisco-Ortega et al. (1996a) with permission from the *Proceedings of the National Academy of Sciences U.S.A.*

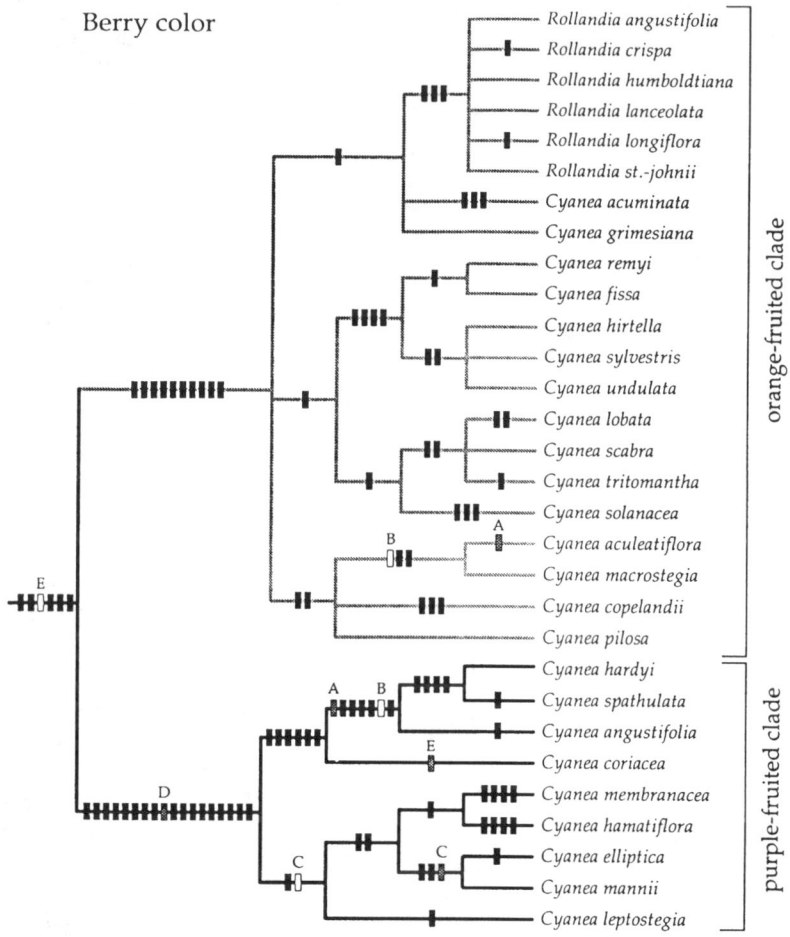

Figure 14.9. Phylogenetic hypothesis for *Cyanea* and *Rollandia* (Campanulaceae-Lobelioideae) in the Hawaiian Islands based on cpDNA restriction-site variation (from Givnish et al., 1995). Letters A–E designate pairs of homoplastic mutations. Vertical bars represent synapomorphic mutations; hollow bars = convergent losses, gray bars = convergent gains, and solid bars = unique gains or losses. The branches for the orange-fruited species are stippled and those for the purple-fruited plants are solid. Note that each fruit color is restricted to one of the two basally divergent clades. Used with permission of the authors.

through simultaneous analysis of all combinable characters (see Hillis et al., 1996), a strategy that may prove particularly successful for island plants (e.g., Mes and 'T Hart, 1996; Baldwin, 1997). Strong conflict between molecular phylogenies of island groups, however, may demand restriction of combined analyses to particular subsets of characters or taxa.

Sampling of Extinct Taxa

PCR technology now offers the potential for extensive sampling of molecular characters from recently extinct island plants. Amplification of short genomic regions (e.g., from *rbcL, ndhF, matK,* and ITS) from small-scale DNA extractions of herbarium specimens has become a widely adopted method of obtaining template for sequencing or restriction-site analysis, with minimal removal of tissue (ca. 100 mg) from collections (see Chapter 1; Jansen et al., 1997). Success at amplifying DNA extracted from herbarium specimens appears to be dependent on a variety of factors, including the taxonomic group in question, treatment of the specimen, and age of the specimen. Successful use of

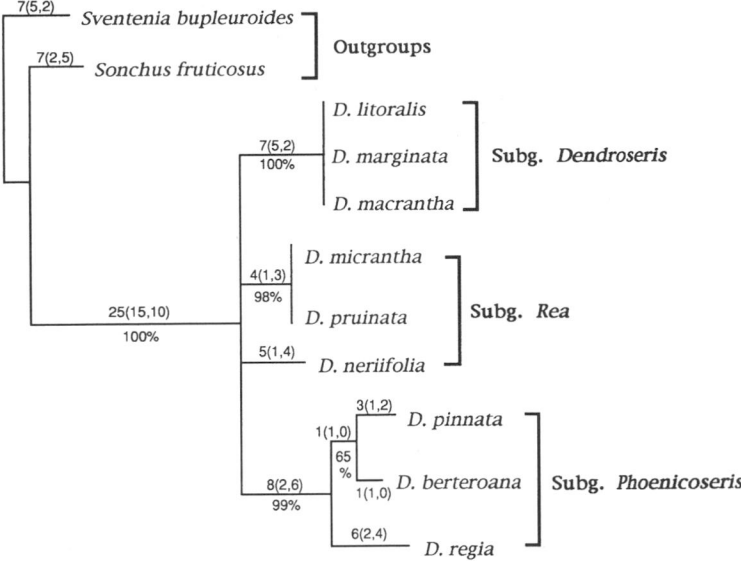

Figure 14.10. Phylogenetic hypothesis for *Dendroseris* (Asteraceae-Lactuceae); a strict consensus tree of the four maximally parsimonious trees produced from ITS sequence variation. Numbers above branches are nucleotide substitutions with the transitions and transversions in parentheses. Numbers below branches are the percentage that a group occurred in 1,000 replicates. Note that subg. *Dendroseris* and subg. *Phoenicoseris* are monophyletic whereas subg. *Rea* is not. From Sang et al. (1994), with permission from the *American Journal of Botany*.

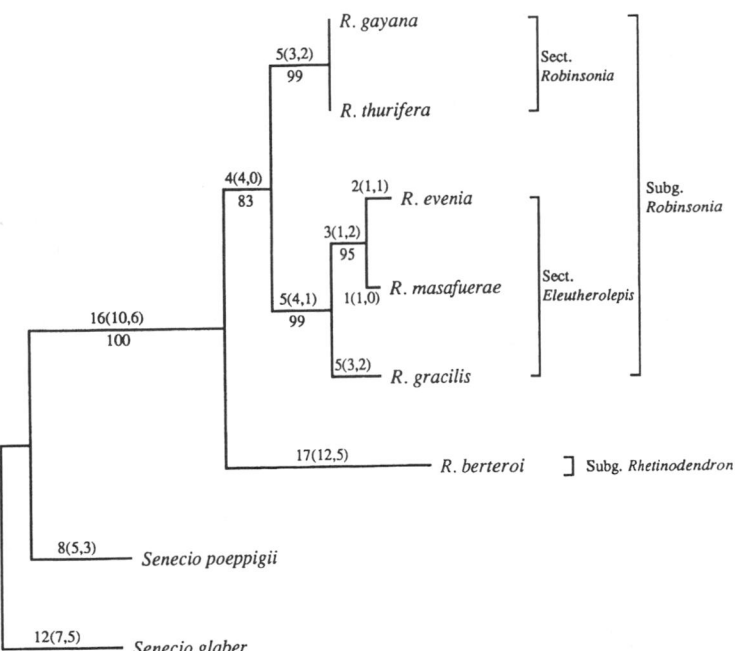

Figure 14.11. Phylogenetic hypothesis for *Robinsonia* (Asteraceae-Senecioneae); the single most parsimonious tree generated from ITS sequences. Numbers above branches are nucleotide substitutions, with numbers of transitions and transversions in parentheses. Note that each subgenus and section of *Robinsonia* is monophyletic. Reproduced from Sang et al. (1995) with permission from *Systematic Botany*.

tissue up to 120 years old by some workers offers hope for new understanding of relationships of long-vanished island plants. To date, we are aware of only one example of genetic analysis of extinct island plants using herbarium material, in Hawaiian *Schiedea* (P. Soltis et al., in prep.), but additional results will no doubt be forthcoming.

DIVERSIFICATION WITHIN AND ACROSS ISLANDS

Taxa endemic to different islands of an archipelago may show more pronounced ecological or even morphological similarity to one another than to coendemics of the same island. Phylogenetic data may be necessary to discriminate between the alternative hypotheses of (a) independent origin of similar characteristics on different islands through parallel intraisland radiations and (b) retention of homologous characteristics (e.g., particular habitat requirements) following dispersal between islands with subsequent, modest divergence in other characters. The extent to which interisland dispersal or intraisland radiation account for observed diversity in oceanic island plants has been assessed from phylogenetic patterns in large lineages from the Hawaiian, Macaronesian, and Juan Fernandez archipelagos, as discussed in the following paragraphs.

Reconstructions of founder events and radiation of the Hawaiian silversword alliance (*Argyroxiphium, Dubautia, Wilkesia*) demonstrate that diversification in the group proceeded primarily by intraisland radiations marked by major ecological transformations within lineages. Baldwin and Robichaux (1995) used parsimony to map interisland dispersal and habitat shifts on the minimum-length ITS trees of the silversword alliance, a group in which all but five (of 28) species are single-island endemics. The reconstructions show that as few as five interisland founder events led to all diversification, which was associated with shifts between wet and dry habitats on most or all of the four islands or island groups (Kauai, Oahu, Maui Nui, and Hawaii) (Fig. 14.12). The authors pointed out that using only the conservative wet-dry categories underestimates the number of eco-

logical shifts that have occurred during evolution on each island. For example, taxa restricted to bog habitats, which pose unique and extreme challenges to plant growth, belong to at least three of the four major ITS lineages in the silversword alliance and occur on all four islands or island groups. Multiple shifts to bog habitats have certainly occurred in the silversword alliance; resolution of the precise number and geographic site of shifts awaits additional sampling of bog-dwelling taxa for molecular phylogenetic analysis.

Reconstructions of historical biogeography and ecology in the Hawaiian endemic lineage comprising *Alsinidendron* and *Schiedea* (Caryophyllaceae) contrast sharply with the patterns of extensive within-island radiation and minimal interisland dispersal evident in the silversword alliance. Phylogenetic evidence from cpDNA restriction sites, ITS sequences, and morphology accords with only one or two shifts between wet and dry habitats during diversification of *Alsinidendron* and *Schiedea* (Sakai et al., 1997; Soltis et al., 1997). Based on the sampling of taxa included in the analyses, interisland dispersal between similar habitats and subsequent speciation appears to account for most interspecific variation in habitat occupancy among endemics of a single island.

A recent phylogenetic study of cpDNA restriction sites in *Argyranthemum* indicates that interisland dispersal between similar habitats has been much more common than single-island radiation into different habitats (Francisco-Ortega et al., 1996a). The two major, basally divergent cpDNA clades in the genus include a large lineage comprising plants from the Canary archipelago and a smaller sister lineage with species from the Madeira Islands (Madeira and Deserta Grande) and the Selvagen Islands (Selvagem Pequeña) (Fig. 14.8). This pattern suggests that dispersal between archipelagos preceded diversification early in the evolution of *Argyranthemum*. The Canary Islands, with seven ecological zones, are much more diverse ecologically than Madeira, Deserta Grande, and Selvagem Pequeña; species of *Argyranthemum* occur in at least three zones on each island of the Canaries. Membership of the two major cpDNA clades of species in the Canaries does

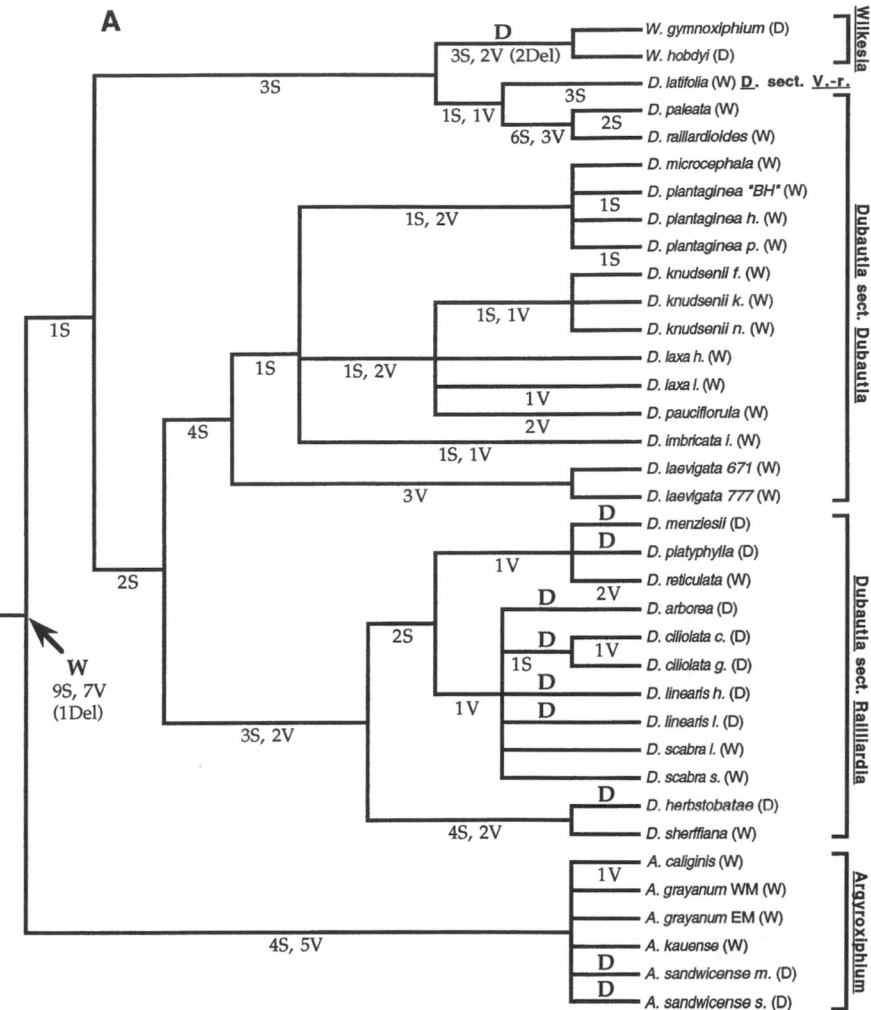

Figure 14.12. Phylogenetic hypothesis of the Hawaiian silversword alliance based on ITS sequences with habitat shifts mapped. The tree shown is one of the eight maximally parsimonious trees from analysis of ITS sequence variation. Deletions not included in the analyses are indicated in parentheses. Number of transitions (S) and transversions (V) are given below the branches. Letters in parentheses following species names and above branches indicate habitat preferences: W = wet habitat, D = dry habitat, and EQ = wet or dry habitat. *A*. Reconstruction of ecological shifts using MacClade 3.0 (Maddison and Maddison, 1992) assuming that the immediate ancestor of the silversword alliance occurred in a wet habitat. *B*. Reconstruction of ecological shifts assuming that the immediate ancestor of the silversword alliance occupied a dry habitat. If the wet habitat is considered ancestral, then at least five shifts have occurred to the dry habitat. Assuming a dry habitat is ancestral requires six or more shifts to wet habitats. Reproduced from Baldwin and Robichaux (1995) with permission of the authors.

not correlate well with taxonomy (species or sectional assignment) or geography (insular occurrence), but rather with the ecological zones in which the plants occur. Most taxa in one clade are from the drier habitats, where the trade winds have no effect; the other clade contains predominantly taxa from more mesic zones influenced by the trade winds (Fig. 14.8). Taxa adapted to the different ecological zones of the Canaries apparently differentiated early in the evolution of *Argyranthemum*. Subsequent interisland dispersal appears to account for the present distribution of taxa in different ecological settings.

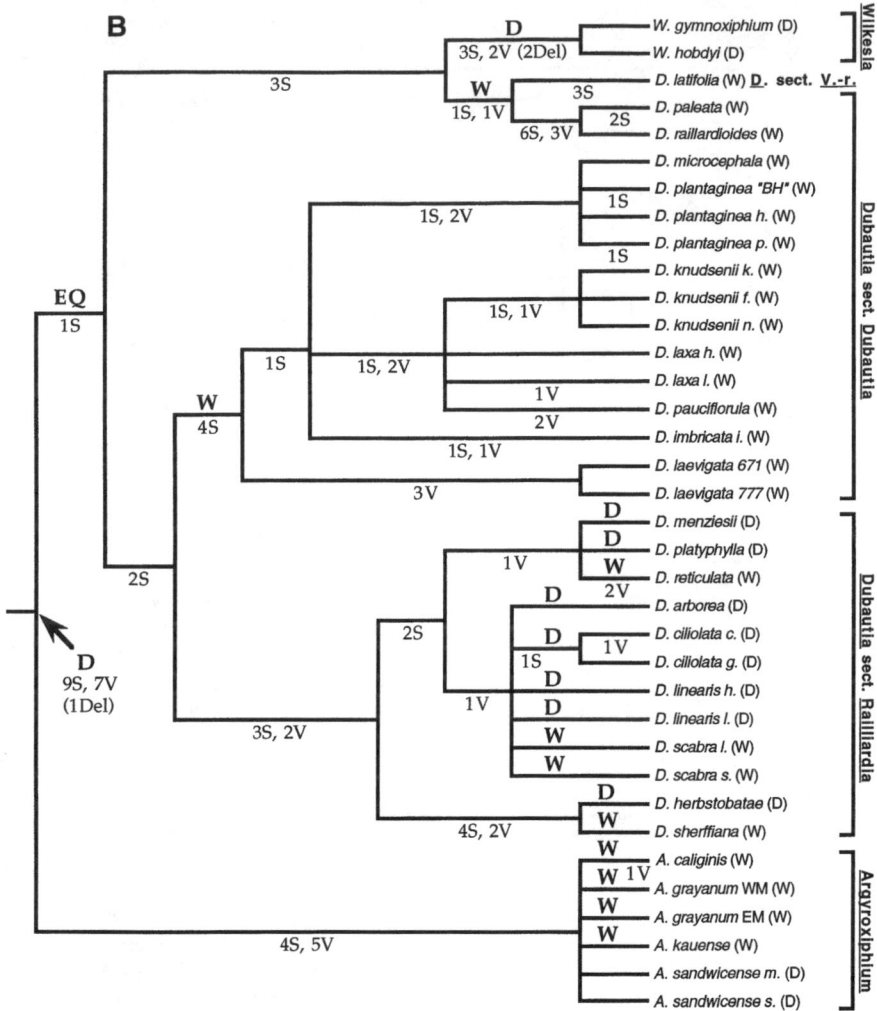

Figure 14.12. Continued

In the small Madeiran clade of *Argyranthemum,* the cpDNA phylogenetic pattern does not provide unequivocal evidence of multiple colonizations of islands, in contrast with the common pattern seen in the Canarian cpDNA clade (Francisco-Ortega et al., 1996a) (Fig. 14.8). Restriction of at least one taxon of *Argyranthemum* to each of the four ecological zones on Madeira suggests that multiple ecological shifts have occurred during speciation on the island. Furthermore, the basal position of taxa from the coastal desert in the Madeiran clade shows that radiation began from a common ancestor of coastal desert, with subsequent occupation of the three other zones. By contrast, extensive radiation is not expected or seen on the other two islands (Deserta Grande and Selvagem Pequeña) occupied by members of the Madeiran clade; both islands are very small, with only one or two ecological zones represented on each.

S.-C. Kim et al. (1996b) used an ITS tree of the woody *Sonchus* alliance in Macaronesia to infer the relative importance of single-island radiations and interisland colonizations in diversification of the group. The results closely paralleled those from *Argyranthemum,* also endemic to Macaronesia. As in *Argyranthemum* from the same islands, the taxa restricted to Madeira constitute one clade; dispersal to the island was apparently followed by diversification there (Fig. 14.6). The strongest case for diversification on one island of the Canary archipelago is in the clade containing *Taeckholmia, Sonchus ortunoi,* and *S. gonzalezpadronii* (Fig. 14.6), but interisland dispersal has apparently been important in that lineage, as well. Seven species of the *Sonchus* alliance occur on more than one island; dispersal has occurred frequently without speciation. Overall, interisland dispersal appears to have been more important to speciation in Macaronesian *Sonchus* than radiation on individual islands.

A third example illustrating the importance of dispersal to biogeographic patterns of ecological diversity in Macaronesian plant groups is provided by Mes and 'T Hart's (1996) phylogeny of *Aeonium* (Crassulaceae) based on cpDNA restriction sites and morphology. The authors concluded that only one unique growth form is represented by each of the clades, despite occurrence of different growth forms on the same islands. The phylogenetic evidence appears to indicate that Macaronesian *Aeonium* underwent an initial diversification into different growth forms and habitats, followed by extensive interisland dispersal and subsequent speciation without major ecological shifts (species sharing the same growth form are apparently restricted to similar habitats). There is no evidence of adaptive radiation occurring repeatedly on different islands in *Aeonium*. The same pattern suggesting repeated interisland dispersal without ecological shifts is also found in *Crambe* sect. *Dendrocrambe* in the Canaries (Francisco-Ortega et al., 1996b).

The Juan Fernandez archipelago represents a biogeographically simple situation compared to the Hawaiian Islands or Macaronesia; there are only two islands and no known seamounts (Sanders et al., 1987). The endemic *Dendroseris* and *Robinsonia* are the two largest genera of plants on the islands and the only lineages that have radiated to any extent there. In both genera, single-island radiation on the older island of Masatierra has been the major pattern of speciation, with eight of the 11 species of *Dendroseris* and six of the seven species of *Robinsonia* present on Masatierra. In *Dendroseris,* one species

from each subgenus is endemic to the younger island of Masafuera.

The ITS strict-consensus tree of *Dendroseris* shows that one of the Masafuera endemics, *D. regia* (subg. *Phoenicoseris*), shared a most recent common ancestor with the two Masatierra species *D. berteroana* and *D. pinnata* (Fig. 14.10; Sang et al., 1994). This pattern suggests that origin of the *D. regia* lineage occurred prior to an instance of speciation on the older island. Most importantly, the phylogenetic topology may indicate that speciation occurred on Masatierra within the last million years or so, that is, since Masafuera was formed (Stuessy et al., 1984). An alternative hypothesis is that the *D. regia* lineage once occurred on Masatierra but became extinct there sometime after its dispersal to Masafuera. Further study of the relative timing of speciation on the two islands has been impeded by difficulty in obtaining DNA from representatives of the other two subgenera of *Dendroseris* on Masafuera. In *Robinsonia,* the sole species from Masafuera, *R. masafuerae,* is sister to a single species on Masatierra in the ITS tree (Fig. 14.11), indicating that dispersal and colonization probably occurred after the last speciation event in sect. *Eleutherolepis* on Masatierra (Sang et al., 1995b).

Based on available data, within-island radiation appears to be more prevalent in plants of the Hawaiian Islands and Juan Fernandez Islands than in the Canary archipelago, but more taxa must be examined before accepting the generalization. Carlquist (1966a, 1996b) has argued cogently that loss of dispersibility has occurred repeatedly in island endemics, and it would be interesting to test for an association between lack of dispersibility and single-island radiation from a phylogenetic perspective. Also, a positive correlation between the isolation of an island by distance from other land masses and the frequency of single-island radiation needs to be tested from phylogenetic data. The correlation seems to hold in at least two groups from Macaronesia: single-island radiation has apparently occurred in both *Argyranthemum* and *Sonchus* on Madeira; diversification of both genera in the less isolated Canaries appears to have primarily followed interisland dispersal (Francisco-Ortega et al., 1996a; S.-C. Kim et al., 1996). We predict that well-resolved trees from molecular data will prove increasingly valuable for assessing the biogeographical and ecological dynamics of diversification in island plants.

INCONGRUITIES BETWEEN MOLECULAR PHYLOGENIES: INSIGHTS INTO EVOLUTIONARY PROCESSES

The general topic of incongruities between phylogenetic data sets is considered in detail in Chapters 10 and 11; discussion is limited here to island plants. The desirability of combining congruent morphological and molecular data sets for obtaining the best estimate of organismal phylogeny was mentioned earlier in this chapter. Lack of congruence between trees based on data sets from different genomes may also offer important insights into biological and evolutionary processes in insular endemics. The value of comparing multiple molecular data sets for achieving an understanding of phylogeny cannot be overemphasized; results from a particular gene or genome may be merely misleading or, alternatively, critical to phylogenetic understanding, depending on the availability of other lines of evidence for comparison (e.g., Sang et al., 1995a; Wendel et al., 1995).

One process responsible for conflict between phylogenies is hybridization. Documentation of interfertility and natural hybrids between species of insular plant lineages has led some botanists to conclude that hybridization has been an important factor in the evolution of island plant groups (e.g., Gillett, 1972; Carr, 1995). Systematists can gain insights into the importance of hybridization in island plants by taking advantage of the different modes of inheritance of cpDNA and nuclear DNA in phylogenetic studies. Chloroplast DNA is usually maternally or uniparentally inherited in angiosperms and has been shown to be particularly susceptible to transfer between species through introgression in a wide diversity of plant groups (see Rieseberg and Soltis, 1991; Rieseberg and Wendel, 1993). Nuclear DNA markers are often more resistant to introgression than cpDNA, although the reverse has been documented in some plants (see Rieseberg and Wendel, 1993). Differential fixation of nuclear markers in the descendants of

plants affected by hybridization can result in patterns of variation that may mislead systematists in the absence of other lines of evidence (see Doyle, 1992; Wendel et al., 1995).

Conflicts among cpDNA, nuclear DNA, and morphological phylogenies can provide evidence of ancient hybridization, if other possible explanations for incongruent patterns (e.g., lineage sorting, homoplasy) can be ruled out (Rieseberg and Soltis, 1991; Doyle, 1992). The best phylogenetic demonstration of a long-term impact of hybridization in plant evolution is evidence that diversification occurred from hybrids or introgressants, as shown in subtribe Helianthinae (Schilling and Panero, 1996).

Conflict of cytogenetic and ITS phylogenetic evidence with cpDNA trees from the Hawaiian silversword alliance provides strong evidence for hybridization-mediated cpDNA acquisition or "capture" across different nuclear lineages of *Dubautia* on Kauai (Baldwin, 1997). Detailed cytogenetic work by Carr and Kyhos (1981, 1986; Carr and Kyhos, pers. comm.) revealed that different species or species groups in the silversword alliance have become genomically differentiated by one or more reciprocal translocation(s) between nonhomologous chromosomes. Chromosome interchange relationships, that is, the number of interchange steps that structurally differentiate pairs of genomes, are understood for the five genomic arrangements known in *Dubautia,* four of which are found among the species on Kauai. Phylogenetic relationships based on ITS sequences among the taxa in *Dubautia* that bear structural genomes DG1, DG3, and DG5 (as abbreviated by Carr and Kyhos, 1986) are perfectly congruent with the cytogenetic data (Fig. 14.13; Baldwin, 1997); single origins of each arrangement are supported by the ITS tree, the topology of which provides evidence of the directionality of chromosome evolution through single evolutionary steps.

Chloroplast DNA relationships among representatives of species in *Dubautia* with genomic arrangements D1 and D3 are in strong conflict with relationships based on cytogenetic and ITS data (Fig. 14.13; Baldwin, 1997): (1) High bootstrap support exists for contrasting phylogenetic placement of the DG1 and DG3 species in sepa-

nrDNA ITS   cpDNA

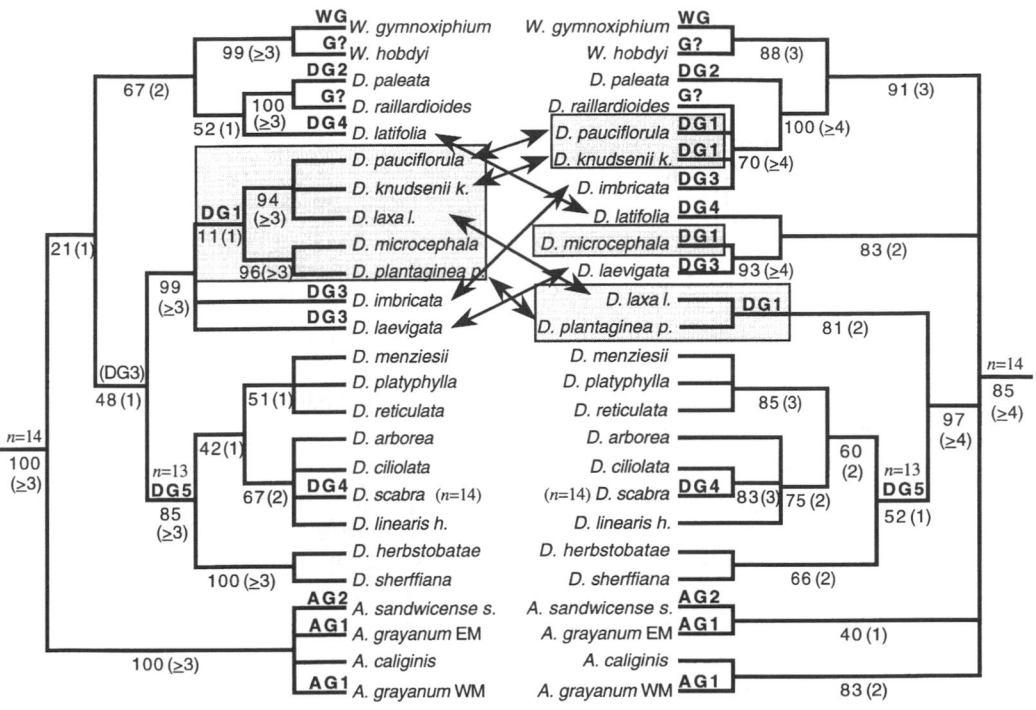

Figure 14.13. Comparison of phylogenetic hypotheses for the Hawaiian silversword alliance obtained from parsimony analyses of ITS sequences (left, Baldwin and Robichaux, 1995) and cpDNA restriction-site variation (right; Baldwin et al., 1990). The two analyses involved the same set of DNA samples. Both trees are semistrict consensuses of the maximally parsimonious trees. Bootstrap values appear below phylogeny branches. Decay index values appear in parentheses below phylogeny branches. Generic abbreviations: *A.* = *Argyroxiphium*, *D.* = *Dubautia*, *W.* = *Wilkesia*. EM = East Maui; WM = West Maui. Nuclear genomic arrangement abbreviations and haploid chromosome numbers (from Carr and Kyhos 1986; Carr and Kyhos, pers. comm.) appear above branches where the occurrence of that arrangement is unequivocal, based on MacClade 3.01 (Maddison and Maddison, 1992) reconstructions (additional data may resolve an earlier occurrence of a transformation). G? = unknown genomic arrangement. Placement of the parenthetical D3 in the ITS tree is based on a MacClade 3.01 reconstruction of chromosome evolution that involved undirected ordering of *Dubautia* and *Wilkesia* genomic arrangements (*Argyroxiphium* not included in analysis). Shaded blocks highlight lineages that unequivocally possess the *Dubautia* Genome 1 arrangement (Carr and Kyhos 1986; Carr and Kyhos, pers. comm.). Arrows point to species that are not in directly opposing positions in the figure; conflicting placement of taxa in the ITS and cpDNA trees is not necessarily implied by arrows. Reproduced from Baldwin (1997).

rate analyses of cpDNA and ITS characters; (2) congruence of signal in the cpDNA and ITS data sets can be rejected (p = 0.01) in a partition homogeneity test (beta version of PAUP* 4.0, provided by D. L. Swofford) if combined cpDNA and ITS data from more than two of the seven DG1/D3 species (i.e., DG1/DG3 species in addition to *D. plantaginea/D. laxa, D. laevigata/D. microcephala,* or *D. knudsenii/D. pauciflorula*) are analyzed with data from the other species in the silversword alliance; and (3) en-

forcement of the cpDNA tree topology as representative of nuclear DNA relationships demands multiple origins of DG1, DG3, or both genomic arrangements. Hybridization across lineages, rather than cpDNA lineage sorting, is apparently responsible for the different cpDNA and nuclear DNA histories; three subgroups of DG1/DG3 *Dubautia* species are deeply separated on the cpDNA tree, where each is placed with sympatric, interfertile species (Fig. 14.13; Baldwin, 1997). A lineage-sorting explanation (see Doyle,

1992) for the cpDNA patterns requires postulating persistence of multiple cpDNA types (each carried by different individuals) through repeated speciation events, a difficult scenario to envision in island plant populations prone to genetic bottlenecks.

In the endemic Macaronesian genus *Argyranthemum,* the cpDNA phylogeny (Fig. 14.8; Francisco-Ortega et al., 1996a) is incongruent with taxonomic (sectional, species) assignments. Given the documentation of hybrid zones between species of *Argyranthemum* (Borger, 1976; Brochmann, 1984, 1987), it is not unexpected that the maternal phylogeny is different from relationships implicit in taxonomic circumscriptions, but the degree of conflict is surprising; hybridization appears to have been much more pervasive in *Argyranthemum* than previously suspected.

CHARACTER EVOLUTION DURING RADIATION IN ARCHIPELAGOS

Reconstructions of character-state transformations from phylogenies can potentially resolve the directionality, sequence, and number of evolutionary changes in morphological, physiological, and ecological characters (see Maddison and Maddison, 1992), evidence that can be invaluable to studies of adaptation. Simple tabulation of life-forms or habitat occurrences outside a phylogenetic context may underestimate evolutionary activity in an island group (e.g., repeated origin of similar morphological characteristics and independent radiations into similar habitats). The following examples illustrate how molecular phylogenies have aided understanding of morphological evolution and ecological shifts in island lineages.

Baldwin and Robichaux (1995) and Baldwin (1998) found that life-form evolution and shifts among wet, dry, and bog habitats in the Hawaiian silverswood alliance (Asteraceae) were more extensive than a simple listing of morphological and ecological types in the group reveals. In *Dubautia* sect. *Raillardia,* ecological shifts between wet and dry habitats occurred in association with diversification on Oahu, Maui Nui, and, perhaps, Hawaii despite high interspecific genetic identities in allozymes (Witter and Carr,

1988), cpDNAs (Baldwin et al., 1990), and ITS sequences (Baldwin and Robichaux, 1995) and genomic structural uniformity among species of the section (Carr and Kyhos, 1986). The unexpected sister group relationship between *Wilkesia* and a lineage of *Dubautia* indicated by cpDNA and ITS studies suggests that shifts among wet, dry, and bog habitats and between rosette-plant and shrub life-forms occurred on Kauai within one of two major lineages of the silversword alliance found on the island (Baldwin and Robichaux, 1995).

Givnish et al. (1995) mapped morphological features on the cpDNA tree of *Cyanea* (including *Rollandia*) and found unanticipated lability and stability in the evolution of certain characteristics. For example, the evolution of fruit color has been surprisingly conservative in the genus and correlates with the two largest cpDNA clades; the purple- and orange-fruited conditions apparently originated only once in the islands during the early diversification of *Cyanea* (Fig. 14.9). The precise evolutionary significance of this finding is unclear, but, as Givnish et al. point out, the pattern may merely reflect an early chance divergence in fruit color. Other features such as thornlike prickles and heterophylly have apparently evolved several times; the adaptive significance of these features is a matter of debate (Givnish et al., 1994; Carlquist, 1995).

Woodiness: An Indicator of Relictualism or Insular Evolution?

A controversial question in island biology is whether endemic island plants are relatives of similar or identical species that once occurred in continental settings or are highly modified products of insular evolution. The woody habit of many Atlantic island endemics has been widely regarded as an ancestral (or primitive) condition that identifies the plants as survivors of ancient lineages that became extinct in continental areas (Bramwell, 1972, 1976; see Cronk, 1992). Woodiness in Pacific island endemics (e.g., in the Hawaiian archipelago) has been more commonly viewed as derived in the island setting via rapid evolution from an ancestral herbaceous condition (Carlquist, 1965, 1974, 1995). Carl-quist (1995) suggested that the above-stated "relictual" hypothesis for Atlantic island endemics would give way to an interpretation of more recent, autochthonous evolution of insular woodiness, once DNA data became available and cladistic analyses were applied. Molecular phylogenetic data are now in hand for several endemic groups in Atlantic and Pacific islands, allowing preliminary evaluation of Carlquist's recent suggestion.

The distinction between stem-based groups and crown groups (see Doyle and Donoghue, 1993) is useful for discussing relictualism. The same lineage of island species might be interpreted as (1) paleoendemic, if divergence of the island group and the extant continental sister group from a common ancestor (i.e., into two stem-based lineages—groups that include all extant and extinct taxa that arose following divergence from a common ancestor) was ancient, or (2) neoendemic, if diversification of the extant island group (i.e., the crown group—the descendents of the most recent common ancestor of extant species) occurred in the island setting. Species-rich Atlantic-island plant groups have been widely interpreted as consistent with both conditions (see Bramwell, 1972, 1976; Aldridge, 1979; Cronk, 1992). Common use of the term *relictual* for such Atlantic island plant groups assumes that similar, extinct continental species belonged to the island stem-based group and that the island group has retained plesiomorphic (ancestral) features that predate dispersal of the original founder(s) from continent to island.

Determination that an insular woody plant group is monophyletic and nested within an extant group of continental herbaceous taxa could be interpreted as consistent with the relictual hypothesis under the assumption that changing climatic conditions selectively purged continental lineages of woody members, leaving only the herbs extant (e.g., Bramwell, 1976; Aldridge, 1979). In the following discussion, we must adhere to the falsifiable null hypothesis that diversity in the plant groups in question is known; conclusions may need to be modified as new fossil evidence or time-calibrated phylogenies become available. A lack of data on actual timing of divergence of insular lineages from their extant sister groups necessitates restriction of most of the discussion to evidence of character evolution.

In the Hawaiian Islands, the silversword alliance offers an example of origin of woodiness in an insular setting. Parsimony-based reconstructions of ancestral habit in cpDNA and ITS trees indicate that the wide array of woody lifeforms in the Hawaiian silversword alliance was derived from an ancestral herbaceous condition found throughout continental *Madia/Raillardiopsis* (Baldwin, 1997), the group within which the Hawaiian lineage is nested. Givnish et al. (1996) briefly reported that the highly diverse Hawaiian lobelioids are part of a larger clade of woody lobelioids from the Bonin Islands, Society Islands, and east Africa. The woody lobelioids may represent an unusual example of a group that developed woodiness in an insular situation and then managed to disperse widely to islands or islandlike settings elsewhere, including the Hawaiian archipelago.

In Macaronesia, two of the most conspicuous radiations of angiosperms appear to represent good examples of acquired woodiness on islands. The genus *Echium* (Boraginaceae) includes some of the most spectacular woody growth forms in Macaronesia, and, like other plants from the archipelago, the woody island species have been viewed as relicts compared to the continental annuals and herbaceous perennials (Bramwell, 1972, 1976; Cronk, 1992). Boehle et al.'s (1996) phylogenetic data from sequences of the ITS region and three noncoding regions of cpDNA in island and continental *Echium* indicate that the insular plants are of recent origin relative to the continental annual group and were derived from an herbaceous island colonist. The phylogenetic position of woody lineages of the Macaronesian *Sonchus* alliance in the ITS tree of Sonchinae (Figs. 14.3, 14.6; S.-C. Kim et al., 1996a) also suggests that woodiness was derived in situ from an ancestral herbaceous condition.

The origin of the woody condition seen in *Argyranthemum* is more difficult to interpret than in the foregoing examples. A cpDNA restriction-site tree (Fig. 14.4) suggests that *Argyranthemum* is sister to a group of three continental, annual genera (*Chrysanthemum, Heteranthemis,* and *Ismelia*) in a well-supported clade corresponding to Chrysantheminae (Fig. 14.4). Although existing data are consistent with the origin of woodiness in *Argyranthemum* from an herbaceous insular ancestor, additional resolution of the relationships of Chrysantheminae to other Anthemideae (L. E. Watson, in prep.) is needed to allow a more reliable analysis of life-form evolution in the group.

An unequivocal example of an oceanic island "relict," from the criterion of stem-based lineage age, is *Lactoris fernandeziana,* the only species of Lactoridaceae. Anatomical and developmental data (Donoghue and Doyle, 1989; Carlquist, 1990; Tucker and Douglas, 1996) as well as an *rbcL* phylogeny (Qiu et al., 1993) place the species deep in the paleoherbs, a group of early origin in the evolution of angiosperms. *Lactoris* is the only known representative of a stem lineage that must be much older than Masatierra (ca. 4 Ma), the oldest island of the Juan Fernandez archipelago (Stuessy et al., 1984). The species or its ancestor must have dispersed to the island long after the origin of Lactoridaceae. In turn, the family must not have become extinct outside of the Juan Fernandez Islands before approximately 4 Ma.

In summary, molecular phylogenetic data provide examples of acquired woodiness on both Pacific and Atlantic islands, reinforcing Carlquist's (1965, 1974, 1995) hypothesis that some of the most striking woody growth forms on oceanic archipelagos are attributable to in situ evolution rather than persistence of relict taxa. The bulk of molecular phylogenetic data on oceanic island plants from Macaronesia and elsewhere (see refs. in Table 14.1) appears unfavorable to the hypothesis that the endemic element of Atlantic island floras includes a large proportion of relicts (Bramwell, 1972, 1976; Cronk, 1992). Additional molecular phylogenetic data on timing of diversifications and the divergence times of insular and continental sister lineages should allow more precise resolution of these issues.

TIMING OF ISLAND RADIATIONS

Molecular phylogenetic data can provide refined estimates of the actual timing of island plant radiations to allow fine-scale insights into historical biogeography and evolutionary rates. Understanding of the phylogeny alone may allow unequivocal parsimony-based reconstruction of the historical pattern of dispersal between islands, for example, if a lineage endemic to one island is nested phylogenetically within a group of species endemic to a different island (Fig. 14.11; see Funk and Wagner, 1995).

Knowledge of island age(s), from potassium-argon dating, allows placement of an outer limit on the timing of an endemic radiation or the age of a lineage that descended from an ancestor that can be unequivocally assigned to an island or archipelago. For example, the most recent common ancestor of endemic lineages of the Juan Fernandez Islands, such as *Dendroseris* and *Robinsonia,* can be assumed to be no older than approximately 4 Ma, the age of the older of the two major islands, Masatierra (Stuessy et al., 1984). Estimates of maximal age for older, hotspot archipelagos, for example, the Hawaiian Islands, are less useful for placing outer limits on the antiquity of insular lineages. Recent reconstructions of island history in the Hawaiian archipelago, for example, leave open the possibility of continual exposure of terrestrial habitats in the chain throughout the last 29 million years (Carson and Clague, 1995), a time frame that greatly exceeds the conceivable age of endemic Compositae (see Graham, 1996), such as the silversword alliance and Hawaiian *Bidens.*

Information about the relationships of island lineages to continental plant groups may offer insights into the timing of insular diversification. Phylogenetic resolution of the relationships of the Hawaiian silversword alliance to continental members of Madiinae has allowed placement of a biogeographically meaningful outer limit on the age of the most recent common ancestor of the modern Hawaiian species. As mentioned above, evidence from cpDNA and ITS sequences has shown that the Hawaiian silversword alliance is deeply nested phylogenetically within the continental tarweed lineage (Baldwin et al., 1991; Baldwin, 1992, 1996; Baldwin and Robichaux, 1995), a group that is nearly restricted to the Mediterranean climatic region of western North America.

Considerable paleoclimatic and fossil evidence for an onset of summer drying conditions in western North America at mid-Miocene (Axelrod, 1992; Flower and Kennett, 1994) dictates that a date no older than 15 Ma can be placed on the most recent common ancestor of the silversword alliance and the tarweed sister lineage in *Madia/Raillardiopsis;* acceptance of a date older than 15 Ma would assume that adaptations to a summer dry climate were acquired independently by multiple continental tarweed lineages. Based on a 13 to 15 Ma calibration point, maximum age of the most recent common ancestor of the silversword alliance was calculated at 4.5 to 6.0 Ma from a well-resolved, rate-constant ITS tree of the silversword alliance/*Madia/Raillardiopsis* lineage (B. G. Baldwin and M. J. Sanderson, in prep.; see Baldwin, 1997). The dates obtained suggest that the silversword alliance diversification occurred within the history of modern high islands of the Hawaiian archipelago, that is, probably not before formation of Kauai (5.1 Ma).

Absolute dating of clades within an insular phylogeny can be subject to considerable error (see Hillis et al., 1996), even with unequivocal reconstruction of an older-to-younger island dispersal event to a particular branch (e.g., in a "progressive grades" topology; Funk and Wagner, 1995) in a well-resolved and supported, rate-constant gene tree. Under such seemingly ideal circumstances, uncertainty about where along the phylogeny branch to place the young island age (e.g., at the point of divergence of the young island lineage from the sister group or at the most recent common ancestor of the young island species) can lead to significant inaccuracy; minor error in choice of a calibration point will be compounded at deeper nodes in the tree. Calibration of phylogeny branches that are deeper than those to be estimated is preferable for maximizing accuracy of lineage age estimates. This approach was followed in seeking the conservative maximum age estimate of the silversword alliance discussed previously.

RATES OF DIVERSIFICATION

A concern related to lineage age is the rate of diversification or lineage-splitting in island plants. Of particular interest to systematists is the question of whether accelerated diversification has occurred in groups commonly regarded as ex-

amples of insular adaptive radiation. Precise phylogenetic data on relationships of island groups to nearest continental relatives is needed for assessing any possible correlation between insular setting and diversification rate. A time-calibrated, rate-constant tree including all lineages within the island ingroup, the sister group, and additional close outgroups would allow for the most sensitive and powerful test of accelerated diversification in an insular clade.

At present, the only phylogenetic data on an island plant group and close continental relatives that provide a rate-constant, time-calibrated tree is the ITS phylogeny of the Hawaiian silversword alliance and continental perennial members of *Madia/Raillardiopsis* (B. G. Baldwin and M. J. Sanderson, in prep.; see Baldwin, 1997). Unfortunately, inclusion of additional, annual continental species from the *Madia/Raillardiopsis* sister group of the silversword alliance in the analysis leads to rejection of rate-constancy of ITS evolution (B. G. Baldwin and M. J. Sanderson, in prep.) using either of two global rate tests based on maximum likelihood (Langley and Fitch, 1974; Felsenstein, 1993). Further investigation is necessary to discern whether a biologically explicable rate difference exists between annual and perennial species in the group. If so, this might allow for use of the time estimates of lineage branching in a test for differences in rates of diversification between the insular and noninsular lineages.

In the absence of information about absolute times of lineage divergences, tests for differences in diversification rates require large disparities in sister lineage sizes to allow for significant results, and directionality of change is difficult to establish (Sanderson and Donoghue, 1996). Comparison of ITS lineage size between the Hawaiian silversword alliance and closest relatives in *Madia/Raillardiopsis* reveals a statistically insignificant difference in numbers of terminal taxa between the two groups (less than 40:1, the ratio of sister group diversities, r and $s,$ necessary for $p = 0.05$ where $p = 2r/(r + s - 1)$; see Slowinski and Guyer, 1989), although the Hawaiian assemblage comprises nearly twice as many species as the continental ITS sister clade. A potentially higher natural rate of extinction in the silversword alliance compared to that of the western American sister group (as might be expected based on the more extreme geographic constriction of populations and greater geologic instability in the insular situation) could mask a significantly higher rate of diversification in the Hawaiian lineage compared to that in continental *Madia/Raillardiopsis*. As phylogenetic data on relationships of insular and noninsular species become available from a wider diversity of plant groups, comparisons of additional replicate insular/continental sister lineages may ultimately allow for a more sensitive test for accelerated diversification in island plants.

CONCLUSIONS

Molecular phylogenetic methods have proved valuable for discerning ancestry and resolving lineages of oceanic island plants. In turn, estimation of phylogenetic relationships among members of island groups and close relatives has allowed insight into the ancestral characteristics of founder species, morphological and ecological change through insular lineages, dispersal to and within archipelagos, and even timing and rates of diversification.

Most molecular evidence supports the conclusion from studies of hybrids and allozymes that oceanic island lineages are younger than may be evident from morphological and ecological divergence between species. Most oceanic island plant lineages that have been successfully placed in a broader phylogenetic framework are deeply nested within larger groups of continental taxa that are relatively conservative morphologically. Molecular phylogenetic data also suggest that woody and semiwoody life-forms of some of the most diverse groups of Atlantic and Pacific island plants arose from an ancestral herbaceous condition. In sum, the weight of molecular evidence is contrary to a relictual hypothesis for most examined oceanic island plant groups, including those from Atlantic archipelagos.

Comparisons of molecular phylogenetic, biogeographic, and ecological patterns across island groups has revealed a striking contrast between (1) lineages marked primarily by within-island diversification in ecological characteristics (i.e., adaptive radiation in the classic sense) and (2) lineages wherein diversification has been commonly associated with dispersal between similar habitats on different islands. Apparent examples of within-island ecological radiation are better represented in the Hawaiian and Juan Fernandez islands than in the Atlantic Islands, where extensive dispersal followed by speciation appears to have occurred in multiple plant lineages of the Canary archipelago.

Promising directions for future molecular phylogenetic work on oceanic island plants include examining rates and timing of diversifications, as better algorithms are developed for dating lineages and discriminating shifts in the tempo of lineage splitting. On a more fundamental level, identification of more rapidly evolving gene regions useful for phylogenetic studies is needed before robust resolution of relationships among insular species can be readily achieved. Progress in pushing back the frontier at the lower limits of phylogenetic resolution (see Chapters 1 and 2) will go far toward achieving a better understanding of plant diversification on islands.

LITERATURE CITED

Aldridge, A. 1979. Evolution within a single genus: *Sonchus* in Macaronesia. *In* Plants and Islands, ed. D. Bramwell, pp. 279–291. Academic Press, London.

Axelrod, D. A. 1992. Miocene floristic change at 15 Ma, Nevada to Washington, U.S.A. Palaeobotanist **41:** 234–239.

Badr, A., W. Martin, and U. Jensen. 1994. Chloroplast DNA restriction site polymorphism in Genisteae (Leguminosae) suggests a common origin for European and American lupines. Plant Systematics and Evolution **193:**95–106.

Baldwin, B. G. 1992. Phylogenetic utility of the internal transcribed spacers of nuclear ribosomal DNA in plants: an example from the Compositae. Molecular Phylogenetics and Evolution **1:**3–16.

Baldwin, B. G. 1996. Phylogenetics of the California tarweeds and the Hawaiian silversword alliance (Madiinae; Heliantheae *sensu lato*). **In** Compositae: Systematics, Proceedings of the International Compositae Conference, Kew, 1994 (D. J. N. Hind, Editor-in-Chief), Vol. 1, eds. D. J. N. Hind and H. Beentje, pp. 377–391. Royal Botanic Gardens, Kew, England.

Baldwin, B. G. 1997. Adaptive radiation of the Hawaiian silversword alliance: congruence and conflict of phylogenetic evidence from molecular and non-molecular investigations. **In** Molecular Evolution and Adaptive

Radiation, eds. T. J. Givnish and K. J. Sytsma, pp. 103-128. Cambridge University Press, Cambridge.

Baldwin, B. G. 1998. Evolution in the endemic Hawaiian Compositae. In Evolution and Speciation in Island Plants, eds. T. F. Stuessy and M. Ono, pp. 49–73. Cambridge University Press, Cambridge, in press.

Baldwin, B. G., and R. H. Robichaux. 1995. Historical biogeography and ecology of the Hawaiian silversword alliance (Asteraceae): new molecular phylogenetic perspectives. In Hawaiian Biogeography: Evolution on a Hotspot Archipelago, eds. W. L. Wagner and V. A. Funk, pp. 257–287. Smithsonian Institution Press, Washington, D.C.

Baldwin, B. G., D. W. Kyhos, and J. Dvorak. 1990. Chloroplast DNA evolution and adaptive radiation in the Hawaiian silversword alliance (Asteraceae-Madiinae). Annals of the Missouri Botanical Garden 77:96–109.

Baldwin, B. G., D. W. Kyhos, J. Dvorak, and G. D. Carr. 1991. Chloroplast DNA evidence for a North American origin of the Hawaiian silversword alliance (Asteraceae). Proceedings of the National Academy of Sciences U.S.A. 88:1840–1843.

Ballard, H. E., Jr., S. Carlquist, and K. J. Sytsma. 1996. ITS sequence data illuminate the biogeographic origin and adaptive radiation of Hawaiian Viola. American Journal of Botany (suppl.) 83:139–140.

Barrett, M., M. J. Donoghue, and E. Sober. 1991. Against consensus. Systematic Zoology 40:486–493.

Boehle, U.-R., H. H. Hilger, and W. F. Martin. 1996. Island colonization and evolution of the insular woody habit in Echium L. (Boraginaceae). Proceedings of the National Academy of Sciences U.S.A. 93:11740–11745.

Borger, L. 1976. Analysis of a hybrid swarm between Argyranthemum adauctum and A. fififolium in the Canary Islands. Norwegian Journal of Botany 23:121–137.

Bramwell, D. 1972. Endemism in the flora of the Canary Islands. In Taxonomy, Phytogeography and Evolution, ed. D. H. Valentine, pp. 141–159. Academic Press, London.

Bramwell, D. 1976. The endemic flora of the Canary Islands; distribution, relationships and phytogeography. In Biogeography and Ecology in the Canary Islands, ed. G. Kunkel, pp. 207–240. Dr. W. Junk, The Hague.

Bremer, K., and C. J. Humphries. 1993. Generic monograph of the Asteraceae-Anthemideae. Bulletin of the Natural History Museum of London (Botany) 23:71–177.

Brochmann, C. 1984. Hybridization and distribution of Argyranthemum coronopifolium (Asteraceae-Anthemideae) in the Canary Islands. Nordic Journal of Botany 4:729–736.

Brochmann, C. 1987. Evaluation of some methods for hybrid analysis, exemplified by hybridization in Argyranthemum (Asteraceae). Nordic Journal of Botany 7:609–630.

Carlquist, S. 1959. Studies on Madiinae: anatomy, cytology, and evolutionary relationships. Aliso 4:171–236.

Carlquist, S. 1965. Island Life. Natural History Press, New York.

Carlquist, S. 1966a. The biota of long distance dispersal. II. Loss of dispersibility in Pacific Compositae. Evolution 20:433–455.

Carlquist, S. 1966b. The biota of long distance dispersal. III. Loss of dispersibility in the Hawaiian flora. Brittonia 18:310–335.

Carlquist, S. 1967. Anatomy and systematics of Dendroseris (sensu lato). Brittonia 19:99–121.

Carlquist, S. 1974. Island Biology. Columbia University Press, New York.

Carlquist, S. 1980. Hawaii, A Natural History: Geology, Climate, Native Flora and Fauna Above the Shoreline, second edition. Pacific Tropical Botanical Garden, Lawai, Hawaii.

Carlquist, S. 1990. Wood anatomy and relationships of Lactoridaceae. American Journal of Botany 77:1498–1505.

Carlquist, S. 1995. Introduction. In Hawaiian Biogeography: Evolution on a Hotspot Archipelago, eds. W. L. Wagner and V. A. Funk, pp. 1–13. Smithsonian Institution Press, Washington, D.C.

Carr, G. D. 1985. Monograph of the Hawaiian Madiinae (Asteraceae): Argyroxiphium, Dubautia, and Wilkesia. Allertonia 4:1–123.

Carr, G. D. 1995. A fully fertile intergeneric hybrid derivative from Argyroxiphium sandwicense ssp. macrocephalum X Dubautia menziesii (Asteraceae) and its relevance to plant evolution in the Hawaiian Islands. American Journal of Botany 82:1574–1581.

Carr, G. D., and D. W. Kyhos. 1981. Adaptive radiation in the Hawaiian silversword alliance (Compositae-Madiinae). I. Cytogenetics of spontaneous hybrids. Evolution 35:543–556.

Carr, G. D., and D. W. Kyhos. 1986. Adaptive radiation in the Hawaiian silversword alliance (Compositae-Madiinae). II. Cytogenetics of artificial and natural hybrids. Evolution 40:959–976.

Carson, H. L., and D. A. Clague. 1995. Geology and biogeography of the Hawaiian Islands. In Hawaiian Biogeography: Evolution on a Hotspot Archipelago, eds. W. L. Wagner and V. A. Funk, pp. 14–29. Smithsonian Institution Press, Washington, D.C.

Carvalho, J. A., and A. Culham. Conservation status and phylogenetics of Isoplexis (Lindl.) Benth. (Scrophulariaceae): an endemic Macaronesian genus. Bulletin do Museo Municipal do Funchal, in press.

Chan, R. K. G. 1994. Molecular Phylogenetic Relationship of Hawaiian Tetramolopium Species (Compositae) using DNA Sequence Data from the Internal Transcribed Spacer (ITS) Regions. Ph.D. dissertation. University of New Mexico, Albuquerque.

Chan, R. K. G., T. K. Lowrey, D. Natvig, and R. Whitkus. 1994. Phylogenetic analysis of internal transcribed spacer (ITS) sequences from nuclear ribosomal DNA of Hawaiian, Cook Island and New Guinea Tetramolopium (Compositae; Astereae). American Journal of Botany (suppl.) 82:118–119.

Charmet, G., C. Ravel, and F. Balfourier. 1997. Phylogenetic analysis in the Festuca-Lolium complex using molecular markers and ITS rDNA. Theoretical and Applied Genetics 94:1038–1040.

Clague, D. A., and G. B. Dalrymple. 1987. The Hawaiian-Emperor volcanic chain, part I: geologic evolution. In

Volcanism in Hawaii, U.S. Geological Survey Professional Paper 1350, eds. R. W. Decker, T. L. Wright, and P. H. Stauffer, pp. 5–54. U.S. Government Printing Office, Washington, D.C.

Crawford, D. J., T. F. Stuessy, M. B. Cosner, D. W. Haines, and M. Silva O. 1993. Ribosomal and chloroplast DNA restriction site mutation and the radiation of *Robinsonia* (Asteraceae: Senecioneae) on the Juan Fernandez Islands. Plant Systematics and Evolution **184**:233–239.

Cronk, Q. C. B. 1992. Relict floras of Atlantic islands: patterns assessed. Biological Journal of the Linnean Society **46**:91–103.

Darwin, C. 1859. On the Origin of Species by Means of Natural Selection (reprint of first edition, 1950). Watts & Co., London.

Dejoode, D. R., and J. F. Wendel. 1992. Genetic diversity and origin of the Hawaiian Islands cotton, *Gossypium tomentosum.* American Journal of Botany **79**: 1311–1319.

de Queiroz, A., M. J. Donoghue, and J. Kim. 1995. Separate versus combined analysis of phylogenetic evidence. Annual Review of Ecology and Systematics **26**:657–681.

Donoghue, M. J., and J. A. Doyle. 1989. Phylogenetic studies of seed plants and angiosperms based on morphological characters. **In** The Hierarchy of Life: Molecules, Morphology, and Phylogenetic Analysis, eds. B. Fernholm, K. Bremer, and H. Jornvall, pp. 181–193, Elsevier Science, Amsterdam.

Donoghue, M. J., J. A. Doyle, J. Gauthier, A. G. Kluge, and T. Rowe. 1989. The importance of fossils in phylogeny reconstruction. Annual Review of Ecology and Systematics **20**:431–460.

Doyle, J. A., and M. J. Donoghue. 1993. Phylogenies and angiosperm diversification. Paleobiology **19**:141–167.

Doyle, J. J. 1992. Gene trees and species trees: molecular systematics as one-character taxonomy. Systematic Botany **17**:144–163.

Felsenstein, J. 1978. Cases in which parsimony and compatibility methods will be positively misleading. Systematic Zoology **27**:401–410.

Felsenstein, J. 1993. PHYLIP (Phylogeny Inference Package), version 3.5. University of Washington, Seattle.

Flower, B. P., and J. P. Kennett. 1994. The middle Miocene climatic transition: East Antarctic ice sheet development, deep ocean circulation and global carbon cycling. Palaeogeography, Palaeoclimatology, Palaeoecology **108**:537–555.

Fosberg, F. R. 1948. Derivation of the flora of the Hawaiian Islands. Insects of Hawaii **1**:107–119.

Francisco-Ortega, J., R. K. Jansen, D. J. Crawford, and A. Santos-Guerra. 1995. Chloroplast DNA evidence for intergeneric relationships of the Macaronesian endemic genus *Argyranthemum* (Asteraceae). Systematic Botany **20**:413–422.

Francisco-Ortega, J., R. K. Jansen, and A. Santos-Guerra. 1996a. Chloroplast DNA evidence of colonization, adaptive radiation, and hybridization in the evolution of the Macaronesian flora. Proceedings of the National Academy of Sciences U.S.A. **93**:4085–4090.

Francisco-Ortega, J., J. Fuertes-Aguilar, S.-C. Kim, D. J. Crawford, A. Santos-Guerra, and R. K. Jansen. 1996b. Molecular evidence for the origin, evolution and dispersal of *Crambe* (Brassicaceae) in the Macaronesian islands. **In** Abstracts. II. Symposium "Fauna and Flora of the Atlantic Islands," p. 41. Universidad de Las Palmas de Gran Canaria, Las Palmas de Gran Canaria, Canary Islands.

Francisco-Ortega, J., A. Santos-Guerra, A. Hines, and R. K. Jansen. 1997a. Molecular evidence for a Mediterranean origin of the Macaronesian endemic genus *Argyranthemum* (Asteraceae). American Journal of Botany **84**:1595–1613.

Francisco-Ortega, J., D. J. Crawford, A. Santos-Guerra, and R. K. Jansen. 1997b. Origin and evolution of *Argyranthemum* (Asteraceae: Anthemideae) in Macaronesia. **In** Molecular Evolution and Adaptive Radiation, eds. T. J. Givnish and K. J. Sytsma, pp. 407–431. Cambridge University Press, Cambridge.

Fuertes-Aguilar, J., M. F. Ray, J. Francisco-Ortega, and R. K. Jansen. 1996. Systematics and evolution of the Macaronesian endemic Malvaceae based on morphological and molecular evidence. **In** Abstracts. II. Symposium "Fauna and Flora of the Atlantic Islands," p. 51. Universidad de Las Palmas de Gran Canaria, Canary Islands.

Funk, V. A., and W. L. Wagner. 1995. Biogeographic patterns in the Hawaiian Islands. **In** Hawaiian Biogeography: Evolution on a Hot Spot Archipelago, eds. W. L. Wagner and V. A. Funk, pp. 379–419. Smithsonian Institution Press, Washington, D.C.

Garnatje, M. T., R. Messenger, and A. Susanna. 1996. New molecular data on the phylogeny of the genus *Cheirolophus* Cass. (Compositae: Cardueae). **In** IV Conference on Plant Taxonomy, eds. C. Blanché, J. Simon, and J. Vallés, p. 98. Editorial Gráficas Signo, Barcelona.

Gillett, G. W. 1972. The role of hybridization in the evolution of the Hawaiian flora. **In** Taxonomy, Phytogeography, and Evolution, ed. D. H. Valentine, pp. 205–219. Academic Press, London.

Gillett, G. W., and E. K. S. Lim. 1970. An experimental study of the genus *Bidens* in the Hawaiian Islands. University of California Publications in Botany **56**:1–6.

Givnish, T. J., K. J. Sytsma, J. F. Smith, and W. J. Hahn. 1994. Thorn-like prickles and heterophylly in *Cyanea:* adaptations to extinct avian browsers on Hawaii? Proceedings of the National Academy of Sciences U.S.A. **91**:2810–2814.

Givnish, T. J., K. J. Sytsma, J. F. Smith, and W. J. Hahn. 1995. Molecular evolution, adaptive radiation, and geographic speciation in *Cyanea* (Campanulaceae, Lobelioideae). **In** Hawaiian Biogeography: Evolution on a Hotspot Archipelago, eds. W. L. Wagner and V. A. Funk, pp. 288–337. Smithsonian Institution Press, Washington, D.C.

Givnish, T. J., E. Knox, T. B. Patterson, J. R. Hapeman, J. D. Palmer, and K. J. Sytsma. 1996. The Hawaiian lobelioids are monophyletic and underwent a rapid initial radiation roughly 15 million years ago. American Journal of Botany (suppl.) **83**:159.

Graham, A. 1996. A contribution to the geological history of the Compositae. In Compositae: Systematics, Proceedings of the International Compositae Conference, Kew, 1994, vol. 1, eds. D. J. N. Hind and H. J. Beentje, pp. 123–140. Royal Botanic Gardens, Kew, England.

Hillis, D. M., B. K. Mable, and C. Moritz. 1996. Applications of molecular systematics: the state of the field and a look to the future. In Molecular Systematics, second edition, eds. D. M. Hillis, C. Moritz, and B. K. Mable, pp. 515–543. Sinauer Associates, Sunderland, Massachusetts.

Huelsenbeck, J. P. 1991. When are fossils better than extant taxa in phylogenetic analysis? Systematic Zoology **40**:458–469.

Huelsenbeck, J. P., and D. M. Hillis. 1993. Success of phylogenetic methods in the four-taxon case. Systematic Biology **42**:247–264.

Humphries, C. J. 1976a. A revision of the Macaronesian genus Argyranthemum Webb ex Schultz Bip. (Compositae: Anthemideae). Bulletin of the British Museum (Natural History) Botany **5**:145–240.

Humphries, C. J. 1976b. Evolution and endemism in Argyranthemum Webb ex Schultz Bip. (Compositae: Anthemideae). Botanica Macaronesica **1**:25–50.

Humphries, C. J. 1979. Endemism and evolution in Macaronesia. In Plants and Islands, ed. D. Bramwell, pp. 171–199. Academic Press, London.

Hutchinson, J. 1917. Notes on African Compositae: IV. Matricaria Linn, and Chrysanthemum DC. Kew Bulletin, pp. 111–118.

Jansen, R. K., D. J. Loockerman, and H.-G. Kim. DNA sampling from herbarium material: a current perspective. In Managing the Modern Herbarium: Dealing with Issues for the 21st Century, eds. D. A. Metsger and S. C. Byers. Society for the Preservation of Natural History Collections, in press.

Jeffrey, C. 1966. Notes on the Compositae: I. The Cichorieae in East Tropical Africa. Kew Bulletin **18**:427–486.

Kadereit, J. W. and C. Jeffrey. 1996. A preliminary analysis of cpDNA variation in the tribe Senecioneae (Compositae) In Compositae: Systematics, Proceedings of the International Compositae Conference, Kew, 1994 (D. J. N. Hind, Editor-in-Chief), Vol. 1, eds. D. J. N. Hind and H. Beentje, pp. 349–360. Royal Botanic Gardens, Kew, England.

Kaess, E., and M. Wink. 1995. Molecular phylogeny of the Papilionoideae (family Leguminosae): rbcL gene sequences versus chemical taxonomy. Botanica Acta **108**:149–162.

Keck, D. D. 1936. The Hawaiian silverswords: systematics, affinities, and phytogeographic problems of the genus Argyroxiphium. Bernice P. Bishop Museum Occasional Papers **11**:1–38.

Kim. H.-G., S. C. Keeley, and R. K. Jansen. 1996. Phylogenetic position of the Hawaiian endemic Hesperomannia (Mutisieae) based on ndhF sequence data. American Journal of Botany (suppl.) **83**:169.

Kim, S.-C., D. J. Crawford, and R. K. Jansen. 1996a. Phylogenetic relationships among the genera of the subtribe

Sonchinae (Asteraceae): evidence from ITS sequences. Systematic Botany **21**:417–432.

Kim S.-C., D. J. Crawford, J. Francisco-Ortega, and A. Santos-Guerra. 1996b. A common origin for woody Sonchus and five related genera in the Macaronesian islands: molecular evidence for extensive radiation. Proceedings of the National Academy of Sciences U.S.A. **93**:7743–7748.

Kim, S.-C., D. J. Crawford, and R. K. Jansen, 1996c. Phylogeny of the subtribe Sonchinae (Asteraceae:Lactuceae): additional information from a noncoding region of cpDNA. American Journal of Botany (suppl.) **83**:168.

Kitamura, S. 1978. Dendranthema and Nipponanthemum. Acta Phytotaxonomica et Geobotanica, Kyoto **29**:165–170.

Knox, E. B., and J. D. Palmer. 1995. The origin of Dendrosenecio within the Senecioneae (Asteraceae) based on chloroplast DNA evidence. American Journal of Botany **82**:1567–1573.

Krupin, A. B., A. Liston, and S. V. Strauss. 1996. Phylogenetic analysis of the hard pines (Pinus subgenus Pinus, Pinaceae) from chloroplast DNA restriction site analysis. American Journal of Botany **83**:489–498.

Langley, C. H., and W. M. Fitch. 1974. An estimation of the constancy of the rate of molecular evolution. Journal of Molecular Evolution **3**:161–177.

Lowe, A. J. and R. J. Abbott. 1996. Origins of the new allopolyploid species Senecio cambrensis (Asteraceae) and its relationship to the Canarian Island endemic Senecio teneriffae. American Journal of Botany **83**:1365–1372.

Lowrey, T. K. 1986. A biosystematic revision of Hawaiian Tetramolopium (Compositae: Astereae). Allertonia **4**:203–265.

Maddison, W. P., and D. R. Maddison. 1992. MacClade, Analysis of Phylogeny and Character Evolution, version 3.01. Sinauer Associates, Sunderland, Massachusetts.

Mes, T. M. H. 1995. Phylogenetic and systematic implications of chloroplast and nuclear spacer sequence variation in the Macaronesian Sempervivoideae and related Sedoideae. In Evolution and Systematics of the Crassulaceae, eds. H. T' Hart and U. Eggli, pp. 30–44. Backhuys Publishers, Leiden.

Mes, T. H. M., and H. T' Hart. 1994. Sedum surculosum and S. jaccardianum (Crassulaceae) share a unique 70 bp deletion in the chloroplast DNA trnL (UAA)-trnF (GAA) intergenic spacer. Plant Systematics and Evolution **193**:213–221.

Mes, T. H. M., and H. T' Hart. 1996. The evolution of growth forms in the Macaronesian genus Aeonium (Crassulaceae) inferred from chloroplast DNA RFLPs and morphology. Molecular Ecology **5**:351–363.

Mes, T. H. M., J. Van Brederode, and H. T' Hart. 1996. Origin of the woody Macaronesian Sempervivoideae and the phylogenetic position of the East African species of Aeonium. Botanica Acta **109**:477–491.

Mes, T. H. M., G. J. Wijers, and H. T' Hart. 1997. Phylogenetic relationships in Monanthes (Crassulaceae) based on morphological, chloroplast and nuclear DNA variation. Journal of Evolutionary Biology **10**:193–216.

Nepokroeff, M., and K. J. Sytsma. 1996. Systematics and patterns of speciation and colonization in Hawaiian *Psychotria* and relatives based on phylogenetic analysis of ITS sequence data. American Journal of Botany (suppl.) **83**:181–182.

Okada, M., R. Whitkus, and T. K. Lowrey. 1995. RFLP diversity and species relationships in *Tetramolopium* (Asteraceae) in Hawaii and the Cook Islands. American Journal of Botany (suppl.) **82**:153.

Olmstead, R., and J. D. Palmer. 1997. Implications for the phylogeny, classification, and biogeography of *Solanum* from cpDNA restriction site variation. Systematic Botany **22**:19–31.

Qiu, Y.-L., M. W. Chase, D. H. Les, and C. R. Parks. 1993. Molecular phylogenetics of the Magnoliidae: cladistic analyses of nucleotide sequences of the plant gene *rbc*L. Annals of the Missouri Botanical Garden **80**:587–606.

Pax, D. L., R. A. Price, and H. J. Michaels. 1997. Phylogenetic position of the Hawaiian geraniums based on *rbc*L sequences. American Journal of Botany **84**:72–78.

Rabakonandrianina, E., and G. D. Carr. 1981. Intergeneric hybridization, induced polyploidy, and the origin of the Hawaiian endemic *Lipochaeta* from *Wedelia* (Compositae). American Journal of Botany **68**: 206–215.

Ray, M. F. 1995. Systematics of *Lavatera* and *Malva* (Malvaceae, Malveae): a new perspective. Plant Systematics and Evolution **198**:29–53.

Rieseberg, L. H., and D. E. Soltis. 1991. Phylogenetic consequences of cytoplasmic gene flow in plants. Evolutionary Trends in Plants **5**:68–84.

Rieseberg, L. H., and J. F. Wendel. 1993. Introgression and its consequences in plants. **In** Hybrid Zones and the Evolutionary Process, ed. R. Harrison, pp. 70–109. Oxford University Press, Oxford.

Sakai, A. K., S. G. Weller, W. L. Wagner, P. S. Soltis, and D. E. Soltis. 1997. Phylogenetic perspectives on the evolution of dioecy: adaptive radiation in the endemic Hawaiian genera *Schiedea* and *Alsinidendron* (Caryophyllaceae: Alsinoideae). **In** Molecular Evolution and Adaptive Radiation, eds. T. J. Givnish and K. J. Sytsma, pp. 455–473. Cambridge University Press, Cambridge.

Sanders, R. W., T. F. Stuessy, C. Marticorena, and M. Silva O. 1987. Phytogeography and evolution of *Dendroseris* and *Robinsonia,* tree-compositae of the Juan Fernandez Islands. Opera Botanica **92**:195–215.

Sanderson, M. J. 1996. How many taxa must be sampled to identify the root node of a large clade? Systematic Biology **45**:168–173.

Sanderson, M. J., and M. J. Donoghue. 1996. Reconstructing shifts in diversification rates on phylogenetic trees. Trends in Ecology and Evolution **11**:15–20.

Sang, T., D. J. Crawford, S.-C. Kim, and T. F. Stuessy. 1994. Radiation of the endemic genus *Dendroseris* (Asteraceae) on the Juan Fernandez Islands: evidence from

sequences of the ITS regions of nuclear ribosomal DNA. American Journal of Botany **81**:1494–1501.

Sang, T., D. J. Crawford, and T. F. Stuessy. 1995a. Documentation of reticulate evolution in peonies (*Paeonia*) using internal transcribed spacer sequences of nuclear ribosomal DNA: implications for biogeography and concerted evolution. Proceedings of the National Academy of Sciences U.S.A. **92**:6813–6817.

Sang, T., D. J. Crawford, T. F. Stuessy, and M. Silva O. 1995b. ITS sequences and the phylogeny of the genus *Robinsonia* (Asteraceae). Systematic Botany **20**: 55–64.

Santoni, S., and A. Berville. 1992. Characterization of the nuclear ribosomal DNA units and phylogeny of *Beta* L., wild forms and cultivated beets. Theoretical and Applied Genetics **83**:533–542.

Schilling, E. E., and J. L. Panero. 1996. Phylogenetic reticulation in subtribe Helianthinae. American Journal of Botany **83**:939–948.

Schilling, E. E., J. L. Panero, and U. H. Eliasson. 1994. Evidence from chloroplast DNA restriction site analysis on the relationships of *Scalesia* (Asteraceae: Heliantheae). American Journal of Botany **8**:248–254.

Schwarzbach, A. E., and J. W. Kadereit. 1996. ITS1 evidence for rapid radiation and allopatric and sympatric modes of speciation in prickly poppies (*Argemone:* Papaveraceae). American Journal of Botany (suppl.) **83**:190–191.

Slowinski, J. B., and C. Guyer. 1989. Testing the stochasticity of patterns of organismal diversity: an improved null model. American Naturalist **134**:907–921.

Smith, A. B. 1994. Rooting molecular trees: problems and strategies. Biological Journal of the Linnean Society **51**:279–292.

Smith, F. D. M., R. M. May, R. Pellew, T. H. Johnson, and K. R. Walter. 1993. How much do we know about the current extinction rate? Trends in Ecology and Evolution **8**:375–378.

Soltis, P. S., D. E. Soltis, S. G. Weller, A. K. Sakai, and W. L. Wagner. 1997. Molecular phylogenetic analysis of the Hawaiian endemics *Schiedea* and *Alsinidendron* (Caryophyllaceae). Systematic Botany **22**:1–15.

Stebbins, G. L. 1953. A new classification of the tribe Cichorieae, family Compositae. Madroño **12**:65–81.

Stebbins, G. L. 1977. Developmental and comparative anatomy of the Compositae. **In** The Biology and Chemistry of the Compositae, eds. V. H. Heywood, J. B. Harborne, and B. L. Turner, pp. 91–109. Academic Press, London.

Strauss, S. H. and A. H. Doerksen. 1990. Restriction fragment analysis of pine phylogeny. Evolution **44**:1081–1096.

Stuessy, T. D., K. A. Foland, J. F. Sutter, and M. Silva O. 1984. Botanical and geological significance of potassium-argon dates from the Juan Fernandez Islands. Science **225**:49–51.

Swofford, D. L., G. J. Olsen, P. J. Waddell, and D. M. Hillis. 1996. Phylogenetic inference. **In** Molecular Systematics, second edition, eds. D. M. Hillis, C. Moritz, and B. K. Mable, pp. 407–514. Sinauer Associates, Sunderland, Massachusetts.

Takhtajan, A. 1986. Floristic Regions of the World (translation by C. Jeffrey). Oliver & Boyd, Edinburgh.

Templeton, A. R. 1981. Mechanisms of speciation—a population genetic approach. Annual Review of Ecology and Systematics 12:23–48.

Tucker, S. C., and A. W. Douglas. 1996. Floral structure, development and relationships of paleoherbs: *Saruma, Cabomba, Lactoris,* and selected Piperales. In Flowering Plant Origin, Evolution and Phylogeny, eds. D. W. Taylor and L. J. Hickey, pp. 141–175. Chapman & Hall, New York.

Vargas, P., B. G. Baldwin, and L. Constance. 1998. Nuclear DNA evidence for a western North American origin of Hawaiian and South American species of *Sanicula* (Apiaceae). Proceedings of the National Academy of Sciences U.S.A. 95:235–240.

Wallace, A. R. 1880. Island Life. Macmillan & Co., London.

Warwick, S. I., and L. D. Black. 1993. Molecular relationships in subtribe Brassicinae (Cruciferae, tribe Brassiceae). Canadian Journal of Botany 71:906–918.

Wendel, J. F. 1989. New World tetraploid cottons contain Old World cytoplasm. Proceedings of the National Academy of Sciences U.S.A. 86:4132–4136.

Wendel, J. F., and A. E. Percival. 1990. Molecular divergence in the Galapagos Islands [Pacific Ocean]: Baja California species pair, *Gossypium klotzschianum* and *Gossypium davidsonii* (Malvaceae). Plant Systematics and Evolution 171:99–116.

Wendel, J. F., A. Schnabel, and T. Seelanan. 1995. Bidirectional interlous concerted evolution following allopolyploid speciation in cotton (*Gossypium*). Proceedings of the National Academy of Sciences U.S.A. 92:280–284.

Wheeler, W. C. 1990. Nucleic acid sequence phylogeny and random outgroups. Cladistics 6:363–368.

Whitton, J. R., R. S. Wallace, and R. K. Jansen. 1995. Phylogenetic relationships and patterns of character change in the tribe Lactuceae (Asteraceae) based on chloroplast DNA restriction site variation. Canadian Journal of Botany 73:1058–1073.

Witter, M. S., and G. D. Carr. 1988. Adaptive radiation and genetic differentiation in the Hawaiian silversword alliance (Compositae: Madiinae). Evolution 42:1278–1287.

World Conservation Monitoring Centre. 1992. Global Biodiversity: Status of the Earth's Living Resources. Chapman & Hall, London.

15

Molecular Markers, Gene Flow, and Natural Selection

Michael L. Arnold and Simon K. Emms

Studies of genetic variation in wild populations usually are designed to identify the processes causing microevolutionary change. The availability of easily accessible, locus-specific genetic markers is crucial to this endeavor. Such markers have been widely available for only 30 years, but since the mid-1960s several major technological advances have given rise to a variety of new methods for measuring genetic variation. Accompanying these advances has been an exponential growth in the number of studies reporting population genetic variation in a wide variety of organisms. Our goal in this paper is to discuss the utility of various molecular markers for addressing microevolutionary questions. We have chosen to emphasize the questions themselves rather than the techniques per se. We hope that this will demonstrate how each technique can be applied successfully to a variety of studies, and that the question being addressed will to some extent dictate which technique is employed. Just as the choice of gene sequence in molecular phylogenetic analyses depends on the age of the lineage under investigation, the appropriate molecular marker for microevolutionary studies depends on the amount of genetic variation needed to elucidate a particular phenomenon. We have concentrated on recently developed DNA markers rather than on isozymes,

not because we believe the former are more valuable, but because some readers may be less familiar with DNA methods than they are with protein electrophoresis. Familiarity with isozyme methods is due in large part to previous reviews of the application of protein electrophoresis in evolutionary studies (e.g., see Soltis and Soltis, 1989).

HISTORY

The application of molecular markers to population biology began in earnest with the benchmark studies of Harris, Hubby, and Lewontin (Harris, 1966; Hubby and Lewontin, 1966; Lewontin and Hubby, 1966). These studies, particularly those of Hubby and Lewontin, used isozyme variation to test two models of natural selection: the classical (purifying selection) model and the balancing selection model (Lewontin, 1974). Although these models make different predictions about the amount and pattern of standing genetic variation in populations, Lewontin (1974, 1991) recently argued that isozyme studies actually have not adequately distinguished between them. However, he also argued that these early studies had a crucially important influence on the development of evolutionary genetics as a discipline. One of the effects he emphasized was the role played by pro-

We thank M. Bulger, J. Burke, S. Carney, J. Williams, and a not so anonymous reviewer for comments on a draft of this manuscript. This work was supported by National Science Foundation grant DEB 9317654 and an American Iris Society Foundation grant.

tein electrophoresis in paving the way for the application of DNA analyses. In some ways the widespread use of protein electrophoresis can be viewed as a preadaptation for the incorporation of newer technologies into evolutionary laboratories. However, this is to understate its value. The technique has allowed a remarkable degree of understanding of population-level processes (Brown and Allard, 1970; Levin, 1975; Hamrick and Holden, 1979) and continues today to be a powerful diagnostic tool for the study of plant (and animal) populations (e.g., Thomson et al., 1991; Husband and Barrett, 1993; Broyles and Wyatt, 1995).

After isozymes, the next major technological advance came in the form of DNA reassociation kinetics (Britten and Koehn, 1968). This technique was used to address macroevolutionary questions such as the origin of "evolutionary novelties" (e.g., Britten and Davidson, 1971) and the effect of genomic evolution on the process of speciation (e.g., Mazrimas and Hatch, 1972), but has not been incorporated into the armory of population biologists. The extension of DNA analyses to microevolutionary studies began with the application of restriction endonuclease techniques to animal systems (e.g., Brown, 1980), for which DNA restriction fragment length polymorphism (RFLP) data accumulated rapidly in numerous taxa. This largely reflected the utility of mitochondrial DNA for within-species comparisons: Portions of the mtDNA molecule were found to incorporate mutations at an evolutionarily rapid rate, leading to measurable sequence divergence among populations (Brown et al., 1979). In contrast, cytoplasmic DNA data were not immediately exploited by plant population biologists. There were two reasons for this. First, the cytoplasmic genomes of plant cells (mitochondrial and chloroplast (cp) DNA) typically were found to demonstrate relatively slow sequence evolution. Second, plant mtDNA showed extraordinarily complex patterns of rearrangement, making it difficult to interpret RFLP data (Palmer, 1990). The cpDNA molecule was found to be extremely useful for a wide range of systematic studies (Jansen and Palmer, 1987), but these tended not to examine more than a handful of individuals from a given taxon (Soltis et al., 1992; McCauley, 1995). Recent studies have shown that some

species have extensive intraspecific cpDNA variation (e.g., Soltis et al., 1991, 1997) whereas others are largely invariant (e.g., Jordan et al., 1996). Empirical analyses of this type (as well as theoretical models of nuclear-cytoplasmic genetic structure: Schnabel and Asmussen, 1989, 1992; Asmussen and Schnabel, 1991) show that cpDNA variation sometimes can provide a great deal of useful information to plant population biologists.

In contrast to cpDNA variants, restriction fragment length variants at nuclear ribosomal DNA (rDNA) loci have been used extensively as markers in microevolutionary studies of plant populations. Because rDNA loci are highly repetitive and contain large, nontranscribed regions (Appels and Dvorak, 1982), they are relatively easy to detect and possess high levels of intraspecific sequence variation. Consequently, they have been used to infer population structure, evolutionary divergence, gene flow, and the effects of natural selection (Saghai-Maroof et al., 1984; Learn and Schaal, 1987; Schaal et al., 1987; Rieseberg et al., 1988; King and Schaal, 1989; Arnold et al., 1990; Whittemore and Schaal, 1991; Dorado et al., 1992; Kim et al., 1992; King, 1993).

The most recent technological advance was the development of methods allowing amplification of specific regions of DNA. As with other new techniques, the polymerase chain reaction (PCR) was quickly incorporated into evolutionary studies. In plant biology, PCR-based approaches have been used mainly for phylogenetic/systematic studies (e.g., Chase, Soltis, Olmstead et al., 1993). However, the technique also has proved to be ideal for addressing many microevolutionary questions. In particular, many studies of gene flow, mating patterns, and natural selection were made possible by the availability of numerous cytoplasmic and nuclear markers derived from PCR-based techniques (e.g., Arnold et al., 1991; Cruzan and Arnold, 1993, 1994; Philbrick, 1993; Powell et al., 1995; Rieseberg and Gerber, 1995; Watano et al., 1995; Chase et al., 1996; McCauley et al., 1996).

ISOZYME ANALYSES

Protein electrophoresis continues to be an invaluable tool for measuring genetic variation and inferring evolutionary process (Hamrick and Godt, 1989, 1996b). Unfortunately, the allure of DNA technologies and their unquestioned power in phylogenetic studies have contributed to an impression that isozyme data are less informative and less accurate than DNA data in all situations. Isozyme data may be less informative, and possibly more subject to error, in phylogenetic analyses because of the problems associated with inferring homology and polarizing the variants. However, they have a great deal of power to measure population substructuring within species and to discriminate between populations of closely related taxa. Moreover, comparisons of isozyme variation in distantly related taxa have also identified common patterns associated with common life histories (Hamrick and Godt, 1989). Thus, for many applications, the strengths of the technique (its low cost, technical ease, the number of available loci, and the codominant inheritance pattern) greatly outweigh its weaknesses (the relatively low variability within loci, the potential for undetected variation, and the difficulty of interpreting some banding patterns). It remains one of the most useful procedures for describing the genetic constitution of natural populations and identifying the factors controlling genotypic structure (see Cruzan, 1998 and Hamrick and Nason, 1996 for recent reviews).

Genetic studies of introduced wild oats (*Avena barbata*) illustrate the manner in which isozyme data can be used to identify microevolutionary processes. Hamrick and Allard (1972) and Hamrick and Holden (1979) found that the multilocus genotypes of wild oat plants varied spatially in natural populations. This genetic variability was correlated with environmental heterogeneity, such that an a priori classification of habitat types could accurately predict the genotypic makeup of plants associated with different habitats (Fig. 15.1; Hamrick and Holden, 1979). Furthermore, Hamrick and Allard (1975) used a common garden experiment to demonstrate that some of the morphological differences between plants with the "xeric" and "mesic" genotypes were (1) genetically determined and (2) possibly adapted to the different environments in which they were found.

Isozyme variation also has been employed to study seed paternity in a variety of plant species (e.g., Meagher, 1986, 1991; Schoen and Stewart, 1986; Stanton et al., 1986; Devlin et al., 1992; Harder and Barrett, 1995). Using both natural populations and experimental arrays of plants, Broyles and Wyatt (1990, 1995) used the technique to determine whether the evolution of floral traits was being driven by fitness returns on male reproductive allocation. They concluded that selection through both male *and* female functions was shaping the evolution of inflorescence size in poke milkweed, *Asclepias exaltata*. In a separate study, Broyles et al. (1994) estimated the degree of long-distance pollen movement and found gene flow to be remarkably extensive: 29%–50% of the seeds were sired by plants outside the study sites. They also demonstrated the existence of interspecific gene flow between *A. exaltata* and *A. syriaca*, despite considerable spatial separation of the two taxa. The high frequency of gene flow between *Asclepias* populations should have been sufficient to counter any divergence resulting from natural selection and genetic drift (Broyles et al., 1994).

The major weakness of isozyme markers is that levels of variation often are inadequate for the task at hand. For example, in a study of seed paternity in the dioecious lily *Chamaelirium luteum*, Meagher (1986) was able to assign only 575 of 2,255 seeds to a most likely male parent using log likelihood ratio (LOD) scores. Of these 575 seeds, only 47 could be unequivocally assigned to one of 273 potential male parents. In addition, there can be interpretive problems due to intra- and interlocus associations of protein molecules, as well as the possibility that band (i.e., electromorph) comigration does not reflect true homology. We should emphasize that the latter is a problem with all techniques to some extent. Nonetheless, for all the reasons just discussed, protein electrophoresis will continue to be a popular and useful tool for plant evolutionary biologists.

NUCLEAR DNA ANALYSES

rDNA RFLPs

The most widely used nuclear DNA markers for measuring population genetic variation in plants have been obtained from ribosomal loci. These

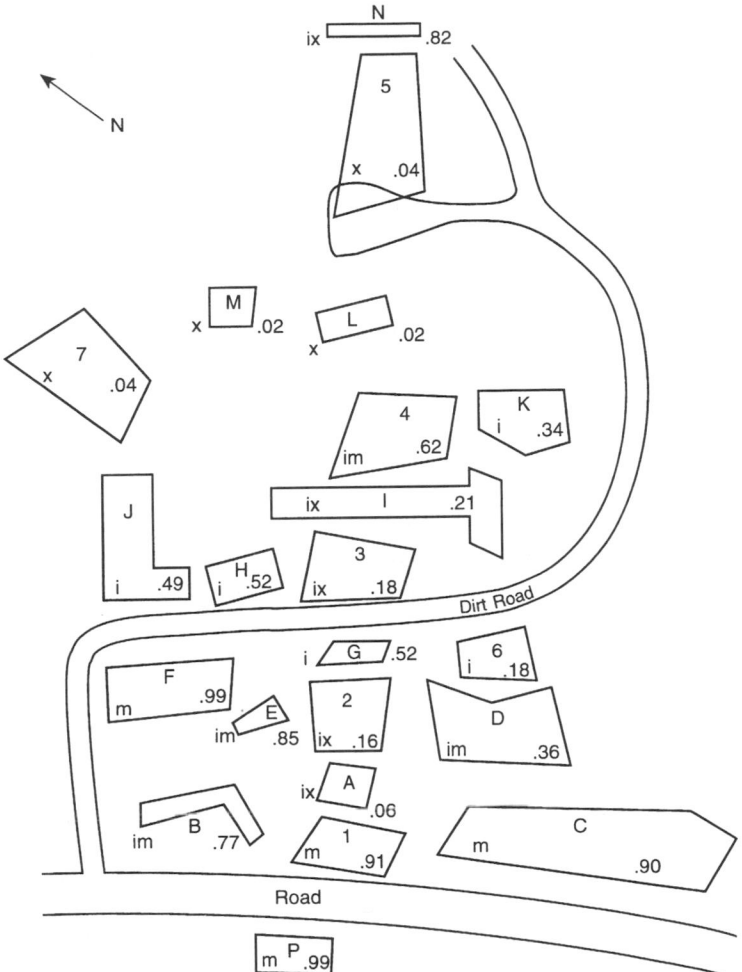

Figure 15.1. Map of collection sites for a study of genetic variation in *Avena barbata*. Sample sizes ranged from 34–250 individuals per site. Nei's genetic identity values (Nei, 1972) are given within or alongside each site. The identity values were calculated by comparing the allele frequencies of each site with a population monomorphic for the alleles characteristic of mesic sites. Numbers 1–7 and upper case letters A–P are site designations. The lower case letters indicate habitat type: m = mesic; im = intermediate-mesic; i = intermediate; ix = intermediate-xeric; x = xeric. Note that genetic identity values are strongly correlated with habitat type. Redrawn from Hamrick and Holden (1979).

markers have the advantages of being moderately variable and easily detected. The amount of variability typically observed makes them ideal for measuring such processes as gene flow within and between taxa.

Saghai-Maroof et al. (1984, 1992) and Schaal and her colleagues (Learn and Schaal, 1987; Schaal et al., 1987; King and Schaal, 1989; Whittemore and Schaal, 1991) have used rDNA intergenic spacer length and restriction site variation to study both intra- and interspecific varia-

tion. The studies of Saghai-Maroof et al. (1984, 1992) involved examining rDNA spacer length variation in hundreds of individual barley plants from accessions of the cultivar and its wild progenitor. One conclusion reached from these studies was that natural selection was acting on the spacer length variants, or on closely associated loci. Their most recent work included a comparison of the patterns of rDNA and cpDNA variation. This analysis also implicated natural selection as a factor determining the observed cytonuclear associations (Saghai-Maroof et al., 1992). However, other processes, including the predominantly selfing mating system, genetic drift, and founder events, were also thought to be affecting the observed population genetic structure.

Schaal and her coworkers analyzed rDNA variation both within and between taxa. The patterns of variation included (1) large amounts of length variation but little, if any, restriction site variation within and between individuals and populations (*Phlox, Quercus, Clematis*); (2) no length variation but significant restriction site variation in *Rudbeckia missouriensis* populations (Learn and Schaal, 1987; Schaal et al., 1987; King and Schaal, 1989; Whittemore and Schaal, 1991). The patterns of rDNA variation in these species allowed inferences to be made about the processes that have shaped present-day population structures. These included such disparate factors as the historical formation of glades inhabited by *R. missouriensis* and interspecific gene flow between the *Quercus* species. In *Quercus,* hybridization between taxa resulted in cpDNA, but not rDNA, introgression (Whittemore and Schaal, 1991), a common pattern when plants hybridize (Rieseberg and Soltis, 1991).

rDNA RFLP data suffer from the same potential major weakness as do isozyme data—a lack of sufficient variation for some purposes. In addition, RFLP techniques are quite expensive and take a considerable amount of time, owing to the use of restriction digests and Southern hybridization. The expense is particularly high given that RFLP digests typically provide information for only one locus. Consequently, relatively few individuals can be analyzed in any given study. PCR-based techniques have largely replaced rDNA RFLP analyses, although the codominant inheritance pattern of RFLP markers can usefully be employed in studies of hybridization (e.g., Emms et al., 1996).

M13 DNA Fingerprint Analysis

DNA "fingerprinting" using minisatellite markers has been a popular technique with animal ecologists, particularly in studies of paternity (see Avise, 1994 for references). Schaal and

coworkers have applied the same technique to several plant systems, using an M13 bacteriophage-derived probe to study genetic variation within and between natural populations of pawpaw *Asimina triloba* (Rogstad et al., 1991) and two species of *Rubus:* blackberry, *R. pensilvanicus,* and raspberry, *R. occidentalis* (Nybom and Schaal, 1990). These studies illustrate how useful the extreme variability detected by the M13 probe sometimes can be. For example, each population of pawpaw was characterized by its own set of unique fingerprint fragments. However, the amount of variation among individuals differed greatly among populations, apparently as a result of variation in the degree of inbreeding and/or clonal propagation (Rogstad et al., 1991). DNA fingerprints of co-occurring blackberry and raspberry plants also reflected differences in growth patterns and mating system. The apomictic blackberry population consisted of a few large clones (20 plants, five genotypes). In contrast, the sexually reproducing raspberry population possessed numerous, spatially nonoverlapping clones (20 plants, 15 genotypes; Nybom and Schaal, 1990). These differences in population structure could be ascribed to biological factors because all plants were collected from the same site.

The major strength of M13 fingerprinting—identification of very high levels of variability within populations—is, for other purposes, a liability. Thus, it is unlikely that the approach generally will be useful for comparing among populations, even within a species. A second weakness is the difficulty of determining band homology. A third, shared with rDNA RFLP analysis, is that the dependence on restriction digests and Southern hybridization limits the number of individuals that can be sampled. M13 fingerprinting probably will have a limited appeal to plant population biologists. Finally, it is not clear that the probes used will be universal in their ability to detect homologous sequences in all species of interest.

RAPD Analysis

The large sample sizes needed for many population genetic studies limited the use of DNA-level variation until the polymerase chain reaction was invented. However, the past decade has seen an increasing application of PCR-based techniques to microevolutionary questions. The random amplified polymorphic DNA (RAPD; Williams et al., 1990) or arbitrarily primed PCR (AP-PCR; Welsh and McClelland, 1990) procedure has proved especially popular. This technique uses short, arbitrary primer sequences to amplify anonymous regions of DNA and produces a series of one to many bands that can be visualized on agarose or polyacrylamide gels. The large number of primers facilitates identification of useful markers, and RAPD screening can be much faster and cheaper than that required to identify RFLPs (but see Ragot and Hoisington, 1993). These factors make it possible to identify numerous polymorphic loci and to sample many individuals. RAPD analysis originally was described as a method for genome mapping (Williams et al., 1990) and it has been widely used for this purpose (see Chapter 16). Genome maps have been used to identify quantitative trait loci affecting both agronomically important traits (e.g., Chunwongse et al., 1994) and phenotypic traits thought to be important in speciation (Bradshaw et al., 1995). Another widespread use of RAPD markers has been the quantification of popula-

tion genetic variability in crop-related species (e.g., Chalmers et al., 1992; Dawson et al., 1993).

The application of RAPD markers to microevolutionary questions began shortly after its description. Arnold et al. (1991) used a combination of RAPD markers and restriction digests of PCR-amplified cpDNA loci to distinguish between pollen- and seed-mediated gene flow in the Louisiana irises *Iris fulva* and *I. hexagona* (see Cytoplasmic DNA Analyses). Using species-specific markers, including RAPDs, Arnold (1993) also was able to determine that *I. nelsonii* (Randolph, 1966) most likely was a hybrid derivative of *I. fulva, I. hexagona,* and a third species, *I. brevicaulis.*

Cruzan and Arnold (1993, 1994) used species-specific RAPD markers for *I. fulva, I. hexagona,* and *I. brevicaulis* to test several hypotheses about the evolution of hybrid zones. Figure 15.2 shows the results of an analysis that tested for associations between RAPD/cpDNA genotypes and habitat in one hybrid population. Hybrid and parental genotypes occurred in different habitats, suggesting that the genetic structure of the hybrid zone was influenced by the environment (Cruzan and Arnold, 1993). Subsequent work, again using RAPD and cpDNA

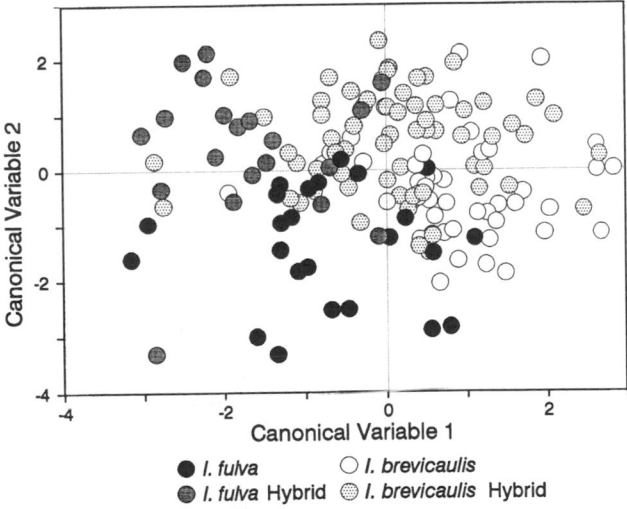

Figure 15.2. Results of a canonical discriminant analysis showing associations between genotype and environment in an iris hybrid zone. The distribution of four genotypic classes is shown on two canonical axes that summarize ambient light levels, elevation above the bayou, and the type of vegetation associated with each iris plant. From Cruzan and Arnold (1993).

markers to define genotypic classes, demonstrated that mating patterns and selection on hybrid viability also affected the structure of this zone. Gene flow between parental species was asymmetric, with *I. fulva* acting as a pollen parent, but not an egg parent, in crosses with *I. brevicaulis* (Cruzan and Arnold, 1994). Hybrids with "intermediate" genotypes were significantly less viable than parental classes and hybrids with genotypes similar to their parents (Fig. 15.3; Cruzan and Arnold, 1994).

RAPD markers also were used to study hybrid zone structure in columbines. Hodges and Arnold (1994) tested whether gene flow between *Aquilegia formosa* and *A. pubescens* was limited by adaptation to different habitats, by different pollination syndromes, or a combination of these factors. Figure 15.4 shows the RAPD and floral character variation across both elevational and ecological transects. Only one of the floral traits introgressed across the transects, but RAPD markers were exchanged to a much greater extent. Hodges and Arnold (1994) concluded that both habitat preferences and pollinator behavior were important in restricting gene flow between the species.

Rieseberg and his colleagues have used RAPD markers to study the evolution of annual sunflowers. Their analyses have focused on genome mapping to identify the origins of hybrid species and the pattern of genomic exchange between sunflower taxa (Rieseberg et al., 1993, 1995a, 1995b, 1996). A detailed description of their findings is provided elsewhere in this volume (see Chapter 16). Briefly, RAPD mapping demonstrated that epistatic interactions, chromosomal rearrangements, and selection among genotypes all were involved in the formation of hybrid species. Furthermore, these same processes also appeared to be involved in determining the genetic structure of contemporary hybrid populations (Rieseberg et al., 1996).

Several investigators have concluded that the major weakness of RAPD technology is its lack of reproducibility (Riedy et al., 1992; Ellsworth et al., 1993). Fragments may not be amplified consistently from one reaction to another, and bands can appear in progeny that were not present in parents. Such results reflect the dependence of the RAPD protocol on initial PCR con-

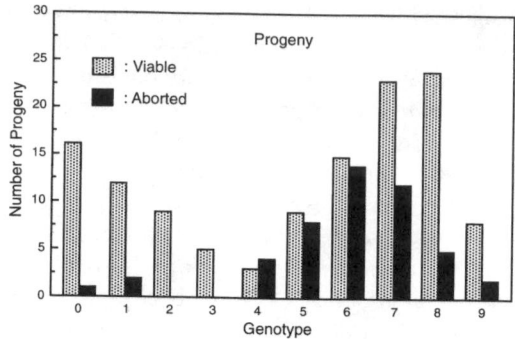

Figure 15.3. The genotypic distribution of viable and inviable (aborted) seed progeny from an *Iris fulva × I. brevicaulis* hybrid population. The genotype scores are derived from RAPD and cpDNA markers. A score of 0 indicates an *I. fulva* genotype; a score of 9 indicates an *I. brevicaulis* genotype. Scores of 1–8 indicate hybrid genotypes that resembled the two parental species to different degrees. From Cruzan and Arnold (1994).

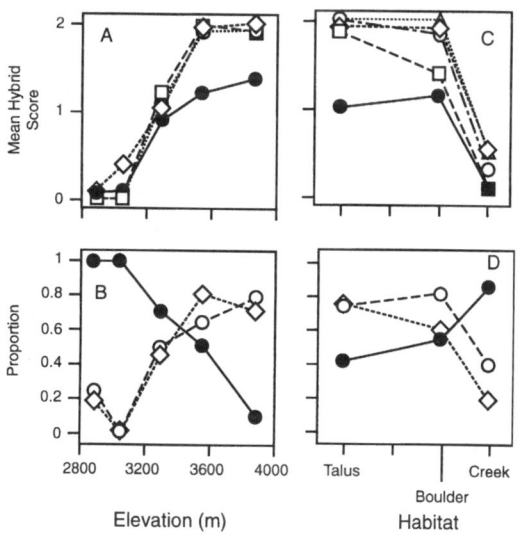

Figure 15.4. Morphological and genetic variation in a hybrid zone between *Aquilegia formosa* and *A. pubescens* across an elevational (A and B) and a habitat (C and D) transect. Variation in floral characters (A and C) ranges from 0 for *A. formosa* to 2 for *A. pubescens*. Floral characters (A and C) are spur color (filled circles), blade color (diamonds), spur length (open circles), blade length (triangles), and flower orientation (squares). Genetic (RAPD) variation (B and D) was measured for one *A. formosa* marker (closed circles) and two *A. pubescens* markers (diamonds and open circles). At the ends of the transects, RAPD genotypes tended to have intermediate hybrid scores (indicative of introgression) whereas floral traits had more extreme scores (indicative of parental genotypes). From Hodges and Arnold (1994).

ditions (Williams et al., 1990). Notwithstanding these problems, many workers have found that RAPD analysis gives rise to consistent, reproducible results. It is also worth noting that "artifactual" banding patterns can occur with every technique discussed in this chapter, not just with RAPDs. A second potential weakness of RAPD markers is that homology is inferred from comigration. This could lead to the erroneous interpretation of banding patterns, although Rieseberg (1996) found that 90% of bands presumed to be homologous on the basis of comigration truly were homologous when examined in more detail using restriction digests or Southern hybridization. Of course, this result also indicates that 10% of the bands that comigrate are *not* homologous. The power of this approach compared to, for example, isozyme analyses that also depend upon comigration to infer homology, is that homology can be tested through Southern analysis, sequencing, or restriction digestion of the RAPD products (Rieseberg, 1996). A third limitation of the RAPD methodology is that the majority of markers are inherited in a dominant fashion, which reduces the information content compared to codominant markers such as isozymes or nuclear RFLPs.

AFLP Analysis

The most recently developed molecular markers have been called AFLPs (amplified fragment length polymorphisms; Vos et al., 1995). AFLP methodology includes digestion of total genomic DNA with two endonucleases (normally one that has only a few recognition sequences present in a given genome compared to the second endonuclease), ligation of short DNA oligonucleotides to the ends of the cut fragments, and two successive PCR amplifications using different sets of primers. The first amplification uses a set of primers that incorporate a core sequence, the recognition sequences of the two endonucleases plus an additional base pair. This additional base pair adds a measure of selectivity to the amplification. The second amplification step uses oligonucleotides that have three base pairs added to the 3' end to make the amplification even more selective. The DNA fragments produced by this analysis normally are separated on denatur-

ing polyacrylamide gels. As with RAPDs, most AFLPs are dominantly inherited, but unlike RAPD protocols (or any other method), a single AFLP reaction can survey up to 100–200 loci (Meksem et al., 1995).

We located only two studies applying AFLPs to the population biology of non-animal species. The first involved an analysis of symbiotic fungi from nests of the fungus-growing ant *Cyphomyrmex minutus* (Mueller et al., 1996). Fourteen fungal isolates were collected from 13 nests at four sites. Four AFLP fingerprint classes were detected from the 14 isolates, with different variants being found both among and within nests. This pattern of variation suggested that the ants might be passing fungal clones from parent to offspring or assimilating free-living clones into their colonies (Mueller et al., 1996).

The second study examined the population genetic structure of *Astragalus cremnophylax* var. *cremnophylax* on the rims of the Grand Canyon (Travis et al., 1996). With nine primer combinations 220 different DNA fragments were scored. Table 15.1 shows the amount of genetic differentiation between samples from the north and south rims and between subsamples from the north rim population. Surprisingly, the two south rim samples were more clearly differentiated than was

Table 15.1. Patterns of genetic differentiation between populations and subpopulations of *Astragalus cremnophylax* var. *cremnophylax* estimated from an AFLP analysis (from Travis et al., 1996).

Population Comparisons	Subpopulation Comparisons	Theta (Corrected F_{ST})	Nm
South Rim Site 1/ South Rim site 2	—	0.51	0.2
South Rim Site 1/ North Rim	—	0.41	0.4
South Rim Site 2/ North Rim	—	0.27	0.7
—	A/B	0.11	2.1
—	A/C	0.13	1.7
—	A/D	0.15	1.4
—	B/C	0.13	1.8
—	B/D	0.14	1.5
—	C/D	0.05	4.5

The letters A–D indicate subpopulations from the north rim population. All F_{ST} values are significantly greater than 0 (Travis et al., 1996).

the south rim sample 2 from the north rim population. Travis et al. (1996) suggested that airborne transport of seeds and pollen was a likely explanation. Overall, gene flow (*Nm*) between populations was estimated to involve less than one migrant per generation (Table 15.1; Travis et al., 1996), and as expected the F_{ST} values for subpopulations from the north rim were significantly lower than those for north and south rim populations. However, the estimates were still significantly greater than zero, indicating differentiation among subpopulations (Table 15.1; Travis et al., 1996).

Because of the recent development of AFLP technology, it is difficult at present to assess its strengths and weaknesses. Two obvious weaknesses, shared with RAPDs, are the dominant mode of inheritance of most AFLP loci, and the fact that homology is inferred from band comigration. The latter problem is likely to be of less concern than for RAPDs because AFLP fragments normally are resolved on denaturing polyacrylamide gels. This should allow resolution down to single base-pair length differences. Another potential problem could arise if partial digests occurred during the first step of the protocol. This could lead to the production of artifactual (nonallelic) fragments during the amplification steps. The strength of the method lies in its ability to resolve tens to hundreds of loci from a single reaction. Thus, it should be possible to identify many polymorphic loci for a given taxon and to screen for these polymorphisms in numerous populations.

Microsatellite Analysis

As with RAPDs and AFLPs, the development of microsatellites (also referred to as short tandem repeats, STRs [Craig et al., 1988], simple sequence length polymorphisms, SSLPs [Rassmann et al., 1991], or simple sequence repeats, SSRs [Jacob et al., 1991]) has been motivated by the need for numerous markers for genome mapping (e.g., Zhao and Kochert, 1993; Roder et al., 1995). The potential uses of microsatellite technology also have been recognized by population biologists (see Ashley and Dow, 1994 for review). As with the application of most molecular genetic techniques, evolutionary zoologists have

been the first to capitalize on microsatellite techniques in studies of natural populations. The application of this methodology to animal taxa is yielding significant contributions to such diverse areas as conservation genetics (Bruford et al., 1996), mating system evolution (Jones and Avise, 1996), and the evolution of polyembryony (Prodohl et al., 1996). The great promise of the technique is based on its potential for identifying numerous loci with many moderately frequent alleles. This type of allelic variation should prove invaluable for addressing such within-population questions as the factors controlling male fitness in plants. The increasing use of microsatellite technology has been accompanied by the development of statistical models for analyzing genetic data derived from microsatellite loci (Valdes et al., 1993; Slatkin, 1995).

Several studies have used the highly polymorphic nature of these loci to track gene movement within natural populations. Figure 15.5 illustrates the allelic variation found at four microsatellite loci in the tropical tree *Pithecellobium elegans*. The variation at these loci theo-

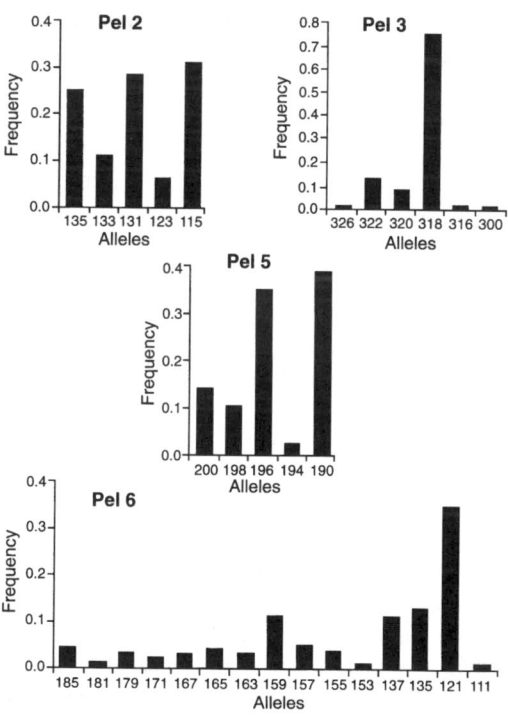

Figure 15.5. Allele frequencies at four microsatellite loci in *Pithecelobium elegans*. Redrawn from Chase et al. (1996).

retically should have allowed a 94% exclusion probability for paternity analysis (Chase et al., 1996). Indeed, the probability of exclusion using only one of the loci (Pel 6) was estimated to be 0.69. Chase et al. (1996) found that seeds within fruits were sired by only a few fathers, but that many more fathers were represented in the fruit crop of individual mother trees. This pattern probably reflected the biology of the pollinators as well as the demography and spatial distribution of the trees themselves.

Microsatellites also were used to map the spatial distribution and infer parentage in a population of 62 adult and 100 bur oak saplings (Dow et al., 1995; Dow and Ashley, 1996). The variation at four microsatellite loci allowed identification of at least one of the two parents for 94 of the 100 saplings. Dow and Ashley (1996) concluded that a majority of the saplings came from seed parents located within the study site, but that 71% were sired by pollen parents growing elsewhere. A paternity analysis on 282 acorns from the same stand found that 174 (62%) were sired by imported pollen, a frequency not significantly different from the 71% reported for saplings (Dow and Ashley, 1996).

Saghai-Maroof et al. (1994) examined the population variability of microsatellite loci in barley. Four loci, each located on a different chromosome, were identified. Two of the loci possessed three alleles; the remaining two possessed 28 and 37 alleles. More allelic diversity was found in wild populations of barley than in cultivars. Moreover, changes in allele frequency in experimental populations were thought to be due to selection on the chromosome segments carrying these markers (Saghai-Maroof et al., 1994).

As with AFLPs, microsatellite techniques are too new for their strengths and weaknesses to be evaluated at present. Potential weaknesses include (1) the relatively large investment of time and money needed to identify appropriate markers; (2) the difficulty of interpreting some banding patterns, especially across gels; (3) limits on the number of samples that can easily be analyzed; (4) the potential for confusion caused by null alleles (Pemberton et al., 1995; Dow and Ashley, 1996); and (5) the likelihood that variation will not be species-specific (although this is

only of concern if one is interested in studying interspecific gene flow or hybrid zone dynamics). A concern for population biologists is that the genetic model normally assumed for the evolution of allelic diversity (the infinite alleles model) is not appropriate for microsatellite data (Slatkin, 1995). However, new statistical techniques have been developed that take into account the elevated mutation rates and different patterns of mutation of many microsatellite loci (Valdes et al., 1993; Slatkin, 1995).

CYTOPLASMIC DNA ANALYSES

An intriguing aspect of plant population biology is that gene flow can occur through two avenues—seeds and pollen. If the initial gene flow between populations is by pollen alone, maternally inherited genomic elements (the chloroplast and mitochondrial DNA of many angiosperms) from the pollen parent are left behind. In contrast, if gene flow is through seeds, both biparentally and maternally inherited genes are transmitted. Thus, cytoplasmic DNA markers can be used to identify the mechanisms of gene flow between populations.

As with nuclear markers, the value of chloroplast and mitochondrial markers depends on the existence of appropriate levels of variation and the ease with which large samples of individuals can be surveyed. RFLP surveys demonstrated that cpDNA sequence variation did indeed exist within species (e.g., Soltis et al., 1989). Moreover, detailed analysis of hybrid zones using cytoplasmic RFLP markers have revealed the importance of gene flow and natural selection in shaping these zones (e.g., Wagner et al., 1987; Rieseberg et al., 1990; Paige et al., 1991). However, the time-consuming nature of RFLP protocols limits the number of individuals that can be sampled in a given study.

One way around this limitation is to perform RFLP surveys on PCR-amplified products, thereby eliminating the time and expense of Southern hybridization. Such an approach was used in studies of cpDNA variation within and between populations of the Louisiana irises *Iris fulva, I. hexagona,* and *I. brevicaulis* (Arnold et al., 1991, 1992). One goal of these studies was to determine whether the initial gene flow between species had arisen by seed immigration followed by pollen exchange or by pollen transfer between populations. cpDNA inheritance in these species is almost exclusively maternal (Cruzan et al., 1993). Thus, seed movement between populations could be monitored by studying the distribution of species-specific cpDNA haplotypes (Arnold et al., 1991, 1992). Figure 15.6 illustrates the pattern of nuclear (RAPD) and cpDNA variation found in one of the *I. fulva* × *I. hexagona* × *I. brevicaulis* hybrid populations. cpDNA haplotypes and nuclear markers of all three species were present in site A, but only *I. fulva* cpDNA haplotypes were present in site B, even though the two sites were less than 30 m apart. In contrast, nuclear markers from both *I. fulva* and *I. brevicaulis* were present in site B. These results suggested that most gene flow between sites was through pollen rather than seeds, and that seed movement could be limited even over very short distances (see Arnold, 1992, 1993 for additional examples).

A similar result was found in studies of the population genetics of *Silene alba* (McCauley, 1994; McCauley et al., 1996). These studies surveyed isozyme and cpDNA variation both within and between populations. The distribution of isozyme variation suggested that plants were mating nearly at random and that gene flow was occurring as a result of pollen movement (McCauley et al., 1996). In contrast, the clustering of cpDNA haplotypes suggested that seed-mediated gene flow was limited even at very small spatial scales (Fig. 15.7; McCauley et al., 1996).

In contrast to the studies in *Iris* and *Silene,* Ferris et al. (1993) found a geographic pattern of cpDNA variation in European oak trees that indicated seed-mediated gene flow. *Quercus rober* and *Q. petraea* shared cpDNA haplotypes in areas of sympatry. Moreover, these shared haplotypes differed between regions. The geographic limits of the haplotype distributions suggested that both species had spread from two or more

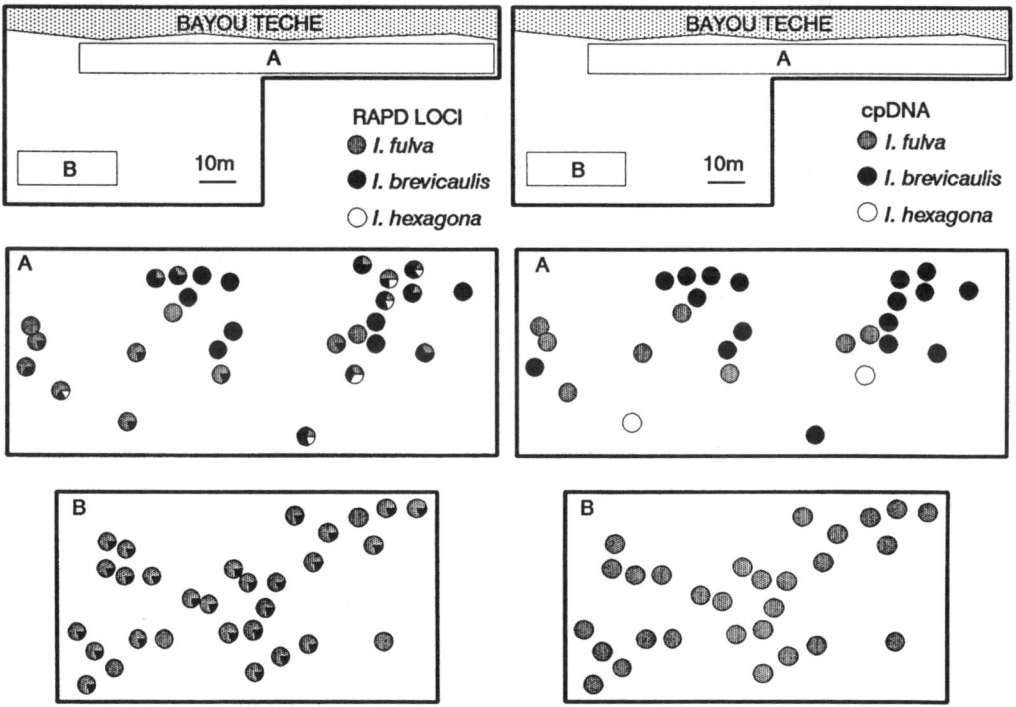

Figure 15.6. Maps of RAPD (left panels) and cpDNA (right panels) genotypes for the Bayou Teche Louisiana iris hybrid population. Each circle represents a single plant within the mapped population. Shading patterns within the circles indicate the fraction of species-specific RAPD markers from *I. fulva, I. hexagona,* and *I. brevicaulis* in each plant. cpDNA haplotype data are missing for two of the individuals sampled for RAPD variation. From Arnold (1993).

S	S	S	S	S
S	-	S	S	S
L	S	S	S	S
-	L	S	S	S
S	S	S	S	S
S	S	L	L	S
S	S	L	L	S
S	S	L	L	L
-	L	L	-	S
-	-	L	S	L
S	-	-	L	-
-	L	L	L	S
-	L	L	-	L
-	-	-	-	S

Figure 15.7. Spatial distribution of two cpDNA haplotypes (S and L) in a population of *Silene alba*. Plants were sampled at 5 m intervals. A dash indicates that no plant was present within 2.5 m. Note the clustering of cpDNA haplotypes. Redrawn from McCauley et al. (1996).

glacial refugia to occupy their current ranges. Thus, seed dispersal and interspecific hybridization apparently are responsible for the genetic structure of contemporary populations of these two species (Ferris et al., 1993).

Three other methods recently have been used to survey cpDNA variation in plant populations. Powell et al. (1995) identified cpDNA microsatellites and studied population diversity in *Pinus leucodermis*. Within-population diversities (G_{ST} values) of four variants ranged from 0 to 0.629. This degree of population subdivision was four times greater than that detected using isozyme loci (Powell et al., 1995) and contrasted with previous cpDNA RFLP data that did not detect any variation (Boscherini et al., 1994). In another analysis of natural populations of pines, Watano et al. (1995) used single-strand conformation polymorphisms (SSCPs) to infer cpDNA sequence variation in *P. pumila, P. parviflora* var. *pentaphylla,* and their putative hybrid derivative, *P. hakkodensis.* (SSCP methods exploit the fact that single-stranded DNA sequence variants have different conformations, and so mi-

grate at different rates through polyacrylamide gels: see Dowling et al., 1996 and Chapter 2 for a description of this and the denaturing gradient gel electrophoresis [DGGE] methodology). They found that individuals assigned to *P. hakkodensis* on the basis of morphology possessed the cpDNA haplotype characteristic of *P. parviflora* var. *pentaphylla*. Furthermore, the SSCP variant of *P. parviflora* var. *pentaphylla* also was present in individuals that possessed the morphology of *P. pumila*. This finding is consistent with the occurrence of asymmetric gene flow from *P. parviflora* var. *pentaphylla* into *P. pumila* (Watano et al., 1995). Finally, Strand et al. (1996) used DGGE to resolve sequence variation in PCR products of a noncoding region of *Aquilegia* cpDNA. In DGGE gels, double-stranded DNA sequences denature, and hence move through the gel, at rates that depend on their base sequence. The technique is capable of identifying sequences that differ by a single nucleotide substitution (Myers et al., 1988). Strand et al. (1996) found that the amount of differentiation between populations was unrelated to geographic separation, which suggested that the populations had been divided from each other during the late Pleistocene or early Holocene, rather than being connected by contemporary gene flow.

For some purposes, the uniparental inheritance of most cytoplasmic markers is a weakness. In addition, techniques that involve Southern hybridization are expensive and time consuming. The incorporation of PCR-based approaches largely alleviates the problem of speed, if not of expense (given that restriction digests are still used), and cpDNA markers should continue to be valuable tools for addressing microevolutionary questions. However, as with techniques such as microsatellites, the paucity of studies applying some of the methodologies (e.g., SSCP and DGGE) limits conclusions concerning their relative strengths and weaknesses.

FUTURE DEVELOPMENTS

The perfect genetic marker would be inexpensive, quickly obtained, easily interpreted, and codominantly inherited (except for those needed to track cytoplasmic DNA movement). The optimal amount of allelic diversity depends on the question being addressed; however, numerous alleles with approximately equivalent frequencies give the greatest power for most population-level analyses. None of the systems discussed above meet all these criteria. However, with careful selection, appropriate techniques already exist for studying a myriad of population-level questions. In fields close to our own research interests, we believe that paternity analyses should flourish in the near future, leading to a better understanding of the factors controlling male fitness and its role in the evolution of floral traits. The identification of multilocus genotypes using species-specific markers will make it possible to measure the relative fitness of various hybrid and parental classes in a range of habitats, thereby elucidating the role of natural hybridization in plant evolution (see Arnold, 1997 for a review). Genome mapping and the identification of quantitative trait loci will allow us to uncover many details of hybridization and speciation (Bradshaw et al., 1995; Rieseberg et al., 1995b; 1996). Studies of population structure and diversity will be of particular importance for the field of conservation biology (Chase et al., 1996; Hamrick and Godt, 1996a; Rieseberg and Swensen, 1996). Such studies will also lead to a greater appreciation of the effect of shared biogeographic history on the evolution of plant lineages (Avise, 1994). The questions are endless, and it seems certain that future technological developments will give rise to as many new and unforeseen questions as have developments since the first application of isozyme markers only 30 years ago.

LITERATURE CITED

Appels, R., and J. Dvorak. 1982. The wheat ribosomal DNA spacer region: its structure and variation in populations and among species. Theoretical and Applied Genetics **63**:337–348.

Arnold, M. L. 1992. Natural hybridization as an evolutionary process. Annual Review of Ecology and Systematics **23**:237–261.

Arnold, M. L. 1993. *Iris nelsonii:* origin and genetic composition of a homoploid hybrid species. American Journal of Botany **80**:577–583.

Arnold, M. L. 1997. Natural Hybridization and Evolution. Oxford University Press, Oxford.

Arnold, M. L., B. D. Bennett, and E. A. Zimmer. 1990. Natural hybridization between *Iris fulva* and *I. hexagona:* pattern of ribosomal DNA variation. Evolution **44**:1512–1521.

Arnold, M. L., C. M. Buckner, and J. J. Robinson. 1991. Pollen mediated introgression and hybrid speciation in Louisiana irises. Proceedings of the National Academy of Sciences U.S.A. **88**:1398–1402.

Arnold, M. L., J. J. Robinson, C. M. Buckner, and B. D. Bennett. 1992. Pollen dispersal and interspecific gene flow in Louisiana irises. Heredity **68**:399–404.

Ashley, M. V., and B. D. Dow. 1994. The use of microsatellite analysis in population biology: background, methods and potential applications. In Molecular Ecology and Evolution: Approaches and Applications, eds. B. Schierwater, B. Streit, G. P. Wagner, and R. DeSalle, pp. 185–201. Birkhäuser Verlag, Basel.

Asmussen, M. A., and A. Schnabel. 1991. Comparative effects of pollen and seed migration on the cytonuclear structure of plant populations. I. Maternal cytoplasmic inheritance. Genetics **128**:639–654.

Avise, J. C. 1994. Molecular Markers, Natural History and Evolution. Chapman & Hall, New York.

Boscherini, G., M. Morgante, P. Rossi, and G. G. Vendramin. 1994. Allozyme and chloroplast DNA variation in Italian and Greek populations of *Pinus leucodermis.* Heredity **73**:284–290.

Bradshaw, H. D., Jr., S. M. Wilbert, K. G. Otto, and D. W. Schemske. 1995. Genetic mapping of floral traits associated with reproductive isolation in monkeyflowers (*Mimulus*). Nature **376**:762–765.

Britten, R. J., and E. H. Davidson. 1971. Repetitive and non-repetitive DNA sequences and a speculation on the origins of evolutionary novelty. Quarterly Review of Biology **46**:111–133.

Britten, R. J., and D. E. Koehn. 1968. Repeated sequences in DNA. Science **161**:529–540.

Brown, A. H. D., and R. W. Allard. 1970. Estimation of the mating system in open-pollinated maize populations using isozyme polymorphisms. Genetics **66**:113–145.

Brown, W. M. 1980. Polymorphism in mitochondrial DNA of humans as revealed by restriction endonuclease analysis. Proceedings of the National Academy of Sciences U.S.A. **77**:3605–3609.

Brown, W. M., M. George, and A. C. Wilson. 1979. Rapid evolution of animal mitochondrial DNA. Proceedings of the National Academy of Sciences U.S.A. **76**:1967–1971.

Broyles, S. B., and R. Wyatt. 1990. Paternity analysis in a natural population of *Asclepias exaltata:* multiple paternity, functional gender, and the "pollen-donation hypothesis." Evolution **44**:1454–1468.

Broyles, S. B., and R. Wyatt. 1995. A reexamination of the pollen-donation hypothesis in an experimental population of *Asclepias exaltata.* Evolution **49**:89–99.

Broyles, S. B., A. Schnabel, and R. Wyatt. 1994. Evidence for long-distance pollen dispersal in milkweeds (*Asclepias exaltata*). Evolution **48**:1032–1040.

Bruford, M. W., D. J. Cheesman, T. Coote, H. A. A. Green, S. A. Haines, C. O'Ryan, and T. R. Williams. 1996.

Microsatellites and their application to conservation genetics. **In** Molecular Genetic Approaches in Conservation, eds. T. B. Smith and R. K. Wayne, pp. 278–297. Oxford University Press, Oxford.

Chalmers, K. J., R. Waugh, J. I. Sprent, A. J. Simons, and W. Powell. 1992. Detection of genetic variation between and within populations of *Gliricidia sepium* and *G. maculata* using RAPD markers. Heredity **69**:465–472.

Chase, M. W., D. E. Soltis, R. G. Olmstead, D. Morgan, D. H. Les, B. D. Mishler, M. R. Duvall, R. A. Price, H. G. Hills, Y.-L. Qiu, K. A. Kron, J. H. Rettig, E. Conti, J. D. Palmer, J. R. Manhart, K. J. Sytsma, H. J. Michaels, W. J. Kress, K. G. Karol, W. D. Clark, M. Hedren, B. S. Gaut, R. K. Jansen, K.-J. Kim, C. F. Wimpee, J. F. Smith, G. R. Furnier, S. H. Strauss, Q.-Y. Xiang, G. M. Plunkett, P. S. Soltis, S. M. Swensen, S. E. Williams, P. A. Gadek, C. J. Quinn, L. E. Eguiarte, E. Golenberg, G. H. Learn, Jr., S. W. Graham, S. C. H. Barrett, S. Dayanandan, and V. A. Albert. 1993. Phylogenetics of seed plants: an analysis of nucleotide sequences from the plastid gene *rbc*L. Annals of the Missouri Botanical Garden **80**:525–580.

Chase, M., R. Kesseli, and K. Bawa. 1996. Microsatellite markers for population and conservation genetics of tropical trees. American Journal of Botany **83**:51–57.

Chunwongse, J., T. B. Bunn, C. Crossman, J. Jiang, and S. D. Tanksley. 1994. Chromosomal localization and molecular marker tagging of the powdery mildew resistance gene (LV) in tomato. Theoretical and Applied Genetics **89**:76–79.

Craig, J., S. Fowler, L. A. Burgoyne, A. C. Scott, and H. W. J. Harding. 1988. Repetitive deoxyribonucleic acid (DNA) human genome variation. A concise review relevant to forensic biology. Journal of Forensic Science **33**:1111–1126.

Cruzan, M. B. 1998. Genetic markers in plant evolutionary ecology. Ecology 79: in press.

Cruzan, M. B., and M. L. Arnold. 1993. Ecological and genetic associations in an *Iris* hybrid zone. Evolution **47**:1432–1445.

Cruzan, M. B., and M. L. Arnold. 1994. Assortative mating and natural selection in an *Iris* hybrid zone. Evolution **48**:1946–1958.

Cruzan, M. B., M. L. Arnold, S. E. Carney, and K. R. Wollenberg. 1993. cpDNA inheritance in interspecific crosses and evolutionary inference in Louisiana irises. American Journal of Botany **80**:344–350.

Dawson, I. K., K. J. Chalmers, R. Waugh, and W. Powell. 1993. Detection and analysis of genetic variation in *Hordeum spontaneum* populations from Israel using RAPD markers. Molecular Ecology **2**:151–159.

Devlin, B., J. Clegg, and N. C. Ellstrand. 1992. The effect of flower production on male reproductive success in wild radish populations. Evolution **46**:1030–1042.

Dorado, O., L. H. Rieseberg, and D. M. Arias. 1992. Chloroplast DNA introgression in southern California sunflowers. Evolution **46**:566–572.

Dow, B. D., and M. V. Ashley. 1996. Microsatellite analysis of seed dispersal and parentage of saplings in bur oak, *Quercus macrocarpa*. Molecular Ecology **5**:615–627.

Dow, B. D., M. V. Ashley, and H. F. Howe. 1995. Characterization of highly variable (GA/CT)$_n$ microsatellites in the bur oak, *Quercus macrocarpa*. Theoretical and Applied Genetics **91**:137–141.

Dowling, T. E., C. Moritz, J. D. Palmer, and L. H. Rieseberg. 1996. Nucleic acids III. Analysis of fragments and restriction sites. **In** Molecular Systematics, second edition, eds. D. M. Hillis, C. Moritz, and B. K. Mable, pp. 249–320. Sinauer Associates, Sunderland, Massachusetts.

Ellsworth, D. L., K. D. Rittenhouse, and R. L. Honeycutt. 1993. Artifactual variation in randomly amplified polymorphic DNA banding patterns. BioTechniques **14**:214–217.

Emms, S. K., S. A. Hodges, and M. L. Arnold. 1996. Pollen-tube competition, siring success, and consistent asymmetric hybridization in Louisiana irises. Evolution **50**:2201–2206.

Ferris, C., R. P. Oliver, A. J. Davy, and G. M. Hewitt. 1993. Native oak chloroplasts reveal an ancient divide across Europe. Molecular Ecology **2**:337–344.

Hamrick, J. L., and R. W. Allard. 1972. Microgeographic variation in allozyme frequencies in *Avena barbata*. Proceedings of the National Academy of Sciences U.S.A. **69**:2100–2104.

Hamrick, J. L., and R. W. Allard. 1975. Correlations between quantitative characters and enzyme genotypes in *Avena barbata*. Evolution **29**:438–442.

Hamrick, J. L., and M. J. W. Godt. 1989. Allozyme diversity in plant species. **In** Plant Population Genetics, Breeding and Genetic Resources, eds. A. H. D. Brown, M. T. Clegg, A. L. Kahler, and B. S. Weir, pp. 43–63. Sinauer Associates, Sunderland, Massachusetts.

Hamrick, J. L., and M. J. W. Godt. 1996a. Conservation genetics of endemic plant species. **In** Conservation Genetics, Case Histories from Nature, eds. J. C. Avise and J. L. Hamrick, pp. 281–304. Chapman & Hall, New York.

Hamrick, J. L., and M. J. W. Godt. 1996b. Effects of life history traits on genetic diversity in plant species. Philosophical Transactions of the Royal Society B **34**: 988–992.

Hamrick, J. L., and L. R. Holden. 1979. Influence of microhabitat heterogeneity on gene frequency distribution and gametic phase disequilibrium in *Avena barbata*. Evolution **33**:521–533.

Hamrick, J. L., and J. D. Nason. 1996. Consequences of dispersal in plants. **In** Population Dynamics in Ecological Space and Time, eds. O. E. Rhodes, R. K. Chesser, and M. H. Smith, pp. 203–236. University of Chicago Press, Chicago.

Harder, L. D., and S. C. H. Barrett. 1995. Mating cost of large floral displays in hermaphrodite plants. Nature **373**:512–515.

Harris, H. 1966. Enzyme polymorphisms in man. Proceedings of the Royal Society of London, B **164**:298–310.

Hodges, S. A., and M. L. Arnold. 1994. Floral and ecological isolation and hybridization between *Aquilegia formosa* and *Aquilegia pubescens*. Proceedings of the National Academy of Sciences U.S.A. **91**:2493–2496.

Hubby, J. L., and R. C. Lewontin. 1966. A molecular approach to the study of genic heterozygosity in natural populations. I. The number of alleles at different loci in *Drosophila pseudoobscura.* Genetics **54:**577–594.

Husband, B. C., and S. C. H. Barrett. 1993. Multiple origins of self-fertilization in tristylous *Eichhornia paniculata* (Pontederiaceae): inferences from style morph and isozyme variation. Journal of Evolutionary Biology **6:**591–608.

Jacob, H. J., K. Linderpaintner, S. E. Lincoln, K. Kusumi, R. K. Bunker, Y. -P. Mao, D. Ganten, V. J. Dzau, and E. S. Lander. 1991. Genetic mapping of a gene causing hypertension in the stroke-prone spontaneously hypertensive rat. Cell **67:**213–224.

Jansen, R. K., and J. D. Palmer. 1987. A chloroplast DNA inversion marks an ancient split in the sunflower family (Asteraceae). Proceedings of the National Academy of Sciences U.S.A. **84:**5818–5822.

Jones, A. G., and J. C. Avise. 1996. Microsatellite analysis of maternity and the mating system in the gulf pipefish *Syngnathus scovelli,* a species with male pregnancy and sex-role reversal. Molecular Ecology **6:**203–213.

Jordan, W. C., M. W. Courtney, and J. E. Neigel. 1996. Low levels of intraspecific genetic variation at a rapidly evolving chloroplast DNA locus in North American duckweeds (Lemnaceae). American Journal of Botany **83:**430–439.

Kim, K. -J., R. K. Jansen, and B. L. Turner. 1992. Evolutionary implications of intraspecific chloroplast DNA variation in dwarf dandelions (*Krigia:* Asteraceae). American Journal of Botany **79:**708–715.

King, L. M. 1993. Origins of genotypic variation in North American dandelions inferred from ribosomal DNA and chloroplast DNA restriction enzyme analysis. Evolution **47:**136–151.

King, L. M., and B. A. Schaal. 1989. Ribosomal-DNA variation and distribution in *Rudbeckia missouriensis.* Evolution **43:**1117–1119.

Learn, G. H., Jr., and B. A. Schaal. 1987. Population subdivision for ribosomal DNA repeat variants in *Clematis fremontii.* Evolution **41:**433–438.

Levin, D. A. 1975. Interspecific hybridization, heterozygosity and gene exchange in *Phlox.* Evolution **29:**37–56.

Lewontin, R. C. 1974. The Genetic Basis of Evolutionary Change. Columbia University Press, New York.

Lewontin, R. C. 1991. Twenty-five years ago in genetics. Electrophoresis in the development of evolutionary genetics: milestone or millstone? Genetics **128:**657–662.

Lewontin, R. C., and J. L. Hubby. 1966. A molecular approach to the study of genic heterozygosity in natural populations. II. Amount of variation and degree of heterozygosity in natural populations of *Drosophila pseudoobscura.* Genetics **54:**595–609.

Mazrimas, J. A., and F. T. Hatch. 1972. A possible relationship between satellite DNA and the evolution of Kangaroo Rat species (genus *Dipodomys*). Nature New Biology **240:**102–105.

McCauley, D. E. 1994. Contrasting the distribution of chloroplast DNA and allozyme polymorphism among local populations of *Silene alba:* Implications for studies of gene flow in plants. Proceedings of the National Academy of Sciences U.S.A. **91:**8127–8131.

McCauley, D. E. 1995. The use of chloroplast DNA polymorphism in studies of gene flow in plants. Trends in Ecology and Evolution **10:**198–202.

McCauley, D. E., J. E. Stevens, P. A. Peroni, and J. A. Raveill. 1996. The spatial distribution of chloroplast DNA and allozyme polymorphisms within a population of *Silene alba* (Caryophyllaceae). American Journal of Botany **83:**727–731.

Meagher, T. R. 1986. Analysis of paternity within a natural population of *Chamaelirium luteum.* 1. Identification of most-likely male parents. American Naturalist **128:** 199–215.

Meagher, T. R. 1991. Analysis of paternity within a natural population of *Chamaelirium luteum.* II. Patterns of male reproductive success. American Naturalist **137:** 738–752.

Meksem, K., D. Leister, J. Peleman, M. Zabeau, F. Salamini, and C. Gebhardt. 1995. A high-resolution map of the vicinity of the *R1* locus on chromosome V of potato based on RFLP and AFLP markers. Molecular and General Genetics **249:**74–81.

Mueller, U. G., S. E. Lipari, and M. G. Milgroom. 1996. Amplified fragment length polymorphism (AFLP) fingerprinting of symbiotic fungi cultured by the fungus-growing ant *Cyphomyrmex minutus.* Molecular Ecology **5:**119–122.

Myers, R. M., V. C. Sheffield, and D. R. Cox. 1988. Detection of single base changes in DNA: ribonuclease cleavage and denaturing gel gradient electrophoresis. In Genome Analysis: A Practical Approach, ed. K. Davies, pp. 95–139. IRL Press, Oxford.

Nei, M. 1972. Genetic distance between populations. American Naturalist **106:**283–292.

Nybom, H., and B. A. Schaal. 1990. DNA "fingerprints" reveal genotypic distributions in natural populations of blackberries and raspberries (*Rubus,* Rosaceae). American Journal of Botany **77:**883–888.

Paige, K. N., W. C. Capman, and P. Jennetten. 1991. Mitochondrial inheritance patterns across a cottonwood hybrid zone: cytonuclear disequilibria and hybrid zone dynamics. Evolution **45:**1360–1369.

Palmer, J. D. 1990. Contrasting modes and tempo of genome evolution in land plant organelles. Trends in Genetics **6:**115–120.

Pemberton, J. M., J. Slate, D. R. Bancroft, and J. A. Barrett. 1995. Nonamplifying alleles at microsatellite loci: a caution for parentage and population studies. Molecular Ecology **4:**249–252.

Philbrick, C. T. 1993. Underwater cross-pollination in *Callitriche hermaphroditica* (Callitrichaceae): evidence from random amplified polymorphic DNA markers. American Journal of Botany **80:**391–394.

Powell, W., M. Morgante, R. McDevitt, G. G. Vendramin, and J. A. Rafalski. 1995. Polymorphic simple sequence repeat regions in chloroplast genomes: applications to the population genetics of pines. Proceedings of the National Academy of Sciences U.S.A. **92:**7759–7763.

Prodohl, P. A., W. J. Loughry, C. M. McDonough, W. S. Nelson, and J. C. Avise. 1996. Molecular documentation of polyembryony and the micro-spatial dispersion of clonal sibships in the nine-banded armadillo, *Dasypus novemcinctus*. Proceedings of the Royal Society of London B **263**:1643–1649.

Ragot, M., and D. A. Hoisington. 1993. Molecular markers for plant-breeding—comparisons of RFLP and RAPD genotyping costs. Theoretical and Applied Genetics **86**:975–984.

Randolph, L. F. 1966. *Iris nelsonii*, a new species of Louisiana iris of hybrid origin. Baileya **14**:143–169.

Rassmann, K., C. Schlotterer, and D. Lautz. 1991. Isolation of simple-sequence loci for use in polymerase chain reaction-based DNA fingerprinting. Electrophoresis **12**:113–118.

Riedy, M. F., W. J. Hamilton, and C. F. Aquadro. 1992. Excess of non-parental bands in offspring from known pedigrees assayed using RAPD PCR. Nucleic Acids Research **20**:918.

Rieseberg, L. H. 1996. Homology among RAPD fragments in interspecific comparisons. Molecular Ecology **5**:99–105.

Rieseberg, L. H., and D. Gerber. 1995. Hybridization in the Catalina Island mountain mahogany (*Cercocarpus traskiae*): RAPD evidence. Conservation Biology **9**:199–203.

Rieseberg, L. H., and D. E. Soltis. 1991. Phylogenetic consequences of cytoplasmic gene flow in plants. Evolutionary Trends in Plants **5**:65–84.

Rieseberg, L. H., and S. M. Swensen. 1996. Conservation genetics of endangered island plants. **In** Conservation Genetics, Case Histories from Nature, eds. J. C. Avise and J. L. Hamrick, pp. 305–334. Chapman & Hall, New York.

Rieseberg, L. H., D. E. Soltis, and J. D. Palmer. 1988. A molecular reexamination of introgression between *Helianthus annuus* and *H. bolanderi* (Compositae). Evolution **42**:227–238.

Rieseberg, L. H., S. Beckstrom-Sternberg, and K. Doan. 1990. *Helianthus annuus* ssp. *texanus* has chloroplast DNA and nuclear ribosomal RNA genes of *Helianthus debilis* ssp. *cucumerifolius*. Proceedings of the National Academy of Sciences U.S.A. **87**:593–597.

Rieseberg, L. H., C. Hichang, R. Chan, and C. Spore. 1993. Genomic map of a diploid hybrid species. Heredity **70**:285–293.

Rieseberg, L. H., C. R. Linder, and G. J. Seiler. 1995a. Chromosomal and genic barriers to introgression in *Helianthus*. Genetics **141**:1163–1171.

Rieseberg, L. H., C. Van Fossen, and A. M. Desrochers. 1995b. Hybrid speciation accompanied by genomic reorganization in wild sunflowers. Nature **375**:313–316.

Rieseberg, L. H., B. Sinervo, C. R. Linder, M. C. Ungerer, and D. M. Arias. 1996. Role of gene interactions in hybrid speciation: evidence from ancient and experimental hybrids. Science **272**:741–745.

Roder, M. S., J. Plaschke, S. U. Konig, A. Borner, M. E. Sorrells, S. D. Tanksley, and M. W. Ganal. 1995. Abundance, variability and chromosomal location of microsatellites in wheat. Molecular and General Genetics **246**:327–333.

Rogstad, S. H., K. Wolff, and B. A. Schaal. 1991. Geographical variation in *Asimina triloba* Dunal (Annonaceae) revealed by the M13 "DNA fingerprinting" probe. American Journal of Botany **78**:1391–1396.

Saghai-Maroof, M. A., K. M. Soliman, R. A. Jorgensen, and R. W. Allard. 1984. Ribosomal DNA spacer-length polymorphisms in barley: Mendelian inheritance, chromosomal location, and population dynamics. Proceedings of the National Academy of Sciences U.S.A. **81**:8014–8018.

Saghai-Maroof, M. A., Q. Zhang, D. B. Neale, and R. W. Allard. 1992. Associations between nuclear loci and chloroplast DNA genotypes in wild barley. Genetics **131**:225–231.

Saghai-Maroof, M. A., R. M. Biyashev, G. P. Yang, Q. Zhang, and R. W. Allard. 1994. Extraordinarily polymorphic microsatellite DNA in barley: species diversity, chromosomal locations, and population dynamics. Proceedings of the National Academy of Sciences U.S.A. **91**:5466–5470.

Schaal, B. A., W. J. Leverich, and J. Nieto-Sotelo. 1987. Ribosomal DNA variation in the native plant *Phlox divaricata*. Molecular Biology and Evolution **4**:611–621.

Schnabel, A., and M. A. Asmussen. 1989. Definition and properties of disequilibria within nuclear-mitochondrial-chloroplast and other nuclear-dicytoplasmic systems. Genetics **123**:199–215.

Schnabel, A., and M. A. Asmussen. 1992. Comparative effects of pollen and seed migration on the cytonuclear structure of plant populations. II. Paternal cytoplasmic inheritance. Genetics **132**:253–267.

Schoen, D. J., and S. C. Stewart. 1986. Variation in male reproductive investment and male reproductive success in white spruce. Evolution **40**: 1109–1120.

Slatkin, M. 1995. A measure of population subdivision based on microsatellite allele frequencies. Genetics **139**:457–462.

Soltis, D. E., and P. S. Soltis (eds.). 1989. Isozymes in Plant Biology. Dioscorides Press, Portland, Oregon.

Soltis, D. E., P. S. Soltis, T. A. Ranker, and B. D. Ness. 1989. Chloroplast DNA variation in a wild plant, *Tolmiea menziesii*. Genetics **121**:819–826.

Soltis, D. E., M. S. Mayer, P. E. Soltis, and M. Edgerton. 1991. Chloroplast-DNA variation in *Tellima grandiflora* (Saxifragaceae). American Journal of Botany **78**:1379–1390

Soltis, D. E., P. S. Soltis, and B. G. Milligan. 1992. Intraspecific chloroplast DNA variation: systematic and phylogenetic implications. **In** Molecular Systematics of Plants, eds. P. S. Soltis, D. E. Soltis, and J. J. Doyle, pp. 117–150. Chapman & Hall, New York.

Soltis, D. E., M. A. Gitzendanner, D. D. Strenge, and P. S. Soltis. 1997. Chloroplast DNA intraspecific phylogeography of plants from the Pacific Northwest of North America. Plant Systematics and Evolution **206**: 353–373.

Stanton, M. L., A. A. Snow, and S. N. Handel. 1986. Floral evolution: attractiveness to pollinators increases male fitness. Science **232**:1625–1627.

Strand, A. E., B. G. Milligan, and C. M. Pruitt. 1996. Are populations islands? Analysis of chloroplast DNA variation in *Aquilegia.* Evolution **50:**1822–1829.

Thomson, J. D., E. A. Herre, J. L. Hamrick, and J. L. Stone. 1991. Genetic mosaics in strangler fig trees: implications for tropical conservation. Science **254:** 1214–1216.

Travis, S. E., J. Maschinski, and P. Keim. 1996. An analysis of genetic variation in *Astragalus cremnophylax* var. *cremnophylax,* a critically-endangered plant, using AFLP markers. Molecular Ecology **5:**735–745.

Valdes, A. M., M. Slatkin, and N. B. Freimer. 1993. Allele frequencies at microsatellite loci: the stepwise mutation model revisited. Genetics **133:**737–749.

Vos, P., R. Hogers, M. Bleeker, M. Reijans, T. van de Lee, M. Hornes, A. Frijters, J. Pot, J. Peleman, M. Kuiper, and M. Zabeau. 1995. AFLP: a new technique for DNA fingerprinting. Nucleic Acids Research **23:**4407–4414.

Wagner, D. B., G. R. Furnier, M. A. Saghai-Maroof, S. M. Williams, B. P. Dancik, and R. W. Allard. 1987. Chloroplast DNA polymorphisms in lodgepole and jack pines and their hybrids. Proceedings of the National Academy of Sciences U.S.A. **84:**2097–2100.

Watano, Y., M. Imazu, and T. Shimizu. 1995. Chloroplast DNA typing by PCR-SSCP in the *Pinus pumila-P. parviflora* var. *pentaphylla* complex (Pinaceae). Journal of Plant Research **108:**493–499.

Welsh, J., and M. McClelland. 1990. Fingerprinting genomes using PCR with arbitrary primers. Nucleic Acids Research **18:**7213–7218.

Whittemore, A. T., and B. A. Schaal. 1991. Interspecific gene flow in sympatric oaks. Proceedings of the National Academy of Sciences U.S.A. **88:**2540–2544.

Williams, J. G. K., A. R. Kubelik, K. J. Livak, J. A. Rafalski, and S. V. Tingey. 1990. DNA polymorphisms amplified by arbitrary primers are useful as genetic markers. Nucleic Acids Research **18:**6531–6535.

Zhao, X., and G. Kochert. 1993. Phylogenetic distribution and genetic mapping of a $(GGC)_n$ microsatellite from rice (*Oryza sativa* L.). Plant Molecular Biology **21:**607–614.

16

Genetic Mapping as a Tool for Studying Speciation

Loren H. Rieseberg

The field of molecular systematics has developed rapidly in response to innovations in molecular biology such as recombinant DNA technology, the polymerase chain reaction (PCR), and efficient and cost-effective methods of DNA sequencing. As revealed by the contents of this book, these methods have been particularly useful when applied to the reconstruction of phylogeny. However, phylogenetic reconstruction represents only one of several contributions molecular biology has made to studies of systematics and evolution.

Another important contribution has been the ability to detect and genetically map large numbers of molecular polymorphisms. Maps based on these molecular markers have contributed a great deal to our understanding of genome structure and evolution. By comparing maps generated for closely related species, the kinds of genomic changes that accompany or facilitate speciation can be inferred (Rieseberg, 1995; Song et al., 1995). For more distantly related species, map comparisons allow insights into mode and tempo of chromosomal evolution (Tanksley et al., 1988a; Whitkus et al., 1992). Completed maps also can be used to monitor the sizes and distribution of parental chromosomal segments in hybrid and introgressive populations (Young and Tanksley, 1989a & b) and allopolyploid species (Reinisch et al., 1994).

In addition to the study of genome evolution, mapped markers provide a means for dissecting the genetic basis of complex traits (Paterson et al., 1988; Tanksley, 1993; Knapp, 1994; Lander and Schork, 1994; Falconer and Mackay, 1996) such as morphological differences (Doebley and Stec, 1993; Hombergen and Bachmann, 1995), life history trade-offs (Mitchell-Olds, 1996), and reproductive barriers (Bradshaw et al., 1995). These kinds of traits are often controlled by many individual genes, which are referred to as polygenes or quantitative trait loci (QTL). The marker-based approach to analyzing QTL is relatively straightforward; in early segregating generations from any given cross, physical linkage leads to linkage disequilibrium between mapped markers and QTL. Given large enough progeny arrays, this makes it possible to detect and estimate the effects of almost all genes or QTL affecting a character (Paterson et al., 1988; Tanksley, 1993; Knapp, 1994; Lander and Schork, 1994; Falconer and Mackay, 1996). Students of speciation are interested in the genetic architecture of complex traits such as reproductive barriers for several reasons. For example, we can begin to address the fundamental question of how many genes are required for speciation (Coyne, 1992). If only one or two genes are necessary for reproductive barrier formation, then speciation may occur rapidly. Moreover, by

The author thanks Mike Arnold, Doug Soltis, Mark Ungerer, and Jeannette Whitton for helpful comments on the manuscript. The sunflower mapping studies discussed in this paper have been supported by grants from NSF and USDA.

mapping, counting, and identifying "speciation" genes, we can begin to reconstruct the sequence of genetic changes resulting in reproductive isolation and speciation (Coyne, 1992; Bradshaw et al., 1995).

This chapter reviews some of the major findings and opportunities resulting from the application of genetic mapping tools to the study of speciation. A discussion of technical issues relevant to mapping, such as marker choice, crossing design, and data analysis strategies, is followed by a detailed exploration of the uses of these maps for studying speciation. Specifically, I examine the genomic structure and composition of diploid hybrids, the role of gene interactions in reproductive isolation and speciation, the evolution of polyploid genomes, and the genetic basis of reproductive barriers. This is followed by a discussion of promising areas for future research and possible approaches that may facilitate studies in these areas. The review is largely restricted to land plants, but includes animal or fungal examples when there are no comparable studies in the plant literature. Likewise, the discussion of mapping is restricted to eukaryotic nuclear chromosomes rather than bacterial or cytoplasmic chromosomes.

Focusing this chapter on the evolution of reproductive barriers implies adherence to the biological species concept (Mayr, 1963). This is not the case, and, in fact, I have previously expressed concern about the limitations of this concept (Rieseberg and Brouillet, 1994). Nonetheless, I do believe that reproductive isolation, in combination with selection and drift, is responsible for generating gaps between otherwise morphologically homogeneous assemblages. Thus, it often plays a critical role in speciation.

APPROACHES TO CHROMOSOME MAPPING

Types of Maps

There are two broad categories of chromosome maps: physical maps and genetic maps. Physical maps are generated by the ordering of distinguishable DNA fragments along a chromosome. A wide variety of physical maps exist, including cytological maps, radiation hybrid maps, STS (sequence-tagged sites) maps, ordered clone collections, restriction maps, and ultimately the DNA sequence of an entire chromosome (Xiong et al., 1996). For large genomes such as those found in plants, the generation of ordered clone collections or contig maps appears to be the physical mapping method of choice (Schmidt et al., 1995). To date, complete contig maps are not available for any plant species, although a physical map for chromosome 4 from *Arabidopsis thaliana* was recently described (Schmidt et al., 1995). With the exception of cytological maps, physical mapping has not been applied to studies of speciation in plants.

Genetic maps are based on the principle that the degree of gene linkage, as measured by recombination frequencies, is correlated with differences in the physical distance separating genes on chromosomes (Lander et al., 1987; Tanksley et al., 1988b). Recombination frequencies between genes or molecular markers can be determined by monitoring meiotic segregation in crosses between individuals that are divergent at these loci. Genetic map distances are based on the frequency of crossing over between markers, where one map unit or centimorgan (cM) is equal to 1% recombination. The average ratio of genetic to physical distance varies widely across chromosomal regions within species, as well as across taxonomic groups. For example, 1 cM corresponds to 2.5–3.0 kb in yeast (Link and Olsen, 1991), ≈140 kb in *Arabidopsis* (Chang et al., 1988), and 350 kb in *Eucalyptus* (Byrne et al., 1995). Genetic maps that cover most of the nuclear genome can be rapidly generated for almost any species (Lander et al., 1987; Tanksley et al., 1988b), and maps now exist for most crop plants (Helentjaris and Burr, 1989; Gresshoff, 1994), as well as many wild plant species. Genetic maps have frequently been used to study speciation and are the focus of this review.

Marker Choice

Over the past two decades a remarkable variety of molecular markers has been developed. These markers differ widely in terms of variability, inheritance, and conservation over taxonomic distances (Table 16.1). Large differences also exist in the laboriousness, expense, and technical difficulty of the methods used to assay these markers (Table 16.1). Advantages and limitations of five classes of markers that are commonly employed for mapping in plants are discussed in the following paragraphs. Additional general information on these methods is provided in Chapters 2 and 15.

Isozymes   Many characteristics of isozymes are favorable for mapping. Isozyme loci exhibit intermediate levels of allelic variability, alleles are typically codominant, and most loci are conserved over land plants (Table 16.1). From a technical standpoint, isozymes are easily and inexpensively assayed, and only small amounts of tissue are required. Nonetheless, the use of isozymes is limited by the small number of polymorphic loci available for mapping, typically fewer than 20.

Restriction Fragment Length Polymorphisms (RFLPs)   RFLPs have been the workhorse of genetic mapping studies since their initial description almost two decades ago (Botstein et al., 1980). Probes for RFLP analysis of anonymous nuclear loci typically are obtained from either cDNA or genomic libraries. From these libraries, clones are selected that hybridize to single or low-copy number sequences. In practice, this requires selecting clones that hybridize to only one or two restriction fragments. There appears to be no correlation between clone size and polymorphism levels (McCouch et al., 1988; Miller and Tanksley, 1990), but cDNA clones typically reveal considerably

more polymorphism than genomic clones, regardless of the enzyme used to generate the library (Landry et al., 1987; McCouch et al., 1988; Miller and Tanksley, 1990).

Like isozymes, RFLP loci are codominant, exhibit intermediate levels of allelic variability, and often can be successfully assayed throughout a plant family (Tanksley et al., 1988a; Whitkus et al., 1992; Teutonico and Osborn, 1994). Although RFLP analysis is more challenging technically than isozyme or random amplified polymorphic DNA (RAPD) analyses (see next section), RFLP methods are well-worked out and should not pose significant difficulties if proper techniques are employed. Although RFLPs are commonly viewed as more laborious and expensive than RAPDs, relative cost and labor actually appear to vary with sample size (Ragot and Hoisington, 1993). According to Ragot and Hoisington (1993), RFLPs are more time- and cost-efficient for large samples (although see following), whereas RAPDs have the advantage for smaller sample sizes. Both methods appear to be more expensive and time-consuming than isozymes and less expensive than microsatellite loci.

Unlike isozymes, RFLPs are not limited by locus number. For example, over 1,000 RFLP loci have been placed on the genomic maps of tomato (Tanksley et al., 1992), potato (Tanksley et al., 1992), rice (Causse et al., 1994; Harushima et al., 1996), and sorghum (Chittenden et al., 1994; Paterson, 1996), and comparable mapping efforts are ongoing in many other crop plants. On the other hand, RFLP analysis does require large quantities of DNA. This is alleviated somewhat by nylon filters that can be

Table 16.1.   Characteristics of molecular markers frequently employed for genetic mapping.

Marker →	AFLPs	Isozymes	Microsatellites	RAPDs	RFLPs
Development time	Low	Low	High	Low	Moderate
Development cost	Low	Low	High	Low	Moderate
Technical difficulty	Moderate	Low	Moderate	Low	Moderate
Post development:					
Efficiency	High	Moderate	Low	Moderate	Moderate
Cost	Moderate	Low	Moderate	Moderate	Moderate
Inheritance	Dominant	Codominant	Codominant	Dominant	Codominant
Variability	Moderate	Moderate	High	Moderate	Moderate
Conservation	Low	High	Moderate	Low	High

reused many times. Nonetheless, DNA amount can be a limitation for wild plants that are small in size or from which large quantities of DNA are difficult to extract.

Random Amplified Polymorphic DNA

The RAPD method is deceptively simple; arbitrary 10-mer primers are deployed in a polymerase chain reaction to amplify random anonymous genomic sequences (Williams et al., 1990) (see Chapter 2 for technical details). Although developed only recently, RAPDs are now widely employed in genetic mapping of wild plant species (Rieseberg et al., 1993, 1995a; Bradshaw et al., 1995; Bachmann and Hombergen, 1996) and crops.

Major advantages of the RAPD method over RFLPs or microsatellites include the technical simplicity of the RAPD method and the large number of loci that can be screened quickly (Williams et al., 1990; Rieseberg and Ellstrand, 1993; Kesseli et al., 1994). The use of RFLPs or microsatellites requires the construction and screening of a genomic library for the appropriate markers, whereas RAPDs require no developmental steps prior to their use. In addition, RAPD primers tend to be consistent across flowering plant genera in terms of amplification strength. This "universality" of RAPD primers, combined with many loci amplified by each, greatly increases the speed with which markers can be screened. Moreover, methods such as bulked segregate analysis (Michelmore et al., 1991) allow rapid identification of markers associated with desired traits (Michelmore et al., 1991) or sex chromosomes (Mulcahy et al., 1992), or that differentiate varieties or species. These methods can greatly increase the efficiency of RAPD searches for character- or species-specific markers relative to RFLPs or microsatellites.

RAPDs also have several potential limitations for mapping studies. First, the majority of RAPD loci is dominant, which decreases linkage sensitivity and the accuracy of ordering in F_2 mapping populations. This problem can be overcome by increasing the size of the mapping population, progeny testing, developing codominant Sequence Characterized Amplified Regions (SCAR) markers (Kesseli et al., 1994), or employing a backcross breeding design (Rieseberg et al., 1993, 1995a). A second potential problem is the rapid loss of homology among RAPD products generated by the same primer as genetic distance increases (Williams et al., 1993; Thormann et al., 1994; Rieseberg, 1996). As a result, comparative mapping studies are difficult with RAPDs, and homology testing is necessary for species-level comparisons (Rieseberg et al., 1995a; Rieseberg, 1996). A third potential problem relates to reports of spurious bands and lack of reproducibility in RAPD surveys of both plants (Ellsworth et al., 1993; Ayliffe et al., 1994) and animals (Riedy et al., 1992). Under appropriate conditions, however, the vast majority of RAPD products exhibits strict Mendelian inheritance and complete reproducibility (Carlson et al., 1991; Heun and Helentjaris, 1993; Kesseli et al., 1994). It should also be pointed out that poor technique can lead to spurious bands and lack of reproducibility for the other types of markers discussed here (see Chapter 15).

Like RFLPs, RAPDs exhibit intermediate levels of variability; however, little DNA is required for RAPD assays. Although RAPDs appear to be roughly comparable to RFLPs in cost (see foregoing), much of the expense associated with RAPDs is due to the high cost of commercially prepared *Taq* polymerase needed for DNA amplification (Fritsch and Rieseberg, 1996). Fortunately, simple methods have been developed for in-house *Taq* production (Engelke et al., 1990; Pluthero, 1993). If the cost of commercial *Taq* is eliminated, then RAPDs are less expensive than RFLPs and roughly comparable in cost to isozymes (Fritsch and Rieseberg, 1996).

Microsatellites

Microsatellites consist of arrays of short tandemly repeated sequences that are assayed by the polymerase chain reaction; alleles differ in the number of repeat units in an array (Edwards et al., 1991). Microsatellites are highly variable, codominant, and require little DNA for analysis, all favorable properties for mapping (Edwards et al., 1991). However, plant microsatellites show little conservation over large evolutionary distances (Lagercrantz et al., 1993; Whitton et al., 1997), suggesting that their use in comparative mapping studies will be restricted to closely related species or genera.

The primary negative attribute of microsatellites relates to the laboriousness and expense of locus development. The development of an adequate number of loci for mapping (>100) requires several years and a substantial financial investment (>$100,000). This cost of development will substantially limit their use in wild plant species or in studies in which large numbers of species-specific markers are required.

Amplified Fragment Length Polymorphisms   AFLP analysis is a very recently developed technique based on the selective amplification of restriction fragments from digested genomic DNA (Vos et al., 1995). As with RAPDs, only small amounts of DNA are required, and DNA fingerprints can be generated without prior sequence knowledge using a limited number of generic primers. Studies published to date suggest that the AFLP technique is more repeatable between labs than RAPDs because reaction conditions used for primer annealing are much more stringent. Moreover, a much larger number of fragments (50–100) can be amplified and detected from a sigle AFLP reaction than for RAPDs (5–10).

Like RAPDs, however, most AFLP loci are dominant, exhibiting intermediate levels of variability, and homologous fragments typically are not conserved over large evolutionary distances (Table 16.1). However, comparative mapping of closely related species does appear feasible. Of the methods described here, AFLP analysis is probably the most challenging technically, but may represent the most efficient and cost-effective approach to mapping.

Mapping Strategy

The choice of plant materials is crucial to the success of mapping studies. Parental individuals should be from sufficiently divergent sources (e.g., geographic races or closely related species) to ensure adequate levels of polymorphism within the mapping population. On the other hand, wide species crosses should be avoided due to the reduced recombination rates and high levels of segregation distortion associated with divergent crosses. Because two generations of crossing are typically required to generate a mapping population, organisms with short generation times and ineffective seed dormancy mechanisms will enhance mapping efficiency. Likewise, self-compatible species are preferable to self-incompatible species for mapping because either self or backcross mapping populations can be generated.

Before the development of molecular markers, backcross populations were typically used for mapping because they allowed all genotypes of dominant morphological characters to be scored. Backcrosses are still the design of choice for mapping dominant molecular markers such as RAPDs or AFLPs for the same basic reason: All molecular genotypes can be scored. Likewise, backcrosses may represent the best approach for mapping in self-incompatible species, where selfing is not possible. If codominant markers are available, F_2 or F_3 populations are superior to backcrosses because two recombination events are detectable (i.e., recombination in both the mega- and microgametophyte can be detected) (Tanksley et al., 1988b). The additional recombination events increase both the maximum distance at which linkage can be detected, as well as the minimum resolvable map distance (Tanksley et al., 1988b). Likewise, F_2 or F_3 populations require fewer progeny for detection of QTL with additive effects, and QTL with partial or complete dominance can be efficiently mapped in only one rather than both backcross directions (Lander and Botstein, 1989).

The number of individuals employed for mapping varies depending on the planned use of the map. As few as 40 individuals are sufficient to detect linkage between markers separated by 30 cM (Tanksley et al., 1988b). However, map distances are more accurately estimated and map resolution is enhanced by larger progenies. Thus, most maps are based on populations ranging from 60 to 120 individuals.

Detection of polygenes or quantitative trait loci with small effects requires even larger populations, particularly for backcrosses (Lander and Botstein, 1989; Tanksley, 1993). Thus, populations for QTL analyses typically range between 100 and 500 individuals. However, even populations at the upper end of this range will fail to detect minor QTL and to differentiate between single QTL and separate linked genes. Because of the laboriousness and expense of surveying large populations, Lander and Botstein (1989) proposed the use of selective genotyping to reduce the number of progeny to be scored for DNA markers. Essentially, only progeny with extreme phenotypes need to be scored, and markers showing significant frequency differences in the extreme subpopulations are assumed to be linked to a QTL affecting the phenotype. However, Tanksley (1993) points out that this method is inefficient for estimating individual QTL effects and for mapping more than one trait, and requires collecting phenotypic information on a larger number of individuals.

The appropriate number of markers employed for mapping also depends on purpose and genome size. For comparative mapping, introgression analyses, and QTL studies, maps with markers every 10–20 cM are sufficient. Because most maps are less than 2,000 cM in length, between 100 and 200 evenly spaced markers would be required. Unfortunately, markers typically are not evenly spaced, and several hundred are usually required to achieve adequate resolution on most maps. However, after the maps have been completed, 100 to 200 evenly spaced markers can be chosen for QTL analyses. For analysis of chromosomal fragment sizes and distributions in synthesized and natural hybrids, the number of markers needed will vary with the age of the hybrid populations. Introgressed fragments are anticipated to be large in primary segregating populations (Fisher, 1949, 1953), thus requiring relatively low marker densities (10–20 cM) for accurate size estimates, whereas higher marker densities (<10 cM) will be needed for accurate fragment size determinations in natural hybrid species or hybrid zones.

Data Analysis

Accurate estimates of gene order require the analysis of several markers simultaneously or "multipoint" linkage analysis (Lander et al., 1987), which is typically accomplished using maximum likelihood methods (Lathrop et al., 1986). A variety of computer programs exists to implement the maximum likelihood approach,

including LINKAGE (Lathrop and Lalouel, 1984), LINKAGE-1 (Suiter et al., 1983), MAPMAKER (Lander et al., 1987), GMENDEL (Liu and Knapp, 1990), and MAPL (Ukai et al., 1995), and most of these are available to academic researchers without charge.

As with linkage analysis, the most powerful approach to QTL mapping requires the simultaneous analysis of the effects of multiple linked markers on a given character or interval mapping (Lander and Botstein, 1989). This basic approach has been further generalized by a number of authors (Haley and Knott, 1992; Zeng, 1993; Jansen, 1994; Hyne and Kearsey, 1995; Xu, 1995; Xu and Atchley, 1995) to allow genome-wide detection of QTLs in a variety of segregating populations, and several computer programs now exist to implement these strategies, including MAPMAKER-QTL (Lander and Botstein, 1989), MAPL (Ukai et al., 1995), and QTL Cartographer (Basten et al., 1996). Recent attention (Feingold et al., 1993; Dupuis, 1994; Kruglyak and Lander, 1995) has focused on the determination of appropriate threshold values for declaring significant QTL effects, and a discussion of various methods for obtaining threshold values is provided by Doerge and Rebai (1996).

Methods for estimating the sizes of introgressed fragments in hybrid or introgressive populations are fairly straightforward. Typically, lengths of introgressed fragments are estimated as the distance spanned by all consecutive introgressed markers, plus half the distance between the terminal introgressed markers and nearest nonintrogressed markers (Wang et al., 1995; Rieseberg et al., 1996b). In theory, these data can be used to estimate the age of hybrid zones or hybrid species, and the number, location, and selection coefficients of loci contributing to reduced hybrid fitness. Unfortunately, strategies for analyzing and interpreting this type of data are poorly developed. To interpret observed patterns of introgression correctly, we need to be able to predict the sizes and distribution of introgressing chromosomal fragments under neutral conditions, as well as under different selection intensities and migration rates. This is computationally difficult because of the need to keep track of large populations of chromosomes, each with very many loci.

Fisher (1949, 1953) developed a method that greatly simplifies the problem. Fisher realized that it was unnecessary to monitor all loci in the genome; rather, knowledge of the location of recombination points within chromosomally heterozygous regions would be sufficient to track parental chromosomal fragments in breeding populations. Fisher referred to these recombination breakpoints between heterogeneous chromosomal segments as "junctions." The junctions can be tracked over multiple generations like point mutations. Some will be lost, whereas others will proceed to fixation. Using this "method of junctions," Fisher (1949, 1953) was able to predict the sizes of parental linkage blocks in several breeding designs including sib-mating and selfing progenies. The theory was extended to tetrasomic plants by Bennett (1953) and to more complex mating designs by Hanson (1959a, 1959b).

Several computer programs are currently being developed that use junction theory to simulate the distribution of chromosomal block sizes in breeding populations (C. Linder, pers. comm.) and natural hybrid zones (Baird, 1995). In addition to the simulations, an analytical solution for estimating chromosomal block sizes in hybrid zones has been provided by Barton (1995). These approaches should be applicable to many multilocus problems, particularly the analysis of introgressive hybridization and hybrid speciation.

GENOMIC STRUCTURE AND COMPOSITION OF DIPLOID HYBRIDS

Primary Hybrids

Hybrid swarms and hybrid zones are often observed in nature, indicating that reproductive isolation is not complete in many plant and animal species (Harrison, 1990; Arnold, 1992; Rieseberg and Wendel, 1993). Nonetheless, many hybrid zones appear to represent substantial barriers to gene exchange (Barton and Hewitt, 1985, 1989). That is, the zones are often very narrow with little gene exchange outside of the area of contact. The genetic structure of these hybrid zones has been of interest to evolutionists, because observed patterns of introgres-

sion can be used to make inferences about the genetic architecture of species barriers (Barton and Hewitt, 1985, 1989). By genetic architecture, I mean the number, effects, interactions, and genomic distribution of factors contributing to reproductive isolation.

Another important issue that has been difficult to test to date is whether coadapted gene complexes represent an important component of species architecture. A long-held view is that species genes are highly integrated or coadapted and that combinations with alien genes or gene complexes will result in progeny of inferior viability that will be eliminated by selection (Harlan, 1936; Dobzhansky, 1941; Mayr, 1954; Rick, 1963; Carson and Templeton, 1984). Thus, coadapted gene complexes can be viewed as groups of genes that interact favorably, or a type of epistasis. Indirect evidence for coadapted gene complexes includes observations of hybrid breakdown in several plant genera (Stephens, 1949; Wall, 1970). The differential elimination of the donor parent genome in interspecific backcrosses also is suggestive of genome coadaptation (Stephens, 1949, 1950; Mangelsdorf, 1958; Rick, 1963; Wall, 1970). However, there is little direct genetic evidence for coadaptation or for its role in resisting introgression (although see Cabot et al., 1994).

Introgression mapping represents a powerful approach to the study of the genetic architecture of species barriers. The first step toward implementing this method is the generation of genetic linkage maps for the hybridizing species (Rieseberg et al., 1995a). The maps can then be compared to identify changes in gene order, and by inference, the structural changes differentiating the genomes of the parental species. After completion of these maps, hybrid and introgressive individuals can be surveyed for the presence of mapped parental molecular markers (Rieseberg et al., 1995b). The resulting "graphical genotypes" can be used to determine rates of introgression as well as the genomic distribution of introgressed chromosomal fragments (Young and Tanksley, 1989a & b). Likewise, the minimum number of genetic intervals contributing to reproductive isolation can be estimated from the distribution of introgressed chromosomal fragments (Rieseberg et al., 1996a). Finally, coadapted gene complexes can be detected by searching for associations among markers that are not physically linked (Rieseberg et al., 1996b).

Perhaps the best example of the utility of this approach comes from studies of two annual sunflower species, *Helianthus annuus* and *H. petiolaris* (Rieseberg et al., 1995a, 1995b, 1996a, 1996b). Both species are self-incompatible, share the same chromosome number ($n = 17$), and are abundant throughout much of the western United States. Nonetheless, they are distinguished by a number of morphological and chromosomal features, and have different ecological requirements. In general, *H. annuus* occurs in heavy, clay soils, whereas *H. petiolaris* is restricted to dry, sandy soils. However, these habitats often are found in close proximity throughout the central and southwestern United States (Heiser, 1947), resulting in the production of literally thousands of hybrid swarms or mosaic hybrid zones.

Comparative genetic linkage mapping using RAPD markers (Rieseberg et al., 1995a) indicates that seven linkage groups are collinear between the two species, whereas the remaining ten linkages differ structurally due to a minimum of seven interchromosomal translocations and three inversions (Fig. 16.1). These structural changes generate multivalent formations and bridges and fragments in hybrids (Heiser, 1947; Chandler et al., 1986), apparently leading to semisterility; F_1 pollen viabilities are typically less than 10%, and seed set is less than 1% (Heiser, 1947). Nonetheless, fertility is rapidly restored in backcross generations (Heiser, 1947).

To reveal the genetic architecture of the reproductive barrier between these two species, two kinds of studies have been performed. In the first experiment, 56 or 58 individuals from three greenhouse synthesized hybrid lineages (lineage I, P-F_1-BC_1-BC_2-F_2-F_3; lineage II, P-F_1-F_2-BC_1-BC_2-F_3; and lineage III, P-F_1-F_2-F_3-BC_1-BC_2) were surveyed for 197 mapped RAPD markers (Rieseberg et al., 1995b, 1996b), whereas the second study, which is ongoing, involves an analysis of genomic patterns of introgression in three natural hybrid zones from western Nebraska (L. Rieseberg and J. Whitton, unpubl.).

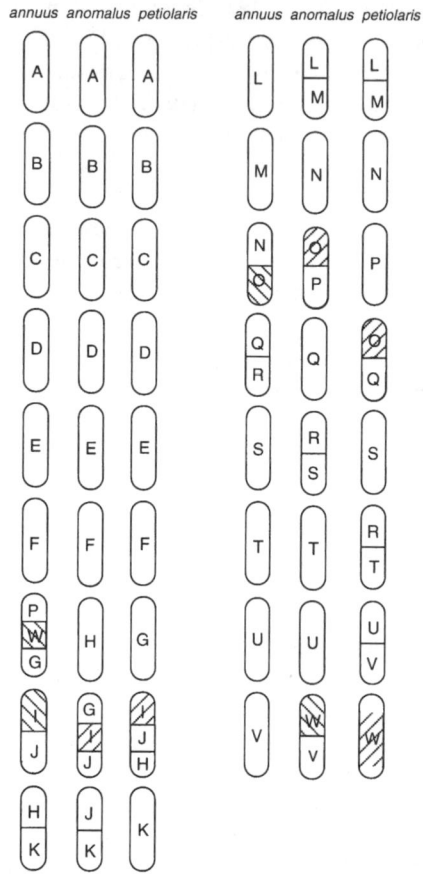

Figure 16.1. Linkage relationships between *Helianthus annuus*, *H. petiolaris*, and their putative hybrid derivative, *H. anomalus*, as inferred from comparative linkage mapping (Rieseberg et al., 1995a). Lines of shading within linkage groups indicate inversions.

The patterns of introgression observed in the synthetic hybrid lineages and natural hybrid zones were used to address several important problems in speciation. The first of these concerns the role of chromosomal structural differences as barriers to interspecific gene flow. Most plant and animal species that have been examined karyotypically differ in chromosome structure (White, 1978; Jackson, 1985). Because these structural changes often appear to be associated with meiotic abnormalities, karyotypic evolution has long been thought to play a major role in reproductive isolation and speciation (Stebbins, 1971; White, 1978; Jackson, 1985). However, this assumption has been questioned recently because individuals heterozygous for chromosomal rearrangements often show little meiotic impairment or loss of fertility (Sites and Moritz, 1987; Coyne et al., 1993).

Introgression analyses can be used to test directly the effects of chromosomal rearrangements on interspecific gene flow. If the chromosomal rearrangements contribute to reduced hybrid fitness, then linkage blocks carrying these rearrangements will be selected against in introgressed progeny. Analysis of the distribution of introgressed linkage blocks in both the synthetic hybrid lineages (Fig. 16.2) and in the natural hybrid zones (Fig. 16.3) revealed that introgression is significantly reduced in the rearranged versus collinear part of the genome (Rieseberg et al., 1995b, 1996a, 1996b). Thus, these data support the view that chromosomal rearrangements do provide significant barriers to gene exchange, particularly within rearranged linkage groups. They also suggest that species genomes are often differentially permeable to introgression, where certain portions of the genome are open to interspecific gene flow, but gene flow is restricted in other parts of the genome (Key, 1968; Carson, 1975; Barton and Hewitt, 1985).

These introgression patterns also can be used to address long-standing questions concerning the number of genes that contribute to reduced hybrid fitness or postmating reproductive barriers (Fisher, 1930; Carson and Templeton, 1984; Coyne, 1992). The rationale for this approach is similar to that described above for the chromosomal rearrangements; introgressed chromosomal fragments that contain sterility genes or genes that reduce hybrid fitness in some way should be selected against in introgressed progeny. Thus, genetic intervals from collinear linkage groups that exhibit reduced rates of introgression (i.e., "resist" introgression), most likely contain one or more genes that contribute to reduced hybrid fitness. By counting up the number of distinct intervals that resist introgression in both the synthetic and natural hybrids, minimum estimates of the number of genes contributing to reduced hybrid fitness can be obtained.

This approach worked very well in *Helianthus* (Rieseberg et al., 1996b). The same ge-

netic intervals resisted introgression in all three synthetic hybrid lineages (Fig. 16.2), and the data from the three natural hybrids, although incomplete, are concordant (Fig. 16.3). Within the collinear portion of the genome, 14 genomic intervals appear to resist introgression, or a minimum of 14 genes (Fig. 16.2). If we assume a similar distribution of sterility genes in the rearranged portion of the genome, then we can extrapolate that as many as 40 genes contribute to reduced hybrid fitness, in addition to the 10 chromosomal rearrangements described earlier. Of course this is a minimum estimate of gene number because each genetic interval may contain more than one gene affecting hybrid fitness. Nonetheless, these data do support a very complex genetic basis for postmating reproductive isolation in sunflower.

It is difficult to conceive of a plausible alternative explanation for these results. Because we obtained almost identical results in all three hybrid lineages, the patterns observed seem unlikely to be a result of stochastic factors such as genetic drift. A mechanical explanation based on physical interactions between chromosomes seems more likely, but does not account well for the complexity of epistasis observed among introgressed chromosomal fragments (below). Possibly, both mechanistic and selective forces have shaped the genomic composition of these lineages.

The multifactorial basis of reduced hybrid fitness reported between these two sunflower species is consistent with estimates of gene numbers from several animal studies. For example, based on the application of cline theory to well-characterized toad (Barton and Hewitt, 1981) and grasshopper (Szymura and Barton, 1991) hybrid zones, 55 and 150 genes, respectively, were estimated to affect hybrid fitness. Likewise, detailed studies of interspecific introgression in *Drosophila* indicate that more than 100 genes contribute to hybrid male sterility (Wu and Palopoli, 1994).

On the other hand, there are many examples in both plants and animals where one or two genes can have major effects on hybrid sterility or inviability; for example, cotton (Stephens, 1950; Gerstel, 1954), rice (Oka, 1974; Wan et al., 1996), wheat (Hermson, 1963), *Mimulus*

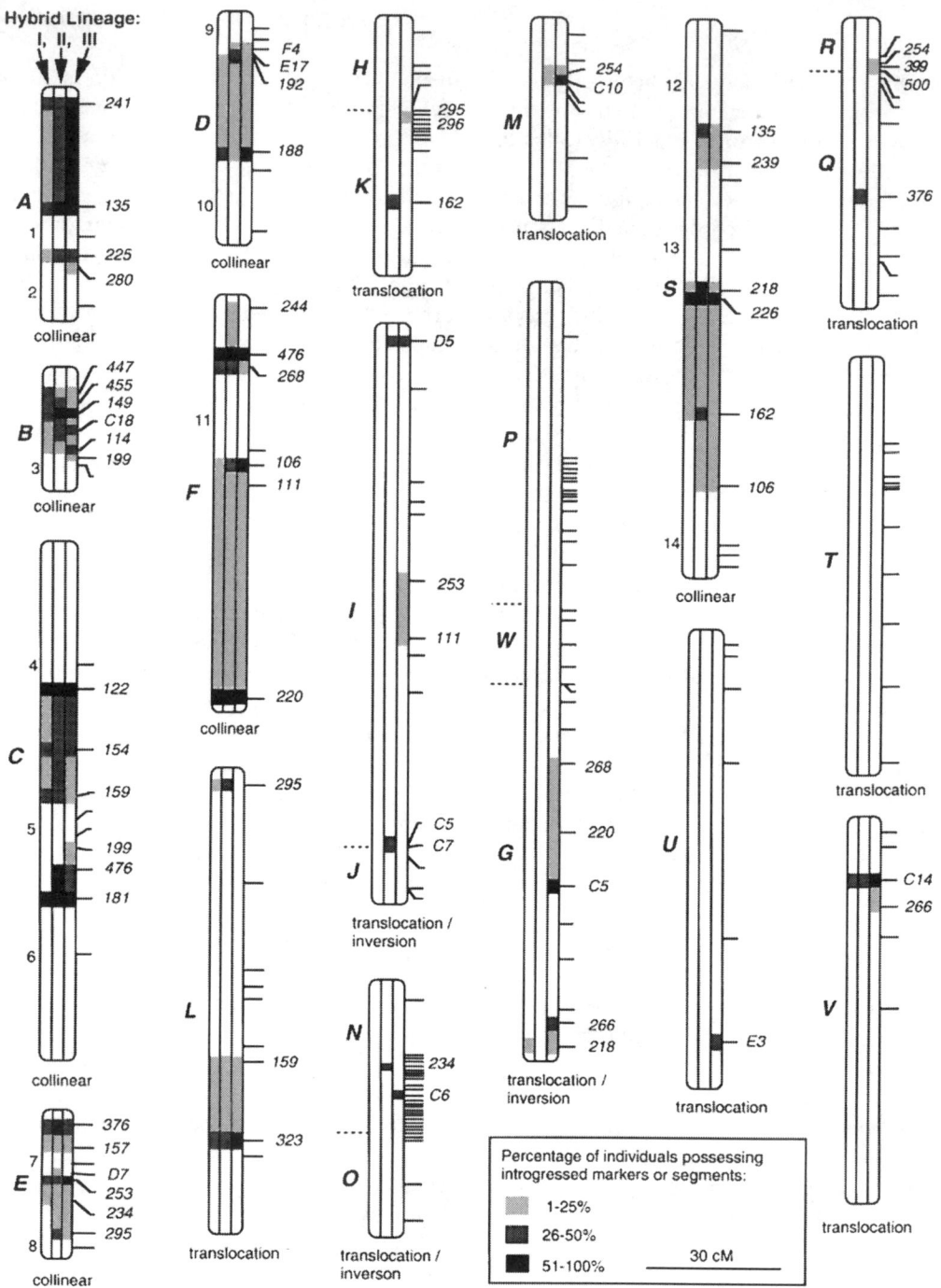

Figure 16.2. Composite graphical genotypes of three hybrid lineages generated between *Helianthus annuus* and *H. petiolaris:* lineage I, P-F₁-BC₂-F₂-F₃; lineage II, P-F₁-F₂-BC₁-BC₂-F₃; and lineage III, P-F₁-F₂-F₃-BC₁-BC₂. Graphical genotypes for lineages I, II, and III are in the left, center, and right side, respectively, of each linkage group. The graphical genotypes are based on the 1,084-cM map of *H. annuus* extended here by approximately 290 cM because several *H. petiolaris* markers occur outside currently mapped regions in *H. annuus* (Rieseberg et al., 1995a). Letters at the left of each linkage group designate major linkage blocks and indicate their relationship to homologous linkages in *H. petiolaris* (cf. Fig. 16.1). Horizontal lines extending to the right of linkage groups indicate the genomic location of the 197 *H. petiolaris* RAPD markers surveyed (Rieseberg et al., 1995b); introgressed markers are identified by primer number. Black or gray bars indicate the frequency of introgressed *H. petiolaris* markers. Numbers at the left of linkage groups indicate linkage blocks that resist introgression in all three lineages and thus probably contribute to hybrid unfitness.

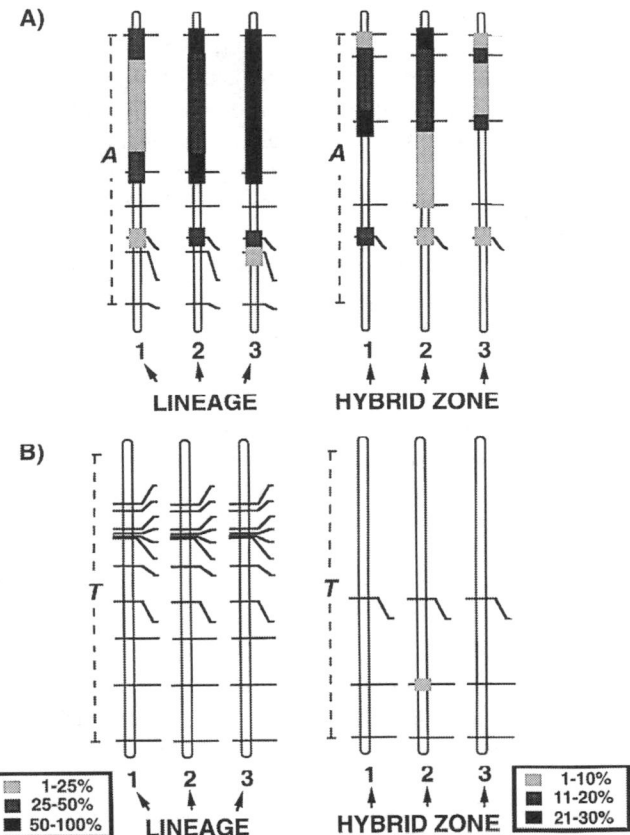

Figure 16.3. Comparison of introgression patterns between the three synthetic hybrid lineages and the three natural hybrid zones in *Helianthus:* A) collinear linkage group; B) rearranged linkage group. The identity of the two linkage groups is indicated by bold letters (cf. Fig. 16.1). Horizontal lines extending through the linkage groups indicate the genomic location of the *H. petiolaris* RAPD markers surveyed (Rieseberg et al., 1995b). Black or gray bars indicate the frequency of introgressed *H. petiolaris* markers.

(Macnair and Christie, 1983; Christie and Macnair, 1984), *Crepis* (Hollingshead, 1930), *Melilotus* (Sano and Kita, 1978), cowpea (Saunders, 1952), barley (Wiebe, 1934), and zebrafish (Wittbrodt et al., 1989). However, these observations do not rule out the possibility that many additional genes may affect hybrid fitness in these species as well. Hybrid fitness is influenced by factors acting throughout the life cycle of an organism, and most of the studies just listed focused on a single factor. Furthermore, the sunflower interspecific cross involves reasonably divergent species, whereas many of the above examples involve intraspecific crosses or crosses between sister taxa. Finally, the experimental designs employed for most of these studies were not powerful enough to detect genic factors with small effects.

There also is substantial indirect evidence from introgression mapping studies that suggests that postmating reproductive isolation in plants often has a complex genetic basis. For instance, the presence of numerous genomic intervals where introgression is reduced or absent is often reported in map-based studies of introgression between crop plants and their wild relatives (Jena et al., 1992; Williams et al., 1993; McGrath et al., 1994, 1996; Wang et al., 1995; Garcia et al., 1995). Presumably, many of these genomic regions harbor genes that contribute to reproductive isolation. Likewise, strong segregation distortion is often observed in interspecific crosses, suggesting that many genes are negatively selected in hybrids. For example, Zamir and Tadmor (1986) report segregation distortion at 54% of loci from interspecific crosses of lentil, pepper, and tomato, compared to only 13% in intraspecific crosses.

Another interesting observation from introgression mapping studies in plants is that certain chromosomal regions often appear to introgress preferentially relative to other fragments. For example, comparisons of observed patterns of introgression with simulations of unrestricted marker introgression in sunflower revealed that 5–6% of *H. petiolaris* markers introgressed at significantly higher than expected rates (Rieseberg et al., 1995a, 1996a, 1996b). Analyses of introgression lines in rice (Jena et al., 1992), peanut (Garcia et al., 1995), and cotton (Wang et al., 1995) reveal similar patterns. In each case, certain fragments appear to be preferentially retained in the introgression lines, even when the lines were derived from different breeding programs. Presumably, these fragments carry genes that interact favorably in a new genetic background. As pointed out by Rieseberg et al. (1996b), this result is important because it contradicts the traditional view that interspecific gene interactions are uniformly negative (Dobzhansky, 1941; Mayr, 1963). Rieseberg et al. (1996b) further speculate that these favorable interspecific gene interactions might provide a source of genetic material for adaptive evolution, as was first postulated by Anderson (1949).

Although the signature of selection is most clearly revealed by the rate and distribution of introgressed chromosomal fragments, selection can also be detected by its effect on the sizes of introgressed fragments. The rationale for this approach is that large fragments are more likely to carry genic or chromosomal factors that contribute to reduced hybrid fitness than small ones. Thus, selection against alien genes should lead to a reduction in overall fragment length. This prediction is borne out in the *Helianthus* example described above (Rieseberg et al., 1996b). Introgressed fragment sizes from the collinear portion of the genome ($\bar{x} = 13.1$ cM) were about one half the length (26–33 cM) predicted by theory (Hanson, 1959a, 1959b). Introgressed fragments within rearranged linkages were even

smaller ($\bar{x} = 9.0$ cM). These observations are suggestive of strong genomewide selection against introgressed chromosomal fragments, particularly in the rearranged portion of the genome. Presumably, junction theory (Baird, 1995) will allow us to use these data to derive selection coefficients at individual loci. Shorter than expected fragment sizes have also been reported for introgressed segments from a wild rice species (*Oryza officinalis*) into domesticated rice (*O. sativa*) and were attributed to a "nonconventional recombination mechanism" (Jena et al., 1992).

Although introgressed fragment lengths in *Helianthus* were shorter than predicted by theory, the expected increase in fragment size with increasing chromosome map length was observed (Fig. 16.4). Introgressed fragment sizes in hybrids and their offspring will be controlled in large part by interactions between selection, recombination rate, and chromosome map length.

Another important aspect of species genetic architecture that can be addressed through intro-

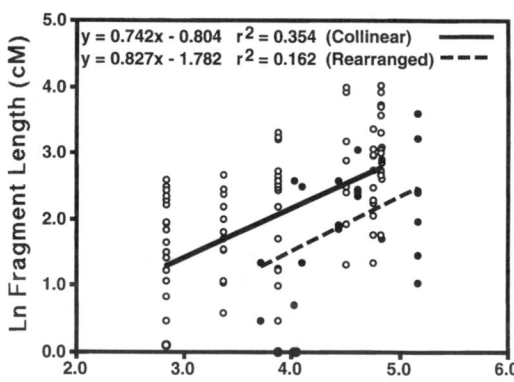

Figure 16.4. Linear regressions of the relationship between ln(chromosome map length) and ln(introgressed fragment length) in the collinear and rearranged portions of the *Helianthus* genome. Solid regression lines are for the collinear linkage groups and dashed lines are for the rearranged linkages. Regression lines have not been extrapolated beyond the data. Note that fragment lengths are significantly longer for collinear linkage groups and that chromosome map length does have a significant effect on fragment length, with longer fragments occurring on longer chromosomes.

gression mapping studies is the role of genome coadaptation or epistatic interactions in reproductive isolation. If coadapted genes represent an important component of species architecture, then significant associations or linkage disequilibria should be observed among loci that are not physically linked (Rieseberg et al., 1996b). That is, selection should retain coadapted gene complexes. In the sunflower example described above, the existence and strength of these epistatic interactions were tested by analyzing all unlinked introgressed markers for significant two-way and three-way associations (Rieseberg et al., 1996b).

The results from these analyses are compelling. Ten or more significant two-way associations were observed in each of the three synthetic hybrid lineages. In the more powerful three-way analysis, 21, 29, and 15 significant three-way associations were observed, generating the complex epistatic webs shown in Fig. 16.5. Even though very stringent significance levels were employed in this analysis ($\alpha \leq 0.0001$), many of the same two- and three-way associations were observed in multiple hybrid lineages (Rieseberg et al., 1996b). Because the hybrid lineages were generated independently, selection rather than random population bottle-

necks must account for these associations. Moreover, markers with epistatic interactions were more often found in all three lineages than markers lacking epistasis, suggesting that these interactions influence hybrid genomic composition (Fig. 16.1). These data therefore strongly support the view that coadapted gene complexes do represent an important component of species architecture.

HYBRID SPECIES

Genetic mapping also has much to offer studies of hybrid speciation. By comparing maps of parental species and their hybrid derivatives, the genomic processes that accompany or facilitate hybrid speciation can be inferred. For example, changes in gene order between the parental and hybrid species can be used to infer the kinds of chromosomal structural changes that accompany hybrid speciation. Likewise, analysis of the genomic distribution of species-specific markers in the hybrid species can be used to determine how the genomes of the parental species merged to form the genome of a new species. The genomic composition of the hybrid species can also be compared with that of synthetic hybrids to provide further insights regarding the processes that constrain hybrid genomic composition and ultimately hybrid speciation.

As with the primary hybrids, most of the relevant mapping data for the study of hybrid species comes from work on the annual sunflowers of the genus *Helianthus* (Rieseberg et al., 1993, 1995a, 1996b). Earlier phylogenetic studies of the group suggested that at least three of 11 species comprising *Helianthus* sect. *Helianthus* are stabilized diploid hybrid derivatives (Rieseberg, 1991). Moreover, all three hybrid derivatives, *H. anomalus, H. paradoxus,* and *H. deserticola,* appeared to be derived from the same two parents, *H. annuus* and *H. petiolaris.* Like their parents, the three putative hybrid species are self-incompatible and have a haploid chromosome number of 17. The three hybrids differ from their parents, however, in terms of geographic distribution and habitat preferences. *Helianthus paradoxus* is endemic to saline brackish marshes in west Texas, whereas *H. anomalus* and *H. deserticola* are xeric species

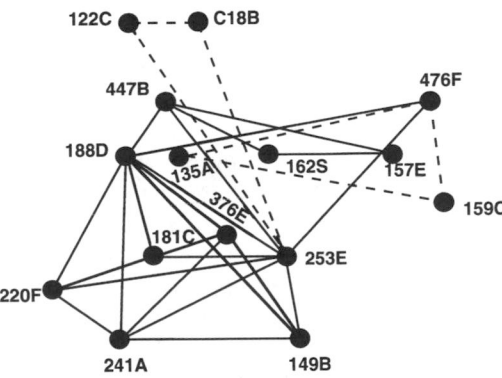

Figure 16.5. Genetic web of three-way epistatic interactions among introgressed chromosomal segments in hybrid lineage I. Marker designations indicate primer number and linkage block (cf. Fig. 16.2). Any three markers that interacted epistatically are connected as a triangle. Positive associations are indicated by solid lines and negative by dashed lines. For example, the three unlinked markers 241_A, 149_B (web base), and 376_E (web interior) were positively associated as indicated by the connecting triangle.

restricted to the Great Basin desert of the southwestern United States. As described in the preceding section, the parental species are widespread throughout the central and western portion of the United States, with *H. annuus* found primarily in mesic soils and *H. petiolaris* found in dry, sandy soils.

Currently, genetic linkage maps based on RAPDs are available for the two parental species, *H. annuus* and *H. petiolaris,* and one of the hybrid species, *H. anomalus* (Rieseberg et al., 1993; 1995a). A partial map has been generated for *H. paradoxus* (L. Rieseberg and A. Desrochers, unpubl.), but the polymorphism levels in the mapping population are low, and the map is poorly resolved.

The first question addressed by these maps was the extent of karyotypic change in *H. anomalus* relative to its parents (Rieseberg et al., 1995a). The extent of change is important because genetic models of hybrid speciation indicate that rapid karyotypic evolution can facilitate the development of reproductive isolation between the new hybrid genotype and its parents (Stebbins, 1957; Grant, 1958; Templeton, 1981), allowing hybrid speciation to occur sympatrically (McCarthy et al., 1995). Gene order comparisons revealed that six of the 17 linkage groups were collinear among all three species, whereas the remaining 11 linkages were not conserved in terms of gene order (Fig. 16.1). The two parental species, *H. annuus* and *H. petiolaris,* differ by at least 10 separate structural rearrangements, including three inversions, and a minimum of seven interchromosomal translocations. The genome of the hybrid species, *H. anomalus,* was extensively rearranged relative to its parents (Fig. 16.1). For four of the 11 rearranged linkages, *H. anomalus* shared the linkage arrangement of one parent or the other. For the remaining seven linkages, however, unique linkage arrangements were displayed. In fact, a minimum of three chromosomal breakages, three fusions, and one duplication are required to achieve the *H. anomalus* genome from its parents. All seven novel rearrangements in *H. anomalus* involve linkage groups that are structurally divergent in the parental species, suggesting that structural differences may induce additional chromosomal rearrangements upon

recombination. Similar increases in chromosomal mutation rates following hybridization have previously been reported in grasshoppers (Shaw et al., 1983).

To reduce gene flow, chromosomal structural differences must enhance reproductive isolation. This does appear to be the case in *H. anomalus,* in which first generation hybrids with its parents are partially sterile, with pollen stainabilities of 1.8–4.1% (*H. annuus*) and 2–58.4% (*H. petiolaris*) (Heiser, 1958; Chandler et al., 1986). Meiotic analyses revealed multivalent formations and bridges and fragments suggesting that chromosomal structural differences are largely responsible for hybrid semisterility (Chandler et al., 1986). Thus, the rapid karyotypic evolution inferred from these mapping data does satisfy genetic models for speciation through hybrid recombination (Stebbins, 1957; Grant, 1958; Templeton, 1981).

In addition to the gene order comparisons, Rieseberg et al. (1995a) analyzed the genomic distribution of species-specific parental markers in the *H. anomalus* genome. Sixteen of 17 linkage groups contained markers from both parental species, indicating that these linkage groups were recombinant (Fig. 16.6). However, the remaining linkage group (linkage group T) contained *H. annuus* markers only, as did two large blocks on linkage groups JK and VW. Rieseberg et al. (1995a) speculated that a combination of epistasis and protection from recombination by structural differences in the parental genomes may have been responsible for the preservation of these linkage blocks.

To provide further insights into the genetic processes affecting hybrid speciation in *Helianthus,* Rieseberg et al. (1996b) compared the genomic composition of the ancient hybrid, *H. anomalus,* with that of individuals from the three synthetic hybrid lineages described in the preceding section (Fig. 16.6). The results were striking. Although generated independently, all three synthesized hybrid lineages converged to nearly identical gene combinations, and this set of gene combinations was recognizably similar to that found in *H. anomalus.* The remarkable concordance in genomic composition between the synthetic and ancient hybrids suggests that selection rather than chance largely governs

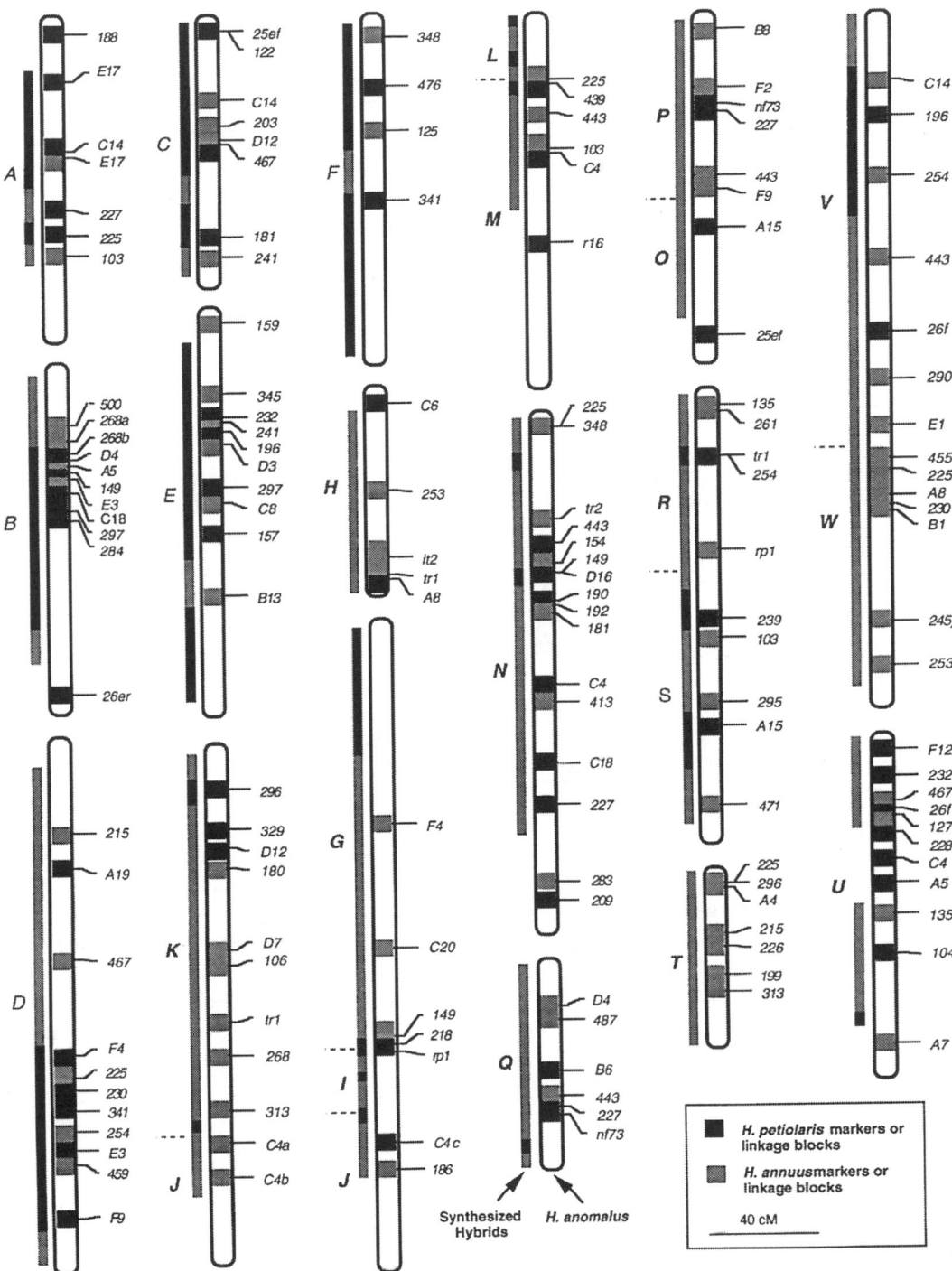

Figure 16.6. Genomic composition of experimental and ancient hybrids. Letters at the left of each linkage group designate major linkage blocks in *H. anomalus* and indicate their relationship to homologous linkages in the parental species (cf. Fig. 16.1). These letters also indicate whether the linkage groups are collinear (bold type) or rearranged (normal type) between the parental species (cf. Fig. 16.1). The distribution of parental markers within the *H. anomalus* genome is indicated by black or gray bars within linkage groups, whereas bars at the left of each linkage group indicate the distribution of parental genomic regions in the synthesized hybrids (cf. Fig. 16.2). Areas lacking information regarding parental origin are white.

hybrid species formation, that the origin of *H. anomalus* probably involved one or two initial backcross generations to *H. annuus,* and that hybrid speciation may be largely repeatable. Because the synthetic hybrid lineages were generated in the greenhouse rather than under natural conditions, congruence in genomic composition appears to result from fertility selection rather than selection for adaptation to a xeric habitat. Moreover, nonrandom rates of introgression and significant associations among unlinked markers in the synthesized hybrid lineages (discussed above) imply that interactions between coadapted parental species' genes constrain the genomic composition of hybrid species.

This study also addressed an ongoing debate about the repeatability of evolution and the relative importance of macroevolutionary versus microevolutionary studies (Coyne, 1996; Yoon, 1996). Macroevolutionists often emphasize the importance of historical contingency and the unpredictability of evolution. This view tends to reduce the importance of microevolutionary studies, which has as one goal to make evolutionary biology a predictive science. Clearly, the sunflower work supports the view that evolution is perhaps more predictable than has previously been thought and demonstrates the utility of genetic mapping for testing the repeatability of microevolutionary change. However, as pointed out by Stephen Jay Gould in Yoon (1996), the argument over predictability may largely concern the scale over which the predictions are made. He argues, and I think correctly, that microevolutionary events such as speciation are more deterministic or predictable than major macroevolutionary changes.

ALLOPOLYPLOIDY

Ancient Polyploids

Genetic mapping also has had a significant impact on our understanding of allopolyploidy primarily because it allows the extent and distribution of duplicated loci to be determined. The potential of this approach for studying allopolyploid speciation was first revealed in a classic study of maize by Helentjaris et al. (1988). Maize is considered to be an ancient polyploid,

and Helentjaris et al. (1988) used RFLP markers to determine the number and location of duplicated sequences in the maize genome. Of the loci examined, 29% were duplicated, supporting the idea that polyploidy may have been involved in the evolution of maize. However, the duplicated loci were distributed in a complex manner not consistent with simple allopolyploidy. To account for this pattern, Helentjaris et al. (1988) suggested that either the duplicated sequences were produced internally (segmental duplication) in the absence of polyploidy or that numerous translocations and inversions had occurred in maize following polyploidization. Whitkus et al. (1992) argue that polyploidy is the more parsimonious explanation for this pattern because of its agreement with cytotaxonomic evidence and because numerous independent segmental duplications are required to explain the data if the polyploidy hypothesis is rejected. Furthermore, follow-up studies of RFLP variation in maize suggest that a much higher proportion of loci (>70%) are duplicated (Ahn and Tanksley, 1993) than was estimated by Helentjaris et al. (1998).

The distribution of duplicated loci has been assayed in other putative ancient polyploids such as sorghum (Whitkus et al., 1992; Chittenden et al., 1994), soybean (Keim et al., 1990), *Brassica* (Slocum, 1989; Slocum et al., 1990; McGrath et al., 1990; McGrath and Quiros, 1991; Kianian and Quiros, 1992), and sunflower (Berry et al., 1995; Gentzbittel et al., 1995). Although all of these studies found evidence of extensive gene duplication in the putative ancient polyploids, only in *Brassica* was the density of duplicated loci high enough for confident identification of "homoeologous" chromosomal regions. In neotetraploid cotton ($n = 26$), patterns of genome duplication not only allow identification of the subgenomic origin of most linkage groups, but also are suggestive of an earlier polyploidization event that generated the diploid ($n = 13$) cottons over 25 million years ago (Reinisch et al., 1994). For example, in several cases, pairs of linked duplicated loci were observed within an individual tetraploid cotton subgenome. Reinisch et al. (1994) speculate that these loci may represent conserved linkage blocks from an ancient polyploidization event.

Neopolyploids

As alluded to in the preceding paragraph, the number and distribution of duplicated sequences have also been determined in several neopolyploid taxa. In tetraploid cotton, for example, mapping of duplicate loci allowed identification of 11 of the 13 expected homoeologous pairs between the A and D subgenomes (Reinisch et al., 1994). Two of these pairs are collinear, four differ by at least one inversion, two differ by at least two inversions, and at least one pair shows a translocation. To determine whether the extensive chromosomal reorganization between the A and D genomes occurred before or after polyploidization, Brubaker et al. (1996) mapped the genomes of the A and D diploid progenitors. Comparison of the diploid and polyploid genomes revealed that the A and D genomes of the diploid species were most similar in gene order and the A and D subgenomes of the tetraploid were most divergent, suggesting that rates of structural evolution increased following polyploidization.

Polyploid genome evolution also has been studied extensively in *Brassica*. *Brassica* provides an ideal system for the study of allopolyploid speciation for several reasons. First, the genome relationships among the three naturally occurring *Brassica* neotetraploids, *B. juncea* ($n = 19$), *B. napus* ($n = 19$), and *B. carinata* ($n = 17$), and their diploid progenitors, *B. rapa* ($n = 10$), *B. nigra* ($n = 8$), and *B. oleracea* ($n = 9$), are well understood. Second, the maternal parent of the *Brassica* tetraploids has been determined by analysis of chloroplast and mitochondrial DNA variation (Erickson et al., 1983; Palmer et al., 1983; Palmer and Herbon, 1988). Third, *in vitro* culture techniques have been developed for *Brassica* that make it possible to synthesize polyploids artificially, which can then be compared to those in nature.

As in cotton, considerable evidence has been compiled for rapid genome evolution following polyploidization in *Brassica*. For example, comparison of synthetic and natural *B. napus* revealed intragenomic rearrangements in three chromosomes in the C genome (Quiros et al., 1993). Likewise, comparison of the *B. napus* A subgenome with its A-genome progenitor, *B. rapa*, revealed the presence of at least one inversion (Teutonico and Osborn, 1994). Quiros et al. (1993) suggests that these changes result from intergenomic recombination, although the possibility that they are due to genome divergence in the diploid progenitor species cannot be ruled out.

Phylogenetic studies of RFLP variation also were suggestive of enhanced rates of genome evolution in polyploid *Brassica* (Song et al., 1988). Two findings were of particular importance. First, the polyploid genomes exhibited significant alterations in RFLP patterns relative to their progenitors, suggesting that the polyploids had undergone considerable genome evolution since their origins. Second, the RFLP composition of the polyploid species was more similar to that of the maternal parent than that of the paternal parent. This led Song et al. (1988) to propose that interactions between the maternal cytoplasmic genome and the paternal nuclear subgenome may cause alterations in genome structure or gene expression patterns. However, it was not clear whether the RFLP divergence observed resulted from directional subgenomic change following polyploidization or from unequal rates of genome divergence in the diploid progenitor species. Nor was it clear whether these changes occur early in polyploid evolution or whether mutation rates in polyploids are enhanced relative to diploids for long periods of evolutionary time.

To address these concerns, Song et al. (1995) generated the following reciprocal synthetic allotetraploids based on interspecific hybridizations between the diploid progenitor species, *B. rapa* (A genome), *B. nigra* (B genome), and *B. oleracea* (C genome): AB (A × B), BA (B × A), AC (C × A), and CA (C × A). The AB and BA synthetic tetraploids have the same nuclear genome complement as the natural tetraploid, *B. juncea,* and the AC and CA synthetic tetraploids have the same nuclear genome complement as *B. napus.* The synthetic tetraploids were then self-pollinated for five generations, and RFLP analysis was conducted on individuals from the F_2 and F_5 generations using 89 nuclear markers.

Comparisons between F_2 and F_5 plants revealed evidence for rapid genome change. On average, 9.6% of restriction fragments was dissimilar between F_2 and F_5 plants from the AB line, 8.2% from the BA line, 4.1% from the AC line, and 3.7% from the CA line. Three types of changes were observed: (1) the loss of fragments found in the F_2 plants; (2) the gain of diploid parental fragments that were absent in the F_2 but present in the F_5 plants; and (3) the gain of novel fragments in the F_5 plants that were not present in either diploid parent or in F_2 plants. The loss of fragments can be easily understood in terms of segregation and deletion. Likewise, the gain of novel fragments in the F_5 plants, although unexpected, can be interpreted to result from an increased mutation rate and/or intergenomic recombination. On the other hand, the regain of parental fragments in the F_5 generation that had been lost in the F_2 generation is difficult to explain in terms of conventional models of Mendelian inheritance and random mutation. Song et al. (1995) suggest that a variety of processes contribute to the high rate of changes observed, including chromosomal re-arrangement, point mutation, DNA methylation, and both intergenomic and intragenic recombination. However, little evidence for changes in DNA methylation was reported, and in my opinion, none of the remaining processes adequately accounts for the convergent regain of many diploid parental species fragments. Clearly, much remains to be understood concerning the genetic processes responsible for rapid genome change in polyploids.

Although the genetic basis of these changes is difficult to understand, important generalizations can be made about polyploid genome evolution by comparing patterns of genome change in the different synthetic lineages. For example, the paternal subgenome in the AB/BA tetraploids showed greater differentiation from its diploid parent than did the maternal subgenome, but no significant differences in subgenome evolution were observed in the AC/CA tetraploids (Fig. 16.7). This finding suggests that cytoplasmic-

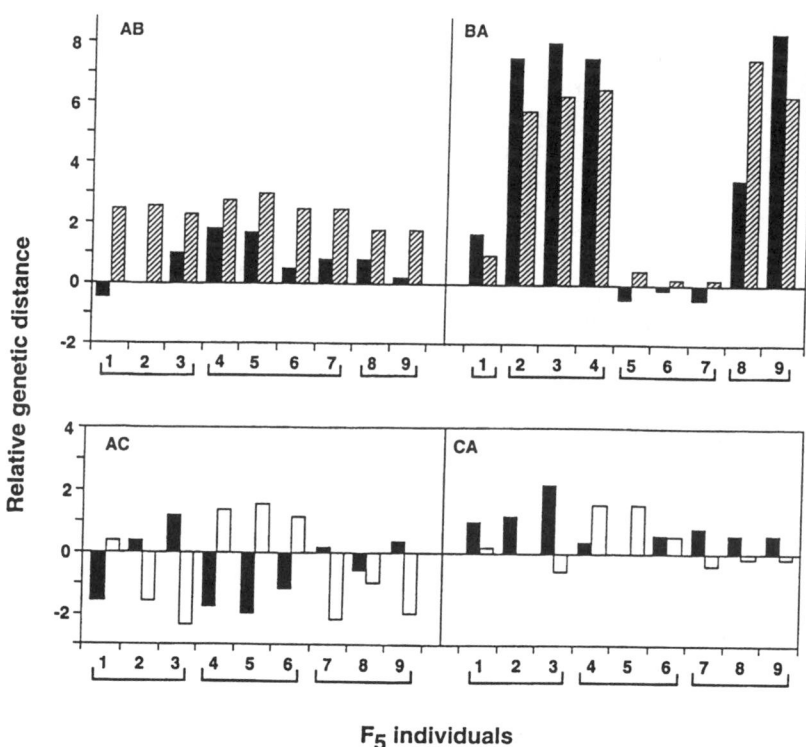

F₅ individuals

Figure 16.7. Relative genetic distances between F_5 individuals of the four synthetic *Brassica* polyploids (AC, CA, AB, and BA) and the diploid parental genomes A (black bars), B (hatched bars), and C (open bars). Bracketed F_5 individuals are full sibs. Positive values indicate that the F_5 plant is more distant from its diploid parent than the F_2, whereas negative values indicate that the F_5 is more similar to its parent than is the F_2. Reprinted with permission from Song et al. (1995).

nuclear interactions influenced genomic change in the AB/BA tetraploids but not in the AC/CA tetraploids. Presumably, the lack of significant cytoplasmic effects in the AC/CA tetraploids is due to higher levels of cytonuclear compatibility in the AC/CA polyploids. Also, the A and C subgenomes were more closely related phylogenetically than were the A and B subgenomes.

As is stressed elsewhere (Soltis and Soltis, 1995; Song et al. 1995), these data have significant implications for our understanding of polyploid evolution. First, they demonstrate that much of the genomic change found in natural polyploids probably occurs in the early generations after polyploid formation. Second, it seems probable that rapid genomic evolution following polyploidy could provide a source of genetic variation for adaptive evolution and diversification within polyploid lineages. Finally, if the kinds of changes observed here are arbitrary and random, it seems likely that polyploid lineages independently derived from the same parental material may rapidly diverge and achieve reproductive isolation. Thus, it would be of interest to test the compatibility of polyploid populations that are known on the basis of molecular evidence to have independent origins from the same parental species.

MAPPING SPECIATION GENES

Several methods have been employed for studying the genetic basis of reproductive barriers (reviewed in Wu and Palopoli, 1994). One of these (low-resolution introgression mapping) was described earlier in this chapter. Other methods include high-resolution introgression mapping such as that performed in *Drosophila* (reviewed in Wu and Palopoli, 1994), classical genetic analyses of F_2 or backcross hybrids (Macnair and Christie, 1983), map-based QTL analyses (Bradshaw et al., 1995), and hybrid viability or sterility rescue (Sawamura et al., 1993a, 1993b).

Each of these approaches has advantages and limitations. Low-resolution introgression mapping provides a crude estimate of the number of genetic intervals contributing to reduced hybrid fitness and the types of interactions between these intervals. However, this method does not distinguish between the effects of closely adjacent loci or allow dissection of the traits that contribute to reduced hybrid fitness. By contrast, high-resolution introgression mapping focuses on a single trait and microdissects a small portion of the genome. Thus, much more detail is obtained, but global coverage is sacrificed.

Classical genetic studies have provided much of our existing knowledge of the genetics of reproductive barriers (Macnair and Cumbes, 1989), but this approach lacks sufficient power to explain the complex segregation patterns resulting from polygenic traits. Map-based QTL studies are much more powerful, and given sufficiently large progenies, can detect the effects of almost all genes affecting a given trait. Nonetheless, small progeny sizes have reduced the effectiveness of most QTL studies that have been conducted to date.

In contrast to the other methods, hybrid viability or sterility rescue takes advantage of mutations within a species that restore hybrid viability or fertility (Wu and Palopoli, 1994). This enables the study of specific viability or sterility genes, but provides little information regarding other genes that contribute to reproductive isolation.

Of the approaches described above, only low-resolution introgression mapping, classical genetic analyses, and map-based QTL studies have been performed in plants. Results from classical genetic studies have been reviewed elsewhere (Gottlieb, 1984; Coyne, 1992). Thus, this section will focus on the handful of studies that has employed QTL analyses to map speciation genes in plants.

The first use of QTL mapping methods for studying reproductive isolation in plants apparently involved an analysis of hybrids between two lentil species, *Lens culinaris* and *L. ervoides* (Tadmor et al., 1987). The two species appear to differ by a single translocation, and it was thought that the translocation was responsible for reduced fertility in first-generation hybrids. To test this hypothesis, Tadmor et al. (1987) used a segregating F_2 population between the two species to generate a map based on 18 isozyme markers. Correlations between four of the isozyme markers and quadrivalent formation in meiosis allowed precise identification of the translocation end points. Moreover, all plants with pollen viability lower than 65% were heterozygous for the translocation, whereas plants with pollen viability higher than 85% were homozygous for the translocation. Thus, the chromosomal translocation does appear to represent the primary postmating reproductive barrier between these two species.

A similar study was conducted recently in *Helianthus*. Most species in the genus appear to differ by one or more chromosomal translocations, and these chromosomal rearrangements are generally well correlated with pollen viability of first-generation hybrids (Chandler et al., 1986). To provide a quantitative estimate of the influence of chromosomal rearrangements on pollen viability, Quillet et al. (1995) analyzed the segregation of 48 isozyme, RAPD, and morphological markers in a BC_1 progeny of an interspecific hybrid between *H. argophyllus* and the common sunflower, *H. annuus*. *Helianthus argophyllus* is the sister species of *H. annuus* (Rieseberg, 1991), and cytogenetic analyses indicate that the two species differ by two reciprocal translocations (Chandler et al., 1986). As predicted by cytogenetic studies, a wide range of variability in pollen viability was observed in the mapping family (27% to 93%). Over 80% of this variation was explained by three genetic intervals located on linkage groups 1, 2, and 3, respectively. Analyses of meiosis in the backcross hybrids revealed that meiotic abnormalities were also tightly correlated with the markers, indicating the chromosomal rearrangements are largely responsible for reducing fertility in hybrids between these species.

QTL mapping has also been used to determine the genetic basis of premating reproductive barriers. The best example in plants involves two closely related monkey flower species, *Mimulus lewisii* and *M. cardinalis* (Bradshaw et al., 1995). Artificial crosses between the two species produce hybrids that are vigorous and fertile. However, hybrids have never been reported in nature even though the species are sympatric over large portions of their geographic distribution and overlap in flowering time. The lack of natural hybrids appears to result from very effective premating reproductive barriers. One of these appears to be pollinator specificity: *M. lewisii* has

pale pink flowers with yellow nectar guides, a wide corolla opening, low volumes of concentrated nectar, and inserted anthers and stigma, and is pollinated by bumblebees. By contrast, *M. cardinalis* has red flowers with no nectar guides, a narrow corolla opening, a large volume of dilute nectar, and exserted anthers and stigma, and is pollinated by hummingbirds. The hummingbird pollination syndrome is thought to be derived from bee-pollinated ancestors in *Mimulus* (Bradshaw et al., 1995).

To determine the genetic basis of these floral differences, Bradshaw et al. (1995) analyzed the segregation of approximately 160 RAPD and isozyme markers in an F_2 mapping population of *M. lewisii* and *M. cardinalis*. Because most of the markers were dominant, two intraspecific maps were created rather than a single integrated map. However, it was possible to identify homologous linkages between the two maps by mapping ten codominant isozyme and RAPD loci. Upon completion of the maps, the number and locations of genes contributing to this premating reproductive barrier were determined by calculating associations between the segregating markers and eight floral traits likely to be involved in pollinator attraction (petal anthocyanins, petal carotenoids, corolla width, and petal width), reward (nectar volume and nectar concentration), and efficiency (stamen length and pistil length).

The results were remarkable. All eight floral traits had at least one major QTL that explained greater than 25% of the variation in that character (Table 16.2). Furthermore, for three of the eight floral traits, a single locus accounted for more than half of the variance (Table 16.2). These data indicate that genes with major effects played an important role in floral evolution in these *Mimulus* species and suggest that speciation may have occurred rapidly (although see Coyne, 1995). Based on these data, Bradshaw et al. (1995) suggested a possible scenario for the evolution of hummingbird pollination requiring a sequence of only three major mutations. In their model, the first mutation would reduce the attractiveness of the flower to insect pollinators by inducing uniform carotenoid deposition throughout the flower and thus eliminating the

Table 16.2. Location and effect of floral QTLs in *Mimulus*

Class	Trait	Linkage Group[a]	Effect (Phen%)[b]
Attraction	Petal anthocyanin	*lewisii-B*	33.5
		cardinalis-B	21.5
	Petal carotenoids	*cardinalis-D*	88.3
	Corolla width	*cardinalis-A*	25.7
		lewisii-C	68.7
		cardinalis-C	33.0
	Petal width	*lewisii-B*	42.4
		cardinalis-B	41.2
		cardinalis-H	25.2
Reward	Nectar volume	*lewisii-B*	48.9
		cardinalis-B	53.1
	Nectar concentration	*lewisii-B*	23.9
		cardinalis-B	28.5
Efficiency	Stamen length	*lewisii-B*	27.7
		cardinalis-B	27.5
		cardinalis-D	21.3
		lewisii-E	18.7
	Pistil length	*cardinalis-D*	43.9
		lewisii-E	51.9

Source: Modified from Bradshaw et al., 1995.

[a]Additive QTL mapped to homologous linkage groups in both species, whereas dominant QTL mapped to a single linkage group.

[b]Phen% is the percentage of phenotypic variance explained by a QTL.

nectar guides. A second mutation would enhance nectar volume to provide a greater reward for hummingbird visitors, whereas the third mutation would increase the length of the pistil to increase the efficacy of hummingbird pollination. The authors plan to test this scenario by introgressing *M. cardinalis* alleles at each of these major QTLs into the background of *M. lewisii* and then measuring the consequences of this sequence of mutations on fitness and pollinator visitation.

QTL mapping also has been employed to analyze the genetic basis of phenotypic differences between closely related plant species (Vlot et al., 1992; van Houten et al., 1994; Hombergen and Bachmann, 1995; Whitkus et al., 1996). Although many of these differences probably have little to do with speciation, the studies do address an old argument in speciation theory: Is speciation a gradual by-product of adaptive differentiation (Dobzhansky, 1941; Mayr, 1942), or is it the outcome of a revolutionary restructuring of the genome (Carson and Templeton, 1984)? The former view predicts that morphological differences between species will be based on a large number of mutations, each with small effect (Fisher, 1930; Barton and Charlesworth, 1984), whereas the latter view predicts a simple genetic basis for most phenotypic differences (Hilu, 1983; Marx, 1983; Gottlieb, 1984; Battjes et al., 1994).

The most detailed QTL studies of morphological differences among closely related wild plant species have been conducted by Bachman and colleagues in the Composite genus, *Microseris* (Vlot et al., 1992; van Houten et al., 1994; Hombergen and Bachmann, 1995; Bachmann and Hombergen, 1996). Characters examined to date include pappus part numbers (Vlot et al., 1992), trichome formation (Hombergen and Bachmann, 1995), and microsporangium numbers (Battjes et al., 1994). Although typically considered to represent classic quantitative traits, all exhibited a pattern of inheritance consistent with a single major gene with large effects plus several modifiers. This pattern of inheritance is similar to that described above for floral differences in *Mimulus,* and both examples appear to support an important role for major genes in speciation.

Future Directions

Speciation remains one of the most poorly understood areas in evolutionary biology (Futuyma, 1983). Although most biologists accept the existence of species as biologically real entities (Mayr, 1988; Gould, 1992), there is surprisingly little agreement about what species are and how they are formed (Otte and Endler, 1989). Numerous speciation theories have been developed, but these have been difficult to test rigorously. Likewise, attempts to generate realistic mathematical models based on these theories have foundered on our lack of understanding of the genetic architecture of reproductive barriers. Differing assumptions of genetic architecture often lead to widely divergent estimates of the plausibility of these theories.

As has been illustrated by the examples presented in this chapter, genetic mapping technology, in combination with phylogenetic and ecological studies, provides an opportunity to answer many previously intractable questions about speciation as well as to test certain models of speciation. Clearly, the primary application of this approach in the future will be to elucidate the genetic architecture of species barriers in a variety of plant and animal groups. By determining the number of genes required for speciation, as well as their genomic locations and interactions, we should be able to start making generalizations about speciation. Presumably, certain genetic architectures will be associated with particular modes of speciation, and possibly may reflect the tempo of the speciation event. Also, it will be of interest to determine whether the same genes are involved in speciation in many different lineages.

Once "speciation" genes have been mapped and identified, the experimental vista will expand exponentially. In addition to our obvious interest in determining what kinds of genes typically are associated with speciation, we can begin to determine whether the same genes are important in speciation in different lineages. For example, will the major QTLs detected by Bradshaw et al. (1995) in *Mimulus* control the shift from insect to bird pollination in other closely related genera and families? Will genes for sterility and inviability be conserved over large evolutionary distances? Is reproductive isolation an inherently convergent process as has been reported for other important QTLs during the evolution of domesticated cereals (Paterson et al., 1995)?

An even more exiting prospect relates to our ability to reconstruct the genetic changes that accompany and facilitate speciation and to study the ecological consequences of these changes. This can be accomplished by sexually transferring important speciation genes from one species into the genetic background of an ancestral species (at least in terms of phenotype) as proposed by Bradshaw et al. (1995) for *Mimulus*. This approach has already been used to study the domestication of maize from teosinte (Roush, 1996), and it seems likely that it will be employed widely to study the evolution and diversification of wild plants as well. Moreover, with genetic engineering, it may ultimately be possible to transfer alleles of interest directly, and thus eliminate the many generations of backcrossing required to remove unwanted alien genetic material. This latter approach will be particularly useful for species in which hybrids cannot be formed.

The experimental reconstruction of past evolutionary events will probably be most powerful and directly relevant to the study of hybrid and allopolyploid speciation events. Because these modes of speciation are both rapid (Soltis and Soltis, 1993; McCarthy et al., 1995) and reticulate, the majority of genetic processes associated with them can be accurately replicated by experimental studies. Numerous allopolyploid species have already been synthesized, and the use of genetic mapping to track genome evolution following polyploidization has already provided surprising data regarding the instability and rapid evolution of polyploid genomes (Song et al., 1995). Clearly, these experiments need to be extended to analyze polyploid speciation events that involve species pairs that vary in terms of genomic divergence, mating system, and life history. In addition, now that numerous duplicated genes have been identified, studies that follow their evolutionary fate become critical. How do genes become silenced, and how do genomes become diploidized?

Progress has also been made in studying diploid hybrid species, but much remains to be done. Although several diploid hybrid species have been experimentally synthesized, only one has been analyzed in detail genetically. Rieseberg et al. (1996b) attempted to replicate the early generations of hybrid speciation, and the genomic analysis of the resulting hybrid lineages was instructive. However, these experiments were conducted in the greenhouse, and it is not clear how selection for xeric conditions might have affected hybrid genomic composition. Clearly, it would be instructive to see whether a hybrid species could be replicated in the wild simply by allowing natural selection to take its course. Genomic composition of the end product could be compared with that of the natural hybrid species to provide insights regarding the repeatability of speciation.

Similarly, it would be of interest to reconstruct the genome of introgressive races. This feat could be accomplished in the greenhouse using a combination of fertility and marker-based selection or by natural selection in field experiments. In both cases, graphical genotype analysis could be used to determine how closely the synthesized introgressive races match those produced historically in nature. Because introgression has often been thought to provide an avenue for the transfer of genetic adaptations, it would be of interest to compare the fitness and ecological amplitude of the synthesized introgressants with the native species and its introgressive races. Moreover, by determining the location of important QTL that differentiate the hybridizing species, it would be possible to verify the transmission of these QTL in both the natural and synthetic introgressive races, as well as to test the adaptive significance of individual QTL.

Clearly, the studies described above are contingent upon a robust phylogenetic framework. Studies of reproductive barriers are most informative in sister species that have only recently diverged, because the genes identified in these studies are more likely to represent those actually involved in speciation rather than genes that have diverged following speciation. Likewise, experimental studies of speciation will be most informative if the genetic background of the "ancestral" species is highly similar to the true ancestor. Studies of hybrid or allopolyploid species require genetic data for precise identifi-

cation of parental species. Moreover, due to the presence of substantial genetic variation within most plant species, it often is necessary to have some knowledge of intraspecific phylogeography (Avise, 1994) to interpret accurately genomic patterns in natural hybrid and polyploid species, as well as to reconstruct the genomes of these species.

In conclusion, genetic mapping tools have the potential to revolutionize the study of speciation. As pointed out by Coyne, pp. 701 (1996), the study of speciation has been "sidetracked . . . by the joys of gel electrophoresis and DNA sequencing." However, these molecular markers also allow us to unravel the genome in an unprecedented fashion and may very well provide the impetus for a whole new round of experimental studies of speciation.

LITERATURE CITED

Ahn, S., and S. D. Tanksley. 1993. Comparative linkage maps of the rice and maize genomes. Proceedings of the National Academy of Sciences U.S.A. **90:** 7980–7984.

Anderson, E. 1949. Introgressive Hybridization. Wiley, New York.

Arnold, M. L. 1992. Natural hybridization as an evolutionary process. Annual Review of Ecology and Systematics **23:**237–261.

Avise, J. C. 1994. Molecular Markers, Natural History, and Evolution. Chapman & Hall, New York.

Ayliffe, M. A., G. J. Lawrence, J. G. Ellis, and A. J. Pryor. 1994. Heteroduplex molecules formed between allelic sequences cause nonparental RAPD bands. Nucleic Acids Research **22:**1632–1636.

Bachmann, K., and E. J. Hombergen. 1996. Mapping genes for phenotypic variation in *Microseris* (Lactuceae) with molecular markers. **In** Proceedings of the International Compositae Conference, Kew, 1994, Vol. 2, Biology and Utilization, eds. P. D. S. Caligari and D. J. N. Hind, pp. 23–43. Royal Botanic Gardens, Kew, England.

Baird, S. J. E. 1995. A simulation study of multilocus clines. Evolution **49:**1038–1045.

Barton, N. H. 1995. Appendix: the fate of introgressing chromosomes. Evolution **49:**1044–1045.

Barton, N. H., and B. Charlesworth. 1984. Genetic revolutions, founder effects, and speciation. Annual Review of Ecology and Systematics **15:**97–131.

Barton, N. H., and G. M. Hewitt. 1981. The genetic basis of hybrid inviability in the grasshopper *Podisma pedestris*. Heredity **47:**367–383.

Barton, N. H., and G. M. Hewitt. 1985. Analysis of hybrid zones. Annual Review of Ecology and Systematics **16:**113–148.

Barton, N. H., and G. M. Hewitt. 1989. Adaptation, speciation and hybrid zones. Nature **341**:497–503.

Basten, C. J., B. S. Weir, and Z.-B. Zeng. 1996. QTL Cartographer. North Carolina State University, Raleigh.

Battjes, J., K. L. Chambers, and K. Bachmann. 1994. Evolution of microsporangium numbers in *Microseris* (Asteraceae: Lactuceae). American Journal of Botany **81**:641–647.

Bennett, J. H. 1953. Junctions in inbreeding. Genetica **26**:392–406.

Berry, S. T., A. J. Leon, C. C. Hanfrey, P. Challis, A. Burkholz, S. R. Barnes, G. K. Rufener, M. Lee, and P. D. S. Caligari. 1995. Molecular marker analysis of *Helianthus annuus* L. 2. Construction of a RFLP linkage map for cultivated sunflower. Theoretical and Applied Genetics **91**:195–199.

Botstein, D., R. L. White, M. H. Skolnick, and R. W. Davis. 1980. Construction of a genetic linkage map in man using restriction fragment length polymorphisms. American Journal of Human Genetics **32**:314–331.

Bradshaw, H. D., S. M. Wilbert, K. G. Otto, and D. W. Schemske. 1995. Genetic mapping of floral traits associated with reproductive isolation in monkeyflowers (*Mimulus*). Nature **376**:762–765.

Brubaker, C. L., A. Paterson, and J. F. Wendel. 1996. Evolution of homoeologous chromosomes in polyploids: evidence from comparative mapping of allopolyploid and diploid *Gossypium*. American Journal of Botany (suppl.) **83**:93.

Byrne, M., J. C. Murrell, B. Allen, and G. F. Moran. 1995. An integrated genetic linkage map for eucalypts using RFLP, RAPD and isozyme markers. Theoretical and Applied Genetics **91**:869–875.

Cabot, E. L., A. W. Davis, N. A. Johnson, and C.-I. Wu. 1994. Genetics of reproductive isolation in the *Drosophila simulans* clade: complex epistasis underlying hybrid male sterility. Genetics **137**:175–189.

Carlson, J. E., L. K. Tulsieram, J. C. Glaubitz, V. W. K. Luk, C. Kauffeldt, and R. Rutledge. 1991. Segregation of random amplified DNA markers in F_1 progeny of conifers. Theoretical and Applied Genetics **83**:194–200.

Carson, H. L. 1975. The genetics of speciation at the diploid level. American Naturalist **109**:73–92.

Carson, H. L., and A. R. Templeton. 1984. Genetic revolutions in relation to speciation phenomena: the founding of new populations. Annual Review of Ecology and Systematics **15**:97–131.

Causse, M. A., M. T. Fulton, Y. G. Cho, S. N. Ahn, J. Chunwongse, F. S. Wu, J. H. Xiao, P. C. Ronald, S. E. Harrington, G. Second, S. R. McCouch, and S. D. Tanksley. 1994. Saturated molecular map of the rice genome based on an interspecific backcross population. Genetics **138**:1251–1274.

Chandler, J. M., C. Jan, and B. H. Beard. 1986. Chromosomal differentiation among the annual *Helianthus* species. Systematic Botany **11**:353–371.

Chang, C., J. L. Bowman, A. W. DeJohn, E. S. Landers, and E. M. Meyerowitz. 1988. Restriction fragment length polymorphism map for *Arabidopsis thaliana*. Proceedings of the National Academy of Sciences U.S.A. **85**:6856–6860.

Chittenden, L. M., K. F. Schertz, Y.-R. Lin, R. A. Wing, and A. H. Paterson. 1994. A detailed RFLP map of *Sorghum bicolor* × *S. pronpinquum,* suitable for high-density mapping, suggests ancestral duplication of *Sorghum* chromosomes or chromosomal segments. Theoretical and Applied Genetics **87**:925–933.

Christie, P., and M. R. Macnair. 1984. Complementary lethal factors in two North American populations of the yellow monkey flower. Journal of Heredity **75**:510–511.

Coyne, J. A. 1992. Genetics and speciation. Nature **355**:511–515.

Coyne, J. A. 1995. Speciation in monkeyflowers. Nature **376**:726–727.

Coyne, J. A. 1996. Speciation in action. Science **272**:700–701.

Coyne, J. A., W. Meyers, A. P. Crittenden, and P. Sniegowski, 1993. The fertility effects of pericentric inversions in *Drosophila melanogaster.* Genetics **134**:487–496.

Dobzhansky, T. H. 1941. Genetics and the Origin of Species. Columbia University Press, New York.

Doebley, J. F., and A. Stec. 1993. Inheritance of the morphological differences between maize and teosinte. Genetics **129**:285–295.

Doerge, R. W., and A. Rebai. 1996. Significance thresholds for QTL interval mapping tests. Heredity **76**:459–464.

Dupuis, J. 1994. Statistical Problems Associated with Mapping Complex and Quantitative Traits from Genomic Mismatch Scanning Data. Ph.D. thesis, Stanford University, Stanford, California.

Edwards, A., A. Civitello, H. A. Hammond, and C. T. Caskey. 1991. DNA typing and genetic mapping with trimeric and tetrameric tandem repeats. American Journal of Human Genetics **49**:746–756.

Ellsworth, D. L., K. D. Rittenhouse, and R. L. Honeycutt. 1993. Artifactual variation in randomly amplified polymorphic DNA banding patterns. BioTechniques **14**:214–217.

Engelke, D. R., A. Krikos, M. E. Bruck, and D. Ginsburg. 1990. Purification of *Thermus aquaticus* DNA polymerase expressed in *Escherichia coli*. Analytical Biochemistry **191**:396–400.

Erickson, L. R., N. A. Strauss, and W. D. Beversdorf. 1983. Restriction patterns reveal origins of chloroplast genomes in *Brassica* amphidiploids. Theoretical and Applied Genetics **65**:201–206.

Falconer, D. S., and T. F. C. Mackay. 1996. Introduction to Quantitative Genetics. Longman, Essex, England.

Feingold, E., P. O. Brown, and D. Siegmund. 1993. Gaussian models for genetic linkage analysis using complete high-resolution maps of identity by descent. American Journal of Human Genetics **53**:234–251.

Fisher, R. A. 1930. The Genetical Theory of Natural Selection. Oxford University Press, Oxford.

Fisher, R. A. 1949. The Theory of Inbreeding. Oliver & Boyd, Edinburgh.

Fisher, R. A. 1953. A fuller theory of junctions in inbreeding. Heredity **8**:187–197.

Fritsch, P. F., and L. H. Rieseberg. 1996. RAPD in conservation biology. In Molecular Genetic Approaches in Conservation, eds. T. B. Smith and B. Wayne, pp. 54–73. Oxford University Press, New York.

Futuyma, D. J. 1983. Speciation. Science 219:1059–1060.

Garcia, G. M., H. T. Stalker, and G. Kochert. 1995. Introgression analysis of an interspecific hybrid population in peanuts (Arachis hypogaea L.) using RFLP and RAPD markers. Genome 38:166–176.

Gentzbittel, F. Vear, Y.-X. Zhang, A. Bervillé, and P. Nicolas. 1995. Development of a consensus linkage RFLP map of cultivated sunflower (Helianthus annuus L.). Theoretical and Applied Genetics 90:1079–1086.

Gerstel, D. U. 1954. A new lethal combination of interspecific cotton hybrids. Genetics 39:628–639.

Gottlieb, L. D. 1984. Genetics and morphological evolution in plants. American Naturalist 123:681–709.

Gould, S. J. 1992. What is a species? Discover 40:40–42.

Grant, V. 1958. The regulation of recombination in plants. Cold Spring Harbor Symposia on Quantitative Biology 23:337–363.

Gresshoff, P. (ed.). 1994. Plant Genome Analysis. CRC Press, Boca Raton, Florida.

Haley, C. S., and S. A. Knott. 1992. A simple regression method for mapping quantitative trait loci in line crosses using flanking markers. Heredity 69:315–324.

Hanson, W. D. 1959a. Early generation analysis of lengths of chromosome segments around a locus held heterozygous with backcrossing or selfing. Genetics 44:833–837.

Hanson, W. D. 1959b. The breakup of initial linkage blocks under selected mating systems. Genetics 44:857–868.

Harlan, S. C. 1936. The genetical conception of the species. Biological Review 11:83–112.

Harrison, R. G. 1990. Hybrid zones: windows on evolutionary process. Oxford Surveys in Evolutionary Biology 7:69–128.

Harushima, Y., N. Kurata, M. Yano, Y. Nagamura, T. Sasaki, Y. Minobe, and M. Nakagahra. 1996. Detection of segregation distortions in an indica-japonica rice cross using a high-resolution map. Theoretical and Applied Genetics 92:145–150.

Heiser, C. B. 1947. Hybridization between the sunflower species Helianthus annuus and H. petiolaris. Evolution 1:249–262.

Heiser, C. B. 1958. Three new annual sunflowers (Helianthus) from the southwestern United States. Rhodora 60:271–283.

Helentjaris, T., and B. Burr (eds.). 1989. Development and Application of Molecular Markers to Problems in Plant Genetics. Cold Spring Harbor Laboratory, Cold Spring Harbor, New York.

Helentjaris, T., D. Weber, and S. Wright. 1988. Identification of the genomic locations of duplicate nucleotide sequences in maize by analysis of restriction fragment length polymorphsms. Genetics 118:353–363.

Hermson, J. G. T. 1963. The genetic basis of hybrid necrosis in wheat. Genetica 33:245–287.

Heun, M., and T. Helentjaris. 1993. Inheritance of RAPDs in F₁ hybrids of corn. Theoretical and Applied Genetics 85:961–968.

Hilu, K. W. 1983. The role of single-gene mutations in the evolution of flowering plants. Evolutionary Biology 16:97–128.

Hollingshead, L. 1930. A lethal factor in Crepis effective only in an interspecific hybrid. Genetics 15:114–140.

Hombergen, E.-J., and K. Bachmann. 1995. RAPD mapping of three QTLs determining trichome formation in Microseris hybrid H27 (Asteraceae: Lactuceae). Theoretical and Applied Genetics 90:853–858.

Hyne, V., and M. J. Kearsey. 1995. QTL analysis: further uses of "marker regression." Theoretical and Applied Genetics 91:471–476.

Jackson, R. C. 1985. Genomic differentiation and its effect on gene flow. Systematic Botany 10:391–404.

Jansen, R. C. 1994. Controlling the type I and type II errors in mapping quantitative trait loci. Genetics 138:871–881.

Jena, K. K., G. S. Khush, and G. Kochert. 1992. RFLP analysis of rice (Oryza sativa L.) introgression lines. Theoretical and Applied Genetics 84:608–616.

Keim, P., B. W. Diers, T. C. Olson, and R. C. Shoemaker. 1990. RFLP mapping in soybean: association between marker loci and variation in quantitative traits. Genetics 126:735–742.

Kesseli, R. V., I. Paran, and R. W. Michelmore. 1994. Analysis of a detailed genetic linkage map of Lactuca sativa (lettuce) constructed from RFLP and RAPD markers. Genetics 136:1435–1446.

Key, K. H. L. 1968. The concept of stasipatric speciation. Systematic Zoology 17:14–22.

Kianian, S. F., and C. F. Quiros. 1992. Generation of a Brassica oleracea composite RFLP map: linkage arrangements among various populations and evolutionary implications. Theoretical and Applied Genetics 84:544–554.

Knapp, S. J. 1994. Mapping quantitative trait loci. In DNA-Based Markers in Plants, eds. R. L. Phillips and I. K. Vasil, pp. 58–96. Kluwer, Dordrecht, The Netherlands.

Kruglyak, L., and E. S. Lander. 1995. A nonparametric approach for mapping quantitative trait loci. Genetics 139:1421–1428.

Lagercrantz, U., H. Ellegren, and L. Andersson. 1993. The abundance of various polymorphic microsatellite motifs differs between plants and vertebrates. Nucleic Acids Research 21:1111–1115.

Lander, E. S., and D. Botstein. 1989. Mapping Mendelian factors underlying quantitative traits using RFLP linkage maps. Genetics 121:185–199.

Lander, E. A., and N. J. Schork. 1994. Genetic dissection of complex traits. Science 265:2037–2049.

Lander, E. S., P. Green, J. Abrahamson, A. Barlow, M. J. Daly, S. E. Lincoln, and L. Newburg. 1987. MAPMAKER: an interactive computer package for constructing primary genetic linkage maps of experimental and natural populations. Genomics 1:174–181.

Landry, B. S., R. Kesseli, H. Leung, and R. W. Michelmore. 1987. Comparison of restriction endonucleases and sources of probes for their efficiency in detecting restriction fragment length polymorphisms in lettuce (Lactuca sativa L.). Theoretical and Applied Genetics 74:646–653.

Lathrop, G. M., and J. M. Lalouel. 1984. Easy calculation of lod scores and genetic risks on small computers. American Journal of Human Genetics **36**:460–465.

Lathrop, G. M., J. M. Lalouel, and R. L. White. 1986. Construction of human linkage maps: likelihood calculations for multilocus linkage analysis. Genetic Epidemiology **3**:39–52.

Link, A. J., and M. V. Olsen. 1991. Physical map of the *Saccharomyces cerevisiae* genome at 110-kilobase resolution. Genetics **127**:681–698.

Liu, B.-H., and S. J. Knapp. 1990. G-MENDEL: a program for Mendelian segregation and linkage analysis of individual or multiple progeny populations using log-likelihood ratios. Journal of Heredity **81**:407.

Macnair, M. R., and P. Christie. 1983. Reproductive isolation as a pleiotropic effect of copper tolerance in *Mimulus guttatus*. American Naturalist **106**:351–372.

Macnair, M. R., and Q. J. Cumbes. 1989. The genetic architecture of interspecific variation in *Mimulus*. Genetics **122**:211–222.

Mangelsdorf, P. C. 1958. The mutagenic effect of hybridizing maize and teosinte. Cold Spring Harbor Symposium on Quantitative Biology **23**:409–421.

Marx, G. A. 1983. Developmental mutations in some annual seed plants. Annual Review of Plant Physiology **34**:389–417.

Mayr, E. 1942. Systematics and the Origin of Species. Columbia University Press, New York.

Mayr, E. 1954. Change of genetic environment and evolution. In Evolution as a Process, eds. J. Huxley, A. C. Hardy, and E. B. Ford, pp. 157–180. Allen & Unwin, London.

Mayr, E. 1963. Animal Species and Evolution. Harvard University Press, Cambridge, Massachusetts.

Mayr, E. 1988. Toward a New Philosophy of Biology. Harvard University Press, Cambridge, Massachusetts.

McCarthy, E. M., M. A. Asmussen, and W. W. Anderson. 1995. A theoretical assessment of recombinational speciation. Heredity **74**:502–209.

McCouch, S. R., G. Kochert, Z. H. Yu, Z. Y. Wang, G. S. Khush, W. R. Coffman, and S. D. Tanksley. 1988. Molecular mapping of rice chromosomes. Theoretical and Applied Genetics **76**:815–829.

McGrath, J. M., and C. F. Quiros. 1991. Inheritance of isozyme and RFLP markers in *Brassica campestris* and comparison with *B. oleracea*. Theoretical and Applied Genetics **82**:668–673.

McGrath, J. M., C. F. Quiros, J. J. Harada, and B. S Landry. 1990. Identification of *Brassica oleracea* monosomic alien chromosome addition lines with molecular markers reveals extensive gene duplication. Molecular and General Genetics **223**:198–204.

McGrath, J. M., S. M. Wielgus, T. G. Uchytil, H. Kim-Lee, G. T. Haberlach, C. E. Williams, and J. P. Helgeson. 1994. Recombination of *Solanum brevidens* chromosomes in the second backcross generation from a somatic hybrid with *S. tuberosum*. Theoretical and Applied Genetics **88**:917–924.

McGrath, J. M., S. M. Wielgus, and J. P. Helgeson. 1996. Segregation and recombination of *Solanum brevidens* synteny groups in progeny of somatic hybrids with *S. tuberosum:* intragenomic equals or exceeds intergenomic recombination. Genetics **142**:1335–1348.

Michelmore, R. W., I. Paran, and R. V. Kesseli. 1991. Identification of markers linked to disease-resistance genes by bulked segregant analysis: a rapid method to detect markers in specific genomic regions by using segregating populations. Proceedings of the National Academy of Sciences U.S.A. **88**:9828–9832.

Miller, J. C., and S. D. Tanksley. 1990. Effect of different restriction enzymes, probe source, and probe length on detecting restriction fragment length polymorphism in tomato. Theoretical and Applied Genetics **80**:385–389.

Mitchell-Olds, T. 1996. Genetic constraints on life-history evolution: quantitative trait loci influencing growth and flowering in *Arabidopsis thaliana*. Evolution **50**:140–145.

Mulcahy, D. L., N. F. Weeden, R. Kesseli, and S. B. Carroll. 1992. DNA probes for the Y-chromosome of *Silene latifolia,* a dioecious angiosperm. Sexual Plant Reproduction **5**:86–88.

Oka, H.-I. 1974. Analysis of genes controlling F_1 sterility in rice by the use of isogenic lines. Genetics **77**:521–534.

Otte, D., and J. A. Endler. 1989. Speciation and Its Consequences. Sinauer Associates, Sunderland, Massachusetts.

Palmer, J. D., and L. A. Herbon. 1988. Plant mitochondrial DNA evolves rapidly in structure but slowly in sequence. Journal of Molecular Evolution **28**:87–97.

Palmer, J. D., C. R. Shields, D. B. Cohen, and T. J. Orton. 1983. Chloroplast DNA evolution and the origin of amphidiploid *Brassica* species. Theoretical and Applied Genetics **65**:181–189.

Paterson, A. H. 1996. The sorghum-Johnson grass ecosystem: risks of gene flow from sorghum to Johnson grass. American Journal of Botany (suppl.) **83**:86.

Paterson, A. H., E. S. Lander, J. D. Hewitt, S. Peterson, S. E. Lincoln, and S. D. Tanksley. 1988. Resolution of quantitative traits into Mendelian factors by using a complete linkage map of restriction fragment length polymorphism. Nature **335**:721–726.

Paterson, A. H., T.-R. Lin, Z. Li, K. F. Schertz, J. F. Doebley, S. R. M. Pinson, S.-C. Liu, J. W. Stansel, and J. E. Irvine. 1995. Convergent domestications of cereal crops by independent mutations at corresponding genetic loci. Science **269**:1714–1718.

Pluthero, F. G. 1993. Rapid purification of high-activity *Taq* DNA polymerase. Nucleic Acids Research **21**:4850–4851.

Quillet, M. C., N. Madjidian, T. Griveau, H. Serieys, M. Tersac, M. Lorieus, and A. Bervillé. 1995. Mapping genetic factors controlling pollen viability in an interspecific cross in *Helianthus* section *Helianthus*. Theoretical and Applied Genetics **91**:1195–1202.

Quiros, C. F., J. Hu, and M. J. Truco. 1993. DNA-based marker *Brassica* maps. In DNA-Based Markers in

Plants, eds. R. L. Phillips and I. K. Vasil, pp. 58–96. Kluwer, Dordrecht, The Netherlands.

Ragot, M., and D. A. Hoisington. 1993. Molecular markers for plant breeding: comparisons of RFLP and RAPD genotyping costs. Theoretical and Applied Genetics 86:975–984.

Reinisch, A. J., J. Dong, C. L. Brubaker, D. M. Stelly, J. F. Wendel, and A. H. Paterson. 1994. A detailed RFLP map of cotton, Gossypium hirsutum × Gossypium barbadense: chromosome organization and evolution in a disomic polyploid genome. Genetics 138: 829–847.

Rick, C. M. 1963. Differential zygotic lethality in a tomato species hybrid. Genetics 48:1497–1507.

Riedy, M. F., W. J. Hamilton III, and C. F. Aquadro. 1992. Excess of non-parental bands in offspring from known primate pedigrees assayed using RAPD PCR. Nucleic Acids Research 20:918.

Rieseberg, L. H. 1991. Homoploid reticulate evolution in Helianthus: evidence from ribosomal genes. American Journal of Botany 78:1218–1237.

Rieseberg, L. H. 1995. The role of hybridization in evolution: old wine in new skins. American Journal of Botany 82:944–953.

Rieseberg, L. H. 1996. Homology among RAPD fragments in interspecific comparisons. Molecular Ecology 5:99–105.

Rieseberg, L. H., and L. Broulliet. 1994. Are many plant species paraphyletic? Taxon 43:21–32.

Rieseberg, L. H., and N. C. Ellstrand. 1993. What can morphological and molecular markers tell us about plant hybridization? Critical Reviews in Plant Sciences 12:213–241.

Rieseberg, L. H., and J. Wendel. 1993. Introgression and its consequences in plants. In Hybrid Zones and the Evolutionary Process, ed. R. Harrison, pp. 70–109. Oxford University Press, New York.

Rieseberg, L. H., H. Choi, R. Chan, and C. Spore. 1993. Genomic map of a diploid hybrid species. Heredity 70:285–293.

Rieseberg, L. H., C. Van Fossen, and A. Desrochers. 1995a. Genomic reorganization accompanies hybrid speciation in wild sunflowers. Nature 375:313–316.

Rieseberg, L. H., C. R. Linder, and G. Seiler. 1995b. Chromosomal and genic barriers to introgression in Helianthus. Genetics 141:1163–1171.

Rieseberg, L. H., B. Sinervo, C. R. Linder, M. Ungerer, and D. M. Arias. 1996a. Role of gene interactions in hybrid speciation: evidence from ancient and experimental hybrids. Science 272:741–745.

Rieseberg, L. H., D. M. Arias, M. Ungerer, C. R. Linder, and B. Sinervo. 1996b. The effects of mating design on introgression between chromosomally divergent sunflower species. Theoretical and Applied Genetics 93:633–644.

Roush, W. 1996. Corn: a lot of change from a little DNA. Science 272:1873.

Sano, Y., and F. Kita. 1978. Reproductive barriers distributed in Melilotus species and their genetic basis. Canadian Journal of Cytology and Genetics 20:275–289.

Saunders, A. R. 1952. Complementary lethal genes in the cowpea. South African Journal of Science 48: 195–197.

Sawamura, K., T. Taira, and T. K. Watanabe. 1993a. Hybrid lethal systems in the Drosophila melanogaster species complex. I. The maternal hybrid rescue (mhr) gene of Drosophila simulans. Genetics 133:299–305.

Sawamura, K., T. Taira, and T. K. Watanabe. 1993b. Hybrid lethal systems in the Drosophila melanogaster species complex. II. The Zygotic hybrid rescue (Zhr) gene of Drosophila simulans. Genetics 133:307–313.

Schmidt, R., J. West, K. Love, Z. Lenehan, C. Lister, H. Thompson, D. Bouchez, and C. Dean. 1995. Physical map and organization of Arabidopsis thaliana chromosome 4. Science 270:480–483.

Shaw, D. D., P. Wilkinson, and D. J. Coates. 1983. Increased chromosomal mutation rate after hybridization between two subspecies of grasshoppers. Science 220:1165–1167.

Sites, J. W., and C. Moritz. 1987. Chromosomal evolution and speciation revisited. Systematic Zoology 36: 153–174.

Slocum, M. K. 1989. Analyzing the genomic structure of Brassica species and subspecies using RFLP analysis. In Development and application of molecular markers to problems in plant genetics, eds. T. Helentjaris and B. Burr, pp. 73–80. Cold Spring Harbor Laboratory Publications, Cold Spring Harbor, New York.

Slocum, M. K., S. S. Figdore, W. C. Kennard, J. Y. Suzuki, and T. C. Osborn, 1990. Linkage arrangement of restriction fragment length polymorphism loci in Brassica oleracea. Theoretical and Applied Genetics 80:57–64.

Soltis, D. E., and P. S. Soltis. 1993. Molecular data and the dynamic nature of polyploidy. Critical Reviews in Plant Sciences 12:243–275.

Soltis, D. E., and P. S. Soltis. 1995. The dynamic nature of polyploid genomes. Proceedings of the National Academy of Sciences U.S.A. 92:8089–8091.

Song, K. M., T. C. Osborn, and P. H. Williams. 1988. Brassica taxonomy based on nuclear restriction fragment length polymorphisms (RFLPs). 1. Genome evolution of diploid and amphidiploid species. Theoretical and Applied Genetics 75:784–794.

Song, K., P. Lu, K. Tang, and T. Osborn. 1995. Rapid genome change in synthetic polyploids of Brassica and its implications for polyploid evolution. Proceedings of the National Academy of Sciences U.S.A. 92:7719–7723.

Stebbins, G. L. 1957. The hybrid origin of microspecies in the Elymus glaucus complex. Cytologia Supplemental Volume 36:336–340.

Stebbins, G. L. 1971. Chromosomal Evolution in Higher Plants. Addison-Wesley, Reading, Massachusetts.

Stephens, S. G. 1949. The cytogenetics of speciation in Gossypium. I. Selective elimination of the donor parent genotype in interspecific backcrosses. Genetics 34:627–637.

Stephens, S. G. 1950. The internal mechanism of speciation in Gossypium. Botanical Review 16:115–149.

Suiter, K. A., J. F. Wendel, and J. S. Case. 1983. Linkage-1: a Pascal computer program for the detection and analysis of genetic linkage. Journal of Heredity **74:**203–204.

Szymura, J. M., and N. H. Barton. 1991. The genetic structure of the hybrid zone between the fire-bellied toads *Bombina bombina* and *B. variegata:* comparisons between transects and between loci. Evolution **45:**237–291.

Tadmor, Y., D. Zamir, and G. Ladizinsky. 1987. Genetic mapping of an ancient translocation in the genus *Lens.* Theoretical and Applied Genetics **73:**883–892.

Tanksley, S. D. 1993. Mapping polygenes. Annual Review of Genetics **27:**205–233.

Tanksley, S. D., R. Bernatsky, N. L. Lapitan, and J. P. Prince. 1988a. Conservation of gene repertoire but not gene order in pepper and tomato. Proceedings of the National Academy of Sciences U.S.A. **85:**6419–6423.

Tanksley, S. D., J. C. Miller, A. H. Paterson, and R. Bernatsky. 1988b. Molecular mapping of plant chromosomes. **In** Chromosome Structure and Function, eds. J. P. Gustafson and R. Appels, pp. 157–173. Plenum Press, New York.

Tanksley, S. D., M. W. Ganal, J. P. Prince, M. C. De Vicente, M. W. Bonierbale, P. Broun, T. M. Fulton, J. J. Giovannoni, S. Grandillo, G. B. Martin, R. Messeguer, J. C. Miller, L. Miller, A. H. Paterson, O. Pineda, M. S. Röder, R. A. Wing, W. Wu, and N. D. Young. 1992. High density molecular linkage maps of the tomato and potato genomes. Genetics **132:**1141–1160.

Templeton, A. R. 1981. Mechanisms of speciation—a population genetic approach. Annual Review of Ecology and Systematics **12:**23–48.

Teutonico, R. A., and T. C. Osborn. 1994. Mapping of RFLP and qualitative trait loci in *Brassica rapa* and comparison to the linkage maps of *B. napus, B. oleracea, Arabidopsis thaliana.* Theoretical and Applied Genetics **89:**885–894.

Thormann, C. E., M. E. Ferreira, L. E. A. Camargo, J. G. Tivang, and T. C. Osborn. 1994. Comparison of RFLP and RAPD markers to estimating genetic relationships within and among cruciferous species. Theoretical and Applied Genetics **88:**973–980.

Ukai, Y., R. Osawa, A. Saito, and T. Hayashi. 1995. MAPL: a package of computer programs for construction of DNA polymorphism linkage maps and analysis of QTL. Breeding Science **45:**139–142.

van Houten, W., L. van Raamsdonk, and K. Bachmann. 1994. Intraspecific evolution in *Microseris pygmaea* (Asteraceae, Lactuceae) analyzed by cosegregation of phenotypic characters (QTLs) and molecular markers (RAPDs). Plant Systematics and Evolution **190:** 49–67.

Vlot, E. C., W. H. J. van Houten, S. Mauthe, and K. Bachmann. 1992. Genetic and nongenetic factors influencing deviations from five pappus parts in a hybrid between *Microseris douglasii* and *M. bigelovii* (Asteraceae, Lactuceae). International Journal of Plant Sciences **153:**89–97.

Vos, P., R. Hogers, M. Bleeker, M. Reijans, T. Van De Lee, M. Hornes, A. Frijters, J. Pot, J. Peleman, M. Kuiper, and M. Zabeau. 1995. AFLP: a new technique for DNA fingerprinting. Nucleic Acids Research **23:** 4407–4414.

Wall, J. R. 1970. Experimental introgression in the genus *Phaseolus.* I. Effect of mating systems on interspecific gene flow. Evolution **24:**356–366.

Wan, J., Y. Yamaguchi, H. Kato, and H. Ikehashi. 1996. Two new loci for hybrid sterility in cultivated rice (*Oryza sativa* L.). Theoretical and Applied Genetics **92:** 183–190.

Wang, G.-L., J.-M. Dong, and A. H. Paterson. 1995. The distribution of *Gossypium hirsutum* chromatin in *G. barbadense* germ plasm: molecular analysis of introgressive hybridization. Theoretical and Applied Genetics **91:**1153–1161.

White, M. J. D. 1978. Modes of Speciation. W. H. Freeman, San Francisco.

Whitkus, R., J. Doebley, and M. Lee. 1992. Comparative genome mapping of sorghum and maize. Genetics **132:**1119–1130.

Whitkus, R., H. Doan, and T. Lowrey. 1996. Genetic control of female sterility in Hawaiian *Tetramolopium* (Asteraceae). American Journal of Botany (suppl.) **83:**100.

Whitton, J., L. H. Rieseberg, and M. C. Ungerer. 1997. Microsatellite loci are not conserved across the Asteraceae. Molecular Biology and Evolution **14:**204–209.

Wiebe, G. A. 1934. Complementary factors in barley giving a lethal progeny. Journal of Heredity **25:**273–275.

Williams, C. E., S. M. Wielgus, G. T. Harberlach, C. Guenther, H. Kim-Lee, and J. P. Helgeson. 1993. RFLP analysis of chromosomal segregation in progeny from an interspecific hexaploid hybrid between *Solanum brevidens* and *Solanum tuberosum.* Genetics **135:**1167–1173.

Williams, J. G. K., A. R. Kubelik, K. J. Livak, J. A. Rafalsky, and S. V. Tingey. 1990. DNA polymorphisms amplified by arbitrary primers are useful as genetic markers. Nucleic Acids Research **18:**6531–6535.

Wittbrodt, J., D. Adam, B. Malitschek, W. Maueler, and F. Raulf. 1989. Novel putative receptor tyrosine kinase encoded by the melanoma-inducing *Tu* locus in *Xiphophorus.* Nature **341:**415–421.

Wu, C.-I., and M. Palopoli. 1994. Genetics of postmating reproductive isolation in animals. Annual Review of Genetics **27:**283–308.

Xiong, M., H. J. Chen, R. A. Prade, Y. Wang, J. Griffith, W. E. Timberlake, and J. Arnold. 1996. On the consistency of a physical mapping method to reconstruct a chromosome *in vitro.* Genetics **142:**267–284.

Xu, S. 1995. A comment on the simple regression method for interval mapping. Genetics **141:**1657–1659.

Xu, S., and W. R. Atchley. 1995. A random model method for interval mapping of quantitative trait loci. Genetics **141:**1189–1197.

Yoon, C. K. 1996. First ever recreation of a new species birth. The New York Times, volume CXLV, number 50, 420, p. B5.

Young, N. D., and S. D. Tanksley. 1989a. Restriction fragment length polymorphism maps and the concept of graphical genotypes. Genetics **77**:95–101.

Young, N. D., and S. D. Tanksley. 1989b. RFLP analysis of the size of chromosomal segments retained around the *Tm-2* locus of tomato during backcross breeding. Theoretical and Applied Genetics **77**:353–359.

Zamir, C., and Y. Tadmor. 1986. Unequal segregation of nuclear genes in plants. Botanical Gazette **147**: 355–358.

Zeng, Z. B. 1993. Theoretical basis for separation of multiple linked genes in mapping quantitative trait loci. Proceedings of the National Academy of Sciences U.S.A. **90**:10972–10976.

17

A Perspective on the Contribution of Plastid *rbcL* DNA Sequences to Angiosperm Phylogenetics

Mark W. Chase and Victor A. Albert

The large *rbcL* analysis published in 1993 (Chase, Soltis, Olmstead et al., 1993) ranks as the largest phylogenetic analysis, molecular or otherwise, ever produced. These data represented over ten years of effort, but significantly the advent of the polymerase chain reaction, PCR, had greatly expanded the numbers of sequences available in the four years just prior to publication. Strategies for production of DNA sequences had become dramatically easier and faster. At the same time that sequencing was becoming more practical, the United States National Science Foundation was approving a large number of molecular systematics proposals, thus making substantial funding available (examine the number of NSF grant numbers listed in the footnotes of the 1993 paper), while Gerard Zurawski (DNAX Corporation) was making available without cost *rbcL* PCR and internal sequencing primers. Consequently, a large number of laboratories began working on this same plastid locus. To a very large extent and until just recently, protein-coding plastid DNA sequences have been the almost exclusive focus of vascular plant molecular systematists, whereas nearly all published work on other organisms has focused

on ribosomal DNA (rDNA; see Chapter 1). By late 1991, the stage had been set for a dramatic increase in the amount of molecular data available on a wide range of seed plants, and this put workers focusing on *rbcL* in a position to discuss a large-scale analysis. Publication was not necessarily their goal; most simply felt that everyone would benefit from the interaction and that potentially we could use this sort of unpublished, but widely circulated, result as a way to focus better the many individual projects underway.

Some authors of the 1993 *rbcL* paper were told by several well-known molecular phylogenists that our efforts were doomed simply because we wished to include so many sequences (see Graur et al., 1996, for a more recent discussion of this topic). We must, we were told, strip our matrices down to a few place holders so that more rigorous analyses could be performed. Because we could and had produced so many *rbcL* sequences, we wanted to explore how sampling was affecting the topologies obtained. It was evident to many workers that smaller matrices did not produce the same overall tree topologies that were obtained when sampling was more extensive. Equally evident

We thank Inger Andersson of the Swedish University of Agricultural Sciences, Uppsala, for her time and interest in analyzing structural correlates of *rbcL* variation. We also acknowledge the many people who provided unpublished sequences of *rbcL*, 18S rDNA, and *atpB*, without which we would have been unable to discuss the issue of congruence (one of the most important aspects of the 1993 *rbcL* tree that previously could not have been considered).

was that the increased sampling produced results that more closely resembled relationships indicated by some classifications and other existing lines of evidence. Our early results with *rbcL* resembled those of Giannasi et al. (1992), but our first larger trees were also in conflict with those published by Boulter (1974), Martin et al. (1985), Martin and Dowd (1991), Troitsky et al. (1991), and Hamby and Zimmer (1992). In 1992, published molecular studies examining angiosperm relationships had a dismally discordant record, and the 1993 *rbcL* tree seemed to many just a further example of an additional gene providing yet another divergent picture (see Syvanen et al., 1989 and Syvanen, 1994, for their views of this incongruence). Many of the individuals carrying out the early *rbcL* work were traditionally trained plant taxonomists who knew a great deal about their research subjects, and they found that in spite of the theoretical problems of working with more than 20 taxa (the greatest number of taxa for most matrices that can be handled with branch and bound algorithms), these analyses seemed to be producing reasonable patterns of relationships. Investigators continued to accumulate large matrices and work with them as best they could, despite having to use only heuristic methods.

It was also clear that projects being pursued independently, given that no classification systems placed the component taxa near each other, were not necessarily independent; for example, Saxifragaceae s. l. were expected to be polyphyletic, but Morgan, Soltis, and Soltis, examining these genera, Chase working on families of Polygalales sensu Cronquist (1981), and Olmstead focusing on Asteridae sensu Cronquist (1981) found in preliminary analyses that their taxa were interdigitating. Combining data seemed the logical solution. It was also evident that we were at an impasse; the results we obtained with small matrices of *rbcL* sequences were not satisfactory, but it seemed extremely intemperate to try to work with, what we estimated at the time, to be 200 or so existing *rbcL* sequences. With encouragement from several other experts, particularly Dave Swofford and Michael Clegg, we persisted in the notion that an attempt would be worthwhile and circulated an informal invitation to anyone working on

seed plant *rbcL* to send sequences to the involved parties, Chase, Soltis, and Olmstead. We were amazed to find that we rapidly accumulated over 400 sequences. At this point, we still had no clear intention to publish our findings.

SEARCH STRATEGIES AND CHOICE OF METHOD

Our first efforts to analyze the 475-taxon matrix on a Macintosh Quadra 800 at Duke University failed because that version of PAUP 3.0 (3.0r; Swofford, 1991) could not handle this number of terminals due to "integer overflow." Coincidentally, Swofford was visiting Duke University and played a major role in the first analysis by making available an uncirculated UNIX trial version of PAUP 3.0, which was loaded on a SPARC II workstation (Sun, Inc.) at Duke University. By the time we were ready to alter the search strategy (see Chase, Soltis, Olmstead, et al., 1993, for a description of these strategies), sequences for several additional families were available, and Swofford had fixed the problem with PAUP (then version 3.0s) so that it could be run on the Macintosh Quadra 800. As the amount of data available to us was expanding, the hardware and software to handle this enormous matrix were improving to the point that an attempt could be made. Had we attempted such an analysis even two years previously, we would have failed because the analytical tools did not exist.

Although putting together the matrices was in the hands of the three first authors, the search strategy was worked out by Brent Mishler, who spent many days modifying the search strategy for Search II (the 500-sequence matrix with 499 taxa; two reasonably divergent *rbcL* sequences had been cloned from the PCR product of *Canella* of Canellaceae). Search II (the B-series figures in Chase, Soltis, Olmstead et al., 1993) was the more rigorous strategy, and it remains our preferred result. Trees from both searches were published to demonstrate that different methods (Search I used the weighting scheme of Albert et al., 1993) and even slightly greater sampling did not have a great effect upon the topology produced. Once we saw the results, it was apparent to us that we should proceed with publication, although several ways of presenting the results were discussed, including just describing the groupings identified and not including figures with trees. In the end, it was decided to illustrate trees.

We also had to face a question of whether an external review of the manuscript was warranted. So many molecular systematists were involved in the project that finding reviewers was difficult (Michael Clegg, Michael Donoghue, and Jim Rodman were among the authors until near the end of manuscript preparation but asked merely to be thanked for their contributions); several of the authors felt that an external review was not necessary due to the wide diversity of opinion held by such a diverse set of contributors. Chase received useful criticisms from Robert Thorne and Peter Martin on a draft of the manuscript. Jeffrey Palmer in particular felt that a formal peer review was important, and ultimately the editors of the *Annals of the Missouri Botanical Gardens* agreed and asked Jeff Doyle to organize some additional reviews, as well as to arbitrate the review process. Doyle solicited two outside anonymous reviewers, one a botanist with acknowledged cladistic expertise, the other a well-known animal molecular systematist. The process of manuscript preparation had been the chief responsibility of the first author (MWC), and a series of seemingly never-ending negotiations took place until all authors approved the document. The external review added some additional helpful perspectives to what by that time had already become a widely discussed and debated text, but some of the major recommendations of these reviewers were rejected. One of the strong recommendations was that the first set of analyses (the A series in the published version) be eliminated from the paper. This suggestion was rejected because the coauthors had debated the issue before the external reviews were solicited and had decided to include both sets of trees. Before publication could then proceed, each coauthor was asked by the editors at the *Annals* to communicate his or her approval of the paper.

Overall, the process had been largely an amicable one, especially considering that so many unpublished data were included. Various issues were debated, and in the end some of the coauthors decided that, in spite of their misgivings, if they did not join in the process, they would lose complete control of what they perceived as their area of interest. Other sets of authors found that they were in fact focusing on the same taxa, even though it had initially appeared that their studies were isolated. In a sense, many sequences appeared "out of the woodwork," and the number and diversity surprised everyone. A certain degree of "dynamic tension" was maintained throughout the whole period of analysis, manuscript preparation, and review, but all authors continued to discuss their misgivings and the perceived shortcomings of the project without resorting to arguments.

One of the major concerns for everyone involved in the *rbcL* project was how to design a robust search strategy. We had few guidelines that could help us in this area because no one had ever before tried to work with so many taxa. The interactive nature of Search II on the Quadra was a benefit, despite the loss of computing power. We have continued to use smaller computers for most large analyses, and we find doing analyses on PCs relatively tractable. The search on the 499-taxon (500 terminal) matrix attempted to use several starting trees and to swap these down to the point at which we seemed unable to find shorter trees. A decision was made *not* to spend several weeks or months more on the search; we were certain that we could have eventually found trees several steps shorter, as have Rice et al. (1997). Prior to publication, we knew that somewhat shorter trees had been found, but these differed little in overall patterns. Indeed, publication could have been delayed for years while searching for shorter trees, but we decided that this was not reasonable; if years were required, then the whole process was wasted because such a lengthy period was not available to us. We sought instead to provide systematists with the database of sequences and trees because of the enormous interest in *rbcL* sequencing and angiosperm phylogeny that was burgeoning at that time.

We suspected that a "reasonably short" tree had been obtained because working on a smaller 100-taxon matrix focusing on hamamelid relationships (Chase, Soltis, Olmstead et al., 1993, fig. 16), we produced a topology similar in nearly all respects to the larger trees for the eudicots. Likewise, Qiu et al. (1993) found that they could duplicate the general patterns found with the larger matrices; they had reduced the matrix to 82 taxa by removing most of the two largest components of the 500-terminal matrix: eudicots (reduced to only 12 genera) and monocots (reduced to 11 genera). From this we also suspected that all of the well-supported groupings had been found. The branching patterns that differed between the larger trees and these two smaller analyses (the hamamelid and Qiu et al. matrices) were weakly supported; lengthy swapping on weakly supported clades is not likely suddenly to produce well-supported groupings. The answer to the question concerning the robustness of the 1993 search strategy was that, if it was not the best, it still seemed to provide a topology that was representative of the overall pattern present in the *rbcL* matrix and that we could duplicate this general topology with smaller matrices if we used the "big tree" (from the 499-taxon analysis) to include taxa from each major lineage.

In the 1993 analyses, parsimony methods alone were used, and this too has been criticized (although not in print). Other methods were available at the time, but the only other apparently viable option would have been neighbor joining (Saitou and Nei, 1987). An attempt to run the matrix on a PC version of neighbor joining in the PHYLIP package (Felsenstein, 1991) failed entirely, probably again due to the size of the matrix. Maximum likelihood (Felsenstein, 1981) was rejected simply because of the size of the matrix, but other papers using a subset of these data demonstrated that the same general result could be produced by this method. For example, a copy of the 500-terminal matrix was sent to M. Nei in 1992, and he produced a neighbor-joining tree that was nearly identical to its parsimony counterpart (M. Nei, pers. comm.). The decision to illustrate only a parsimony tree was based on two practical factors: (1) we did not wish to get into a debate over which methods were best at the same time we were struggling just to push one method reasonably far, and (2) at the time, no other programs appeared to be able to handle the matrix. Many of the authors also believed that parsimony represented the method with the fewest ad hoc assumptions and that because it produced output that allowed us to examine in an unadulterated manner how the characters, the individual base positions, were changing, it permitted us to examine patterns of change in the gene with the greatest ease and accuracy (see Chapters 4, 6, and 7). All other methods involve certain "corrections" for unobserved substitutions, and many of these assumptions lack empirical support. For example, we were told by molecular biologists that third positions were almost certainly saturated and so we should discard them or, better yet, convert the sequences into amino acids. However, our experiences with the *rbcL* data convinced us that third positions contained useful information. This prejudiced many of us to be skeptical of methods that in any way altered the data prior to producing trees.

Parsimony is also the only method available for workers who are interested in combining in a single matrix molecular and morphological/anatomical data (Albert et al., 1994a, 1994b; Doyle et al., 1994; Chase et al., 1995b; Sheahan and Chase, 1996; Morton et al., 1997b; Nandi et al., 1998). Recent simulations by Hillis (1996) using the large 18S rDNA data set of Soltis, Soltis, Nickrent et al. (1997) have demonstrated a higher rate of reproducing the known tree with parsimony and weighted parsimony than with neighbor joining, even when the pairwise distances matched the substitution model exactly. It is safe to conclude that parsimony methods were appropriate for analyzing the *rbcL* data, and no extensive distortions were introduced by using only a single method of analysis, parsimony.

PATTERNS OF *rbcL* EVOLUTION AND FUNCTIONAL CONSTRAINTS

One of the factors that had been suggested to have potentially distorting effects on the *rbcL* topology was the occurrence of lineage-specific rate variation. For example, Duvall et al. (1993) used maximum likelihood because rate variation had been found among lineages in the monocotyledons (Gaut et al., 1992), and it has been stated by Felsenstein (1981) that maximum likelihood is less biased by rate differences, although he has never indicated the relative scale of differences in rate that could cause such distortion (Felsenstein, 1978). Albert et al. (1992b, 1993) calculated the range of rate differences that would be needed to put an analysis into the "Felsenstein zone," and they demonstrated that rate differences required to introduce spuriousness into results were far beyond what would be considered appropriate for use in phylogenetic studies. The differences among lineages in evolutionary rate for *rbcL* are all within a narrow range of low values. Albert et al. (1994a) examined rates in 19 woody pairs of taxa and concluded that none of these was sufficiently out of range of the others to cause problems in analysis. The molecular clock for *rbcL* appeared "quasi-ultrametric" and therefore indicative of near neutrality. That neutrality is desirable for characters used in phylogenetic analysis has been contentious, but the argument that rate differences are of sufficient magnitude to present problems for parsimony analysis is not valid. This has been borne out as well by the agreement among trees produced by parsimony, neighbor joining, and maximum likelihood: If there is spuriousness in the *rbcL* trees, it equally affects all methods of analysis, and therefore requires more data from other sources to evaluate the robustness of the results, not simply a different method of analysis.

Further assessment of the reliability of the *rbcL* tree published in 1993 must involve other sources of data; particularly useful are analyses of information from genomes other than the plastid. We believe it is safe to assume that the 1993 *rbcL* tree was a good representation of the patterns present in *rbcL* sequences, which is a much firmer conclusion than we were able to reach at that time. The next questions that must be addressed are: (1) Does *rbcL* evolve in any peculiar way that would provide a nonhistorical tree (perhaps due to its functional constraints)? and (2) Do other genes (particularly those from the nucleus) provide similar patterns of relationships?

Before examining the patterns of variation in *rbcL*, we would like to present the bootstrap tree of the 500-terminal matrix (Fig. 17.1) because it was not included in the 1993 paper. At that time, a bootstrap or other analysis of internal support was not deemed feasible. This consensus tree was produced with 1,000 replicates of the "fast" bootstrap method available on PAUP 3.1.1. (Swofford, 1993). This method is far faster than standard methods of bootstrapping, but when compared to these more "standard methods" on smaller data matrices, both provide comparable levels of support. It makes use of the first-pass length calculation that other algorithms use to evaluate starting trees for branch swapping; patterns that are well-supported will be evident in the starting trees as a percentage of replicates without swapping (i.e., frequency of presence in the starting trees is what is then being measured, and amount of internal support is inferred from this percentage rather than group presence in the final trees found in each replicate after swapping).

There are two perspectives relevant to the bootstrap consensus tree for *rbcL*. The first is to focus on the number of supported groups, of which there are many. The often-cited reason for why *rbcL* was selected in the first place (Palmer et al., 1988) is that it is a relatively conserved gene that could be expected to be more useful at higher taxonomic levels. It is quite evident from Fig. 17.1 that *rbcL* is more useful at the level of family and below; at higher levels it simply lacks a strong phylogenetic signal, with the exceptions of eudicots, monocotyledons, and all angiosperms. The "spine" of the tree has no bootstrap support. There is a consistent pattern in the most parsimonious trees (and this is true for the shorter trees found by Rice et al., 1997), but only more terminal groupings of genera receive high bootstraps (i.e., better than 80%). From the standpoint of elucidating relationships among archaic angiosperms, the *rbcL* tree of 1993 was inconclusive. The patterns of relationships recently revealed by broad analyses of 18S rDNA (Soltis, Soltis, Nickrent et al., 1997) and *atpB* (Savolainen et al., 1996) are highly congruent with those retrieved in the 1993 *rbcL* analyses, lending additional support to the 1993 tree. By itself, the *rbcL* analysis left more questions than it answered, which is of course worthwhile on its own merits (i.e., it provided a focus for other studies).

Perhaps one of the most valuable aspects of the existence of a large *rbcL* matrix has been the ability to use these strongly supported terminal

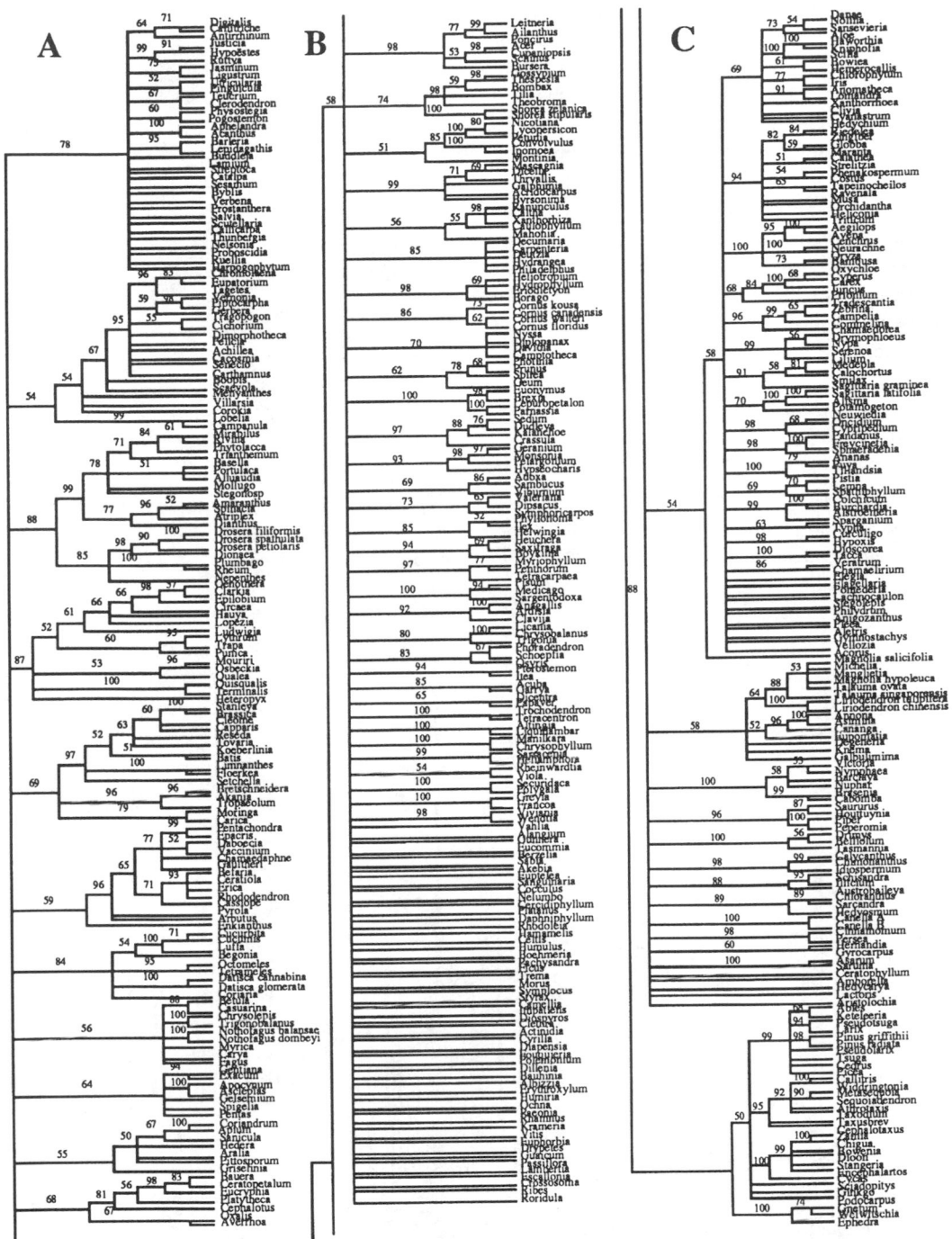

Figure 17.1. Bootstrap consensus tree (saving all groups with greater than 50% occurrence) produced with 1,000 replicates of "fast" bootstrapping with PAUP version 3.1.1 (Swofford, 1993; see text for description and rationale for this procedure) on the 500-sequence matrix of Chase, Soltis, Olmstead et al. (1993). The tree is broken into three sections; the eudicots (triaperturate-pollen families) are shown in parts A and B, whereas the outgroups, "primitive" dicots, and monocots are in part C.

relationships to place taxa that have long been problematic. If these taxa have close relationships to clades already in the matrix, it is thus a simple task to find where they fit. This use has been of enormous benefit to the systematics community, and it is one of the reasons why workers continue to produce *rbcL* sequences. Because of the existence of this now enormous database, doing a few *rbcL* sequences allows a worker to develop a nearly instantaneous focus on the scale of the problems faced, and a great number of follow-up projects, both molecular and otherwise, have benefited greatly from this focus.

Variable sites are more or less evenly distributed along the length of the exon (Fig. 17.2). There are no particular hot spots, although some sites do experience quite frequent substitutions. The character transformation patterns (Fig. 17.3) demonstrate no particular bias, and the transi-

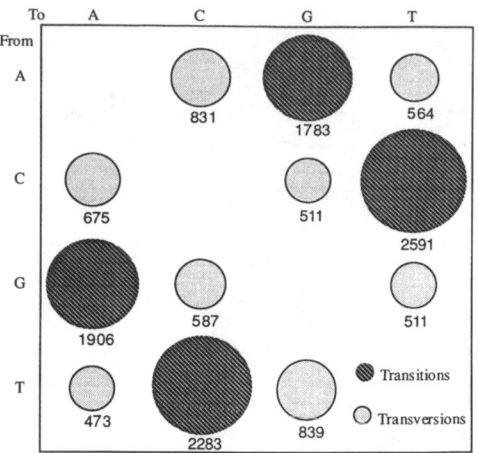

Figure 17.3. Frequency of unambiguous transformations on the same tree as in Figure 17.2 from Chase, Soltis, and Olmstead (1993). The overall transition/transversion ratio for this tree is 1.72:1. (This figure was produced by the software program MacClade; Maddison and Maddison, 1992.)

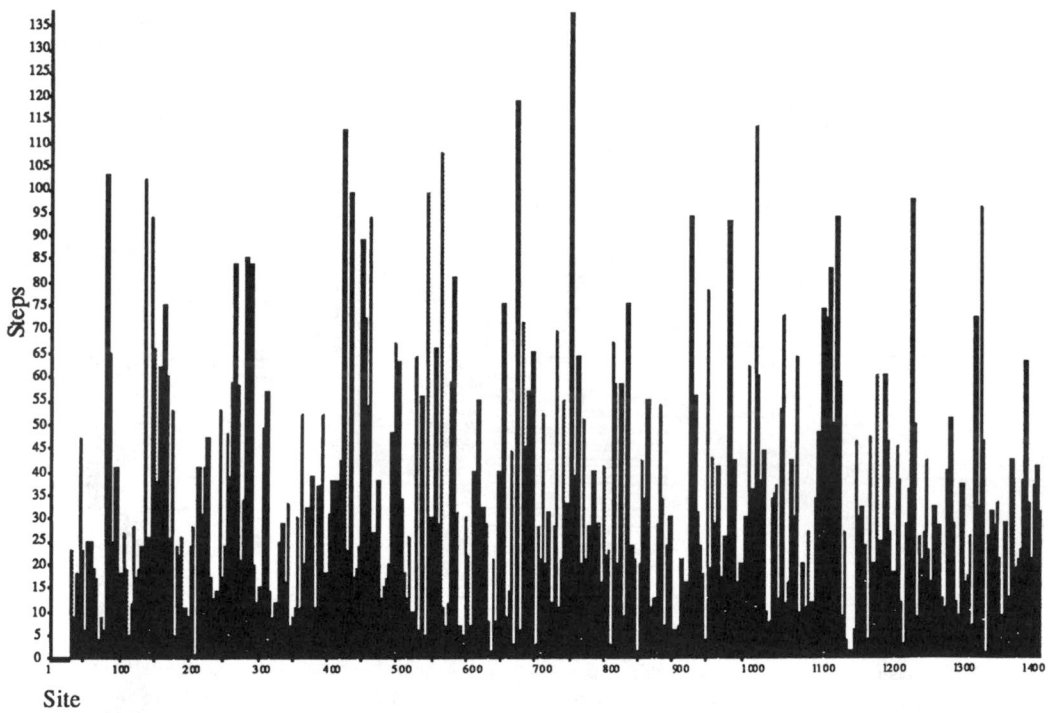

Site

Figure 17.2. Number of steps at each base position along the length of the *rbcL* exon for the tree from the 500-sequence analysis illustrated in Chase, Soltis, Olmstead et al. (1993). The first 26 base pairs are not included because this region corresponds to the forward PCR primer, which is incorporated into the amplified products, thus disguising the native sequence that may differ somewhat from that of the primer. Note that the variation is not strongly clumped in certain regions. (This figure was produced by the software program MacClade; Maddison and Maddison, 1992.)

tion/transversion ratio is 1.72:1, which is also within the expected range (see Albert et al., 1992b, 1993). As expected (and similar to Chase et al., 1995a, for monocotyledons), most steps occur at third positions (72%; Table 17.1; Fig. 17.4), and first-position changes follow this (18%). Third positions have the lowest consistency indices (0.09; Table 17.1), although not

Table 17.1. Codon position information.

Codon Position	CI	RI	Steps	Percent of Total
First	0.16	0.58	2,992	18
Second	0.25	0.54	1,661	10
Third	0.09	0.66	11,885	72

Note: CI, consistency index; RI, retention index.

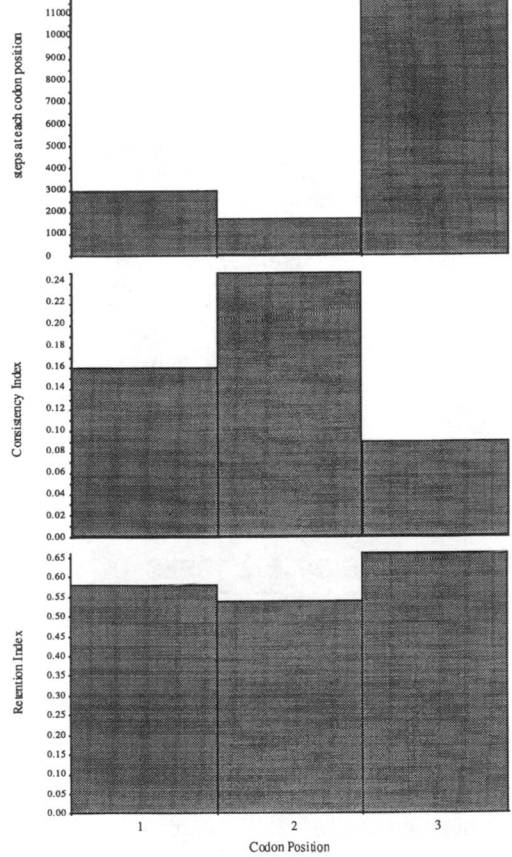

Figure 17.4. For each codon position in *rbcL,* the number of steps, consistency index (CI), and retention index (RI), for the same tree as in Figs. 17.2 and 17.3. (This figure was produced by the software program MacClade; Maddison and Maddison, 1992). These data are also summarized in Table 17.1.

vastly lower than the other two positions (0.16 and 0.25, respectively). The retention indices for third positions are the highest and are unexpectedly higher than those for either first or second positions. Because the retention index measures the extent of character congruence at maximum parsimony, third positions as a class appear to show the greatest overall agreement in supporting the tree topology. Many researchers have experimented with using only first and second positions or converting the DNA sequence into amino acids (see comparisons in Albert et al., 1994a; Manhart, 1994), but neither of these provides a vastly different pattern or, more importantly, better support. Among land plants and certainly within angiosperms, third positions supply information as reliable as that provided by first and second positions; because substitutions at the third position are the most abundant, their elimination simply provides less resolution and weaker measures of internal support.

Asymmetries in codon usage have been suggested (Ritland and Clegg, 1987) between some groups of organisms, and differential availability of transfer RNAs has been suggested for angiosperms (Morton, 1993, 1994; Morton and Clegg, 1995). Another decisive factor in codon usage appears to be a bias toward A and T based on neighboring base composition (Morton, 1995; Morton and Clegg, 1995). Other phenomena become apparent, however, when a greater phylogenetic range of land plants is considered. Albert et al. (1994b) produced a comparison of codon frequencies for a 40-taxon matrix that included representatives of all extant land plant lineages, and this information is repeated here (Table 17.2). Frequencies are highly asymmetric, and this could be explained in many instances by an AT bias in synonymous codons (i.e., as with valine in which GTT and GTA are much more frequent than GTC and GTG; Table 17.2). In other cases, (e.g., leucine and serine), codon usage is more complex, and Albert et al. (1994b) have suggested that high frequency codons are linked by additional substitutions to other frequent codons and that those with lower frequencies are linked by subsequent substitutions to lower-frequency, possibly functionally deleterious, codons or to stop codons. Frequency patterns in which third-position

Table 17.2. Codon frequencies for an *rbcL* land plant matrix (Albert et al., 1994a), from Albert et al. (1994b).

Codon	Amino acid	Number	Σ Amino Acid	Proportions	Proportion Σ	Codon	Amino acid	Number	Σ Amino Acid	Proportions	Proportion Σ
TGT	Cys	211		0.0116		CAG	Glm	94	408	0.0052	0.0225
TGC	Cys	116	0327	0.0064	0.0189	CAT	Hla	358		0.0197	
TCT	Ser	242		0.0133		CAC	Hla	213	571	0.0117	0.0315
TCC	Ser	165		0.0091		CGT	Arg	478		0.0263	
TCA	Ser	44		0.0024		CGC	Arg	112		0.0062	
TCG	Ser	30		0.0017		CGA	Arg	248		0.0137	
AGT	Ser	111		0.0061		CGG	Arg	42		0.0023	
AGC	Ser	60	0652	0.0033	0.0360	AGA	Arg	237		0.0131	
ACT	Thr	613		0.0338		AGG	Arg	49	1166	0.0027	0.0643
ACC	Thr	280		0.0154		AAA	Lys	687		0.0379	
ACA	Thr	162		0.0089		AAG	Lys	175	862	0.0096	0.0475
ACG	Thr	58	1113	0.0032	0.0613	ATG	Met	362	362	0.0199	0.0199
CCT	Pro	438		0.0241		ATT	Ile	472		0.0260	
CCC	Pro	160		0.0088		ATC	Ile	295		0.0163	
CCA	Pro	161		0.0089		ATA	Ile	51	818	0.0028	0.0451
CCG	Pro	68	0827	0.0038	0.0456	TTA	Leu	379		0.0209	
GCT	Ala	881		0.0486		TTG	Leu	339		0.0187	
GCC	Ala	220		0.0121		CTT	Leu	334		0.0184	
GCA	Ala	470		0.0299		CTC	Leu	47		0.0026	
GCG	Ala	135	1706	0.0074	0.0940	CTA	Leu	279		0.0154	
GGT	Gly	826		0.0455		CTG	Leu	164	1542	0.0090	0.0850
GGC	Gly	134		0.0074		GTT	Val	476		0.0262	
GGA	Gly	609		0.0336		GTC	Val	93		0.0051	
GGG	Gly	247	1816	0.0136	0.1001	GTA	Val	549		0.0303	
AAT	Asm	401		0.0221		GTG	Val	148	1266	0.0082	0.0698
AAC	Asm	182	583	0.0100	0.0321	TTT	Phe	444		0.0245	
GAT	Asp	804		0.0443		TTC	Phe	370	814	0.0204	0.0449
GAC	Asp	224	1028	0.0123	0.0566	TAT	Tyr	474		0.0261	
GAA	Glu	947		0.0522		TAC	Tyr	218	692	0.0120	0.0381
GAG	Glu	314	1261	0.0173	0.0695	TGG	Trp	307	307	0.0169	0.0169
CAA	Glm	314		0.0173							

Source: From Albert et al., 1994b.

nucleotide identity does not correlate with AT bias (e.g., serine TCC is 3.8 times more frequent than TCA; Table 17.2) suggest that, at least in some cases, functional conservation may be as important for third positions as for first and second positions. However, it may be usual that third-position nucleotides are more consistent characters on average than first or second positions because they are in fact less functionally constrained (and thus more subject to AT bias; Morton, 1995; Morton and Clegg, 1995). We would expect other protein-coding genes to exhibit similar patterns; such comparisons can now be made because other gene data sets now exist.

SIMILARITIES OF *rbcL* TREES TO THOSE FROM OTHER PHYLOGENETIC STUDIES

In 1993, the *rbcL* tree could only be compared to the various taxonomic schemes for angiosperms. Although such comparisons were made difficult due to the vague use of concepts such as "linking groups" (e.g., Cronquist, 1981), the general impression was that the molecular tree was dissimilar to any previously proposed taxonomic scheme. Many of the unusual groupings in the *rbcL* tree were present in some of Dahlgren's schemes, particularly in monocotyledons (many of the families and superorders of Dahlgren et al., 1985, can be overlaid easily onto the monocotyledons in the *rbcL* topology; Duvall et al., 1993; Chase et al., 1995a).

We have constructed representative cladograms like that of fig. 2B of Chase, Soltis, Olmstead et al. (1993) for both 18S rDNA (Fig. 17.5; Soltis, Soltis, Nickrent et al., 1997) and *atpB* (Fig. 17.6; Savolainen et al., 1996), and both of these are highly congruent with that of *rbcL*. The issue of congruence between *rbcL* and 18S rDNA trees has been reviewed in Soltis et al. (1997). To be sure, there are differences, for example the positions of Caryophyllidae s. l. and *Ceratophyllum,* but none of these differences has any degree of internal support (less than 50% with the bootstrap), and thus cannot be seriously considered as incongruent. When there is strong support from one, the others provide either a similar pattern or simply lack support; there are no strongly supported incongruent groupings. In particular, the same major clades are common to all three gene trees, and we assume that all three have recovered the same pattern. This pattern can be interpreted as historical

because they collectively represent evidence from both the nuclear (18S rDNA) and plastid (*rbcL, atpB*) genomes. In the only published comparison of several gene trees, Hoot, Culham, and Crane (1995) demonstrated that all three genes examined (*rbcL, atpB,* and 18S rDNA) also recovered the same topology within Lardizabalaceae. It would appear that worries over producing incongruent results will not be a major problem above the genus level.

Recently, a broad phylogenetic analysis of nonmolecular data was conducted for angiosperms using 161 taxa (Nandi et al., 1998). Although the trees recovered in this study clearly differ from those of the above-noted molecular analyses, the patterns for many groups are highly congruent with those observed in the gene trees. For example, Caryophyllidae sensu Cronquist are clustered with families such as Droseraceae, Nepenthaceae, Tamaricaceae, and Frankeniaceae; Malvales s. l. include such families as Bixaceae, Cistaceae, Cochlospermaceae, Sarcolaenaceae, and Sphaerosepalaceae; rosid families such as Aquifoliaceae, Cornaceae, and Escalloniaceae are among Asteridae sensu

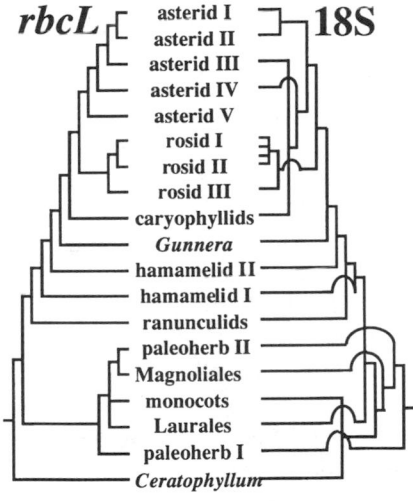

Figure 17.5. Schematic comparison of the general topology found for the 500-taxon search of Chase, Soltis, Olmstead et al. (1993) compared with that for 18S rDNA nuclear ribosomal matrix of Soltis, Soltis, Nickrent et al. (1997). Nearly the same terminal groupings were found in both studies, but neither of these trees is well-supported except within the terminal sets of families identified. In this figure, the paleoherb I group in the 18S rDNA tree is a paraphyletic grade rather than a clade as illustrated.

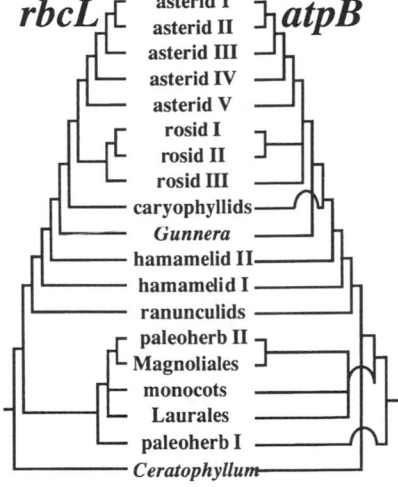

Figure 17.6. Schematic comparison of the general topology found for the 500-taxon search of Chase, Soltis, Olmstead et al. (1993) compared with that for plastid *atpB* of Savolainen et al. (1996). Identical terminal groupings were found in both studies, but internal support is again weak for the main spine of both trees. These two plastid gene analyses are much more congruent than either is to the 18S rDNA tree (Fig. 17.5). In this figure, the paleoherb I group in the *atpB* tree is a paraphyletic grade rather than a clade as illustrated.

Cronquist; and Primulales, Ebenales, and Lecythidales are associated with each other and some Theales such as Theaceae, Marcgraviaceae, and Scytopetalaceae, and all of these are sister to Asteridae. Whereas it is the most dissimilar of these broad analyses, it has many patterns congruent with the gene trees. All of these are weakly supported due to the large number of unknown data points in the matrix, an inevitable outcome of assembling a matrix covering all angiosperms.

HYBRIDIZATION AND HORIZONTAL GENE TRANSFER

Many studies at the species/genus levels have found evidence for hybridization (see Chapter 10), but this should not be expected to affect the sort of interfamilial studies addressed by Chase, Soltis, Olmstead et al. (1993) and others. Syvanen (1994) has argued that the history of angiosperm classification has been made difficult by horizontal transfer of genes and even parts of genes between highly divergent genomes, and that classifications have differed because the emphasis (weighting) applied to the various characters produced different conclusions. The assumption is that these different characters have been inherited idiosyncratically, and this would be expected to give different ideas of relationships because they have had different evolutionary histories. Likewise, gene trees for the angiosperms published until only recently have appeared to be incongruent, and this too was cited as a reflection of the same unconnected evolutionary histories as were responsible for different angiosperm classifications. The 1993 *rbcL* tree was included in this summary of the "angiosperm paradox." Syvanen (1994) did not consider that previous taxonomic schemes differed simply because the authors focused on subsets of characters and tried to sift through these in a subjective manner rather than synthesizing all characters objectively using a phylogenetic (i.e., cladistic) approach. In fact, few have attempted phylogenetic analysis of morphological/anatomical and chemical characters broadly within the angiosperms; the two notable exceptions are Hufford's (1992) treatment of rosids and the recent investigation of all an-

giosperms (Nandi et al., 1998). Neither of these supports any previous classification, but both provide many similar groupings found in the *rbcL* and the other well-sampled gene trees. No one would disagree with Syvanen's assertion that "the problem in plant taxonomy lies in the widespread existence of parallel traits," but we think that it is unlikely that this parallelism is caused by widespread horizontal transfer among angiosperms. In fact, the high degree of concordance between well-sampled plastid (*rbcL* and *atpB*) and nuclear gene trees (Soltis et al., 1997) suggests that horizontal gene transfer has rarely occurred.

Chase, Soltis, Olmstead et al. (1993) and Olmstead et al. (1992, 1993) stressed the importance of sufficient taxonomic sampling or "taxon density"; the importance of this can be seen in the differences between the *rbcL* trees of Albert et al. (1992a), Giannasi et al. (1992), Soltis et al. (1990), and Chase, Soltis, Olmstead et al. (1993). These trees have the appearance of incongruence for the taxa shared, but because these are data from the same gene, then sampling differences, not horizontal transfer, must be the cause. Syvanen (1994) stated that all published gene trees disagree, thereby supporting his hypothesis of their nonhistorical pattern of descent. If published *rbcL* trees disagree because of different levels of sampling and the more thoroughly sampled studies produce more reasonable and similar results, as argued by Chase, Soltis, Olmstead et al. (1993) and Olmstead et al. (1992, 1993), then this appearance of incongruence could also be generated by different levels of sampling. The cytochrome c data (Syvanen et al., 1989) that produced "four highly incongruent sets of trees" simply represented superficial sampling or were not suitable for phylogenetic analysis in the first place. The congruence of the three well-sampled gene trees for angiosperms, *rbcL,* 18S rDNA, and *atpB* (Figs. 17.5 and 17.6), supports the hypothesis that there is a common pattern in all these data.

CONTINUED SEQUENCING OF *rbcL*

Since the publication of the 1993 *rbcL* paper, numerous additional *rbcL* sequences have been produced. The original purpose for which *rbcL* sequencing began was not to elucidate higher order relationships among angiosperms, but rather to evaluate interfamilial patterns within a restricted set of purportedly closely related groups. We would be surprised if addition of any of these "new" sequences to the 499-taxon matrix would add substantially to our knowledge of higher-level relationships within the angiosperms. This is not to say that increased taxon-density within specific large clades would be similarly unproductive. For example, we could now learn a great deal about relationships within Rosidae s. l., a clade underrepresented in the 1993 analysis, but most of the clades revealed by the 1993 study were adequately sampled. As a case in point, sampling of those families considered potentially to be "first-branching" was extensive in the 1993 paper. Little would be gained via additional sequencing of *rbcL* because in this case there are apparently no other unsampled lineages. Continued *rbcL* sequencing has to have another focus.

By continuing to sequence *rbcL* for additional representatives of angiosperm lineages already in the matrix, we can begin to identify those clades to which problematic taxa belong. Examples of this type of use of the large *rbcL* data base include: Fernando et al. (1993) on Surianaceae; Shinwari et al. (1994) on *Prosartes;* Crayn et al. (1995) on the genus *Recchia;* Fay et al. (1997) on *Medusagyne;* Morton et al. (1997) on *Physena* and *Asteropeia;* Fay et al. (1998) on *Diegodendron;* Fay et al. (1997) on *Rhabdodendron;* and Hibsch-Jetter et al. (1997) on *Eremosyne.* These studies are made robust because, although the main spine of the *rbcL* tree does not have measurable internal support, more terminal groupings are strongly supported (Fig. 17.1), thus making it possible to place accurately taxa that are additional members of these groups. Conversely, taxa without close relatives, such as Zygophyllaceae and Krameriaceae (Sheahan and Chase, 1996), are difficult to place and will require the combination of multiple data sets before a well-supported position can be determined. These isolated taxa are almost certainly sisters of the major clades of angiosperms (i.e., all of the rosid I families), but because these cases are plagued by both low internal support and short

branch lengths, their positions are not yet final. Obtaining strongly supported placements of such taxa will have to await the results of studies that combine several genes.

In addition to the placement of problematic taxa, a second reason for continuing to sequence *rbcL* is the refinement of infrafamilial and intraordinal relationships; the level of variation and support in *rbcL* is appropriate to address these levels of relationships. Examples of the former use are: Morgan and Soltis (1993) on Saxifragaceae s. l.; Soltis et al. (1993) on Saxifragaceae s. s.; Price and Palmer (1993) on Geraniaceae; Kron and Chase (1993) on Ericaceae; Michaels et al. (1993) on Asteraceae; Conti et al. (1993) on Onagraceae; Xiang et al. (1993) on Cornaceae; Morgan et al. (1994) on Rosaceae; Williams et al. (1994) on Droseraceae; Fernando et al. (1995) on Simaroubaceae; Soltis et al. (1995) on Hydrangeaceae; Hedrén et al. (1995) on Acanthaceae; Morton et al. (1997b) on Lecythidaceae. Ordinal-level studies have also been successful in sorting out which families belong to higher-level groupings; these include: Rettig et al. (1992) on Caryophyllales; Rodman et al. (1993) on Capparales; Smith et al. (1993) on Zingiberales; Chase et al. (1995a) on Lilianae; Conti et al. (1996) on Myrtales; Plunkett et al. (1996) on Apiales; Gadek et al. (1996) on Sapindales; and Morton et al. (1997a) on Ebenales. These studies have led to refined infrafamilial and infraordinal classification, and to further studies to evaluate more carefully the often novel patterns detected with *rbcL* phylogenies.

One last purpose for which additional *rbcL* sequences are now being generated is in the context of large combined matrices. We now know that large numbers of characters improve both resolution and support as well as decrease time spent in analysis (Hillis, 1996; Soltis et al., 1997; Chase and Cox, 1998). Thus, data sets that combine evidence from multiple genes are more likely to provide strong estimates of phylogeny than are those based on a single gene. In only a few cases was another gene sequenced before *rbcL* (Nickrent and Franchina, 1990; Hamby and Zimmer, 1992). In many instances, initial sequencing of *rbcL* within a group (particularly within a family) provided too little sequence variation on its own to be usefully extended to many taxa, but with the need to combine *rbcL* with other data, comparable sampling for all loci is required, regardless of the relatively low variation present in any of the loci individually. Although it has been assumed that *rbcL* is relatively conserved, many studies are finding that plastid spacers are not that much more variable than *rbcL* that they obviate the need to produce *rbcL* data as well (Fay et al., 1996). It is safe to assume that sequencing of *rbcL* continues to fulfill a useful role in synthetic phylogenetic studies.

STUDIES OF PLANT LIFE HISTORY TRAITS AND ECOLOGY

The extensive taxon coverage of the 1993 *rbcL* tree has guaranteed that this tree has had a wide application outside the field of systematics. Botanists and ecologists interested in the evolution of life history traits have seized upon this most extensive analysis of angiosperms to ask questions that require a well-developed phylogenetic context (i.e., that need a real tree). These workers have accepted the caveats made in the 1993 paper and generally not pushed evolutionary/phylogenetic interpretations beyond what seems reasonable in the context of these limitations. One such application has involved attempts to assess whether phylogenetic correlations exist for two or more features. Several such examples of this approach were recently published in the *Proceedings of the Royal Society of London* (B series, volume 262, 1996). As one example, Barraclough et al. (1996) examined the relationship between sequence divergence and species diversification; they concluded that species-rich groups are much more likely to be sequence-divergent than species-poor groups. This approach is an important shift in the attitude that ecological characteristics do not require a phylogenetic context to be accurately understood.

STRUCTURAL EVOLUTION OF RUBISCO

The expanding *rbcL* database has also facilitated detailed studies of molecular and structural evolution. As suggested by Clegg (1993), the large

data set of *rbcL* sequences can be applied to existing detailed information on the crystal structure of the RUBISCO holoenzyme (Chapman et al., 1988; Andersson et al., 1989; Knight et al., 1990; Schneider et al., 1990; Taylor and Andersson, 1996). Comparative analysis of the spinach and tobacco enzymes revealed that most amino acid differences (54 out of 61) were to be found on the protein surfaces (Schreuder et al., 1993).

These observations have been extended across a large sample of sequences representing all clades of land plants (Kellogg and Juliano, 1997; V. A. Albert and I. Andersson, unpubl.). Here again, the vast majority of amino acid differences among approximately 600 land plant *rbcL* sequences was found to be localized on the surface of the large-subunit protein. As one might predict, the active site and associated amino acid residues remained nearly constant over all species studied. At a lower taxonomic level, the single genus *Drosera* (Droseraceae) exemplifies the insights that may be garnered via the careful analysis of the RUBISCO holoenzyme based on *rbcL* sequence data. One species that varied quite close to the active site was *Drosera peltata* (V. A. Albert and I. Andersson, unpubl.), a carnivorous plant from Australia. At residue 174 of the large subunit, proximal to active site residue 175, threonine (T) was substituted for isoleucine (I). This change seemed rather odd because it would likely affect the critical structural conformation around residue 175. Because no other Droseraceae showed this change, it may be ascribable to sequencing error, but the intense analysis of Droseraceae *rbcL* sequences that ensued uncovered some additional potential structural differences compared to most other angiosperms, including all other genera of carnivorous plants. This suggests that the unusual single change in *D. peltata* cited above is part of a syndrome of potentially altered RUBISCO function that occurs throughout Droseraceae.

RUBISCO large subunits form four interacting dimers in the holoenzyme (Knight et al., 1990). Residue R258 (arginine) is at the L-dimer interfaces of the spinach enzyme and interacts ionically with residue E259 (glutamic acid) of neighboring dimers. This salt-link is disrupted in four *Drosera* species, in which basic arginine is substituted for uncharged serine (S, in *Drosera gigantea, D. peltata,* and *D. stolonifera*) or polar tyrosine (Y, in *D. binata*). Four other land plants had altered residues at position 258, but only one of these would have broken the salt bridge. Another disrupted salt-link at the L-dimer interface occurs at D286 (aspartic acid), which in spinach links to K252 (lysine); several Droseraceae (*Drosera binata, D. gigantea, D. peltata, D. stolonifera,* and *Dionaea muscipula*) have polar asparagine (N) substituted for acidic aspartic acid. These same species have N287 substituted with basic histidine (H), which may form a compensatory salt-link elsewhere. Only two other unrelated land plants had altered amino acids at positions 286 and 287. Interestingly, although less relevant structurally, I301 (isoleucine) at the C-terminal interface within L dimers is changed to L (leucine) in *Drosera filiformis* and *D. indica.* Only two other unrelated land plants had amino acid substitutions at this site. A nearly *Drosera*-specific site of change was R435, which lies between two alpha helices. This basic residue is changed to polar Q (glutamine) in *Drosera burmanni, D. capensis, D. filiformis, D. peltata,* and *D. spathulata;* to nonpolar L (leucine) in *D. dichrosepala;* to basic H in *D. gigantea;* and to polar Y (tyrosine) in *D. petiolaris.* Only *Eupomatia* (Magnoliidae) has an altered R435 (to H). This concentration of unusual amino acid substitutions near active sites in Droseraceae suggests that these carnivorous plants may be the single instance of altered RUBISCO function found among all photosynthetic organisms. The most interesting aspect of this finding is its uneven pattern within *Drosera;* these changes are clustered in *Drosera* but not uniform (i.e., they are not always synapomorphies for clades of species), suggesting an uneven pattern that is tending toward altered function.

The nearly universal clustering of amino acid substitutions away from the functionally conserved active site of RUBISCO is not a surprising result. However, changes in external residues could also have functional consequences, if not catastrophic ones. For example, it is possible that variation in protein surface attributes could affect intermolecular interactions between RUBISCO and other factors. Alteration of chaperone specificity is one possibility. Molecular chaperones are known to function in the assembly of the RUBISCO holoenzyme (Ellis, 1990; Gatenby and Ellis, 1990; Ellis and Hartl, 1996). Some research on cloning and understanding chaperones is directed toward crop improvement, because the RUBISCO of higher plants will not assemble properly using bacterial chaperonins. Indeed, genetic engineering of the many variable sites that face out into solution could be directed toward modifying interactions between RUBISCO and its environment, including its chaperone-mediated assembly.

The unusual amino acid changes among Droseraceae may partly reflect altered functional properties; others may be, in addition, lineage-specific markers of phylogenetic history (cf. Albert et al., 1994a, 1994b), but a more thorough study of Droseraceae would be useful to resolve these questions. At present, only a few of the more than 100 species in *Drosera* have been studied. The disrupted L-dimer salt bridges could have structural implications for the holoenzyme, but this can be adequately studied only by solving the crystal structure of RUBISCO of Droseraceae. It cannot be inferred by comparisons to structure of RUBISCO from *Nicotiana.* There appears to be no functional correlation with angiosperm carnivory (Albert et al., 1992a) because partial heterotrophy is observed in many taxa that do not share these unique substitutions. If these preliminary observations are any indication, further analyses of variation in RUBISCO structure using the vast database of *rbcL* sequences will likely uncover as many questions as answers.

STABILITY OF PHYLOGENETIC PATTERNS DISCOVERED USING *rbcL* DATA

Chase, Soltis, Olmstead et al. is a tribute to the spirit of cooperation that exists among plant systematists; so many were willing to contribute previously unpublished sequences to facilitate the advancement of their discipline. Had individual authors first published their data separately, the publication of this work would have been delayed by at least two years. Plant systematists were clearly interested more in the progress of their discipline than in personal accomplishment. Significantly, this same spirit of cooperation has now been carried to the next stage of investigation, that of the production of additional large matrices of gene sequences that can be compared and ultimately combined with that from *rbcL* (such as has been done already in the context of examining the congruence of 18S rDNA and *rbcL;* Soltis et al., 1997).

In addition to fostering large-scale collaboration, the 1993 *rbcL* analysis demonstrated that large matrices could be successfully analyzed, although this "proof" has only recently emerged with the analysis of two additional large gene matrices, that of 18S rDNA (Soltis, Soltis, Nickrent et al., 1997) and *atpB* (Savolainen et al., 1996). Although none of these broad searches has swapped to completion, the "estimates" of phylogenetic relationships obtained using each gene are highly similar. This advancement occurred due to increased computer capacity and speed, both in hardware and software. At the time of publication, the *rbcL* tree was considered by some molecular phylogenists to be a "grand folly" because it was obvious to any "knowledgeable" researcher that matrices of this dimension were intractable due to the astronomical number of tree topologies that would have to be evaluated (comments made to M. Chase by more than one recognized authority, unnamed here; see Graur et al., 1996, for an alternative approach). We thought at the time that it would be unlikely that the 1993 *rbcL* tree would gain wide acceptance, but we also knew that analyses with smaller sets of taxa produced topologies that were not only less meaningful, but in some instances clearly erroneous. Our experiments with larger matrices on the other hand convinced many of us that more thorough sampling produced better results. Thus, our decision was the one stated in the 1993 paper: if fewer taxa produced better phylogenetic analyses in terms of completeness, whereas increased numbers of taxa, which were impractical in terms of methods, produced results that seemed more reasonable (i.e., were congruent with other kinds of information), we felt obliged to experiment with more taxa to determine the limits of what was feasible. Rice et al. (1997) have con-

tinued to find shorter trees than those reported for the 500-sequence matrix in 1993, but they have not recovered any new patterns with great support, nor are they likely to. It is perfectly clear that the 1993 analysis was adequate: It found all of the many clades supported by the bootstrap (Fig. 17.1).

Rice et al. (1997) have continued to work with the weakly supported groupings, and many of us believe that effort would be better directed toward the corroboration of the *rbcL* tree with data from other genes. The recently produced 18S rDNA (Soltis, Soltis, Nickrent et al., 1997) and *atpB* (Savolainen et al., 1996) studies have taken the same intensive sampling approach as the 1993 *rbcL* project. The high degree of congruence and the increased level of support for major groupings (reviewed in Soltis et al., 1997) provide clear evidence that large analyses can be meaningful. Such high levels of congruence cannot be due to chance. The same well-supported major groupings are present in all three analyses, and the weakly supported interrelationships of these major groups are also highly similar. Whereas it is undoubtedly true that an adequate search of all possible tree patterns is impossible for such large numbers of terminals, it has become equally clear that the structured nature of the variation in these matrices precludes having to evaluate the overwhelming majority of the potential tree topologies. Many trees obviously need not be considered simply on the basis of crude distance calculations, and of course several heuristic parsimony programs (e.g., "mhennig" of Henning86 [Farris, 1988]; and "simple" data addition of PAUP, Swofford, 1993) use such calculations to generate the starting trees upon which swapping then begins. Other programs, such as neighbor joining, consist entirely of distance calculations that have been corrected for multiple, undetected substitutions, and these can be expected to produce nearly identical topologies to those produced by parsimony, especially when the underlying signal is strong.

Soltis et al. (1997) have already demonstrated that combining these gene sequence matrices produces greater levels of internal support for larger numbers of groupings, and this would be expected if the matrices are compatible. They used the partition homogeneity test of PAUP 4.0* (after Farris et al., 1995; used with permission of the author, D. Swofford) to evaluate 18S rDNA and *rbcL* matrices, but they found that they were significantly incongruent ($P < 0.05$). Nevertheless, combining the two matrices provided fewer trees and greater levels of internal support, both of which would not be expected if the underlying signals were incongruent. Chase et al. (1995b) found a similar synergistic situation when combining apparently incongruent morphological and molecular data for monocotyledons. Another important point noted by Soltis et al. (1997) is that replicate searches of the combined 18S rDNA/*rbcL* matrix for 232 taxa swapped to completion in a matter of days. With what is by now several years' worth of computer time, not even one search of the 1993 500-sequence *rbcL* matrix has been completed; likewise the 232-taxon 18S rDNA matrix (Soltis, Soltis, Nickrent et al., 1997) was swapped on for more than 2 years (D. Soltis, pers. comm.) without completing the search. By combining *rbcL* and 18S rDNA with *atpB* (D. Soltis, P. Soltis, M. Chase, M. Mort, and V. Savolainen unpubl.), even greater levels of internal support and more resolution are obtained, and these matrices also swap to completion in only a few days (Chase and Cox, 1998, examine one of the reasons why this occurs).

A NEW ANGIOSPERM CLASSIFICATION

The robustness of the combined 18S rDNA/*rbcL*/*atpB* tree puts angiosperm systematists in the unexpected position of having a well-supported phylogenetic tree for the angiosperms. This phylogenetic hypothesis will serve as the basis for constructing a new classification for the angiosperms. Nonetheless, questions will remain about the appropriateness of basing a classification solely on molecular data. It is noteworthy, therefore, that the shortest trees recovered from a large (161-taxon) analysis of nonmolecular data (Nandi et al., 1998) are remarkably similar in many groupings to the molecular trees (even though the trees themselves are modestly different). It is clear that other lines of evidence, including data from phytochemistry and micromorphology, lend additional support to the topologies obtained from the three genes. This observation should encourage us to begin the process of reclassification. We would suggest that this could best be carried out by the establishment of an electronic "homepage" based at one institution that would be responsible for the updating of the classification as new, well-supported evidence, molecular or otherwise, comes to light (such a "homepage" effort already exists at Uppsala University, under the aegis of K. Bremer, B. Bremer, and M. Thulin). Few new names would be required; nearly every conceivable category and combination exists in the literature (J. Reveal, University of Maryland at College Park, has a database, *Indices Nominum Supragenericorum Plantarum Vascularium,* available on-line). There is no reason not to reclassify based on a well-supported phylogenetic analysis, regardless of the type of data that were analyzed. The alternative is the continued use of classifications that nearly all would now agree are hopelessly flawed and out of date. The real problem is that publication of this new system is impractical, and the rate of change is already rapid and only bound to increase in the coming years. The only viable way to produce this new collaborative system is to make it available electronically and update it as soon as evidence emerges.

Those interested in teaching plant taxonomy have a challenge to teach their students in this rapidly changing environment, and this electronically formatted classification could be periodically published and made available in hard copy, complete with cladograms, illustrations, and descriptions. This publication would both document the expanding scope of what is known, as well as highlight those taxa for which information is still lacking or inconclusive with presently available data. Printed classifications have certain advantages of course and, if for nothing else, they would serve as a permanent record of the developing body of knowledge of angiosperm relationships.

It is unlikely that another 42-author effort like that of 1993 *rbcL* analysis will again be necessary. With the advent of automated sequencing, the rate at which new data are being produced is going to continue to escalate, and increasingly computerized methods of data management are going to be required. Large matrices will be created, but the number of investigators required to produce these data will shrink. The 1993 *rbcL* paper had 42 coauthors, the 1997 18S rDNA paper had only 16, and the 1998 *atpB* paper will have fewer than ten. This is not to indicate that fewer people are involved overall in systematic studies, but rather it is a reflection of the diversity of approaches that is being employed to evaluate angiosperm relationships. The publication of the 1993 *rbcL* tree set the stage for an entirely novel and collaborative approach to angiosperm systematics, and that approach has begun to produce enormous change in the way that classification of angiosperms is being assessed and modified.

LITERATURE CITED

Albert, V. A., S. E. Williams, and M. W. Chase. 1992a. Carnivorous plants: phylogeny and structural evolution. Science **257**:1491–1495.

Albert, V. A., B. A. Mishler, and M. W. Chase. 1992b. Character-state weighting for restriction site data in phylogenetic reconstruction, with an example from chloroplast DNA. In Molecular Systematics of Plants, eds. P. S. Soltis, D. E. Soltis, and J. J. Doyle, pp. 369–401. Chapman & Hall, New York.

Albert, V. A., M. W. Chase, and B. A. Mishler. 1993. Character-state weighting for cladistic analysis of protein-coding DNA sequences. Annals of the Missouri Botanical Garden **80**:752–766.

Albert, V. A., A. Backlund, K. Bremer, M. W. Chase, J. R. Manhart, B. D. Mishler, and K. C. Nixon. 1994a. Functional constraints and *rbcL* evidence for land plant phylogeny. Annals of the Missouri Botanical Garden **81**:534–567.

Albert, V. A., A. Backlund, and K. Bremer. 1994b. DNA characters and cladistics: the optimization of functional history. In Models in Phylogeny Reconstruction, eds. R. W. Scotland, D. J. Siebert, and D. M. Williams, pp. 251–272. Clarendon Press, Oxford.

Andersson, I., S. Knight, G. Schneider, Y. Lindqvist, T. Lundqvist, C.-I. Brändén, and G. H. Lorimer. 1989. Crystal structure of the active site of ribulose-bisphosphate carboxylase. Nature **337**:229–234.

Barraclough, T. G., P. H. Harvey, and S. Nee. 1996. Rate of *rbcL* gene sequence evolution and species diversification in flowering plants (angiosperms). Proceedings of the Royal Society of London B **263**:589–591.

Boulter, D. 1974. The evolution of plant proteins with special reference to higher plant cytochrome c. Current Advances in Plant Science **4**:1–16.

Chapman M. S., S. W. Suh, P. M. Curmi, D. Cascio, W. W. Smith, and D. S. Eisenberg. 1988. Tertiary structure of plant RuBisCO: domains and their contacts. Science **241**:71–74.

Chase, M. W., and A. V. Cox. 1998. Gene sequences, collaboration, and analysis of large data sets. Australian Systematic Botany: in press.

Chase, M. W., D. E. Soltis, R. G. Olmstead, D. Morgan, D. H. Les, B. Mishler, M. R. Duvall, R. A. Price, H. G. Hills, Y.-L. Qiu, K. A. Kron, J. H. Rettig, E. Conti, J. D. Palmer, J. R. Manhart, K. J. Sytsma, H. J. Michaels, W. J. Kress, K. G. Karol, W. D. Clark, M. Hedrén, B. S. Gaut, R. K. Jansen, K.-J. Kim, C. F. Wimpee, J. F. Smith, G. R. Furnier, S. H. Straus, Q.-Y. Xiang, G. M. Plunkett, P. S. Soltis, S. M. Swensen, S. E. Williams, P. A. Gadek, C. J. Quinn, L. Eguiarte, E. Golenberg, G. H. Learn, S. W. Graham, S. C. H. Barrett, S. Dayanandan, and V. A. Albert. 1993. Phylogenetics of seed plants: an analysis of nucleotide sequences from the plastic gene *rbcL*. Annals of the Missouri Botanical Garden **80**:528–580.

Chase, M. W., M. R. Duvall, H. G. Hills, J. G. Conran, A. V. Cox, L. E. Eguiarte, J. Hartwell, M. F. Fay, L. R. Caddick, K. M. Cameron, and S. Hoot. 1995a. Molecular phylogenetics of Lilianae. In Monocotyledons: Systematics and Evolution, eds. P. Rudall, P. J. Cribb, D. F. Cutler, and C. J. Humphries, pp. 109–137. Royal Botanic Gardens, Kew, London.

Chase, M. W., D. W. Stevenson, P. Wilkin, and P. J. Rudall. 1995b. Monocot systematics: a combined analysis. In Monocotyledons: Systematics and Evolution, eds. P. Rudall, P. J. Cribb, D. F. Cutler, and C. J. Humphries, pp. 685–730. Royal Botanic Gardens, Kew, London.

Clegg, M. T. 1993. Chloroplast gene sequences and the study of plant evolution. Proceedings of the National Academy of Sciences U.S.A. **90**:363–367.

Conti, E., A. Fishbach, and K. J. Sytsma. 1993. Tribal relationships in Onagraceae: implications from *rbcL* sequence data. Annals of the Missouri Botanical Garden **80**:672–685.

Conti, E., A. Litt, and K. J. Sytsma. 1996. Circumscription of Myrtales and their relationships to other rosids: evidence from *rbcL* sequence data. American Journal of Botany **83**:221–233.

Crayn, D. M., E. S. Fernando, P. A. Gadek, and C. J. Quinn. 1995. A reassessment of the familial affinities of the Mexican genus *Recchia* Mocino and Sesse ex DC. Brittonia **47**:397–402.

Cronquist, A. 1981. An Integrated System of Classification of Flowering Plants. Columbia University Press, New York.

Dahlgren, R. T., H. T. Clifford, and P. F. Yeo. 1985. The Families of the Monocotyledons: Structure, Evolution, and Taxonomy. Springer-Verlag, Berlin.

Doyle, J. A., M. J. Donoghue, and E. A. Zimmer. 1994. Integration of morphological and ribosomal RNA data on the origin of the angiosperms. Annals of the Missouri Botanical Garden **81**:419–450.

Duvall, M. R., M. T. Clegg, M. W. Chase, W. D. Clark, W. J. Kress, H. G. Hills, L. E. Eguiarte, J. F. Smith, B. S. Gaut, E. A. Zimmer, and G. H. Learn, Jr. 1993. Phylogenetic hypotheses for the monocotyledons constructed from *rbcL* sequence data. Annals of the Missouri Botanical Garden **80**:607–619.

Ellis, R. J. 1990. Molecular chaperones: the plant connection. Science **250**:954–959.

Ellis, R. J., and F. U. Hartl. 1996. Protein folding in the cell: competing models of chaperonin function. FASEB Journal **10**:20–26.

Farris, J. S. 1988. HENNIG86 reference, version 1.5. Published by the author, Port Jefferson, New York.

Farris, J. S., M. Källersjo, A. G. Kluge, and C. Bult. 1995. Testing significance of incongruence. Cladistics **10**:315–319.

Fay, M. F., W. S. Davis, L. Hufford, and M. W. Chase. 1996. A combined cladistic analysis of Themidaceae. American Journal of Botany (suppl.) **83**:155.

Fay, M. F., S. M. Swensen, and M. W. Chase. 1997. Taxonomic affinities of *Medusagyne oppositifolia*. Kew Bulletin **52**:111–120. 1997.

Fay, M. F., C. Bayer, W. Alverson, A. Y. de Bruijn, S. M. Swensen, and M. W. Chase. 1998. Plastid *rbcL* sequences indicate a close affinity between *Diegodendron* and *Bixa*. Taxon **47**: in press.

Fay, M. F., K. M. Cameron, G. T. Prance, D. Lledó, and M. W. Chase. 1997. Familial relationships of *Rhabdodendron* (Rhabdodendraceae): plastid *rbcL* sequences indicate a caryophyllid placement. Kew Bulletin **52**: 923–932.

Felsenstein, J. 1978. Cases in which parsimony or compatibility methods will be positively misleading. Systematic Zoology **20**:406–416.

Felsenstein, J. 1981. Evolutionary trees from DNA sequences: a maximum likelihood approach. Journal of Molecular Evolution **17**:368–376.

Felsenstein, J. 1991. PHYLIP (Phylogeny Inference Package) version 3.4. University of Washington, Seattle.

Fernando, E. S., P. A. Gadek, D. M. Crayn, and C. J. Quinn. 1993. Rosid affinities of Surianaceae: molecular evidence. Molecular Phylogenetics and Evolution **2**:344–350.

Fernando, E. S., P. A. Gadek, and C. J. Quinn. 1995. Simaroubaceae, an artificial construct: evidence from *rbcL* sequence variation. American Journal of Botany **82**:92–103.

Gadek, P. A., E. S. Fernando, C. J. Quinn, S. B. Hoot, T. Terrazas, M. C. Sheahan, and M. W. Chase. 1996. Sapindales: molecular delimitation and infraordinal groups. American Journal of Botany **83**:802–811.

Gatenby, A. A., and R. J. Ellis. 1990. Chaperone function: the assembly of ribulose bisphosphate carboxylase-oxygenase. Annual Review of Cell Biology **6**:125–149.

Gaut, B. S., S. V. Muse, W. D. Clark, and M. T. Clegg. 1992. Relative rates of nucleotide substitution at the *rbcL* locus of monocotyledonous plants. Journal of Molecular Evolution **35**:292–303.

Giannasi, D. E., G. Zurawski, G. Learn, and M. T. Clegg. 1992. Evolutionary relationships of the Caryophyllidae based on comparative *rbcL* sequences. Systematic Botany **17**:1–15.

Graur, D., L. Duret, and M. Gouy. 1996. Phylogenetic position of the order Lagomorpha (rabbits, hares and allies). Nature **379**:333–334.

Hamby, R. K., and E. A. Zimmer. 1992. Ribosomal RNA as a tool in plant systematics. **In** Molecular Systematics of Plants, eds. P. S. Soltis, D. E. Soltis, and J. J. Doyle, pp. 50–91. Chapman & Hall, New York.

Hedrén, M., M. W. Chase, and R. G. Olmstead. 1995. Relationships in the Acanthaceae and related families as suggested by cladistic analysis of *rbcL* nucleotide sequences. Plant Systematics and Evolution **194**:93–109.

Hibsch-Jetter, C. D., D. E. Soltis, and T. D. MacFarlane. 1997. Phylogenetic relationships of *Eremosyne pectinata* Endl. (Saxifragaceae sensu lato) based on *rbcL* sequence data. Plant Systematics and Evolution **204**:225–232.

Hillis, D. M. 1996. Inferring complex phylogenies. Nature **383**:130.

Hoot, S. B., A. Culham, and P. R. Crane. 1995. The utility of *atpB* gene sequences in resolving phylogenetic relationships: comparisons with *rbcL* and 18S ribosomal DNA sequences in Lardizabalaceae. Annals of the Missouri Botanical Garden **82**:194–207.

Hufford, L. 1992. Rosidae and their relationships to other nonmagnoliid dicotyledons: a phylogenetic analysis using morphological and chemical data. Annals of the Missouri Botanical Garden **79**:218–248.

Kellogg, E. A., and N. D. Juliano. 1997. The structure and function of RUBISCO and their implications for systematic studies. American Journal of Botany **84**: 413–428.

Knight, S., I. Andersson, and C.-I. Brändén. 1990. Crystallographic analysis of ribulose 1,5-bisphosphate carboxylase from spinach at 2.4 Å resolution. Subunit interactions and active site. Journal of Molecular Biology **215**:113–160.

Kron, K. A., and M. W. Chase. 1993. Systematics of the Ericaceae, Empetraceae, Epacridaceae and related taxa based upon *rbcL* sequence data. Annals of the Missouri Botanical Garden **80**:735–741.

Maddison, W. P., and D. R. Maddison. 1992. MacClade: Analysis of Phylogeny and Character Evolution, version 3.0. Sinauer Associates, Sunderland, Massachusetts.

Manhart, J. R. 1994. Phylogenetic analysis of green plant *rbcL* sequences. Molecular Phylogenetics and Evolution **3**:114–127.

Martin, P. G., and J. M. Dowd. 1991. A comparison of 18S ribosomal RNA and RUBISCO large subunit sequences for studying angiosperm phylogeny. Journal of Molecular Evolution **33**:274–282.

Martin, P. G., D. Boulter, and D. Penny. 1985. Angiosperm phylogeny studied using sequences of five macromolecules. Taxon **34**:393–400.

Michaels, H. J., K. M. Scott, R. G. Olmstead, T. Szaro, R. K. Jansen, and J. D. Palmer. 1993. Interfamilial relationships of the Asteraceae: insights from *rbcL* sequence variation. Annals of the Missouri Botanical Garden **80**:742–765.

Morgan, D. R., and D. E. Soltis. 1993. Phylogenetic relationships among members of Saxifragaceae *sensu lato* based on *rbcL* sequence data. Annals of the Missouri Botanical Garden **80**:631–660.

Morgan, D. R., D. E. Soltis, and K. R. Robertson. 1994. Systematic and evolutionary implications of *rbcL* sequence variation in Rosaceae. American Journal of Botany **81**:890–903.

Morton, B. R. 1993. Chloroplast DNA codon use: evidence for selection at the *psbA* locus based on tRNA availability. Journal of Molecular Evolution **37**:273–280.

Morton, B. R. 1994. Codon use and the rate of divergence of land plant chloroplast genes. Molecular Biology and Evolution **11**:231–238.

Morton, B. R. 1995. Neighboring base composition and transversion/transition bias in a comparison of rice and maize chloroplast noncoding regions. Proceedings of the National Academy of Sciences U.S.A. **92**: 9717–9721.

Morton, B. R., and M. T. Clegg. 1995. Neighboring base composition is strongly correlated with base substitution bias in a region of the chloroplast genome. Journal of Molecular Evolution **41**:597–603.

Morton, C. M., M. W. Chase, and K. G. Karol. 1997. Taxonomic affinities of *Physena* (Physenaceae) and *Asteropeia* (Asteropeiaceae) Botanical Review 231–239.

Morton, C. M., M. W. Chase, K. A. Kron, and S. A. Swensen. 1997. A molecular evaluation of the monophyly of the order Ebenales based upon *rbcL* sequence data. Systematic Botany **21**:567–586.

Morton, C. M., S. A. Mori, G. T. Prance, K. G. Karol, and M. W. Chase. 1997. Phylogenetic relationships of Lecythidaceae: a cladistic analysis using *rbcL* sequences and morphological data. American Journal of Botany **84**: 530–540.

Nandi, O., M. W. Chase, and P. K. Endress. 1998. A combined cladistic analysis of angiosperms using *rbcL* and non-molecular data sets. Plant Systematics and Evolution: in press.

Nickrent, D. L., and C. R. Franchina. 1990. Phylogenetic relationships of Santalales and relatives. Journal of Molecular Evolution **31**:294–301.

Olmstead, R. G., H. J. Michaels, K. M. Scott, and J. D. Palmer. 1992. Monophyly of the Asteridae and identification of their major lineages inferred from DNA sequences of *rbcL*. Annals of the Missouri Botanical Garden **79**:249–265.

Olmstead, R. G., B. Bremer, K. M. Scott, and J. D. Palmer. 1993. A parsimony analysis of the Asteridae *sensu lato* based on *rbcL* sequences. Annals of the Missouri Botanical Garden **80**:700–722.

Palmer, J. D., R. K. Jansen, H. J. Michaels, M. W. Chase, and J. M. Manhart. 1988. Analysis of chloroplast DNA variation. Annals of the Missouri Botanical Garden **75**:1180–1206.

Plunkett, G. M., D. E. Soltis, and P. S. Soltis, 1996. Higher level relationships of Apiales (Apiaceae and Araliaceae) based on phylogenetic analysis of *rbcL* sequences. American Journal of Botany **83**:499–515.

Price, R. A., and J. D. Palmer. 1993. Phylogenetic relationships of the Geraniaceae and Geraniales from *rbcL* sequence comparisons. Annals of the Missouri Botanical Garden **80**:661–671.

Qiu, Y.-L., M. W. Chase, D. H. Les, and C. R. Parks. 1993. Molecular phylogenetics of the Magnoliidae: cladistic analyses of nucleotide sequences of the plastid gene *rbcL*. Annals of the Missouri Botanical Garden **80**: 587–606.

Rettig, J. H., H. D. Wilson, and J. M. Manhart. 1992. Phylogeny of the Caryophyllales—gene sequence data. Taxon **41**:201–209.

Rice, K. A., M. J. Donoghue, and R. G. Olmstead. 1997. A reanalysis of the large *rbcL* dataset: implications for future phylogenetic studies. Systematic Biology **46**:554–563.

Ritland, K., and M. T. Clegg. 1987. Evolutionary analysis of plant DNA sequences. American Naturalist **130**: S74–100.

Rodman, J., R. Price, K. Karol, E. Conti, K. Sytsma, and J. D. Palmer. 1993. Nucleotide sequences of the *rbcL* gene indicate monophyly of mustard oil plants. Annals of the Missouri Botanical Garden **80**:686–699.

Saitou, N., and M. Nei. 1987. The neighbor-joining method: a new method for reconstructing phylogenetic trees. Molecular Biology and Evolution **4**:406–425.

Savolainen, V., C. M. Morton, S. B. Hoot, and M. W. Chase. 1996. An examination of phylogenetic patterns of plastid *atpB* gene sequences among eudicots. American Journal of Botany (suppl.) **83**:190.

Schneider G., S. Knight, I. Andersson, C.-I. Brändén, Y. Lindqvist, and T. Lundqvist. 1990. Comparison of the crystal structures of L2 and L8S8 RUBISCO suggests a functional role for the small subunit. EMBO Journal **9**:2045–2050.

Schreuder H. A., S. Knight, P. M. Curmi, I. Andersson, D. Cascio, R. M. Sweet, C.-I. Brändén, and D. Eisenberg. 1993. Crystal structure of activated tobacco RUBISCO complexed with the reaction-intermediate analogue 2-carboxy-arabinitol 1,5-bisphosphate. Protein Science **2**:1136–1146.

Sheahan, M. C., and M. W. Chase. 1996. A phylogenetic analysis of Zygophyllaceae R. Br. based on morphological, anatomical, and *rbcL* DNA sequence data. Botanical Journal of the Linnean Society **122**: 279–300.

Shinwari, Z. K., R. Terauchi, F. H. Utech, and S. Kawano. 1994. Recognition of the New World *Disporum* section *Prosartes* as *Prosartes* (Liliaceae) based on the sequence data of the *rbcL* gene. Taxon **43**:353–366.

Smith, J. F., W. J. Kress, and E. A. Zimmer. 1993. Phylogenetic analysis of the Zingiberales based on *rbcL* sequences. Annals of the Missouri Botanical Garden **80**:620–630.

Soltis, D. E., P. S. Soltis, M. T. Clegg, and M. Durbin. 1990. *rbcL* sequence divergence and phylogenetic relationships in Saxifragaceae sensu lato. Proceedings of the National Academy of Sciences U.S.A. **87**:4640–4644.

Soltis, D. E., D. R. Morgan, A. Grable, P. S. Soltis, and R. Kuzoff. 1993. Molecular systematics of Saxifragaceae sensu stricto. American Journal of Botany **80**:1056–1081.

Soltis, D. E., Q.-Y. Xiang, and L. Hufford. 1995. Relationships and evolution of Hydrangeaceae based on *rbcL* sequence data. American Journal of Botany **82**: 504–514.

Soltis, D. E., C. Hibsch-Jetter, P. S. Soltis, M. W. Chase, and J. S. Farris. 1997. Molecular phylogenetic relationships among angiosperms: an overview based on *rbcL* and 18S rDNA sequences. Journal of Plant Research, in press.

Soltis, D. E., P. S. Soltis, D. L. Nickrent, L. A. Johnson, W. J. Hahn, S. B. Hoot, J. A. Sweere, R. K. Kuzoff, K. A. Kron, M. W. Chase, S. M. Swensen, E. A. Zimmer, S.-M. Chaw, L. J. Gillespie, W. J. Kress, and K. J. Sytsma. 1997. Angiosperm phylogeny inferred from 18S ribosomal DNA sequences. Annals of the Missouri Botanical Garden **84:**1–49.

Swofford, D. L. 1991. PAUP: Phylogenetic Analysis Using Parsimony, version 3.0q. Illinois Natural History Survey, Champaign, Illinois.

Swofford, D. L. 1993. PAUP: Phylogenetic Analysis Using Parsimony, version 3.1.1. Illinois Natural History Survey, Champaign, Illinois.

Syvanen, M. 1994. Horizontal gene transfer: evidence and possible consequences. Annual Review of Genetics **28:**237–261.

Syvanen, M., H. Hartman, and P. F. Stevens. 1989. Classical plant taxonomic ambiguities extend to the molecular level. Journal of Molecular Evolution **28:**536–544.

Taylor T. C., and I. Andersson. 1996. Structural transitions during activation and ligand binding in hexadecameric RUBISCO inferred from the crystal structure of the activated unliganded spinach enzyme. Natural Structural Biology **3:**95–101.

Troitsky, A. V., Y. F. Malakhovets, G. M. Rakhimova, V. K. Bobrova, K. M. Valiejo-Roman, and A. S. Antonov. 1991. Angiosperm origin and early stages of seed plant evolution deduced from rRNA sequence comparisons. Journal of Molecular Evolution **32:**253–261.

Williams, S. E., V. A. Albert, and M. W. Chase. 1994. Relationships of Droseraceae: a cladistic analysis of *rbcL* sequence and morphological data. American Journal of Botany **81:**1027–1037.

Xiang, Q.-Y., D. E. Soltis, D. R. Morgan, and P. S. Soltis. 1993. Phylogenetic relationships of *Cornus* L. sensu lato and putative relatives inferred from *rbcL* sequence data. Annals of the Missouri Botanical Garden **80:**723–734.

18

Molecular Systematics of the Green Algae

Russell L. Chapman, Mark A. Buchheim, Charles F. Delwiche,
Thomas Friedl, Volker A. R. Huss, Kenneth G. Karol,
Louise A. Lewis, Jim Manhart, Richard M. McCourt,
Jeanine L. Olsen, and Debra A. Waters

Ranging from unicells to complex "plantlike" organisms that are adapted to habitats from subaerial or terrestrial to freshwater or marine, the green algae represent a diversity of life forms that offer a daunting challenge in the search for shared morphological characters. The ultrastuctural techniques that fueled the 1970s and early 1980s revolution in algal systematics revealed a suite of new morphological characters, but many were not global (i.e., present in all of the taxa). Controversy over the interpretation of the importance of ultrastructural features (e.g., of cell division versus flagellar apparatus) led to conflicting hypotheses. Also, different researchers studied different details of different taxa, and thus a data matrix reporting a complete set of morphological and ultrastructural characters over a wide range of algal taxa was not available. Thus, it is no wonder that many researchers interested in unraveling the mystery of green algal phylogeny embraced molecular systematics, hoping that its early promise of relatively simple access to ample global characters would lead, finally, to a "true" phylogeny. The extent to which this promise has been fulfilled, or is likely to be fulfilled, is the subject of this chapter.

Beginning with a brief history of green algal systematics prior to the advent of molecular techniques, we then move on to the impact that molecular tools have made in algal systematics, emphasizing results of sequence data from genomic ribosomal DNA and the chloroplast genome. We offer some analyses completed specifically for this chapter, as well as a review of the current literature, and attempt to present as complete a picture of each group discussed as is currently available. Unfortunately, not all algal groups are well studied. This is a product of the tremendous diversity of algae and the relatively small number of scientists studying this diversity. Thus, the chapter attempts to present a snapshot of the present understanding of algal systematics in its current state of flux.

As alluded to above, classification of the algae has been a challenge, partly because the

We thank Karen LeBlanc for assistance with preparing the manuscript. This work was supported by NSF grants DEB-9408057 and DEB-9107389 to RLC and an NSF EPSCoR-Louisiana Board of Regents LaSER grant in support of molecular evolution studies at LSU; NSF grant DEB-9220834 to MAB and RLC; DOE/NSF/USDA Joint Program on Collaborative Research in Plant Biology (USDA grant 94-37105-0173) to MAB and RLC; NSF grant DEB-9103673 to CFD; NSF grant DEB-9306462 to JM; BSR 9020117 to RMM and R. W. Hoshaw and DEB 9407606 to RMM.

500–900 million year old group of organisms traditionally called "algae" comprises such a wide variety of taxa that even defining the term "algae" succinctly is impossible. There are unicellular, filamentous, parenchymatous, and macroscopic forms of algae (Figures 18.1, 18.2), and many of these forms exhibit environmentally induced phenotypic plasticity (Norton et al., 1996). This diverse morphology led to the proposal of at least 16 classification schemes by 14 authors from 1935 to 1980 (Bold and Wynne, 1985). These schemes, based largely on morphology and biochemistry, considered features such as pigments, cell wall composition, and storage products. Some of these characters were based on chloroplast features, and, until the reality of endosymbiotic events was considered, this was a confusing factor in some studies. Beginning in the 1970s, electron microscopy revealed features of the flagellar apparatus and cell division, among other things, that were extremely useful in comparative cell ultrastructure.

By the 1980s, concepts of a natural group of green algae, Chlorophyta, had been refined. The euglenoids were recognized to be a group separate from Chlorophyta, and the concept of Charales as a separate phylum was dropped. The algae included in Chlorophyta share the following characteristics: chlorophylls *a* and *b;* chloroplast features such as a double membrane, lack of endoplasmic reticulum (ER), storage of true starch as a reserve polysaccharide, and stacked thylakoids; production of cellulose; and flagellated cells (if produced) containing a unique stellate structure (referred to as a "star" or "H-piece" [Fig. 18.3]) in the flagellar transition region (Kessler, 1984; Mattox and Stewart, 1984). Generally, Chlorophyta share these characteristics with the "higher plants," forming an arguably monophyletic group termed Viridiplantae by Cavalier-Smith (1981) and Chlorobionta by Bremer (1985).

In 1984, Mattox and Stewart published a classification of Chlorophyta that is still a standard against which hypotheses of algal phylogeny are tested by many investigators, although alternative hypotheses have subsequently been proposed (e.g., van den Hoek et al., 1992). The five-class scheme of Mattox and Stewart comprises Micromonadophyceae (including *Micromonas,*

Pedinomonas, Pyraminonas, Mesostigma, Scourfieldia, etc.), Charophyceae (Chlorokybales, Klebsormidiales, Zygnematales, Coleochaetales, and Charales), Ulvophyceae (including *Ulothrix, Cladophora, Ctenocladus,* and genera in the Trentepohliaceae), Pleurastrophyceae (including the Tetraselmidales and Pleurastrales), and Chlorophyceae (consisting of Chlamydomonadales, Volvocales, Chlorococcales, Sphaeropleales, Chlorosarcinales, Chaetophorales, and Oedogoniales). The most important aspect of this scheme is "the degree to which characteristics correlate and predictions can be made," leading to a classification "more natural than any previous system" (Mattox and Stewart, pp. 62, 1984). Even so, Mattox and Stewart (1984) considered the Micromonadophyceae to be an unnatural, catch-all class of unicellular taxa, based solely on shared primitive characteristics. Mattox and Stewart intuitively interpreted the characters used in their classification scheme, rather than using explicit data analyses. However, as pointed out by Mishler and Churchill (1985), the "logical processes followed by contemporary phycologists (e.g., Pickett-Heaps, 1979; Mattox and Stewart, 1984)" were "remarkably akin . . . to those advocated by Hennigians," and subsequent explicit phylogenetic analyses (Mishler and Churchill, 1985; Bremer et al., 1987) did support some, but not all, of the Mattox and Stewart hypotheses. A discussion of the tremendous impact of phylogenetic systematics is outside the scope of this paper. (Please see Hennig, 1965, 1966 for the original concepts, and Theriot, 1989, 1992 for discussions of "phylogenetic systematics for phycologists.")

With green algal phylogeny unresolved by organismal characters, the tools of molecular biology that became available in the mid-1980s were welcomed by phycologists as a possible panacea. Gene sequencing especially, enhanced by polymerase chain reaction (PCR) amplification of DNA (Li et al., 1988; Pääbo, 1989; Innis et al., 1990), offers a great many characters that are relatively easy to obtain, for which positional homology is generally easy to determine. Many genes are more global within a taxon than many morphological characters. Initially, of course, gathering molecular data was more difficult and costly than traditional methods (as was

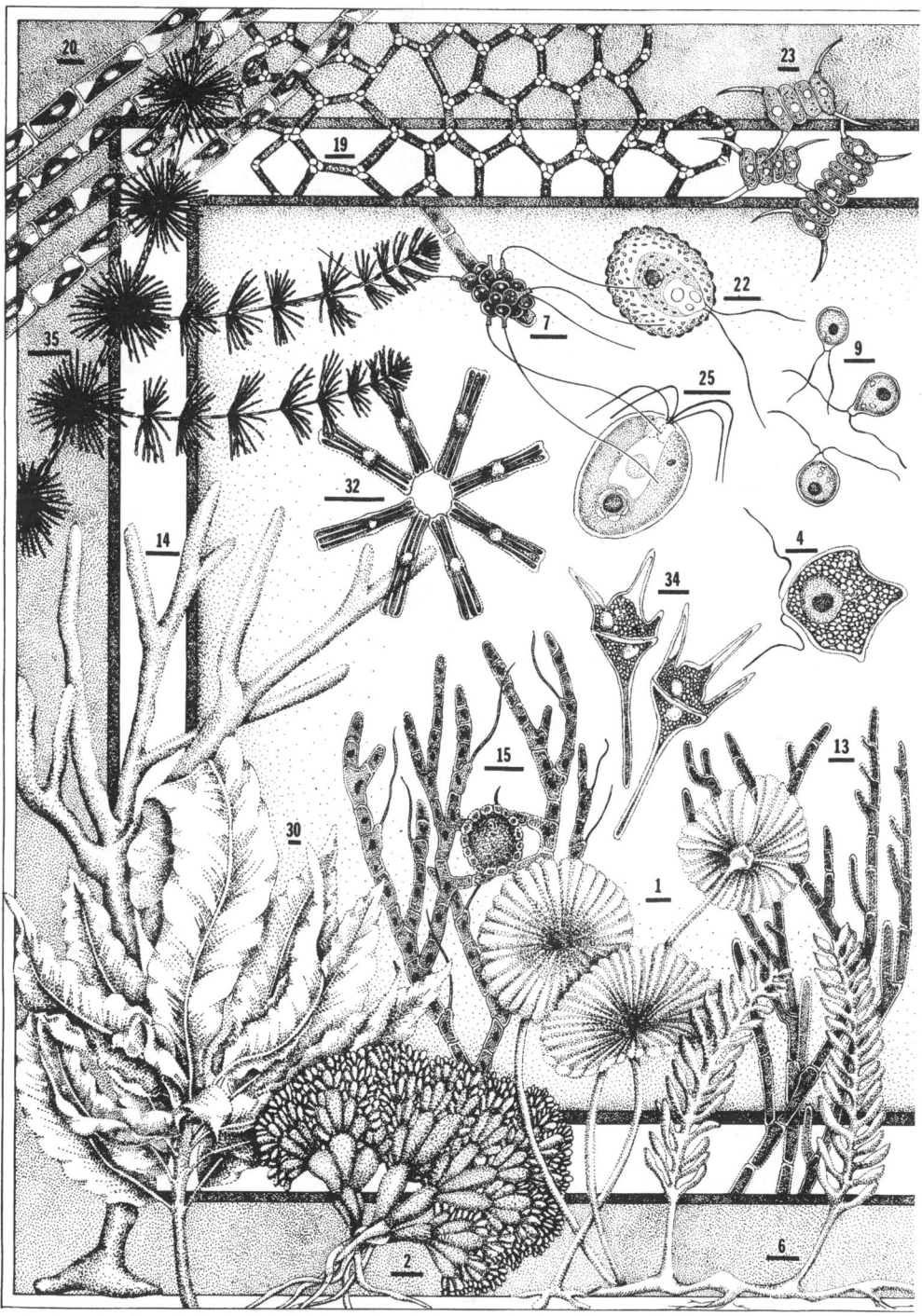

Figure 18.1. This "Algal Fantasia" depicts a potpourri that would never be found in nature. To illustrate the beauty and morphological diversity of the algae, not only are microscopic taxa shown with the same stature as macroscopic ones (see scale bars), but freshwater (F) algae cavort happily with marine (M) species. Genera (although not necessarily the same species) 1–28 are mentioned in this chapter and/or included in the research projects covered; however, 29–34 are included simply to acquaint the non-phycologist with a sampling of algal diversity. 1. *Acetabularia sp.* 1 mm (M) 2. *Anadyomene stellata* 250 μm (M) 3. *Ankistrodesmus sp.* 5 μm (F) 4. *Brachiomonas sp.* 5 μm (F) 5. *Carteria crucifera* 1 μm (F) 6. *Caulerpa floridana* 1 cm (M) 7. *Chaetosphaeridium sp.* 25 μm (F) 8. *Chara zeylanica* 1 mm (F) 9. *Chlamydomonas sp.* 5 μm (F) 10. *Chlorella minutissima* 1 μm (F) 11. *Chlorococcum sp.* 5 μm (F) 12. *Chlorococcum sp.* (nonmotile) 5 μm (F)

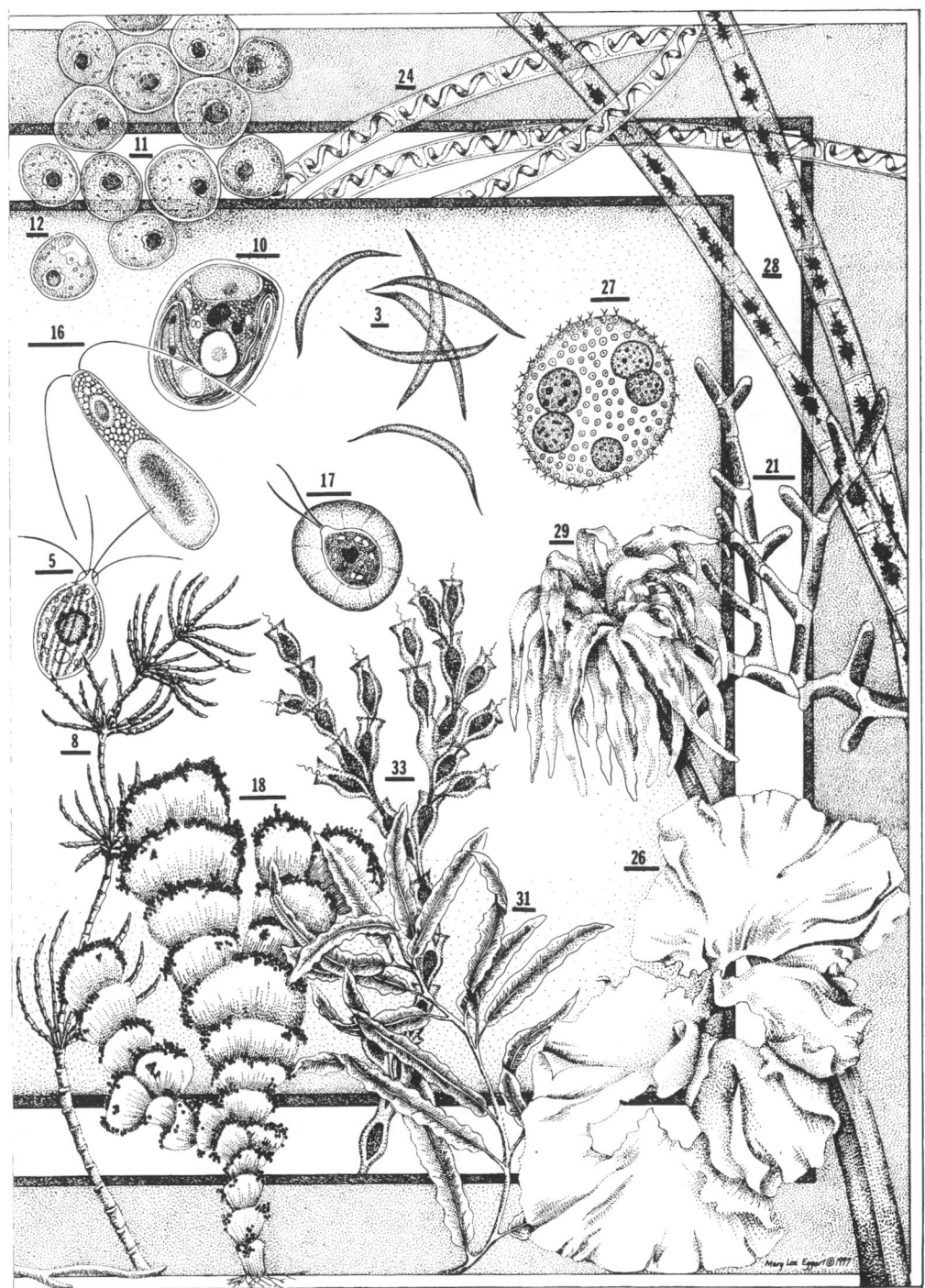

13. *Cladophora sp.* 10 μm (F or M) 14. *Codium fragile* 1 cm (M) 15. *Coleochaete pulvinata* 50 μm (F) 16. *Dunaliella sp.* 5 μm (F) 17. *Haematococcus sp.* 10 μm (F) 18. *Halimeda tuna* 1 cm (M) 19. *Hydrodictyon sp.* 10 μm (F) 20. *Klebsormidium flaccidum* 25 μm (F) 21. *Microthamnion küttzingianum* 25 μm (F) 22. *Phacotus lenticularis* 10 μm (M) 23. *Scenedesmus sp.* 10 μm (F) 24. *Spirogyra sp.* 1 μm (F) 25. *Tetraselmis suecica* 5 μm (F) 26. *Ulva rotundata* 1 cm (M) 27. *Volvox aureus* 5 μm (F) 28. *Zygnema sp.* 10 μm (F) 29. *Postelsia palmaeformis* 1 cm (M) 30. *Alaria esculenta* 1 cm (M) 31. *Sargassum filipendula* 1 cm (M) 32. *Tabellaria sp.* 5 μm (F) 33. *Dinobryon sp.* 10 μm (F) 34. *Ceratium sp.* 25 μm (F) 35. *Batrachospermum sp.* 50 μm (F). Artist: Mary Lee Eggart.

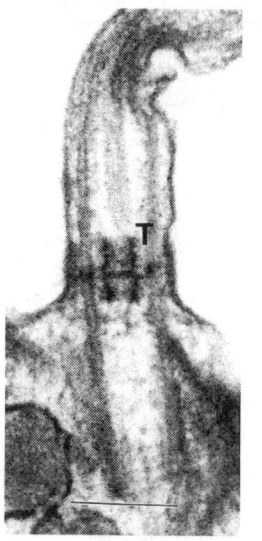 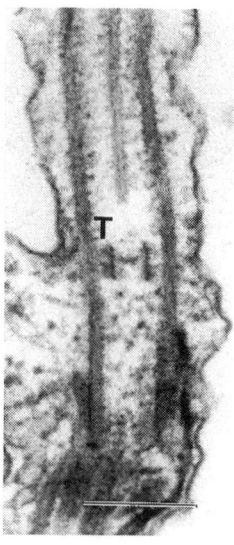

Figure 18.2. Different ultrastructure of transition regions in the flagellar apparatus of zoospores from two species of *Myrmecia* (Trebouxiophyceae). Unpublished micrographs by T. Friedl. (**Left**) *Myrmecia bisecta*. Note prominent H-like structure in the transition region. Scale bar = 0.2 μm. (**Right**) *Myrmecia biatorellae*. Note reduced H-like structure in the transition region. Scale bar = 0.2 μm. Comparisons of the complete 18S rRNA sequences from species of *Myrmecia* (Friedl 1995) showed the genus being paraphyletic in its previous state, and that *M. bisecta* actually lies outside of the monophyletic cluster of other *Myrmecia* spp. Therefore, the phylogenetic importance of ultrastructural differences seen in the transition region of *Myrmecia* zoospores has been substantiated by the rRNA sequence analyses. T = transition region between basal body and emergent flagellum.

the situation in the early days of electron microscopy).

Some of the first molecular studies involved nuclear 5S ribosomal RNA (Hori et al., 1985; Hori and Osawa, 1987; Devereux et al., 1990; Krishnan et al., 1990; van de Peer et al., 1990), but, as discussed elsewhere (Halanych, 1991; Steele et al., 1991), this gene proved largely unsuitable for algal systematics. Most research has centered on nuclear genomic ribosomal DNA, mostly of the small subunit (18S or SSU rDNA), and chloroplast DNA, mostly from the large subunit of the RUBISCO enzyme (ribulose-1, 5-bisphosphate carboxylase oxygenase, or *rbcL*). Many of these studies have focused on smaller

groups (corresponding to the class level), which are discussed later in this chapter, though some have taken more of a big-picture approach.

The most comprehensive attempts at resolving algal phylogenies have included both molecular and morphological data (Mishler et al., 1994; Watanabe et al., 1995). One consistent result of these studies has been support for the original Pickett-Heaps and Marchant (1972) suggestion that there are two major lineages of green algae: the charophycean lineage, which gave rise to the land plants and with them forms a clade, the Streptophyta; and the chlorophycean lineage, which is a monophyletic assemblage of the taxa that comprise the Chlorophyceae, Pleurastrophyceae, Micromonadophyceae, and Ulvophyceae sensu Mattox and Stewart (1984). This chlorophycean lineage comprises the classes Chlorophyceae, Trebouxiophyceae, Prasinophyceae, and Ulvophyceae sensu Friedl (1995). Figure 18.4 (see also Friedl, 1995; Melkonian and Surek, 1995; Tree of Life Web Site) illustrates the consensus of current opinion about the divergences among major groups within the chlorophycean and charophycean lineages. Only the major divergence between the chlorophycean and charophycean taxa is robustly supported. Relationships among groups within these two lineages are not resolved, although the relationships shown among the Ulvophyceae, Chlorophyceae, and Trebouxiophyceae (Microthamniales) are supported by some studies, and the paraphyly of the Prasinophyceae is strongly supported.

CHLOROPHYCEAN LINEAGE (SENSU PICKETT-HEAPS AND MARCHANT, 1972)

In this section, we discuss six examples of recent 18S rDNA research emphasizing different chlorophycean green algal groups: the prasinophytes, *Chlamydomonas* and its allies, the coccoid green algae, Dasycladales, Cladophorales, and Trebouxiophyceae. Some of the research is overlapping, and slight differences in tree topologies demonstrate the effect of using different taxa and different methods of data analysis (see also Wilcox et al., 1993; Ragan et al., 1994; Kranz et al., 1995) in cases in which the phylogenetic signal is not strong. However, the trees

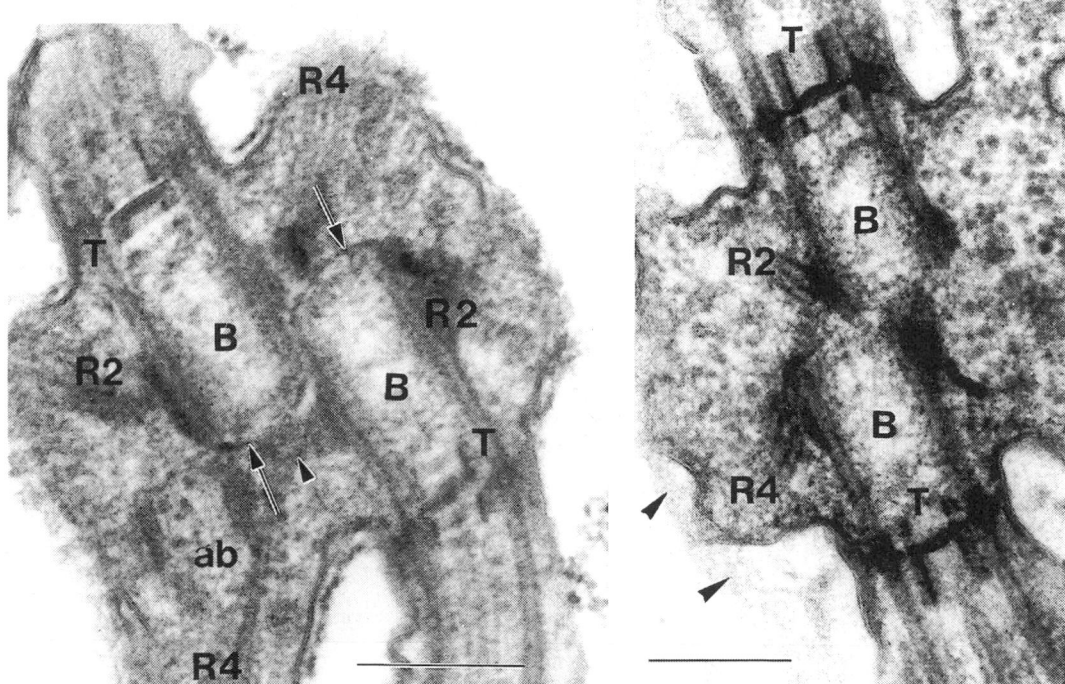

Figure 18.3. Two types of basal body orientation in flagellated stages of green algae, anterior view. (Both micrographs were published in *Phycologia* 35:456–469, 1996, and are reprinted by permission.) (**Left**) Counter clockwise (CCW or 11/5 o'clock) orientation in zoospores of *Gloeotilopsis paucicellulare* (Ulvophyceae). Note accessory body connected with R4 microtubular root. Arrows = terminal caps at proximal ends of basal bodies; arrowheads = electron-dense material that connects the two basal bodies. Scale bar = 0.2 μm. (**Right**) Clockwise (CW or 1/7 o'clock) orientation in zoospores of *Pleurastrum insigne* (Chlorophyceae). Only one microtubular root of each basal body is visible. Note textured thin wall (arrows). Scale bar = 0.2 μm. **Abbreviations:** B = basal body; ab = accessory basal body; R2 = two-membered microtubular root; R4 = four-membered microtubular root; T = transition region between basal body and emergent flagellum.

for the *Chlamydomonas* group (Fig. 18.5) and the coccoid green algae (Fig. 18.6) show remarkable similarities in the topologies of overlapping chlamydomonad taxa (relative position of *C. humicola* is the only difference), despite different methods of analysis and different sets of taxa.

Prasinophytes

The prasinophytes comprise a diverse assemblage of motile chlorophytes with cell bodies and flagella covered by nonmineralized organic scales. They are considered by some to be descendants of the earliest group of chlorophytes, which presumably gave rise to all other groups of green algae (Melkonian, 1990a). Steinkötter et al. (1994) demonstrated a paraphyletic origin of some representatives of the prasinophytes using

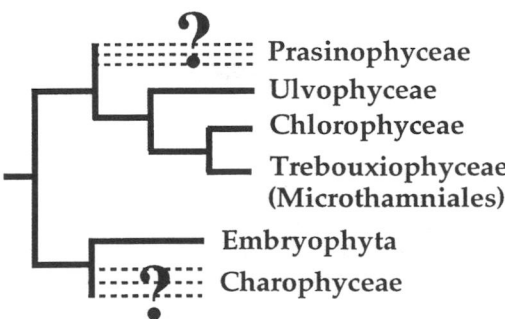

Figure 18.4. Green algae and land plants: current concepts of phylogeny.

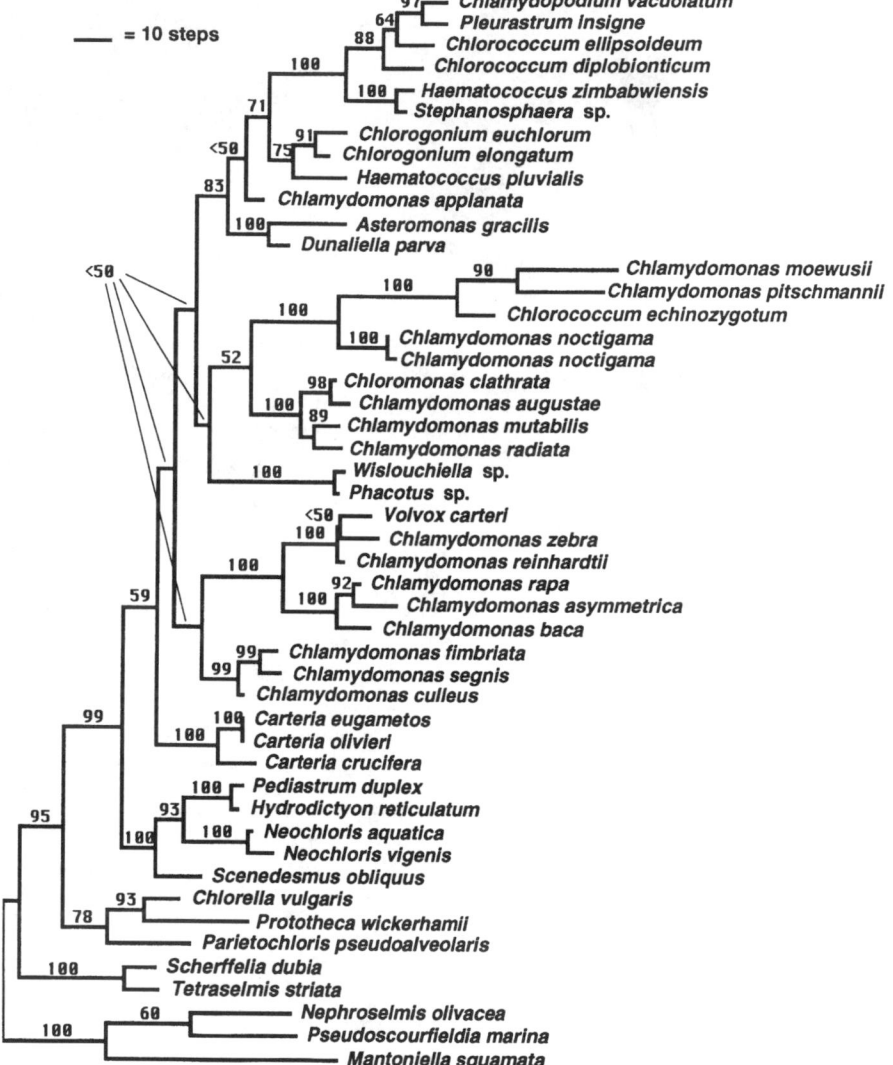

Figure 18.5. Phylogram from maximum parsimony analysis of 18S rRNA data. The aligned data (1,648 unambiguously aligned sites from the nuclear 18S rRNA gene) were analyzed heuristically (PAUP 3.1.1, Swofford, 1993) under MULPARS and using the TBR search algorithm. *Mantoniella, Nephroselmis,* and *Pseudoscourfieldia* were used as outgroups, following Mattox and Stewart (1984). Bootstrap values (Felsenstein, 1985) are presented in association with the corresponding node. The topologies from a neighbor-joining analysis of a K2P (Kimura, 1980) matrix (constructed using PHYLIP 3.57, Felsenstein, 1993) and from a maximum likelihood analysis (constructed using fastDNAML, Olsen et al., 1994a) are identical except for the position of the phacotacean clade (see text). This analysis includes a number of published 18S rDNA sequences (Gunderson et al., 1987; Rausch et al., 1989; Huss and Sogin, 1990; Lewis et al., 1992; Wilcox et al., 1992, 1993; Steinkötter et al., 1994; Friedl, 1995; Gordon et al., 1995; Buchheim et al., 1996b).

18S rRNA sequence analyses. The prasinophyte taxa in this study (*Tetraselmis striata, Scherffelia dubia, Pseudoscourfieldia marina,* and *Nephroselmis olivacea*) were basal to the ulvophycean and chlorophycean green algal taxa used, but not to the charophycean green algal taxa.

An ancient lineage sister to all green algae might be represented by *Mantoniella squamata,* as suggested in Figure 18.5. However, the exact position of *Mantoniella* cannot be conclusively deduced from this, or from a previous study by Kranz et al. (1995), which yielded inconsistent

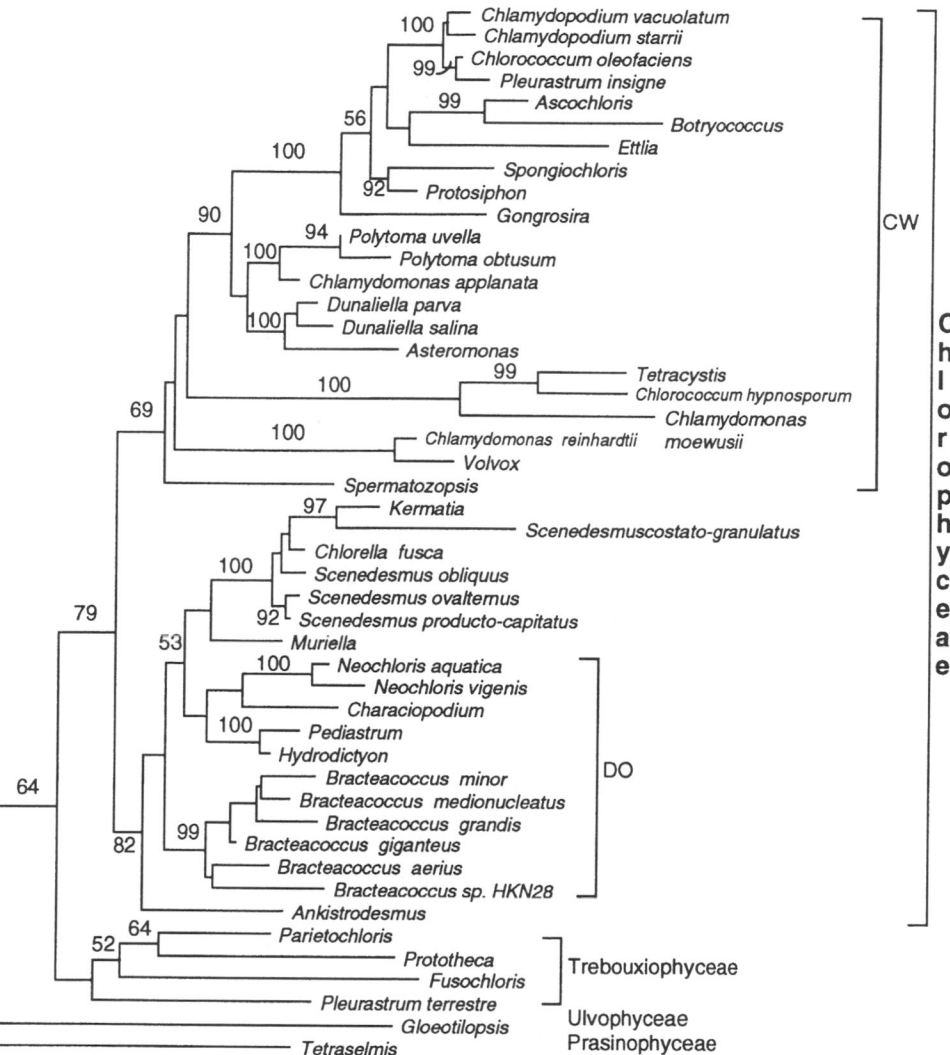

Figure 18.6. Maximum likelihood tree using 18S ribosomal RNA gene sequence data, using the HKY85 model (Hasegawa et al., 1985) plus discrete rate heterogeneity (alpha shape = 0.5), ts/tv set to 2.0, as implemented in a prerelease version of PAUP* 4.0 (Swofford, 1998, used with permission of author). Branch lengths are drawn proportional to expected number of substitutions. Bootstrap values (100 replicates) are shown above the nodes; bootstrap values were obtained using F84 (plus gamma) distances and heuristic searches. Alignment of the 18S rRNA gene among the chlorophycean green algae is relatively straightforward. In those instances in which homology of characters could not be ascertained with certainty, the sites were excluded from the analysis.

results with respect to the phylogenetic position of *Mantoniella*. There is a critical need for more sampling of these unicellular flagellates.

Chlamydomonas and Allies

Chlamydomonas, a biflagellate, unicellular genus that comprises more than 400 recognized species (Ettl, 1976), is placed with most other walled, unicellular, green flagellates in the order Chlamydomonadales (sensu Mattox and Stewart, 1984). *Chlamydomonas* and its putative allies traditionally have been regarded as a sister group to the colonial flagellates (Volvocales, sensu Mattox and Stewart, 1984). Molecular phylogenetic investigations have revealed a number of insights regarding relationships among selected species of these groups. The earliest molecular phylogenetic investigations of volvocalean and chlamydomonadalean flagellates demonstrated that *Chlamydomonas* is not only highly diverse, but also fails tests for monophyly (Buchheim et al., 1990—partial 18S rRNA). Subsequent investigations have also noted that most of the colonial flagellates, such as *Volvox,* are closely allied with *Chlamydomonas reinhardtii* (Buchheim and Chapman, 1991—partial 18S and 26S rRNA; Larson et al., 1991—partial 18S and 26S rRNA; Buchheim et al., 1994a—partial 18S and 26S rRNA; Nozaki et al., 1995—*rbcL* data). However, *Stephanosphaera* represents an exception to this characterization since this colonial flagellate (which shows a number of cytological differences from other Volvocales) has been shown to be a close ally of *Chlamydomonas applanata* (Buchheim and Chapman, 1991—partial 18S and 26S rRNA; Buchheim et al., 1994a—partial 18S and 26S rRNA; Rumpf et al., 1996—partial 18S rRNA). *Chlamydomonas moewusii* has been grouped with several marine species, including *Chlamydomonas parkeae* (Kim et al., 1994—chloroplast 16S rRNA; Buchheim et al., 1995—18S rRNA); and *Chlamydomonas euryale* (Buchheim et al., 1995—18S rRNA).

Molecular phylogenetic investigations also have focused on variability in *Carteria,* a quadriflagellate, unicellular alga, with some results challenging the concept of monophyly for the genus (Buchheim and Chapman, 1992; Buchheim et al., 1996b). However, phylogenetic analyses of sequence data from the chloroplast 23S rRNA gene are ambiguous in supporting either monophyly or paraphyly of *Carteria* (Buchheim et al., 1996b).

Recent 18S rRNA sequence data (Rumpf et al., 1996) suggest that, as long suspected, *Polytoma,* a colorless, biflagellate unicell, is closely related to *Chlamydomonas.* However, two distinct clades of *Polytoma* taxa were detected, suggesting that the genus is not monophyletic and that photosynthetic capability has been lost at least twice within the chlamydomonad flagellates (Rumpf et al., 1996).

Results of a new analysis of 18S rRNA sequence data for a broad sampling of *Chlamydomonas* taxa and their unicellular or coccoid allies are presented in Figure 18.5 and discussed below under the subheadings *Chloromonas, Phacotaceae,* and *Chlorococcum* (see also discussion below on The Chlorophycean Coccoid Green Algae). The chlamydomonad flagellates are regarded as members of the class Chlorophyceae (sensu Mattox and Stewart, 1984). Their sister group within the Chlorophyceae is considered to be the Sphaeropleales (sensu Deason et al., 1991); several taxa from this sister group have been included in this analysis. To place this set of taxa in an even broader context, several basal lineages also have been included (e.g., Trebouxiophyceae sensu Friedl, 1995).

Chloromonas The genus *Chloromonas* is a biflagellate, unicellular member of Chlamydomonadales, distinguished from *Chlamydomonas* by the apparent absence of a pyrenoid in its vegetative cells (Ettl, 1970, 1983; Hoham, 1980; Iyengar and Desikachary, 1981). Phylogenetic analysis of 18S rRNA data (Fig. 18.5) suggests that the snow alga, *Chloromonas clathrata* (formerly listed as *Chlamydomonas yellowstonensis,* Starr and Zeikus, 1993), is most closely related to another snow alga, *Chlamydomonas augustae* (formerly listed as *Chlamydomonas nivalis,* Starr and Zeikus, 1993). These two snow algae are part of a recently identified, robustly supported lineage (Buchheim et al., 1996a) that also includes *Chlamydomonas radiata* and *Chlamydomonas mutabilis.* A taxonomically broader study of 18S rRNA variability in *Chloromonas* (Buchheim et al., 1997) demonstrated that the genus is not monophyletic. In other words, the molecular data indicate that the absence of a vegetative pyrenoid is not a good indicator of natural history, whereas habitat (e.g., snow vs. temperate, nonfrozen substrates) correlates well with phylogenetic pattern (Buchheim et al., 1997).

Phacotaceae Members of Phacotaceae, although closely linked to *Chlamydomonas* and its traditional allies (Mattox and Stewart, 1984; Bold and Wynne, 1985), are distinguished from other chlamydomonadalean taxa by the presence of a lorica or specialized outer investment (Iyengar and Desikachary, 1981; Ettl, 1983). Two representative members of Phacotaceae, *Phacotus* sp. and *Wislouchiella* sp., are resolved as a monophyletic group within the chlamydomonad flagellates (Fig. 18.5). The results of the analysis place them as a basal lineage in a clade that includes *Chlamydomonas moewusii, Chlorococcum echinozygotum, Chlamydomonas mutabilis,* and *Chloromonas clathrata.* As the low bootstrap value implies, the position of the phacotacean clade is not robust to taxon extinction experiments nor to different methods of phylogenetic reconstruction. Topologies from neighbor joining (Saitou and Nei, 1987) and maximum likelihood (Felsenstein, 1993) analyses, which are otherwise identical to the parsimony topology, place the phacotacean alliance at the base of the clade that includes *Haematococcus* (unpubl.).

Previous studies of partial 18S rRNA sequences (Buchheim et al., 1994b) indicated that *Dysmorphococcus globosus,* another member of Phacotaceae, was not resolved with other members of the family including *Wislouchiella* and *Phacotus,* but instead was grouped with *Haematococcus* and *Chlorogonium.* Unfortunately, a complete 18S rRNA sequence for *D. globosus* was not available for this new analysis, which means that although the data presented here robustly support monophyly of Phacotaceae, the one taxon that could challenge this conclusion was not included. Additional taxon sampling also may help resolve questions regarding the position of Phacotaceae within the chlamydomonad lineage.

Chlorococcum Members of the genus *Chlorococcum* produce a motile vegetative stage that strongly resembles vegetative cells of *Chlamydomonas.* However, this motile vegetative stage is short-lived and is replaced by a nonmotile stage that possesses a cellulosic cell wall (Bold and Wynne, 1985). Based on this distinguishing nonmotile stage, traditional classifications that are constructed almost exclusively from gross morphological evidence place *Chlorococcum* in a separate order, Chlorococcales (e.g., Bold and Wynne, 1985). However, as has been demonstrated previously (Buchheim et al., 1996b), the nuclear 18S rRNA data place *Chlorococcum echinozygotum* in a clade with *Chlamydomonas moewusii* and other species of *Chlamydomonas* (Fig. 18.5). Furthermore, *Chlorococcum ellipsoideum* and *Chlorococcum diplobionticum,* which have been added to the present analysis, represent part of an alliance (separate from *Chlorococcum echinozygotum*) with *Pleurastrum, Characium, Dunaliella, Asteromonas, Chlamydomonas applanata,* and the Haematococcaceae (*Haematococcus, Stephanosphaera, Chlorogonium,* sensu Ettl, 1983). Thus, the 18S data do not support the monophyly of the genus *Chlorococcum* (see also Nakayama et al., 1996 and discussion of the coccoid taxa, below).

As mentioned in Buchheim et al. (1996b), the inclusion of chlorococcalean taxa (sensu lato) such as *Chlorococcum, Characium,* and *Pleurastrum* among the chlamydomonad flagellates challenges the traditional, morphology-based concepts of both groups. On the other hand, these observations corroborate the primary diagnostic criterion of Chlamydophyceae (sensu Ettl, 1981)—the presence of a chlamydomonad cell wall (the "chlamys") surrounding the motile vegetative stage versus Chlorophyceae with wall-less motile cells (see Melkonian, 1982b and Deason, 1984 for dissenting opinions). Each of the chlorococcalean taxa (sensu lato) included in this analysis possesses a walled motile stage. Ettl (1981) considered neither *Asteromonas* nor *Dunaliella,* which traditionally have been regarded as "naked" flagellates, to be members of Chlamydophyceae. However, ultrastructural investigations of *Asteromonas* (Chappell et al., 1989) have demonstrated that this ostensibly naked flagellate does indeed have striplike wall elements that resemble the wall elements observed in *Chlorogonium.* Regardless of the actual walled or wall-less status of these naked flagellates, one is forced to make the ad hoc assumption, within the context of the molecular phylogeny presented here, that both *Asteromonas* and *Dunaliella* have lost a typical chlamydomonad cell wall. The rank at which this assemblage of walled flagellates is placed may be debated; however, the molecular data support Ettl's (1981) basic criterion for the circumscription of this group of organisms with a walled, motile vegetative stage.

The Chlorophycean Coccoid Green Algae

Some of the most dramatic examples of the tremendous impact molecular data have had on the understanding of evolutionary relationships among green algae come from studies of the coccoid green algae. These vegetatively nonmotile cells that produce zoospores or autospores are traditionally placed in the order Chlorococcales sensu lato (Chlorophyceae). The taxa are morphologically simple and are usually unicells with simplified shapes, coenobic colonies, or loose clusters of cells. Because of their morphology, they have been considered derived, by loss of the motile vegetative stage, from a *Chlamydomonas*-like flagellated ancestor, and an intermediate form between the flagellated to the filamentous habit. Many are plankton of freshwater lakes, others are found as common elements in soils, and still others exist as epiphytes or endophytes associated with other plants and animals (Bold and Wynne, 1985; Melkonian, 1990b).

Acknowledged as an unnatural group by many, the coccoid green algae have a long history of attempts to establish a standard by which the few morphological characters present could be used in their classification. Much progress in describing the phenotypically plastic coccoids was made by Starr, Bold, and their students, who undertook detailed examinations of life cycles, cellular structures, antibiotic studies, and the macroscopic appearance of colonies under standardized conditions (Starr, 1955; Deason and Bold, 1960; Bold and Parker, 1962; Chantanachat and Bold, 1962; Bischoff and Bold, 1964; Brown and Bold, 1964; Archibald and Bold, 1970; Bold, 1970; Lee and Bold, 1974).

Morphological characters that were used at the class level were discussed by Ettl (1981) and Ettl and Komárek (1982) and are mentioned in the foregoing discussion of *Chlamydomonas* and its allies. At the ordinal or family level, the classification of coccoids has also been subject to numerous revisions. Ettl and Komárek's (1982) proposal was based on the nuclear condition of vegetative cells and was later disputed because using this character would tend to split taxa that were considered to be closely related (Kouwets, 1991).

Ultrastructural research, which opened up a new set of characters for studies on other groups of green algae, also provided significant insights into the coccoid green algae. Three different flagellar apparatus orientations, counterclockwise (CCW), directly opposite (DO), and clockwise (CW) (Fig. 18.3), were found within the coccoid genera (Watanabe and Floyd, 1992; Floyd et al., 1993), which provided evidence that despite the similarities in vegetative morphology Chlorococcales were not a monophyletic assemblage. These groups were considered to correspond to different classes and orders of green algae according to the scheme of Mattox and Stewart (1984). Taxa possessing DO flagellar orientations were transferred to Sphaeropleales (Chlorophyceae), and those taxa with CCW flagellar orientations were treated in Trebouxiophyceae (see Friedl, 1995). The remainder were left in Chlorococcales sensu stricto (Deason et al., 1991).

The first molecular studies done on coccoid green algae were DNA reassociation studies on *Chlorella* (see Huss et al., 1989), which proved useful in differentiating species, but not in determining relationships between *Chlorella* species and other green algae. Analysis of 18S rDNA sequences for eight species of *Chlorella* (Huss and Sogin, 1990) further demonstrated that the genus was not monophyletic, but consisted of two groups corresponding to the traditional families Oocystaceae and Scenedesmaceae. Molecular comparisons between *Chlorella* species were especially useful because motile stages are completely lacking in this genus.

Through ultrastructural studies, the zoospores of different species of *Neochloris* and *Characium* were found to have three distinct flagellar apparatus types (Watanabe and Floyd, 1989, 1992; Floyd et al., 1993), with taxa in the CCW, CW, and DO clades. Lewis et al. (1992) obtained 18S rDNA data for representatives of each flagellar apparatus type in each genus.

These data verified that, within this group, flagellar apparatus configuration is indicative of evolutionary history, but that characters such as nuclear condition or absence of zoospore walls are not by themselves indicators of natural groups. In addition to placing the zoosporic taxa in different groups, the 18S sequence data allowed the phylogenetic position of autosporic taxa (such as *Chlorella*) to be assessed more fully (Wilcox et al., 1992). The coccoid genus, *Chlorococcum,* has been found to be paraphyletic (Buchheim et al., 1996a; Nakayama et al., 1996), whereas another, *Bracteacoccus,* seems to be monophyletic (Lewis, 1997).

A summary analysis of the currently available 18S rDNA sequences from chlorophycean coccoid green algae is shown in Figure 18.6. To present as thorough a picture as is currently possible of the diversity of this group, most of the 18S rDNA sequences available in Genbank (http://www.ncbi.nlm.nih.gov) for coccoid green algae and related taxa were included in the analysis (although this does not guarantee uniform taxon coverage). Because Trebouxiophyceae are addressed elsewhere in this chapter, only four taxa from this class were included to help in orientation of the tree. Results of this analysis place the chlorophycean algae into two groups, corresponding to taxa with either clockwise or directly opposed flagellar apparatus orientation. The first clade includes diverse taxa traditionally placed in Chlorococcales, Volvocales, and Chlorosarcinales and representing motile and nonmotile taxa, uninucleate and multinucleate types, and taxa with both walled and thin-walled zoospores. All of these contain (or are assumed to contain) taxa with the CW flagellar orientation. Despite the diversity of vegetative cell types, patterns emerge within this clade. There is a well-supported group of vegetatively nonmotile taxa with thin-walled or "naked" zoospores (*Gongrosira* to *Chlamydopodium* plus the autosporic taxon *Botryococcus.*). A second group is composed of the flagellates *Polytoma, Chlamydomonas applanata, Asteromonas,* and *Dunaliella.* Finally, there is a grade of both flagellated and nonflagellated vegetative cells which have thick-walled zoospores. The relationship between taxa with thick- and thin-walled zoospores was recently discussed by

Nakayama et al. (1996) and points to the studies by Ettl (1981) addressing the importance of the presence of zoospore walls. However, the condition of wall-less zoospores does not define a clade in this analysis. Likewise, for these taxa, the nuclear condition of cells and the loss of motility do not seem to be appropriate diagnostic characters at the ordinal level, but will perhaps be useful to distinguish families.

The second clade corresponds to taxa included in the order Sphaeropleales (sensu Deason et al., 1991), plus *Bracteacoccus* (Lewis, 1997) and autosporic chlorococcalean taxa such as *Scenedesmus* and *Ankistrodesmus.* Sphaeropleales include taxa (both unicells and colonies) with naked zoospores, and the DO flagellar apparatus orientation. The close association of the autosporic taxa with the zoosporic DO taxa (especially *Hydrodictyon*) is not particularly surprising given similarities in cell structure and coenobial growth forms. A number of additional taxa, such as *Atractomorpha* and *Sphaeroplea* (Sphaeropleaceae), and taxa producing both zoospores and autospores (*Tetraedron* and *Chlorotetraedron*), need to be examined in future studies.

What can be concluded from analyses of these taxa? First, Chlorophyceae are those green algae producing motile cells with CW or DO flagellar apparatus orientations, and each type (at least with the presently included taxa) forms its own lineage (discussed by O'Kelly, 1992). Thus, flagellar apparatus orientation is a reliable and easily interpreted character in the group. Second, the wall thickness of zoospores seems to be a distinguishing character among examined lineages of CW algae, but not a useful character outside of the CW clade. Third, nuclear condition (whether uni- or multinucleate), the loss of motility, and the colonial condition are characters that are distributed throughout the different lineages of Chlorophyceae and, therefore, should be reserved for analyses at lower taxonomic levels (i.e., family and order).

Formal revisions of the above groups await further ultrastructural and sequence data. Representatives of several orders (Chaetophorales, except for *Gongrosira* [but see Watanabe et al., 1992], Oedogoniales, and Chaetopeltidales) could not be included here because 18S rDNA sequences are not yet available. Inclusion of these additional taxa will no doubt help to resolve the relationship between the DO and CW clades.

Phylogeny of the Class Trebouxiophyceae (Chlorophyta)

The class Trebouxiophyceae has been described recently on the basis of rDNA sequence comparisons (Friedl, 1995). Phylogenetic inferences from complete 18S rDNA sequences (Fig. 18.7) demonstrate a single origin of many coccoid green algae that completely lack motile stages (autosporic coccoids e.g., *Chlorella* spp.) with those that are defined by the ultrastructure of their zoospores as members of Microthamniales sensu Melkonian (1982a, 1990c) (Pleurastrales sensu Mattox and Stewart, 1984). No known morphological characters available would reveal these relationships independent from the rDNA data. The specific tree topology suggests a zoosporic ancestry of autosporic coccoids. For example, the zoosporic *Dictyochloropsis reticulata* shares a sister group relationship with two strains of *Chlorella saccharophila* and *Chl. luteoviridis* (Fig. 18.7; Friedl, 1995). The loss of flagellated stages may have occurred independently several times within the Trebouxiophyceae and within Chlorophyceae (indicated by arrows in Fig. 18.7). Autosporic coccoids define additional lineages within Trebouxiophyceae (e.g., *Chl. ellipsoidea/Chl. mirabilis, Nannochloris* sp./*Choricystis minor,* and *Chlorella* ssp., *Nanochlorum eucaryotum, Prototheca wickerhamii;* Fig. 18.7). These analyses clearly indicate the multiple origins of coccoid green algae (e.g., Huss and Sogin, 1990; Lewis et al., 1992; Wilcox et al., 1992).

Members of Trebouxiophyceae share various common features such as ecology and growth morphology. Most known trebouxiophytes are coccoids; they live in terrestrial habitats or occur in symbioses with lichen fungi (e.g., *Trebouxia* spp.) or invertebrates (e.g., *Chlorella* sp. HvT). The rDNA analyses indicate that the capacity to exist in lichen associations has multiple independent origins (marked by a gray dot in Fig. 18.7) and that lichen algae are derived from non-symbiotic terrestrial green algae. For example,

the lichen symbionts *Myrmecia biatorellae* and *Trebouxia* spp. share a common origin with the terrestrial *M. israeliensis*. The lichen photobiont *Dictyochloropsis reticulata* has a sister group relationship with species of *Chlorella* that have been isolated from tree bark (Friedl, 1995).

The specific topology of the rDNA phylogeny (Fig. 18.7) suggests that the filamentous organization has several independent origins within Trebouxiophyceae, which is in sharp contrast to traditional taxonomic concepts of the green algae (e.g., Fritsch, 1935; Fott, 1971). *Leptosira terrestris* and *Microthamnion keutzingianum* are filamentous algae; *M. kuetzingianum* is evolutionarily very closely related to the coccoid *Fusochloris perforata* (Fig. 18.7; Friedl and Zeltner, 1994). *Leptosira terrestris* constitutes an independent lineage whose position among other trebouxiophycean taxa is unresolved (Friedl, 1996).

The rDNA analyses substantiate the significance of motile cell features over vegetative cell morphology for delineating Trebouxiophycean genera. Despite their very similar vegetative morphology, the rDNA analyses resolve a monophyletic origin only for those species of *Myrmecia* (*M. biatorellae* and *M. israeliensis*) that share identical zoospore characters, that is, lacking an eyespot and having an elongated, considerably flattened and asymmetrical shape (Friedl, 1995). *Myrmecia bisecta,* with zoospores that have an eyespot and are more compact and droplike in shape, forms an independent lineage in the rDNA phylogeny (Fig. 18.7; Friedl, 1995). A grouping of *Trebouxia* spp. with species of *Myrmecia* (to the exclusion of *M. bisecta*) is well resolved; the order Trebouxiales has been assigned to this particular clade (Fig. 18.7). This assignment is in concordance with the presence of unique motile cell features that separate these taxa from other zoosporic members of Trebouxiophyceae; these include the pronounced flattening of the cell, a reduced transition zone in the flagellar region, and a reduced system II fiber in the flagellar apparatus (Melkonian and Berns, 1983; Melkonian and Peveling, 1988). The rDNA analyses provide an even finer resolution of evolutionary relationships than motile cell ultrastructure. A paraphyletic origin of the genus *Trebouxia* is

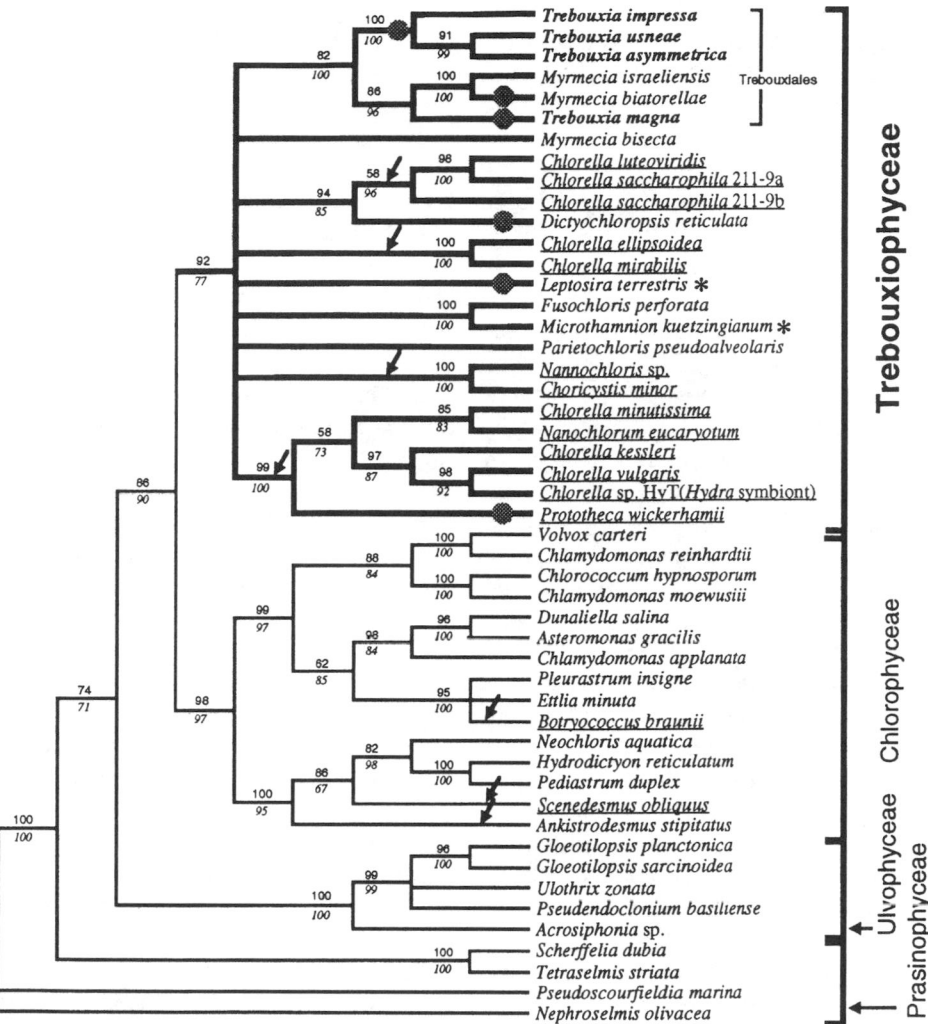

Figure 18.7. Strict consensus tree of three independently inferred phylogenies from distance, weighted maximum parsimony, and maximum likelihood methods. A total of 1,719 nucleotide positions was considered for the phylogenetic analyses. The prasinophytes *Nephroselmis olivacea* and *Pseudoscourfieldia marina* were employed as outgroup taxa (Steinkötter et al., 1994). For the distance method analysis, pairwise similarities between sequences were converted to evolutionary distances using the correction of Kimura (1980) and taken as input for a neighbor-joining phylogenetic reconstruction (Saitou and Nei, 1987) using PHYLIP 3.5c (Felsenstein, 1993). Maximum-parsimony analyses were performed using PAUP 3.1.1 (Swofford, 1993) with the nucleotide positions weighted (rescaled consistency index over an interval of 1–1,000) with a random addition of sequences with 10 replicates and a branch swapping algorithm (TBR, or tree bisection-reconnection). The maximum-parsimony method is improved when sites that have relatively higher rates of change are given less weight in the phylogenetic analysis (Hillis et al., 1994). Parsimony analysis of the weighted data resulted in a single most parsimonious tree with a consistency index of 0.708. For maximum likelihood analyses, the program fastDNAml (Olsen et al., 1994a) was used with the global search option, rearrangements of partial trees crossing one branch, and rearrangements of the full tree crossing 46 branches. A jumbled taxon addition and a transition:transversion ratio of 2.0 were used in searching for the best log-likelihood score. The maximum likelihood phylogeny had a Ln likelihood of −14659.83. Bootstrap values were computed independently for 500 resamplings of neighbor-joining analysis (above line) and for 100 resamplings of weighted parsimony analysis (below line) in order to estimate support for internal branches in the neighbor-joining and parsimony trees (Felsenstein, 1985).

Autosporic coccoid green algae, formerly classified in a single order "Chlorococcales," are indicated in the tree by underlined species names. Arrows on branch lengths indicate putative independent losses of flagellated stages; asterisks beside species names indicate filamentous taxa. Lineages that lead to lichen photobionts are marked by a gray dot. Species of *Trebouxia* are shown in bold-face letters.

suggested in the rDNA phylogenies, i.e., *T. magna* groups with *Myrmecia* spp. and not with other *Trebouxia* spp. (Fig. 18.7; Friedl and Zeltner, 1994), although all members of Trebouxiales share identical zoospore characters.

The monophyletic origin of Trebouxiophyceae is well resolved in most neighbor-joining analyses (e.g., 92% in Fig. 18.7). That the topologies within the maximum parsimony trees are less robust (e.g., 77% in Fig. 18.7) than those in the neighbor-joining analyses may reflect the greater sensitivity of parsimony methods to unequal rates of change within lineages (Felsenstein, 1978, 1981). Actual branch lengths are not seen in the consensus tree of the three independent analyses in Figure 18.7, but the internal branches in these phylogenies may be considerably different in length (e.g., the branches leading to *Trebouxia impressa* and *Chlorella luteoviridis* in the neighbor-joining analysis Figure 1B of Friedl, 1995). Trebouxiophyceae appear as an array of several independent lineages whose interrelationships are not resolved in the rDNA analyses. The relative branching order for lineages within the Trebouxiophyceae obtained from three independent phylogeny reconstruction methods is different; these divergences are also not resolved in bootstrap resamplings using neighbor-joining and parsimony methods.

Zoosporic taxa within Trebouxiophyceae and Ulvophyceae (represented by members of the order Ulotrichales in Fig. 18.7) share the counterclockwise (CCW) basal body orientation; they have been regarded as closely related orders within one class, the Ulvophyceae sensu Sluiman (1989) (e.g., Lokhorst and Rongen, 1994). However, the rDNA phylogenies provide strong evidence that Trebouxiophyceae and Ulotrichales are not evolutionarily close to one another. Instead, they represent two independent lineages of green algae (Steinkötter et al., 1994; Friedl, 1996). The rDNA analysis (Fig. 18.7) suggests that the CCW basal body orientation is a symplesiomorphic character. Also, the "phycoplast"-type of cytokinesis that has been found in a few members of Trebouxiophyceae and in Ulvophyceae/Ulotrichales (Sluiman, 1991) is a symplesiomorphy in rDNA phylogenies. Most analyses show Trebouxiophyceae and Chloro-

phyceae as sister groups with Ulvophyceae positioned at the base (Fig. 18.7). These relationships, however, are not unequivocally resolved. Analyses of user-defined trees (Friedl, 1996) show that a forced sister-group relationship of Ulvophyceae with Chlorophyceae (with Trebouxiophyceae basal) is not significantly worse than the best tree. The forced topology wherein Trebouxiophyceae and Ulvophyceae are sister groups, assuming that both were closely related orders of one class, is, however, significantly worse.

Zoosporic members of Trebouxiophyceae have been grouped within the class Pleurastrophyceae (order Pleurastrales, Mattox and Stewart, 1984). This class also included *Tetraselmis/ Scherffelia*, a clade that diverges prior to the radiation of the Trebouxiophyceae/Chlorophyceae/Ulvophyceae (Fig. 18.7). Pleurastrophyceae are, therefore, polyphyletic in the rDNA analyses (Kantz et al., 1990; Steinkötter et al., 1994). The "metacentric" type of phycoplast (Molnar et al., 1975) shared by Trebouxiophyceae and *Scherffelia/Tetrasemis* (Friedl, 1996) is a shared primitive character. Species previously assigned to *Pleurastrum* are of polyphyletic origin as indicated by their zoospore ultrastructure and by 18S rDNA sequence analyses (Friedl, 1996). The type species of *Pleurastrum, P. insigne,* lies outside the monophyletic clade that contains Pleurastrales sensu Mattox and Stewart (Fig. 18.7). Hence, "Pleurastrophyceae," permanently tied to *P. insigne,* becomes a later synonym of Chlorophyceae.

Dasycladales

Dasycladales are a tropical cosmopolitan order of benthic marine green algae with a fossil record that extends back to the Precambrian-Cambrian boundary (approximately 550 Ma) (Schopf, 1970; Berger and Kaever, 1992). Of some 175 known fossil genera, only 11 are extant and, among the 38 described species, approximately half belong to the "*Acetabularia*" group. Dasycladales are in fact living fossils. As defined by Stanley (1979), living fossils include extant clades that have survived for long intervals of geologic time at low numerical diversity and exhibit primitive morphological characters that have undergone little evolutionary change. The unmistakable radial symmetry of dasyclads forms a basic architectural groundplan from which many elaborations are possible. However, establishing phylogenetic relationships among genera and species has remained problematic, because, despite the distinctive morphology and complex developmental patterns documented in micrographic and ultrastructural detail (Berger and Kaever, 1992), there is a lack of morphological transitions between taxa. Moreover, differences in emphasis and approach used by paleobotanists (Deloffre and Granier, 1991) and phycologists with regard to morphological characters have further confounded efforts to establish phylogenetic classifications. As a result, the classifications of Solms-Laubach (1895), Valet (1979), and Berger and Kaever (1992) are all phenetic, though all discuss "phylogenetic affinities." Parsimony analysis based on morphological characters (Olsen, unpubl.) confirms the problem of convergent and autapomorphous characters at the expense of synapomorphous ones. Branching orders within families cannot be resolved.

18S rDNA sequences analyzed with parsimony and distance methods (Olsen et al.,

1994b) provided the first phylogenetic tree for 14 dasyclad species representing eight of the 11 extant genera. An 18S molecular clock calibrated against four fossil dating points supports a divergence time of 1%/25 Ma. Results from that study (Fig. 18.8) support the two main familial lineages but call into question the longstanding arguments for the recognition of *Acicularia, Acetabularia,* and *Polyphysa* as separate genera. In the unrooted analyses, the genera within the Dasycladaceae form a well-resolved clade. In the rooted analyses, this clade collapses, because all potentially suitable outgroups (Caulerpales and Siphonocladales-Cladophorales) are too distant. Long branch effects are clearly a problem. Based on morphological data, Acetabulariaceae are recognized on the basis of transient whorls and the presence of a corona superior, whereas the Dasycladaceae are supported by permanent whorls. All other morphological characters are sympleisomorphous, having arisen at least twice and often more times. For this reason, genera within the families cannot be resolved. They possess autapomorphies or are defined by absence of the character as compared with the sister taxon. Monophyly of the Dasycladales has never been

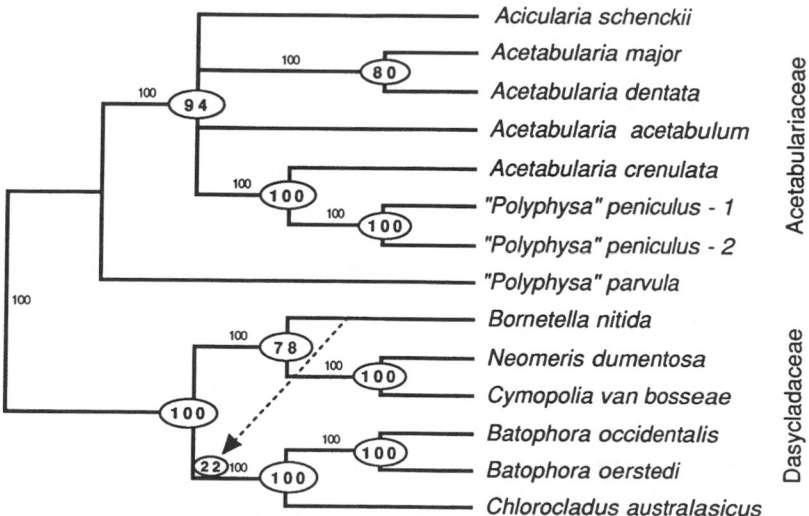

Figure 18.8. 18S rDNA tree for Dasycladales. Strict consensus of the two MPTs (L = 732 steps, RI = 0.707) found in an unweighted, Branch and Bound analysis (PAUP) of complete sequences which included secondary structure in the alignment (alignment length 1,733, variable positions 411, informative positions 251 [=61%]). Bootstrap values circled. Arrow from *Bornetella* shows that in 22% of the bootstrap replicates it groups with the *Batophora-Chlorocladus* clade. From Olsen et al. (1994b). Reprinted by permission.

questioned and is unlikely to be. The absence of the E10-1 stemloop is a synapomorphy for the order as is strong radial symmetry; it remains a unique lineage.

Phylogenetic analyses of the chloroplast encoded *rbcL* gene, for the same isolates used in the 18S rDNA work, are currently in progress. So far, the results follow the same general topology recognizing the two major families (Zechman, unpubl.). The main difference (which may not hold) is the position of *Cymopolia* as sister genus to *Neomeris* vis-à-vis as basal taxon to all of the Dasycladales. Taxon sampling and long branches play a significant role in the analysis. *Cymopolia van bosseae* (an Indo-West Pacific endemic) was used in the 18S analysis, whereas *C. barbata* (a tropical Atlantic endemic) was used in the *rbcL* tree. The two do not appear to be sister taxa. The degree to which this split of the genus *Cymopolia* reflects "cryptic genera," oddities in G + C content and mutation rate, and/or the possible presence of pseudogenes is being studied.

Still missing from all molecular analyses are *Halicoryne, Chalmasia,* and *Dasycladus. Halicoryne* is the putatively most primitive of the extant Acetabulariaceae, as is *Dasycladus* for the Dasycladaceae. *Chalmasia* is predicted to fall within *Acetabularia.* A total evidence analysis based on 18S and *rbcL* data combined with morphology (Olsen and Zechman, unpubl.) awaits the inclusion of the three missing genera and both putative species of *Cymopolia.*

An exhaustive review of the entire body of dasycladalean literature (more than 450 scientific articles) can be found in Bonotto (1988) and in Berger and Kaever (1992). For a review of *Acetabularia* as a model system for cell and molecular biology, see Zeller and Mandoli (1993) and Hunt and Mandoli (1996). Molecular phylogenetics of Dasycladales are reviewed by Olsen et al. (1994b).

Siphonocladales-Cladophorales Complex

Molecular phylogenetic studies of the Siphonocladales-Cladophorales Complex (SCC) are so far limited. At present only about 10 of the 20 or so recognized genera have been sampled

(Fig. 18.9), and among those usually only a single representative species has been sequenced. Rampant paraphyly has been demonstrated in the SCC (Kooistra et al., 1993; Bakker et al., 1994, 1995a, 1995b); inclusion of species throughout the range of the genus is essential for assessment of correct sister-group relationships. Nuclear rDNA internal transcribed spacer (ITS) sequences have been very useful for species-level studies (Kooistra et al., 1992; Bakker et al., 1995a, 1995b). Zechman et al. (1990) compared partial, reverse transcriptase-based 18S and 26S rRNA sequences from *Anadyomene stellata, Chaetomorpha linum, Cladophora albida, Microdictyon boergesenii, Cladophoropsis membranacea, Dictyosphaeria versluysii,* and *Blastophysa rhizopus.* Bakker et al. (1994) compared complete 18S rDNA sequences from eleven species of *Cladophora, Chaetomorpha* sp., *Cladophoropsis zollingerii, C. membranacea, Chamaedoris peniculum, Valonia utricularis, Microdictyon boergesenii, Ernodesmis verticillata,* and *Siphonocladus tropicus.* Unfortunately, it is not possible to combine these data sets because of taxon sampling and differences between the reverse transcriptase-based 18S rRNA and the 18S rDNA sequences. Virtually all of the reverse transcriptase-based 18S sequences are presently being resequenced (Zechman, pers. comm.) and *rbcL* sequences added. Specific genera and complexes are being investigated with one or more genes by different international groups to incorporate better biogeographic sampling (i.e., *Valonia/Ventricaria/ Ernodesmis/Boergesenia* complex, Olsen and Stam, pers. comm.; and freshwater representatives, particularly *Cladophora* species, Ueda et al., pers. comm.). Key windows to the molecular, morphological, and ultrastructural literature on the SCC include O'Kelly and Floyd (1984), Olsen-Stojkovich et al. (1986), Zechman et al. (1990), and Bakker et al. (1994).

The Eusiphonous Lineages

The eusiphonous lineages include about 30 genera of predominantly tropical species. Some of the more common genera are *Halimeda, Caulerpa, Bryopsis,* and *Codium.* At present, only several species in this diverse assemblage

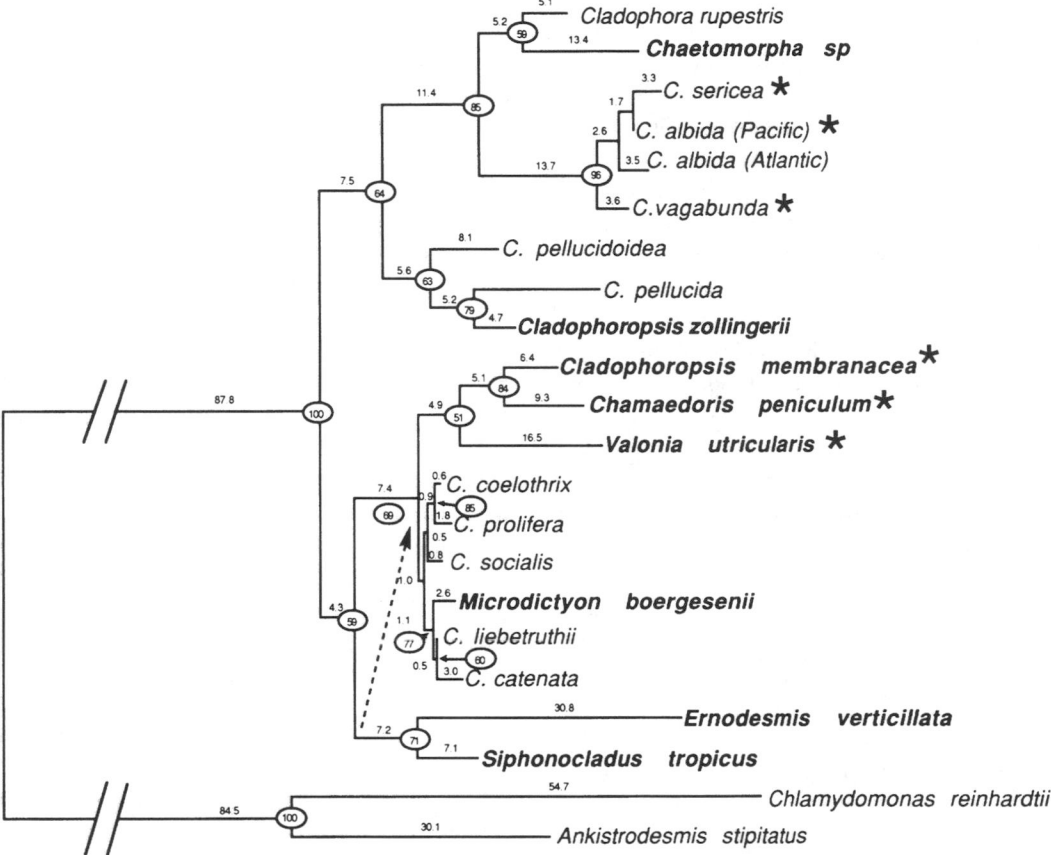

Figure 18.9. 18S rDNA tree for the *Cladophora* complex. Single most parsimonious tree found in general heuristic search with combinatorial weighting and double-stranded positions weighted 0.66. Sequences are complete. Bootstrap values circled. Siphonocladalean representatives are indicated in boldface. A star indicates that rDNA-ITS studies have been published for these species. Arrow shows position of *Ernodesmis/Siphonocladus* clade in 30% of the bootstrap replicates. From Bakker et al. (1994). Reprinted by permission.

have been sequenced and so the placement of these probable ancient lineages is still not known with certainty. Preliminary analyses based on 18S rDNA data (Zechman, pers. comm.) support Caulerpales as sister group to Dasycladales, but completion of this research awaits further taxon sampling. Both *rbcL* and 18S rDNA sequence studies are in early stages by several research groups (Australian and Indo-West Pacific taxa—especially *Caulerpa,* King et al., University of Melbourne; Caribbean *Caulerpa* species, Manhart et al., Texas A & M; Udoteaceae, Zechman, Duke University; Caribbean *Halimeda,* Hillis and Kooistra; *Bryopsis,* Ciaster, Duke University).

THE CHAROPHYCEAN LINEAGE (SENSU PICKETT-HEAPS AND MARCHANT, 1972)

Undoubtedly, the closest algal relatives of the land plants are from the charophycean lineage of green algae. After a general discussion of this lineage, we present the findings of recent research on three charophycean groups, Charales, Coleochaetales, and Zygnematales, and their relationships to each other and to the higher land plants.

With unusual and relatively complex thalli and reproductive organs, Charales sensu Bold and Wynne (1985) have been considered to be either a highly derived member of Chlorophyta or a separate division, Charophyta (Bold and Wynne, 1985). Their potentially close relationship to higher plants has long been suspected, and this hypothesis has received strong support from both morphological (primarily ultrastructural) and molecular evidence (McCourt et al., 1996b). Mattox and Stewart (1984) viewed Charales, along with Chlorokybales, Klebsormidiales, Zygnematales, and Coleochaetales, as members of the class Charophyceae, which formed one of two major clades of green algae. In their scheme, this class, although paraphyletic (Mishler and Churchill, 1985), contains the extant taxa of green algae that are most closely related to higher plants (Graham, 1993). Although taxonomic treatments may vary, all of these algae can be called collectively the "charophycean lineage" or "Charophyceae," and individually, the "charophytes" or "charophycean green algae."

In the debate over which charophyte (if any) can be regarded as ancestral to the land plants, several authors suggest that *Coleochaete* rather than Charales is the sister group of the land plants (Pickett-Heaps, 1979; Taylor, 1982; Graham, 1982, 1983; Graham and Wilcox, 1983; Mishler and Churchill, 1985; Graham et al., 1991). Among other characteristics, the multicellular development of the male gametangia and retention of the zygote on the gametophyte, as well as the presence of a structure resembling the gametophytic placental transfer cells of the land plants, have served as arguments for a possible close relationship of *Coleochaete* and the embryophytes. However, a considerable problem with this argument is that only some of the species (e.g., *C. orbicularis*) possess the characters listed above (Bremer, 1985).

Mattox and Stewart (1984) based their original classification on ultrastructural features of mitosis and motile cells and biochemical features. Molecular data from both the rDNA and *rbcL* genes have provided additional strong support for the placement of the charophycean algae on the evolutionary lineage leading to land plants.

Sequences from the 18S rRNA gene have been used to reconstruct the phylogeny of eukaryotic cells at the deepest levels (Hillis and Dixon, 1991), and some of the first strong molecular evidence supporting the relationship of Charales to land plants came from such studies. In general, the 18S data support the monophyly of the charophycean green algae and land plants, although the precise topology of this relationship is not clear (McCourt, 1995; Melkonian and Surek, 1995). In several reports Charales were found to be the sister taxon to land plants (Wilcox et al., 1992; Surek et al., 1993), whereas in another all the orders of the charophycean green algae formed a monophyletic sister group to land plants (Bhattacharya et al., 1994). Ragan et al. (1994) found support for several sister-group relationships among charophycean green algae depending on the method of analysis employed. All of these studies used some of the same sequences, so we assume that the conflicts between topologies were due to taxon sampling (that is, the trees changed with the addition or omission of particular taxa) and analytical method.

Data from *rbcL* sequences also support the monophyly of the charophycean green algae and land plants (Manhart, 1994; McCourt et al., 1995, 1996a). These data place two monophyletic orders, Charales and Coleochaetales, in a clade that is sister taxon to the land plants, with the other orders basal to the charophyte-embryophyte clade (Streptophytes sensu Bremer, 1985). However, bootstrap and decay support for this topology are weak and three lineages (two algal orders and land plants) form an unresolved trichotomy in such statistical analyses. Again, expanded sampling of taxa from the charophycean green algae, as well as additional sequence and morphological characters, may lead to a more robust phylogeny.

Below we discuss the findings of recent research on three charophycean groups, Charales, Coleochaetales, and Zygnematales, and their relationships to each other and to the higher land plants.

Charales

Characeae are the sole extant family of Charales, a large group of aquatic green algae known as stoneworts or brittleworts (Wood and Imahori, 1965). The group is well represented in the fossil record in the form of fossilized oogonia or remnants of their contained zygotes (Grambast, 1974; Feist and Grambast-Fessard, 1991). The group is characterized by relatively complex thalli, consisting of axial filaments of giant internodal cells alternating with shorter node regions composed of many smaller cells and whorled branchlets.

For Charales, *rbcL* sequence data support several conventional hypotheses of generic relationships (McCourt et al., 1996a) (Fig. 18.10). One of the two tribes proposed by Wood and Imahori (1965) (Chareae, composed of four genera) is strongly supported as monophyletic, whereas the other (Nitelleae, composed of two genera) is not. The latter decomposes into a grade with *Tolypella* sister to the entire Charales, and *Nitella* sister to the Chareae clade. These relationships are, however, the most weakly supported of any within the *rbcL* topology for the family and may be contradicted by additional studies. The *rbcL* sequence data are generally congruent with morphological analyses, although the latter suffer from a lack of sufficient phylogenetically informative characters

to provide robust support for a morphology-based tree (McCourt et al., 1996a). Enlarging the number of taxa external to Charales to include other charophycean green algae, as well as taxa from the chlorophycean algal lineage, does not lead to a change in the inferred phylogeny within the Charales (see *Zygnematales,* below).

In recent 18S rDNA data analyses (Figs. 18.11 and 18.12), Charales, often regarded as the more advanced of the charophycean green algae, are shown to constitute a distinct and ancient lineage within this class. The remaining Charophyceae are most closely related to the bryophytes which emerged first in the evolution of the land plants as inferred from the 18S rRNA tree (Figs. 18.11 and 18.12) (see also discussion below on the phylogenetic position of the charophytes within the green plants [Viridiplantae/Chlorobionta]).

Coleochaetales

Coleochaetales, including the genera *Coleochaete* and *Chaetosphaeridium*, are structurally complex microscopic green algae with a life

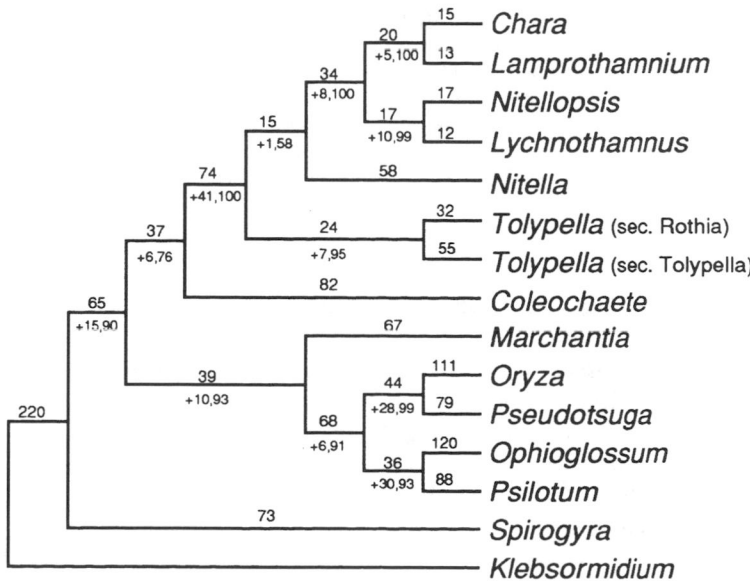

Figure 18.10.   Phylogeny of Charales based on *rbcL* sequences. The topology represents the single most parsimonious tree found with branch and bound search, equal weights for all nucleotide positions (PAUP 3.1.1, Swofford, 1993). Tree length = 1,515 steps, CI = 0.547 (0.497 excluding uninformative characters). Values above branches are branch lengths; decay (+) and bootstrap values (percent of 1,000 replicates) shown below branches. (Modified from McCourt et al., 1996a).

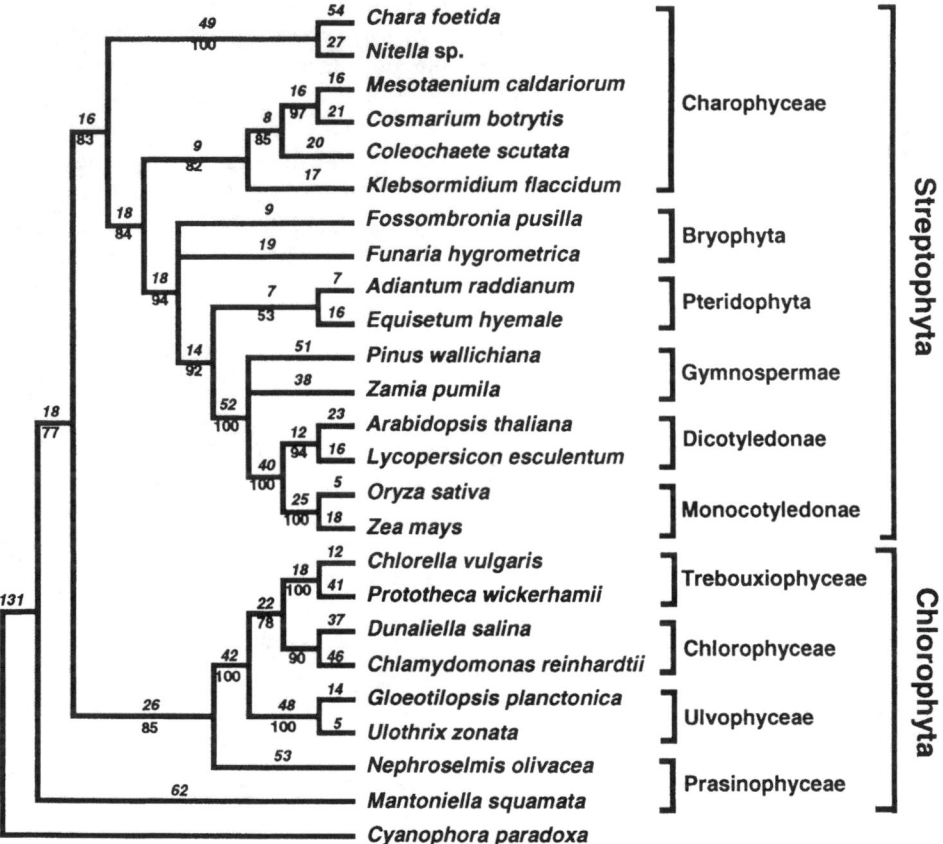

Figure 18.11. 18S rDNA phylogeny of green plants. The tree shown is a maximum parsimony bootstrap analysis based on 1,676 nucleotide positions. The number of steps separating two nodes is taken from one of four most parsimonious trees that resulted from a heuristic search with random addition of sequences, and shown in italics above branches. Overall tree length was 1,236 steps with a consistency index (CI) of 0.569 and a retention index (RI) of 0.647. Bootstrap values based on 500 resamplings with 10 heuristic searches each are given below branches. The tree includes all major lineages of green plants and demonstrates that chlorophytes are sister group to streptophytes (= land plants + charophytes). The glaucocystophyte *Cyanophora paradox* (EMBL Acc. No.: X68483) was used as outgroup. Figure by Huss and Kranz, 1998; reprinted by permission.

history that (although haplontic) is reminiscent of the alternation of generations seen in land plants (the embryophytes). Largely on the basis of the perceived similarity of life histories, Bower (1908) proposed that *Coleochaete* might be a close relative of land plants, but this idea fell into disrepute later in the twentieth century as the full diversity of green algal life histories was discovered.

Cladistic analyses supported the idea that land plants are closely related to the charophycean green algae, specifically Coleochaetales. Mishler and Churchill (1985) performed a phylogenetic analysis that indicated that the genus *Coleochaete* was the sister taxon to the land plants, but this study was controversial because some of the characters that linked *Coleochaete* to land plants were not uniform throughout the genus *Coleochaete* (Whittemore, 1987). Mishler and Churchill (1987) acknowledged this problem and proposed that *Coleochaete* may be paraphyletic, with the species *C. orbicularis* the putative sister taxon to the land plants. Subsequent cladistic analyses of structural information lent some support to this hypothesis (Graham et al., 1991), but the number of characters in these analyses was small, and although the data supported a charophycean origin of plants,

A)

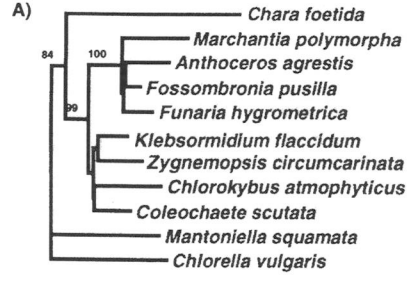

B)

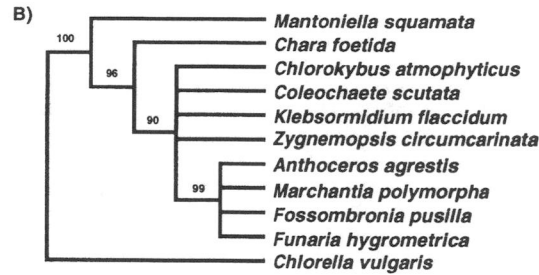

C)

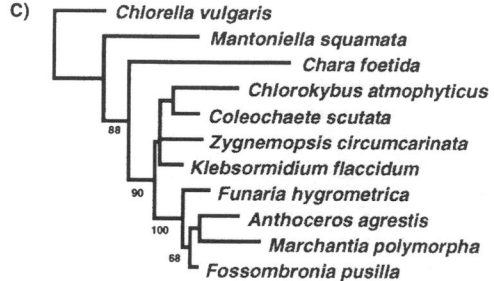

Figure 18.12. Comparison of the three most widely used methods for phylogenetic reconstruction to show the relationships between charophytes and their closest relatives, the bryophytes, based on 18S rDNA sequence data. All analyses are based on 1,575 nucelotide positions. Only bootstrap values >50% are shown. *Chlorella vulgaris* was used as outgroup. *A.* Neighbor Joining (1,000 bootstrap resamplings). *B.* Maximum Parsimony (1,000 bootstrap resamplings). *C.* Maximum likelihood (250 bootstrap resamplings). Figure by Huss and Kranz, 1998; reprinted by permission.

several taxa, including the orders Coleochaetales and Charales, plausibly could be interpreted as the sister taxon to the land plants.

The hypothesis that *Coleochaete orbicularis* is the sister taxon to land plants could not be tested until molecular data were available from several taxa in the order Coleochaetales. C. Delwiche, K. Karol, and K. Sytsma (unpubl.) have recently completed an analysis of *rbcL* sequences from a number of species in Coleochaetales. These analyses support Mattox and Stewart's (1984) concept of Charophyceae (Fig. 18.13). Coleochaetales and Charales are each monophyletic groups closely related to the land plants, but the branching order among these three groups is not well resolved. Bootstrap support for the monophyly of Coleochaetales is strong (93%), as is support for the branching order among taxa within the order. This is inconsistent with the hypothesis that the sister taxon to land plants is the species *Coleochaete orbicularis,* but provides further support for a close relationship between land plants and the green algal orders Coleochaetales and Charales.

Zygnematales

Zygnematales sensu Bold and Wynne (1985) comprise the conjugating green algae, including several groups of unicellular, filamentous, and colonial taxa. The classification system that is widely used is that of Mix (1972, 1973; see also Brook, 1981), who delineated families based on ultrastructural features of the multilayered walls of vegetative cells. Two families (approximately 18 genera), Mesotaeniaceae (saccoderm desmids) and Zygnemataceae, possess the simplest walls: one homogeneous piece with no (or weak) encrustations and lacking pores in any of the three layers (outer, primary, and secondary). The other four families in the Mix (1972, 1973) scheme are commonly called "placoderm desmids" and have been placed by some authors in a separate order, Desmidiales (Gerrath, 1993). Three of these families, Gonatozygaceae (two genera), Peniaceae (one genus), and Closteriaceae (two genera), exhibit a moderate degree of wall ornamentation (warts, spines, and ridges) and pores in the outer (middle) layer. (Kouwets

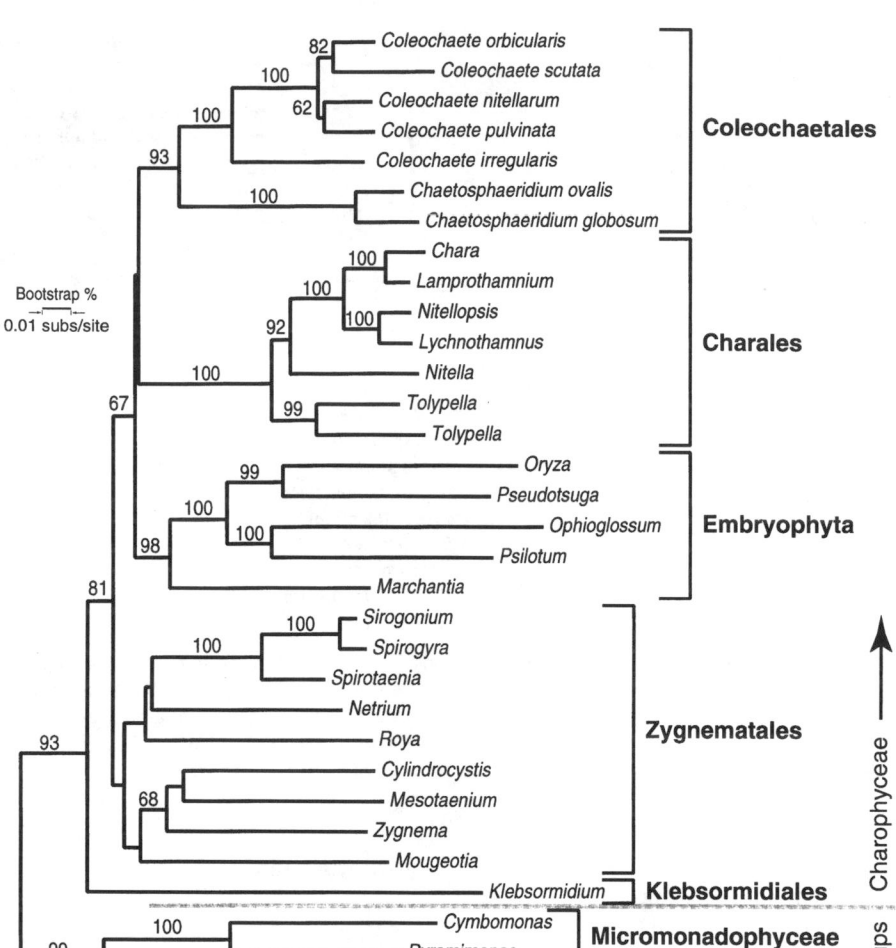

Figure 18.13.   Neighbor-joining tree calculated with Kimura three-parameter distances using *rbcL* nucleotide data showing that Coleochaetales and Charales are monophyletic groups closely related to land plants. Bootstrap values above 60% are shown above branches.

and Coesel, 1984, merged Gonatozygaceae with Peniaceae). The last family, Desmidiaceae (approximately 31 genera), exhibits the most marked degree of wall ornamentation, with spines and other protuberances arising in the innermost secondary wall. Elaborate pores are located in the secondary wall as well. The primary wall is shed.

18S rDNA sequence data have provided support for the monophyly of Zygnematales and for the close relationship of this group and other charophycean green algae to higher plants

(Surek et al., 1993; Bhattacharya et al., 1994, 1996a, 1996b). These 18S studies also suggested that Zygnemataceae and Mesotaeniaceae form a paraphyletic assemblage, whereas the placoderm desmids as a whole appear to be monophyletic. The taxon sampling, however, did not include many genera (only 10 genera of the approximately 54), and the findings warrant further sampling in the group. In addition to 18S rDNA sequences, Bhattacharya et al. (1994, 1996a) showed that Zygnematales possessed a Group I intron within the 18S rDNA, which does not occur in other charophycean green algae and probably represents a phylogenetically informative lateral transfer of the intron. The intron exhibits several interesting features related to self-splicing; sequences for this intron may also contain phylogenetic signal useful in reconstructing the phylogeny of the group.

McCourt et al. (unpubl.) and Park et al. (1996) used *rbcL* sequences to reconstruct phylogeny within Zygnematales. This gene exhibits a faster divergence rate than does 18S rDNA, and Manhart (1994) suggested that *rbcL* is not informative at the deepest levels in the phylogeny of green plants. However, within Zygnematales and Charales (McCourt et al., 1996a), *rbcL* appears to provide useful data for testing hypotheses of relationships. As with many molecular studies dealing with groups that diverged probably several hundred million years ago, adequate taxon sampling and choice of outgroup affect the topology of resulting trees. Figure 18.13 shows a composite parsimony analysis of *rbcL* data for Zygnematales, Charales, and Coleochaetales (Delwiche, unpubl.). The topology of Charales in this expanded data set is identical to that obtained with a smaller data set of charophycean taxa (McCourt et al., 1996a, 1996b). The topology of Zygnemataceae and Mesotaeniaceae is somewhat different from that obtained with a smaller taxon set (McCourt et al., 1995). The more diverse Zygnematales are apparently more sensitive to taxon sampling than Charales (all six extant genera were included).

Figure 18.14 refutes the monophyly of Zygnemataceae and Mesotaeniaceae, the two families exhibiting the simplest wall structure. Two of the other four families (Peniaceae and Desmidiaceae) are supported as monophyletic, both individually and as sister taxa. The topology of the most parsimonious tree obtained is consistent with monophyly of the four families, although bootstrap support is moderate to weak. Gonatozygaceae are strongly supported as monophyletic, although the curious placement of the saccoderm desmid *Roya* as sister to this clade remains an enigma (Park et al., 1996). Along with Gonatozygaceae, the Peniaceae (one described genus) and Closteriaceae (two described genera, one included in this analysis) occupy a basal position in the placoderms (Fig.

18.14). This placement does not refute the monophyly of any of Mix's (1972, 1973) families. Bootstrap support for these nodes is moderate to weak, however. Desmidiaceae (sensu Mix, 1972, 1973) also received moderate support as a monophyletic assemblage, with several clades internal to the group receiving strong support (e.g., *Sphaerozosma* and *Groenbladia*).

Comparison of the *rbcL* tree with the distribution of wall structural characters provides support for the monophyly of the families exhibiting moderate to strong wall ornamentation (Fig. 18.14). The two families with the simplest structure (Mesotaeniaceae and Zygnemataceae) are paraphyletic. McCourt et al. (unpubl.) suggest that this is due to the probable plesiomorphic nature of the smooth, one-piece, unornamented wall of these two families. Outgroups lack the derived pore structures and segmented walls of the placoderm desmids; it is therefore not surprising that groups erected on shared plesiomorphic wall characters may be nonmonophyletic. The delineation of the other families based on the derived characters (synapomorphies) of moderate wall ornamentation and other features does result in monophyletic taxa.

The *rbcL* topology is congruent with that of the 18S rDNA trees (Surek et al., 1993; Bhattacharya et al., 1994; McCourt et al., 1995; Park et al., 1996). However, taxon sampling did not overlap sufficiently across studies to predict whether further 18S rDNA sampling will exhibit the same topology as that found in the *rbcL* analyses.

When the *rbcL* sequence data for Zygnematales are combined with those of other charophycean algae (Fig. 18.13), the order is not strongly supported as monophyletic, and the topology within the order changes significantly. This result suggests that *rbcL* sequence data alone do not provide a robust phylogeny for this group.

Phylogenetic Position of the Charophytes within the Green Plants (Viridiplantae/Chlorobionta)— 18S rRNA Data

Unfortunately, as with other analyses presented in this chapter, the rDNA sequence data currently available are unable to resolve the question of which extant green alga is most closely related

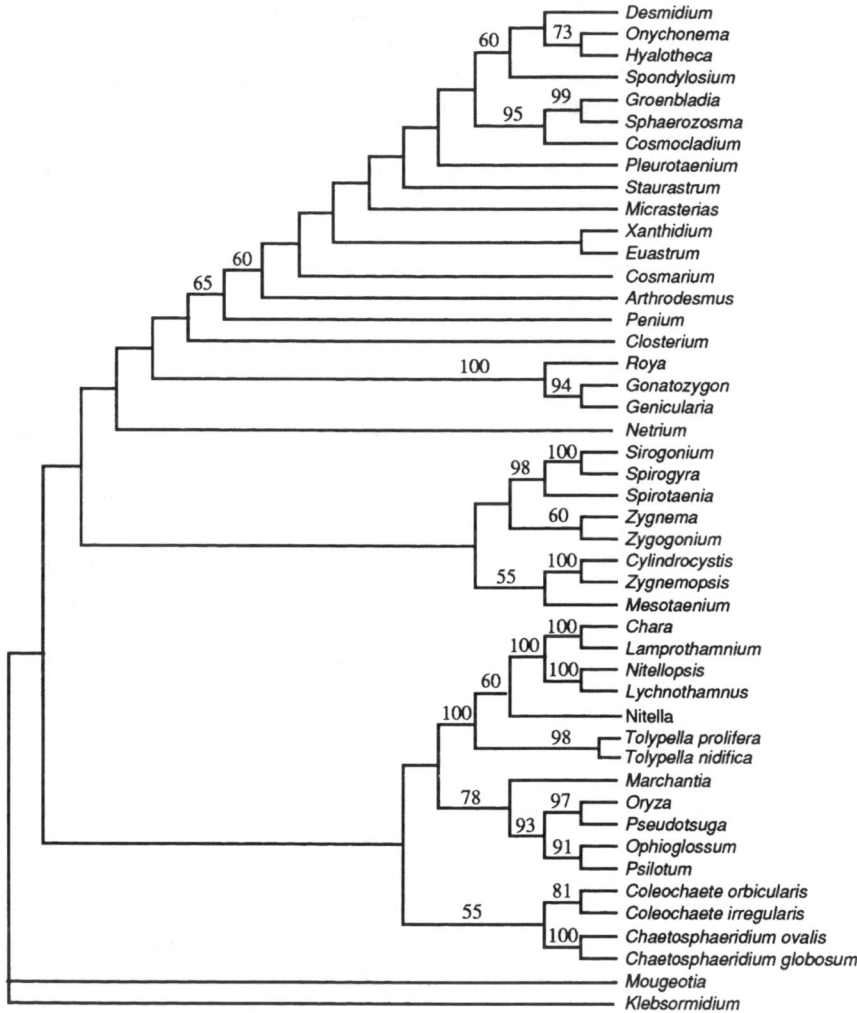

Figure 18.14.   Phylogeny of Zygnematales based on *rbcL* sequences. Strict consensus of three equally parsimonious trees found with heuristic search, 10 random taxon addition replicates, steepest descent, equal weights for all nucleotide positions (Paup* 4.0d53, used with permission of D. Swofford). Tree length = 3,683 steps, CI = 0.278 (0.262 excluding uninformative characters). Numbers above branches are bootstrap values (percent of 1,000 replicates). (McCourt and Karol, unpubl.).

to the land plants. Extensive bootstrap analyses with the three most widely used methods for phylogenetic reconstruction, neighbor joining, maximum parsimony, and maximum likelihood (trees A, B, and C, respectively, in Fig. 18.12), all indicate that the different charophyte lineages (except Charales) diverged within a short period of time and should, together, be regarded as a sister group to the bryophytes. There is no indication that any one of the charophytes investigated is significantly more closely related to the bryophytes than are the others. There is also no indication that some bryophytes might

have independently evolved from different green algae, such as the hornworts from a *Coleochaete*-like alga, as suggested by Sluiman (1983). Thus, features of *C. orbicularis* such as zygote-associated cells that resemble the placental transfer cells of embryophytes might well be the product of parallel evolution. Lack of resolution within the charophytes and also within the different lineages of bryophytes is indicated by low bootstrap values (<50%) in the distance and the maximum likelihood analyses, and by multifurcation in the maximum parsimony analysis.

Why do these results contradict previous molecular studies even with complete 18S rRNA sequences, which favored Charales as closest relatives to the land plants? The large evolutionary gap between the charophytes and the land plants, together with the fact that Charales and land plants have an accelerated mutation rate compared with the remaining charophytes, bryophytes, and pteridophytes (Kranz et al., 1995; Kranz and Huss, 1996), might have been responsible for artificially grouping Charales with the land plants, a phenomenon known as long branch attraction (cf. Li et al., 1987; Felsenstein, 1988). Including sequences of organisms much more closely related to the charophytes, such as bryophytes and pteridophytes, is a means of avoiding such artifacts. Turning to an independent data set from a different molecule, for example, a chloroplast gene, could clarify relationships, and, in the next section, such additional clues are discussed.

ADDITIONAL CHLOROPLAST CLUES TO GREEN ALGAL PHYLOGENY

Several characteristics of land plant chloroplast DNAs (cpDNAs) distinguish them from those of other chloroplast-containing taxa, including the green algae (reviewed in Palmer, 1991). These characteristics include a highly stable genome that is divided by inverted repeats (present in some green algae) into large and small single-copy regions, Group II introns in specific genes, and several unique, apparently derived operons. The genome organization and gene order characteristic of land plants have been found only in that group and may represent a unique derived condition. Only a few green algal cpDNAs have been characterized, and the only ones studied in detail have been several species of *Chlamydomonas,* but it is clear that green algal cpDNAs are highly variable in terms of size and genome structure. It is impossible to rationalize the many changes necessary to reconcile land plant cpDNA structures with those of the green algae at this point. However, investigations of finer levels of genome organization such as the presence/absence of genes and introns and operon structures have

revealed additional apparent synapomorphies of land plants and charophytes.

Investigations of the *tufA* gene suggest, in contrast to many of the results presented earlier in this chaper, that Zygnematales might be the closest sister group to the land plants (Baldauf et al., 1990). The *tufA* gene, encoding protein synthesis elongation factor Tu, was transferred from the chloroplast to the nucleus within the charophyte lineage. This gene is present in the cpDNA of all Chlorophyceae and Ulvophyceae, but was not detected in the chloroplast genomes of some members of the charophycean green algae and was absent in all land plant chloroplasts examined. Within the charophycean green algae, only the cpDNA of Charales showed strong hybridization signals with a *tufA*-specific probe (consistent with the view that Charales are the most basal charophyte lineage); a weak signal resulting from a very divergent and probably nonfunctional *tufA* gene was obtained with *Coleochaete,* whereas no signal was detected with chloroplasts from Zygnematales (*Spirogyra* and *Sirogonium*).

One other similarity of the chloroplast genomes of land plants and *Spirogyra* has been revealed. The organization of the *rps12* and *rps7* genes is unusual in land plants. These two genes are apparently the remnants of the *str* operon in prokaryotes, which is composed of *rps12, rps7, fusA,* and *tufA*. In the cpDNAs of land plants, the *fusA* (*fusA* has not been found in any chloroplast genome) and *tufA* genes are absent and *rps12* is split into three pieces: exon 1 is located on another part of the genome and exons 2 and 3 are separated by an intron. Exons 2 and 3 of *rps12* are cotranscribed with *rps7*. In *Spirogyra maxima* cpDNA, the organization of *rps12* and *rps7* is the same as that in land plants (Lew and Manhart, 1993). Thus, although the overall genome structure of *Spirogyra maxima* cpDNA is very different from that of land plants (Manhart et al., 1990), an examination of genome organization in finer detail has revealed two similarities that support a close relationship of *Spirogyra* with land plants.

There are 17 genes in land plant cpDNAs that contain Group II introns (Palmer, 1991), and none of these introns has been found in any other taxonomic group except the charophycean green algae. Group II introns in *trnI* and *trnA* have been documented in *Coleochate orbicularis, Nitella translucens,* and in the *trnA* intron in *Spirogyra maxima* (Manhart and Palmer, 1990). The absence of the intron in the *trnI* gene of *Spirogyra* indicates that *Coleochaete* and *Nitella* are closer to the land plant lineage than is *Spirogyra,* a relationship supported by some phylogenetic analyses of DNA sequences and morphological characters (see previous sections of this chapter), but at variance with the distribution of *tufA.*

Additional operons unique to land plants are currently being investigated in Charophyceae. The *rpl23* operon is the largest operon in land plant cpDNAs and contains genes of the S10, spc, and alpha operons of prokaryotes (Tanaka et al., 1986; Sugiura, 1992). The gene order of the *rpl23* operon is (S10)*rpl23 rpl2-rps19-rpl22-rps3-rpl16-*(spc)*rpl14-rps8-infA-rpl36-*(alpha)*rpl11-rpoA* and the *rpl23* end is next to inverted repeat B or included in it. Among green algae, detailed information on these genes is only available for *Chlamydomonas* (Boudreau et al., 1994). In *Chlamydomonas reinhardtii,* there is a gene cluster that has some of the same genes of the S10 and spc operons as in land plants, but in *C. moewusii,* this cluster is split.

The *petB* and *petD* genes are members of the *petBD* operon in cyanobacteria (Brand et al., 1992), and these genes are clustered together in all known chloroplast genomes with the exception of *Chlamydomonas* (Buschlen et al., 1991), where they are located in two different operons. The *petBD* operon is clearly a plesiomorphic character in chloroplast genomes, and the operon organization of *Chlamydomonas* is derived. In land plants *petB* and *petD* are members of the *psbB* operon. This operon, which is unique to land plants, is composed of *psbB-psbT-*(*psbN*)*-psbH-petB-petD* (Freyer et al., 1993). The operon is a pentacistronic transcriptional unit; *psbN,* about 100 bp, is transcribed independently on the other strand (Kanno and Hirai, 1993). The characterization of these gene clusters at the sequence level in the charophycean green algae should add to our understanding of phylogenetic relationships of these algae and land plants and also of the evolution of chloroplast genomes.

CONCLUSIONS

Although molecular systematics has provided some answers to questions about green algal phylogeny, many remain. In one example presented in this chapter, rDNA analyses corroborate the significance of motile cell features for delineating trebouxiophycean algae, but do not resolve the interrelationships of lineages within Trebouxiophyceae. In another, there is a general consensus that there are two major lineages of green algae, and that one of these, the charophycean green algae, forms a monophyletic group (Viridiplantae/Chlorobionta) with the land plants. However, which charophycean green alga is the sister taxon to the land plants is not yet resolved. If there was a rapid, ancient radiation of these lineages, there is reason to believe no stochastically changing molecule will ever provide a resolution (Lanyon, 1988).

Molecular data, which generate a large pool of global characters, can provide an independent test of phylogenetic hypotheses. These results are used to test phylogenies based on morphology and other "organismal" characters (e.g., habitat, biochemistry, life cycles). Indeed, the mapping of organismal characters on a robust molecular tree is a means of assessing the usefulness of characters at various taxonomic levels. This approach assumes that molecular data provide the true phylogeny. However, to some extent there is a slight circularity: The usefulness of molecular data is assessed by the congruence of molecular phylogenies with phylogenies inferred from the organismal data, the usefulness of which is being tested. A total evidence approach (sensu Kluge, 1989), involving the analysis of combined, homogeneous (Bull et al., 1993), multiple data sets based on different genes and well-defined morphological characters, is likely to provide the most robust phylogenies.

Such a total evidence approach can only be achieved through coordinated acquisition of data. Results of research on green algal phylogeny, including databases showing which genes have been studied for which taxa, are available through the Green Plant Phylogeny Research Coordination Group (http://ucjeps. berkeley.edu/bryolab/greenplantpage.html).

Similarly, the Tree of Life Project (http://phylogeny.arizona.edu/tree/phylogeny.html) provides a continually updated summary of green algal phylogenies. (The Tree of Life project provides information on all organisms, not just algae, and is "designed to contain information about the phylogenetic relationships and characteristics of organisms, to illustrate the diversity and unity of living organisms, and to link biological information available on the Internet in the form of a phylogenetic navigator.")

We anticipate that as more sequences of more genes are obtained for more taxa, better resolution of phylogenies will result, and these results can be used to test hypotheses regarding character evolution, biogeography, and gene sequence changes as well as the origin and evolution of green algae and early land plants. Though not a panacea or quick fix, molecular systematics continues to add new information to a growing understanding of relationships among the first green plants—the green algae.

LITERATURE CITED

Archibald, P. A., and H. C. Bold. 1970. Phycological Studies XI. The Genus *Chlorococcum* Meneghini, Publication 7015. University of Texas, Austin.

Bakker, F. T., J. L. Olsen, W. T. Stam, and C. van den Hoek. 1994. The *Cladophora* complex (Chlorophyta): new views based on 18S rRNA gene sequences. Molecular Phylogenetics and Evolution 3:365–382.

Bakker, F. T., J. L. Olsen, and W. T. Stam. 1995a. Evolution of nuclear rDNA ITS sequences in the *Cladophora bida/sericea* clade (Chlorophyta). Journal of Molecular Evolution 40:640–651.

Bakker, F. T., J. L. Olsen, and W. T. Stam. 1995b. Global phylogeography in the cosmopolitan species *Cladophora vagabunda* (Chlorophyta) based on nuclear rDNA internal transcribed spacer sequences. European Journal of Phycology 30:197–208.

Baldauf, S. L., J. R. Manhart, and J. D. Palmer. 1990. Different fates of the chloroplast *tufA* gene following its transfer to the nucleus in green algae. Proceedings of the National Academy of Sciences U.S.A. 87:5317–5321.

Berger, S., and M. J. Kaever. 1992. Dasycladales: An Illustrated Monograph of a Fascinating Algal Order. Thieme, Stuttgart.

Bhattacharya, D., B. Surek, M. Rüsing, S. Damberger, and M. Melkonian. 1994. Group I introns are inherited through common ancestry in the nuclear-encoded rRNA of Zygnematales (Charophyceae). Proceedings of the National Academy of Sciences U.S.A. 91:9916–9920.

Bhattacharya, D., S. Damberger, B. Surek, and M. Melkonian. 1996a. Primary and secondary structure analyses of the rDNA group-I introns of the Zygnematales (Charophyta). Current Genetics 29:282–286.

Bhattacharya, D., T. Friedl, and S. Damberger. 1996b. Nuclear-encoded rDNA group-I introns: origin and phylogenetic relationships of insertion site lineages in the green algae. Molecular Biology and Evolution 13:978–989.

Bischoff, H. W., and H. C. Bold. 1964. Phycological Studies. IV. Some Soil Algae from Enchanted Rock and Related Algal Species, Publication 6318. University of Texas, Austin.

Bold, H. C. 1970. Some aspects of the taxonomy of soil algae. Annals of the New York Academy of Science 17:601–616.

Bold, H. C., and B. C. Parker. 1962. Some supplementary attributes in the classification of *Chlorococcum* species. Archiv für Mikrobiologie 42:167–288.

Bold, H. C., and M. J. Wynne. 1985. Introduction to the Algae, second edition. Prentice-Hall, Englewood Cliffs, New Jersey.

Bonotto, S. 1988. Recent progress in research on *Acetabularia* and related Dasycladales. Progress in Phycological Research 6:59–235.

Boudreau, E., C. Otis, and M. Turmel. 1994. Conserved gene clusters in the highly rearranged chloroplast genomes of *Chlamydomonas moewussi* and *Chlamydomonas reinhardtii*. Plant Molecular Biology 24:585–602.

Bower, F. O. 1908. The Origin of a Land Flora: A Theory Based Upon the Facts of Alternation. Macmillan, London.

Brand, S. N., X. Tan, and W. R. Widger. 1992. Cloning and sequencing of the *petBD* operon from the cyanobacterium *Synechococcus* sp. PCC 7002. Plant Molecular Biology 20:481–491.

Bremer K. 1985. Summary of green plant phylogeny and classification. Cladistics 1:369–385.

Bremer, K., C. J. Humphries, B. D. Mishler, and S. P. Churchill. 1987. On cladistic relationships in green plants. Taxon 36:339–349.

Brook, A. J. 1981. The Biology of Desmids. Blackwell Scientific, Oxford.

Brown, R. M., Jr., and H. C. Bold. 1964. Phycological Studies. V. Comparative Studies of the Algal Genera *Tetracystis* and *Chlorococcum*, Publication 6417. University of Texas, Austin.

Buchheim, M. A., and R. L. Chapman. 1991. Phylogeny of the colonial green flagellates: a study of 18S and 26S rRNA sequence data. BioSystems 25:85–100.

Buchheim, M. A., and R. L. Chapman. 1992. Phylogeny of the genus *Carteria* (Chlorophyceae) inferred from organismal and molecular evidence. Journal of Phycology 28:362–374.

Buchheim, M. A., M. Turmel, E. A. Zimmer, and R. L. Chapman. 1990. Phylogeny of *Chlamydomonas* (Chlorophyta) based on cladistic analysis of nuclear 18S rRNA sequence data. Journal of Phycology 26:689–699.

Buchheim, M. A., J. A. Buchheim, and R. L. Chapman. 1994a. Molecular and morphological approaches to phylogenetic problems in the Phacotaceae (Chlorophyceae). Journal of Phycology (suppl.) **30**:15.

Buchheim, M. A., M. A. McAuley, E. A. Zimmer, E. C. Theriot, and R. L. Chapman. 1994b. Multiple origins of colonial green flagellates from unicells: evidence from molecular and organismal characters. Molecular Phylogenetics and Evolution **3**:322–343.

Buchheim, M. A., J. A. Buchheim, and R. L. Chapman. 1995. Systematics of the Chlamydomonadales: phylogeny of brackish water and marine taxa. Journal of Phycology (suppl.) **30**:17.

Buchheim, M. A., J. A. Buchheim, and R. L. Chapman. 1996a. Molecular phylogenetics of *Chloromonas* (Chlorophyceae). Journal of Phycology (suppl.) **32**:9.

Buchheim, M. A., C. Lemieux, C. Otis, R. R. Gutell, R. L. Chapman, and M. Turmel. 1996b. Phylogeny of the Chlamydomonadales (Chlorophyceae): a comparison of ribosomal RNA gene sequences from the nucleus and the chloroplast. Molecular Phylogenetics and Evolution **5**:391–402.

Buchheim, M. A., J. A. Buchheim, and R. L. Chapman. 1997. Phylogeny of *Chloromonas* (Chlorophyceae): a study of 18S rRNA gene sequences. Journal of Phycology **33**:286–293.

Bull, J. J., J. P. Huelsenbeck, C. W. Cunningham, D. L. Swofford, and P. J. Waddell. 1993. Partitioning and combining data in phylogenetic analysis. Systematic Biology **42**:384–397.

Buschlen, S., Y. Choquet, R. Kuras, and F.-A. Wollman. 1991. Nucleotide sequences of the continuous and separated *petA, petB,* and *petD* chloroplast genes in *Chlamydomonas reinhardtii.* FEBS Letters **284**:257–262.

Cavalier-Smith, T. 1981. Eukaryote kingdoms: seven or nine? BioSystems **14**:461–481.

Chantanachat, S., and H. C. Bold. 1962. Phycological Studies II. Some Algae from Arid Soils, Publication 6218. University of Texas, Austin.

Chappell, D. F., H. J. Hoops, and G. L. Floyd. 1989. Strip-like covering revealed on "wall-less" *Asteromonas* (Chlorophyceae). Journal of Phycology **25**:197–199.

Deason, T. R. 1984. A discussion of the classes Chlamydophyceae and Chlorophyceae and their subordinate taxa. Plant Systematics and Evolution **146**:75–86.

Deason, T. R., and H. C. Bold. 1960. Phycological Studies I. Exploratory Studies of Texas Soil Algae, Publication 6022. University of Texas, Austin.

Deason, T. R., P. C. Silva, S. Watanabe, and G. L. Floyd. 1991. Taxonomic status of the species of the green algal genus *Neochloris.* Plant Systematics and Evolution **177**:213–219.

Deloffre, R., and B. Granier. 1991. Hypothèse phylogénique des algues Dasycladales, Comptes Rendu Academie Science Paris (ser. II) **312**:1673–1676.

Devereux, R., A. R. Loehlich III, and G. E. Fox. 1990. Higher plant origins and the phylogeny of green algae. Journal of Molecular Evolution **31**:18–24.

Ettl, H. 1970. Die Gattung *Chloromonas* Gobi emend. Wille. Nova Hedwigia **34**:1–283.

Ettl, H. 1976. Die Gattung *Chlamydomonas* Ehrenberg. Nova Hedwigia **49**:1–1122.

Ettl, H. 1981. Die neue Klasse Chlamydophyceae, eine natürliche Gruppe der Grünalgen (Chlorophyta). Plant Systematics and Evolution **137**:107–126.

Ettl, H. 1983. Süßwasserflora von Mitteleuropa, Band 9: Chlorophyta I. Phytomonadina. VEB Gustav Fischer Verlag, Jena, Germany.

Ettl, H., and J. Komárek. 1982. Was versteht man unter dem Begriff "coccale Grünalgen"? (Systematische Bemerkungen zu den Grünalgen II). Archiv für Hydrobiologie (Supplement 60) Algological Studies **27**:345–374.

Feist, M., and N. Grambast-Fessard. 1991. The genus concept in Charophyta: evidence from Palaeozoic to Recent. In Calcareous Algae and Stromatolites, ed. R. Riding, pp. 189–203. Springer-Verlag, New York.

Felsenstein, J. 1978. Cases in which parsimony and compatibility methods will be positively misleading. Systematic Zoology **27**:401–410.

Felsenstein, J. 1981. Evolutionary trees from DNA sequences: a maximum likelihood approach. Journal of Molecular Evolution **17**:368–376.

Felsenstein, J. 1985. Confidence limits on phylogenies: an approach using the bootstrap. Evolution **39**:66–70.

Felsenstein, J. 1988. Phylogenies from molecular sequences: inference and reliability. Annual Review of Genetics **22**:521–565.

Felsenstein, J. 1993. PHYLIP (Phylogeny Inference Package), version 3.5c. University of Washington, Seattle.

Floyd, G. L., S. Watanabe, and T. R. Deason. 1993. Comparative ultrastructure of the zoospores of eight species of *Characium* (Chlorophyta). Archiv für Protistenkunde **143**:63–73.

Fott, B. 1971. Algenkunde, second edition. VEB Gustuv Fischer Verlag, Jena, Germany.

Freyer, R., B. Hoch, K. Neckermann, R. M. Maier, and H. Kossel. 1993. RNA editing in maize chloroplasts is a processing step independent of splicing and cleavage to monocistronic mRNAS. Plant Journal **4**:621–629.

Friedl, T. 1995. Inferring taxonomic positions and testing genus level assignments in coccoid green lichen algae: a phylogenetic analysis of 18S ribosomal RNA sequences from *Dictyochloropsis reticulata* and from members of the genus *Myrmecia* (Chlorophyta, Trebouxiophyceae *cl. nov.*). Journal of Phycology **31**:632–639.

Friedl, T. 1996. Evolution of the polyphyletic genus *Pleurastrum* (Chlorophyta): inferences from nuclear-encoded ribosomal DNA sequences and motile cell ultrastructure. Phycologia **34**:456–469.

Friedl, T., and C. Zeltner. 1994. Assessing the relationships of some coccoid green lichen algae and the Microthamniales (Chlorophyta) with 18S ribosomal RNA gene sequence comparisons. Journal of Phycology **30**:500–506.

Fritsch, F. E. 1935. The Structure and Reproduction of the Algae, Vol. 1. Cambridge University Press, London.

Gerrath, J. F. 1993. The biology of desmids: a decade of progress. In Progress in Phycological Research, Vol. 9, eds. F. E. Round and D. J. Chapman, pp. 79–193. Biopress, Bristol, England.

Gordon, J., R. Rumpf, S. L. Shank, D. Vernon, and C. W. Birky, Jr. 1995. Sequences of the rrn 18 genes of *Chlamydomonas humicola* and *C. dysosmos* are identical, in agreement with their combination in the species *C. applanata* (Chlorophyta). Journal of Phycology **31**:312–313.

Graham, L. E. 1982. The occurrence, evolution, and phylogenetic significance of parenchyma in *Coleochaete* Breb. American Journal of Botany **69**:447–454.

Graham, L. E. 1983. *Coleochaete:* advanced green alga or primitive embryophyte? American Journal of Botany (suppl.) **70**:5.

Graham, L. E. 1993. Origin of Land Plants. Wiley, New York.

Graham, L. E., and L. W. Wilcox. 1983. The occurrence and phylogenetic significance of putative placental transfer cells in the green alga *Coleochaete*. American Journal of Botany **70**:113–120.

Graham, L. E., C. F. Delwiche, and B. Mishler. 1991. Phylogenetic connections between the 'green algae' and the 'bryophytes'. Advances in Bryology **4**:213–244.

Grambast, L. 1974. Phylogeny of the Charophyta. Taxon **23**:463–481.

Gunderson, J. H., H. Elwood, A. Ingold, K. Kindle, and M. L. Sogin. 1987. Phylogenetic relationships between chlorophytes, chrysophytes, and oomycetes. Proceedings of the National Academy of Sciences U.S.A. **84**:5823–5827.

Halanych, K. M. 1991. 5S ribosomal RNA sequences inappropriate for phylogenetic reconstruction. Molecular Biology and Evolution **22**:160–174.

Hasegawa, M., H. Kishino, and T. Yano. 1985. Dating the ape-human split by a molecular clock of mitochondrial DNA. Journal of Molecular Evolution **22**:160–174.

Hennig, W. 1965. Phylogenetic systematics. Annual Review of Entomology **10**:97–110.

Hennig, W. 1966. Phylogenetic Systematics. University of Illinois Press, Urbana.

Hillis, D. M., and M. T. Dixon. 1991. Ribosomal DNA: molecular evolution and phylogenetic inference. Quarterly Review of Biology **66**:411–453.

Hillis, D. M., J. P. Huelsenbeck, and D. Swofford. 1994. Hobgoblin of phylogenetics. Nature **369**:363–364.

Hoham, R. W. 1980. Unicellular chlorophytes-snow algae. **In** Phytoflagellates, ed. E. R. Cox, pp. 61–84. Elsevier, North Holland, Amsterdam.

Hori, H., and S. Osawa. 1987. Origin and evolution of organisms as deduced from 5S ribosomal RNA sequences. Molecular Biology and Evolution **4**:445.

Hori, H., B.-L. Lim, and S. Osawa. 1985. Evolution of green plants as deduced from 5S rRNA sequences. Proceedings of the National Academy of Sciences U.S.A. **82**:820.

Hunt, B. E., and D. F. Mandoli. 1996. A new, artificial seawater that facilitates growth of large numbers of cells of *Acetabularia acetabulum* (Chlorophyta) and reduces the labor inherent in cell culture. Journal of Phycology **32**:483–495.

Huss, V. A. R., and M. L. Sogin. 1990. Phylogenetic position of some *Chlorella* species within the Chlorococcales based upon complete small-subunit ribosomal RNA sequences. Journal of Molecular Evolution **31**:432–442.

Huss, V. A. R., G. Huss, and E. Kessler. 1989. Deoxyribonucleic acid reassociation and interspecies relationships of the genus *Chlorella* (Chlorophyceae). Plant Systematics and Evolution **168**:71–82.

Huss, V. A. R., and H. D. Kranz. 1998. Charophyte evolution and the origin of land plants. **In** The Origins of the Algae and Their Plastids, ed. D. Bhattacharya, in press. Springer-Verlag, Wien, Austria.

Innis, M. A., D. H. Gefland, J. J. Sninsky, and T. J. White, eds. 1990. PCR Protocols: A Guide to Methods and Applications. Academic Press, New York.

Iyengar, M. O. P., and T. V. Desikachary. 1981. Volvocales. Indian Council of Agricultural Research, New Dehli, India.

Kanno, A., and A. Hirai. 1993. A transcription map of the chloroplast genome from rice (*Oryza sativa*). Current Genetics **23**:166–174.

Kantz, T. S., E. C. Theriot, E. A. Zimmer, and R. L. Chapman. 1990. The Pleurastrophyceae and Micromonadophyceae: a cladistic analysis of nuclear rRNA sequence data. Journal of Phycology **23**:633–638.

Kessler, E. 1984. A general review on the contribution of chemotaxonomy to the systematics of green algae. **In** Systematics of the Green Algae, eds. D. E. G. Irvine and J. M. John, pp. 391–408. Academic Press, London.

Kim, Y.-S., H. Oyaizu, S. Matsumoto, M. M. Watanabe, and H. Nozaki. 1994. Chloroplast small-subunit ribosomal RNA gene sequence from *Chlamydomonas parkeae* (Chlorophyta): molecular phylogeny of a green alga with a peculiar pigment composition. European Journal of Phycology **29**:213–217.

Kimura, M. 1980. A simple model for estimating evolutionary rates of base substitutions through comparative studies of nucleotide sequences. Journal of Molecular Evolution **16**:111–120.

Kluge, A. J. 1989. A concern for evidence and a phylogenetic hypothesis of relationships among *Epicrates* (Boidae, Serpentes). Systematic Zoology **38**:7–25.

Kooistra, W. H. C. F., W. T. Stam, J. L. Olsen, and C. van den Hoek. 1992. Biogeography of *Cladophoropsis membranacea* (Chlorophyta) based on comparisons of nuclear rDNA ITS sequences. Journal of Phycology **28**:660–668.

Kooistra, W. H. C. F., J. L. Olsen, W. T. Stam, and C. van den Hoek. 1993. Problems related to species sampling in phylogenetic studies: an example of non-monophyly in *Cladophoropsis* and *Struvea* (Siphonocladales, Chlorophyta). Phycologia **32**:419–428.

Kouwets, F. A. 1991. Mitosis in the coenocytic coccoid green alga *Dictyochloris fragrans* Vischer ex Starr (Chlorellales, Chlorophyta). Cryptogamic Botany **2/3**:104–114.

Kouwets, F. A. C., and P. F. M. Coesel. 1984. Taxonomic revision of the conjugatophycean family Peniaceae on the basis of cell wall ultrastructure. Journal of Phycology **20**:555–562.

Kranz, H. D., and V. A. R. Huss. 1996. Molecular evolution of pteridophytes and their relationship to seed plants: evidence from complete 18S rRNA gene sequences. Plant Systematics and Evolution **202:**1–11.

Kranz, H. D., D. Mikš, M.-L. Siegler, I. Capesius, C. W. Sensen, and V. A. R. Huss. 1995. The origin of land plants: phylogenetic relationships among charophytes, bryophytes, and vascular plants inferred from complete small-subunit ribosomal RNA gene sequences. Journal of Molecular Evolution **41:**74–84.

Krishnan, S., S. Barnabas, and J. Barnabas. 1990. Interrelationships among major protistan groups based on a parsimony network of 5S rRNA sequences. Biosystems **24:**135–144.

Lanyon, S. M. 1988. The stochastic mode of molecular evolution: what consequences for systematic investigations? The Auk **105:**565–573.

Larson, A., M. M. Kirk, and D. L. Kirk. 1991. Molecular phylogeny of the volvocine flagellates. Molecular Biology and Evolution **9:**85–102.

Lee, K. W., and H. C. Bold. 1974. Phycological Studies XII. *Characium* and Some *Characium*-like Algae, Publication 7403. University of Texas, Austin.

Lew, K. A., and J. R. Manhart. 1993. The *rps12* gene in *Spirogyra maxima* (Chlorophyta): its evolutionary significance. Journal of Phycology **29:**500–505.

Lewis, L. A. 1997. Diversity and phylogenetic placement of *Bracteacoccus* Tereg (Chlorophyceae, Chlorophyta) based on 18S rRNA gene sequence data. Journal of Phycology **33:**279–285.

Lewis, L. A., L. W. Wilcox, P. A. Fuerst, and G. L. Floyd. 1992. Concordance of molecular and ultrastructural data in the study of zoosporic chlorococcalean green algae. Journal of Phycology **28:**375–380.

Li, H., U. B. Gyllensten, X. Cui, X. F. R. K. Saiki, H. A. Erlich, and N. Arnheim. 1988. Amplification and analysis of DNA sequences in single human sperm and diploid cells. Nature **335:**414–417.

Li, W.-H., K. H. Wolfe, J. Sourdis, and P. M. Sharp. 1987. Reconstruction of phylogenetic trees and estimation of divergence times under nonconstant rates of evolution. Cold Spring Harbor Symposia on Quantitative Biology **52:**847–856.

Lokhorst, G. M., and G. P. J. Rongen. 1994. Comparative ultrastructural studies of division processes in the terrestrial green alga *Leptosira erumpens* (Deason et Bold) Lukeshová confirm the ordinal status of the Pleurastrales. Cryptogamic Botany **4:**394–409.

Manhart, J. 1994. Phylogeny of green plants based on *rbcL* sequences. Molecular Biology and Evolution **3:**114–127.

Manhart, J. R., and J. D. Palmer. 1990. The gain of two chloroplast tRNA introns marks the green algal ancestors of land plants. Nature **345:**268–270.

Manhart, J. R., R. W. Hoshaw, and J. D. Palmer. 1990. Physical and gene mapping of *Spirogyra* cpDNA reveals a unique chloroplast genome. Journal of Phycology **26:**490–494.

Mattox, K. R., and K. D. Stewart. 1984. The classification of the green algae: a concept based on comparative cytology. In The Systematics of the Green Algae, eds. D. E. G. Irvine and D. M. John, pp. 29–72. Academic Press, London.

McCourt, R. M. 1995. Green algal phylogeny. Trends in Ecology and Evolution **10:**159–163.

McCourt, R. M., K. G. Karol, S. Kaplan, and R. W. Hoshaw. 1995. Using *rbcL* sequences to test hypotheses of chloroplast and thallus evolution in conjugating green algae (Zygnematales, Charophyceae). Journal of Phycology **31:**989–995.

McCourt, R. M., K. G. Karol, M. Guerlisquine, and M. Feist. 1996a. Phylogeny of extant genera in the family Characeae (Division Charophyta) based on *rbcL* sequences and morphology. American Journal of Botany **83:**125–131.

McCourt, R. M., S. Meiers, K. G. Karol, and R. L. Chapman. 1996b. Molecular systematics of the Charales. In Cytology, Genetics and Molecular Biology of Algae, eds. D. B. Chaudhary and S. B. Agrawal, pp. 323–336. SPB Publishing, Amsterdam.

Melkonian, M. 1982a. Two different types of motile cells within the Chlorococcales and Chlorosarcinales: taxonomic implications. British Phycological Journal **17:**236.

Melkonian, M. 1982b. Systematics and evolution of the algae. In Progress in Botany, Vol. 44, eds. H. Ellenberg, K. Esser, K. Kubitzki, E. Schnepf, and H. Ziegler, pp. 315–344. Springer-Verlag, New York.

Melkonian, M. 1990a. Phylum Chlorophyta; Class Prasinophyceae. In Handbook of Protoctista, eds. L. Margulis, J. O. Corliss, M. Melkonian, and D. J. Chapman, pp. 641–648. Jones & Bartlett Publishers, Boston.

Melkonian, M. 1990b. Phylum Chlorophyta, Class Chlorophyceae. In Handbook of Protoctista, eds. L. Margulis, J. O. Corliss, M. Melkonian, and D. J. Chapman, pp. 608–616. Jones & Bartlett Publishers, Boston.

Melkonian, M. 1990c. Chlorophyte orders of uncertain affinities: Order Microthamniales. In Handbook of Protoctista, eds. L. Margulis, J. O. Corliss, M. Melkonian, and D. J. Chapman, pp. 652–654. Jones & Bartlett Publishers, Boston.

Melkonian M., and B. Berns. 1983. Zoospore ultrastructure in the green alga *Friedmannia israelensis:* an absolute configuration analysis. Protoplasma **114:**67–84.

Melkonian, M., and E. Peveling. 1988. Zoospore ultrastructure in species of *Trebouxia* and *Pseudotrebouxia* (Chlorophyta). Plant Systematics and Evolution **158:**183–210.

Melkonian, M., and B. Surek. 1995. Phylogeny of the Chlorophyta: congruence between ultrastructural and molecular evidence. Bulletin de la Société de Zoologique de France **120:**191–208.

Mishler, B. D., and S. P. Churchill. 1985. Transition to a land flora: phylogenetic relationships of the green algae and bryophytes. Cladistics **1:**305–328.

Mishler, B. D., and S. P. Churchill. 1987. Transition to a land flora: a reply. Cladistics **3:**66–71.

Mishler, B. D., L. A. Lewis, M. A. Buchheim, K. S. Renzaglia, D. J. Garbary, C. F. Delwiche, F. W. Zechman, T. S. Kantz, and R. L. Chapman. 1994. Phylogenetic relationships of the "green algae" and "bryophytes." Annals of the Missouri Botanical Garden **81**:451–483.

Mix, M. 1972. Die Feinstruktur der Zellwände über Mesotaeniaceae und Gonatozygaceae mit einer vergleichenden Betrachtung der verschiedenen Wandtypen der Conjugatophyceae und ber deren systematischen Wert. Archives für Microbiologie **81**:197–220.

Mix, M. 1973. Die Feinstruktur der Zellwände der Conjugaten und ihre systematische Bedeutung. Nova Hedwigia Beihefte **42**:179–194.

Molnar, K. E., K. D. Stewart, and K. R. Mattox. 1975. Cell division in the filamentous *Pleurastrum* and its comparison with the unicellular *Platymonas* (Chlorophyceae). Journal of Phycology **11**:287–296.

Nakayama, T., S. Watanabe, K. Mitsui, H. Uchida, and I. Inouye. 1996. The phylogenetic relationships between the Chlamydomonadales and Chlorococcales inferred from 18S rDNA sequence data. Phycological Research **44**:47–56.

Norton, T. A., M. Melkonian, and R. A. Andersen. 1996. Algal biodiversity. Phycologia **35**:308–326.

Nozaki, H., M. M. Watanabe, and K. Aizawa. 1995. Morphology and paedogamous sexual reproduction in *Chlorogonium capillatum* sp. nov. (Volvocales, Chlorophyta). Journal of Phycology **31**:655–663.

O'Kelly, C. J. 1992. Flagellar apparatus architecture and the phylogeny of "green" algae: chlorophytes, euglenoids, glaucophytes. In The Cytoskeleton of the Algae, ed. D. Menzel, pp. 315–345. CRC Press, Boca Raton, Florida.

O'Kelly, C. J., and G. L. Floyd. 1984. Correlations among patterns of sporangial structure and development, life histories and ultrastructural features in the Ulvophyceae. In Systematics of the Green Algae, eds. D. E. G. Irvine and D. M. John, pp. 121–156. Academic Press, London.

Olsen, G. J., H. Matsuda, R. Hagstrom, and R. Overbeek. 1994a. fastDNAml: a tool for construction of phylogenetic trees of DNA sequences using maximum likelihood. CABIOS **10**:41–48.

Olsen, J. L., W. T. Stam, S. Berger, and D. Menzel. 1994b. 18S rDNA and evolution in the Dasycladales (Chlorophyta): modern living fossils. Journal of Phycology **30**:729–744.

Olsen-Stojkovich, J. L., J. A. West, and J. M. Lowenstein. 1986. Phylogenetics and biogeography in the Cladophorales complex (Chlorophyta): some insights from immunological distance data. Botanica Marina **29**:239–249.

Pääbo, S. 1989. Ancient DNA: extraction, characterization, molecular cloning, and enzymatic amplification. Proceedings of the National Academy of Sciences U.S.A. **86**:1939.

Palmer, J. D. 1991. Plastid chromosomes: structure and evolution. In The Molecular Biology of Plastids, eds. L. Bogorad and I. K. Vasil, pp. 5–53. Academic Press, San Diego.

Park, N. E., K. G. Karol, R. W. Hoshaw, and R. M. McCourt. 1996. Phylogeny of *Gonatozygon* and *Genicularia* (Gonatozygaceae, Desmidiales) based on *rbcL* sequences. European Journal of Phycology **31**:309–313.

Pickett-Heaps, J. D. 1979. Electron microscopy and the phylogeny of green algae and land plants. American Zoologist **19**:545–554.

Pickett-Heaps, J. D., and H. J. Marchant. 1972. The phylogeny of the green algae; a new proposal. Cytobios **6**:255–264.

Ragan, M. A., T. J. Parsons, T. Sawa, and N. A. Straus. 1994. 18S ribosomal DNA sequences indicate a monophyletic origin of Charophyceae. Journal of Phycology **30**:490–500.

Rausch, H., N. Larsen, and R. Schmitt. 1989. Phylogenetic relationships of the green alga *Volvox carteri* deduced from small-subunit ribosomal RNA comparisons. Journal of Molecular Evolution **29**:255–265.

Rumpf, R., D. Vernon, D. Schreiber, and C. W. Birky, Jr. 1996. Evolutionary consequences of the loss of photosynthesis in Chlamydomonadaceae: phylogenetic analysis of Rrn 18 (18S rDNA) in 13 *Polytoma* strains (Chlorophyta). Journal of Phycology **32**:119–126.

Saitou, N., and M. Nei. 1987. The neighbor-joining method: a new method for reconstructing phylogenetic trees. Molecular Biology and Evolution **4**:406–425.

Schopf, J. W. 1970. Pre-Cambrian micro-organisms and evolutionary events prior to the origin of vascular plants. Biological Review **45**:319–352.

Sluiman, H. J. 1983. The flagellar apparatus of the zoospore of the filamentous green alga *Coleochaete pulvinata:* absolute configuration and phylogenetic significance. Protoplasma **115**:160–175.

Sluiman, H. J. 1989. The green algal class Ulvophyceae—an ultrastructural survey and classification. Cryptogamic Botany **1**:83–94.

Sluiman, H. J. 1991. Cell division in *Gloeotilopsis planctonica,* a newly identified ulvophycean alga (Chlorophyta) studied by freeze fixation and freeze substitution. Journal of Phycology **27**:291–298.

Solms-Laubach, H. 1895. Graf zu: Monograph of the Acetabularieae. Transactions of the Linnean Society of Botany London, 2nd Ser. Bot. **5**:1–39 (pls. 1–4).

Stanley, S. M. 1979. Macroevolution. W. H. Freeman, San Francisco.

Starr, R. 1955. A Comparative Study of *Chlorococcum meneghini* and Other Spherical, Zoospore-Producing Genera of the Chlorococcales. Indiana University Press, Bloomington.

Starr, R. C., and J. A. Zeikus. 1993. UTEX—the culture collection of algae at the University of Texas at Austin 1993 list of cultures. Journal of Phycology (suppl.) **29**:1–106.

Steele, K. P., K. E. Holsinger, R. K. Jansen, and D. W. Taylor. 1991. Assessing the reliability of 5S rRNA sequence data for phylogenetic analysis in green plants. Molecular Biology and Evolution **8**:240–248.

Steinkötter, J., D. Bhattacharya, I. Semmelroth, C. Bibeau, and M. Melkonian. 1994. Prasinophytes form independent lineages within the Chlorophyta: evidence from ribosomal RNA sequence comparisons. Journal of Phycology **30**:340–345.

Sugiura, M. 1992. The chloroplast genome. Plant Molecular Biology **19**:149–168.

Surek, B., U. Beemelmanns, M. Melkonian, and D. Bhattacharya. 1993. Ribosomal RNA sequence comparisons demonstrate an evolutionary relationship between Zygnematales and charophytes. Plant Systematics and Evolution **191**:171–181.

Swofford, D. L. 1993. PAUP: Phylogenetic Analysis Using Parsimony, version 3.1. Computer program distributed by the Illinois Natural History Survey, Champaign, Illinois.

Swofford, D. L. 1998. PAUP*: Phylogenetic Analysis Using Parsimony, version 4.0. Sinauer, Sunderland, Massachusetts, in press.

Tanaka, M., T. Wakasugi, M. Sugita, K. Shinozaki, and M. Sugiura. 1986. Genes for the eight ribosomal proteins are clustered on the chloroplast genome of tobacco (*Nicotiana tabacum*): similarity to the S10 and spc operons of *Escherichia coli*. Proceedings of the National Academy of Sciences U.S.A. **83**:6030–6034.

Taylor, T. N. 1982. The origin of land plants: a paleobotanical perspective. Taxon **31**:155–177.

Theriot, E. 1989. Phylogenetic systematics for phycology. Journal of Phycology **25**:407–411.

Theriot, E. 1992. Phylogenetic systematics, the theory and practice of taxonomy. In Progress in Phycological Research, Vol. 8, eds. F. E. Round and D. J. Chapman, pp. 179–207. Biopress, Bristol, England.

Valet, B. 1979. Essai evolutif et phylogénétique des Dasycladales actuelles. Bulletin des Centres de Recherches Exploration—Production Elf-Aquitaine **3**:855–857.

van den Hoek, C., W. T. Stam, and J. L. Olsen. 1992. Phylogenetic changes in peroxisomes of algae: phylogeny of plant peroxisomes. In The Chlorophyta: Systematics and Phylogeny, ed. H. Stabenau, pp. 330–362. University of Oldenburg, Germany.

van de Peer, Y., R. De Baere, J. Cauwenberghs, and R. DeWachter. 1990. Evolution of green plants and their relationship with other photosynthetic eukaryotes as deduced from 5S ribosomal RNA sequences. Plant Systematics and Evolution **170**:85–96.

Watanabe, S., and G. L. Floyd. 1989. Comparative ultrastructure of the zoospores of nine species of *Neochloris* (Chlorophyta). Plant Systematics and Evolution **168**:195–219.

Watanabe, S., and G. L. Floyd. 1992. Comparative ultrastructure of zoospores with parallel basal bodies from the green algae *Dictyochloris fragrans* and *Bracteacoccus* sp. American Journal of Botany **79**:551–555.

Watanabe, S., D. F. Chapell, and G. L. Floyd. 1992. Ultrastructure of the flagellar apparatus of the gametes of *Gongrosira papuasica* (Chlorophyta). British Phycological Journal **27**:21–28.

Watanabe, S., T. Nakayama, K. Mitsui, I. Inouye, and G. L. Floyd. 1995. Systematics of the coccoid green algae: a review based primarily on ultrastructural features. Journal of Phycology (suppl.) **31**:5.

Whittemore, A. T. 1987. Transition to a land flora: a critique. Cladistics **3**:60–65.

Wilcox, L. W., L. A. Lewis, P. A. Fuerst, and G. L. Floyd. 1992. Assessing the relationships of autosporic and zoosporic chlorococcalean green algae with 18S rDNA sequence data. Journal of Phycology **28**:381–386.

Wilcox, L. W., P. A. Fuerst, and G. L. Floyd. 1993. Phylogenetic relationships of four charophycean green algae inferred from complete nuclear-encoded small subunit rRNA gene sequences. American Journal of Botany **80**:1028–1033.

Wood, R. D., and K. Imahori. 1965. Monograph of the Characeae. First Part of a Revision of the Characeae. Verlag von J. Cramer, Weinheim, Germany.

Zechman, F. W., E. C. Theriot, E. A. Zimmer, and R. L. Chapman. 1990. Phylogeny of the Ulvophyceae (Chlorophyta): cladistic analysis of nuclear-encoded rRNA sequence data. Journal of Phycology **26**:700–710.

Zeller, A., and D. F. Mandoli. 1993. Growth of *Acetabularia acetabulum* (Dasycladales, Chlorophyta) on solid substrata at specific cell densities. Phycologia **32**:136–142.

19

Phylogenetic Studies
of Extant Pteridophytes

*Paul G. Wolf, Kathleen M. Pryer, Alan R. Smith,
and Mitsuyasu Hasebe*

Pteridophytes represent the most poorly understood group of vascular plants from a phylogenetic perspective (Stewart and Rothwell, 1993). The group is probably polyphyletic and includes four extant divisions (following Cronquist et al., 1966): Polypodiophyta (ferns), Psilotophyta (Psilotaceae, or whisk ferns), Lycopodiophyta (lycopods), and Equisetophyta (horsetails). Estimating phylogenetic relationships among these groups, and their relationship to seed plants and to many extinct groups of land plants, remains one of the greatest challenges in plant systematics. In this chapter we review some of the literature bearing on relationships among pteridophytes, focusing on studies of ferns. We also present an exploratory analysis, using nucleotide sequences from three genes and data from 77 morphological characters, to examine the feasibility of a combined approach to inferring pteridophyte phylogeny. We then discuss the problems associated with resolving ancient divergence events and with analyzing large and diverse data sets. We conclude with what we believe to be the most fertile directions for future research on pteridophyte phylogeny.

An improved phylogenetic framework of pteridophytes is required for developing classifications of land plants that reflect evolutionary history. Phylogenetic hypotheses are also necessary for understanding the sequence of events associated with major changes in vegetative morphology (Kenrick and Crane, 1991), reproductive characters (Crane, 1990; Lugardon, 1990; Kenrick, 1994), ecology (Brooks and McLennan, 1991), life histories (Stearns, 1992), and habitat (Mishler and Churchill, 1985). Robust phylogenies also allow us to learn more about evolution at the molecular level (Avise et al., 1994; Sharp et al., 1995), which in turn can provide more realistic models for inferring phylogeny using molecular data (Swofford et al., 1996). Moreover, some authors have argued that phylogenetic hypotheses permit inferences on aspects of evolutionary processes themselves such as patterns of speciation, biogeography, and adaptation (Harvey and Pagel, 1991; Philippe and Adoutte, 1996). Thus, improving the resolution of pteridophyte phylogeny will enhance our understanding of vascular plant diversification and the evolution of terrestrial ecosystems.

Much of our understanding of vascular plant relationships comes from studies of the fossil record. For example, characteristics of pteridosperm fossils strongly suggest that ferns and seed plants share a common ancestor (Stewart,

Thanks to M. Chase and V. Savolainen for an unpublished *atpB* sequence from *Ginkgo*. This work was supported financially by grants from the National Science Foundation (DEB-9615533 to KMP, DEB-9707087 to PGW, and DEB-9616260 to ARS).

1981). Incorporating data from both extant and fossil taxa should provide the most informative approach to resolving relationships (Rothwell, 1994; Rothwell and Stockey, 1994; Smith and Littlewood, 1994). However, our goal here is to focus initially on developing better phylogenetic hypotheses for extant taxa. Because this synthesis will be based partly on morphological characters, subsequent studies that include fossil taxa should then become more feasible (Donoghue et al., 1989; Doyle et al., 1994; Huelsenbeck, 1994).

NONMOLECULAR PHYLOGENETIC HYPOTHESES FOR PTERIDOPHYTES

Detailed overviews of pteridophyte phylogenetic studies and taxonomic treatments can be found elsewhere (Takhtajan, 1953; Foster and Gifford, 1974; Crane, 1990; Kramer, 1991; Stewart and Rothwell, 1993; Kenrick, 1994; Kenrick and Friis, 1995; Kenrick and Crane, 1997). The view that pteridophytes are not monophyletic developed long before the application of formal cladistic approaches. For example, Jeffrey (1902) recognized two phyla of vascular plants: the Lycopsida (including Psilotaceae, lycopods, and horsetails) and the Pteropsida (ferns and seed plants). Although Jeffrey (1902) acknowledged that his Lycopsida was unnatural (not monophyletic), it was not until the work of Eames (1936), Smith (1938), and Zimmerman (1959) that the current four-taxon system for pteridophytes (sensu Cronquist et al., 1966; Kramer and Green, 1990) was adopted. Many variations on this basic four-group theme have emerged, the most evident (and perhaps phylogenetically accurate) being the inclusion of ferns and seed plants as a taxon (e.g., Jeffrey, 1902). More recently, Bierhorst (1968, 1977) suggested that Psilotaceae (*Psilotum* and *Tmesipteris*) are closely related to some leptosporangiate ferns, rather than representing a separate and old lineage of vascular plants. Bierhorst's evidence came from the structure and development of gametophytes, embryos, "fronds," and stems, and the similarities of these characters in Psilotaceae and the fern genera *Stromatopteris* and *Actinostachys*. However, other pteridologists have disagreed with these interpretations (Kaplan, 1977; Wagner, 1977); as a result, Bierhorst's phylogenetic hypotheses were never incorporated into taxonomic treatments other than his own (Bierhorst, 1971). Another phylogenetic hypothesis for pteridophytes was proposed by Kato (1983) in which the lycopods and Psilotaceae together form a subdivision, as do ferns plus horsetails. This "biphyletic" classification was also based on a reevaluation of comparative morphology, but again without a formal cladistic analysis of characters. Given the wide range of treatments for pteridophytes it is no wonder that the most recent classification does not propose relationships among the four classes (Kramer and Green, 1990). Alternatively, if the four groups all diverged more or less simultaneously in geologic time, then lack of resolution may be an accurate reflection of relationships.

Somewhat independent of taxonomic treatments, several phylogenetic analyses of vascular plants have been conducted. We discuss studies using nonmolecular data, followed by a review of molecular studies. Parenti (1980) used 24 cellular, morphological, and anatomical characters to generate a cladogram of the major groups of land plants, including fossil taxa. The branching order among extant vascular plants was (Psilotaceae (lycopods (horsetails (ferns (seed plants))))). Thus, ferns are sister to extant seed plants and Psilotaceae are sister to all other extant vascular plants. Bremer et al. (1987) used 88 morphological characters and obtained the same branching order of extant vascular plant groups. Crane (1990) built on the data of Bremer et al. (1987), but presented an unresolved tree for the taxa of concern here. In Crane's analysis, Psilotaceae were included with true ferns (as noted above, a classification proposed by Bierhorst, 1971), and ferns, horsetails, and seed plants formed an unresolved trichotomy, leaving lycopods as sister to other extant vascular plants. Similar trees were developed by Kenrick (1994) and Kenrick and Friis (1995), where the focus was the inclusion of extinct and extant taxa, as it was by Crane (1990). The most notable departure from hypotheses discussed so far was presented by Garbary et al. (1993) in a cladistic study of land plants using strictly characters of male gametogenesis. Their results are controversial because they infer polyphyly of lycopods (*Selaginella* is sister to a bryophyte clade), with the remaining vascular plants branching as follows: (*Lycopodium* (*Marsilea* ((horsetails + ferns) seed plants))). *Psilotum* was not included in their analysis.

Ongoing studies by Rothwell (1998) involve cladistic analysis of 101 morphological characters of ferns and fernlike plants (including many fossil groups). The results suggest that ferns sensu lato are polyphyletic, but that extant ferns represent a distinct clade. Also, Psilotaceae appear near the base of the vascular plant tree. Perhaps Rothwell's most significant finding is that there are some changes in the extant fern tree (e.g., positions of *Osmunda* and *Schizaea*) when the extinct taxa are excluded.

In summary, nonmolecular phylogenetic hypotheses for vascular plants depict rather varied phylogenetic relationships. Yet, there appears to be a general consensus on two points: that extant lycopods are the sister to other extant vascular plants (Banks, 1975; DiMichele and Skog, 1992) and that ferns and seed plants form a monophyletic group.

NONMOLECULAR PHYLOGENETIC HYPOTHESES FOR FERNS

Many evolutionary trees (phyletic schemes), as well as evolutionarily-based classifications, have been proposed for the approximately 30 families of ferns (e.g., Ching, 1940; Wagner, 1969; Holttum, 1973; Mickel, 1974; Crabbe et al., 1975; Lovis, 1977; Pichi Sermolli, 1977; Kramer and Green, 1990). These schemes were based mostly on morphological (but also cytological) characters, and despite the many differences among schemes, there is consensus on some points. Details of the characters used and areas of concordance among schemes were most recently provided by Smith (1995). Seventeen categories of characters were discussed; the most widely used have been sorus position; indusial presence, shape, and orientation; spore structure; rhizome structure; root anatomy; and stipe vasculature. Many of the differences among phyletic schemes arise from emphases on different characters. Nevertheless, four areas of agreement were noted among these systems: (1) Ophioglossaceae

and Marattiaceae are distant relatives of leptosporangiate ferns. (2) About ten families are regarded as "primitive," i.e., branching early with respect to other families. These include: Osmundaceae, Schizaeaceae, Gleicheniaceae, Matoniaceae, Dipteridaceae, Plagiogyriaceae, Loxomataceae, Hymenophyllaceae, Dicksoniaceae, and Cyatheaceae. (3) Families considered to have arisen more recently include Dennstaedtiaceae, Pteridaceae, Vittariaceae, Polypodiaceae, Grammitidaceae, Thelypteridaceae, Dryopteridaceae, Aspleniaceae, and Blechnaceae. (4) The "higher" leptosporangiate ferns (see foregoing discussion) have had more than one origin, with Pteridaceae "derived" from schizaeoid stock, Polypodiaceae and Grammitidaceae from gleichenioid progenitors, and most of the remaining families from dennstaedtioid ancestors.

In addition to the partly intuitive "phyletic schemes" mentioned above, a few studies have attempted to estimate fern phylogeny more objectively using morphological data. The first was by Wagner (1969), who used his ground plan/divergence method. This approach examined character distributions within groups to identify "primitive" versus "specialized" conditions, and then summed across characters to develop a relative score for each group, from which a tree was generated. The first study to use parsimony methods to analyze morphological data in ferns was by Pryer et al. (1995). That study was based on 77 parsimony-informative characters, and the results of one of their analyses (analysis 1A of Pryer et al., 1995) is reproduced in Figure 19.1. Unlike Wagner's (1969) analysis, the more recent study made no a priori assumptions about ontogeny or whether characters are ancestral or derived. Character polarity was only inferred later from the most parsimonious trees. Several aspects of Pryer et al.'s (1995) tree are consistent with that of Wagner's and with the general consensus of previous works discussed by Smith (1995). Ophioglossaceae and Marattiaceae diverged near the base of the fern tree. The general positions of the ten "primitive" families were basal, and the "advanced families" were sister to a clade that included the tree ferns. Thus, although the "higher" fern families were more recently derived than the "primitive" families, the former

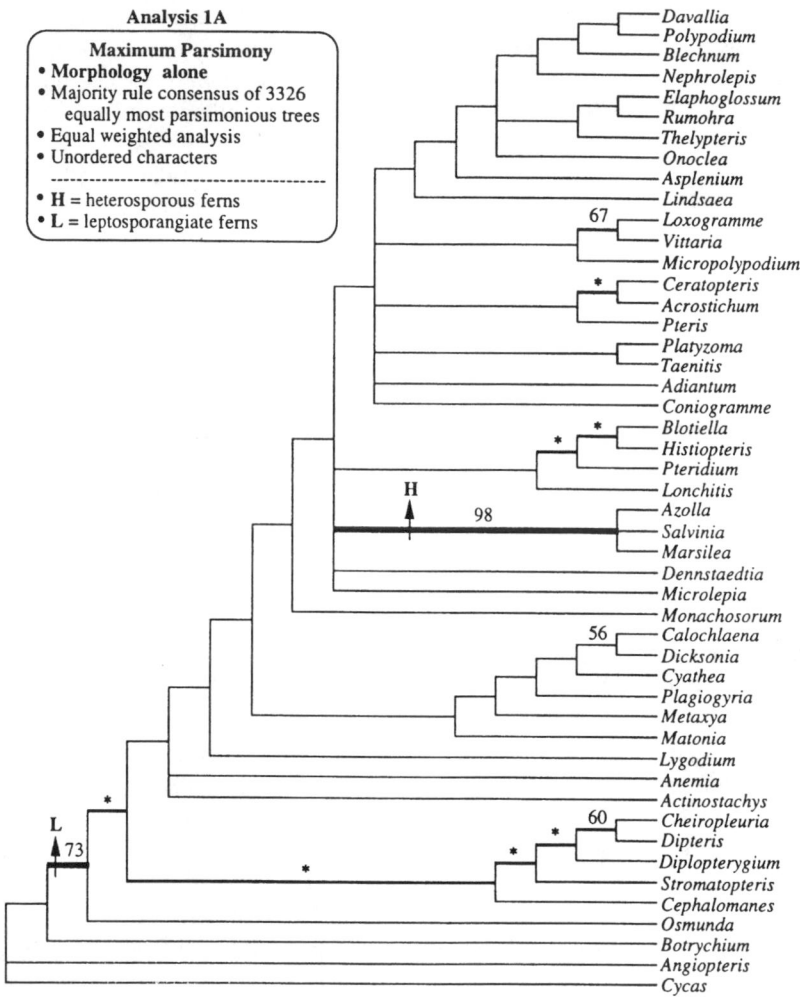

Figure 19.1. Analysis 1A (fig. 2, p. 222) of Pryer et al. (1995). Majority rule consensus of 3,326 most parsimonious trees at 449 steps based on 77 parsimony-informative characters that were equally weighted and unordered. Numbers above branches indicate bootstrap percent values. Asterisks denote branches also found in the strict consensus, but not the bootstrap majority rule tree. Tree was rooted with *Cycas;* CI = 0.325; RI = 0.541. Figure reproduced with permission from the American Fern Society.

were monophyletic rather than polyphyletic, as suggested by previous workers. Also, the heterosporous ferns appeared to be monophyletic, and Dryopteridaceae were paraphyletic (Pryer et al., 1995). As will be seen later, this phylogenetic analysis based on morphology agreed more than it conflicted with analyses based on molecular data.

Another recent cladistic analysis of morphological data used 146 characters, including 22 root characters, to infer relationships among all major groups of ferns using parsimony methods

(Schneider, 1996). Several aspects of Schneider's results are different from those of Pryer et al. (1995). For example, in Schneider's (1996) analysis, *Lindsaea* is in a pteridoid clade, *Asplenium* is sister to *Polypodium, Platyzoma* is sister to *Azolla + Marsilea,* and *Plagiogyria* is at the base of the tree, above *Osmunda.* Clades that are similar in the two studies include a tree fern clade and some aspects of the dryopteroid group. Schneider (1996) used some characters not studied by Pryer et al. (1995), most notably several additional root characters. Future studies might benefit from evaluating and integrating characters from the two data matrices.

MOLECULAR SYSTEMATICS OF PTERIDOPHYTES

Stein (1985) reviewed the early work that used nucleic acid data for fern systematics, and in the same paper used reassociation kinetic data to test the hybrid origin of *Osmunda* species. In addition to these DNA hybridization studies, Stein's lab pioneered the use of restriction site data for phylogenetic studies of ferns (Yatskievych et al., 1988; Stein et al., 1989) using heterologous chloroplast DNA probes from *Lactuca* and *Petunia*. Later, probes from *Adiantum* chloroplast DNA (Hasebe and Iwatsuki, 1990a, 1990b) permitted more restriction site studies in ferns (e.g., Gastony et al., 1992; Murakami and Schaal, 1994; Conant et al., 1995; Haufler et al., 1995). The restriction site approach has provided robust data for inferring relationships among species within genera and among closely related genera, but it has not been used extensively for studies at the family level or higher. At these higher levels two approaches have been used: characterization of chloroplast DNA structure and variation of nucleotide sequences in coding genes (both chloroplast and nuclear). Structural rearrangements are relatively rare events and therefore can be powerful phylogenetic markers (Downie and Palmer, 1992; see Chapter 1). Several rearrangements have been detected in ferns and used to make phylogenetic inferences (Stein et al., 1992; Raubeson and Stein, 1995). The colinearity of gene order in *Marchantia* and lycopods supports the hypothesis (e.g., Crane, 1990) that lycopods are the sister group to all other vascular plants (Raubeson and Jansen, 1992). Although there is clear phylogenetic potential for such cpDNA structural characters, to date the number detected and the number of pteridophyte taxa surveyed are too low to make robust conclusions.

With the incorporation of nucleotide sequence data into plant systematics (e.g., Ritland and Clegg, 1987; Hamby and Zimmer, 1988), it was not long before such approaches were used by fern systematists. The first gene to be sequenced from a sufficient number of pterido-

phyte taxa for phylogenetic studies was *rbcL* (Hasebe et al., 1993). Attempts have been made to use *rbcL* sequence data to infer relationships among groups of vascular plants (Hasebe et al., 1993) and to examine relationships among all major groups of green plants (Manhart, 1994). However, at these higher levels, sites are generally saturated, producing only weak phylogenetic signal. Using amino acid sequences (Hasebe et al., 1993) or focusing on first and second codon positions (Manhart, 1994) did not solve this problem. Better phylogenetic resolution was achieved with sequences from nuclear 18S ribosomal RNA genes (Kranz et al., 1995; Kranz and Huss, 1996). In the most recent analysis, Kranz and Huss (1996) examined 22 complete sequences using maximum parsimony and maximum likelihood, comparing support for alternative topologies. The best supported trees had Lycopsida as the sister group to the remaining vascular plants and *Psilotum* as sister to the seed plants: (lycopods (ferns (horsetails (Psilotaceae (seed plants))))). However, short branches suggest that the ferns, horsetails, Psilotaceae, and seed plants radiated more or less simultaneously, perhaps from a trimerophyte ancestor (Kranz and Huss, 1996).

Hiesel et al. (1994) used mitochondrial *coxIII* (cytochrome oxidase subunit III) sequences to examine the phylogeny of land plants. Parsimony analysis resulted in *Lycopodium* as sister to a clade that included bryophytes and vascular plants. However, maximum likelihood trees were more congruent with inferences from other data: (lycopods (horsetails (Psilotaceae (ferns + seed plants)))). An alternative topology was inferred by Kolukisaoglu et al. (1995) using nuclear encoded phytochrome genes: ((lycopods + horsetails) (ferns (Psilotaceae + seed plants))). Two trees were presented by Kolukisaoglu et al. (1995), one of which includes a basal clade with bryophytes, *Selaginella,* and *Equisetum.* Many aspects of the phytochrome gene tree are consistent with other studies. However, phytochrome genes are part of a gene family (Schneider-Poetsch et al., 1994; Mathews et al., 1995), and sequences from different taxa may not be orthologous, potentially resulting in conflict between the gene tree and organismal trees.

Boivin et al. (1996) used sequences of the chloroplast *chlB* gene (which encodes a subunit of the light-independent protochlorophyllide reductase) to examine relationships among diverse land plants. The gene could not be PCR-amplified from *Ephedra* or *Psilotum.* Neighbor-joining and parsimony analyses of an internal fragment of approximately 350 bp resulted in optimal trees that were consistent with many previous analyses; bryophytes were paraphyletic, lycopods (except *Isoetes*) were the most basal vascular plant clade, and ferns appeared as a monophyletic group: (lycopods (*Equisetum* (ferns + seed plants))). However, unlike in most previous studies, *Isoetes* appeared as a sister to all land plants. The *chlB* gene appears to contain good phylogenetic signal, but failure to detect the gene in plastids of various angiosperms and critical taxa such as *Ephedra* and *Psilotum* may limit its use as a phylogenetic tool at the level of vascular plants.

Manhart (1995) examined chloroplast 16S rRNA gene sequences from 23 land plants using maximum parsimony. Phylogenetic resolution appeared to be better for the deeper nodes than with *rbcL* sequences (Manhart, 1994), but the base of the vascular plant clade was still poorly resolved; the topology was dependent on the inclusion of *Selaginella,* which was connected by a long branch. With *Selaginella* included, the vascular plant tree was (horsetails (ferns ((Psilotaceae + Ophioglossaceae) (lycopods and seed plants)))). Lycopods did not form a clade: ((*Lycopodium* + *Selaginella*) (*Isoetes* + seed plants)). With *Selaginella* removed, *Lycopodium* became the sister to the remaining vascular plants, and *Isoetes* remained sister to seed plants. The clade that included Psilotaceae and Ophioglossaceae (the eusporangiate ferns *Ophioglossum* and *Botrychium*) was reasonably well supported with a bootstrap of at least 80%. This clade had also been detected (with less support) in Manhart's (1994) *rbcL* study. More recently, Wolf (1996) detected the same clade using parsimony analysis of nucleotide data from the chloroplast gene *atpB.* The *atpB* tree with the greatest support was (*Selaginella* (horsetails (*Angiopteris* (Psilotaceae + Ophioglossaceae) (other lycopods + ferns)))). Seed plants were not included in the *atpB* analysis, and lycopods

were polyphyletic as in the 16S rDNA analysis. The Psilotaceae + Ophioglossaceae clade has not been proposed on the basis of data other than nucleotide sequences. Congruence among the three data sets (*rbcL,* 16S rDNA, and *atpB*) could be interpreted as strong support for this clade. An alternative explanation for this congruence could be long-branch attraction (Felsenstein, 1978; Hendy and Penny, 1989). Long branches on a tree can result from multiple extinctions along a lineage, increased substitution rate, or ancient divergence. Successive long branches can cause parsimony methods to converge on an incorrect tree that unites taxa with long branches together in a false clade. Under these conditions, adding more data actually increases support for the false clade. Maximum likelihood approaches should be less susceptible to these effects (Swofford et al., 1996). Preliminary maximum likelihood analyses using the *atpB* data set (P. Wolf, unpubl.) still support the Psilotaceae + Ophioglossaceae clade, suggesting that long-branch attraction may not be the explanation. Further analyses are needed for *rbcL,* 16S rDNA, and 18S rDNA to test the support for this possibly spurious clade.

MOLECULAR SYSTEMATICS OF FERNS

Although *rbcL* nucleotide data alone were not highly informative for inferring relationships among groups of vascular plants, these data have provided evidence for many well-supported clades within the ferns (Hasebe et al., 1994, 1995; Wolf et al., 1994). Pryer et al. (1995) used parsimony analysis of *rbcL* data for the same set of taxa for which they analyzed 77 morphological characters. The *rbcL* tree is reproduced in Figure 19.2, which can be compared more directly (than the other *rbcL* studies above) with the morphological analysis (Fig. 19.1). Most genera from the more derived families formed well-resolved clades, whereas basal clades were poorly supported by bootstrap analysis. The consensus *rbcL* tree is not significantly different from the consensus tree based on morphological data, including some results that differed from most (but not all) previous phylogenetic hypotheses. For example, *Polypodium* appeared closely related to *Davallia* (Davalliaceae) in

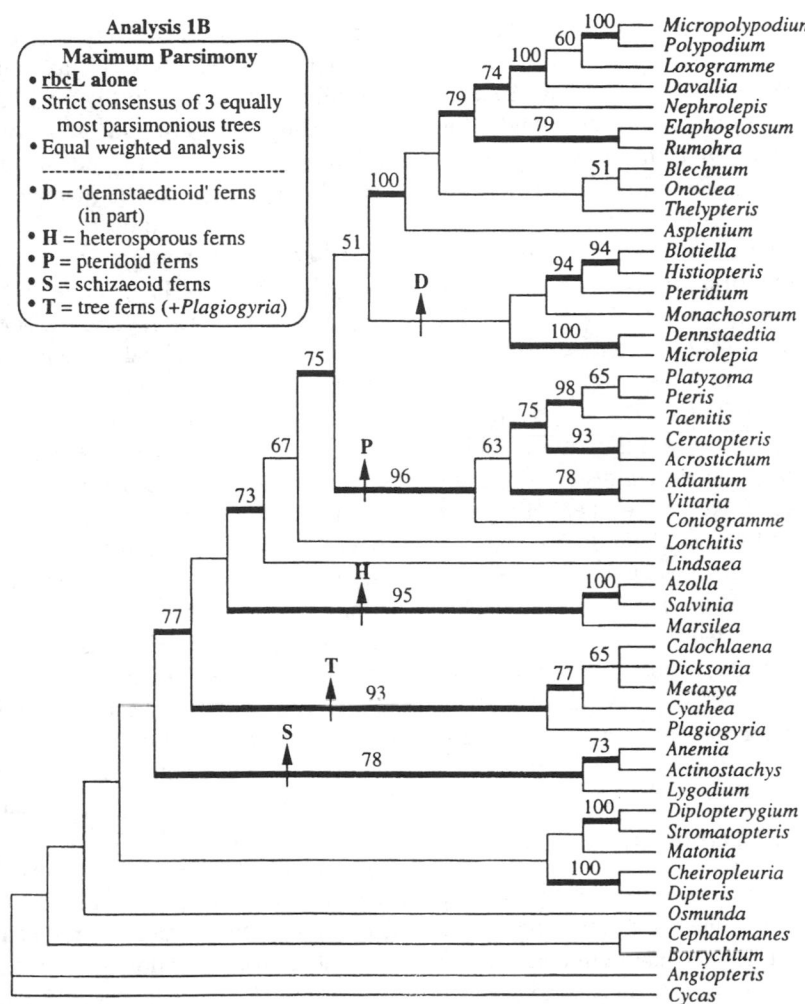

Analysis 1B

Maximum Parsimony
- **rbcL alone**
- Strict consensus of 3 equally
  most parsimonious trees
- Equal weighted analysis

- D = 'dennstaedtioid' ferns
  (in part)
- H = heterosporous ferns
- P = pteridoid ferns
- S = schizaeoid ferns
- T = tree ferns (+*Plagiogyria*)

Figure 19.2. Analysis 1B (fig. 3, p. 223) of Pryer et al. (1995). Strict consensus of three most parsimonious trees at 3,639 steps based on 489 parsimony-informative characters. See Fig. 19.1 for additional notation. CI = 0.235; RI = 0.464. Figure reproduced with permission from the American Fern Society.

both morphological and *rbcL* analyses. This result does not support the traditional view of the Polypodiaceae being more closely related to gleichenioid ferns (Wagner, 1969; Lovis, 1977), but is consistent with evidence from sporangial and gametophyte structures, as suggested by Jarrett (1980). Most traditionally recognized fern families (sensu Kramer and Green, 1990) appear consistently as monophyletic groups based on analysis of *rbcL,* for example, Polypodiaceae (with *Micropolypodium* included), Oleandraceae, Blechnaceae, Aspleniaceae, Vittariaceae, Marsileaceae, Schizaeaceae, Gleiche-

niaceae, Hymenophyllaceae, and Ophioglossaceae (Hasebe et al., 1995). Consistently nonmonophyletic families include Dennstaedtiaceae, Dryopteridaceae (by inclusion of at least eight other families), and Pteridaceae (by inclusion of Vittariaceae).

Although *rbcL* appears to lack signal for inferring basal clades of ferns and pteridophytes, the gene seems to be extremely informative for inferring relationships among genera within families. In a parsimony analysis of *rbcL* sequences in Vittariaceae (Crane et al., 1995), most clades were highly resolved with 100% (or nearly 100%) bootstrap support, and seven clades had decay indices greater than 10. Crane et al.'s (1995) results indicate that *Vittaria* is probably polyphyletic. Gastony and Rollo (1995) analyzed *rbcL* from 25 species of cheilanthoid ferns, also finding strongly supported clades. The resulting trees suggest that both *Pellaea* and *Cheilanthes* are polyphyletic, and the segregation of *Argyrochosma* from *Notholaena* is supported. Wolf (1995) used sequence data from *rbcL* and nuclear 18S rRNA genes to infer relationships among genera of Dennstaedtiaceae. The results suggest that the family is polyphyletic with the lindsaeoid genera representing one main lineage and Dennstaedtiaceae sensu stricto another. The genera *Orthiopteris* and *Tapeinidium* do not appear to be supported within either clade. *rbcL* sequence data have also been used to infer relationships among closely related genera (e.g., onocleoid ferns, Gastony and Ungerer, 1997) and among species within genera (e.g., *Polypodium*, Haufler and Ranker, 1995; *Botrychium* subg. *Botrychium*, Hauk, 1995).

ANALYSES OF COMBINED DATA SETS

It has become increasingly clear that progress in systematics rarely comes from focusing on only one type of data. Rather, taking into account information from fossils, extant taxa, morphology, and molecules has resulted in many of the more robust insights into phylogeny (Patterson, 1987). Thus, several studies have used approaches that combine extinct and extant taxa (Rothwell, 1987; Donoghue et al., 1989; Doyle and Donoghue, 1992; Nixon et al., 1994; Rothwell and Serbet, 1994; Nixon, 1996; Rothwell, 1998), other studies have combined morphological and molecular data (e.g., Baldwin, 1993; Doyle et al., 1994; Mishler et al., 1994; Lutzoni and Vilgalys, 1995; Pryer et al., 1995), and some studies have combined different molecular data sets (e.g., Olmstead and Sweere, 1994; Hoot, 1995; Hoot and Crane, 1995; Mason-Gamer and Kellogg, 1996; Soltis et al., 1996). Pryer et al. (1995) combined data from 77 parsimony-informative, morphological characters with 1,206 bp for *rbcL* (of which 490 were parsimony-informative) for 49 pteridophyte taxa. The strict

consensus of 34 most parsimonious trees for a smaller "fern" data set (47 taxa plus one outgroup) is reproduced in Figure 19.3. The most significant finding of this study was that the combined analysis provided better resolution than the separate analyses (compare Figs. 19.1, 19.2, and 19.3). For almost all clades detected, bootstrap support was greatest in the combined analysis (table 4 of Pryer et al., 1995). For example, a leptosporangiate fern clade (including *Osmunda* at the base) did not appear on the strict consensus *rbcL* tree, the clade had 73% bootstrap support on the morphology tree, and 89% support on the combined tree. Apparently, *rbcL* sequence data and morphology provided optimal information at different levels in the phylogeny: *rbcL* for more recent divergences and morphology for older events, and together, the information is complementary. Increased phylogenetic resolution and increased internal support in analyses of combined (versus separate) data sets have also been observed in other studies (e.g., Kim et al., 1992; Doyle et al., 1994; Mishler et al., 1994; Olmstead and Sweere, 1994; Soltis et al., 1996, 1997; see also Chapters 11 and 17). The approach used by Pryer et al. (1995) suggests that increasing the number of gene sequences and the number of morphological characters may help to resolve further the base of both the fern and pteridophyte phylogeny, where most of the phylogenetic uncertainty remains.

A tree based on nucleotide sequence data estimates a gene phylogeny, which may not necessarily agree with the organismal phylogeny, because, for example, of introgression, lineage sorting, or gene duplication (e.g., Hillis, 1987; Pamilo and Nei, 1988; Doyle, 1992; Lutzoni and Vilgalys, 1995). In addition, different data sets can provide evidence for different trees because of sampling error (Rodrigo et al., 1993) or use of an inappropriate evolutionary model for the data (Bull et al., 1993). The issue of whether to combine molecular and morphological data (or multiple molecular data sets) is not trivial. Arguments have been presented for combining data sets to analyze the "total evidence" (Kluge, 1989; Barrett et al., 1991), for analyzing data sets separately and then comparing trees and testing for congruence (Miyamoto and Fitch, 1995), and for first testing for homogeneity among data sets

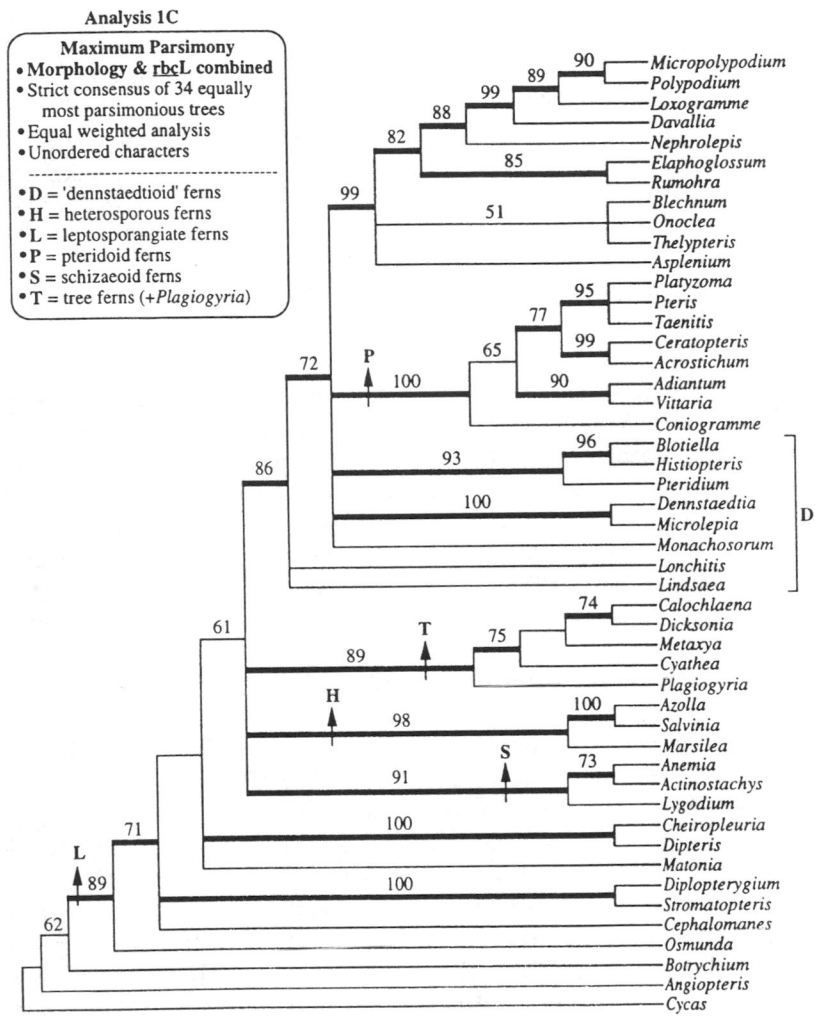

Analysis 1C

Maximum Parsimony
- **Morphology & rbcL combined**
- Strict consensus of 34 equally
  most parsimonious trees
- Equal weighted analysis
- Unordered characters
- -
- D = 'dennstaedtioid' ferns
- H = heterosporous ferns
- L = leptosporangiate ferns
- P = pteridoid ferns
- S = schizaeoid ferns
- T = tree ferns (+*Plagiogyria*)

Figure 19.3. Analysis 1C (fig. 4, p. 225) of Pryer et al. (1995). Strict consensus of 34 most parsimonious trees at 4,128 steps based on 564 parsimony-informative characters. See Fig. 19.1 for additional notation. CI = 0.242; RI = 0.466. Figure reproduced with permission from the American Fern Society.

to see if combining is appropriate (Bull et al., 1993; de Queiroz, 1993; Rodrigo et al., 1993; Lutzoni and Vilgalys, 1995; for reviews, see de Queiroz et al., 1995, Huelsenbeck et al., 1996, and Chapter 11). Regardless of how multiple data sets should be analyzed, the use of several independent sources of phylogenetic data should greatly increase the chance of correctly inferring evolutionary relationships (Penny et al., 1991). This is especially important for ancient divergence events (such as the base of the vascular plant phylogeny) where any single data set is likely to have a weak phylogenetic signal.

EXPLORATORY ANALYSIS OF PTERIDOPHYTE PHYLOGENY USING MULTIPLE DATA SETS

To explore the feasibility of combining several diverse data sets for phylogenetic analysis of pteridophytes, we chose 12 representative taxa for which we could gather data on 77 morphological characters and nucleotide sequence data from *rbcL* (1,206 bp), *atpB* (588 bp), and the 18S nuclear rRNA gene (1,617 bp). Previous studies have examined the effect of combining two data sets in ferns. Our goal was to see if combining four data sets could provide even better phylogenetic resolution.

Our taxon sample was limited because few pteridophytes have been examined for many genes. However, restricting the number of taxa to 12 had the advantage of allowing for thorough searches of tree space. We chose one representative of each class of fern allies (*Huperzia, Equisetum,* and *Psilotum*), eight ferns (*Ophioglossum, Vandenboschia, Osmunda, Dicksonia, Adiantum, Lonchitis, Pteridium,* and *Blechnum*), and one seed plant (*Ginkgo*). We used *Huperzia* to root the trees. Obviously, as data sets grow, more outgroups need to be included, especially bryophytes. We used an exemplar approach, with each taxon represented by the same genus for all data sets. For nine of the 12 taxa we used a single exemplar species (Table 19.1). The exceptions (where no specific epithet is given in Table 19.1) were *Ophioglossum—O. reticulatum* for *atpB, O. petiolatum* for 18S rDNA, *O. engelmannii* for *rbcL,* all three for morphology; *Equisetum— E. hyemale* for 18S rDNA, *E. arvense* for the other three data sets; *Huperzia—H. campiana* for *rbcL, H. lucidula* for the other three data sets.

Complete matrices for all data sets are available upon request from any of the authors. The Pryer et al. (1995) morphology data matrix (with the four new taxa included here) is also available at http://ucjeps.berkeley.edu/bryolab/greenplant-page.html. Nuclear 18S rDNA, *atpB,* and *rbcL* sequences were found in GenBank for all 12 taxa, except for the *atpB* sequence of *Ginkgo,* which was kindly provided by M. Chase and V. Savolainen. Morphological data for 77 char-

acters are available for eight of the 12 taxa (Pryer et al., 1995), and this data matrix was extended with information for *Huperzia, Ginkgo, Vandenboschia,* and *Ophioglossum.* Sources of information, additional to those cited in Pryer et al. (1995), used for the extended morphological data matrix were: Arnott (1959), Fryns-Claussens and Van Cotthem (1973), and Rohr (1977) for *Ginkgo;* and Gewirtz and Fahn (1960) and Pant and Khare (1969) for *Ophioglossum.*

We used maximum parsimony to infer phylogeny from each data set separately and from a combined data set of 3,488 characters. We used PAUP version 3.1.1 (Swofford, 1993), searching for shortest trees using the branch-and-bound algorithm. Multistate morphological characters were treated as polymorphic; ambiguous nucleotides were treated as uncertain in the separate analyses. Bootstrapping (100 branch-and-bound replicates) was used to assess the support of each branch. Because current versions of PAUP do not allow for mixing different treatments of multistate characters, we used the "uncertain" coding for the combined analysis (see Swofford and Begle, 1993; Doyle et al., 1994, for further discussion).

Numbers of shortest trees, tree lengths, and other tree statistics are included in Figures 19.4, 19.5, and 19.6. There were several areas of congruence among the resultant trees. A "higher indusiate" clade (*Dicksonia, Lonchitis, Pteridium, Blechnum,* and *Adiantum*) was seen in all three molecular-based trees, and the morphology-based tree differed slightly with the inclusion of *Vandenboschia.* The positions of basal clades

Table 19.1. Taxa used and GenBank accession numbers for nucleotide sequence data.

Taxon	18S rDNA	*rbcL*	*atpB*
Osmunda cinnamomea	U18516	D14882	U93827
Vandenboschia davallioides	U18629	U05948	U93828
Dicksonia antarctica	U18624	U05919	U93829
Pteridium aquilinum	U18628	U05939	U93835
Adiantum raddianum	X78889	U05906	U93840
Blechnum occidentale	U18622	U05909	U93838
Lonchitis hirsuta	U18632	U05929	U93830
Psilotum nudum	X81963	U30835	U93822
Ophioglossum	U18515	L11058	U93825
Huperzia	U18505	X98282	U93819
Equisetum	X78890	L11053	U93824
Ginkgo biloba	D16448	D10733	Chase unpubl.

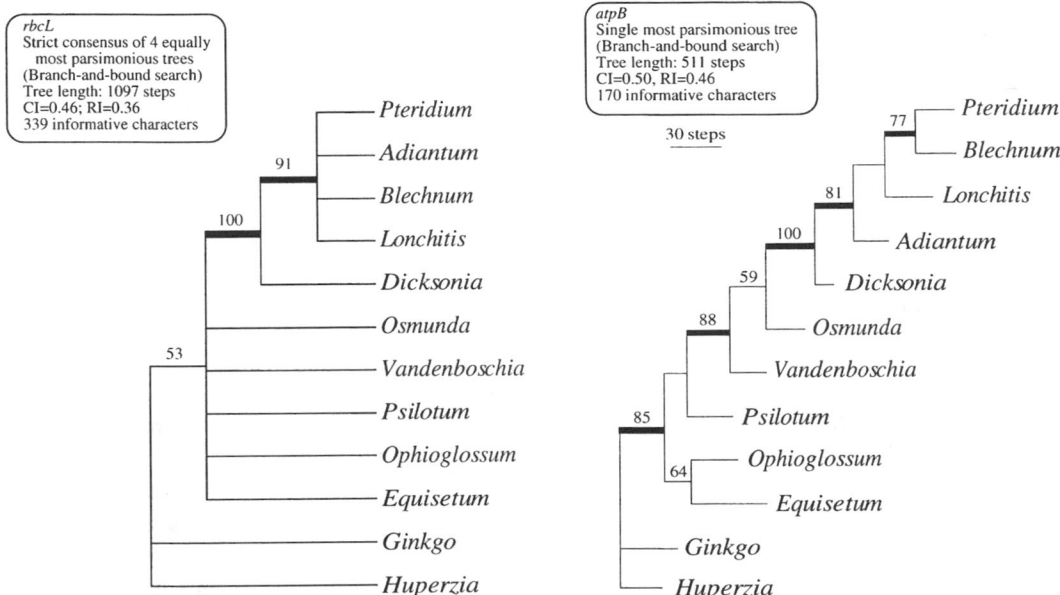

Figure 19.4. Strict consensus and single most parsimonious trees resulting, respectively, from phylogenetic analysis of *rbcL* (left) and *atpB* (right). Numbers above branches denote bootstrap percent values. Trees were rooted with *Huperzia*. For *rbcL* the strict consensus of four most parsimonious trees is shown; tree length of 1,097 steps; CI = 0.46; RI = 0.36. For *atpB* the single most parsimonious tree is depicted; tree length of 511 steps; CI = 0.50; RI = 0.46.

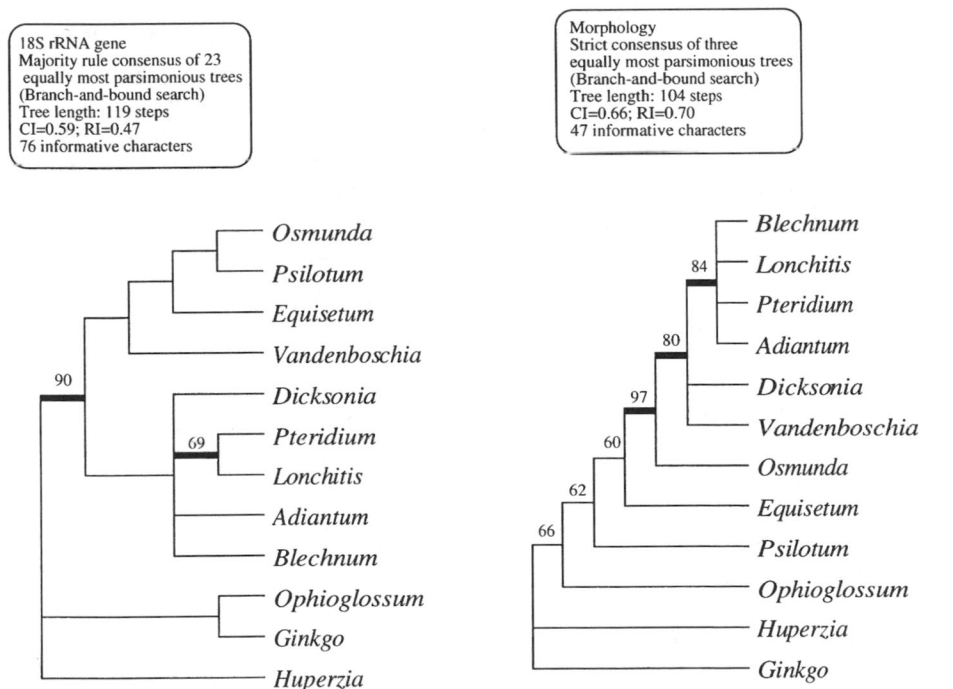

Figure 19.5. Majority rule and strict consensus trees resulting, respectively, from analysis of 18S rRNA gene (left) and morphology (right). Numbers above branches denote bootstrap percent values. Trees were rooted with *Huperzia*. For the 18S rRNA gene the majority rule consensus of 23 shortest trees is shown; tree length of 119 steps; CI = 0.59; RI = 0.47. For morphology the strict consensus of three most parsimonious trees is shown; tree length of 104 steps; CI = 0.66; RI = 0.70.

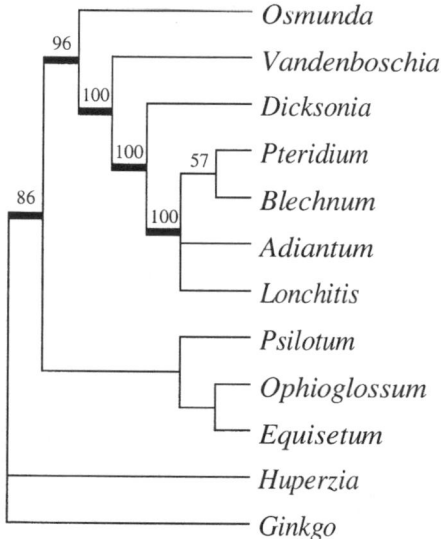

Analysis of combined data sets
Strict consensus of 2 equally
most parsimonious trees
(Branch-and-bound search)
Tree length: 1872 steps
CI=0.47; RI=0.39
634 informative characters

Figure 19.6. Strict consensus tree resulting from analysis of four combined data sets (*rbcL, atpB,* 18S rRNA gene, and morphology) at 1,872 steps. Numbers above branches denote bootstrap percent values. Tree was rooted with *Huperzia;* CI = 0.47; RI = 0.39.

varied more among trees. For example, the *atpB* tree had an (*Equisetum + Ophioglossum*) clade, whereas the 18S rDNA tree had a (*Psilotum + Osmunda*) clade. The *atpB* and morphology trees had resolution spread relatively evenly across the tree, as indicated by high bootstrap values. The *rbcL* and 18S rDNA trees were not as well resolved. However, the combined analysis using all four data sets had branches with the highest bootstrap support at deep nodes of the tree (Fig. 19.6).

Because our taxon sampling was limited, we avoid making strong statements about vascular plant phylogeny. Rather, this exploratory exercise illustrates that combining several data sets can result in trees with better support than those from single data sets. This finding is not new; it has been noted previously from combining two or more data sets in other studies (e.g., Marshall, 1992; Doyle et al., 1994; Olmstead and Sweere, 1994; Hoot and Crane, 1995; Pryer et al., 1995;

Soltis et al., 1996, 1997). Currently, seven genes appear to have some phylogenetic utility for vascular plant studies: *rbcL, atpB,* 16S rDNA, 18S rDNA, 26S rDNA, phytochrome, and the mitochondrial *coxIII;* this list is growing continually (see Chapter 1). If combining several molecular data sets plus morphology increases the phylogenetic signal, then the potential exists to improve significantly the resolution at the base of the extant vascular plant tree using such a combined approach. Such a task will require coordination among workers to ensure accuracy of data, use of the same vouchered taxa (preferably same DNA sources), and careful consideration of taxon sampling to achieve good representation of all major extant groups. Hillis (1996) showed that, up to a point, increasing the length of variable molecular data sets should improve phylogenetic accuracy, and that large data sets (in terms also of taxa) can be used effectively to infer phylogeny. Hillis's (1996) simulations (based on 18S rDNA data) suggest that phylogenies can be accurately reconstructed for over 200 taxa using 5,000 informative characters. The total length of the seven genes listed above is approximately 9,000 bp.

Several studies have found that genomic rearrangements, which by their nature tend to be rare, can be more useful for unraveling ancient phylogenetic patterns (Raubeson and Jansen, 1992; Stein et al., 1992; Raubeson and Stein, 1995; Manhart and McCourt, 1996; Qiu and Palmer, 1996). A limitation with the current genomic mapping strategies is that they involve one-by-one screening of taxa for intron positions and large-scale rearrangements, using probing techniques. An alternative strategy would be to use complete chloroplast genome sequences. Through automated DNA sequencing, this approach is now feasible and not completely out of reach in terms of cost. Complete sequences of the chloroplast genomes of several pteridophytes (150–180 kb each) could be added to the six cpDNA sequences already published. Genomic mapping algorithms, developed for the human genome sequencing project, can be applied to large sequence data sets to search for rearrangements and large-scale losses and gains of fragments. In the long run, this should be a faster and more systematic approach to looking for needles in the haystack. An added advantage over traditional probing methods is that evolutionarily homologous losses and gains of regions could be identified by sequence identity in the splicing regions. Also, by generating complete chloroplast genome sequences, we can search for new regions where the sequences themselves contain phylogenetic signal.

The accuracy of our knowledge of relationships of pteridophytes, and vascular plants in general, will depend on continued study in all areas: single gene sequences, genomic sequences, developmental patterns, and morphological characters. These large-scale studies are feasible only for extant taxa. Many of the phylogenetically key taxa are from extinct lineages; hence, the incorporation of fossil data is also essential for continued progress in estimating vascular plant phylogenies.

LITERATURE CITED

Arnott, H. J. 1959. Anastomoses in the venation of *Ginkgo biloba*. American Journal of Botany **46**:405–411.

Avise, J. C., W. S. Nelson, and H. Sugita. 1994. A speciational history of "living fossils": molecular evolutionary patterns in horseshoe crabs. Evolution **48**:1986–2001.

Baldwin, B. G. 1993. Molecular phylogenetics of *Calycadenia* (Compositae) based on ITS sequences of nuclear ribosomal DNA: chromosomal and morphological evolution reexamined. American Journal of Botany **80**:222–238.

Banks, H. P. 1975. Reclassification of Psilophyta. Taxon **24**:401–413.

Barrett, M., M. J. Donoghue, and E. Sober. 1991. Against consensus. Systematic Zoology **40**:486–493.

Bierhorst, D. W. 1968. On the Stromatopteridaceae (fam. nov.) and on the Psilotaceae. Phytomorphology **18**:232–268.

Bierhorst, D. W. 1971. Morphology of Vascular Plants. Macmillan, New York.

Bierhorst, D. W. 1977. The systematic position of *Psilotum* and *Tmesipteris*. Brittonia **29**:3–13.

Boivin, M. R. R., D. Beauseigle, J. Bousquet, and G. Bellemare. 1996. Phylogenetic inferences from chloroplast *chlB* gene sequences of *Nephrolepsis exalta* (Filicopsida), *Ephedra altissima* (Gnetopsida) and diverse land plants. Molecular Phylogenetics and Evolution **6**:19–29.

Bremer, K., C. J. Humphries, B. D. Mishler, and S. P. Churchill. 1987. On cladistic relationships of green plants. Taxon **36**:339–349.

Brooks, D. R., and D. A. McLennan. 1991. Phylogeny, Ecology, and Behavior. A Research Program in Comparative Biology. University of Chicago Press, Chicago.

Bull, J. J., J. P. Huelsenbeck, C. W. Cunningham, D. L. Swofford, and P. J. Waddell. 1993. Partitioning and combining data in phylogenetic analysis. Systematic Biology **42**:384–397.

Ching, R. C. 1940. On natural classification of the family "Polypodiaceae". Sunyatsenia **5**:201–268.

Conant, D. S., L. A. Raubeson, D. K. Attwood, and D. B. Stein. 1995. The relationships of Papuasian Cyatheaceae to New World tree ferns. American Fern Journal **85**:328–340.

Crabbe, J. A., A. C. Jermy, and J. T. Mickel. 1975. A new generic sequence for the pteridophyte herbarium. Fern Gazette **11**:141–162.

Crane, E. H., D. R. Farrar, and J. F. Wendel. 1995. Phylogeny of the Vittariaceae: convergent simplification leads to a polyphyletic *Vittaria*. American Fern Journal **85**:283–305.

Crane, P. 1990. The phylogenetic context of microsporogenesis. **In** Microspores: Evolution and Ontogeny, eds. S. Blackmore and R. B. Knox, pp. 11–41. Academic Press, London.

Cronquist, A., A. L. Takhtajan, and W. Zimmerman. 1966. On the higher taxa of Embryobionta. Taxon **15**:129–134.

de Queiroz, A. 1993. For consensus (sometimes). Systematic Biology **42**:368–372.

de Queiroz, A., M. J. Donoghue, and J. Kim. 1995. Separate versus combined analysis of phylogenetic evidence. Annual Review of Ecology and Systematics **26**:657–681.

DiMichele, W. A., and J. E. Skog. 1992. The Lycopsida: a symposium. Annals of the Missouri Botanical Garden **79**:447–449.

Donoghue, M. J., J. A. Doyle, J. Gauthier, A. G. Kluge, and T. Rowe. 1989. The importance of fossils in phylogeny reconstruction. Annual Review of Systematics and Ecology **20**:431–460.

Downie, S. R., and J. D. Palmer. 1992. The use of chloroplast DNA rearrangements in reconstructing plant phylogeny. **In** Molecular Systematics of Plants, eds. P. S. Soltis, D. E. Soltis, and J. J. Doyle, pp. 14–35. Chapman & Hall, New York.

Doyle, J. A., and M. J. Donoghue. 1992. Fossils and seed plant phylogeny reanalyzed. Brittonia **44**:89–106.

Doyle, J. A., M. J. Donoghue, and E. A. Zimmer. 1994. Integration of morphological and ribosomal RNA data on the origin of angiosperms. Annals of the Missouri Botanical Garden **81**:419–450.

Doyle, J. J. 1992. Gene trees and species trees: molecular systematics as one-character taxonomy. Systematic Botany **17**:144–163.

Eames, A. J. 1936. Morphology of the Vascular Plants: Lower Groups. McGraw-Hill, New York.

Felsenstein, J. 1978. Cases in which parsimony or compatibility methods will be positively misleading. Systematic Zoology **27**:401–410.

Foster, A. S., and E. M. Gifford, Jr. 1974. Comparative Morphology of Vascular Plants. W. H. Freeman, San Francisco.

Fryns-Claussens, E., and W. R. J. Van Cotthem. 1973. A new classification of ontogenetic types of stomata. Botanical Review **39**:71–138.

Garbary, D. J., K. S. Renzaglia, and J. G. Duckett. 1993. The phylogeny of land plants—a cladistic analysis based on male gametogenesis. Plant Systematics and Evolution **188:**237–269.

Gastony, G. J., and D. R. Rollo. 1995. Phylogeny and generic circumscriptions of cheilanthoid ferns (Pteridaceae: Cheilanthoideae) inferred from *rbcL* nucleotide sequences. American Fern Journal **85:**341–360.

Gastony, G. J., and M. C. Ungerer. 1997. Molecular systematics and a revised taxonomy of the onocleoid ferns (Dryopteridaceae: Onocleeae). American Journal of Botany **84:**840–849.

Gastony, G. J., G. Yatskievych, and C. K. Dixon. 1992. Chloroplast DNA restriction site variation in the fern genus *Pellaea:* phylogenetic relationships of the *Pellaea glabella* complex. American Journal of Botany **79:**1072–1080.

Gewirtz, M., and A. Fahn. 1960. The anatomy of the sporophyte and gametophyte of *Ophioglossum lusitanicum* L. Phytomorphology **10:**342–351.

Hamby, R. K., and E. A. Zimmer. 1988. Ribosomal RNA sequences for inferring phylogeny within the grass family (Poaceae). Plant Systematics and Evolution **160:**29–37.

Harvey, P. H., and M. D. Pagel. 1991. The Comparative Method in Evolutionary Biology. Oxford University Press, Oxford.

Hasebe, M., and K. Iwatsuki. 1990a. *Adiantum capillus-veneris* chloroplast DNA clone bank: as useful heterologous probes in the systematics of the leptosporangiate ferns. American Fern Journal **80:**20–25.

Hasebe, M., and K. Iwatsuki. 1990b. Chloroplast DNA from *Adiantum capillus-veneris* L., a fern species (Adiantaceae); clone bank, physical map and unusual gene localization in comparison with angiosperm chloroplast DNA. Current Genetics **17:**359–364.

Hasebe, M., M. Ito, R. Kofuji, K. Ueda, and K. Iwatsuki. 1993. Phylogenetic relationships of ferns deduced from *rbcL* gene sequence. Journal of Molecular Evolution **37:**476–482.

Hasebe, M., T. Omori, M. Nakazawa, T. Sano, M. Kato, and K. Iwatsuki. 1994. *rbcL* gene sequences provide evidence for the evolutionary lineages of leptosporangiate ferns. Proceedings of the National Academy of Sciences U.S.A. **91:**5730–5734.

Hasebe, M., P. G. Wolf, K. M. Pryer, K. Ueda, M. Ito, R. Sano, G. J. Gastony, J. Yokoyama, J. R. Manhart, N. Murakami, E. H. Crane, C. H. Haufler, and W. D. Hauk. 1995. Fern phylogeny based on *rbcL* nucleotide sequences. American Fern Journal **85:**134–181.

Haufler, C. H., and T. A. Ranker. 1995. *rbcL* sequences provide phylogenetic insights among sister species of the fern genus *Polypodium.* American Fern Journal **85:**361–374.

Haufler, C. H., D. E. Soltis, and P. S. Soltis. 1995. Phylogeny of the *Polypodium vulgare* complex: insights from chloroplast DNA restriction site data. Systematic Botany **20:**110–119.

Hauk, W. D. 1995. A molecular assessment of relationships among cryptic species of *Botrychium* subgenus *Botrychium* (Ophioglossaceae). American Fern Journal **85:**375–394.

Hendy, M. D., and D. Penny. 1989. A framework for the quantitative study of evolutionary trees. Systematic Zoology **38:**297–309.

Hiesel, R., A. von Haeseler, and A. Brennicke. 1994. Plant mitochondrial nucleic acid sequences as a tool for phylogenetic analysis. Proceedings of the National Academy of Sciences U.S.A. **91:**634–638.

Hillis, D. M. 1987. Molecular versus morphological approaches to systematics. Annual Review of Ecology and Systematics **18:**23–42.

Hillis, D. M. 1996. Inferring complex phylogenies. Nature **383:**130–131.

Holttum, R. E. 1973. Posing the problems. In The Phylogeny and Classification of the Ferns, eds. A. C. Jermy, J. A. Crabbe, and B. A. Thomas, pp. 1–10. Academic Press, London.

Hoot, S. B. 1995. Phylogeny of the Ranunculaceae based on preliminary *atpB, rbcL* and 18S nuclear ribosomal DNA sequence data. Plant Systematics and Evolution **9:**241–251.

Hoot, S. B., and P. R. Crane. 1995. Inter-familial relationships in the Ranunculidae based on molecular systematics. Plant Systematics and Evolution (suppl.) **9:**119–131.

Huelsenbeck, J. P. 1994. Comparing the stratigraphic record to estimates of phylogeny. Paleobiology **20:**470–483.

Huelsenbeck, J. P., J. J. Bull, and C. W. Cunningham. 1996. Combining data in phylogenetic analysis. Trends in Ecology and Evolution **11:**152–158.

Jarrett, F. M. 1980. Studies in the classification of the leptosporangiate ferns: I. The affinities of the Polypodiaceae sensu stricto and the Grammitidaceae. Kew Bulletin **34:**825–833.

Jeffrey, E. C. 1902. The structure and development of the stem in pteridophytes and gymnosperms. Philosophical Transactions of the Royal Society of London, Series B **195:**119–146.

Kaplan, D. R. 1977. Morphological status of the shoot systems of Psilotaceae. Brittonia **29:**30–53.

Kato, M. 1983. The classification of the major groups of pteridophytes. Journal of the Faculty of Science, University of Tokyo, Sect. III **13:**263–283.

Kenrick, P. 1994. Alternation of generations in land plants: new phylogenetic and palaeobotanical evidence. Biological Reviews of the Cambridge Philosophical Society **69:**293–330.

Kenrick, P., and P. R. Crane. 1991. Water-conducting cells in early fossil plants: implications for the evolution of early tracheophytes. Botanical Gazette **152:**335–356.

Kenrick, P., and P. R. Crane. 1997. The Origin and Early Diversification of Land Plants: A Cladistic Study. Smithsonian Press, Washington, D.C.

Kenrick, P., and E. M. Friis. 1995. Paleobotany of land plants. In Progress in Botany, Vol. 56, eds. H. D. Behnke, U. Luttge, K. Esser, J. W. Kadereit, and M. Runge, pp. 372–395. Springer-Verlag, Berlin.

Kim, K., R. K. Jansen, R. S. Wallace, H. J. Michaels, and J. D. Palmer. 1992. Phylogenetic implications of *rbcL* sequence variation in the *Asteraceae.* Annals of the Missouri Botanical Garden **79:**428–445.

Kluge, A. G. 1989. A concern for evidence and a phylogenetic hypothesis of relationships among *Epicrates* (Boidae, Serpentes). Systematic Zoology **38**:7–25.

Kolukisaoglu, H. Ü., S. Marx, C. Wiegmann, S. Hanelt, and H. A. W. Schneider-Poetsch. 1995. Divergence of the phytochrome gene family predates angiosperm evolution and suggests that *Selaginella* and *Equisetum* arose prior to *Psilotum*. Journal of Molecular Evolution **41**:329–337.

Kramer, K. U. 1991. Systematics of the pteridophytes. In Progress in Botany, Vol. 52, eds. H. D. Behnke, K. Esser, K. Kubitzki, M. Runge, and H. Ziegles, pp. 342–358. Springer-Verlag, Berlin.

Kramer, K. U., and P. S. Green, eds. 1990. Pteridophytes and gymnosperms. In The Families and Genera of Vascular Plants, Vol. 1 (series ed. K. Kubitzki). Springer-Verlag, Berlin.

Kranz, H. D., and V. A. R. Huss. 1996. Molecular evolution of pteridophytes and their relationship to seed plants: evidence from complete 18S rRNA gene sequences. Plant Systematics and Evolution **202**:1–11.

Kranz, H. D., D. Miks, M. Siegler, I. Capesius, C. W. Sensen, and V. A. R. Huss. 1995. The origin of land plants: phylogenetic relationships among charophytes, bryophytes, and vascular plants inferred from complete small-subunit ribosomal RNA gene sequences. Journal of Molecular Evolution **41**:74–84.

Lovis, J. D. 1977. Evolutionary patterns and processes in ferns. Advances in Botanical Research **4**:229–415.

Lugardon, B. 1990. Pteridophyte sporogenesis: a survey of spore wall ontogeny and fine structure in a polyphyletic plant group. In Microspores: Evolution and Ontogeny, eds. S. Blackmore and R. B. Knox, pp. 95–112. Academic Press, London.

Lutzoni, F., and R. Vilgalys. 1995. Integration of morphological and molecular data sets in estimating fungal phylogenies. Canadian Journal of Botany **73**: S649–S659.

Manhart, J. R. 1994. Phylogenetic analysis of green plant *rbcL* sequences. Molecular Phylogenetics and Evolution **3**:114–127.

Manhart, J. R. 1995. Chloroplast 16S rDNA sequences and phylogenetic relationships of fern allies and ferns. American Fern Journal **85**:182–192.

Manhart, J. R., and R. M. McCourt. 1996. Utility of introns and other genomic features in the phylogeny of green algae and land plants. American Journal of Botany (suppl.) **86**:108.

Marshall, C. R. 1992. Character analysis and the integration of molecular and morphological data in an understanding of sand dollar phylogeny. Molecular Biology and Evolution **9**:309–322.

Mason-Gamer, R. J., and E. A. Kellogg. 1996. Testing for phylogenetic conflict among molecular data sets in the tribe Triticeae (Gramineae). Systematic Biology **45**:524–545.

Mathews, S., M. Lavin, and R. A. Sharrock. 1995. Evolution of the phytochrome gene family and its utility for phylogenetic analyses of angiosperms. Annals of the Missouri Botanical Garden **82**:296–321.

Mickel, J. T. 1974. Phyletic lines in the modern ferns. Annals of the Missouri Botanical Garden **61**:474–482.

Mishler, B. D., and S. P. Churchill. 1985. Transition to a land flora: phylogenetic relationships of the green algae and bryophytes. Cladistics **1**:305–328.

Mishler, B. D., L. A. Lewis, M. A. Buchheim, K. S. Renzaglia, D. J. Garbary, C. F. Delwiche, F. W. Zechman, T. S. Kantz, and R. L. Chapman. 1994. Phylogenetic relationships of the "green algae" and "bryophytes". Annals of the Missouri Botanical Garden **81**:451–483.

Miyamoto, M. M., and W. M. Fitch. 1995. Testing species phylogenies and phylogenetic methods with congruence. Systematic Biology **44**:64–76.

Murakami, N., and B. A. Schaal. 1994. Chloroplast DNA variation and the phylogeny of *Asplenium* sect. *Hymenasplenium* (Aspleniaceae) in the New World tropics. Journal of Plant Research **107**:245–251.

Nixon, K. C. 1996. Paleobotany in cladistics and cladistics in paleobotany: enlightenment and uncertainty. Review of Paleobotany and Palynology **90**:361–373.

Nixon, K. C., W. L. Crepet, D. Stevenson, and E. M. Friis. 1994. A reevaluation of seed plant phylogeny. Annals of the Missouri Botanical Garden **81**:484–533.

Olmstead, R. G., and J. A. Sweere. 1994. Combining data in phylogenetic systematics: an empirical approach using three molecular data sets in the Solanaceae. Systematic Biology **43**:467–481.

Pamilo, P., and M. Nei. 1988. Relationships between gene trees and species trees. Molecular Biology and Evolution **5**:568–583.

Pant, D. D., and P. K. Khare. 1969. Epidermal structure and stomatal ontogeny in some eusporangiate ferns. Annals of Botany **33**:795–805.

Parenti, L. R. 1980. A phylogenetic analysis of the land plants. Biological Journal of the Linnean Society **13**:225–242.

Patterson, C. 1987. Molecules and Morphology in Evolution: Conflict or Compromise? Cambridge University Press, Cambridge.

Penny, D., M. D. Hendy, and M. A. Steel. 1991. Testing the theory of descent. In Phylogenetic Analysis of DNA Sequences, eds. M. M. Miyamoto and J. Cracraft, pp. 155–183. Oxford University Press, Oxford.

Philippe, H., and A. Adoutte. 1996. What can phylogenetic patterns tell us about the evolutionary processes generating biodiversity? In Aspects of the Genesis and Maintenance of Biological Diversity, eds. M. E. Hochberg, J. Clobert, and R. Barbault, pp. 41–59. Oxford University Press, Oxford.

Pichi Sermolli, R. E. G. 1977. Tentamen Pteridophytorum genera in taxonomicum ordinem redigendi. Webbia **31**:313–512.

Pryer, K. M., A. R. Smith, and J. E. Skog. 1995. Phylogenetic relationships of extant ferns based on evidence from morphology and *rbcL* sequences. American Fern Journal **85**:205–282.

Qiu, Y.-L., and J. D. Palmer. 1996. Intron evolution and angiosperm phylogeny. American Journal of Botany (suppl.) **86**:188.

Raubeson, L. A., and R. K. Jansen. 1992. Chloroplast DNA evidence on the ancient evolutionary split in vascular land plants. Science **255**:1697–1699.

Raubeson, L. A., and D. B. Stein. 1995. Insights into fern evolution from mapping chloroplast genomes. American Fern Journal **85**:193–204.

Ritland, K., and M. T. Clegg. 1987. Evolutionary analysis of plant DNA sequences. The American Naturalist **130**:74–100.

Rodrigo, A. G., M. Kellyborges, P. R. Bergquist, and P. L. Bergquist. 1993. A randomisation test of the null hypothesis that two cladograms are sample estimates of a parametric phylogenetic tree. New Zealand Journal of Botany **31**:257–268.

Rohr, R. 1977. Etude comparée de la formation de l'exine au cours de la microsporogénèse chez une gymnosperme (*Taxus baccata*) et une préphanerogame (*Ginkgo biloba*). Cytologia **42**:156–167.

Rothwell, G. W. 1987. Complex Paleozoic Filicales in the evolutionary radiation of ferns. American Journal of Botany **74**:458–461.

Rothwell, G. W. 1994. Phylogenetic relationships among ferns and gymnosperms: an overview. Journal of Plant Research **107**:411–416.

Rothwell, G. W. 1998. Fossils and ferns in the resolution of land plant phylogeny. Botanical Review, in press.

Rothwell, G. W., and R. Serbet. 1994. Lignophyte phylogeny and the evolution of spermatophytes—a numerical cladistic analysis. Systematic Botany **19**: 443–482.

Rothwell, G. W., and R. A. Stockey. 1994. Pteridophytes and gymnosperms: current concepts of structure, evolutionary history, and phylogeny. Journal of Plant Research **107**:409.

Schneider, H. 1996. Vergleichende Wurzelanatomie der Farne. Ph.D. dissertation. Universität Zürich, Zürich.

Schneider-Poetsch, H. A. W., S. Marx, H. U. Kolukisaoglu, S. Hanelt, and B. Braun. 1994. Phytochrome evolution: phytochrome genes in ferns and mosses. Phycologia **91**:241–250.

Sharp, P. M., M. Averof, A. T. Lloyd, G. Matassi, and J. F. Peden. 1995. DNA sequence evolution: the sounds of silence. Philosophical Transactions of the Royal Society London (Biology) **349**:241–247.

Smith, A. B., and D. T. J. Littlewood. 1994. Paleontological data and molecular phylogenetic analysis. Paleobiology **20**:259–273.

Smith, A. R. 1995. Non-molecular phylogenetic hypotheses for ferns. American Fern Journal **85**:104–122.

Smith, G. M. 1938. Cryptogamic Botany, Vol. 2, Bryophytes and Pteridophytes. McGraw-Hill, London.

Soltis, D. E., R. K. Kuzoff, E. Conti, R. Gornall, and K. Ferguson. 1996. *matK* and *rbcL* gene sequence data indicate that *Saxifraga* (Saxifragaceae) is polyphyletic. American Journal of Botany **83**:371–382.

Soltis, D. E., C. Hibsch-Jetter, P. S. Soltis, M. W. Chase, and J. S. Farris. 1997. Molecular phylogenetic relationships among angiosperms: an overview based on *rbcL* and 18S rDNA sequences. Journal of Plant Research, in press.

Stearns, S. C. 1992. The Evolution of Life Histories. Oxford University Press, Oxford.

Stein, D. B. 1985. Nucleic acid comparisons as a tool in understanding species interrelationships and phylogeny. Proceedings of the Royal Society Edinburgh **86B**: 283–288.

Stein, D. B., G. Yatskievych, and G. J. Gastony. 1989. Chloroplast DNA evolution and phylogeny of some polystichoid ferns. Biochemical Systematics and Ecology **17**:93–101.

Stein, D. B., D. S. Conant, M. E. Ahearn, E. T. Jordan, S. A. Kirch, M. Hasebe, K. Iwatsuki, M. K. Tan, and J. A. Thompson. 1992. Structural rearrangements of the chloroplast genome provide an important phylogenetic link in ferns. Proceedings of the National Academy of Sciences U.S.A. **89**:1856–1860.

Stewart, W. N. 1981. The progymnospermopsida: the construction of a concept. Canadian Journal of Botany **59**:1539–1542.

Stewart, W. W., and G. W. Rothwell. 1993. Paleobotany and the Evolution of Plants. Cambridge University Press, Cambridge.

Swofford, D. L. 1993. PAUP: Phylogenetic Analysis Using Parsimony, version 3.1.1. Illinois Natural History Survey, Champaign, Illinois.

Swofford, D. L., and D. P. Begle. 1993. PAUP: Phylogenetic Analysis Using Parsimony, Version 3.1.1, User's Manual. Laboratory of Molecular Systematics, Smithsonian Institution, Washington, D.C.

Swofford, D. L., G. J. Olsen, P. J. Waddell, and D. M. Hillis. 1996. Phylogenetic inference. **In** Molecular Systematics, eds. D. M. Hillis, C. Moritz, and B. K. Mable, pp. 407–514. Sinauer Associates, Sunderland, Massachusetts.

Takhtajan, A. L. 1953. Phylogenetic principles of the system of higher plants. Botanical Review **19**:1–45.

Wagner, W. H., Jr. 1969. The construction of a classification. **In** Systematic Biology, U.S. National Academy of Sciences Publications No. 1692, pp. 67–90. National Academy Press, Washington, D.C.

Wagner, W. H., Jr. 1977. Systematic implications of the Psilotaceae. Brittonia **29**:54–63.

Wolf, P. G. 1995. Phylogenetic analyses of *rbcL* and nuclear ribosomal RNA gene sequences in Dennstaedtiaceae. American Fern Journal **85**:306–327.

Wolf, P. G. 1996. Nucleotide sequences from *atpB* provide useful data for fern systematics. American Journal of Botany (suppl.) **83**:133.

Wolf, P. G., P. S. Soltis, and D. S. Soltis. 1994. Phylogenetic relationships of dennstaedtioid ferns: Evidence from *rbcL* sequences. Molecular Phylogenetics and Evolution **3**:383–392.

Yatskievych, G., D. B. Stein, and G. J. Gastony. 1988. Chloroplast DNA evolution and systematics of *Phanerophlebia* (Dryopteridaceae) and related fern genera. Proceedings of the National Academy of Sciences U.S.A. **85**:2589–2593.

Zimmerman, W. 1959. Die Phylogenie der Pflanzen. Gustav Fischer Verlag, Stuttgart.

Index